CRC Handbook of Proximate Analysis Tables of Higher Plants

Authors

James A. Duke
Botanist
Germplasm Introduction and Evaluation Laboratory
United States Department of Agriculture
Agricultural Research Service
Beltsville, Maryland

Alan A. Atchley
Research Botanist
Germplasm Resources Laboratory
United States Department of Agriculture
Agricultural Research Service
Beltsville, Maryland

CRC

CRC Press, Inc.
Boca Raton, Florida

Library of Congress Cataloging-in-Publication Data

Duke, James, A., 1929—
CRC handbook of proximate analysis tables of higher plants.

Bibliography: p.
1. Plants, Edible--Composition--Tables. 2. Food crops--Composition--Tables. 3. Nutrition--Tables.
I. Atchley, Alan A. II. Title. III. Title: Handbook of proximate analysis tables of higher plants.
TX557.D85 1986 641.3′3 86-4251
ISBN 0-8493-3634-1

Direct all inquiries to CRC Press, Inc., 2000 Corporate Blvd., N.W., Boca Raton, Florida, 33431.

International Standard Book Number 0-8493-3634-1

Library of Congress Card Number 86-4251
Printed in the United States

INTRODUCTION

When CRC Press asked about the possibility of publishing a nutritional table we were pleased, believing that our Germplasm Introduction and Evaluation Laboratory has one of the largest computerized collections of proximate analyses, at least as far as numbers of species are concerned. That was in 1983, when the data base contained most plant species analyzed in Food Composition Tables for Africa,[1] East Asia,[2] and Latin America,[3] to complement data from Agriculture Handbook No. 8,[4] the Wealth of India[5] and Forage Plant Analyses.[6]

With the CRC stimulus, we were encouraged to incorporate selected entries from other sources,[7-17] especially Gohl's 1981 compendium[18] devoted largely to tropical forages. Sensing renewed interest in exotic oilseeds as potential energy and feed sources, we further decided to enter many of the analyses provided by a very significant seed screening program[19-21] launched years ago by Q. Jones, F. R. Earle, and A. S. Barclay.

Although we have thousands of entries, covering hundreds of species, our data are still very incomplete; for example, as far as we can determine there was no complete proximate analysis of the groundnut *(Apios americana* L.) — the roots which helped our forefather Pilgrims survive their first American winters, thanks to the teachings of the American Indians. We have only recently gathered enough groundnuts to satisfy requirements for that proximate analysis. Many of our native American Food species, some considered endangered or threatened, have never been subjected to a proximate nutritional analysis.

Similar data gaps prompted Duke to get a proximate analysis of the leaves of coca.[14] These leaves proved to be extremely rich in iron and calcium, more so than any other plant in the Food Composition Table for Latin America, on an "as purchased" basis (APB).

If you find our Nutrition Tables as useful as we have for information on esoteric economic plants, we hope you will send us references or reprints on species not included in our tables. Since our data base is on computer tape, we can augment and edit it. We plan an annual update. With certain rather stringent specifications, we can make copies of the computer tape available to responsible agencies where mutual benefits would accrue.

Readers should be advised that the inclusion of a species in a nutritional table does not necessarily imply that it is edible. As a matter of fact, some of the species are quite poisonous, e.g., *Abrus precatorius*. However, in some developing countries the seeds are consumed after elaborate processing.

In spite of painstaking and tedious work, we fear that some errors will have escaped editorial eyes. Via the computer, corrections are easily made. Users are urged to report any errors they uncover. We will not travel these trails (and trials) again very often.

Since one of the major functions of this publication is to compare nutritional chemistry of as many plant species as possible, the reader is referred to Table 1. In Table 1, rather conventional food species from Watt and Merrill's Composition of Foods[4] are grouped into ten arbitrary categories, ranging from the wettest, lowest-calorie category to the driest, highest-calorie category. Five or ten elements from each category are presented so that the reader can compare less conventional foods with these conventional foods. These are averaged, based on the APB moisture content. To enable other comparisons, the ZMB (Zero-Moisture Basis) has been calculated as well, by multiplying each column except water by 100/100-X where X is the percentage of water. This reduces all the entries to the common denominator ZMB (with some distortion since fiber is included twice). Clearly there is gradation from the APB to ZMB. A fresh grape contains *ca.* 81.5% moisture but only 4 ppm Fe, whereas a dried grape or raisin contains (by analysis) less than 20% water and closer to 35 ppm iron. We hear the raisin described as iron-rich, but on a ZMB, our aerial "vegetables" contain 79 ppm iron, our leafy vegetables 253 ppm, our shoots 109, wild greens 222, other fruits 40 ppm (about on par with raisins), underground "vegetables" 55, ''roots'' 49, cereals 38, pulses 72, and nuts 43. Are raisins then really rich in iron? While we have prepared Table 1 manually, and converted our averages to ZMB, our computerized data can be mass converted to ZMB.

Table 2 represents the direct transcription of data from one of each species from the source indicated. Sometimes these represent fresh weight analyses on an APB. APB data are most useful for converting conventional yields (kg/ha) to, e.g., protein yield. One shortcoming in this ''as purchased'' table is that it is difficult to compare protein contents of greens, hay, and seeds, for example, when their moisture contents may vary from 95% to 5% more or less. Yet many prefer to compare on an APB; others prefer the ZMB.

To prepare Table 3, a ZMB table, we instructed the computer to select credible entries [those in which water, plus protein, plus, oil, plus total carbohydrate (-fiber), plus fiber, plus ash equals 100 ± 1]. Then we instructed the computer to multiply all entries (except water) in such analyses by 100/100-X where X is the percentage of water. Incomplete entries that already had zero in the water column were retained as is for the ZMB table. Data within the same column here are probably more closely comparable than in the APB table. ZMB tables, if computerized, allow us to calculate readily how many calories will accompany the RDA of a nutritional element in a given food.

The format for Tables 2 and 3 is as follows: PLNA (species name); Part (plant part); Calories (per 100g); Water (%); Protein (%); Fat (%); Total carbohydrate (%); Fiber (%); Ash (%); Calcium (mg/100 g); Phosphorus (mg/100 g); Iron (mg/100 g); Sodium (mg/100 g); Potassium (mg/100 g); Carotene (μg/100g — Vitamin A units multiplied by 2 to convert to beta-carotene equivalent); Thiamine (mg/100 g); Riboflavin (mg/100 g); Niacin (mg/100 g); Ascorbic acid (mg/100 g). Ref Co = Reference Code (see references). Note that plant parts are coded as follows: A = aril; BU = bulb; C = twig; CA = callus; D = pod; E = epidermis, bark; F = fruit, mature fruit; G = green fruit, green pulp; H = hay, dry; HE = heart; I = inflorescence, flower; IS = immature seed, green seed; J = juice; K = dry fruit; L = leaf, fresh; N = dry leaf; O = oil; P = pulp of fruit, mature pulp; Q = green root; R = root; S = seed, mature seed; SH = shoot with fresh leaves, bud; T = tuber; U = cooked seed; V = dry seed; W = wet hay; X = seedling; Y = stem; YC = Corm; Z = leaf stalk.

THE AUTHORS

James A. Duke was born in Birmingham, Alabama, in 1929. He is a Phi Beta Kappa graduate of the University of North Carolina, where he took his AB (1952) and BS (1955) in Botany. Following a $2^1/_2$ year tour of military duty, which included a mycological assignment at Fort Dietrick, Maryland, he took his PhD in Botany, in 1961, at the University of North Carolina, moving on to postdoctoral activities at Washington University and the Missouri Botanical Garden, in St. Louis, Missouri, where he assumed professorial and curatorial duties, respectively. At the Missouri Botanical Garden, Duke began intensive studies of neotropical ethnobotany, which is his overriding interest to this day. From 1963 to 1965, Duke served with the U.S. Department of Agriculture at Beltsville, Maryland, devoting much of his time to neotropical ecology, especially seedling ecology. In 1965, he joined Battelle Columbus Laboratories, for ecological and ethnological studies in Panama and Colombia, returning to the USDA in 1971 for crop diversification and medicinal plant studies in developing countries. As a collaborator with the Smithsonian Institution, he has lectured and taught there, with emphasis on neotropical ethnobotany and folk medicine. Considered a key figure in the ''Herbal Renaissance'', Duke received the Cutty Sark Science Award in 1981. In 1985, he received the Maharishi University Award and was awarded Honorary Membership in the Oriental Healing Arts Institute. Currently he is an economic botanist with the Germplasm Introduction and Evaluation Laboratory, Agricultural Research Service, U.S. Department of Agriculture, Beltsville, Maryland.

Dr. Duke is an active member in the American Herb Association (Life), American Society of Pharmacognosy, Association for Tropical Biology, Herb Research Foundation, International Association of Plant Taxonomists (Life), International Society for Tropical Ecology, International Weed Science Society (Life), National Association for Professional Bureaucrats, Organization for Tropical Studies (Life), Society for Economic Botany (Life), Southern Appalachian Botanical Club (Life), Tri-State Bluegrass Association (Life), and Weed Science Society of America.

Widely travelled in his ethnobotanical studies, Dr. Duke has more than 100 scientific publications to his credit. His books include *Isthmian Ethnobotanical Dictionary,* Harrod and Company, Baltimore, 1972 (being reissued with illustrations by Scientific Publishers, Jodphur, India); *A Handbook of Legumes of World Economic Importance,* Plenum Press, New York, 1981; *Medicinal Plants of the Bible,* Trado-Medic Books, Buffalo, New York, 1983; with E. S. Ayensu, *Medicinal Plants of China,* 2 volumes, Reference Publications, Algonac, Michigan, 1985; *A Culinary Herbal,* Trado-Medic Books, Buffalo, New York, 1985; and the *CRC Handbook of Medicinal Herbs,* 1985, CRC Press, Boca Raton, Florida. Also in press for 1986: CRC Handbook of Agricultural Energy Potential of Developing Countries, CRC Press, Boca Raton, Florida; *Gilding of the Ginseng,* Reference Publications, Algonac, Michigan; and *Handbook of Northeastern Indian Medicinal Plants,* Quarterman Publications, Lawrence, Massachusetts. In early 1986, Dr. Duke's first LP album was produced, entitled ''Herbalbum'' with herbal country music.

Alan A. Atchley was born in 1946 in Akron, Ohio, but grew up mostly elsewhere in the Midwest, the South, Appalachia, and southern Arizona. He believes this background may have "pre-adapted" him to the study of ecogeography, which tries to discover why plant and animal species occupy only certain areas. The availability of plant food, for example, can have crucial bearing on the habitability of an area.

Dr. Atchley received his bachelor's degree *cum laude* from Harvard College in 1967 and his doctorate from the University of Maryland in 1976. Since then he has worked for the United States Department of Agriculture's Agricultural Research Service in a variety of roles. From 1976 to 1980 he participated in ecological field work in and climatological research on various Central American, Caribbean, and Middle Eastern countries. Since 1980 he has been in charge of the Agricultural, Ecological, and Geographic Information System (AEGIS), now part of the Germplasm Introduction and Evaluation Laboratory. AEGIS aims, in part, to evaluate potential new and replacement crops on the basis of ecological suitability and nutritional value. Dr. Atchley has built upon this facility to investigate a new range of germplasm and crop replacement questions using Artificial Intelligence.

TABLE OF CONTENTS

ACKNOWLEDGMENTS

It was with great relief that we entrusted Michael Woodbridge with transcribing recent handwritten notes and new material to the computer. Kazue Wain transcribed our manual entries from the Food Composition Tables,[1-3] Agriculture Handbook No. 8,[4] and The Wealth of India.[5] Concetta Pirrone proofed the latter entries while Michael proofed his own entries. Thanks to all who have helped. It has been tough and trying. We hope the utility of the tables will be a monument to the monumental efforts of our skilled technicians.

Table 1
PROXIMATE ANALYSES OF CONVENTIONAL PLANT FOODS (PER 100 g)[a]

"VEGETABLE" (aerial)

Food item	H_2O (g)	Cal	Prot (g)	Fat (g)	Total carb (g)	Fiber (g)	Ash (g)	Ca (mg)	P (mg)	Fe (mg)	Na (mg)	K (mg)	β-Car (μg)	Thia (mg)	Rib (mg)	Nia (mg)	Vit C (mg)
Eggplant	92.4	25	1.2	.2	5.6	.9	.6	12	26	.7	2	214	6	.05	.05	.6	5
Okra	88.9	36	2.4	.3	7.6	1.0	.8	92	51	.6	3	249	312	.17	.21	1.0	31
Pepper	93.4	22	1.2	.2	4.8	1.4	.4	9	22	.7	13	213	252	.08	.08	.5	128
Squash	94.0	19	1.1	.1	4.2	.6	.6	28	29	.4	1	202	246	.05	.09	1.0	22
Tomato	93.5	22	1.1	.2	4.7	.5	.5	13	27	.5	3	244	540	.06	.04	.7	23
Average																	
(APB	92.44	25	1.4	.2	5.4	.9	.6	31	31	.6	4	224	271	.08	.09	.8	42
(ZMB)	0	331	18.5	2.6	71.4	11.9	7.9	410	410	7.9	53	2,963	3,585	1.06	1.19	10.6	556

LEAFY VEGETABLE (cultivated)

Food item	H_2O (g)	Cal	Prot (g)	Fat (g)	Total carb (g)	Fiber (g)	Ash (g)	Ca (mg)	P (mg)	Fe (mg)	Na (mg)	K (mg)	β-Car (μg)	Thia (mg)	Rib (mg)	Nia (mg)	Vit C (mg)
Beet greens	90.9	24	2.2	.3	4.6	1.3	2.0	119	40	3.3	130	570	3,660	.10	.22	.4	30
Cabbage	92.4	24	1.3	.2	5.4	.8	.7	49	29	.4	20	233	78	.05	.05	.3	47
Chard	91.1	25	2.4	.3	4.6	.8	1.6	88	39	3.2	147	550	3,900	.06	.17	.5	32
Kale	87.5	38	4.2	.8	6.0	1.3	1.5	179	73	2.2	75	378	5,340	—	—	—	125
Spinach	90.7	26	3.2	.3	4.3	.6	1.5	93	51	3.1	71	470	4,860	.10	.20	.6	51
Average																	
(APB)	90.5	27.4	2.7	.38	5.0	.96	1.46	106	46	2.4	89	440	3,568	.08	.16	.45	57
(ZMB)		289	28.4	4.0	54.7	10.5	15.8	1,116	484	25.3	937	4,633	37,571	.84	1.68	4.74	600

Table 1 (continued)
PROXIMATE ANALYSES OF CONVENTIONAL PLANT FOODS (PER 100 g)[a]

Food item	H_2O (g)	Cal	Prot (g)	Fat (g)	Total carb (g)	Fiber (g)	Ash (g)	Ca (mg)	P (mg)	Fe (mg)	Na (mg)	K (mg)	β-Car (μg)	Thia (mg)	Rib (mg)	Nia (mg)	Vit C (mg)
SHOOTS																	
Asparagus	91.7	26	2.5	.2	5.0	.7	.6	22	62	1.0	2	278	540	.18	.20	1.5	33
Bamboo	91.0	27	2.6	.3	5.2	.7	.9	13	59	.5	—	533	12	.15	.07	.6	4
Mungbean	88.8	35	3.8	.2	6.6	.7	.6	19	64	1.3	5	223	12	.13	.13	.8	19
Pokeweed	91.6	23	2.6	.4	3.7	—	1.7	53	44	1.7	—	—	5,220	.08	.33	1.2	136
Soybean	86.3	46	6.2	1.4	5.3	.8	.8	48	67	1.0	—	—	48	.23	.20	.8	13
Average																	
(APB)	89.9	31	3.5	.5	5.2	.6	.9	31	59	1.1	4	345	1,166	.15	.19	1.0	41
(ZMB)	0	307	34.6	5.0	51.5	5.9	8.9	307	584	10.9	40	3,416	11,543	1.48	1.88	9.9	405
GREENS (wild)																	
Amaranth	86.9	36	3.5	.5	6.5	1.3	2.6	267	67	3.9	—	411	3,660	.08	.16	1.4	80
Dandelion	85.6	45	2.7	.7	9.2	1.6	1.8	187	66	3.1	76	397	8,400	.19	.26	—	35
Dock	90.9	28	2.1	.3	5.6	.8	1.1	66	41	1.6	5	338	7,740	.09	.22	.5	119
Lambsquarters	84.3	43	4.2	.8	7.3	2.1	3.4	309	72	1.2	—	—	6,960	.16	.44	1.2	80
Purslane	92.5	21	1.7	.4	3.8	.9	1.6	103	39	3.5	—	—	1,500	.03	.10	.5	25
Average																	
(APB)	88	35	2.84	.54	6.48	1.34	2.1	186.4	57	2.66	40.5	382	5,652	.11	.24	.9	68
(ZMB)		292	23.7	4.5	54.0	11.2	17.5	1,553	475	22.2	337	3,183	47,433	.9	2.0	7.5	567

Table 1 (continued)
PROXIMATE ANALYSES OF CONVENTIONAL PLANT FOODS (PER 100 g)[a]

Food item	H_2O (g)	Cal	Prot (g)	Fat (g)	Total carb (g)	Fiber (g)	Ash (g)	Ca (mg)	P (mg)	Fe (mg)	Na (mg)	K (mg)	β-Car (μg)	Thia (mg)	Rib (mg)	Nia (mg)	Vit C (mg)
FRUIT																	
Apples	84.4	58	.2	.6	14.5	1.0	.3	7	10	.3	1	110	54	.03	.02	.1	4
Apricots	85.3	51	1.0	.2	12.8	.6	.7	17	23	.5	1	281	1,620	.03	.04	.6	10
Bananas	75.7	85	1.1	.2	22.2	.5	.8	8	26	.7	1	370	114	.05	.06	.7	10
Blackberries	84.5	58	1.2	.9	12.9	4.1	.5	32	19	.9	1	170	120	.03	.04	.4	21
Blueberries	83.2	62	.7	.5	15.3	1.5	.3	15	13	1.0	1	81	60	.03	.06	.5	14
Cherries	83.7	58	1.2	.3	14.3	.2	.5	22	19	.4	2	191	600	.05	.06	.4	10
Cranberries	87.9	46	.4	.7	0.8	1.4	.2	14	10	.5	2	82	24	.03	.02	.1	11
Grapefruit	88.4	41	.5	.1	10.6	.2	.4	16	16	.4	1	135	48	.04	.02	.2	38
Orange	86.0	49	1.0	.2	12.2	.5	.6	41	20	.4	1	200	120	.10	.04	.4	50
Strawberries	89.9	37	.7	.5	8.4	1.3	.5	21	21	1.0	1	164	36	.03	.07	.6	59
Average																	
(APB)	84.9	54.5	.8	.4	13.4	1.1	.5	19.3	17.7	.61	1.2	178.4	280	.042	.043	.4	22.7
(ZMB)	0	361	5.2	2.6	88.7	7.3	3.3	128	117	4.0	7.9	1,181	1,854	.28	.28	2.65	150
"VEGETABLE" (underground)																	
Beets	87.3	43	1.6	.1	9.9	.8	1.1	16	3 3	.7	60	335	12	.03	.05	.4	10
Carrots	88.2	42	1.1	.2	9.7	1.0	.8	37	36	.7	47	341	6,600	.06	.05	.6	8
Garlic	61.3	137	6.2	.2	30.8	1.5	1.5	29	202	1.5	19	529	Trace	.25	.08	.5	15
Onions	89.1	38	1.5	.1	8.7	.6	.6	27	36	.5	10	157	24	.03	.04	.2	10
Radishes	94.5	17	1.0	.1	3.6	.7	.8	30	31	1.0	18	322	6	.03	.03	.3	26
Average																	
(APB)	84.1	55	2.3	.16	12.5	.9	.96	28	68	.88	31	337	1,328	.08	.05	.4	14
(ZMB)	0	346	14.5	1.0	78.6	5.7	6.0	176	428	5.5	194	2,119	8,352	.50	.31	2.5	88

Table 1 (continued)
PROXIMATE ANALYSES OF CONVENTIONAL PLANT FOODS (PER 100 g)[a]

Food item	H_2O (g)	Cal	Prot (g)	Fat (g)	Total carb (g)	Fiber (g)	Ash (g)	Ca (mg)	P (mg)	Fe (mg)	Na (mg)	K (mg)	β-Car (μg)	Thia (mg)	Rib (mg)	Nia (mg)	Vit C (mg)
ROOTS																	
Jerusalem-artichoke	79.8	—	2.3	.1	16.7	.8	1.1	14	78	3.4	—	—	12	.20	.06	1.3	4
Parsnip	79.1	76	1.7	.5	17.5	2.0	1.2	50	77	.7	12	541	18	.08	.09	.2	16
Potato	79.8	76	2.1	.1	17.1	.5	.9	7	53	.6	3	407	Trace	.10	.04	1.5	20
Rutabaga	87.0	46	1.1	.1	11.0	1.1	.8	66	39	.4	5	239	348	.07	.07	1.1	43
Salsify	77.6	—	2.9	.6	18.0	1.8	.9	47	66	1.5	—	380	6	.04	.04	.3	11
Sweetpotato	70.6	114	1.7	.4	26.3	.7	1.0	32	47	.7	10	243	5,280	.10	.06	.6	21
Taro	73.0	98	1.9	.2	23.7	.8	1.2	28	61	1.0	7	514	12	.13	.04	1.1	4
Turnip	91.5	30	1.0	.2	6.6	.9	.7	39	30	.5	49	268	Trace	.04	.07	.6	36
Yam	73.5	101	2.1	.2	23.2	.9	1.0	20	69	.6	—	600	Trace	.10	.04	.5	9
Yambean	85.1	55	1.4	.2	12.8	.7	.5	15	18	.6	—	—	Trace	.04	.03	.3	20
Average																	
(APB)	79.7	74	1.8	.3	17.3	1.0	.9	32	54	1.0	14	399	568	.09	.05	.8	18
(ZMB)	0	365	8.9	1.5	85.2	4.9	4.4	158	266	4.9	69	1,966	2,798	.44	.25	3.9	89
CEREALS																	
Barley	11.1	349	8.2	1.0	78.8	.5	.9	16	189	2.0	3	160	0	.12	.05	3.1	0
Buckwheat	11.0	335	11.7	2.4	72.9	9.9	2.0	114	282	3.1	—	448	0	.60	—	4.4	0
Millet	11.8	327	9.9	2.9	72.9	3.2	2.5	20	311	6.8	—	430	0	.73	.38	2.3	0
Popcorn	9.8	362	11.9	4.7	72.1	2.1	1.5	10	264	2.5	3	—	—	.39	.11	2.1	0
Rice	12.0	360	7.5	1.9	77.4	.9	1.2	32	221	1.6	9	214	0	.34	.05	4.7	0
Rye	11.	334	12.1	1.7	73.4	2.0	1.8	38	376	3.7	1	467	0	.43	.22	1.6	0
Sorghum	11.	332	11.0	3.3	73.0	1.7	1.7	28	287	4.4	—	350	0	.38	.15	3.9	0
Wheat, red	13.0	330	14.0	2.2	69.1	2.3	1.7	36	383	3.1	3	370	0	.57	.12	4.3	0
Wheat, white	11.5	335	9.4	2.0	75.4	1.9	1.7	36	394	3.0	3	390	0	.53	.12	5.3	0
Wildrice	8.5	353	14.1	.7	75.3	1.0	1.4	19	339	4.2	7	220	0	.45	.63	6.2	0
Average																	
(APB)	11.1	342	11.0	2.3	74.0	2.6	1.6	35	305	3.4	4	339	0	.45	.20	3.8	0
(ZMB)	0	385	12.4	2.6	83.2	2.9	1.8	39	343	3.8	4	381	0	.51	.22	4.3	0

Table 1 (continued)
PROXIMATE ANALYSES OF CONVENTIONAL PLANT FOODS (PER 100 g)[a]

Food item	H_2O (g)	Cal	Prot (g)	Fat (g)	Total carb (g)	Fiber (g)	Ash (g)	Ca (mg)	P (mg)	Fe (mg)	Na (mg)	K (mg)	β-Car (μg)	Thia (mg)	Rib (mg)	Nia (mg)	Vit C (mg)
PULSES (dry)																	
Bean (lima)	10.3	345	20.4	1.6	64.0	4.3	3.7	72	385	7.8	4	1,529	Trace	.48	.17	1.9	—
Bean (mung)	10.7	340	24.2	1.3	60.3	4.4	3.5	118	340	7.7	6	1,028	48	.38	.21	2.6	—
Bean (red)	10.4	343	22.5	1.5	61.9	4.2	3.7	110	406	6.9	10	984	12	.51	.20	2.3	—
Broadbean	11.9	338	25.1	1.7	58.2	6.7	3.1	102	391	7.1	—	—	42	.50	.30	2.5	—
Chickpea	10.7	360	20.5	4.8	61.0	5.0	3.0	150	331	6.9	26	797	30	.31	.15	2.0	—
Cowpea	10.5	343	22.8	1.5	61.7	4.4	3.5	74	426	5.8	35	1,024	18	1.05	.21	2.2	—
Lentil	11.1	340	24.7	1.1	60.1	3.9	3.0	79	377	6.8	30	790	36	.37	.22	2.0	—
Peanut	5.6	564	26.0	47.5	18.6	2.4	2.3	69	401	2.1	5	674	—	1.14	.13	17.2	0
Pea	11.7	340	24.1	1.3	60.3	4.9	2.6	64	340	5.1	35	1,005	72	.74	.29	3.0	—
Soybean	10.0	403	34.1	17.7	33.5	4.9	4.7	226	554	8.4	5	1,677	48	1.10	.31	2.2	—
Average																	
(APB)	10.3	372	24.4	8.0	54.0	4.5	3.3	106	395	6.5	17	1,056	34	.66	.22	3.8	0
(ZMB)	0	415	27.2	8.9	60.2	5.0	3.7	118	440	7.2	19	1.177	37	.73	.25	4.2	0

Table 1 (continued)
PROXIMATE ANALYSES OF CONVENTIONAL PLANT FOODS (PER 100 g)[a]

Food item	H_2O (g)	Cal	Prot (g)	Fat (g)	Total carb (g)	Fiber (g)	Ash (g)	Ca (mg)	P (mg)	Fe (mg)	Na (mg)	K (mg)	β-Car (μg)	Thia (mg)	Rib (mg)	Nia (mg)	Vit C (mg)
NUTS																	
Almond	4.7	598	18.6	54.2	19.5	2.6	3.0	234	504	4.7	4	773	0	.24	.92	3.5	Trace
Brazilnut	4.6	654	14.3	66.9	10.9	3.1	3.3	186	693	3.4	1	715	Trace	.96	.12	1.6	—
Cashew	5.2	561	17.2	45.7	29.3	1.4	2.6	38	373	3.8	15	464	60	.43	.25	1.8	—
Hickory	3.3	673	13.2	68.7	12.8	1.9	2.0	Trace	360	2.4	—	—	—	—	—	—	—
Macadamia	3.0	691	7.8	71.6	15.9	2.5	1.7	48	161	2.0	—	264	0	.34	.11	1.3	0
Pecan	3.4	687	9.2	71.2	14.6	2.3	1.6	73	289	2.4	Trace	603	78	.86	.13	.9	2
Pilinut	6.3	669	11.4	71.1	8.4	2.7	2.8	140	554	3.4	3	489	24	.88	.09	.5	Trace
Pinon	3.1	635	13.0	60.5	20.5	1.1	2.9	12	604	5.2	—	—	18	1.28	.23	4.5	Trace
Pistachio	5.3	594	19.3	53.7	19.0	1.9	2.7	131	500	7.3	—	972	138	.67	—	1.4	0
Walnut	3.1	628	20.5	59.3	14.8	1.7	2.3	Trace	570	6.0	3	460	180	.22	.11	.7	—
Average																	
(APB)	4.2	639	14.4	62.3	16.6	1.7	2.5	86	461	4.1	4	592	55	.65	.24	.18	Trace
(ZMB)	0	667	15.0	65.0	17.3	1.8	2.6	90	481	4.3	4	618	57	.68	.25	1.9	Trace

[a] See introduction.

TABLE 2 - DATA DIRECTLY TRANSCRIBED FROM ORIGINAL SOURCES.

OBS	PLNA	PART	CALORIES	WATER	PROTEIN	FAT	TOTCARB	FIBER	ASH	CALCIUM	PHOSPHON	IRON	SODIUM	POTASS	CAROT	THIAMINE	RIBOFLAV	NIACIN	ASCORBIC	REFCO
1			.	.	.	.	.	.	.	.	.	.	.	.	.	.	.	.	.	.
2	ABELMOSCHUS ESCULENTUS	F	36	88.9	2.4	0.3	7.6	1.0	0.8	92.0	51	0.6	3	249	312	0.17	0.21	1	31	4
3	ABELMOSCHUS ESCULENTUS	S	.	9.9	19.4	16.8	26.0	21.0	4.7	210.0	790	.	.	820	.	.	.	.	.	5
4	ABRUS PRECATORIUS	L	62	79.0	5.6	.	13.7	3.7	1.7	266.0	65	.	.	.	.	.	.	.	.	1
5	ABRUS PRECATORIUS	L	295	0.0	26.7	.	65.2	17.6	8.1	126.6	309	.	.	.	.	.	.	.	.	10
6	ABRUS PRECATORIUS	S	.	0.0	19.4	2.8	.	.	.	.	.	.	.	.	.	.	.	.	.	20
7	ABUTILON AMERICANUM	S	.	0.0	23.1	20.2	.	.	.	.	.	.	.	.	.	.	.	.	.	20
8	ABUTILON BIDENTATUM	S	.	0.0	20.6	14.6	.	.	.	.	.	.	.	.	.	.	.	.	.	21
9	ABUTILON INCANUM	S	.	0.0	24.4	14.7	.	.	.	.	.	.	.	.	.	.	.	.	.	19
10	ABUTILON INCANUM	L	.	0.0	19.6	1.8	68.7	22.5	9.9	2320.0	280	.	.	2444	.	.	.	.	.	6
11	ABUTILON MOLLE	S	.	0.0	23.8	14.7	.	.	.	.	.	.	.	.	.	.	.	.	.	21
12	ABUTILON THEOPHRASTI	S	.	0.0	23.1	15.3	.	.	4.9	.	.	.	.	.	.	.	.	.	.	19
13	ABUTILON THEOPHRASTI	S	.	0.0	20.9	17.7	.	.	.	.	.	.	.	.	.	.	.	.	.	21
14	ABUTILON TRISULCATUM	S	.	0.0	16.8	12.7	.	.	.	.	.	.	.	.	.	.	.	.	.	21
15	ACACIA ALBIDA	L	.	0.0	17.1	2.3	72.2	12.4	8.4	.	.	.	.	.	.	.	.	.	.	18
16	ACACIA ALBIDA	F	.	0.0	14.3	1.5	77.9	24.7	6.3	1110.0	140	.	.	.	.	.	.	.	.	18
17	ACACIA ALBIDA	L	.	0.0	19.7	1.6	71.5	19.6	7.2	1000.0	230	.	.	.	.	.	.	.	.	18
18	ACACIA ANEURA	SH	.	0.0	9.2	4.6	80.6	30.1	5.6	870.0	70	.	.	.	.	.	.	.	.	18
19	ACACIA ANGUSTISSIMA	S	.	0.0	47.5	4.4	.	.	.	.	.	.	.	.	.	.	.	.	.	20
20	ACACIA ANGUSTISSIMA	L	.	0.0	26.6	3.8	64.7	16.9	4.9	970.0	290	.	.	1280	.	.	.	.	.	6
21	ACACIA AURICULAEFORMIS	S	.	0.0	33.6	7.4	.	.	.	.	.	.	.	.	.	.	.	.	.	21
22	ACACIA BAILEYANA	S	.	0.0	18.8	13.6	.	.	.	.	.	.	.	.	.	.	.	.	.	20
23	ACACIA BREVISPICA	L	.	0.0	18.8	5.7	69.0	18.0	6.5	1860.0	310	.	.	.	.	.	.	.	.	18
24	ACACIA BREVISPICA	F	.	0.0	17.5	1.1	76.8	24.0	4.6	460.0	270	.	.	.	.	.	.	.	.	18
25	ACACIA CATECHU	L	.	0.0	13.0	4.6	72.6	22.6	9.8	2740.0	170	.	.	.	.	.	.	.	.	18
26	ACACIA CAVEN	S	.	0.0	24.5	4.7	.	.	.	.	.	.	.	.	.	.	.	.	.	21
27	ACACIA CONSTRICTA	S	.	0.0	40.6	9.5	.	.	3.9	.	.	.	.	.	.	.	.	.	.	19
28	ACACIA CONSTRICTA	S	.	0.0	41.9	11.0	.	.	.	.	.	.	.	.	.	.	.	.	.	20
29	ACACIA CONSTRICTA VAR. PAUCISPINA	S	.	0.0	31.9	12.6	.	.	.	.	.	.	.	.	.	.	.	.	.	19
30	ACACIA CYANOPHYLLA	S	.	0.0	28.1	13.6	.	.	.	.	.	.	.	.	.	.	.	.	.	20
31	ACACIA CYANOPHYLLA	S	.	0.0	26.8	12.8	.	.	.	.	.	.	.	.	.	.	.	.	.	21
32	ACACIA DECURRENS	S	.	0.0	21.2	12.3	.	.	3.8	.	.	.	.	.	.	.	.	.	.	19
33	ACACIA ELATA	S	.	0.0	25.0	9.0	.	.	.	.	.	.	.	.	.	.	.	.	.	20
34	ACACIA FARNESIANA	SH	.	0.0	17.2	2.1	75.0	17.9	5.2	.	.	.	.	.	.	.	.	.	.	18
35	ACACIA FARNESIANA	F	.	0.0	17.2	1.6	77.3	19.4	4.1	550.0	200	.	.	.	.	.	.	.	.	18
36	ACACIA FARNESIANA	S	.	0.0	20.9	2.3	73.0	18.3	3.8	390.0	270	.	.	.	.	.	.	.	.	18
37	ACACIA FARNESIANA	S	.	0.0	55.0	7.8	.	.	4.3	.	.	.	.	.	.	.	.	.	.	19
38	ACACIA FARNESIANA	S	.	0.0	4.0	3.5	.	.	.	.	.	.	.	.	.	.	.	.	.	20
39	ACACIA FARNESIANA	S	.	0.0	22.4	3.0	.	.	.	.	.	.	.	.	.	.	.	.	.	21
40	ACACIA GALPINII	F	.	0.0	12.7	5.4	78.1	26.3	3.8	.	.	.	.	.	.	.	.	.	.	18
41	ACACIA GERRARDII	F	.	0.0	20.9	1.3	72.7	27.4	5.1	.	.	.	.	.	.	.	.	.	.	18
42	ACACIA GERRARDII	SH	.	0.0	17.7	1.6	74.7	24.2	6.0	630.0	250	.	.	.	.	.	.	.	.	18
43	ACACIA GIRAFFAE	F	.	0.0	13.8	1.7	80.9	32.9	3.6	710.0	110	.	.	.	.	.	.	.	.	18
44	ACACIA GIRAFFAE	S	.	0.0	26.3	4.7	.	.	.	.	.	.	.	.	.	.	.	.	.	21
45	ACACIA GREGGII	S	.	0.0	22.0	11.6	.	.	2.0	.	.	.	.	.	.	.	.	.	.	19

TABLE 2 - DATA DIRECTLY TRANSCRIBED FROM ORIGINAL SOURCES.

OBS	PLNA	PART	CALORIES	WATER	PROTEIN	FAT	TOTLCARB	FIBER	ASH	CALCIUM	PHOSPHO	IRON	SODIUM	POTASS	CAROTE	THIAMINE	RIBOFLAV	NIACIN	ASCORBIC	REF
46	ACACIA GREGGII	S	.	0	13.2	2.2	.	.	3.5	.	.	.	.	.	.	.	.	.	.	19
47	ACACIA GREGGII	S	.	0	14.4	9.2	.	.	.	.	.	.	.	.	.	.	.	.	.	19
48	ACACIA GREGGII	S	.	0	17.5	8.8	.	.	.	.	.	.	.	.	.	.	.	.	.	20
49	ACACIA GREGGII	S	.	0	18.1	8.4	.	.	.	.	.	.	.	.	.	.	.	.	.	20
50	ACACIA LAETA	L	.	0	26.4	3.0	61.4	18.9	9.2	.	.	.	.	.	.	.	.	.	.	18
51	ACACIA LINEARIS	S	.	0	23.1	9.4	.	.	.	.	.	.	.	.	.	.	.	.	.	20
52	ACACIA LONGIFOLIA	S	.	0	26.9	13.2	.	.	.	.	.	.	.	.	.	.	.	.	.	20
53	ACACIA MELANOXYLON	S	.	0	39.4	7.6	.	.	3.9	.	.	.	.	.	.	.	.	.	.	19
54	ACACIA MELLIFERA	L	.	0	42.8	2.6	48.4	16.2	6.2	510	580	.	.	.	.	.	.	.	.	18
55	ACACIA MELLIFERA	L	.	0	21.3	2.6	67.5	14.1	8.6	1140	180	.	.	.	.	.	.	.	.	18
56	ACACIA MILLEFOLIA	S	.	0	38.8	13.1	.	.	.	.	.	.	.	.	.	.	.	.	.	19
57	ACACIA MODESTA	S	.	0	38.9	8.4	.	.	.	.	.	.	.	.	.	.	.	.	.	21
58	ACACIA NILOTICA	L	.	0	12.9	12.6	68.1	11.3	6.4	1090	280	.	.	.	.	.	.	.	.	18
59	ACACIA NILOTICA	S	.	0	18.2	4.2	.	.	.	.	.	.	.	.	.	.	.	.	.	21
60	ACACIA NILOTICA	S	.	0	23.4	10.6	.	.	.	.	.	.	.	.	.	.	.	.	.	21
61	ACACIA NILOTICA SUBSP. SUBALATA	S	.	0	13.6	5.0	77.6	10.2	3.8	.	.	.	.	.	.	.	.	.	.	18
62	ACACIA NUBICA	L	.	0	32.4	1.7	57.1	15.8	8.8	1710	440	.	.	.	.	.	.	.	.	18
63	ACACIA NUBICA	F	.	0	15.2	1.9	75.4	37.4	7.5	1460	230	.	.	.	.	.	.	.	.	18
64	ACACIA OCCIDENTALIS	S	.	0	18.9	7.4	.	.	2.9	.	.	.	.	.	.	.	.	.	.	19
65	ACACIA OCCIDENTALIS	S	.	0	22.5	8.6	.	.	.	.	.	.	.	.	.	.	.	.	.	19
66	ACACIA PENNATULA	S	.	0	24.4	3.8	.	.	.	.	.	.	.	.	.	.	.	.	.	20
67	ACACIA PYCNANTHA	S	.	0	26.9	13.1	.	.	3.4	.	.	.	.	.	.	.	.	.	.	19
68	ACACIA RADDIANA	L	.	0	17.9	4.7	69.7	17.2	7.7	1550	210	.	.	.	.	.	.	.	.	18
69	ACACIA RADDIANA	C	.	.	16.2	2.4	75.4	28.2	6.0	1020	190	.	.	.	.	.	.	.	.	18
70	ACACIA RADDIANA	F	.	0	17.5	1.8	74.5	20.4	6.2	900	260	.	.	.	.	.	.	.	.	18
71	ACACIA ROEMERIANA	L	.	0	20.5	3.6	70.6	20.2	5.3	1510	200	.	.	1050	.	.	.	.	.	6
72	ACACIA SALINGA	S	.	0	26.9	12.7	.	.	4.8	.	.	.	.	.	.	.	.	.	.	19
73	ACACIA SCHAFFNERI	S	.	0	27.5	3.5	.	.	.	.	.	.	.	.	.	.	.	.	.	20
74	ACACIA SENEGAL	L	.	0	18.2	6.7	66.7	11.2	8.4	1320	140	.	.	.	.	.	.	.	.	18
75	ACACIA SENEGAL	F	.	0	22.0	1.0	69.9	39.0	7.1	.	.	.	.	.	.	.	.	.	.	18
76	ACACIA SENEGAL	S	.	0	40.1	5.8	49.1	13.7	5.0	750	310	.	.	.	.	.	.	.	.	18
77	ACACIA SEYAL	L	.	0	22.6	4.0	64.4	8.4	9.0	3830	210	.	.	.	.	.	.	.	.	18
78	ACACIA SEYAL	F	.	0	20.9	1.9	67.9	20.2	9.3	1330	370	.	.	.	.	.	.	.	.	18
79	ACACIA SEYAL	E	.	0	10.6	0.9	79.7	22.2	8.8	2500	70	.	.	.	.	.	.	.	.	18
80	ACACIA SIEBERIANA	F	.	0	8.4	1.9	83.0	25.1	6.7	100	140	.	.	.	.	.	.	.	.	18
81	ACACIA SIEBERIANA	L	.	0	15.8	6.3	66.3	23.6	11.6	2500	100	.	.	.	.	.	.	.	.	18
82	ACACIA SP.	L	307	0	43.0	3.2	48.4	30.1	5.4	500	452	19.9	.	.	.	1.08	0.91	45.73	264	10
83	ACACIA SP.	S	.	0	44.8	4.7	.	.	4.4	.	.	.	.	.	.	.	.	.	.	19
84	ACACIA SUAVEOLENS	S	.	0	22.5	12.4	.	.	.	.	.	.	.	.	.	.	.	.	.	20
85	ACACIA TORTILIS	L	.	0	19.2	6.1	66.0	11.6	8.7	2270	170	.	.	.	.	.	.	.	.	18
86	ACACIA TORTILIS	F	.	0	17.3	3.1	73.9	24.8	5.7	790	340	.	.	.	.	.	.	.	.	18
87	ACACIA TORTILIS	S	.	0	37.8	6.0	50.6	10.9	5.9	560	730	.	.	.	.	.	.	.	.	18
88	ACACIA WILLARDIANA	S	.	0	35.2	21.3	.	.	4.9	.	.	.	.	.	.	.	.	.	.	19
89	ACALYPHA FRUTICOSA	SH	.	0	14.1	3.8	69.3	18.1	12.8	.	.	.	.	.	.	.	.	.	.	18
90	ACANTHOPANAX SPINOSUM	S	.	0	30.6	36.6	.	.	3.5	.	.	.	.	.	.	.	.	.	.	19

TABLE 2 - DATA DIRECTLY TRANSCRIBED FROM ORIGINAL SOURCES.

OBS	PLNA	PART	CALORIES	WATER	PROTEIN	FAT	TOTLCARB	FIBER	ASH	CALCIUM	PHOSPHO	IRON	SODIUM	POTASS	CAROT	THIAMINE	RIBOFLAV	NIACIN	ASCORBIC	REFCO
91	ACANTHOPANAX SPINOSUM	S	.	0.0	5.6	5.6	.	.	2.1	.	.	.	.	.	.	.	.	.	.	19
92	ACANTHUS HIRSUTUS	S	.	0.0	22.1	2.5	.	.	.	.	.	.	.	.	.	.	.	.	.	20
93	ACANTHUS MOLLIS	W	.	0.0	25.3	2.3	56.7	11.2	15.7	830	340	.	.	.	.	.	.	.	.	18
94	ACANTHUS MOLLIS	S	.	0.0	26.1	1.4	69.7	2.8	2.8	.	.	.	.	.	.	.	.	.	.	18
95	ACANTHUS MOLLIS	S	.	0.0	27.5	1.2	.	.	.	.	.	.	.	.	.	.	.	.	.	20
96	ACANTHUS SYRIACUS	S	.	0.0	22.4	2.2	.	.	.	.	.	.	.	.	.	.	.	.	.	21
97	ACANTHUS SYRIACUS	S	.	0.0	24.9	2.4	.	.	.	.	.	.	.	.	.	.	.	.	.	21
98	ACER ARGUTUM	S	.	0.0	7.2	2.3	.	.	.	.	.	.	.	.	.	.	.	.	.	21
99	ACER BUERGERIANUM	S	.	0.0	15.6	9.5	.	.	.	.	.	.	.	.	.	.	.	.	.	20
100	ACER CAMPESTRE	S	.	0.0	7.2	1.3	.	.	.	.	.	.	.	.	.	.	.	.	.	21
101	ACER CISSIFOLIUM	S	.	0.0	23.8	17.4	.	.	.	.	.	.	.	.	.	.	.	.	.	20
102	ACER GLABRUM	S	.	0.0	8.3	5.8	.	.	.	.	.	.	.	.	.	.	.	.	.	21
103	ACER JAPONICUM	S	.	0.0	12.9	10.6	.	.	.	.	.	.	.	.	.	.	.	.	.	21
104	ACER NEGUNDO	S	.	0.0	48.8	28.2	.	.	.	.	.	.	.	.	.	.	.	.	.	20
105	ACER PALMATUM	S	.	0.0	12.9	5.3	.	.	.	.	.	.	.	.	.	.	.	.	.	21
106	ACER RUBRUM	S	.	0.0	31.8	3.9	.	.	4.3	.	.	.	.	.	.	.	.	.	.	19
107	ACER RUBRUM	L	.	0.0	.	.	.	.	.	850	220	29.0	.	.	.	.	.	.	.	6
108	ACER TATARICUM	S	.	0.0	17.2	8.9	.	.	.	.	.	.	.	.	.	.	.	.	.	21
109	ACER TRUNCATUM	S	.	0.0	17.5	33.9	.	.	.	.	.	.	.	.	.	.	.	.	.	20
110	ACER TRUNCATUM	S	.	0.0	15.6	36.6	.	.	.	.	.	.	.	.	.	.	.	.	.	20
111	ACHILLEA AEGYPTIACA	S	.	0.0	33.1	30.4	.	.	.	.	.	.	.	.	.	.	.	.	.	21
112	ACHILLEA FILIPENDULINA	S	.	0.0	33.8	27.6	.	.	.	.	.	.	.	.	.	.	.	.	.	20
113	ACHILLEA LANULOSA	W	.	0.0	12.8	3.9	71.8	23.6	10.4	1330	360	.	.	.	.	.	.	.	.	6
114	ACHILLEA MILLEFOLIUM	S	.	0.0	33.8	25.4	.	.	.	.	.	.	.	.	.	.	.	.	.	19
115	ACHILLEA MILLEFOLIUM	S	.	0.0	36.9	22.3	.	.	.	.	.	.	.	.	.	.	.	.	.	20
116	ACHILLEA MILLEFOLIUM	S	.	0.0	28.6	33.4	.	.	.	.	.	.	.	.	.	.	.	.	.	21
117	ACHILLEA MILLEFOLIUM	W	.	0.0	14.4	1.8	71.3	20.1	12.5	.	.	.	.	.	.	.	.	.	.	6
118	ACHILLEA SANTOLINA	S	.	0.0	31.9	24.6	.	.	.	.	.	.	.	.	.	.	.	.	.	21
119	ACHILLEA SANTOLINA	S	.	0.0	40.9	27.5	.	.	.	.	.	.	.	.	.	.	.	.	.	21
120	ACHRAS ZAPOTA	S	.	0.0	9.6	22.5	.	.	4.7	.	.	.	.	.	.	.	.	.	.	19
121	ACHRAS ZAPOTA	S	.	0.0	1.4	0.7	.	.	0.3	.	.	.	.	.	.	.	.	.	.	19
122	ACHRYANTHES ASPERA	S	.	0.0	20.5	3.5	.	.	.	.	.	.	.	.	.	.	.	.	.	21
123	ACHYRANTHES ASPERA	W	.	0.0	26.4	1.4	53.6	20.2	18.6	1790	310	.	.	.	.	.	.	.	.	18
124	ACHYRANTHES ASPERA	I	46	85.0	6.4	1.0	6.2	0.8	1.4	158	59	0.0	.	.	6555	0.13	0.48	1.0	30	2
125	ACLEISANTHES SP.	L	.	0.0	24.4	2.0	62.4	24.7	11.2	2650	150	.	.	3370	.	.	.	.	.	6
126	ACNIDA CANNABINA	S	.	0.0	16.9	12.7	.	.	.	.	.	.	.	.	.	.	.	.	.	19
127	ACONITUM NAPELLUS	S	.	0.0	28.1	37.4	.	.	.	.	.	.	.	.	.	.	.	.	.	20
128	ACROCERAS AMPLECTENS	F	332	10.4	6.5	5.3	70.0	.	7.8	10	.	4.0	.	.	.	0.12	0.23	3.5	.	1
129	ACROCERAS MACRUM	H	.	0.0	8.5	1.8	83.6	31.8	6.1	.	.	.	.	.	.	.	.	.	.	18
130	ACROCERAS MACRUM	L	.	0.0	21.3	3.8	66.5	30.0	8.4	.	.	.	.	.	.	.	.	.	.	18
131	ACROCLINIUM ROSEUM	S	.	0.0	37.5	25.9	.	.	.	.	.	.	.	.	.	.	.	.	.	19
132	ACROCOMIA MEXICANA	SH	39	87.6	2.4	0.4	8.4	0.7	1.2	157	59	1.0	.	.	0	0.04	0.08	1.3	20	3
133	ACROCOMIA MEXICANA	F	231	51.8	4.4	13.8	27.9	13.4	2.1	199	57	0.2	.	.	140	0.14	0.09	1.0	28	3
134	ACTAEA RUBRA	S	.	0.0	14.4	26.0	.	.	.	.	.	.	.	.	.	.	.	.	.	20
135	ACTINIDIA ARGUTA	S	.	0.0	16.2	22.4	.	.	3.0	.	.	.	.	.	.	.	.	.	.	19

TABLE 2 - DATA DIRECTLY TRANSCRIBED FROM ORIGINAL SOURCES.

OBS	PLANA	PART	CALORIES	WATER	PROTEIN	FAT	TOTCARB	FIBER	ASH	CALCIUM	PHOSPHO	IRON	SODIUM	POTASS	CAROT	THIAMINE	RIBOFLAV	NIACIN	ASCORBIC	REFCO
136	ACTINIDIA CHINENSIS	S	.	0.0	15.0	34.1	.	.	.	.	.	.	.	.	.	.	.	.	.	19
137	ACTINOLEMA MACROLEMA	S	.	0.0	21.3	48.4	.	.	.	.	.	.	.	.	.	.	.	.	.	21
138	ACTINOMERIS ALTERNIFOLIA	S	.	0.0	43.1	38.8	.	.	.	.	.	.	.	.	.	.	.	.	.	19
139	ACTINOPHLOEUS MACARTHURII	S	.	0.0	5.9	1.6	.	.	1.4	.	.	.	.	.	.	.	.	.	.	19
140	ADANSONIA DIGITATA	P	290	16.0	2.2	0.8	76.7	6.8	4.3	284	118	7.4	.	.	70	0.37	0.06	2.1	270	1
141	ADANSONIA DIGITATA	L	282	11.8	12.3	3.1	63.2	9.7	9.6	2241	275	24.0	.	.	9710	0.13	0.82	4.4	0	1
142	ADANSONIA DIGITATA	S	452	7.8	30.0	29.6	24.7	3.2	7.9	263	1494	13.9	.	.	0	1.38	0.13	1.4	.	1
143	ADENANTHERA MICROSPERMA	S	.	0.0	22.5	14.3	.	.	.	.	.	.	.	.	.	.	.	.	.	19
144	ADENANTHERA PAVONINA	SH	.	0.0	24.8	3.5	65.9	14.0	5.8	190	240	.	.	.	.	.	.	.	.	18
145	ADENANTHERA PAVONINA	S	.	0.0	40.0	26.1	.	.	4.2	.	.	.	.	.	.	.	.	.	.	19
146	ADENANTHERA PAVONINA	S	.	0.0	4.0	1.0	.	.	4.5	.	.	.	.	.	.	.	.	.	.	19
147	ADENANTHERA PAVONINA	S	.	0.0	21.2	13.5	.	.	.	.	.	.	.	.	.	.	.	.	.	20
148	ADENIA GUMMIFERA	L	59	81.0	3.6	1.1	11.5	1.9	2.8	186	48	4.2	.	.	.	.	.	.	.	1
149	ADENOCARPUS HISPANICUS	S	.	0.0	38.9	6.2	.	.	.	.	.	.	.	.	.	.	.	.	.	21
150	ADENOCARPUS SP.	S	.	0.0	39.2	7.4	.	.	.	.	.	.	.	.	.	.	.	.	.	21
151	ADENOPHORA LILIFOLIA	S	.	0.0	19.4	35.1	.	.	.	.	.	.	.	.	.	.	.	.	.	20
152	ADENOPHORA POTANINII	S	.	0.0	21.2	37.0	.	.	.	.	.	.	.	.	.	.	.	.	.	20
153	ADLUMIA FUNGOSA	S	.	0.0	14.4	31.0	.	.	.	.	.	.	.	.	.	.	.	.	.	20
154	ADONIDIA MERRILLII	S	.	0.0	4.1	1.3	.	.	1.5	.	.	.	.	.	.	.	.	.	.	19
155	ADONIS AESTIVALIS	S	.	0.0	10.0	17.8	.	.	.	.	.	.	.	.	.	.	.	.	.	20
156	ADONIS AESTIVALIS	S	.	0.0	11.9	20.4	.	.	.	.	.	.	.	.	.	.	.	.	.	20
157	ADONIS AUTUMNALIS	S	.	0.0	11.2	16.4	.	.	.	.	.	.	.	.	.	.	.	.	.	20
158	ADONIS FLAMMEA	S	.	0.0	14.4	20.5	.	.	.	.	.	.	.	.	.	.	.	.	.	20
159	AEGILOPS CYLINDRICA	S	.	0.0	21.6	2.1	.	.	.	.	.	.	.	.	.	.	.	.	.	21
160	AEGILOPS TRIUNCIALIS	S	.	0.0	23.1	2.7	.	.	.	.	.	.	.	.	.	.	.	.	.	21
161	AEGLE MARMELOS	F	133	61.5	1.8	0.3	34.7	2.9	1.7	85	50	0.6	.	600	55	1.30	1.19	1.1	8	2
162	AELUROPUS LITTORALIS	S	.	0.0	15.5	3.1	.	.	.	.	.	.	.	.	.	.	.	.	.	21
163	AERVA LANATA	L	48	83.1	3.9	0.6	9.4	2.1	3.0	493	86	.	.	.	.	.	.	.	.	1
164	AERVA PERSICA	W	.	0.0	10.3	0.8	73.0	41.7	8.4	.	.	.	.	.	.	.	.	.	.	18
165	AESCHYNOMENE AMERICANA	S	.	0.0	37.3	7.0	.	.	4.5	.	.	.	.	.	.	.	.	.	.	19
166	AESCHYNOMENE AMERICANA	S	.	0.0	35.8	6.8	.	.	4.7	.	.	.	.	.	.	.	.	.	.	19
167	AESCHYNOMENE AMORPHOIDES	S	.	0.0	44.4	10.8	.	.	.	.	.	.	.	.	.	.	.	.	.	21
168	AESCHYNOMENE INDICA	S	.	0.0	33.1	9.1	.	.	.	.	.	.	.	.	.	.	.	.	.	20
169	AESCHYNOMENE RUDIS	S	.	0.0	33.4	11.5	.	.	.	.	.	.	.	.	.	.	.	.	.	21
170	AESCHYNOMENE SCABRA	S	.	0.0	31.4	7.2	.	.	.	.	.	.	.	.	.	.	.	.	.	21
171	AESCULUS OCTANDRA	S	.	0.0	16.2	3.7	.	.	.	.	.	.	.	.	.	.	.	.	.	20
172	AESCULUS TURBINATA	S	369	14.3	2.6	6.1	75.9	3.9	1.1	240	.	.	.	.	0	.	.	2.1	0	2
173	AESCULUS TURBINATA	S	.	0.0	6.9	6.6	.	.	.	.	.	.	.	.	.	.	.	.	.	20
174	AFRAEGLE PANICULATA	L	74	76.5	6.4	0.7	14.7	3.5	1.7	470	54	.	.	.	.	.	.	.	.	1
175	AFRAMOMUM MELEGUETA	F	.	21.0	7.5	.	27.8	.	4.2	750	0	.	.	.	.	.	.	.	.	1
176	AGASTACHE URTICIFOLIA	S	.	0.0	21.1	32.0	.	.	.	.	.	.	.	.	.	.	.	.	.	20
177	AGASTACHE URTICIFOLIA	S	.	0.0	17.7	31.1	.	.	.	.	.	.	.	.	.	.	.	.	.	21
178	AGASTACHE URTICIFOLIA	W	.	0.0	22.8	3.0	62.3	15.8	11.9	1490	420	.	.	.	.	.	.	.	.	6
179	AGAVE AMERICANA	H	.	0.0	6.1	1.3	82.2	15.6	12.4	.	.	.	.	.	.	.	.	.	.	18
180	AGAVE ATROVIRENS	L	.	0.0	2.3	2.8	86.0	32.4	8.9	.	.	.	.	.	.	.	.	.	.	18

TABLE 2 - DATA DIRECTLY TRANSCRIBED FROM ORIGINAL SOURCES.

OBS	PLANA	PART	CALORIES	WATER	PROTEIN	FAT	TOTLCARB	FIBER	ASH	CALCIUM	PHOSPHO	IRON	SODIUM	POTASS	CAROT	THIAMINE	RIBOFLAV	NIACIN	ASCORBIC	REFCO
181	AGAVE ATROVIRENS	I	30	91	0.9	0.2	7.3	.	0.6	114	27	0.9	.	.	320	0.11	0.05	0.2	59	3
182	AGAVE GEMINIFLORA	S	.	0	32.5	12.8	.	.	.	.	.	.	.	.	.	.	.	.	.	20
183	AGAVE LECHEGUILLA	S	.	0	24.9	17.4	.	.	.	.	.	.	.	.	.	.	.	.	.	21
184	AGAVE PERRYI	S	.	0	26.9	25.1	.	.	.	.	.	.	.	.	.	.	.	.	.	19
185	AGAVE PERRYI	S	.	0	8.1	0.8	.	.	.	.	.	.	.	.	.	.	.	.	.	19
186	AGAVE POTRERANA	S	.	0	27.5	18.1	.	.	.	.	.	.	.	.	.	.	.	.	.	20
187	AGAVE SCHOTTII	S	.	0	29.4	21.4	.	.	0.3	.	.	.	.	.	.	.	.	.	.	19
188	AGERATUM HOUSTONIANUM	S	.	0	25.2	30.8	.	.	7.4	.	.	.	.	.	.	.	.	.	.	19
189	AGERATUM HOUSTONIANUM	S	.	0	24.4	31.4	.	.	5.7	.	.	.	.	.	.	.	.	.	.	19
190	AGRIMONIA EUPATORIA	S	.	0	35.3	5.8	.	.	.	.	.	.	.	.	.	.	.	.	.	21
191	AGRIMONIA GRYPOSEPALA	S	.	0	16.8	10.1	.	.	.	.	.	.	.	.	.	.	.	.	.	21
192	AGROPYRON CRISTATUM	S	.	0	24.4	2.6	.	.	.	.	.	.	.	.	.	.	.	.	.	20
193	AGROPYRON CRISTATUM	S	.	0	18.6	2.3	.	.	.	.	.	.	.	.	.	.	.	.	.	21
194	AGROPYRON CRISTATUM	W	.	0	12.1	2.9	76.1	30.4	8.9	410	210	.	.	.	25	0.33	0.86	.	.	6
195	AGROPYRON DASYSTACHYUM	S	.	0	13.8	2.5	.	.	6.9	.	.	.	.	.	.	.	.	.	.	19
196	AGROPYRON INERME	W	.	0	11.3	3.7	77.3	29.7	7.7	580	240	.	.	.	32	.	.	.	.	6
197	AGROPYRON INTERMEDIUM	W	.	0	19.3	5.9	62.3	26.6	12.5	.	.	.	.	.	.	.	.	.	.	18
198	AGROPYRON INTERMEDIUM	H	.	0	8.7	2.3	79.8	34.8	9.2	360	.	.	.	.	.	.	.	.	.	6
199	AGROPYRON JUNCEUM	S	.	0	20.9	3.1	.	.	.	.	.	.	.	.	.	.	.	.	.	21
200	AGROPYRON REPENS	S	.	0	21.2	12.1	.	.	.	.	.	.	.	.	.	.	.	.	.	20
201	AGROPYRON REPENS	S	.	0	16.8	1.8	.	.	.	.	.	.	.	.	.	.	.	.	.	21
202	AGROPYRON REPENS	H	.	0	10.6	2.8	79.7	33.9	6.9	.	.	.	.	.	.	.	.	.	.	6
203	AGROPYRON REPENS	W	.	0	11.2	3.8	77.2	31.7	7.8	330	280	.	.	.	12	.	.	.	.	6
204	AGROPYRON SMITHII	H	.	0	7.5	3.0	80.9	33.5	8.6	300	140	11.0	.	.	13	.	.	.	.	6
205	AGROPYRON SPICATUM	W	.	0	10.2	3.1	78.8	31.4	7.9	420	240	.	.	.	23	.	.	.	.	6
206	AGROPYRON SUBSECUNDUM	H	.	0	3.4	4.1	80.9	35.0	11.6	.	.	.	.	.	.	.	.	.	.	6
207	AGROPYRON SUBSECUNDUM	W	.	.	.	.	.	.	.	570	340	.	.	.	.	.	.	.	.	6
208	AGROPYRON TRACHYCAULUM	S	.	0	14.5	2.0	.	.	5.1	.	.	.	.	.	.	.	.	.	.	19
209	AGROPYRON TRACHYCAULUM	H	.	0	9.9	2.6	78.6	34.2	8.9	360	240	.	.	2680	.	0.51	1.06	.	.	6
210	AGROPYRON X. TRITICUM	S	.	0	26.2	2.6	.	.	.	.	.	.	.	.	.	.	.	.	.	20
211	AGROPYRON X. TRITICUM	S	.	0	26.2	2.0	.	.	.	.	.	.	.	.	.	.	.	.	.	20
212	AGROSTEMMA GITHAGO	S	.	0	15.4	6.4	.	.	4.1	.	.	.	.	.	.	.	.	.	.	19
213	AGROSTEMMA GITHAGO	S	.	0	16.2	6.8	.	.	.	.	.	.	.	.	.	.	.	.	.	20
214	AGROSTEMMA GITHAGO	S	.	0	14.4	6.4	.	.	.	.	.	.	.	.	.	.	.	.	.	21
215	AGROSTIS ALBA	H	.	0	8.6	2.7	81.4	31.7	7.3	450	230	27.0	.	1690	0	.	.	.	.	6
216	AGROSTIS ALBA	W	.	0	13.3	3.8	74.5	25.0	8.4	620	370	25.0	.	2350	20	.	.	.	.	6
217	AGROSTIS CANINA	W	.	0	19.3	3.0	67.5	20.6	10.2	.	490	45.0	.	.	.	.	.	.	.	6
218	AGROSTIS HIEMALIS	W	.	0	9.6	2.8	78.8	34.4	8.8	390	260	.	.	.	.	.	.	.	.	6
219	AGROSTIS PALUSTRIS	H	.	0	4.3	1.4	88.5	30.8	5.8	.	.	.	.	.	.	.	.	.	.	6
220	AGROSTIS SCHIMPERANA	W	.	0	17.9	2.1	69.3	28.2	10.7	.	.	.	.	.	.	.	.	.	.	18
221	AGROSTIS TENUIS	H	.	0	7.1	2.3	83.5	31.4	7.1	.	.	.	.	.	.	.	.	.	.	6
222	AGROSTIS TENUIS	W	.	0	17.7	3.5	68.6	23.5	10.2	660	390	.	.	2270	13	.	.	.	.	6
223	AGROSTIS VERTICILLATA	W	.	0	9.3	2.4	73.8	29.7	14.5	530	180	.	.	.	.	.	.	.	.	6
224	AILANTHUS ALTISSIMA	S	.	0	27.6	55.5	.	.	4.1	.	.	.	.	.	.	.	.	.	.	19
225	AILANTHUS ALTISSIMA	S	.	0	27.5	59.1	.	.	.	.	.	.	.	.	.	.	.	.	.	19

TABLE 2 - DATA DIRECTLY TRANSCRIBED FROM ORIGINAL SOURCES.

OBS	PLNA	PART	CALORIES	WATER	PROTEIN	FAT	TOTCARB	FIBER	ASH	CALCIUM	PHOSPHO	IRON	SODIUM	POTASS	CAROT	THIAMINE	RIBOFLAV	NIACIN	ASCORBIC	REF
226	AJUGA CHIA	S	.	0.0	11.8	20.8	.	.	.	.	.	.	.	.	.	.	.	.	.	20
227	AKEBIA QUINATA	S	.	0.0	15.6	38.3	.	.	.	.	.	.	.	.	.	.	.	.	.	20
228	AKEBIA TRIFOLIATA	S	.	0.0	15.6	38.9	.	.	.	.	.	.	.	.	.	.	.	.	.	20
229	ALBIZIA AMARA	F	.	0.0	26.8	1.7	64.5	21.4	7.0	1020	280	.	.	.	.	.	.	.	.	18
230	ALBIZIA AMARA	SH	.	0.0	26.8	1.9	65.2	26.8	6.1	470	250	.	.	.	.	.	.	.	.	18
231	ALBIZIA AMARA	F	.	0.0	13.7	7.5	74.5	35.2	4.3	.	.	.	.	.	.	.	.	.	.	18
232	ALBIZIA CHINENSIS	L	.	0.0	15.1	4.4	75.2	31.6	5.3	1170	140	.	.	.	.	.	.	.	.	18
233	ALBIZIA JULIBRISSIN	S	.	0.0	33.0	11.1	.	.	4.7	.	.	.	.	.	.	.	.	.	.	19
234	ALBIZIA JULIBRISSIN	S	.	0.0	32.1	8.5	.	.	.	.	.	.	.	.	.	.	.	.	.	21
235	ALBIZIA LEBBEK	L	.	0.0	18.1	4.7	69.2	26.5	8.0	2020	140	.	.	.	.	.	.	.	.	18
236	ALBIZIA LEBBEK	F	.	0.0	21.1	4.6	69.7	23.0	4.6	.	.	.	.	.	.	.	.	.	.	18
237	ALBIZIA LEBBEK	S	.	0.0	30.2	5.3	.	.	.	.	.	.	.	.	.	.	.	.	.	21
238	ALBIZIA LOPHANTHA	S	.	0.0	32.5	8.0	.	.	1.3	.	.	.	.	.	.	.	.	.	.	19
239	ALBIZIA SP.	L	62	80.0	7.0	0.2	12.2	3.2	0.6	420	62	.	.	.	.	.	.	.	.	1
240	ALBIZZIA SP.	L	310	0.0	35.0	1.0	61.0	16.0	3.0	2100	310	.	.	.	.	.	.	.	.	10
241	ALCEA ACAULIS	S	.	0.0	12.2	6.6	.	.	.	.	.	.	.	.	.	.	.	.	.	21
242	ALCEA PALLIDA	S	.	0.0	16.4	12.8	.	.	.	.	.	.	.	.	.	.	.	.	.	21
243	ALCEA SETOSA	S	.	0.0	15.1	6.9	.	.	.	.	.	.	.	.	.	.	.	.	.	21
244	ALCEA SP.	S	.	0.0	16.4	7.0	.	.	.	.	.	.	.	.	.	.	.	.	.	21
245	ALCEA SP.	S	.	0.0	15.2	3.2	.	.	.	.	.	.	.	.	.	.	.	.	.	21
246	ALCEA STRIATA SUBSP. STRIATA	S	.	0.0	33.4	17.4	.	.	.	.	.	.	.	.	.	.	.	.	.	21
247	ALEURITES MOLUCCANA	S	.	0.0	23.1	64.0	.	.	3.6	.	.	.	.	.	.	.	.	.	.	19
248	ALEURITES MOLUCCANA	S	626	7.0	19.0	63.0	8.0	.	3.0	80	200	2.0	.	.	0	0.06	.	.	0	2
249	ALEURITES TRISPERMA	S	.	0.0	4.9	65.4	.	.	.	.	.	.	.	.	.	.	.	.	.	20
250	ALHAGI CAMELORUM	S	.	0.0	21.8	4.0	.	.	.	.	.	.	.	.	.	.	.	.	.	21
251	ALHAGI CAMELORUM	S	.	0.0	27.7	5.0	.	.	.	.	.	.	.	.	.	.	.	.	.	21
252	ALISMA PLANTAGO-AQUATICA	S	.	0.0	13.2	4.2	.	.	.	.	.	.	.	.	.	.	.	.	.	21
253	ALLAEANTHUS LUZONICUS	I	70	77.6	7.7	1.2	11.4	2.4	2.1	282	124	6.8	5	784	1170	0.15	0.35	1.6	24	2
254	ALLANBLACKIA FLORIBUNDA	S	648	6.0	3.6	64.4	23.8	3.1	2.2	122	169	.	.	.	.	.	.	.	.	1
255	ALLANBLACKIA FLORIBUNDA	L	66	79.0	5.9	1.1	12.0	2.4	2.0	273	85	3.7	5	479	2980	0.14	0.35	1.2	95	2
256	ALLANBLACKIA FLORIBUNDA	F	49	86.7	0.4	0.5	12.2	1.0	0.2	5	13	0.4	.	.	50	0.06	0.04	0.3	5	3
257	ALLIUM AMPELOPRASUM	R	45	86.3	2.2	0.3	10.3	.	0.9	52	50	1.1	.	.	25	0.11	.	.	17	2
258	ALLIUM AMPELOPRASUM	L	35	89.4	2.3	0.5	7.0	1.0	0.8	58	48	2.7	5	316	2435	0.09	0.10	0.6	32	2
259	ALLIUM AMPELOPRASUM	I	55	83.1	5.5	0.5	10.5	1.1	0.4	23	38	0.9	.	.	2550	0.14	0.19	0.9	40	2
260	ALLIUM AMPELOPRASUM	L	52	85.4	2.2	0.3	11.2	1.3	0.9	52	50	1.1	5	347	24	0.11	0.06	0.5	17	4
261	ALLIUM CANADENSE	BU	.	0.0	13.1	1.9	.	.	.	.	.	.	.	.	.	.	.	.	.	19
262	ALLIUM CEPA	R	48	86.6	1.9	0.3	10.4	0.8	0.8	26	43	0.7	8	297	0	0.04	0.05	1.0	6	2
263	ALLIUM CEPA	R	38	88.6	1.6	0.2	9.0	0.7	0.6	30	44	1.0	9	166	0	0.06	0.04	0.2	9	2
264	ALLIUM CEPA	Q	39	89.6	1.5	0.2	8.3	1.0	0.4	34	42	1.4	.	.	0	0.04	0.04	0.4	13	3
265	ALLIUM CEPA	L	26	92.2	1.8	0.6	4.7	1.1	0.7	42	43	3.4	.	.	1230	0.05	0.11	0.7	39	3
266	ALLIUM CEPA	R	38	89.1	1.5	0.1	8.7	0.6	0.6	27	36	0.5	10	157	24	0.03	0.04	0.2	10	4
267	ALLIUM CF. DRUMMONDII	S	.	0.0	24.4	17.7	.	.	.	.	.	.	.	.	.	.	.	.	.	20
268	ALLIUM FISTULOSUM	R	30	91.4	1.6	0.3	6.1	0.8	0.6	55	41	1.1	15	192	630	0.06	0.08	0.5	19	2
269	ALLIUM ODORUM	L	34	88.4	3.0	0.4	6.8	0.8	1.4	59	66	2.6	6	234	1020	0.08	0.15	0.9	36	2
270	ALLIUM PANICULATUM	S	.	0.0	30.5	11.3	.	.	.	.	.	.	.	.	.	.	.	.	.	21

TABLE 2 - DATA DIRECTLY TRANSCRIBED FROM ORIGINAL SOURCES.

OBS	PLANA	PART	CALORIES	WATER	PROTEIN	FAT	TOTLCARB	FIBER	ASH	CALCIUM	PHOSPHO	IRON	SODIUM	POTASS	CAROT	THIAMINE	RIBOFLAV	NIACIN	ASCORBIC	REF
271	ALLIUM PORRUM	S	.	0.0	25.9	15.9	.	.	4.0	.	.	.	.	.	.	.	.	.	.	19
272	ALLIUM PORRUM	S	.	0.0	28.2	13.9	.	.	4.4	.	.	.	.	.	.	.	.	.	.	19
273	ALLIUM ROTUNDUM	BU	.	0.0	15.8	2.2	.	.	.	.	.	.	.	.	.	.	.	.	.	21
274	ALLIUM ROTUNDUM	S	.	0.0	28.5	15.3	.	.	.	.	.	.	.	.	.	.	.	.	.	21
275	ALLIUM SATIVUM	R	117	67.8	3.5	0.3	27.4	0.7	1.0	18	88	1.5	18	373	0	0.24	0.05	0.4	10	2
276	ALLIUM SATIVUM	L	44	86.4	2.6	0.5	9.5	1.8	1.0	58	46	0.6	4	326	920	0.11	0.14	0.6	39	2
277	ALLIUM SATIVUM	SH	76	77.7	1.2	0.3	20.1	1.7	0.7	12	52	1.7	.	273	200	0.14	0.06	0.5	42	2
278	ALLIUM SATIVUM	I	39	88.4	1.4	0.2	9.4	0.8	0.6	25	46	0.9	.	.	60	0.11	0.06	0.4	44	2
279	ALLIUM SATIVUM	R	137	61.3	6.2	0.2	30.8	1.5	1.5	29	202	1.5	19	529	0	0.25	0.08	0.5	15	4
280	ALLIUM SCHOENOPRASUM	L	27	92.0	2.7	0.6	4.3	0.7	0.4	83	41	0.8	.	.	.	0.10	0.06	0.5	32	2
281	ALLIUM SCHOENOPRASUM	L	28	91.3	1.8	0.3	5.8	1.1	0.8	69	44	1.7	.	250	3480	0.08	0.13	0.5	56	4
282	ALLIUM SP.	BU	.	0.0	14.4	1.8	.	.	2.9	.	.	.	.	.	.	.	.	.	.	19
283	ALLIUM SP.	S	.	0.0	31.8	10.5	.	.	.	.	.	.	.	.	.	.	.	.	.	21
284	ALLIUM VINEALE	BU	.	0.0	27.8	2.8	.	.	3.9	.	.	.	.	.	.	.	.	.	.	19
285	ALLOTEROPSIS SEMIALATA	W	.	0.0	15.6	1.1	74.3	14.7	9.0	.	.	.	.	.	.	.	.	.	.	18
286	ALNUS CORDIFOLIA	S	.	0.0	24.4	23.2	.	.	.	.	.	.	.	.	.	.	.	.	.	20
287	ALNUS CORDIFOLIA	S	.	0.0	15.6	16.8	.	.	.	.	.	.	.	.	.	.	.	.	.	20
288	ALNUS GLUTINOSA	S	.	0.0	19.4	16.1	.	.	.	.	.	.	.	.	.	.	.	.	.	19
289	ALNUS GLUTINOSA	S	.	0.0	11.1	11.7	.	.	.	.	.	.	.	.	.	.	.	.	.	21
290	ALNUS INCANA	S	.	0.0	16.2	9.8	.	.	.	.	.	.	.	.	.	.	.	.	.	20
291	ALOCASIA MACRORRHIZA	L	.	0.0	24.0	6.3	52.0	12.5	17.7	1570	140	.	.	.	.	.	.	.	.	18
292	ALOCASIA MACRORRHIZA	R	61	84.0	0.6	0.3	14.8	.	0.3	30	50	1.0	.	.	0	0.05	.	.	5	2
293	ALONSOA WARSCEWICZII	S	.	0.0	20.0	26.2	.	.	4.3	.	.	.	.	.	.	.	.	.	.	19
294	ALOPECURUS ARUNDINACEUS	S	.	0.0	24.4	13.9	.	.	.	.	.	.	.	.	.	.	.	.	.	20
295	ALOPECURUS ARUNDINACEUS	S	.	0.0	21.0	14.0	.	.	.	.	.	.	.	.	.	.	.	.	.	21
296	ALOPECURUS PRATENSIS	S	.	0.0	20.6	15.8	.	.	.	.	.	.	.	.	.	.	.	.	.	20
297	ALOPECURUS PRATENSIS	W	.	0.0	17.2	4.6	67.5	21.5	10.7	570	460	.	.	.	.	.	.	.	.	6
298	ALPINIA GALANGA	R	51	85.9	1.0	0.4	11.7	3.1	1.0	31	25	2.1	.	.	1520	0.05	0.02	1.0	26	2
299	ALTERNANTHERA MARITIMA	L	55	78.5	5.0	0.9	10.0	2.2	5.6	142	35	.	.	.	.	.	.	.	.	1
300	ALTERNANTHERA PHILOXEROIDES	SH	.	0.0	7.3	1.6	77.5	8.5	13.6	140	.	.	.	.	.	.	.	.	.	18
301	ALTERNANTHERA SESSILIS	L	60	80.2	4.7	0.8	11.8	2.1	2.5	146	45	.	.	.	.	.	.	.	.	1
302	ALTERNANTHERA SESSILIS	S	.	0.0	18.2	11.4	.	.	.	.	.	.	.	.	.	.	.	.	.	21
303	ALTERNANTHERA SP.	L	43	85.7	1.8	0.2	10.3	.	2.0	202	72	16.0	115	468	845	0.02	0.28	1.5	12	2
304	ALTERNANTHERA SP.		.	0.0	21.1	1.6	55.9	14.8	21.4	3060	300	.	.	3750	.	.	.	.	.	6
305	ALTHAEA OFFICINALIS	S	.	0.0	23.7	12.8	.	.	.	.	.	.	.	.	.	.	.	.	.	21
306	ALTHAEA OFFICINALIS	S	.	0.0	22.7	17.2	.	.	.	.	.	.	.	.	.	.	.	.	.	21
307	ALTHAEA ROSEA	S	.	0.0	25.6	12.8	.	.	.	.	.	.	.	.	.	.	.	.	.	19
308	ALVARADOA AMORPHOIDES	S	.	0.0	25.0	58.8	.	.	.	.	.	.	.	.	.	.	.	.	.	20
309	ALYSICARPUS RUGOSUS	SH	.	0.0	14.7	1.8	73.5	39.1	10.0	1260	230	.	.	.	.	.	.	.	.	18
310	ALYSICARPUS VAGINALIS	L	.	0.0	14.0	6.3	69.6	29.3	10.1	.	.	.	.	.	.	.	.	.	.	18
311	ALYSICARPUS VAGINALIS	SH	.	0.0	13.9	2.5	70.7	21.4	12.9	.	.	.	.	.	.	.	.	.	.	18
312	ALYSICARPUS VAGINALIS	S	.	0.0	12.0	1.4	75.6	48.9	11.0	.	.	.	.	.	.	.	.	.	.	18
313	ALYSICARPUS VAGINALIS	S	.	0.0	34.2	6.6	.	.	4.1	.	.	.	.	.	.	.	.	.	.	19
314	ALYSICARPUS VAGINALIS	S	.	0.0	36.6	6.2	.	.	4.0	.	.	.	.	.	.	.	.	.	.	19
315	ALYSICARPUS VAGINALIS	S	.	0.0	35.6	5.6	.	.	.	.	.	.	.	.	.	.	.	.	.	20

TABLE 2 - DATA DIRECTLY TRANSCRIBED FROM ORIGINAL SOURCES.

OBS	PLANA	PART	CALORIES	WATER	PROTEIN	FAT	TOTCARB	FIBER	ASH	CALCIUM	PHOSPHO	IRON	SODIUM	POTASS	CAROTENE	THIAMINE	RIBOFLAV	NIACIN	ASCORBIC	REFCO
316	ALYSICARPUS VAGINALIS	H	.	0.0	12.6	1.7	79.2	39.4	6.5	.	.	.	.	.	.	.	.	.	.	6
317	ALYSSUM CAMPESTRE	S	.	0.0	25.8	18.1	.	.	.	.	.	.	.	.	.	.	.	.	.	20
318	ALYSSUM CONSTELLATUM	S	.	0.0	26.6	9.5	.	.	.	.	.	.	.	.	.	.	.	.	.	21
319	ALYSSUM DASYCARPUM	S	.	0.0	33.7	29.2	.	.	.	.	.	.	.	.	.	.	.	.	.	20
320	ALYSSUM MINIMUM	S	.	0.0	25.0	20.9	.	.	.	.	.	.	.	.	.	.	.	.	.	20
321	ALYSSUM SAXATILE	S	.	0.0	26.9	18.3	.	.	.	.	.	.	.	.	.	.	.	.	.	20
322	ALYSSUM SAXATILE	S	.	0.0	27.5	16.8	.	.	.	.	.	.	.	.	.	.	.	.	.	20
323	ALYSSUM SAXATILE VAR. LUTEUM	S	.	0.0	28.8	15.7	.	.	.	.	.	.	.	.	.	.	.	.	.	20
324	ALYSSUM TORTUOSUM	S	.	0.0	31.8	25.4	.	.	.	.	.	.	.	.	.	.	.	.	.	20
325	AMARANTHUS ALBUS	S	.	0.0	16.3	7.6	.	.	.	.	.	.	.	.	.	.	.	.	.	21
326	AMARANTHUS BLITOIDES	S	.	0.0	16.1	6.8	.	.	.	.	.	.	.	.	.	.	.	.	.	21
327	AMARANTHUS BLITOIDES	W	.	0.0	27.0	1.4	50.4	11.7	21.2	4190	490	.	.	5200	.	.	.	.	.	6
328	AMARANTHUS BLITOIDES	W	.	0.0	18.5	1.3	53.9	13.1	26.3	3200	180	.	.	4560	.	.	.	.	.	6
329	AMARANTHUS CAUDATUS	L	265	0.0	26.7	2.8	49.5	8.7	21.0	2341	304	.	11	263	.	.	.	4.74	.	12
330	AMARANTHUS CAUDATUS	S	.	0.0	17.5	6.5	.	.	2.8	.	.	.	.	.	.	.	.	.	.	19
331	AMARANTHUS CAUDATUS	S	.	0.0	16.9	6.9	.	.	.	.	.	.	.	.	.	.	.	.	.	20
332	AMARANTHUS CAUDATUS	S	358	12.3	12.9	7.2	65.1	6.7	2.5	247	500	3.4	.	.	0	0.14	0.32	1.00	3	3
333	AMARANTHUS DEFLEXUS	S	.	0.0	17.2	6.5	.	.	.	.	.	.	.	.	.	.	.	.	.	21
334	AMARANTHUS GRACILIS	L	256	0.0	31.9	3.0	43.0	9.2	22.2	2877	559	.	14	403	.	.	.	7.19	.	12
335	AMARANTHUS GRAECIZANS	L	269	0.0	23.2	3.2	52.0	13.5	21.6	1827	455	.	11	226	.	.	.	6.23	.	12
336	AMARANTHUS GRAECIZANS	SH	.	0.0	19.9	1.5	61.6	21.0	17.0	1810	540	.	.	.	.	.	.	.	.	18
337	AMARANTHUS HYBRIDUS	S	.	0.0	16.2	6.6	.	.	.	.	.	.	.	.	.	.	.	.	.	19
338	AMARANTHUS HYBRIDUS	S	.	0.0	16.7	4.8	.	.	.	.	.	.	.	.	.	.	.	.	.	21
339	AMARANTHUS LIVIDUS	S	.	0.0	15.4	7.6	.	.	.	.	.	.	.	.	.	.	.	.	.	21
340	AMARANTHUS PALMERI	S	.	0.0	15.8	8.1	.	.	.	.	.	.	.	.	.	.	.	.	.	21
341	AMARANTHUS RETROFLEXUS	S	.	0.0	17.5	7.2	.	.	3.4	.	.	.	.	.	.	.	.	.	.	19
342	AMARANTHUS RETROFLEXUS	S	.	0.0	16.8	6.9	.	.	.	.	.	.	.	.	.	.	.	.	.	21
343	AMARANTHUS RETROFLEXUS	S	.	0.0	16.9	7.4	.	.	.	.	.	.	.	.	.	.	.	.	.	21
344	AMARANTHUS RETROFLEXUS	S	.	0.0	16.9	6.6	.	.	.	.	.	.	.	.	.	.	.	.	.	21
345	AMARANTHUS RETROFLEXUS	S	.	0.0	17.5	7.4	.	.	.	.	.	.	.	.	.	.	.	.	.	21
346	AMARANTHUS RETROFLEXUS	W	.	0.0	25.7	1.5	54.0	10.8	18.8	3950	510	.	.	5150	10	.	.	.	.	6
347	AMARANTHUS SP.	L	42	84.0	4.6	0.2	8.3	1.8	2.9	410	103	8.9	.	.	5716	0.05	0.42	1.20	64	1
348	AMARANTHUS SP.	S	356	12.7	14.0	6.0	63.1	9.4	4.2	490	455	.	.	.	.	.	.	.	.	1
349	AMARANTHUS SP.	L	36	86.9	3.5	0.5	6.5	1.3	2.6	267	67	3.9	.	411	3660	0.08	0.16	1.40	80	4
350	AMARANTHUS SPINOSUS	L	267	0.0	28.4	4.5	45.0	10.4	22.1	1795	430	.	13	337	.	.	.	7.73	.	12
351	AMARANTHUS SPINOSUS	L	45	83.7	5.6	0.6	7.4	1.6	2.7	476	75	2.2	6	575	6655	0.01	0.33	1.40	82	2
352	AMARANTHUS SPINOSUS	L	.	85.0	3.0	0.3	8.1	.	3.6	800	50	22.9	.	.	.	.	.	.	.	5
353	AMARANTHUS TRICOLOR	L	32	88.4	3.0	0.5	5.7	1.3	2.4	237	67	5.6	4	416	3085	0.06	0.20	1.20	53	2
354	AMARANTHUS TRICOLOR	L	33	87.8	3.3	0.3	6.2	1.6	2.4	288	123	5.4	.	546	4590	0.08	0.16	1.40	89	2
355	AMARANTHUS TRICOLOR	L	34	88.2	3.2	0.4	6.3	1.5	1.9	288	80	6.1	.	.	65	0.08	0.28	0.30	27	2
356	AMARANTHUS VIRIDIS	L	43	84.8	5.2	0.8	6.7	1.0	2.5	341	76	4.1	51	442	7715	0.01	0.37	1.80	120	2
357	AMARANTHUS VIRIDIS	S	.	0.0	16.2	7.0	.	.	.	.	.	.	.	.	.	.	.	.	.	20
358	AMARANTHUS VIRIDIS	S	.	0.0	14.1	4.7	.	.	.	.	.	.	.	.	.	.	.	.	.	21
359	AMBLYOLEPIS SETIGERA	S	.	0.0	16.9	13.6	.	.	.	.	.	.	.	.	.	.	.	.	.	20
360	AMBLYOPYRUM MUTICUM	S	.	0.0	24.5	3.1	.	.	.	.	.	.	.	.	.	.	.	.	.	21

TABLE 2 - DATA DIRECTLY TRANSCRIBED FROM ORIGINAL SOURCES.

OBS	PLNA	PART	CALORIES	WATER	PROTEIN	FAT	TOTCARB	FIBER	ASH	CALCIUM	PHOSPHO	IRON	SODIUM	POTASS	CAROT	THIAMINE	RIBOFLAV	NIACIN	ASCORBIC	REF
361	AMBLYOPYRUM MUTICUM	S	.	0.0	26.3	3.4	.	.	.	.	.	.	.	.	.	.	.	.	.	21
362	AMBROSIA ARTEMISIAEFOLIA	W	.	0.0	18.7	9.8	66.4	34.4	5.1	.	.	.	.	.	.	.	.	.	.	6
363	AMBROSIA PSILOSTACHYA	W	.	0.0	23.8	5.5	54.9	12.0	15.8	.	.	.	.	.	.	.	.	.	.	6
364	AMBROSIA TRIFIDA	S	.	0.0	21.6	24.7	.	.	3.7	.	.	.	.	.	.	.	.	.	.	19
365	AMBROSIA TRIFIDA	S	.	0.0	15.2	19.9	.	.	.	.	.	.	.	.	.	.	.	.	.	21
366	AMBURANA SP.	S	.	0.0	25.6	35.2	.	.	.	.	.	.	.	.	.	.	.	.	.	20
367	AMELANCHIER ALNIFOLIA	S	.	0.0	18.1	13.0	.	.	.	.	.	.	.	.	.	.	.	.	.	20
368	AMELANCHIER ALNIFOLIA	S	.	0.0	15.0	10.2	.	.	.	.	.	.	.	.	.	.	.	.	.	20
369	AMELANCHIER ALNIFOLIA	S	.	0.0	14.2	10.3	.	.	.	.	.	.	.	.	.	.	.	.	.	20
370	AMELANCHIER ALNIFOLIA	L	.	0.0	10.9	6.2	74.4	14.8	8.5	2320	620	.	.	.	.	.	.	.	.	6
371	AMELANCHIER ALNIFOLIA	L	.	0.0	18.0	3.9	71.4	13.4	6.7	1600	480	.	.	.	.	.	.	.	.	6
372	AMELANCHIER ALNIFOLIA	C	.	0.0	5.5	3.4	88.3	34.8	2.8	.	.	.	.	.	.	.	.	.	.	6
373	AMELANCHIER OVALIS	S	.	0.0	11.8	7.8	.	.	.	.	.	.	.	.	.	.	.	.	.	21
374	AMICIA ZYGOMERIS	S	.	0.0	22.5	20.6	.	.	.	.	.	.	.	.	.	.	.	.	.	20
375	AMMANIA BACCIFERA	L	25	90.9	3.6	0.0	4.5	.	1.0	60	68	6.4	.	.	3875	0.07	0.05	.	53	2
376	AMMI MAJUS	S	.	0.0	18.8	11.6	.	.	.	.	.	.	.	.	.	.	.	.	.	19
377	AMMI MAJUS	S	.	0.0	17.8	8.9	.	.	.	.	.	.	.	.	.	.	.	.	.	20
378	AMMI MAJUS	S	.	0.0	18.9	12.5	.	.	.	.	.	.	.	.	.	.	.	.	.	21
379	AMMI MAJUS	S	.	0.0	19.1	19.4	.	.	.	.	.	.	.	.	.	.	.	.	.	21
380	AMMI VISNAGA	S	.	0.0	18.1	19.2	.	.	8.9	.	.	.	.	.	.	.	.	.	.	19
381	AMMI VISNAGA	S	.	0.0	17.3	15.3	.	.	.	.	.	.	.	.	.	.	.	.	.	21
382	AMMI VISNAGA	S	.	0.0	19.9	16.9	.	.	.	.	.	.	.	.	.	.	.	.	.	21
383	AMMI VISNAGA	S	.	0.0	18.4	27.0	.	.	.	.	.	.	.	.	.	.	.	.	.	21
384	AMMI VISNAGA	S	.	0.0	10.9	16.9	.	.	.	.	.	.	.	.	.	.	.	.	.	21
385	AMMOBIUM ALATUM	S	.	0.0	28.1	27.0	.	.	.	.	.	.	.	.	.	.	.	.	.	20
386	AMMOBIUM ALATUM VAR. GRANDIFLORUM	S	.	0.0	30.0	25.0	.	.	.	.	.	.	.	.	.	.	.	.	.	20
387	AMMOPHILA ARENARIA	S	.	0.0	23.7	12.5	.	.	.	.	.	.	.	.	.	.	.	.	.	21
388	AMMOSELINUM POPEI	S	.	0.0	13.1	7.8	.	.	.	.	.	.	.	.	.	.	.	.	.	20
389	AMMOSELINUM POPEI	S	.	0.0	16.3	9.9	.	.	.	.	.	.	.	.	.	.	.	.	.	21
390	AMOREUXIA PALMATIFIDA	S	.	0.0	11.9	14.0	.	.	.	.	.	.	.	.	.	.	.	.	.	20
391	AMORPHA CANESCENS	S	.	0.0	17.6	3.4	.	.	4.9	.	.	.	.	.	.	.	.	.	.	19
392	AMORPHA CANESCENS	S	.	0.0	39.8	6.6	.	.	4.3	.	.	.	.	.	.	.	.	.	.	19
393	AMORPHA CANESCENS	S	.	0.0	12.4	7.0	.	.	5.3	.	.	.	.	.	.	.	.	.	.	19
394	AMORPHA CANESCENS	W	.	0.0	8.2	8.5	72.9	21.5	10.4	2800	920	.	.	.	.	.	.	.	.	6
395	AMORPHA FRUTICOSA	S	.	0.0	32.3	11.0	.	.	4.8	.	.	.	.	.	.	.	.	.	.	19
396	AMORPHA FRUTICOSA	S	.	0.0	21.2	7.2	.	.	5.9	.	.	.	.	.	.	.	.	.	.	19
397	AMORPHA FRUTICOSA	S	.	0.0	28.4	10.0	.	.	4.5	.	.	.	.	.	.	.	.	.	.	19
398	AMORPHA FRUTICOSA	S	.	0.0	10.5	9.4	.	.	6.0	.	.	.	.	.	.	.	.	.	.	19
399	AMORPHA FRUTICOSA	S	.	0.0	30.7	8.6	.	.	4.1	.	.	.	.	.	.	.	.	.	.	19
400	AMORPHA FRUTICOSA	S	.	0.0	11.5	10.1	.	.	6.1	.	.	.	.	.	.	.	.	.	.	19
401	AMORPHOPHALLUS APHYLLUS	R	108	71.2	2.4	0.4	24.3	1.4	1.7	152	109	.	.	.	.	.	.	.	.	1
402	AMORPHOPHALLUS CAMPANULATUS	SH	.	0.0	15.3	3.3	65.1	14.2	16.3	1520	810	.	.	.	.	.	.	.	.	18
403	AMORPHOPHALLUS CAMPANULATUS	R	.	0.0	19.8	1.6	72.0	2.3	6.6	.	.	.	.	.	.	.	.	.	.	18
404	AMORPHOPHALLUS CAMPANULATUS	R	81	78.5	2.0	0.2	18.4	0.7	0.9	38	38	2.4	5	416	0	0.06	0.02	1.7	6	2
405	AMORPHOPHALLUS CONJAC	L	8	97.4	0.1	0.0	2.3	0.1	0.2	17	7	0.3	.	.	.	0.00	0.00	0.0	0	2

TABLE 2 - DATA DIRECTLY TRANSCRIBED FROM ORIGINAL SOURCES.

OBS	PLANA	PART	CALORIES	WATER	PROTEIN	FAT	TOTLCARB	FIBER	ASH	CALCIUM	PHOSPHO	IRON	SODIUM	POTASS	CAROT	THIAMINE	RIBOFLAV	NIACIN	ASCORBIC	REFCO
406	AMSINCKIA DOUGLASIANA		.	0.0	13.9	.	.	23.0	16.2	2280	520	.	.	4420	.	.	.	.	.	6
407	ANACARDIUM OCCIDENTALE	S	542	7.6	17.4	43.4	29.2	1.4	2.4	76	578	18.0	.	.	.	0.65	0.25	1.60	7	1
408	ANACARDIUM OCCIDENTALE	F	53	85.6	1.0	0.7	12.3	0.6	0.4	12	45	1.4	.	.	760	0.03	0.24	0.30	252	1
409	ANACARDIUM OCCIDENTALE	L	100	69.9	5.2	0.6	23.1	.	1.2	.	.	.	.	.	615	0.01	0.01	1.40	89	2
410	ANACARDIUM OCCIDENTALE	S	533	2.7	15.2	37.0	42.0	1.4	3.1	24	580	1.8	.	.	0	0.85	0.32	2.10	.	3
411	ANACARDIUM OCCIDENTALE	S	561	5.2	17.2	45.7	29.3	1.4	2.6	38	373	3.8	15	464	60	0.43	0.25	1.80	.	4
412	ANACOLOSA LUZONIENSIS	P	101	74.3	3.2	0.4	21.3	0.8	0.8	27	50	1.3	.	.	725	0.07	0.08	1.10	20	2
413	ANACOLOSA LUZONIENSIS	S	229	44.9	3.6	2.6	47.9	1.8	1.0	23	127	1.0	.	.	.	0.14	0.14	0.90	9	2
414	ANACYCLUS CLAVATUS	S	.	0.0	21.3	19.8	.	.	.	.	.	.	.	.	.	.	.	.	.	21
415	ANAGALLIS ARVENSIS	S	.	0.0	14.4	17.7	.	.	.	.	.	.	.	.	.	.	.	.	.	20
416	ANAGALLIS ARVENSIS	S	.	0.0	15.3	20.7	.	.	.	.	.	.	.	.	.	.	.	.	.	21
417	ANAGYRIS FOETIDA	S	.	0.0	21.8	10.3	.	.	.	.	.	.	.	.	.	.	.	.	.	21
418	ANAGYRIS FOETIDA	S	.	0.0	22.8	9.6	.	.	.	.	.	.	.	.	.	.	.	.	.	21
419	ANAGYRIS FOETIDA	S	.	0.0	24.8	10.1	.	.	.	.	.	.	.	.	.	.	.	.	.	21
420	ANAGYRIS FOETIDA	S	.	0.0	26.4	11.4	.	.	.	.	.	.	.	.	.	.	.	.	.	21
421	ANAGYRIS FOETIDA	S	.	0.0	24.7	11.2	.	.	.	.	.	.	.	.	.	.	.	.	.	21
422	ANANAS COMOSUS	F	47	87.0	0.7	0.3	11.6	0.5	0.4	17	12	0.5	2	125	35	0.06	0.03	0.30	22	2
423	ANANAS COMOSUS	F	52	85.4	0.4	0.2	13.7	0.4	0.3	18	8	0.5	.	.	32	0.08	0.04	0.20	61	3
424	ANANAS COMOSUS	F	52	85.3	0.4	0.2	13.7	0.4	0.4	17	8	0.5	1	146	42	0.09	0.03	0.20	17	4
425	ANANAS COMOSUS	W	.	0.0	11.9	3.4	75.9	20.3	8.8	.	.	.	.	.	.	.	.	.	.	6
426	ANCHOMANES DIFFORMIS	R	101	72.4	3.4	0.1	22.6	1.4	1.5	201	69	.	.	.	.	.	.	.	.	1
427	ANCHUSA AZUREA	S	.	0.0	16.4	21.4	.	.	16.8	.	.	.	.	.	.	.	.	.	.	19
428	ANCHUSA CAPENSIS	S	.	0.0	19.4	29.9	.	.	15.7	.	.	.	.	.	.	.	.	.	.	19
429	ANCHUSA HYBRIDA	S	.	0.0	20.4	20.5	.	.	.	.	.	.	.	.	.	.	.	.	.	20
430	ANDROPOGON BARBINODIS	W	.	0.0	7.2	2.1	82.2	33.0	8.5	690	390	.	.	.	.	.	.	.	.	6
431	ANDROPOGON CANALICULATUS	S	.	0.0	13.3	2.0	74.9	31.9	9.8	.	.	.	.	.	.	.	.	.	.	18
432	ANDROPOGON CIRRATUS	W	.	0.0	3.7	1.8	89.7	36.5	4.8	290	70	.	.	520	.	.	.	.	.	6
433	ANDROPOGON GAYANUS	W	.	0.0	6.5	3.1	80.5	29.6	9.9	.	.	.	.	.	.	.	.	.	.	18
434	ANDROPOGON GAYANUS	H	.	0.0	6.1	1.7	84.3	35.1	7.9	.	.	.	.	.	.	.	.	.	.	18
435	ANDROPOGON GERARDII	S	.	0.0	31.9	8.3	.	.	.	.	.	.	.	.	.	.	.	.	.	20
436	ANDROPOGON GERARDII	W	.	0.0	7.0	2.3	82.9	33.2	7.8	360	160	17.0	.	1220	13	.	.	.	.	6
437	ANDROPOGON GLOMERATUS(IMMATURE)	W	.	0.0	10.6	2.1	76.7	31.3	10.6	.	.	.	.	.	.	.	.	.	.	6
438	ANDROPOGON GLOMERATUS(MATURE)	W	.	0.0	4.6	1.5	86.5	34.1	7.4	.	.	.	.	.	.	.	.	.	.	6
439	ANDROPOGON HALLII	S	.	0.0	26.2	6.0	.	.	.	.	.	.	.	.	.	.	.	.	.	20
440	ANDROPOGON HALLII	W	.	0.0	5.8	2.3	84.3	33.5	7.6	.	.	.	.	.	.	.	.	.	.	6
441	ANDROPOGON ISCHAEMUM	S	.	0.0	30.0	9.8	.	.	.	.	.	.	.	.	.	.	.	.	.	19
442	ANDROPOGON ISCHAEMUM	W	.	0.0	8.0	2.0	81.5	33.5	8.5	.	.	.	.	.	.	.	.	.	.	6
443	ANDROPOGON PERTUSUS	S	.	0.0	26.2	8.6	.	.	.	.	.	.	.	.	.	.	.	.	.	19
444	ANDROPOGON SACCHAROIDES	W	.	0.0	7.1	1.8	80.3	33.8	10.8	.	.	.	.	.	.	.	.	.	.	6
445	ANDROPOGON SCOPARIUS	S	.	0.0	33.8	7.2	.	.	.	.	.	.	.	.	.	.	.	.	.	20
446	ANDROPOGON SCOPARIUS	W	.	0.0	6.2	2.2	83.9	32.8	7.7	350	120	.	.	.	7	.	.	.	.	6
447	ANDROPOGON TENER	W	.	0.0	6.3	2.4	82.3	33.8	9.0	490	100	.	.	.	.	.	.	.	.	6
448	ANDROPOGON TERNARIUS	S	.	0.0	37.5	9.2	.	.	.	.	.	.	.	.	.	.	.	.	.	20
449	ANDROPOGON VIRGINICUS	W	.	0.0	13.5	1.8	73.8	30.9	10.9	.	.	.	.	.	.	.	.	.	.	6
450	ANDROSACE MAXIMA	S	.	0.0	11.1	8.9	.	.	.	.	.	.	.	.	.	.	.	.	.	21

TABLE 2 - DATA DIRECTLY TRANSCRIBED FROM ORIGINAL SOURCES.

OBS	PLANA	PART	CALORIES	WATER	PROTEIN	FAT	TOTCARB	FIBER	ASH	CALCIUM	PHOSPHO	IRON	SODIUM	POTASS	CAROT	THIAMINE	RIBOFLAV	NIACIN	ASCORBIC	REFCO
451	ANEILEMA PEDUNCULOSUM	L	24	91.3	2.1	0.4	4.3	1.3	1.9	153	53	.	.	.	.	.	.	.	.	1
452	ANEMONE CORONARIA	S	.	0.0	25.6	26.6	.	.	4.5	.	.	.	.	.	.	.	.	.	.	19
453	ANEMONE CORONARIA (ST.BRIGID)	S	.	0.0	26.0	20.7	.	.	4.1	.	.	.	.	.	.	.	.	.	.	19
454	ANEMONE CYLINDRICA	S	.	0.0	20.6	27.9	.	.	.	.	.	.	.	.	.	.	.	.	.	21
455	ANEMONE DECAPETALA	S	.	0.0	22.5	32.9	.	.	.	.	.	.	.	.	.	.	.	.	.	20
456	ANETHUM GRAVEOLENS	S	305	7.7	16.0	14.5	55.2	21.1	6.6	1516	277	16.3	20	1186	32	0.42	0.28	2.81	.	16
457	ANETHUM GRAVEOLENS	H	253	7.3	20.0	4.4	55.8	11.9	12.6	1784	543	48.8	208	3308	.	0.42	0.28	2.81	.	16
458	ANETHUM GRAVEOLENS	S	.	0.0	21.3	17.7	.	.	7.6	.	.	.	.	.	.	.	.	.	.	19
459	ANETHUM GRAVEOLENS	S	.	0.0	22.6	9.9	.	.	8.4	.	.	.	.	.	.	.	.	.	.	19
460	ANETHUM GRAVEOLENS	S	.	0.0	18.6	6.8	.	.	7.3	.	.	.	.	.	.	.	.	.	.	19
461	ANGELICA AMPLA	S	.	0.0	29.5	26.0	.	.	.	.	.	.	.	.	.	.	.	.	.	20
462	ANGELICA ARCHANGELICA	S	.	0.0	31.9	18.9	.	.	.	.	.	.	.	.	.	.	.	.	.	20
463	ANGELICA ARCHANGELICA	S	.	0.0	32.5	20.3	.	.	.	.	.	.	.	.	.	.	.	.	.	20
464	ANISOMELES INDICA	S	.	0.0	14.6	27.5	.	.	.	.	.	.	.	.	.	.	.	.	.	21
465	ANNONA CHERIMOLA	F	110	68.7	1.5	0.1	29.0	2.5	0.7	9	24	0.2	.	.	0	0.11	0.11	1.00	12	2
466	ANNONA CHERIMOLA	F	82	76.6	1.1	0.2	21.3	1.9	0.8	34	35	0.6	.	.	0	0.09	0.13	0.90	17	3
467	ANNONA CHERIMOLA	F	94	73.5	1.3	0.4	24.0	2.2	0.8	23	40	0.5	.	.	6	0.10	0.11	1.30	9	4
468	ANNONA DIVERSIFOLIA	F	98	71.5	2.8	0.2	24.1	1.3	1.4	32	52	0.7	.	.	0	0.02	0.03	2.20	14	3
469	ANNONA GLABRA	F	52	86.8	0.4	1.6	10.3	3.8	0.9	52	24	2.3	.	.	310	0.04	0.07	0.60	21	3
470	ANNONA MURICATA	F	59	83.2	1.0	0.2	15.1	0.6	0.5	14	21	0.5	8	293	0	0.08	0.10	1.30	24	2
471	ANNONA MURICATA	F	60	83.1	1.0	0.4	14.9	1.1	0.6	24	28	0.5	.	.	10	0.07	0.05	0.90	26	3
472	ANNONA MURICATA	F	65	81.7	1.0	0.3	16.3	1.1	0.7	14	27	0.6	14	265	6	0.07	0.05	0.90	20	4
473	ANNONA PURPUREA	F	101	85.0	0.7	10.5	2.9	1.2	0.9	19	22	1.0	720	.	.	0.05	0.07	0.80	28	3
474	ANNONA RETICULATA	F	76	78.3	1.5	0.3	19.0	1.4	0.9	27	32	0.5	6	495	20	0.11	0.07	0.60	21	2
475	ANNONA RETICULATA	F	75	78.6	1.9	0.2	18.5	2.2	0.8	24	26	1.0	.	.	0	0.07	0.12	0.70	30	3
476	ANNONA RETICULATA	F	101	71.5	1.7	0.6	25.2	3.4	1.0	27	20	0.8	.	.	0	0.08	0.10	0.50	22	4
477	ANNONA SQUAMOSA	F	78	77.5	1.4	0.2	20.0	1.6	0.9	30	36	0.6	5	299	5	0.11	0.10	0.80	36	2
478	ANNONA SQUAMOSA	F	96	72.8	1.6	0.2	24.6	1.6	0.8	28	36	1.8	.	.	0	0.11	0.15	0.90	35	3
479	ANNONA SQUAMOSA	F	94	73.3	1.8	0.3	23.7	1.7	0.9	22	41	0.6	11	275	6	0.10	0.14	1.00	34	4
480	ANODA CRISTATA	S	.	0.0	19.9	18.1	.	.	.	.	.	.	.	.	.	.	.	.	.	20
481	ANOGEISSUS LATIFOLIA	L	.	.	7.5	3.6	69.4	24.2	9.5	3030	340	.	.	.	.	.	.	.	.	18
482	ANTHEMIS SANCTI-JOHANNIS	S	.	0.0	17.7	24.4	.	.	6.8	.	.	.	.	.	.	.	.	.	.	19
483	ANTHEMIS TINCTORIA	S	.	0.0	18.6	19.2	.	.	.	.	.	.	.	.	.	.	.	.	.	20
484	ANTHEMIS TINCTORIA	S	.	0.0	33.7	38.1	.	.	.	.	.	.	.	.	.	.	.	.	.	21
485	ANTHEMIS TINCTORIA VAR. KELWAYI	S	.	0.0	16.2	22.6	.	.	.	.	.	.	.	.	.	.	.	.	.	19
486	ANTHEMIS TINCTORIA VAR. KELWAYI	S	.	0.0	16.2	22.3	.	.	.	.	.	.	.	.	.	.	.	.	.	20
487	ANTHEMIS TINCTORIA VAR. KELWAYI	S	.	0.0	15.0	23.0	.	.	.	.	.	.	.	.	.	.	.	.	.	20
488	ANTHEPHORA NIGRITANA	W	.	0.0	14.1	1.8	79.4	38.0	8.4	.	.	.	.	.	.	.	.	.	.	18
489	ANTHOXANTHUM ODORATUM	W	.	0.0	20.0	4.6	65.4	17.8	10.0	635	400	61.0	.	.	.	.	.	.	.	6
490	ANTHRISCUS CEREFOLIUM	L	237	7.2	23.2	3.9	49.1	11.3	16.6	1346	450	32.0	83	4740	.	.	.	.	.	16
491	ANTHRISCUS CEREFOLIUM	S	.	0.0	18.8	15.6	.	.	.	.	.	.	.	.	.	.	.	.	.	19
492	ANTHRISCUS CEREFOLIUM	S	.	0.0	19.4	10.5	.	.	.	.	.	.	.	.	.	.	.	.	.	20
493	ANTHYLLIS CORNICINA	S	.	0.0	45.3	9.0	.	.	.	.	.	.	.	.	.	.	.	.	.	21
494	ANTHYLLIS LOTOIDES	S	.	0.0	43.8	7.3	.	.	.	.	.	.	.	.	.	.	.	.	.	21
495	ANTIDESMA BUNIUS	F	36	89.9	0.6	0.6	8.1	0.8	0.8	30	24	0.8	.	.	25	0.01	0.04	0.40	6	2

TABLE 2 - DATA DIRECTLY TRANSCRIBED FROM ORIGINAL SOURCES.

OBS	PLNA	PART	CALORIES	WATER	PROTEIN	FAT	TOTCARB	FIBER	ASH	CALCIUM	PHOSPHO	IRON	SODIUM	POTASS	CAROTE	THIAMINE	RIBOFLAV	NIACIN	ASCORBIC	REFCO
496	ANTIDESMA PLATYPHYLLUM	F	67	80.7	0.4	0.1	18.0	0.1	0.8	16	42	1.8	.	.	130	0.01	0.03	0.5	17	3
497	ANTIGONON LEPTOPUS	S	.	0.0	10.6	1.7	.	.	.	.	.	.	.	.	.	.	.	.	.	20
498	ANTIRRHINUM MAJUS	S	.	0.0	16.2	42.9	.	.	5.0	.	.	.	.	.	.	.	.	.	.	19
499	ANTIRRHINUM MAJUS	S	.	0.0	.	36.2	.	.	.	.	.	.	.	.	.	.	.	.	.	21
500	ANTIRRHINUM ORONTIUM	S	.	0.0	18.6	39.1	.	.	.	.	.	.	.	.	.	.	.	.	.	21
501	ANTIRRHINUM SICULUM	S	.	0.0	15.1	37.6	.	.	.	.	.	.	.	.	.	.	.	.	.	21
502	ANTROCARYON MICRASTER	S	683	3.7	10.8	70.0	14.6	2.5	0.9	67	558	.	.	.	.	.	.	.	.	1
503	APHANOSTEPHUS ARIZONICUS	S	.	0.0	12.5	11.8	.	.	.	.	.	.	.	.	.	.	.	.	.	20
504	APHANOSTEPHUS SP.	W	.	0.0	14.4	3.1	73.9	22.6	8.6	1050	200	.	.	2540	.	.	.	.	.	6
505	APIUM GRAVEOLENS	S	392	6.0	18.1	25.3	41.4	11.8	9.3	1767	547	44.9	160	1400	31	.	.	.	17	16
506	APIUM GRAVEOLENS	L	.	0.0	25.6	.	.	.	16.0	1310	.	.	2340	2050	.	.	.	.	.	17
507	APIUM GRAVEOLENS	L	.	0.0	27.2	6.9	46.9	3.5	19.0	2700	1700	.	.	.	.	.	.	.	.	18
508	APIUM GRAVEOLENS	S	.	0.0	14.8	32.9	.	.	6.2	.	.	.	.	.	.	.	.	.	.	19
509	APIUM GRAVEOLENS	S	.	0.0	23.3	30.4	.	.	7.3	.	.	.	.	.	.	.	.	.	.	19
510	APIUM GRAVEOLENS	S	.	0.0	20.5	36.2	.	.	.	.	.	.	.	.	.	.	.	.	.	21
511	APIUM GRAVEOLENS	L	.	0.0	28.2	6.2	.	7.9	.	.	.	.	.	.	.	.	.	.	.	6
512	APIUM GRAVEOLENS VAR. DULCE	L	20	92.7	1.1	0.1	4.5	1.0	1.6	87	50	0.9	.	.	.	0.03	0.25	.	17	1
513	APIUM GRAVEOLENS VAR. DULCE	L	21	92.8	1.4	0.3	4.2	1.0	1.3	62	37	2.5	96	326	1040	0.06	0.07	0.4	20	2
514	APIUM GRAVEOLENS VAR. DULCE	L	17	94.1	0.9	0.1	3.9	0.6	1.0	39	28	0.3	126	341	144	0.03	0.03	0.3	9	4
515	APIUM SELLOWIANUM	S	.	0.0	20.2	25.6	.	.	.	.	.	.	.	.	.	.	.	.	.	21
516	APOCYNUM CANNABINUM	S	.	0.0	28.7	22.6	.	.	4.7	.	.	.	.	.	.	.	.	.	.	19
517	APOCYNUM CANNABINUM	S	.	0.0	33.1	23.8	.	.	.	.	.	.	.	.	.	.	.	.	.	19
518	APODANTHERA UNDULATA	S	.	0.0	41.9	47.7	.	.	.	.	.	.	.	.	.	.	.	.	.	19
519	AQUILEGIA ALPINA	S	.	0.0	21.2	27.6	.	.	.	.	.	.	.	.	.	.	.	.	.	20
520	AQUILEGIA CAERULEA	S	.	0.0	23.8	30.2	.	.	.	.	.	.	.	.	.	.	.	.	.	20
521	AQUILEGIA SP.	S	.	0.0	20.6	26.4	.	.	.	.	.	.	.	.	.	.	.	.	.	20
522	AQUILEGIA VULGARIS	S	.	0.0	19.6	26.9	.	.	.	.	.	.	.	.	.	.	.	.	.	21
523	ARABIS ALPINA	S	.	0.0	24.4	30.6	.	.	.	.	.	.	.	.	.	.	.	.	.	20
524	ARABIS ALPINA	S	.	0.0	21.2	32.6	.	.	.	.	.	.	.	.	.	.	.	.	.	20
525	ARABIS ALPINA	S	.	0.0	20.6	32.5	.	.	.	.	.	.	.	.	.	.	.	.	.	20
526	ARABIS GLABRA	S	.	0.0	24.9	37.7	.	.	.	.	.	.	.	.	.	.	.	.	.	20
527	ARABIS LAEVIGATA	S	.	0.0	23.8	38.7	.	.	.	.	.	.	.	.	.	.	.	.	.	20
528	ARABIS SP.	S	.	0.0	28.6	38.0	.	.	.	.	.	.	.	.	.	.	.	.	.	20
529	ARACHIS HYPOGAEA	L	69	78.5	4.4	0.6	14.9	4.6	1.6	262	82	4.2	.	.	7735	0.23	0.58	1.6	98	1
530	ARACHIS HYPOGAEA	SH	.	0.0	17.5	2.2	71.7	20.1	8.6	930	200	.	.	.	.	.	.	.	.	18
531	ARACHIS HYPOGAEA	H	.	0.0	26.2	2.7	59.5	27.9	11.6	3150	810	.	.	.	.	.	.	.	.	18
532	ARACHIS HYPOGAEA	S	303	42.2	15.0	19.4	21.8	1.1	1.6	56	245	2.1	4	421	20	0.97	0.18	9.7	11	2
533	ARACHIS HYPOGAEA	S	564	5.6	26.0	47.5	18.6	2.4	2.3	69	401	2.1	5	674	.	1.14	0.13	17.2	0	4
534	ARACHIS HYPOGAEA	W	.	0.0	16.4	9.5	66.8	25.1	7.3	1710	290	.	.	970	.	.	.	.	.	6
535	ARALIA CORDATA	L	14	95.3	1.0	0.2	2.9	0.5	0.6	13	25	0.3	2	310	0	0.06	0.02	0.8	5	2
536	ARALIA ELATA	L	48	84.0	5.5	0.9	7.6	2.2	2.0	81	139	7.4	.	.	.	0.10	0.05	1.0	10	2
537	ARALIA SPINOSA	S	.	0.0	17.5	46.3	.	.	3.7	.	.	.	.	.	.	.	.	.	.	19
538	ARALIA SPINOSA	S	.	0.0	5.8	4.2	.	.	5.7	.	.	.	.	.	.	.	.	.	.	19
539	ARBUTUS ANDRACHNE	S	.	0.0	10.6	1.6	.	.	.	.	.	.	.	.	.	.	.	.	.	21
540	ARBUTUS ANDRACHNE	S	.	0.0	11.1	0.7	.	.	.	.	.	.	.	.	.	.	.	.	.	21

TABLE 2 - DATA DIRECTLY TRANSCRIBED FROM ORIGINAL SOURCES.

OBS	PLNA	PART	CALORIES	WATER	PROTEIN	FAT	TOTLCARB	FIBER	ASH	CALCIUM	PHOSPHON	IRON	SODIUM	POTASS	CAROTE	THIAMINE	RIBOFLAV	NIACIN	ASCORBIC	REF
541	ARBUTUS MENZIESII	S	.	0.0	15.6	22.2	.	.	.	.	.	.	.	.	.	.	.	.	.	20
542	ARBUTUS UNEDO	S	.	0.0	33.1	36.8	.	.	.	.	.	.	.	.	.	.	.	.	.	20
543	ARBUTUS UNEDO	S	.	0.0	3.1	3.4	.	.	.	.	.	.	.	.	.	.	.	.	.	20
544	ARBUTUS UNEDO	S	.	0.0	23.5	16.5	.	.	.	.	.	.	.	.	.	.	.	.	.	21
545	ARCTIUM LAPPA	R	89	76.5	2.5	0.1	20.1	1.7	0.8	50	58	1.2	30	180	0	0.25	0.08	0.3	2	2
546	ARCTIUM LAPPA	S	.	0.0	16.8	21.4	.	.	.	.	.	.	.	.	.	.	.	.	.	21
547	ARCTIUM MINUS	S	.	0.0	16.9	20.8	.	.	4.2	.	.	.	.	.	.	.	.	.	.	19
548	ARCTIUM MINUS	S	.	0.0	18.8	16.6	.	.	.	.	.	.	.	.	.	.	.	.	.	20
549	ARCTIUM MINUS	S	.	0.0	21.3	16.1	.	.	.	.	.	.	.	.	.	.	.	.	.	21
550	ARCTIUM MINUS	S	.	0.0	17.6	16.5	.	.	.	.	.	.	.	.	.	.	.	.	.	21
551	ARCTIUM MINUS	S	.	0.0	18.6	13.8	.	.	.	.	.	.	.	.	.	.	.	.	.	21
552	ARCTIUM MINUS	S	.	0.0	21.6	21.4	.	.	.	.	.	.	.	.	.	.	.	.	.	21
553	ARCTIUM MINUS	S	.	0.0	19.9	15.4	.	.	.	.	.	.	.	.	.	.	.	.	.	21
554	ARCTOSTAPHYLOS GLAUCA	S	.	0.0	24.9	55.7	.	.	3.2	.	.	.	.	.	.	.	.	.	.	19
555	ARCTOSTAPHYLOS PATULA	S	.	0.0	3.1	1.8	.	.	.	.	.	.	.	.	.	.	.	.	.	21
556	ARCTOSTAPHYLOS PUNGENS	S	.	0.0	2.5	3.6	.	.	.	.	.	.	.	.	.	.	.	.	.	19
557	ARCTOSTAPHYLOS PUNGENS	S	.	0.0	2.5	3.7	.	.	.	.	.	.	.	.	.	.	.	.	.	19
558	ARCTOSTAPHYLOS TOMENTOSA	F	111	69.7	1.2	1.1	27.1	1.4	0.9	.	33	.	.	.	1020	0.03	0.00	0.4	78	3
559	ARCTOSTAPHYLOS UVA-URSI	S	.	0.0	25.6	50.0	.	.	3.3	.	.	.	.	.	.	.	.	.	.	19
560	ARCTOSTAPHYLOS UVA-URSI	S	.	0.0	2.5	2.0	.	.	0.6	.	.	.	.	.	.	.	.	.	.	19
561	ARCTOTIS GRANDIS	S	.	0.0	36.2	47.7	.	.	6.3	.	.	.	.	.	.	.	.	.	.	19
562	ARCTOTIS GRANDIS	S	.	0.0	12.5	10.2	.	.	7.1	.	.	.	.	.	.	.	.	.	.	19
563	ARDISIA CRENULATA	S	.	0.0	7.5	0.7	.	.	2.7	.	.	.	.	.	.	.	.	.	.	19
564	ARDISIA ELLIPTCA	F	100	72.1	1.3	0.9	24.5	3.1	1.2	130	38	0.8	.	.	0	0.00	0.01	1.6	10	3
565	ARECA CATECHU	SH	43	86.4	3.3	0.3	9.0	.	1.0	6	89	2.0	.	.	.	.	.	.	.	2
566	ARECA CATECHU	S	394	12.3	6.0	10.8	69.4	15.9	1.5	542	63	5.7	76	446	0	0.17	0.69	0.6	0	2
567	ARECA CATECHU	S	.	31.3	4.9	4.4	47.2	.	1.0	50	130	1.5	.	.	.	.	.	.	.	5
568	ARECASTRUM ROMANZOFFIANUM	S	.	0.0	12.8	64.7	.	.	.	.	.	.	.	.	.	.	.	.	.	19
569	ARENGA PINNATA	SH	19	94.7	0.1	0.2	4.9	0.5	0.1	21	3	0.5	2	7	.	0.00	0.01	0.1	.	2
570	ARGANIA SPINOSA	S	.	0.0	9.0	71.7	17.2	6.8	2.1	.	.	.	.	.	.	.	.	.	.	18
571	ARGEMONE ALBIFLORA	S	.	0.0	17.5	33.0	.	.	.	.	.	.	.	.	.	.	.	.	.	20
572	ARGEMONE ALBIFLORA	S	.	0.0	16.9	33.0	.	.	.	.	.	.	.	.	.	.	.	.	.	20
573	ARGEMONE HISPIDA	S	.	0.0	17.0	31.5	.	.	7.3	.	.	.	.	.	.	.	.	.	.	19
574	ARGEMONE INTERMEDIA	S	.	0.0	18.2	34.0	.	.	6.8	.	.	.	.	.	.	.	.	.	.	19
575	ARGEMONE INTERMEDIA	S	.	0.0	19.4	40.1	.	.	5.4	.	.	.	.	.	.	.	.	.	.	19
576	ARGEMONE INTERMEDIA	S	.	0.0	18.1	38.8	.	.	.	.	.	.	.	.	.	.	.	.	.	20
577	ARGEMONE MEXICANA	S	.	0.0	18.1	36.4	.	.	.	.	.	.	.	.	.	.	.	.	.	19
578	ARGEMONE MEXICANA	S	.	0.0	17.6	39.3	.	.	.	.	.	.	.	.	.	.	.	.	.	21
579	ARGEMONE MUNITA SUBSP. ROTUNDATA	S	.	0.0	15.4	34.4	.	.	.	.	.	.	.	.	.	.	.	.	.	21
580	ARGEMONE PLATYCERAS	S	.	0.0	24.4	33.7	.	.	6.6	.	.	.	.	.	.	.	.	.	.	19
581	ARGEMONE POLYANTHEMOS	S	.	0.0	18.6	36.2	.	.	.	.	.	.	.	.	.	.	.	.	.	21
582	ARGEMONE SP.	S	.	0.0	19.4	36.8	.	.	.	.	.	.	.	.	.	.	.	.	.	19
583	ARGYROLOBIUM UNIFLORUM	S	.	0.0	30.8	4.6	.	.	.	.	.	.	.	.	.	.	.	.	.	21
584	ARISAEMA JACQUEMONTII	S	.	0.0	14.0	1.8	.	.	.	.	.	.	.	.	.	.	.	.	.	21
585	ARISAEMA TRIPHYLLUM	S	.	0.0	15.0	2.8	.	.	.	.	.	.	.	.	.	.	.	.	.	19

TABLE 2 - DATA DIRECTLY TRANSCRIBED FROM ORIGINAL SOURCES.

OBS	PLNA	PART	CALORIES	WATER	PROTEIN	FAT	TOTLCARB	FIBER	ASH	CALCIUM	PHOSPHO	IRON	SODIUM	POTASS	CAROT	THIAMINE	RIBOFLAV	NIACIN	ASCORBIC	REFCO
586	ARISTIDA ADSCENSIONIS	W	.	0.0	10.5	1.6	78.4	33.6	9.5	.	.	.	.	.	.	.	.	.	.	18
587	ARISTIDA ADSCENSIONIS	H	.	0.0	2.0	0.9	89.2	45.7	7.9	.	.	.	.	.	.	.	.	.	.	18
588	ARISTIDA LONGISETA	W	.	0.0	5.6	1.4	81.9	36.1	11.1	760	90	.	.	.	.	.	.	.	.	6
589	ARISTIDA LONGSPICA	W	.	0.0	6.8	2.0	80.6	31.9	10.6	.	.	.	.	.	.	.	.	.	.	6
590	ARISTIDA OLIGANTHA	W	.	0.0	5.2	2.2	82.8	31.4	9.8	.	.	.	.	.	.	.	.	.	.	6
591	ARISTIDA PURPURASCENS	W	.	0.0	5.2	1.2	82.0	35.9	11.6	.	.	.	.	.	.	.	.	.	.	6
592	ARISTIDA PURPUREA	W	.	0.0	6.8	1.4	80.0	33.9	11.8	.	.	.	.	.	.	.	.	.	.	6
593	ARISTIDA WRIGHTII	W	.	0.0	7.8	1.7	79.1	32.4	11.4	.	.	.	.	.	.	.	.	.	.	6
594	ARISTOLOCHIA MAURORUM	S	.	0.0	21.1	23.4	.	.	.	.	.	.	.	.	.	.	.	.	.	20
595	ARMERIA PSEUD-ARMERIA	S	.	0.0	37.5	6.5	.	.	.	.	.	.	.	.	.	.	.	.	.	20
596	ARMERIA PSEUD-ARMERIA	S	.	0.0	32.5	7.4	.	.	.	.	.	.	.	.	.	.	.	.	.	20
597	ARMORACIA RUSTICANA	R	69	80.2	2.7	0.2	15.5	1.4	1.4	24	65	2.1	.	.	5	0.03	0.07	0.40	22	2
598	ARMORACIA RUSTICANA	R	87	74.6	3.2	0.3	19.7	2.4	2.2	140	64	1.4	8	564	.	0.07	.	.	81	4
599	ARRABIDAEA SP.	S	.	0.0	20.0	22.8	.	.	.	.	.	.	.	.	.	.	.	.	.	20
600	ARRACACIA XANTHORRHIZA	L	102	73.4	0.8	0.2	24.4	1.0	1.2	26	52	0.9	.	.	0	0.07	0.06	2.80	23	3
601	ARRHENATHERUM ELATIUS	S	.	0.0	25.0	8.3	.	.	.	.	.	.	.	.	.	.	.	.	.	20
602	ARRHENATHERUM ELATIUS	S	.	0.0	23.8	8.7	.	.	.	.	.	.	.	.	.	.	.	.	.	20
603	ARRHENATHERUM ELATIUS	W	.	0.0	9.5	3.1	80.4	31.2	7.0	.	.	.	.	.	47	.	.	.	.	6
604	ARTEDIA SQUAMATA	S	.	0.0	11.2	9.5	.	.	.	.	.	.	.	.	.	.	.	.	.	21
605	ARTEDIA SQUAMATA	S	.	0.0	8.3	4.8	.	.	.	.	.	.	.	.	.	.	.	.	.	21
606	ARTEMESIA ABSINTHIUM	S	.	0.0	25.8	33.4	.	.	6.6	.	.	.	.	.	.	.	.	.	.	19
607	ARTEMISIA ABSINTHIUM	S	.	0.0	28.1	38.1	.	.	.	.	.	.	.	.	.	.	.	.	.	20
608	ARTEMISIA ARBUSCULA	W	.	0.0	8.6	7.3	77.3	20.2	6.8	.	.	.	.	.	.	.	.	.	.	6
609	ARTEMISIA CAMPESTRIS	S	.	0.0	33.1	35.6	.	.	.	.	.	.	.	.	.	.	.	.	.	21
610	ARTEMISIA DRACUNCULUS	H	295	7.7	22.8	7.2	50.2	7.4	12.0	1139	313	32.3	62	3020	2520	0.25	1.34	8.95	.	16
611	ARTEMISIA DRACUNCULUS	S	.	0.0	31.5	38.1	.	.	.	.	.	.	.	.	.	.	.	.	.	21
612	ARTEMISIA FILIFOLIA	W	.	0.0	11.2	8.2	75.4	23.5	5.2	750	230	.	.	.	.	.	.	.	.	6
613	ARTEMISIA FRIGIDA	S	.	0.0	30.4	31.4	.	.	.	.	.	.	.	.	.	.	.	.	.	21
614	ARTEMISIA FRIGIDA	W	.	0.0	9.4	4.0	72.3	30.8	14.3	600	290	.	.	.	.	.	.	.	.	6
615	ARTEMISIA GNAPHALODES	W	.	0.0	11.1	3.1	77.7	26.9	8.1	.	.	.	.	.	.	.	.	.	.	6
616	ARTEMISIA HERBA-ALBA	S	.	0.0	30.1	30.6	.	.	.	.	.	.	.	.	.	.	.	.	.	21
617	ARTEMISIA LACTIFLORA	L	40	88.1	2.8	0.8	7.3	1.4	1.0	45	48	2.2	.	.	2700	0.12	0.26	0.40	36	2
618	ARTEMISIA MARITIMA	S	.	0.0	43.8	22.0	.	.	.	.	.	.	.	.	.	.	.	.	.	21
619	ARTEMISIA NOVA	W	508	0.0	8.6	10.0	75.3	27.5	6.1	650	200	.	.	.	2	.	.	.	.	6
620	ARTEMISIA PEDATIFIDA	W	.	0.0	7.0	2.4	62.8	23.7	27.8	.	.	.	.	.	.	.	.	.	.	6
621	ARTEMISIA RIGIDA	W	.	0.0	21.1	3.7	68.6	22.0	6.6	.	.	.	.	.	.	.	.	.	.	6
622	ARTEMISIA SPINESCENS	W	424	0.0	18.5	3.3	61.8	20.1	16.4	1570	420	.	.	.	2	.	.	.	.	6
623	ARTEMISIA TRIDENTATA	W	.	0.0	9.9	13.0	71.1	25.3	6.0	720	180	.	.	.	2	.	.	.	.	6
624	ARTEMISIA TRIPARTITA	W	501	0.0	13.1	.	.	.	.	790	210	.	.	.	.	.	.	.	.	6
625	ARTEMISIA VARIABILIS	S	.	0.0	24.8	38.7	.	.	.	.	.	.	.	.	.	.	.	.	.	21
626	ARTEMISIA VULGARIS	L	35	87.3	5.2	0.8	4.5	3.4	2.2	82	40	1.5	.	.	2140	0.15	0.16	3.00	72	2
627	ARTEMISIA VULGARIS	W	.	0.0	11.4	6.0	76.3	33.5	6.3	860	190	.	.	2200	.	.	.	.	.	6
628	ARTOCARPUS ALTILIS	F	96	72.9	1.3	0.3	24.7	1.3	0.8	29	40	0.7	13	396	10	0.08	0.06	1.20	12	2
629	ARTOCARPUS ALTILIS	L	75	75.5	5.0	0.0	2.0	.	2.0	170	60	17.5	.	.	.	0.10	.	.	70	2
630	ARTOCARPUS ALTILIS	S	434	20.2	15.1	29.0	34.0	2.5	1.7	66	320	6.7	41	380	280	0.88	0.55	0.80	12	2

TABLE 2 - DATA DIRECTLY TRANSCRIBED FROM ORIGINAL SOURCES.

OBS	PLANA	PART	CALORIES	WATER	PROTEIN	FAT	TOTCARB	FIBER	ASH	CALCIUM	PHOSPHO	IRON	SODIUM	POTASS	CAROTENE	THIAMINE	RIBOFLAV	NIACIN	ASCORBIC	REFCO
631	ARTOCARPUS ALTILIS	F	103	70.8	1.7	0.3	26.2	1.2	1.0	33	32	1.2	15	439	24	0.11	0.03	0.9	29	4
632	ARTOCARPUS HETEROPHYLLUS	L	.	0.0	18.5	5.0	66.3	26.2	10.2	2000	110	.	.	.	.	.	.	.	.	18
633	ARTOCARPUS HETEROPHYLLUS	F	94	72.9	1.7	0.3	23.7	0.9	1.4	27	38	0.6	2	407	235	0.09	0.11	0.7	9	2
634	ARTOCARPUS HETEROPHYLLUS	G	53	84.7	2.1	0.4	11.9	2.8	0.9	48	25	0.3	3	323	10	0.10	0.04	0.3	16	2
635	ARTOCARPUS ODORATISSIMA	S	151	60.9	4.3	0.4	32.6	1.5	1.8	35	126	1.2	22	841	25	0.18	0.05	0.5	17	2
636	ARTOCARPUS ODORATISSIMA	L	75	75.5	5.0	0.0	17.5	.	2.0	170	60	17.5	.	.	.	0.10	.	.	70	2
637	ARTOCARPUS ODORATISSIMA	F	104	70.6	0.5	0.2	27.9	0.8	0.8	12	33	3.5	.	.	.	0.06	0.05	0.6	30	2
638	ARTOCARPUS ODORATISSIMA	U	232	48.5	2.6	6.1	42.2	0.7	0.6	27	117	0.6	.	.	.	0.26	0.04	0.6	.	2
639	ARUM ORIENTALE	S	.	0.0	14.6	2.0	.	.	.	.	.	.	.	.	.	.	.	.	.	21
640	ARUNDINARIA CANNAVIEIRA	L	.	0.0	7.8	2.4	86.5	45.0	3.3	.	.	.	.	.	.	.	.	.	.	18
641	ARUNDINARIA JAPONICA	H	.	0.0	2.4	2.3	91.6	33.9	3.7	370	150	.	.	680	.	.	.	.	.	6
642	ARUNDINARIA JAPONICA	H	.	0.0	2.2	2.3	92.2	34.1	3.3	.	.	.	.	.	.	.	.	.	.	6
643	ARUNDINARIA JAPONICA	C	.	0.0	2.7	1.1	92.6	28.0	3.6	.	.	.	.	.	.	.	.	.	.	6
644	ARUNDINARIA TECTA	W	.	0.0	15.1	3.5	70.8	29.4	10.6	210	190	20.0	.	.	.	.	.	.	.	6
645	ARUNDINARIA TECTA	L	.	0.0	12.1	.	.	.	.	240	60	.	.	.	.	.	.	.	.	6
646	ARUNDO DONAX	W	.	0.0	13.2	1.9	69.9	28.5	15.0	.	.	.	.	.	.	.	.	.	.	18
647	ARUNDO DONAX	H	.	0.0	8.8	1.1	77.7	33.0	12.4	.	.	.	.	.	.	.	.	.	.	18
648	ARUNDO DONAX	W	.	0.0	6.9	1.3	82.6	37.5	9.2	490	110	.	.	2550	.	.	.	.	.	6
649	ASCLEPIAS CAPRICORNUS	S	.	0.0	33.8	22.3	.	.	.	.	.	.	.	.	.	.	.	.	.	19
650	ASCLEPIAS CURASSAVICA	S	.	0.0	22.4	25.9	.	.	.	.	.	.	.	.	.	.	.	.	.	21
651	ASCLEPIAS ENGELMANNIANA	S	.	0.0	41.5	38.4	.	.	6.2	.	.	.	.	.	.	.	.	.	.	19
652	ASCLEPIAS ENGELMANNIANA	S	.	0.0	32.6	18.1	.	.	3.7	.	.	.	.	.	.	.	.	.	.	19
653	ASCLEPIAS INCARNATA	S	.	0.0	30.9	26.8	.	.	3.9	.	.	.	.	.	.	.	.	.	.	19
654	ASCLEPIAS LINARIA	S	.	0.0	26.2	24.4	.	.	.	.	.	.	.	.	.	.	.	.	.	19
655	ASCLEPIAS SPECIOSA	S	.	0.0	31.8	25.2	.	.	.	.	.	.	.	.	.	.	.	.	.	21
656	ASCLEPIAS SUBVERTICILLATA	S	.	0.0	26.1	20.1	.	.	.	.	.	.	.	.	.	.	.	.	.	20
657	ASCLEPIAS SYRIACA	S	.	0.0	33.8	31.8	.	.	3.7	.	.	.	.	.	.	.	.	.	.	19
658	ASCLEPIAS SYRIACA	S	.	0.0	37.2	28.2	.	.	4.0	.	.	.	.	.	.	.	.	.	.	19
659	ASCLEPIAS SYRIACA	S	.	0.0	17.6	9.1	.	.	4.2	.	.	.	.	.	.	.	.	.	.	19
660	ASCLEPIAS TUBEROSA	S	.	0.0	32.5	24.1	.	.	.	.	.	.	.	.	.	.	.	.	.	20
661	ASCOPHYLLUM NODOSUM	W	.	0.0	8.4	3.2	.	.	20.9	1160	110	.	.	.	.	.	.	.	.	6
662	ASIMINA TRILOBA	F	85	76.6	5.2	0.9	16.8	.	0.5	.	.	.	.	.	.	.	.	.	.	4
663	ASPALATHUS CONTAMINATUS	L	313	10.1	5.6	1.5	80.3	.	2.5	133	220	46.3	.	.	.	.	.	.	11	1
664	ASPALATHUS SP.	S	.	0.0	41.0	4.5	.	.	.	.	.	.	.	.	.	.	.	.	.	21
665	ASPARAGUS ACUTIFOLIUS	S	.	0.0	18.5	19.8	.	.	.	.	.	.	.	.	.	.	.	.	.	21
666	ASPARAGUS OFFICINALIS	S	.	0.0	16.2	15.4	.	.	2.4	.	.	.	.	.	.	.	.	.	.	19
667	ASPARAGUS OFFICINALIS	SH	21	92.7	2.5	0.3	3.6	0.9	0.9	16	59	1.4	5	221	380	0.15	0.18	1.8	20	2
668	ASPARAGUS OFFICINALIS	SH	23	91.8	2.4	0.3	4.1	0.8	1.4	25	84	0.9	3	.	30	0.16	0.30	3.0	10	2
669	ASPARAGUS OFFICINALIS	SH	26	91.7	2.5	0.2	5.0	0.7	0.6	22	62	1.0	2	278	540	0.18	0.20	1.5	33	4
670	ASPARAGUS OFFICINALIS	SH	.	93.0	2.2	0.2	3.9	0.7	0.7	25	39	1.0	.	.	840	0.18	0.13	.	40	5
671	ASPARAGUS THUNBERGIANUS	S	.	0.0	13.5	8.9	.	.	.	.	.	.	.	.	.	.	.	.	.	21
672	ASPHODELINE ISTHMOCARPA	S	.	0.0	28.4	24.6	.	.	.	.	.	.	.	.	.	.	.	.	.	21
673	ASPHODELUS FISTULOSUS	S	.	0.0	21.4	23.8	.	.	.	.	.	.	.	.	.	.	.	.	.	21
674	ASPHODELUS MICROCARPUS	S	.	0.0	38.1	13.6	.	.	.	.	.	.	.	.	.	.	.	.	.	21
675	ASPHODELUS TENUIFOLIUS	S	.	0.0	18.6	16.8	.	.	.	.	.	.	.	.	.	.	.	.	.	21

TABLE 2 - DATA DIRECTLY TRANSCRIBED FROM ORIGINAL SOURCES.

OBS	PLNA	PART	CALORIES	WATER	PROTEIN	FAT	TOTLCARB	FIBER	ASH	CALCIUM	PHOSPHON	IRON	SODIUM	POTASS	CAROT	THIAMINE	RIBOFLAV	NIACIN	ASCORBIC	REFCO
676	ASPHODELUS TENUIFOLIUS	S	.	0.0	15.2	28.8	.	.	.	.	.	.	.	.	.	.	.	.	.	21
677	ASTER ALPINUS	S	.	0.0	19.3	22.7	.	.	5.6	.	.	.	.	.	.	.	.	.	.	19
678	ASTER AMELLUS	L	39	87.2	4.2	0.7	6.4	1.1	1.5	42	76	4	.	533	3345	0.18	0.36	1.1	88	2
679	ASTER ERICOIDES	S	.	0.0	28.1	30.4	.	.	5.5	.	.	.	.	.	.	.	.	.	.	19
680	ASTER ERICOIDES	W	.	0.0	7.6	5.0	81.0	21.1	6.4	2470	100	.	.	940	.	.	.	.	.	6
681	ASTER FREMONTII	C	.	0.0	5.3	1.5	.	.	5.8	910	730	.	.	.	.	.	.	.	.	6
682	ASTER FREMONTII	W	.	0.0	9.1	3.1	.	.	8.8	1260	350	.	.	.	.	.	.	.	.	6
683	ASTER FREMONTII	L	.	0.0	12.2	4.8	.	.	11.7	1600	510	.	.	.	.	.	.	.	.	6
684	ASTER NOVAE-ANGLIAE	S	.	0.0	32.5	28.7	.	.	.	.	.	.	.	.	.	.	.	.	.	20
685	ASTER NOVI-BELGII	S	.	0.0	27.5	24.3	.	.	.	.	.	.	.	.	.	.	.	.	.	20
686	ASTER SQUAMATUS	S	.	0.0	27.9	29.8	.	.	.	.	.	.	.	.	.	.	.	.	.	21
687	ASTER TANACETIFOLIUS	S	.	0.0	15.8	8.4	.	.	.	.	.	.	.	.	.	.	.	.	.	20
688	ASTER TONGOLENSIS	S	.	0.0	20.0	19.3	.	.	.	.	.	.	.	.	.	.	.	.	.	20
689	ASTER WASATCHENSIS	S	.	0.0	24.9	25.2	.	.	.	.	.	.	.	.	.	.	.	.	.	21
690	ASTRAGALUS ADSURGENS	S	.	0.0	42.1	5.3	.	.	3.7	.	.	.	.	.	.	.	.	.	.	19
691	ASTRAGALUS ANGUSTIFOLIUS	S	.	0.0	43.5	5.7	.	.	.	.	.	.	.	.	.	.	.	.	.	21
692	ASTRAGALUS ANGUSTIFOLIUS	S	.	0.0	43.1	5.4	.	.	.	.	.	.	.	.	.	.	.	.	.	21
693	ASTRAGALUS BOETICUS	S	.	0.0	29.5	1.2	.	.	.	.	.	.	.	.	.	.	.	.	.	21
694	ASTRAGALUS BRAZOENSIS	S	.	0.0	43.8	1.6	.	.	.	.	.	.	.	.	.	.	.	.	.	19
695	ASTRAGALUS CANADENSIS	S	.	0.0	40.2	2.8	.	.	3.0	.	.	.	.	.	.	.	.	.	.	19
696	ASTRAGALUS CICER	S	.	0.0	39.9	3.2	.	.	4.6	.	.	.	.	.	.	.	.	.	.	19
697	ASTRAGALUS CICER	S	.	0.0	37.0	3.6	.	.	6.2	.	.	.	.	.	.	.	.	.	.	19
698	ASTRAGALUS CICER	S	.	0.0	40.1	3.5	.	.	4.3	.	.	.	.	.	.	.	.	.	.	19
699	ASTRAGALUS CICER	S	.	0.0	38.8	4.2	.	.	4.2	.	.	.	.	.	.	.	.	.	.	19
700	ASTRAGALUS CRASSICARPUS	S	.	0.0	47.5	5.7	.	.	.	.	.	.	.	.	.	.	.	.	.	20
701	ASTRAGALUS FALCATUS	S	.	0.0	41.0	3.2	.	.	4.7	.	.	.	.	.	.	.	.	.	.	19
702	ASTRAGALUS FLEXUOSUS	S	.	0.0	43.8	4.0	.	.	5.6	.	.	.	.	.	.	.	.	.	.	19
703	ASTRAGALUS FLEXUOSUS	S	.	0.0	42.9	4.3	.	.	4.9	.	.	.	.	.	.	.	.	.	.	19
704	ASTRAGALUS GALEGIFORMIS	S	.	0.0	43.7	5.2	.	.	4.5	.	.	.	.	.	.	.	.	.	.	19
705	ASTRAGALUS GALEGIFORMIS	S	.	0.0	43.9	5.4	.	.	3.8	.	.	.	.	.	.	.	.	.	.	19
706	ASTRAGALUS GALEGIFORMIS	S	.	0.0	44.5	6.0	.	.	.	.	.	.	.	.	.	.	.	.	.	21
707	ASTRAGALUS GLYCYPHYLLOS	S	.	0.0	36.1	2.2	.	.	4.3	.	.	.	.	.	.	.	.	.	.	19
708	ASTRAGALUS GLYCYPHYLLOS	S	.	0.0	40.0	3.0	.	.	.	.	.	.	.	.	.	.	.	.	.	21
709	ASTRAGALUS HAMOSUS	S	.	0.0	34.2	1.4	.	.	.	.	.	.	.	.	.	.	.	.	.	20
710	ASTRAGALUS HAMOSUS	S	.	0.0	26.4	1.2	.	.	.	.	.	.	.	.	.	.	.	.	.	21
711	ASTRAGALUS HAMOSUS	S	.	0.0	34.8	1.7	.	.	.	.	.	.	.	.	.	.	.	.	.	21
712	ASTRAGALUS LONGIDENTATUS	S	.	0.0	30.4	2.0	.	.	.	.	.	.	.	.	.	.	.	.	.	21
713	ASTRAGALUS MACROCEPHALUS	S	.	0.0	50.9	5.5	.	.	.	.	.	.	.	.	.	.	.	.	.	21
714	ASTRAGALUS MACROCEPHALUS	S	.	0.0	31.2	2.7	.	.	.	.	.	.	.	.	.	.	.	.	.	21
715	ASTRAGALUS MENZIESII	S	.	0.0	36.5	4.9	.	.	2.0	.	.	.	.	.	.	.	.	.	.	19
716	ASTRAGALUS MEXICANUS	S	.	0.0	38.9	4.6	.	.	4.4	.	.	.	.	.	.	.	.	.	.	19
717	ASTRAGALUS MOLLISSIMUS	S	.	0.0	37.6	6.0	.	.	4.0	.	.	.	.	.	.	.	.	.	.	19
718	ASTRAGALUS NERVULOSUS	S	.	0.0	52.6	9.1	.	.	.	.	.	.	.	.	.	.	.	.	.	21
719	ASTRAGALUS NUTTALLIANUS	S	.	0.0	35.0	2.6	.	.	.	.	.	.	.	.	.	.	.	.	.	20
720	ASTRAGALUS NUTTALLIANUS VAR. TRICHOCARPUS	S	.	0.0	35.0	1.8	.	.	.	.	.	.	.	.	.	.	.	.	.	19

TABLE 2 - DATA DIRECTLY TRANSCRIBED FROM ORIGINAL SOURCES.

OBS	PLANA	PART	CALORIES	WATER	PROTEIN	FAT	TOTCARB	FIBER	ASH	CALCIUM	PHOSPHO	IRON	SODIUM	POTASS	CAROT	THIAMINE	RIBOFLAV	NIACIN	ASCORBIC	REFCO
766	ATRIPLEX LENTIFORMIS	S	.	0.0	12.5	2.1	.	.	13.6	.	.	.	.	.	.	.	.	.	.	19
767	ATRIPLEX NUMMULARIA	SH	.	0.0	21.3	2.6	53.9	10.3	22.2	1220	170	.	.	.	.	.	.	.	.	18
768	ATRIPLEX NUTTALLII	W	.	.	.	.	.	.	.	3270	130	.	.	.	.	.	.	.	.	6
769	ATRIPLEX ROSEA	L	.	0.0	15.5	1.8	59.7	14.5	23.0	.	.	.	.	.	.	.	.	.	.	6
770	ATRIPLEX ROSEA	L	.	0.0	23.7	2.1	42.5	8.1	31.7	.	.	.	.	.	.	.	.	.	.	6
771	ATRIPLEX SEMIBACCATA	W	.	0.0	16.6	1.6	58.2	18.7	23.6	1810	140	.	.	.	.	.	.	.	.	18
772	ATRIPLEX SEMIBACCATA	SH	.	0.0	21.4	1.4	57.9	17.0	19.3	.	.	.	.	.	.	.	.	.	.	18
773	ATRIPLEX SEMIBACCATA	W	.	0.0	17.5	1.6	59.2	18.8	21.7	.	.	.	.	.	.	.	.	.	.	6
774	ATRIPLEX SP.	S	.	0.0	23.8	1.9	.	.	.	.	.	.	.	.	.	.	.	.	.	19
775	ATRIPLEX TATARICA	S	.	0.0	20.6	11.1	.	.	.	.	.	.	.	.	.	.	.	.	.	21
776	ATRIPLEX TATARICA	S	.	0.0	19.9	6.7	.	.	.	.	.	.	.	.	.	.	.	.	.	21
777	ATROPA ACUMINATA	S	.	0.0	18.4	41.3	.	.	.	.	.	.	.	.	.	.	.	.	.	21
778	ATROPA BELLADONA	S	.	0.0	21.0	27.6	.	.	.	.	.	.	.	.	.	.	.	.	.	21
779	AUCUBA JAPONICA	S	.	0.0	8.8	0.8	.	.	3.8	.	.	.	.	.	.	.	.	.	.	19
780	AULAX PALLASIA	S	.	0.0	71.1	23.7	.	.	.	.	.	.	.	.	.	.	.	.	.	21
781	AVENA BARBATA	S	.	0.0	16.2	11.3	.	.	2.0	.	.	.	.	.	.	.	.	.	.	19
782	AVENA BARBATA	S	.	0.0	17.5	9.7	.	.	2.6	.	.	.	.	.	.	.	.	.	.	19
783	AVENA BARBATA	W	.	0.0	8.6	.	.	34.1	8.5	320	310	.	.	2620	.	.	.	.	.	6
784	AVENA BYZANTINA	W	.	0.0	26.1	5.3	56.9	18.1	11.7	.	.	.	.	.	.	.	.	.	.	18
785	AVENA BYZANTINA	W	.	0.0	8.5	3.0	81.7	18.4	6.8	.	.	.	.	.	.	.	.	.	.	18
786	AVENA BYZANTINA	S	.	0.0	12.8	5.6	77.6	13.9	4.0	.	.	.	.	.	.	.	.	.	.	6
787	AVENA FATUA	S	.	0.0	14.8	11.5	.	.	2.8	.	.	.	.	.	.	.	.	.	.	19
788	AVENA FATUA	H	.	0.0	7.5	2.9	82.0	34.1	7.6	.	.	.	.	.	.	.	.	.	.	6
789	AVENA SATIVA		.	0.0	34.1	6.5	49.4	21.4	10.0	.	.	.	.	.	.	.	.	.	.	18
790	AVENA SATIVA	H	.	0.0	7.1	1.9	86.1	30.9	4.9	.	.	.	.	.	.	.	.	.	.	18
791	AVENA SATIVA	S	374	11.0	13.1	6.1	67.4	5.8	2.4	59	425	4.6	10	.	0	0.35	0.09	2.20	0	2
792	AVENA SATIVA	W	.	0.0	18.0	3.9	67.4	26.3	10.7	400	370	57.0	.	3230	43	0.13	1.10	4.94	.	6
793	AVENA SATIVA	H	454	0.0	9.2	3.1	80.2	31.0	7.5	260	240	53.0	.	970	10	0.35	1.01	.	.	6
794	AVENA SATIVA	S	471	0.0	13.3	5.1	77.5	12.0	4.1	110	390	8.0	.	420	0	0.70	0.17	1.79	.	6
795	AVERRHOA BILIMBI	F	27	92.5	0.6	0.3	6.3	0.6	0.3	5	13	0.6	4	130	105	0.02	0.04	0.20	35	2
796	AVERRHOA BILIMBI	F	20	94.2	0.8	0.1	4.6	0.6	0.3	3	11	1.0	.	.	40	0.01	0.03	0.30	16	3
797	AVERRHOA CARAMBOLA	F	45	87.4	0.4	0.3	11.5	0.8	0.4	6	16	0.9	.	.	210	0.03	0.02	0.40	39	1
798	AVERRHOA CARAMBOLA	F	28	92.3	0.3	0.4	6.7	1.0	0.3	8	15	0.9	2	181	160	0.05	0.04	0.40	38	2
799	AVERRHOA CARAMBOLA	F	36	90.6	0.5	0.3	8.8	0.6	0.4	5	18	0.4	.	.	180	0.04	0.02	0.30	35	3
800	AVERRHOA CARAMBOLA	F	35	90.4	0.7	0.5	8.0	0.9	0.4	4	17	1.5	2	192	720	0.04	0.02	0.30	35	4
801	AVICENNIA GERMINANS	S	354	9.8	5.6	0.5	81.3	4.0	2.8	207	117	.	.	.	.	.	.	.	.	1
802	AVICENNIA NITIDA	L	.	0.0	10.7	4.4	69.2	23.9	15.7	.	.	.	.	.	.	.	.	.	.	6
803	AXONOPUS AFFINIS	W	.	0.0	8.3	1.7	84.1	30.4	5.9	.	.	.	.	.	.	.	.	.	.	18
804	AXONOPUS COMPRESSUS	W	.	0.0	11.4	1.8	76.4	42.4	10.4	.	.	.	.	.	.	.	.	.	.	18
805	AXONOPUS SP.	W	.	0.0	9.3	1.7	77.9	27.7	11.1	460	190	44.0	.	850	14	.	.	.	.	6
806	AZADIRACHTA INDICA	L	.	0.0	15.4	4.2	69.2	12.7	11.2	2650	240	.	.	.	.	.	.	.	.	18
807	AZADIRACHTA INDICA	F	67	81.9	1.3	1.0	15.1	.	0.7	25	41	.	.	.	.	.	.	.	.	2
808	BACCAUREA MOTLEYANA	F	65	79.0	1.7	0.2	16.1	.	3.0	13	20	0.8	.	.	0	0.00	.	.	5	2
809	BACCHARIS HALIMIFOLIA	S	.	0.0	28.8	35.3	.	.	.	.	.	.	.	.	.	.	.	.	.	20
810	BACCHARIS SAROTHROIDES	S	.	0.0	23.8	31.3	.	.	.	.	.	.	.	.	.	.	.	.	.	20

TABLE 2 - DATA DIRECTLY TRANSCRIBED FROM ORIGINAL SOURCES.

OBS	PLANA	PART	CALORIES	WATER	PROTEIN	FAT	TOTCARB	FIBER	ASH	CALCIUM	PHOSPHO	IRON	SODIUM	POTASS	CAROT	THIAMINE	RIBOFLAV	NIACIN	ASCORBIC	REF
721	ASTRAGALUS ODORATUS	S	.	0.0	41.0	3.2	.	.	.	.	.	.	.	.	.	.	.	.	.	21
722	ASTRAGALUS OVALIS	S	.	0.0	49.8	6.8	.	.	.	.	.	.	.	.	.	.	.	.	.	21
723	ASTRAGALUS PANDURATUS	S	.	0.0	52.4	8.5	.	.	.	.	.	.	.	.	.	.	.	.	.	20
724	ASTRAGALUS PANDURATUS	S	.	0.0	50.9	7.1	.	.	.	.	.	.	.	.	.	.	.	.	.	21
725	ASTRAGALUS PARISHII	S	.	0.0	38.1	5.3	.	.	.	.	.	.	.	.	.	.	.	.	.	21
726	ASTRAGALUS PODOCARPUS	S	.	0.0	37.0	2.6	.	.	4.4	.	.	.	.	.	.	.	.	.	.	19
727	ASTRAGALUS PONTICUS	S	.	0.0	46.3	5.7	.	.	.	.	.	.	.	.	.	.	.	.	.	21
728	ASTRAGALUS PONTICUS	S	.	0.0	40.5	6.3	.	.	.	.	.	.	.	.	.	.	.	.	.	21
729	ASTRAGALUS PYRRHOTRICHUS	S	.	0.0	41.7	7.4	.	.	.	.	.	.	.	.	.	.	.	.	.	21
730	ASTRAGALUS RACEMOSUS	S	.	0.0	40.9	6.5	.	.	.	.	.	.	.	.	.	.	.	.	.	20
731	ASTRAGALUS RUBYI	S	.	0.0	35.4	5.1	.	.	4.5	.	.	.	.	.	.	.	.	.	.	19
732	ASTRAGALUS SEMIBILOCULARIS	S	.	0.0	41.9	5.3	.	.	4.7	.	.	.	.	.	.	.	.	.	.	19
733	ASTRAGALUS SETULOSUS	S	.	0.0	43.4	4.8	.	.	.	.	.	.	.	.	.	.	.	.	.	21
734	ASTRAGALUS SINICUS	S	.	0.0	36.6	5.3	.	.	3.3	.	.	.	.	.	.	.	.	.	.	19
735	ASTRAGALUS SP.	S	.	0.0	33.8	4.7	.	.	4.0	.	.	.	.	.	.	.	.	.	.	19
736	ASTRAGALUS SP.	S	.	0.0	44.5	4.0	.	.	3.9	.	.	.	.	.	.	.	.	.	.	19
737	ASTRAGALUS SP.	S	.	0.0	36.7	4.2	.	.	4.8	.	.	.	.	.	.	.	.	.	.	19
738	ASTRAGALUS SP.	S	.	0.0	61.1	7.6	.	.	.	.	.	.	.	.	.	.	.	.	.	21
739	ASTRAGALUS SP.	W	.	0.0	21.2	2.2	63.5	20.9	13.1	1180	300	.	.	2800	.	.	.	.	.	6
740	ASTRAGALUS SP.	W	.	0.0	23.2	2.7	65.4	19.1	8.7	1650	390	.	.	2800	.	.	.	.	.	6
741	ASTRAGALUS STIPULATUS	S	.	0.0	33.3	5.2	.	.	4.8	.	.	.	.	.	.	.	.	.	.	19
742	ASTRAGALUS TENELLUS	S	.	0.0	32.1	5.0	.	.	3.7	.	.	.	.	.	.	.	.	.	.	19
743	ASTRAGALUS VULNERARIAE	S	.	0.0	45.6	5.1	.	.	.	.	.	.	.	.	.	.	.	.	.	21
744	ASTRANTIA MAXIMA	S	.	0.0	28.9	39.2	.	.	.	.	.	.	.	.	.	.	.	.	.	21
745	ASTROCARYUM STANDLEYANUM	F	99	71.9	1.7	0.7	24.3	5.7	1.4	62	41	1.5	.	.	12290	0.06	0.07	0.5	4	3
746	ASTRODAUCUS ORIENTALIS	S	.	0.0	17.1	11.8	.	.	.	.	.	.	.	.	.	.	.	.	.	21
747	ASYSTASIA GANGETICA	L	56	82.6	3.7	1.2	10.4	1.2	2.1	226	30	4.7	.	.	6250	0.19	0.21	1.0	42	2
748	ATHANASIA TOMENTOSA	S	.	0.0	21.7	5.7	.	.	.	.	.	.	.	.	.	.	.	.	.	21
749	ATRAPHAXIS BILLARDIERI	S	.	0.0	13.6	2.0	.	.	.	.	.	.	.	.	.	.	.	.	.	20
750	ATRIPLEX ARGENTEA	W	.	0.0	10.3	1.5	67.8	28.9	20.4	.	.	.	.	.	.	.	.	.	.	6
751	ATRIPLEX CANESCENS	S	.	0.0	5.4	1.0	.	.	6.5	.	.	.	.	.	.	.	.	.	.	19
752	ATRIPLEX CANESCENS	S	.	0.0	5.6	1.5	.	.	.	.	.	.	.	.	.	.	.	.	.	19
753	ATRIPLEX CANESCENS	S	.	0.0	23.6	10.6	.	.	.	.	.	.	.	.	.	.	.	.	.	21
754	ATRIPLEX CANESCENS	S	.	0.0	22.1	11.5	.	.	.	.	.	.	.	.	.	.	.	.	.	21
755	ATRIPLEX CANESCENS	W	.	0.0	14.5	2.1	71.6	16.2	11.8	1190	150	10.0	.	810	.	.	.	.	.	6
756	ATRIPLEX CONFERTIFOLIA	W	389	0.0	7.9	2.2	68.2	30.7	21.7	2610	200	.	.	.	2	.	.	.	.	6
757	ATRIPLEX CONFERTIFOLIA	W	.	0.0	8.0	2.2	67.9	31.1	21.9	2270	140	.	.	.	.	.	.	.	.	6
758	ATRIPLEX CONFERTIFOLIA	L	.	0.0	14.9	1.9	52.1	10.8	31.1	.	.	.	.	.	.	.	.	.	.	6
759	ATRIPLEX GARDNERI	W	.	0.0	8.4	3.0	73.0	30.6	15.6	.	.	.	.	.	.	.	.	.	.	6
760	ATRIPLEX GARDNERI	L	.	0.0	16.9	2.7	56.3	11.3	24.1	.	.	.	.	.	.	.	.	.	.	6
761	ATRIPLEX GARDNERI	C	.	0.0	7.7	5.0	79.6	35.7	7.7	.	.	.	.	.	.	.	.	.	.	6
762	ATRIPLEX HASTATA	S	.	0.0	21.8	9.0	.	.	.	.	.	.	.	.	.	.	.	.	.	21
763	ATRIPLEX HORTENSIS	S	.	0.0	30.3	5.9	.	.	.	.	.	.	.	.	.	.	.	.	.	21
764	ATRIPLEX HORTENSIS	S	.	0.0	31.2	5.6	.	.	.	.	.	.	.	.	.	.	.	.	.	21
765	ATRIPLEX LAEVIS	S	.	0.0	20.8	8.4	.	.	.	.	.	.	.	.	.	.	.	.	.	21

TABLE 2 - DATA DIRECTLY TRANSCRIBED FROM ORIGINAL SOURCES.

OBS	PLANA	PART	CALORIES	WATER	PROTEIN	FAT	TOTCARB	FIBER	ASH	CALCIUM	PHOSPHO	IRON	SODIUM	POTASS	CAROTE	THIAMINE	RIBOFLAV	NIACIN	ASCORBIC	REF
811	BACCHARIS VACCINIOIDES	S	.	0.0	25.0	22.5	.	.	.	.	.	.	.	.	.	.	.	.	.	20
812	BACTRIS MINOR	F	70	79.6	1.2	0.2	17.8	2.1	1.2	14	28	1.2	.	.	170	0.03	0.06	0.4	8	3
813	BAHIA GLANDULOSA	S	.	0.0	19.4	22.7	.	.	.	.	.	.	.	.	.	.	.	.	.	20
814	BAILEYA PLENIRADIATA	S	.	0.0	20.0	13.0	.	.	.	.	.	.	.	.	.	.	.	.	.	21
815	BALANITES AEGYPTIACA	F	122	64.0	2.2	0.0	31.9	.	1.9	47	.	.	.	.	.	.	.	.	14	1
816	BALANITES AEGYPTIACA	L	.	0.0	11.6	4.2	71.5	13.6	12.7	.	.	.	.	.	.	.	.	.	.	18
817	BALANITES AEGYPTIACA	SH	.	0.0	27.5	1.5	64.4	23.3	6.6	480	380	.	.	.	.	.	.	.	.	18
818	BALANITES AEGYPTIACA	F	.	0.0	11.1	1.7	79.1	10.2	8.1	380	400	.	.	.	.	.	.	.	.	18
819	BALANITES AEGYPTIACA	S	.	0.0	21.9	45.7	.	.	3.3	.	.	.	.	.	.	.	.	.	.	19
820	BALSAMORHIZA SAGITTATA		.	0.0	15.1	3.4	.	20.2	.	1790	260	.	.	.	.	.	.	.	.	6
821	BAMBUSA ARUNDINACEA	L	.	0.0	18.6	4.1	65.5	24.1	11.8	.	.	.	.	.	.	.	.	.	.	18
822	BAMBUSA ARUNDINACEA	SH	29	90.7	2.3	0.2	6.1	0.5	0.7	33	41	0.4	.	.	20	0.15	0.07	0.6	4	3
823	BAMBUSA ARUNDINACEA	S	.	11.0	11.8	0.6	75.4	1.7	1.2	.	.	.	.	.	.	.	.	.	.	5
824	BAMBUSA SP.	SH	32	89.6	2.7	0.3	6.4	0.9	1.0	10	58	0.6	9	486	15	0.13	0.10	0.7	10	2
825	BAMBUSA SP.	SH	23	92.4	2.0	0.2	4.6	1.9	0.8	7	47	1.0	.	398	0	.	.	.	.	2
826	BAMBUSA SP.	SH	35	88.7	3.2	0.5	6.4	0.7	1.2	33	48	0.2	.	694	50	0.07	0.09	0.6	9	2
827	BAMBUSA SP.	SH	27	91.0	2.6	0.3	5.2	0.7	0.9	13	59	0.5	.	533	12	0.15	0.07	0.6	4	4
828	BAMBUSA VULGARIS	L	.	0.0	19.0	0.5	63.6	28.8	16.9	.	.	.	.	.	.	.	.	.	.	18
829	BAPTISIA AUSTRALIS	S	.	0.0	34.8	10.4	.	.	3.4	.	.	.	.	.	.	.	.	.	.	19
830	BAPTISIA LEUCANTHA	S	.	0.0	37.7	10.2	.	.	.	.	.	.	.	.	.	.	.	.	.	20
831	BAPTISIA VIRIDIS	S	.	0.0	38.2	11.4	.	.	.	.	.	.	.	.	.	.	.	.	.	20
832	BARBAREA VULGARIS	S	.	0.0	15.6	31.0	.	.	.	.	.	.	.	.	.	.	.	.	.	20
833	BARLERIA OPACA	L	55	81.0	3.6	0.5	11.7	2.7	3.2	874	38	.	.	.	.	.	.	.	.	1
834	BAROSMA BETULINA	S	.	0.0	27.1	7.8	.	.	.	.	.	.	.	.	.	.	.	.	.	21
835	BAROSMA BETULINA	S	.	0.0	29.9	33.8	.	.	.	.	.	.	.	.	.	.	.	.	.	21
836	BARRINGTONIA ACUTANGULA	L	42	86.8	2.2	0.2	9.7	.	1.1	40	62	2.7	12	374	1300	0.02	.	.	.	2
837	BASELLA ALBA	L	19	93.4	1.6	0.3	3.5	0.6	1.2	106	39	1.6	23	488	3490	0.06	0.17	0.6	86	2
838	BASELLA ALBA	L	19	93.0	1.2	0.2	4.1	0.6	1.5	346	11	3.9	.	.	1720	0.03	0.08	0.4	69	3
839	BASELLA ALBA	L	19	93.1	1.8	0.3	3.4	0.7	1.4	109	52	1.2	.	.	4800	0.05	.	0.5	102	4
840	BASSIA HYSSOPIFOLIA	S	.	0.0	26.8	14.0	.	.	.	.	.	.	.	.	.	.	.	.	.	21
841	BASSIA HYSSOPIFOLIA	S	.	0.0	24.3	17.6	.	.	.	.	.	.	.	.	.	.	.	.	.	21
842	BAUHINIA ESCULENTA	L	82	72.3	4.3	0.1	19.9	6.8	2.4	436	86	.	.	.	.	.	.	.	68	1
843	BAUHINIA ESCULENTA	S	554	1.1	29.2	42.4	24.1	20.3	3.2	194	474	6.5	.	.	.	0.08	0.97	1.9	.	1
844	BAUHINIA GALPINII	F	.	0.0	17.2	3.1	76.1	20.1	3.6	200	200	.	.	.	.	.	.	.	.	18
845	BAUHINIA HERRERAE	S	.	0.0	19.4	2.4	.	.	4.0	.	.	.	.	.	.	.	.	.	.	19
846	BAUHINIA PURPUREA	S	.	0.0	27.5	20.0	.	.	.	.	.	.	.	.	.	.	.	.	.	19
847	BAUHINIA RETICULATA	L	295	0.0	22.1	0.5	66.4	31.3	11.1	2005	369	.	.	.	.	.	.	.	.	10
848	BAUHINIA SP.	S	.	0.0	26.9	7.5	.	.	.	.	.	.	.	.	.	.	.	.	.	20
849	BAUHINIA VARIEGATA	L	.	0.0	18.5	9.0	62.5	10.5	10.0	1730	160	.	.	.	.	.	.	.	.	18
850	BAUHINIA VARIEGATA	S	.	0.0	32.5	19.3	.	.	.	.	.	.	.	.	.	.	.	.	.	21
851	BECKMANNIA ERUCAEFORMIS	S	.	0.0	17.7	14.8	.	.	.	.	.	.	.	.	.	.	.	.	.	21
852	BECKMANNIA SYZIGACHNE		.	0.0	5.0	1.3	82.1	37.9	11.6	.	.	.	.	.	.	.	.	.	.	6
853	BECKMANNIA SYZIGACHNE		.	0.0	3.3	1.1	83.2	39.2	12.4	.	.	.	.	.	.	.	.	.	.	6
854	BEILSCHMIEDIA MANNII	V	333	14.6	5.9	0.5	75.8	1.6	3.2	220	100	.	.	.	.	.	.	.	.	1
855	BELLARDIA TRIXAGO	S	.	0.0	30.1	33.8	.	.	.	.	.	.	.	.	.	.	.	.	.	21

TABLE 2 - DATA DIRECTLY TRANSCRIBED FROM ORIGINAL SOURCES.

OBS	PLANA	PART	CALORIES	WATER	PROTEIN	FAT	TOTCARB	FIBER	ASH	CALCIUM	PHOSPHON	IRON	SODIUM	POTASS	CAROT	THIAMINE	RIBOFLAV	NIACIN	ASCORBIC	REFCO
856	BELLARDIA TRIXAGO	S	.	0.0	30.0	31.6	.	.	.	.	.	.	.	.	.	.	.	.	.	21
857	BENINCASA HISPIDA	F	12	96.2	0.5	0.1	2.9	0.6	0.3	17	19	0.4	5	111	5	0.03	0.03	0.20	20	2
858	BENINCASA HISPIDA	F	14	95.8	0.5	0.1	3.3	0.8	0.3	21	7	0.5	.	.	10	0.00	0.01	0.30	27	3
859	BENINCASA HISPIDA	F	13	96.1	0.4	0.2	3.0	0.5	0.3	19	19	0.4	6	111	0	0.04	0.11	0.40	13	4
860	BEQUAERTIODENDRON MAGALISMONTANUM	S	.	0.0	5.9	5.8	.	.	.	.	.	.	.	.	.	.	.	.	.	21
861	BERBERIS CRATAEGINA	S	.	0.0	17.2	17.9	.	.	.	.	.	.	.	.	.	.	.	.	.	21
862	BERBERIS NERVOSA	F	379	0.0	18.0	8.0	71.0	.	4.0	191	.	3.0	.	.	.	.	.	.	145	15
863	BERTHOLLETIA EXCELSA	S	648	5.3	14.8	66.8	9.2	.	3.9	.	.	.	.	.	.	.	.	.	.	1
864	BERTHOLLETIA EXCELSA	S	644	4.7	17.4	65.0	9.6	3.9	3.3	169	620	3.6	2	.	5	0.20	0.69	0.20	2	2
865	BERTHOLLETIA EXCELSA	S	640	2.6	13.2	60.3	20.5	1.2	3.4	166	660	3.2	.	.	40	1.09	0.12	7.70	10	3
866	BERTHOLLETIA EXCELSA	S	654	4.6	14.3	66.9	10.9	3.1	3.3	186	693	3.4	1	715	0	0.96	0.12	1.60	.	4
867	BETA LOMATOGONA	S	.	0.0	11.7	2.6	.	.	.	.	.	.	.	.	.	.	.	.	.	21
868	BETA VULGARIS	R	46	86.7	1.9	0.1	10.5	1.2	0.8	24	43	0.8	.	.	.	.	.	.	3	1
869	BETA VULGARIS	L	21	92.4	2.0	0.2	3.6	0.4	1.8	132	34	0.7	.	.	.	0.00	0.20	.	.	1
870	BETA VULGARIS	S	.	0.0	15.4	6.3	.	.	5.9	.	.	.	.	.	.	.	.	.	.	19
871	BETA VULGARIS	S	.	0.0	14.3	6.4	.	.	5.1	.	.	.	.	.	.	.	.	.	.	19
872	BETA VULGARIS	S	.	0.0	15.4	7.1	.	.	5.6	.	.	.	.	.	.	.	.	.	.	19
873	BETA VULGARIS	S	.	0.0	12.6	4.0	.	.	5.5	.	.	.	.	.	.	.	.	.	.	19
874	BETA VULGARIS	S	.	0.0	14.6	3.4	.	.	6.8	.	.	.	.	.	.	.	.	.	.	19
875	BETA VULGARIS	S	.	0.0	13.2	5.8	.	.	6.5	.	.	.	.	.	.	.	.	.	.	19
876	BETA VULGARIS	S	.	0.0	11.1	2.8	.	.	6.6	.	.	.	.	.	.	.	.	.	.	19
877	BETA VULGARIS	R	44	87.4	1.6	0.2	10.0	0.9	0.8	23	35	1.1	36	330	0	0.02	0.04	0.30	6	2
878	BETA VULGARIS	L	28	89.6	2.4	0.6	4.8	1.1	2.6	96	38	0.9	.	438	2665	0.07	0.24	0.60	12	2
879	BETA VULGARIS	L	21	92.6	1.8	0.4	3.7	0.9	1.5	70	44	2.9	.	.	1720	0.10	0.19	0.50	22	2
880	BETA VULGARIS	S	.	0.0	9.3	4.6	.	.	.	.	.	.	.	.	.	.	.	.	.	21
881	BETA VULGARIS	R	44	87.8	1.7	0.1	9.5	1.0	0.9	14	38	0.8	.	.	0	0.01	0.04	0.20	5	3
882	BETA VULGARIS	L	45	86.4	3.2	0.4	8.1	3.8	1.9	114	34	3.1	.	.	3150	0.07	0.22	0.60	50	3
883	BETA VULGARIS	L	27	90.8	1.6	0.4	5.6	1.0	1.6	110	29	3.6	.	.	1750	0.03	0.09	0.40	34	3
884	BETA VULGARIS	R	43	87.3	1.6	0.1	9.9	0.8	1.1	16	33	0.7	60	335	12	0.03	0.05	0.40	10	4
885	BETA VULGARIS	L	24	90.9	2.2	0.3	4.6	1.3	2.0	119	40	3.3	130	500	3660	0.10	0.22	0.40	30	4
886	BETA VULGARIS	L	25	91.1	2.4	0.3	4.6	0.8	1.6	88	39	3.2	147	550	3900	0.06	0.17	0.50	32	4
887	BETA VULGARIS	W	.	0.0	16.0	2.7	59.8	10.9	21.5	1030	230	.	.	5700	.	0.59	0.61	4.73	.	6
888	BETULA LENTA	L	.	0.0	28.1	8.6	55.6	16.9	7.7	.	.	.	.	.	.	.	.	.	.	6
889	BETULA PLATYPHYLLA	S	.	0.0	12.4	19.0	.	.	.	.	.	.	.	.	.	.	.	.	.	20
890	BETULA POPULIFOLIA	S	.	0.0	6.2	5.8	.	.	.	.	.	.	.	.	.	.	.	.	.	21
891	BIDENS BITERNATA	S	.	0.0	23.1	14.5	.	.	.	.	.	.	.	.	.	.	.	.	.	21
892	BIDENS FRONDOSA	S	.	0.0	33.8	34.3	.	.	.	.	.	.	.	.	.	.	.	.	.	19
893	BIDENS GENTRYI	S	.	0.0	22.9	14.4	.	.	.	.	.	.	.	.	.	.	.	.	.	21
894	BIDENS PILOSA	L	43	85.1	3.8	0.5	8.4	3.9	2.2	340	67	.	.	.	1800	.	.	.	.	1
895	BIDENS PILOSA	L	295	0.0	24.5	4.0	56.4	12.1	15.1	1721	273	.	11	267	.	.	.	.	.	12
896	BIDENS PILOSA	S	.	0.0	17.5	17.1	.	.	.	.	.	.	.	.	.	.	.	.	.	20
897	BIDENS PILOSA F. RADIATA	S	.	0.0	20.6	18.9	.	.	.	.	.	.	.	.	.	.	.	.	.	20
898	BIDENS SP.	S	.	0.0	35.6	33.1	.	.	7.1	.	.	.	.	.	.	.	.	.	.	19
899	BIDENS SP.	S	.	0.0	26.4	27.0	.	.	.	.	.	.	.	.	.	.	.	.	.	21
900	BIDENS TRIPARTITA	S	.	0.0	20.6	6.7	.	.	.	.	.	.	.	.	.	.	.	.	.	21

TABLE 2 - DATA DIRECTLY TRANSCRIBED FROM ORIGINAL SOURCES.

OBS	PLNA	PART	CALORIES	WATER	PROTEIN	FAT	TOTCARB	FIBER	ASH	CALCIUM	PHOSPHO	IRON	SODIUM	POTASS	CAROT	THIAMINE	RIBOFLAV	NIACIN	ASCORBIC	REFCO
901	BIFORA AMERICANA	S	.	0.0	9.1	17.9	.	.	.	.	.	.	.	.	.	.	.	.	.	20
902	BIGNONIA UNGUIS-CATI	S	.	0.0	20.0	27.9	.	.	4.20	.	.	.	.	.	.	.	.	.	.	19
903	BIGNONIA UNGUIS-CATI	S	.	0.0	26.3	38.9	.	.	.	.	.	.	.	.	.	.	.	.	.	21
904	BISERRULA PELECINUS	S	.	0.0	32.8	1.9	.	.	.	.	.	.	.	.	.	.	.	.	.	21
905	BIXA ORELLANA	S	.	0.0	13.1	5.0	.	.	5.40	.	.	.	.	.	.	.	.	.	.	19
906	BIXA ORELLANA	S	.	0.0	13.2	3.8	.	.	.	.	.	.	.	.	.	.	.	.	.	20
907	BIXA ORELLANA	P	54	84.4	0.0	0.3	14.3	0.5	1.00	7	10	0.8	.	.	90	0.00	0.05	0.3	2	3
908	BLEPHAROCALYX TWEEDIEI	S	.	0.0	7.4	24.3	.	.	.	.	.	.	.	.	.	.	.	.	.	21
909	BLIGHIA SAPIDA	A	196	69.2	5.0	20.0	4.6	1.6	1.20	40	16	2.7	.	.	555	0.13	0.14	1.4	26	1
910	BLIGHIA SAPIDA	S	64	79.8	5.4	1.2	11.5	3.1	2.10	20	.	3.5	.	.	.	0.15	0.09	1.7	.	1
911	BLIGHIA SAPIDA	S	180	71.5	4.2	17.4	5.5	0.3	1.40	37	70	1.7	.	.	.	0.07	0.17	1.6	78	3
912	BOCCONIA ARBOREA	S	.	0.0	14.4	35.1	.	.	.	.	.	.	.	.	.	.	.	.	.	20
913	BOEHMERIA CYLINDRICA	S	.	0.0	20.6	25.9	.	.	.	.	.	.	.	.	.	.	.	.	.	19
914	BOEHMERIA NIVEA	W	.	0.0	30.1	4.4	47.3	13.1	18.20	.	.	.	.	.	.	.	.	.	.	18
915	BOEHMERIA NIVEA	H	.	0.0	22.4	3.2	56.7	11.9	17.70	450	140	.	.	.	.	.	.	.	.	18
916	BOEHMERIA NIVEA	L	38	87.0	3.3	0.5	7.1	2.4	2.10	459	45	1.6	.	.	5240	0.03	0.28	0.7	73	3
917	BOEHMERIA NIVEA	W	.	0.0	21.9	4.1	60.6	25.9	13.40	4600	240	.	.	.	30	0.73	1.65	6.7	.	6
918	BOERHAAVIA DIFFUSA	K	326	10.2	8.3	5.3	70.6	20.1	5.60	730	240	.	.	.	.	.	.	.	.	1
919	BOERHAAVIA DIFFUSA	L	52	82.0	4.5	0.5	10.3	2.2	2.70	432	72	.	.	.	.	.	.	.	.	1
920	BOISSIERA SQUARROSA	S	.	0.0	29.6	2.2	.	.	.	.	.	.	.	.	.	.	.	.	.	21
921	BOLTONIA ASTEROIDES	S	.	0.0	19.1	21.1	.	.	6.20	.	.	.	.	.	.	.	.	.	.	19
922	BOMBAX BUONOPOZENSE	L	62	80.0	3.4	0.6	14.5	3.5	1.50	276	32	.	.	.	.	.	.	.	.	1
923	BOMBAX BUONOPOZENSE	I	64	82.0	1.2	0.8	14.8	2.4	1.20	174	27	.	.	.	.	.	.	.	.	1
924	BOMBAX BUONOPOZENSE	S	465	6.0	19.3	27.5	41.3	12.2	5.90	348	676	.	.	.	.	.	.	.	.	1
925	BORAGO OFFICINALIS	S	.	0.0	20.9	38.3	.	.	11.00	.	.	.	.	.	.	.	.	.	.	19
926	BORASSUS FLABELLIFER	F	43	87.6	0.8	0.1	10.9	2.0	0.60	27	30	1.0	.	.	0	0.04	0.02	0.3	5	1
927	BORASSUS FLABELLIFER	IS	71	82.3	0.9	0.1	16.4	1.4	0.30	30	32	.	.	.	.	.	.	.	.	1
928	BORASSUS FLABELLIFER	L	.	0.0	13.3	4.6	74.7	38.0	7.40	.	.	.	.	.	.	.	.	.	.	18
929	BORASSUS FLABELLIFER	F	.	0.0	6.5	0.8	87.9	16.1	4.80	210	240	.	.	.	.	.	.	.	.	18
930	BORASSUS FLABELLIFER	SH	103	69.5	2.7	0.2	26.6	2.2	1.00	18	140	.	.	.	.	0.05	0.18	0.9	8	2
931	BORASSUS FLABELLIFER	F	.	77.2	0.7	0.2	21.3	.	0.70	.	.	.	.	.	.	.	.	.	24	5
932	BORASSUS FLABELLIFER	S	.	92.6	0.6	0.1	6.3	.	0.26	.	.	.	.	.	.	.	.	.	13	5
933	BORASSUS FLABELLIFER	J	.	85.9	0.2	0.0	26.1	.	0.30	.	.	.	.	.	.	.	.	.	6	5
934	BOREAVA ORIENTALIS	S	.	0.0	36.2	41.2	.	.	.	.	.	.	.	.	.	.	.	.	.	20
935	BORONIA MEGASTIGMA	S	.	0.0	20.0	32.3	.	.	.	.	.	.	.	.	.	.	.	.	.	20
936	BOSCIA ALBITRUNCA	L	.	0.0	15.8	.	.	30.3	7.40	1440	120	.	.	.	.	.	.	.	.	18
937	BOSCIA ANGUSTIFOLIA	SH	.	0.0	19.3	1.3	72.3	39.2	7.10	610	180	.	.	.	.	.	.	.	.	18
938	BOSCIA FOETIDA	S	.	0.0	25.2	23.3	.	.	.	.	.	.	.	.	.	.	.	.	.	21
939	BOSCIA SALICIFOLIA	L	62	82.0	3.1	1.6	11.4	3.7	1.90	200	13	3.2	.	.	.	.	.	.	.	1
940	BOSCIA SENEGALENSIS	S	341	11.4	21.2	1.6	62.4	1.7	3.40	123	323	6.8	.	.	165	0.02	0.03	8.8	5	1
941	BOTHRIOCHLOA CAUCASICA	S	.	0.0	21.9	7.0	.	.	.	.	.	.	.	.	.	.	.	.	.	20
942	BOTHRIOCHLOA INSCULPTA	W	.	0.0	9.7	2.6	75.0	32.5	12.70	.	.	.	.	.	.	.	.	.	.	18
943	BOTHRIOCHLOA INSCULPTA	H	.	0.0	4.6	1.8	83.7	35.2	9.90	.	.	.	.	.	.	.	.	.	.	18
944	BOTHRIOCHLOA INTERMEDIA	W	.	0.0	3.9	1.3	85.3	38.9	9.50	.	.	.	.	.	.	.	.	.	.	18
945	BOTHRIOCHLOA INTERMEDIA	H	.	0.0	4.2	1.5	85.1	43.2	9.20	.	.	.	.	.	.	.	.	.	.	18

TABLE 2 - DATA DIRECTLY TRANSCRIBED FROM ORIGINAL SOURCES.

OBS	PLNA	PART	CALORIES	WATER	PROTEIN	FAT	TOTLCARB	FIBER	ASH	CALCIUM	PHOSPHO	IRON	SODIUM	POTASS	CAROT	THIAMINE	RIBOFLAV	NIACIN	ASCORBIC	REFCO
946	BOTHRIOCHLOA ISCHAEMUM	S	.	0.0	28.8	9.6	.	.	.	.	.	.	.	.	.	.	.	.	.	20
947	BOTHRIOCHLOA PERTUSA	H	.	0.0	3.0	1.2	85.1	39.2	10.7	.	.	.	.	.	.	.	.	.	.	18
948	BOTHRIOCHLOA PERTUSA	W	.	0.0	5.7	1.0	81.6	34.9	11.7	.	.	.	.	.	.	.	.	.	.	18
949	BOTHRIOCHLOA RADICANS	W	.	0.0	10.2	3.7	77.5	28.8	8.6	.	.	.	.	.	.	.	.	.	.	18
950	BOTHRIOCHLOA SACCHAROIDES	S	.	0.0	27.5	10.8	.	.	.	.	.	.	.	.	.	.	.	.	.	20
951	BOUEA MACROPHYLLA	F	52	85.2	0.7	0.0	13.9	0.6	0.2	6	11	0.3	.	.	40	0.03	0.03	0.3	75	3
952	BOUEA SP.	F	60	83.0	0.8	0.1	15.7	.	0.4	12	22	0.9	.	148	240	0.04	0.04	0.4	107	2
953	BOUEA SP.	L	60	81.0	3.1	0.3	14.0	.	1.6	40	45	4.7	.	.	360	0.00	.	.	61	2
954	BOURGAEA TOURNEFORTII	S	.	0.0	42.5	22.6	.	.	.	.	.	.	.	.	.	.	.	.	.	21
955	BOUTELOUA CURTIPENDULA		.	0.0	5.8	1.7	79.9	32.1	12.6	390	90	.	.	710	3	.	.	.	.	6
956	BOUTELOUA ERIOPODA		.	0.0	8.2	1.7	82.7	31.9	7.4	240	120	.	.	.	4	.	.	.	.	6
957	BOUTELOUA FILIFORMIS		.	0.0	9.0	2.0	73.1	30.4	14.9	.	.	.	.	.	.	.	.	.	.	6
958	BOUTELOUA GRACILIS		.	0.0	8.1	1.9	77.8	29.8	12.2	570	240	.	.	.	10	.	.	.	.	6
959	BOUTELOUA HIRSUTA		.	0.0	7.2	1.6	73.6	29.0	17.6	580	90	.	.	870	.	.	.	.	.	6
960	BOUTELOUA RIGIDISETA		.	0.0	8.0	1.9	75.7	30.9	14.4	620	100	.	.	.	.	.	.	.	.	6
961	BOUTELOUA TRIFIDA		.	0.0	7.6	2.3	81.4	33.0	8.7	610	100	.	.	600	.	.	.	.	.	6
962	BRACHIARIA BRIZANTHA	W	.	0.0	4.0	1.3	83.6	31.1	11.1	.	.	.	.	.	.	.	.	.	.	18
963	BRACHIARIA BRIZANTHA	H	.	0.0	8.6	1.9	78.9	31.3	10.6	.	.	.	.	.	.	.	.	.	.	18
964	BRACHIARIA DECUMBENS	W	.	0.0	11.2	2.8	76.1	28.0	9.9	.	.	.	.	.	.	.	.	.	.	18
965	BRACHIARIA DISTACHYA	W	.	0.0	10.1	3.0	73.4	26.6	13.5	.	.	.	.	.	.	.	.	.	.	18
966	BRACHIARIA LEERSIOIDES	W	.	0.0	11.8	1.9	72.8	31.5	13.5	.	.	.	.	.	.	.	.	.	.	18
967	BRACHIARIA MUTICA	L	.	0.0	14.0	2.2	75.7	31.0	8.1	.	.	.	.	.	.	.	.	.	.	18
968	BRACHIARIA MUTICA	Y	.	0.0	5.9	0.8	87.4	41.5	6.3	.	.	.	.	.	.	.	.	.	.	18
969	BRACHIARIA MUTICA	H	.	0.0	12.4	2.0	74.9	30.3	10.7	.	.	.	.	.	.	.	.	.	.	18
970	BRACHIARIA NIGROPEDATA	W	.	0.0	8.5	1.6	80.9	30.7	9.0	.	.	.	.	.	.	.	.	.	.	18
971	BRACHIARIA RAMOSA	S	.	0.0	15.8	4.6	.	.	.	.	.	.	.	.	.	.	.	.	.	21
972	BRACHIARIA RUZIZIENSIS	W	.	0.0	13.9	2.3	73.8	27.2	9.0	.	.	.	.	.	.	.	.	.	.	18
973	BRACHYCHITON ACERIFOLIUM	S	.	0.0	36.2	29.9	.	.	5.7	.	.	.	.	.	.	.	.	.	.	19
974	BRACHYCHITON ACERIFOLIUM	S	.	0.0	3.4	1.5	.	.	2.6	.	.	.	.	.	.	.	.	.	.	19
975	BRACHYCOME IBERIDIFOLIA	S	.	0.0	26.2	27.2	.	.	7.4	.	.	.	.	.	.	.	.	.	.	19
976	BRACHYPODIUM DISTACHYON	S	.	0.0	10.2	1.2	.	.	.	.	.	.	.	.	.	.	.	.	.	21
977	BRACHYPODIUM PHOENICOIDES	S	.	0.0	21.6	2.5	.	.	.	.	.	.	.	.	.	.	.	.	.	21
978	BRACHYPODIUM PINNATUM	S	.	0.0	16.3	2.3	.	.	.	.	.	.	.	.	.	.	.	.	.	21
979	BRASENIA SCHREBERI	L	10	92.6	0.7	0.2	1.8	0.1	4.7	9	23	2.0	.	.	10	0.03	0.03	0.3	0	2
980	BRASSICA CAMPESTRIS	S	.	0.0	21.6	43.7	28.7	6.3	6.0	490	710	.	.	.	.	.	.	.	.	18
981	BRASSICA CAMPESTRIS	S	.	0.0	22.6	42.9	.	.	4.3	.	.	.	.	.	.	.	.	.	.	19
982	BRASSICA CAMPESTRIS	S	.	0.0	28.8	38.8	.	.	.	.	.	.	.	.	.	.	.	.	.	20
983	BRASSICA CAMPESTRIS	S	.	0.0	22.5	44.2	.	.	.	.	.	.	.	.	.	.	.	.	.	20
984	BRASSICA CAMPESTRIS	S	.	0.0	21.2	44.8	.	.	.	.	.	.	.	.	.	.	.	.	.	20
985	BRASSICA CAMPESTRIS	S	.	0.0	26.8	39.2	.	.	.	.	.	.	.	.	.	.	.	.	.	21
986	BRASSICA CAMPESTRIS	S	.	0.0	27.4	29.1	.	.	.	.	.	.	.	.	.	.	.	.	.	21
987	BRASSICA CARINATA	S	.	0.0	34.4	26.4	.	.	5.5	.	.	.	.	.	.	.	.	.	.	19
988	BRASSICA CARINATA	S	.	0.0	32.6	29.7	.	.	5.6	.	.	.	.	.	.	.	.	.	.	19
989	BRASSICA JAPONICA	L	23	92.0	2.5	0.4	3.7	1.4	1.4	140	42	2.7	120	340	500	0.08	0.13	1.0	70	2
990	BRASSICA JUNCEA	L	24	91.8	2.4	0.4	4.3	1.0	1.1	160	48	2.7	24	297	1825	0.06	0.14	0.8	73	2

TABLE 2 - DATA DIRECTLY TRANSCRIBED FROM ORIGINAL SOURCES.

OBS	PLNA	PART	CALORIES	WATER	PROTEIN	FAT	TOTLCARB	FIBER	ASH	CALCIUM	PHOSPHO	IRON	SODIUM	POTASS	CAROTE	THIAMINE	RIBOFLAV	NIACIN	ASCORBIC	REFCO
991	BRASSICA JUNCEA	R	38	85.2	1.9	0.3	8.8	2.0	3.8	111	65	1.6	.	447	45	0.05	0.12	0.7	21	2
992	BRASSICA JUNCEA	I	16	94.6	1.3	0.3	2.8	0.5	1.0	144	35	1.0	.	257	1815	0.03	0.08	.	28	2
993	BRASSICA JUNCEA	SH	21	93.5	1.8	0.5	3.4	0.8	0.8	71	18	2.2	.	222	.	.	.	.	.	2
994	BRASSICA JUNCEA	S	.	0.0	28.8	34.2	.	.	.	.	.	.	.	.	.	.	.	.	.	20
995	BRASSICA JUNCEA	L	27	90.6	2.6	0.4	4.8	1.0	1.6	80	40	4.0	.	.	1220	0.07	0.21	6.2	62	3
996	BRASSICA NAPUS	S	.	0.0	19.3	43.4	33.8	27.2	3.5	320	650	.	.	.	.	.	.	.	.	18
997	BRASSICA NAPUS	S	.	0.0	23.8	45.9	.	.	.	.	.	.	.	.	.	.	.	.	.	20
998	BRASSICA NAPUS	S	.	0.0	26.9	41.6	.	.	.	.	.	.	.	.	.	.	.	.	.	20
999	BRASSICA NAPUS	S	.	0.0	23.1	43.9	.	.	.	.	.	.	.	.	.	.	.	.	.	20
1000	BRASSICA NAPUS	S	.	0.0	21.2	47.4	.	.	.	.	.	.	.	.	.	.	.	.	.	20
1001	BRASSICA NAPUS	S	.	0.0	20.0	47.8	.	.	.	.	.	.	.	.	.	.	.	.	.	20
1002	BRASSICA NAPUS	S	.	0.0	21.9	44.6	.	.	.	.	.	.	.	.	.	.	.	.	.	20
1003	BRASSICA NAPUS	S	.	0.0	29.9	36.8	.	.	.	.	.	.	.	.	.	.	.	.	.	21
1004	BRASSICA NAPUS	S	.	0.0	29.8	34.5	.	.	.	.	.	.	.	.	.	.	.	.	.	21
1005	BRASSICA NAPUS	S	.	0.0	30.1	36.4	.	.	.	.	.	.	.	.	.	.	.	.	.	21
1006	BRASSICA NAPUS	I	37	87.4	4.3	0.4	6.4	.	1.5	117	97	3.4	.	.	3200	0.04	0.02	1.2	42	3
1007	BRASSICA NAPUS	L	61	83.3	2.9	1.7	11.2	1.8	0.9	136	38	4.6	.	.	2680	0.08	0.15	0.5	120	3
1008	BRASSICA NAPUS	R	22	93.2	1.7	0.1	4.2	0.8	0.8	20	20	1.5	.	.	0	0.04	0.04	0.7	26	3
1009	BRASSICA NAPUS	R	46	87.0	1.1	0.1	11.0	1.1	0.8	66	39	0.4	5	239	348	0.07	0.07	1.1	43	4
1010	BRASSICA NAPUS	W	.	0.0	16.0	3.6	67.2	17.2	13.2	1190	340	18.0	.	2580	.	.	.	.	.	6
1011	BRASSICA NAPUS NAPOBRASSICA GROUP	R	46	87.0	1.1	0.1	11.0	1.1	0.8	66	39	0.4	5	239	350	0.07	0.07	1.1	43	2
1012	BRASSICA NAPUS NAPOBRASSICA GROUP	W	.	0.0	16.4	2.6	55.4	12.3	25.6	.	.	.	.	.	.	.	.	.	.	6
1013	BRASSICA NAPUS NAPOBRASSICA GROUP	Y	.	0.0	18.5	.	.	14.9	.	.	.	.	.	.	1	.	0.86	.	.	6
1014	BRASSICA NARINOSA	L	23	91.4	2.8	0.3	4.0	0.9	1.6	200	58	3.8	.	599	2040	0.08	0.15	0.6	58	2
1015	BRASSICA NIGRA	S	.	0.0	36.2	31.7	.	.	.	.	.	.	.	.	.	.	.	.	.	20
1016	BRASSICA NIGRA	S	.	0.0	31.9	32.0	.	.	.	.	.	.	.	.	.	.	.	.	.	21
1017	BRASSICA NIGRA	S	.	0.0	29.8	30.1	.	.	.	.	.	.	.	.	.	.	.	.	.	21
1018	BRASSICA NIGRA	L	31	89.5	3.0	0.5	5.6	1.1	1.4	183	50	3.0	32	377	4200	0.11	0.22	0.8	97	4
1019	BRASSICA NIGRA	S	.	7.6	29.1	28.2	30.2	11.0	0.5	.	.	.	.	.	.	.	.	.	.	5
1020	BRASSICA OLERACEA ACEPHALA GROUP	L	43	84.6	3.5	0.8	7.7	1.6	3.4	132	77	1.3	.	.	900	.	.	.	110	1
1021	BRASSICA OLERACEA ACEPHALA GROUP	L	35	89.0	3.0	0.4	6.8	1.2	0.8	230	56	2.0	.	.	270	0.10	0.13	0.4	93	2
1022	BRASSICA OLERACEA ACEPHALA GROUP	L	44	85.4	4.5	0.7	7.5	1.3	1.9	252	66	2.2	.	.	4030	0.16	0.24	1.2	125	3
1023	BRASSICA OLERACEA ACEPHALA GROUP	L	45	85.3	4.8	0.8	7.5	1.2	1.6	250	82	1.5	.	450	5580	0.16	0.31	1.7	152	4
1024	BRASSICA OLERACEA ACEPHALA GROUP	Y	.	0.0	17.6	3.7	.	11.2	.	.	.	.	.	.	.	.	.	.	.	6
1025	BRASSICA OLERACEA BOTRYTIS GROUP	L	29	90.5	2.8	0.2	5.7	0.9	0.8	30	58	1.0	20	349	55	0.07	0.08	0.7	72	2
1026	BRASSICA OLERACEA BOTRYTIS GROUP	L	39	87.3	4.5	0.6	6.4	1.6	1.2	116	81	1.3	.	.	1120	0.12	0.18	1.1	94	3
1027	BRASSICA OLERACEA BOTRYTIS GROUP	L	27	91.0	2.7	0.2	5.2	1.0	0.9	25	56	1.1	13	295	36	0.11	0.10	0.7	78	4
1028	BRASSICA OLERACEA BOTRYTIS GROUP	L	32	89.1	3.6	0.3	5.9	1.5	1.1	103	78	1.1	15	382	1500	0.10	0.23	0.9	113	4
1029	BRASSICA OLERACEA BOTRYTIS GROUP	Y	.	0.0	21.1	3.3	.	16.6	.	.	.	.	.	.	.	.	.	.	.	6
1030	BRASSICA OLERACEA BOTRYTIS GROUP	Y	.	0.0	17.1	.	.	17.3	.	.	.	.	.	.	.	.	.	.	.	6
1031	BRASSICA OLERACEA CAPITATA GROUP	L	26	91.4	1.7	0.1	6.0	1.2	0.8	47	40	0.7	.	.	100	0.04	0.04	0.3	54	1
1032	BRASSICA OLERACEA CAPITATA GROUP	L	43	86.4	3.0	0.3	9.2	1.6	1.1	71	60	1.5	4	266	35	0.12	0.08	0.6	85	2
1033	BRASSICA OLERACEA CAPITATA GROUP	L	22	93.0	1.6	0.3	4.4	0.8	0.7	55	31	0.8	10	238	280	0.06	0.06	0.3	46	2
1034	BRASSICA OLERACEA CAPITATA GROUP	L	28	91.4	1.7	0.2	6.1	1.0	0.6	43	36	0.7	.	.	60	0.06	0.04	0.3	43	3
1035	BRASSICA OLERACEA CAPITATA GROUP	L	31	90.2	2.0	0.2	6.9	1.0	0.7	42	35	0.8	26	268	24	0.09	0.06	0.4	61	4

TABLE 2 - DATA DIRECTLY TRANSCRIBED FROM ORIGINAL SOURCES.

OBS	PLANA	PART	CALORIES	WATER	PROTEIN	FAT	TOTCARB	FIBER	ASH	CALCIUM	PHOSPHO	IRON	SODIUM	POTASS	CAROTE	THIAMINE	RIBOFLAV	NIACIN	ASCORBIC	REFCO
1036	BRASSICA OLERACEA CAPITATA GROUP	L	24	92.4	1.3	0.2	5.4	0.8	0.7	49	29	0.4	20	233	78	0.05	0.05	0.3	47	4
1037	BRASSICA OLERACEA CAPITATA GROUP	L	.	0.0	24.0	4.0	62.5	12.6	9.5	630	210	.	.	.	.	.	0.89	.	.	6
1038	BRASSICA OLERACEA GEMMIFERA GROUP	SH	47	85.2	3.9	0.4	9.5	1.6	1.1	40	74	1.4	14	480	50	0.16	0.13	0.5	71	2
1039	BRASSICA OLERACEA GEMMIFERA GROUP	SH	50	83.3	5.2	0.3	9.9	1.9	1.3	47	92	1.7	.	.	290	0.17	0.16	1.1	82	3
1040	BRASSICA OLERACEA GEMMIFERA GROUP	SH	45	85.2	4.9	0.4	8.3	1.6	1.2	36	80	1.5	14	390	330	0.10	0.16	0.9	102	4
1041	BRASSICA OLERACEA GONGYLODES GROUP	L	26	91.3	2.3	0.2	5.1	1.4	1.1	36	36	1.6	.	.	.	.	.	.	.	1
1042	BRASSICA OLERACEA GONGYLODES GROUP	L	31	89.9	2.6	0.4	6.0	1.0	1.1	52	48	1.3	8	249	2000	0.06	0.10	0.6	68	2
1043	BRASSICA OLERACEA GONGYLODES GROUP	L	31	91.0	2.0	0.1	6.1	1.0	0.8	32	48	0.3	.	.	0	0.05	0.03	0.4	60	3
1044	BRASSICA OLERACEA GONGYLODES GROUP	L	29	90.3	2.0	0.1	6.6	1.0	1.0	41	51	0.5	8	372	12	0.06	0.04	0.3	66	4
1045	BRASSICA OLERACEA VAR. CAPITATA	S	.	0.0	33.8	25.9	.	.	5.8	.	.	.	.	.	.	.	.	.	.	19
1046	BRASSICA RAPA	R	21	93.5	1.0	0.0	4.7	0.9	0.8	38	42	0.5	.	.	.	0.03	0.05	.	75	1
1047	BRASSICA RAPA	L	29	90.4	2.9	0.4	5.3	0.4	1.0	158	131	1.9	.	.	.	0.02	0.25	.	.	1
1048	BRASSICA RAPA	L	25	91.5	2.5	0.3	4.5	0.8	1.2	127	53	1.2	.	388	2600	0.14	0.09	0.7	45	2
1049	BRASSICA RAPA	R	21	92.8	1.0	0.2	4.4	0.8	0.8	43	33	0.9	60	281	20	0.04	0.04	0.5	22	2
1050	BRASSICA RAPA	L	23	92.7	1.9	0.2	4.6	1.0	0.6	168	52	2.6	78	420	1330	0.10	0.18	0.7	47	2
1051	BRASSICA RAPA	S	.	0.0	26.2	30.7	.	.	.	.	.	.	.	.	.	.	.	.	.	20
1052	BRASSICA RAPA	R	27	92.5	0.8	0.2	5.7	0.8	0.8	32	26	0.4	.	.	0	0.03	0.03	0.5	28	3
1053	BRASSICA RAPA	L	61	83.3	2.9	1.7	11.2	1.8	0.9	136	38	4.6	.	.	2680	0.08	0.15	0.5	120	3
1054	BRASSICA RAPA	R	30	91.5	1.0	0.2	6.6	0.9	0.7	39	30	0.5	49	268	0	0.04	0.07	0.6	36	4
1055	BRASSICA RAPA	L	28	90.3	3.0	0.3	5.0	0.8	1.4	246	58	1.8	.	.	4560	0.21	0.39	0.8	139	4
1056	BRASSICA RAPA CHINENSIS GROUP	L	25	90.9	2.0	0.3	5.0	1.4	1.8	58	12	0.0	.	.	2280	0.00	0.18	6.5	.	1
1057	BRASSICA RAPA CHINENSIS GROUP	S	416	7.5	22.2	18.3	45.6	15.9	6.4	.	.	.	.	.	.	0.54	0.40	.	.	1
1058	BRASSICA RAPA CHINENSIS GROUP	L	17	94.2	1.7	0.2	3.1	0.7	0.8	102	46	2.6	22	279	2305	0.07	0.13	0.8	53	2
1059	BRASSICA RAPA CHINENSIS GROUP	L	16	94.3	1.6	0.2	2.9	0.6	1.0	165	44	0.8	26	306	1860	0.05	0.10	0.8	25	4
1060	BRASSICA RAPA CHINENSIS GROUP	L	14	95.0	1.2	0.1	3.0	0.6	0.7	43	40	0.6	23	253	90	0.05	0.04	0.6	25	4
1061	BRASSICA RAPA PEKINENSIS GROUP	L	13	94.5	1.3	0.2	2.3	0.9	1.7	35	39	0.5	.	.	.	.	.	.	.	1
1062	BRASSICA RAPA RAPIFERA GROUP	Y	.	0.0	18.0	.	.	10.3	.	.	.	.	.	.	.	.	.	.	.	6
1063	BRASSICA RAPA RAPIFERA GROUP	W	.	0.0	17.5	2.4	61.9	11.2	18.2	2870	580	38.0	.	3790	5	0.14	0.24	.	.	6
1064	BRASSICA SP.	S	.	0.0	26.3	33.1	.	.	5.2	.	.	.	.	.	.	.	.	.	.	19
1065	BRASSICA SP.	S	.	0.0	26.2	40.3	.	.	.	.	.	.	.	.	.	.	.	.	.	19
1066	BRASSICA SP.	S	.	0.0	26.2	39.3	.	.	.	.	.	.	.	.	.	.	.	.	.	19
1067	BRASSICA SP.	S	.	0.0	31.2	28.1	.	.	.	.	.	.	.	.	.	.	.	.	.	19
1068	BRASSICA SP.	L	40	87.8	2.7	0.7	7.8	1.2	1.0	192	58	2.4	.	363	1200	0.04	0.14	0.5	40	2
1069	BRASSICA SP.	S	.	0.0	35.6	32.9	.	.	.	.	.	.	.	.	.	.	.	.	.	20
1070	BRASSICA SP.	S	.	0.0	17.5	51.4	.	.	.	.	.	.	.	.	.	.	.	.	.	20
1071	BRASSICA SP. X SINAPIS SP.	S	.	0.0	30.4	28.2	.	.	.	.	.	.	.	.	.	.	.	.	.	21
1072	BRASSICA TOURNEFORTII	S	.	0.0	26.6	28.1	.	.	.	.	.	.	.	.	.	.	.	.	.	21
1073	BRASSICA TOURNEFORTII	S	.	0.0	25.4	29.3	.	.	.	.	.	.	.	.	.	.	.	.	.	21
1074	BRASSICA TOURNEFORTII	S	.	0.0	29.0	22.6	.	.	.	.	.	.	.	.	.	.	.	.	.	21
1075	BRASSICA TOURNEFORTII	S	.	0.0	22.1	29.6	.	.	.	.	.	.	.	.	.	.	.	.	.	21
1076	BRASSICA TOURNEFORTII	S	.	0.0	21.6	32.2	.	.	.	.	.	.	.	.	.	.	.	.	.	21
1077	BRASSICA TOURNEFORTII	S	.	0.0	27.2	29.8	.	.	.	.	.	.	.	.	.	.	.	.	.	21
1078	BRASSICA TOURNEFORTII	S	.	0.0	27.6	30.2	.	.	.	.	.	.	.	.	.	.	.	.	.	21
1079	BRAZORIA SCUTELLARIOIDES	S	.	0.0	18.5	36.1	.	.	.	.	.	.	.	.	.	.	.	.	.	20
1080	BREYNIA RHAMNOIDES	S	.	0.0	16.4	27.2	.	.	.	.	.	.	.	.	.	.	.	.	.	21

TABLE 2 - DATA DIRECTLY TRANSCRIBED FROM ORIGINAL SOURCES.

OBS	PLANA	PART	CALORIES	WATER	PROTEIN	FAT	TOTCARB	FIBER	ASH	CALCIUM	PHOSPHO	IRON	SODIUM	POTASS	CAROTE	THIAMINE	RIBOFLAV	NIACIN	ASCORBIC	REFCO
1081	BRICKELLIA CALIFORNICA	S	.	0.0	15.4	5.6	.	.	.	.	.	.	.	.	.	.	.	.	.	21
1082	BRICKELLIA SCOPARIA VAR. SUBAURICULATA	S	.	0.0	32.5	21.1	.	.	.	.	.	.	.	.	.	.	.	.	.	20
1083	BRIDELIA STIPULARIA	S	.	0.0	15.2	18.4	.	.	.	.	.	.	.	.	.	.	.	.	.	21
1084	BRIZA FUSCA	S	.	0.0	31.8	17.2	.	.	.	.	.	.	.	.	.	.	.	.	.	21
1085	BRIZA MAXIMA	S	.	0.0	21.2	5.5	.	.	.	.	.	.	.	.	.	.	.	.	.	21
1086	BRIZA SPICATA	S	.	0.0	15.4	18.3	.	.	.	.	.	.	.	.	.	.	.	.	.	20
1087	BRIZA SPICATA	S	.	0.0	15.9	21.0	.	.	.	.	.	.	.	.	.	.	.	.	.	21
1088	BROMELIA PINGUIN	L	25	92.0	0.9	0.3	5.7	0.6	1.1	116	24	0.6	.	.	0	0.03	0.04	0.4	21	3
1089	BROMELIA PLUMIERI	L	43	87.1	1.2	0.2	10.7	0.4	0.8	48	6	0.6	.	.	10	0.06	0.02	0.3	47	3
1090	BROMUS ANOMALUS		.	.	.	.	.	.	.	.	.	.	.	.	.	.	.	.	.	6
1091	BROMUS ARENARIUS		.	0.0	8.0	.	.	31.4	5.5	360	270	.	.	1410	.	.	.	.	.	6
1092	BROMUS ARVENSIS	S	.	0.0	11.2	1.3	.	.	3.0	.	.	.	.	.	.	.	.	.	.	19
1093	BROMUS ARVENSIS	S	.	0.0	15.1	1.3	.	.	.	.	.	.	.	.	.	.	.	.	.	21
1094	BROMUS BREVIS	S	.	0.0	14.8	1.4	.	.	2.1	.	.	.	.	.	.	.	.	.	.	19
1095	BROMUS BREVIS	S	.	0.0	6.9	2.1	.	.	6.7	.	.	.	.	.	.	.	.	.	.	19
1096	BROMUS CAPPADOCICUS	S	.	0.0	13.9	1.2	.	.	.	.	.	.	.	.	.	.	.	.	.	21
1097	BROMUS CARINATUS		.	0.0	7.6	1.9	83.2	37.6	7.3	390	240	.	.	.	.	.	.	.	.	6
1098	BROMUS CATHARTICUS	S	.	0.0	17.5	0.8	.	.	.	.	.	.	.	.	.	.	.	.	.	19
1099	BROMUS CATHARTICUS	S	.	0.0	18.8	0.8	.	.	.	.	.	.	.	.	.	.	.	.	.	19
1100	BROMUS CATHARTICUS		.	0.0	16.3	3.2	67.0	24.1	13.5	560	290	.	.	3850	.	.	.	.	.	6
1101	BROMUS INERMIS	S	.	0.0	19.4	1.5	.	.	3.0	.	.	.	.	.	.	.	.	.	.	19
1102	BROMUS INERMUS		.	0.0	11.4	3.0	77.0	28.4	8.6	440	250	.	.	2700	32	0.31	0.77	.	.	6
1103	BROMUS JAPONICUS		.	0.0	16.1	3.1	70.7	28.7	10.1	400	260	.	.	.	.	.	.	.	.	6
1104	BROMUS MOLLIS		.	0.0	11.2	2.4	77.4	29.3	9.0	506	366	.	.	2480	.	.	.	.	.	6
1105	BROMUS RIGIDUS		.	0.0	14.0	.	.	26.6	8.5	480	390	.	.	.	.	.	.	.	.	6
1106	BROMUS RUBENS	S	.	0.0	17.9	2.0	.	.	2.8	.	.	.	.	.	.	.	.	.	.	19
1107	BROMUS RUBENS		.	0.0	8.5	.	.	31.6	7.2	280	230	.	.	1450	.	.	.	.	.	6
1108	BROMUS SCOPARIUS	S	.	0.0	16.1	1.4	.	.	.	.	.	.	.	.	.	.	.	.	.	21
1109	BROMUS SP.	S	.	0.0	16.9	1.6	.	.	.	.	.	.	.	.	.	.	.	.	.	20
1110	BROMUS TECTORUM	S	.	0.0	14.1	1.2	.	.	.	.	.	.	.	.	.	.	.	.	.	21
1111	BROMUS TECTORUM		.	0.0	9.7	2.0	78.9	24.4	9.4	440	260	.	.	.	.	.	.	.	.	6
1112	BROMUS TEXENSIS		.	.	.	.	.	.	.	600	200	.	.	.	.	.	.	.	.	6
1113	BROMUS VALDIVIANUS	S	.	0.0	8.1	2.5	.	.	.	.	.	.	.	.	.	.	.	.	.	19
1114	BROMUS WILLDENOWII	S	.	0.0	14.5	1.0	.	.	.	.	.	.	.	.	.	.	.	.	.	21
1115	BRONGNIARTIA ALAMOSSANA	S	.	0.0	26.9	33.9	.	.	.	.	.	.	.	.	.	.	.	.	.	20
1116	BROSIMUM ALICASTRUM	L	.	0.0	13.9	3.4	74.8	26.5	7.9	.	.	.	.	.	.	.	.	.	.	18
1117	BROSIMUM ALICASTRUM	S	.	0.0	9.9	1.1	.	.	4.1	.	.	.	.	.	.	.	.	.	.	19
1118	BROSIMUM ALICASTRUM	L	127	62.0	3.2	1.2	30.6	8.9	3.0	530	68	5.4	.	.	820	0.24	0.51	1.4	55	3
1119	BROSIMUM ALICASTRUM	F	56	84.0	2.5	0.5	12.1	1.2	0.9	45	36	0.8	.	.	560	0.05	1.52	0.8	28	3
1120	BROSIMUM ALICASTRUM	S	363	6.5	11.4	1.6	76.1	6.2	4.4	211	142	4.6	.	.	130	0.03	0.14	2.1	.	3
1121	BROUSSONETIA PAPYRIFERA	S	.	0.0	18.5	30.1	.	.	7.7	.	.	.	.	.	.	.	.	.	.	19
1122	BROWALLIA AMERICANA	S	.	0.0	25.0	20.8	.	.	.	.	.	.	.	.	.	.	.	.	.	20
1123	BRYONIA SP.	S	.	0.0	19.1	23.3	.	.	.	.	.	.	.	.	.	.	.	.	.	21
1124	BUCHLOE DACTYLOIDES	H	.	0.0	9.2	1.8	78.6	27.7	12.4	520	160	.	.	710	9	.	.	.	.	6
1125	BUDDLEIA DAVIDII	S	.	0.0	18.8	27.7	.	.	4.2	.	.	.	.	.	.	.	.	.	.	19

TABLE 2 - DATA DIRECTLY TRANSCRIBED FROM ORIGINAL SOURCES.

OBS	PLNA	PART	CALORIES	WATER	PROTEIN	FAT	TOTCARB	FIBER	ASH	CALCIUM	PHOSPHO	IRON	SODIUM	POTASS	CAROT	THIAMINE	RIBOFLAV	NIACIN	ASCORBIC	REFCO
1126	BUMELIA TENAX	L	.	0.0	13.0	10.3	70.1	25.3	6.6	3040	120	.	.	570	.	.	.	.	.	6
1127	BUNCHOSIA PALMERI	S	.	0.0	19.3	1.8	.	.	.	.	.	.	.	.	.	.	.	.	.	21
1128	BUNIUM PERSICUM	S	.	0.0	19.9	13.6	.	.	.	.	.	.	.	.	.	.	.	.	.	21
1129	BUPLEURUM CROECEUM	S	.	0.0	14.9	9.5	.	.	.	.	.	.	.	.	.	.	.	.	.	21
1130	BUPLEURUM HELDREICHII	S	.	0.0	15.1	8.2	.	.	.	.	.	.	.	.	.	.	.	.	.	21
1131	BUPLEURUM SUBOVATUM	S	.	0.0	10.3	6.8	.	.	.	.	.	.	.	.	.	.	.	.	.	21
1132	BURSERA CONFUSA	S	.	0.0	11.9	16.7	.	.	.	.	.	.	.	.	.	.	.	.	.	19
1133	BURSERA CONFUSA	S	.	0.0	6.2	16.9	.	.	.	.	.	.	.	.	.	.	.	.	.	19
1134	BURSERA CONFUSA	S	.	0.0	3.8	6.7	.	.	.	.	.	.	.	.	.	.	.	.	.	20
1135	BURSERA GRANDIFOLIA	S	.	0.0	1.2	0.4	.	.	.	.	.	.	.	.	.	.	.	.	.	19
1136	BURSERA GRANDIFOLIA	S	.	0.0	3.2	2.7	.	.	.	.	.	.	.	.	.	.	.	.	.	20
1137	BURSERA INOPINATA	S	.	0.0	7.5	7.6	.	.	.	.	.	.	.	.	.	.	.	.	.	20
1138	BUTIA CAPITATA	S	.	0.0	15.7	56.5	.	.	.	.	.	.	.	.	.	.	.	.	.	21
1139	BUTIA CAPITATA X EREOSPATHA	S	.	0.0	12.9	44.1	.	.	1.8	.	.	.	.	.	.	.	.	.	.	19
1140	BUTYROSPERMUM PARADOXUM	S	579	6.9	6.8	49.0	35.6	5.2	1.7	100	40	3.0	.	.	.	0.52	.	.	.	1
1141	BUXUS SEMPERVIRENS	S	.	0.0	21.4	41.6	.	.	1.6	.	.	.	.	.	.	.	.	.	.	19
1142	BUXUS SEMPERVIRENS	S	.	0.0	21.8	44.3	.	.	.	.	.	.	.	.	.	.	.	.	.	21
1143	BYRSONIMA CRASSIFOLIA	F	66	82.8	0.9	1.3	14.4	2.2	0.6	33	17	2.0	.	.	40	0.02	0.04	0.4	84	3
1144	CACALIA DECOMPOSITA	S	.	0.0	33.8	34.2	.	.	.	.	.	.	.	.	.	.	.	.	.	20
1145	CACALIA PLANTAGINEA	S	.	0.0	26.2	36.5	.	.	.	.	.	.	.	.	.	.	.	.	.	19
1146	CACALIA TUBEROSA	S	.	0.0	27.9	28.3	.	.	.	.	.	.	.	.	.	.	.	.	.	20
1147	CACHRYS GONIOCARPA	S	.	0.0	37.5	21.5	.	.	.	.	.	.	.	.	.	.	.	.	.	21
1148	CADABA FARINOSA	L	85	70.0	9.2	1.3	14.6	1.9	4.9	738	36	3.8	.	.	.	.	.	.	.	1
1149	CADABA FARINOSA	L	.	0.0	30.6	1.8	52.1	9.3	15.4	920	160	.	.	.	.	.	.	.	.	18
1150	CADABA FARINOSA	SH	.	0.0	18.2	1.8	73.0	27.3	7.0	390	170	.	.	.	.	.	.	.	.	18
1151	CAESALPINIA PLATYLOBA	S	.	0.0	19.4	12.8	.	.	.	.	.	.	.	.	.	.	.	.	.	19
1152	CAESALPINIA SPINOSA	F	.	0.0	8.8	1.6	81.4	9.6	8.2	220	200	.	.	.	.	.	.	.	.	18
1153	CAJANUS CAJAN	V	345	9.9	19.5	1.3	65.5	7.3	3.8	161	285	15.0	.	.	55	0.72	0.14	2.9	0	1
1154	CAJANUS CAJAN	W	.	0.0	21.4	6.0	66.8	30.8	5.8	890	240	.	.	.	.	.	.	.	.	18
1155	CAJANUS CAJAN	L	.	0.0	11.0	6.9	63.6	18.3	18.5	.	.	.	.	.	.	.	.	.	.	18
1156	CAJANUS CAJAN	F	.	0.0	20.3	1.7	74.7	35.2	3.3	.	.	.	.	.	.	.	.	.	.	18
1157	CAJANUS CAJAN	S	.	0.0	23.4	0.9	71.4	10.6	4.3	140	450	.	.	.	.	.	.	.	.	18
1158	CAJANUS CAJAN	S	.	0.0	21.8	1.4	.	.	4.0	.	.	.	.	.	.	.	.	.	.	19
1159	CAJANUS CAJAN	V	339	11.5	20.4	1.2	63.4	4.4	3.5	103	224	4.9	26	654	160	0.49	0.21	2.2	0	2
1160	CAJANUS CAJAN	G	114	64.4	8.7	0.6	24.5	3.6	1.8	72	174	2.0	5	622	145	0.44	0.16	1.8	32	2
1161	CAJANUS CAJAN	IS	119	68.9	7.5	0.6	21.6	.	1.4	29	135	1.3	5	563	145	0.40	0.25	2.4	26	2
1162	CAJANUS CAJAN	S	.	0.0	22.5	.	.	.	.	.	.	.	.	.	.	.	.	.	.	20
1163	CAJANUS CAJAN	S	.	0.0	21.5	.	.	.	.	.	.	.	.	.	.	.	.	.	.	21
1164	CAJANUS CAJAN	V	337	12.2	19.2	1.5	63.3	8.1	3.8	137	322	5.0	.	.	40	0.72	0.17	2.6	0	3
1165	CAJANUS CAJAN	F	118	69.4	7.0	0.6	21.7	3.5	1.3	35	124	1.7	.	.	40	0.34	0.21	2.3	49	3
1166	CAJANUS CAJAN	IS	117	69.5	7.2	0.6	21.3	3.3	1.4	42	127	1.6	5	552	84	0.40	0.17	2.2	39	4
1167	CAJANUS CAJAN	W	.	0.0	14.5	4.4	76.1	37.8	5.0	.	.	.	.	.	.	.	.	.	.	6
1168	CAKILE EDENTULA	S	.	0.0	31.2	48.7	.	.	.	.	.	.	.	.	.	.	.	.	.	19
1169	CAKILE MARITIMA	S	.	0.0	33.4	43.8	.	.	.	.	.	.	.	.	.	.	.	.	.	21
1170	CAKILE MARITIMA	S	.	0.0	32.7	45.6	.	.	.	.	.	.	.	.	.	.	.	.	.	21

TABLE 2 - DATA DIRECTLY TRANSCRIBED FROM ORIGINAL SOURCES.

OBS	PLNA	PART	CALORIES	WATER	PROTEIN	FAT	TOTLCARB	FIBER	ASH	CALCIUM	PHOSPHO	IRON	SODIUM	POTASS	CAROT	THIAMINE	RIBOFLAV	NIACIN	ASCORBIC	REFCO
1171	CALAMOGROSTIS CANADENSIS	W	.	0.0	9.2	2.70	78.9	34.1	9.2	.	.	.	.	.	.	.	.	.	.	6
1172	CALAMOVILFA LONGIFOLIA	W	.	0.0	5.9	1.90	86.8	37.8	5.4	.	.	.	.	.	.	.	.	.	.	6
1173	CALAMUS ORNATUS	F	79	79.0	0.6	1.20	18.6	0.5	0.6	19	10	1.7	.	.	.	0.06	0.01	0.9	5	2
1174	CALANDRINIA MICRANTHA	L	19	93.0	2.2	0.30	3.1	0.9	1.4	48	31	3.2	.	.	960	0.08	0.09	0.4	17	3
1175	CALATHEA MACROSEPALA		22	92.3	1.8	0.20	4.4	1.5	1.3	20	49	1.0	.	.	0	0.04	0.10	0.7	7	3
1176	CALEA URTICAEFOLIA	S	.	0.0	18.8	21.00	.	.	.	.	.	.	.	.	.	.	.	.	.	20
1177	CALEA URTICAEFOLIA	S	.	0.0	12.9	7.50	.	.	.	.	.	.	.	.	.	.	.	.	.	21
1178	CALENDULA ARVENSIS	S	.	0.0	41.1	40.60	.	.	.	.	.	.	.	.	.	.	.	.	.	21
1179	CALENDULA OFFICINALIS	S	.	0.0	33.1	42.20	.	.	.	.	.	.	.	.	.	.	.	.	.	19
1180	CALENDULA OFFICINALIS	S	.	0.0	8.1	3.30	.	.	.	.	.	.	.	.	.	.	.	.	.	19
1181	CALENDULA OFFICINALIS	S	.	0.0	30.0	45.30	.	.	.	.	.	.	.	.	.	.	.	.	.	20
1182	CALENDULA OFFICINALIS	S	.	0.0	33.8	42.10	.	.	.	.	.	.	.	.	.	.	.	.	.	20
1183	CALENDULA OFFICINALIS	S	.	0.0	31.2	43.60	.	.	.	.	.	.	.	.	.	.	.	.	.	20
1184	CALENDULA OFFICINALIS	S	.	0.0	36.9	40.80	.	.	.	.	.	.	.	.	.	.	.	.	.	20
1185	CALENDULA OFFICINALIS	S	.	0.0	33.8	44.20	.	.	.	.	.	.	.	.	.	.	.	.	.	20
1186	CALENDULA OFFICINALIS	S	.	0.0	31.9	43.60	.	.	.	.	.	.	.	.	.	.	.	.	.	20
1187	CALENDULA OFFICINALIS	S	.	0.0	30.6	45.10	.	.	.	.	.	.	.	.	.	.	.	.	.	20
1188	CALENDULA OFFICINALIS	S	.	0.0	30.6	45.80	.	.	.	.	.	.	.	.	.	.	.	.	.	20
1189	CALENDULA OFFICINALIS	S	.	0.0	32.5	44.70	.	.	.	.	.	.	.	.	.	.	.	.	.	20
1190	CALENDULA OFFICINALIS	S	.	0.0	30.6	45.20	.	.	.	.	.	.	.	.	.	.	.	.	.	20
1191	CALENDULA OFFICINALIS	S	.	0.0	32.5	43.20	.	.	.	.	.	.	.	.	.	.	.	.	.	20
1192	CALENDULA OFFICINALIS	S	.	0.0	32.5	44.30	.	.	.	.	.	.	.	.	.	.	.	.	.	20
1193	CALENDULA OFFICINALIS	S	.	0.0	32.5	44.00	.	.	.	.	.	.	.	.	.	.	.	.	.	20
1194	CALENDULA OFFICINALIS	S	.	0.0	33.1	44.10	.	.	.	.	.	.	.	.	.	.	.	.	.	20
1195	CALENDULA OFFICINALIS	S	.	0.0	33.8	43.10	.	.	.	.	.	.	.	.	.	.	.	.	.	20
1196	CALENDULA OFFICINALIS	S	.	0.0	31.9	43.60	.	.	.	.	.	.	.	.	.	.	.	.	.	20
1197	CALENDULA OFFICINALIS	S	.	0.0	33.8	43.50	.	.	.	.	.	.	.	.	.	.	.	.	.	20
1198	CALENDULA OFFICINALIS	S	.	0.0	32.5	43.70	.	.	.	.	.	.	.	.	.	.	.	.	.	20
1199	CALENDULA OFFICINALIS	S	.	0.0	.	43.10	.	.	.	.	.	.	.	.	.	.	.	.	.	21
1200	CALLIANDRA ERIOPHYLLA	S	.	0.0	38.6	15.90	.	.	4.1	.	.	.	.	.	.	.	.	.	.	19
1201	CALLIANDRA ERIOPHYLLA	S	.	0.0	38.9	16.40	.	.	4.4	.	.	.	.	.	.	.	.	.	.	19
1202	CALLIANDRA GRACILIS	S	.	0.0	26.7	2.20	.	.	.	.	.	.	.	.	.	.	.	.	.	21
1203	CALLICARPA AMERICANA	S	.	0.0	6.2	9.80	.	.	.	.	.	.	.	.	.	.	.	.	.	19
1204	CALLIOPSIS DRUMMONDII	S	.	0.0	18.1	22.40	.	.	.	.	.	.	.	.	.	.	.	.	.	19
1205	CALLIRHOE PEDATA	W	.	0.0	10.4	2.70	77.6	30.2	9.3	2461	450	223.0	.	.	.	.	.	.	.	6
1206	CALOCHORTUS SP.	S	.	0.0	14.6	6.00	.	.	.	.	.	.	.	.	.	.	.	.	.	20
1207	CALOPOGONIUM MUCUNOIDES	W	.	0.0	15.6	2.30	75.9	31.5	6.2	1420	190	.	.	.	.	.	.	.	.	18
1208	CALOTROPIS GIGANTEA	S	.	0.0	28.8	36.10	.	.	.	.	.	.	.	.	.	.	.	.	.	20
1209	CALOTROPIS PROCERA	S	.	0.0	27.5	36.60	.	.	.	.	.	.	.	.	.	.	.	.	.	21
1210	CALOTROPIS PROCERA	S	.	0.0	30.1	34.80	.	.	.	.	.	.	.	.	.	.	.	.	.	21
1211	CALYCANTHUS FLORIDUS	S	.	0.0	24.4	48.30	.	.	2.6	.	.	.	.	.	.	.	.	.	.	19
1212	CALYCOTOME VILLOSA	S	.	0.0	31.4	3.00	.	.	.	.	.	.	.	.	.	.	.	.	.	21
1213	CAMASSIA QUAMASH	TU	.	70.0	0.7	0.23	27.1	1.3	2.0	.	.	.	.	.	.	0.07	0.05	.	4	9
1214	CAMELINA MICROCARPA	S	.	0.0	25.6	34.10	.	.	.	.	.	.	.	.	.	.	.	.	.	20
1215	CAMELINA RUMELICA	S	.	0.0	29.1	35.20	.	.	.	.	.	.	.	.	.	.	.	.	.	20

TABLE 2 - DATA DIRECTLY TRANSCRIBED FROM ORIGINAL SOURCES.

OBS	PLNA	PART	CALORIES	WATER	PROTEIN	FAT	TOTLCARB	FIBER	ASH	CALCIUM	PHOSPHO	IRON	SODIUM	POTASS	CAROT	THIAMINE	RIBOFLAV	NIACIN	ASCORBIC	REFCO
1216	CAMELINA SATIVA	S	.	0.0	33.1	33.2	.	.	.	.	.	.	.	.	.	.	.	.	.	19
1217	CAMELLIA JAPONICA	S	.	0.0	8.8	66.4	.	.	.	.	.	.	.	.	.	.	.	.	.	20
1218	CAMELLIA OLEIFERA	S	.	0.0	9.4	41.6	.	.	.	.	.	.	.	.	.	.	.	.	.	19
1219	CAMPANULA CARPATICA	S	.	0.0	16.9	38.4	.	.	.	.	.	.	.	.	.	.	.	.	.	20
1220	CAMPANULA PERSICIFOLIA	S	.	0.0	21.2	37.8	.	.	.	.	.	.	.	.	.	.	.	.	.	20
1221	CAMPSIS RADICANS	S	.	0.0	22.5	24.4	.	.	.	.	.	.	.	.	.	.	.	.	.	19
1222	CAMPSIS RADICANS	S	.	0.0	22.5	20.2	.	.	.	.	.	.	.	.	.	.	.	.	.	19
1223	CAMPSIS RADICANS	S	.	0.0	24.4	28.8	.	.	.	.	.	.	.	.	.	.	.	.	.	19
1224	CAMPTOSEMA RUBICUNDUM	S	.	0.0	21.9	4.1	.	.	.	.	.	.	.	.	.	.	.	.	.	21
1225	CANARIUM ALBUM	F	72	80.2	1.2	1.2	16.2	3.4	1.2	18	29	2.1	47	416	330	0.02	0.11	0.4	20	2
1226	CANARIUM OVATUM	S	669	6.3	11.4	71.1	8.4	2.7	2.8	140	554	3.4	3	489	24	0.88	0.09	0.5	0	
1227	CANARIUM OVATUM	S	645	8.9	14.2	68.5	5.5	3.2	2.9	119	508	2.6	3	489	25	0.95	0.12	0.4	0	2
1228	CANARIUM OVATUM	S	600	14.2	10.6	62.8	9.3	2.0	3.1	154	61	4.1	.	.	30	0.64	0.06	0.5	21	3
1229	CANARIUM SCHWEINFURTHII	S	546	6.0	18.3	45.8	24.4	3.4	5.5	160	1260	.	.	.	.	.	.	.	.	1
1230	CANARIUM SCHWEINFURTHII	F	239	49.3	1.5	13.5	33.5	.	2.2	85	630	5.0	.	.	.	.	.	.	.	1
1231	CANAVALIA ENSIFORMIS	G	37	88.6	2.4	0.3	8.1	1.8	0.6	60	40	2.0	.	.	.	.	.	.	.	1
1232	CANAVALIA ENSIFORMIS	S	347	10.7	24.5	2.6	59.0	7.4	3.2	158	298	7.0	.	.	.	0.77	0.15	1.8	.	1
1233	CANAVALIA ENSIFORMIS	W	.	0.0	22.5	2.1	63.8	27.4	11.6	.	.	.	.	.	.	.	.	.	.	18
1234	CANAVALIA ENSIFORMIS	S	.	0.0	36.3	2.5	58.3	9.9	2.9	190	270	.	.	.	.	.	.	.	.	18
1235	CANAVALIA ENSIFORMIS	F	.	0.0	4.5	1.5	90.2	48.1	3.8	300	10	.	.	.	.	.	.	.	.	18
1236	CANAVALIA ENSIFORMIS	S	.	0.0	28.1	2.3	.	.	.	.	.	.	.	.	.	.	.	.	.	19
1237	CANAVALIA ENSIFORMIS	S	.	0.0	29.4	2.8	.	.	.	.	.	.	.	.	.	.	.	.	.	19
1238	CANAVALIA ENSIFORMIS	S	.	0.0	27.5	3.2	.	.	.	.	.	.	.	.	.	.	.	.	.	19
1239	CANAVALIA ENSIFORMIS	S	.	0.0	30.0	2.4	.	.	.	.	.	.	.	.	.	.	.	.	.	19
1240	CANAVALIA ENSIFORMIS	G	43	88.6	2.7	0.2	7.9	1.8	0.6	60	40	2.0	.	.	25	0.10	.	.	0	2
1241	CANAVALIA ENSIFORMIS	S	348	11.2	21.0	3.2	61.0	7.6	3.6	134	300	8.6	35	750	.	0.65	0.13	3.1	2	2
1242	CANAVALIA ENSIFORMIS	S	.	0.0	27.2	2.8	.	.	.	.	.	.	.	.	.	.	.	.	.	21
1243	CANAVALIA ENSIFORMIS	S	.	0.0	28.7	2.8	.	.	.	.	.	.	.	.	.	.	.	.	.	21
1244	CANAVALIA ENSIFORMIS	G	82	78.5	6.9	0.5	13.3	3.3	0.8	33	66	1.2	.	.	30	0.22	0.10	2.0	32	3
1245	CANAVALIA ENSIFORMIS	S	331	12.0	25.4	1.3	57.1	4.9	4.2	96	343	4.9	.	.	20	0.46	0.15	2.1	.	3
1246	CANAVALIA GLADIATA	S	318	14.9	27.1	0.6	53.8	11.6	3.6	.	.	.	.	.	.	.	.	.	.	1
1247	CANAVALIA GLADIATA	G	34	89.2	2.8	0.2	7.3	1.5	0.5	.	.	.	.	.	.	.	.	.	.	1
1248	CANAVALIA GLADIATA	S	.	0.0	33.2	2.6	60.6	8.0	3.6	.	.	.	.	.	.	.	.	.	.	18
1249	CANAVALIA GLADIATA	G	.	88.6	2.7	0.2	7.9	1.5	0.6	.	.	.	.	.	24	.	.	.	.	5
1250	CANAVALIA SP.	S	.	0.0	33.1	1.4	.	.	.	.	.	.	.	.	.	.	.	.	.	20
1251	CANAVALIA SP.	W	.	0.0	22.4	2.2	63.8	27.6	11.6	.	.	.	.	.	.	.	.	.	.	6
1252	CANNA EDULIS	L	.	0.0	14.1	2.5	71.1	24.6	12.3	.	.	.	.	.	.	.	.	.	.	18
1253	CANNA EDULIS	SH	.	0.0	10.2	5.1	68.2	19.6	16.5	.	.	.	.	.	.	.	.	.	.	18
1254	CANNA EDULIS	R	.	0.0	3.6	0.8	88.2	3.4	7.4	.	.	.	.	.	.	.	.	.	.	18
1255	CANNA EDULIS	R	130	66.8	0.9	0.1	31.3	0.5	0.9	15	63	1.4	.	.	0	0.03	0.01	0.4	7	3
1256	CANNA FLACCIDA	S	.	0.0	9.4	0.8	.	.	.	.	.	.	.	.	.	.	.	.	.	20
1257	CANNA INDICA	R	.	0.0	3.6	0.7	90.2	4.0	5.3	130	220	.	.	.	.	.	.	.	.	18
1258	CANNA INDICA	S	.	0.0	10.0	0.8	.	.	.	.	.	.	.	.	.	.	.	.	.	20
1259	CANNA INDICA	S	.	0.0	8.7	0.5	.	.	.	.	.	.	.	.	.	.	.	.	.	20
1260	CANNABIS SATIVA	S	.	0.0	20.0	35.8	39.6	16.5	4.6	.	.	.	.	.	.	.	.	.	.	18

TABLE 2 - DATA DIRECTLY TRANSCRIBED FROM ORIGINAL SOURCES.

OBS	PLNA	PART	CALORIES	WATER	PROTEIN	FAT	TOTCARB	FIBER	ASH	CALCIUM	PHOSPHON	IRON	SODIUM	POTASS	CAROTENE	THIAMINE	RIBOFLAV	NIACIN	ASCORBIC	REFCO
1261	CANNABIS SATIVA	S	421	13.6	27.1	25.6	27.6	20.3	6.1	120	970	12.0	.	.	5	0.32	0.17	2.10	0	2
1262	CANNABIS SATIVA	S	.	8.8	21.5	30.4	34.7	18.8	4.6	.	.	.	.	.	.	.	.	.	.	5
1263	CANNOMOIS VIRGATA	S	.	0.0	7.6	0.8	.	.	.	.	.	.	.	.	.	.	.	.	.	21
1264	CAPPARIS CORYMBOSA	L	84	72.0	8.4	1.2	15.0	4.1	3.4	926	30	5.7	.	.	.	.	.	.	.	1
1265	CAPPARIS OVATA	S	.	0.0	22.4	30.6	.	.	.	.	.	.	.	.	.	.	.	.	.	21
1266	CAPPARIS SPINOSA	S	.	0.0	22.0	34.6	.	.	4.8	.	.	.	.	.	.	.	.	.	.	19
1267	CAPPARIS SPINOSA	S	.	0.0	19.0	31.6	.	.	.	.	.	.	.	.	.	.	.	.	.	21
1268	CAPSELLA BURSA-PASTORIS	L	33	88.2	4.2	0.5	5.2	1.2	1.9	208	86	4.8	.	394	2590	0.25	0.17	0.40	36	2
1269	CAPSELLA BURSA-PASTORIS	S	.	0.0	32.5	25.9	.	.	.	.	.	.	.	.	.	.	.	.	.	20
1270	CAPSELLA BURSA-PASTORIS	S	.	0.0	24.4	39.7	.	.	.	.	.	.	.	.	.	.	.	.	.	20
1271	CAPSELLA BURSA-PASTORIS	S	.	0.0	27.8	26.0	.	.	.	.	.	.	.	.	.	.	.	.	.	21
1272	CAPSICUM ABYSSINICUM	F	100	72.8	3.0	2.4	20.2	9.2	1.6	19	110	5.6	.	.	1410	.	0.30	1.60	147	1
1273	CAPSICUM ANNUUM	F	289	9.5	14.8	13.0	55.7	20.9	7.0	177	345	23.6	34	2344	36362	0.64	1.74	15.32	71	16
1274	CAPSICUM ANNUUM VAR. ANNUUM	F	45	86.9	2.0	0.8	9.5	1.7	0.8	11	47	0.9	.	374	4770	0.09	0.12	0.40	86	2
1275	CAPSICUM ANNUUM VAR. ANNUUM	F	26	92.0	1.3	0.2	6.0	1.4	0.5	12	34	0.9	5	274	1750	0.07	0.08	0.80	103	2
1276	CAPSICUM ANNUUM VAR. ANNUUM	F	296	13.2	13.2	5.1	61.9	10.4	6.6	150	206	7.7	.	.	6460	0.22	1.04	15.00	79	3
1277	CAPSICUM ANNUUM VAR. ANNUUM	F	56	84.8	2.8	1.7	9.9	4.0	0.8	29	49	1.4	.	.	980	0.14	0.09	1.60	103	3
1278	CAPSICUM ANNUUM VAR. ANNUUM	F	82	77.2	3.3	2.0	16.1	8.0	1.4	26	69	1.1	.	.	940	0.19	0.16	2.60	87	3
1279	CAPSICUM ANNUUM VAR. ANNUUM	F	38	88.6	1.6	0.3	8.9	1.8	0.6	16	34	0.8	.	.	760	0.07	0.06	1.30	181	3
1280	CAPSICUM ANNUUM VAR. ANNUUM	F	31	90.8	1.2	0.3	7.1	1.3	0.6	8	27	0.6	.	.	290	0.06	0.06	1.00	114	3
1281	CAPSICUM ANNUUM VAR. ANNUUM	S	382	8.4	16.9	18.3	52.6	17.9	3.8	87	450	5.6	.	.	380	0.55	0.67	13.60	26	3
1282	CAPSICUM ANNUUM VAR. ANNUUM	S	309	7.4	16.1	1.8	71.3	35.0	3.4	57	466	7.0	.	.	300	0.64	0.29	11.80	29	3
1283	CAPSICUM ANNUUM VAR. ANNUUM	S	394	6.7	16.7	19.4	53.6	35.2	3.6	38	460	7.8	.	.	260	0.12	0.21	8.40	23	3
1284	CAPSICUM ANNUUM VAR. ANNUUM	S	416	6.5	16.6	23.2	50.7	24.6	3.0	51	448	7.2	.	.	2212	0.26	0.24	10.10	31	3
1285	CAPSICUM ANNUUM VAR. ANNUUM	F	93	74.3	3.7	2.3	18.1	9.0	1.6	29	78	1.2	.	.	12960	0.22	0.36	4.40	369	4
1286	CAPSICUM ANNUUM VAR. ANNUUM	F	37	88.8	1.3	0.2	9.1	1.8	0.6	10	25	0.7	.	.	462	0.09	0.06	1.70	235	4
1287	CAPSICUM FRUTESCENS	F	42	87.4	1.1	0.1	10.7	1.4	0.7	5	18	1.2	.	.	330	0.04	0.05	2.00	40	
1288	CAPSICUM FRUTESCENS	F	94	74.2	4.1	2.3	18.0	6.0	1.4	58	101	2.9	.	.	7140	0.25	0.20	2.40	121	1
1289	CAPSICUM FRUTESCENS	F	318	8.0	12.0	17.3	55.6	24.9	6.0	148	293	7.8	30	2014	24966	0.33	0.92	8.70	76	16
1290	CAPSICUM FRUTESCENS	S	.	0.0	16.2	18.2	.	.	.	.	.	.	.	.	.	.	.	.	.	19
1291	CAPSICUM FRUTESCENS	S	.	0.0	20.0	21.0	.	.	.	.	.	.	.	.	.	.	.	.	.	19
1292	CAPSICUM FRUTESCENS	S	.	0.0	16.2	17.5	.	.	.	.	.	.	.	.	.	.	.	.	.	19
1293	CAPSICUM FRUTESCENS	S	.	0.0	20.0	24.8	.	.	4.6	.	.	.	.	.	.	.	.	.	.	19
1294	CAPSICUM FRUTESCENS	S	.	0.0	20.0	26.1	.	.	4.6	.	.	.	.	.	.	.	.	.	.	19
1295	CAPSICUM FRUTESCENS	P	49	85.9	1.6	0.7	11.1	1.5	0.7	22	32	1.9	.	.	720	0.07	0.09	1.10	125	3
1296	CAPSICUM PUBESCENS	F	36	89.3	1.2	0.2	8.7	1.0	0.6	6	20	0.7	.	.	300	0.06	0.08	2.10	35	3
1297	CAPSICUM SP.	F	347	10.2	13.8	14.0	54.9	16.4	7.1	130	190	.	.	.	1060	0.63	0.27	8.30	180	1
1298	CAPSICUM SP.	L	53	82.1	5.8	1.0	8.5	1.9	2.6	246	56	1.4	5	714	6210	0.40	0.33	1.90	68	2
1299	CAPSICUM SP.	F	38	88.8	1.9	0.6	8.0	2.2	0.7	20	28	1.7	.	.	940	0.09	0.13	1.50	91	3
1300	CARAGANA ARBORESCENS	S	.	0.0	35.5	13.6	.	.	3.4	.	.	.	.	.	.	.	.	.	.	19
1301	CARAGANA ARBORESCENS	S	.	0.0	36.4	13.2	.	.	3.6	.	.	.	.	.	.	.	.	.	.	19
1302	CARDARIA DRABA	S	.	0.0	26.4	19.3	.	.	.	.	.	.	.	.	.	.	.	.	.	21
1303	CARDIOSPERMUM HALICACABUM	S	.	0.0	16.8	18.6	.	.	.	.	.	.	.	.	.	.	.	.	.	21
1304	CARDIOSPERMUM MICROCARPUM	S	.	0.0	16.6	25.9	.	.	.	.	.	.	.	.	.	.	.	.	.	21
1305	CARDOPATIUM CORYMBOSUM	S	.	0.0	43.4	34.2	.	.	.	.	.	.	.	.	.	.	.	.	.	21

TABLE 2 - DATA DIRECTLY TRANSCRIBED FROM ORIGINAL SOURCES.

OBS	PLNA	PART	CALORIES	WATER	PROTEIN	FAT	TOTCARB	FIBER	ASH	CALCIUM	PHOSPHO	IRON	SODIUM	POTASS	CAROT	THIAMINE	RIBOFLAV	NIACIN	ASCORBIC	REFCO
1306	CARDUUS NUTANS	S	.	0.0	17.3	29.1	.	.	.	.	.	.	.	.	.	.	.	.	.	20
1307	CARDUUS SP.	W	.	0.0	9.3	4.2	75.7	30.4	10.8	.	.	.	.	.	.	.	.	.	.	6
1308	CAREX AQUATILIS	W	.	0.0	13.2	4.0	74.4	30.2	8.4	400	310	.	.	.	.	.	.	.	.	6
1309	CAREX ATHROSTACHYA	W	.	0.0	8.0	3.1	82.6	32.8	6.3	.	.	.	.	.	.	.	.	.	.	6
1310	CAREX BICKNELLII	S	.	0.0	7.7	3.5	.	.	2.2	.	.	.	.	.	.	.	.	.	.	19
1311	CAREX CRINATA	S	.	0.0	14.8	5.7	.	.	.	.	.	.	.	.	.	.	.	.	.	20
1312	CAREX CRUS-CORVI	S	.	0.0	10.6	8.4	.	.	.	.	.	.	.	.	.	.	.	.	.	19
1313	CAREX DOUGLASII	W	.	0.0	10.1	4.0	79.1	29.7	6.8	.	.	.	.	.	.	.	.	.	.	6
1314	CAREX EBENEA	S	.	0.0	23.8	16.7	.	.	.	.	.	.	.	.	.	.	.	.	.	21
1315	CAREX FESTIVELLA	W	.	0.0	12.9	2.5	79.7	29.6	4.9	.	.	.	.	.	.	.	.	.	.	6
1316	CAREX FILIFOLIA	W	.	0.0	7.5	3.2	78.5	32.9	10.8	.	.	.	.	.	.	.	.	.	.	6
1317	CAREX HELIOPHILA	W	.	0.0	15.9	2.4	73.9	22.5	7.8	.	.	.	.	.	.	.	.	.	.	6
1318	CAREX LANUGINOSA	W	.	0.0	9.4	2.3	81.0	32.3	7.3	.	.	.	.	.	.	.	.	.	.	6
1319	CAREX NEBRASKENSIS	S	.	0.0	6.7	4.6	.	.	2.0	.	.	.	.	.	.	.	.	.	.	19
1320	CAREX NEBRASKENSIS	W	.	0.0	9.6	3.2	79.5	30.2	7.7	.	.	.	.	.	.	.	.	.	.	6
1321	CAREX PRESLII	W	.	0.0	22.8	2.9	65.6	20.9	8.7	.	.	.	.	.	.	.	.	.	.	6
1322	CAREX ROSTRATA	S	.	0.0	11.3	4.8	.	.	.	.	.	.	.	.	.	.	.	.	.	20
1323	CAREX ROSTRATA	W	.	0.0	11.2	3.4	76.6	28.9	8.8	380	280	.	.	.	.	.	.	.	.	6
1324	CAREX ROSTRATA	W	.	0.0	11.7	2.8	76.9	29.1	8.6	570	310	.	.	.	.	.	.	.	.	6
1325	CAREX SICCATA	W	.	0.0	8.9	2.8	81.4	30.6	6.9	.	.	.	.	.	.	.	.	.	.	6
1326	CAREX SP.	S	.	0.0	13.1	6.1	.	.	.	.	.	.	.	.	.	.	.	.	.	20
1327	CAREX ULTRA	S	.	0.0	13.0	12.9	.	.	6.6	.	.	.	.	.	.	.	.	.	.	19
1328	CAREX VESICARIA	S	.	0.0	12.9	8.5	.	.	.	.	.	.	.	.	.	.	.	.	.	21
1329	CARICA CANDAMARCENSIS	F	24	93.0	0.8	0.2	5.4	1.2	0.6	12	14	0.5	.	.	70	0.02	0.03	1.00	47	3
1330	CARICA PAPAYA	F	32	90.8	0.4	0.1	8.3	0.9	0.4	21	15	0.6	.	.	950	0.03	0.03	0.40	52	1
1331	CARICA PAPAYA	L	.	0.0	32.6	0.8	55.6	7.3	11.0	.	.	.	.	.	.	.	.	.	.	18
1332	CARICA PAPAYA	L	.	0.0	20.9	13.6	50.1	14.5	15.4	2380	220	.	.	.	.	.	.	.	.	18
1333	CARICA PAPAYA	F	.	0.0	11.1	1.2	78.3	11.7	9.4	230	160	.	.	.	.	.	.	.	.	18
1334	CARICA PAPAYA	F	45	87.1	0.5	0.1	11.8	0.5	0.5	24	22	0.7	4	221	710	0.03	0.05	0.40	73	2
1335	CARICA PAPAYA	L	74	77.5	7.0	2.0	11.3	1.8	2.2	344	142	0.8	16	652	11565	0.09	0.48	2.10	140	2
1336	CARICA PAPAYA	G	26	92.1	1.0	0.1	6.2	0.9	0.6	38	20	0.3	7	215	15	0.02	0.03	0.30	40	2
1337	CARICA PAPAYA	F	39	88.7	0.6	0.1	10.0	0.9	0.6	20	16	0.3	3	234	1050	0.04	0.04	0.30	56	4
1338	CARICA PENTAGONA	F	23	93.2	0.9	0.1	5.4	0.5	0.4	11	14	0.4	.	.	60	0.02	0.06	1.00	29	3
1339	CARISSA MACROCARPA	F	68	81.9	0.4	0.9	16.4	0.8	0.4	11	7	1.3	.	.	25	0.04	0.06	0.20	56	2
1340	CARISSA MACROCARPA	F	70	80.8	0.5	1.3	16.0	0.9	0.4	.	.	.	.	.	24	0.04	0.06	0.20	38	4
1341	CARPHOCHAETE WISLIZENII	S	.	0.0	20.0	13.3	.	.	.	.	.	.	.	.	.	.	.	.	.	20
1342	CARRICHTERA ANNUA	S	.	0.0	31.4	18.6	.	.	.	.	.	.	.	.	.	.	.	.	.	21
1343	CARRICHTERA ANNUA	S	.	0.0	35.1	13.4	.	.	.	.	.	.	.	.	.	.	.	.	.	21
1344	CARTHAMUS ARBORESCENS	S	.	0.0	8.5	12.7	.	.	.	.	.	.	.	.	.	.	.	.	.	21
1345	CARTHAMUS FLAVESCENS	S	.	0.0	14.9	19.9	.	.	.	.	.	.	.	.	.	.	.	.	.	21
1346	CARTHAMUS GLAUCUS	S	.	0.0	16.9	20.4	.	.	.	.	.	.	.	.	.	.	.	.	.	21
1347	CARTHAMUS LANATUS	S	.	0.0	12.4	16.9	.	.	.	.	.	.	.	.	.	.	.	.	.	21
1348	CARTHAMUS LANATUS	S	.	0.0	13.8	21.3	.	.	.	.	.	.	.	.	.	.	.	.	.	21
1349	CARTHAMUS NITIDUS	S	.	0.0	16.8	14.0	.	.	.	.	.	.	.	.	.	.	.	.	.	21
1350	CARTHAMUS OXYACANTHA	S	.	0.0	15.1	26.9	.	.	.	.	.	.	.	.	.	.	.	.	.	21

TABLE 2 - DATA DIRECTLY TRANSCRIBED FROM ORIGINAL SOURCES.

OBS	PLNA	PART	CALORIES	WATER	PROTEIN	FAT	TOTCARB	FIBER	ASH	CALCIUM	PHOSPHO	IRON	SODIUM	POTASS	CAROTE	THIAMINE	RIBOFLAV	NIACIN	ASCORBIC	REFCO
1351	CARTHAMUS TINCTORIUS	S	482	4.8	12.6	27.8	50.5	25.1	4.3	126	310	9.7	.	.	0	0.59	0.14	0.50	0	1
1352	CARTHAMUS TINCTORIUS	W	.	0.0	17.0	1.6	69.2	10.9	12.2	1700	340	.	.	.	.	.	.	.	.	18
1353	CARTHAMUS TINCTORIUS	H	.	0.0	12.2	2.4	76.9	31.1	8.5	.	.	.	.	.	.	.	.	.	.	18
1354	CARTHAMUS TINCTORIUS	S	615	5.0	19.1	59.5	12.4	.	4.0	.	.	.	.	.	.	.	.	.	.	4
1355	CARTHAMUS TINCTORIUS	H	.	0.0	4.2	5.1	89.2	58.2	1.5	.	.	.	.	.	.	.	.	.	.	6
1356	CARUM CARVI	S	333	9.9	19.8	14.6	49.9	12.6	5.9	689	568	16.2	17	1351	218	0.38	0.38	3.61	.	16
1357	CARYA ILLINOENSIS	S	691	3.2	9.3	72.0	13.9	2.2	1.6	73	290	2.4	3	604	80	0.86	0.13	0.90	2	2
1358	CARYA ILLINOENSIS	S	696	3.0	9.4	73.0	13.0	2.2	1.6	74	324	2.4	.	.	20	0.72	0.11	0.90	2	3
1359	CARYA ILLINOENSIS	S	687	3.4	9.2	71.2	14.6	2.3	1.6	73	289	2.4	0	603	78	0.86	0.13	0.90	2	4
1360	CASIMIROA EDULIS	S	.	0.0	16.3	0.8	.	.	5.8	.	.	.	.	.	.	.	.	.	.	19
1361	CASIMIROA EDULIS	S	.	0.0	2.8	0.4	.	.	1.1	.	.	.	.	.	.	.	.	.	.	19
1362	CASIMIROA EDULIS	F	65	82.0	1.4	0.4	15.7	1.7	0.5	8	18	0.2	.	.	30	0.04	0.07	0.50	23	3
1363	CASIMIROA TETRAMERIA	S	.	0.0	12.5	0.9	.	.	5.1	.	.	.	.	.	.	.	.	.	.	19
1364	CASIMIROA TETRAMERIA	S	.	0.0	5.6	0.7	.	.	1.8	.	.	.	.	.	.	.	.	.	.	19
1365	CASSIA ALATA	S	.	0.0	18.1	2.9	.	.	.	.	.	.	.	.	.	.	.	.	.	20
1366	CASSIA ALATA	S	.	0.0	16.1	3.5	.	.	.	.	.	.	.	.	.	.	.	.	.	21
1367	CASSIA ARMATA	S	.	0.0	23.6	12.4	.	.	.	.	.	.	.	.	.	.	.	.	.	21
1368	CASSIA BAUHINIOIDES	S	.	0.0	23.1	3.4	.	.	.	.	.	.	.	.	.	.	.	.	.	19
1369	CASSIA BIFLORA	S	.	0.0	23.1	4.4	.	.	.	.	.	.	.	.	.	.	.	.	.	20
1370	CASSIA BONARIENSIS	S	.	0.0	20.9	3.3	.	.	.	.	.	.	.	.	.	.	.	.	.	21
1371	CASSIA BONARIENSIS	S	.	0.0	20.0	3.1	.	.	.	.	.	.	.	.	.	.	.	.	.	21
1372	CASSIA CF. LINDHEIMERIANA	S	.	0.0	19.4	5.1	.	.	.	.	.	.	.	.	.	.	.	.	.	20
1373	CASSIA CF. UNIFLORA	S	.	0.0	24.4	5.3	.	.	.	.	.	.	.	.	.	.	.	.	.	20
1374	CASSIA CORYMBOSA	S	.	0.0	20.0	4.3	.	.	.	.	.	.	.	.	.	.	.	.	.	20
1375	CASSIA COVESII	S	.	0.0	20.6	3.6	.	.	.	.	.	.	.	.	.	.	.	.	.	20
1376	CASSIA DENSIFLORA	S	.	0.0	19.7	6.6	.	.	.	.	.	.	.	.	.	.	.	.	.	21
1377	CASSIA DIDYMOBOTRYA	S	.	0.0	23.9	4.0	.	.	6.0	.	.	.	.	.	.	.	.	.	.	19
1378	CASSIA DURANGENSIS	S	.	0.0	18.7	3.6	.	.	.	.	.	.	.	.	.	.	.	.	.	20
1379	CASSIA EMARGINATA	S	.	0.0	25.0	2.5	.	.	.	.	.	.	.	.	.	.	.	.	.	19
1380	CASSIA FASCICULATA	S	.	0.0	37.2	4.3	.	.	4.0	.	.	.	.	.	.	.	.	.	.	19
1381	CASSIA FASCICULATA	S	.	0.0	37.4	5.2	.	.	4.4	.	.	.	.	.	.	.	.	.	.	19
1382	CASSIA FASCICULATA	S	.	0.0	33.8	4.6	.	.	3.7	.	.	.	.	.	.	.	.	.	.	19
1383	CASSIA FASCICULATA	S	.	0.0	32.2	4.4	.	.	3.9	.	.	.	.	.	.	.	.	.	.	19
1384	CASSIA FASCICULATA	S	.	0.0	35.1	4.2	.	.	.	.	.	.	.	.	.	.	.	.	.	20
1385	CASSIA FISTULA	L	.	0.0	17.6	7.8	66.8	30.2	7.8	3270	330	.	.	.	.	.	.	.	.	18
1386	CASSIA FISTULA	S	.	0.0	19.8	3.6	.	.	.	.	.	.	.	.	.	.	.	.	.	21
1387	CASSIA HIRSUTA	S	.	0.0	17.5	2.4	.	.	.	.	.	.	.	.	.	.	.	.	.	20
1388	CASSIA JAVANICA	S	.	0.0	11.9	1.4	.	.	.	.	.	.	.	.	.	.	.	.	.	20
1389	CASSIA LAEVIGATA	SH	.	0.0	17.9	6.2	65.8	24.7	7.4	.	.	.	.	.	.	.	.	.	.	18
1390	CASSIA LEPTADENIA	S	.	0.0	31.3	3.5	.	.	.	.	.	.	.	.	.	.	.	.	.	21
1391	CASSIA LEPTOCARPA	S	.	0.0	20.3	3.6	.	.	4.0	.	.	.	.	.	.	.	.	.	.	19
1392	CASSIA LEPTOCARPA	S	.	0.0	21.5	4.4	.	.	4.1	.	.	.	.	.	.	.	.	.	.	19
1393	CASSIA LEPTOCARPA	S	.	0.0	21.5	4.4	.	.	4.1	.	.	.	.	.	.	.	.	.	.	19
1394	CASSIA LEPTOCARPA	S	.	0.0	18.8	4.8	.	.	.	.	.	.	.	.	.	.	.	.	.	19
1395	CASSIA LEPTOCARPA	S	.	0.0	22.5	3.7	.	.	.	.	.	.	.	.	.	.	.	.	.	20

TABLE 2 - DATA DIRECTLY TRANSCRIBED FROM ORIGINAL SOURCES.

OBS	PLNA	PART	CALORIES	WATER	PROTEIN	FAT	TOTLCARB	FIBER	ASH	CALCIUM	PHOSPHO	IRON	SODIUM	POTASS	CAROT	THIAMINE	RIBOFLAV	NIACIN	ASCORBIC	REFCO
1396	CASSIA LEPTOCARPA	S	.	0.0	18.9	4.7	.	.	.	.	.	.	.	.	.	.	.	.	.	21
1397	CASSIA LINDHEIMERIANA	S	.	0.0	18.8	4.6	.	.	.	.	.	.	.	.	.	.	.	.	.	20
1398	CASSIA MARILANDICA	S	.	0.0	16.9	3.6	.	.	.	.	.	.	.	.	.	.	.	.	.	19
1399	CASSIA MARILANDICA	S	.	0.0	21.9	3.2	.	.	3.4	.	.	.	.	.	.	.	.	.	.	21
1400	CASSIA OBTUSIFOLIA	L	60	79.7	5.6	0.2	12.5	2.3	2.0	589	96	5.9	.	.	7940	0.23	0.71	1.5	113	1
1401	CASSIA OBTUSIFOLIA	S	.	0.0	19.0	6.5	.	.	.	.	.	.	.	.	.	.	.	.	.	21
1402	CASSIA OCCIDENTALIS	S	.	0.0	22.5	3.4	.	.	4.4	.	.	.	.	.	.	.	.	.	.	19
1403	CASSIA OCCIDENTALIS	S	.	0.0	18.6	3.2	.	.	.	.	.	.	.	.	.	.	.	.	.	21
1404	CASSIA OCCIDENTALIS	S	.	0.0	18.1	.	.	.	.	.	.	.	.	.	.	.	.	.	.	21
1405	CASSIA OCCIDENTALIS	S	.	0.0	18.2	3.6	.	.	.	.	.	.	.	.	.	.	.	.	.	21
1406	CASSIA OCCIDENTALIS	S	.	0.0	24.9	3.6	.	.	.	.	.	.	.	.	.	.	.	.	.	21
1407	CASSIA OCCIDENTALIS	S	.	0.0	19.1	3.9	.	.	.	.	.	.	.	.	.	.	.	.	.	21
1408	CASSIA OCCIDENTALIS	S	.	0.0	20.6	2.7	.	.	.	.	.	.	.	.	.	.	.	.	.	21
1409	CASSIA PURPUREA	S	.	0.0	19.1	5.5	.	.	.	.	.	.	.	.	.	.	.	.	.	21
1410	CASSIA ROEMERIANA	S	.	0.0	21.5	3.7	.	.	.	.	.	.	.	.	.	.	.	.	.	20
1411	CASSIA SIAMEA	C	.	0.0	20.0	5.6	69.1	16.5	5.3	1140	140	.	.	.	.	.	.	.	.	18
1412	CASSIA SIAMEA	L	.	0.0	16.8	11.2	66.6	19.8	5.4	.	.	.	.	.	.	.	.	.	.	18
1413	CASSIA SIAMEA	L	114	65.6	7.4	1.0	24.4	3.7	1.6	100	155	5.3	.	.	6640	0.04	0.69	1.3	56	2
1414	CASSIA SKINNERI	S	.	0.0	23.2	7.6	.	.	.	.	.	.	.	.	.	.	.	.	.	21
1415	CASSIA SP.	S	346	10.3	18.0	1.3	66.9	6.5	3.5	205	184	.	.	.	.	1.04	.	.	.	1
1416	CASSIA SP.	S	.	0.0	17.5	3.9	.	.	.	.	.	.	.	.	.	.	.	.	.	19
1417	CASSIA SP.	S	.	0.0	30.0	6.2	.	.	.	.	.	.	.	.	.	.	.	.	.	20
1418	CASSIA SP.	S	.	0.0	31.2	5.3	.	.	.	.	.	.	.	.	.	.	.	.	.	20
1419	CASSIA SP.	S	.	0.0	16.3	5.0	.	.	.	.	.	.	.	.	.	.	.	.	.	21
1420	CASSIA SPECTABILIS	S	.	0.0	22.9	2.8	.	.	5.3	.	.	.	.	.	.	.	.	.	.	19
1421	CASSIA TOMENTOSA	S	.	0.0	17.4	3.5	.	.	.	.	.	.	.	.	.	.	.	.	.	21
1422	CASSIA TORA	S	.	0.0	18.2	7.4	65.3	4.6	9.1	720	200	.	.	.	.	.	.	.	.	18
1423	CASSIA TORA	SH	.	0.0	13.2	4.4	72.6	19.9	9.9	2810	570	.	.	.	.	.	.	.	.	18
1424	CASSIA TORA	S	.	0.0	21.9	3.3	.	.	.	.	.	.	.	.	.	.	.	.	.	19
1425	CASSIA TORA	S	.	0.0	18.8	5.7	.	.	.	.	.	.	.	.	.	.	.	.	.	20
1426	CASSIA TORA	S	.	0.0	36.0	3.8	.	.	.	.	.	.	.	.	.	.	.	.	.	21
1427	CASSIA UNIFLORA	S	.	0.0	21.2	6.3	.	.	.	.	.	.	.	.	.	.	.	.	.	20
1428	CASSIA UNIFLORA	S	.	0.0	25.0	4.6	.	.	.	.	.	.	.	.	.	.	.	.	.	20
1429	CASSIA WISLIZENII	S	.	0.0	33.1	5.3	.	.	5.4	.	.	.	.	.	.	.	.	.	.	19
1430	CASSINE EUCLEAEFORMIS	S	.	0.0	14.5	47.9	.	.	.	.	.	.	.	.	.	.	.	.	.	21
1431	CASSINE EUCLEAEFORMIS	S	.	0.0	5.8	65.0	.	.	.	.	.	.	.	.	.	.	.	.	.	21
1432	CASSINE EUCLEAEFORMIS	S	.	0.0	11.3	51.0	.	.	.	.	.	.	.	.	.	.	.	.	.	21
1433	CASSINE EUCLEAEFORMIS	S	.	0.0	5.4	70.4	.	.	.	.	.	.	.	.	.	.	.	.	.	21
1434	CASSINE PAPILLOSA	S	.	0.0	15.8	60.3	.	.	.	.	.	.	.	.	.	.	.	.	.	21
1435	CASTALIS NUDICAULIS VAR. GRAMINIFOLIA	S	.	0.0	.	7.2	.	.	.	.	.	.	.	.	.	.	.	.	.	21
1436	CASTALIS NUDICAULIS VAR. NUDICAULIS	S	.	0.0	.	10.8	.	.	.	.	.	.	.	.	.	.	.	.	.	21
1437	CASTANEA CRENATA	S	170	57.4	3.0	0.6	38.0	1.0	1.0	34	80	1.6	16	.	.	0.38	0.18	1.6	29	2
1438	CASTANEA MOLLISSIMA	S	223	44.6	6.6	1.5	46.1	.	1.2	20	93	2.1	.	.	120	0.16	0.18	0.8	36	2
1439	CASTANEA SATIVA	S	201	50.5	3.7	1.4	43.4	1.2	1.0	15	91	1.7	16	844	.	0.23	0.23	0.6	.	2
1440	CASTANEA SATIVA	S	194	52.5	2.9	1.5	42.1	1.1	1.0	27	88	1.7	6	454	.	0.22	0.22	0.6	.	4

TABLE 2 - DATA DIRECTLY TRANSCRIBED FROM ORIGINAL SOURCES.

OBS	PLNA	PART	CALORIES	WATER	PROTEIN	FAT	TOTCARB	FIBER	ASH	CALCIUM	PHOSPHO	IRON	SODIUM	POTASS	CAROT	THIAMINE	RIBOFLAV	NIACIN	ASCORBIC	REFCO
1441	CASTILLEJA SP.	W	.	0.0	9.9	2.6	75.0	14.4	12.5	910	490	.	.	.	2	.	.	.	.	6
1442	CASUARINA TORULOSA	S	.	0.0	48.8	41.9	.	.	5.2	.	.	.	.	.	.	.	.	.	.	19
1443	CATALPA BIGNONIOIDES	S	.	0.0	36.3	33.0	.	.	4.7	.	.	.	.	.	.	.	.	.	.	19
1444	CATANANCHE CAERULEA	S	.	0.0	33.8	27.1	.	.	.	.	.	.	.	.	.	.	.	.	.	19
1445	CATHA EDULIS	L	.	.	5.2	.	.	2.7	1.6	290	.	18.0	.	.	1800	0.05	0.05	14.8	161	11
1446	CAUCALIS DAUCOIDES	S	.	0.0	14.7	11.6	.	.	.	.	.	.	.	.	.	.	.	.	.	20
1447	CAULANTHUS INFLATUS	S	.	0.0	28.8	30.5	.	.	.	.	.	.	.	.	.	.	.	.	.	21
1448	CAYAPONIA MAXIMOWICZII	S	.	0.0	30.2	40.3	.	.	.	.	.	.	.	.	.	.	.	.	.	21
1449	CAYAPONIA SP.	S	.	0.0	16.2	17.9	.	.	2.9	.	.	.	.	.	.	.	.	.	.	19
1450	CEDRELA ODORATA	S	.	0.0	25.0	21.4	.	.	8.3	.	.	.	.	.	.	.	.	.	.	19
1451	CEDRELA SINENSIS	SH	51	83.6	6.0	1.0	7.9	1.3	1.5	30	102	3.2	.	576	.	.	.	.	.	2
1452	CEDRUS ATLANTICA	S	.	0.0	14.4	43.0	.	.	.	.	.	.	.	.	.	.	.	.	.	20
1453	CEDRUS DEODARA	S	.	0.0	13.1	44.7	.	.	.	.	.	.	.	.	.	.	.	.	.	20
1454	CEIBA ACUMINATA	S	.	0.0	31.2	32.0	.	.	3.2	.	.	.	.	.	.	.	.	.	.	19
1455	CEIBA ACUMINATA	S	.	0.0	39.4	42.7	.	.	.	.	.	.	.	.	.	.	.	.	.	19
1456	CEIBA PENTANDRA	S	494	6.8	30.9	36.5	20.1	1.5	5.7	214	1183	.	.	.	.	.	.	.	.	1
1457	CEIBA PENTANDRA	S	.	0.0	30.4	23.1	38.3	19.6	8.2	470	970	.	.	.	.	.	.	.	.	18
1458	CEIBA SP.	S	.	0.0	35.6	43.5	.	.	.	.	.	.	.	.	.	.	.	.	.	19
1459	CELASTRUS ORBICULATUS	S	.	0.0	21.9	46.1	.	.	3.6	.	.	.	.	.	.	.	.	.	.	19
1460	CELOSIA ANTHELMINTICA	SH	.	0.0	15.0	2.1	76.4	32.9	6.5	560	180	.	.	.	.	.	.	.	.	18
1461	CELOSIA ARGENTEA	L	44	83.8	4.7	0.7	7.6	1.8	3.2	260	43	7.8	.	.	.	.	.	.	.	1
1462	CELOSIA ARGENTEA	S	.	0.0	14.9	11.2	.	.	3.3	.	.	.	.	.	.	.	.	.	.	19
1463	CELOSIA ARGENTEA	S	.	0.0	13.0	9.3	.	.	.	.	.	.	.	.	.	.	.	.	.	21
1464	CELOSIA CRISTATA	S	.	0.0	18.1	13.4	.	.	.	.	.	.	.	.	.	.	.	.	.	20
1465	CELOSIA LAXA	L	33	88.9	2.7	0.4	6.4	1.4	1.6	154	32	5.0	.	.	1925	.	0.10	.	10	1
1466	CELOSIA TRIGYNA	L	33	88.9	2.7	0.4	6.4	1.4	1.6	154	32	5.0	.	.	1925	.	0.10	.	10	1
1467	CELTIS AUSTRALIS	L	.	0.0	15.5	4.6	69.8	22.0	10.1	2830	110	.	.	.	.	.	.	.	.	18
1468	CELTIS AUSTRALIS	S	.	0.0	11.9	14.5	.	.	.	.	.	.	.	.	.	.	.	.	.	20
1469	CELTIS MISSISSIPPIENSIS	L	253	10.8	8.0	4.8	54.1	.	22.3	.	166	19.7	.	.	.	.	.	.	.	1
1470	CELTIS OCCIDENTALIS	S	.	0.0	11.2	9.5	.	.	.	.	.	.	.	.	.	.	.	.	.	20
1471	CELTIS PALLIDA	F	72	79.5	1.0	0.3	18.3	0.9	0.9	63	22	2.8	.	.	30	0.00	0.00	0.3	51	3
1472	CELTIS RETICULATA	L	.	0.0	15.8	1.8	66.6	18.5	15.8	4910	180	.	.	1380	.	.	.	.	.	6
1473	CENCHRUS BIFLORUS	S	370	9.8	17.8	8.5	62.3	0.6	1.6	144	270	22.0	.	.	.	.	.	.	.	1
1474	CENCHRUS BIFLORUS	W	.	0.0	11.6	2.8	77.9	33.9	7.7	.	.	.	.	.	.	.	.	.	.	18
1475	CENCHRUS BIFLORUS	H	.	0.0	9.0	1.0	78.4	32.9	11.6	.	.	.	.	.	.	.	.	.	.	18
1476	CENCHRUS CILIARIS	W	.	0.0	11.0	2.6	73.2	31.9	13.2	.	.	.	.	.	.	.	.	.	.	18
1477	CENCHRUS CILIARIS	H	.	0.0	7.4	1.7	79.2	35.2	11.7	.	.	.	.	.	.	.	.	.	.	18
1478	CENCHRUS SETIGERAS	W	.	0.0	18.6	1.9	67.6	28.3	11.9	.	.	.	.	.	.	.	.	.	.	18
1479	CENCHRUS SP.	W	.	0.0	10.7	2.3	80.0	36.5	7.0	.	.	.	.	.	.	.	.	.	.	6
1480	CENCHRUS TRIBULOIDES	S	.	0.0	25.6	8.8	.	.	2.2	.	.	.	.	.	.	.	.	.	.	19
1481	CENTAUREA AMERICANA	S	.	0.0	20.2	14.3	.	.	3.2	.	.	.	.	.	.	.	.	.	.	19
1482	CENTAUREA AMERICANA	S	.	0.0	15.1	10.6	.	.	.	.	.	.	.	.	.	.	.	.	.	20
1483	CENTAUREA AMERICANA	S	.	0.0	19.6	16.9	.	.	.	.	.	.	.	.	.	.	.	.	.	20
1484	CENTAUREA ASPERA	S	.	0.0	15.4	20.5	.	.	.	.	.	.	.	.	.	.	.	.	.	21
1485	CENTAUREA CALCITRAPA	S	.	0.0	24.6	21.2	.	.	.	.	.	.	.	.	.	.	.	.	.	21

TABLE 2 - DATA DIRECTLY TRANSCRIBED FROM ORIGINAL SOURCES.

OBS	PLANA	PART	CALORIES	WATER	PROTEIN	FAT	TOTLCARB	FIBER	ASH	CALCIUM	PHOSPHO	IRON	SODIUM	POTASS	CAROT	THIAMINE	RIBOFLAV	NIACIN	ASCORBIC	REFCO
1486	CENTAUREA CALCITRAPA	S	.	0.0	20.0	24.1	.	.	.	.	.	.	.	.	.	.	.	.	.	21
1487	CENTAUREA CYANUS	S	.	0.0	15.8	23.6	.	.	3.4	.	.	.	.	.	.	.	.	.	.	19
1488	CENTAUREA CYANUS	S	.	0.0	15.8	25.4	.	.	.	.	.	.	.	.	.	.	.	.	.	20
1489	CENTAUREA CYANUS	S	.	0.0	18.1	25.0	.	.	.	.	.	.	.	.	.	.	.	.	.	20
1490	CENTAUREA DEALBATA	S	.	0.0	23.8	19.4	.	.	.	.	.	.	.	.	.	.	.	.	.	20
1491	CENTAUREA DEPRESSA	S	.	0.0	17.4	24.8	.	.	.	.	.	.	.	.	.	.	.	.	.	21
1492	CENTAUREA GYMNOCARPA	S	.	0.0	18.8	23.5	.	.	.	.	.	.	.	.	.	.	.	.	.	20
1493	CENTAUREA IBERICA	S	.	0.0	21.6	24.8	.	.	.	.	.	.	.	.	.	.	.	.	.	21
1494	CENTAUREA INTYBACEA	S	.	0.0	13.2	18.5	.	.	.	.	.	.	.	.	.	.	.	.	.	21
1495	CENTAUREA MACROCEPHALA	S	.	0.0	35.0	19.9	.	.	.	.	.	.	.	.	.	.	.	.	.	20
1496	CENTAUREA MACROCEPHALA	S	.	0.0	33.2	18.3	.	.	.	.	.	.	.	.	.	.	.	.	.	21
1497	CENTAUREA MONTANA	S	.	0.0	21.2	22.0	.	.	.	.	.	.	.	.	.	.	.	.	.	20
1498	CENTAUREA MOSCHATA	S	.	0.0	22.5	28.4	.	.	.	.	.	.	.	.	.	.	.	.	.	20
1499	CENTAUREA ORNATA	S	.	0.0	17.4	8.5	.	.	.	.	.	.	.	.	.	.	.	.	.	21
1500	CENTAUREA REPENS	S	.	0.0	12.4	9.0	.	.	.	.	.	.	.	.	.	.	.	.	.	21
1501	CENTAUREA REPENS	S	.	0.0	16.7	9.1	.	.	.	.	.	.	.	.	.	.	.	.	.	21
1502	CENTAUREA RIGIDA	S	.	0.0	25.9	20.7	.	.	.	.	.	.	.	.	.	.	.	.	.	21
1503	CENTAUREA SQUARROSA	S	.	0.0	25.3	11.6	.	.	.	.	.	.	.	.	.	.	.	.	.	21
1504	CENTAUREA TRIUMFETTII	S	.	0.0	13.6	23.2	.	.	.	.	.	.	.	.	.	.	.	.	.	21
1505	CENTAUREA VERUTUM	S	.	0.0	18.9	20.4	.	.	.	.	.	.	.	.	.	.	.	.	.	21
1506	CENTELLA ASIATICA	L	34	89.3	1.6	0.6	6.9	2.0	1.6	170	30	3.1	.	414	6580	0.15	0.14	1.2	4	2
1507	CENTROSEMA PUBESCENS	W	.	0.0	25.3	3.2	63.7	31.8	7.3	1040	320	.	.	.	.	.	.	.	.	18
1508	CEPHALANTHUS GLABRATUS	S	.	0.0	8.9	8.5	.	.	.	.	.	.	.	.	.	.	.	.	.	21
1509	CEPHALANTHUS OCCIDENTALIS	S	.	0.0	9.8	9.4	.	.	3.2	.	.	.	.	.	.	.	.	.	.	19
1510	CEPHALARIA JOPPICA	S	.	0.0	16.4	23.9	.	.	.	.	.	.	.	.	.	.	.	.	.	21
1511	CEPHALARIA LEUCANTHA	S	.	0.0	17.0	13.7	.	.	.	.	.	.	.	.	.	.	.	.	.	21
1512	CEPHALARIA MICROCEPHALA	S	.	0.0	31.8	19.3	.	.	.	.	.	.	.	.	.	.	.	.	.	21
1513	CEPHALARIA SETOSA	S	.	0.0	17.1	26.0	.	.	.	.	.	.	.	.	.	.	.	.	.	21
1514	CEPHALARIA TRANSYLVANICA	S	.	0.0	21.4	24.1	.	.	.	.	.	.	.	.	.	.	.	.	.	21
1515	CEPHALOTAXUS HARRINGTONIA VAR. DRUPACEAE	S	.	0.0	12.5	65.2	.	.	.	.	.	.	.	.	.	.	.	.	.	20
1516	CERASTIUM DICHOTOMUM	S	.	0.0	15.7	4.1	.	.	.	.	.	.	.	.	.	.	.	.	.	21
1517	CERASTIUM PERFOLIATUM	S	.	0.0	17.4	3.8	.	.	.	.	.	.	.	.	.	.	.	.	.	20
1518	CERASTIUM TOMENTOSUM	S	.	0.0	18.8	5.2	.	.	3.8	.	.	.	.	.	.	.	.	.	.	19
1519	CERASTIUM VISCOSUM	S	.	0.0	16.9	5.9	.	.	.	.	.	.	.	.	.	.	.	.	.	20
1520	CERATONIA SILIQUA	F	.	0.0	6.9	1.2	88.2	12.0	3.7	560	120	.	.	.	.	.	.	.	.	18
1521	CERATONIA SILIQUA	S	.	0.0	18.5	2.0	76.0	8.0	3.5	.	.	.	.	.	.	.	.	.	.	18
1522	CERATONIA SILIQUA	S	.	0.0	17.5	1.5	.	.	.	.	.	.	.	.	.	.	.	.	.	19
1523	CERATONIA SILIQUA		180	11.2	4.5	1.4	80.7	7.7	2.2	352	81	.	.	.	.	.	.	.	.	4
1524	CERATONIA SILIQUA	S	.	8.9	14.4	1.8	62.5	6.9	2.3	.	.	.	.	.	.	.	.	.	.	5
1525	CERATONIA SILIQUA	S	.	13.6	19.7	3.1	70.9	8.3	3.6	.	.	.	.	.	.	.	.	.	.	5
1526	CERATOPETALUM GUMMIFERUM	S	.	0.0	8.8	8.6	.	.	.	.	.	.	.	.	.	.	.	.	.	20
1527	CERATOTHECA SESAMOIDES	L	54	81.0	4.2	0.5	11.0	2.3	3.3	300	86	3.2	.	.	.	.	.	.	28	1
1528	CERATOTHECA SESAMOIDES	S	550	7.0	14.2	46.5	27.5	17.3	4.8	887	.	38.0	.	.	.	0.75	0.30	4.4	.	1
1529	CERCIDIUM FLORIDUM	S	.	0.0	54.4	15.7	.	.	5.1	.	.	.	.	.	.	.	.	.	.	19
1530	CERCIDIUM FLORIDUM	S	.	0.0	6.6	1.1	.	.	2.6	.	.	.	.	.	.	.	.	.	.	19

TABLE 2 - DATA DIRECTLY TRANSCRIBED FROM ORIGINAL SOURCES.

OBS	PLANA	PART	CALORIES	WATER	PROTEIN	FAT	TOTLCARB	FIBER	ASH	CALCIUM	PHOSPHO	IRON	SODIUM	POTASS	CAROT	THIAMINE	RIBOFLAV	NIACIN	ASCORBIC	REF
1531	CERCIDIUM FLORIDUM	S	.	0.0	27.5	5.0	.	.	.	.	.	.	.	.	.	.	.	.	.	20
1532	CERCIDIUM TORREYANUM	S	.	0.0	27.5	4.3	.	.	.	.	.	.	.	.	.	.	.	.	.	20
1533	CERCIS CANADENSIS	S	.	0.0	22.9	8.2	.	.	3.1	.	.	.	.	.	.	.	.	.	.	19
1534	CERCIS CANADENSIS	S	.	0.0	23.1	8.8	.	.	2.9	.	.	.	.	.	.	.	.	.	.	19
1535	CERCIS CANADENSIS	S	.	0.0	27.3	8.3	.	.	3.1	.	.	.	.	.	.	.	.	.	.	19
1536	CERCIS CANADENSIS	S	.	0.0	27.5	7.7	.	.	.	.	.	.	.	.	.	.	.	.	.	19
1537	CERCIS CHINENSIS	S	.	0.0	16.6	9.2	.	.	2.8	.	.	.	.	.	.	.	.	.	.	19
1538	CERCIS OCCIDENTALIS	S	.	0.0	18.6	8.8	.	.	2.6	.	.	.	.	.	.	.	.	.	.	19
1539	CERCIS SILIQUASTRUM	S	.	0.0	20.4	8.9	.	.	3.0	.	.	.	.	.	.	.	.	.	.	19
1540	CERCIS SILIQUASTRUM	S	.	0.0	23.1	8.0	.	.	.	.	.	.	.	.	.	.	.	.	.	21
1541	CERCOCARPUS MONTANUS	S	.	0.0	16.9	8.3	.	.	.	.	.	.	.	.	.	.	.	.	.	20
1542	CEREUS HEXAGONUS	I	29	91.0	.	.	8.2	.	0.8	92	27	4.4	.	.	260	0.03	0.01	0.5	40	3
1543	CEREUS HEXAGONUS	F	74	79.6	2.1	1.0	16.3	.	1.0	44	23	.	.	.	70	0.04	0.03	0.5	32	3
1544	CERINTHE MAJOR	S	.	0.0	13.1	25.3	.	.	.	.	.	.	.	.	.	.	.	.	.	21
1545	CERINTHE MINOR	S	.	0.0	14.1	10.4	.	.	.	.	.	.	.	.	.	.	.	.	.	20
1546	CEVALLIA SINUATA	S	.	0.0	20.9	21.7	.	.	.	.	.	.	.	.	.	.	.	.	.	21
1547	CHAENACTIS ALPINA	S	.	0.0	21.8	18.8	.	.	.	.	.	.	.	.	.	.	.	.	.	20
1548	CHAENACTIS ALPINA	S	.	0.0	21.2	21.0	.	.	.	.	.	.	.	.	.	.	.	.	.	20
1549	CHAENACTIS SP.	S	.	0.0	18.1	27.8	.	.	.	.	.	.	.	.	.	.	.	.	.	20
1550	CHAEROPHYLLUM AUREUM	S	.	0.0	24.1	12.8	.	.	.	.	.	.	.	.	.	.	.	.	.	21
1551	CHAEROPHYLLUM TAINTURIERI	S	.	0.0	20.6	10.5	.	.	.	.	.	.	.	.	.	.	.	.	.	20
1552	CHAEROPHYLLUM TAINTURIERI	S	.	0.0	14.9	9.1	.	.	.	.	.	.	.	.	.	.	.	.	.	20
1553	CHAEROPHYLLUM TAINTURIERI	S	.	0.0	13.4	10.6	.	.	.	.	.	.	.	.	.	.	.	.	.	21
1554	CHAMAECYPARIS LAWSONIANA	S	.	0.0	11.9	24.7	.	.	.	.	.	.	.	.	.	.	.	.	.	20
1555	CHAMAECYTISUS PYGMAEUS	S	.	0.0	34.5	8.9	.	.	.	.	.	.	.	.	.	.	.	.	.	21
1556	CHAMAEDOREA SP.		45	85.0	4.0	0.7	8.3	1.2	2.0	369	106	1.4	.	.	10	0.08	0.10	0.9	14	3
1557	CHAMAEPEUCE HISPANICA	S	.	0.0	19.9	24.1	.	.	.	.	.	.	.	.	.	.	.	.	.	21
1558	CHAMAEROPS HUMILIS	S	.	0.0	5.0	8.7	.	.	.	.	.	.	.	.	.	.	.	.	.	20
1559	CHAMAESARACHA CORONOPUS	S	.	0.0	20.6	20.3	.	.	.	.	.	.	.	.	.	.	.	.	.	19
1560	CHARDINIA ORIENTALIS	S	.	0.0	23.2	26.1	.	.	.	.	.	.	.	.	.	.	.	.	.	21
1561	CHASCANUM CARIPENSE	F	.	0.0	14.3	2.4	74.5	27.3	8.8	1910	110	.	.	.	.	.	.	.	.	18
1562	CHEIRANTHUS CHEIRI	S	.	0.0	27.5	29.5	.	.	.	.	.	.	.	.	.	.	.	.	.	20
1563	CHELONE BARBATA	S	.	0.0	18.9	24.2	.	.	3.3	.	.	.	.	.	.	.	.	.	.	19
1564	CHENOPODIUM ALBUM	L	61	75.9	7.4	1.0	9.8	1.6	5.9	371	80	.	.	.	.	.	.	.	.	1
1565	CHENOPODIUM ALBUM	L	.	0.0	21.7	3.7	41.8	8.3	32.8	.	.	.	.	.	.	.	.	.	.	18
1566	CHENOPODIUM ALBUM	S	.	0.0	19.4	9.3	.	.	.	.	.	.	.	.	.	.	.	.	.	19
1567	CHENOPODIUM ALBUM	L	32	88.0	2.1	0.7	6.0	2.6	3.2	160	442	3.0	.	.	3790	0.08	0.19	0.3	.	3
1568	CHENOPODIUM ALBUM	S	.	0.0	16.1	7.0	69.0	20.0	5.9	.	.	.	.	.	.	.	.	.	.	5
1569	CHENOPODIUM ALBUM	W	.	0.0	26.9	1.7	43.8	11.2	27.6	3330	510	.	.	8710	.	.	.	.	.	6
1570	CHENOPODIUM AMBROSIOIDES	S	.	0.0	15.6	9.2	.	.	.	.	.	.	.	.	.	.	.	.	.	19
1571	CHENOPODIUM AMBROSIOIDES	S	.	0.0	17.5	2.5	.	.	.	.	.	.	.	.	.	.	.	.	.	19
1572	CHENOPODIUM AMBROSIOIDES	S	.	0.0	12.1	8.5	.	.	.	.	.	.	.	.	.	.	.	.	.	21
1573	CHENOPODIUM AMBROSIOIDES	L	42	85.5	3.8	0.7	7.6	1.3	2.4	304	52	5.2	.	.	2420	0.06	0.28	0.6	11	3
1574	CHENOPODIUM BERLANDIERI	L	44	83.0	7.2	0.9	5.4	2.5	3.5	156	68	3.0	.	.	5530	0.17	0.47	1.2	109	3
1575	CHENOPODIUM BOTRYS	S	.	0.0	15.2	5.2	.	.	.	.	.	.	.	.	.	.	.	.	.	21

TABLE 2 - DATA DIRECTLY TRANSCRIBED FROM ORIGINAL SOURCES.

OBS	PLANA	PART	CALORIES	WATER	PROTEIN	FAT	TOTCARB	FIBER	ASH	CALCIUM	PHOSPHO	IRON	SODIUM	POTASS	CAROT	THIAMINE	RIBOFLAV	NIACIN	ASCORBIC	REFCO
1576	CHENOPODIUM FOLIOSUM	S	.	0.0	13.5	2.0	.	.	.	.	.	.	.	.	.	.	.	.	.	21
1577	CHENOPODIUM LEPTOPHYLLUM	S	.	0.0	16.7	6.9	.	.	.	.	.	.	.	.	.	.	.	.	.	20
1578	CHENOPODIUM MURALE	S	.	0.0	15.4	8.6	.	.	.	.	.	.	.	.	.	.	.	.	.	21
1579	CHENOPODIUM OPULIFOLIUM	S	.	0.0	17.2	9.3	.	.	.	.	.	.	.	.	.	.	.	.	.	21
1580	CHENOPODIUM OPULIFOLIUM	S	.	0.0	18.6	9.3	.	.	.	.	.	.	.	.	.	.	.	.	.	21
1581	CHENOPODIUM PALLIDICAULE	S	.	0.0	16.9	8.8	71.5	3.9	2.8	.	.	.	.	.	.	.	.	.	.	18
1582	CHENOPODIUM PALLIDICAULE	S	327	12.5	14.2	2.7	66.0	11.0	4.6	134	386	10.5	.	.	.	0.86	0.83	1.1	2	3
1583	CHENOPODIUM PALLIDICAULE	S	334	12.4	13.9	4.5	64.2	9.8	5.0	110	375	13.3	.	.	.	0.47	0.65	1.1	1	3
1584	CHENOPODIUM PALLIDICAULE	S	320	13.4	13.8	1.5	67.0	11.8	4.3	131	363	10.0	.	.	.	0.86	0.30	1.3	1	3
1585	CHENOPODIUM QUINOA	S	.	0.0	12.1	7.3	.	.	3.1	.	.	.	.	.	.	.	.	.	.	19
1586	CHENOPODIUM QUINOA	S	.	0.0	12.5	7.2	.	.	.	.	.	.	.	.	.	.	.	.	.	19
1587	CHENOPODIUM QUINOA	S	351	11.0	12.3	6.1	67.7	4.6	2.9	112	286	7.5	.	.	0	0.36	0.42	1.4	3	3
1588	CHENOPODIUM RUBRUM	S	.	0.0	14.8	9.1	.	.	.	.	.	.	.	.	.	.	.	.	.	21
1589	CHENOPODIUM SP.	S	.	0.0	14.3	2.7	.	.	3.1	.	.	.	.	.	.	.	.	.	.	19
1590	CHENOPODIUM SP.	S	.	0.0	12.7	3.7	.	.	1.6	.	.	.	.	.	.	.	.	.	.	19
1591	CHENOPODIUM SP.	L	43	84.3	4.2	0.8	7.3	2.1	3.4	309	72	1.2	.	.	6960	0.16	0.44	1.2	80	4
1592	CHILOPSIS LINEARIS	S	.	0.0	17.3	29.0	.	.	3.3	.	.	.	.	.	.	.	.	.	.	19
1593	CHILOPSIS LINEARIS	S	.	0.0	22.1	35.0	.	.	4.0	.	.	.	.	.	.	.	.	.	.	19
1594	CHILOPSIS LINEARIS	S	.	0.0	22.5	31.9	.	.	.	.	.	.	.	.	.	.	.	.	.	19
1595	CHIMONANTHUS PRAECOX	S	.	0.0	34.4	36.8	.	.	2.7	.	.	.	.	.	.	.	.	.	.	19
1596	CHIMONANTHUS PRAECOX	S	.	0.0	1.6	5.9	.	.	1.0	.	.	.	.	.	.	.	.	.	.	19
1597	CHLORIS CUCULLATA	W	.	0.0	9.9	2.2	71.7	28.6	16.2	4800	160	.	.	.	.	.	.	.	.	6
1598	CHLORIS GAYANA	W	.	0.0	8.0	2.0	76.9	37.2	13.1	.	.	.	.	.	.	.	.	.	.	18
1599	CHLORIS GAYANA	H	.	0.0	9.9	27.0	78.5	33.4	8.9	.	.	.	.	.	.	.	.	.	.	18
1600	CHLORIS GAYANA	S	.	0.0	21.2	4.7	.	.	.	.	.	.	.	.	.	.	.	.	.	19
1601	CHLORIS GAYANA	S	.	0.0	21.9	4.6	.	.	.	.	.	.	.	.	.	.	.	.	.	19
1602	CHLORIS GAYANA	S	.	0.0	23.1	5.5	.	.	.	.	.	.	.	.	.	.	.	.	.	20
1603	CHLORIS GAYANA	W	.	0.0	7.4	1.6	80.5	37.5	10.5	.	.	.	.	.	.	.	.	.	.	6
1604	CHLORIS PRIEURII	W	.	0.0	9.2	1.8	79.8	37.1	9.2	.	.	.	.	.	.	.	.	.	.	18
1605	CHLORIS ROXBURGHIANA	W	.	0.0	16.1	0.9	74.2	30.8	8.8	.	.	.	.	.	.	.	.	.	.	18
1606	CHLORIS VERTICILLATA	W	.	0.0	8.2	2.0	74.0	29.2	15.8	640	180	.	.	.	.	.	.	.	.	6
1607	CHLORIS VIRGATA	W	.	0.0	12.9	1.8	71.5	31.1	13.8	.	.	.	.	.	.	.	.	.	.	18
1608	CHLORIS VIRGATA	H	.	0.0	10.3	1.8	75.2	34.2	12.7	.	.	.	.	.	.	.	.	.	.	18
1609	CHLORIS VIRGATA	S	.	0.0	28.1	10.7	.	.	.	.	.	.	.	.	.	.	.	.	.	20
1610	CHONDROPETALUM MACROCARPUM	S	.	0.0	34.4	3.1	.	.	.	.	.	.	.	.	.	.	.	.	.	21
1611	CHONDROPETALUM MICROCARPUM	S	.	0.0	38.4	1.0	.	.	.	.	.	.	.	.	.	.	.	.	.	21
1612	CHROZOPHORA HIEROSOLYMITANA	S	.	0.0	22.2	28.1	.	.	.	.	.	.	.	.	.	.	.	.	.	21
1613	CHROZOPHORA TINCTORIA	S	.	0.0	18.3	25.3	.	.	.	.	.	.	.	.	.	.	.	.	.	21
1614	CHROZOPHORA TINCTORIA	S	.	0.0	37.2	48.6	.	.	.	.	.	.	.	.	.	.	.	.	.	21
1615	CHROZOPHORA TINCTORIA	S	.	0.0	34.7	56.5	.	.	.	.	.	.	.	.	.	.	.	.	.	21
1616	CHRYSALIDOCARPUS LUCUBENSIS	S	.	0.0	2.9	8.2	.	.	.	.	.	.	.	.	.	.	.	.	.	20
1617	CHRYSALIDOCARPUS LUTESCENS	S	.	0.0	6.9	7.2	.	.	.	.	.	.	.	.	.	.	.	.	.	20
1618	CHRYSANTHEMOIDES INCANA	S	.	0.0	.	34.9	.	.	.	.	.	.	.	.	.	.	.	.	.	21
1619	CHRYSANTHEMOIDES MONILIFERA	S	.	0.0	.	50.4	.	.	.	.	.	.	.	.	.	.	.	.	.	20
1620	CHRYSANTHEMOIDES MONILIFERA	S	.	0.0	.	49.0	.	.	.	.	.	.	.	.	.	.	.	.	.	21

TABLE 2 - DATA DIRECTLY TRANSCRIBED FROM ORIGINAL SOURCES.

OBS	PLNA	PART	CALORIES	WATER	PROTEIN	FAT	TOTCARB	FIBER	ASH	CALCIUM	PHOSPHO	IRON	SODIUM	POTASS	CAROT	THIAMINE	RIBOFLAV	NIACIN	ASCORBIC	REFCO
1621	CHRYSANTHEMOIDES MONILIFERA	S	.	0.0	.	50.0	.	.	.	.	.	.	.	.	.	.	.	.	.	21
1622	CHRYSANTHEMUM BALSAMITA	S	.	0.0	24.6	20.0	.	.	.	.	.	.	.	.	.	.	.	.	.	21
1623	CHRYSANTHEMUM CARINATUM	S	.	0.0	10.2	13.6	.	.	4.5	.	.	.	.	.	.	.	.	.	.	19
1624	CHRYSANTHEMUM CINERARIIFOLIUM	SH	.	0.0	13.0	0.5	79.4	23.6	7.1	530	240	.	.	.	.	.	.	.	.	18
1625	CHRYSANTHEMUM CORONARIUM	S	.	0.0	13.2	21.9	.	.	4.6	.	.	.	.	.	.	.	.	.	.	19
1626	CHRYSANTHEMUM CORONARIUM	S	.	0.0	17.6	19.3	.	.	5.7	.	.	.	.	.	.	.	.	.	.	19
1627	CHRYSANTHEMUM CORONARIUM	S	.	0.0	15.0	19.5	.	.	.	.	.	.	.	.	.	.	.	.	.	19
1628	CHRYSANTHEMUM CORONARIUM	L	19	93.5	1.8	0.3	3.3	0.9	1.1	63	34	2.5	106	256	3160	0.09	0.19	0.6	27	2
1629	CHRYSANTHEMUM CORONARIUM	S	.	0.0	13.8	12.8	.	.	.	.	.	.	.	.	.	.	.	.	.	20
1630	CHRYSANTHEMUM CORONARIUM	S	.	0.0	11.4	4.4	.	.	.	.	.	.	.	.	.	.	.	.	.	21
1631	CHRYSANTHEMUM CORONARIUM	S	.	0.0	6.9	4.4	.	.	.	.	.	.	.	.	.	.	.	.	.	21
1632	CHRYSANTHEMUM CORONARIUM	S	.	0.0	33.6	44.6	.	.	.	.	.	.	.	.	.	.	.	.	.	21
1633	CHRYSANTHEMUM CORONARIUM	S	.	0.0	12.8	7.1	.	.	.	.	.	.	.	.	.	.	.	.	.	21
1634	CHRYSANTHEMUM CORONARIUM	S	.	0.0	9.2	8.7	.	.	.	.	.	.	.	.	.	.	.	.	.	21
1635	CHRYSANTHEMUM LEUCANTHEMUM	S	.	0.0	24.3	23.4	.	.	5.3	.	.	.	.	.	.	.	.	.	.	19
1636	CHRYSANTHEMUM LEUCANTHEMUM	S	.	0.0	19.4	22.9	.	.	.	.	.	.	.	.	.	.	.	.	.	20
1637	CHRYSANTHEMUM LEUCANTHEMUM	S	.	0.0	16.4	18.7	.	.	.	.	.	.	.	.	.	.	.	.	.	21
1638	CHRYSANTHEMUM LEUCANTHEMUM CV. ALASKA	S	.	0.0	23.8	27.9	.	.	.	.	.	.	.	.	.	.	.	.	.	19
1639	CHRYSANTHEMUM PARTHENIUM	S	.	0.0	22.2	31.2	.	.	.	.	.	.	.	.	.	.	.	.	.	21
1640	CHRYSANTHEMUM SEGETUM	L	19	93.2	1.2	0.3	3.9	0.7	1.4	124	50	3.0	.	.	1160	0.13	0.10	0.7	9	3
1641	CHRYSOBALANUS ICACO	F	56	84.5	0.6	0.3	14.2	1.1	0.4	29	16	1.2	.	.	78	0.02	0.03	0.8	22	
1642	CHRYSOBALANUS ICACO	F	47	86.3	0.4	0.1	12.4	1.0	0.8	38	17	0.6	.	.	0	0.04	0.03	0.3	9	3
1643	CHRYSOBALANUS ICACO	S	.	2.4	10.7	.	63.8	.	1.7	.	.	.	.	.	.	.	.	.	.	5
1644	CHRYSOPHYLLUM CAINITO	F	68	81.5	1.0	0.7	16.4	2.0	0.4	17	16	0.4	5	140	0	0.01	0.02	0.9	8	2
1645	CHRYSOPHYLLUM CAINITO	F	68	82.8	0.8	1.6	14.5	1.0	0.3	21	17	0.8	.	.	10	0.04	0.03	1.0	11	3
1646	CHRYSOPHYLLUM CAINITO	G	84	80.1	1.4	3.2	14.6	3.3	0.7	22	21	0.6	.	.	40	0.02	0.01	0.9	3	3
1647	CHRYSOPOGON FULVUS	W	.	0.0	9.0	2.1	78.0	32.6	11.0	.	.	.	.	.	.	.	.	.	.	18
1648	CHRYSOPOGON FULVUS	H	.	0.0	4.7	1.1	85.5	36.6	8.7	.	.	.	.	.	.	.	.	.	.	18
1649	CHRYSOPOGON GRYLLUS	S	.	0.0	40.5	6.0	.	.	.	.	.	.	.	.	.	.	.	.	.	21
1650	CHRYSOPOGON LANCEARIUS	L	.	0.0	4.7	.	.	27.6	16.3	1860	620	.	2230	.	.	.	.	.	.	5
1651	CHRYSOPOGON MONTANUS	L	.	0.0	4.5	1.8	84.1	26.5	9.6	.	.	.	.	.	.	.	.	.	.	5
1652	CHRYSOPOGON ORIENTALIS	L	.	7.8	4.4	1.1	78.2	35.6	8.4	.	.	.	.	.	.	.	.	.	.	5
1653	CHRYSOPSIS FOLIOSA	S	.	0.0	24.8	21.2	.	.	.	.	.	.	.	.	.	.	.	.	.	21
1654	CHRYSOTHAMNUS LANCEOLATUS	W	.	0.0	6.1	5.0	82.6	28.6	6.3	.	.	.	.	.	.	.	.	.	.	6
1655	CHRYSOTHAMNUS NAUSEOSUS	W	.	0.0	12.8	3.2	75.2	22.6	8.8	950	380	.	.	.	.	.	.	.	.	6
1656	CHRYSOTHAMNUS PARRYI SUBSP. NEVADENSIS	S	.	0.0	32.2	28.0	.	.	.	.	.	.	.	.	.	.	.	.	.	21
1657	CHRYSOTHAMNUS STENOPHYLLUS	W	.	0.0	6.6	8.9	76.6	29.8	7.9	2210	110	.	.	.	.	.	.	.	.	6
1658	CHRYSOTHAMNUS VISCIDIFLORUS	W	.	0.0	13.4	5.4	73.2	23.8	8.0	1400	340	.	.	.	.	.	.	.	.	6
1659	CHUSQUEA BACULIFERA	L	.	0.0	8.8	3.1	77.6	34.4	10.5	.	.	.	.	.	.	.	.	.	.	18
1660	CHUSQUEA PINIFOLIA	L	.	0.0	7.6	5.0	80.1	41.2	7.3	.	.	.	.	.	.	.	.	.	.	18
1661	CHUSQUEA QUITA	L	.	0.0	10.0	3.6	75.0	34.5	11.3	.	.	.	.	.	.	.	.	.	.	18
1662	CICER ANATOLICUM	S	.	0.0	24.1	3.4	.	.	.	.	.	.	.	.	.	.	.	.	.	21
1663	CICER ARIETINUM	S	357	10.0	19.6	3.7	63.5	6.7	3.2	252	271	11.1	.	.	60	0.48	0.16	1.8	.	1
1664	CICER ARIETINUM	SH	.	0.0	11.3	2.2	75.1	27.2	11.4	1410	250	.	.	.	.	.	.	.	.	18
1665	CICER ARIETINUM	H	.	0.0	12.9	1.5	74.4	35.3	11.2	1530	200	.	.	.	.	.	.	.	.	18

TABLE 2 - DATA DIRECTLY TRANSCRIBED FROM ORIGINAL SOURCES.

OBS	PLNA	PART	CALORIES	WATER	PROTEIN	FAT	TOTLCARB	FIBER	ASH	CALCIUM	PHOSPHO	IRON	SODIUM	POTASS	CAROT	THIAMINE	RIBOFLAV	NIACIN	ASCORBIC	REFCO
1666	CICER ARIETINUM	F	.	0.0	17.0	2.9	70.5	25.0	9.6	1390	240	.	.	.	.	.	.	.	.	18
1667	CICER ARIETINUM	S	.	0.0	18.1	4.9	73.5	9.8	3.5	260	410	.	.	.	.	.	.	.	.	18
1668	CICER ARIETINUM	S	.	0.0	19.4	5.5	.	.	2.7	.	.	.	.	.	.	.	.	.	.	19
1669	CICER ARIETINUM	S	364	11.5	18.2	6.2	61.1	3.4	3.0	134	324	7.3	.	.	30	0.46	0.16	1.7	1	3
1670	CICER ARIETINUM	S	360	10.7	20.5	4.8	61.0	5.0	3.0	150	331	6.9	26	797	30	0.31	0.15	2.0	.	4
1671	CICHORIUM ENDIVIA	L	21	92.8	1.5	0.2	4.3	1.1	1.2	60	36	1.2	.	.	.	.	.	.	.	1
1672	CICHORIUM ENDIVIA	L	20	93.1	1.7	0.2	4.1	0.9	0.9	79	.	1.7	.	.	1980	0.07	0.12	0.4	11	3
1673	CICHORIUM ENDIVIA	L	20	93.1	1.7	0.1	4.1	0.9	1.0	81	54	1.7	14	294	1980	0.07	0.14	0.5	10	4
1674	CICHORIUM ENDIVIA	SH	15	95.1	1.0	0.1	3.2	.	0.6	18	21	0.5	7	182	0	.	.	.	.	4
1675	CICHORIUM ENDIVIA	L	20	92.8	1.8	0.3	3.8	0.8	1.3	86	40	0.9	.	420	2400	0.06	0.10	0.5	22	4
1676	CICHORIUM INTYBUS	S	.	0.0	18.9	17.4	.	.	.	.	.	.	.	.	.	.	.	.	.	20
1677	CICHORIUM INTYBUS	S	.	0.0	26.8	21.4	.	.	.	.	.	.	.	.	.	.	.	.	.	20
1678	CICHORIUM INTYBUS	S	.	0.0	23.2	23.2	.	.	.	.	.	.	.	.	.	.	.	.	.	21
1679	CICHORIUM INTYBUS	S	.	0.0	21.8	28.1	.	.	.	.	.	.	.	.	.	.	.	.	.	21
1680	CICHORIUM INTYBUS	L	20	93.1	1.7	0.2	4.1	0.9	0.9	79	.	1.7	.	.	1580	0.07	0.12	0.4	11	3
1681	CICHORIUM INTYBUS	W	.	0.0	14.4	2.2	65.4	15.3	18.0	1890	220	.	.	.	.	.	.	.	.	6
1682	CICUTA DOUGLASII	S	.	0.0	25.4	16.8	.	.	.	.	.	.	.	.	.	.	.	.	.	20
1683	CICUTA DOUGLASII	S	.	0.0	26.2	16.5	.	.	.	.	.	.	.	.	.	.	.	.	.	20
1684	CICUTA MACULATA	S	.	0.0	23.8	15.9	.	.	.	.	.	.	.	.	.	.	.	.	.	20
1685	CICUTA MEXICANA	S	.	0.0	16.2	32.6	.	.	.	.	.	.	.	.	.	.	.	.	.	20
1686	CIMICIFUGA RACEMOSA	S	.	0.0	23.8	45.1	.	.	.	.	.	.	.	.	.	.	.	.	.	19
1687	CINCHONA SP.	S	.	0.0	18.4	16.0	.	.	6.1	.	.	.	.	.	.	.	.	.	.	19
1688	CINNAMOMUM CASSIA & CINNAMOMUM VERUM	E	261	9.5	3.9	3.2	79.8	24.4	3.6	1228	61	38.1	26	500	156	0.08	0.14	1.3	28	16
1689	CINNAMOMUM VERUM	E	305	11.8	3.1	1.2	80.6	.	3.3	470	.	.	.	.	.	0.02	0.07	0.7	.	1
1690	CIRCAEA QUADRISULCATA	S	.	0.0	17.5	18.5	.	.	.	.	.	.	.	.	.	.	.	.	.	19
1691	CIRSIUM ACARNA	S	.	0.0	18.5	29.2	.	.	.	.	.	.	.	.	.	.	.	.	.	21
1692	CIRSIUM ALTISSIMUM	S	.	0.0	19.1	22.2	.	.	3.4	.	.	.	.	.	.	.	.	.	.	19
1693	CIRSIUM ARVENSE	S	.	0.0	17.6	15.7	.	.	.	.	.	.	.	.	.	.	.	.	.	21
1694	CIRSIUM CAROLINIANUM	S	.	0.0	15.9	17.4	.	.	.	.	.	.	.	.	.	.	.	.	.	20
1695	CIRSIUM DIPSACOLEPSII	L	63	80.9	3.0	0.1	15.3	1.5	0.7	20	79	8.5	.	.	0	0.02	0.06	0.2	0	2
1696	CIRSIUM MEGACEPHALUM	S	.	0.0	16.8	12.3	.	.	.	.	.	.	.	.	.	.	.	.	.	20
1697	CIRSIUM SP.	W	.	0.0	24.9	2.7	57.6	14.4	14.8	4260	210	.	.	3180	2	.	.	.	.	6
1698	CISSUS POPULNEA	G	72	80.0	1.3	1.8	15.2	3.2	1.7	392	.	.	.	.	.	.	.	.	.	1
1699	CISSUS STRIATA	SH	.	0.0	20.8	2.5	68.6	14.2	8.1	2500	350	.	.	.	.	.	.	.	.	18
1700	CISTUS LAURIFOLIUS	S	.	0.0	20.6	10.3	.	.	.	.	.	.	.	.	.	.	.	.	.	21
1701	CISTUS PARVIFLORUS	S	.	0.0	17.7	7.2	.	.	.	.	.	.	.	.	.	.	.	.	.	21
1702	CISTUS PARVIFLORUS	S	.	0.0	21.6	7.4	.	.	.	.	.	.	.	.	.	.	.	.	.	21
1703	CISTUS SALVIFOLIUS	S	.	0.0	13.3	7.2	.	.	.	.	.	.	.	.	.	.	.	.	.	21
1704	CISTUS VILLOSUS	S	.	0.0	18.9	7.4	.	.	.	.	.	.	.	.	.	.	.	.	.	21
1705	CITRULLUS COLOCYNTHIS	S	556	6.7	23.6	47.2	19.5	1.5	3.0	46	580	.	.	.	.	0.11	.	.	.	1
1706	CITRULLUS COLOCYNTHIS	S	.	0.0	12.2	18.5	.	.	.	.	.	.	.	.	.	.	.	.	.	21
1707	CITRULLUS COLOCYNTHIS	S	.	0.0	16.6	22.7	.	.	.	.	.	.	.	.	.	.	.	.	.	21
1708	CITRULLUS LANATUS	F	22	93.6	0.5	0.1	5.5	0.4	0.3	8	9	0.3	.	.	250	0.04	0.05	0.1	8	1
1709	CITRULLUS LANATUS	S	518	7.3	18.4	38.7	32.0	22.4	3.6	120	770	.	.	.	.	.	.	.	.	1
1710	CITRULLUS LANATUS	K	567	5.7	25.8	49.7	15.1	4.0	3.7	53	.	7.4	.	.	.	0.10	0.12	1.4	.	1

TABLE 2 - DATA DIRECTLY TRANSCRIBED FROM ORIGINAL SOURCES.

OBS	PLNA	PART	CALORIES	WATER	PROTEIN	FAT	TOTCARB	FIBER	ASH	CALCIUM	PHOSPHO	IRON	SODIUM	POTASS	CAROT	THIAMINE	RIBOFLAV	NIACIN	ASCORBIC	REF
1711	CITRULLUS LANATUS	SH	.	0.0	20.0	2.2	59.4	17.1	18.4	1960	460	.	.	.	.	.	.	.	.	18
1712	CITRULLUS LANATUS	F	.	0.0	10.0	8.9	73.3	25.7	7.8	340	290	.	.	.	.	.	.	.	.	18
1713	CITRULLUS LANATUS	S	.	0.0	24.4	35.4	36.0	31.6	4.2	.	.	.	.	.	.	.	.	.	.	18
1714	CITRULLUS LANATUS	F	26	92.6	0.5	0.2	6.4	0.3	0.3	7	10	0.5	1	100	354	0.03	0.03	0.2	7	4
1715	CITRULLUS VULGARIS	S	.	0.0	37.5	51.2	.	.	3.7	.	.	.	.	.	.	.	.	.	.	19
1716	CITRULLUS VULGARIS	S	.	0.0	37.5	51.4	.	.	3.8	.	.	.	.	.	.	.	.	.	.	19
1717	CITRULLUS VULGARIS	S	.	0.0	38.1	51.8	.	.	3.6	.	.	.	.	.	.	.	.	.	.	19
1718	CITRULLUS VULGARIS	S	.	0.0	37.5	52.0	.	.	3.6	.	.	.	.	.	.	.	.	.	.	19
1719	CITRULLUS VULGARIS	S	.	0.0	12.3	17.5	.	.	1.4	.	.	.	.	.	.	.	.	.	.	19
1720	CITRULLUS VULGARIS	S	.	0.0	11.5	17.0	.	.	.	.	.	.	.	.	.	.	.	.	.	21
1721	CITRUS AURANTIFOLIA	F	32	89.5	0.6	0.8	8.7	0.7	0.4	19	21	0.7	.	.	10	0.03	0.02	0.3	45	1
1722	CITRUS AURANTIFOLIA	F	36	91.0	0.5	2.4	5.9	0.3	0.2	13	11	0.0	2	82	10	0.03	0.02	0.1	46	2
1723	CITRUS AURANTIFOLIA	F	32	91.0	0.4	1.4	7.0	0.3	0.2	24	14	0.4	.	.	10	0.03	0.02	0.2	40	3
1724	CITRUS AURANTIFOLIA	F	28	89.3	0.7	0.2	9.5	0.5	0.3	33	18	0.6	2	102	6	0.03	0.02	0.2	37	4
1725	CITRUS AURANTIUM	F	37	89.2	0.6	0.0	9.7	0.4	0.5	18	12	.	.	.	290	.	.	.	48	1
1726	CITRUS AURANTIUM	F	44	87.5	0.7	0.1	11.2	2.0	0.5	42	20	0.4	.	.	70	0.07	0.03	0.3	43	2
1727	CITRUS AURANTIUM	F	50	85.7	0.7	0.1	13.0	0.9	0.5	43	17	0.6	.	.	40	0.06	0.04	0.3	42	3
1728	CITRUS GRANDIS	F	25	92.4	0.5	0.1	6.3	0.5	0.7	26	22	0.3	.	.	.	0.07	0.02	.	30	1
1729	CITRUS GRANDIS	F	39	88.9	0.7	0.3	9.5	0.4	0.6	27	22	0.5	1	235	30	0.05	0.02	0.3	53	2
1730	CITRUS GRANDIS	F	34	90.3	0.6	0.2	8.5	0.8	0.4	26	26	0.5	.	.	0	0.04	0.02	0.2	35	3
1731	CITRUS HYSTRIX	F	38	90.8	0.6	1.3	7.0	0.4	0.3	23	8	0.6	.	.	.	0.04	0.02	0.1	43	2
1732	CITRUS IYO	F	40	88.6	0.6	0.3	10.0	0.4	0.5	21	17	0.3	.	.	30	0.09	0.06	0.8	30	2
1733	CITRUS JAPONICA	F	28	93.5	0.2	1.2	4.8	0.0	0.3	16	12	0.7	3	800	0	0.01	0.01	0.2	43	2
1734	CITRUS LIMETTA	F	30	89.9	0.7	0.6	8.4	1.0	0.4	28	20	0.5	.	.	0	0.05	0.03	0.2	48	3
1735	CITRUS LIMON	F	29	89.8	0.7	0.4	8.8	0.9	0.3	25	18	0.5	.	.	.	.	0.02	.	.	1
1736	CITRUS LIMON	F	28	90.4	0.7	0.8	7.6	0.7	0.5	32	19	0.3	4	137	15	0.04	0.02	0.1	50	2
1737	CITRUS LIMON	F	29	90.3	0.6	0.6	8.1	0.6	0.4	41	15	0.7	.	.	10	0.06	0.02	0.1	51	3
1738	CITRUS LIMON	F	27	90.1	1.1	0.3	8.2	0.4	0.3	26	16	0.6	2	138	12	0.04	0.02	0.1	53	4
1739	CITRUS MEDICA	L	56	81.8	3.2	0.0	13.5	.	1.5	34	77	4.4	.	.	11935	0.04	.	.	.	2
1740	CITRUS MEDICA	F	40	88.7	0.6	0.1	10.2	1.4	0.4	42	20	0.4	.	.	10	0.06	0.04	0.2	32	3
1741	CITRUS PARADISI	F	34	90.0	0.8	0.1	8.6	0.6	0.5	21	18	0.6	.	.	25	0.05	0.03	0.2	44	1
1742	CITRUS PARADISI	F	39	88.6	0.6	0.1	10.1	0.5	0.6	12	22	1.0	2	135	60	0.07	0.06	0.5	38	2
1743	CITRUS PARADISI	F	38	89.2	0.6	0.2	9.6	0.2	0.4	18	21	0.5	.	.	20	0.05	0.02	0.2	43	3
1744	CITRUS PARADISI	F	41	88.4	0.5	0.1	10.6	0.2	0.4	16	16	0.4	1	135	48	0.04	0.02	0.2	38	4
1745	CITRUS POONENSIS	F	41	88.4	0.8	0.2	10.2	0.4	0.4	21	19	0.3	.	167	340	0.10	0.04	0.6	54	2
1746	CITRUS RETICULATA	F	49	86.1	0.8	0.3	12.3	0.6	0.5	38	17	1.1	.	.	230	0.08	0.05	0.2	28	1
1747	CITRUS RETICULATA	F	41	88.6	0.7	0.2	10.2	0.3	0.3	26	14	0.2	1	149	465	0.09	0.04	0.4	42	2
1748	CITRUS RETICULATA	F	43	87.8	0.7	0.2	10.9	0.4	0.4	30	16	0.4	.	.	80	0.08	0.03	0.3	33	3
1749	CITRUS RETICULATA	F	46	87.0	0.8	0.2	11.6	0.5	0.4	40	18	0.4	2	126	252	0.06	0.02	0.1	31	4
1750	CITRUS RETICULATA X C. SINENSIS	F	44	87.6	0.8	0.2	11.0	0.2	0.4	37	16	0.2	.	.	330	0.10	0.03	0.2	36	3
1751	CITRUS SINENSIS	F	43	88.0	0.6	0.4	10.5	0.6	0.5	28	17	0.1	.	.	75	0.02	0.03	0.2	46	1
1752	CITRUS SINENSIS	F	40	88.6	0.8	0.2	9.9	0.4	0.5	21	20	0.3	2	162	150	0.07	0.04	0.4	43	2
1753	CITRUS SINENSIS	F	42	87.7	0.8	0.2	10.5	0.4	0.8	34	20	0.7	.	.	80	0.09	0.03	0.2	59	3
1754	CITRUS SINENSIS	F	49	86.0	1.0	0.2	12.2	0.5	0.6	41	20	0.4	1	200	120	0.10	0.04	0.4	50	4
1755	CITRUS TANKAN	F	38	89.4	0.6	0.4	9.1	0.6	0.5	33	20	0.3	.	323	840	0.08	0.05	0.5	57	2

TABLE 2 - DATA DIRECTLY TRANSCRIBED FROM ORIGINAL SOURCES.

OBS	PLANA	PART	CALORIES	WATER	PROTEIN	FAT	TOTCARB	FIBER	ASH	CALCIUM	PHOSPHO	IRON	SODIUM	POTASS	CAROT	THIAMINE	RIBOFLAV	NIACIN	ASCORBIC	REFCO
1756	CLADIUM JAMAICENSE	S	.	0.0	7.2	5.4	.	.	.	.	.	.	.	.	.	.	.	.	.	20
1757	CLARKIA AMOENA	S	.	0.0	31.2	34.6	.	.	.	.	.	.	.	.	.	.	.	.	.	20
1758	CLARKIA ELEGANS	S	.	0.0	28.7	35.5	.	.	7.0	.	.	.	.	.	.	.	.	.	.	19
1759	CLARKIA ELEGANS	S	.	0.0	28.8	36.7	.	.	7.9	.	.	.	.	.	.	.	.	.	.	19
1760	CLARKIA ELEGANS	S	.	0.0	29.4	37.1	.	.	.	.	.	.	.	.	.	.	.	.	.	20
1761	CLAUSENA LANSIUM	F	55	84.0	0.9	0.1	14.1	0.8	0.9	15	19	0.0	16	281	0	0.02	0.11	3.3	148	2
1762	CLEISTOPHOLIS SP	F	57	84.0	3.0	0.5	12.0	.	0.5	23	40	2.2	.	.	.	.	.	.	10	1
1763	CLEMATIS DRUMMONDI	W	.	0.0	15.8	2.0	76.8	27.6	5.4	1470	210	.	.	1550	.	.	.	.	.	6
1764	CLEMATIS FLAMMULA	S	.	0.0	13.1	15.4	.	.	.	.	.	.	.	.	.	.	.	.	.	20
1765	CLEMATIS FLAMMULA	S	.	0.0	13.1	13.8	.	.	.	.	.	.	.	.	.	.	.	.	.	20
1766	CLEMATIS ORIENTALIS	S	.	0.0	22.8	25.3	.	.	.	.	.	.	.	.	.	.	.	.	.	21
1767	CLEMATIS ORIENTALIS	S	.	0.0	16.3	18.7	.	.	.	.	.	.	.	.	.	.	.	.	.	21
1768	CLEMATIS PANICULATA	S	.	0.0	19.4	17.9	.	.	.	.	.	.	.	.	.	.	.	.	.	20
1769	CLEMATIS RECTA	S	.	0.0	13.8	18.1	.	.	.	.	.	.	.	.	.	.	.	.	.	20
1770	CLEMATIS VIRGINIANA	S	.	0.0	20.6	14.0	.	.	4.6	.	.	.	.	.	.	.	.	.	.	19
1771	CLEMATIS VITALBA	S	.	0.0	15.1	14.0	.	.	.	.	.	.	.	.	.	.	.	.	.	21
1772	CLEMATIS VITICELLA	S	.	0.0	15.0	19.6	.	.	.	.	.	.	.	.	.	.	.	.	.	20
1773	CLEMATIS VITICELLA	S	.	0.0	15.0	16.7	.	.	.	.	.	.	.	.	.	.	.	.	.	20
1774	CLEMATIS VITICELLA	S	.	0.0	15.1	19.2	.	.	.	.	.	.	.	.	.	.	.	.	.	21
1775	CLEOME ARABICA	S	.	0.0	17.0	21.4	.	.	.	.	.	.	.	.	.	.	.	.	.	21
1776	CLEOME DIANDRA	S	.	0.0	20.8	27.8	.	.	.	.	.	.	.	.	.	.	.	.	.	21
1777	CLEOME PILOSA	S	.	0.0	18.8	30.1	.	.	.	.	.	.	.	.	.	.	.	.	.	21
1778	CLEOME SERRULATA	S	.	0.0	23.1	28.6	.	.	3.1	.	.	.	.	.	.	.	.	.	.	19
1779	CLEOME SERRULATA	S	.	0.0	20.8	33.6	.	.	2.7	.	.	.	.	.	.	.	.	.	.	19
1780	CLEOME SERRULATA	S	.	0.0	18.1	23.7	.	.	.	.	.	.	.	.	.	.	.	.	.	20
1781	CLEOME SONORAE	S	.	0.0	22.9	28.9	.	.	.	.	.	.	.	.	.	.	.	.	.	21
1782	CLEOME SP.	L	57	81.0	5.5	0.9	10.1	1.7	2.5	454	59	2.7	.	.	.	.	.	.	.	1
1783	CLEOME SPINOSA	S	.	0.0	19.8	33.0	.	.	2.6	.	.	.	.	.	.	.	.	.	.	19
1784	CLEOME SPINOSA	S	.	0.0	19.4	35.7	.	.	2.6	.	.	.	.	.	.	.	.	.	.	19
1785	CLEOME SPINOSA	S	.	0.0	18.8	36.6	.	.	2.6	.	.	.	.	.	.	.	.	.	.	19
1786	CLEOME TENUIS	S	.	0.0	18.8	30.0	.	.	.	.	.	.	.	.	.	.	.	.	.	19
1787	CLEOME VISCOSA	S	.	0.0	14.8	23.4	.	.	.	.	.	.	.	.	.	.	.	.	.	21
1788	CLERODENDRUM INDICUM	L	82	73.1	5.7	0.0	19.0	.	2.2	53	85	.	.	.	7906	.	.	.	18	2
1789	CLERODENDRUM TRICHOTOMUM	S	.	0.0	25.0	60.7	.	.	.	.	.	.	.	.	.	.	.	.	.	20
1790	CLITORIA TERNATEA	SH	.	0.0	19.5	4.5	67.4	11.8	8.6	3290	170	.	.	.	.	.	.	.	.	18
1791	CLITORIA TERNATEA	S	.	0.0	46.9	12.1	.	.	4.3	.	.	.	.	.	.	.	.	.	.	19
1792	CLITORIA TERNATEA	S	.	0.0	46.3	12.3	.	.	4.7	.	.	.	.	.	.	.	.	.	.	19
1793	CLITORIA TERNATEA	F	66	80.0	3.8	0.4	15.0	4.8	0.8	40	24	0.4	4	309	670	0.04	0.18	1.4	248	2
1794	CLITORIA TERNATEA	S	.	0.0	46.9	.	.	.	.	.	.	.	.	.	.	.	.	.	.	20
1795	CLUTIA AFFINIS	S	.	0.0	23.2	41.7	.	.	.	.	.	.	.	.	.	.	.	.	.	21
1796	CNICUS BENEDICTUS	S	.	0.0	17.5	27.1	.	.	4.1	.	.	.	.	.	.	.	.	.	.	19
1797	CNICUS BENEDICTUS	S	.	0.0	13.8	25.8	.	.	.	.	.	.	.	.	.	.	.	.	.	21
1798	CNIDOSCOLUS ACONITIFOLIUM	L	64	79.8	6.2	1.3	10.7	2.0	2.0	234	76	2.8	.	.	4740	0.20	0.40	1.6	194	3
1799	CNIDOSCOLUS ANGUSTIDENS	S	.	0.0	42.5	43.7	.	.	.	.	.	.	.	.	.	.	.	.	.	19
1800	CNIDOSCOLUS ELASTICUS	S	.	0.0	12.7	26.2	.	.	.	.	.	.	.	.	.	.	.	.	.	21

TABLE 2 - DATA DIRECTLY TRANSCRIBED FROM ORIGINAL SOURCES.

OBS	PLNA	PART	CALORIES	WATER	PROTEIN	FAT	TOTLCARB	FIBER	ASH	CALCIUM	PHOSPHO	IRON	SODIUM	POTASS	CAROT	THIAMINE	RIBOFLAV	NIACIN	ASCORBIC	REFCO
1801	CNIDOSCOLUS SP.	S	.	0.0	19.4	30.8	.	.	.	.	.	.	.	.	.	.	.	.	.	19
1802	CNIDOSCOLUS TEPIQUENSIS	S	.	0.0	13.4	19.8	.	.	.	.	.	.	.	.	.	.	.	.	.	21
1803	COBAEA SCANDENS	S	.	0.0	18.2	22.1	.	.	3.6	.	.	.	.	.	.	.	.	.	.	19
1804	COCCINEA CORDIFOLIA	L	28	90.6	4.1	0.4	4.2	1.0	0.7	126	30	4.6	.	.	10845	0.17	0.13	3.80	48	2
1805	COCCINEA CORDIFOLIA	G	.	93.1	1.2	0.1	5.1	1.6	0.5	40	30	1.4	.	.	156	.	.	.	28	5
1806	COCCOLOBA BARBADENSIS	S	.	0.0	4.8	1.2	.	.	.	.	.	.	.	.	.	.	.	.	.	21
1807	COCCOLOBA UVIFERA	F	73	79.1	1.4	0.3	18.2	2.1	1.0	54	33	1.2	.	.	0	0.03	0.03	0.60	17	3
1808	COCCULUS CAROLINUS	S	.	0.0	16.9	18.4	.	.	.	.	.	.	.	.	.	.	.	.	.	19
1809	COCHLOSPERMUM RELIGIOSUM	S	.	9.2	20.9	14.2	50.4	14.6	5.2	.	.	.	.	.	.	.	.	.	.	5
1810	COCHLOSPERMUM VITIFOLIUM	S	.	0.0	14.6	15.6	.	.	5.2	.	.	.	.	.	.	.	.	.	.	19
1811	COCOS NUCIFERA	S	388	42.6	3.6	39.0	13.8	6.6	1.0	21	103	2.5	.	.	25	0.03	0.03	0.60	2	1
1812	COCOS NUCIFERA	IS	200	68.0	2.0	17.4	11.7	3.7	0.9	.	.	.	.	.	.	.	.	.	.	1
1813	COCOS NUCIFERA	S	312	51.7	3.2	28.2	16.0	3.0	0.9	23	112	2.5	7	555	0	0.05	0.03	0.60	3	2
1814	COCOS NUCIFERA	IS	77	84.0	1.4	3.6	10.3	0.4	0.7	42	56	1.0	51	257	0	0.04	0.03	0.80	6	2
1815	COCOS NUCIFERA	SH	48	86.4	1.8	1.3	9.1	1.0	1.4	27	60	0.5	23	665	.	0.03	0.04	1.20	9	2
1816	COCOS NUCIFERA	S	296	54.6	3.5	27.2	13.7	3.8	1.0	13	83	1.8	.	.	0	0.04	0.03	0.60	4	3
1817	COCOS NUCIFERA	IS	122	81.4	1.9	11.9	4.0	0.7	0.8	11	42	1.1	.	.	0	0.05	0.03	0.80	7	3
1818	COCOS NUCIFERA	S	346	50.9	3.5	35.3	9.4	4.0	0.9	13	95	1.7	23	256	0	0.05	0.02	0.50	3	4
1819	COCOS NUCIFERA	Z	.	11.3	.	.	.	.	5.2	270	170	.	.	260	.	.	.	.	.	5
1820	CODON ROYENII	S	.	0.0	14.2	8.4	.	.	.	.	.	.	.	.	.	.	.	.	.	21
1821	CODONOPSIS LANCEOLATA	L	74	82.2	2.3	3.5	10.9	6.4	1.1	90	121	2.1	.	.	0	0.12	0.22	0.80	0	2
1822	COFFEA ARABICA	L	306	6.4	9.3	5.5	66.6	17.5	12.2	1910	170	96.6	.	.	2360	0.00	0.21	5.20	.	1
1823	COFFEA ARABICA	L	.	0.0	9.9	5.9	71.2	18.7	13.0	.	.	.	.	.	.	.	.	.	.	18
1824	COFFEA ARABICA	S	203	6.3	11.7	10.8	68.2	22.9	3.0	120	178	2.9	.	.	20	0.22	0.06	1.30	.	3
1825	COIX LACRYMA-JOBI	S	333	8.9	10.4	5.3	66.5	10.5	8.9	.	.	.	.	.	.	.	.	.	.	1
1826	COIX LACRYMA-JOBI	W	.	0.0	8.5	2.8	79.8	27.9	8.9	.	.	.	.	.	.	.	.	.	.	18
1827	COIX LACRYMA-JOBI	S	.	0.0	8.1	3.6	82.4	17.0	5.9	.	.	.	.	.	.	.	.	.	.	18
1828	COIX LACRYMA-JOBI	S	.	0.0	21.9	7.9	.	.	1.8	.	.	.	.	.	.	.	.	.	.	19
1829	COIX LACRYMA-JOBI	S	.	0.0	2.0	6.4	.	.	5.8	.	.	.	.	.	.	.	.	.	.	19
1830	COIX LACRYMA-JOBI	S	.	0.0	26.2	5.0	.	.	.	.	.	.	.	.	.	.	.	.	.	21
1831	COIX LACRYMA-JOBI	S	380	11.2	15.4	6.2	65.3	0.8	1.9	25	435	5.0	.	.	0	0.28	0.19	4.30	0	3
1832	COIX LACRYMA-JOBI	S	.	10.1	10.3	3.1	73.6	0.3	0.7	5	30	.	.	.	.	.	.	.	.	5
1833	COIX LACRYMA-JOBI	S	.	11.3	12.1	3.8	74.6	0.3	1.0	6	50	.	.	.	.	.	.	.	.	5
1834	COLA ACUMINATA	F	148	62.9	2.2	0.4	33.7	1.4	0.8	58	86	2.0	.	.	25	0.03	0.03	0.60	54	1
1835	COLA ACUMINATA	A	62	83.3	1.5	0.6	14.4	0.8	0.2	3	17	1.4	.	.	30	0.01	0.05	0.70	10	3
1836	COLA CORDIFOLIA	P	91	73.8	1.2	0.0	24.2	0.4	0.8	55	47	2.5	.	.	460	0.01	.	0.40	10	1
1837	COLCHICUM BORNMULLERI	S	.	0.0	8.9	1.1	.	.	.	.	.	.	.	.	.	.	.	.	.	21
1838	COLEUS AMBOINICUS	L	17	94.4	0.9	0.4	3.1	0.5	1.2	232	8	3.9	.	.	980	0.03	0.07	0.20	12	3
1839	COLLADONIA CRENATA	S	.	0.0	22.9	6.0	.	.	.	.	.	.	.	.	.	.	.	.	.	21
1840	COLLADONIA CRENATA	S	.	0.0	30.7	13.1	.	.	.	.	.	.	.	.	.	.	.	.	.	21
1841	COLLINSIA HETEROPHYLLA	S	.	0.0	16.9	8.2	.	.	3.9	.	.	.	.	.	.	.	.	.	.	19
1842	COLLINSIA HETEROPHYLLA	S	.	0.0	13.2	9.5	.	.	.	.	.	.	.	.	.	.	.	.	.	21
1843	COLLINSIA PARVIFLORA	S	.	0.0	14.7	13.0	.	.	.	.	.	.	.	.	.	.	.	.	.	21
1844	COLLINSIA TINCTORIA	S	.	0.0	15.7	5.6	.	.	.	.	.	.	.	.	.	.	.	.	.	21
1845	COLOCASIA ANTIQUORUM	L	281	0.0	26.8	8.0	41.6	14.6	15.5	993	401	.	11	213	.	.	.	7.12	.	12

TABLE 2 - DATA DIRECTLY TRANSCRIBED FROM ORIGINAL SOURCES.

OBS	PLANA	PART	CALORIES	WATER	PROTEIN	FAT	TOTCARB	FIBER	ASH	CALCIUM	PHOSPHO	IRON	SODIUM	POTASS	CAROTENE	THIAMINE	RIBOFLAV	NIACIN	ASCORBIC	REFCO
1846	COLOCASIA ESCULENTA	R	102	73.1	1.8	0.1	23.8	1.0	1.2	51	88	1.2	.	.	0	0.10	0.03	0.80	8	1
1847	COLOCASIA ESCULENTA	L	31	90.0	2.4	0.6	5.7	1.5	1.3	98	49	2.0	.	.	1800	0.17	0.35	0.80	11	1
1848	COLOCASIA ESCULENTA	L	.	0.0	25.0	10.7	52.9	12.1	12.4	1740	580	.	.	.	.	.	.	.	.	18
1849	COLOCASIA ESCULENTA	R	.	0.0	9.5	0.1	86.3	1.7	4.1	.	.	.	.	.	.	.	.	.	.	18
1850	COLOCASIA ESCULENTA	L	61	81.4	4.1	1.0	11.9	1.2	1.6	162	69	1.0	9	963	5535	0.13	0.34	1.50	63	2
1851	COLOCASIA ESCULENTA	Z	24	92.7	0.5	0.2	5.8	0.9	0.8	49	25	0.9	4	334	180	0.02	0.04	0.40	13	2
1852	COLOCASIA ESCULENTA	R	94	75.4	2.2	0.4	21.0	0.8	1.0	34	62	1.2	10	448	0	0.12	0.04	1.00	8	2
1853	COLOCASIA ESCULENTA	R	92	74.6	1.6	0.2	22.4	0.8	1.2	96	88	1.2	.	.	10	0.08	0.04	0.70	7	3
1854	COLOCASIA ESCULENTA	R	98	73.0	1.9	0.2	23.7	0.8	1.2	28	61	1.0	7	514	20	0.13	0.04	1.10	4	4
1855	COLOCASIA ESCULENTA	L	40	87.2	3.0	0.8	7.4	1.4	1.6	76	59	1.0	.	.	.	.	.	.	31	4
1856	COLOCASIA ESCULENTA	R	.	73.1	3.0	0.1	22.1	.	1.7	40	140	2.1	.	.	24	.	.	.	0	5
1857	COLOCASIA ESCULENTA	Z	.	93.4	0.3	0.3	4.7	0.6	1.2	60	20	0.5	.	.	.	.	.	.	.	5
1858	COLUBRINA TEXENSIS	S	.	0.0	20.0	17.9	.	.	.	.	.	.	.	.	.	.	.	.	.	19
1859	COLUBRINA TEXENSIS	W	.	0.0	19.5	4.5	67.3	11.8	8.7	3290	170	.	.	1280	.	.	.	.	.	6
1860	COLUTEA ARBORESCENS	S	.	0.0	40.6	3.0	.	.	4.7	.	.	.	.	.	.	.	.	.	.	19
1861	COLUTEA ARBORESCENS	S	.	0.0	40.5	4.1	.	.	.	.	.	.	.	.	.	.	.	.	.	21
1862	COLUTEA CILICICA	S	.	0.0	45.4	2.6	.	.	.	.	.	.	.	.	.	.	.	.	.	21
1863	COLUTEA ISTRIA	S	.	0.0	39.2	2.3	.	.	.	.	.	.	.	.	.	.	.	.	.	21
1864	COMANDRA PALLIDA	S	.	0.0	8.1	24.5	.	.	.	.	.	.	.	.	.	.	.	.	.	20
1865	COMBRETUM APICULATUM	L	.	0.0	12.2	3.8	74.0	21.6	10.0	2840	190	.	.	.	.	.	.	.	.	18
1866	COMMELINA BENGHALENSIS	L	43	86.2	2.3	0.4	9.5	2.4	1.6	210	52	.	.	.	.	.	.	.	.	1
1867	CONDALIA LYCIOIDES VAR. CANESCENS	S	.	0.0	38.1	17.8	.	.	.	.	.	.	.	.	.	.	.	.	.	19
1868	CONDALIA OBTUSIFOLIA	W	.	0.0	22.5	2.4	67.4	10.1	7.7	1500	220	.	.	2390	.	.	.	.	.	6
1869	CONDALIA SP.	S	.	0.0	11.2	68.6	.	.	.	.	.	.	.	.	.	.	.	.	.	20
1870	CONIUM MACULATUM	S	.	0.0	29.2	16.4	.	.	.	.	.	.	.	.	.	.	.	.	.	20
1871	CONIUM MACULATUM	S	.	0.0	23.5	15.5	.	.	.	.	.	.	.	.	.	.	.	.	.	21
1872	CONIUM MACULATUM	S	.	0.0	22.5	17.7	.	.	.	.	.	.	.	.	.	.	.	.	.	21
1873	CONIUM MACULATUM	S	.	0.0	21.8	14.3	.	.	.	.	.	.	.	.	.	.	.	.	.	21
1874	CONIUM MACULATUM	S	.	0.0	20.6	17.5	.	.	.	.	.	.	.	.	.	.	.	.	.	21
1875	CONIUM MACULATUM	S	.	0.0	24.9	11.6	.	.	.	.	.	.	.	.	.	.	.	.	.	21
1876	CONIUM MACULATUM	S	.	0.0	23.7	17.9	.	.	.	.	.	.	.	.	.	.	.	.	.	21
1877	CONRINGIA ORIENTALIS	S	.	0.0	30.6	27.3	.	.	.	.	.	.	.	.	.	.	.	.	.	20
1878	CONRINGIA PLANISILIQUA	S	.	0.0	29.9	34.4	.	.	.	.	.	.	.	.	.	.	.	.	.	20
1879	CONSOLIDA ORIENTALIS	S	.	0.0	21.1	45.7	.	.	.	.	.	.	.	.	.	.	.	.	.	21
1880	CONVOLVULUS ARVENSIS	S	.	0.0	22.0	8.8	.	.	.	.	.	.	.	.	.	.	.	.	.	21
1881	CONVOLVULUS INCANUS	S	.	0.0	20.1	6.8	.	.	.	.	.	.	.	.	.	.	.	.	.	20
1882	CONVOLVULUS SECUNDUS	S	.	0.0	24.2	8.0	.	.	.	.	.	.	.	.	.	.	.	.	.	21
1883	CONVOLVULUS TRICOLOR	S	.	0.0	22.2	7.7	.	.	5.7	.	.	.	.	.	.	.	.	.	.	19
1884	CONVOLVULUS TRICOLOR	S	.	0.0	16.9	10.7	.	.	.	.	.	.	.	.	.	.	.	.	.	20
1885	CONYZA BONARIENSIS	S	.	0.0	26.4	24.3	.	.	.	.	.	.	.	.	.	.	.	.	.	21
1886	CORCHORUS ACUTANGULUS	S	.	0.0	22.6	12.8	.	.	.	.	.	.	.	.	.	.	.	.	.	21
1887	CORCHORUS CAPSULARIS	S	.	0.0	21.8	11.7	.	.	7.1	.	.	.	.	.	.	.	.	.	.	19
1888	CORCHORUS CLITORIUS	S	.	0.0	21.0	14.1	.	.	7.8	.	.	.	.	.	.	.	.	.	.	19
1889	CORCHORUS OLITORIUS	L	58	80.4	4.5	0.3	12.4	2.0	2.4	360	122	7.2	.	.	6410	0.15	0.53	1.20	80	1
1890	CORCHORUS OLITORIUS	L	43	84.1	5.6	0.3	7.6	1.7	2.4	266	97	7.7	12	444	7850	0.13	0.26	1.10	53	2

TABLE 2 - DATA DIRECTLY TRANSCRIBED FROM ORIGINAL SOURCES.

OBS	PLANA	PART	CALORIES	WATER	PROTEIN	FAT	TOTCARB	FIBER	ASH	CALCIUM	PHOSPHO	IRON	SODIUM	POTASS	CAROTE	THIAMINE	RIBOFLAV	NIACIN	ASCORBIC	REF
1891	CORCHORUS TRIDENS	L	308	0.0	23.6	3.4	62.3	8.4	10.8	1912	308	.	14	198	.	.	.	5.31	.	12
1892	CORCHORUS TRILOCULARIS	L	318	0.0	30.4	4.8	57.2	7.9	7.7	1428	292	.	135	270	.	.	.	7.00	.	12
1893	CORDEAUXIA EDULIS	S	398	11.1	10.8	12.0	63.9	1.4	2.2	32	185	6.4	.	.	.	.	.	.	.	1
1894	CORDIA ALBA	F	45	86.7	1.2	0.4	10.5	0.1	1.2	7	41	1.3	.	.	10	0.00	0.04	0.50	21	3
1895	CORDIA MYXA	S	503	5.6	34.8	37.0	17.9	2.0	4.7	.	.	.	.	.	.	.	.	.	.	1
1896	CORDIA MYXA	L	.	0.0	15.8	6.8	64.3	14.7	13.1	2560	220	.	.	.	.	.	.	.	.	18
1897	CORDIA MYXA	L	65	64.0	5.1	0.3	14.0	3.6	16.6	81	67	7.2	.	.	330	0.10	0.29	0.20	0	2
1898	CORDYLA PINNATA	G	67	80.9	2.3	0.2	16.0	1.5	0.6	23	126	.	.	.	.	.	.	.	93	1
1899	CORDYLA PINNATA	F	70	80.0	1.4	0.1	17.8	1.1	0.7	29	142	1.8	.	.	310	0.02	.	0.80	74	1
1900	CORDYLA PINNATA	F	67	80.9	2.3	0.2	16.0	1.5	0.6	23	126	.	.	.	.	.	.	.	93	1
1901	CORDYLINE AUSTRALIS	S	.	0.0	19.9	48.3	.	.	2.8	.	.	.	.	.	.	.	.	.	.	19
1902	CORDYLINE INDIVISA	S	.	0.0	15.6	29.1	.	.	.	.	.	.	.	.	.	.	.	.	.	20
1903	CORDYLINE TERMINALIS	L	.	0.0	13.9	3.1	73.9	30.9	9.1	.	.	.	.	.	.	.	.	.	.	18
1904	COREOPSIS BASILIS	S	.	0.0	14.4	21.6	.	.	.	.	.	.	.	.	.	.	.	.	.	20
1905	COREOPSIS GRANDIFLORA	S	.	0.0	12.5	13.2	.	.	.	.	.	.	.	.	.	.	.	.	.	20
1906	COREOPSIS LANCEOLATA	S	.	0.0	14.6	14.6	.	.	4.1	.	.	.	.	.	.	.	.	.	.	19
1907	COREOPSIS TINCTORIA	S	.	0.0	20.6	24.2	.	.	4.7	.	.	.	.	.	.	.	.	.	.	19
1908	CORIANDRUM SATIVUM	L	279	7.3	21.8	4.8	52.1	10.4	14.0	1246	481	42.5	211	4466	.	1.25	1.50	10.70	567	16
1909	CORIANDRUM SATIVUM	S	298	8.9	12.4	17.8	55.0	29.1	6.0	709	409	16.3	35	1267	0	0.24	0.29	2.13	.	16
1910	CORIANDRUM SATIVUM	L	37	87.6	2.6	0.6	7.3	1.6	1.9	133	80	4.5	94	560	2860	0.11	0.15	1.30	78	2
1911	CORIANDRUM SATIVUM	S	.	0.0	13.8	22.2	.	.	.	.	.	.	.	.	.	.	.	.	.	20
1912	CORIANDRUM SATIVUM	S	.	0.0	14.4	18.1	.	.	.	.	.	.	.	.	.	.	.	.	.	20
1913	CORIANDRUM SATIVUM	S	.	0.0	11.1	11.5	.	.	.	.	.	.	.	.	.	.	.	.	.	21
1914	CORIANDRUM SATIVUM	L	42	86.0	3.3	0.7	8.0	1.7	2.0	188	72	3.0	.	.	3200	0.15	0.28	1.60	75	3
1915	CORIANDRUM SATIVUM	F	.	11.2	14.1	16.1	54.2	32.6	4.4	630	370	17.9	.	.	5200	.	.	.	250	5
1916	CORNUS FLORIDA	S	.	0.0	34.9	49.4	.	.	3.7	.	.	.	.	.	.	.	.	.	.	19
1917	CORNUS FLORIDA	S	.	0.0	2.3	2.5	.	.	3.6	.	.	.	.	.	.	.	.	.	.	19
1918	CORNUS FLORIDA	S	.	0.0	38.8	43.8	.	.	3.6	.	.	.	.	.	.	.	.	.	.	19
1919	CORNUS FLORIDA	S	.	0.0	2.6	2.4	.	.	2.8	.	.	.	.	.	.	.	.	.	.	19
1920	CORNUS FLORIDA	S	.	0.0	35.3	43.3	.	.	3.6	.	.	.	.	.	.	.	.	.	.	19
1921	CORNUS FLORIDA	L	.	0.0	12.1	5.4	73.3	16.0	9.2	2590	150	.	.	.	.	.	.	.	.	6
1922	CORNUS MAS	S	.	0.0	31.1	49.5	.	.	3.1	.	.	.	.	.	.	.	.	.	.	19
1923	CORNUS MAS	S	.	0.0	1.8	2.3	.	.	1.0	.	.	.	.	.	.	.	.	.	.	19
1924	CORNUS MAS	S	.	0.0	4.1	7.3	.	.	.	.	.	.	.	.	.	.	.	.	.	21
1925	CORNUS MAS	S	.	0.0	3.8	4.5	.	.	.	.	.	.	.	.	.	.	.	.	.	21
1926	CORNUS SANGUINEA SUBSP. AUSTRALIS	S	.	0.0	6.9	22.2	.	.	.	.	.	.	.	.	.	.	.	.	.	21
1927	CORONILLA CRETICA	S	.	0.0	28.8	4.0	.	.	3.5	.	.	.	.	.	.	.	.	.	.	19
1928	CORONILLA CRETICA	S	.	0.0	24.9	3.9	.	.	3.8	.	.	.	.	.	.	.	.	.	.	19
1929	CORONILLA CRETICA	S	.	0.0	28.4	4.8	.	.	3.7	.	.	.	.	.	.	.	.	.	.	19
1930	CORONILLA EMEROIDES	S	.	0.0	36.9	7.2	.	.	.	.	.	.	.	.	.	.	.	.	.	21
1931	CORONILLA JUNCEA	S	.	0.0	17.4	6.3	.	.	.	.	.	.	.	.	.	.	.	.	.	21
1932	CORONILLA VARIA	S	.	0.0	31.5	8.3	.	.	3.5	.	.	.	.	.	.	.	.	.	.	19
1933	CORONILLA VARIA	S	.	0.0	33.9	6.7	.	.	3.8	.	.	.	.	.	.	.	.	.	.	19
1934	CORONILLA VARIA	S	.	0.0	35.8	6.8	.	.	3.5	.	.	.	.	.	.	.	.	.	.	19
1935	CORONILLA VARIA	S	.	0.0	31.7	8.2	.	.	3.5	.	.	.	.	.	.	.	.	.	.	19

TABLE 2 - DATA DIRECTLY TRANSCRIBED FROM ORIGINAL SOURCES.

OBS	PLNA	PART	CALORIES	WATER	PROTEIN	FAT	TOTLCARB	FIBER	ASH	CALCIUM	PHOSPHO	IRON	SODIUM	POTASS	CAROT	THIAMINE	RIBOFLAV	NIACIN	ASCORBIC	REFCO
1936	CORONOPUS DIDYMUS	S	.	0.0	27.4	40.9	.	.	.	.	.	.	.	.	.	.	.	.	.	21
1937	CORYDALIS AUREA	L	.	0.0	38.9	3.4	43.2	9.3	14.5	2500	720	.	.	4250	.	.	.	.	.	6
1938	CORYLUS AVELLANA	S	588	5.2	19.9	51.5	21.7	5.6	1.7	.	.	.	.	.	.	.	.	.	.	1
1939	CORYLUS AVELLANA	S	587	7.4	15.2	54.2	19.8	3.1	3.4	186	428	4.2	.	967	10	0.16	0.41	5.0	2	2
1940	CORYLUS SP.	S	647	3.4	10.8	63.2	19.8	2.3	2.8	254	319	3.6	.	.	128	0.46	0.55	5.0	7	3
1941	CORYLUS SP.	S	634	5.8	12.6	62.4	16.7	3.0	2.5	209	337	3.4	2	704	.	.	0.46	.	9	4
1942	CORYPHA UTAN	F	62	81.0	0.7	0.1	17.8	1.3	0.4	14	17	0.2	.	.	.	0.01	0.02	0.6	11	2
1943	COSMOS BIPINNATUS	S	.	0.0	19.0	20.2	.	.	4.1	.	.	.	.	.	.	.	.	.	.	19
1944	COSMOS BIPINNATUS	S	.	0.0	20.0	17.4	.	.	.	.	.	.	.	.	.	.	.	.	.	20
1945	COSMOS BIPINNATUS	S	.	0.0	15.6	21.4	.	.	.	.	.	.	.	.	.	.	.	.	.	20
1946	COSMOS SEEMANNII	S	.	0.0	21.2	11.6	.	.	.	.	.	.	.	.	.	.	.	.	.	20
1947	COSMOS SULPHUREUS	S	.	0.0	18.6	19.2	.	.	.	.	.	.	.	.	.	.	.	.	.	21
1948	COTONEASTER ACUMINATA	S	.	0.0	8.8	5.7	.	.	.	.	.	.	.	.	.	.	.	.	.	20
1949	COTONEASTER BACILLARIS	S	.	0.0	5.4	5.0	.	.	.	.	.	.	.	.	.	.	.	.	.	21
1950	COTONEASTER NUMMULARIA	S	.	0.0	6.3	5.3	.	.	.	.	.	.	.	.	.	.	.	.	.	21
1951	COUEPIA SP.	S	.	0.0	12.5	73.9	.	.	.	.	.	.	.	.	.	.	.	.	.	20
1952	COULA EDULIS	S	463	8.3	7.2	23.6	59.0	2.2	1.9	165	247	.	.	.	.	.	.	.	.	1
1953	COUMEROUNA ALATA	S	.	0.0	15.1	45.3	.	.	2.6	.	.	.	.	.	.	.	.	.	.	19
1954	COUMEROUNA ALATA	S	.	0.0	1.2	9.3	.	.	0.4	.	.	.	.	.	.	.	.	.	.	19
1955	COURSETIA GLANDULOSA	S	.	0.0	55.6	17.1	.	.	.	.	.	.	.	.	.	.	.	.	.	19
1956	COUSINIA BRACHYPTERA	S	.	0.0	25.4	17.5	.	.	.	.	.	.	.	.	.	.	.	.	.	21
1957	COWANIA STANSBURIANA	S	.	0.0	14.4	7.6	.	.	.	.	.	.	.	.	.	.	.	.	.	20
1958	CRAMBE ABYSSINICA	L	32	89.0	4.2	0.6	4.6	1.6	1.6	176	70	2.9	.	.	85	0.00	0.09	0.7	.	1
1959	CRAMBE ABYSSINICA	S	.	0.0	15.5	55.0	26.5	10.8	3.0	.	.	.	.	.	.	.	.	.	.	18
1960	CRAMBE ABYSSINICA	S	.	0.0	31.5	35.5	.	.	12.6	.	.	.	.	.	.	.	.	.	.	19
1961	CRAMBE ABYSSINICA	S	.	0.0	22.6	17.9	.	.	8.5	.	.	.	.	.	.	.	.	.	.	19
1962	CRAMBE ABYSSINICA	S	.	0.0	25.0	48.1	.	.	4.2	.	.	.	.	.	.	.	.	.	.	19
1963	CRAMBE ABYSSINICA	S	.	0.0	9.8	12.7	.	.	8.7	.	.	.	.	.	.	.	.	.	.	19
1964	CRAMBE ABYSSINICA	S	.	0.0	25.0	33.8	.	.	.	.	.	.	.	.	.	.	.	.	.	20
1965	CRAMBE ABYSSINICA	S	.	0.0	29.9	40.5	.	.	.	.	.	.	.	.	.	.	.	.	.	21
1966	CRAMBE ABYSSINICA	S	.	0.0	27.9	46.3	.	.	.	.	.	.	.	.	.	.	.	.	.	21
1967	CRAMBE ABYSSINICA	S	.	0.0	23.0	36.8	.	.	.	.	.	.	.	.	.	.	.	.	.	21
1968	CRAMBE ABYSSINICA	S	.	0.0	30.4	43.3	.	.	.	.	.	.	.	.	.	.	.	.	.	21
1969	CRAMBE ABYSSINICA	S	.	0.0	22.0	31.1	.	.	.	.	.	.	.	.	.	.	.	.	.	21
1970	CRAMBE ABYSSINICA	S	.	0.0	30.2	43.1	.	.	.	.	.	.	.	.	.	.	.	.	.	21
1971	CRAMBE ABYSSINICA	S	.	0.0	22.2	30.7	.	.	.	.	.	.	.	.	.	.	.	.	.	21
1972	CRAMBE ABYSSINICA	S	.	0.0	30.7	43.3	.	.	.	.	.	.	.	.	.	.	.	.	.	21
1973	CRAMBF ABYSSINICA	S	.	0.0	22.0	30.8	.	.	.	.	.	.	.	.	.	.	.	.	.	21
1974	CRAMBE ABYSSINICA	S	.	0.0	30.4	43.9	.	.	.	.	.	.	.	.	.	.	.	.	.	21
1975	CRAMBE ABYSSINICA	S	.	0.0	23.3	30.7	.	.	.	.	.	.	.	.	.	.	.	.	.	21
1976	CRAMBE ABYSSINICA	S	.	0.0	31.1	42.5	.	.	.	.	.	.	.	.	.	.	.	.	.	21
1977	CRAMBE ABYSSINICA	S	.	0.0	22.6	30.3	.	.	.	.	.	.	.	.	.	.	.	.	.	21
1978	CRAMBE CORDIFOLIA	S	.	0.0	45.2	25.7	.	.	.	.	.	.	.	.	.	.	.	.	.	21
1979	CRAMBE CORDIFOLIA	S	.	0.0	29.4	16.4	.	.	.	.	.	.	.	.	.	.	.	.	.	21
1980	CRAMBE HISPANICA	S	.	0.0	29.2	44.8	.	.	.	.	.	.	.	.	.	.	.	.	.	21

TABLE 2 - DATA DIRECTLY TRANSCRIBED FROM ORIGINAL SOURCES.

OBS	PLNA	PART	CALORIES	WATER	PROTEIN	FAT	TOTLCARB	FIBER	ASH	CALCIUM	PHOSPHO	IRON	SODIUM	POTASS	CAROT	THIAMINE	RIBOFLAV	NIACIN	ASCORBIC	REFCO
2026	CROTALARIA INCANA	S	.	0.0	33.1	.	.	.	.	.	.	.	.	.	.	.	.	.	.	21
2027	CROTALARIA INCANA	S	.	0.0	28.6	.	.	.	.	.	.	.	.	.	.	.	.	.	.	21
2028	CROTALARIA INCANA	S	.	0.0	29.1	.	.	.	.	.	.	.	.	.	.	.	.	.	.	21
2029	CROTALARIA INTERMEDIA	S	.	0.0	30.5	0.4	.	.	3.8	.	.	.	.	.	.	.	.	.	.	19
2030	CROTALARIA INTERMEDIA	S	.	0.0	34.8	3.2	.	.	3.5	.	.	.	.	.	.	.	.	.	.	19
2031	CROTALARIA INTERMEDIA	S	.	0.0	30.6	2.7	.	.	.	.	.	.	.	.	.	.	.	.	.	19
2032	CROTALARIA INTERMEDIA	S	.	0.0	31.8	2.6	.	.	.	.	.	.	.	.	.	.	.	.	.	20
2033	CROTALARIA INTERMEDIA	S	.	0.0	33.5	3.2	.	.	.	.	.	.	.	.	.	.	.	.	.	20
2034	CROTALARIA INTERMEDIA	S	.	0.0	31.4	.	.	.	.	.	.	.	.	.	.	.	.	.	.	21
2035	CROTALARIA INTERMEDIA	S	.	0.0	31.8	.	.	.	.	.	.	.	.	.	.	.	.	.	.	21
2036	CROTALARIA INTERMEDIA	S	.	0.0	31.6	.	.	.	.	.	.	.	.	.	.	.	.	.	.	21
2037	CROTALARIA INTERMEDIA	W	.	0.0	14.5	2.1	78.3	40.3	5.1	1820	590	.	.	.	.	.	.	.	.	6
2038	CROTALARIA INTERMEDIA(BUD)	W	.	0.0	13.7	2.2	79.2	40.1	4.9	2930	810	.	.	.	.	.	.	.	.	6
2039	CROTALARIA INTERMEDIA(POD)	W	.	0.0	11.9	1.5	82.7	47.9	3.9	620	190	.	.	.	.	.	.	.	.	6
2040	CROTALARIA JUNCEA	W	.	0.0	15.0	5.8	71.0	30.0	8.2	2080	510	.	.	.	.	.	.	.	.	18
2041	CROTALARIA JUNCEA	H	.	0.0	12.5	1.0	81.0	45.7	5.5	.	.	.	.	.	.	.	.	.	.	18
2042	CROTALARIA JUNCEA	S	.	0.0	40.3	1.4	52.7	10.0	5.6	.	.	.	.	.	.	.	.	.	.	18
2043	CROTALARIA JUNCEA	S	.	0.0	41.1	3.5	.	.	4.6	.	.	.	.	.	.	.	.	.	.	19
2044	CROTALARIA JUNCEA	S	.	0.0	39.0	4.1	.	.	4.6	.	.	.	.	.	.	.	.	.	.	19
2045	CROTALARIA JUNCEA	S	.	0.0	39.8	3.7	.	.	4.7	.	.	.	.	.	.	.	.	.	.	19
2046	CROTALARIA JUNCEA	S	.	0.0	38.7	2.7	.	.	1.2	.	.	.	.	.	.	.	.	.	.	19
2047	CROTALARIA JUNCEA	S	.	0.0	37.8	3.8	.	.	4.5	.	.	.	.	.	.	.	.	.	.	19
2048	CROTALARIA JUNCEA	S	.	0.0	40.0	3.2	.	.	.	.	.	.	.	.	.	.	.	.	.	20
2049	CROTALARIA JUNCEA	S	.	0.0	24.7	.	.	.	.	.	.	.	.	.	.	.	.	.	.	21
2050	CROTALARIA JUNCEA	S	.	0.0	27.8	.	.	.	.	.	.	.	.	.	.	.	.	.	.	21
2051	CROTALARIA LABURNIFOLIA	S	.	0.0	31.4	3.8	.	.	.	.	.	.	.	.	.	.	.	.	.	21
2052	CROTALARIA LANCEOLATA	S	.	0.0	31.4	2.5	.	.	3.5	.	.	.	.	.	.	.	.	.	.	19
2053	CROTALARIA LANCEOLATA	S	.	0.0	30.9	.	.	.	.	.	.	.	.	.	.	.	.	.	.	21
2054	CROTALARIA LEIOLOBA	S	.	0.0	26.5	2.0	.	.	.	.	.	.	.	.	.	.	.	.	.	21
2055	CROTALARIA LONGIROSTRATA	S	.	0.0	33.9	3.4	.	.	.	.	.	.	.	.	.	.	.	.	.	21
2056	CROTALARIA LONGIROSTRATA	S	.	0.0	33.0	4.4	.	.	.	.	.	.	.	.	.	.	.	.	.	21
2057	CROTALARIA LONGIROSTRATA	L	56	81.6	7.0	0.8	9.1	2.0	1.5	287	72	4.7	.	.	6130	1.33	0.49	2	100	3
2058	CROTALARIA MACROCALYX	W	.	0.0	15.1	2.4	76.3	37.7	6.2	560	120	.	.	.	.	.	.	.	.	18
2059	CROTALARIA MEDICAGINEA	S	.	0.0	35.2	3.1	.	.	.	.	.	.	.	.	.	.	.	.	.	21
2060	CROTALARIA MUCRONATA	S	.	0.0	30.5	2.7	.	.	3.5	.	.	.	.	.	.	.	.	.	.	19
2061	CROTALARIA MUCRONATA	S	.	0.0	25.0	2.1	.	.	.	.	.	.	.	.	.	.	.	.	.	20
2062	CROTALARIA MUCRONATA	S	.	0.0	33.5	1.9	.	.	.	.	.	.	.	.	.	.	.	.	.	20
2063	CROTALARIA MUCRONATA	S	.	0.0	28.0	.	.	.	.	.	.	.	.	.	.	.	.	.	.	21
2064	CROTALARIA MUCRONATA	S	.	0.0	27.1	.	.	.	.	.	.	.	.	.	.	.	.	.	.	21
2065	CROTALARIA MUCRONATA	S	.	0.0	24.1	.	.	.	.	.	.	.	.	.	.	.	.	.	.	21
2066	CROTALARIA MUCRONATA	S	.	0.0	24.4	.	.	.	.	.	.	.	.	.	.	.	.	.	.	21
2067	CROTALARIA MUCRONATA	S	.	0.0	23.8	.	.	.	.	.	.	.	.	.	.	.	.	.	.	21
2068	CROTALARIA MUCRONATA	S	.	0.0	28.0	.	.	.	.	.	.	.	.	.	.	.	.	.	.	21
2069	CROTALARIA MUCRONATA	S	.	0.0	29.5	.	.	.	.	.	.	.	.	.	.	.	.	.	.	21
2070	CROTALARIA MUCRONATA	S	.	0.0	24.6	.	.	.	.	.	.	.	.	.	.	.	.	.	.	21

TABLE 2 - DATA DIRECTLY TRANSCRIBED FROM ORIGINAL SOURCES.

OBS	PLANA	PART	CALORIES	WATER	PROTEIN	FAT	TOTCARB	FIBER	ASH	CALCIUM	PHOSPHO	IRON	SODIUM	POTASS	CAROT	THIAMINE	RIBOFLAV	NIACIN	ASCORBIC	REFCO
1981	CRAMBE ORIENTALIS	S	.	0.0	31.0	43.1	.	.	.	.	.	.	.	.	.	.	.	.	.	20
1982	CRAMBE TATARICA	S	.	0.0	40.0	33.4	.	.	.	.	.	.	.	.	.	.	.	.	.	20
1983	CRAMBE TATARICA	S	.	0.0	40.8	33.3	.	.	.	.	.	.	.	.	.	.	.	.	.	21
1984	CRASSOCEPHALUM RUBENS	L	64	79.9	3.2	0.7	14.0	1.9	2.2	260	52	.	.	.	.	.	.	.	.	1
1985	CRATAEGUS AZAROLUS	S	.	0.0	3.2	0.6	.	.	.	.	.	.	.	.	.	.	.	.	.	21
1986	CRATAEGUS AZAROLUS	S	.	0.0	4.1	0.7	.	.	.	.	.	.	.	.	.	.	.	.	.	21
1987	CRATAEGUS CHRYSOCARPA	S	.	0.0	4.4	2.0	.	.	.	.	.	.	.	.	.	.	.	.	.	21
1988	CRATAEGUS CRUS-GALLI	S	.	0.0	41.9	35.4	.	.	.	.	.	.	.	.	.	.	.	.	.	20
1989	CRATAEGUS MOLLIS	S	.	0.0	46.9	31.3	.	.	5.1	.	.	.	.	.	.	.	.	.	.	19
1990	CRATAEGUS MOLLIS	S	.	0.0	1.9	0.8	.	.	0.9	.	.	.	.	.	.	.	.	.	.	19
1991	CRATAEGUS MONOGYNA	S	.	0.0	4.6	0.9	.	.	.	.	.	.	.	.	.	.	.	.	.	21
1992	CRATAEGUS MONOGYNA	S	.	0.0	5.1	2.4	.	.	.	.	.	.	.	.	.	.	.	.	.	21
1993	CRATAEGUS ORIENTALIS	S	.	0.0	5.1	3.7	.	.	.	.	.	.	.	.	.	.	.	.	.	21
1994	CRATAEGUS PINNATIFIDA	F	92	73.9	0.4	1.0	23.9	1.8	0.8	85	25	2.1	.	327	.	.	.	.	.	2
1995	CRATAEGUS PUBESCENS	F	89	74.8	0.7	0.3	23.5	2.4	0.7	82	28	1.6	.	.	1670	0.03	0.04	0.4	79	3
1996	CRATAEGUS SP.	S	.	0.0	44.2	39.2	.	.	.	.	.	.	.	.	.	.	.	.	.	21
1997	CRATEAGUS SP.	F	68	87.1	0.4	0.3	17.8	0.6	0.4	6	13	0.3	1	110	24	0.03	0.02	0.1	8	4
1998	CRATEVA RELIGIOSA	L	70	76.0	5.9	0.1	15.4	3.7	2.6	405	155	8.4	.	.	4960	0.39	0.71	1.4	192	1
1999	CRATEVA RELIGIOSA	L	93	73.4	3.4	1.3	20.6	4.9	1.3	124	20	5.3	.	.	3650	0.08	0.25	1.5	5	2
2000	CREPIS ACUMINATA	W	.	0.0	14.1	.	.	.	.	.	.	.	.	.	.	.	.	.	.	6
2001	CREPIS FOETIDA	S	.	0.0	24.3	23.9	.	.	.	.	.	.	.	.	.	.	.	.	.	21
2002	CREPIS OCCIDENTALIS	S	.	0.0	.	22.4	.	.	.	.	.	.	.	.	.	.	.	.	.	21
2003	CREPIS RUBRA	S	.	0.0	.	22.2	.	.	.	.	.	.	.	.	.	.	.	.	.	21
2004	CRESCENTIA ALATA	S	530	3.4	30.2	39.7	22.9	2.4	3.8	50	968	9.4	.	.	20	0.73	0.12	0.9	.	3
2005	CROCUS SATIVUS	I	310	11.9	11.4	5.8	65.4	3.9	5.4	111	252	11.1	148	1724	.	.	.	.	.	16
2006	CROTALARIA ALATA	S	.	0.0	28.2	2.8	.	.	.	.	.	.	.	.	.	.	.	.	.	21
2007	CROTALARIA ANAGYROIDES	SH	.	0.0	23.0	2.2	67.6	28.0	7.2	570	280	.	.	.	.	.	.	.	.	18
2008	CROTALARIA ANAGYROIDES	S	.	0.0	35.0	4.3	.	.	.	.	.	.	.	.	.	.	.	.	.	20
2009	CROTALARIA ANAGYROIDES	S	.	0.0	34.9	6.1	.	.	.	.	.	.	.	.	.	.	.	.	.	21
2010	CROTALARIA ARENARIA	SH	.	0.0	14.3	1.7	71.9	37.5	12.1	1230	220	.	.	.	.	.	.	.	.	18
2011	CROTALARIA AUSTRALIS	S	.	0.0	23.5	2.0	.	.	.	.	.	.	.	.	.	.	.	.	.	21
2012	CROTALARIA BREVIDENS	SH	.	0.0	28.9	3.3	57.9	22.1	9.9	440	270	.	.	.	.	.	.	.	.	18
2013	CROTALARIA BREVIDENS VAR. INTERMEDIA	L	.	74.5	8.8	.	.	.	1.6	222	.	0.8	.	.	.	.	.	.	.	1
2014	CROTALARIA CAPENSIS	S	.	0.0	27.4	8.0	.	.	.	.	.	.	.	.	.	.	.	.	.	21
2015	CROTALARIA CF. INCANA	S	.	0.0	31.9	4.5	.	.	.	.	.	.	.	.	.	.	.	.	.	20
2016	CROTALARIA CF. LONGIROSTRATA	S	.	0.0	31.2	1.7	.	.	.	.	.	.	.	.	.	.	.	.	.	20
2017	CROTALARIA ERIOCARPA	S	.	0.0	33.8	4.4	.	.	.	.	.	.	.	.	.	.	.	.	.	20
2018	CROTALARIA FALCATA	S	.	0.0	27.8	2.2	.	.	.	.	.	.	.	.	.	.	.	.	.	21
2019	CROTALARIA GRANTIANA	S	.	0.0	30.3	1.9	.	.	.	.	.	.	.	.	.	.	.	.	.	21
2020	CROTALARIA GRANTIANA	S	.	0.0	30.4	.	.	.	.	.	.	.	.	.	.	.	.	.	.	21
2021	CROTALARIA INCANA	SH	.	0.0	23.8	2.9	63.4	27.1	9.9	810	260	.	.	.	.	.	.	.	.	18
2022	CROTALARIA INCANA	S	.	0.0	30.2	3.0	.	.	3.3	.	.	.	.	.	.	.	.	.	.	19
2023	CROTALARIA INCANA	S	.	0.0	29.5	2.7	.	.	3.4	.	.	.	.	.	.	.	.	.	.	19
2024	CROTALARIA INCANA	S	.	0.0	33.5	3.0	.	.	3.7	.	.	.	.	.	.	.	.	.	.	19
2025	CROTALARIA INCANA	S	.	0.0	27.9	.	.	.	.	.	.	.	.	.	.	.	.	.	.	21

TABLE 2 - DATA DIRECTLY TRANSCRIBED FROM ORIGINAL SOURCES.

OBS	PLNA	PART	CALORIES	WATER	PROTEIN	FAT	TOTLCARB	FIBER	ASH	CALCIUM	PHOSPHO	IRON	SODIUM	POTASS	CAROT	THIAMINE	RIBOFLAV	NIACIN	ASCORBIC	REFCO
2071	CROTALARIA MUCRONATA	S	.	0.0	31.5	2.9	.	.	.	.	.	.	.	.	.	.	.	.	.	21
2072	CROTALARIA OCHROLEUCA	S	.	0.0	31.0	2.6	.	.	.	.	.	.	.	.	.	.	.	.	.	21
2073	CROTALARIA PUMILA	S	.	0.0	24.4	3.2	.	.	4.1	.	.	.	.	.	.	.	.	.	.	19
2074	CROTALARIA PUMILA	S	.	0.0	31.8	4.4	.	.	3.3	.	.	.	.	.	.	.	.	.	.	19
2075	CROTALARIA PUMILA	S	.	0.0	36.2	3.0	.	.	.	.	.	.	.	.	.	.	.	.	.	20
2076	CROTALARIA PUMILA	S	.	0.0	31.9	2.9	.	.	.	.	.	.	.	.	.	.	.	.	.	20
2077	CROTALARIA PUMILA	S	.	0.0	31.8	3.4	.	.	.	.	.	.	.	.	.	.	.	.	.	21
2078	CROTALARIA PURPUREA	S	.	0.0	31.4	3.8	.	.	.	.	.	.	.	.	.	.	.	.	.	21
2079	CROTALARIA QUINQUEFOLIA	S	.	0.0	24.2	2.6	.	.	.	.	.	.	.	.	.	.	.	.	.	20
2080	CROTALARIA RETUSA	S	.	0.0	27.5	.	.	.	.	.	.	.	.	.	.	.	.	.	.	20
2081	CROTALARIA RETUSA	S	.	0.0	21.6	.	.	.	.	.	.	.	.	.	.	.	.	.	.	21
2082	CROTALARIA RETUSA	S	.	0.0	24.3	.	.	.	.	.	.	.	.	.	.	.	.	.	.	21
2083	CROTALARIA RETUSA	S	.	0.0	22.2	.	.	.	.	.	.	.	.	.	.	.	.	.	.	21
2084	CROTALARIA RHODESIAE	S	.	0.0	29.4	2.2	.	.	.	.	.	.	.	.	.	.	.	.	.	21
2085	CROTALARIA SAGITTALIS	S	.	0.0	33.8	4.5	.	.	.	.	.	.	.	.	.	.	.	.	.	19
2086	CROTALARIA SALTIANA	S	.	0.0	29.0	2.1	.	.	.	.	.	.	.	.	.	.	.	.	.	21
2087	CROTALARIA SERICAE	S	.	0.0	29.2	2.3	.	.	.	.	.	.	.	.	.	.	.	.	.	21
2088	CROTALARIA SP.	S	.	0.0	36.8	2.0	.	.	3.6	.	.	.	.	.	.	.	.	.	.	19
2089	CROTALARIA SP.	S	.	0.0	26.0	.	.	.	.	.	.	.	.	.	.	.	.	.	.	21
2090	CROTALARIA SP.	S	.	0.0	33.2	.	.	.	.	.	.	.	.	.	.	.	.	.	.	21
2091	CROTALARIA SP.	S	.	0.0	32.1	.	.	.	.	.	.	.	.	.	.	.	.	.	.	21
2092	CROTALARIA SP.	S	.	0.0	28.5	3.2	.	.	.	.	.	.	.	.	.	.	.	.	.	21
2093	CROTALARIA SPECTABILIS	S	.	0.0	30.8	2.6	.	.	3.9	.	.	.	.	.	.	.	.	.	.	19
2094	CROTALARIA SPECTABILIS	S	.	0.0	29.4	.	.	.	.	.	.	.	.	.	.	.	.	.	.	20
2095	CROTALARIA SPECTABILIS	S	.	0.0	32.5	2.3	.	.	.	.	.	.	.	.	.	.	.	.	.	20
2096	CROTALARIA SPECTABILIS	S	.	0.0	28.1	2.6	.	.	.	.	.	.	.	.	.	.	.	.	.	20
2097	CROTALARIA SPECTABILIS	S	.	0.0	30.2	3.4	.	.	.	.	.	.	.	.	.	.	.	.	.	20
2098	CROTALARIA SPECTABILIS	S	.	0.0	28.6	2.7	.	.	.	.	.	.	.	.	.	.	.	.	.	20
2099	CROTALARIA USARAMOENSIS	S	.	0.0	30.2	2.2	.	.	.	.	.	.	.	.	.	.	.	.	.	21
2100	CROTALARIA USARAMOENSIS	S	.	0.0	32.9	.	.	.	.	.	.	.	.	.	.	.	.	.	.	21
2101	CROTALARIA USARAMOENSIS	S	.	0.0	29.8	.	.	.	.	.	.	.	.	.	.	.	.	.	.	21
2102	CROTALARIA USARAMOENSIS	S	.	0.0	32.3	2.8	.	.	.	.	.	.	.	.	.	.	.	.	.	21
2103	CROTALARIA VERRUCOSA	S	.	0.0	33.3	3.0	.	.	.	.	.	.	.	.	.	.	.	.	.	21
2104	CROTALARIA VERRUCOSA	S	.	0.0	25.4	.	.	.	.	.	.	.	.	.	.	.	.	.	.	21
2105	CROTALARIA VITELLINA	S	.	0.0	32.5	3.7	.	.	.	.	.	.	.	.	.	.	.	.	.	21
2106	CROTALARIA VITELLINA	S	.	0.0	30.6	5.0	.	.	.	.	.	.	.	.	.	.	.	.	.	21
2107	CROTON CAPITATUS	S	.	0.0	20.9	14.8	.	.	2.9	.	.	.	.	.	.	.	.	.	.	19
2108	CROTON CORYMBULOSUS	S	.	0.0	22.9	23.6	.	.	.	.	.	.	.	.	.	.	.	.	.	21
2109	CROTON DICHOGAMUS	W	.	0.0	25.1	2.0	62.4	20.1	10.5	1860	490	.	.	.	.	.	.	.	.	18
2110	CROTON FRAGILIS	S	.	0.0	26.9	25.4	.	.	.	.	.	.	.	.	.	.	.	.	.	20
2111	CROTON SP.	W	.	0.0	16.0	4.4	72.2	23.0	7.4	2120	160	.	.	1510	.	.	.	.	.	6
2112	CROTON SP.(FRUITING)	W	.	0.0	16.3	4.2	72.8	23.8	6.7	2290	180	.	.	1370	.	.	.	.	.	6
2113	CROTON SP.(IMMATURE)	W	.	0.0	19.0	3.5	69.6	20.5	7.9	1890	160	.	.	1750	.	.	.	.	.	6
2114	CROTON TEXENSIS	S	.	0.0	24.9	26.4	.	.	3.2	.	.	.	.	.	.	.	.	.	.	19
2115	CRUCIANELLA SP.	S	.	0.0	19.7	10.0	.	.	.	.	.	.	.	.	.	.	.	.	.	21

TABLE 2 - DATA DIRECTLY TRANSCRIBED FROM ORIGINAL SOURCES.

OBS	PLNA	PART	CALORIES	WATER	PROTEIN	FAT	TOTCARB	FIBER	ASH	CALCIUM	PHOSPHO	IRON	SODIUM	POTASS	CAROT	THIAMINE	RIBOFLAV	NIACIN	ASCORBIC	REFCO
2116	CRYPTANTHA BRADBURIANA	S	.	0.0	21.1	27.2	.	.	.	.	.	.	.	.	.	.	.	.	.	21
2117	CRYPTANTHA SP.	W	.	0.0	7.5	.	.	27.7	13.3	2060	300	.	.	2520	.	.	.	.	.	6
2118	CRYPTOMERIA JAPONICA	S	.	0.0	10.6	10.6	.	.	2.3	.	.	.	.	.	.	.	.	.	.	19
2119	CRYPTOSTEMMA CALENDULACEUM	S	.	0.0	15.3	24.0	.	.	.	.	.	.	.	.	.	.	.	.	.	21
2120	CRYPTOTAENIA JAPONICA	L	18	93.5	2.0	0.1	3.4	1.3	1.0	81	45	1.8	7	490	780	0.15	0.20	0.5	60	2
2121	CUCUMEROPSIS MANNII	S	523	5.8	15.9	37.9	37.1	15.8	3.3	141	678	.	.	.	.	.	.	.	.	1
2122	CUCUMIS ANGURIA	S	.	0.0	25.8	25.3	.	.	.	.	.	.	.	.	.	.	.	.	.	21
2123	CUCUMIS DIPSACEUS	L	35	89.9	1.0	0.7	7.6	0.0	0.8	14	25	1.0	.	.	.	0.03	0.05	.	21	1
2124	CUCUMIS MELO	F	18	94.8	0.5	0.1	4.2	0.5	0.4	20	28	1.7	.	.	30	0.04	0.02	0.6	12	1
2125	CUCUMIS MELO	L	41	85.0	4.2	0.2	8.1	.	2.5	.	.	.	.	.	.	.	.	.	.	1
2126	CUCUMIS MELO	SH	.	0.0	27.4	2.8	65.3	16.9	4.5	2980	680	.	.	.	.	.	.	.	.	18
2127	CUCUMIS MELO	F	.	0.0	21.2	3.0	66.8	21.2	9.0	450	300	.	.	.	.	.	.	.	.	18
2128	CUCUMIS MELO	F	13	95.6	1.0	0.1	2.8	0.6	0.5	30	14	0.4	20	.	0	0.02	0.02	0.2	14	2
2129	CUCUMIS MELO	F	24	93.2	0.3	0.2	5.8	0.4	0.5	8	19	0.3	.	30	1705	0.04	0.02	0.7	38	2
2130	CUCUMIS MELO	F	26	92.4	1.0	0.3	5.6	0.5	0.7	18	14	0.5	14	207	0	0.05	0.02	0.4	25	2
2131	CUCUMIS MELO	F	26	92.4	1.0	0.3	5.6	0.5	0.7	13	14	0.5	14	207	90	0.05	0.02	0.4	25	2
2132	CUCUMIS MELO	F	25	92.8	0.5	0.1	6.2	0.5	0.4	15	15	1.2	.	.	700	0.04	0.03	0.6	29	3
2133	CUCUMIS MELO	F	44	87.4	0.6	0.3	11.1	0.4	0.6	32	32	0.7	.	.	0	0.03	0.02	0.6	23	3
2134	CUCUMIS MELO	F	30	91.2	0.7	0.1	7.5	0.3	0.5	14	16	0.4	12	251	2040	0.04	0.03	0.6	33	4
2135	CUCUMIS MELO	F	27	91.5	1.2	0.0	6.5	0.5	0.8	14	16	0.4	12	251	18	0.04	0.03	0.6	13	4
2136	CUCUMIS MELO	F	33	90.6	0.8	0.3	7.7	0.6	0.6	14	16	0.4	12	251	24	0.04	0.03	0.6	23	4
2137	CUCUMIS MELO	S	.	0.0	35.8	44.6	.	.	5.6	.	.	.	.	.	.	.	.	.	.	5
2138	CUCUMIS MELO	F	.	.	0.6	0.2	5.9	0.5	0.6	16	15	3.9	.	.	1440	0.06	0.07	.	25	5
2139	CUCUMIS SATIVUS	F	15	95.1	0.8	0.1	3.4	0.8	0.6	13	30	0.5	.	.	0	0.02	0.01	0.3	14	1
2140	CUCUMIS SATIVUS	F	12	96.2	0.6	0.1	2.7	0.5	0.4	21	24	0.4	13	154	85	0.03	0.04	0.2	11	2
2141	CUCUMIS SATIVUS	S	.	0.0	27.5	33.0	.	.	.	.	.	.	.	.	.	.	.	.	.	20
2142	CUCUMIS SATIVUS	S	.	0.0	26.9	34.0	.	.	.	.	.	.	.	.	.	.	.	.	.	20
2143	CUCUMIS SATIVUS	S	.	0.0	26.2	27.2	.	.	.	.	.	.	.	.	.	.	.	.	.	20
2144	CUCUMIS SATIVUS	S	.	0.0	26.9	33.5	.	.	.	.	.	.	.	.	.	.	.	.	.	20
2145	CUCUMIS SATIVUS	S	.	0.0	31.2	33.0	.	.	.	.	.	.	.	.	.	.	.	.	.	20
2146	CUCUMIS SATIVUS	F	15	95.4	0.7	0.1	3.4	0.4	0.4	16	24	0.6	.	.	10	0.03	0.04	0.2	14	3
2147	CUCUMIS SATIVUS	F	15	95.1	0.9	0.1	3.4	0.6	0.5	25	27	1.1	6	160	150	0.03	0.04	0.2	11	4
2148	CUCUMIS SATIVUS	S	.	0.0	42.0	42.5	.	.	.	.	620	.	.	.	.	.	.	.	.	5
2149	CUCURBITA DIGITATA	S	.	0.0	27.5	22.8	.	.	4.3	.	.	.	.	.	.	.	.	.	.	19
2150	CUCURBITA DIGITATA	S	.	0.0	36.9	48.2	.	.	.	.	.	.	.	.	.	.	.	.	.	19
2151	CUCURBITA DIGITATA	S	.	0.0	22.5	27.5	.	.	.	.	.	.	.	.	.	.	.	.	.	19
2152	CUCURBITA FICIFOLIA	S	.	0.0	29.4	19.5	.	.	.	.	.	.	.	.	.	.	.	.	.	19
2153	CUCURBITA FICIFOLIA	G	24	92.7	0.8	0.1	6.0	0.3	0.4	12	41	0.2	.	.	0	0.06	0.04	0.7	46	3
2154	CUCURBITA FICIFOLIA	F	21	93.6	0.8	0.1	5.1	0.4	0.4	15	19	0.4	.	.	20	0.04	0.03	0.3	11	3
2155	CUCURBITA FOETIDISSIMA	S	.	0.0	32.5	33.9	.	.	.	.	.	.	.	.	.	.	.	.	.	19
2156	CUCURBITA FOETIDISSIMA	S	.	0.0	33.8	31.1	.	.	.	.	.	.	.	.	.	.	.	.	.	20
2157	CUCURBITA FOETIDISSIMA	L	.	0.0	25.3	4.0	49.2	10.2	21.5	7760	260	.	.	2530	.	.	.	.	.	6
2158	CUCURBITA LUNDELLIANA	S	.	0.0	28.8	23.0	.	.	.	.	.	.	.	.	.	.	.	.	.	20
2159	CUCURBITA MAXIMA	L	27	89.2	4.0	0.2	4.4	2.4	2.2	477	136	0.8	.	.	3600	0.06	0.32	.	80	1
2160	CUCURBITA MAXIMA	S	.	0.0	35.6	51.4	.	.	.	.	.	.	.	.	.	.	.	.	.	19

TABLE 2 - DATA DIRECTLY TRANSCRIBED FROM ORIGINAL SOURCES.

OBS	PLNA	PART	CALORIES	WATER	PROTEIN	FAT	TOTLCARB	FIBER	ASH	CALCIUM	PHOSPHO	IRON	SODIUM	POTASS	CAROT	THIAMINE	RIBOFLAV	NIACIN	ASCORBIC	REFCO
2161	CUCURBITA MAXIMA	S	.	0.0	39.4	50.7	.	.	.	.	.	.	.	.	.	.	.	.	.	19
2162	CUCURBITA MAXIMA	S	.	0.0	41.9	43.6	.	.	.	.	.	.	.	.	.	.	.	.	.	19
2163	CUCURBITA MAXIMA	S	.	0.0	38.8	51.0	.	.	.	.	.	.	.	.	.	.	.	.	.	19
2164	CUCURBITA MAXIMA	S	.	0.0	36.2	51.5	.	.	.	.	.	.	.	.	.	.	.	.	.	19
2165	CUCURBITA MAXIMA	S	.	0.0	38.1	51.9	.	.	.	.	.	.	.	.	.	.	.	.	.	19
2166	CUCURBITA MAXIMA	S	.	0.0	40.0	50.3	.	.	.	.	.	.	.	.	.	.	.	.	.	19
2167	CUCURBITA MAXIMA	L	35	89.0	2.8	0.9	5.8	1.4	1.5	96	82	1.7	6	454	1460	0.13	0.18	1.10	17	2
2168	CUCURBITA MAXIMA	F	50	85.2	1.4	0.2	12.5	0.8	0.7	27	43	0.6	7	351	1475	0.09	0.06	1.00	14	2
2169	CUCURBITA MAXIMA	L	15	94.8	1.3	0.2	2.9	0.6	0.8	88	40	3.1	5	240	140	0.05	0.08	0.80	27	2
2170	CUCURBITA MAXIMA	G	24	92.8	1.0	0.2	5.5	0.4	0.5	19	32	0.6	.	.	30	0.05	0.04	0.50	19	3
2171	CUCURBITA MAXIMA	F	35	89.1	1.7	0.2	8.1	0.9	0.9	32	24	2.3	.	.	2290	0.07	0.05	0.80	11	3
2172	CUCURBITA MAXIMA	S	547	4.9	30.3	45.8	14.7	2.2	4.6	38	1064	9.2	.	.	30	0.23	0.16	2.90	0	3
2173	CUCURBITA MAXIMA	F	50	85.1	1.4	0.3	12.4	1.4	0.8	22	38	0.6	1	369	2200	0.05	0.11	0.60	13	4
2174	CUCURBITA MAXIMA	F	.	92.6	1.4	0.1	5.3	.	0.6	10	30	0.7	.	.	50	.	.	.	2	5
2175	CUCURBITA MOSCHATA	S	.	0.0	40.0	47.7	.	.	.	.	.	.	.	.	.	.	.	.	.	19
2176	CUCURBITA MOSCHATA	S	.	0.0	40.0	48.7	.	.	.	.	.	.	.	.	.	.	.	.	.	19
2177	CUCURBITA MOSCHATA	S	.	0.0	41.2	48.1	.	.	.	.	.	.	.	.	.	.	.	.	.	19
2178	CUCURBITA MOSCHATA	S	.	0.0	36.9	49.4	.	.	.	.	.	.	.	.	.	.	.	.	.	19
2179	CUCURBITA MOSCHATA	S	.	0.0	38.1	52.9	.	.	.	.	.	.	.	.	.	.	.	.	.	19
2180	CUCURBITA MOSCHATA	S	.	0.0	36.9	51.1	.	.	.	.	.	.	.	.	.	.	.	.	.	19
2181	CUCURBITA MOSCHATA	F	28	91.4	1.1	0.2	6.6	1.2	0.8	33	48	1.7	.	.	310	0.11	0.09	0.70	24	2
2182	CUCURBITA MOSCHATA	L	30	89.7	3.6	0.6	4.5	.	1.6	138	99	3.7	.	.	1650	0.14	.	.	36	2
2183	CUCURBITA MOSCHATA	F	40	88.3	1.2	0.3	9.8	0.6	0.4	12	27	0.7	.	.	2110	0.05	0.04	0.60	42	3
2184	CUCURBITA MOSCHATA	S	465	7.7	30.5	27.9	30.8	4.9	3.1	35	725	1.1	.	.	0	0.13	0.12	2.20	0	3
2185	CUCURBITA OKEECHOBEENSIS	S	.	0.0	38.1	52.6	.	.	.	.	.	.	.	.	.	.	.	.	.	19
2186	CUCURBITA PALMERI	S	.	0.0	32.4	39.4	.	.	.	.	.	.	.	.	.	.	.	.	.	21
2187	CUCURBITA PALMERI	S	.	0.0	30.6	43.5	.	.	.	.	.	.	.	.	.	.	.	.	.	21
2188	CUCURBITA PEPO	G	22	92.4	1.5	0.1	5.0	1.0	1.0	24	38	0.4	.	.	.	0.00	0.17	0.80	.	1
2189	CUCURBITA PEPO	F	23	92.6	1.0	0.1	5.5	0.8	0.8	25	32	1.4	.	.	3565	0.05	0.02	0.50	8	1
2190	CUCURBITA PEPO	L	27	89.2	4.0	0.2	4.4	2.4	2.2	477	136	0.8	.	.	3600	.	0.06	0.32	80	1
2191	CUCURBITA PEPO	S	.	0.0	38.1	48.1	.	.	.	.	.	.	.	.	.	.	.	.	.	19
2192	CUCURBITA PEPO	S	.	0.0	40.6	48.3	.	.	.	.	.	.	.	.	.	.	.	.	.	19
2193	CUCURBITA PEPO	S	.	0.0	39.4	49.8	.	.	.	.	.	.	.	.	.	.	.	.	.	19
2194	CUCURBITA PEPO	S	.	0.0	38.8	50.7	.	.	.	.	.	.	.	.	.	.	.	.	.	19
2195	CUCURBITA PEPO	S	.	0.0	36.9	48.7	.	.	.	.	.	.	.	.	.	.	.	.	.	19
2196	CUCURBITA PEPO	S	.	0.0	39.3	46.2	.	.	4.8	.	.	.	.	.	.	.	.	.	.	19
2197	CUCURBITA PEPO	S	.	0.0	38.1	50.4	.	.	.	.	.	.	.	.	.	.	.	.	.	19
2198	CUCURBITA PEPO	S	.	0.0	40.0	48.7	.	.	5.2	.	.	.	.	.	.	.	.	.	.	19
2199	CUCURBITA PEPO	S	.	0.0	27.5	5.1	.	.	1.6	.	.	.	.	.	.	.	.	.	.	19
2200	CUCURBITA PEPO	L	21	92.6	3.0	0.4	3.0	.	1.0	37	99	2.1	6	472	1940	0.09	0.13	0.90	11	2
2201	CUCURBITA PEPO	S	542	3.7	29.4	40.4	25.1	2.0	1.4	33	714	9.9	444	293	235	0.40	0.14	3.20	0	2
2202	CUCURBITA PEPO	F	27	91.9	0.7	0.2	6.6	0.8	0.6	24	33	0.7	8	350	785	0.03	0.04	0.50	14	2
2203	CUCURBITA PEPO	I	16	94.8	1.4	0.3	2.7	0.6	0.8	47	86	1.0	.	.	400	0.02	0.11	0.60	18	3
2204	CUCURBITA PEPO	L	26	90.4	4.2	0.4	3.4	1.5	1.6	127	96	5.8	.	.	1630	0.14	0.17	1.80	58	3
2205	CUCURBITA PEPO	G	24	92.8	1.0	0.2	5.5	0.4	0.5	19	32	0.6	.	.	30	0.05	0.04	0.50	19	3

TABLE 2 - DATA DIRECTLY TRANSCRIBED FROM ORIGINAL SOURCES.

OBS	PLANA	PART	CALORIES	WATER	PROTEIN	FAT	TOTLCARB	FIBER	ASH	CALCIUM	PHOSPHON	IRON	SODIUM	POTASS	CAROT	THIAMINE	RIBOFLAV	NIACIN	ASCORBIC	REFCO
2206	CUCURBITA PEPO	F	30	91.0	0.6	0.2	7.6	0.7	0.6	19	22	0.5	.	.	1840	0.04	0.04	0.50	15	3
2207	CUCURBITA PEPO	F	16	94.8	1.4	0.3	2.7	0.6	0.8	47	86	1.0	.	.	400	0.02	0.11	0.60	18	3
2208	CUCURBITA PEPO	L	26	90.4	4.2	0.4	3.4	1.5	1.6	127	96	5.8	.	.	1630	0.14	0.17	1.80	58	3
2209	CUCURBITA PEPO	F	15	95.2	0.8	0.1	3.5	0.4	0.4	18	14	2.3	.	.	0	0.03	0.03	0.40	17	3
2210	CUCURBITA PEPO	F	19	94.0	1.1	0.1	4.2	0.6	0.6	28	29	0.4	1	202	246	0.05	0.09	1.00	22	4
2211	CUCURBITA PEPO	F	26	91.6	1.0	0.1	6.5	1.1	0.8	21	44	0.8	1	340	960	0.05	0.11	0.60	9	4
2212	CUCURBITA PEPO	S	553	4.4	29.0	46.7	15.0	1.9	4.9	51	1144	11.2	.	.	42	0.24	0.19	2.40	.	4
2213	CUCURBITA PEPO	S	.	.	30.3	38.4	27.3	18.1	3.4	.	.	.	.	.	.	.	.	.	.	5
2214	CUMINUM CYMINUM	S	375	8.1	17.8	22.3	44.2	10.5	7.6	931	499	66.4	168	1788	762	0.63	0.33	4.58	7	16
2215	CUMINUM CYMINUM	S	332	12.0	16.6	13.4	50.2	.	7.8	1365	.	24.3	.	.	520	.	.	.	0	2
2216	CUMINUM CYMINUM	S	.	11.9	18.7	15.0	48.6	12.0	5.8	1080	490	31.0	.	.	522	.	.	.	3	5
2217	CUNNINGHAMIA LANCEOLATA	S	.	0.0	4.4	1.6	.	.	.	.	.	.	.	.	.	.	.	.	.	20
2218	CUPHEA CARTHAGENENSIS	S	.	0.0	16.2	32.7	.	.	.	.	.	.	.	.	.	.	.	.	.	21
2219	CUPHEA HOOKERIANA	S	.	0.0	18.6	15.8	.	.	.	.	.	.	.	.	.	.	.	.	.	21
2220	CUPHEA IGNEA	S	.	0.0	18.1	33.6	.	.	.	.	.	.	.	.	.	.	.	.	.	20
2221	CUPHEA LLAVEA	S	.	0.0	21.2	30.0	.	.	.	.	.	.	.	.	.	.	.	.	.	20
2222	CUPHEA LLAVEA	S	.	0.0	15.2	17.0	.	.	.	.	.	.	.	.	.	.	.	.	.	21
2223	CUPHEA LLAVEA	S	.	0.0	15.2	22.8	.	.	.	.	.	.	.	.	.	.	.	.	.	21
2224	CUPHEA LLAVEA VAR. MINIATA	S	.	0.0	16.9	20.9	.	.	6.8	.	.	.	.	.	.	.	.	.	.	19
2225	CUPHEA PAINTERI	S	.	0.0	23.4	35.8	.	.	.	.	.	.	.	.	.	.	.	.	.	21
2226	CUPRESSUS ARIZONICA	S	.	0.0	4.4	4.8	.	.	.	.	.	.	.	.	.	.	.	.	.	20
2227	CUPRESSUS SEMPERVIRENS	S	.	0.0	9.8	9.4	.	.	.	.	.	.	.	.	.	.	.	.	.	21
2228	CURCUMA AROMATICA	R	.	13.3	30.6	12.1	31.9	8.4	4.5	.	.	.	.	.	.	.	.	.	.	5
2229	CURCUMA DOMESTICA	R	354	11.4	7.8	9.9	64.9	6.7	6.0	182	268	41.4	38	2525	0	0.15	0.23	5.14	26	16
2230	CURCUMA DOMESTICA	R	46	88.2	1.2	1.8	7.9	0.9	0.9	27	64	2.3	3	487	0	0.02	0.03	0.50	0	2
2231	CURCUMA DOMESTICA	R	53	82.4	4.1	0.0	10.9	1.2	2.6	51	111	.	.	.	0	0.15	0.21	1.10	5	3
2232	CURCUMA DOMESTICA	R	.	13.1	6.3	5.1	72.0	2.6	3.5	.	.	.	.	30	.	.	.	.	.	5
2233	CUSCUTA GRONOVII	S	.	0.0	13.8	4.7	.	.	.	.	.	.	.	.	.	.	.	.	.	19
2234	CUSCUTA SP.	W	.	0.0	12.0	3.4	82.1	19.6	2.5	.	.	.	.	.	.	.	.	.	.	6
2235	CUSCUTA SP.	Y	.	0.0	7.6	1.7	87.8	24.4	2.9	110	130	.	.	1240	.	.	.	.	.	6
2236	CYAMOPSIS TETRAGONOLOBA	C	.	0.0	6.8	1.0	87.7	48.0	4.5	750	120	.	.	.	.	.	.	.	.	18
2237	CYAMOPSIS TETRAGONOLOBA	L	.	0.0	22.5	3.5	59.6	9.7	14.4	.	.	.	.	.	.	.	.	.	.	18
2238	CYAMOPSIS TETRAGONOLOBA	H	.	0.0	25.2	0.9	57.4	13.8	16.5	2420	270	.	.	.	.	.	.	.	.	18
2239	CYAMOPSIS TETRAGONOLOBA	SH	.	0.0	16.0	1.9	65.1	22.7	17.0	1800	100	.	.	.	.	.	.	.	.	18
2240	CYAMOPSIS TETRAGONOLOBA	S	.	0.0	28.0	3.7	64.3	5.0	4.0	.	420	.	.	.	.	.	.	.	.	18
2241	CYAMOPSIS TETRAGONOLOBA	S	.	0.0	32.1	3.1	.	.	4.2	.	.	.	.	.	.	.	.	.	.	19
2242	CYAMOPSIS TETRAGONOLOBA	S	.	0.0	36.2	3.3	.	.	4.3	.	.	.	.	.	.	.	.	.	.	19
2243	CYAMOPSIS TETRAGONOLOBA	S	.	0.0	35.4	2.9	.	.	3.8	.	.	.	.	.	.	.	.	.	.	19
2244	CYAMOPSIS TETRAGONOLOBA	S	.	0.0	36.1	2.8	.	.	3.8	.	.	.	.	.	.	.	.	.	.	19
2245	CYAMOPSIS TETRAGONOLOBA	S	.	0.0	35.5	3.0	.	.	4.0	.	.	.	.	.	.	.	.	.	.	19
2246	CYAMOPSIS TETRAGONOLOBA	S	.	0.0	35.1	2.7	.	.	3.8	.	.	.	.	.	.	.	.	.	.	19
2247	CYAMOPSIS TETRAGONOLOBA	S	.	0.0	34.9	2.8	.	.	3.9	.	.	.	.	.	.	.	.	.	.	19
2248	CYAMOPSIS TETRAGONOLOBA	S	.	0.0	28.4	2.8	.	.	.	.	.	.	.	.	.	.	.	.	.	19
2249	CYAMOPSIS TETRAGONOLOBA	S	.	0.0	29.4	2.7	.	.	3.6	.	.	.	.	.	.	.	.	.	.	19
2250	CYAMOPSIS TETRAGONOLOBA	S	.	0.0	25.1	2.9	.	.	3.9	.	.	.	.	.	.	.	.	.	.	19

TABLE 2 - DATA DIRECTLY TRANSCRIBED FROM ORIGINAL SOURCES.

OBS	PLNA	PART	CALORIES	WATER	PROTEIN	FAT	TOTCARB	FIBER	ASH	CALCIUM	PHOSPHO	IRON	SODIUM	POTASS	CAROT	THIAMINE	RIBOFLAV	NIACIN	ASCORBIC	REF
2251	CYAMOPSIS TETRAGONOLOBA	S	.	0.0	24.2	3.7	.	.	3.3	.	.	.	.	.	.	.	.	.	.	19
2252	CYAMOPSIS TETRAGONOLOBA	S	.	0.0	31.2	3.5	.	.	3.4	.	.	.	.	.	.	.	.	.	.	19
2253	CYAMOPSIS TETRAGONOLOBA	S	.	0.0	34.4	2.9	.	.	3.5	.	.	.	.	.	.	.	.	.	.	19
2254	CYAMOPSIS TETRAGONOLOBA	S	.	0.0	33.8	3.2	.	.	.	.	.	.	.	.	.	.	.	.	.	20
2255	CYAMOPSIS TETRAGONOLOBA	G	.	82.5	3.7	0.2	12.2	2.3	1.4	130	50	5.8	.	.	198	.	.	.	49	5
2256	CYAMOPSIS TETRAGONOLOBA	L	.	.	25.2	0.9	57.4	13.8	16.5	.	.	.	.	.	.	.	.	.	.	5
2257	CYAMOPSIS TETRAGONOLOBA	W	.	0.0	14.6	2.5	70.7	22.8	12.2	.	.	.	.	.	.	.	.	.	.	6
2258	CYAMOPSIS TETRAGONOLOBA (FRUITING)	W	.	0.0	11.9	2.4	78.2	25.8	7.5	1980	160	.	.	1910	.	.	.	.	.	6
2259	CYCAS REVOLUTA	S	.	0.0	15.4	0.9	.	.	.	.	.	.	.	.	.	.	.	.	.	21
2260	CYCAS REVOLUTA	S	.	0.0	13.9	1.0	.	.	.	.	.	.	.	.	.	.	.	.	.	21
2261	CYCLANTHERA PEDATA	F	17	94.5	0.6	0.1	4.0	0.7	0.8	13	26	0.6	.	.	30	0.04	0.04	0.3	14	3
2262	CYCLOPIA SUBTERNATA	S	.	0.0	24.8	4.6	.	.	.	.	.	.	.	.	.	.	.	.	.	21
2263	CYDONIA OBLONGA	F	62	82.6	0.4	0.3	16.2	3.1	0.5	14	15	0.9	.	.	.	.	.	.	.	1
2264	CYDONIA OBLONGA	F	63	82.4	0.6	0.3	16.3	2.2	0.4	6	15	0.6	.	.	20	0.03	0.03	0.4	17	3
2265	CYDONIA OBLONGA	F	57	83.8	0.4	0.1	15.3	1.7	0.4	11	17	0.7	4	197	24	0.02	0.03	0.2	15	4
2266	CYDONIA OBLONGA	F	.	85.7	0.3	0.1	13.6	1.7	0.3	.	.	.	.	.	.	.	.	.	10	5
2267	CYMBOPOGON CITRATUS	L	92	74.3	1.0	1.4	21.9	4.2	1.4	32	30	1.8	.	.	425	0.05	0.02	2.2	1	2
2268	CYMBOPOGON GIGANTEUS	W	.	0.0	5.0	4.1	84.4	33.2	6.5	.	.	.	.	.	.	.	.	.	.	18
2269	CYNANCHUM ACUTUM	S	.	0.0	28.0	24.1	.	.	.	.	.	.	.	.	.	.	.	.	.	21
2270	CYNANCHUM ACUTUM	S	.	0.0	16.0	11.9	.	.	.	.	.	.	.	.	.	.	.	.	.	21
2271	CYNARA ALBA	S	.	0.0	27.4	17.1	.	.	.	.	.	.	.	.	.	.	.	.	.	21
2272	CYNARA CARDUNCULUS	SH	.	0.0	16.9	6.2	62.3	26.9	14.6	1450	330	.	.	.	.	.	.	.	.	18
2273	CYNARA CARDUNCULUS	S	.	0.0	27.1	42.2	.	.	5.3	.	.	.	.	.	.	.	.	.	.	19
2274	CYNARA CARDUNCULUS	S	.	0.0	7.4	4.6	.	.	2.5	.	.	.	.	.	.	.	.	.	.	19
2275	CYNARA CARDUNCULUS	S	.	0.0	27.5	43.0	.	.	.	.	.	.	.	.	.	.	.	.	.	20
2276	CYNARA CARDUNCULUS	S	.	0.0	21.4	21.5	.	.	.	.	.	.	.	.	.	.	.	.	.	21
2277	CYNARA CARDUNCULUS	SH	18	93.8	0.5	0.2	4.1	1.0	1.4	114	10	1.5	.	.	0	0.01	0.03	0.2	1	3
2278	CYNARA HUMILIS	S	.	0.0	21.9	21.0	.	.	.	.	.	.	.	.	.	.	.	.	.	21
2279	CYNARA SCOLYMUS	I	42	85.9	3.4	0.2	9.0	0.8	1.5	64	79	0.2	.	.	.	.	.	.	.	1
2280	CYNARA SCOLYMUS	I	73	77.4	4.5	0.3	16.6	3.8	1.2	80	138	2.0	67	673	150	0.12	0.08	1.6	0	2
2281	CYNARA SCOLYMUS	S	.	0.0	21.9	17.1	.	.	.	.	.	.	.	.	.	.	.	.	.	20
2282	CYNARA SCOLYMUS	I	29	90.2	2.7	0.2	5.9	2.2	1.0	44	58	0.8	.	.	190	0.06	0.07	0.8	5	3
2283	CYNARA SCOLYMUS	I	28	85.5	2.9	0.2	10.6	2.4	0.8	51	88	1.3	43	430	96	0.08	0.05	1.0	12	4
2284	CYNARA SCOLYMUS	I	.	77.3	3.6	0.1	17.2	1.2	1.8	120	100	2.3	.	.	36	.	.	.	0	5
2285	CYNARA SCOLYMUS	W	.	0.0	5.1	1.1	86.1	18.0	7.7	.	.	.	.	.	.	.	.	.	.	6
2286	CYNODON AETHIOPICUS	W	.	0.0	12.2	1.7	74.2	26.6	11.9	.	.	.	.	.	.	.	.	.	.	18
2287	CYNODON AETHIOPICUS	H	.	0.0	6.4	0.6	86.2	35.6	6.8	.	.	.	.	.	.	.	.	.	.	18
2288	CYNODON DACTYLON	W	.	0.0	14.2	1.9	71.5	26.6	12.4	.	.	.	.	.	.	.	.	.	.	18
2289	CYNODON DACTYLON	H	.	0.0	11.1	1.4	75.0	18.4	12.5	.	.	.	.	.	.	.	.	.	.	18
2290	CYNODON DACTYLON	L	.	0.0	10.5	1.8	76.0	28.2	11.8	.	.	.	.	.	.	.	.	.	.	5
2291	CYNODON DACTYLON	W	.	0.0	11.6	2.1	75.9	25.9	10.4	530	220	112.0	.	1630	28	.	.	.	.	6
2292	CYNODON PLECTOSTACHYUS	W	.	0.0	19.1	2.9	65.0	22.6	13.0	.	.	.	.	.	.	.	.	.	.	18
2293	CYNODON PLECTOSTACHYUS	H	.	0.0	5.2	0.9	85.8	39.7	8.1	.	.	.	.	.	.	.	.	.	.	18
2294	CYNOGLOSSUM AMABILE	S	.	0.0	17.6	22.6	.	.	11.6	.	.	.	.	.	.	.	.	.	.	19
2295	CYNOGLOSSUM CRETICUM	S	.	0.0	24.8	42.2	.	.	.	.	.	.	.	.	.	.	.	.	.	21

TABLE 2 - DATA DIRECTLY TRANSCRIBED FROM ORIGINAL SOURCES.

OBS	PLNA	PART	CALORIES	WATER	PROTEIN	FAT	TOTCARB	FIBER	ASH	CALCIUM	PHOSPHO	IRON	SODIUM	POTASS	CAROT	THIAMINE	RIBOFLAV	NIACIN	ASCORBIC	REFCO
2296	CYNOGLOSSUM OFFICINALE	S	.	0.0	20.4	21.2	.	.	.	.	.	.	.	.	.	.	.	.	.	21
2297	CYNOGLOSSUM PICTUM	S	.	0.0	16.7	22.3	.	.	.	.	.	.	.	.	.	.	.	.	.	21
2298	CYNOSURUS CRISTATUS	S	.	0.0	15.1	4.7	.	.	.	.	.	.	.	.	.	.	.	.	.	21
2299	CYPERUS ESCULENTUS	R	302	36.5	3.5	12.7	46.1	7.4	1.2	25	204	8.0	.	.	.	0.28	0.09	1.3	3	1
2300	CYPERUS ESCULENTUS	R	.	0.0	6.1	27.4	.	.	2.4	.	.	.	.	.	.	.	.	.	.	19
2301	CYPERUS ESCULENTUS	R	311	32.6	4.4	17.2	43.9	7.1	1.9	59	155	2.4	.	.	.	0.09	.	.	.	3
2302	CYPERUS ESCULENTUS	W	.	.	.	.	.	.	.	670	150	.	.	740	.	.	.	.	.	6
2303	CYPERUS KALLI	S	.	0.0	17.4	20.0	.	.	.	.	.	.	.	.	.	.	.	.	.	21
2304	CYPERUS VEGETUS	S	.	0.0	10.0	8.7	.	.	.	.	.	.	.	.	.	.	.	.	.	21
2305	CYPERUS VIRENS	S	.	0.0	11.5	8.2	.	.	6.9	.	.	.	.	.	.	.	.	.	.	19
2306	CYPERUS VIRENS	S	.	0.0	8.8	2.6	.	.	8.3	.	.	.	.	.	.	.	.	.	.	19
2307	CYPHOLEPSIS YEMINICA	W	.	0.0	15.5	1.8	74.9	37.9	7.8	.	.	.	.	.	.	.	.	.	.	18
2308	CYPHOMANDRA BETACEA	F	57	82.7	1.5	0.2	14.5	4.2	1.1	71	30	0.7	.	.	.	.	.	.	.	1
2309	CYPHOMANDRA BETACEA	F	48	85.9	1.5	0.3	11.3	2.2	1.0	13	24	0.8	.	.	460	0.04	0.04	1.0	17	2
2310	CYPHOMANDRA BETACEA	F	50	85.9	2.2	0.9	10.3	1.6	0.7	9	48	0.8	.	.	460	0.10	0.04	1.2	29	3
2311	CYPHOMANDRA BETACEA	F	.	82.7	1.5	0.2	14.5	4.2	1.1	100	30	0.7	.	.	324	.	.	.	.	5
2312	CYTISUS ALBUS	S	.	0.0	33.8	6.9	.	.	.	.	.	.	.	.	.	.	.	.	.	20
2313	CYTISUS CANTABRICUS	S	.	0.0	30.1	7.7	.	.	.	.	.	.	.	.	.	.	.	.	.	21
2314	CYTISUS MULTIFLORUS	S	.	0.0	34.0	7.0	.	.	3.2	.	.	.	.	.	.	.	.	.	.	19
2315	CYTISUS MULTIFLORUS	S	.	0.0	32.9	7.5	.	.	.	.	.	.	.	.	.	.	.	.	.	21
2316	CYTISUS NIGRICANS	S	.	0.0	27.6	7.7	.	.	3.0	.	.	.	.	.	.	.	.	.	.	19
2317	CYTISUS PROLIFERUS	W	.	0.0	25.0	4.0	66.0	22.0	5.0	570	190	.	.	.	.	.	.	.	.	18
2318	CYTISUS SCOPARIUS	S	.	0.0	32.5	6.0	.	.	3.8	.	.	.	.	.	.	.	.	.	.	19
2319	DACRYODES EDULIS	F	263	56.2	4.6	23.2	14.9	8.2	1.1	43	60	0.8	.	.	.	.	.	.	19	2
2320	DACTYLIS GLOMERATA	W	.	0.0	12.6	3.5	74.9	33.1	9.0	.	.	.	.	.	.	.	.	.	.	18
2321	DACTYLIS GLOMERATA	H	.	0.0	16.1	3.3	68.4	37.7	12.2	.	.	.	.	.	.	.	.	.	.	18
2322	DACTYLIS GLOMERATA	S	.	0.0	29.6	11.6	.	.	.	.	.	.	.	.	.	.	.	.	.	19
2323	DACTYLIS GLOMERATA	S	.	0.0	6.9	2.3	.	.	10.3	.	.	.	.	.	.	.	.	.	.	19
2324	DACTYLIS GLOMERATA	S	.	0.0	26.2	10.8	.	.	2.7	.	.	.	.	.	.	.	.	.	.	19
2325	DACTYLIS GLOMERATA	S	.	0.0	27.5	10.9	.	.	.	.	.	.	.	.	.	.	.	.	.	20
2326	DACTYLIS GLOMERATA	S	.	0.0	25.3	10.9	.	.	.	.	.	.	.	.	.	.	.	.	.	21
2327	DACTYLIS GLOMERATA	L	.	77.3	4.1	1.3	27.7	11.1	2.9	.	.	.	.	.	.	.	.	.	.	5
2328	DACTYLIS GLOMERATA	L	.	66.9	1.9	0.7	15.7	5.8	1.6	.	.	.	.	.	.	.	.	.	.	5
2329	DACTYLIS GLOMERATA	W	.	0.0	13.8	4.3	72.7	27.9	9.2	530	510	23.0	.	3440	32	0.17	.	.	.	6
2330	DACTYLOCTENIUM AEGYPTIUM	S	344	11.4	15.7	1.9	68.1	2.8	2.9	820	460	.	.	.	.	.	.	.	.	1
2331	DACTYLOCTENIUM AEGYPTIUM	W	.	0.0	15.8	1.8	72.4	26.8	10.5	.	.	.	.	.	.	.	.	.	.	18
2332	DACTYLOCTENIUM AEGYPTIUM	H	.	0.0	8.3	0.8	84.4	36.9	6.5	.	.	.	.	.	.	.	.	.	.	18
2333	DAHLIA PINNATA	S	.	0.0	31.9	17.8	.	.	.	.	.	.	.	.	.	.	.	.	.	20
2334	DALBERGIA CULTRATA	L	79	74.4	5.5	0.0	18.4	.	1.7	28	78	.	.	.	.	.	.	.	26	1
2335	DALBERGIA SISSOO	L	.	0.0	24.1	2.0	67.3	12.5	6.6	840	420	.	.	.	.	.	.	.	.	18
2336	DALEA ALBIFLORA	S	.	0.0	33.8	5.8	.	.	.	.	.	.	.	.	.	.	.	.	.	19
2337	DALEA ALBIFLORA	S	.	0.0	35.0	7.0	.	.	.	.	.	.	.	.	.	.	.	.	.	20
2338	DALEA ALOPECUROIDES	S	.	0.0	36.6	5.7	.	.	3.5	.	.	.	.	.	.	.	.	.	.	19
2339	DALEA ALOPECUROIDES	S	.	0.0	30.2	6.5	.	.	4.7	.	.	.	.	.	.	.	.	.	.	19
2340	DALEA ALOPECUROIDES	S	.	0.0	31.2	4.7	.	.	.	.	.	.	.	.	.	.	.	.	.	21

TABLE 2 - DATA DIRECTLY TRANSCRIBED FROM ORIGINAL SOURCES.

OBS	PLNA	PART	CALORIES	WATER	PROTEIN	FAT	TOTLCARB	FIBER	ASH	CALCIUM	PHOSPHO	IRON	SODIUM	POTASS	CAROT	THIAMINE	RIBOFLAV	NIACIN	ASCORBIC	REFCO
2341	DALEA CITRIODORA	S	.	0	31.2	7.1	.	.	.	.	.	.	.	.	.	.	.	.	.	20
2342	DALEA DIFFUSA	S	.	0	32.5	6.4	.	.	.	.	.	.	.	.	.	.	.	.	.	19
2343	DALEA DIFFUSA & DALEA NUTANS	S	.	0	28.8	11.9	.	.	.	.	.	.	.	.	.	.	.	.	.	21
2344	DALEA GRAYI	S	.	0	34.0	11.8	.	.	3.2	.	.	.	.	.	.	.	.	.	.	19
2345	DALEA LACHNOSTACHYS	S	.	0	45.0	11.9	.	.	.	.	.	.	.	.	.	.	.	.	.	19
2346	DALEA LAGOPINA	S	.	0	33.1	6.3	.	.	.	.	.	.	.	.	.	.	.	.	.	20
2347	DALEA LUMHOLTZII	S	.	0	33.4	6.9	.	.	4.5	.	.	.	.	.	.	.	.	.	.	19
2348	DALEA MUCRONATA	S	.	0	37.5	4.9	.	.	.	.	.	.	.	.	.	.	.	.	.	20
2349	DALEA NEGLECTA	S	.	0	41.2	13.7	.	.	.	.	.	.	.	.	.	.	.	.	.	20
2350	DALEA NUTANS	S	.	0	26.9	10.9	.	.	.	.	.	.	.	.	.	.	.	.	.	20
2351	DALEA NUTANS	S	.	0	31.5	9.7	.	.	.	.	.	.	.	.	.	.	.	.	.	21
2352	DALEA OAXACANA	S	.	0	36.6	6.8	.	.	6.6	.	.	.	.	.	.	.	.	.	.	19
2353	DALEA PSORALEOIDES	S	.	0	27.5	2.7	.	.	.	.	.	.	.	.	.	.	.	.	.	20
2354	DALEA SP.	S	.	0	28.0	3.4	.	.	.	.	.	.	.	.	.	.	.	.	.	21
2355	DALEA SP.	S	.	0	33.9	3.3	.	.	.	.	.	.	.	.	.	.	.	.	.	21
2356	DALEA SP.	W	.	0	19.0	6.5	67.8	14.5	6.7	2270	140	.	.	1210	.	.	.	.	.	6
2357	DALEA TOMENTOSA	S	.	0	26.4	3.4	.	.	.	.	.	.	.	.	.	.	.	.	.	20
2358	DALEA TUBERCULINA	S	.	0	36.9	8.2	.	.	.	.	.	.	.	.	.	.	.	.	.	20
2359	DALEA VIRIDIFLORA	S	.	0	36.2	2.7	.	.	.	.	.	.	.	.	.	.	.	.	.	20
2360	DANAE RACEMOSA	S	.	0	13.8	0.4	.	.	.	.	.	.	.	.	.	.	.	.	.	20
2361	DAPHNE MEZEREUM	S	.	0	24.4	65.4	.	.	.	.	.	.	.	.	.	.	.	.	.	19
2362	DAPHNE MEZEREUM	S	.	0	7.5	6.0	.	.	.	.	.	.	.	.	.	.	.	.	.	19
2363	DAPHNE PONTICA	S	.	0	7.8	28.0	.	.	.	.	.	.	.	.	.	.	.	.	.	21
2364	DAPHNIPHYLLUM HUMILE	S	.	0	15.0	34.8	.	.	2.3	.	.	.	.	.	.	.	.	.	.	19
2365	DAPHNIPHYLLUM MACROPODUM	S	.	0	18.1	37.4	.	.	.	.	.	.	.	.	.	.	.	.	.	20
2366	DASYLIRION TEXANUM	S	.	0	27.5	23.0	.	.	.	.	.	.	.	.	.	.	.	.	.	19
2367	DASYLIRION TEXANUM	W	.	0	7.1	2.2	87.7	39.0	3.0	690	80	.	.	760	.	.	.	.	.	6
2368	DASYLIRION WHEELERI	S	.	0	26.3	22.1	.	.	2.1	.	.	.	.	.	.	.	.	.	.	19
2369	DASYLIRION WHEELERI	S	.	0	7.1	1.5	.	.	6.0	.	.	.	.	.	.	.	.	.	.	19
2370	DASYLIRION WHEELERI	S	.	0	26.2	23.8	.	.	.	.	.	.	.	.	.	.	.	.	.	19
2371	DASYLIRION WHEELERI	S	.	0	42.5	28.3	.	.	.	.	.	.	.	.	.	.	.	.	.	19
2372	DASYLIRION WHEELERI	S	.	0	50.0	21.7	.	.	.	.	.	.	.	.	.	.	.	.	.	19
2373	DASYLIRION WHEELERI	Y	.	0	3.7	.	.	.	11.6	890	210	19	2100	.	.	.	.	.	.	6
2374	DASYLIRION WHEELERI	L	.	0	4.1	.	.	.	7.7	1440	260	13	.	1630	.	.	.	.	.	6
2375	DATISCA CANNABINA	S	.	0	25.8	40.9	.	.	.	.	.	.	.	.	.	.	.	.	.	21
2376	DATURA CF. INOXIA	S	.	0	17.5	22.7	.	.	.	.	.	.	.	.	.	.	.	.	.	20
2377	DATURA DISCOLOR	S	.	0	21.2	20.8	.	.	.	.	.	.	.	.	.	.	.	.	.	20
2378	DATURA FEROX	S	.	0	18.1	17.6	.	.	.	.	.	.	.	.	.	.	.	.	.	21
2379	DATURA INOXIA	S	.	0	13.8	19.6	.	.	.	.	.	.	.	.	.	.	.	.	.	20
2380	DATURA INOXIA	S	.	0	15.1	17.3	.	.	.	.	.	.	.	.	.	.	.	.	.	21
2381	DATURA INOXIA	S	.	0	20.0	12.0	.	.	.	.	.	.	.	.	.	.	.	.	.	21
2382	DATURA METAL	S	.	0	13.4	18.8	.	.	4.5	.	.	.	.	.	.	.	.	.	.	19
2383	DATURA METAL	S	.	0	12.5	17.2	.	.	5.6	.	.	.	.	.	.	.	.	.	.	19
2384	DATURA METALOIDES	S	.	0	17.5	21.8	.	.	.	.	.	.	.	.	.	.	.	.	.	19
2385	DATURA QUERCIFOLIA	S	.	0	21.2	16.1	.	.	.	.	.	.	.	.	.	.	.	.	.	19

TABLE 2 - DATA DIRECTLY TRANSCRIBED FROM ORIGINAL SOURCES.

OBS	PLNA	PART	CALORIES	WATER	PROTEIN	FAT	TOTCARB	FIBER	ASH	CALCIUM	PHOSPHO	IRON	SODIUM	POTASS	CAROT	THIAMINE	RIBOFLAV	NIACIN	ASCORBIC	REFCO
2386	DATURA REBURRA	S	.	0.0	12.3	19.5	.	.	.	.	.	.	.	.	.	.	.	.	.	21
2387	DATURA SP.	S	.	0.0	14.4	22.3	.	.	.	.	.	.	.	.	.	.	.	.	.	19
2388	DATURA STRAMONIUM	S	.	0.0	18.3	25.8	.	.	2.7	.	.	.	.	.	.	.	.	.	.	19
2389	DATURA STRAMONIUM	S	.	0.0	18.2	22.8	.	.	2.7	.	.	.	.	.	.	.	.	.	.	19
2390	DATURA STRAMONIUM	S	.	0.0	18.1	27.1	.	.	.	.	.	.	.	.	.	.	.	.	.	20
2391	DATURA STRAMONIUM	S	.	0.0	19.3	24.9	.	.	.	.	.	.	.	.	.	.	.	.	.	21
2392	DATURA STRAMONIUM	S	.	0.0	14.0	18.4	.	.	.	.	.	.	.	.	.	.	.	.	.	21
2393	DATURA STRAMONIUM	S	.	0.0	17.2	19.6	.	.	.	.	.	.	.	.	.	.	.	.	.	21
2394	DATURA STRAMONIUM	S	.	0.0	19.4	28.5	.	.	.	.	.	.	.	.	.	.	.	.	.	21
2395	DATURA WRIGHTII	S	.	0.0	13.4	24.2	.	.	.	.	.	.	.	.	.	.	.	.	.	21
2396	DAUCUS AUREUS	S	.	0.0	15.6	15.0	.	.	.	.	.	.	.	.	.	.	.	.	.	21
2397	DAUCUS BROTERI	S	.	0.0	18.9	18.6	.	.	.	.	.	.	.	.	.	.	.	.	.	21
2398	DAUCUS CAROTA	R	40	88.6	0.9	0.1	9.6	1.4	0.8	35	38	0.7	.	.	5480	0.04	0.04	0.6	8	1
2399	DAUCUS CAROTA	SH	.	0.0	12.1	2.3	67.4	16.7	18.2	.	.	.	.	.	.	.	.	.	.	18
2400	DAUCUS CAROTA	R	.	0.0	12.7	1.5	77.6	10.4	8.2	.	.	.	.	.	.	.	.	.	.	18
2401	DAUCUS CAROTA	S	.	0.0	28.4	20.1	.	.	6.6	.	.	.	.	.	.	.	.	.	.	19
2402	DAUCUS CAROTA	S	.	0.0	28.9	28.3	.	.	7.4	.	.	.	.	.	.	.	.	.	.	19
2403	DAUCUS CAROTA	S	.	0.0	26.9	28.2	.	.	6.7	.	.	.	.	.	.	.	.	.	.	19
2404	DAUCUS CAROTA	S	.	0.0	28.5	19.8	.	.	7.4	.	.	.	.	.	.	.	.	.	.	19
2405	DAUCUS CAROTA	S	.	0.0	23.9	16.3	.	.	7.6	.	.	.	.	.	.	.	.	.	.	19
2406	DAUCUS CAROTA	S	.	0.0	35.8	16.4	.	.	4.1	.	.	.	.	.	.	.	.	.	.	19
2407	DAUCUS CAROTA	S	.	0.0	22.5	18.2	.	.	6.5	.	.	.	.	.	.	.	.	.	.	19
2408	DAUCUS CAROTA	S	.	0.0	25.4	26.5	.	.	8.4	.	.	.	.	.	.	.	.	.	.	19
2409	DAUCUS CAROTA	S	.	0.0	29.8	30.2	.	.	6.7	.	.	.	.	.	.	.	.	.	.	19
2410	DAUCUS CAROTA	S	.	0.0	13.4	8.7	.	.	11.8	.	.	.	.	.	.	.	.	.	.	19
2411	DAUCUS CAROTA	S	.	0.0	26.2	19.6	.	.	.	.	.	.	.	.	.	.	.	.	.	19
2412	DAUCUS CAROTA	S	.	0.0	20.0	19.2	.	.	.	.	.	.	.	.	.	.	.	.	.	19
2413	DAUCUS CAROTA	R	37	89.6	1.1	0.3	8.3	0.9	0.7	36	38	1.2	70	245	7000	0.06	0.05	0.7	8	2
2414	DAUCUS CAROTA	S	.	0.0	20.0	12.9	.	.	.	.	.	.	.	.	.	.	.	.	.	20
2415	DAUCUS CAROTA	S	.	0.0	25.6	15.6	.	.	.	.	.	.	.	.	.	.	.	.	.	20
2416	DAUCUS CAROTA	S	.	0.0	22.4	27.2	.	.	.	.	.	.	.	.	.	.	.	.	.	21
2417	DAUCUS CAROTA	R	41	89.1	0.8	0.4	8.9	0.8	0.8	34	26	0.9	.	.	7060	0.06	0.04	0.6	5	3
2418	DAUCUS CAROTA	R	42	88.2	1.1	0.2	9.7	1.0	0.8	37	36	0.7	47	341	6600	0.06	0.05	0.6	8	4
2419	DAUCUS CAROTA	R	.	86.0	0.9	0.1	11.9	1.2	1.1	80	30	1.5	.	.	1860	.	.	.	3	5
2420	DAUCUS CAROTA	L	.	0.0	16.4	4.6	64.0	11.1	15.0	.	.	.	.	.	.	.	.	.	.	6
2421	DAUCUS CAROTA	Y	.	0.0	10.4	5.5	.	18.1	.	.	.	.	.	.	.	.	.	.	.	6
2422	DAUCUS CAROTA VAR. MAXIMUS	S	.	0.0	19.1	18.9	.	.	.	.	.	.	.	.	.	.	.	.	.	21
2423	DAUCUS GUTTATUS	S	.	0.0	20.9	23.3	.	.	.	.	.	.	.	.	.	.	.	.	.	21
2424	DAUCUS PUSILLUS	S	.	0.0	20.0	27.1	.	.	6.7	.	.	.	.	.	.	.	.	.	.	19
2425	DAUCUS PUSILLUS	S	.	0.0	21.4	22.7	.	.	.	.	.	.	.	.	.	.	.	.	.	20
2426	DAUCUS PUSILLUS	W	.	0.0	11.7	6.4	72.6	18.9	9.3	2350	350	.	.	2590	.	.	.	.	.	6
2427	DELONIX REGIA	F	.	0.0	11.1	.	.	49.7	4.7	230	140	.	.	.	.	.	.	.	.	18
2428	DELONIX REGIA	S	.	0.0	17.5	3.3	.	.	.	.	.	.	.	.	.	.	.	.	.	19
2429	DELONIX REGIA	S	.	0.0	16.9	3.0	.	.	.	.	.	.	.	.	.	.	.	.	.	19
2430	DELONIX REGIA	S	65	82.8	6.1	0.9	8.9	.	1.3	116	66	1.1	1	395	300	.	0.20	0.8	.	2

TABLE 2 - DATA DIRECTLY TRANSCRIBED FROM ORIGINAL SOURCES.

OBS	PLNA	PART	CALORIES	WATER	PROTEIN	FAT	TOTLCARB	FIBER	ASH	CALCIUM	PHOSPHO	IRON	SODIUM	POTASS	CAROT	THIAMINE	RIBOFLAV	NIACIN	ASCORBIC	REFCO
2431	DELONIX REGIA	S	.	6.4	60.3	9.7	16.2	.	7.4	.	.	.	.	.	.	.	.	.	.	5
2432	DELPHINIUM GRANDIFLORUM	S	.	0.0	29.4	27.2	.	.	.	.	.	.	.	.	.	.	.	.	.	20
2433	DELPHINIUM HANSENII	S	.	0.0	25.1	27.3	.	.	.	.	.	.	.	.	.	.	.	.	.	21
2434	DELPHINIUM HYBRIDUM	S	.	0.0	31.2	28.2	.	.	.	.	.	.	.	.	.	.	.	.	.	20
2435	DELPHINIUM OCCIDENTALE SSP. CUCULLATUM	S	.	0.0	27.6	42.3	.	.	.	.	.	.	.	.	.	.	.	.	.	20
2436	DELPHINIUM SP.	S	.	0.0	22.5	6.6	.	.	5.5	.	.	.	.	.	.	.	.	.	.	19
2437	DELPHINIUM SP.	W	.	.	.	.	.	.	.	1400	350	.	.	.	.	.	.	.	.	6
2438	DELPHINIUM TENUISECTUM	S	.	0.0	28.8	33.6	.	.	.	.	.	.	.	.	.	.	.	.	.	20
2439	DENDROCALAMUS STRICTUS	L	.	0.0	20.5	7.5	57.2	26.0	14.8	.	.	.	.	.	.	.	.	.	.	18
2440	DESCURAINIA PINNATA	S	.	0.0	24.4	38.4	.	.	.	.	.	.	.	.	.	.	.	.	.	20
2441	DESCURAINIA PINNATA	W	.	.	.	.	.	.	.	1970	690	.	.	4450	.	.	.	.	.	6
2442	DESCURAINIA SOPHIA	S	.	0.0	29.5	35.3	.	.	3.9	.	.	.	.	.	.	.	.	.	.	19
2443	DESCURAINIA SOPHIA	S	.	0.0	25.5	39.7	.	.	3.6	.	.	.	.	.	.	.	.	.	.	19
2444	DESCURAINIA SOPHIA	S	.	0.0	28.1	38.3	.	.	.	.	.	.	.	.	.	.	.	.	.	20
2445	DESCURAINIA SOPHIA	S	.	0.0	29.2	39.4	.	.	.	.	.	.	.	.	.	.	.	.	.	20
2446	DESCURAINIA SOPHIA	S	.	0.0	28.8	38.9	.	.	.	.	.	.	.	.	.	.	.	.	.	20
2447	DESCURAINIA SOPHIA	S	.	0.0	27.6	37.6	.	.	.	.	.	.	.	.	.	.	.	.	.	20
2448	DESCURAINIA SOPHIA	S	.	0.0	29.9	26.9	.	.	.	.	.	.	.	.	.	.	.	.	.	21
2449	DESCURAINIA SOPHIA	S	.	0.0	29.3	38.1	.	.	.	.	.	.	.	.	.	.	.	.	.	21
2450	DESCURAINIA SP.	S	.	0.0	36.9	33.6	.	.	.	.	.	.	.	.	.	.	.	.	.	21
2451	DESCURAINIA SP.	W	.	0.0	7.6	2.4	.	.	5.6	1490	490	.	.	4450	.	.	.	.	.	6
2452	DESMANTHUS ACUMINATUS	S	.	0.0	33.8	2.6	.	.	.	.	.	.	.	.	.	.	.	.	.	20
2453	DESMANTHUS ILLINOENSIS	S	.	0.0	34.9	2.3	.	.	3.7	.	.	.	.	.	.	.	.	.	.	19
2454	DESMANTHUS ILLINOENSIS	S	.	0.0	34.4	2.2	.	.	3.8	.	.	.	.	.	.	.	.	.	.	19
2455	DESMANTHUS ILLINOENSIS	S	.	0.0	34.4	2.3	.	.	3.6	.	.	.	.	.	.	.	.	.	.	19
2456	DESMANTHUS SUBULATUS	S	.	0.0	32.6	3.3	.	.	.	.	.	.	.	.	.	.	.	.	.	21
2457	DESMANTHUS VELUTINUS	S	.	0.0	31.2	3.9	.	.	.	.	.	.	.	.	.	.	.	.	.	19
2458	DESMANTHUS VIRGATUS	SH	.	0.0	11.5	2.4	80.3	39.3	5.8	.	.	.	.	.	.	.	.	.	.	18
2459	DESMANTHUS VIRGATUS	S	.	0.0	31.9	4.4	.	.	.	.	.	.	.	.	.	.	.	.	.	20
2460	DESMANTHUS VIRGATUS	S	.	0.0	35.0	4.3	.	.	.	.	.	.	.	.	.	.	.	.	.	20
2461	DESMANTHUS VIRGATUS	W	.	0.0	12.6	2.2	79.5	38.8	5.7	1750	350	.	.	1810	.	.	.	.	.	6
2462	DESMODIUM AFFINE	S	.	0.0	27.5	14.0	.	.	.	.	.	.	.	.	.	.	.	.	.	21
2463	DESMODIUM BARBATUM	SH	.	0.0	10.4	4.0	81.7	34.7	3.9	.	.	.	.	.	.	.	.	.	.	18
2464	DESMODIUM BELLUM	S	.	0.0	34.4	8.5	.	.	.	.	.	.	.	.	.	.	.	.	.	20
2465	DESMODIUM CF. DISTORTUM	S	.	0.0	28.5	8.2	.	.	.	.	.	.	.	.	.	.	.	.	.	21
2466	DESMODIUM CINERASCENS	S	.	0.0	37.7	12.0	.	.	4.2	.	.	.	.	.	.	.	.	.	.	19
2467	DESMODIUM CINEREUM	L	88	74.1	5.8	1.8	16.5	6.1	1.8	227	108	3.8	.	.	3560	0.25	0.11	1.50	164	3
2468	DESMODIUM DILLENII	S	.	0.0	39.4	14.1	.	.	.	.	.	.	.	.	.	.	.	.	.	20
2469	DESMODIUM DISCOLOR	SH	.	0.0	18.7	0.3	71.0	33.4	10.0	.	.	.	.	.	.	.	.	.	.	18
2470	DESMODIUM DISCOLOR	SH	.	0.0	12.1	2.8	78.4	35.1	6.7	1250	190	.	.	.	.	.	.	.	.	18
2471	DESMODIUM GYROIDES	SH	.	0.0	11.5	0.7	79.3	45.2	8.5	970	400	.	.	.	.	.	.	.	.	18
2472	DESMODIUM HETEROPHYLLUM	SH	.	0.0	8.9	1.4	77.2	27.1	12.5	740	100	.	.	.	.	.	.	.	.	18
2473	DESMODIUM INTORTUM	SH	.	0.0	20.7	3.3	66.6	25.4	9.4	1410	520	.	.	.	.	.	.	.	.	18
2474	DESMODIUM NICARAQUENSE	S	.	0.0	31.9	9.0	.	.	4.3	.	.	.	.	.	.	.	.	.	.	19
2475	DESMODIUM OVALIFOLIUM	S	.	0.0	32.7	3.7	.	.	.	.	.	.	.	.	.	.	.	.	.	21

TABLE 2 - DATA DIRECTLY TRANSCRIBED FROM ORIGINAL SOURCES.

OBS	PLNA	PART	CALORIES	WATER	PROTEIN	FAT	TOTLCARB	FIBER	ASH	CALCIUM	PHOSPHO	IRON	SODIUM	POTASS	CAROT	THIAMINE	RIBOFLAV	NIACIN	ASCORBIC	REFCO
2476	DESMODIUM PABULARIS	SH	.	0.0	23.8	2.3	63.7	23.4	10.2	.	.	.	.	.	.	.	.	.	.	18
2477	DESMODIUM SCORPIARUS	SH	.	0.0	15.5	3.0	71.8	29.3	9.7	.	.	.	.	.	.	.	.	.	.	18
2478	DESMODIUM SP.	W	.	0.0	15.0	2.5	71.7	26.1	10.8	1240	280	47.0	.	.	18	0.37	0.77	5.25	.	6
2479	DESMODIUM TILIAEFOLIUM	S	.	0.0	44.2	13.1	.	.	.	.	.	.	.	.	.	.	.	.	.	21
2480	DESMODIUM TRIFLORUM	SH	.	0.0	8.8	2.8	79.6	29.2	3.8	740	120	.	.	.	.	.	.	.	.	18
2481	DESMODIUM UNCINATUM	SH	.	0.0	12.8	3.4	74.7	29.7	9.1	1060	410	.	.	.	.	.	.	.	.	18
2482	DESMOSTACHYA BIPINNATA	L	.	0.0	6.8	1.6	82.5	40.3	9.1	.	.	.	.	.	.	.	.	.	.	5
2483	DETARIUM MICROCARPUM	F	311	11.1	4.9	0.4	81.0	6.6	2.6	82	84	1.8	.	.	.	0.03	.	.	32	1
2484	DEUTZIA SCABRA	S	.	0.0	26.2	33.8	.	.	.	.	.	.	.	.	.	.	.	.	.	19
2485	DIALIUM GUINEENSE	F	294	17.1	3.4	0.1	78.4	2.2	1.0	196	77	3.7	.	.	.	.	.	.	.	1
2486	DIANTHUS BARBATUS	S	.	0.0	18.8	6.1	.	.	.	.	.	.	.	.	.	.	.	.	.	20
2487	DIANTHUS CHINENSIS	S	.	0.0	18.1	6.0	.	.	.	.	.	.	.	.	.	.	.	.	.	20
2488	DICENTRA OCHROLEUCA	S	.	0.0	13.6	30.9	.	.	2.8	.	.	.	.	.	.	.	.	.	.	19
2489	DICHANTHIUM ANNULATUM	W	.	0.0	10.4	1.7	75.8	34.9	12.1	.	.	.	.	.	.	.	.	.	.	18
2490	DICHANTHIUM ANNULATUM	H	.	0.0	2.7	1.2	84.6	39.1	11.5	.	.	.	.	.	.	.	.	.	.	18
2491	DICHANTHIUM ANNULATUM	L	.	69.9	2.1	1.6	22.6	9.2	3.8	.	.	.	.	.	.	.	.	.	.	5
2492	DICHANTHIUM ANNULATUM	L	.	65.9	2.2	1.7	26.2	11.6	3.7	.	.	.	.	.	.	.	.	.	.	5
2493	DICHANTHIUM ANNULATUM	L	.	65.4	2.0	1.7	27.1	14.3	3.8	.	.	.	.	.	.	.	.	.	.	5
2494	DICHANTHIUM ANNULATUM	L	.	.	2.7	1.2	84.7	39.1	11.5	560	110	.	350	500	.	.	.	.	.	5
2495	DICHANTHIUM CARICOSUM	W	.	0.0	7.0	1.5	79.8	36.7	11.7	.	.	.	.	.	.	.	.	.	.	18
2496	DICHANTHIUM CARICOSUM	H	.	0.0	6.1	1.7	80.4	40.3	11.8	.	.	.	.	.	.	.	.	.	.	18
2497	DICHROSTACHYS CINEREA	F	.	0.0	16.6	5.4	72.4	22.1	5.6	920	120	.	.	.	.	.	.	.	.	18
2498	DICHROSTACHYS GLOMERATA	F	.	0.0	10.8	1.4	82.4	26.6	5.4	510	200	.	.	.	.	.	.	.	.	18
2499	DICLIPTERA ROXBURGHIANA	S	.	0.0	23.4	21.2	.	.	.	.	.	.	.	.	.	.	.	.	.	21
2500	DICTAMNUS ALBUS	S	.	0.0	24.4	36.2	.	.	.	.	.	.	.	.	.	.	.	.	.	20
2501	DIDISCUS CAERULEUS	S	.	0.0	22.5	37.4	.	.	3.0	.	.	.	.	.	.	.	.	.	.	19
2502	DIGITALIS FERRUGINEA	S	.	0.0	19.1	28.2	.	.	.	.	.	.	.	.	.	.	.	.	.	21
2503	DIGITALIS LANATA	S	.	0.0	17.1	34.7	.	.	.	.	.	.	.	.	.	.	.	.	.	21
2504	DIGITALIS ORIENTALIS	S	.	0.0	17.9	35.8	.	.	.	.	.	.	.	.	.	.	.	.	.	21
2505	DIGITALIS PURPUREA	S	.	0.0	16.8	36.1	.	.	3.6	.	.	.	.	.	.	.	.	.	.	19
2506	DIGITALIS PURPUREA	S	.	0.0	17.5	36.7	.	.	.	.	.	.	.	.	.	.	.	.	.	20
2507	DIGITALIS PURPUREA	S	.	0.0	14.8	38.9	.	.	.	.	.	.	.	.	.	.	.	.	.	21
2508	DIGITALIS PURPUREA VAR. GLOXINIAEFLORA	S	.	0.0	16.2	40.4	.	.	.	.	.	.	.	.	.	.	.	.	.	20
2509	DIGITALIS THASPI	S	.	0.0	14.9	35.0	.	.	.	.	.	.	.	.	.	.	.	.	.	21
2510	DIGITARIA DECUMBENS	W	.	0.0	14.9	3.0	70.7	31.0	11.4	.	.	.	.	.	.	.	.	.	.	18
2511	DIGITARIA DECUMBENS	H	.	0.0	3.2	1.3	89.6	33.5	5.9	.	.	.	.	.	.	.	.	.	.	18
2512	DIGITARIA DECUMBENS	W	.	0.0	10.8	2.0	77.4	29.8	9.8	450	350	.	.	.	.	.	.	.	.	6
2513	DIGITARIA EXILIS	S	332	11.2	7.1	3.0	74.4	7.4	4.3	41	191	8.5	.	.	.	0.24	0.10	1.90	.	1
2514	DIGITARIA EXILIS	H	.	0.0	5.4	1.7	85.9	30.1	7.0	.	.	.	.	.	.	.	.	.	.	18
2515	DIGITARIA GAZENSIS	W	.	0.0	13.7	3.7	74.5	28.2	8.1	.	.	.	.	.	.	.	.	.	.	18
2516	DIGITARIA GAZENSIS	H	.	0.0	11.5	2.4	72.8	30.1	13.3	.	.	.	.	.	.	.	.	.	.	18
2517	DIGITARIA IBURUA	S	343	10.3	8.9	3.0	75.6	6.2	2.2	.	234	10.0	.	.	.	.	.	.	.	1
2518	DIGITARIA MACROBLEPHERA	W	.	0.0	12.3	1.8	74.7	32.2	11.2	.	.	.	.	.	.	.	.	.	.	18
2519	DIGITARIA MILANJIANA	W	.	0.0	8.1	2.3	79.9	36.8	9.7	.	.	.	.	.	.	.	.	.	.	18
2520	DIGITARIA MILANJIANA	H	.	0.0	12.5	2.3	76.0	29.2	9.2	.	.	.	.	.	.	.	.	.	.	18

TABLE 2 - DATA DIRECTLY TRANSCRIBED FROM ORIGINAL SOURCES.

OBS	PLNA	PART	CALORIES	WATER	PROTEIN	FAT	TOTLCARB	FIBER	ASH	CALCIUM	PHOSPHON	IRON	SODIUM	POTASS	CAROT	THIAMINE	RIBOFLAV	NIACIN	ASCORBIC	REFCO
2521	DIGITARIA PENTZII	W	.	0.0	14.0	3.0	68.7	28.4	14.3	.	.	.	.	.	.	.	.	.	.	18
2522	DIGITARIA PENTZII	H	.	0.0	10.1	2.0	80.1	32.9	7.8	.	.	.	.	.	.	.	.	.	.	18
2523	DIGITARIA SANGUINALIS	W	.	0.0	13.6	2.9	71.7	30.9	11.8	680	250	.	.	4220	.	.	.	.	.	6
2524	DIGITARIA SCALARUM	W	.	0.0	14.7	3.8	72.0	29.0	9.5	.	.	.	.	.	.	.	.	.	.	18
2525	DIGITARIA SETIGERA	L	.	.	14.5	2.2	72.0	28.0	11.3	820	460	.	.	.	.	.	.	.	.	5
2526	DIGITARIA SETIVALVA	H	.	0.0	13.2	2.6	72.8	27.7	11.4	.	.	.	.	.	.	.	.	.	.	18
2527	DIGITARIA SETIVALVA	W	.	0.0	10.4	2.4	75.8	29.8	11.3	.	.	.	.	.	.	.	.	.	.	18
2528	DIGITARIA SWAZILANDENSIS	H	.	0.0	13.2	1.3	75.8	34.6	9.7	.	.	.	.	.	.	.	.	.	.	18
2529	DIGITARIA SWAZILANDENSIS	W	.	0.0	7.5	1.9	79.0	30.5	11.6	.	.	.	.	.	.	.	.	.	.	18
2530	DILLENIA BURBIDGEI	F	29	91.7	0.3	0.2	7.4	0.9	0.4	28	5	0.1	10	115	25	0.02	0.04	0.2	2	2
2531	DILLENIA PHILIPPINENSIS	F	29	91.7	0.3	0.2	7.4	0.9	0.4	28	5	0.1	10	115	25	0.02	0.04	0.2	2	2
2532	DIMOCARPUS LONGAN	S	.	0.0	6.9	4.4	.	.	1.5	.	.	.	.	.	.	.	.	.	.	19
2533	DIMOCARPUS LONGAN	F	71	81.0	1.0	1.4	15.6	0.3	1.0	23	36	0.4	.	.	.	0.03	0.14	0.3	56	2
2534	DIMOCARPUS LONGAN	F	61	82.4	1.0	0.1	15.8	0.4	0.7	10	42	1.2	.	.	.	.	.	.	6	4
2535	DIMOCARPUS LONGAN	F	.	82.4	1.4	0.4	9.0	0.6	.	.	.	.	.	.	.	.	.	.	.	5
2536	DIMORPHOTHECA ANNUA VAR. RINGENS	S	.	0.0	35.6	40.9	.	.	.	.	.	.	.	.	.	.	.	.	.	19
2537	DIMORPHOTHECA AURANTIACA	S	.	0.0	37.8	34.0	.	.	5.4	.	.	.	.	.	.	.	.	.	.	19
2538	DIMORPHOTHECA AURANTIACA	S	.	0.0	14.4	12.5	.	.	5.5	.	.	.	.	.	.	.	.	.	.	19
2539	DIMORPHOTHECA AURANTIACA	S	.	0.0	40.0	33.6	.	.	.	.	.	.	.	.	.	.	.	.	.	19
2540	DIMORPHOTHECA AURANTIACA	S	.	0.0	12.5	4.9	.	.	.	.	.	.	.	.	.	.	.	.	.	19
2541	DIMORPHOTHECA AURANTIACA	S	.	0.0	36.2	39.2	.	.	.	.	.	.	.	.	.	.	.	.	.	19
2542	DIMORPHOTHECA AURANTIACA	S	.	0.0	38.8	37.9	.	.	.	.	.	.	.	.	.	.	.	.	.	19
2543	DIMORPHOTHECA AURANTIACA	S	.	0.0	37.5	35.0	.	.	.	.	.	.	.	.	.	.	.	.	.	19
2544	DIMORPHOTHECA CALENDULACEA	S	.	0.0	38.1	35.5	.	.	.	.	.	.	.	.	.	.	.	.	.	19
2545	DIMORPHOTHECA CHRYSANTHEMIFOLIA	S	.	0.0	28.2	48.0	.	.	.	.	.	.	.	.	.	.	.	.	.	20
2546	DIMORPHOTHECA CUNEATA	S	.	0.0	38.2	32.2	.	.	.	.	.	.	.	.	.	.	.	.	.	20
2547	DIMORPHOTHECA PLUVIALIS	S	.	0.0	38.1	36.2	.	.	.	.	.	.	.	.	.	.	.	.	.	20
2548	DIMORPHOTHECA PLUVIALIS	S	.	0.0	27.1	44.6	.	.	.	.	.	.	.	.	.	.	.	.	.	20
2549	DIMORPHOTHECA PLUVIALIS	S	.	0.0	32.4	41.5	.	.	.	.	.	.	.	.	.	.	.	.	.	20
2550	DIMORPHOTHECA PLUVIALIS	S	.	0.0	25.4	32.9	.	.	.	.	.	.	.	.	.	.	.	.	.	20
2551	DIMORPHOTHECA PLUVIALIS	S	.	0.0	35.3	41.7	.	.	.	.	.	.	.	.	.	.	.	.	.	20
2552	DIMORPHOTHECA SINUATA	S	.	0.0	41.9	34.4	.	.	.	.	.	.	.	.	.	.	.	.	.	19
2553	DIMORPHOTHECA SINUATA	S	.	0.0	32.5	43.7	.	.	.	.	.	.	.	.	.	.	.	.	.	19
2554	DIMORPHOTHECA SINUATA	S	.	0.0	22.5	20.3	.	.	.	.	.	.	.	.	.	.	.	.	.	20
2555	DIMORPHOTHECA SINUATA	S	.	0.0	39.4	42.5	.	.	.	.	.	.	.	.	.	.	.	.	.	20
2556	DIMORPHOTHECA SINUATA	S	.	0.0	35.6	39.2	.	.	.	.	.	.	.	.	.	.	.	.	.	20
2557	DIMORPHOTHECA SINUATA	S	.	0.0	33.0	42.1	.	.	.	.	.	.	.	.	.	.	.	.	.	20
2558	DIMORPHOTHECA SINUATA	S	.	0.0	32.3	42.4	.	.	.	.	.	.	.	.	.	.	.	.	.	20
2559	DIMORPHOTHECA SINUATA	S	.	0.0	38.8	38.2	.	.	.	.	.	.	.	.	.	.	.	.	.	20
2560	DIMORPHOTHECA SINUATA	S	.	0.0	.	33.4	.	.	.	.	.	.	.	.	.	.	.	.	.	21
2561	DIMORPHOTHECA SINUATA	S	.	0.0	.	32.0	.	.	.	.	.	.	.	.	.	.	.	.	.	21
2562	DIMORPHOTHECA SINUATA	S	.	0.0	.	30.7	.	.	.	.	.	.	.	.	.	.	.	.	.	21
2563	DIMORPHOTHECA SP.	S	.	0.0	36.2	41.2	.	.	.	.	.	.	.	.	.	.	.	.	.	20
2564	DIMORPHOTHECA SP.	S	.	0.0	36.9	41.4	.	.	.	.	.	.	.	.	.	.	.	.	.	20
2565	DIMORPHOTHECA ZEYHERI	S	.	0.0	42.4	36.2	.	.	.	.	.	.	.	.	.	.	.	.	.	20

TABLE 2 - DATA DIRECTLY TRANSCRIBED FROM ORIGINAL SOURCES.

OBS	PLANA	PART	CALORIES	WATER	PROTEIN	FAT	TOTCARB	FIBER	ASH	CALCIUM	PHOSPHO	IRON	SODIUM	POTASS	CAROT	THIAMINE	RIBOFLAV	NIACIN	ASCORBIC	REFCO
2566	DINEBRA RETROFLEXA	L	.	72.3	1.0	0.8	23.9	6.3	2.1	.	.	.	.	.	.	.	.	.	.	5
2567	DINEBRA RETROFLEXA	L	.	69.9	1.6	0.9	25.4	6.3	2.3	.	.	.	.	.	.	.	.	.	.	5
2568	DINEBRA RETROFLEXA	L	.	66.9	1.1	1.3	27.0	12.3	3.8	.	.	.	.	.	.	.	.	.	.	5
2569	DIODIA TERES VAR. SETIFERA	S	.	0.0	13.6	9.1	.	.	.	.	.	.	.	.	.	.	.	.	.	21
2570	DIOSCOREA ALATA	R	135	65.1	2.3	0.1	31.8	1.5	0.7	28	52	1.6	.	.	10	0.05	0.03	0.50	12	1
2571	DIOSCOREA ALATA	W	.	0.0	12.0	2.3	77.8	25.3	7.9	950	160	.	.	.	.	.	.	.	.	18
2572	DIOSCOREA ALATA	R	.	0.0	8.1	0.8	85.9	2.6	5.2	120	180	.	.	.	.	.	.	.	.	18
2573	DIOSCOREA ALATA	R	.	.	8.0	0.6	73.9	2.2	4.2	.	440	.	.	.	.	.	.	.	.	5
2574	DIOSCOREA ALATA	R	.	.	15.7	1.3	91.1	6.1	7.3	.	980	.	.	.	.	.	.	.	.	5
2575	DIOSCOREA BULBIFERA	R	112	71.0	1.5	0.1	26.5	0.9	0.9	69	29	.	.	.	.	.	.	.	.	1
2576	DIOSCOREA CAYENENSIS	R	71	80.8	1.5	0.1	16.4	0.6	1.2	36	17	5.2	.	.	.	.	.	.	.	1
2577	DIOSCOREA CAYENENSIS	R	.	0.0	6.2	0.4	90.2	2.4	3.2	.	.	.	.	.	.	.	.	.	.	18
2578	DIOSCOREA DUMETORUM	R	124	67.3	3.2	0.1	28.3	0.8	1.1	52	45	.	.	.	.	.	.	.	.	1
2579	DIOSCOREA ESCULENTA	R	102	73.6	1.5	0.2	23.9	0.6	0.8	12	35	0.8	8	366	.	0.10	0.01	0.80	15	
2580	DIOSCOREA ESCULENTA	R	112	70.2	3.5	0.1	25.2	0.5	1.0	62	53	.	.	.	.	.	.	.	.	1
2581	DIOSCOREA ESCULENTA	R	.	0.0	8.7	0.5	86.6	3.5	4.2	.	.	.	.	.	.	.	.	.	.	18
2582	DIOSCOREA ESCULENTA	R	95	75.0	1.5	0.1	22.4	.	1.0	14	49	0.8	.	.	0	0.05	.	.	4	2
2583	DIOSCOREA ESCULENTA	R	.	0.0	10.8	1.7	77.8	6.5	9.6	.	940	.	.	.	.	.	.	.	.	5
2584	DIOSCOREA LUZONENSIS	R	97	74.7	2.5	0.1	22.1	0.4	0.6	60	41	0.7	8	305	0	0.11	0.02	0.70	18	2
2585	DIOSCOREA MINUTIFLORA	R	118	68.4	3.0	0.1	27.0	10.0	1.5	69	41	.	.	.	.	.	.	.	.	1
2586	DIOSCOREA MINUTIFLORA	R	118	68.4	3.0	0.1	27.0	10.0	1.5	69	41	.	.	.	.	.	.	.	.	1
2587	DIOSCOREA PREUSII	R	116	68.8	3.0	0.1	26.4	1.3	1.7	60	50	.	.	.	.	.	.	.	.	1
2588	DIOSCOREA ROTUNDATA	R	.	0.0	5.9	0.5	89.3	2.4	4.3	.	.	.	.	.	.	.	.	.	.	18
2589	DIOSCOREA SP.	R	108	71.8	2.0	0.1	25.0	0.5	1.0	22	39	1.0	10	294	0	0.10	0.04	0.70	8	2
2590	DIOSCOREA SP.	L	.	86.0	0.4	0.0	.	.	.	201	.	1.1	.	.	6520	0.15	.	.	31	2
2591	DIOSCOREA SP.	R	100	72.6	2.0	0.2	24.3	0.6	0.9	14	43	1.3	.	.	0	0.13	0.02	0.40	3	3
2592	DIOSCOREA SP.	R	101	73.5	2.1	0.2	23.2	0.9	1.0	20	69	0.6	.	600	0	0.10	0.04	0.50	9	4
2593	DIOSPYROS ANISANDRA	S	.	0.0	7.5	3.0	.	.	.	.	.	.	.	.	.	.	.	.	.	19
2594	DIOSPYROS DISCOLOR	F	49	86.0	0.6	0.2	12.6	1.6	0.6	38	19	0.8	3	354	15	0.00	0.04	0.30	28	2
2595	DIOSPYROS DISCOLOR	F	.	83.0	2.8	0.2	13.5	1.8	0.4	.	.	.	.	.	.	.	.	.	.	5
2596	DIOSPYROS DISCOLOR	F	.	72.0	0.9	0.2	25.7	1.7	1.2	.	.	.	.	.	.	.	.	.	.	5
2597	DIOSPYROS EBENUM	F	113	77.8	0.7	7.5	13.3	0.2	0.7	15	25	0.3	.	.	20	0.00	0.05	0.30	4	2
2598	DIOSPYROS EBENUM	F	66	82.0	0.7	1.2	15.0	1.6	1.1	18	26	1.2	.	.	80	0.02	0.03	0.20	29	3
2599	DIOSPYROS KAKI	F	.	80.9	.	.	.	.	0.5	24	.	.	.	.	1420	.	.	.	5	1
2600	DIOSPYROS KAKI	F	56	83.7	0.6	0.2	14.6	1.2	0.9	9	24	0.4	2	176	600	0.03	0.04	0.30	20	2
2601	DIOSPYROS KAKI	F	78	78.2	0.8	0.4	20.0	1.9	0.6	6	26	0.3	.	.	1500	0.05	0.05	0.00	11	3
2602	DIOSPYROS KAKI	F	77	78.6	0.7	0.4	19.7	1.6	0.6	6	26	0.3	6	174	1626	0.03	0.02	0.10	11	4
2603	DIOSPYROS KAKI	F	.	79.6	0.8	0.2	19.0	.	0.4	10	10	3.0	.	.	1026	0.03	0.03	0.05	10	5
2604	DIOSPYROS LOTUS	S	.	0.0	9.3	2.6	.	.	.	.	.	.	.	.	.	.	.	.	.	21
2605	DIOSPYROS LYCIOIDES SUBSP. SERICEA	F	.	63.5	.	.	.	.	.	.	.	.	.	.	.	.	.	.	32	1
2606	DIOSPYROS MESPILIFORMIS	F	125	64.5	1.0	0.0	33.9	0.6	0.6	42	46	2.0	.	.	.	0.01	.	0.20	13	1
2607	DIOSPYROS TEXANA	S	.	0.0	8.1	2.8	.	.	.	.	.	.	.	.	.	.	.	.	.	19
2608	DIOSPYROS TEXANA	L	.	0.0	13.0	7.4	69.2	21.0	10.4	4110	120	.	.	530	.	.	.	.	.	6
2609	DIOSPYROS VIRGINIANA	S	.	0.0	10.0	2.6	.	.	.	.	.	.	.	.	.	.	.	.	.	19
2610	DIOSPYROS VIRGINIANA	F	127	64.4	0.8	0.4	33.5	1.5	0.9	27	26	2.5	1	310	.	.	.	.	66	4

TABLE 2 - DATA DIRECTLY TRANSCRIBED FROM ORIGINAL SOURCES.

OBS	PLNA	PART	CALORIES	WATER	PROTEIN	FAT	TOTLCARB	FIBER	ASH	CALCIUM	PHOSPHO	IRON	SODIUM	POTASS	CAROT	THIAMINE	RIBOFLAV	NIACIN	ASCORBIC	REFCO
2611	DIOTIS MARITIMA	S	.	0.0	8.9	7.0	.	.	.	.	.	.	.	.	.	.	.	.	.	21
2612	DIPCADI GLAUCUM	S	.	0.0	27.1	10.3	.	.	.	.	.	.	.	.	.	.	.	.	.	21
2613	DIPLOKNEMA BUTYRACEA	S	.	5.0	5.2	55.9	30.0	.	3.8	.	.	.	.	.	.	.	.	.	.	5
2614	DIPLOTAXIS CATHOLICA	S	.	0.0	25.4	31.4	.	.	.	.	.	.	.	.	.	.	.	.	.	21
2615	DIPLOTAXIS ERUCOIDES	S	.	0.0	32.1	32.1	.	.	.	.	.	.	.	.	.	.	.	.	.	21
2616	DIPLOTAXIS ERUCOIDES	S	.	0.0	26.7	35.9	.	.	.	.	.	.	.	.	.	.	.	.	.	21
2617	DIPLOTAXIS SIIFOLIA	S	.	0.0	27.8	33.0	.	.	.	.	.	.	.	.	.	.	.	.	.	21
2618	DIPSACUS SYLVESTRIS	S	.	0.0	23.6	27.7	.	.	.	.	.	.	.	.	.	.	.	.	.	21
2619	DIPSACUS SYLVESTRIS	S	.	0.0	22.6	28.3	.	.	.	.	.	.	.	.	.	.	.	.	.	21
2620	DIPSACUS SYLVESTRIS		.	.	.	.	.	.	.	2090	540	19.0	.	2740	.	.	.	.	.	6
2621	DISANTHUS CERCIDIFOLIUS	S	.	0.0	12.5	16.2	.	.	.	.	.	.	.	.	.	.	.	.	.	20
2622	DISTICHLIS SPICATA	W	.	0.0	6.9	1.7	82.1	29.8	9.3	330	120	19.0	.	240	.	.	.	.	.	6
2623	DISTICHLIS STRICTA	W	.	0.0	5.9	1.7	85.6	29.7	6.8	160	90	.	.	.	.	.	.	.	.	6
2624	DITHYREA WISLIZENII	S	.	0.0	16.9	3.9	.	.	.	.	.	.	.	.	.	.	.	.	.	19
2625	DODONAEA VISCOSA	S	.	0.0	21.4	19.0	.	.	.	.	.	.	.	.	.	.	.	.	.	21
2626	DODONAEA VISCOSA	S	.	0.0	20.4	7.6	.	.	.	.	.	.	.	.	.	.	.	.	.	21
2627	DOLICHOS AXILLARIS	S	.	0.0	28.8	0.6	.	.	.	.	.	.	.	.	.	.	.	.	.	19
2628	DOLICHOS BIFLORUS	S	.	0.0	33.1	0.6	.	.	6.3	.	.	.	.	.	.	.	.	.	.	19
2629	DOLICHOS BIFLORUS	W	.	0.0	16.4	0.2	72.3	30.2	11.1	.	.	.	.	.	.	.	.	.	.	6
2630	DOLICHOS LABLAB	S	340	10.9	22.8	1.0	62.1	8.6	3.2	90	328	9.0	.	.	.	0.54	0.14	2.3	0	1
2631	DOLICHOS LABLAB	SH	.	0.0	14.2	3.5	67.5	28.1	14.8	1980	260	.	.	.	.	.	.	.	.	18
2632	DOLICHOS LABLAB	H	.	0.0	16.6	2.9	71.2	37.1	9.3	.	.	.	.	.	.	.	.	.	.	18
2633	DOLICHOS LABLAB	F	.	0.0	10.1	0.8	82.9	36.2	6.2	900	120	.	.	.	.	.	.	.	.	18
2634	DOLICHOS LABLAB	S	.	0.0	28.0	1.2	66.6	8.6	4.2	490	360	.	.	.	.	.	.	.	.	18
2635	DOLICHOS LABLAB	S	.	0.0	24.9	1.0	.	.	4.2	.	.	.	.	.	.	.	.	.	.	19
2636	DOLICHOS LABLAB	S	.	0.0	21.8	1.2	.	.	3.7	.	.	.	.	.	.	.	.	.	.	19
2637	DOLICHOS LABLAB	S	340	9.2	24.6	0.8	61.0	.	4.4	178	388	3.9	.	.	.	0.17	.	.	.	2
2638	DOLICHOS LABLAB	S	334	12.1	21.5	1.2	61.4	6.8	3.8	98	345	3.9	.	.	.	0.40	0.12	1.8	0	2
2639	DOLICHOS LABLAB	S	334	13.1	22.1	1.0	61.2	7.9	2.6	46	375	7.3	.	.	.	0.44	0.12	1.5	.	3
2640	DOLICHOS LABLAB	S	338	11.8	22.2	1.5	61.0	6.9	3.5	73	418	5.1	.	.	.	0.62	0.18	2.1	.	4
2641	DOLICHOS LABLAB	F	35	88.8	2.8	0.3	7.3	1.8	0.8	57	53	1.0	2	285	348	0.09	0.11	0.9	20	4
2642	DOLICHOS LABLAB	F	.	82.4	4.5	0.1	12.0	2.0	1.0	50	60	1.7	.	.	.	.	.	0.8	9	5
2643	DOLICHOS LIGNOSUS	S	374	0.0	27.1	0.9	67.0	.	4.8	196	196	.	.	.	.	.	.	.	.	10
2644	DOLICHOS SP.	S	.	0.0	28.8	0.5	.	.	.	.	.	.	.	.	.	.	.	.	.	19
2645	DORONICUM CAUCASICUM	S	.	0.0	20.6	38.4	.	.	.	.	.	.	.	.	.	.	.	.	.	20
2646	DORYCNIUM ANATOLICUM	S	.	0.0	42.2	5.7	.	.	.	.	.	.	.	.	.	.	.	.	.	21
2647	DORYCNIUM INTERMEDIUM	S	.	0.0	31.2	5.3	.	.	.	.	.	.	.	.	.	.	.	.	.	21
2648	DOVYALIS HEBECARPA	F	63	82.8	1.2	0.8	14.6	1.8	0.6	13	26	1.2	.	.	210	0.02	0.04	0.3	98	3
2649	DRABA SP.	W	.	0.0	15.8	2.1	67.7	17.6	14.4	3310	330	.	.	1780	.	.	.	.	.	6
2650	DRACOCEPHALUM MOLDAVICA	S	.	0.0	21.2	25.3	.	.	.	.	.	.	.	.	.	.	.	.	.	20
2651	DRACOPIS AMPLEXICAULIS	S	.	0.0	19.3	29.9	.	.	.	.	.	.	.	.	.	.	.	.	.	20
2652	DUCROSIA ANETHIFOLIA	S	.	0.0	13.4	11.6	.	.	.	.	.	.	.	.	.	.	.	.	.	21
2653	DURANTA REPENS	S	.	0.0	13.1	1.4	.	.	.	.	.	.	.	.	.	.	.	.	.	21
2654	DURIO ZIBETHINUS	F	124	66.8	2.5	1.6	28.3	1.4	0.8	20	63	0.9	1	601	10	0.27	0.29	1.2	37	2
2655	DURIO ZIBETHINUS	S	190	51.5	2.6	0.4	43.6	.	1.9	17	68	1.0	3	962	250	.	0.05	0.9	.	2

TABLE 2 - DATA DIRECTLY TRANSCRIBED FROM ORIGINAL SOURCES.

OBS	PLNA	PART	CALORIES	WATER	PROTEIN	FAT	TOTLCARB	FIBER
2656	DURIO ZIBETHINUS	F	67	81.1	2.2	0.8	14.8	1.6
2657	DURIO ZIBETHINUS	P	.	58.0	2.8	3.9	34.1	.
2658	DYSSODIA CANCELLATA	S	.	0.0	20.6	15.2	.	.
2659	DYSSODIA CANCELLATA	S	.	0.0	23.8	19.4	.	.
2660	DYSSODIA HARTWEGI & DYSSODIA PENTACHAETA	S	.	0.0	21.2	18.0	.	.
2661	DYSSODIA PINNATA	S	.	0.0	18.1	11.8	.	.
2662	DYSSODIA ROSEATA	S	.	0.0	18.8	20.3	.	.
2663	DYSSODIA TAGETIFLORA	S	.	0.0	23.8	21.4	.	.
2664	EBENUS HIRSUTA	S	.	0.0	53.7	10.7	.	.
2665	ECBALLIUM ELATERIUM	S	.	0.0	28.2	57.0	.	.
2666	ECBALLIUM ELATERIUM	S	.	0.0	29.7	57.6	.	.
2667	ECHINACEA ANGUSTIFOLIA	S	.	0.0	49.4	37.1	.	.
2668	ECHINACEA ANGUSTIFOLIA	S	.	0.0	11.0	6.7	.	.
2669	ECHINACEA ANGUSTIFOLIA	S	.	0.0	23.2	9.4	.	.
2670	ECHINACEA SP.	S	.	0.0	30.6	16.0	.	.
2671	ECHINARIA CAPITATA	S	.	0.0	17.2	1.6	.	.
2672	ECHINOCHLOA COLONUM	SH	.	0.0	12.5	2.4	72.7	28.6
2673	ECHINOCHLOA COLONUM	S	.	0.0	12.5	7.0	.	.
2674	ECHINOCHLOA COLONUM	W	.	0.0	17.4	2.3	66.7	29.6
2675	ECHINOCHLOA CRUS-GALLI	SH	.	0.0	7.4	2.9	81.1	31.1
2676	ECHINOCHLOA CRUS-GALLI	H	.	0.0	13.5	2.5	73.6	22.7
2677	ECHINOCHLOA CRUS-GALLI	S	.	0.0	15.6	4.3	.	.
2678	ECHINOCHLOA CRUS-GALLI	S	342	11.1	9.8	3.9	72.6	7.2
2679	ECHINOCHLOA CRUS-GALLI	S	.	0.0	12.1	7.1	.	.

OBS	ASH	CALCIUM	PHOSPHO	IRON	SODIUM	POTASS	CAROT	THIAMINE	RIBOFLAV	NIACIN	ASCORBIC	REFCO
2656	1.1	8	38	0.7	.	.	20	C.35	0.20	0.70	24	3
2657	1.2	10	50	1.0	.	.	12	.	.	.	25	5
2658	.	.	.	.	.	.	.	.	.	.	.	20
2659	.	.	.	.	.	.	.	.	.	.	.	20
2660	.	.	.	.	.	.	.	.	.	.	.	20
2661	.	.	.	.	.	.	.	.	.	.	.	20
2662	.	.	.	.	.	.	.	.	.	.	.	20
2663	.	.	.	.	.	.	.	.	.	.	.	21
2664	.	.	.	.	.	.	.	.	.	.	.	21
2665	.	.	.	.	.	.	.	.	.	.	.	21
2666	.	.	.	.	.	.	.	.	.	.	.	21
2667	2.7	.	.	.	.	.	.	.	.	.	.	19
2668	4.3	.	.	.	.	.	.	.	.	.	.	19
2669	.	.	.	.	.	.	.	.	.	.	.	20
2670	.	.	.	.	.	.	.	.	.	.	.	20
2671	.	.	.	.	.	.	.	.	.	.	.	21
2672	12.4	.	.	.	.	.	.	.	.	.	.	18
2673	.	.	.	.	.	.	.	.	.	.	.	19
2674	13.6	830	240	.	.	3670	.	.	.	.	.	6
2675	8.6	.	.	.	.	.	.	.	.	.	.	18
2676	10.4	.	.	.	.	.	.	.	.	.	.	18
2677	4.2	.	.	.	.	.	.	.	.	.	.	19
2678	2.6	41	330	4.3	.	.	0	0.33	0.10	4.00	0	2
2679	.	.	.	.	.	.	.	.	.	.	.	21

TABLE 2 - DATA DIRECTLY TRANSCRIBED FROM ORIGINAL SOURCES.

OBS	PLNA	PART	CALORIES	WATER	PROTEIN	FAT	TOTLCARB	FIBER
2680	ECHINOCHLOA CRUS-GALLI	S	.	11.9	6.2	2.2	75.3	9.8
2681	ECHINOCHLOA CRUS-GALLI	W	.	0.0	13.7	2.9	72.2	22.0
2682	ECHINOCHLOA CRUS-GALLI	W	.	0.0	9.0	2.7	79.9	29.9
2683	ECHINOCHLOA CRUSGALLI VAR. FRUMENTACEA	S	.	0.0	11.9	5.9	.	.
2684	ECHINOCHLOA HAPLOCLADA	SH	.	0.0	13.7	3.0	72.0	30.7
2685	ECHINOCHLOA POLYSTACHYA	SH	.	0.0	13.8	4.2	69.3	32.0
2686	ECHINOCHLOA POLYSTACHYA	H	.	0.0	7.9	2.1	79.5	35.5
2687	ECHINOCHLOA PYRAMIDALIS	SH	.	0.0	7.0	1.1	83.3	31.4
2688	ECHINOCHLOA PYRAMIDALIS	H	.	0.0	15.6	2.4	71.8	33.5
2689	ECHINOCHLOA SP.	S	.	0.0	18.8	6.2	.	.
2690	ECHINOCHLOA SP.	S	.	0.0	10.0	3.1	.	.
2691	ECHINOCHLOA SP.	S	.	0.0	18.8	6.4	.	.
2692	ECHINOCHLOA STAGNINA	SH	.	0.0	11.3	2.2	76.6	32.5
2693	ECHINOCYSTIS FABACEA	S	.	0.0	31.2	51.1	.	.
2694	ECHINOCYSTIS OREGANA	S	.	0.0	33.5	47.0	.	.
2695	ECHINOPEPON HORRIDUS	S	.	0.0	20.2	21.4	.	.
2696	ECHINOPS EXALTATUS	S	.	0.0	37.4	34.7	.	.
2697	ECHINOPS STRIGOSUS	S	.	0.0	36.9	38.8	.	.
2698	ECHINOPS VISCOSUS	S	.	0.0	37.6	27.8	.	.
2699	ECHIUM ITALICUM	S	.	0.0	16.1	17.4	.	.
2700	ECHIUM PLANTAGINEUM	S	.	0.0	19.4	29.9	.	.
2701	ECLIPTA ALBA	S	.	0.0	15.6	13.1	.	.
2702	EHRHARTA CALYCINA	S	.	0.0	13.1	2.2	.	.
2703	EHRHARTA CALYCINA	S	.	0.0	8.1	2.1	.	.

OBS	ASH	CALCIUM	PHOSPHO	IRON	SODIUM	POTASS	CAROT	THIAMINE	RIBOFLAV	NIACIN	ASCORBIC	REFCO
2680	4.4	20	28	2.9	.	.	0	.	.	.	.	5
2681	11.2	.	.	.	.	.	.	.	.	.	.	6
2682	8.4	.	.	.	.	.	.	.	.	.	.	6
2683	3.2	.	.	.	.	.	.	.	.	.	.	19
2684	11.3	.	.	.	.	.	.	.	.	.	.	18
2685	12.7	.	.	.	.	.	.	.	.	.	.	18
2686	10.5	.	.	.	.	.	.	.	.	.	.	18
2687	8.6	.	.	.	.	.	.	.	.	.	.	18
2688	10.2	.	.	.	.	.	.	.	.	.	.	18
2689	.	.	.	.	.	.	.	.	.	.	.	19
2690	.	.	.	.	.	.	.	.	.	.	.	19
2691	.	.	.	.	.	.	.	.	.	.	.	20
2692	9.9	.	.	.	.	.	.	.	.	.	.	18
2693	3.2	.	.	.	.	.	.	.	.	.	.	19
2694	3.7	.	.	.	.	.	.	.	.	.	.	19
2695	.	.	.	.	.	.	.	.	.	.	.	21
2696	5.1	.	.	.	.	.	.	.	.	.	.	19
2697	.	.	.	.	.	.	.	.	.	.	.	21
2698	.	.	.	.	.	.	.	.	.	.	.	21
2699	.	.	.	.	.	.	.	.	.	.	.	21
2700	.	.	.	.	.	.	.	.	.	.	.	20
2701	.	.	.	.	.	.	.	.	.	.	.	21
2702	2.0	.	.	.	.	.	.	.	.	.	.	19
2703	6.3	.	.	.	.	.	.	.	.	.	.	19

TABLE 2 - DATA DIRECTLY TRANSCRIBED FROM ORIGINAL SOURCES.

OBS	PLANA	PART	CALORIES	WATER	PROTEIN	FAT	TOTCARB	FIBER	ASH	CALCIUM	PHOSPHO	IRON	SODIUM	POTASS	CAROTENE	THIAMINE	RIBOFLAV	NIACIN	ASCORBIC	REFCO
2704	EICHHORNIA CRASSIPES	L	.	0.0	22.3	2.0	60.6	19.5	15.1	745	890	158.0	1860	2780	245	0.59	3.07	7.94	.	13
2705	EICHHORNIA CRASSIPES	L	.	0.0	23.4	2.2	54.0	17.1	20.4	560	850	256.0	1630	3450	.	.	.	.	.	13
2706	EICHHORNIA CRASSIPES	L	.	0.0	17.1	1.6	70.2	18.6	11.1	1720	450	212.0	1660	1640	0	.	.	.	.	13
2707	EICHHORNIA CRASSIPES	L	.	0.0	9.7	1.7	68.7	19.2	19.9	792	310	615.0	2020	2660	.	.	.	.	.	13
2708	EICHHORNIA CRASSIPES	W	.	0.0	13.1	1.3	70.3	18.2	15.3	2160	410	.	.	.	.	.	.	.	.	18
2709	EICHHORNIA CRASSIPES	H	.	0.0	11.6	0.7	69.9	24.2	17.8	2190	640	.	.	.	.	.	.	.	.	18
2710	EICHHORNIA CRASSIPES	R	.	0.0	5.8	0.9	89.6	20.5	3.7	.	.	.	.	.	.	.	.	.	.	18
2711	EICHHORNIA CRASSIPES	L	30	89.8	0.5	0.1	7.5	2.4	2.1	.	.	.	.	.	.	.	.	.	.	2
2712	ELAEAGNUS ANGUSTIFOLIA	S	.	0.0	41.9	26.3	.	.	2.4	.	.	.	.	.	.	.	.	.	.	19
2713	ELAEAGNUS ANGUSTIFOLIA	S	.	0.0	9.2	1.8	.	.	.	.	.	.	.	.	.	.	.	.	.	21
2714	ELAEAGNUS ANGUSTIFOLIA	S	.	0.0	9.5	5.3	.	.	.	.	.	.	.	.	.	.	.	.	.	21
2715	ELAEAGNUS COMMUTATA	F	51	86.2	1.3	0.9	10.9	0.5	0.7	7	20	0.4	.	.	20	0.03	0.05	0.40	10	2
2716	ELAEAGNUS PUNGENS	S	.	0.0	42.2	23.1	.	.	.	.	.	.	.	.	.	.	.	.	.	21
2717	ELAEAGNUS UMBELLATA	S	.	0.0	26.0	14.5	.	.	.	.	.	.	.	.	.	.	.	.	.	21
2718	ELAEIS GUINEENSIS	F	540	26.2	1.9	58.4	12.5	3.2	1.0	82	47	4.5	.	.	42420	0.20	0.10	1.40	12	1
2719	ELAEIS GUINEENSIS	F	.	0.0	7.9	54.0	36.4	3.9	1.7	90	310	.	.	.	.	.	.	.	.	18
2720	ELAEIS GUINEENSIS	S	.	0.0	9.9	54.4	.	.	1.6	.	.	.	.	.	.	.	.	.	.	19
2721	ELAEIS GUINEENSIS	F	537	28.0	1.6	59.0	10.5	2.7	0.9	98	44	4.0	.	.	36490	0.25	0.12	1.30	9	3
2722	ELAEIS GUINEENSIS	O	878	0.5	0.0	99.1	0.4	0.0	0.0	7	8	5.5	.	.	27280	0.00	0.03	0.00	.	3
2723	ELAEOCARPUS SERRATUS	F	91	74.2	1.1	0.1	23.9	3.1	0.7	44	36	1.6	.	.	160	0.02	0.01	0.40	101	3
2724	ELAEOCARPUS SERRATUS	F	.	77.2	0.7	0.5	19.5	1.5	0.6	.	.	.	.	.	.	.	.	.	47	5
2725	ELEGIA GALPINII	S	.	0.0	3.1	0.3	.	.	.	.	.	.	.	.	.	.	.	.	.	21
2726	ELEGIA PARVIFLORA	S	.	0.0	17.6	0.7	.	.	.	.	.	.	.	.	.	.	.	.	.	21
2727	ELEOCHARIS DULCIS	R	68	81.1	1.4	0.3	16.1	0.6	1.1	5	77	0.7	10	481	0	0.03	0.02	1.00	6	2
2728	ELEOCHARIS DULCIS	R	79	78.3	1.4	0.2	19.0	0.8	1.1	4	65	0.6	20	500	0	0.14	0.20	1.00	4	4
2729	ELEOCHARIS DULCIS	R	.	77.3	1.5	0.2	35.5	0.9	1.2	.	.	.	.	.	.	.	.	.	.	5
2730	ELEOCHARIS DULCIS	R	.	4.3	8.1	0.7	81.2	3.7	5.7	.	.	.	.	.	.	.	.	.	.	5
2731	ELEOCHARIS PALUSTRIS	W	.	0.0	10.3	.	.	29.0	13.9	450	210	.	.	2720	.	.	.	.	.	6
2732	ELEPHANTOPUS SCABER	S	.	0.0	26.9	21.5	.	.	.	.	.	.	.	.	.	.	.	.	.	20
2733	ELEPHANTORRHIZA ELEPHANTINA	S	.	0.0	22.2	39.0	.	.	.	.	.	.	.	.	.	.	.	.	.	21
2734	ELETTARIA CARDAMOMUM	S	311	8.3	10.8	6.7	68.5	11.3	5.8	383	178	14.0	18	1179	.	0.20	0.18	1.10	.	16
2735	ELETTARIA CARDAMOMUM	F	.	20.0	10.2	2.2	42.1	20.1	5.4	130	160	5.0	.	.	.	.	.	.	.	5
2736	ELETTARIA CARDAMOMUM	F	.	20.0	10.2	2.2	62.2	20.1	5.4	130	160	5.0	.	.	.	.	.	.	.	5
2737	ELEUSINE CORACANA	S	329	10.9	7.4	1.3	77.7	4.3	2.7	397	244	17.1	.	.	0	0.18	0.11	0.80	.	1
2738	ELEUSINE CORACANA	W	.	0.0	7.6	1.1	76.2	33.6	15.1	.	.	.	.	.	.	.	.	.	.	18
2739	ELEUSINE CORACANA	H	.	0.0	3.2	1.3	87.6	34.2	7.9	.	.	.	.	.	.	.	.	.	.	18
2740	ELEUSINE CORACANA	S	332	11.7	6.2	1.4	78.2	1.2	2.5	329	254	5.3	.	.	0	0.33	0.10	1.30	0	2
2741	ELEUSINE CORACANA	S	.	13.1	7.1	1.3	76.3	.	2.2	330	270	5.4	.	.	42	0.40	.	1.10	.	5
2742	ELEUSINE CORACANA	L	.	80.8	1.9	0.5	13.2	5.4	3.5	.	.	.	.	.	.	.	.	.	.	5
2743	ELEUSINE INDICA	W	.	0.0	13.3	2.8	71.9	28.5	12.0	.	.	.	.	.	.	.	.	.	.	18
2744	ELEUSINE INDICA	H	.	0.0	3.5	1.1	85.5	35.8	9.9	.	.	.	.	.	.	.	.	.	.	18
2745	ELEUSINE INDICA	H	.	0.0	3.5	1.1	85.5	35.8	9.9	.	.	.	.	.	.	.	.	.	.	6
2746	ELIONURUS TRIPSACOIDES	L	.	0.0	5.3	1.7	87.6	39.1	5.4	250	100	.	.	.	.	.	.	.	.	6
2747	ELYMUS AMBIGUUS	H	.	0.0	7.9	1.8	84.5	36.5	5.8	.	.	.	.	.	.	.	.	.	.	6
2748	ELYMUS CANADENSIS	S	.	0.0	26.9	2.2	.	.	2.9	.	.	.	.	.	.	.	.	.	.	19

TABLE 2 - DATA DIRECTLY TRANSCRIBED FROM ORIGINAL SOURCES.

OBS	PLANA	PART	CALORIES	WATER	PROTEIN	FAT	TOTCARB	FIBER	ASH	CALCIUM	PHOSPHO	IRON	SODIUM	POTASS	CAROT	THIAMINE	RIBOFLAV	NIACIN	ASCORBIC	REFCO
2749	ELYMUS CANADENSIS	S	.	0.0	16.2	3.2	.	.	.	.	.	.	.	.	.	.	.	.	.	19
2750	ELYMUS CANADENSIS	W	.	0.0	8.4	2.3	77.8	30.6	11.5	760	210	.	.	.	.	.	.	.	.	6
2751	ELYMUS CAPUT-MEDUSAE	S	.	0.0	14.2	1.9	.	.	.	.	.	.	.	.	.	.	.	.	.	21
2752	ELYMUS CONDENSATUS	W	.	0.0	9.3	3.2	77.4	35.9	10.1	390	170	.	.	.	.	.	.	.	.	6
2753	ELYMUS GIGANTEUS	S	.	0.0	16.3	2.1	.	.	3.0	.	.	.	.	.	.	.	.	.	.	19
2754	ELYMUS GLAUCUS	W	.	0.0	7.3	2.3	.	.	5.5	.	.	.	.	.	.	.	.	.	.	6
2755	ELYMUS JUNCEUS	S	.	0.0	17.1	1.9	.	.	6.2	.	.	.	.	.	.	.	.	.	.	19
2756	ELYMUS JUNCEUS	W	.	0.0	14.1	2.6	75.3	22.4	8.0	.	.	.	.	.	.	.	.	.	.	6
2757	ELYMUS VILLOSUS	W	.	0.0	21.5	3.0	59.9	21.7	15.6	530	350	.	.	.	.	.	.	.	.	6
2758	ELYMUS VIRGINICUS	W	.	0.0	5.9	2.3	85.5	34.4	6.3	400	210	.	.	.	.	.	.	.	.	6
2759	ELYONURUS ARGENTEUS	W	.	0.0	8.0	1.9	82.6	37.8	7.5	.	.	.	.	.	.	.	.	.	.	18
2760	EMILIA COCCINEA	L	64	79.9	3.2	0.7	14.0	1.9	2.2	260	52	.	.	.	.	.	.	.	.	1
2761	EMILIA SONCHIFOLIA	L	56	81.8	4.0	0.6	11.7	2.0	1.9	398	118	.	.	.	.	.	.	.	.	1
2762	ENARTHROCARPUS STRANGULATUS	S	.	0.0	35.8	39.6	.	.	.	.	.	.	.	.	.	.	.	.	.	21
2763	ENCELIA FARINOSA	S	.	0.0	23.1	23.9	.	.	4.5	.	.	.	.	.	.	.	.	.	.	19
2764	ENDOPLEURA UCHI	F	240	50.0	0.3	13.1	35.9	2.8	0.7	20	17	2.0	.	.	330	0.01	0.01	2.7	6	3
2765	ENGELMANNIA PINNATIFIDA	S	.	0.0	26.6	25.5	.	.	.	.	.	.	.	.	.	.	.	.	.	20
2766	ENGELMANNIA PINNATIFIDA	S	.	0.0	30.6	26.9	.	.	.	.	.	.	.	.	.	.	.	.	.	21
2767	ENGELMANNIA PINNATIFIDA	L	.	0.0	23.2	1.2	58.2	9.9	17.4	2630	370	.	.	4730	.	.	.	.	.	6
2768	ENKIANTHUS PERULATUS	S	.	0.0	14.2	18.6	.	.	.	.	.	.	.	.	.	.	.	.	.	21
2769	ENSETE VENTRICOSUM	HE	171	56.3	1.2	0.2	41.3	1.6	1.0	120	.	5.3	.	.	.	0.03	0.04	0.1	.	1
2770	ENSETE VENTRICOSUM	HE	225	43.7	0.2	0.1	55.4	0.1	0.6	44	.	7.7	.	.	.	.	0.02	0.0	.	1
2771	ENSETE VENTRICOSUM	S	352	12.4	13.3	1.4	72.2	0.5	0.7	.	.	.	.	.	.	.	.	.	.	1
2772	ENTEROLOBIUM CONTORTISILIQUA	S	.	0.0	25.6	2.3	.	.	.	.	.	.	.	.	.	.	.	.	.	21
2773	ENTEROLOBIUM CYCLOCARPUM	C	.	0.0	16.8	5.3	72.9	12.2	5.0	670	140	.	.	.	.	.	.	.	.	18
2774	ENTEROLOBIUM CYCLOCARPUM	S	.	0.0	21.8	1.7	72.5	13.5	4.1	.	.	.	.	.	.	.	.	.	.	18
2775	ENTEROLOBIUM CYCLOCARPUM	S	.	0.0	43.4	2.5	.	.	3.3	.	.	.	.	.	.	.	.	.	.	19
2776	ENTEROLOBIUM CYCLOCARPUM	S	.	0.0	3.9	0.3	.	.	5.0	.	.	.	.	.	.	.	.	.	.	19
2777	ENTEROLOBIUM CYCLOCARPUM	S	.	0.0	26.0	1.6	.	.	.	.	.	.	.	.	.	.	.	.	.	20
2778	ENTEROLOBIUM CYCLOCARPUM	S	386	1.4	31.3	3.9	60.2	2.0	3.2	30	547	4.6	.	.	30	2.74	0.23	1.8	7	3
2779	ENTEROPOGON MONOSTACHYOS	L	.	8.4	4.3	1.4	77.8	33.5	8.0	.	.	.	.	.	.	.	.	.	.	5
2780	ENTOLASIA IMBRICATA	W	.	0.0	19.5	1.1	72.9	14.8	8.5	.	.	.	.	.	.	.	.	.	.	18
2781	EPHEDRA ANTISYPHILITICA	Y	.	0.0	7.0	1.5	85.2	33.7	6.3	3060	80	.	.	620	.	.	.	.	.	6
2782	EPHEDRA NEVADENSIS	L	.	0.0	5.0	.	.	.	.	5810	500	.	.	.	.	.	.	.	.	6
2783	EPHEDRA TRIFURCA	L	.	0.0	6.0	.	.	.	7.7	3310	80	13.0	.	420	.	.	.	.	.	6
2784	EPHEDRA TWEEDIANA	S	.	0.0	15.1	14.0	.	.	.	.	.	.	.	.	.	.	.	.	.	21
2785	EPILOBIUM ANGUSTIFOLIUM	L	.	0.0	2.8	7.1	82.5	12.7	7.6	.	.	.	.	.	.	.	.	.	.	6
2786	EPILOBIUM HIRSUTUM	S	.	0.0	25.2	31.7	.	.	.	.	.	.	.	.	.	.	.	.	.	21
2787	EQUISETUM ARVENSE	L	20	93.7	1.0	0.2	4.4	1.1	0.7	58	93	4.4	.	.	300	0.00	0.07	5.6	50	2
2788	EQUISETUM ARVENSE	W	.	0.0	5.3	2.4	73.8	23.5	18.5	.	.	.	.	.	.	.	.	.	.	6
2789	ERAGROSTIS BEYRICHII	W	.	0.0	7.2	1.6	79.5	32.7	11.7	.	.	.	.	.	.	.	.	.	.	6
2790	ERAGROSTIS CAESPITOSA	W	.	0.0	8.7	1.8	81.4	37.5	8.1	.	.	.	.	.	.	.	.	.	.	18
2791	ERAGROSTIS CHLOROMELAS	W	.	0.0	10.4	3.0	79.8	34.5	7.8	.	.	.	.	.	.	.	.	.	.	6
2792	ERAGROSTIS CILIANENSIS	W	.	0.0	15.3	2.4	71.7	29.0	10.6	.	.	.	.	.	.	.	.	.	.	18
2793	ERAGROSTIS CILIANENSIS	W	.	0.0	12.2	1.6	75.3	27.8	10.9	950	240	.	.	1610	.	.	.	.	.	6

TABLE 2 - DATA DIRECTLY TRANSCRIBED FROM ORIGINAL SOURCES.

OBS	PLNA	PART	CALORIES	WATER	PROTEIN	FAT	TOTCARB	FIBER	ASH	CALCIUM	PHOSPHO	IRON	SODIUM	POTASS	CAROT	THIAMINE	RIBOFLAV	NIACIN	ASCORBIC	REFCO
2794	ERAGROSTIS CURVULA	W	.	0.0	13.1	2.1	78.8	31.5	6.0	.	.	.	.	.	.	.	.	.	.	18
2795	ERAGROSTIS CURVULA	W	.	0.0	9.3	3.0	82.1	32.4	5.6	310	180	.	.	.	28	.	.	.	.	6
2796	ERAGROSTIS DIFFUSA	W	.	0.0	11.8	3.7	76.1	30.8	8.4	380	200	.	.	.	.	.	.	.	.	6
2797	ERAGROSTIS INTERMEDIA	W	.	0.0	9.9	2.0	76.6	32.8	11.5	410	170	.	.	.	.	.	.	.	.	6
2798	ERAGROSTIS LEHMANNIANA	H	.	0.0	10.4	1.7	79.2	31.9	8.7	.	.	.	.	.	.	.	.	.	.	18
2799	ERAGROSTIS LEHMANNIANA	W	.	0.0	6.7	1.7	84.0	35.0	7.6	.	.	.	.	.	.	.	.	.	.	6
2800	ERAGROSTIS LUGENS	W	.	0.0	13.2	2.1	73.9	29.4	10.8	370	160	.	.	.	.	.	.	.	.	6
2801	ERAGROSTIS PLUMOSA	L	.	0.0	9.9	2.2	79.7	33.6	8.2	550	870	.	.	.	.	.	.	.	.	5
2802	ERAGROSTIS SECUNDIFLORA	W	.	0.0	6.5	2.0	81.4	37.1	10.1	340	120	.	.	.	.	.	.	.	.	6
2803	ERAGROSTIS SUPERBA	W	.	0.0	11.3	1.9	80.2	32.3	6.6	.	.	.	.	.	.	.	.	.	.	18
2804	ERAGROSTIS SUPERBA	S	.	0.0	21.2	5.0	.	.	3.4	.	.	.	.	.	.	.	.	.	.	19
2805	ERAGROSTIS TEF	S	328	12.5	9.6	2.6	72.0	2.7	3.3	172	313	75.5	.	.	5	0.13	0.12	1.4	88	1
2806	ERAGROSTIS TEF	S	332	11.3	8.7	1.9	75.5	2.3	2.6	146	380	20.9	.	.	0	0.38	0.14	1.9	.	1
2807	ERAGROSTIS TEF	H	.	0.0	10.5	1.1	83.1	34.2	5.3	.	.	.	.	.	.	.	.	.	.	18
2808	ERAGROSTIS TEF	S	.	8.8	10.4	2.2	.	4.3	2.6	.	.	.	.	.	.	.	.	.	.	5
2809	ERAGROSTIS TEF	L	.	6.4	8.4	1.9	74.1	33.7	9.6	.	.	.	.	.	.	.	.	.	.	5
2810	ERAGROSTIS TENELLA	L	.	0.0	.	.	.	.	4.0	493	613	0.0	162	1441	.	.	.	.	.	5
2811	ERAGROSTIS TREMULA	L	.	6.2	3.1	1.7	84.1	38.5	4.8	.	.	.	.	.	.	.	.	.	.	5
2812	ERAGROSTIS TRICHODES	W	.	0.0	9.8	2.5	79.3	27.8	8.4	490	180	.	.	.	.	.	.	.	.	6
2813	ERAGROSTIS TRICHODES	W	.	0.0	4.8	2.6	89.6	34.3	3.0	380	180	.	.	.	.	.	.	.	.	6
2814	ERAGROSTIS TRICHODES	W	.	0.0	3.1	2.1	89.5	36.3	5.3	370	120	.	.	.	.	.	.	.	.	6
2815	EREMIA TOTTA	S	.	0.0	2.1	3.1	.	.	.	.	.	.	.	.	.	.	.	.	.	21
2816	EREMOCARPUS SETIGERUS	S	.	0.0	28.1	28.0	.	.	.	.	.	.	.	.	.	.	.	.	.	20
2817	EREMOCHLOA OPHIUROIDES	W	.	0.0	11.3	2.8	80.3	31.6	5.6	580	260	13.0	.	.	.	.	.	.	.	6
2818	EREMOPOGON FOVEOLATUS	L	.	7.2	2.3	1.3	81.8	37.8	7.4	.	.	.	.	.	.	.	.	.	.	5
2819	EREMOSTACHYS LACINIATA	S	.	0.0	26.4	10.4	.	.	.	.	.	.	.	.	.	.	.	.	.	21
2820	EREMOSTACHYS LACINIATA	S	.	0.0	21.1	12.2	.	.	.	.	.	.	.	.	.	.	.	.	.	21
2821	EREMOSTACHYS SPECIOSA	S	.	0.0	24.2	21.1	.	.	.	.	.	.	.	.	.	.	.	.	.	21
2822	EREMOSTACHYS VICARYI	S	.	0.0	25.0	19.0	.	.	.	.	.	.	.	.	.	.	.	.	.	21
2823	EREMURUS AUCHERIANUS	S	.	0.0	30.6	29.5	.	.	.	.	.	.	.	.	.	.	.	.	.	21
2824	EREMURUS AURANTIACUS	S	.	0.0	34.1	27.3	.	.	.	.	.	.	.	.	.	.	.	.	.	21
2825	ERIANTHUS LONGESETOSUS	L	.	0.0	8.0	2.2	80.6	35.0	9.2	210	160	.	.	.	.	.	.	.	.	5
2826	ERIANTHUS RAVENNAE	H	.	0.0	1.9	1.6	89.6	43.9	6.9	.	.	.	.	.	.	.	.	.	.	18
2827	ERIANTHUS SP.	W	.	0.0	6.1	3.5	88.5	48.9	1.9	610	100	.	.	.	1	.	.	.	.	6
2828	ERIGERON CANADENSIS	W	.	0.0	14.9	1.8	75.1	26.1	8.2	1010	280	.	.	2610	.	.	.	.	.	6
2829	ERIGERON SPECIOSUS	S	.	0.0	26.1	21.8	.	.	.	.	.	.	.	.	.	.	.	.	.	21
2830	ERIOBOTRYA JAPONICA	F	40	88.6	0.5	0.2	10.2	0.5	0.5	18	25	0.2	4	315	775	0.02	0.04	0.2	4	2
2831	ERIOBOTRYA JAPONICA	F	44	88.1	0.2	0.6	10.7	0.8	0.4	18	14	0.8	.	.	260	0.02	0.05	0.3	10	3
2832	ERIOBOTRYA JAPONICA	F	48	86.5	0.4	0.2	12.4	0.5	0.5	20	36	0.4	.	348	402	.	.	.	1	4
2833	ERIOBOTRYA JAPONICA	F	.	87.4	0.7	0.3	11.1	0.9	0.5	30	20	0.7	.	.	.	.	.	.	.	5
2834	ERIOCHLOA GRACILIS	W	.	0.0	13.3	2.4	73.5	29.5	10.8	680	220	.	.	3040	.	.	.	.	.	6
2835	ERIOCHLOA NUBICA	W	.	0.0	10.2	1.4	72.2	33.5	16.2	.	.	.	.	.	.	.	.	.	.	18
2836	ERIOCHLOA POLYSTACHYA	SH	.	0.0	10.3	1.9	77.2	32.9	10.6	.	.	.	.	.	.	.	.	.	.	18
2837	ERIOCHLOA POLYSTACHYA	C	.	0.0	7.1	1.3	81.0	37.1	10.6	.	.	.	.	.	.	.	.	.	.	18
2838	ERIOCHLOA POLYSTACHYA	L	.	0.0	14.4	2.8	72.1	27.5	10.7	.	.	.	.	.	.	.	.	.	.	18

TABLE 2 - DATA DIRECTLY TRANSCRIBED FROM ORIGINAL SOURCES.

OBS	PLNA	PART	CALORIES	WATER	PROTEIN	FAT	TOTLCARB	FIBER	ASH	CALCIUM	PHOSPHO	IRON	SODIUM	POTASS	CAROT	THIAMINE	RIBOFLAV	NIACIN	ASCORBIC	REFCO
2839	ERIOGONUM ALATUM	S	.	0.0	18.8	2.6	.	.	.	.	.	.	.	.	.	.	.	.	.	19
2840	ERIOGONUM ALATUM	S	.	0.0	7.5	1.9	.	.	.	.	.	.	.	.	.	.	.	.	.	19
2841	ERIOGONUM ALATUM	S	.	0.0	18.8	2.9	.	.	3.2	.	.	.	.	.	.	.	.	.	.	19
2842	ERIOGONUM ALATUM	S	.	0.0	6.2	1.7	.	.	5.0	.	.	.	.	.	.	.	.	.	.	19
2843	ERIOGONUM ANNUUM	S	.	0.0	19.2	5.9	.	.	.	.	.	.	.	.	.	.	.	.	.	21
2844	ERIOGONUM SP.	W	.	0.0	6.6	.	.	.	13.7	1030	90	.	.	.	1	.	.	.	.	6
2845	ERIOGONUM SP.	I	.	0.0	17.0	2.2	74.3	13.5	6.5	.	.	.	.	.	.	.	.	.	.	6
2846	ERIOSEMA GRANDIFLORUM	S	.	0.0	23.2	1.9	.	.	.	.	.	.	.	.	.	.	.	.	.	21
2847	ERODIUM BOTRYS	W	.	0.0	17.0	4.8	65.7	22.0	12.5	2400	540	.	.	.	.	.	.	.	.	18
2848	ERODIUM BOTRYS	W	.	0.0	13.5	2.8	81.1	21.2	12.6	1950	330	.	.	2950	.	.	.	.	.	6
2849	ERODIUM CICONIUM	S	.	0.0	21.0	12.0	.	.	.	.	.	.	.	.	.	.	.	.	.	21
2850	ERODIUM CICUTARIUM	W	.	0.0	16.5	.	.	18.2	13.1	2510	470	.	.	3370	.	.	.	.	.	6
2851	ERODIUM GRUINUM	S	.	0.0	20.6	10.7	.	.	.	.	.	.	.	.	.	.	.	.	.	20
2852	ERODIUM MOSCHATUM	W	.	0.0	16.9	5.1	65.2	22.1	12.8	2490	530	.	.	.	.	.	.	.	.	18
2853	ERODIUM MOSCHATUM	W	.	0.0	16.2	.	.	18.1	15.0	2210	440	.	.	3450	.	.	.	.	.	6
2854	ERODIUM TEXANUM	S	.	0.0	25.6	15.8	.	.	.	.	.	.	.	.	.	.	.	.	.	19
2855	ERODIUM TEXANUM	W	.	0.0	24.4	2.4	62.0	8.1	11.2	1960	260	.	.	2280	.	.	.	.	.	6
2856	ERUCA LONGIROSTRIS	S	.	0.0	31.6	25.5	.	.	.	.	.	.	.	.	.	.	.	.	.	21
2857	ERUCA SATIVA	S	.	0.0	37.4	22.5	.	.	4.8	.	.	.	.	.	.	.	.	.	.	19
2858	ERUCA SATIVA	S	.	0.0	30.6	34.6	.	.	.	.	.	.	.	.	.	.	.	.	.	20
2859	ERUCA SATIVA	S	.	0.0	29.6	35.1	.	.	.	.	.	.	.	.	.	.	.	.	.	21
2860	ERUCA SATIVA	S	.	0.0	35.1	24.2	.	.	.	.	.	.	.	.	.	.	.	.	.	21
2861	ERUCA VESICARIA	S	.	0.0	33.8	27.2	.	.	.	.	.	.	.	.	.	.	.	.	.	21
2862	ERUCARIA HISPANICA	S	.	0.0	35.1	31.9	.	.	.	.	.	.	.	.	.	.	.	.	.	21
2863	ERUCASTRUM STRIGOSUM	S	.	0.0	25.6	28.2	.	.	.	.	.	.	.	.	.	.	.	.	.	21
2864	ERUCASTRUM STRIGOSUM	S	.	0.0	28.8	25.2	.	.	.	.	.	.	.	.	.	.	.	.	.	21
2865	ERYNGIUM AQUATICUM VAR. RAVENELII	L	44	85.7	2.3	0.7	9.2	2.0	2.1	295	46	5.7	.	.	3320	0.09	0.22	1.00	28	3
2866	ERYNGIUM CRETICUM	S	.	0.0	15.9	9.1	.	.	.	.	.	.	.	.	.	.	.	.	.	21
2867	ERYNGIUM GIGANTEUM	S	.	0.0	20.9	26.7	.	.	7.1	.	.	.	.	.	.	.	.	.	.	19
2868	ERYNGIUM MARITIMUM	S	.	0.0	13.8	23.2	.	.	.	.	.	.	.	.	.	.	.	.	.	21
2869	ERYNGIUM PLANUM	S	.	0.0	30.0	27.9	.	.	.	.	.	.	.	.	.	.	.	.	.	20
2870	ERYNGIUM SP.	S	.	0.0	27.6	31.6	.	.	.	.	.	.	.	.	.	.	.	.	.	21
2871	ERYSIMUM CUSPIDATUM	S	.	0.0	35.8	32.8	.	.	.	.	.	.	.	.	.	.	.	.	.	21
2872	ERYSIMUM PEROFSKIANUM	S	.	0.0	28.0	35.3	.	.	5.4	.	.	.	.	.	.	.	.	.	.	19
2873	ERYTHEA SP.	S	.	0.0	5.8	6.6	.	.	.	.	.	.	.	.	.	.	.	.	.	19
2874	ERYTHEA SP.	S	.	0.0	5.6	9.2	.	.	.	.	.	.	.	.	.	.	.	.	.	21
2875	ERYTHRINA BERTEROANA	SH	48	84.2	4.4	0.2	10.0	2.4	1.2	108	80	2.2	.	.	220	0.19	0.19	1.20	37	3
2876	ERYTHRINA CRISTA-GALLI	S	.	0.0	30.3	16.7	.	.	.	.	.	.	.	.	.	.	.	.	.	21
2877	ERYTHRINA FLABELLIFORMIS	S	.	0.0	36.2	26.1	.	.	.	.	.	.	.	.	.	.	.	.	.	19
2878	ERYTHRINA FLABELLIFORMIS	S	.	0.0	27.5	20.1	.	.	.	.	.	.	.	.	.	.	.	.	.	20
2879	ERYTHRINA FLABELLIFORMIS	S	.	0.0	25.6	19.6	.	.	.	.	.	.	.	.	.	.	.	.	.	20
2880	ERYTHRINA FUSCA	L	60	81.5	4.6	0.8	11.7	4.1	1.4	57	40	1.8	.	.	2300	0.24	0.17	4.70	3	2
2881	ERYTHRINA HERBACEA	S	.	0.0	26.0	21.3	.	.	.	.	.	.	.	.	.	.	.	.	.	21
2882	ERYTHRINA POEPPIGIANA	S	.	0.0	36.8	12.4	.	.	5.3	.	.	.	.	.	.	.	.	.	.	19
2883	ERYTHRINA SP.	S	.	0.0	42.5	19.0	.	.	6.2	.	.	.	.	.	.	.	.	.	.	19

TABLE 2 - DATA DIRECTLY TRANSCRIBED FROM ORIGINAL SOURCES.

OBS	PLANA	PART	CALORIES	WATER	PROTEIN	FAT	TOTCARB	FIBER	ASH	CALCIUM	PHOSPHO	IRON	SODIUM	POTASS	CAROT	THIAMINE	RIBOFLAV	NIACIN	ASCORBIC	REFCO
2884	ERYTHRINA SP.	S	.	0.0	5.4	1.4	.	.	2.4	.	.	.	.	.	.	.	.	.	.	19
2885	ERYTHROCOCCA BONGESIS	L	106	68.0	7.2	2.7	18.5	4.3	3.6	678	107	.	.	.	.	.	.	.	.	1
2886	ERYTHROXYLUM COCA	L	305	6.5	18.9	5.0	60.6	14.4	9.0	1540	911	45.8	41	2020	11000	0.35	1.91	1.29	1	14
2887	ERYTHROXYLUM MEXICANUM	S	.	0.0	15.6	6.2	.	.	.	.	.	.	.	.	.	.	.	.	.	20
2888	ESCHSCHOLZIA CALIFORNICA	S	.	0.0	25.0	46.8	.	.	.	.	.	.	.	.	.	.	.	.	.	20
2889	ESCOBEDIA CF. CRASSIPES	S	.	0.0	20.6	15.0	.	.	.	.	.	.	.	.	.	.	.	.	.	20
2890	ESCOBEDIA PEDUNCULARIS	S	.	0.0	21.2	12.5	.	.	.	.	.	.	.	.	.	.	.	.	.	20
2891	EUADENIA SP	L	77	75.0	5.4	0.9	15.7	3.1	3.0	420	40	.	.	.	.	.	.	.	.	1
2892	EUCHLAENA MEXICANA	S	.	0.0	10.4	3.8	.	.	5.3	.	.	.	.	.	.	.	.	.	.	19
2893	EUCHLAENA SP.	W	.	0.0	8.4	2.3	79.5	31.1	9.8	.	.	.	.	.	.	.	.	.	.	6
2894	EUCLEA ACUTIFOLIA	S	.	0.0	7.0	5.1	.	.	.	.	.	.	.	.	.	.	.	.	.	21
2895	EUCLEA POLYANDRA	S	.	0.0	5.8	3.8	.	.	.	.	.	.	.	.	.	.	.	.	.	21
2896	EUGENIA DOMBEYI	S	.	0.0	5.3	0.8	.	.	2.7	.	.	.	.	.	.	.	.	.	.	19
2897	EUGENIA DOMBEYI	F	53	85.3	0.6	0.3	13.4	0.6	0.4	40	14	0.5	.	.	40	0.04	0.03	0.30	19	3
2898	EUGENIA JAMBOLANA	S	.	0.0	6.6	1.4	.	.	.	.	.	.	.	.	.	.	.	.	.	21
2899	EUGENIA JAMBOS	S	.	0.0	5.8	3.8	.	.	2.0	.	.	.	.	.	.	.	.	.	.	19
2900	EUGENIA SP.	F	31	91.3	0.5	0.1	7.9	0.6	0.2	11	11	0.4	8	60	0	0.02	0.02	0.40	13	2
2901	EUGENIA SP.	F	51	85.8	0.8	0.4	12.5	0.6	0.5	9	11	0.2	.	.	1270	0.03	0.06	0.30	14	3
2902	EUGENIA UNIFLORA	S	.	0.0	5.6	0.5	.	.	2.9	.	.	.	.	.	.	.	.	.	.	19
2903	EUGENIA UNIFLORA	F	38	89.0	0.5	0.1	10.2	0.3	0.2	7	9	0.1	.	.	1120	0.02	0.05	0.20	19	2
2904	EUGENIA UNIFLORA	F	51	85.8	0.8	0.4	12.5	0.6	0.5	9	11	0.2	.	.	900	0.03	0.04	0.30	30	4
2905	EULALIOPSIS BINATA	L	.	.	2.3	2.3	89.0	32.1	6.4	650	50	.	310	260	.	.	.	.	.	5
2906	EUONYMUS ALATUS	S	.	0.0	20.5	44.4	.	.	3.9	.	.	.	.	.	.	.	.	.	.	19
2907	EUONYMUS EUROPAEUS	S	.	0.0	20.7	51.5	.	.	.	.	.	.	.	.	.	.	.	.	.	21
2908	EUONYMUS HAMILTONIANUS	S	.	0.0	16.1	51.8	.	.	.	.	.	.	.	.	.	.	.	.	.	21
2909	EUONYMUS LATIFOLIUS	S	.	0.0	20.8	47.9	.	.	.	.	.	.	.	.	.	.	.	.	.	21
2910	EUONYMUS VERRUCOSUS	S	.	0.0	27.2	36.2	.	.	.	.	.	.	.	.	.	.	.	.	.	21
2911	EUONYMUS VERRUCOSUS	S	.	0.0	19.2	52.5	.	.	.	.	.	.	.	.	.	.	.	.	.	21
2912	EUPATORIUM HYSSOPINUM & EUPATORIUM SPP.	S	.	0.0	30.6	21.7	.	.	.	.	.	.	.	.	.	.	.	.	.	20
2913	EUPATORIUM LIEBMANNII	S	.	0.0	29.4	30.8	.	.	.	.	.	.	.	.	.	.	.	.	.	20
2914	EUPATORIUM ODORATUM	S	.	0.0	20.6	15.7	.	.	.	.	.	.	.	.	.	.	.	.	.	19
2915	EUPATORIUM ODORATUM	S	.	0.0	21.2	27.5	.	.	.	.	.	.	.	.	.	.	.	.	.	20
2916	EUPATORIUM RUGOSUM	S	.	0.0	25.6	39.6	.	.	3.8	.	.	.	.	.	.	.	.	.	.	19
2917	EUPATORIUM SEROTINUM	S	.	0.0	23.3	34.2	.	.	.	.	.	.	.	.	.	.	.	.	.	21
2918	EUPATORIUM SP.	S	.	0.0	15.6	26.5	.	.	.	.	.	.	.	.	.	.	.	.	.	20
2919	EUPATORIUM SP.	S	.	0.0	23.8	24.9	.	.	.	.	.	.	.	.	.	.	.	.	.	20
2920	EUPHORBIA AKENOCARPA	S	.	0.0	23.8	42.0	.	.	.	.	.	.	.	.	.	.	.	.	.	21
2921	EUPHORBIA AMYGDALOIDES	S	.	0.0	21.6	40.0	.	.	.	.	.	.	.	.	.	.	.	.	.	21
2922	EUPHORBIA BICOLOR	S	.	0.0	23.1	26.8	.	.	.	.	.	.	.	.	.	.	.	.	.	21
2923	EUPHORBIA COERULESCENS	SH	.	0.0	15.1	22.6	50.5	15.3	11.8	170	240	.	.	.	.	.	.	.	.	18
2924	EUPHORBIA CUPHOSPERMA	S	.	0.0	7.5	7.1	.	.	.	.	.	.	.	.	.	.	.	.	.	20
2925	EUPHORBIA HETEROPHYLLA	S	.	0.0	25.2	37.4	.	.	5.7	.	.	.	.	.	.	.	.	.	.	19
2926	EUPHORBIA HETEROPHYLLA	S	.	0.0	26.2	37.9	.	.	6.5	.	.	.	.	.	.	.	.	.	.	19
2927	EUPHORBIA HETEROPHYLLA	S	.	0.0	26.9	29.6	.	.	.	.	.	.	.	.	.	.	.	.	.	19
2928	EUPHORBIA HETEROPHYLLA	S	.	0.0	23.1	40.1	.	.	.	.	.	.	.	.	.	.	.	.	.	20

TABLE 2 - DATA DIRECTLY TRANSCRIBED FROM ORIGINAL SOURCES.

OBS	PLNA	PART	CALORIES	WATER	PROTEIN	FAT	TOTLCARB	FIBER	ASH	CALCIUM	PHOSPHO	IRON	SODIUM	POTASS	CAROT	THIAMINE	RIBOFLAV	NIACIN	ASCORBIC	REFCO
2929	EUPHORBIA KOTSCHYANA	S	.	0.0	39.4	36.0	.	.	.	.	.	.	.	.	.	.	.	.	.	21
2930	EUPHORBIA LATHYRUS	S	.	0.0	15.8	44.3	.	.	2.6	.	.	.	.	.	.	.	.	.	.	19
2931	EUPHORBIA LATHYRUS	S	.	0.0	8.4	6.5	.	.	9.3	.	.	.	.	.	.	.	.	.	.	19
2932	EUPHORBIA LATHYRUS	S	.	0.0	15.0	47.5	.	.	.	.	.	.	.	.	.	.	.	.	.	20
2933	EUPHORBIA MACROCLADA	S	.	0.0	32.2	24.5	.	.	.	.	.	.	.	.	.	.	.	.	.	21
2934	EUPHORBIA MARGINATA	S	.	0.0	21.4	32.0	.	.	3.2	.	.	.	.	.	.	.	.	.	.	19
2935	EUPHORBIA MARGINATA	S	.	0.0	20.9	31.7	.	.	3.2	.	.	.	.	.	.	.	.	.	.	19
2936	EUPHORBIA MARGINATA	S	.	0.0	16.4	17.9	.	.	3.2	.	.	.	.	.	.	.	.	.	.	19
2937	EUPHORBIA MAURITANICA	S	.	0.0	17.6	45.0	.	.	.	.	.	.	.	.	.	.	.	.	.	21
2938	EUPHORBIA MYRSINITES	S	.	0.0	27.2	32.3	.	.	.	.	.	.	.	.	.	.	.	.	.	20
2939	EUPHORBIA PARALIAS	S	.	0.0	21.7	41.0	.	.	.	.	.	.	.	.	.	.	.	.	.	21
2940	EUPHORBIA PARALIAS	S	.	0.0	20.4	49.4	.	.	.	.	.	.	.	.	.	.	.	.	.	21
2941	EUPHORBIA PARRYI	S	.	0.0	21.9	44.4	.	.	.	.	.	.	.	.	.	.	.	.	.	20
2942	EUPHORBIA SALICIFOLIA	S	.	0.0	24.2	30.6	.	.	.	.	.	.	.	.	.	.	.	.	.	21
2943	EUPHORBIA SEGETALIS	S	.	0.0	22.6	39.8	.	.	.	.	.	.	.	.	.	.	.	.	.	21
2944	EUPHORBIA SERRATA	S	.	0.0	21.3	42.1	.	.	.	.	.	.	.	.	.	.	.	.	.	21
2945	EUPHORBIA SP.	S	.	0.0	20.0	24.3	.	.	.	.	.	.	.	.	.	.	.	.	.	20
2946	EUPHORBIA SP.	S	.	0.0	9.3	13.8	.	.	.	.	.	.	.	.	.	.	.	.	.	21
2947	EUPHORBIA SP.	W	.	0.0	14.4	4.1	70.5	13.0	11.0	1190	260	.	.	1940	.	.	.	.	.	6
2948	EUPHORBIA TERRACINA	S	.	0.0	17.6	39.8	.	.	.	.	.	.	.	.	.	.	.	.	.	21
2949	EUPHORBIA TERRACINA	S	.	0.0	17.8	32.5	.	.	.	.	.	.	.	.	.	.	.	.	.	21
2950	EUPHORBIA THAMNOIDES	S	.	0.0	25.6	34.5	.	.	.	.	.	.	.	.	.	.	.	.	.	21
2951	EUPTELEA POLYANDRA	S	.	0.0	7.5	11.0	.	.	.	.	.	.	.	.	.	.	.	.	.	19
2952	EUROTIA LANATA	W	.	0.0	11.3	2.3	72.5	31.5	13.9	1900	120	.	.	.	2	.	.	.	.	6
2953	EURYALE FEROX	S	346	13.6	8.3	0.3	77.2	0.2	0.6	39	86	1.2	.	57	.	.	.	.	.	2
2954	EURYALE FEROX	S	.	12.8	9.7	0.1	76.9	.	0.5	20	90	1.4	.	.	0	.	.	.	.	5
2955	EURYOPS ATHANASIAE	S	.	0.0	22.8	40.4	.	.	.	.	.	.	.	.	.	.	.	.	.	21
2956	EUSTACHYS PASPALOIDES	W	.	0.0	9.7	2.6	77.7	33.8	10.0	.	.	.	.	.	.	.	.	.	.	18
2957	EUTERPE LONGEPETIOLATA	SH	26	91.0	2.2	0.2	5.2	0.6	1.4	86	79	0.8	.	.	0	0.04	0.09	0.7	17	3
2958	EUTERPE OLERACEA	F	265	41.0	3.4	12.2	42.2	18.0	1.2	.	.	.	.	.	.	0.36	.	.	9	3
2959	EUTREMA WASABI	R	74	76.7	5.1	0.2	16.7	1.4	1.3	93	72	0.8	.	.	30	0.15	0.10	0.5	80	
2960	EXOCHORDA RACEMOSA	S	.	0.0	48.1	27.7	.	.	4.1	.	.	.	.	.	.	.	.	.	.	19
2961	EXOCHORDA RACEMOSA	S	.	0.0	10.6	5.2	.	.	2.2	.	.	.	.	.	.	.	.	.	.	19
2962	FAGARA CHALYBEA	L	.	0.0	10.3	3.3	79.6	19.3	6.8	1700	470	.	.	.	.	.	.	.	.	18
2963	FAGOPYRUM ESCULENTUM	S	.	0.0	12.9	2.6	.	.	1.9	.	.	.	.	.	.	.	.	.	.	19
2964	FAGOPYRUM ESCULENTUM	S	336	11.3	10.3	2.4	73.8	6.6	2.2	44	320	3.0	.	446	0	0.36	0.18	2.0	0	2
2965	FAGOPYRUM ESCULENTUM	S	335	11.0	11.7	2.4	72.9	9.9	2.0	114	282	3.1	.	448	0	0.60	.	4.4	0	4
2966	FAGOPYRUM ESCULENTUM	S	.	11.3	10.3	2.4	73.6	8.6	2.4	70	300	13.2	.	.	.	0.90	.	4.4	.	5
2967	FAGOPYRUM SP.	W	.	0.0	12.6	2.5	75.1	21.9	9.8	.	.	.	.	.	.	.	.	.	.	6
2968	FAGOPYRUM SP.	H	.	0.0	4.8	1.1	84.3	40.2	9.8	1420	200	.	.	3220	.	.	.	.	.	6
2969	FAGOPYRUM TARTARICUM	S	.	0.0	11.7	2.7	.	.	2.3	.	.	.	.	.	.	.	.	.	.	19
2970	FAGOPYRUM TARTARICUM	S	.	0.0	11.2	2.4	.	.	5.3	.	.	.	.	.	.	.	.	.	.	19
2971	FAGUS GRANDIFOLIA	S	568	6.6	19.4	50.0	20.3	3.7	3.7	.	.	.	.	.	.	.	.	.	.	4
2972	FAGUS ORIENTALIS	S	.	0.0	31.8	44.6	.	.	.	.	.	.	.	.	.	.	.	.	.	21
2973	FALCARIA VULGARIS	S	.	0.0	24.8	11.4	.	.	.	.	.	.	.	.	.	.	.	.	.	20

TABLE 2 - DATA DIRECTLY TRANSCRIBED FROM ORIGINAL SOURCES.

OBS	PLANA	PART	CALORIES	WATER	PROTEIN	FAT	TOTCARB	FIBER	ASH	CALCIUM	PHOSPHO	IRON	SODIUM	POTASS	CAROT	THIAMINE	RIBOFLAV	NIACIN	ASCORBIC	REFCO
2974	FALLUGIA PARADOXA	S	.	0.0	30.6	36.8	.	.	.	.	.	.	.	.	.	.	.	.	.	20
2975	FEIJOA SELLOWIANA	F	.	79.2	.	.	.	.	0.5	36	.	.	.	.	.	.	.	.	34	1
2976	FELICIA AMELLOIDES	S	.	0.0	18.8	28.1	.	.	.	.	.	.	.	.	.	.	.	.	.	20
2977	FERNALDIA PANDURATA	L	32	89.2	2.6	0.2	6.8	1.4	1.2	58	46	1.1	.	.	110	0.64	0.11	2.3	12	3
2978	FEROCACTUS SP.	S	.	0.0	18.8	17.2	.	.	.	.	.	.	.	.	.	.	.	.	.	19
2979	FEROCACTUS SP.	S	.	0.0	17.5	17.4	.	.	.	.	.	.	.	.	.	.	.	.	.	19
2980	FERONIA LIMONIA	F	.	69.5	7.3	0.6	20.7	5.2	1.9	130	110	0.6	.	.	.	.	0.17	.	2	5
2981	FERULA COMMUNIS	S	.	0.0	28.3	9.2	.	.	.	.	.	.	.	.	.	.	.	.	.	21
2982	FERULA COPODA	S	.	0.0	25.4	19.1	.	.	.	.	.	.	.	.	.	.	.	.	.	21
2983	FERULA GALVANIFLUA	S	.	0.0	13.1	46.9	.	.	.	.	.	.	.	.	.	.	.	.	.	20
2984	FERULA OVINA	S	.	0.0	21.4	19.1	.	.	.	.	.	.	.	.	.	.	.	.	.	21
2985	FERULAGO SP.	S	.	0.0	18.9	6.9	.	.	.	.	.	.	.	.	.	.	.	.	.	21
2986	FESTUCA ARUNDINACEA	S	.	0.0	11.9	1.7	.	.	6.1	.	.	.	.	.	.	.	.	.	.	19
2987	FESTUCA ARUNDINACEA	S	.	0.0	22.5	1.4	.	.	.	.	.	.	.	.	.	.	.	.	.	20
2988	FESTUCA ARUNDINACEA	W	.	0.0	11.6	3.3	76.8	29.6	8.3	.	.	.	.	.	.	.	.	.	.	6
2989	FESTUCA ELATIOR	S	.	0.0	21.9	1.6	.	.	2.4	.	.	.	.	.	.	.	.	.	.	19
2990	FESTUCA ELATIOR	W	.	0.0	15.1	4.1	71.8	27.1	9.0	510	380	.	.	2000	9	1.52	0.55	.	.	6
2991	FESTUCA IDAHOENSIS	W	.	0.0	7.8	3.8	80.9	35.9	7.5	390	220	.	.	.	.	.	.	.	.	6
2992	FESTUCA MEGALURA	W	.	0.0	17.2	4.6	67.5	21.5	10.7	570	460	.	.	.	.	.	.	.	.	6
2993	FESTUCA OCTOFLORA	W	.	0.0	13.0	2.5	74.5	27.4	10.0	440	220	.	.	1720	.	.	.	.	.	6
2994	FESTUCA OVINA	S	.	0.0	26.1	2.2	.	.	.	.	.	.	.	.	.	.	.	.	.	21
2995	FESTUCA OVINA	W	.	0.0	16.3	3.2	73.2	27.1	8.3	490	360	43.0	.	1850	.	.	.	.	.	6
2996	FESTUCA RUBRA	W	.	0.0	13.9	3.1	73.5	27.8	9.5	540	440	.	.	.	.	.	.	.	.	6
2997	FESTUCA RUBRA SUBSP. COMMUTATA	H	.	0.0	13.9	.	.	.	.	500	380	.	.	1580	.	.	.	.	.	6
2998	FIBIGIA CLYPEATA	S	.	0.0	22.8	11.9	.	.	.	.	.	.	.	.	.	.	.	.	.	20
2999	FIBIGIA CLYPEATA	S	.	0.0	27.8	12.4	.	.	.	.	.	.	.	.	.	.	.	.	.	21
3000	FICUS BENGHALENSIS	L	.	0.0	9.7	2.9	73.0	22.6	14.4	2560	190	.	.	.	.	.	.	.	.	18
3001	FICUS BENGHALENSIS	F	34	90.3	0.5	0.0	9.0	0.0	0.2	11	11	0.4	.	.	0	0.00	0.00	0.3	15	3
3002	FICUS CARICA	F	52	85.2	0.8	0.2	13.2	1.3	0.6	44	20	0.5	.	.	30	.	.	.	15	1
3003	FICUS CARICA	L	.	0.0	14.2	5.9	63.2	17.1	16.7	3160	160	.	.	.	.	.	.	.	.	18
3004	FICUS CARICA	F	59	83.6	1.0	0.4	14.5	1.5	0.5	41	29	0.5	6	177	140	0.03	0.03	0.4	2	2
3005	FICUS CARICA	F	62	82.2	1.2	0.2	15.6	1.6	0.8	50	30	0.5	.	.	60	0.04	0.05	0.4	4	3
3006	FICUS CARICA	G	42	87.7	1.6	0.2	9.8	1.9	0.7	52	34	0.7	.	.	200	0.04	0.07	0.4	20	3
3007	FICUS CARICA	F	80	77.5	1.2	0.3	20.3	1.2	0.7	35	22	0.6	2	194	80	0.06	0.05	0.4	2	4
3008	FICUS CUNIA	F	.	13.5	8.7	5.7	60.8	17.7	11.3	.	1120	.	.	.	.	.	.	.	.	5
3009	FICUS ELASTICA	L	.	0.0	13.0	4.4	75.3	32.2	7.3	1110	130	.	.	.	.	.	.	.	.	18
3010	FICUS GLABRATA	F	33	90.3	0.5	0.1	8.4	1.7	0.7	62	7	2.3	.	.	50	0.03	0.02	0.2	0	3
3011	FICUS GLOMERATA	L	.	0.0	11.2	2.4	71.3	12.3	15.1	2680	310	.	.	.	.	.	.	.	.	18
3012	FICUS GLOMERATA	F	.	13.6	7.4	5.6	66.9	17.9	6.5	.	910	.	.	.	.	.	.	.	.	5
3013	FICUS INFECTORIA	L	.	0.0	12.5	3.5	70.9	20.0	13.1	2210	290	.	.	.	.	.	.	.	.	18
3014	FICUS ITEOPHYLLA	F	76	77.1	3.0	0.2	17.9	5.8	1.8	290	75	10.0	.	.	235	0.04	.	.	25	1
3015	FICUS PLATYPHYLLA	F	56	83.0	1.9	0.1	13.7	3.6	1.3	205	66	4.0	.	.	.	0.04	0.04	0.3	22	1
3016	FICUS RELIGIOSA	L	.	0.0	9.0	2.7	68.3	15.9	20.0	2970	210	.	.	.	.	.	.	.	.	18
3017	FICUS RELIGIOSA	SH	.	0.0	11.7	2.9	70.1	26.1	15.3	3690	270	.	.	.	.	.	.	.	.	18
3018	FICUS RELIGIOSA	F	.	9.9	7.9	5.3	34.9	.	8.3	.	690	.	.	.	.	.	.	.	.	5

TABLE 2 - DATA DIRECTLY TRANSCRIBED FROM ORIGINAL SOURCES.

OBS	PLANA	PART	CALORIES	WATER	PROTEIN	FAT	TOTCARB	FIBER	ASH	CALCIUM	PHOSPHON	IRON	SODIUM	POTASS	CAROTE	THIAMINE	RIBOFLAV	NIACIN	ASCORBIC	REFCO
3019	FICUS SYCOMORUS	F	77	78.5	1.6	1.2	17.1	5.1	1.6	250	65	6.0	.	.	220	0.03	0.06	0.50	15	1
3020	FICUS SYCOMORUS	L	57	80.5	6.1	0.2	11.3	2.4	1.9	74	163	15.8	.	.	.	.	.	.	35	1
3021	FILAGO GALLICA	W	.	0.0	10.5	.	.	28.1	9.8	1050	330	.	.	2310	.	.	.	.	.	6
3022	FINGERHUTHIA SESLERIAEFORMIS	S	.	0.0	33.1	4.0	.	.	3.6	.	.	.	.	.	.	.	.	.	.	19
3023	FIRMIANA SIMPLEX	S	.	0.0	25.6	36.9	.	.	5.0	.	.	.	.	.	.	.	.	.	.	19
3024	FIRMIANA SIMPLEX	S	533	2.5	20.0	38.8	34.2	2.2	4.5	27	212	4.8	.	1206	.	.	.	.	.	2
3025	FIRMIANA SIMPLEX	S	.	0.0	24.2	37.8	.	.	.	.	.	.	.	.	.	.	.	.	.	21
3026	FLACOURTIA INDICA	F	.	72.3	.	.	.	.	.	.	.	.	.	.	.	.	.	.	0	1
3027	FLACOURTIA INDICA	S	.	0.0	6.8	12.0	.	.	.	.	.	.	.	.	.	.	.	.	.	21
3028	FLACOURTIA INERMIS	F	94	74.2	0.5	0.6	24.2	1.2	0.5	33	17	0.7	1.0	171	30	0.01	0.02	0.40	5	2
3029	FLACOURTIA INERMIS	F	.	86.6	0.4	0.2	12.5	1.1	0.4	.	.	.	.	.	.	.	.	.	.	5
3030	FLACOURTIA JANGOMAS	F	78	77.7	0.5	0.1	20.9	1.0	0.8	43	25	.	.	.	.	.	.	.	.	2
3031	FLACOURTIA JANGOMAS	F	108	69.5	0.5	0.0	29.5	0.5	0.5	20	44	1.2	.	.	30	0.02	0.03	0.60	14	3
3032	FLACOURTIA MONTANA	S	.	0.0	10.8	1.1	.	.	.	.	.	.	.	.	.	.	.	.	.	21
3033	FLAVERIA TRINERVIA	S	.	0.0	26.9	27.7	.	.	.	.	.	.	.	.	.	.	.	.	.	20
3034	FLAVERIA TRINERVIA	S	.	0.0	26.9	27.5	.	.	.	.	.	.	.	.	.	.	.	.	.	20
3035	FLEMINGIA CONGESTA	S	.	0.0	24.2	0.7	.	.	.	.	.	.	.	.	.	.	.	.	.	21
3036	FLEMINGIA SP.	S	.	0.0	22.5	0.5	.	.	.	.	.	.	.	.	.	.	.	.	.	20
3037	FLOURENSIA CERNUS	S	.	0.0	16.2	6.6	.	.	.	.	.	.	.	.	.	.	.	.	.	20
3038	FLOURENSIA PRINGLEI	S	.	0.0	46.2	30.6	.	.	.	.	.	.	.	.	.	.	.	.	.	20
3039	FOENICULUM VULGARE	L	31	89.2	2.9	0.5	5.6	0.5	1.8	114	54	2.9	.	338	2610	0.12	0.15	0.70	34	1
3040	FOENICULUM VULGARE	S	345	8.8	15.8	14.9	52.3	15.7	8.2	1196	487	18.5	8.8	1694	81	0.41	0.35	6.05	.	16
3041	FOENICULUM VULGARE	S	.	0.0	20.1	26.6	.	.	7.1	.	.	.	.	.	.	.	.	.	.	19
3042	FOENICULUM VULGARE	S	.	0.0	20.0	22.9	.	.	.	.	.	.	.	.	.	.	.	.	.	19
3043	FOENICULUM VULGARE	S	.	0.0	25.0	21.5	.	.	.	.	.	.	.	.	.	.	.	.	.	21
3044	FOENICULUM VULGARE	L	28	90.0	2.8	0.4	5.1	0.5	1.7	100	51	2.7	.	397	2100	.	.	.	31	4
3045	FOENICULUM VULGARE VAR. DULCE	S	.	0.0	19.7	24.0	.	.	8.2	.	.	.	.	.	.	.	.	.	.	19
3046	FORESTIERA NEOMEXICANA	L	.	0.0	19.7	2.9	71.0	9.4	6.4	1000	250	.	.	2550	.	.	.	.	.	6
3047	FORESTIERA PUBESCENS	S	.	0.0	12.5	18.9	.	.	.	.	.	.	.	.	.	.	.	.	.	20
3048	FORTUNELLA MARGARITA	F	48	86.7	0.4	0.5	11.9	1.5	0.5	16	65	0.8	111.0	156	25	0.09	0.06	0.50	40	2
3049	FORTUNELLA MARGARITA	F	62	82.4	1.3	0.2	15.6	1.5	0.5	62	22	0.8	.	.	320	0.14	0.06	0.80	55	3
3050	FORTUNELLA SP.	F	65	81.3	0.9	0.1	17.1	3.7	0.6	63	23	0.4	7.0	236	360	0.08	0.10	.	36	4
3051	FOUQUIERIA SPLENDENS	S	.	0.0	28.8	18.6	.	.	.	.	.	.	.	.	.	.	.	.	.	19
3052	FOUQUIERIA SPLENDENS	W	.	0.0	3.7	.	.	.	3.5	630	70	17.0	.	500	.	.	.	.	.	6
3053	FRAGARIA SP.	F	37	89.0	0.8	0.3	8.9	1.2	1.0	18	18	0.7	.	.	80	.	.	1.90	53	1
3054	FRAGARIA SP.	F	34	90.6	0.8	0.5	7.6	1.7	0.5	25	30	0.9	1.0	193	10	0.04	0.03	0.40	53	1
3055	FRAGARIA SP.	F	36	90.0	0.8	0.3	8.5	1.3	0.4	29	29	1.0	.	.	20	0.03	0.04	0.40	70	3
3056	FRAGARIA SP.	F	37	89.9	0.7	0.5	8.4	1.3	0.5	21	21	1.0	1.0	164	36	0.03	0.07	0.60	59	4
3057	FRANGULA ALNUS	S	.	0.0	12.9	15.8	.	.	.	.	.	.	.	.	.	.	.	.	.	21
3058	FRANGULA ALNUS	S	.	0.0	15.7	10.7	.	.	.	.	.	.	.	.	.	.	.	.	.	21
3059	FRANSERIA ACANTHICARPA	S	.	0.0	14.9	20.7	.	.	.	.	.	.	.	.	.	.	.	.	.	21
3060	FRASERA PARRYI	S	.	0.0	15.6	27.0	.	.	.	.	.	.	.	.	.	.	.	.	.	20
3061	FRAXINUS AMERICANA	S	.	0.0	25.2	33.4	.	.	3.5	.	.	.	.	.	.	.	.	.	.	19
3062	FRAXINUS AMERICANA	S	.	0.0	3.9	2.7	.	.	5.6	.	.	.	.	.	.	.	.	.	.	19
3063	FRAXINUS ORNUS	S	.	0.0	15.2	26.5	.	.	.	.	.	.	.	.	.	.	.	.	.	21

TABLE 2 - DATA DIRECTLY TRANSCRIBED FROM ORIGINAL SOURCES.

OBS	PLANA	PART	CALORIES	WATER	PROTEIN	FAT	TOTCARB	FIBER	ASH	CALCIUM	PHOSPHO	IRON	SODIUM	POTASS	CAROT	THIAMINE	RIBOFLAV	NIACIN	ASCORBIC	REFCO
3064	FRAXINUS OXYCARPA	S	.	0.0	18.1	29.9	.	.	.	.	.	.	.	.	.	.	.	.	.	21
3065	FRAXINUS XANTHOXYLOIDES	S	.	0.0	16.6	26.7	.	.	.	.	.	.	.	.	.	.	.	.	.	21
3066	FROELICHIA FLORIDANA	S	.	0.0	21.8	9.0	.	.	.	.	.	.	.	.	.	.	.	.	.	20
3067	FUMARIA DENSIFLORA	S	.	0.0	20.0	32.7	.	.	.	.	.	.	.	.	.	.	.	.	.	21
3068	GAILLARDIA ARISTATA	S	.	0.0	45.9	27.6	.	.	5.3	.	.	.	.	.	.	.	.	.	.	19
3069	GAILLARDIA ARISTATA	S	.	0.0	15.0	8.5	.	.	10.3	.	.	.	.	.	.	.	.	.	.	19
3070	GAILLARDIA PINNATIFIDA	S	.	0.0	29.1	19.2	.	.	.	.	.	.	.	.	.	.	.	.	.	20
3071	GAILLARDIA PULCHELLA	S	.	0.0	31.5	18.3	.	.	7.2	.	.	.	.	.	.	.	.	.	.	19
3072	GAILLARDIA PULCHELLA	S	.	0.0	22.6	17.2	.	.	.	.	.	.	.	.	.	.	.	.	.	20
3073	GAILLARDIA PULCHELLA	W	.	0.0	13.0	4.7	71.0	28.6	11.3	2680	280	.	.	3270	.	.	.	.	.	6
3074	GALACTIA WRIGHTII	S	.	0.0	25.0	1.6	.	.	3.6	.	.	.	.	.	.	.	.	.	.	19
3075	GALACTITES TOMENTOSA	S	.	0.0	20.2	23.6	.	.	.	.	.	.	.	.	.	.	.	.	.	21
3076	GALEGA CORONILLOIDES	S	.	0.0	40.7	3.7	.	.	.	.	.	.	.	.	.	.	.	.	.	21
3077	GALEGA OFFICINALIS	S	.	0.0	41.8	2.8	.	.	.	.	.	.	.	.	.	.	.	.	.	21
3078	GALENIA AFRICANA	C	.	0.0	10.2	4.9	75.0	34.2	9.9	1540	60	.	.	.	.	.	.	.	.	18
3079	GALINSOGA PARVIFLORA	L	42	86.1	3.2	0.5	8.3	1.3	1.9	245	45	7.1	.	.	2240	0.11	0.27	2.1	30	3
3080	GALIUM APARINE	S	.	0.0	12.7	3.1	.	.	.	.	.	.	.	.	.	.	.	.	.	20
3081	GALIUM ARTICULATUM	S	.	0.0	17.0	10.7	.	.	.	.	.	.	.	.	.	.	.	.	.	20
3082	GALIUM CORONATUM	S	.	0.0	13.3	9.3	.	.	.	.	.	.	.	.	.	.	.	.	.	21
3083	GALIUM PARISIENSE	S	.	0.0	17.8	9.6	.	.	.	.	.	.	.	.	.	.	.	.	.	21
3084	GAMOLEPIS CHRYSANTHEMOIDES	S	.	0.0	14.6	25.6	.	.	.	.	.	.	.	.	.	.	.	.	.	21
3085	GARCINIA HUILLENSIS	F	.	75.9	.	.	.	.	.	.	.	.	.	.	.	.	.	.	8	1
3086	GARCINIA MANGOSTANA	F	57	84.3	0.5	0.3	14.7	5.0	0.2	10	10	0.5	1	135	0	0.03	0.02	0.6	4	2
3087	GARCINIA MANGOSTANA	P	.	84.9	0.5	0.1	14.3	.	0.2	10	20	0.2	.	.	.	.	.	.	.	5
3088	GARCINIA TINCTORIA	S	.	0.0	4.2	26.1	.	.	1.8	.	.	.	.	.	.	.	.	.	.	19
3089	GARCINIA TINCTORIA	S	.	0.0	5.1	7.7	.	.	0.2	.	.	.	.	.	.	.	.	.	.	19
3090	GARDENIA JASMINOIDES	S	.	0.0	14.4	20.0	.	.	2.5	.	.	.	.	.	.	.	.	.	.	19
3091	GAULTHERIA SHALLON	F	371	0.0	13.0	5.0	79.0	.	3.0	377	.	4.0	.	.	.	.	.	.	487	15
3092	GAULTHERIA SHALLON	F	351	0.0	6.0	1.0	88.0	.	4.0	344	.	4.0	.	.	.	.	.	.	710	15
3093	GAURA NEOMEXICANA	S	.	0.0	31.2	48.3	.	.	.	.	.	.	.	.	.	.	.	.	.	21
3094	GAURA PARVIFLORA	S	.	0.0	11.4	13.7	.	.	4.6	.	.	.	.	.	.	.	.	.	.	19
3095	GAURA PARVIFLORA	S	.	0.0	30.6	41.2	.	.	.	.	.	.	.	.	.	.	.	.	.	19
3096	GAURA SP.	W	.	0.0	25.4	2.0	59.8	15.7	12.8	.	.	.	.	.	.	.	.	.	.	6
3097	GAURA SUFFULTA	S	.	0.0	9.9	8.9	.	.	.	.	.	.	.	.	.	.	.	.	.	20
3098	GAZANIA KREBSIANA SUBSP. KREBSIANA	S	.	0.0	45.1	27.5	.	.	.	.	.	.	.	.	.	.	.	.	.	21
3099	GAZANIA PINNATA	S	.	0.0	45.8	27.2	.	.	.	.	.	.	.	.	.	.	.	.	.	21
3100	GAZANIA PINNATA	S	.	0.0	42.7	26.6	.	.	.	.	.	.	.	.	.	.	.	.	.	21
3101	GENISTA AETHNENSIS	S	.	0.0	36.9	8.5	.	.	3.0	.	.	.	.	.	.	.	.	.	.	19
3102	GENISTA CARINALIS	S	.	0.0	36.2	4.8	.	.	.	.	.	.	.	.	.	.	.	.	.	21
3103	GENISTA JAUBERTII	S	.	0.0	43.9	11.0	.	.	.	.	.	.	.	.	.	.	.	.	.	21
3104	GENISTA LYDIA	S	.	0.0	40.9	11.3	.	.	.	.	.	.	.	.	.	.	.	.	.	21
3105	GENISTA MONOSPERMA	S	.	0.0	28.1	5.2	.	.	.	.	.	.	.	.	.	.	.	.	.	19
3106	GENISTA TINCTORIA	S	.	0.0	41.2	9.7	.	.	.	.	.	.	.	.	.	.	.	.	.	21
3107	GENTIANA LUTEA	S	.	0.0	31.0	26.0	.	.	.	.	.	.	.	.	.	.	.	.	.	21
3108	GEONOMA EDULIS	SH	35	88.2	3.2	0.3	7.0	1.5	1.3	41	82	1.8	.	.	0	0.07	0.13	1.2	18	3

TABLE 2 - DATA DIRECTLY TRANSCRIBED FROM ORIGINAL SOURCES.

OBS	PLANA	PART	CALORIES	WATER	PROTEIN	FAT	TOTLCARB	FIBER	ASH	CALCIUM	PHOSPHO	IRON	SODIUM	POTASS	CAROT	THIAMINE	RIBOFLAV	NIACIN	ASCORBIC	REFCO
3109	GEOPRUMNON CRASSICARPUM	S	.	0.0	45.0	5.4	.	.	3.9	.	.	.	.	.	.	.	.	.	.	19
3110	GERANIUM CAROLINIANUM	S	.	0.0	22.9	17.0	.	.	.	.	.	.	.	.	.	.	.	.	.	20
3111	GERANIUM CAROLINIANUM	S	.	0.0	22.3	19.5	.	.	.	.	.	.	.	.	.	.	.	.	.	21
3112	GERANIUM CAROLINIANUM	W	.	0.0	13.3	2.0	71.7	16.6	13.0	1340	570	.	.	2160	.	.	.	.	.	6
3113	GERANIUM FREMONTII	W	.	0.0	18.8	3.3	70.7	8.4	7.2	1250	480	.	.	.	.	.	.	.	.	6
3114	GERANIUM VISCOSISSIMUM	W	.	0.0	28.9	2.4	60.4	8.2	8.3	.	.	.	.	.	.	.	.	.	.	6
3115	GEROPOGON GLABER	S	.	0.0	19.4	16.5	.	.	.	.	.	.	.	.	.	.	.	.	.	21
3116	GEUM CHILOENSE	S	.	0.0	29.6	24.4	.	.	4.7	.	.	.	.	.	.	.	.	.	.	19
3117	GEUM CHILOENSE	S	.	0.0	13.1	7.4	.	.	8.4	.	.	.	.	.	.	.	.	.	.	19
3118	GEUM MACROPHYLLUM	S	.	0.0	15.4	21.7	.	.	.	.	.	.	.	.	.	.	.	.	.	21
3119	GEUM SP.	S	.	0.0	14.4	18.8	.	.	.	.	.	.	.	.	.	.	.	.	.	19
3120	GEUM SP.	H	.	0.0	17.4	17.9	60.8	21.2	3.9	.	.	.	.	.	.	.	.	.	.	6
3121	GEUM URBANUM	S	.	0.0	13.9	21.0	.	.	.	.	.	.	.	.	.	.	.	.	.	21
3122	GILIA CANDIDA	S	.	0.0	16.8	16.1	.	.	.	.	.	.	.	.	.	.	.	.	.	21
3123	GILIA CAPITATA	S	.	0.0	15.6	22.7	.	.	.	.	.	.	.	.	.	.	.	.	.	21
3124	GILIA PINNATA	S	.	0.0	18.1	21.0	.	.	.	.	.	.	.	.	.	.	.	.	.	20
3125	GILIA RIGIDULA	S	.	0.0	14.1	17.5	.	.	.	.	.	.	.	.	.	.	.	.	.	20
3126	GILIA SP.	S	.	0.0	26.9	30.7	.	.	.	.	.	.	.	.	.	.	.	.	.	20
3127	GILIA TRICOLOR		.	0.0	12.0	.	.	23.1	8.8	1440	450	.	.	2600	.	.	.	.	.	6
3128	GILIA VISCIDULA	W	.	0.0	8.0	.	.	24.9	8.6	1550	180	.	.	1190	.	.	.	.	.	6
3129	GINGKO BILOBA	S	.	0.0	10.2	3.1	.	.	3.8	.	.	.	.	.	.	.	.	.	.	19
3130	GINKGO BILOBA	S	185	54.1	4.8	1.6	38.1	0.6	1.4	5	150	1.2	7	523	180	0.24	0.12	2.8	25	2
3131	GLADIOLUS UNGUICULATUS	R	135	60.2	2.0	0.3	35.7	3.1	1.8	215	.	19.0	.	.	0	0.10	0.08	1.1	3	1
3132	GLAUCIUM ARABICUM	S	.	0.0	23.4	34.7	.	.	.	.	.	.	.	.	.	.	.	.	.	21
3133	GLAUCIUM CORNICULATUM	S	.	0.0	20.6	33.5	.	.	.	.	.	.	.	.	.	.	.	.	.	21
3134	GLAUCIUM CORNICULATUM	S	.	0.0	22.8	31.3	.	.	.	.	.	.	.	.	.	.	.	.	.	21
3135	GLEDITSIA JAPONICA	S	.	0.0	22.4	2.8	.	.	.	.	.	.	.	.	.	.	.	.	.	21
3136	GLEDITSIA TRIACANTHOS	F	.	0.0	23.1	4.6	66.9	12.7	5.4	.	.	.	.	.	.	.	.	.	.	18
3137	GLEDITSIA TRIACANTHOS	S	.	0.0	10.6	0.8	84.7	21.1	3.9	280	320	.	.	.	.	.	.	.	.	18
3138	GLEDITSIA TRIACANTHOS	S	.	0.0	20.6	2.6	.	.	.	.	.	.	.	.	.	.	.	.	.	19
3139	GLEDITSIA TRIACANTHOS	S	.	0.0	21.8	2.3	.	.	3.8	.	.	.	.	.	.	.	.	.	.	19
3140	GLEDITSIA TRIACANTHOS	S	.	0.0	22.1	1.7	.	.	4.0	.	.	.	.	.	.	.	.	.	.	19
3141	GLEDITSIA TRIACANTHOS	S	.	0.0	24.1	4.3	.	.	.	.	.	.	.	.	.	.	.	.	.	21
3142	GLEDITSIA TRIACANTHOS	F	.	.	23.1	4.6	96.9	12.7	.	.	.	.	.	.	.	.	.	.	.	5
3143	GLIRICIDIA MACULATA	C	.	0.0	18.8	3.7	71.2	15.5	6.3	660	110	.	.	.	.	.	.	.	.	18
3144	GLIRICIDIA MACULATA	L	.	0.0	30.0	4.3	57.7	14.1	8.0	.	.	.	.	.	.	.	.	.	.	18
3145	GLIRICIDIA SEPIUM	L	52	84.7	2.4	0.5	11.9	1.3	0.5	17	34	0.8	.	.	.	0.14	0.08	1.0	12	2
3146	GLIRICIDIA SEPIUM	I	46	86.0	2.4	0.2	10.8	1.6	0.6	22	37	1.8	.	.	70	0.11	0.08	0.6	44	3
3147	GLIRICIDIA SEPIUM	I	.	85.5	3.7	1.5	8.4	2.4	1.0	.	.	.	.	.	.	.	.	.	.	5
3148	GLIRICIDIA SEPIUM	S	.	13.2	53.8	16.1	3.4	.	3.5	.	.	.	.	.	.	.	.	.	.	5
3149	GLIRICIDIA SP.	S	.	0.0	47.5	20.4	.	.	4.6	.	.	.	.	.	.	.	.	.	.	19
3150	GLOBULARIA ORIENTALIS	S	.	0.0	27.5	39.0	.	.	.	.	.	.	.	.	.	.	.	.	.	21
3151	GLOBULARIA TRICHOSANTHA	S	.	0.0	36.1	35.2	.	.	.	.	.	.	.	.	.	.	.	.	.	21
3152	GLYCERIA GRANDIS	W	.	0.0	12.5	2.4	76.7	32.8	8.4	260	550	.	.	.	14	.	.	.	.	6
3153	GLYCERIA OBTUSA	S	.	0.0	23.1	3.4	.	.	.	.	.	.	.	.	.	.	.	.	.	20

TABLE 2 - DATA DIRECTLY TRANSCRIBED FROM ORIGINAL SOURCES.

OBS	PLNA	PART	CALORIES	WATER	PROTEIN	FAT	TOTLCARB	FIBER	ASH	CALCIUM	PHOSPHO	IRON	SODIUM	POTASS	CAROT	THIAMINE	RIBOFLAV	NIACIN	ASCORBIC	REFCO
3154	GLYCERIA STRIATA	S	.	0.0	22.5	2.1	.	.	.	.	.	.	.	.	.	.	.	.	.	20
3155	GLYCINE GRACILIS	S	.	0.0	44.0	14.3	.	.	5.9	.	.	.	.	.	.	.	.	.	.	19
3156	GLYCINE MAX	S	405	9.5	33.7	17.9	33.9	4.7	5.0	183	541	6.1	.	.	55	0.71	0.25	2.0	.	1
3157	GLYCINE MAX	W	.	0.0	16.1	6.2	65.8	29.6	11.9	1220	240	.	.	.	.	.	.	.	.	18
3158	GLYCINE MAX	H	.	0.0	13.9	2.5	78.1	35.5	5.5	940	130	.	.	.	.	.	.	.	.	18
3159	GLYCINE MAX	S	.	0.0	31.2	21.8	.	.	5.5	.	.	.	.	.	.	.	.	.	.	19
3160	GLYCINE MAX	S	385	12.3	33.3	15.0	35.4	4.3	4.0	213	509	9.5	.	410	10	0.65	0.23	2.8	0	2
3161	GLYCINE MAX	S	139	68.2	13.0	5.7	11.4	1.9	1.7	78	158	3.8	15	607	360	0.40	0.17	1.5	27	2
3162	GLYCINE MAX	X	62	81.5	7.7	1.8	8.0	0.7	1.0	52	58	1.1	30	279	25	0.19	0.15	0.8	10	2
3163	GLYCINE MAX	S	398	9.2	33.4	16.4	35.5	5.7	5.5	222	730	11.5	.	.	0	0.88	0.27	2.2	.	3
3164	GLYCINE MAX	G	134	69.2	10.9	5.1	13.2	1.4	1.6	67	225	2.8	.	.	414	0.44	0.16	1.4	29	4
3165	GLYCINE MAX	X	46	86.3	6.2	1.4	5.3	0.8	0.8	48	67	1.0	.	.	48	0.23	0.20	0.8	13	4
3166	GLYCINE MAX	W	.	0.0	16.6	4.0	68.8	28.1	10.6	1300	530	21.0	.	1060	29	.	.	.	.	6
3167	GLYCINE SP.	S	.	0.0	32.4	4.5	.	.	5.0	.	.	.	.	.	.	.	.	.	.	19
3168	GLYCINE WIGHTII	W	.	0.0	10.1	0.7	80.6	32.7	8.6	530	330	.	.	.	.	.	.	.	.	18
3169	GLYCINE WIGHTII	H	.	0.0	14.4	2.6	74.5	34.2	8.5	990	310	.	.	.	.	.	.	.	.	18
3170	GLYCYRRHIZA ECHINATA	S	.	0.0	42.2	3.1	.	.	.	.	.	.	.	.	.	.	.	.	.	21
3171	GLYCYRRHIZA GLABRA	S	.	0.0	26.8	1.3	.	.	.	.	.	.	.	.	.	.	.	.	.	21
3172	GLYCYRRHIZA GLABRA	S	.	0.0	31.3	9.0	.	.	.	.	.	.	.	.	.	.	.	.	.	21
3173	GLYCYRRHIZA GLABRA	S	.	0.0	21.1	11.2	.	.	.	.	.	.	.	.	.	.	.	.	.	21
3174	GLYCYRRHIZA LEPIDOTA	S	.	0.0	38.5	6.4	.	.	3.6	.	.	.	.	.	.	.	.	.	.	19
3175	GLYCYRRHIZA LEPIDOTA	S	.	0.0	27.6	2.1	.	.	.	.	.	.	.	.	.	.	.	.	.	21
3176	GNETUM BUCHHOLZIANUM	L	103	69.8	6.0	1.3	21.6	5.5	1.3	130	54	5.6	.	.	.	.	.	.	100	1
3177	GNETUM GNEMON	L	82	75.4	3.8	0.8	18.4	2.6	1.6	128	100	2.7	1	83	7110	0.17	0.24	1.6	113	2
3178	GNETUM GNEMON	F	66	80.0	5.0	0.7	13.3	.	1.0	163	75	2.8	.	.	600	0.10	.	.	100	2
3179	GNETUM GNEMON	S	345	13.0	12.0	1.5	71.5	.	2.0	100	400	5.0	.	.	0	0.20	.	.	0	2
3180	GNETUM GNEMON	L	121	65.6	5.9	2.2	24.6	11.8	1.7	336	73	2.7	.	.	8990	0.11	0.33	1.5	215	3
3181	GNETUM GNEMON	L	.	81.9	.	.	.	.	1.3	110	240	10.0	.	.	.	.	.	.	.	5
3182	GNETUM GNEMON	S	.	30.0	10.9	1.6	54.9	0.9	1.7	.	.	.	.	.	.	.	.	.	.	5
3183	GODETIA AMOENA	W	.	0.0	6.1	.	.	19.1	8.0	1865	265	.	.	1445	.	.	.	.	.	6
3184	GODETIA GRANDIFLORA	S	.	0.0	32.5	33.1	.	.	.	.	.	.	.	.	.	.	.	.	.	19
3185	GODETIA QUADRIVULNERA	W	.	0.0	4.0	.	.	32.6	6.9	1760	240	.	.	1140	.	.	.	.	.	6
3186	GODETIA QUADRIVULNERA	W	.	0.0	6.8	.	.	23.8	7.5	1840	380	.	.	1460	.	.	.	.	.	6
3187	GOLDMANIA FOETIDA	S	.	0.0	28.9	7.1	.	.	.	.	.	.	.	.	.	.	.	.	.	21
3188	GONOLOBUS SALVINII	F	29	92.2	1.1	0.1	6.2	0.8	0.4	12	17	0.4	.	.	10	0.10	0.02	0.2	120	3
3189	GOSSYPIUM HIRSUTUM	S	.	0.0	37.1	38.4	.	.	5.2	.	.	.	.	.	.	.	.	.	.	19
3190	GOSSYPIUM HIRSUTUM	S	398	8.1	32.9	16.1	36.7	4.8	6.2	149	1022	10.2	.	.	0	0.14	0.03	3.6	.	3
3191	GOSSYPIUM SP.	I	.	84.9	2.2	0.9	9.2	1.2	1.6	60	50	.	.	.	.	.	.	.	.	
3192	GOSSYPIUM SP.	S	437	8.0	20.2	21.2	46.6	18.5	4.0	158	824	.	.	.	.	.	.	.	.	1
3193	GOSSYPIUM SP.	L	.	0.0	16.8	7.8	57.8	11.2	17.6	3510	290	.	.	1170	.	.	.	.	.	6
3194	GOSSYPIUM SP.	Y	.	0.0	6.2	0.9	86.1	47.2	6.8	.	.	.	.	.	.	.	.	.	.	6
3195	GRAYIA SPINOSA	W	.	0.0	8.9	2.0	84.0	38.3	5.1	1100	190	.	.	.	.	.	.	.	.	6
3196	GRAYIA SPINOSA	L	.	0.0	10.2	1.7	57.5	9.2	30.6	.	.	.	.	.	.	.	.	.	.	6
3197	GREWIA ASIATICA	S	.	0.0	9.3	6.1	.	.	.	.	.	.	.	.	.	.	.	.	.	21
3198	GREWIA BICOLOR	SH	.	0.0	12.7	3.5	77.8	23.0	6.0	1600	300	.	.	.	.	.	.	.	.	18

TABLE 2 - DATA DIRECTLY TRANSCRIBED FROM ORIGINAL SOURCES.

OBS	PLNA	PART	CALORIES	WATER	PROTEIN	FAT	TOTLCARB	FIBER	ASH	CALCIUM	PHOSPHO	IRON	SODIUM	POTASS	CAROT	THIAMINE	RIBOFLAV	NIACIN	ASCORBIC	REFCO
3199	GREWIA BICOLOR	L	.	0.0	5.8	7.4	73.3	17.5	13.5	3020	140	.	.	.	.	.	.	.	.	18
3200	GREWIA KAKOTHAMNOS	SH	.	0.0	8.7	2.5	81.1	32.6	7.7	1530	360	.	.	.	.	.	.	.	.	18
3201	GREWIA OPPOSITIFOLIA	L	.	0.0	16.4	8.4	70.3	16.6	14.9	.	.	.	.	.	.	.	.	.	.	18
3202	GREWIA OPTIVA	S	.	0.0	33.4	20.6	.	.	.	.	.	.	.	.	.	.	.	.	.	21
3203	GREWIA OPTIVA	L	.	0.0	10.1	6.8	68.9	14.1	14.2	4180	250	.	.	.	.	.	.	.	.	5
3204	GREWIA TENAX	S	.	0.0	29.4	23.0	.	.	.	.	.	.	.	.	.	.	.	.	.	21
3205	GRIELUM SP.	S	.	0.0	18.6	2.5	.	.	.	.	.	.	.	.	.	.	.	.	.	21
3206	GRINDELIA LANCEOLATA	S	.	0.0	16.0	12.1	.	.	.	.	.	.	.	.	.	.	.	.	.	21
3207	GRINDELIA OXYLEPIS	S	.	0.0	13.1	9.0	.	.	.	.	.	.	.	.	.	.	.	.	.	20
3208	GRINDELIA ROBUSTA	S	.	0.0	17.4	14.5	.	.	.	.	.	.	.	.	.	.	.	.	.	21
3209	GRINDELIA SQUARROSA	S	.	0.0	13.8	19.6	.	.	3.4	.	.	.	.	.	.	.	.	.	.	19
3210	GRINDELIA SUBALPINA VAR. ERECTA	S	.	0.0	16.5	19.1	.	.	.	.	.	.	.	.	.	.	.	.	.	21
3211	GUAIACUM COULTERI	S	.	0.0	14.4	3.4	.	.	.	.	.	.	.	.	.	.	.	.	.	19
3212	GUARA VILLOSA	S	.	0.0	37.2	43.9	.	.	.	.	.	.	.	.	.	.	.	.	.	21
3213	GUETTARDA URUGUENSIS	S	.	0.0	6.2	3.6	.	.	.	.	.	.	.	.	.	.	.	.	.	21
3214	GUIZOTIA ABYSSINICA	S	483	6.2	17.3	33.9	34.2	13.5	8.4	470	800	.	.	.	0	0.43	0.55	3	0	1
3215	GUIZOTIA ABYSSINICA	H	.	0.0	12.8	10.9	64.8	24.0	11.5	.	.	.	.	.	.	.	.	.	.	18
3216	GUIZOTIA ABYSSINICA	S	.	0.0	24.2	40.0	30.5	16.8	5.3	.	.	.	.	.	.	.	.	.	.	18
3217	GUIZOTIA ABYSSINICA	S	.	0.0	21.1	39.0	.	.	10.0	.	.	.	.	.	.	.	.	.	.	19
3218	GUIZOTIA ABYSSINICA	S	.	7.8	19.4	31.3	39.7	.	1.8	50	180	.	.	.	.	.	.	.	.	5
3219	GUNDELIA TOURNEFORTII	S	.	0.0	44.2	39.3	.	.	.	.	.	.	.	.	.	.	.	.	.	21
3220	GUTIERREZIA MICROCEPHALA	W	.	0.0	10.4	15.8	69.6	22.5	4.2	740	310	.	.	1350	.	.	.	.	.	6
3221	GUTIERREZIA SAROTHRAE	S	.	0.0	32.3	31.8	.	.	.	.	.	.	.	.	.	.	.	.	.	21
3222	GUTIERREZIA SAROTHRAE	W	.	0.0	6.7	11.2	76.8	29.3	5.3	1320	200	.	.	.	.	.	.	.	.	6
3223	GUTIERREZIA TEXANA	W	.	0.0	14.7	7.2	72.1	19.0	6.0	650	170	.	.	2980	.	.	.	.	.	6
3224	GYMNOCLADUS DIOICA	S	.	0.0	29.6	24.2	.	.	4.0	.	.	.	.	.	.	.	.	.	.	19
3225	GYMNOCLADUS DIOICA	S	.	0.0	4.2	0.5	.	.	5.4	.	.	.	.	.	.	.	.	.	.	19
3226	GYMNOCLADUS DIOICA	S	.	0.0	30.7	21.7	.	.	3.8	.	.	.	.	.	.	.	.	.	.	19
3227	GYNANDROPSIS GYNANDRA	L	34	86.6	4.8	0.4	5.2	1.2	3.0	288	111	6	.	.	.	.	.	.	13	1
3228	GYNANDROPSIS GYNANDRA	SH	.	0.0	25.7	2.7	52.5	18.6	19.1	1690	590	.	.	.	.	.	.	.	.	18
3229	GYNANDROPSIS GYNANDRA	S	.	0.0	18.5	25.5	.	.	.	.	.	.	.	.	.	.	.	.	.	21
3230	GYPSOPHILA ANATOLICA	S	.	0.0	19.3	7.1	.	.	.	.	.	.	.	.	.	.	.	.	.	21
3231	GYPSOPHILA ELEGANS	S	.	0.0	16.9	6.6	.	.	5.0	.	.	.	.	.	.	.	.	.	.	19
3232	GYPSOPHILA ELEGANS	S	.	0.0	16.3	5.7	.	.	4.6	.	.	.	.	.	.	.	.	.	.	19
3233	GYPSOPHILA ELEGANS	S	.	0.0	16.9	6.9	.	.	.	.	.	.	.	.	.	.	.	.	.	20
3234	GYPSOPHILA PORRIGENS	S	.	0.0	19.1	8.4	.	.	.	.	.	.	.	.	.	.	.	.	.	20
3235	GYPSOPHILA RUSCIFOLIA	S	.	0.0	19.8	7.6	.	.	.	.	.	.	.	.	.	.	.	.	.	21
3236	HALOCNEMUM STROBILACEUM	S	.	0.0	27.1	12.6	.	.	.	.	.	.	.	.	.	.	.	.	.	21
3237	HAMAMELIS JAPONICA	S	.	0.0	14.9	16.4	.	.	.	.	.	.	.	.	.	.	.	.	.	21
3238	HAPLOPAPPUS CILIATUS	S	.	0.0	12.8	11.8	.	.	.	.	.	.	.	.	.	.	.	.	.	21
3239	HAPLOPAPPUS CROCEUS	S	.	0.0	23.9	13.6	.	.	.	.	.	.	.	.	.	.	.	.	.	21
3240	HAYNALDIA VILLOSA	S	.	0.0	26.2	2.7	.	.	.	.	.	.	.	.	.	.	.	.	.	21
3241	HEDEOMA DRUMMONDII	S	.	0.0	8.8	6.9	.	.	.	.	.	.	.	.	.	.	.	.	.	20
3242	HEDEOMA DRUMMONDII	W	.	0.0	10.8	3.8	77.3	30.1	8.1	1990	190	.	.	1730	.	.	.	.	.	6
3243	HEDERA HELIX	S	.	0.0	16.2	35.1	.	.	.	.	.	.	.	.	.	.	.	.	.	19

TABLE 2 - DATA DIRECTLY TRANSCRIBED FROM ORIGINAL SOURCES.

OBS	PLNA	PART	CALORIES	WATER	PROTEIN	FAT	TOTCARB	FIBER	ASH	CALCIUM	PHOSPHO	IRON	SODIUM	POTASS	CAROTENE	THIAMINE	RIBOFLAV	NIACIN	ASCORBIC	REFCO
3244	HEDYPNOIS CRETICA	S	.	0.0	15.0	12.1	.	.	.	.	.	.	.	.	.	.	.	.	.	21
3245	HEDYSARUM BOREALE	S	.	0.0	43.0	8.8	.	.	5.1	.	.	.	.	.	.	.	.	.	.	19
3246	HEDYSARUM CORONARIUM	W	.	0.0	19.6	2.4	64.0	24.0	14.0	1120	340	.	.	.	.	.	.	.	.	18
3247	HEDYSARUM CORONARIUM	H	.	0.0	5.8	0.3	86.4	51.8	7.5	.	.	.	.	.	.	.	.	.	.	18
3248	HEDYSARUM CORONARIUM	F	.	0.0	13.8	2.1	68.0	23.6	15.5	.	.	.	.	.	.	.	.	.	.	18
3249	HEDYSARUM CORONARIUM	L	.	83.0	2.3	0.4	12.6	5.2	1.8	.	.	.	.	.	.	.	.	.	.	5
3250	HEDYSARUM HUMILE	S	.	0.0	55.8	10.0	.	.	.	.	.	.	.	.	.	.	.	.	.	21
3251	HEDYSARUM SPINOSISSIMUM	S	.	0.0	61.6	11.1	.	.	.	.	.	.	.	.	.	.	.	.	.	21
3252	HEDYSARUM VARIUM	S	.	0.0	50.1	8.3	.	.	.	.	.	.	.	.	.	.	.	.	.	20
3253	HEIMIA SALICIFOLIA	S	.	0.0	21.6	28.7	.	.	.	.	.	.	.	.	.	.	.	.	.	21
3254	HELENIUM AUTUMNALE	S	.	0.0	25.2	22.8	.	.	7.8	.	.	.	.	.	.	.	.	.	.	19
3255	HELENIUM AUTUMNALE	S	.	0.0	27.5	26.3	.	.	.	.	.	.	.	.	.	.	.	.	.	20
3256	HELENIUM LACINIATUM	S	.	0.0	23.1	28.6	.	.	.	.	.	.	.	.	.	.	.	.	.	20
3257	HELENIUM OOCLINIUM	S	.	0.0	30.3	26.3	.	.	.	.	.	.	.	.	.	.	.	.	.	20
3258	HELENIUM SP.	W	.	0.0	18.7	4.8	64.2	11.0	12.3	2440	610	.	.	3950	.	.	.	.	.	6
3259	HELIANTHELLA CALIFORNICA	S	.	0.0	30.8	21.3	.	.	.	.	.	.	.	.	.	.	.	.	.	21
3260	HELIANTHELLA UNIFLORA	W	.	0.0	7.4	4.3	.	.	9.4	2030	310	.	.	.	.	.	.	.	.	6
3261	HELIANTHEMUM LEDIFOLIUM	S	.	0.0	19.8	5.0	.	.	.	.	.	.	.	.	.	.	.	.	.	21
3262	HELIANTHEMUM NUMMULARIUM VAR. MUTABILE	S	.	0.0	17.5	4.8	.	.	.	.	.	.	.	.	.	.	.	.	.	20
3263	HELIANTHEMUM STIPULATUM	S	.	0.0	22.8	8.9	.	.	.	.	.	.	.	.	.	.	.	.	.	21
3264	HELIANTHUS ANNUUS	S	486	5.5	13.0	27.7	51.3	25.9	2.5	.	.	.	.	.	.	.	.	.	.	1
3265	HELIANTHUS ANNUUS	SH	.	0.0	13.0	1.9	70.3	20.4	14.8	1670	370	.	.	.	.	.	.	.	.	18
3266	HELIANTHUS ANNUUS	S	.	0.0	12.3	42.0	42.6	27.9	3.1	210	590	.	.	.	.	.	.	.	.	18
3267	HELIANTHUS ANNUUS	I	.	0.0	12.7	13.7	64.3	23.9	9.3	630	80	.	.	.	.	.	.	.	.	18
3268	HELIANTHUS ANNUUS	S	.	0.0	34.4	48.8	.	.	4.4	.	.	.	.	.	.	.	.	.	.	19
3269	HELIANTHUS ANNUUS	S	.	0.0	26.9	54.4	.	.	4.0	.	.	.	.	.	.	.	.	.	.	19
3270	HELIANTHUS ANNUUS	S	.	0.0	41.2	42.4	.	.	4.6	.	.	.	.	.	.	.	.	.	.	19
3271	HELIANTHUS ANNUUS	S	.	0.0	35.0	45.3	.	.	5.0	.	.	.	.	.	.	.	.	.	.	19
3272	HELIANTHUS ANNUUS	S	.	0.0	36.9	42.2	.	.	4.6	.	.	.	.	.	.	.	.	.	.	19
3273	HELIANTHUS ANNUUS	S	.	0.0	30.0	49.0	.	.	4.8	.	.	.	.	.	.	.	.	.	.	19
3274	HELIANTHUS ANNUUS	S	.	0.0	29.4	48.8	.	.	5.0	.	.	.	.	.	.	.	.	.	.	19
3275	HELIANTHUS ANNUUS	S	.	0.0	32.5	44.1	.	.	4.9	.	.	.	.	.	.	.	.	.	.	19
3276	HELIANTHUS ANNUUS	S	.	0.0	36.2	43.5	.	.	5.4	.	.	.	.	.	.	.	.	.	.	19
3277	HELIANTHUS ANNUUS	S	.	0.0	31.2	46.0	.	.	4.6	.	.	.	.	.	.	.	.	.	.	19
3278	HELIANTHUS ANNUUS	S	.	0.0	33.8	45.4	.	.	4.6	.	.	.	.	.	.	.	.	.	.	19
3279	HELIANTHUS ANNUUS	S	.	0.0	27.5	50.8	.	.	4.2	.	.	.	.	.	.	.	.	.	.	19
3280	HELIANTHUS ANNUUS	S	.	0.0	26.2	51.6	.	.	4.9	.	.	.	.	.	.	.	.	.	.	19
3281	HELIANTHUS ANNUUS	S	.	0.0	31.2	44.3	.	.	5.0	.	.	.	.	.	.	.	.	.	.	19
3282	HELIANTHUS ANNUUS	S	.	0.0	26.2	54.5	.	.	4.3	.	.	.	.	.	.	.	.	.	.	19
3283	HELIANTHUS ANNUUS	S	.	0.0	34.4	42.1	.	.	5.6	.	.	.	.	.	.	.	.	.	.	19
3284	HELIANTHUS ANNUUS	S	.	0.0	33.8	49.5	.	.	5.1	.	.	.	.	.	.	.	.	.	.	19
3285	HELIANTHUS ANNUUS	S	.	0.0	40.0	43.8	.	.	5.0	.	.	.	.	.	.	.	.	.	.	19
3286	HELIANTHUS ANNUUS	S	.	0.0	40.6	43.8	.	.	5.1	.	.	.	.	.	.	.	.	.	.	19
3287	HELIANTHUS ANNUUS	S	.	0.0	41.9	39.9	.	.	5.2	.	.	.	.	.	.	.	.	.	.	19
3288	HELIANTHUS ANNUUS	S	.	0.0	43.8	40.2	.	.	5.2	.	.	.	.	.	.	.	.	.	.	19

TABLE 2 - DATA DIRECTLY TRANSCRIBED FROM ORIGINAL SOURCES.

OBS	PLNA	PART	CALORIES	WATER	PROTEIN	FAT	TOTLCARB	FIBER	ASH	CALCIUM	PHOSPHO	IRON	SODIUM	POTASS	CAROT	THIAMINE	RIBOFLAV	NIACIN	ASCORBIC	REFCO
3289	HELIANTHUS ANNUUS	S	.	0.0	40.0	40.0	.	.	4.6	.	.	.	.	.	.	.	.	.	.	19
3290	HELIANTHUS ANNUUS	S	.	0.0	40.0	44.5	.	.	4.6	.	.	.	.	.	.	.	.	.	.	19
3291	HELIANTHUS ANNUUS	S	.	0.0	37.5	45.1	.	.	4.5	.	.	.	.	.	.	.	.	.	.	19
3292	HELIANTHUS ANNUUS	S	.	0.0	19.3	28.2	.	.	3.1	.	.	.	.	.	.	.	.	.	.	19
3293	HELIANTHUS ANNUUS	S	.	0.0	18.8	28.6	.	.	.	.	.	.	.	.	.	.	.	.	.	19
3294	HELIANTHUS ANNUUS	S	490	8.5	16.7	32.8	38.6	3.7	3.4	92	632	5.8	.	815	0	.	0.07	2.4	0	2
3295	HELIANTHUS ANNUUS	S	.	0.0	15.1	24.4	.	.	.	.	.	.	.	.	.	.	.	.	.	20
3296	HELIANTHUS ANNUUS	S	.	0.0	18.1	32.4	.	.	.	.	.	.	.	.	.	.	.	.	.	21
3297	HELIANTHUS ANNUUS	S	.	0.0	16.8	33.1	.	.	.	.	.	.	.	.	.	.	.	.	.	21
3298	HELIANTHUS ANNUUS	S	.	0.0	16.5	34.4	.	.	.	.	.	.	.	.	.	.	.	.	.	21
3299	HELIANTHUS ANNUUS	S	.	0.0	16.7	34.3	.	.	.	.	.	.	.	.	.	.	.	.	.	21
3300	HELIANTHUS ANNUUS	S	.	0.0	17.3	45.7	.	.	.	.	.	.	.	.	.	.	.	.	.	21
3301	HELIANTHUS ANNUUS	S	.	0.0	18.9	46.7	.	.	.	.	.	.	.	.	.	.	.	.	.	21
3302	HELIANTHUS ANNUUS	S	.	0.0	18.2	43.7	.	.	.	.	.	.	.	.	.	.	.	.	.	21
3303	HELIANTHUS ANNUUS	S	.	0.0	19.4	40.4	.	.	.	.	.	.	.	.	.	.	.	.	.	21
3304	HELIANTHUS ANNUUS	S	.	0.0	18.7	43.6	.	.	.	.	.	.	.	.	.	.	.	.	.	21
3305	HELIANTHUS ANNUUS	S	.	0.0	19.9	45.7	.	.	.	.	.	.	.	.	.	.	.	.	.	21
3306	HELIANTHUS ANNUUS	S	.	0.0	17.2	43.6	.	.	.	.	.	.	.	.	.	.	.	.	.	21
3307	HELIANTHUS ANNUUS	S	.	0.0	19.1	43.9	.	.	.	.	.	.	.	.	.	.	.	.	.	21
3308	HELIANTHUS ANNUUS	S	.	0.0	19.6	44.2	.	.	.	.	.	.	.	.	.	.	.	.	.	21
3309	HELIANTHUS ANNUUS	S	.	0.0	20.2	41.2	.	.	.	.	.	.	.	.	.	.	.	.	.	21
3310	HELIANTHUS ANNUUS	S	.	0.0	16.3	39.0	.	.	.	.	.	.	.	.	.	.	.	.	.	21
3311	HELIANTHUS ANNUUS	S	.	0.0	18.6	37.9	.	.	.	.	.	.	.	.	.	.	.	.	.	21
3312	HELIANTHUS ANNUUS	S	575	5.6	22.4	51.4	16.5	5.0	4.1	105	868	7.7	.	.	10	2.00	0.19	7.6	0	3
3313	HELIANTHUS ANNUUS	S	560	4.8	24.0	47.3	19.9	3.8	4.0	120	837	7.1	30	920	30	1.96	0.23	5.4	.	4
3314	HELIANTHUS ANNUUS	W	.	0.0	8.8	2.9	77.2	30.3	11.1	.	.	.	.	.	.	.	.	.	.	6
3315	HELIANTHUS ANNUUS(IMMATURE)	W	.	0.0	20.5	6.9	55.7	10.7	16.9	2130	220	.	.	4470	.	.	.	.	.	6
3316	HELIANTHUS CILIARIA	S	.	0.0	15.7	15.6	.	.	4.4	.	.	.	.	.	.	.	.	.	.	19
3317	HELIANTHUS CILIARIS	S	.	0.0	15.6	23.6	.	.	.	.	.	.	.	.	.	.	.	.	.	20
3318	HELIANTHUS MAXIMILIANI	S	.	0.0	30.8	30.4	.	.	4.3	.	.	.	.	.	.	.	.	.	.	19
3319	HELIANTHUS MAXIMILIANI	S	.	0.0	30.6	29.9	.	.	.	.	.	.	.	.	.	.	.	.	.	20
3320	HELIANTHUS MOLLIS	S	.	0.0	25.6	28.6	.	.	.	.	.	.	.	.	.	.	.	.	.	19
3321	HELIANTHUS NUTTALLII	S	.	0.0	30.6	23.4	.	.	.	.	.	.	.	.	.	.	.	.	.	21
3322	HELIANTHUS PETIOLARIS	S	.	0.0	19.3	29.9	.	.	2.8	.	.	.	.	.	.	.	.	.	.	19
3323	HELIANTHUS PETIOLARIS	S	.	0.0	16.9	28.4	.	.	.	.	.	.	.	.	.	.	.	.	.	19
3324	HELIANTHUS PETIOLARIS	S	.	0.0	21.2	24.1	.	.	.	.	.	.	.	.	.	.	.	.	.	21
3325	HELIANTHUS PETIOLARIS	W	.	0.0	10.2	6.9	71.4	35.2	11.5	.	.	.	.	.	.	.	.	.	.	6
3326	HELIANTHUS PETIOLARIS SUBSP. FALLAX	S	.	0.0	22.6	32.5	.	.	.	.	.	.	.	.	.	.	.	.	.	21
3327	HELIANTHUS PUMILIS	S	.	0.0	24.2	27.7	.	.	.	.	.	.	.	.	.	.	.	.	.	21
3328	HELIANTHUS RYDBERGII	S	.	0.0	26.1	30.3	.	.	3.8	.	.	.	.	.	.	.	.	.	.	19
3329	HELIANTHUS SCABERRIMUS	S	.	0.0	24.3	29.5	.	.	3.6	.	.	.	.	.	.	.	.	.	.	19
3330	HELIANTHUS SP.	S	.	0.0	32.2	32.2	.	.	4.3	.	.	.	.	.	.	.	.	.	.	19
3331	HELIANTHUS SP.	S	.	0.0	22.5	37.8	.	.	.	.	.	.	.	.	.	.	.	.	.	20
3332	HELIANTHUS TUBEROSUS	R	67	79.8	2.0	0.1	17.1	1.2	1.0	32	88	0.4	.	.	.	.	.	.	.	1
3333	HELIANTHUS TUBEROSUS	L	.	0.0	20.7	2.3	60.9	12.4	16.1	2040	360	.	.	.	.	.	.	.	.	18

TABLE 2 - DATA DIRECTLY TRANSCRIBED FROM ORIGINAL SOURCES.

OBS	PLNA	PART	CALORIES	WATER	PROTEIN	FAT	TOTCARB	FIBER	ASH	CALCIUM	PHOSPHO	IRON	SODIUM	POTASS	CAROT	THIAMINE	RIBOFLAV	NIACIN	ASCORBIC	REFCO
3334	HELIANTHUS TUBEROSUS	SH	.	0.0	10.5	3.4	70.6	16.7	15.5	.	.	.	.	.	.	.	.	.	.	18
3335	HELIANTHUS TUBEROSUS	R	.	0.0	8.3	0.6	85.1	4.8	6.0	210	350	.	.	.	.	.	.	.	.	18
3336	HELIANTHUS TUBEROSUS	S	.	0.0	30.6	26.5	.	.	4.8	.	.	.	.	.	.	.	.	.	.	19
3337	HELIANTHUS TUBEROSUS	R	69	80.4	1.8	0.1	16.5	0.8	1.2	29	66	1.6	.	382	0	0.20	0.05	0.2	6	2
3338	HELIANTHUS TUBEROSUS	R	65	80.0	2.2	0.1	16.5	0.8	1.2	32	.	0.6	.	.	10	0.20	0.05	0.1	6	3
3339	HELIANTHUS TUBEROSUS	R	41	79.8	2.3	0.1	16.7	0.8	1.1	14	78	3.4	.	.	12	0.20	0.06	1.3	4	4
3340	HELIANTHUS TUBEROSUS	SH	.	72.8	1.4	0.3	23.4	4.9	2.1	440	30	.	.	370	.	.	.	.	.	5
3341	HELICHRYSUM BRACTEATUM	S	.	0.0	21.9	24.2	.	.	5.7	.	.	.	.	.	.	.	.	.	.	19
3342	HELICHRYSUM BRACTEATUM	S	.	0.0	21.9	27.8	.	.	.	.	.	.	.	.	.	.	.	.	.	20
3343	HELICHRYSUM FULGIDUM	S	.	0.0	22.2	25.6	.	.	.	.	.	.	.	.	.	.	.	.	.	21
3344	HELICHRYSUM MONSTROSUM	S	.	0.0	21.9	29.0	.	.	.	.	.	.	.	.	.	.	.	.	.	20
3345	HELICONIA BREVISPATHA	L	26	91.9	0.4	0.1	6.7	1.6	0.9	45	22	1.0	85	159	.	0.01	0.03	0.1	.	2
3346	HELICTERES ISORA	S	.	0.0	19.5	9.4	.	.	.	.	.	.	.	.	.	.	.	.	.	21
3347	HELIOPSIS HELIANTHOIDES VAR. PITCHERIANA	S	.	0.0	18.9	22.6	.	.	3.5	.	.	.	.	.	.	.	.	.	.	19
3348	HELIOPSIS HELIANTHOIDES VAR. PITCHERIANA	S	.	0.0	22.5	23.0	.	.	.	.	.	.	.	.	.	.	.	.	.	20
3349	HELIOPSIS LAEVIS	S	.	0.0	18.1	11.3	.	.	.	.	.	.	.	.	.	.	.	.	.	20
3350	HELIOPSIS SCABRA VAR. ZINNIAEFLORA	S	.	0.0	25.0	21.3	.	.	.	.	.	.	.	.	.	.	.	.	.	20
3351	HELIOTROPIUM EUROPAEUM	S	.	0.0	20.9	23.8	.	.	.	.	.	.	.	.	.	.	.	.	.	21
3352	HELIOTROPIUM INDICUM	S	.	0.0	15.0	11.6	.	.	.	.	.	.	.	.	.	.	.	.	.	19
3353	HELIPTERUM MANGLESII	S	.	0.0	35.6	25.5	.	.	.	.	.	.	.	.	.	.	.	.	.	20
3354	HELIPTERUM MANGLESII VAR. MACULATUM	S	.	0.0	33.8	25.9	.	.	.	.	.	.	.	.	.	.	.	.	.	20
3355	HELIPTERUM ROSEUM	S	.	0.0	37.5	26.5	.	.	.	.	.	.	.	.	.	.	.	.	.	20
3356	HELLEBORUS NIGER	S	.	0.0	21.2	35.2	.	.	.	.	.	.	.	.	.	.	.	.	.	20
3357	HEMEROCALLIS FULVA	S	.	0.0	38.8	27.2	.	.	3.5	.	.	.	.	.	.	.	.	.	.	19
3358	HEMEROCALLIS LILIO-ASPHODELUS	I	42	87.2	2.0	0.4	9.6	1.2	0.8	87	175	1.2	24	170	3045	0.16	0.21	0.8	88	2
3359	HEMEROCALLIS LILIO-ASPHODELUS	I	.	85.5	1.7	0.4	11.7	1.2	0.8	.	.	.	.	.	.	.	.	.	.	5
3360	HERACLEUM LANATUM	S	.	0.0	28.2	15.0	.	.	5.2	.	.	.	.	.	.	.	.	.	.	19
3361	HERACLEUM LANATUM	S	.	0.0	10.9	1.0	.	.	7.6	.	.	.	.	.	.	.	.	.	.	19
3362	HERACLEUM LANATUM	S	.	0.0	21.5	17.8	.	.	8.1	.	.	.	.	.	.	.	.	.	.	19
3363	HERACLEUM LANATUM	S	.	0.0	20.0	14.0	.	.	.	.	.	.	.	.	.	.	.	.	.	20
3364	HERACLEUM LANATUM	I	.	0.0	25.3	5.2	57.9	16.4	11.6	.	.	.	.	.	.	.	.	.	.	6
3365	HERACLEUM LANATUM	L	.	0.0	20.2	3.6	61.7	12.4	14.5	.	.	.	.	.	.	.	.	.	.	6
3366	HERACLEUM LANATUM	H	.	0.0	24.5	3.1	56.6	12.8	15.8	.	.	.	.	.	.	.	.	.	.	6
3367	HERACLEUM MAXIMUM	S	.	0.0	29.3	21.0	.	.	.	.	.	.	.	.	.	.	.	.	.	20
3368	HERACLEUM SPHONDYLIUM	S	.	0.0	15.8	15.2	.	.	.	.	.	.	.	.	.	.	.	.	.	21
3369	HERNIARIA GLABRA	S	.	0.0	16.9	8.6	.	.	.	.	.	.	.	.	.	.	.	.	.	21
3370	HERNIARIA INCANA	S	.	0.0	8.4	2.7	.	.	.	.	.	.	.	.	.	.	.	.	.	21
3371	HESPERIS MATRONALIS	S	.	0.0	24.9	31.8	.	.	4.6	.	.	.	.	.	.	.	.	.	.	19
3372	HESPEROCHLOA KINGII	W	.	0.0	11.5	4.2	76.9	33.9	7.4	440	300	.	.	.	.	.	.	.	.	6
3373	HESPEROCHLOA KINGII(IMMATURE)	W	.	0.0	18.8	4.4	69.2	29.2	7.6	490	380	.	.	.	.	.	.	.	.	6
3374	HESPEROCHLOA KINGII(MATURE)	W	.	0.0	6.6	3.8	82.1	37.2	7.5	440	200	.	.	.	.	.	.	.	.	6
3375	HETERANTHERA RENIFORMIS	L	18	93.8	0.9	0.2	4.0	0.8	1.1	63	37	18.5	.	.	1460	0.03	0.08	0.3	16	3
3376	HETEROPHRAGMA ADENOPHYLLUM	S	.	0.0	62.5	20.2	.	.	.	.	.	.	.	.	.	.	.	.	.	20
3377	HETEROPOGON CONTORTUS	W	.	0.0	9.4	1.0	83.3	28.9	6.3	.	.	.	.	.	.	.	.	.	.	18
3378	HETEROPOGON CONTORTUS	H	.	0.0	3.5	1.0	87.3	40.1	8.2	.	.	.	.	.	.	.	.	.	.	18

TABLE 2 - DATA DIRECTLY TRANSCRIBED FROM ORIGINAL SOURCES.

OBS	PLNA	PART	CALORIES	WATER	PROTEIN	FAT	TOTLCARB	FIBER	ASH	CALCIUM	PHOSPHO	IRON	SODIUM	POTASS	CAROT	THIAMINE	RIBOFLAV	NIACIN	ASCORBIC	REFCO
3379	HETEROPOGON CONTORTUS	L	.	0.0	7.5	3.2	78.4	28.9	.	950	400	.	.	.	.	.	.	.	.	5
3380	HETEROPOGON CONTORTUS	L	.	0.0	4.4	3.2	83.7	26.9	.	460	320	.	.	.	.	.	.	.	.	5
3381	HETEROPOGON CONTORTUS	L	.	0.0	2.3	2.5	87.5	28.9	.	340	370	.	.	.	.	.	.	.	.	5
3382	HETEROPOGON CONTORTUS(IMMATURE)	W	.	0.0	9.6	2.0	78.4	31.0	10.0	470	160	.	.	.	.	.	.	.	.	6
3383	HETEROPOGON CONTORTUS(MATURE)	W	.	0.0	3.4	1.9	88.1	34.4	6.6	220	50	.	.	.	.	.	.	.	.	6
3384	HETEROTHECA LATIFOLIA VAR. LATIFOLIA	S	.	0.0	17.2	11.8	.	.	.	.	.	.	.	.	.	.	.	.	.	21
3385	HETEROTHECA SUBAXILLARIS	S	.	0.0	13.9	13.4	.	.	2.8	.	.	.	.	.	.	.	.	.	.	19
3386	HEVEA BRASILIENSIS	F	.	0.0	18.0	52.3	27.1	1.9	2.6	.	.	.	.	.	.	.	.	.	.	18
3387	HEVEA BRASILIENSIS	S	.	8.5	17.6	48.5	22.9	.	2.5	120	430	.	.	.	.	.	.	.	.	5
3388	HIBISCUS ABELMOSCHUS	S	.	0.0	27.8	21.5	.	.	.	.	.	.	.	.	.	.	.	.	.	20
3389	HIBISCUS ASPER	L	.	0.0	31.1	3.5	55.6	13.5	9.8	1040	90	.	.	.	.	.	.	.	.	18
3390	HIBISCUS CANNABINUS	L	67	79.0	5.5	1.2	12.2	2.3	2.1	484	18	12.1	.	.	.	.	.	.	75	1
3391	HIBISCUS CANNABINUS	C	.	0.0	15.0	5.8	71.0	19.9	8.2	2080	500	.	.	.	.	.	.	.	.	18
3392	HIBISCUS CANNABINUS	L	.	0.0	13.1	2.1	73.0	11.6	11.8	3310	350	.	.	.	.	.	.	.	.	18
3393	HIBISCUS CANNABINUS	S	.	0.0	27.2	15.4	51.3	25.1	6.1	600	630	.	.	.	.	.	.	.	.	18
3394	HIBISCUS CANNABINUS	S	.	0.0	27.9	18.5	.	.	5.6	.	.	.	.	.	.	.	.	.	.	19
3395	HIBISCUS CANNABINUS	S	.	0.0	28.8	22.9	.	.	.	.	.	.	.	.	.	.	.	.	.	19
3396	HIBISCUS CANNABINUS	S	.	0.0	32.5	17.5	.	.	.	.	.	.	.	.	.	.	.	.	.	19
3397	HIBISCUS CANNABINUS	S	.	0.0	29.8	19.8	.	.	5.5	.	.	.	.	.	.	.	.	.	.	19
3398	HIBISCUS CANNABINUS	S	.	0.0	31.4	19.1	.	.	6.0	.	.	.	.	.	.	.	.	.	.	19
3399	HIBISCUS CANNABINUS	S	.	0.0	27.4	20.7	.	.	4.9	.	.	.	.	.	.	.	.	.	.	19
3400	HIBISCUS CANNABINUS	S	.	0.0	37.0	21.7	.	.	4.3	.	.	.	.	.	.	.	.	.	.	19
3401	HIBISCUS CANNABINUS	S	.	0.0	37.0	22.9	.	.	4.6	.	.	.	.	.	.	.	.	.	.	19
3402	HIBISCUS CANNABINUS	S	.	0.0	28.1	18.8	.	.	6.1	.	.	.	.	.	.	.	.	.	.	19
3403	HIBISCUS CANNABINUS	S	.	0.0	30.1	17.9	.	.	6.0	.	.	.	.	.	.	.	.	.	.	19
3404	HIBISCUS CANNABINUS	S	.	0.0	32.7	18.4	.	.	6.0	.	.	.	.	.	.	.	.	.	.	19
3405	HIBISCUS CANNABINUS	S	.	0.0	30.6	18.3	.	.	5.8	.	.	.	.	.	.	.	.	.	.	19
3406	HIBISCUS CANNABINUS	S	.	0.0	28.8	18.7	.	.	5.8	.	.	.	.	.	.	.	.	.	.	19
3407	HIBISCUS CANNABINUS	S	.	0.0	28.9	19.0	.	.	5.6	.	.	.	.	.	.	.	.	.	.	19
3408	HIBISCUS CANNABINUS	S	.	0.0	26.6	14.2	.	.	5.5	.	.	.	.	.	.	.	.	.	.	19
3409	HIBISCUS CANNABINUS	S	.	0.0	28.9	19.0	.	.	6.3	.	.	.	.	.	.	.	.	.	.	19
3410	HIBISCUS CANNABINUS	S	.	0.0	32.1	20.7	.	.	5.6	.	.	.	.	.	.	.	.	.	.	19
3411	HIBISCUS CANNABINUS	S	.	0.0	27.5	20.9	.	.	4.6	.	.	.	.	.	.	.	.	.	.	19
3412	HIBISCUS CANNABINUS	S	.	0.0	26.2	20.7	.	.	4.6	.	.	.	.	.	.	.	.	.	.	19
3413	HIBISCUS CANNABINUS	S	.	0.0	25.6	20.4	.	.	4.5	.	.	.	.	.	.	.	.	.	.	19
3414	HIBISCUS CANNABINUS	S	.	0.0	27.5	21.5	.	.	4.3	.	.	.	.	.	.	.	.	.	.	19
3415	HIBISCUS CANNABINUS	S	.	0.0	26.2	20.9	.	.	4.6	.	.	.	.	.	.	.	.	.	.	19
3416	HIBISCUS CANNABINUS	S	.	0.0	26.2	20.6	.	.	4.8	.	.	.	.	.	.	.	.	.	.	19
3417	HIBISCUS CANNABINUS	S	.	0.0	28.8	20.7	.	.	4.8	.	.	.	.	.	.	.	.	.	.	19
3418	HIBISCUS CANNABINUS	S	.	0.0	25.6	20.5	.	.	4.3	.	.	.	.	.	.	.	.	.	.	19
3419	HIBISCUS CANNABINUS	S	.	0.0	26.9	21.3	.	.	4.4	.	.	.	.	.	.	.	.	.	.	19
3420	HIBISCUS CANNABINUS	S	.	0.0	28.1	21.5	.	.	4.4	.	.	.	.	.	.	.	.	.	.	19
3421	HIBISCUS CANNABINUS	S	.	0.0	26.2	21.2	.	.	4.2	.	.	.	.	.	.	.	.	.	.	19
3422	HIBISCUS CANNABINUS	S	.	0.0	25.6	21.5	.	.	4.6	.	.	.	.	.	.	.	.	.	.	19
3423	HIBISCUS CANNABINUS	S	.	0.0	28.8	20.2	.	.	4.8	.	.	.	.	.	.	.	.	.	.	19

TABLE 2 - DATA DIRECTLY TRANSCRIBED FROM ORIGINAL SOURCES.

OBS	PLNA	PART	CALORIES	WATER	PROTEIN	FAT	TOTLCARB	FIBER	ASH	CALCIUM	PHOSPHO	IRON	SODIUM	POTASS	CAROT	THIAMINE	RIBOFLAV	NIACIN	ASCORBIC	REFCO
3424	HIBISCUS CANNABINUS	S	.	0.0	28.1	21.1	.	.	4.9	.	.	.	.	.	.	.	.	.	.	19
3425	HIBISCUS CANNABINUS	S	.	0.0	27.5	19.6	.	.	5.4	.	.	.	.	.	.	.	.	.	.	19
3426	HIBISCUS CANNABINUS	S	.	0.0	28.1	20.2	.	.	4.8	.	.	.	.	.	.	.	.	.	.	19
3427	HIBISCUS CANNABINUS	S	.	0.0	26.2	21.1	.	.	4.3	.	.	.	.	.	.	.	.	.	.	19
3428	HIBISCUS CANNABINUS	L	42	86.5	3.1	0.0	9.6	.	0.8	62	38	1.3	.	.	.	0.07	0.06	.	.	2
3429	HIBISCUS CANNABINUS	S	.	0.0	24.2	21.7	.	.	.	.	.	.	.	.	.	.	.	.	.	20
3430	HIBISCUS CANNABINUS	S	.	0.0	22.5	15.1	.	.	.	.	.	.	.	.	.	.	.	.	.	20
3431	HIBISCUS ESCULENTUS	S	.	0.0	23.8	14.7	.	.	5.0	.	.	.	.	.	.	.	.	.	.	19
3432	HIBISCUS ESCULENTUS	S	.	0.0	25.6	15.4	.	.	.	.	.	.	.	.	.	.	.	.	.	20
3433	HIBISCUS ESCULENTUS	S	.	0.0	22.5	15.1	.	.	.	.	.	.	.	.	.	.	.	.	.	20
3434	HIBISCUS GIBSONII	S	.	0.0	23.2	15.1	.	.	.	.	.	.	.	.	.	.	.	.	.	21
3435	HIBISCUS GRANDIFLORUS	S	.	0.0	24.4	11.7	.	.	.	.	.	.	.	.	.	.	.	.	.	20
3436	HIBISCUS LASIOCARPUS	S	.	0.0	21.2	12.1	.	.	.	.	.	.	.	.	.	.	.	.	.	20
3437	HIBISCUS LUNARIFOLIUS	W	.	0.0	17.8	1.6	71.4	37.4	9.2	1930	330	.	.	.	.	.	.	.	.	18
3438	HIBISCUS MOSCHEUTOS	S	.	0.0	22.5	11.8	.	.	3.4	.	.	.	.	.	.	.	.	.	.	19
3439	HIBISCUS MOSCHEUTOS	S	.	0.0	21.9	12.7	.	.	.	.	.	.	.	.	.	.	.	.	.	19
3440	HIBISCUS MOSCHEUTOS	S	.	0.0	23.8	11.6	.	.	.	.	.	.	.	.	.	.	.	.	.	20
3441	HIBISCUS PALUSTRIS	S	.	0.0	21.9	12.4	.	.	.	.	.	.	.	.	.	.	.	.	.	19
3442	HIBISCUS PLATANIFOLIUS	S	.	0.0	13.2	8.9	.	.	2.1	.	.	.	.	.	.	.	.	.	.	19
3443	HIBISCUS ROSA-SINENSIS	W	.	0.0	15.4	3.5	69.7	15.5	11.4	1670	520	.	.	.	.	.	.	.	.	18
3444	HIBISCUS ROSA-SINENSIS	F	36	89.8	0.4	0.4	8.8	1.6	0.6	4	27	1.7	.	.	.	0.03	0.05	0.6	4	2
3445	HIBISCUS ROSA-SINENSIS	I	.	89.8	0.4	0.4	.	1.6	.	4	27	1.7	.	.	.	.	.	.	.	5
3446	HIBISCUS SABDARIFFA	F	49	84.5	1.9	0.1	12.3	2.3	1.2	172	57	2.9	.	.	300	.	.	.	14	1
3447	HIBISCUS SABDARIFFA	L	43	85.6	3.3	0.3	9.2	1.6	1.6	213	93	4.8	.	.	4135	0.17	0.45	1.2	54	1
3448	HIBISCUS SABDARIFFA	I	44	86.2	1.6	0.1	11.1	2.5	1.0	160	60	3.8	.	.	285	0.04	0.06	0.5	14	1
3449	HIBISCUS SABDARIFFA	S	.	0.0	21.4	17.5	55.7	12.0	5.4	610	320	.	.	.	.	.	.	.	.	18
3450	HIBISCUS SABDARIFFA	L	.	0.0	12.3	7.8	72.5	7.8	7.4	1310	280	.	.	.	.	.	.	.	.	18
3451	HIBISCUS SABDARIFFA	SH	.	0.0	30.1	1.9	58.9	10.9	9.1	.	.	.	.	.	.	.	.	.	.	18
3452	HIBISCUS SABDARIFFA	L	.	0.0	25.7	3.3	60.4	11.4	10.6	.	.	.	.	.	.	.	.	.	.	18
3453	HIBISCUS SABDARIFFA	F	39	90.0	0.7	1.1	7.6	1.4	0.6	174	18	0.1	.	.	110	0.01	0.02	0.4	10	2
3454	HIBISCUS SABDARIFFA	L	44	86.4	1.9	0.3	10.4	1.3	1.0	116	48	1.5	.	.	7550	0.02	0.15	1.8	34	2
3455	HIBISCUS SABDARIFFA	I	55	84.5	1.7	1.0	12.0	1.0	0.8	110	39	2.2	.	.	20	0.04	0.06	0.4	18	3
3456	HIBISCUS SCHIZOPETALUS	L	.	0.0	17.4	8.5	62.8	14.9	11.3	.	.	.	.	.	.	.	.	.	.	18
3457	HIBISCUS SYRIACUS	S	.	0.0	28.8	29.4	.	.	.	.	.	.	.	.	.	.	.	.	.	19
3458	HIBISCUS SYRIACUS	S	.	0.0	30.6	28.0	.	.	.	.	.	.	.	.	.	.	.	.	.	19
3459	HIBISCUS SYRIACUS	S	.	0.0	28.8	15.2	.	.	.	.	.	.	.	.	.	.	.	.	.	19
3460	HIBISCUS SYRIACUS	S	.	0.0	28.8	27.7	.	.	.	.	.	.	.	.	.	.	.	.	.	20
3461	HIBISCUS SYRIACUS	S	.	0.0	27.5	24.4	.	.	.	.	.	.	.	.	.	.	.	.	.	20
3462	HIBISCUS TILIACEUS	S	.	0.0	18.2	6.3	.	.	.	.	.	.	.	.	.	.	.	.	.	21
3463	HIBISCUS TRIONUM	S	.	0.0	18.6	14.0	.	.	5.7	.	.	.	.	.	.	.	.	.	.	19
3464	HILARIA BELANGERI(ALL)	W	.	0.0	7.5	2.2	74.6	28.4	15.7	560	120	.	.	670	.	.	.	.	.	6
3465	HILARIA BELANGERI(IMMATURE)	W	.	0.0	17.2	2.5	64.6	25.6	15.7	1040	260	.	.	790	.	.	.	.	.	6
3466	HILARIA BELANGERI(MATURE)	W	.	0.0	6.0	2.2	74.7	28.8	17.1	550	90	.	.	390	.	.	.	.	.	6
3467	HILARIA BELANGERI(VERY MATURE)	W	.	0.0	4.9	2.0	79.6	29.3	13.5	330	60	.	.	420	.	.	.	.	.	6
3468	HILARIA JAMESII	W	.	0.0	4.9	1.5	79.3	32.9	14.3	670	110	.	.	.	.	.	.	.	.	6

TABLE 2 - DATA DIRECTLY TRANSCRIBED FROM ORIGINAL SOURCES.

OBS	PLNA	PART	CALORIES	WATER	PROTEIN	FAT	TOTLCARB	FIBER	ASH	CALCIUM	PHOSPHO	IRON	SODIUM	POTASS	CAROT	THIAMINE	RIBOFLAV	NIACIN	ASCORBIC	REFCO
3469	HILARIA JAMESII(MATURE)	W	.	0.0	6.2	2.6	.	.	19.1	1160	80	.	.	.	.	.	.	.	.	6
3470	HILARIA JAMESII(VERY MATURE)	W	.	0.0	3.8	1.2	81.0	33.0	14.0	590	80	.	.	.	.	.	.	.	.	6
3471	HILARIA MUTICA(ALL)	W	.	0.0	7.0	1.5	80.6	33.4	10.9	250	130	.	.	.	.	.	.	.	.	6
3472	HILLERIA LATIFOLIA	L	44	84.3	4.1	0.8	7.8	.	3.0	349	.	4.1	.	.	.	.	.	.	22	1
3473	HIPPOCREPIS UNISILIQUOSA	S	.	0.0	22.1	1.8	.	.	.	.	.	.	.	.	.	.	.	.	.	21
3474	HIPPOMARATHRUM BOISSIERI	S	.	0.0	15.4	20.2	.	.	.	.	.	.	.	.	.	.	.	.	.	21
3475	HIPPOMARATHRUM PAUCIRADIATUM	S	.	0.0	36.9	31.3	.	.	.	.	.	.	.	.	.	.	.	.	.	21
3476	HIPPOPHAE RHAMNOIDES	S	.	0.0	28.1	9.5	.	.	.	.	.	.	.	.	.	.	.	.	.	20
3477	HIPPOPHAE RHAMNOIDES	S	.	0.0	22.6	9.2	.	.	.	.	.	.	.	.	.	.	.	.	.	21
3478	HIRPICIUM INTEGRIFOLIUM	S	.	0.0	46.3	22.6	.	.	.	.	.	.	.	.	.	.	.	.	.	21
3479	HIRSCHFELDIA INCANA	S	.	0.0	28.1	30.6	.	.	.	.	.	.	.	.	.	.	.	.	.	21
3480	HIRSCHFELDIA INCANA	S	.	0.0	32.8	32.7	.	.	.	.	.	.	.	.	.	.	.	.	.	21
3481	HOFFMANNSEGGIA SP.	W	.	0.0	14.7	2.5	74.9	17.1	7.9	3180	180	.	.	1210	.	.	.	.	.	6
3482	HOLCUS LANATUS	S	.	0.0	20.6	17.9	.	.	5.4	.	.	.	.	.	.	.	.	.	.	19
3483	HOLCUS LANATUS	S	.	0.0	23.4	20.5	.	.	.	.	.	.	.	.	.	.	.	.	.	21
3484	HORDEUM BULBOSUM	S	.	0.0	27.5	2.6	.	.	4.3	.	.	.	.	.	.	.	.	.	.	19
3485	HORDEUM BULBOSUM	S	.	0.0	23.2	2.0	.	.	.	.	.	.	.	.	.	.	.	.	.	21
3486	HORDEUM BULBOSUM	S	.	0.0	21.6	3.1	.	.	.	.	.	.	.	.	.	.	.	.	.	21
3487	HORDEUM MURINUM	S	.	0.0	23.5	1.4	.	.	.	.	.	.	.	.	.	.	.	.	.	21
3488	HORDEUM VULGARE	S	337	9.9	10.0	1.6	76.2	6.0	2.3	61	251	17.9	.	.	.	0.36	.	.	.	1
3489	HORDEUM VULGARE	F	.	0.0	10.3	2.4	83.8	6.2	3.5	80	290	.	.	.	.	.	.	.	.	18
3490	HORDEUM VULGARE	W	.	0.0	22.7	5.1	59.8	20.2	12.4	.	.	.	.	.	.	.	.	.	.	18
3491	HORDEUM VULGARE	H	.	0.0	2.5	1.4	78.5	44.6	17.6	.	.	.	.	.	.	.	.	.	.	18
3492	HORDEUM VULGARE	S	327	13.7	10.5	2.1	71.8	2.5	1.9	50	378	6.0	4	562	0	0.31	0.10	5.2	0	2
3493	HORDEUM VULGARE	S	348	10.5	9.7	1.9	75.4	6.5	2.5	55	341	4.5	.	.	10	0.38	0.20	7.2	0	3
3494	HORDEUM VULGARE	S	349	11.1	8.2	1.0	78.8	0.5	0.9	16	189	2.0	3	160	0	0.12	0.05	3.1	0	4
3495	HORDEUM VULGARE(ALL)	W	.	0.0	21.2	3.9	62.0	20.8	12.9	480	370	.	.	3060	.	.	.	.	.	6
3496	HOSTA VENTRICOSA	S	.	0.0	25.6	36.4	.	.	.	.	.	.	.	.	.	.	.	.	.	19
3497	HOVENIA DULCIS	S	.	0.0	15.0	7.8	.	.	.	.	.	.	.	.	.	.	.	.	.	20
3498	HULSEA HETEROCHROMA	S	.	0.0	20.0	21.9	.	.	4.6	.	.	.	.	.	.	.	.	.	.	19
3499	HUMULUS SCANDENS	S	.	0.0	23.9	31.0	.	.	10.3	.	.	.	.	.	.	.	.	.	.	19
3500	HYACINTHUS SP.	W	.	0.0	8.0	3.3	76.2	18.0	12.5	.	.	.	.	.	.	.	.	.	.	6
3501	HYDRANGEA PANICULATA	S	.	0.0	25.0	31.3	.	.	.	.	.	.	.	.	.	.	.	.	.	20
3502	HYDRANGEA PETIOLARIS	S	.	0.0	29.1	38.0	.	.	.	.	.	.	.	.	.	.	.	.	.	21
3503	HYDROCOTYLE BONARIENSIS	S	.	0.0	10.0	13.4	.	.	.	.	.	.	.	.	.	.	.	.	.	20
3504	HYGROPHILA SPINOSA	L	56	82.2	4.1	.	12.8	.	0.9	55	73	.	.	.	25670	.	.	.	100	2
3505	HYLOCEREUS UNDATUS	F	54	84.4	1.4	0.4	13.2	1.4	0.6	10	26	1.3	.	.	0	0.04	0.04	0.3	8	3
3506	HYMENACHNE AMPLEXICAULIS	S	.	0.0	8.9	1.0	78.6	36.7	11.5	.	.	.	.	.	.	.	.	.	.	18
3507	HYMENACHNE AMPLEXICAULIS	L	.	0.0	22.6	2.8	67.4	32.4	7.2	.	.	.	.	.	.	.	.	.	.	18
3508	HYMENACHNE AMPLEXICAULIS	W	.	0.0	9.4	2.3	76.1	22.1	12.2	.	.	.	.	.	.	.	.	.	.	18
3509	HYMENACHNE AMPLEXICAULIS	H	.	0.0	7.5	1.4	78.2	29.2	12.9	.	.	.	.	.	.	.	.	.	.	18
3510	HYMENACHNE AMPLEXICAULIS	L	.	0.0	9.4	2.3	76.1	22.1	12.2	130	206	.	.	.	.	.	.	.	.	5
3511	HYMENACHNE AMPLEXICAULIS	L	.	0.0	7.5	1.4	79.1	29.2	12.9	141	186	.	.	.	.	.	.	.	.	5
3512	HYMENAEA COURBARIL	F	361	0.0	6.9	2.5	88.1	15.7	2.3	32	167	3.7	.	.	1	0.27	0.20	4.8	13	10
3513	HYMENOCARPUS CIRCINNATA	S	.	0.0	46.0	12.8	.	.	.	.	.	.	.	.	.	.	.	.	.	21

TABLE 2 - DATA DIRECTLY TRANSCRIBED FROM ORIGINAL SOURCES.

OBS	PLNA	PART	CALORIES	WATER	PROTEIN	FAT	TOTLCARB	FIBER	ASH	CALCIUM	PHOSPHO	IRON	SODIUM	POTASS	CAROT	THIAMINE	RIBOFLAV	NIACIN	ASCORBIC	REFCO
3514	HYMENOCARPUS CIRCINNATA	S	.	0.0	39.4	17.4	.	.	.	.	.	.	.	.	.	.	.	.	.	21
3515	HYMENOCLEA MONOGYRA	S	.	0.0	38.8	37.1	.	.	.	.	.	.	.	.	.	.	.	.	.	19
3516	HYMENOPAPPUS ARTEMISIAEFOLIUS	S	.	0.0	16.9	19.6	.	.	.	.	.	.	.	.	.	.	.	.	.	19
3517	HYMENOPAPPUS ARTEMISIAEFOLIUS	S	.	0.0	15.6	22.5	.	.	.	.	.	.	.	.	.	.	.	.	.	20
3518	HYMENOXYS ODORATA(IMMATURE)	W	.	0.0	33.3	3.3	51.0	11.0	12.4	1840	460	.	.	4590	.	.	.	.	.	6
3519	HYMENOXYS SP.(PAST FLOWERING)	W	.	0.0	12.3	6.1	73.4	23.9	8.2	1040	180	.	.	2540	.	.	.	.	.	6
3520	HYOSCYAMUS AUREUS	S	.	0.0	19.4	31.7	.	.	.	.	.	.	.	.	.	.	.	.	.	21
3521	HYOSCYAMUS NIGER	S	.	0.0	21.2	38.9	.	.	.	.	.	.	.	.	.	.	.	.	.	21
3522	HYOSCYAMUS RETICULATUS	S	.	0.0	20.2	36.1	.	.	.	.	.	.	.	.	.	.	.	.	.	20
3523	HYOSCYAMUS RETICULATUS	S	.	0.0	22.0	34.8	.	.	.	.	.	.	.	.	.	.	.	.	.	21
3524	HYPARRHENIA CYMBARIA	W	.	0.0	7.3	2.6	75.0	30.3	15.1	.	.	.	.	.	.	.	.	.	.	18
3525	HYPARRHENIA DISSOLUTA	W	.	0.0	12.9	3.0	75.3	33.7	8.8	.	.	.	.	.	.	.	.	.	.	18
3526	HYPARRHENIA DISSOLUTA	H	.	0.0	6.8	1.5	85.9	36.9	5.8	.	.	.	.	.	.	.	.	.	.	18
3527	HYPARRHENIA FILIPENDULA	W	.	0.0	6.6	1.8	85.8	36.3	5.8	.	.	.	.	.	.	.	.	.	.	18
3528	HYPARRHENIA LINTONII	W	.	0.0	11.4	5.0	75.0	31.8	8.6	.	.	.	.	.	.	.	.	.	.	18
3529	HYPARRHENIA RUFA	W	.	0.0	9.2	2.6	73.3	28.9	14.9	.	.	.	.	.	.	.	.	.	.	18
3530	HYPARRHENIA RUFA	H	.	0.0	6.5	2.3	73.3	35.0	17.9	.	.	.	.	.	.	.	.	.	.	18
3531	HYPARRHENIA RUFA	S	.	0.0	11.2	2.0	.	.	9.4	.	.	.	.	.	.	.	.	.	.	19
3532	HYPERICUM ANDROSAEMUM	S	.	0.0	14.7	28.9	.	.	.	.	.	.	.	.	.	.	.	.	.	21
3533	HYPERICUM CALYCINUM	S	.	0.0	.	31.6	.	.	.	.	.	.	.	.	.	.	.	.	.	21
3534	HYPERICUM CALYCINUM	S	.	0.0	14.9	44.9	.	.	.	.	.	.	.	.	.	.	.	.	.	21
3535	HYPERICUM CISTIFOLIUM	S	.	0.0	23.1	38.6	.	.	.	.	.	.	.	.	.	.	.	.	.	20
3536	HYPERICUM PERFORATUM	S	.	0.0	20.7	33.2	.	.	.	.	.	.	.	.	.	.	.	.	.	21
3537	HYPERICUM PERFORATUM	S	.	0.0	18.1	32.8	.	.	.	.	.	.	.	.	.	.	.	.	.	21
3538	HYPERICUM TETRAPTERUM	S	.	0.0	24.2	34.9	.	.	.	.	.	.	.	.	.	.	.	.	.	20
3539	HYPERICUM TRIQUETRIFOLIUM	S	.	0.0	16.2	32.6	.	.	.	.	.	.	.	.	.	.	.	.	.	21
3540	HYPHAENE SP.	S	.	0.0	9.0	7.0	81.2	7.3	2.8	.	.	.	.	.	.	.	.	.	.	18
3541	HYPHAENE SP.	F	.	0.0	4.5	2.6	89.3	24.7	3.6	.	.	.	.	.	.	.	.	.	.	18
3542	HYPHAENE THEBAICA	S	395	6.0	3.9	6.4	80.6	9.4	3.1	144	226	.	.	.	.	.	.	.	.	1
3543	HYPHAENE TURBINATA	S	.	0.0	8.1	13.4	.	.	2.3	.	.	.	.	.	.	.	.	.	.	19
3544	HYPOCALYPTUS OBCORDATUS	S	.	0.0	24.1	4.1	.	.	.	.	.	.	.	.	.	.	.	.	.	21
3545	HYPTIS CF. AMERICANA	S	.	0.0	13.1	14.4	.	.	.	.	.	.	.	.	.	.	.	.	.	20
3546	HYPTIS DECURRENS	S	.	0.0	15.0	19.5	.	.	.	.	.	.	.	.	.	.	.	.	.	20
3547	HYPTIS SPICIGERA	S	479	4.4	17.0	29.1	43.4	17.2	6.1	214	638	.	.	.	.	.	.	.	.	1
3548	HYPTIS SPICIGERA	S	.	0.0	14.4	18.8	.	.	.	.	.	.	.	.	.	.	.	.	.	19
3549	HYPTIS STELLULATA	S	.	0.0	19.7	22.3	.	.	.	.	.	.	.	.	.	.	.	.	.	21
3550	HYPTIS SUAVEOLENS	S	.	0.0	21.9	24.4	.	.	.	.	.	.	.	.	.	.	.	.	.	19
3551	HYPTIS SUAVEOLENS	S	.	0.0	18.8	14.3	.	.	.	.	.	.	.	.	.	.	.	.	.	20
3552	IBERIS AMARA	S	.	0.0	28.1	26.8	.	.	.	.	.	.	.	.	.	.	.	.	.	20
3553	IBERIS UMBELLATA	S	.	0.0	34.4	25.6	.	.	.	.	.	.	.	.	.	.	.	.	.	20
3554	ICACINA SENEGALENSIS	R	335	11.7	10.3	0.7	74.5	.	2.8	150	.	7	.	.	.	0.04	0.18	1.4	.	1
3555	IDESIA POLYCARPA	S	.	0.0	11.9	29.4	.	.	.	.	.	.	.	.	.	.	.	.	.	20
3556	ILEX AQUIFOLIUM	S	.	0.0	6.6	1.9	.	.	.	.	.	.	.	.	.	.	.	.	.	21
3557	ILEX CRENATA	S	.	0.0	24.2	48.0	.	.	.	.	.	.	.	.	.	.	.	.	.	21
3558	ILEX LATIFOLIA	S	.	0.0	7.0	9.9	.	.	.	.	.	.	.	.	.	.	.	.	.	21

TABLE 2 - DATA DIRECTLY TRANSCRIBED FROM ORIGINAL SOURCES.

OBS	PLNA	PART	CALORIES	WATER	PROTEIN	FAT	TOTLCARB	FIBER	ASH	CALCIUM	PHOSPHO	IRON	SODIUM	POTASS	CAROT	THIAMINE	RIBOFLAV	NIACIN	ASCORBIC	REFCO
3559	ILEX PARAGUARIENSIS	S	.	7.2	6.4	16.1	.	52.0	1.3	.	.	.	.	.	.	.	.	.	.	5
3560	ILEX PURPUREA	S	.	0.0	13.0	8.5	.	.	2.2	.	.	.	.	.	.	.	.	.	.	19
3561	ILEX SERRATA	S	.	0.0	21.0	30.4	.	.	.	.	.	.	.	.	.	.	.	.	.	20
3562	ILEX VERTICILLATA	S	.	0.0	12.5	15.9	.	.	.	.	.	.	.	.	.	.	.	.	.	19
3563	IMPATIENS BALSAMINA	S	.	0.0	16.2	21.8	.	.	3.3	.	.	.	.	.	.	.	.	.	.	19
3564	IMPATIENS GLANDULIFERA	S	.	6.5	21.0	50.0	.	.	.	.	.	.	.	.	.	.	.	.	.	5
3565	IMPERATA CYLINDRICA	W	.	0.0	11.8	1.9	79.2	32.1	7.1	.	.	.	.	.	.	.	.	.	.	18
3566	IMPERATA CYLINDRICA	H	.	0.0	3.8	0.7	87.7	39.7	7.8	.	.	.	.	.	.	.	.	.	.	18
3567	IMPERATA CYLINDRICA	SH	.	0.0	6.6	3.3	82.2	34.6	7.9	390	220	.	.	.	.	.	.	.	.	5
3568	INDIGOFERA ARRECTA	W	.	0.0	26.2	2.2	62.7	25.5	8.9	1360	350	.	.	.	.	.	.	.	.	18
3569	INDIGOFERA BRACHYSTACHYA	S	.	0.0	34.1	3.1	.	.	.	.	.	.	.	.	.	.	.	.	.	21
3570	INDIGOFERA DENSIFLORA	S	.	0.0	35.0	3.0	.	.	.	.	.	.	.	.	.	.	.	.	.	20
3571	INDIGOFERA DIPHYLLA	W	.	0.0	15.8	5.4	72.9	38.8	5.9	570	150	.	.	.	.	.	.	.	.	18
3572	INDIGOFERA ECHINATA	S	.	0.0	36.2	4.0	.	.	2.9	.	.	.	.	.	.	.	.	.	.	19
3573	INDIGOFERA ENDECAPHYLLA	S	.	0.0	28.9	3.5	.	.	4.0	.	.	.	.	.	.	.	.	.	.	19
3574	INDIGOFERA ENDECAPHYLLA	W	.	0.0	13.0	2.2	79.4	33.5	5.4	.	.	.	.	.	.	.	.	.	.	6
3575	INDIGOFERA FRUTESCENS	S	.	0.0	32.1	4.8	.	.	.	.	.	.	.	.	.	.	.	.	.	21
3576	INDIGOFERA GLANDULOSA	S	.	8.2	31.9	2.2	54.5	7.8	3.2	.	.	.	.	.	.	.	.	.	.	5
3577	INDIGOFERA GLANDULOSA	S	.	8.2	37.1	4.9	45.3	9.3	4.1	.	.	.	.	.	.	.	.	.	.	5
3578	INDIGOFERA HIRSUTA	W	.	0.0	19.5	2.9	67.3	22.7	10.3	1590	200	.	.	.	.	.	.	.	.	18
3579	INDIGOFERA HIRSUTA	S	.	0.0	27.1	1.9	.	.	4.0	.	.	.	.	.	.	.	.	.	.	19
3580	INDIGOFERA HIRSUTA	S	.	0.0	30.1	1.5	.	.	3.8	.	.	.	.	.	.	.	.	.	.	19
3581	INDIGOFERA HIRSUTA	S	.	0.0	30.2	1.6	.	.	.	.	.	.	.	.	.	.	.	.	.	21
3582	INDIGOFERA JALISCENSIS	S	.	0.0	15.6	1.4	.	.	.	.	.	.	.	.	.	.	.	.	.	20
3583	INDIGOFERA LEPTOSEPALA	S	.	0.0	39.9	5.0	.	.	3.7	.	.	.	.	.	.	.	.	.	.	19
3584	INDIGOFERA LINIFOLIA	S	.	0.0	37.9	3.1	.	.	.	.	.	.	.	.	.	.	.	.	.	21
3585	INDIGOFERA PILOSA	S	.	0.0	28.9	3.3	.	.	3.8	.	.	.	.	.	.	.	.	.	.	19
3586	INDIGOFERA RETROFLEXA	W	.	0.0	16.8	1.1	72.6	31.8	9.5	.	.	.	.	.	.	.	.	.	.	18
3587	INDIGOFERA SECUNDIFLORA	W	.	0.0	16.9	1.9	64.0	19.1	17.2	4260	230	.	.	.	.	.	.	.	.	18
3588	INDIGOFERA SP.	H	.	0.0	12.0	2.5	79.8	24.9	5.7	1500	210	.	.	.	.	.	.	.	.	6
3589	INDIGOFERA SPHAEROCARPA	S	.	0.0	31.2	3.0	.	.	.	.	.	.	.	.	.	.	.	.	.	19
3590	INDIGOFERA SPICATA	W	.	0.0	12.3	0.8	75.2	38.4	11.7	1230	400	.	.	.	.	.	.	.	.	18
3591	INDIGOFERA SPICATA	L	.	80.5	4.1	0.6	12.6	4.2	2.2	.	.	.	.	.	.	.	.	.	.	5
3592	INDIGOFERA SUFFRUTICOSA	S	.	0.0	27.8	1.6	.	.	3.2	.	.	.	.	.	.	.	.	.	.	19
3593	INDIGOFERA SULCATA	S	.	0.0	27.2	4.7	.	.	.	.	.	.	.	.	.	.	.	.	.	21
3594	INDIGOFERA ZOLLINGERIANA	S	.	0.0	28.8	1.3	.	.	.	.	.	.	.	.	.	.	.	.	.	20
3595	INGA EDULIS	S	118	63.3	10.7	0.7	24.0	1.6	1.3	192	128	3.1	.	.	0	0.19	0.13	1.2	13	3
3596	INGA SP.	P	60	83.0	1.0	0.1	15.5	1.2	0.4	21	20	0.9	.	.	0	0.04	0.06	0.4	9	3
3597	INGA SP.	S	339	12.6	18.9	2.1	62.9	3.4	3.5	127	256	4.8	.	.	0	0.32	0.03	1.4	3	3
3598	INOCARPUS FAGIFERUS	S	243	43.0	3.8	4.5	47.2	2.5	1.5	.	.	.	.	.	0	0.26	.	.	2	2
3599	INULA GRAVEOLENS	S	.	0.0	27.8	29.8	.	.	.	.	.	.	.	.	.	.	.	.	.	21
3600	INULA VISCOSA	S	.	0.0	20.8	23.9	.	.	.	.	.	.	.	.	.	.	.	.	.	21
3601	IOSTEPHANE HETEROPHYLLA	S	.	0.0	26.2	14.9	.	.	.	.	.	.	.	.	.	.	.	.	.	20
3602	IPOMOEA AQUATICA	L	44	85.0	3.6	0.4	9.0	1.9	2.0	180	42	5.4	.	.	.	.	.	.	.	1
3603	IPOMOEA AQUATICA	W	.	0.0	18.8	2.1	60.9	20.9	18.2	710	320	.	.	.	.	.	.	.	.	18

TABLE 2 - DATA DIRECTLY TRANSCRIBED FROM ORIGINAL SOURCES.

OBS	PLNA	PART	CALORIES	WATER	PROTEIN	FAT	TOTLCARB	FIBER	ASH	CALCIUM	PHOSPHO	IRON	SODIUM	POTASS	CAROT	THIAMINE	RIBOFLAV	NIACIN	ASCORBIC	REFCO
3604	IPOMOEA AQUATICA	L	.	0.0	24.0	2.7	60.0	12.7	13.3	1200	280	.	.	.	.	.	.	.	.	18
3605	IPOMOEA AQUATICA	SH	.	0.0	34.3	3.9	48.9	10.2	12.9	.	.	.	.	.	.	.	.	.	.	18
3606	IPOMOEA AQUATICA	L	30	90.0	2.7	0.4	5.6	1.1	1.3	60	42	2.5	44	469	2865	0.09	0.16	1.1	47	2
3607	IPOMOEA AQUATICA	L	29	89.7	3.0	0.3	5.4	1.1	1.6	73	51	2.5	.	150	3780	0.07	0.12	0.7	32	4
3608	IPOMOEA BATATAS	R	121	68.8	1.6	0.2	28.5	1.0	0.9	33	38	2.0	.	.	75	0.09	0.04	0.7	37	1
3609	IPOMOEA BATATAS	R	121	68.8	1.6	0.2	28.5	1.0	0.9	33	38	2.0	.	.	1255	0.09	0.04	0.7	37	1
3610	IPOMOEA BATATAS	L	49	83.0	4.6	0.2	10.2	2.4	2.0	158	84	6.2	.	.	5870	0.10	0.28	0.9	70	1
3611	IPOMOEA BATATAS	L	.	0.0	19.4	3.7	51.0	10.2	25.9	.	.	.	.	.	.	.	.	.	.	18
3612	IPOMOEA BATATAS	W	.	0.0	21.9	3.4	56.7	15.0	18.0	1790	240	.	.	.	.	.	.	.	.	18
3613	IPOMOEA BATATAS	H	.	0.0	16.4	5.2	65.8	27.4	12.6	.	.	.	.	.	.	.	.	.	.	18
3614	IPOMOEA BATATAS	R	.	0.0	8.8	0.6	88.5	2.4	2.1	230	200	.	.	.	.	.	.	.	.	18
3615	IPOMOEA BATATAS	R	108	72.3	1.0	0.3	25.6	0.8	0.7	21	50	0.9	31	210	35	0.14	0.05	0.7	21	2
3616	IPOMOEA BATATAS	R	115	70.7	1.2	0.3	27.1	0.8	0.7	36	56	0.9	36	304	1680	0.12	0.05	0.6	30	2
3617	IPOMOEA BATATAS	L	42	86.7	3.2	0.7	8.0	1.6	1.4	86	81	4.5	5	562	2700	0.13	0.26	0.9	21	2
3618	IPOMOEA BATATAS	L	48	86.3	3.2	0.8	8.5	1.6	1.2	111	30	2.9	.	.	2920	0.08	0.26	0.7	58	3
3619	IPOMOEA BATATAS	R	114	70.6	1.7	0.4	26.3	0.7	1.0	32	47	0.7	10	243	5280	0.10	0.06	0.6	21	4
3620	IPOMOEA BATATAS	R	.	68.5	1.2	0.3	31.0	.	1.0	.	.	.	.	.	.	.	.	.	.	5
3621	IPOMOEA BATATAS		.	0.0	.	.	.	.	.	.	.	.	.	.	.	.	.	.	.	6
3622	IPOMOEA CARDIOPHYLLA	S	.	0.0	31.4	10.1	.	.	.	.	.	.	.	.	.	.	.	.	.	21
3623	IPOMOEA CARDIOPHYLLA	S	.	0.0	23.5	10.0	.	.	.	.	.	.	.	.	.	.	.	.	.	21
3624	IPOMOEA CARDIOPHYLLA	S	.	0.0	24.4	10.7	.	.	.	.	.	.	.	.	.	.	.	.	.	21
3625	IPOMOEA CARDIOPHYLLA	S	.	0.0	26.6	12.7	.	.	.	.	.	.	.	.	.	.	.	.	.	21
3626	IPOMOEA CARNEA	S	.	0.0	19.9	9.8	.	.	.	.	.	.	.	.	.	.	.	.	.	21
3627	IPOMOEA CISSOIDES	S	.	0.0	17.4	6.3	.	.	.	.	.	.	.	.	.	.	.	.	.	21
3628	IPOMOEA HEDERACEA	S	.	0.0	25.0	11.8	.	.	.	.	.	.	.	.	.	.	.	.	.	20
3629	IPOMOEA HISPIDA	S	.	0.0	24.8	9.5	.	.	.	.	.	.	.	.	.	.	.	.	.	21
3630	IPOMOEA LEPTOPHYLLA	S	.	0.0	31.2	11.1	.	.	.	.	.	.	.	.	.	.	.	.	.	19
3631	IPOMOEA MURICATA	S	.	0.0	15.0	18.8	.	.	.	.	.	.	.	.	.	.	.	.	.	21
3632	IPOMOEA NIL	S	.	0.0	25.4	11.3	.	.	.	.	.	.	.	.	.	.	.	.	.	21
3633	IPOMOEA PARASITICA	S	.	0.0	25.6	18.6	.	.	.	.	.	.	.	.	.	.	.	.	.	20
3634	IPOMOEA PARASITICA	S	.	0.0	24.3	20.1	.	.	.	.	.	.	.	.	.	.	.	.	.	21
3635	IPOMOEA PENDURATA	S	.	0.0	16.2	7.1	.	.	3.9	.	.	.	.	.	.	.	.	.	.	19
3636	IPOMOEA PENTAPHYLLA	S	.	0.0	19.9	8.2	.	.	.	.	.	.	.	.	.	.	.	.	.	21
3637	IPOMOEA PES-TIGRIDIS	S	.	0.0	26.7	9.7	.	.	.	.	.	.	.	.	.	.	.	.	.	21
3638	IPOMOEA PES-TIGRIDIS	L	.	0.0	12.3	3.0	74.4	26.2	10.3	917	479	.	.	.	.	.	.	.	.	5
3639	IPOMOEA PES-TIGRIDIS	L	.	0.0	13.6	3.5	71.0	27.0	11.9	1480	344	.	.	.	.	.	.	.	.	5
3640	IPOMOEA PILOSA	S	.	0.0	31.6	9.3	.	.	.	.	.	.	.	.	.	.	.	.	.	21
3641	IPOMOEA PURPUREA	S	.	0.0	22.5	11.3	.	.	3.4	.	.	.	.	.	.	.	.	.	.	19
3642	IPOMOEA SP.	S	.	0.0	25.6	17.6	.	.	.	.	.	.	.	.	.	.	.	.	.	19
3643	IPOMOEA SP.	S	.	0.0	18.1	5.8	.	.	.	.	.	.	.	.	.	.	.	.	.	20
3644	IPOMOEA SP.	S	.	0.0	23.8	15.6	.	.	.	.	.	.	.	.	.	.	.	.	.	21
3645	IPOMOEA SP.	S	.	0.0	19.9	8.6	.	.	.	.	.	.	.	.	.	.	.	.	.	21
3646	IPOMOEA SP.	S	.	0.0	25.6	17.7	.	.	.	.	.	.	.	.	.	.	.	.	.	21
3647	IPOMOEA TRILOBA	S	.	0.0	20.4	11.8	.	.	.	.	.	.	.	.	.	.	.	.	.	21
3648	IRIS GERMANICA	S	.	0.0	14.8	17.3	.	.	5.1	.	.	.	.	.	.	.	.	.	.	19

TABLE 2 - DATA DIRECTLY TRANSCRIBED FROM ORIGINAL SOURCES.

OBS	PLNA	PART	CALORIES	WATER	PROTEIN	FAT	TOTCARB	FIBER	ASH	CALCIUM	PHOSPHO	IRON	SODIUM	POTASS	CAROT	THIAMINE	RIBOFLAV	NIACIN	ASCORBIC	REFCO
3649	IRIS GERMANICA	S	.	0.0	15.8	19.2	.	.	.	.	.	.	.	.	.	.	.	.	.	21
3650	IRIS KERNERIANA	S	.	0.0	21.9	11.6	.	.	.	.	.	.	.	.	.	.	.	.	.	21
3651	IRIS MISSOURIENSIS	S	.	0.0	13.1	13.9	.	.	.	.	.	.	.	.	.	.	.	.	.	20
3652	IRIS MISSOURIENSIS	S	.	0.0	16.9	14.3	.	.	.	.	.	.	.	.	.	.	.	.	.	20
3653	IRIS MISSOURIENSIS	S	.	0.0	16.2	14.7	.	.	.	.	.	.	.	.	.	.	.	.	.	20
3654	IRIS MISSOURIENSIS	S	.	0.0	11.2	13.9	.	.	.	.	.	.	.	.	.	.	.	.	.	21
3655	IRIS OCHROLEUCA	S	.	0.0	14.4	11.6	.	.	.	.	.	.	.	.	.	.	.	.	.	21
3656	IRIS SP.	S	.	0.0	14.2	16.3	.	.	.	.	.	.	.	.	.	.	.	.	.	21
3657	IRIS SPURIA	S	.	0.0	12.6	14.9	.	.	.	.	.	.	.	.	.	.	.	.	.	21
3658	IRVINGIA GABONENSIS	S	670	5.1	7.5	68.9	16.6	2.2	1.9	126	202	3.4	.	.	.	0.16	0.08	0.7	.	1
3659	ISACHNE GLOBOSA	L	.	0.0	10.5	1.8	72.0	32.0	15.7	.	.	.	.	.	.	.	.	.	.	5
3660	ISACHNE MILIACEA	L	.	0.0	10.4	1.5	70.8	29.0	17.3	.	.	.	.	.	.	.	.	.	.	5
3661	ISATIS AUCHERI	S	.	0.0	36.1	33.2	.	.	.	.	.	.	.	.	.	.	.	.	.	20
3662	ISATIS LUSITANICA	S	.	0.0	11.6	8.8	.	.	.	.	.	.	.	.	.	.	.	.	.	21
3663	ISATIS TINCTORIA	S	.	0.0	32.9	38.4	.	.	.	.	.	.	.	.	.	.	.	.	.	21
3664	ISATIS TINCTORIS	S	.	0.0	12.5	12.6	.	.	.	.	.	.	.	.	.	.	.	.	.	20
3665	ISATIS TINCTORIS	S	.	0.0	34.0	37.4	.	.	.	.	.	.	.	.	.	.	.	.	.	20
3666	ISCHAEMUM ARISTATUM	W	.	0.0	14.2	2.5	72.7	31.6	10.6	.	.	.	.	.	.	.	.	.	.	18
3667	ISCHAEMUM ARISTATUM	W	.	0.0	8.2	1.3	81.6	35.0	8.9	.	.	.	.	.	.	.	.	.	.	18
3668	ISCHAEMUM ARISTATUM	L	.	5.4	4.6	1.5	83.6	41.5	5.0	.	.	.	.	.	5	.	.	.	.	5
3669	ISCHAEMUM MUTICUM	L	.	74.0	2.4	0.4	21.3	6.4	1.9	50	30	.	.	.	.	.	.	.	.	5
3670	ISCHAEMUM PILOSUM	L	.	70.0	2.2	1.0	23.6	9.0	3.1	.	.	.	.	.	.	.	.	.	.	5
3671	ISCHAEMUM TIMORENSE	L	.	0.0	15.3	1.6	73.4	30.0	9.7	.	.	.	.	.	.	.	.	.	.	18
3672	ISCHAEMUM TIMORENSE	C	.	0.0	6.2	0.8	84.6	38.3	8.4	.	.	.	.	.	.	.	.	.	.	18
3673	ISCHAEMUM TIMORENSE	W	.	0.0	10.7	2.0	77.0	32.7	10.3	.	.	.	.	.	.	.	.	.	.	18
3674	ISEILEMA ANTHEPHOROIDES	L	.	7.2	4.3	2.0	75.4	32.2	11.1	.	.	.	.	.	.	.	.	.	.	5
3675	ISEILEMA LAXUM	H	.	0.0	6.4	0.6	86.2	35.6	6.8	.	.	.	.	.	.	.	.	.	.	18
3676	ISEILEMA LAXUM	W	.	0.0	5.1	1.4	81.9	34.2	11.6	.	.	.	.	.	.	.	.	.	.	18
3677	ISEILEMA LAXUM	L	.	0.0	5.7	2.7	79.8	28.0	11.8	1300	630	.	300	1450	.	.	.	.	.	5
3678	ISOMERIS ARBOREA	S	.	0.0	37.2	44.8	.	.	3.5	.	.	.	.	.	.	.	.	.	.	19
3679	ISOMERIS ARBOREA	S	.	0.0	6.9	2.8	.	.	1.2	.	.	.	.	.	.	.	.	.	.	19
3680	ISOMERIS ARBOREA	S	.	0.0	40.6	42.0	.	.	.	.	.	.	.	.	.	.	.	.	.	20
3681	IVA AXILLARIS	S	.	0.0	3.6	8.5	.	.	.	.	.	.	.	.	.	.	.	.	.	21
3682	IVA FRUTESCENS	S	.	0.0	18.8	11.9	.	.	.	.	.	.	.	.	.	.	.	.	.	19
3683	IVA XANTHIFOLIA	S	.	0.0	19.4	22.8	.	.	5.2	.	.	.	.	.	.	.	.	.	.	19
3684	IVA XANTHIFOLIA	S	.	0.0	19.6	31.8	.	.	.	.	.	.	.	.	.	.	.	.	.	21
3685	IXOPHORUS UNISETUS	W	.	0.0	14.4	3.0	67.9	30.0	14.8	.	.	.	.	.	.	.	.	.	.	18
3686	IXOPHORUS UNISETUS	H	.	0.0	5.4	0.9	80.4	35.3	13.3	.	.	.	.	.	.	.	.	.	.	18
3687	JACARANDA MIMOSIFOLIA	S	.	0.0	33.1	32.7	.	.	4.3	.	.	.	.	.	.	.	.	.	.	19
3688	JACQUINIA PUNGENS	S	.	0.0	11.2	11.5	.	.	.	.	.	.	.	.	.	.	.	.	.	19
3689	JACQUINIA PUNGENS	S	.	0.0	11.2	11.6	.	.	.	.	.	.	.	.	.	.	.	.	.	20
3690	JACQUINIA PUNGENS	S	.	0.0	10.6	11.1	.	.	.	.	.	.	.	.	.	.	.	.	.	20
3691	JASMINUM FRUTICANS	S	.	0.0	10.8	22.5	.	.	.	.	.	.	.	.	.	.	.	.	.	20
3692	JATROPHA CORDATA	S	.	0.0	15.6	39.1	.	.	.	.	.	.	.	.	.	.	.	.	.	20
3693	JATROPHA CURCAS	S	.	0.0	18.6	31.1	.	.	.	.	.	.	.	.	.	.	.	.	.	21

TABLE 2 - DATA DIRECTLY TRANSCRIBED FROM ORIGINAL SOURCES.

OBS	PLNA	PART	CALORIES	WATER	PROTEIN	FAT	TOTCARB	FIBER	ASH	CALCIUM	PHOSPHO	IRON	SODIUM	POTASS	CAROT	THIAMINE	RIBOFLAV	NIACIN	ASCORBIC	REFCO
3694	JATROPHA CURCAS	S	.	6.6	18.2	38.0	33.5	15.5	4.5	.	.	.	.	.	.	.	.	.	.	5
3695	JATROPHA MACRORHIZA	S	.	0.0	36.2	52.8	.	.	.	.	.	.	.	.	.	.	.	.	.	20
3696	JATROPHA SPATHULATA	S	.	0.0	26.2	58.2	.	.	.	.	.	.	.	.	.	.	.	.	.	19
3697	JUBAEA CHILENSIS	S	622	12.9	7.1	65.6	13.5	5.9	0.9	32	176	5.3	.	.	0	0.26	0.05	0.40	0	3
3698	JUGLANS CINEREA	S	629	3.8	23.7	61.2	8.4	.	2.9	.	.	6.8	.	.	.	.	.	.	.	4
3699	JUGLANS NIGRA	S	628	3.1	20.5	59.3	14.8	1.7	2.3	0	570	6.0	3	460	180	0.22	0.11	0.70	.	4
3700	JUGLANS REGIA	S	657	2.5	15.6	64.4	15.8	.	1.7	.	.	.	.	.	.	.	.	.	.	1
3701	JUGLANS REGIA	S	.	0.0	15.1	58.2	24.8	9.2	1.9	.	280	.	.	.	.	.	.	.	.	18
3702	JUGLANS REGIA	S	647	3.6	18.2	63.6	12.6	1.6	2.0	106	326	3.0	3	536	0	0.50	0.08	3.00	5	2
3703	JUGLANS REGIA	S	664	4.2	13.7	67.2	13.2	2.0	1.7	92	379	3.3	.	.	50	0.27	0.51	3.00	0	3
3704	JUGLANS REGIA	S	651	3.5	14.8	64.0	15.8	2.1	1.9	99	380	3.1	2	450	18	0.33	0.13	0.90	2	4
3705	JUNCUS ACUTUS	S	.	0.0	20.5	5.8	.	.	.	.	.	.	.	.	.	.	.	.	.	21
3706	JUNCUS ACUTUS	S	.	0.0	20.9	6.8	.	.	.	.	.	.	.	.	.	.	.	.	.	21
3707	JUNCUS BALTICUS	W	.	0.0	10.0	1.8	82.5	33.0	5.7	.	.	.	.	.	.	.	.	.	.	6
3708	JUNCUS BUFONIUS	W	.	0.0	10.0	.	.	27.7	5.7	280	390	.	.	1910	.	.	.	.	.	6
3709	JUNCUS EFFUSUS	S	.	0.0	13.1	5.0	.	.	.	.	.	.	.	.	.	.	.	.	.	20
3710	JUNCUS GERARDII	H	.	0.0	8.4	2.8	80.7	28.0	8.1	.	.	.	.	.	.	.	.	.	.	6
3711	JUNCUS INFLEXUS	S	.	0.0	15.7	6.4	.	.	.	.	.	.	.	.	.	.	.	.	.	21
3712	JUNCUS INFLEXUS	S	.	0.0	18.6	6.0	.	.	.	.	.	.	.	.	.	.	.	.	.	21
3713	JUNCUS LONGISTYLIS	H	.	0.0	10.8	1.4	81.5	33.8	6.3	.	.	.	.	.	.	.	.	.	.	6
3714	JUNCUS MARGINATUS	S	.	0.0	12.8	2.7	.	.	.	.	.	.	.	.	.	.	.	.	.	21
3715	JUNCUS OXYMERIS		.	0.0	10.8	.	.	31.3	8.1	290	210	.	.	2790	.	.	.	.	.	6
3716	JUNCUS SP.	W	.	0.0	10.0	1.9	81.5	31.8	6.6	300	250	.	.	2360	.	.	.	.	.	6
3717	JUNIPERUS COMMUNIS SUBSP. NANA	S	.	0.0	30.9	53.9	.	.	.	.	.	.	.	.	.	.	.	.	.	21
3718	JUNIPERUS EXCELSA	S	.	0.0	4.7	3.4	.	.	.	.	.	.	.	.	.	.	.	.	.	21
3719	JUNIPERUS OSTEOSPERMA	W	.	0.0	6.4	16.3	72.8	22.0	4.5	1590	170	.	.	.	.	.	.	.	.	6
3720	JUNIPERUS OXYCEDRUS	S	.	0.0	3.1	5.1	.	.	.	.	.	.	.	.	.	.	.	.	.	21
3721	JUNIPERUS OXYCEDRUS	S	.	0.0	4.4	10.0	.	.	.	.	.	.	.	.	.	.	.	.	.	21
3722	JUNIPERUS PINCHOTII	W	.	0.0	8.6	10.3	76.3	19.9	4.8	.	.	.	.	.	.	.	.	.	.	6
3723	JUNIPERUS SP.	S	.	0.0	6.4	14.1	.	.	.	.	.	.	.	.	.	.	.	.	.	21
3724	JUNIPERUS THURIFERA	S	.	0.0	3.8	2.3	.	.	.	.	.	.	.	.	.	.	.	.	.	21
3725	JUNIPERUS VIRGINIANA	S	.	0.0	13.8	16.4	.	.	.	.	.	.	.	.	.	.	.	.	.	20
3726	JURINEA ANATOLICA	S	.	0.0	20.9	24.0	.	.	.	.	.	.	.	.	.	.	.	.	.	21
3727	JUSSIAEA LEPTOCARPA	S	.	0.0	13.8	12.5	.	.	.	.	.	.	.	.	.	.	.	.	.	20
3728	JUSTICIA EXIGUA	SH	.	0.0	14.0	3.2	63.9	21.9	18.9	1670	450	.	.	.	.	.	.	.	.	18
3729	JUSTICIA PECTORALIS	L	44	85.0	3.9	0.6	8.2	2.8	2.3	663	35	7.4	.	.	2670	0.04	0.20	2.50	28	3
3730	JUSTICIA SP.	L	33	86.8	3.3	0.4	6.1	1.7	3.4	510	70	.	.	.	.	.	.	.	.	1
3731	KAEMPFERIA CANDIDA	R	38	89.5	1.1	0.8	7.3	.	1.3	28	40	2.0	.	.	3000	0.08	0.02	.	.	2
3732	KALIMERIS YOMENA	L	29	90.2	2.2	0.4	5.7	0.5	1.5	80	66	4.0	.	.	1860	0.15	0.20	2.00	50	2
3733	KALMIA LATIFOLIA	W	.	0.0	8.7	9.5	78.2	14.7	3.6	960	110	5.0	.	.	.	.	.	.	.	6
3734	KALOPANAX SEPTEMLOBUS	S	.	0.0	3.2	21.7	.	.	.	.	.	.	.	.	.	.	.	.	.	21
3735	KERRIA JAPONICA	S	.	0.0	44.9	45.3	.	.	.	.	.	.	.	.	.	.	.	.	.	21
3736	KERSTINGIELLA GEOCARPA	S	386	0.0	21.5	1.2	73.9	6.1	3.6	114	435	16.6	.	.	.	0.84	0.21	2.55	0	10
3737	KIGELIA AFRICANA	F	54	85.4	0.8	0.9	12.2	4.3	0.7	.	.	.	.	.	.	.	.	.	.	1
3738	KIGELIA PINNATA	F	.	0.0	5.8	6.0	83.7	29.4	4.5	.	.	.	.	.	.	.	.	.	.	18

TABLE 2 - DATA DIRECTLY TRANSCRIBED FROM ORIGINAL SOURCES.

OBS	PLNA	PART	CALORIES	WATER	PROTEIN	FAT	TOTLCARB	FIBER	ASH	CALCIUM	PHOSPHO	IRON	SODIUM	POTASS	CAROT	THIAMINE	RIBOFLAV	NIACIN	ASCORBIC	REFCO
3739	KLAINEDOXA GABONESIS	S	612	7.8	8.4	59.8	20.3	1.7	3.7	313	424	.	.	.	.	.	.	.	.	1
3740	KLEINHOVIA HOSPITA	S	.	0.0	21.3	8.5	.	.	.	.	.	.	.	.	.	.	.	.	.	21
3741	KOCHIA PROSTRATA	L	.	0.0	9.8	3.0	74.5	35.3	.	.	.	.	.	.	.	.	.	.	.	5
3742	KOCHIA SCOPARIA	S	.	0.0	23.1	11.2	.	.	4.5	.	.	.	.	.	.	.	.	.	.	19
3743	KOCHIA SCOPARIA	S	.	0.0	20.4	8.8	.	.	9.4	.	.	.	.	.	.	.	.	.	.	19
3744	KOCHIA SCOPARIA	S	.	0.0	27.5	16.0	.	.	3.4	.	.	.	.	.	.	.	.	.	.	19
3745	KOCHIA SCOPARIA	L	.	13.1	18.7	2.1	49.4	17.1	16.7	.	.	.	.	.	4	.	.	.	.	5
3746	KOCHIA SCOPARIA	H	.	0.0	13.8	1.5	71.3	24.9	13.4	.	.	.	.	.	.	.	.	.	.	6
3747	KOCHIA SCOPARIA(FLOWERING)	H	.	0.0	13.1	1.6	71.4	27.6	13.9	.	.	.	.	.	.	.	.	.	.	6
3748	KOCHIA SCOPARIA(IMMATURE)	H	.	0.0	18.8	1.4	66.1	19.7	13.7	.	.	.	.	.	.	.	.	.	.	6
3749	KOCHIA SCOPARIA(MATURE)	H	.	0.0	11.2	1.5	75.1	35.6	12.2	.	.	.	.	.	.	.	.	.	.	6
3750	KOCHIA SP.	H	.	0.0	13.7	1.6	71.2	25.7	13.5	.	.	.	.	.	.	.	.	.	.	6
3751	KOCHIA VESTITA	W	.	0.0	9.0	3.7	62.9	15.5	24.4	2370	120	.	.	.	.	.	.	.	.	6
3752	KOELERIA CRISTATA	L	.	0.0	8.9	2.6	79.2	38.5	8.3	500	600	.	.	.	.	.	.	.	.	5
3753	KOELERIA CRISTATA	W	.	0.0	7.3	2.1	82.5	37.8	8.1	320	240	.	.	.	8	.	.	.	.	6
3754	KOELERIA CRISTATA(MATURE)	W	.	0.0	5.8	2.5	83.6	38.6	8.1	330	310	.	.	.	.	.	.	.	.	6
3755	KOELERIA CRISTATA(VERY MATURE)	W	.	0.0	3.8	1.8	86.7	41.0	7.7	290	140	.	.	.	.	.	.	.	.	6
3756	KOELERIA GLAUCOVIRENS	S	.	0.0	.	5.6	.	.	.	.	.	.	.	.	.	.	.	.	.	21
3757	KOELERIA PHLEOIDES	L	.	.	11.0	2.1	66.4	24.7	18.9	870	690	.	.	.	.	.	.	.	.	5
3758	KOELERIA PHLEOIDES	L	.	.	5.9	2.9	74.9	33.0	14.9	650	740	.	.	.	.	.	.	.	.	5
3759	KOELREUTERIA BIPINNATA	S	.	0.0	28.5	54.0	.	.	3.1	.	.	.	.	.	.	.	.	.	.	19
3760	KOELREUTERIA FORMOSANA	S	.	0.0	28.9	52.7	.	.	5.9	.	.	.	.	.	.	.	.	.	.	19
3761	KOELREUTERIA PANICULATA	S	.	0.0	30.6	40.8	.	.	.	.	.	.	.	.	.	.	.	.	.	19
3762	KOELREUTERIA PANICULATA	S	.	0.0	26.9	41.9	.	.	.	.	.	.	.	.	.	.	.	.	.	19
3763	KOELREUTERIA PANICULATA	S	.	0.0	27.5	45.8	.	.	.	.	.	.	.	.	.	.	.	.	.	19
3764	KOSTELETZKYA SP.	H	.	0.0	12.8	2.2	79.3	36.3	5.7	260	210	.	.	900	.	.	.	.	.	6
3765	KOSTELETZKYA VIRGINICA	S	.	0.0	26.9	18.9	.	.	4.0	.	.	.	.	.	.	.	.	.	.	19
3766	KUHNIA GLUTINOSA	S	.	0.0	33.1	21.5	.	.	4.3	.	.	.	.	.	.	.	.	.	.	19
3767	KUNDMANNIA SICULA	L	.	0.0	19.8	.	.	.	20.3	2270	.	.	1910	3130	.	.	.	.	.	17
3768	LACTUCA CRETICA	S	.	0.0	30.2	31.4	.	.	.	.	.	.	.	.	.	.	.	.	.	21
3769	LACTUCA SATIVA	L	20	93.7	1.2	0.2	4.3	0.6	0.6	26	40	0.7	.	.	.	.	.	.	10	1
3770	LACTUCA SATIVA	L	.	87.6	.	.	.	.	0.8	.	.	1.0	.	.	.	.	.	.	.	1
3771	LACTUCA SATIVA	L	16	94.7	1.2	0.2	3.2	0.4	0.7	56	53	1.1	10	.	720	0.06	0.06	0.60	15	1
3772	LACTUCA SATIVA	S	.	0.0	26.8	41.6	.	.	3.8	.	.	.	.	.	.	.	.	.	.	19
3773	LACTUCA SATIVA	S	.	0.0	31.8	33.8	.	.	4.7	.	.	.	.	.	.	.	.	.	.	19
3774	LACTUCA SATIVA	L	20	93.6	1.4	0.3	3.9	0.6	0.8	56	34	2.1	14	254	2035	0.06	0.12	0.50	17	2
3775	LACTUCA SATIVA	L	14	95.8	1.0	0.4	2.4	0.4	0.4	18	22	0.4	.	.	885	0.04	0.04	0.20	4	2
3776	LACTUCA SATIVA	L	15	94.9	1.3	0.2	2.9	0.7	0.7	43	34	1.3	.	.	520	0.08	0.08	0.40	12	3
3777	LACTUCA SATIVA	L	13	95.8	1.0	0.1	2.7	0.5	0.4	16	23	0.4	.	.	.	0.05	0.03	0.30	7	3
3778	LACTUCA SATIVA	L	18	94.0	1.3	0.3	3.5	0.7	0.9	68	25	1.4	9	264	1140	0.05	0.08	0.40	18	4
3779	LACTUCA SATIVA	L	14	95.1	1.2	0.2	2.5	0.5	1.0	35	26	2.0	9	264	582	0.06	0.06	0.30	8	4
3780	LACTUCA SCARIOLA	S	.	0.0	29.4	40.8	.	.	4.4	.	.	.	.	.	.	.	.	.	.	19
3781	LACTUCA SCARIOLA	S	.	0.0	25.9	39.5	.	.	.	.	.	.	.	.	.	.	.	.	.	21
3782	LACTUCA SP.	W	.	0.0	12.8	6.1	60.3	19.1	20.8	.	.	.	.	.	.	.	.	.	.	6
3783	LACTUCA TARAXAEIFOLIA	L	44	84.3	3.2	0.8	8.3	2.0	3.4	326	58	.	.	.	.	.	.	.	.	1

TABLE 2 - DATA DIRECTLY TRANSCRIBED FROM ORIGINAL SOURCES.

OBS	PLNA	PART	CALORIES	WATER	PROTEIN	FAT	TOTCARB	FIBER	ASH	CALCIUM	PHOSPHON	IRON	SODIUM	POTASS	CAROTE	THIAMINE	RIBOFLAV	NIACIN	ASCORBIC	REFCO
3784	LAGENARIA SICERARIA	F	26	92.2	0.7	0.2	6.3	1.5	0.6	18	21	0.5	.	.	0	0.04	0.03	0.60	19	3
3785	LAGENARIA SICERARIA	S	.	2.5	30.7	52.5	9.9	1.6	4.4	110	2460	.	.	.	.	.	.	.	.	5
3786	LAGENARIA SICERARIA	F	.	96.3	0.2	0.1	2.9	.	0.5	20	.	0.7	11	86	.	0.04	0.02	0.33	13	5
3787	LAGERSTROEMIA INDICA	S	.	0.0	21.2	8.3	.	.	.	.	.	.	.	.	.	.	.	.	.	20
3788	LAGERSTROEMIA SPECIOSA	S	.	0.0	4.4	1.5	.	.	.	.	.	.	.	.	.	.	.	.	.	20
3789	LAGERSTROEMIA TOMENTOSA	S	.	0.0	12.5	3.0	.	.	.	.	.	.	.	.	.	.	.	.	.	20
3790	LALLEMANTIA ROYLEANA	S	.	0.0	17.5	19.7	.	.	.	.	.	.	.	.	.	.	.	.	.	21
3791	LANARIA LANATA	S	.	0.0	16.7	17.0	.	.	.	.	.	.	.	.	.	.	.	.	.	21
3792	LANSIUM DOMESTICUM	F	55	84.2	0.9	0.1	14.2	0.8	0.6	12	24	0.9	1	232	0	0.08	0.05	1.00	2	2
3793	LANSIUM DOMESTICUM	F	.	86.5	0.8	0.3	11.8	2.3	0.6	20	30	0.5	.	.	8	0.09	0.12	0.00	1	5
3794	LANTANA CAMARA	S	.	0.0	35.1	48.0	.	.	.	.	.	.	.	.	.	.	.	.	.	21
3795	LANTANA SP.	S	.	0.0	13.8	5.1	.	.	.	.	.	.	.	.	.	.	.	.	.	20
3796	LAPORTEA AESTUANS	L	53	80.0	5.8	0.4	10.0	3.0	3.8	440	114	1.5	.	.	.	.	.	.	.	1
3797	LAPPULA REDOWSKII	S	.	0.0	17.8	18.8	.	.	.	.	.	.	.	.	.	.	.	.	.	20
3798	LAPPULA SP.	L	.	0.0	22.8	3.3	62.3	10.6	11.6	.	.	.	.	.	.	.	.	.	.	6
3799	LARIX SP	S	662	5.5	15.8	68.2	9.0	0.9	1.5	13	165	4.7	.	.	30	0.33	0.10	7.00	0	2
3800	LARREA DIVARICATA	S	.	0.0	29.0	12.9	.	.	.	.	.	.	.	.	.	.	.	.	.	20
3801	LARREA TRIDENTATA	S	.	0.0	30.0	11.7	.	.	5.8	.	.	.	.	.	.	.	.	.	.	19
3802	LARREA TRIDENTATA(IMMATURE)	W	.	0.0	15.8	11.4	62.5	14.5	10.3	2520	260	.	.	1390	.	.	.	.	.	6
3803	LASIURUS HIRSUTUS	W	.	0.0	10.4	8.2	69.5	34.1	11.9	.	.	.	.	.	.	.	.	.	.	18
3804	LASIURUS HIRSUTUS	H	.	0.0	5.9	0.3	82.7	38.0	11.0	.	.	.	.	.	.	.	.	.	.	18
3805	LATHYRUS CICERA	S	.	0.0	33.3	0.8	.	.	3.4	.	.	.	.	.	.	.	.	.	.	19
3806	LATHYRUS CICERA	S	.	0.0	34.7	0.7	.	.	3.3	.	.	.	.	.	.	.	.	.	.	19
3807	LATHYRUS CICERA	S	.	0.0	30.3	0.7	.	.	3.6	.	.	.	.	.	.	.	.	.	.	19
3808	LATHYRUS CICERA	S	.	0.0	33.7	0.7	.	.	3.9	.	.	.	.	.	.	.	.	.	.	19
3809	LATHYRUS CICERA	S	.	0.0	34.4	0.7	.	.	3.6	.	.	.	.	.	.	.	.	.	.	19
3810	LATHYRUS CICERA	S	.	0.0	31.9	0.8	.	.	3.2	.	.	.	.	.	.	.	.	.	.	19
3811	LATHYRUS CICERA	S	.	0.0	32.0	0.7	.	.	3.0	.	.	.	.	.	.	.	.	.	.	19
3812	LATHYRUS CICERA	S	.	0.0	29.8	0.6	.	.	2.5	.	.	.	.	.	.	.	.	.	.	19
3813	LATHYRUS CICERA	S	.	0.0	32.0	0.8	.	.	3.5	.	.	.	.	.	.	.	.	.	.	19
3814	LATHYRUS CICERA	S	.	0.0	29.3	0.6	.	.	3.5	.	.	.	.	.	.	.	.	.	.	19
3815	LATHYRUS CICERA	S	.	0.0	27.8	1.0	.	.	3.1	.	.	.	.	.	.	.	.	.	.	19
3816	LATHYRUS CICERA	S	.	0.0	28.5	0.7	.	.	3.3	.	.	.	.	.	.	.	.	.	.	19
3817	LATHYRUS CICERA	S	.	0.0	27.8	0.9	.	.	3.3	.	.	.	.	.	.	.	.	.	.	19
3818	LATHYRUS CICERA	S	.	0.0	27.7	0.7	.	.	3.3	.	.	.	.	.	.	.	.	.	.	19
3819	LATHYRUS CICERA	S	.	0.0	27.8	0.8	.	.	3.3	.	.	.	.	.	.	.	.	.	.	19
3820	LATHYRUS CICERA	S	.	0.0	29.6	0.8	.	.	3.3	.	.	.	.	.	.	.	.	.	.	19
3821	LATHYRUS CICERA	S	.	0.0	30.1	0.9	.	.	3.6	.	.	.	.	.	.	.	.	.	.	19
3822	LATHYRUS CICERA	S	.	0.0	30.0	0.7	.	.	3.2	.	.	.	.	.	.	.	.	.	.	19
3823	LATHYRUS DIGITATUS	S	.	0.0	27.8	0.8	.	.	.	.	.	.	.	.	.	.	.	.	.	21
3824	LATHYRUS GORGONII	S	.	0.0	24.4	0.5	.	.	.	.	.	.	.	.	.	.	.	.	.	21
3825	LATHYRUS HIEROSOLYMITANUS	S	.	0.0	27.5	0.8	.	.	.	.	.	.	.	.	.	.	.	.	.	21
3826	LATHYRUS HIRSUTUS	S	.	0.0	28.6	0.7	.	.	3.5	.	.	.	.	.	.	.	.	.	.	19
3827	LATHYRUS HIRSUTUS	S	.	0.0	28.0	0.6	.	.	3.8	.	.	.	.	.	.	.	.	.	.	19
3828	LATHYRUS HIRSUTUS	S	.	0.0	26.9	0.6	.	.	4.4	.	.	.	.	.	.	.	.	.	.	19

TABLE 2 - DATA DIRECTLY TRANSCRIBED FROM ORIGINAL SOURCES.

OBS	PLNA	PART	CALORIES	WATER	PROTEIN	FAT	TOTLCARB	FIBER	ASH	CALCIUM	PHOSPHO	IRON	SODIUM	POTASS	CAROT	THIAMINE	RIBOFLAV	NIACIN	ASCORBIC	REFCO
3829	LATHYRUS INCURVUS	S	.	0.0	27.6	1.1	.	.	.	.	.	.	.	.	.	.	.	.	.	21
3830	LATHYRUS LATIFOLIUS	S	.	0.0	32.1	0.8	.	.	.	.	.	.	.	.	.	.	.	.	.	21
3831	LATHYRUS LEUCANTHUS	H	.	0.0	16.1	2.8	.	.	9.1	1350	310	.	.	.	.	.	.	.	.	6
3832	LATHYRUS ORNATUS	S	.	0.0	32.9	1.4	.	.	2.8	.	.	.	.	.	.	.	.	.	.	19
3833	LATHYRUS PALLESCENS	S	.	0.0	27.2	0.9	.	.	.	.	.	.	.	.	.	.	.	.	.	21
3834	LATHYRUS PSEUDO-CICERA	S	.	0.0	28.8	0.8	.	.	.	.	.	.	.	.	.	.	.	.	.	21
3835	LATHYRUS SATIVUS	W	.	0.0	20.7	3.4	65.0	23.0	10.9	1320	340	.	.	.	.	.	.	.	.	18
3836	LATHYRUS SATIVUS	H	.	0.0	18.2	3.3	67.4	30.5	11.1	.	.	.	.	.	.	.	.	.	.	18
3837	LATHYRUS SATIVUS	S	.	0.0	20.9	2.0	74.5	5.9	2.6	650	300	.	.	.	.	.	.	.	.	18
3838	LATHYRUS SATIVUS	S	.	0.0	28.1	0.9	.	.	3.4	.	.	.	.	.	.	.	.	.	.	19
3839	LATHYRUS SATIVUS	S	.	0.0	33.4	0.9	.	.	3.5	.	.	.	.	.	.	.	.	.	.	19
3840	LATHYRUS SATIVUS	S	.	0.0	33.1	0.8	.	.	3.2	.	.	.	.	.	.	.	.	.	.	19
3841	LATHYRUS SATIVUS	S	.	0.0	30.1	0.8	.	.	4.0	.	.	.	.	.	.	.	.	.	.	19
3842	LATHYRUS SATIVUS	S	.	0.0	33.6	0.8	.	.	3.0	.	.	.	.	.	.	.	.	.	.	19
3843	LATHYRUS SATIVUS	S	.	0.0	30.8	0.7	.	.	3.4	.	.	.	.	.	.	.	.	.	.	19
3844	LATHYRUS SATIVUS	S	.	0.0	.	0.7	.	.	3.6	.	.	.	.	.	.	.	.	.	.	19
3845	LATHYRUS SATIVUS	S	.	0.0	.	0.7	.	.	3.5	.	.	.	.	.	.	.	.	.	.	19
3846	LATHYRUS SATIVUS	S	.	0.0	.	0.5	.	.	2.8	.	.	.	.	.	.	.	.	.	.	19
3847	LATHYRUS SATIVUS	L	.	84.2	6.1	1.0	7.6	.	1.1	160	100	7.3	.	.	3600	.	.	.	.	5
3848	LATHYRUS SP.	S	.	0.0	29.6	0.9	.	.	4.0	.	.	.	.	.	.	.	.	.	.	19
3849	LATHYRUS SP.	W	.	0.0	17.2	3.4	54.5	26.2	24.9	.	.	.	.	.	.	.	.	.	.	6
3850	LATHYRUS SP.	H	.	0.0	16.7	2.7	72.3	29.0	8.3	1390	310	.	.	.	.	.	.	.	.	6
3851	LATHYRUS SP.(LEAFMEAL)	W	.	0.0	21.3	6.6	61.9	14.8	10.2	1870	270	.	.	.	.	.	.	.	.	6
3852	LATHYRUS SP.(STEMMEAL)	W	.	0.0	11.5	2.4	78.3	29.4	7.8	600	180	.	.	.	.	.	.	.	.	6
3853	LATHYRUS SYLVESTRIS	S	.	0.0	34.6	0.8	.	.	3.2	.	.	.	.	.	.	.	.	.	.	19
3854	LATHYRUS SYLVESTRIS	S	.	0.0	32.4	0.8	.	.	3.0	.	.	.	.	.	.	.	.	.	.	19
3855	LATHYRUS SYLVESTRIS(ALL)	H	.	0.0	28.1	3.9	60.9	28.6	7.1	.	.	.	.	.	.	.	.	.	.	6
3856	LATHYRUS SYLVESTRIS(IMMATURE)	H	.	0.0	32.4	3.7	56.9	26.6	7.0	520	490	.	.	.	.	.	.	.	.	6
3857	LATHYRUS SYLVESTRIS(PAST BLOOMING)	H	.	0.0	24.4	4.6	64.1	30.4	6.9	.	300	.	.	.	.	.	.	.	.	6
3858	LATHYRUS TINGITANUS	S	.	0.0	42.9	1.2	.	.	3.7	.	.	.	.	.	.	.	.	.	.	19
3859	LATHYRUS TUBEROSUS	S	.	0.0	27.5	0.9	.	.	3.6	.	.	.	.	.	.	.	.	.	.	19
3860	LATHYRUS TUBEROSUS	S	.	0.0	23.3	1.1	.	.	3.8	.	.	.	.	.	.	.	.	.	.	19
3861	LATHYRUS TUBEROSUS	S	.	0.0	25.0	1.0	.	.	.	.	.	.	.	.	.	.	.	.	.	19
3862	LATIPES SENEGALENSIS	W	.	0.0	9.3	1.3	81.9	38.9	7.5	.	.	.	.	.	.	.	.	.	.	18
3863	LATUS CORNICULATUS	W	.	0.0	21.6	2.5	67.8	22.2	8.1	1760	200	.	.	1840	.	.	.	.	.	6
3864	LAURUS NOBILIS	L	313	5.4	7.6	8.4	75.0	26.3	3.6	834	113	43.0	23	529	3711	0.01	0.42	2.0	46	16
3865	LAURUS NOBILIS	S	.	0.0	10.2	26.9	.	.	.	.	.	.	.	.	.	.	.	.	.	21
3866	LAURUS NOBILIS	S	.	0.0	8.2	25.2	.	.	.	.	.	.	.	.	.	.	.	.	.	21
3867	LAURUS NOBILIS	L	188	45.2	4.2	1.2	47.1	4.6	2.3	187	70	5.3	.	.	1050	0.04	0.21	1.7	54	3
3868	LAVANDULA LATIFOLIA	S	.	0.0	23.0	28.4	.	.	.	.	.	.	.	.	.	.	.	.	.	21
3869	LAVANDULA PEDUNCULATA	S	.	0.0	27.0	20.0	.	.	.	.	.	.	.	.	.	.	.	.	.	21
3870	LAVANDULA SP.	S	.	0.0	29.4	21.8	.	.	.	.	.	.	.	.	.	.	.	.	.	20
3871	LAVANDULA STOECHAS	S	.	0.0	23.3	20.7	.	.	.	.	.	.	.	.	.	.	.	.	.	21
3872	LAVATERA ASSURGENTIFLORA	S	.	0.0	18.8	11.2	.	.	.	.	.	.	.	.	.	.	.	.	.	20
3873	LAVATERA PUNCTATA	S	.	0.0	18.2	12.9	.	.	.	.	.	.	.	.	.	.	.	.	.	21

TABLE 2 - DATA DIRECTLY TRANSCRIBED FROM ORIGINAL SOURCES.

OBS	PLNA	PART	CALORIES	WATER	PROTEIN	FAT	TOTLCARB	FIBER	ASH	CALCIUM	PHOSPHO	IRON	SODIUM	POTASS	CAROT	THIAMINE	RIBOFLAV	NIACIN	ASCORBIC	REFCO
3874	LAVATERA PUNCTATA	S	.	0.0	19.8	14.4	.	.	.	.	.	.	.	.	.	.	.	.	.	21
3875	LAVATERA PUNCTATA	S	.	0.0	18.7	13.6	.	.	.	.	.	.	.	.	.	.	.	.	.	21
3876	LAVATERA SP.	S	.	0.0	17.5	14.0	.	.	.	.	.	.	.	.	.	.	.	.	.	20
3877	LAVATERA TRILOBIA SUBSP. TRILOBA	S	.	0.0	22.0	14.8	.	.	.	.	.	.	.	.	.	.	.	.	.	21
3878	LAVATERA TRIMESTRIS	S	.	0.0	25.4	16.4	.	.	5.3	.	.	.	.	.	.	.	.	.	.	19
3879	LAVATERA TRIMESTRIS	S	.	0.0	25.0	17.1	.	.	.	.	.	.	.	.	.	.	.	.	.	20
3880	LAVATERA TRIMESTRIS	S	.	0.0	26.2	17.0	.	.	.	.	.	.	.	.	.	.	.	.	.	20
3881	LAVATERA TRIMESTRIS	S	.	0.0	23.1	17.2	.	.	.	.	.	.	.	.	.	.	.	.	.	20
3882	LAVATERA TRIMESTRIS	S	.	0.0	23.2	19.0	.	.	.	.	.	.	.	.	.	.	.	.	.	20
3883	LAVATERA TRIMESTRIS	S	.	0.0	26.2	16.9	.	.	.	.	.	.	.	.	.	.	.	.	.	20
3884	LAVATERA TRIMESTRIS	S	.	0.0	20.6	16.4	.	.	.	.	.	.	.	.	.	.	.	.	.	20
3885	LAVATERA TRIMESTRIS	S	.	0.0	22.5	16.4	.	.	.	.	.	.	.	.	.	.	.	.	.	20
3886	LAWSONIA INERMIS	S	.	10.6	5.0	10.5	67.2	33.6	4.8	.	.	.	.	.	.	.	.	.	.	5
3887	LAYIA SP.	W	.	0.0	10.0	.	.	24.0	10.8	1360	410	.	.	2900	.	.	.	.	.	6
3888	LEBECKIA SP.	S	.	0.0	44.3	8.6	.	.	.	.	.	.	.	.	.	.	.	.	.	21
3889	LECHEA TENUIFOLIA	S	.	0.0	22.9	3.6	.	.	.	.	.	.	.	.	.	.	.	.	.	20
3890	LECHEA VILLOSA	S	.	0.0	20.7	2.0	.	.	.	.	.	.	.	.	.	.	.	.	.	20
3891	LEERSIA HEXANDRA	W	.	0.0	10.1	1.8	77.7	25.6	10.4	.	.	.	.	.	.	.	.	.	.	18
3892	LEERSIA HEXANDRA	H	.	0.0	6.3	1.5	77.3	31.4	14.9	.	.	.	.	.	.	.	.	.	.	18
3893	LEERSIA HEXANDRA	L	.	0.0	5.8	2.1	75.4	28.4	16.7	190	140	.	.	.	.	.	.	.	.	5
3894	LEMNA MINOR	L	18	93.2	2.1	0.3	2.8	.	1.6	142	4	.	.	.	1120	0.06	0.13	0.6	5	3
3895	LENS CULINARIS	S	345	9.9	24.9	1.2	61.1	3.9	2.9	64	298	7.0	.	.	60	0.41	0.19	2.2	0	1
3896	LENS CULINARIS	S	.	0.0	26.7	1.2	60.1	0.5	12.0	.	.	.	.	.	.	.	.	.	.	18
3897	LENS CULINARIS	S	.	0.0	27.1	0.8	.	.	2.6	.	.	.	.	.	.	.	.	.	.	19
3898	LENS CULINARIS	S	340	12.0	20.2	0.6	65.0	.	2.1	68	325	7.0	29	780	.	0.46	0.33	1.3	0	2
3899	LENS CULINARIS	S	340	12.2	23.7	1.3	60.7	3.2	2.1	68	353	7.0	.	.	20	0.46	0.33	2.4	5	3
3900	LENS CULINARIS	S	340	11.1	24.7	1.1	60.1	3.9	3.0	79	377	6.8	30	790	36	0.37	0.22	2.0	.	4
3901	LEONOTIS NEPETAEFOLIA	S	.	0.0	26.6	38.0	.	.	.	.	.	.	.	.	.	.	.	.	.	21
3902	LEONOTIS NEPETAEFOLIA	S	.	8.8	23.6	28.0	33.6	14.3	5.9	.	.	.	.	.	.	.	.	.	.	5
3903	LEONOTIS NEPETIFOLIA	S	.	0.0	28.8	35.9	.	.	.	.	.	.	.	.	.	.	.	.	.	20
3904	LEONTICE LEONTOPETALUM	S	.	0.0	23.3	15.0	.	.	.	.	.	.	.	.	.	.	.	.	.	21
3905	LEONTICE LEONTOPETALUM	S	.	0.0	18.7	12.0	.	.	.	.	.	.	.	.	.	.	.	.	.	21
3906	LEONURUS CARDIACA	S	.	0.0	22.3	30.2	.	.	6.4	.	.	.	.	.	.	.	.	.	.	19
3907	LEONURUS SIBIRICUS	S	.	5.4	10.1	37.0	40.0	1.9	7.5	.	.	.	.	.	.	.	.	.	.	5
3908	LEPECHINIA SPICATA	S	.	0.0	14.0	15.2	.	.	.	.	.	.	.	.	.	.	.	.	.	21
3909	LEPECHINIA SPICATA	S	.	0.0	12.1	16.3	.	.	.	.	.	.	.	.	.	.	.	.	.	21
3910	LEPEDIUM SP.(FLOWERING)	W	.	0.0	24.4	2.0	65.0	22.3	8.6	1550	380	.	.	2050	.	.	.	.	.	6
3911	LEPIDAGATHIS CRISTATA	L	.	0.0	9.2	3.9	73.4	32.6	13.4	1970	310	1.3	.	.	.	.	.	.	.	5
3912	LEPIDIUM DENSIFLORUM	S	.	0.0	27.0	23.6	.	.	.	.	.	.	.	.	.	.	.	.	.	21
3913	LEPIDIUM DRABA	S	.	0.0	24.1	15.9	.	.	.	.	.	.	.	.	.	.	.	.	.	20
3914	LEPIDIUM LASIOCARPUM	S	.	0.0	26.3	23.5	.	.	.	.	.	.	.	.	.	.	.	.	.	20
3915	LEPIDIUM LATIFOLIUM	S	.	0.0	25.6	18.7	.	.	.	.	.	.	.	.	.	.	.	.	.	21
3916	LEPIDIUM MONTANUM VAR. ANGUSTIFOLIUM	S	.	0.0	28.1	27.7	.	.	.	.	.	.	.	.	.	.	.	.	.	20
3917	LEPIDIUM PERFOLIATUM	S	.	0.0	29.4	19.2	.	.	.	.	.	.	.	.	.	.	.	.	.	20
3918	LEPIDIUM PERFOLIATUM	S	.	0.0	22.9	21.9	.	.	.	.	.	.	.	.	.	.	.	.	.	21

TABLE 2 - DATA DIRECTLY TRANSCRIBED FROM ORIGINAL SOURCES.

OBS	PLANA	PART	CALORIES	WATER	PROTEIN	FAT	TOTLCARB	FIBER	ASH	CALCIUM	PHOSPHO	IRON	SODIUM	POTASS	CAROTE	THIAMINE	RIBOFLAV	NIACIN	ASCORBIC	REFCO
3919	LEPIDIUM SATIVUM	S	.	0.0	31.2	21.9	.	.	5.7	.	.	.	.	.	.	.	.	.	.	19
3920	LEPIDIUM SATIVUM	L	32	89.4	2.6	0.7	5.5	1.1	1.8	81	76	1.3	14	606	5580	0.08	0.26	1.0	69	2
3921	LEPIDIUM SATIVUM	L	.	82.3	5.8	1.0	8.7	.	2.2	360	110	28.6	.	.	1782	0.11	0.17	1.0	87	5
3922	LEPIDIUM SATIVUM	S	.	5.7	23.5	15.9	.	.	5.7	310	1650	.	.	.	.	.	.	.	.	5
3923	LEPIDIUM SP.(IMMATURE)	W	.	0.0	38.9	1.9	45.7	11.9	13.5	1950	600	.	.	3180	.	.	.	.	.	6
3924	LEPIDIUM VIRGINICUM	S	.	0.0	26.2	19.7	.	.	.	.	.	.	.	.	.	.	.	.	.	19
3925	LEPIDIUM VIRGINICUM	S	.	0.0	28.1	19.2	.	.	.	.	.	.	.	.	.	.	.	.	.	20
3926	LEPTADENIA HASTATA	L	54	81.0	4.9	0.2	11.3	4.7	2.6	417	94	5.4	.	.	4915	0.25	0.35	1.9	78	1
3927	LEPTOCHLOA CHINENSIS	L	.	0.0	9.6	1.5	75.1	31.6	13.8	.	.	.	.	.	.	.	.	.	.	5
3928	LEPTOCHLOA DUBIA	W	.	0.0	11.6	2.1	76.8	32.4	9.5	400	230	.	.	1850	.	.	.	.	.	6
3929	LEPTOCHLOA OBTUSIFLORA	W	.	0.0	18.4	1.7	70.4	28.0	9.5	.	.	.	.	.	.	.	.	.	.	18
3930	LEPTOCHLOA OBTUSIFOLIA	I	.	6.9	4.0	1.2	80.4	36.5	7.6	523	793	.	.	.	.	.	.	.	.	5
3931	LEPTOCHLOA PANICEA	L	.	0.0	6.7	0.9	74.2	30.8	18.2	.	.	.	.	.	.	.	.	.	.	5
3932	LEPTOLOMA COGNATUM	W	.	0.0	9.7	2.3	79.9	37.1	8.1	540	150	.	.	2490	.	.	.	.	.	6
3933	LEPTOLOMA COGNATUM(FULL BLOOM)	W	.	0.0	12.3	1.9	75.9	36.6	9.9	580	190	.	.	1750	.	.	.	.	.	6
3934	LESPEDEZA BICOLOR	S	.	0.0	36.6	11.9	.	.	4.7	.	.	.	.	.	.	.	.	.	.	19
3935	LESPEDEZA BICOLOR	S	.	0.0	39.6	10.6	.	.	5.0	.	.	.	.	.	.	.	.	.	.	19
3936	LESPEDEZA BICOLOR	S	.	0.0	37.9	11.2	.	.	4.9	.	.	.	.	.	.	.	.	.	.	19
3937	LESPEDEZA BICOLOR	S	.	0.0	37.2	10.9	.	.	5.1	.	.	.	.	.	.	.	.	.	.	19
3938	LESPEDEZA BICOLOR	S	.	0.0	36.3	11.0	.	.	5.0	.	.	.	.	.	.	.	.	.	.	19
3939	LESPEDEZA BICOLOR	S	.	0.0	33.4	12.6	.	.	5.1	.	.	.	.	.	.	.	.	.	.	19
3940	LESPEDEZA CAPITATA	S	.	0.0	38.3	8.5	.	.	3.4	.	.	.	.	.	.	.	.	.	.	19
3941	LESPEDEZA CAPITATA	S	.	0.0	5.6	1.6	.	.	1.3	.	.	.	.	.	.	.	.	.	.	19
3942	LESPEDEZA CAPITATA	S	.	0.0	37.5	10.2	.	.	.	.	.	.	.	.	.	.	.	.	.	19
3943	LESPEDEZA CAPITATA	S	.	0.0	35.1	5.1	.	.	.	.	.	.	.	.	.	.	.	.	.	21
3944	LESPEDEZA CUNEATA	W	.	0.0	16.4	2.0	68.8	32.0	12.8	.	.	.	.	.	.	.	.	.	.	18
3945	LESPEDEZA CUNEATA	H	.	0.0	11.3	2.3	83.0	42.6	3.4	.	.	.	.	.	.	.	.	.	.	18
3946	LESPEDEZA CUNEATA	S	.	0.0	40.5	5.9	.	.	3.4	.	.	.	.	.	.	.	.	.	.	19
3947	LESPEDEZA CUNEATA	S	.	0.0	39.9	7.1	.	.	3.7	.	.	.	.	.	.	.	.	.	.	19
3948	LESPEDEZA CUNEATA	S	.	0.0	38.1	7.5	.	.	4.6	.	.	.	.	.	.	.	.	.	.	19
3949	LESPEDEZA CUNEATA	H	.	0.0	15.2	3.0	75.7	27.5	6.1	1460	230	29.0	.	1030	.	.	.	.	.	6
3950	LESPEDEZA CYRTOBOTRYA	S	.	0.0	33.2	14.2	.	.	5.5	.	.	.	.	.	.	.	.	.	.	19
3951	LESPEDEZA HEDYSARIOIDES	S	.	0.0	37.4	8.2	.	.	4.5	.	.	.	.	.	.	.	.	.	.	19
3952	LESPEDEZA HEDYSARIOIDES	S	.	0.0	35.8	7.4	.	.	3.7	.	.	.	.	.	.	.	.	.	.	19
3953	LESPEDEZA HEDYSARIOIDES	S	.	0.0	35.1	7.8	.	.	5.6	.	.	.	.	.	.	.	.	.	.	19
3954	LESPEDEZA JAPONICA VAR. INTERMEDIA	S	.	0.0	41.2	9.9	.	.	4.6	.	.	.	.	.	.	.	.	.	.	19
3955	LESPEDEZA LATISSIMA	S	.	0.0	36.1	8.2	.	.	5.0	.	.	.	.	.	.	.	.	.	.	19
3956	LESPEDEZA SP.	W	.	0.0	16.2	2.7	72.7	30.8	8.4	1140	310	31.0	.	1280	.	.	.	.	.	6
3957	LESPEDEZA STIPULACEA	S	.	0.0	52.0	7.1	.	.	4.4	.	.	.	.	.	.	.	.	.	.	19
3958	LESPEDEZA STIPULACEA	H	.	0.0	14.7	3.5	75.2	30.6	6.6	1000	220	21.0	.	1040	4	0.96	.	.	.	6
3959	LESPEDEZA STRIATA	S	.	0.0	51.9	7.3	.	.	4.5	.	.	.	.	.	.	.	.	.	.	19
3960	LESPEDEZA STRIATA	S	.	0.0	26.4	4.3	.	.	4.6	.	.	.	.	.	.	.	.	.	.	19
3961	LESPEDEZA STRIATA	H	.	0.0	13.7	2.7	78.1	32.0	5.5	1050	200	31.0	.	1050	5	1.00	.	.	.	6
3962	LESPEDEZA STRIATA	W	.	0.0	15.3	2.1	71.4	36.1	11.2	1130	270	32.0	.	1160	.	.	.	.	.	6
3963	LESQUERELLA ANGUSTIFOLIA	S	.	0.0	25.0	26.2	.	.	.	.	.	.	.	.	.	.	.	.	.	20

TABLE 2 - DATA DIRECTLY TRANSCRIBED FROM ORIGINAL SOURCES.

OBS	PLNA	PART	CALORIES	WATER	PROTEIN	FAT	TOTLCARB	FIBER	ASH	CALCIUM	PHOSPHO	IRON	SODIUM	POTASS	CAROT	THIAMINE	RIBOFLAV	NIACIN	ASCORBIC	REFCO
3964	LESQUERELLA ARGYRAEA	S	.	0.0	23.8	26.4	.	.	.	.	.	.	.	.	.	.	.	.	.	20
3965	LESQUERELLA ARGYRAEA	S	.	0.0	20.0	27.4	.	.	.	.	.	.	.	.	.	.	.	.	.	20
3966	LESQUERELLA DENSIPILA	S	.	0.0	20.6	24.4	.	.	.	.	.	.	.	.	.	.	.	.	.	20
3967	LESQUERELLA DENSIPILA	S	.	0.0	.	27.4	.	.	.	.	.	.	.	.	.	.	.	.	.	21
3968	LESQUERELLA ENGELMANNII	S	.	0.0	21.2	11.1	.	.	.	.	.	.	.	.	.	.	.	.	.	20
3969	LESQUERELLA ENGELMANNII	S	.	0.0	21.2	21.2	.	.	.	.	.	.	.	.	.	.	.	.	.	20
3970	LESQUERELLA ENGELMANNII	S	.	0.0	21.2	19.8	.	.	.	.	.	.	.	.	.	.	.	.	.	20
3971	LESQUERELLA FENDLERI	S	.	0.0	21.2	19.8	.	.	.	.	.	.	.	.	.	.	.	.	.	20
3972	LESQUERELLA FENDLERI	S	.	0.0	22.5	28.1	.	.	.	.	.	.	.	.	.	.	.	.	.	20
3973	LESQUERELLA FENDLERI	S	.	0.0	20.6	22.8	.	.	.	.	.	.	.	.	.	.	.	.	.	20
3974	LESQUERELLA FENDLERI	S	.	0.0	21.0	24.6	.	.	.	.	.	.	.	.	.	.	.	.	.	21
3975	LESQUERELLA FENDLERI	S	.	0.0	24.1	21.6	.	.	.	.	.	.	.	.	.	.	.	.	.	21
3976	LESQUERELLA FENDLERI	S	.	0.0	20.9	27.6	.	.	.	.	.	.	.	.	.	.	.	.	.	21
3977	LESQUERELLA FENDLERI	S	.	0.0	25.8	26.2	.	.	.	.	.	.	.	.	.	.	.	.	.	21
3978	LESQUERELLA FENDLERI	S	.	0.0	26.1	25.8	.	.	.	.	.	.	.	.	.	.	.	.	.	21
3979	LESQUERELLA GLOBOSA	S	.	0.0	24.4	39.4	.	.	.	.	.	.	.	.	.	.	.	.	.	20
3980	LESQUERELLA GORDONII	S	.	0.0	23.1	28.8	.	.	.	.	.	.	.	.	.	.	.	.	.	20
3981	LESQUERELLA GORDONII	S	.	0.0	23.1	26.0	.	.	.	.	.	.	.	.	.	.	.	.	.	20
3982	LESQUERELLA GRANDIFLORA	S	.	0.0	19.4	37.2	.	.	.	.	.	.	.	.	.	.	.	.	.	20
3983	LESQUERELLA GRANDIFLORA	S	.	0.0	21.2	26.8	.	.	.	.	.	.	.	.	.	.	.	.	.	20
3984	LESQUERELLA LASIOCARPA	S	.	0.0	19.4	28.6	.	.	.	.	.	.	.	.	.	.	.	.	.	19
3985	LESQUERELLA LASIOCARPA	S	.	0.0	22.5	30.4	.	.	.	.	.	.	.	.	.	.	.	.	.	20
3986	LESQUERELLA LESCURII	S	.	0.0	21.2	28.1	.	.	.	.	.	.	.	.	.	.	.	.	.	20
3987	LESQUERELLA LESCURII	S	.	0.0	18.9	28.7	.	.	.	.	.	.	.	.	.	.	.	.	.	21
3988	LESQUERELLA LINDHEIMERI	S	.	0.0	21.2	25.5	.	.	.	.	.	.	.	.	.	.	.	.	.	19
3989	LESQUERELLA LINDHEIMERI	S	.	0.0	.	21.4	.	.	.	.	.	.	.	.	.	.	.	.	.	20
3990	LESQUERELLA LYRATA	S	.	0.0	18.4	25.9	.	.	.	.	.	.	.	.	.	.	.	.	.	21
3991	LESQUERELLA OVALIFOLIA	S	.	0.0	24.4	24.0	.	.	.	.	.	.	.	.	.	.	.	.	.	20
3992	LESQUERELLA PERFORATA	S	.	0.0	18.0	23.4	.	.	.	.	.	.	.	.	.	.	.	.	.	21
3993	LESQUERELLA PINETORUM	S	.	0.0	22.5	26.6	.	.	.	.	.	.	.	.	.	.	.	.	.	20
3994	LESQUERELLA SP.	W	.	0.0	21.4	2.2	65.8	17.4	10.6	4640	250	.	.	1680	.	.	.	.	.	6
3995	LESQUERELLA SP.	W	.	0.0	27.0	1.6	56.5	15.3	14.9	5020	350	.	.	2060	.	.	.	.	.	6
3996	LESQUERELLA STONENSIS	S	.	0.0	17.2	23.5	.	.	.	.	.	.	.	.	.	.	.	.	.	21
3997	LESSERTIA BRACHYSTACHYA	S	.	0.0	32.8	2.0	.	.	.	.	.	.	.	.	.	.	.	.	.	21
3998	LESSINGIA GERMANORUM	W	.	0.0	7.6	.	.	25.5	7.9	990	280	.	.	2180	.	.	.	.	.	6
3999	LEUCADENDRON AURANTIACUM	S	.	0.0	24.3	13.3	.	.	.	.	.	.	.	.	.	.	.	.	.	21
4000	LEUCADENDRON CORYMBOSUM	S	.	0.0	35.6	16.5	.	.	.	.	.	.	.	.	.	.	.	.	.	21
4001	LEUCADENDRON DREGEI	S	.	0.0	20.5	13.0	.	.	.	.	.	.	.	.	.	.	.	.	.	21
4002	LEUCADENDRON GALPINII	S	.	0.0	25.1	12.0	.	.	.	.	.	.	.	.	.	.	.	.	.	21
4003	LEUCADENDRON PLUMOSUM	S	.	0.0	56.5	21.2	.	.	.	.	.	.	.	.	.	.	.	.	.	21
4004	LEUCADENDRON PSEUDOSPATHULATUM	S	.	0.0	50.5	41.4	.	.	.	.	.	.	.	.	.	.	.	.	.	21
4005	LEUCADENDRON PSEUDOSPATHULATUM	S	.	0.0	53.8	42.0	.	.	.	.	.	.	.	.	.	.	.	.	.	21
4006	LEUCADENDRON TORTUM	S	.	0.0	69.3	28.5	.	.	.	.	.	.	.	.	.	.	.	.	.	21
4007	LEUCADENDRON TRUNCATUM	S	.	0.0	59.1	34.7	.	.	.	.	.	.	.	.	.	.	.	.	.	21
4008	LEUCADENDRON VENOSUM	S	.	0.0	56.3	39.9	.	.	.	.	.	.	.	.	.	.	.	.	.	21

TABLE 2 - DATA DIRECTLY TRANSCRIBED FROM ORIGINAL SOURCES.

OBS	PLANA	PART	CALORIES	WATER	PROTEIN	FAT	TOTCARB	FIBER	ASH	CALCIUM	PHOSPHON	IRON	SODIUM	POTASS	CAROT	THIAMINE	RIBOFLAV	NIACIN	ASCORBIC	REFCO
4009	LEUCAENA GLAUCA	S	.	0.0	30.8	6.8	.	.	3.8	.	.	.	.	.	.	.	.	.	.	19
4010	LEUCAENA GLAUCA	S	.	0.0	32.5	6.4	.	.	.	.	.	.	.	.	.	.	.	.	.	19
4011	LEUCAENA LANCEOLATA	S	.	0.0	38.1	7.6	.	.	.	.	.	.	.	.	.	.	.	.	.	20
4012	LEUCAENA LANCEOLATA	S	.	0.0	36.9	4.7	.	.	.	.	.	.	.	.	.	.	.	.	.	20
4013	LEUCAENA LANCEOLATA	S	.	0.0	36.2	5.2	.	.	.	.	.	.	.	.	.	.	.	.	.	21
4014	LEUCAENA LEUCOCEPHALA	L	.	70.6	5.3	0.6	21.9	9.7	1.8	.	.	.	.	.	.	.	.	.	.	
4015	LEUCAENA LEUCOCEPHALA	L	68	79.5	2.9	0.8	15.3	1.8	1.5	553.0	51	.	.	.	.	.	.	.	.	1
4016	LEUCAENA LEUCOCEPHALA	L	.	0.0	21.0	6.5	64.1	18.1	8.4	.	.	.	.	.	.	.	.	.	.	18
4017	LEUCAENA LEUCOCEPHALA	C	.	0.0	27.8	3.2	65.5	10.4	3.5	.	.	.	.	.	.	.	.	.	.	18
4018	LEUCAENA LEUCOCEPHALA	SH	.	0.0	24.2	2.7	64.2	24.2	8.9	.	.	.	.	.	.	.	.	.	.	18
4019	LEUCAENA LEUCOCEPHALA	F	.	0.0	21.7	1.4	71.1	25.6	5.8	.	.	.	.	.	.	.	.	.	.	18
4020	LEUCAENA LEUCOCEPHALA	S	.	0.0	35.8	7.5	52.3	11.4	4.4	.	.	.	.	.	.	.	.	.	.	18
4021	LEUCAENA LEUCOCEPHALA	SH	59	80.7	8.4	0.9	8.8	3.8	1.2	137.0	11	9.2	.	.	4730	.	0.09	5.4	8	2
4022	LEUCAENA LEUCOCEPHALA	S	.	0.0	28.1	10.0	.	.	.	.	.	.	.	.	.	.	.	.	.	20
4023	LEUCAENA LEUCOCEPHALA	H	.	0.0	14.3	2.1	77.9	33.6	5.7	.	.	.	.	.	.	.	.	.	.	6
4024	LEUCAENA LEUCOCEPHALA(IMMATURE)	L	.	0.0	18.3	2.1	72.9	37.7	6.7	.	.	.	.	.	.	.	.	.	.	6
4025	LEUCAENA LEUCOCEPHALA(MATURE)	L	.	0.0	12.8	1.2	78.4	43.4	7.6	.	.	.	.	.	.	.	.	.	.	6
4026	LEUCAENA RETUSA	S	.	0.0	29.4	7.5	.	.	.	.	.	.	.	.	.	.	.	.	.	19
4027	LEUCAENA RETUSA	S	.	0.0	32.5	6.5	.	.	.	.	.	.	.	.	.	.	.	.	.	20
4028	LEUCAENA RETUSA	S	.	0.0	12.3	9.2	.	.	.	.	.	.	.	.	.	.	.	.	.	20
4029	LEUCAENA RETUSA(IMMATURE)	L	.	0.0	17.6	3.9	72.1	24.4	6.4	2810.0	120	.	.	1300	.	.	.	.	.	6
4030	LEUCAS MICROPHYLLA	SH	.	0.0	11.4	3.3	76.5	34.1	8.8	910.0	620	.	.	.	.	.	.	.	.	18
4031	LEUCOPHYLUM SP.	W	.	0.0	7.0	5.8	83.5	25.8	3.7	1140.0	100	.	.	700	.	.	.	.	.	6
4032	LEUCOSPAERA BAINESII	L	.	0.0	11.7	2.0	76.1	29.0	10.2	1750.0	70	.	.	.	.	.	.	.	.	18
4033	LEUCOSPERMUM ATTENUATUM	S	.	0.0	49.8	44.1	.	.	.	.	.	.	.	.	.	.	.	.	.	21
4034	LEUCOSPERMUM CATHERINAE	S	.	0.0	47.6	46.8	.	.	.	.	.	.	.	.	.	.	.	.	.	21
4035	LEUCOSPERMUM CONOCARPUM	S	.	0.0	56.9	40.8	.	.	.	.	.	.	.	.	.	.	.	.	.	21
4036	LEUCOSPERMUM NUTANS	S	.	0.0	52.4	44.7	.	.	.	.	.	.	.	.	.	.	.	.	.	21
4037	LEVISTICUM OFFICINALE	S	.	0.0	20.0	14.7	.	.	.	.	.	.	.	.	.	.	.	.	.	20
4038	LEWISIA REDIVIVA	R	.	76.6	0.7	0.1	21.6	1.5	1.0	18.7	.	1.4	.	.	.	0.09	0.02	.	27	9
4039	LIATRIS PUNCTATA	S	.	0.0	47.2	29.4	.	.	3.5	.	.	.	.	.	.	.	.	.	.	19
4040	LIATRIS PUNCTATA	S	.	0.0	15.1	9.2	.	.	5.4	.	.	.	.	.	.	.	.	.	.	19
4041	LIATRIS PUNCTATA		.	0.0	7.5	8.8	73.7	29.4	10.0	.	.	.	.	.	.	.	.	.	.	6
4042	LIATRIS PYCNOSTACHYA	S	.	0.0	34.2	29.6	.	.	3.4	.	.	.	.	.	.	.	.	.	.	19
4043	LIATRIS PYCNOSTACHYA	S	.	0.0	15.0	3.7	.	.	.	.	.	.	.	.	.	.	.	.	.	20
4044	LIATRIS SP.	S	.	0.0	46.6	26.4	.	.	4.3	.	.	.	.	.	.	.	.	.	.	19
4045	LIATRIS SPICATA	S	.	0.0	35.6	35.9	.	.	4.5	.	.	.	.	.	.	.	.	.	.	19
4046	LIATRIS SPICATA	S	.	0.0	11.1	10.5	.	.	5.2	.	.	.	.	.	.	.	.	.	.	19
4047	LIBOCEDRUS DECURRENS	S	.	0.0	11.9	45.3	.	.	.	.	.	.	.	.	.	.	.	.	.	20
4048	LICANIA PLATYPUS	F	116	67.4	1.4	0.5	29.7	0.9	1.0	11.0	29	1.7	.	.	250	0.02	0.01	1.5	11	3
4049	LICANIA RIGIDA	S	.	0.0	8.3	49.4	.	.	.	.	.	.	.	.	.	.	.	.	.	21
4050	LIGUSTICUM PORTERI	S	.	0.0	27.5	20.5	.	.	.	.	.	.	.	.	.	.	.	.	.	20
4051	LIGUSTICUM PORTERI	S	.	0.0	26.4	23.9	.	.	.	.	.	.	.	.	.	.	.	.	.	20
4052	LIGUSTICUM PORTERI	I	.	0.0	18.9	2.7	67.2	15.9	11.2	.	.	.	.	.	.	.	.	.	.	6
4053	LIGUSTICUM PORTERI	L	.	0.0	16.1	3.0	68.1	14.4	12.8	.	.	.	.	.	.	.	.	.	.	6

TABLE 2 - DATA DIRECTLY TRANSCRIBED FROM ORIGINAL SOURCES.

OBS	PLANA	PART	CALORIES	WATER	PROTEIN	FAT	TOTCARB	FIBER	ASH	CALCIUM	PHOSPHON	IRON	SODIUM	POTASS	CAROT	THIAMINE	RIBOFLAV	NIACIN	ASCORBIC	REFCO
4054	LIGUSTICUM SP.	W	.	0.0	28.5	3.0	57.0	12.8	11.5	.	.	.	.	.	.	.	.	.	.	6
4055	LIGUSTICUM SP.	L	.	0.0	18.0	3.2	66.1	14.0	12.7	.	.	.	.	.	.	.	.	.	.	6
4056	LIGUSTRUM IBOTA	S	.	0.0	14.1	22.0	.	.	.	.	.	.	.	.	.	.	.	.	.	21
4057	LIGUSTRUM JAPONICUM	S	.	0.0	14.4	25.7	.	.	.	.	.	.	.	.	.	.	.	.	.	20
4058	LIGUSTRUM JAPONICUM	S	.	0.0	12.2	20.1	.	.	.	.	.	.	.	.	.	.	.	.	.	20
4059	LILIUM LANCIFOLIUM	R	113	66.0	4.8	0.6	27.2	1.1	1.4	5	240	0.5	10	.	0	0.05	0.04	0.2	20	2
4060	LILIUM LANCIFOLIUM	R	315	14.5	7.9	1.8	72.5	1.4	3.3	30	122	5.0	88	685	0	0.08	0.08	2.4	17	2
4061	LILIUM PONTICUM	S	.	0.0	14.2	11.8	.	.	.	.	.	.	.	.	.	.	.	.	.	21
4062	LIMNANTHES ALBA VAR. ALBA	S	.	0.0	19.9	26.8	.	.	.	.	.	.	.	.	.	.	.	.	.	21
4063	LIMNANTHES ALBA VAR. VERSICOLOR	S	.	0.0	19.2	30.9	.	.	.	.	.	.	.	.	.	.	.	.	.	21
4064	LIMNANTHES BAKERI	S	.	0.0	15.9	25.6	.	.	.	.	.	.	.	.	.	.	.	.	.	21
4065	LIMNANTHES DOUGLASII	S	.	0.0	24.7	24.5	.	.	7.3	.	.	.	.	.	.	.	.	.	.	19
4066	LIMNANTHES DOUGLASII	S	.	0.0	25.6	25.2	.	.	.	.	.	.	.	.	.	.	.	.	.	19
4067	LIMNANTHES DOUGLASII	S	.	0.0	22.5	27.4	.	.	.	.	.	.	.	.	.	.	.	.	.	19
4068	LIMNANTHES DOUGLASII	S	.	0.0	20.6	29.1	.	.	.	.	.	.	.	.	.	.	.	.	.	19
4069	LIMNANTHES DOUGLASII	S	.	0.0	23.1	25.6	.	.	.	.	.	.	.	.	.	.	.	.	.	20
4070	LIMNANTHES DOUGLASII	S	.	0.0	23.8	24.0	.	.	.	.	.	.	.	.	.	.	.	.	.	20
4071	LIMNANTHES DOUGLASII VAR. DOUGLASII	S	.	0.0	15.8	29.8	.	.	.	.	.	.	.	.	.	.	.	.	.	21
4072	LIMNANTHES DOUGLASII VAR. DOUGLASII	S	.	0.0	17.0	24.5	.	.	.	.	.	.	.	.	.	.	.	.	.	21
4073	LIMNANTHES DOUGLASII VAR. DOUGLASII	S	.	0.0	20.2	28.4	.	.	.	.	.	.	.	.	.	.	.	.	.	21
4074	LIMNANTHES DOUGLASII VAR. NIVEA	S	.	0.0	17.4	26.1	.	.	.	.	.	.	.	.	.	.	.	.	.	21
4075	LIMNANTHES DOUGLASII VAR. NIVEA	S	.	0.0	19.2	29.8	.	.	.	.	.	.	.	.	.	.	.	.	.	21
4076	LIMNANTHES DOUGLASII VAR. NIVEA	S	.	0.0	22.2	30.4	.	.	.	.	.	.	.	.	.	.	.	.	.	21
4077	LIMNANTHES DOUGLASII VAR. NIVEA	S	.	0.0	14.9	33.2	.	.	.	.	.	.	.	.	.	.	.	.	.	21
4078	LIMNANTHES DOUGLASII VAR. ROSEA	S	.	0.0	16.4	20.2	.	.	.	.	.	.	.	.	.	.	.	.	.	21
4079	LIMNANTHES DOUGLASII VAR. ROSEA	S	.	0.0	16.8	21.7	.	.	.	.	.	.	.	.	.	.	.	.	.	21
4080	LIMNANTHES FLOCCOSSA	S	.	0.0	24.6	28.3	.	.	.	.	.	.	.	.	.	.	.	.	.	21
4081	LIMNANTHES GRACILIS VAR. GRACILIS	S	.	0.0	16.6	28.8	.	.	.	.	.	.	.	.	.	.	.	.	.	21
4082	LIMNANTHES GRACILIS VAR. PARISHII	S	.	0.0	18.2	33.3	.	.	.	.	.	.	.	.	.	.	.	.	.	21
4083	LIMNANTHES MONTANA	S	.	0.0	20.7	25.8	.	.	.	.	.	.	.	.	.	.	.	.	.	21
4084	LIMNANTHES STRIATA	S	.	0.0	20.3	29.0	.	.	.	.	.	.	.	.	.	.	.	.	.	21
4085	LIMNOCHARIS FLAVA	L	33	90.0	1.7	0.2	7.7	.	0.4	62	33	2.1	.	.	2280	0.07	.	.	54	2
4086	LIMNODEA ARKANSANA	W	.	0.0	11.4	2.4	79.6	29.2	6.6	340	190	.	.	1770	.	.	.	.	.	6
4087	LIMNOPHILA SP.	L	24	92.2	1.4	0.3	5.2	.	0.9	76	24	3.5	123	137	2000	0.03	0.14	0.6	1	2
4088	LIMNOSCIADIUM PUMILUM	S	.	0.0	12.5	16.2	.	.	.	.	.	.	.	.	.	.	.	.	.	20
4089	LIMONIUM GMELINII	S	.	0.0	10.3	2.3	.	.	.	.	.	.	.	.	.	.	.	.	.	21
4090	LIMONIUM HUMILE	S	.	0.0	33.5	7.8	.	.	.	.	.	.	.	.	.	.	.	.	.	21
4091	LIMONIUM SINUATUM	S	.	0.0	36.6	9.7	.	.	5.9	.	.	.	.	.	.	.	.	.	.	19
4092	LIMONIUM SINUATUM	S	.	0.0	7.8	1.6	.	.	5.0	.	.	.	.	.	.	.	.	.	.	19
4093	LIMONIUM TETARICUM	S	.	0.0	33.8	5.3	.	.	4.4	.	.	.	.	.	.	.	.	.	.	19
4094	LIMONIUM TETARICUM	S	.	0.0	12.3	2.3	.	.	5.3	.	.	.	.	.	.	.	.	.	.	19
4095	LINARIA LANIGERA	S	.	0.0	20.6	43.0	.	.	.	.	.	.	.	.	.	.	.	.	.	21
4096	LINARIA MAROCCANA	S	.	0.0	21.1	43.1	.	.	4.7	.	.	.	.	.	.	.	.	.	.	19
4097	LINARIA NIVEA	S	.	0.0	15.8	50.7	.	.	.	.	.	.	.	.	.	.	.	.	.	21
4098	LINARIA PYRAMIDATA	S	.	0.0	22.4	44.4	.	.	.	.	.	.	.	.	.	.	.	.	.	21

TABLE 2 - DATA DIRECTLY TRANSCRIBED FROM ORIGINAL SOURCES.

OBS	PLNA	PART	CALORIES	WATER	PROTEIN	FAT	TOTCARB	FIBER	ASH	CALCIUM	PHOSPHO	IRON	SODIUM	POTASS	CAROT	THIAMINE	RIBOFLAV	NIACIN	ASCORBIC	REF
4099	LINDERA BENZOIN	S	.	0.0	21.2	61.30	.	.	.	.	.	.	.	.	.	.	.	.	.	19
4100	LINDERA PRAECOX	S	.	0.0	12.2	29.70	.	.	.	.	.	.	.	.	.	.	.	.	.	21
4101	LINUM AUSTRALE	S	.	0.0	22.2	28.80	.	.	.	.	.	.	.	.	.	.	.	.	.	20
4102	LINUM BIENNE	S	.	0.0	24.2	29.90	.	.	.	.	.	.	.	.	.	.	.	.	.	21
4103	LINUM GRANDIFLORUM	S	.	0.0	23.1	28.20	.	.	.	.	.	.	.	.	.	.	.	.	.	20
4104	LINUM HIRSUTUM	S	.	0.0	24.2	34.10	.	.	.	.	.	.	.	.	.	.	.	.	.	21
4105	LINUM PERENNE	S	.	0.0	23.1	28.80	.	.	.	.	.	.	.	.	.	.	.	.	.	20
4106	LINUM SP.	S	.	0.0	16.2	19.40	.	.	.	.	.	.	.	.	.	.	.	.	.	20
4107	LINUM STRICTUM	S	.	0.0	23.2	32.60	.	.	.	.	.	.	.	.	.	.	.	.	.	21
4108	LINUM USITATISSIMUM	S	498	6.3	18.0	34.00	37.2	8.8	4.5	271.0	462	43.8	.	.	0	0.17	0.16	1.4	.	1
4109	LINUM USITATISSIMUM	H	.	0.0	7.6	6.10	80.1	36.5	6.2	.	30	.	.	.	.	.	.	.	.	18
4110	LINUM USITATISSIMUM	S	.	0.0	29.2	36.00	29.9	4.8	4.9	200.0	600	.	.	.	.	.	.	.	.	18
4111	LINUM USITATISSIMUM	S	.	0.0	25.6	40.40	.	.	3.7	.	.	.	.	.	.	.	.	.	.	19
4112	LINUM USITATISSIMUM	S	.	0.0	26.6	4.00	.	.	.	.	.	.	.	.	.	.	.	.	.	21
4113	LINUM USITATISSIMUM	S	.	6.6	20.3	37.10	33.6	4.8	2.4	170.0	370	2.7	.	.	30	.	.	.	.	5
4114	LINUM USITATISSIMUM	H	.	0.0	7.8	3.30	81.7	46.2	7.2	720.0	110	.	.	1740	.	.	.	.	.	6
4115	LIPPIA DURANGENSE	S	.	0.0	18.1	23.80	.	.	.	.	.	.	.	.	.	.	.	.	.	20
4116	LIQUIDAMBAR STYRACIFLUA	S	.	0.0	33.1	29.70	.	.	5.6	.	.	.	.	.	.	.	.	.	.	19
4117	LISAEA PAPYRACEA	S	.	0.0	12.4	17.20	.	.	.	.	.	.	.	.	.	.	.	.	.	21
4118	LITCHI CHINENSIS	F	56	84.2	0.8	0.20	14.3	0.5	0.5	8.0	22	0.4	.	.	.	.	.	.	.	1
4119	LITCHI CHINENSIS	F	65	82.1	0.8	0.40	16.3	0.2	0.4	10.0	29	0.3	3	170	0	0.05	0.06	0.6	50	2
4120	LITCHI CHINENSIS	F	60	83.1	0.8	0.10	15.7	0.1	0.3	5.0	31	1.1	.	.	0	0.02	0.03	0.9	72	3
4121	LITCHI CHINENSIS	F	64	81.9	0.9	0.30	16.4	0.3	0.5	8.0	42	0.4	3	170	.	.	0.05	.	42	4
4122	LITHOSPERMUM OFFICINALE	S	.	0.0	34.6	54.30	7.7	2.8	3.7	.	.	.	.	.	.	.	.	.	.	5
4123	LITHOSPERMUM RUDERALE	W	.	0.0	27.5	3.90	54.3	12.6	14.3	.	.	.	.	.	.	.	.	.	.	6
4124	LITHOSPERMUM RUDERALE	L	.	0.0	17.2	3.00	64.5	15.8	15.3	2520.0	360	.	.	.	.	.	.	.	.	6
4125	LITHOSPERMUM TENUIFLORUM	S	.	0.0	20.1	16.40	.	.	.	.	.	.	.	.	.	.	.	.	.	21
4126	LITHRAEA BRASILIENSIS	S	.	0.0	7.1	11.00	.	.	.	.	.	.	.	.	.	.	.	.	.	21
4127	LITSEA GLAUCESCENS	S	.	0.0	15.0	10.90	.	.	.	.	.	.	.	.	.	.	.	.	.	20
4128	LITSEA GUATEMALENSIS	L	329	8.0	13.7	7.00	66.4	23.7	4.9	803.0	114	15.0	.	.	8300	0.10	0.65	2.5	.	3
4129	LOBELIA ERINUS	S	.	0.0	21.9	46.70	.	.	2.9	.	.	.	.	.	.	.	.	.	.	19
4130	LOBULARIA MARITIMA	S	.	0.0	26.2	35.60	.	.	.	.	.	.	.	.	.	.	.	.	.	20
4131	LOBULARIA MARITIMA	S	.	0.0	26.9	33.10	.	.	.	.	.	.	.	.	.	.	.	.	.	20
4132	LOLIUM MULTIFLORUM	S	.	0.0	15.0	1.60	.	.	3.3	.	.	.	.	.	.	.	.	.	.	19
4133	LOLIUM MULTIFLORUM	S	.	0.0	2.1	2.00	.	.	4.4	.	.	.	.	.	.	.	.	.	.	19
4134	LOLIUM MULTIFLORUM	S	.	0.0	11.4	1.40	.	.	3.9	.	.	.	.	.	.	.	.	.	.	19
4135	LOLIUM MULTIFLORUM	L	.	72.9	3.1	1.30	20.2	6.8	2.5	130.0	80	.	.	400	.	.	.	.	.	5
4136	LOLIUM MULTIFLORUM	L	.	11.4	8.1	1.90	71.1	27.8	7.5	.	.	.	.	.	.	.	.	.	.	5
4137	LOLIUM MULTIFLORUM	S	.	.	9.8	2.40	82.1	10.3	5.7	.	.	.	.	.	.	.	.	.	.	5
4138	LOLIUM MULTIFLORUM	W	.	0.0	18.5	4.20	58.1	20.0	19.2	640.0	410	.	.	2000	.	.	.	.	.	6
4139	LOLIUM MULTIFLORUM	H	.	0.0	8.0	1.80	81.9	32.6	8.3	620.0	410	.	.	1990	.	.	.	.	.	6
4140	LOLIUM PERENNE	S	.	0.0	18.8	1.80	.	.	.	.	.	.	.	.	.	.	.	.	.	20
4141	LOLIUM PERENNE	L	.	73.4	3.0	1.30	19.9	6.7	2.4	120.0	70	.	.	510	.	.	.	.	.	5
4142	LOLIUM PERENNE	L	.	12.0	9.2	3.10	67.6	24.2	8.1	.	.	.	.	.	.	.	.	.	.	5
4143	LOLIUM PERENNE	W	.	0.0	24.7	3.90	55.4	18.7	16.0	600.0	480	.	.	.	.	.	.	.	.	6

TABLE 2 - DATA DIRECTLY TRANSCRIBED FROM ORIGINAL SOURCES.

OBS	PLNA	PART	CALORIES	WATER	PROTEIN	FAT	TOTCARB	FIBER	ASH	CALCIUM	PHOSPHO	IRON	SODIUM	POTASS	CAROT	THIAMINE	RIBOFLAV	NIACIN	ASCORBIC	REFCO
4144	LOLIUM PERENNE	H	.	0.0	11.1	3.40	76.1	29.1	9.4	400.0	260	.	.	1910	.	.	.	.	.	6
4145	LOLIUM SP.	W	.	0.0	16.2	3.90	66.7	23.2	13.2	550.0	450	.	.	3400	40	.	.	.	.	6
4146	LOLIUM TEMULENTUM	S	.	0.0	16.2	1.40	.	.	.	.	.	.	.	.	.	.	.	.	.	20
4147	LOLIUM TEMULENTUM	S	.	.	10.9	2.60	95.2	39.0	11.9	.	.	.	.	.	.	.	.	.	.	5
4148	LOMATIUM CANBYI	R	.	71.9	0.8	0.12	25.9	2.2	1.3	47.7	.	1.5	.	.	.	0.11	0.06	.	20	9
4149	LOMATIUM COUS	R	.	67.1	1.0	0.40	30.0	2.6	1.5	79.0	.	3.6	.	.	.	0.06	0.07	.	17	9
4150	LOMATIUM DAUCIFOLIUM	S	.	0.0	21.0	15.20	.	.	.	.	.	.	.	.	.	.	.	.	.	20
4151	LOMATIUM DAUCIFOLIUM	S	.	0.0	22.1	17.60	.	.	.	.	.	.	.	.	.	.	.	.	.	20
4152	LOMATIUM MACROCARPUM	S	.	0.0	23.9	23.50	.	.	7.0	.	.	.	.	.	.	.	.	.	.	19
4153	LOMATIUM MACROCARPUM	S	.	0.0	8.2	8.40	.	.	11.9	.	.	.	.	.	.	.	.	.	.	19
4154	LOMATIUM NUDICAULE	S	.	0.0	18.1	18.40	.	.	.	.	.	.	.	.	.	.	.	.	.	20
4155	LOMATIUM NUDICAULE	L	.	87.8	.	.	.	.	.	36.6	.	0.7	.	.	.	0.02	0.08	.	66	9
4156	LOMATIUM SP.	W	.	0.0	23.0	5.00	61.5	14.6	10.5	.	.	.	.	.	.	.	.	.	.	6
4157	LONAS ANNUS	S	.	0.0	25.6	27.80	.	.	.	.	.	.	.	.	.	.	.	.	.	20
4158	LONICERA INVOLUCRATA(IMMATURE)	L	.	0.0	18.2	2.40	72.6	10.5	6.8	2730.0	620	.	.	.	.	.	.	.	.	6
4159	LONICERA INVOLUCRATA(MATURE)	L	.	0.0	9.6	5.30	74.1	11.6	11.0	1050.0	420	.	.	.	.	.	.	.	.	6
4160	LONICERA JAPONICA	S	.	0.0	21.0	26.80	.	.	.	.	.	.	.	.	.	.	.	.	.	21
4161	LONICERA ORIENTALIS	S	.	0.0	7.2	16.50	.	.	.	.	.	.	.	.	.	.	.	.	.	21
4162	LONICERA SP.	L	.	0.0	12.0	3.50	76.1	17.0	8.4	1420.0	300	.	.	2390	.	.	.	.	.	6
4163	LONICERA TARTARICA	S	.	0.0	20.9	32.40	.	.	3.4	.	.	.	.	.	.	.	.	.	.	19
4164	LOPHIRA LANCEOLATA	S	547	8.0	13.8	44.10	32.0	1.2	2.1	101.0	156	.	.	.	.	.	.	.	.	1
4165	LOPHOPOGON TRIDENTATUS	L	.	7.8	2.4	1.50	68.8	29.2	19.6	.	.	.	.	.	.	.	.	.	.	5
4166	LOTONONIS BAINESII	W	.	0.0	19.3	4.00	68.6	27.0	8.1	650.0	320	.	.	.	.	.	.	.	.	18
4167	LOTUS AMERICANUS	S	.	0.0	48.3	6.50	.	.	3.2	.	.	.	.	.	.	.	.	.	.	19
4168	LOTUS AMERICANUS	W	.	0.0	15.0	1.90	75.5	22.5	7.6	1650.0	210	.	.	1680	.	.	.	.	.	6
4169	LOTUS ARABICUS	S	.	0.0	36.0	5.30	.	.	6.0	.	.	.	.	.	.	.	.	.	.	19
4170	LOTUS CORNICULATUS	W	.	0.0	13.4	4.30	73.3	26.4	8.8	1130.0	230	.	.	.	.	.	.	.	.	18
4171	LOTUS CORNICULATUS	S	.	0.0	35.5	5.80	.	.	7.8	.	.	.	.	.	.	.	.	.	.	19
4172	LOTUS CORNICULATUS	S	.	0.0	37.4	6.20	.	.	6.9	.	.	.	.	.	.	.	.	.	.	19
4173	LOTUS CORNICULATUS	S	.	0.0	35.4	6.10	.	.	9.9	.	.	.	.	.	.	.	.	.	.	19
4174	LOTUS CORNICULATUS	S	.	0.0	37.3	6.70	.	.	4.1	.	.	.	.	.	.	.	.	.	.	19
4175	LOTUS CORNICULATUS	S	.	0.0	34.9	6.20	.	.	4.3	.	.	.	.	.	.	.	.	.	.	19
4176	LOTUS CORNICULATUS	S	.	0.0	36.0	6.00	.	.	5.1	.	.	.	.	.	.	.	.	.	.	19
4177	LOTUS CORNICULATUS	S	.	0.0	33.8	6.70	.	.	.	.	.	.	.	.	.	.	.	.	.	19
4178	LOTUS CORNICULATUS	S	.	0.0	38.6	7.50	.	.	4.3	.	.	.	.	.	.	.	.	.	.	19
4179	LOTUS CORNICULATUS	S	.	0.0	36.8	7.80	.	.	4.2	.	.	.	.	.	.	.	.	.	.	19
4180	LOTUS CORNICULATUS	S	.	0.0	36.1	6.30	.	.	4.5	.	.	.	.	.	.	.	.	.	.	19
4181	LOTUS CORNICULATUS	L	.	80.0	5.6	1.00	11.9	2.6	1.5	.	.	.	.	.	.	.	.	.	.	5
4182	LOTUS CORNICULATUS(FLOWERING)	W	.	0.0	19.9	2.00	70.5	21.6	7.6	.	.	.	.	.	.	.	.	.	.	6
4183	LOTUS CORNICULATUS(IMMATURE)	W	.	0.0	26.7	2.40	60.9	17.2	10.0	.	.	.	.	.	.	.	.	.	.	6
4184	LOTUS CORNICULATUS(PAST FLOWERING)	W	.	0.0	18.7	1.70	72.2	27.1	7.4	.	.	.	.	.	.	.	.	.	.	6
4185	LOTUS CRASSIFOLIUS	H	.	0.0	13.4	3.00	75.4	24.5	8.2	.	.	.	.	.	.	.	.	.	.	6
4186	LOTUS CRETICUS	S	.	0.0	31.6	7.60	.	.	.	.	.	.	.	.	.	.	.	.	.	21
4187	LOTUS CRETICUS	S	.	0.0	31.8	6.80	.	.	.	.	.	.	.	.	.	.	.	.	.	21
4188	LOTUS GEBELIA VAR. VILLOSUS	S	.	0.0	36.2	7.30	.	.	.	.	.	.	.	.	.	.	.	.	.	21

TABLE 2 - DATA DIRECTLY TRANSCRIBED FROM ORIGINAL SOURCES.

OBS	PLNA	PART	CALORIES	WATER	PROTEIN	FAT	TOTCARB	FIBER	ASH	CALCIUM	PHOSPHO	IRON	SODIUM	POTASS	CAROT	THIAMINE	RIBOFLAV	NIACIN	ASCORBIC	REFCO
4189	LOTUS RIGIDUS	S	.	0.0	38.0	6.8	.	.	4.0	.	.	.	.	.	.	.	.	.	.	19
4190	LOTUS SCOPARIUS	S	.	0.0	35.3	8.4	.	.	3.7	.	.	.	.	.	.	.	.	.	.	19
4191	LOTUS SCOPARIUS	S	.	0.0	38.1	7.6	.	.	.	.	.	.	.	.	.	.	.	.	.	20
4192	LOTUS SCOPARIUS	S	.	0.0	40.0	7.8	.	.	.	.	.	.	.	.	.	.	.	.	.	20
4193	LOTUS SUBPINNATUS(IMMATURE)	W	.	0.0	23.4	.	.	16.7	13.1	2120	340	.	.	2350	.	.	.	.	.	6
4194	LOTUS TENUIS	S	.	0.0	34.2	7.0	.	.	.	.	.	.	.	.	.	.	.	.	.	21
4195	LOTUS ULIGINOSUS	S	.	0.0	35.2	6.4	.	.	8.1	.	.	.	.	.	.	.	.	.	.	19
4196	LOTUS ULIGINOSUS	S	.	0.0	33.9	5.1	.	.	4.9	.	.	.	.	.	.	.	.	.	.	19
4197	LOTUS ULIGINOSUS	S	.	0.0	34.5	5.6	.	.	3.8	.	.	.	.	.	.	.	.	.	.	19
4198	LUDWIGIA PEPLOIDES	L	41	87.0	3.3	0.4	8.3	3.3	1.0	57	300	12.7	.	.	5925	0.00	0.01	2.8	3	2
4199	LUDWIGIA PEPLOIDES	L	43	87.5	2.5	1.0	7.9	0.8	1.1	144	65	8.0	.	.	3450	0.04	0.12	0.8	87	3
4200	LUDWIGIA PILOSA	S	.	0.0	13.8	28.7	.	.	.	.	.	.	.	.	.	.	.	.	.	21
4201	LUFFA ACUTANGULA	S	.	0.0	27.5	23.5	.	.	3.2	.	.	.	.	.	.	.	.	.	.	19
4202	LUFFA ACUTANGULA	S	.	0.0	41.4	44.6	.	.	4.7	.	.	.	.	.	.	.	.	.	.	19
4203	LUFFA ACUTANGULA	F	17	94.6	0.7	0.1	4.1	0.6	0.5	25	39	0.5	1	142	30	0.03	0.03	0.2	6	2
4204	LUFFA ACUTANGULA	L	27	90.1	5.1	0.0	4.1	1.5	0.7	56	140	11.5	.	.	9240	0.05	0.06	.	95	2
4205	LUFFA ACUTANGULA	F	18	94.5	0.8	0.2	4.1	0.5	0.4	19	33	0.9	.	.	228	0.03	0.04	0.4	8	4
4206	LUFFA ACUTANGULA	F	.	95.4	0.5	0.1	3.7	.	0.3	40	40	1.6	.	.	34	.	.	.	.	5
4207	LUFFA ACUTANGULA	S	.	0.0	39.9	48.4	7.7	1.9	4.8	.	.	.	.	.	.	.	.	.	.	5
4208	LUFFA AEGYPTIACA	L	14	94.0	1.6	0.1	2.7	.	1.6	330	33	.	.	.	.	.	.	.	.	1
4209	LUFFA AEGYPTIACA	L	27	90.1	5.1	0.0	4.1	1.5	0.7	56	140	11.5	.	.	9240	0.05	0.06	.	95	2
4210	LUFFA AEGYPTIACA	F	19	94.3	1.1	0.2	4.0	1.0	0.4	2	30	0.7	3	154	170	0.03	0.04	0.3	10	2
4211	LUFFA AEGYPTIACA	F	.	93.2	1.2	0.2	4.9	2.0	0.5	36	19	1.1	.	.	120	0.02	0.06	0.4	0	5
4212	LUFFA AEGYPTIACA	S	.	.	40.9	45.7	8.3	2.9	4.8	.	1830	.	.	.	.	.	.	.	.	5
4213	LUFFA OPERCULATA	S	.	0.0	24.4	28.6	.	.	.	.	.	.	.	.	.	.	.	.	.	21
4214	LUNARIA ANNUA	S	.	0.0	24.6	32.9	.	.	4.2	.	.	.	.	.	.	.	.	.	.	19
4215	LUNARIA ANNUA	S	.	0.0	27.5	47.5	.	.	2.6	.	.	.	.	.	.	.	.	.	.	19
4216	LUNARIA ANNUA	S	.	0.0	28.1	30.3	.	.	.	.	.	.	.	.	.	.	.	.	.	20
4217	LUPINUS ALBUS	S	.	0.0	37.0	9.7	.	.	3.1	.	.	.	.	.	.	.	.	.	.	19
4218	LUPINUS ALBUS	S	.	0.0	36.0	9.0	.	.	3.4	.	.	.	.	.	.	.	.	.	.	19
4219	LUPINUS ALBUS	S	.	0.0	38.8	9.1	.	.	3.2	.	.	.	.	.	.	.	.	.	.	19
4220	LUPINUS ALBUS	S	.	0.0	35.7	9.7	.	.	3.8	.	.	.	.	.	.	.	.	.	.	19
4221	LUPINUS ALBUS	S	.	11.0	32.8	8.9	51.0	.	2.6	.	.	.	.	.	.	.	.	.	.	5
4222	LUPINUS ANGUSTIFOLIUS	S	.	0.0	34.9	5.3	.	.	2.7	.	.	.	.	.	.	.	.	.	.	19
4223	LUPINUS ANGUSTIFOLIUS	S	.	0.0	35.3	3.0	.	.	.	.	.	.	.	.	.	.	.	.	.	19
4224	LUPINUS ANGUSTIFOLIUS	S	.	0.0	35.4	5.4	.	.	2.9	.	.	.	.	.	.	.	.	.	.	19
4225	LUPINUS ANGUSTIFOLIUS	S	.	0.0	36.6	5.4	.	.	3.0	.	.	.	.	.	.	.	.	.	.	19
4226	LUPINUS ANGUSTIFOLIUS	S	.	0.0	31.6	5.5	.	.	2.9	.	.	.	.	.	.	.	.	.	.	19
4227	LUPINUS ANGUSTIFOLIUS	S	.	0.0	36.8	5.2	.	.	3.4	.	.	.	.	.	.	.	.	.	.	19
4228	LUPINUS ANGUSTIFOLIUS	S	.	13.8	31.0	6.4	46.4	11.5	2.4	.	.	.	.	.	.	.	.	.	.	5
4229	LUPINUS ANGUSTIFOLIUS	H	.	0.0	32.5	5.2	58.6	20.1	3.7	590	470	.	.	.	.	.	.	.	.	6
4230	LUPINUS BENTHAMI	W	.	0.0	18.3	.	.	23.9	10.4	1430	310	.	.	3070	.	.	.	.	.	6
4231	LUPINUS BICOLOR	W	.	0.0	17.1	.	.	24.9	9.1	1410	290	.	.	2790	.	.	.	.	.	6
4232	LUPINUS CAUDATUS	L	.	0.0	19.9	3.1	69.7	30.2	7.3	1890	230	.	.	.	.	.	.	.	.	6
4233	LUPINUS HARTWEGII	S	.	0.0	38.0	9.1	.	.	4.8	.	.	.	.	.	.	.	.	.	.	19

TABLE 2 - DATA DIRECTLY TRANSCRIBED FROM ORIGINAL SOURCES.

OBS	PLNA	PART	CALORIES	WATER	PROTEIN	FAT	TOTLCARB	FIBER	ASH	CALCIUM	PHOSPHO	IRON	SODIUM	POTASS	CAROT	THIAMINE	RIBOFLAV	NIACIN	ASCORBIC	REFCO
4234	LUPINUS HIRSUTISSIMUS	S	.	0.0	41.3	10.3	.	.	.	.	.	.	.	.	.	.	.	.	.	21
4235	LUPINUS HISPANICUS	S	.	0.0	44.4	3.2	.	.	.	.	.	.	.	.	.	.	.	.	.	21
4236	LUPINUS HISPANICUS	S	.	0.0	39.1	3.7	.	.	.	.	.	.	.	.	.	.	.	.	.	21
4237	LUPINUS LATIFOLIUS	S	.	0.0	37.4	4.8	.	.	4.5	.	.	.	.	.	.	.	.	.	.	19
4238	LUPINUS LATIFOLIUS	S	.	0.0	38.8	5.9	.	.	.	.	.	.	.	.	.	.	.	.	.	20
4239	LUPINUS LUTEOLUS	S	.	0.0	44.1	9.5	.	.	.	.	.	.	.	.	.	.	.	.	.	21
4240	LUPINUS LUTEOLUS	S	.	0.0	40.7	10.5	.	.	.	.	.	.	.	.	.	.	.	.	.	21
4241	LUPINUS LUTEOLUS	S	.	0.0	54.0	10.0	.	.	.	.	.	.	.	.	.	.	.	.	.	21
4242	LUPINUS LUTEUS	S	.	0.0	44.1	5.4	.	.	5.3	.	.	.	.	.	.	.	.	.	.	19
4243	LUPINUS LUTEUS	S	.	0.0	42.2	5.5	.	.	4.2	.	.	.	.	.	.	.	.	.	.	19
4244	LUPINUS LUTEUS	S	.	11.1	39.8	4.9	39.7	14.0	4.5	230	390	.	.	810	.	.	.	.	.	5
4245	LUPINUS LUTEUS	H	.	0.0	35.6	5.2	54.2	19.6	5.0	640	820	.	.	.	.	.	.	.	.	6
4246	LUPINUS MICRANTHUS	S	.	0.0	46.7	5.3	.	.	5.4	.	.	.	.	.	.	.	.	.	.	19
4247	LUPINUS MUTABILIS	S	407	7.7	44.3	16.5	28.2	7.1	3.3	90	545	6.3	.	.	0	0.28	0.50	2.6	.	3
4248	LUPINUS MUTABILIS	F	126	71.4	13.7	5.1	9.0	0.6	0.8	28	168	1.9	.	.	30	0.09	0.15	0.8	.	3
4249	LUPINUS PALAESTINUS	S	.	0.0	27.7	5.3	.	.	.	.	.	.	.	.	.	.	.	.	.	21
4250	LUPINUS PARVIFLORUS	W	.	0.0	19.4	.	.	27.0	6.2	.	.	.	.	.	.	.	.	.	.	6
4251	LUPINUS PLATTENSIS	S	.	0.0	39.9	8.4	.	.	4.1	.	.	.	.	.	.	.	.	.	.	19
4252	LUPINUS POLYPHYLLUS VAR. CARMINEUS	S	.	0.0	41.5	9.2	.	.	5.0	.	.	.	.	.	.	.	.	.	.	19
4253	LUPINUS RIVULARIS	S	.	0.0	45.9	7.5	.	.	3.6	.	.	.	.	.	.	.	.	.	.	19
4254	LUPINUS SP.	W	.	0.0	26.6	2.6	56.9	19.1	13.9	1280	250	.	.	.	.	.	.	.	.	18
4255	LUPINUS SP.	S	.	0.0	45.0	5.0	45.2	16.2	4.8	370	200	.	.	.	.	.	.	.	.	18
4256	LUPINUS SP.	S	.	0.0	42.6	10.4	.	.	4.3	.	.	.	.	.	.	.	.	.	.	19
4257	LUPINUS SP.	S	.	0.0	36.2	6.3	.	.	4.3	.	.	.	.	.	.	.	.	.	.	19
4258	LUPINUS SP.	S	.	0.0	41.2	8.7	.	.	3.9	.	.	.	.	.	.	.	.	.	.	19
4259	LUPINUS SP.	S	.	0.0	43.1	10.0	.	.	.	.	.	.	.	.	.	.	.	.	.	20
4260	LUPINUS SP.	S	.	0.0	41.2	8.3	.	.	.	.	.	.	.	.	.	.	.	.	.	21
4261	LUPINUS STIPULATUS	S	.	0.0	41.9	9.1	.	.	.	.	.	.	.	.	.	.	.	.	.	20
4262	LUPINUS SUBCARNOSUS	L	.	0.0	16.2	2.3	72.2	22.5	9.3	2570	240	.	.	1710	.	.	.	.	.	6
4263	LUPINUS SUCCULENTUS	S	.	0.0	38.2	5.3	.	.	4.9	.	.	.	.	.	.	.	.	.	.	19
4264	LUPINUS TEXENSIS	S	.	0.0	34.4	2.4	.	.	.	.	.	.	.	.	.	.	.	.	.	19
4265	LUPINUS TEXENSIS	S	.	0.0	30.6	2.7	.	.	.	.	.	.	.	.	.	.	.	.	.	20
4266	LUPINUS VARIUS	S	.	0.0	27.2	3.5	.	.	.	.	.	.	.	.	.	.	.	.	.	21
4267	LUPINUS VARIUS SUBSP. ORIENTALIS	S	.	0.0	26.9	3.6	.	.	.	.	.	.	.	.	.	.	.	.	.	21
4268	LYCHNIS CHALCEDONICA	S	.	0.0	16.9	6.5	.	.	.	.	.	.	.	.	.	.	.	.	.	20
4269	LYCHNIS VISCARIA	S	.	0.0	20.2	6.7	.	.	4.0	.	.	.	.	.	.	.	.	.	.	19
4270	LYCIUM CHINENSE	L	29	89.6	4.1	0.6	4.0	1.3	1.7	148	43	5.4	191	518	4450	0.08	0.31	0.8	8	2
4271	LYCIUM CHINENSE	S	.	0.0	16.2	19.1	.	.	.	.	.	.	.	.	.	.	.	.	.	20
4272	LYCOPERSICON ESCULENTUM	L	40	86.8	2.8	0.6	7.9	0.1	1.9	433	86	1.3	.	.	.	0.08	0.47	.	175	1
4273	LYCOPERSICON ESCULENTUM	F	22	93.2	1.0	0.2	4.9	0.4	0.7	29	62	1.7	.	.	1040	0.05	0.04	0.7	50	1
4274	LYCOPERSICON ESCULENTUM	F	20	93.8	1.2	0.3	4.2	0.7	0.5	7	30	0.6	4	235	505	0.06	0.04	0.6	23	2
4275	LYCOPERSICON ESCULENTUM	F	32	90.4	1.6	0.4	6.8	1.2	0.8	15	35	0.6	.	.	660	0.13	0.08	0.8	36	3
4276	LYCOPERSICON ESCULENTUM	F	21	93.8	0.8	0.3	4.6	0.6	0.5	7	24	0.6	.	.	360	0.06	0.05	0.7	23	3
4277	LYCOPERSICON ESCULENTUM	F	21	93.6	1.2	0.2	4.6	0.6	0.4	6	20	0.6	.	.	90	0.06	0.04	0.4	18	3
4278	LYCOPERSICON ESCULENTUM	G	24	93.0	1.2	0.2	5.1	0.5	0.5	13	27	0.5	3	244	162	0.06	0.04	0.5	20	4

TABLE 2 - DATA DIRECTLY TRANSCRIBED FROM ORIGINAL SOURCES.

OBS	PLANA	PART	CALORIES	WATER	PROTEIN	FAT	TOTCARB	FIBER	ASH	CALCIUM	PHOSPHO	IRON	SODIUM	POTASS	CAROTE	THIAMINE	RIBOFLAV	NIACIN	ASCORBIC	REFCO
4279	LYCOPERSICON ESCULENTUM	F	22	93.5	1.1	0.2	4.7	0.5	0.5	13	27	0.5	3	244	540	0.06	0.04	0.70	23	4
4280	LYCOPERSICON ESCULENTUM	S	.	9.0	27.6	24.4	35.0	13.6	4.0	.	.	.	.	.	.	.	.	.	.	5
4281	LYCOPERSICON ESCULENTUM	W	.	0.0	26.4	1.7	45.3	15.4	26.6	.	.	.	.	.	.	.	.	.	.	6
4282	LYCOPUS ASPER	S	.	0.0	16.6	25.4	.	.	6.7	.	.	.	.	.	.	.	.	.	.	19
4283	LYCOPUS EUROPAEUS	S	.	0.0	19.3	33.0	.	.	.	.	.	.	.	.	.	.	.	.	.	21
4284	LYCOPUS EUROPAEUS	S	.	0.0	15.9	26.2	.	.	.	.	.	.	.	.	.	.	.	.	.	21
4285	LYSILOMA ACAPULCENSIS	S	.	0.0	40.6	9.1	.	.	.	.	.	.	.	.	.	.	.	.	.	20
4286	LYSILOMA BAHAMENSIS	S	.	0.0	25.0	4.9	.	.	.	.	.	.	.	.	.	.	.	.	.	19
4287	LYSILOMA DIVARICATA	S	.	0.0	33.1	12.9	.	.	4.7	.	.	.	.	.	.	.	.	.	.	19
4288	LYTHRUM SALICARIA	S	.	0.0	16.3	10.9	.	.	.	.	.	.	.	.	.	.	.	.	.	21
4289	MACADAMIA INTEGRIFOLIA	F	.	0.0	9.9	70.1	18.6	9.2	1.4	100	250	.	.	.	.	.	.	.	.	18
4290	MACADAMIA TENIFOLIA	S	691	3.0	7.8	71.6	15.9	2.5	1.7	48	161	2.0	.	264	0	0.34	0.11	1.30	0	4
4291	MACADAMIA TENIFOLIA	S	.	3.1	8.7	71.4	15.1	2.5	1.7	.	.	.	.	.	.	.	.	.	.	5
4292	MACHAERANTHERA TANACETIFOLIA	S	.	0.0	17.8	19.0	.	.	.	.	.	.	.	.	.	.	.	.	.	21
4293	MACHAERIUM ROSESCENS	S	.	0.0	41.9	23.1	.	.	6.2	.	.	.	.	.	.	.	.	.	.	19
4294	MACLEYA CORDATA VAR. JAPONICA	S	.	0.0	17.6	40.7	.	.	5.2	.	.	.	.	.	.	.	.	.	.	19
4295	MACLURA POMIFERA	S	.	0.0	38.2	44.5	.	.	3.0	.	.	.	.	.	.	.	.	.	.	19
4296	MACLURA POMIFERA	S	.	0.0	6.3	8.6	.	.	4.5	.	.	.	.	.	.	.	.	.	.	19
4297	MACLURA POMIFERA	S	.	0.0	46.9	35.8	.	.	3.4	.	.	.	.	.	.	.	.	.	.	19
4298	MACLURA POMIFERA	L	.	0.0	23.4	1.9	.	.	10.0	.	.	.	.	.	.	.	.	.	.	5
4299	MACROPTILIUM ATROPURPUREUM	W	.	0.0	23.0	3.1	60.1	30.4	13.7	1420	210	.	.	.	.	.	.	.	.	18
4300	MACROTYLOMA UNIFLORUM	S	333	9.7	22.5	1.0	60.5	4.7	6.3	.	.	.	.	.	.	.	.	.	.	1
4301	MACROTYLOMA UNIFLORUM	S	354	9.0	28.9	4.1	53.8	.	4.2	294	.	8.2	.	.	.	.	.	.	.	1
4302	MACROTYLOMA UNIFLORUM	S	.	11.8	22.0	0.5	62.6	5.3	3.1	280	390	7.6	.	.	71	.	.	1.50	.	5
4303	MADHUCA BUTYRACEA	S	.	5.0	5.2	55.9	30.0	.	3.8	.	.	.	.	.	.	.	.	.	.	5
4304	MADHUCA INDICA	L	.	0.0	9.1	4.1	79.0	18.7	7.8	1530	220	.	.	.	.	.	.	.	.	18
4305	MADHUCA INDICA	I	.	18.6	4.4	0.5	72.9	1.7	2.7	140	140	15.0	.	.	23	0.03	0.88	5.20	7	5
4306	MADHUCA INDICA	F	.	73.6	1.4	1.6	22.7	.	0.7	45	22	1.1	.	.	307	.	.	.	40	5
4307	MADHUCA LONGIFOLIA	I	.	0.0	5.0	1.8	89.0	1.6	4.2	130	120	.	.	.	.	.	.	.	.	18
4308	MADHUCA LONGIFOLIA	L	.	0.0	9.1	3.9	79.4	19.0	7.6	1460	210	.	.	.	.	.	.	.	.	18
4309	MADIA SP.	W	.	0.0	13.0	.	.	18.0	11.1	1540	340	.	.	3270	.	.	.	.	.	6
4310	MADIA SP.(FLOWERING)	W	.	0.0	6.7	.	.	19.3	6.9	1640	470	.	.	4930	.	.	.	.	.	6
4311	MADIA SP.(IMMATURE)	W	.	0.0	22.9	.	.	12.8	15.7	2050	520	.	.	5230	.	.	.	.	.	6
4312	MADIA SP.(MATURE)	W	.	0.0	5.5	.	.	22.2	7.0	1040	110	.	.	1090	.	.	.	.	.	6
4313	MAERUA ANGOLENSIS	L	60	80.0	5.4	0.6	11.7	1.7	2.3	420	84	.	.	.	.	.	.	.	.	1
4314	MAERUA CRASSIFOLIA	L	.	0.0	28.8	2.7	55.1	10.4	13.4	2060	170	.	.	.	.	.	.	.	.	18
4315	MAERUA CRASSIFOLIA	SH	.	0.0	11.3	1.5	77.1	33.7	10.1	2090	130	.	.	.	.	.	.	.	.	18
4316	MAGNOLIA GRANDIFLORA	S	.	0.0	8.8	56.4	.	.	.	.	.	.	.	.	.	.	.	.	.	20
4317	MAGNOLIA GRANDIFLORA	S	.	6.9	9.1	46.2	35.8	19.1	2.0	.	.	.	.	.	.	.	.	.	.	5
4318	MAGNOLIA MACROPHYLLA	S	.	0.0	15.6	46.1	.	.	.	.	.	.	.	.	.	.	.	.	.	20
4319	MAHONIA TRIFOLIATA	L	.	0.0	11.0	2.6	83.2	32.6	3.2	670	160	.	.	800	.	.	.	.	.	6
4320	MAJORANA HORTENSIS	H	271	7.6	12.7	7.0	60.6	18.1	12.1	1990	306	82.7	77	1522	4841	0.29	0.32	4.12	51	16
4321	MAJORANA HORTENSIS	S	.	0.0	31.5	39.7	.	.	5.2	.	.	.	.	.	.	.	.	.	.	19
4322	MALABALIA SEKAKUL	S	.	0.0	16.8	16.0	.	.	.	.	.	.	.	.	.	.	.	.	.	21
4323	MALACHRA ALCEAEFOLIA	S	.	0.0	23.4	17.7	.	.	.	.	.	.	.	.	.	.	.	.	.	21

TABLE 2 - DATA DIRECTLY TRANSCRIBED FROM ORIGINAL SOURCES.

OBS	PLNA	PART	CALORIES	WATER	PROTEIN	FAT	TOTCARB	FIBER	ASH	CALCIUM	PHOSPHO	IRON	SODIUM	POTASS	CAROT	THIAMINE	RIBOFLAV	NIACIN	ASCORBIC	REFCO
4324	MALACHRA CAPITATA	S	.	0.0	22.5	15.5	.	.	.	.	.	.	.	.	.	.	.	.	.	20
4325	MALCOLMIA AFRICANA	S	.	0.0	32.4	30.6	.	.	.	.	.	.	.	.	.	.	.	.	.	21
4326	MALCOLMIA AFRICANA	S	.	0.0	37.0	29.8	.	.	.	.	.	.	.	.	.	.	.	.	.	21
4327	MALCOLMIA MARITIMA	S	.	0.0	23.8	20.4	.	.	.	.	.	.	.	.	.	.	.	.	.	20
4328	MALLOTUS PHILIPPINENSIS	S	.	0.0	18.5	25.8	.	.	.	.	.	.	.	.	.	.	.	.	.	21
4329	MALLOTUS PHILIPPINENSIS	S	.	0.0	20.0	23.7	.	.	.	.	.	.	.	.	.	.	.	.	.	21
4330	MALOPE TRIFIDA	S	.	0.0	23.1	16.2	.	.	.	.	.	.	.	.	.	.	.	.	.	20
4331	MALOPE TRIFIDA	S	.	0.0	26.9	17.5	.	.	.	.	.	.	.	.	.	.	.	.	.	20
4332	MALPIGHIA GLABRA	F	39	83.2	1.8	1.0	6.8	1.2	7.2	27	27	1.2	.	.	20	0.03	0.04	0.50	21	3
4333	MALPIGHIA UMBELLATA	S	.	0.0	41.9	45.8	.	.	.	.	.	.	.	.	.	.	.	.	.	19
4334	MALPIGHIA UMBELLATA	S	.	0.0	10.6	1.4	.	.	.	.	.	.	.	.	.	.	.	.	.	19
4335	MALUS SPECTABILIS	F	89	75.0	0.2	0.2	24.1	1.7	0.5	66	6	1.3	.	.	460	0.01	0.02	0.20	2	2
4336	MALUS SYLVESTRIS	F	58	84.0	0.2	0.4	15.1	0.8	0.3	3	7	0.2	.	.	80	.	.	.	1	1
4337	MALUS SYLVESTRIS	F	51	85.8	0.4	0.3	13.2	0.6	0.3	10	10	0.5	2	130	20	0.02	0.03	0.20	4	2
4338	MALUS SYLVESTRIS	F	58	84.0	0.3	0.3	15.2	0.7	0.2	6	10	0.4	.	.	20	0.03	0.05	0.20	6	3
4339	MALUS SYLVESTRIS	F	58	84.4	0.2	0.6	14.5	1.0	0.3	7	10	0.3	1	110	54	0.03	0.02	0.10	4	4
4340	MALVA AEGYPTIA	S	.	0.0	25.4	17.3	.	.	.	.	.	.	.	.	.	.	.	.	.	21
4341	MALVA MONTANA	S	.	0.0	26.4	18.4	.	.	.	.	.	.	.	.	.	.	.	.	.	21
4342	MALVA NEGLECTA	S	.	0.0	21.0	15.2	.	.	.	.	.	.	.	.	.	.	.	.	.	21
4343	MALVA NICAEENSIS	S	.	0.0	20.8	26.1	.	.	.	.	.	.	.	.	.	.	.	.	.	21
4344	MALVA PARVIFLORA	L	36	86.3	4.8	0.2	6.4	1.5	2.3	324	67	.	.	.	.	.	.	.	65	1
4345	MALVA PARVIFLORA	S	.	0.0	21.2	13.0	.	.	.	.	.	.	.	.	.	.	.	.	.	19
4346	MALVA PARVIFLORA	S	.	0.0	19.0	14.4	.	.	.	.	.	.	.	.	.	.	.	.	.	21
4347	MALVA PARVIFLORA	S	.	0.0	30.4	14.7	.	.	.	.	.	.	.	.	.	.	.	.	.	21
4348	MALVA PUSILLA	L	68	78.6	5.5	1.5	11.8	8.9	2.6	.	.	.	.	.	.	.	.	.	.	1
4349	MALVA SP.	S	.	0.0	15.6	11.0	.	.	.	.	.	.	.	.	.	.	.	.	.	20
4350	MALVA SP.	S	.	0.0	17.5	11.3	.	.	.	.	.	.	.	.	.	.	.	.	.	20
4351	MALVA SP.	S	.	0.0	23.6	16.2	.	.	.	.	.	.	.	.	.	.	.	.	.	21
4352	MALVA SP.	L	37	86.7	4.4	0.6	5.9	1.6	2.4	249	69	12.7	.	.	4380	0.13	0.20	1.00	35	3
4353	MALVA SYLVESTRIS	L	28	91.6	3.6	1.4	2.1	0.9	1.3	90	42	3.7	.	.	3315	0.17	0.29	0.50	24	2
4354	MALVA SYLVESTRIS	S	.	0.0	19.0	14.2	.	.	.	.	.	.	.	.	.	.	.	.	.	21
4355	MALVA SYLVESTRIS	S	.	0.0	19.8	9.7	.	.	.	.	.	.	.	.	.	.	.	.	.	21
4356	MALVA SYLVESTRIS	L	.	92.0	2.1	.	.	.	1.7	221	36	9.7	.	.	5	0.10	0.24	0.44	.	5
4357	MALYASTRUM COROMANDELIANUM	S	.	0.0	26.0	13.7	.	.	.	.	.	.	.	.	.	.	.	.	.	21
4358	MALYASTRUM SPICATUM	S	.	0.0	26.2	15.2	.	.	.	.	.	.	.	.	.	.	.	.	.	21
4359	MAMMEA AFRICANA	S	415	5.8	2.0	9.0	81.8	3.1	1.4	339	169	.	.	.	.	.	.	.	.	1
4360	MAMMEA AMERICANA	F	47	86.8	0.6	0.2	12.1	1.0	0.3	13	12	0.4	.	.	180	0.03	0.05	0.40	16	3
4361	MAMMEA AMERICANA	F	51	86.2	0.5	0.5	12.5	1.0	0.3	11	11	0.7	15	47	138	0.02	0.04	0.40	14	4
4362	MAMMEA AMERICANA	F	.	86.5	0.6	0.3	12.3	1.5	0.3	13	.	0.4	.	.	168	0.02	0.04	0.50	15	5
4363	MANDRAGORA OFFICINARUM	S	.	0.0	22.1	22.6	.	.	.	.	.	.	.	.	.	.	.	.	.	21
4364	MANFREDA SP.	S	.	0.0	20.0	18.4	.	.	.	.	.	.	.	.	.	.	.	.	.	20
4365	MANFREDA SP.	S	.	0.0	20.2	22.4	.	.	.	.	.	.	.	.	.	.	.	.	.	21
4366	MANGIFERA CAESIA	F	48	86.5	1.0	0.2	11.9	.	0.4	10	24	0.0	.	.	5	0.08	.	.	58	2
4367	MANGIFERA INDICA	F	60	82.9	0.6	0.2	15.8	0.9	0.5	24	22	1.2	.	.	3200	0.03	0.05	0.40	42	1
4368	MANGIFERA INDICA	G	48	86.2	1.0	0.1	12.2	1.7	0.5	57	35	1.3	.	.	1000	0.03	0.06	0.30	270	1

TABLE 2 - DATA DIRECTLY TRANSCRIBED FROM ORIGINAL SOURCES.

OBS	PLANT	PART	CALORIES	WATER	PROTEIN	FAT	TOTLCARB	FIBER	ASH	CALCIUM	PHOSPHO	IRON	SODIUM	POTASS	CAROT	THIAMINE	RIBOFLAV	NIACIN	ASCORBIC	REFCO
4369	MANGIFERA INDICA	L	.	0.0	9.5	4.8	72.6	22.6	13.1	3140	160	.	.	.	.	.	.	.	.	18
4370	MANGIFERA INDICA	F	62	82.6	0.6	0.3	15.9	0.5	0.6	10	15	0.3	3	214	1880	0.06	0.05	0.6	36	2
4371	MANGIFERA INDICA	G	60	82.9	0.6	0.4	15.3	0.4	0.8	10	15	0.2	6	139	110	0.06	0.05	0.6	62	2
4372	MANGIFERA INDICA	L	54	82.1	3.9	0.2	12.1	1.6	1.7	19	77	2.8	.	.	610	0.06	0.09	1.5	27	2
4373	MANGIFERA INDICA	F	59	83.5	0.5	0.2	15.4	0.8	0.4	12	12	0.8	.	.	1260	0.05	0.06	0.4	53	3
4374	MANGIFERA INDICA	G	44	87.6	0.4	0.2	11.5	0.8	0.3	14	10	0.4	.	.	80	0.02	0.03	0.2	128	3
4375	MANGIFERA INDICA	L	66	81.7	0.7	0.4	16.8	0.9	0.4	10	13	0.4	7	189	2880	0.05	0.05	1.1	35	4
4376	MANGIFERA INDICA	L	.	78.2	3.0	0.4	16.5	1.6	1.9	29	72	6.2	.	.	894	0.04	0.06	2.2	53	5
4377	MANGIFERA ODORATA	F	70	79.9	0.9	0.1	18.5	0.8	0.6	4	18	0.2	.	.	360	0.04	0.06	0.7	13	3
4378	MANIHOT ESCULENTA	L	91	71.7	7.0	1.0	18.3	4.0	2.0	303	119	7.6	.	.	11775	0.25	0.60	2.4	311	1
4379	MANIHOT ESCULENTA	R	357	8.7	1.3	0.5	86.6	1.8	2.9	121	118	.	.	.	.	.	.	.	.	1
4380	MANIHOT ESCULENTA	R	149	62.0	1.2	0.2	35.7	1.1	0.9	68	42	1.9	.	.	30	0.04	0.05	0.6	31	1
4381	MANIHOT ESCULENTA	L	.	0.0	24.8	5.2	61.5	18.3	8.5	980	520	.	.	.	.	.	.	.	.	18
4382	MANIHOT ESCULENTA	R	.	0.0	2.4	0.7	94.0	2.0	2.9	140	20	.	.	.	.	.	.	.	.	18
4383	MANIHOT ESCULENTA	R	135	65.5	1.0	0.2	32.4	1.0	0.9	26	32	0.9	2	394	0	0.05	0.04	0.6	34	2
4384	MANIHOT ESCULENTA	L	60	81.0	6.9	1.3	9.2	2.1	1.6	144	68	2.8	4	409	8280	0.16	0.32	1.8	82	2
4385	MANIHOT ESCULENTA	R	148	60.6	0.8	0.3	37.4	1.0	0.9	36	48	1.1	.	.	10	0.06	0.04	0.7	40	3
4386	MANIHOT ESCULENTA	R	132	65.2	1.0	0.4	32.8	1.0	0.6	40	34	1.4	.	.	0	0.05	0.04	0.6	19	3
4387	MANIHOT ESCULENTA	L	80	77.2	6.8	1.4	12.8	2.4	1.8	206	86	2.0	.	.	60	0.12	0.27	1.7	290	3
4388	MANIHOT ESCULENTA	L	.	0.0	23.7	3.0	62.5	9.6	10.8	380	380	12.0	.	.	.	.	.	.	.	6
4389	MANIHOT ISOLOBA	S	.	0.0	21.2	32.0	.	.	.	.	.	.	.	.	.	.	.	.	.	20
4390	MANIHOT RUBRICAULIS	S	.	0.0	25.6	28.2	.	.	.	.	.	.	.	.	.	.	.	.	.	20
4391	MANILKARA ZAPOTA	F	76	79.3	0.4	0.7	19.1	2.2	0.5	27	11	0.6	3	181	25	0.00	0.00	0.2	13	2
4392	MANILKARA ZAPOTA	F	94	75.0	0.5	1.1	23.0	1.6	0.4	24	10	1.0	.	.	20	0.01	0.01	0.2	15	3
4393	MARAH GILENSIS	S	.	0.0	27.8	57.1	.	.	2.6	.	.	.	.	.	.	.	.	.	.	19
4394	MARAH GILENSIS	S	.	0.0	3.8	2.8	.	.	1.1	.	.	.	.	.	.	.	.	.	.	19
4395	MARAH GILENSIS	S	.	0.0	30.0	56.7	.	.	.	.	.	.	.	.	.	.	.	.	.	20
4396	MARAH MACROCARPA	S	.	0.0	29.4	54.4	.	.	3.0	.	.	.	.	.	.	.	.	.	.	19
4397	MARANTA ARUNDINACEA	R	125	67.4	1.7	0.2	29.5	2.0	1.2	15	18	1.9	.	.	.	0.13	0.02	0.5	7	2
4398	MARANTA ARUNDINACEA	R	157	57.2	2.4	0.1	39.0	1.9	1.3	20	24	3.2	.	.	0	0.08	0.03	0.7	9	3
4399	MARRUBIUM PESTALOZZAE	S	.	0.0	24.4	34.0	.	.	.	.	.	.	.	.	.	.	.	.	.	21
4400	MARRUBIUM VULGARE	S	.	0.0	25.0	35.1	.	.	.	.	.	.	.	.	.	.	.	.	.	19
4401	MARRUBIUM VULGARE	S	.	0.0	24.3	34.0	.	.	.	.	.	.	.	.	.	.	.	.	.	20
4402	MARRUBIUM VULGARE	S	.	0.0	23.3	37.2	.	.	.	.	.	.	.	.	.	.	.	.	.	20
4403	MARRUBIUM VULGARE	S	.	0.0	18.5	26.4	.	.	.	.	.	.	.	.	.	.	.	.	.	21
4404	MARRUBIUM VULGARE	S	.	0.0	25.6	36.9	.	.	.	.	.	.	.	.	.	.	.	.	.	21
4405	MARSDENIA EDULIS	S	.	0.0	27.4	47.0	.	.	3.4	.	.	.	.	.	.	.	.	.	.	19
4406	MARSHALLIA CAESPITOSE VAR. SIGNATA	S	.	0.0	31.1	25.6	.	.	.	.	.	.	.	.	.	.	.	.	.	20
4407	MARSILEA CRENATA	L	27	93.0	1.0	1.2	4.0	3.3	0.8	48	0	.	.	.	7300	0.10	0.07	3.4	3	2
4408	MARTINEZIA LINDENIANA	S	.	0.0	7.8	55.3	.	.	2.0	.	.	.	.	.	.	.	.	.	.	19
4409	MARTINEZIA LINDENIANA	S	.	0.0	2.2	0.6	.	.	1.0	.	.	.	.	.	.	.	.	.	.	19
4410	MARTINEZIA LINDENIANA	S	.	0.0	7.1	56.5	.	.	2.5	.	.	.	.	.	.	.	.	.	.	19
4411	MARTYNIA LOUISIANICA	S	.	0.0	25.0	59.8	.	.	.	.	.	.	.	.	.	.	.	.	.	19
4412	MARTYNIA PARVIFLORA	S	.	0.0	30.0	47.9	.	.	.	.	.	.	.	.	.	.	.	.	.	19
4413	MARTYNIA PARVIFLORA	S	.	0.0	15.0	3.5	.	.	.	.	.	.	.	.	.	.	.	.	.	19

TABLE 2 - DATA DIRECTLY TRANSCRIBED FROM ORIGINAL SOURCES.

OBS	PLNA	PART	CALORIES	WATER	PROTEIN	FAT	TOTCARB	FIBER	ASH	CALCIUM	PHOSPHO	IRON	SODIUM	POTASS	CAROT	THIAMINE	RIBOFLAV	NIACIN	ASCORBIC	REFCO
4414	MARTYNIA SP.	S	.	0.0	31.9	56.5	.	.	.	.	.	.	.	.	.	.	.	.	.	19
4415	MASTICHODENDRON CAPIRI	L	132	64.6	2.2	1.7	30.7	1.7	0.8	74	24	.	.	.	220	0.08	0.06	0.50	149	3
4416	MATRICARIA CAPENSIS	S	.	0.0	26.2	26.6	.	.	.	.	.	.	.	.	.	.	.	.	.	19
4417	MATTHIOLA BICORNIS	S	.	0.0	35.3	29.4	.	.	4.5	.	.	.	.	.	.	.	.	.	.	19
4418	MATTHIOLA BICORNIS	S	.	0.0	33.1	28.8	.	.	.	.	.	.	.	.	.	.	.	.	.	19
4419	MATTHIOLA FRUTICULOSA	S	.	0.0	33.6	33.4	.	.	.	.	.	.	.	.	.	.	.	.	.	21
4420	MATTHIOLA LONGIPETALA SUBSP. BICORNIS	S	.	0.0	30.0	33.1	.	.	.	.	.	.	.	.	.	.	.	.	.	21
4421	MAURITIA VINIFERA	F	143	72.8	3.0	10.5	12.5	11.4	1.2	113	19	3.5	.	.	24750	0.03	0.23	0.70	26	3
4422	MAXIMOWICZIA SONORAE	S	.	0.0	26.9	28.5	.	.	.	.	.	.	.	.	.	.	.	.	.	20
4423	MEDICAGO ARABICA	S	.	0.0	43.7	8.4	.	.	4.5	.	.	.	.	.	.	.	.	.	.	19
4424	MEDICAGO ARABICA	S	.	0.0	38.2	8.2	.	.	3.6	.	.	.	.	.	.	.	.	.	.	19
4425	MEDICAGO ARABICA	S	.	0.0	11.4	1.6	.	.	6.4	.	.	.	.	.	.	.	.	.	.	19
4426	MEDICAGO CILIARIS	S	.	0.0	38.6	10.3	.	.	.	.	.	.	.	.	.	.	.	.	.	21
4427	MEDICAGO DENTICULATA	L	284	0.0	35.8	2.2	50.0	6.7	11.9	992	462	380.0	.	2342	18015	1.34	2.09	5.22	731	10
4428	MEDICAGO FALCATA	S	.	0.0	42.7	9.9	.	.	3.9	.	.	.	.	.	.	.	.	.	.	19
4429	MEDICAGO FALCATA	S	.	0.0	36.8	10.8	.	.	3.9	.	.	.	.	.	.	.	.	.	.	19
4430	MEDICAGO FALCATA	S	.	0.0	42.5	9.8	.	.	4.3	.	.	.	.	.	.	.	.	.	.	19
4431	MEDICAGO HISPIDA	S	.	0.0	35.9	6.6	.	.	8.5	.	.	.	.	.	.	.	.	.	.	19
4432	MEDICAGO HISPIDA	S	.	0.0	36.5	7.3	.	.	3.0	.	.	.	.	.	.	.	.	.	.	19
4433	MEDICAGO HISPIDA	S	.	0.0	9.1	1.0	.	.	2.5	.	.	.	.	.	.	.	.	.	.	19
4434	MEDICAGO HISPIDA	S	.	0.0	37.9	7.3	.	.	3.8	.	.	.	.	.	.	.	.	.	.	19
4435	MEDICAGO LITTORALIS	S	.	0.0	46.3	12.5	.	.	.	.	.	.	.	.	.	.	.	.	.	21
4436	MEDICAGO LUPULINA	W	.	0.0	25.2	3.3	58.8	21.3	12.7	1610	410	.	.	.	.	.	.	.	.	18
4437	MEDICAGO LUPULINA	S	.	0.0	41.7	6.6	.	.	3.5	.	.	.	.	.	.	.	.	.	.	19
4438	MEDICAGO LUPULINA	S	.	0.0	37.9	5.7	.	.	3.2	.	.	.	.	.	.	.	.	.	.	19
4439	MEDICAGO LUPULINA	S	.	0.0	33.4	5.0	.	.	6.2	.	.	.	.	.	.	.	.	.	.	19
4440	MEDICAGO LUPULINA	S	.	0.0	29.9	4.5	.	.	.	.	.	.	.	.	.	.	.	.	.	21
4441	MEDICAGO LUPULINA	L	.	72.8	7.0	0.7	16.8	4.3	2.8	.	.	.	.	.	.	.	.	.	.	5
4442	MEDICAGO LUPULINA	W	.	0.0	23.3	3.3	63.2	24.7	10.2	960	320	.	.	2280	.	.	.	.	.	6
4443	MEDICAGO MARINA	S	.	0.0	38.6	14.9	.	.	.	.	.	.	.	.	.	.	.	.	.	21
4444	MEDICAGO MARINA	S	.	0.0	36.8	13.4	.	.	.	.	.	.	.	.	.	.	.	.	.	21
4445	MEDICAGO MINIMA	S	.	0.0	48.6	7.8	.	.	.	.	.	.	.	.	.	.	.	.	.	21
4446	MEDICAGO MINIMA	S	.	0.0	46.2	8.6	.	.	.	.	.	.	.	.	.	.	.	.	.	21
4447	MEDICAGO MINIMA	L	.	.	18.0	2.4	69.3	28.1	10.3	1750	710	.	.	.	.	.	.	.	.	5
4448	MEDICAGO ORBICULARIS	S	.	0.0	31.4	5.2	.	.	5.7	.	.	.	.	.	.	.	.	.	.	19
4449	MEDICAGO ORBICULARIS	S	.	0.0	29.1	3.3	.	.	4.6	.	.	.	.	.	.	.	.	.	.	19
4450	MEDICAGO ORBICULARIS	S	.	0.0	36.6	3.4	.	.	.	.	.	.	.	.	.	.	.	.	.	21
4451	MEDICAGO ORBICULARIS	S	.	0.0	29.9	3.8	.	.	.	.	.	.	.	.	.	.	.	.	.	21
4452	MEDICAGO POLYMORPHA	W	.	0.0	26.7	3.6	55.2	21.4	14.5	1620	470	.	.	.	.	.	.	.	.	18
4453	MEDICAGO POLYMORPHA	S	.	0.0	17.4	2.5	72.7	38.6	7.4	1830	190	.	.	.	.	.	.	.	.	18
4454	MEDICAGO POLYMORPHA	L	38	86.6	4.8	0.3	6.7	0.9	1.6	133	62	5.1	.	314	2415	0.18	0.28	0.70	98	2
4455	MEDICAGO POLYMORPHA	L	.	79.2	5.1	1.7	11.7	3.9	2.3	.	.	.	.	.	.	.	.	.	850	5
4456	MEDICAGO POLYMORPHA	F	.	.	22.9	3.1	69.4	31.6	4.7	.	610	.	.	.	.	.	.	.	.	5
4457	MEDICAGO POLYMORPHA	H	.	0.0	19.0	2.8	68.5	25.1	9.7	.	.	.	.	.	.	.	.	.	.	6
4458	MEDICAGO POLYMORPHA(IMMATURE)	H	.	0.0	26.0	2.3	58.3	23.1	13.4	.	.	.	.	.	.	.	.	.	.	6

TABLE 2 - DATA DIRECTLY TRANSCRIBED FROM ORIGINAL SOURCES.

OBS	PLNA	PART	CALORIES	WATER	PROTEIN	FAT	TOTLCARB	FIBER	ASH	CALCIUM	PHOSPHO	IRON	SODIUM	POTASS	CAROT	THIAMINE	RIBOFLAV	NIACIN	ASCORBIC	REFCO
4459	MEDICAGO POLYMORPHA(MATURE)	H	.	0.0	17.7	2.7	71.0	25.2	8.6	.	.	.	.	.	.	.	.	.	.	6
4460	MEDICAGO RIGIDULA	S	.	0.0	47.4	9.4	.	.	.	.	.	.	.	.	.	.	.	.	.	21
4461	MEDICAGO SATIVA	W	.	0.0	25.3	2.9	60.0	23.5	11.8	2410	350	.	.	.	.	.	.	.	.	18
4462	MEDICAGO SATIVA	L	.	0.0	21.7	2.4	58.9	20.1	17.0	1440	200	.	.	.	.	.	.	.	.	18
4463	MEDICAGO SATIVA	H	.	0.0	18.7	2.6	68.9	34.4	9.8	1260	180	.	.	.	.	.	.	.	.	18
4464	MEDICAGO SATIVA	S	.	0.0	37.0	12.3	.	.	3.7	.	.	.	.	.	.	.	.	.	.	19
4465	MEDICAGO SATIVA	S	.	0.0	36.9	11.3	.	.	3.7	.	.	.	.	.	.	.	.	.	.	19
4466	MEDICAGO SATIVA	S	.	0.0	38.5	10.1	.	.	.	.	.	.	.	.	.	.	.	.	.	21
4467	MEDICAGO SATIVA	S	.	0.0	38.0	10.6	.	.	.	.	.	.	.	.	.	.	.	.	.	21
4468	MEDICAGO SATIVA	L	52	82.7	6.0	0.4	9.5	3.1	1.4	12	51	5.4	.	.	6820	0.13	0.14	0.50	162	3
4469	MEDICAGO SATIVA	H	.	0.0	17.3	2.1	71.7	31.4	8.9	1640	260	24.0	.	1770	6	0.35	1.47	4.00	.	6
4470	MEDICAGO SATIVA	W	.	0.0	19.3	3.0	68.7	27.4	9.0	1720	310	33.0	.	2030	20	0.64	1.61	4.42	.	6
4471	MEDICAGO SCUTELLATA	S	.	0.0	33.5	11.5	.	.	4.9	.	.	.	.	.	.	.	.	.	.	19
4472	MEDICAGO SCUTELLATA	S	.	0.0	34.5	12.4	.	.	4.4	.	.	.	.	.	.	.	.	.	.	19
4473	MEDICAGO SCUTELLATA	S	.	0.0	39.2	9.8	.	.	.	.	.	.	.	.	.	.	.	.	.	21
4474	MEDICAGO SP.	S	.	0.0	34.9	10.3	.	.	5.1	.	.	.	.	.	.	.	.	.	.	19
4475	MEDICAGO TRIBULOIDES	S	.	0.0	43.8	15.9	.	.	3.2	.	.	.	.	.	.	.	.	.	.	19
4476	MEDICAGO TRUNCATULA	W	.	0.0	25.8	3.1	55.9	24.0	15.2	2130	460	.	.	.	.	.	.	.	.	18
4477	MEDICAGO TURBINATA	S	.	0.0	44.6	10.6	.	.	.	.	.	.	.	.	.	.	.	.	.	20
4478	MEDICAGO TURBINATA	S	.	0.0	40.8	15.5	.	.	.	.	.	.	.	.	.	.	.	.	.	21
4479	MEDICAGO TURBINATA	S	.	0.0	47.9	15.1	.	.	.	.	.	.	.	.	.	.	.	.	.	21
4480	MELAMPODIUM PERFOLIATUM	S	.	0.0	11.2	15.1	.	.	.	.	.	.	.	.	.	.	.	.	.	20
4481	MELAMPODIUM PERFOLIATUM	S	.	0.0	8.8	3.8	.	.	.	.	.	.	.	.	.	.	.	.	.	20
4482	MELAPODIUM LEUCANTHUM	W	.	0.0	13.5	10.0	66.6	15.9	9.9	1730	180	.	.	2900	.	.	.	.	.	6
4483	MELHANIA FUTTEYPORENSIS	S	.	0.0	19.3	4.8	.	.	.	.	.	.	.	.	.	.	.	.	.	21
4484	MELIA AZEDARACH	L	.	0.0	21.8	8.8	58.6	11.6	10.8	2310	220	.	.	.	.	.	.	.	.	18
4485	MELIA AZEDARACH	S	.	0.0	30.4	44.7	.	.	3.4	.	.	.	.	.	.	.	.	.	.	19
4486	MELIA AZEDARACH	S	.	0.0	28.7	44.0	.	.	.	.	.	.	.	.	.	.	.	.	.	21
4487	MELIA AZEDERACH	S	.	0.0	2.5	2.4	.	.	1.9	.	.	.	.	.	.	.	.	.	.	19
4488	MELICA CILIATA VAR. NEBRODENSIS	S	.	0.0	25.2	3.5	.	.	.	.	.	.	.	.	.	.	.	.	.	21
4489	MELICA IMPERFECTA	W	.	0.0	11.0	2.6	79.9	30.3	6.5	390	260	.	.	1940	.	.	.	.	.	6
4490	MELICOCCUS BIJUGATUS	F	.	0.0	10.1	1.3	85.9	10.8	2.7	.	.	.	.	.	.	.	.	.	.	18
4491	MELICOCCUS BIJUGATUS	S	.	0.0	11.6	1.2	84.5	11.2	2.7	.	.	.	.	.	.	.	.	.	.	18
4492	MELICOCCUS BIJUGATUS	F	77	78.4	1.1	0.2	19.9	1.4	0.4	12	20	0.6	.	.	25	0.05	0.02	0.70	5	2
4493	MELICOCCUS BIJUGATUS	F	59	78.4	1.1	0.2	19.9	1.4	0.4	12	20	0.6	.	.	50	0.05	0.02	0.70	5	3
4494	MELILOTUS ALBA	S	.	0.0	37.3	6.9	.	.	3.9	.	.	.	.	.	.	.	.	.	.	19
4495	MELILOTUS ALBA	S	.	0.0	40.6	5.6	.	.	4.0	.	.	.	.	.	.	.	.	.	.	19
4496	MELILOTUS ALBA	S	.	0.0	42.8	6.2	.	.	4.2	.	.	.	.	.	.	.	.	.	.	19
4497	MELILOTUS ALBA	S	.	0.0	35.7	7.5	.	.	4.5	.	.	.	.	.	.	.	.	.	.	19
4498	MELILOTUS ALBA	S	.	0.0	38.1	6.5	.	.	3.8	.	.	.	.	.	.	.	.	.	.	19
4499	MELILOTUS ALBA	S	.	0.0	39.1	5.9	.	.	.	.	.	.	.	.	.	.	.	.	.	21
4500	MELILOTUS ALBA	S	.	0.0	36.4	8.3	.	.	.	.	.	.	.	.	.	.	.	.	.	21
4501	MELILOTUS ALBA	S	.	0.0	39.4	7.6	26.7	12.9	4.1	.	.	.	.	.	.	.	.	.	.	5
4502	MELILOTUS ALBA VAR. ANNUA	S	.	0.0	39.5	6.7	.	.	4.3	.	.	.	.	.	.	.	.	.	.	19
4503	MELILOTUS INDICA	W	.	0.0	15.3	1.7	69.9	29.4	13.1	.	.	.	.	.	.	.	.	.	.	18

TABLE 2 - DATA DIRECTLY TRANSCRIBED FROM ORIGINAL SOURCES.

OBS	PLNA	PART	CALORIES	WATER	PROTEIN	FAT	TOTCARB	FIBER	ASH	CALCIUM	PHOSPHO	IRON	SODIUM	POTASS	CAROT	THIAMINE	RIBOFLAV	NIACIN	ASCORBIC	REFCO
4504	MELILOTUS INDICA	F	.	0.0	25.30	1.9	64.30	14.90	8.5	1340.0	350	.	.	.	.	.	.	.	.	18
4505	MELILOTUS INDICA	S	.	0.0	29.00	3.2	73.30	16.90	4.5	.	.	.	.	.	.	.	.	.	.	18
4506	MELILOTUS INDICA	S	.	0.0	32.40	4.6	.	.	6.9	.	.	.	.	.	.	.	.	.	.	19
4507	MELILOTUS INDICA	S	.	0.0	32.50	5.4	.	.	3.7	.	.	.	.	.	.	.	.	.	.	19
4508	MELILOTUS INDICUS	S	.	0.0	29.40	4.6	.	.	.	.	.	.	.	.	.	.	.	.	.	20
4509	MELILOTUS ITALICA	S	.	0.0	36.80	6.9	.	.	3.8	.	.	.	.	.	.	.	.	.	.	19
4510	MELILOTUS NEAPOLITANA	S	.	0.0	35.40	1.9	.	.	.	.	.	.	.	.	.	.	.	.	.	21
4511	MELILOTUS OFFICINALIS	S	.	0.0	41.20	6.5	.	.	4.0	.	.	.	.	.	.	.	.	.	.	19
4512	MELILOTUS OFFICINALIS	S	.	0.0	41.60	7.6	.	.	4.7	.	.	.	.	.	.	.	.	.	.	19
4513	MELILOTUS OFFICINALIS	S	.	0.0	37.60	7.3	.	.	4.6	.	.	.	.	.	.	.	.	.	.	19
4514	MELILOTUS OFFICINALIS	S	.	0.0	38.40	4.8	.	.	4.3	.	.	.	.	.	.	.	.	.	.	19
4515	MELILOTUS OFFICINALIS	S	.	0.0	33.90	7.0	.	.	4.4	.	.	.	.	.	.	.	.	.	.	19
4516	MELILOTUS OFFICINALIS	S	.	0.0	12.80	1.9	.	.	14.3	.	.	.	.	.	.	.	.	.	.	19
4517	MELILOTUS OFFICINALIS	S	.	0.0	34.90	7.5	.	.	.	.	.	.	.	.	.	.	.	.	.	20
4518	MELILOTUS SP.	S	.	0.0	37.10	4.6	.	.	3.8	.	.	.	.	.	.	.	.	.	.	19
4519	MELILOTUS SP.	H	.	0.0	16.30	2.2	72.80	32.20	8.7	1770.0	260	15.0	.	1340	12	.	.	.	.	6
4520	MELILOTUS SP.	W	.	0.0	17.80	2.8	71.20	29.40	8.2	1320.0	270	14.0	.	1650	27	0.53	8.38	3.61	.	6
4521	MELILOTUS TAURICA	S	.	0.0	41.20	7.4	.	.	6.1	.	.	.	.	.	.	.	.	.	.	19
4522	MELINIS MINUTIFLORA	W	.	0.0	9.00	3.0	81.20	36.50	7.8	.	.	.	.	.	.	.	.	.	.	18
4523	MELINIS MINUTIFLORA	H	.	0.0	13.80	4.1	74.70	32.30	7.4	.	.	.	.	.	.	.	.	.	.	18
4524	MELINUS MINUTIFLORA	W	.	0.0	4.60	2.5	85.20	38.60	7.7	.	.	.	.	.	.	.	.	.	.	6
4525	MELIOSMA MYRIANTHA	S	.	0.0	8.10	9.6	.	.	.	.	.	.	.	.	.	.	.	.	.	20
4526	MELISSA OFFICINALIS	S	.	0.0	29.30	11.5	.	.	.	.	.	.	.	.	.	.	.	.	.	21
4527	MELOCHIA HIRSUTA	S	.	0.0	19.40	7.3	.	.	.	.	.	.	.	.	.	.	.	.	.	20
4528	MENISPERMUM CANADENSE	S	.	0.0	13.10	16.0	.	.	2.6	.	.	.	.	.	.	.	.	.	.	19
4529	MENTHA ARVENSIS	S	.	0.0	17.80	23.0	.	.	8.9	.	.	.	.	.	.	.	.	.	.	19
4530	MENTHA ARVENSIS		.	81.8	.	.	.	.	.	67.3	.	7.2	.	.	.	0.04	0.12	.	15	9
4531	MENTHA PULEGIUM	L	.	0.0	10.00	.	.	.	11.1	944.0	.	.	541	1531	.	.	.	.	.	17
4532	MENTHA PULEGIUM	S	.	0.0	24.60	26.6	.	.	.	.	.	.	.	.	.	.	.	.	.	21
4533	MENTHA ROTUNDIFOLIA	L	.	0.0	14.40	.	.	.	9.7	1630.0	.	.	252	1710	.	.	.	.	.	17
4534	MENTHA ROTUNDIFOLIA	S	.	0.0	19.60	28.2	.	.	.	.	.	.	.	.	.	.	.	.	.	21
4535	MENTHA SP.	S	.	0.0	25.10	44.4	.	.	3.1	.	.	.	.	.	.	.	.	.	.	19
4536	MENTHA SP.	L	32	89.3	3.00	0.7	5.40	.	1.6	194.0	48	3.8	2	179	2160	0.13	0.16	0.70	64	2
4537	MENTHA SP.	L	45	86.5	2.70	1.3	7.70	1.10	1.8	138.0	41	6.7	.	.	5080	0.13	0.26	0.90	64	3
4538	MENTHA SP.	H	.	0.0	14.80	2.5	73.80	23.70	8.9	1710.0	220	.	.	.	.	.	.	.	.	6
4539	MENTHA SPICATA	L	.	83.0	4.80	0.6	10.00	2.00	1.6	200.0	80	15.6	.	.	1620	0.05	0.08	0.40	.	5
4540	MENTHA TOMENTOSA	S	.	0.0	19.10	28.4	.	.	.	.	.	.	.	.	.	.	.	.	.	21
4541	MENTZELIA ALBICAULIS	S	.	0.0	19.40	34.5	.	.	.	.	.	.	.	.	.	.	.	.	.	19
4542	MENTZELIA DECAPETALA	S	.	0.0	21.30	37.0	.	.	11.1	.	.	.	.	.	.	.	.	.	.	19
4543	MENTZELIA DISPERSA	S	.	0.0	22.50	46.5	.	.	.	.	.	.	.	.	.	.	.	.	.	21
4544	MENTZELIA NUDA	S	.	0.0	19.00	39.1	.	.	.	.	.	.	.	.	.	.	.	.	.	21
4545	MENTZELIA NUDA	S	.	0.0	21.20	39.3	.	.	.	.	.	.	.	.	.	.	.	.	.	21
4546	MENTZELLA DECAPETALA	S	.	0.0	21.40	42.4	.	.	12.5	.	.	.	.	.	.	.	.	.	.	19
4547	MERCURIALIS ANNUA	S	.	0.0	18.90	37.1	.	.	11.0	.	.	.	.	.	.	.	.	.	.	19
4548	MERREMIA PALMERI	S	.	0.0	18.20	9.0	.	.	.	.	.	.	.	.	.	.	.	.	.	21

TABLE 2 - DATA DIRECTLY TRANSCRIBED FROM ORIGINAL SOURCES.

OBS	PLANA	PART	CALORIES	WATER	PROTEIN	FAT	TOTCARB	FIBER	ASH	CALCIUM	PHOSPHO	IRON	SODIUM	POTASS	CAROT	THIAMINE	RIBOFLAV	NIACIN	ASCORBIC	REFC
4549	MERTENSIA SP.	W	.	0.0	25.30	3.3	69.80	17.10	11.6	.	.	.	.	.	.	.	.	.	.	6
4550	MESEMBRYANTHEMUM CRYSTALLINUM	S	.	0.0	31.20	7.8	.	.	5.0	.	.	.	.	.	.	.	.	.	.	19
4551	MESEMBRYANTHEMUM SP.		.	0.0	7.63	.	63.73	19.22	22.5	1350	100	79.3	.	.	.	.	.	.	.	8
4552	MESPILUS GERMANICA	S	.	0.0	1.20	0.7	.	.	.	.	.	.	.	.	.	.	.	.	.	20
4553	MESPILUS GERMANICA	S	.	0.0	45.40	40.7	.	.	.	.	.	.	.	.	.	.	.	.	.	21
4554	MESPILUS GERMANICA	S	.	0.0	3.30	2.3	.	.	.	.	.	.	.	.	.	.	.	.	.	21
4555	METROXYLON SP.	HE	357	13.1	1.40	0.2	85.90	0.20	0.4	15	10	1.4	8	36	0	0.01	.	.	0	2
4556	METROXYLON SP.	HE	.	12.2	0.20	0.2	87.10	.	0.3	10	10	1.3	.	.	.	.	.	0.2	.	5
4557	MICROLONCHUS DURIAEI	S	.	0.0	20.90	23.5	.	.	.	.	.	.	.	.	.	.	.	.	.	21
4558	MICROMERIA SERPYLLIFOLIA	S	.	0.0	20.40	27.9	.	.	.	.	.	.	.	.	.	.	.	.	.	21
4559	MIKANIA CORDATA	L	.	90.9	1.50	0.3	6.20	2.10	1.1	120	80	.	.	.	.	.	.	.	.	5
4560	MILLERIA QUINQUEFLORA	S	.	0.0	12.10	13.4	.	.	.	.	.	.	.	.	.	.	.	.	.	21
4561	MILLETTIA AURICULATA	L	.	0.0	22.70	4.6	63.40	32.50	9.3	1920	290	.	.	.	.	.	.	.	.	18
4562	MILLETTIA OVALIFOLIA	S	.	0.0	26.90	37.0	.	.	4.7	.	.	.	.	.	.	.	.	.	.	19
4563	MILLETTIA OVALIFOLIA	S	.	0.0	32.10	7.8	.	.	.	.	.	.	.	.	.	.	.	.	.	21
4564	MIMOSA ACULEATICARPA	S	.	0.0	34.40	6.2	.	.	.	.	.	.	.	.	.	.	.	.	.	20
4565	MIMOSA HAMATA	L	.	0.0	11.90	7.9	70.40	12.10	9.8	.	.	.	.	.	.	.	.	.	.	5
4566	MIMOSA HIMALAYANA	S	.	0.0	35.90	2.2	.	.	.	.	.	.	.	.	.	.	.	.	.	21
4567	MIMOSA INVISA	W	.	0.0	16.50	2.5	73.70	21.80	7.3	1100	320	.	.	.	.	.	.	.	.	18
4568	MIMOSA PALMERI	S	.	0.0	36.90	6.1	.	.	.	.	.	.	.	.	.	.	.	.	.	19
4569	MIMOSA PUDICA	P	159	54.7	1.80	1.0	40.20	4.30	2.3	212	30	.	.	.	.	.	.	.	.	
4570	MIMOSA PUDICA	S	.	0.0	30.40	2.8	.	.	3.6	.	.	.	.	.	.	.	.	.	.	19
4571	MIMOSA PUDICA	S	.	0.0	11.60	2.4	.	.	5.4	.	.	.	.	.	.	.	.	.	.	19
4572	MIMOSA PUDICA	L	22	93.0	1.40	0.0	5.10	.	0.5	35	53	1.4	.	.	6510	0.12	0.05	.	18	2
4573	MIMOSA PUDICA	L	.	74.9	5.50	0.8	17.20	8.20	1.6	.	.	.	.	.	.	.	.	.	.	5
4574	MIMOSA SP.	L	.	0.0	20.60	3.4	68.90	21.10	7.1	2380	170	.	.	1140	.	.	.	.	.	6
4575	MIMOSA URUGUENSIS	S	.	0.0	31.80	5.2	.	.	.	.	.	.	.	.	.	.	.	.	.	21
4576	MIMULUS GUTTATUS	S	.	0.0	20.20	40.2	.	.	.	.	.	.	.	.	.	.	.	.	.	21
4577	MIMULUS LEWISII	W	.	0.0	15.60	5.9	64.60	17.10	13.9	.	.	.	.	.	.	.	.	.	.	6
4578	MIRABILIS JALAPA	S	.	0.0	17.10	3.9	.	.	3.2	.	.	.	.	.	.	.	.	.	.	19
4579	MIRABILIS JALAPA	S	.	0.0	18.10	4.6	.	.	.	.	.	.	.	.	.	.	.	.	.	21
4580	MIRABILIS NYCTAGINEA	S	.	0.0	21.20	7.0	.	.	3.5	.	.	.	.	.	.	.	.	.	.	19
4581	MNESITHEA LAEVIS	L	.	0.0	3.90	0.8	86.60	44.20	8.7	.	.	.	.	.	.	.	.	.	.	5
4582	MODIOLA CAROLINIANA	S	.	0.0	19.30	18.6	.	.	.	.	.	.	.	.	.	.	.	.	.	20
4583	MODIOLA CAROLINIANA	S	.	0.0	18.60	16.1	.	.	.	.	.	.	.	.	.	.	.	.	.	21
4584	MODIOLA CAROLINIANA	L	15	91.5	2.20	.	2.70	.	3.6	263	39	6.0	.	.	1860	0.08	0.16	0.4	.	3
4585	MOLDAVICA PARVIFLORA	S	.	0.0	15.10	18.3	.	.	.	.	.	.	.	.	.	.	.	.	.	21
4586	MOLLUGO VERTICILLATA	S	.	0.0	13.10	15.6	.	.	.	.	.	.	.	.	.	.	.	.	.	20
4587	MOLTKIA CAERULEA	S	.	0.0	8.90	10.3	.	.	.	.	.	.	.	.	.	.	.	.	.	21
4588	MOLUCCELLA LAEVIS	S	.	0.0	27.50	33.4	.	.	.	.	.	.	.	.	.	.	.	.	.	20
4589	MOLUCCELLA LAEVIS	S	.	0.0	24.30	34.4	.	.	.	.	.	.	.	.	.	.	.	.	.	21
4590	MOLUCCELLA LAEVIS	S	.	0.0	25.10	37.2	.	.	.	.	.	.	.	.	.	.	.	.	.	21
4591	MOLUCCELLA SPINOSA	S	.	0.0	19.50	38.5	.	.	.	.	.	.	.	.	.	.	.	.	.	21
4592	MOMORDICA BALSAMINA	L	259	0.0	26.40	2.9	47.60	14.00	23.1	2157	270	.	19	193	.	.	.	10.3	.	12
4593	MOMORDICA BALSAMINA	F	246	0.0	20.80	2.1	49.80	17.60	9.8	1199	348	.	11	352	.	.	.	7.9	.	12

TABLE 2 - DATA DIRECTLY TRANSCRIBED FROM ORIGINAL SOURCES.

OBS	PLANA	PART	CALORIES	WATER	PROTEIN	FAT	TOTCARB	FIBER	ASH	CALCIUM	PHOSPHO	IRON	SODIUM	POTASS	CAROT	THIAMINE	RIBOFLAV	NIACIN	ASCORBIC	REFCO
4594	MOMORDICA BALSAMINA	W	.	0.0	24.8	1.0	58.5	21.1	15.7	1440	290	.	.	.	.	.	.	.	.	18
4595	MOMORDICA BALSAMINA	S	.	0.0	19.2	22.8	.	.	2.4	.	.	.	.	.	.	.	.	.	.	19
4596	MOMORDICA BALSAMINA	S	.	0.0	28.1	42.7	.	.	3.2	.	.	.	.	.	.	.	.	.	.	19
4597	MOMORDICA BALSAMINA	S	.	0.0	28.1	39.7	.	.	3.3	.	.	.	.	.	.	.	.	.	.	19
4598	MOMORDICA BALSAMINA	S	.	0.0	30.6	38.8	.	.	3.2	.	.	.	.	.	.	.	.	.	.	19
4599	MOMORDICA BALSAMINA	S	.	0.0	20.6	20.4	.	.	.	.	.	.	.	.	.	.	.	.	.	19
4600	MOMORDICA BALSAMINA	S	.	0.0	32.5	37.6	.	.	.	.	.	.	.	.	.	.	.	.	.	20
4601	MOMORDICA CHARANTIA	L	60	80.1	5.1	0.4	12.3	0.5	2.1	264	666	7.1	.	.	.	0.05	0.47	.	247	1
4602	MOMORDICA CHARANTIA	F	25	92.4	1.2	0.2	5.8	1.0	0.4	13	32	0.2	.	.	.	0.02	0.07	.	277	1
4603	MOMORDICA CHARANTIA	S	.	0.0	32.8	46.2	.	.	3.4	.	.	.	.	.	.	.	.	.	.	19
4604	MOMORDICA CHARANTIA	S	.	0.0	5.3	1.0	.	.	1.1	.	.	.	.	.	.	.	.	.	.	19
4605	MOMORDICA CHARANTIA	S	.	0.0	35.6	49.6	.	.	.	.	.	.	.	.	.	.	.	.	.	19
4606	MOMORDICA CHARANTIA	F	19	94.0	0.8	0.1	4.5	1.0	0.6	26	32	2.3	2	270	110	0.06	0.04	0.30	57	2
4607	MOMORDICA CHARANTIA	L	44	84.6	5.6	0.6	7.0	1.6	2.2	288	54	5.0	19	510	5085	0.13	0.46	1.50	170	2
4608	MOMORDICA CHARANTIA	S	.	0.0	36.2	48.6	.	.	.	.	.	.	.	.	.	.	.	.	.	20
4609	MOMORDICA CHARANTIA	S	.	0.0	21.3	31.6	.	.	.	.	.	.	.	.	.	.	.	.	.	21
4610	MOMORDICA CHARANTIA	F	21	93.4	0.9	0.2	4.7	1.2	0.8	26	34	1.7	.	.	50	0.08	0.06	0.30	51	3
4611	MOMORDICA CHARANTIA	L	.	82.5	6.5	0.5	.	1.8	2.6	297	53	3.3	.	.	6	0.14	0.55	1.80	210	5
4612	MOMORDICA CHARANTIA	F	.	83.2	2.9	1.0	11.5	1.7	1.4	50	140	9.4	.	.	126	0.07	0.09	0.50	88	5
4613	MOMORDICA COCHINCHINENSIS	F	.	84.1	2.6	0.7	11.6	5.9	1.0	21	148	2.6	.	.	.	.	.	.	247	5
4614	MOMORDICA DIOICA	F	.	84.1	3.1	1.0	10.7	3.0	1.1	33	42	4.6	.	.	1620	0.05	0.17	0.59	275	5
4615	MONANTHOCHLOE LITTORALIS		.	0.0	6.6	1.5	79.1	29.1	12.8	310	180	.	.	.	.	.	.	.	.	6
4616	MONARDA FISTULOSA	S	.	0.0	26.7	31.8	.	.	6.7	.	.	.	.	.	.	.	.	.	.	19
4617	MONARDA FISTULOSA VAR. MENTHAEFOLIA	S	.	0.0	27.5	36.5	.	.	.	.	.	.	.	.	.	.	.	.	.	21
4618	MONARDA PUNCTATA	S	.	0.0	21.3	31.3	.	.	.	.	.	.	.	.	.	.	.	.	.	20
4619	MONARDELLA SP.	S	.	0.0	19.4	26.8	.	.	.	.	.	.	.	.	.	.	.	.	.	19
4620	MONNINA WRIGHTII	S	.	0.0	10.0	13.7	.	.	.	.	.	.	.	.	.	.	.	.	.	20
4621	MONOCHORIA VAGINALIS	L	18	93.0	1.0	0.2	3.8	.	2.0	80	45	3.7	.	.	600	0.08	.	.	50	2
4622	MONODORA MYRSTICA	S	477	10.4	8.8	29.0	50.0	23.7	1.8	103	.	9.3	.	.	.	0.09	0.10	0.50	.	1
4623	MONTANOA ARBORESCENS	S	.	0.0	13.8	7.3	.	.	.	.	.	.	.	.	.	.	.	.	.	20
4624	MORINDA CITRIFOLIA	L	107	77.6	4.5	7.8	8.6	4.0	1.5	36	15	2.3	.	.	36265	0.16	0.09	7.20	52	2
4625	MORINGA OLEIFERA	L	78	73.9	8.2	0.6	14.7	2.1	2.6	571	149	11.4	.	.	11920	0.22	0.77	2.60	182	1
4626	MORINGA OLEIFERA	I	49	84.2	3.3	0.1	11.2	2.4	1.2	22	79	5.2	.	.	460	0.06	.	1.40	248	1
4627	MORINGA OLEIFERA	L	.	0.0	29.0	5.2	56.7	19.1	9.1	2060	240	.	.	.	.	.	.	.	.	18
4628	MORINGA OLEIFERA	F	.	0.0	20.7	1.0	69.4	27.0	8.9	.	.	.	.	.	.	.	.	.	.	18
4629	MORINGA OLEIFERA	L	72	77.6	7.4	1.5	11.6	1.2	1.9	297	90	3.6	4	473	8855	0.20	0.73	3.40	167	2
4630	MORINGA OLEIFERA	F	42	86.7	2.5	0.2	9.5	.	1.1	58	62	0.8	.	.	75	0.05	0.06	0.60	159	2
4631	MORINGA OLEIFERA	F	.	86.9	2.5	0.1	8.5	4.8	2.0	30	110	5.3	.	.	110	.	.	0.20	120	5
4632	MORINGA OLEIFERA	L	.	75.0	6.7	1.7	14.3	0.9	2.3	440	70	7.0	.	.	6780	0.21	.	0.80	220	5
4633	MORTONIA SCABRELLA	S	.	0.0	8.8	10.5	.	.	3.6	.	.	.	.	.	.	.	.	.	.	19
4634	MORUS ALBA	L	.	0.0	15.0	7.4	63.3	15.3	14.3	2420	240	.	.	.	.	.	.	.	.	18
4635	MORUS ALBA	F	53	85.0	1.7	0.4	12.2	0.9	0.7	30	32	3.7	37	152	10	0.03	0.06	0.70	5	2
4636	MORUS ALBA	L	.	0.0	14.0	6.8	49.7	.	13.8	2740	450	.	.	.	.	.	.	.	.	5
4637	MORUS ALBA	F	.	87.5	1.5	0.4	9.7	.	0.9	80	40	1.9	.	.	104	0.01	0.18	0.80	13	5
4638	MORUS NIGRA	L	.	0.0	17.6	11.5	50.5	7.4	20.4	2150	130	.	.	.	.	.	.	.	.	18

TABLE 2 - DATA DIRECTLY TRANSCRIBED FROM ORIGINAL SOURCES.

OBS	PLNA	PART	CALORIES	WATER	PROTEIN	FAT	TOTCARB	FIBER	ASH	CALCIUM	PHOSPHO	IRON	SODIUM	POTASS	CAROTE	THIAMINE	RIBOFLAV	NIACIN	ASCORBIC	REFCO
4639	MORUS NIGRA	F	42	87.9	1.4	0.3	9.8	0.7	0.6	24	26	3.0	30	123	15	0.04	0.08	0.7	39	2
4640	MORUS NIGRA	F	.	85.5	0.7	0.4	12.2	0.8	0.4	60	20	2.6	.	.	10	0.06	0.09	0.2	10	5
4641	MORUS RUBRA	S	.	0.0	25.6	37.6	.	.	.	.	.	.	.	.	.	.	.	.	.	20
4642	MORUS SP.	F	.	88.4	.	.	.	.	.	38	33	1.5	.	.	.	0.08	.	.	61	1
4643	MORUS SP.	L	.	0.0	18.0	5.7	61.3	11.2	15.0	4640	170	.	.	1980	.	.	.	.	.	6
4644	MUCUNA DEERINGIANA	S	352	11.4	21.2	4.4	59.5	4.6	3.5	132	320	.	.	.	.	0.12	0.10	3.0	2	1
4645	MUCUNA DEERINGIANA	W	.	0.0	15.5	4.3	74.6	34.4	5.6	1210	130	.	.	.	.	.	.	.	.	18
4646	MUCUNA DEERINGIANA	D	.	0.0	21.0	2.6	71.9	15.6	4.5	.	.	.	.	.	.	.	.	.	.	18
4647	MUCUNA DEERINGIANA	S	.	0.0	28.6	0.7	66.7	9.5	4.0	.	.	.	.	.	.	.	.	.	.	18
4648	MUCUNA DEERINGIANA	F	.	10.0	18.1	4.4	63.3	13.0	4.2	.	.	.	.	.	.	.	.	.	.	5
4649	MUCUNA DEERINGIANA	S	.	10.0	23.4	5.7	57.9	6.4	3.0	.	.	.	.	.	.	.	.	.	.	5
4650	MUCUNA DEERINGIANA	L	.	0.0	15.1	2.1	67.8	19.3	14.9	.	.	.	.	.	.	.	.	.	.	5
4651	MUCUNA HOLTONII	S	.	0.0	20.5	4.1	.	.	3.2	.	.	.	.	.	.	.	.	.	.	19
4652	MUCUNA PRURIENS	S	344	10.9	29.5	3.2	52.9	7.8	3.5	.	.	.	.	.	.	0.51	0.24	1.7	.	1
4653	MUCUNA PRURIENS	S	.	9.1	25.0	3.0	.	6.8	4.0	160	470	20.0	.	.	.	.	.	.	.	5
4654	MUCUNA SP.	S	360	10.6	24.9	5.6	55.6	4.4	3.3	.	.	.	.	.	.	.	.	.	.	1
4655	MUCUNA UTILIS	S	332	15.0	24.0	3.0	55.0	.	3.0	130	200	2.0	.	.	40	0.30	0.17	2.4	0	2
4656	MUHLENBERGIA ARENACEA	W	.	0.0	13.3	2.0	70.3	30.5	14.4	1040	330	.	.	.	.	.	.	.	.	6
4657	MUHLENBERGIA CAPILLARIS(IMMATURE)	W	.	0.0	9.9	1.9	78.4	35.0	9.8	430	280	.	.	.	.	.	.	.	.	6
4658	MUHLENBERGIA CAPILLARIS(MATURE)	W	.	0.0	3.9	1.7	87.0	37.1	7.4	260	170	.	.	.	.	.	.	.	.	6
4659	MUHLENBERGIA CUSPIDATA	W	.	0.0	7.4	1.9	80.2	33.5	10.5	.	.	.	.	.	.	.	.	.	.	6
4660	MUHLENBERGIA EMERSLEYI	H	.	0.0	7.5	1.5	82.1	34.1	8.9	.	120	.	.	.	.	.	.	.	.	6
4661	MUHLENBERGIA PORTERI	W	.	0.0	7.3	1.8	85.0	36.9	5.9	270	90	.	.	.	.	.	.	.	.	6
4662	MUHLENBERGIA PUNGENS	W	.	0.0	18.3	1.7	68.9	26.9	11.1	980	450	.	.	.	.	.	.	.	.	6
4663	MUHLENBERGIA SP.(ALL)	W	.	0.0	8.8	1.8	81.4	35.2	8.0	420	170	.	.	.	.	.	.	.	.	6
4664	MUNDIA SPINOSA	S	.	0.0	29.6	52.1	.	.	.	.	.	.	.	.	.	.	.	.	.	21
4665	MUNDIA SPINOSA	S	.	0.0	30.7	44.2	.	.	.	.	.	.	.	.	.	.	.	.	.	21
4666	MUNROA SQUARROSA	W	.	0.0	12.5	1.7	71.9	17.3	13.9	480	160	.	.	.	.	.	.	.	.	6
4667	MUNTINGIA CALABURA	F	87	75.1	2.0	0.4	21.3	2.0	1.2	64	72	1.2	2	390	0	0.03	0.04	0.4	86	2
4668	MUNTINGIA CALABURA	F	91	76.3	2.1	2.3	17.9	6.0	1.4	125	94	1.2	.	.	30	0.06	0.05	0.5	90	3
4669	MURRAYA KOENIGII	L	88	73.0	9.7	1.7	14.1	.	1.5	811	.	3.1	.	.	6000	0.08	0.21	2.3	12	2
4670	MURRAYA KOENIGII	L	.	66.3	6.1	1.0	22.4	6.4	4.2	810	600	3.1	.	.	7560	0.00	0.00	2.3	4	5
4671	MUSA ACUMINATA	F	72	79.2	1.8	0.2	18.0	0.2	0.8	10	24	1.3	18	435	80	0.03	0.04	0.6	8	2
4672	MUSA ACUMINATA	F	107	69.6	2.0	0.2	27.3	1.1	0.9	8	29	0.5	.	.	.	0.06	0.06	0.7	13	3
4673	MUSA NANA	G	108	72.4	1.1	0.3	25.3	1.0	0.9	11	28	0.9	.	.	.	.	.	.	.	1
4674	MUSA NANA	F	111	71.8	1.2	0.4	25.8	0.9	0.8	10	26	0.8	.	.	285	.	.	.	1	1
4675	MUSA NANA	F	97	72.3	1.4	0.2	25.2	0.6	0.9	10	32	0.9	.	.	160	0.05	0.09	0.6	8	3
4676	MUSA PARADISIACA SAPIENTUM	L	.	0.0	16.6	4.3	66.6	19.2	12.5	590	220	17.0	.	.	.	.	.	.	.	6
4677	MUSA TROGLODYTARUM	F	109	68.9	1.0	0.2	29.0	1.1	0.9	6	40	0.4	.	.	100	0.03	0.03	0.5	20	3
4678	MUSA URANOSCOPOS	F	99	71.8	1.1	0.2	25.9	0.9	1.0	15	25	0.4	.	.	560	0.04	0.04	0.4	8	3
4679	MUSA X PARADISIACA	G	110	69.0	1.4	0.2	28.7	0.5	0.7	8	35	0.9	.	.	290	0.04	0.02	0.6	31	
4680	MUSA X PARADISIACA	F	135	65.4	1.2	0.3	32.1	0.5	1.0	8	38	1.3	.	.	780	0.08	0.04	0.6	20	1
4681	MUSA X PARADISIACA	F	88	77.0	1.5	0.1	20.6	0.9	0.8	9	21	1.4	.	.	120	0.03	0.03	0.6	9	1
4682	MUSA X PARADISIACA	L	.	0.0	9.9	11.8	69.5	24.0	8.8	750	240	.	.	.	.	.	.	.	.	18
4683	MUSA X PARADISIACA	SH	.	0.0	2.4	2.3	81.0	20.5	14.3	1160	.	.	.	.	.	.	.	.	.	18

TABLE 2 - DATA DIRECTLY TRANSCRIBED FROM ORIGINAL SOURCES.

OBS	PLNA	PART	CALORIES	WATER	PROTEIN	FAT	TOTCARB	FIBER	ASH	CALCIUM	PHOSPHO	IRON	SODIUM	POTASS	CAROT	THIAMINE	RIBOFLAV	NIACIN	ASCORBIC	REFCO
4684	MUSA X PARADISIACA	F	.	0.0	5.4	0.9	91.4	2.2	3.3	.	.	.	.	.	.	.	.	.	.	18
4685	MUSA X PARADISIACA	F	100	71.6	1.2	0.3	26.1	0.6	0.8	12	32	0.8	4	401	225	0.03	0.04	0.60	14	2
4686	MUSA X PARADISIACA	F	91	74.1	1.3	0.2	23.5	0.6	0.9	9	11	0.6	.	.	70	0.18	0.05	0.50	8	3
4687	MUSA X PARADISIACA	F	122	65.6	1.0	0.3	32.3	0.5	0.8	8	34	0.8	.	.	350	0.06	0.04	0.60	20	3
4688	MUSA X PARADISIACA	F	92	73.6	1.2	0.2	24.1	0.4	0.9	10	20	0.5	.	.	330	0.04	0.04	0.60	8	3
4689	MUSA X PARADISIACA	G	132	62.6	1.2	0.1	35.3	0.4	0.8	8	40	0.8	.	.	760	0.07	0.04	0.50	28	3
4690	MUSA X PARADISIACA	F	119	66.4	1.1	0.4	31.2	0.4	0.9	7	30	0.7	5	385	363	0.06	0.04	0.60	14	4
4691	MUSA X PARADISIACA	F	85	75.7	1.1	0.2	22.2	0.5	0.8	8	26	0.7	1	370	114	0.05	0.06	0.70	10	4
4692	MUSCARI LONGIPES	S	.	0.0	15.4	10.4	.	.	.	.	.	.	.	.	.	.	.	.	.	21
4693	MUSCARI RACEMOSUM	S	.	0.0	14.9	10.0	.	.	.	.	.	.	.	.	.	.	.	.	.	21
4694	MUSCARI SP.	S	.	0.0	19.5	11.4	.	.	.	.	.	.	.	.	.	.	.	.	.	21
4695	MYOSOTIS SYLVATICA	S	.	0.0	16.9	42.8	.	.	.	.	.	.	.	.	.	.	.	.	.	20
4696	MYRCIARIA CAULIFLORA	F	66	83.5	0.5	1.8	13.8	1.0	0.4	61	17	0.2	.	.	0	0.01	0.02	0.30	17	2
4697	MYRCIARIA FLORIBUNDA	F	46	87.1	0.1	0.0	12.6	0.1	0.2	6	9	0.5	.	.	0	0.02	0.02	0.20	23	3
4698	MYRIANTHUS SP.	F	49	85.5	1.9	.	11.8	.	0.8	44	70	1.1	.	.	.	.	.	.	.	1
4699	MYRIANTHUS SP.	S	471	13.5	23.6	33.4	27.0	3.5	2.5	132	371	6.6	.	.	.	.	.	.	.	1
4700	MYRICA CERIFERA	S	.	0.0	4.9	13.3	.	.	.	.	.	.	.	.	.	.	.	.	.	20
4701	MYRICA CERIFERA	L	.	0.0	10.8	2.2	74.3	29.8	12.7	.	.	.	.	.	.	.	.	.	.	6
4702	MYRICA CORDIFOLIA	S	.	0.0	5.5	8.9	.	.	.	.	.	.	.	.	.	.	.	.	.	21
4703	MYRICA HETEROPHYLLA	S	.	0.0	5.6	31.2	.	.	.	.	.	.	.	.	.	.	.	.	.	19
4704	MYRICA PENSYLVANICA	S	.	0.0	6.2	35.3	.	.	.	.	.	.	.	.	.	.	.	.	.	20
4705	MYRICA PENSYLVANICA	S	.	0.0	9.4	4.3	.	.	.	.	.	.	.	.	.	.	.	.	.	20
4706	MYRISTICA CANARICA	S	.	7.2	7.3	49.4	34.8	20.1	1.2	.	.	.	.	.	.	.	.	.	.	5
4707	MYRISTICA FRAGRANS	A	475	8.2	6.7	32.4	50.5	4.8	2.2	252	110	13.9	80	463	480	0.31	0.45	1.35	.	16
4708	MYRISTICA FRAGRANS	S	525	6.2	5.8	36.3	49.3	4.0	2.3	184	213	3.0	16	350	61	0.35	0.06	1.30	.	16
4709	MYRISTICA FRAGRANS	S	.	14.3	7.5	36.4	40.1	11.6	1.7	120	240	4.6	.	.	.	.	.	.	.	5
4710	MYRISTICA FRAGRANS	A	.	15.9	6.5	24.4	51.6	3.8	1.6	180	100	12.6	.	.	.	.	.	.	.	5
4711	MYRISTICA FRAGRANS	F	.	86.8	1.0	0.4	11.2	.	0.6	40	10	2.0	.	.	5	.	.	.	.	5
4712	MYRISTICA MALABARICA	S	.	6.9	6.5	.	44.5	2.3	1.3	.	.	.	.	.	.	.	.	.	.	5
4713	MYRISTICA MALABARICA	A	.	4.1	7.3	.	23.8	3.1	1.5	.	.	.	.	.	.	.	.	.	.	5
4714	MYRSINE AFRICANA	S	.	0.0	10.3	19.7	.	.	.	.	.	.	.	.	.	.	.	.	.	21
4715	MYRTUS COMMUNIS	S	.	0.0	6.6	0.6	.	.	.	.	.	.	.	.	.	.	.	.	.	21
4716	MYRTUS COMMUNIS	S	.	0.0	6.9	6.3	.	.	.	.	.	.	.	.	.	.	.	.	.	21
4717	NAMA HAVARDII	S	.	0.0	22.3	14.5	.	.	.	.	.	.	.	.	.	.	.	.	.	20
4718	NAMA SP.	L	.	0.0	14.1	2.4	65.2	19.2	18.3	5020	120	.	.	2570	.	.	.	.	.	6
4719	NANDINA DOMESTICA	S	.	0.0	20.0	14.2	.	.	2.7	.	.	.	.	.	.	.	.	.	.	19
4720	NASTURTIOPSIS ARABICA	S	.	0.0	30.6	33.4	.	.	.	.	.	.	.	.	.	.	.	.	.	21
4721	NASTURTIUM OFFICINALE	L	18	93.6	1.9	0.2	3.4	0.8	0.9	64	46	1.1	.	.	960	0.03	0.27	.	.	1
4722	NASTURTIUM OFFICINALE	L	.	0.0	28.3	.	.	.	16.4	2060	.	.	335	3520	.	.	.	.	.	17
4723	NASTURTIUM OFFICINALE	L	18	93.8	1.7	0.4	3.0	1.1	1.1	114	42	1.9	41	224	1025	0.07	0.13	0.50	46	2
4724	NASTURTIUM OFFICINALE	S	.	0.0	29.4	27.6	.	.	.	.	.	.	.	.	.	.	.	.	.	20
4725	NASTURTIUM OFFICINALE	S	.	0.0	30.0	27.2	.	.	.	.	.	.	.	.	.	.	.	.	.	20
4726	NASTURTIUM OFFICINALE	L	22	92.2	2.8	0.4	3.3	1.1	1.3	117	76	1.9	.	.	2210	0.12	0.10	1.00	44	3
4727	NASTURTIUM OFFICINALE	L	19	93.3	2.2	0.3	3.0	0.7	1.2	151	54	1.7	52	282	2940	0.08	0.16	0.90	79	4
4728	NASTURTIUM OFFICINALE	L	.	89.2	2.9	0.2	5.5	.	2.2	290	140	4.6	.	.	.	.	.	.	.	5

TABLE 2 - DATA DIRECTLY TRANSCRIBED FROM ORIGINAL SOURCES.

OBS	PLNA	PART	CALORIES	WATER	PROTEIN	FAT	TOTCARB	FIBER	ASH	CALCIUM	PHOSPHO	IRON	SODIUM	POTASS	CAROT	THIAMINE	RIBOFLAV	NIACIN	ASCORBIC	REFCO
4729	NELUMBO NUCIFERA	S	63	80.2	4.8	0.2	13.9	2.6	0.9	59	59	.	.	.	.	.	.	.	.	1
4730	NELUMBO NUCIFERA	S	88	77.0	5.2	0.5	16.1	1.1	1.2	76	164	1.4	4	383	0	0.15	0.06	1.80	10	2
4731	NELUMBO NUCIFERA	S	334	14.3	14.2	2.3	65.3	2.1	3.9	119	466	6.1	42	807	30	0.64	0.15	1.60	0	2
4732	NELUMBO NUCIFERA	R	.	83.8	2.7	0.1	12.0	0.8	1.1	60	.	.	.	.	.	0.22	0.06	2.10	15	5
4733	NEMESIA SUTTONII	S	.	0.0	28.8	47.7	.	.	4.7	.	.	.	.	.	.	.	.	.	.	19
4734	NEMOPHILA MENZIESII	S	.	0.0	11.5	3.3	.	.	3.1	.	.	.	.	.	.	.	.	.	.	19
4735	NEMOPHILA MENZIESII	S	.	0.0	13.1	4.6	.	.	3.4	.	.	.	.	.	.	.	.	.	.	19
4736	NENAX ACEROSA	S	.	0.0	33.2	6.1	.	.	.	.	.	.	.	.	.	.	.	.	.	21
4737	NEPETA CATARIA	S	.	0.0	18.4	21.2	.	.	3.2	.	.	.	.	.	.	.	.	.	.	19
4738	NEPETA CONGESTA	S	.	0.0	18.2	16.9	.	.	.	.	.	.	.	.	.	.	.	.	.	20
4739	NEPETA ITALICA	S	.	0.0	16.9	22.1	.	.	.	.	.	.	.	.	.	.	.	.	.	21
4740	NEPETA MUSSINI	S	.	0.0	21.7	25.3	.	.	8.0	.	.	.	.	.	.	.	.	.	.	19
4741	NEPETA NEPETELLA	S	.	0.0	18.6	26.8	.	.	.	.	.	.	.	.	.	.	.	.	.	21
4742	NEPETA TMOLEA	S	.	0.0	22.2	29.4	.	.	.	.	.	.	.	.	.	.	.	.	.	21
4743	NEPHELIUM LAPPACEUM	F	64	82.0	1.0	0.1	16.5	1.1	0.4	20	15	1.9	1	64	0	0.01	0.06	0.40	53	2
4744	NEPHELIUM LAPPACEUM	A	.	82.3	0.5	0.1	16.0	0.3	0.9	11	13	.	.	.	.	.	.	.	30	5
4745	NEPHELIUM MUTABILE	F	60	82.0	0.8	0.0	15.9	0.3	1.3	14	34	0.4	.	.	.	0.05	0.13	0.30	72	2
4746	NEPHELIUM MUTABILE	F	55	83.0	0.8	0.0	14.5	1.0	1.7	25	14	3.0	.	.	0	0.05	0.06	0.40	48	2
4747	NEPTUNIA OLERACEA	L	34	88.0	5.2	0.2	5.4	1.8	1.2	56	36	3.9	107	277	3555	0.12	0.14	1.10	21	2
4748	NEPTUNIA PUBESCENS	S	.	0.0	26.2	3.9	.	.	.	.	.	.	.	.	.	.	.	.	.	20
4749	NERISYRENIA CAMPORUM	S	.	0.0	27.5	37.6	.	.	.	.	.	.	.	.	.	.	.	.	.	20
4750	NERIUM OLEANDER	S	.	0.0	30.0	29.4	.	.	.	.	.	.	.	.	.	.	.	.	.	20
4751	NESLIA PANICULATA	S	.	0.0	31.0	46.8	.	.	.	.	.	.	.	.	.	.	.	.	.	21
4752	NEUROLAENA LOBATA	S	.	0.0	21.9	34.8	.	.	.	.	.	.	.	.	.	.	.	.	.	20
4753	NICANDRA PHYSALODES	S	.	0.0	16.6	26.2	.	.	.	.	.	.	.	.	.	.	.	.	.	21
4754	NICOTIANA AFFINIS	S	.	0.0	20.0	39.2	.	.	4.0	.	.	.	.	.	.	.	.	.	.	19
4755	NICOTIANA ALATA	S	.	0.0	20.0	38.4	.	.	.	.	.	.	.	.	.	.	.	.	.	20
4756	NICOTIANA TABACUM	S	.	0.0	22.4	45.4	.	.	.	.	.	.	.	.	.	.	.	.	.	21
4757	NIGELLA CILIARIS	S	.	0.0	21.1	30.4	.	.	.	.	.	.	.	.	.	.	.	.	.	21
4758	NIGELLA DAMASCENA	S	.	0.0	25.7	42.7	.	.	5.3	.	.	.	.	.	.	.	.	.	.	19
4759	NIGELLA DAMASCENA	S	.	0.0	25.0	40.5	.	.	.	.	.	.	.	.	.	.	.	.	.	20
4760	NIGELLA HISPANICA	S	.	0.0	23.8	46.0	.	.	5.0	.	.	.	.	.	.	.	.	.	.	19
4761	NOLINA DURANGENSIS	S	.	0.0	14.4	23.3	.	.	.	.	.	.	.	.	.	.	.	.	.	20
4762	NOLINA MATAPENSIS	S	.	0.0	20.4	17.8	.	.	.	.	.	.	.	.	.	.	.	.	.	21
4763	NOLINA MICROCARPA	L	.	0.0	6.3	3.0	86.8	41.9	3.9	.	.	.	.	.	.	.	.	.	.	6
4764	NOLINA SP.	L	.	0.0	6.3	3.1	86.5	42.7	4.1	920	90	.	.	700	.	.	.	.	.	6
4765	NOLINA TEXANA	S	.	0.0	15.6	20.3	.	.	.	.	.	.	.	.	.	.	.	.	.	19
4766	NOLINA TEXANA	S	.	0.0	11.2	10.2	.	.	.	.	.	.	.	.	.	.	.	.	.	19
4767	NOPALEA COCHENILLIFERA	F	60	83.4	1.3	0.3	14.7	2.8	0.3	14	21	0.3	.	.	0	0.01	0.03	0.50	17	3
4768	NOPALEA COCHENILLIFERA	SH	29	90.8	1.3	0.1	6.9	6.8	0.9	.	17	2.7	.	.	440	0.03	0.04	0.40	16	3
4769	NOTOBASIS SYRIACA	S	.	0.0	17.6	24.7	.	.	.	.	.	.	.	.	.	.	.	.	.	21
4770	NYMPHAEA ELEGANS	R	159	55.9	7.9	0.3	35.0	0.4	0.9	55	153	2.8	.	.	0	0.17	0.04	6.50	1	3
4771	NYMPHAEA LOTUS	R	120	64.0	5.2	0.2	29.5	1.0	1.1	93	151	.	.	.	.	.	.	.	.	1
4772	NYMPHAEA LOTUS	I	48	84.2	1.4	0.2	11.9	3.4	2.3	44	16	.	.	.	.	.	.	.	.	1
4773	NYMPHAEA LOTUS	S	379	6.5	8.4	2.2	81.5	6.1	1.4	.	.	13.5	.	.	.	.	.	.	.	1

TABLE 2 - DATA DIRECTLY TRANSCRIBED FROM ORIGINAL SOURCES.

OBS	PLNA	PART	CALORIES	WATER	PROTEIN	FAT	TOTLCARB	FIBER	ASH	CALCIUM	PHOSPHO	IRON	SODIUM	POTASS	CAROT	THIAMINE	RIBOFLAV	NIACIN	ASCORBIC	REFCO
4774	NYSSA OGECHE	S	.	0.0	5.6	7.4	.	.	2.0	.	.	.	.	.	.	.	.	.	.	19
4775	NYSSA OGECHE	S	.	0.0	3.8	1.6	.	.	2.7	.	.	.	.	.	.	.	.	.	.	19
4776	NYSSA OGECHE	S	.	0.0	5.6	8.9	.	.	.	.	.	.	.	.	.	.	.	.	.	21
4777	NYSSA OGECHE	S	.	0.0	2.2	1.2	.	.	.	.	.	.	.	.	.	.	.	.	.	21
4778	OCIMUM AMERICANUM	S	.	0.0	17.5	18.5	.	.	.	.	.	.	.	.	.	.	.	.	.	21
4779	OCIMUM BASILICUM	L	75	77.1	5.5	1.4	13.9	.	2.1	487	.	5.2	.	.	.	.	.	.	37	1
4780	OCIMUM BASILICUM	H	.	6.4	14.4	4.0	61.0	17.8	14.3	2113	490	42.0	34	3433	5625	0.15	0.32	6.95	61	16
4781	OCIMUM BASILICUM	S	.	0.0	21.4	24.2	.	.	7.2	.	.	.	.	.	.	.	.	.	.	19
4782	OCIMUM BASILICUM	L	43	86.5	3.3	1.2	7.0	2.0	2.0	320	38	4.8	12	429	4500	0.08	0.35	0.80	27	2
4783	OCIMUM CANUM	L	40	87.4	3.4	0.8	7.0	2.6	1.4	92	58	2.0	.	.	7950	0.10	0.28	0.60	26	2
4784	OCIMUM KILIMANDSCHARICUM	S	.	6.4	18.8	17.4	50.8	27.0	6.5	.	.	.	.	.	.	.	.	.	.	5
4785	OCIMUM SANCTUM	S	.	0.0	13.8	15.5	.	.	.	.	.	.	.	.	.	.	.	.	.	20
4786	OCIMUM SP.	L	38	88.4	3.8	1.4	4.7	.	1.7	630	.	.	.	.	.	.	.	.	52	1
4787	OCOTEA ACUTIFOLIA	S	.	0.0	8.7	24.2	.	.	.	.	.	.	.	.	.	.	.	.	.	21
4788	OENANTHE GLOBULOSA	S	.	0.0	9.6	8.8	.	.	.	.	.	.	.	.	.	.	.	.	.	21
4789	OENANTHE JAVANICA	L	28	90.6	1.8	0.3	5.9	1.2	1.4	113	55	3.0	18	443	2190	0.06	0.22	1.00	14	2
4790	OENANTHE LACHENALII	S	.	0.0	17.6	22.1	.	.	.	.	.	.	.	.	.	.	.	.	.	21
4791	OENANTHE PIMPINELLOIDES	S	.	0.0	13.8	12.6	.	.	.	.	.	.	.	.	.	.	.	.	.	21
4792	OENOTHERA BIENNIS	S	.	0.0	16.4	24.6	.	.	6.3	.	.	.	.	.	.	.	.	.	.	19
4793	OENOTHERA BIENNIS	S	.	0.0	16.9	27.5	.	.	.	.	.	.	.	.	.	.	.	.	.	19
4794	OENOTHERA DEPRESSA	S	.	0.0	16.1	21.8	.	.	.	.	.	.	.	.	.	.	.	.	.	21
4795	OENOTHERA DEPRESSA	S	.	0.0	17.2	27.7	.	.	.	.	.	.	.	.	.	.	.	.	.	21
4796	OENOTHERA DEPRESSA	S	.	0.0	13.5	26.4	.	.	.	.	.	.	.	.	.	.	.	.	.	21
4797	OENOTHERA DRUMMONDII	S	.	0.0	16.9	25.0	.	.	.	.	.	.	.	.	.	.	.	.	.	20
4798	OENOTHERA DRUMMONDII	S	.	0.0	19.7	29.0	.	.	.	.	.	.	.	.	.	.	.	.	.	21
4799	OENOTHERA LACINIATA	S	.	0.0	11.9	17.3	.	.	.	.	.	.	.	.	.	.	.	.	.	20
4800	OENOTHERA LAMARCKIANA	S	.	0.0	17.2	26.1	.	.	6.8	.	.	.	.	.	.	.	.	.	.	19
4801	OENOTHERA LAMARCKIANA	S	.	0.0	20.0	28.4	.	.	8.7	.	.	.	.	.	.	.	.	.	.	19
4802	OENOTHERA LAMARCKIANA	S	.	0.0	20.0	28.3	.	.	.	.	.	.	.	.	.	.	.	.	.	20
4803	OENOTHERA MISSOURIENSIS VAR. INCANA	S	.	0.0	22.9	20.8	.	.	.	.	.	.	.	.	.	.	.	.	.	20
4804	OENOTHERA RHOMBIPETALA	S	.	0.0	15.6	30.7	.	.	9.3	.	.	.	.	.	.	.	.	.	.	19
4805	OENOTHERA SERRULATA	S	.	0.0	24.9	22.0	.	.	.	.	.	.	.	.	.	.	.	.	.	20
4806	OENOTHERA SP.	S	.	0.0	19.4	30.8	.	.	.	.	.	.	.	.	.	.	.	.	.	20
4807	OENOTHERA SP.	W	.	0.0	14.4	3.0	74.7	20.0	7.9	1930	340	.	.	2940	.	.	.	.	.	6
4808	OLEA EUROPAEA	L	.	0.0	13.1	7.3	73.5	17.7	6.1	1180	90	.	.	.	.	.	.	.	.	18
4809	OLEA EUROPAEA	C	.	0.0	8.9	6.1	76.5	28.9	8.5	.	.	.	.	.	.	.	.	.	.	18
4810	OLEA EUROPAEA	F	116	78.2	1.4	12.7	1.3	1.3	6.4	61	17	1.6	2400	55	180	.	.	.	.	4
4811	ONCOBA SPINOSA	S	.	0.0	12.5	29.1	.	.	.	.	.	.	.	.	.	.	.	.	.	20
4812	ONOBRYCHIS ARENARIA	S	.	0.0	.	6.1	.	.	.	.	.	.	.	.	.	.	.	.	.	20
4813	ONOBRYCHIS ARGYREA	S	.	0.0	49.2	16.8	.	.	.	.	.	.	.	.	.	.	.	.	.	20
4814	ONOBRYCHIS AURANTIACA	S	.	0.0	57.2	13.4	.	.	.	.	.	.	.	.	.	.	.	.	.	21
4815	ONOBRYCHIS CADMEA	S	.	0.0	41.5	5.7	.	.	.	.	.	.	.	.	.	.	.	.	.	21
4816	ONOBRYCHIS HYPARGYREA	S	.	0.0	47.6	13.8	.	.	.	.	.	.	.	.	.	.	.	.	.	21
4817	ONOBRYCHIS MATRITENSIS	S	.	0.0	47.0	7.3	.	.	.	.	.	.	.	.	.	.	.	.	.	21
4818	ONOBRYCHIS SP.	H	.	0.0	22.6	3.8	67.2	28.2	6.4	.	.	.	.	.	.	.	.	.	.	6

TABLE 2 - DATA DIRECTLY TRANSCRIBED FROM ORIGINAL SOURCES.

OBS	PLNA	PART	CALORIES	WATER	PROTEIN	FAT	TOTLCARB	FIBER	ASH	CALCIUM	PHOSPHO	IRON	SODIUM	POTASS	CAROT	THIAMINE	RIBOFLAV	NIACIN	ASCORBIC	REFCO
4819	ONOBRYCHIS SP.	W	.	0.0	17.4	3.3	70.9	25.5	8.4	.	.	.	.	.	.	.	.	.	.	6
4820	ONOBRYCHIS VICIAEFOLIA	S	.	0.0	40.9	7.6	.	.	3.4	.	.	.	.	.	.	.	.	.	.	19
4821	ONOBRYCHIS VICIAEFOLIA	S	.	0.0	33.3	7.9	.	.	4.8	.	.	.	.	.	.	.	.	.	.	19
4822	ONOBRYCHIS VULGARIS	S	.	0.0	40.5	7.0	.	.	3.9	.	.	.	.	.	.	.	.	.	.	19
4823	ONOBRYCHIS VULGARIS	S	.	0.0	40.9	6.8	.	.	3.9	.	.	.	.	.	.	.	.	.	.	19
4824	ONOBRYCHIS VULGARIS	S	.	0.0	39.8	6.7	.	.	4.2	.	.	.	.	.	.	.	.	.	.	19
4825	ONONIS ALOPECUROIDES	S	.	0.0	41.4	6.3	.	.	.	.	.	.	.	.	.	.	.	.	.	21
4826	ONONIS LEIOSPERMA	S	.	0.0	46.2	8.2	.	.	.	.	.	.	.	.	.	.	.	.	.	21
4827	ONONIS PUBESCENS	S	.	0.0	35.7	12.0	.	.	.	.	.	.	.	.	.	.	.	.	.	21
4828	ONOPORDUM ACANTHIUM	S	.	0.0	18.8	20.1	.	.	.	.	.	.	.	.	.	.	.	.	.	20
4829	ONOPORDUM ACANTHIUM	S	.	0.0	16.8	10.1	.	.	.	.	.	.	.	.	.	.	.	.	.	21
4830	ONOPORDUM ACANTHIUM	S	.	0.0	18.6	17.4	.	.	.	.	.	.	.	.	.	.	.	.	.	21
4831	ONOPORDUM ARABICUM	S	.	0.0	16.9	19.2	.	.	.	.	.	.	.	.	.	.	.	.	.	20
4832	ONOPORDUM ARABICUM	S	.	0.0	17.9	12.5	.	.	.	.	.	.	.	.	.	.	.	.	.	21
4833	ONOPORDUM CYNAROCEPHALUM	S	.	0.0	17.8	16.6	.	.	.	.	.	.	.	.	.	.	.	.	.	21
4834	ONOSMA SERICEUM	S	.	0.0	19.4	20.3	.	.	.	.	.	.	.	.	.	.	.	.	.	21
4835	ONOSMA STELLULATUM	S	.	0.0	24.4	22.9	.	.	.	.	.	.	.	.	.	.	.	.	.	21
4836	ONOSMODIUM MOLLE	S	.	0.0	11.9	17.2	.	.	.	.	.	.	.	.	.	.	.	.	.	21
4837	ONOSMODIUM MOLLE	S	.	0.0	33.0	53.6	.	.	.	.	.	.	.	.	.	.	.	.	.	21
4838	ONOSMODIUM OCCIDENTALE	S	.	0.0	35.7	56.4	.	.	3.6	.	.	.	.	.	.	.	.	.	.	19
4839	ONOSMODIUM OCCIDENTALE	S	.	0.0	6.8	7.1	.	.	59.0	.	.	.	.	.	.	.	.	.	.	19
4840	OOTHRINAX ANOMALA	S	.	0.0	4.9	1.8	.	.	1.6	.	.	.	.	.	.	.	.	.	.	19
4841	OOTHRINAX ANOMALA	S	.	0.0	2.8	0.6	.	.	0.6	.	.	.	.	.	.	.	.	.	.	19
4842	OPERCULINA TURPETHUM	L	48	85.0	3.9	0.7	9.0	1.4	1.4	120	82	6.4	.	.	2345	0.18	0.25	1.50	48	2
4843	OPOPANAX HISPIDUM	S	.	0.0	23.8	15.6	.	.	.	.	.	.	.	.	.	.	.	.	.	21
4844	OPUNTIA AMYCLAEA	SH	.	0.0	4.0	2.7	68.9	12.0	24.4	.	.	.	.	.	.	.	.	.	.	18
4845	OPUNTIA BRASILIENSIS	SH	.	0.0	6.5	3.3	.	10.9	17.6	.	.	.	.	.	.	.	.	.	.	18
4846	OPUNTIA CHOLLA	SH	.	0.0	5.5	4.1	76.4	13.8	14.0	.	.	.	.	.	.	.	.	.	.	18
4847	OPUNTIA ELATIOR	SH	.	0.0	6.3	3.3	71.6	13.5	18.8	.	.	.	.	.	.	.	.	.	.	18
4848	OPUNTIA FICUS-INDICA	SH	.	0.0	9.6	1.3	63.2	13.4	25.9	2810	280	.	.	.	.	.	.	.	.	18
4849	OPUNTIA FICUS-INDICA	F	67	81.4	1.1	0.4	16.6	1.1	0.5	57	32	1.2	.	.	.	0.01	0.02	0.30	18	3
4850	OPUNTIA FICUS-INDICA	F	.	90.2	1.5	0.0	5.6	.	0.3	18	140	.	.	.	.	.	.	.	.	5
4851	OPUNTIA FICUS-INDICA	S	.	13.6	12.1	5.4	48.2	11.1	20.8	.	.	.	.	.	.	.	.	.	.	5
4852	OPUNTIA MAXIMA	SH	.	0.0	6.5	1.6	74.2	11.3	17.7	.	.	.	.	.	.	.	.	.	.	18
4853	OPUNTIA MEGACANTHA	F	53	84.5	0.8	0.1	13.8	1.9	0.8	43	9	0.3	.	.	0	0.02	0.03	0.40	25	2
4854	OPUNTIA SP.	F	62	82.9	1.0	0.4	15.4	3.9	0.3	31	27	0.3	.	.	.	.	.	.	20	1
4855	OPUNTIA SP.	F	37	88.8	1.1	0.4	8.8	2.6	0.9	110	20	0.5	.	.	100	0.04	0.04	0.20	19	3
4856	OPUNTIA SP.	F	60	84.6	1.4	1.4	12.1	6.6	0.5	46	32	1.2	.	.	20	0.02	0.03	0.40	22	3
4857	OPUNTIA SP.	F	42	88.0	0.5	0.1	10.9	1.6	0.5	20	28	0.3	2	166	36	0.01	0.03	0.40	22	4
4858	OPUNTIA SP.(ALL)	W	.	0.0	5.0	2.0	73.0	13.3	20.0	6290	80	90.0	.	1210	1	.	.	.	.	6
4859	ORBIGNYA COHUNE	S	.	0.0	6.9	52.2	.	.	.	.	.	.	.	.	.	.	.	.	.	19
4860	ORBIGNYA COHUNE	S	.	0.0	1.2	0.5	.	.	.	.	.	.	.	.	.	.	.	.	.	19
4861	ORBIGNYA SPECIOSA	S	.	0.0	9.4	62.9	.	.	.	.	.	.	.	.	.	.	.	.	.	19
4862	ORBIGNYA SPECIOSA	S	.	0.0	16.2	0.2	.	.	.	.	.	.	.	.	.	.	.	.	.	19
4863	ORCHIS LAXIFLORA	S	.	0.0	10.9	0.7	.	.	.	.	.	.	.	.	.	.	.	.	.	21

TABLE 2 - DATA DIRECTLY TRANSCRIBED FROM ORIGINAL SOURCES.

OBS	PLNA	PART	CALORIES	WATER	PROTEIN	FAT	TOTLCARB	FIBER	ASH	CALCIUM	PHOSPHO	IRON	SODIUM	POTASS	CAROT	THIAMINE	RIBOFLAV	NIACIN	ASCORBIC	REFC
4864	ORIGANUM VULGARE	H	306	7.2	11.0	10.2	64.4	15.0	7.2	1576	200	44.0	15	1669	4142	0.34	.	6.22	.	16
4865	ORMOSIA JAMAICENSIS	S	.	0.0	12.5	9.3	.	.	2.0	.	.	.	.	.	.	.	.	.	.	19
4866	ORMOSIA PANAMENSIS	S	.	0.0	15.0	7.7	.	.	1.8	.	.	.	.	.	.	.	.	.	.	19
4867	ORNITHOGALUM NUTANS	S	.	0.0	17.5	6.3	.	.	.	.	.	.	.	.	.	.	.	.	.	20
4868	ORNITHOPUS COMPRESSUS	S	.	0.0	41.4	15.5	.	.	.	.	.	.	.	.	.	.	.	.	.	21
4869	ORNITHOPUS SATIVUS	W	.	0.0	24.4	3.7	60.4	18.7	11.5	1210	240	.	.	.	.	.	.	.	.	18
4870	ORNITHOPUS SATIVUS	W	.	0.0	14.8	3.7	66.2	23.0	15.3	.	.	.	.	.	.	.	.	.	.	6
4871	ORTHOCARPUS LUTEUS	W	.	0.0	6.7	2.1	.	.	9.2	1030	490	.	.	.	.	.	.	.	.	6
4872	ORYZA GLABERRIMA	S	344	10.9	6.6	1.9	74.1	.	6.5	25	236	2.0	.	.	.	0.31	.	5.00	.	1
4873	ORYZA SATIVA	S	353	10.3	6.2	2.0	76.4	10.3	5.1	27	312	7.6	.	.	.	0.34	0.05	5.80	.	1
4874	ORYZA SATIVA	W	.	0.0	9.0	1.8	73.3	28.3	15.9	.	.	.	.	.	.	.	.	.	.	18
4875	ORYZA SATIVA	H	.	0.0	8.2	1.8	74.3	32.0	15.7	.	.	.	.	.	.	.	.	.	.	18
4876	ORYZA SATIVA	S	341	13.7	5.8	2.3	73.4	10.4	4.8	24	236	1.4	81	156	.	0.33	0.06	5.60	.	2
4877	ORYZA SATIVA	S	357	13.0	7.2	1.5	77.6	0.8	0.7	14	231	2.6	.	.	0	0.22	0.05	4.00	0	3
4878	ORYZA SATIVA	S	360	12.0	7.5	1.9	77.4	0.9	1.2	32	221	1.6	9	214	0	0.34	0.05	4.70	0	4
4879	ORYZA SATIVA	H	.	0.0	4.2	1.4	77.5	35.1	16.9	240	90	.	.	1320	.	.	.	.	.	6
4880	ORYZOPSIS HYMENOIDES	S	.	0.0	11.9	2.2	.	.	.	.	.	.	.	.	.	.	.	.	.	19
4881	ORYZOPSIS HYMENOIDES	W	.	0.0	4.0	1.9	85.4	38.0	8.7	580	50	.	.	.	.	.	.	.	.	6
4882	ORYZOPSIS MILIACEA	S	.	0.0	14.4	2.1	.	.	3.1	.	.	.	.	.	.	.	.	.	.	19
4883	OSMANTHUS HETEROPHYLLUS	S	.	0.0	14.2	25.9	.	.	.	.	.	.	.	.	.	.	.	.	.	21
4884	OSMORHIZA CHILENSIS	S	.	0.0	24.7	8.5	.	.	.	.	.	.	.	.	.	.	.	.	.	21
4885	OSMORHIZA OCCIDENTALIS	S	.	0.0	20.8	10.4	.	.	.	.	.	.	.	.	.	.	.	.	.	20
4886	OSMORHIZA OCCIDENTALIS	W	.	0.0	10.5	2.9	.	.	16.5	2940	450	.	.	.	.	.	.	.	.	6
4887	OSTEOSPERMUM AMPLECTANS	S	.	0.0	13.1	17.1	.	.	.	.	.	.	.	.	.	.	.	.	.	20
4888	OSTEOSPERMUM ASPERULUM	S	.	0.0	.	9.3	.	.	.	.	.	.	.	.	.	.	.	.	.	21
4889	OSTEOSPERMUM CALENDULACEUM	S	.	0.0	.	52.0	.	.	.	.	.	.	.	.	.	.	.	.	.	20
4890	OSTEOSPERMUM CAULESCENS	S	.	0.0	13.4	14.2	.	.	.	.	.	.	.	.	.	.	.	.	.	20
4891	OSTEOSPERMUM CLANDESTINUM	S	.	0.0	12.5	6.6	.	.	.	.	.	.	.	.	.	.	.	.	.	20
4892	OSTEOSPERMUM CLANDESTINUM	S	.	0.0	38.1	42.0	.	.	.	.	.	.	.	.	.	.	.	.	.	20
4893	OSTEOSPERMUM CLANDESTINUM	S	.	0.0	38.0	42.6	.	.	.	.	.	.	.	.	.	.	.	.	.	20
4894	OSTEOSPERMUM CORYMBOSUM	S	.	0.0	.	8.7	.	.	.	.	.	.	.	.	.	.	.	.	.	21
4895	OSTEOSPERMUM DREGEI	S	.	0.0	9.4	14.6	.	.	.	.	.	.	.	.	.	.	.	.	.	20
4896	OSTEOSPERMUM ECKLONIS	S	.	0.0	23.8	48.3	.	.	.	.	.	.	.	.	.	.	.	.	.	20
4897	OSTEOSPERMUM ECKLONIS	S	.	0.0	30.6	50.4	.	.	.	.	.	.	.	.	.	.	.	.	.	20
4898	OSTEOSPERMUM FRUTICOSUM	S	.	0.0	.	12.9	.	.	.	.	.	.	.	.	.	.	.	.	.	21
4899	OSTEOSPERMUM HYOSEROIDES	S	.	0.0	35.1	49.7	.	.	.	.	.	.	.	.	.	.	.	.	.	20
4900	OSTEOSPERMUM IMBRICATUM	S	.	0.0	.	4.7	.	.	.	.	.	.	.	.	.	.	.	.	.	21
4901	OSTEOSPERMUM IMBRICATUM	S	.	0.0	.	6.4	.	.	.	.	.	.	.	.	.	.	.	.	.	21
4902	OSTEOSPERMUM IMBRICATUM SUBSP. NERVATUM	S	.	0.0	.	10.3	.	.	.	.	.	.	.	.	.	.	.	.	.	21
4903	OSTEOSPERMUM JUCUNDUM	S	.	0.0	.	16.6	.	.	.	.	.	.	.	.	.	.	.	.	.	21
4904	OSTEOSPERMUM JUNCEUM	S	.	0.0	25.8	61.2	.	.	.	.	.	.	.	.	.	.	.	.	.	20
4905	OSTEOSPERMUM MURICATUM	S	.	0.0	11.9	9.9	.	.	.	.	.	.	.	.	.	.	.	.	.	20
4906	OSTEOSPERMUM MURICATUM	S	.	0.0	11.8	8.8	.	.	.	.	.	.	.	.	.	.	.	.	.	20
4907	OSTEOSPERMUM MURICATUM	S	.	0.0	11.9	10.6	.	.	.	.	.	.	.	.	.	.	.	.	.	20
4908	OSTEOSPERMUM MURICATUM	S	.	0.0	11.4	8.2	.	.	.	.	.	.	.	.	.	.	.	.	.	20

TABLE 2 - DATA DIRECTLY TRANSCRIBED FROM ORIGINAL SOURCES.

OBS	PLNA	PART	CALORIES	WATER	PROTEIN	FAT	TOTLCARB	FIBER	ASH	CALCIUM	PHOSPHO	IRON	SODIUM	POTASS	CAROT	THIAMINE	RIBOFLAV	NIACIN	ASCORBIC	REFCO
4909	OSTEOSPERMUM MURICATUM	S	.	0.0	.	5.6	.	.	.	.	.	.	.	.	.	.	.	.	.	20
4910	OSTEOSPERMUM SCARIOSUM	S	.	0.0	37.10	40.8	.	.	.	.	.	.	.	.	.	.	.	.	.	20
4911	OSTEOSPERMUM SCARIOSUM	S	.	0.0	.	39.5	.	.	.	.	.	.	.	.	.	.	.	.	.	20
4912	OSTEOSPERMUM SCARIOSUM	S	.	0.0	37.40	42.6	.	.	.	.	.	.	.	.	.	.	.	.	.	20
4913	OSTEOSPERMUM SINUATUM	S	.	0.0	41.20	43.9	.	.	.	.	.	.	.	.	.	.	.	.	.	20
4914	OSTEOSPERMUM SINUATUM	S	.	0.0	41.90	42.6	.	.	.	.	.	.	.	.	.	.	.	.	.	20
4915	OSTEOSPERMUM SINUATUM	S	.	0.0	39.40	43.3	.	.	.	.	.	.	.	.	.	.	.	.	.	20
4916	OSTEOSPERMUM SINUATUM	S	.	0.0	47.00	31.8	.	.	.	.	.	.	.	.	.	.	.	.	.	20
4917	OSTEOSPERMUM SINUATUM	S	.	0.0	48.20	30.8	.	.	.	.	.	.	.	.	.	.	.	.	.	20
4918	OSTEOSPERMUM SINUATUM	S	.	0.0	45.40	39.4	.	.	.	.	.	.	.	.	.	.	.	.	.	21
4919	OSTEOSPERMUM SINUATUM		.	0.0	14.01	.	60.58	20.22	22.04	980	200	18.6	650	1780	.	.	.	.	.	8
4920	OSTEOSPERMUM SPINESCENS	C	.	0.0	9.20	3.2	72.30	26.40	15.30	1040	240	.	.	.	.	.	.	.	.	18
4921	OSTEOSPERMUM SPINESCENS	S	.	0.0	40.00	43.3	.	.	.	.	.	.	.	.	.	.	.	.	.	20
4922	OSTEOSPERMUM SPINESCENS	S	.	0.0	.	41.2	.	.	.	.	.	.	.	.	.	.	.	.	.	20
4923	OSTEOSPERMUM SPINESCENS	S	.	0.0	.	36.8	.	.	.	.	.	.	.	.	.	.	.	.	.	20
4924	OSTEOSPERMUM SPINESCENS	S	.	0.0	.	4.5	.	.	.	.	.	.	.	.	.	.	.	.	.	21
4925	OSTEOSPERMUM SPINESCENS	S	.	0.0	.	7.2	.	.	.	.	.	.	.	.	.	.	.	.	.	21
4926	OSTEOSPERMUM SPINESCENS	S	.	0.0	41.90	42.6	.	.	.	.	.	.	.	.	.	.	.	.	.	21
4927	OSTEOSPERMUM SPINOSUM	S	.	0.0	.	6.6	.	.	.	.	.	.	.	.	.	.	.	.	.	21
4928	OSTEOSPERMUM SPINOSUM	S	.	0.0	.	6.9	.	.	.	.	.	.	.	.	.	.	.	.	.	21
4929	OSTRYA CARPINIFOLIA	S	.	0.0	13.00	12.7	.	.	.	.	.	.	.	.	.	.	.	.	.	21
4930	OSYRIS ALBA	S	.	0.0	10.60	36.1	.	.	.	.	.	.	.	.	.	.	.	.	.	21
4931	OTHONNOPSIS INTERMEDIA	S	.	0.0	31.80	44.1	.	.	.	.	.	.	.	.	.	.	.	.	.	21
4932	OXALIS EUROPAEA	S	.	0.0	21.90	47.6	.	.	.	.	.	.	.	.	.	.	.	.	.	20
4933	OXALIS TUBEROSA	R	63	83.8	1.00	0.6	13.80	0.80	0.80	4	34	0.8	.	.	0	0.05	0.07	0.4	37	3
4934	OXYTROPIS FLORIBUNDA	S	.	0.0	34.40	4.0	.	.	5.10	.	.	.	.	.	.	.	.	.	.	19
4935	OXYTROPIS LAGOPUS	W	.	0.0	17.10	2.0	74.30	29.30	6.60	.	.	.	.	.	.	.	.	.	.	6
4936	OXYTROPIS LAMBERTII	S	.	0.0	36.50	7.2	.	.	.	.	.	.	.	.	.	.	.	.	.	20
4937	OXYTROPIS SERICEA	S	.	0.0	41.20	6.0	.	.	.	.	.	.	.	.	.	.	.	.	.	20
4938	OXYTROPIS URALENSIS	S	.	0.0	27.40	4.3	.	.	3.60	.	.	.	.	.	.	.	.	.	.	19
4939	PACHIRA MACROCARPA	S	530	5.4	16.00	39.2	35.90	12.40	3.50	83	286	3.8	72	663	1230	0.03	0.06	3.8	24	2
4940	PACHYCEREUS PECTEN-ABORIGINUM	S	.	0.0	29.40	32.0	.	.	.	.	.	.	.	.	.	.	.	.	.	19
4941	PACHYRRHIZUS ANGULATUS	R	55	84.6	1.70	0.1	13.00	1.40	0.60	15	41	0.8	.	.	0	0.04	0.04	0.3	10	2
4942	PACHYRRHIZUS EROSUS	R	47	86.9	1.40	0.1	11.00	0.60	0.60	.	.	.	.	.	.	.	.	.	.	1
4943	PACHYRRHIZUS EROSUS	R	46	87.4	1.60	0.2	10.30	1.30	0.50	18	17	0.8	2	88	0	0.08	0.06	0.6	15	2
4944	PACHYRRHIZUS EROSUS	R	45	87.8	1.20	0.1	10.60	0.70	0.30	18	16	0.8	.	.	0	0.03	0.03	0.3	21	3
4945	PACHYRRHIZUS SP.	R	55	85.1	1.40	0.2	12.80	0.70	0.50	15	18	0.6	.	.	0	0.04	0.03	0.3	20	4
4946	PACHYRRHIZUS TUBEROSUS	S	.	6.7	26.20	27.3	27.00	7.00	3.60	.	.	.	.	.	.	.	.	.	.	5
4947	PACHYRRHIZUS TUBEROSUS	F	.	86.4	2.60	0.3	12.90	2.90	0.70	121	39	1.3	.	.	345	0.11	0.09	0.8	.	5
4948	PACHYRRHIZUS TUBEROSUS	R	.	82.4	1.50	0.1	15.60	0.60	0.50	16	.	1.1	.	.	.	0.05	0.02	0.2	14	5
4949	PACHYRRHIZUS VERNALIS	S	.	0.0	28.10	24.5	.	.	.	.	.	.	.	.	.	.	.	.	.	20
4950	PAEONIA BROWNII	S	.	0.0	14.10	18.7	.	.	3.00	.	.	.	.	.	.	.	.	.	.	19
4951	PAEONIA PEREGRINA	S	.	0.0	18.80	38.3	.	.	.	.	.	.	.	.	.	.	.	.	.	20
4952	PALIURIS SPINA-CHRISTI	S	.	0.0	31.50	19.2	.	.	.	.	.	.	.	.	.	.	.	.	.	21
4953	PALIURUS SPINA-CHRISTI	S	.	0.0	32.50	21.0	.	.	.	.	.	.	.	.	.	.	.	.	.	20

TABLE 2 - DATA DIRECTLY TRANSCRIBED FROM ORIGINAL SOURCES.

OBS	PLANA	PART	CALORIES	WATER	PROTEIN	FAT	TOTCARB	FIBER	ASH	CALCIUM	PHOSPHO	IRON	SODIUM	POTASS	CAROT	THIAMINE	RIBOFLAV	NIACIN	ASCORBIC	REFCO
4954	PALLENSIS SPINOSA	S	.	0.0	28.8	25.2	.	.	.	.	.	.	.	.	.	.	.	.	.	21
4955	PANAX QUINQUEFOLIUS	CA	.	.	33.0	0.9	53.6	6.7	12.5	360	520	30.6	480	3380	.	.	.	.	.	7
4956	PANAX QUINQUEFOLIUS	R	.	0.0	8.7	0.6	86.1	6.4	4.6	250	270	36.6	52	660	.	.	.	.	.	7
4957	PANAX QUINQUEFOLIUS	SH	.	.	10.3	1.4	79.8	24.0	8.5	1080	330	10.8	56	1720	.	.	.	.	.	7
4958	PANCRATIUM MARITIMUM	S	.	0.0	22.0	37.2	.	.	.	.	.	.	.	.	.	.	.	.	.	21
4959	PANDA OLEOSA	S	498	26.8	15.3	51.5	3.3	.	3.1	85	174	.	.	.	.	.	.	.	.	1
4960	PANDA OLEOSA	S	553	4.8	23.4	45.2	22.9	6.0	3.7	371	523	.	.	.	.	.	.	.	.	1
4961	PANDANUS SP.	F	150	68.5	4.9	8.2	17.9	6.8	0.5	.	.	.	.	.	10	0.09	.	1.4	9	2
4962	PANICUM ANTIDOTALE	W	.	0.0	18.8	3.0	68.1	28.0	10.1	.	.	.	.	.	.	.	.	.	.	18
4963	PANICUM ANTIDOTALE	S	.	0.0	15.0	5.8	.	.	4.0	.	.	.	.	.	.	.	.	.	.	19
4964	PANICUM ANTIDOTALE	S	.	0.0	9.2	5.4	.	.	.	.	.	.	.	.	.	.	.	.	.	21
4965	PANICUM ANTIDOTALE	W	.	0.0	18.9	2.3	65.7	23.5	13.1	610	260	.	.	.	.	.	.	.	.	6
4966	PANICUM ARIZONICUM	W	.	0.0	8.0	1.6	78.0	36.9	12.4	540	130	.	.	.	.	.	.	.	.	6
4967	PANICUM BULBOSUM	W	.	0.0	20.8	3.4	62.7	24.5	13.1	.	.	.	.	.	.	.	.	.	.	6
4968	PANICUM CAPILLARE	W	.	0.0	16.6	1.9	68.3	27.8	13.2	630	270	.	.	.	.	.	.	.	.	6
4969	PANICUM CAPILLARE OCCIDENTALE	W	.	0.0	9.6	2.2	77.6	33.1	10.6	610	220	.	.	.	.	.	.	.	.	6
4970	PANICUM COLORATUM	W	.	0.0	18.9	2.6	67.5	28.6	11.0	.	.	.	.	.	.	.	.	.	.	18
4971	PANICUM COLORATUM	H	.	0.0	9.0	2.8	80.3	36.7	7.9	.	.	.	.	.	.	.	.	.	.	18
4972	PANICUM COLORATUM	S	.	0.0	16.2	4.2	.	.	.	.	.	.	.	.	.	.	.	.	.	20
4973	PANICUM FASCICULATUM	W	.	0.0	18.3	2.5	65.5	27.6	13.7	.	.	.	.	.	.	.	.	.	.	6
4974	PANICUM GEMINATUM	W	.	0.0	9.9	2.1	73.0	31.1	14.9	490	230	.	.	.	.	.	.	.	.	6
4975	PANICUM HALLII	W	.	0.0	11.0	2.0	77.3	29.3	9.7	520	170	.	.	1520	.	.	.	.	.	6
4976	PANICUM HELLERI	W	.	0.0	10.3	2.5	72.6	28.1	14.6	320	100	.	.	.	.	.	.	.	.	6
4977	PANICUM HEMITOMON	W	.	0.0	16.5	1.9	71.3	29.9	10.3	250	380	11.0	.	1590	.	.	.	.	.	6
4978	PANICUM HIRTICAULE	W	.	0.0	11.9	1.8	75.3	40.8	11.0	480	180	.	.	1690	.	.	.	.	.	6
4979	PANICUM LAEVIFOLIUM	W	.	0.0	5.4	1.5	82.6	40.2	10.5	.	.	.	.	.	.	.	.	.	.	18
4980	PANICUM LAEVIFOLIUM	H	.	0.0	8.7	1.9	82.0	39.9	7.4	.	.	.	.	.	.	.	.	.	.	18
4981	PANICUM MASAIENSE	W	.	0.0	19.7	2.5	67.1	24.0	10.7	.	.	.	.	.	.	.	.	.	.	18
4982	PANICUM MAXIMUM	W	.	0.0	20.5	0.9	67.2	24.1	11.4	.	.	.	.	.	.	.	.	.	.	18
4983	PANICUM MAXIMUM	W	.	0.0	5.3	1.4	82.7	39.6	10.6	.	.	.	.	.	.	.	.	.	.	18
4984	PANICUM MAXIMUM	H	.	0.0	11.9	3.2	72.9	31.7	12.0	.	.	.	.	.	.	.	.	.	.	18
4985	PANICUM MAXIMUM	H	.	0.0	5.5	1.4	82.7	40.1	10.4	.	.	.	.	.	.	.	.	.	.	18
4986	PANICUM MAXIMUM	L	.	.	7.9	1.2	75.4	38.4	15.5	.	.	.	.	.	.	.	.	.	.	5
4987	PANICUM MAXIMUM	L	.	.	4.8	0.7	82.2	42.1	12.3	.	.	.	.	.	.	.	.	.	.	5
4988	PANICUM MAXIMUM	L	.	.	7.6	1.2	75.2	38.1	16.0	.	.	.	.	.	.	.	.	.	.	5
4989	PANICUM MAXIMUM	L	.	.	4.8	1.1	81.8	42.1	12.3	.	.	.	.	.	.	.	.	.	.	5
4990	PANICUM MAXIMUM	W	.	0.0	5.9	1.6	81.9	35.7	10.6	2090	590	.	.	.	.	.	.	.	.	6
4991	PANICUM MILIACEUM	H	.	0.0	12.5	2.5	78.4	33.9	6.6	.	.	.	.	.	.	.	.	.	.	18
4992	PANICUM MILIACEUM	S	326	13.5	12.7	3.5	66.5	9.1	3.8	20	270	3.5	.	.	0	0.40	0.10	4.0	0	2
4993	PANICUM MILIACEUM	S	327	11.8	9.9	2.9	72.9	3.2	2.5	20	311	6.8	.	430	0	0.73	0.38	2.3	0	4
4994	PANICUM MILIACEUM	S	.	11.9	12.5	1.1	71.1	2.2	3.4	10	330	5.7	.	.	0	0.78	0.10	1.0	0	5
4995	PANICUM MILIACEUM	L	.	75.3	2.0	0.6	20.3	7.4	1.8	.	.	.	.	.	.	.	.	.	.	5
4996	PANICUM MILIACEUM	L	.	9.7	9.3	2.2	71.5	23.9	7.3	.	.	.	.	.	.	.	.	.	.	5
4997	PANICUM MILIACEUM	W	.	0.0	8.4	2.6	81.5	29.0	7.5	.	.	.	.	.	.	.	.	.	.	6
4998	PANICUM OBTUSUM	W	.	0.0	5.4	1.8	82.3	35.8	10.5	400	200	.	.	2480	.	.	.	.	.	6

TABLE 2 - DATA DIRECTLY TRANSCRIBED FROM ORIGINAL SOURCES.

OBS	PLNA	PART	CALORIES	WATER	PROTEIN	FAT	TOTCARB	FIBER	ASH	CALCIUM	PHOSPHO	IRON	SODIUM	POTASS	CAROT	THIAMINE	RIBOFLAV	NIACIN	ASCORBIC	REFCO
4999	PANICUM PRIONITIS	S	.	0.0	22.2	6.2	.	.	.	.	.	.	.	.	.	.	.	.	.	21
5000	PANICUM PURPURASCENS	W	.	0.0	7.7	1.7	80.2	32.6	10.4	600	340	15.0	.	1550	.	.	.	.	.	6
5001	PANICUM REPENS	W	.	0.0	14.0	2.1	70.5	32.6	13.4	.	.	.	.	.	.	.	.	.	.	18
5002	PANICUM REPENS	L	.	0.0	8.5	3.2	75.6	27.4	12.8	670	730	.	.	.	.	.	.	.	.	5
5003	PANICUM REPENS	L	.	11.4	8.0	1.5	73.9	21.6	5.1	.	.	.	.	.	.	.	.	.	.	5
5004	PANICUM SUMATRENSIS	S	.	11.0	7.2	4.9	73.4	9.6	3.5	24	320	7.0	.	.	.	0.34	.	.	.	5
5005	PANICUM TEXANUM	W	.	0.0	13.4	2.7	66.3	28.3	17.6	230	230	.	.	.	.	.	.	.	.	6
5006	PANICUM TRICHOCLADUM	W	.	0.0	9.6	2.5	79.4	35.0	8.5	.	.	.	.	.	.	.	.	.	.	18
5007	PANICUM TURGIDUM	S	.	0.0	12.4	4.2	.	.	.	.	.	.	.	.	.	.	.	.	.	21
5008	PANICUM VIRGATUM	W	.	0.0	6.0	2.3	85.5	35.0	6.2	330	120	.	.	.	.	.	.	.	.	6
5009	PAPAVER CLAVATUM	S	.	0.0	25.4	34.9	.	.	.	.	.	.	.	.	.	.	.	.	.	21
5010	PAPAVER MONATHUM	S	.	0.0	25.5	47.7	.	.	4.6	.	.	.	.	.	.	.	.	.	.	19
5011	PAPAVER RHOEAS	S	.	0.0	21.8	47.6	.	.	8.0	.	.	.	.	.	.	.	.	.	.	19
5012	PAPAVER RHOEAS	S	.	0.0	25.0	43.0	.	.	.	.	.	.	.	.	.	.	.	.	.	20
5013	PAPAVER RHOEAS	S	.	0.0	21.7	42.3	.	.	.	.	.	.	.	.	.	.	.	.	.	20
5014	PAPAVER RHOEAS	S	.	0.0	26.9	45.1	.	.	.	.	.	.	.	.	.	.	.	.	.	21
5015	PAPAVER RHOEAS	S	.	0.0	27.3	40.1	.	.	.	.	.	.	.	.	.	.	.	.	.	21
5016	PAPAVER SOMNIFERUM	S	533	6.8	18.0	44.7	23.7	6.3	6.8	1448	848	9.4	21	700	0	0.85	0.17	0.98	.	16
5017	PAPAVER SOMNIFERUM	S	.	0.0	22.5	35.0	34.4	9.3	8.1	.	.	.	.	.	.	.	.	.	.	18
5018	PAPAVER SOMNIFERUM	S	.	5.2	24.4	49.1	20.1	5.8	6.0	1450	890	11.1	.	.	0	1.18	1.20	1.28	.	5
5019	PAPAVER SOMNIFERUM	S	.	4.3	22.3	46.5	16.5	4.8	5.6	1030	790	8.5	.	.	0	0.74	0.76	0.80	.	5
5020	PARACARYUM ANGUSTIFOLIUM	S	.	0.0	29.4	39.0	.	.	.	.	.	.	.	.	.	.	.	.	.	21
5021	PARACARYUM CAELESTINUM	S	.	0.0	18.8	20.7	.	.	.	.	.	.	.	.	.	.	.	.	.	20
5022	PARINARI EXCELSA	F	117	66.8	1.3	0.2	30.9	1.5	0.8	38	27	1.7	.	.	290	.	.	.	31	1
5023	PARINARI GLABERRIMUM	F	23	92.0	0.8	0.2	5.2	.	1.8	30	50	4.6	.	.	0	.	.	.	38	2
5024	PARINARI MACROPHYLLA	F	142	59.6	1.4	0.1	37.9	2.4	1.0	42	54	1.7	.	.	40	0.03	0.08	0.70	95	1
5025	PARINARI MACROPHYLLA	S	652	3.0	15.4	64.5	14.4	5.0	2.7	92	497	6.3	.	.	210	0.57	0.07	0.50	4	1
5026	PARINARIUM SP.	S	.	0.0	9.4	73.6	.	.	.	.	.	.	.	.	.	.	.	.	.	20
5027	PARKIA FILICOIDEA	F	.	0.0	13.7	7.3	72.3	19.4	6.7	.	.	.	.	.	.	.	.	.	.	18
5028	PARKIA FILICOIDEA	S	.	0.0	31.8	17.4	46.4	9.4	4.4	.	.	.	.	.	.	.	.	.	.	18
5029	PARKIA SP.	S	426	7.0	32.3	19.5	37.1	4.1	4.1	291	384	33.2	.	.	.	0.30	0.20	3.00	6	1
5030	PARKIA SP.	P	303	13.2	3.4	0.4	80.1	12.6	2.9	124	160	3.6	.	.	2430	1.05	0.71	1.00	242	1
5031	PARKIA SPECIOSA	F	130	70.7	8.0	8.1	11.9	0.5	1.3	76	83	0.7	.	.	.	0.11	0.01	1.00	6	2
5032	PARKINSONIA ACULEATA	S	.	0.0	24.3	6.8	.	.	3.8	.	.	.	.	.	.	.	.	.	.	19
5033	PARKINSONIA ACULEATA	S	.	0.0	5.9	0.8	.	.	7.0	.	.	.	.	.	.	.	.	.	.	19
5034	PARKINSONIA ACULEATA	S	.	0.0	20.5	3.1	.	.	3.4	.	.	.	.	.	.	.	.	.	.	19
5035	PARKINSONIA ACULEATA	S	.	0.0	21.2	3.8	.	.	.	.	.	.	.	.	.	.	.	.	.	20
5036	PARKINSONIA ACULEATA	L	.	0.0	7.5	1.8	73.8	29.0	16.9	4160	170	.	.	.	.	.	.	.	.	5
5037	PARKINSONIA AFRICANA	S	.	0.0	23.8	4.2	.	.	.	.	.	.	.	.	.	.	.	.	.	21
5038	PARMENTIERA ACULEATA	F	150	60.8	4.5	4.9	26.1	20.9	3.7	55	146	.	.	.	120	0.20	0.10	1.20	26	3
5039	PARMENTIERA ACULEATA	F	44	87.6	1.4	0.1	10.6	0.3	0.3	14	29	3.2	.	.	20	0.06	0.06	2.00	5	3
5040	PARONYCHIA KURDICA	S	.	0.0	22.5	9.5	.	.	.	.	.	.	.	.	.	.	.	.	.	21
5041	PARROTIA JACQUEMONTIANA	S	.	0.0	14.8	16.8	.	.	.	.	.	.	.	.	.	.	.	.	.	21
5042	PARTHENIUM ARGENTATUM	S	.	0.0	34.9	24.0	.	.	6.6	.	.	.	.	.	.	.	.	.	.	19
5043	PARTHENIUM INCANUM	L	.	0.0	21.2	6.8	61.4	19.4	10.6	.	.	.	.	.	.	.	.	.	.	6

TABLE 2 - DATA DIRECTLY TRANSCRIBED FROM ORIGINAL SOURCES.

OBS	PLANA	PART	CALORIES	WATER	PROTEIN	FAT	TOTCARB	FIBER	ASH	CALCIUM	PHOSPHO	IRON	SODIUM	POTASS	CAROT	THIAMINE	RIBOFLAV	NIACIN	ASCORBIC	REF
5044	PARTHENOCISSUS QUINQUEFOLIA	S	.	0.0	15.0	33.1	.	.	.	.	.	.	.	.	.	.	.	.	.	19
5045	PARTHENOCISSUS QUINQUEFOLIA	S	.	0.0	16.2	27.5	.	.	.	.	.	.	.	.	.	.	.	.	.	19
5046	PARTHENOCISSUS QUINQUEFOLIA	S	.	0.0	15.6	25.8	.	.	.	.	.	.	.	.	.	.	.	.	.	20
5047	PARTHENOCISSUS TRICUSPIDATA VAR. VEITCHII	S	.	0.0	14.4	31.3	.	.	.	.	.	.	.	.	.	.	.	.	.	20
5048	PASPALIDIUM DESERTORUM	W	.	0.0	13.8	1.2	72.0	29.4	13.0	.	.	.	.	.	.	.	.	.	.	18
5049	PASPALUM ALMUM		.	0.0	6.8	2.2	82.0	30.7	9.0	800	120	.	.	.	.	.	.	.	.	6
5050	PASPALUM CILIATIFOLIUM	W	.	0.0	7.3	1.8	76.2	33.3	14.7	640	150	.	.	.	.	.	.	.	.	6
5051	PASPALUM COMMERSONII	S	335	9.6	5.4	2.0	79.0	9.5	4.0	170	160	.	.	.	.	.	.	.	.	1
5052	PASPALUM CONJUGATUM	W	.	0.0	13.6	1.9	70.9	26.3	13.6	.	.	.	.	.	.	.	.	.	.	18
5053	PASPALUM CONJUGATUM	C	.	0.0	6.8	2.0	87.4	40.6	3.8	.	.	.	.	.	.	.	.	.	.	18
5054	PASPALUM CONJUGATUM	L	.	0.0	14.9	4.9	75.3	28.0	4.9	.	.	.	.	.	.	.	.	.	.	18
5055	PASPALUM CONJUGATUM	L	.	78.7	2.9	0.4	15.1	5.6	2.9	92	43	.	.	.	.	.	.	.	.	5
5056	PASPALUM DILATATUM	W	.	0.0	9.9	1.8	80.1	35.7	8.2	.	.	.	.	.	.	.	.	.	.	18
5057	PASPALUM DILATATUM	S	.	0.0	11.2	5.0	.	.	5.8	.	.	.	.	.	.	.	.	.	.	19
5058	PASPALUM DILATATUM	L	.	75.0	3.0	0.6	18.2	7.2	3.2	140	50	4.0	.	430	.	.	.	.	.	5
5059	PASPALUM DILATATUM	H	.	0.0	12.8	1.9	77.9	34.1	7.4	.	.	.	.	.	.	.	.	.	.	6
5060	PASPALUM DILATATUM	W	.	0.0	12.1	2.5	74.1	29.4	11.3	550	270	17.0	.	1480	.	.	.	.	.	6
5061	PASPALUM DILATATUM(ALL)	W	.	0.0	10.1	2.1	79.1	33.9	8.7	510	200	12.0	.	.	.	.	.	.	.	6
5062	PASPALUM DISTICHUM	W	.	0.0	9.6	2.3	75.6	28.8	12.5	670	380	.	.	2390	30	.	.	.	.	6
5063	PASPALUM FASCICULATUM	W	.	0.0	19.3	2.0	65.4	25.3	13.3	.	.	.	.	.	.	.	.	.	.	18
5064	PASPALUM FLORIDANUM	W	.	0.0	5.6	1.6	83.3	34.0	9.5	470	90	.	.	.	.	.	.	.	.	6
5065	PASPALUM HARTWEGIANUM(IMMATURE)	W	.	0.0	12.5	2.6	72.7	26.7	12.2	570	170	.	.	.	.	.	.	.	.	6
5066	PASPALUM HARTWEGIANUM(MATURE)	W	.	0.0	3.7	1.4	86.3	35.3	8.6	450	60	.	.	.	.	.	.	.	.	6
5067	PASPALUM LANGEI		.	.	.	.	.	.	.	480	220	.	.	.	.	.	.	.	.	6
5068	PASPALUM LIVIDUM	W	.	0.0	5.4	1.6	83.0	30.2	10.0	440	140	.	.	.	.	.	.	.	.	6
5069	PASPALUM NOTATUM	W	.	0.0	13.0	4.0	76.2	34.5	6.8	.	.	.	.	.	.	.	.	.	.	18
5070	PASPALUM NOTATUM	H	.	0.0	12.3	1.4	74.9	33.5	11.4	.	.	.	.	.	.	.	.	.	.	18
5071	PASPALUM NOTATUM	S	.	0.0	7.4	3.1	.	.	3.1	.	.	.	.	.	.	.	.	.	.	19
5072	PASPALUM NOTATUM	W	.	0.0	8.5	1.6	79.7	31.6	10.2	480	300	.	.	1450	.	.	.	.	.	6
5073	PASPALUM PLICATULUM	W	.	0.0	6.4	2.0	81.6	31.4	10.0	530	90	.	.	.	.	.	.	.	.	6
5074	PASPALUM PUBESCENS	W	.	0.0	4.9	1.4	84.5	35.6	9.2	320	80	.	.	.	.	.	.	.	.	6
5075	PASPALUM PUBIFLORUM	W	.	0.0	9.0	1.5	75.0	31.5	14.5	490	100	.	.	.	.	.	.	.	.	6
5076	PASPALUM SCORBICULATUM	H	.	0.0	3.5	1.5	82.7	34.3	12.3	.	.	.	.	.	.	.	.	.	.	18
5077	PASPALUM SCROBICULATUM	W	.	0.0	11.4	1.4	72.9	23.8	14.3	.	.	.	.	.	.	.	.	.	.	18
5078	PASPALUM SCROBICULATUM	S	.	11.6	10.6	4.2	69.2	10.0	4.4	50	284	6.0	.	.	.	0.40	.	.	.	5
5079	PASPALUM SCROBICULATUM	L	.	9.2	2.2	3.2	74.0	29.7	11.0	.	.	.	.	.	.	.	.	.	.	5
5080	PASPALUM SETACEUM	W	.	0.0	8.3	1.9	75.4	30.6	14.4	680	160	.	.	.	.	.	.	.	.	6
5081	PASPALUM STRAMINEUM	W	.	0.0	8.3	2.3	78.1	34.3	11.3	600	150	.	.	.	.	.	.	.	.	6
5082	PASPALUM UNISPICATUM	W	.	0.0	14.4	1.8	67.9	26.5	15.9	710	180	.	.	.	.	.	.	.	.	6
5083	PASPALUM URVILLEI	W	.	0.0	8.4	2.2	78.7	36.3	10.7	540	130	.	.	.	.	.	.	.	.	6
5084	PASSIFLORA EDULIS	F	.	79.0	.	.	.	.	.	25	48	.	.	.	.	0.03	0.10	.	.	1
5085	PASSIFLORA EDULIS	F	90	75.1	2.2	0.7	21.2	.	0.8	13	64	1.6	28	348	420	0.00	0.13	1.5	30	4
5086	PASSIFLORA EDULIS	S	.	5.4	11.1	23.8	58.8	53.7	1.8	80	640	18.0	.	.	.	.	.	.	.	5
5087	PASSIFLORA FOETIDA	F	64	82.0	1.8	0.4	15.2	.	0.6	20	48	0.7	3	341	0	0.01	0.06	1.4	15	2
5088	PASSIFLORA FOETIDA	S	.	0.0	10.8	24.8	.	.	.	.	.	.	.	.	.	.	.	.	.	21

TABLE 2 - DATA DIRECTLY TRANSCRIBED FROM ORIGINAL SOURCES.

OBS	PLANA	PART	CALORIES	WATER	PROTEIN	FAT	TOTCARB	FIBER	ASH	CALCIUM	PHOSPHO	IRON	SODIUM	POTASS	CAROT	THIAMINE	RIBOFLAV	NIACIN	ASCORBIC	REFCO
5089	PASSIFLORA INCARNATA	F	111	72.5	2.3	3.3	21.0	7.3	0.9	14	43	1.6	.	.	.	.	.	.	.	1
5090	PASSIFLORA LIGULARIS	F	94	76.3	2.4	2.8	17.3	4.2	1.2	10	64	0.9	.	.	10	0.00	0.11	1.6	20	3
5091	PASSIFLORA MOLLISSIMA	F	25	92.6	0.5	0.1	6.2	0.6	0.6	8	18	0.4	.	.	40	0.00	0.04	1.5	52	3
5092	PASSIFLORA QUADRANGULARIS	F	77	80.0	2.6	1.9	14.5	4.9	1.0	9	36	0.6	.	.	.	.	.	.	20	1
5093	PASSIFLORA QUADRANGULARIS	F	20	94.4	0.7	0.2	4.3	0.7	0.4	14	17	0.8	.	.	0	0.00	0.03	3.8	15	3
5094	PASSIFLORA QUADRANGULARIS	S	81	78.4	1.9	1.3	17.6	5.6	0.8	9	39	2.9	.	.	20	0.00	0.12	1.9	15	3
5095	PASTINACA SATIVA	R	65	81.2	1.6	0.3	15.2	1.5	1.7	48	65	2.1	.	.	.	.	.	.	.	1
5096	PASTINACA SATIVA	S	.	0.0	15.9	27.3	.	.	4.9	.	.	.	.	.	.	.	.	.	.	19
5097	PASTINACA SATIVA	R	76	79.1	1.7	0.5	17.5	2.0	1.2	50	77	0.7	12	541	20	0.08	0.09	0.2	16	2
5098	PASTINACA SATIVA	S	.	0.0	15.0	22.0	.	.	.	.	.	.	.	.	.	.	.	.	.	20
5099	PASTINACA SATIVA	S	.	0.0	15.5	28.8	.	.	.	.	.	.	.	.	.	.	.	.	.	21
5100	PASTINACA SATIVA	R	.	72.4	1.3	0.3	24.9	1.7	1.1	50	40	0.4	.	.	18	0.31	0.09	0.4	16	5
5101	PASTINACA SATIVA	L	.	0.0	22.9	5.0	.	8.0	.	.	.	.	.	.	.	.	.	.	.	6
5102	PASTINACA SATIVA	Y	.	0.0	6.0	.	.	17.2	.	.	.	.	.	.	.	.	.	.	.	6
5103	PAULLINIA PINNATA	A	69	83.0	0.7	0.6	15.2	0.3	0.5	4	171	.	.	.	.	.	.	.	.	1
5104	PAULOWNIA TOMENTOSA	S	.	0.0	11.9	23.6	.	.	3.8	.	.	.	.	.	.	.	.	.	.	19
5105	PAULOWNIA TOMENTOSA	S	.	0.0	7.8	13.5	.	.	5.2	.	.	.	.	.	.	.	.	.	.	19
5106	PAVETTA CRASSIPES	L	68	79.5	3.2	0.5	15.8	3.1	1.0	170	70	.	.	.	.	.	.	.	.	1
5107	PEDICULARIS GROENLANDICA	S	.	0.0	30.5	13.9	.	.	.	.	.	.	.	.	.	.	.	.	.	20
5108	PEDICULARIS RACEMOSA	W	.	0.0	15.0	1.8	71.6	17.7	11.6	.	.	.	.	.	.	.	.	.	.	6
5109	PEDILANTHUS MACROCARPUS	S	.	0.0	15.9	61.9	.	.	.	.	.	.	.	.	.	.	.	.	.	21
5110	PEGANUM HARMALA	S	.	0.0	29.4	15.4	.	.	.	.	.	.	.	.	.	.	.	.	.	21
5111	PEGANUM HARMALA	S	.	0.0	33.3	17.5	.	.	.	.	.	.	.	.	.	.	.	.	.	21
5112	PEGANUM HARMALA	S	.	0.0	35.2	13.5	.	.	.	.	.	.	.	.	.	.	.	.	.	21
5113	PEGANUM HARMALA	S	.	0.0	34.3	16.8	.	.	.	.	.	.	.	.	.	.	.	.	.	21
5114	PELTARIA ANGUSTIFOLIA	S	.	0.0	24.6	32.4	.	.	.	.	.	.	.	.	.	.	.	.	.	21
5115	PELTOPHORUM INERME	S	.	0.0	14.7	4.7	.	.	.	.	.	.	.	.	.	.	.	.	.	21
5116	PENNISETUM AMERICANUM	X	.	5.1	.	.	.	.	.	.	.	.	.	.	.	0.21	0.32	3.0	.	
5117	PENNISETUM AMERICANUM	S	341	12.0	10.4	4.0	71.6	1.9	2.0	22	286	20.7	.	.	0	0.30	0.22	1.7	3	1
5118	PENNISETUM AMERICANUM	S	349	10.4	9.6	4.0	74.4	2.1	1.6	23	307	7.5	.	.	.	0.30	0.14	1.6	.	1
5119	PENNISETUM AMERICANUM	S	377	8.5	9.7	5.7	74.2	2.4	1.9	17	353	17.3	.	.	.	0.32	0.19	1.6	.	1
5120	PENNISETUM AMERICANUM	W	.	0.0	17.7	3.2	65.7	21.1	13.4	.	.	.	.	.	.	.	.	.	.	18
5121	PENNISETUM AMERICANUM	H	.	0.0	6.6	1.3	82.1	41.2	10.0	.	.	.	.	.	.	.	.	.	.	18
5122	PENNISETUM AMERICANUM	S	335	11.0	11.4	5.0	70.2	1.5	2.4	50	358	9.0	.	.	.	0.31	0.14	1.8	0	2
5123	PENNISETUM AMERICANUM	S	.	0.0	13.4	5.1	79.3	2.5	2.2	60	430	1.0	.	.	.	.	.	.	.	6
5124	PENNISETUM AMERICANUM	H	.	0.0	9.1	2.1	78.7	35.9	10.1	.	.	.	.	.	.	.	.	.	.	6
5125	PENNISETUM AMERICANUM	W	.	0.0	10.0	2.8	78.1	31.2	9.1	.	.	.	.	.	.	.	.	.	.	6
5126	PENNISETUM CILIARE	S	.	0.0	31.2	8.5	.	.	.	.	.	.	.	.	.	.	.	.	.	19
5127	PENNISETUM CILIARE	S	.	0.0	24.4	9.0	.	.	3.0	.	.	.	.	.	.	.	.	.	.	19
5128	PENNISETUM CILIARE	S	.	0.0	30.6	10.0	.	.	.	.	.	.	.	.	.	.	.	.	.	20
5129	PENNISETUM CILIARE	S	.	0.0	24.4	10.9	.	.	.	.	.	.	.	.	.	.	.	.	.	20
5130	PENNISETUM CILIARE	S	.	0.0	28.8	13.6	.	.	.	.	.	.	.	.	.	.	.	.	.	20
5131	PENNISETUM CILIARE	S	.	0.0	30.6	10.6	.	.	.	.	.	.	.	.	.	.	.	.	.	20
5132	PENNISETUM CILIARE	S	.	0.0	33.8	10.0	.	.	.	.	.	.	.	.	.	.	.	.	.	20
5133	PENNISETUM CLANDESTINUM	W	.	0.0	24.8	4.0	61.7	20.9	9.5	.	.	.	.	.	.	.	.	.	.	18

TABLE 2 - DATA DIRECTLY TRANSCRIBED FROM ORIGINAL SOURCES.

OBS	PLANA	PART	CALORIES	WATER	PROTEIN	FAT	TOTCARB	FIBER	ASH	CALCIUM	PHOSPHON	IRON	SODIUM	POTASS	CAROT	THIAMINE	RIBOFLAV	NIACIN	ASCORBIC	REFCO
5134	PENNISETUM CLANDESTINUM	H	.	0.0	12.10	0.8	76.6	35.20	10.5	.	.	.	.	.	.	.	.	.	.	18
5135	PENNISETUM CLANDESTINUM	W	.	0.0	17.70	1.9	69.5	22.90	10.9	430	250	23.0	.	4590	.	.	.	.	.	6
5136	PENNISETUM GLAUCUM	S	.	0.0	11.50	6.1	.	.	1.8	.	.	.	.	.	.	.	.	.	.	19
5137	PENNISETUM GLAUCUM	S	.	0.0	17.50	6.0	.	.	2.1	.	.	.	.	.	.	.	.	.	.	19
5138	PENNISETUM GLAUCUM	S	.	0.0	15.60	5.3	.	.	1.8	.	.	.	.	.	.	.	.	.	.	19
5139	PENNISETUM GLAUCUM	S	.	0.0	16.20	5.7	.	.	2.0	.	.	.	.	.	.	.	.	.	.	19
5140	PENNISETUM ORIENTALE	W	.	0.0	11.00	7.7	74.0	37.30	7.3	.	.	.	.	.	.	.	.	.	.	18
5141	PENNISETUM ORIENTALE	S	.	0.0	13.90	8.1	.	.	.	.	.	.	.	.	.	.	.	.	.	21
5142	PENNISETUM PEDICELLATUM	W	.	0.0	12.50	1.9	73.6	35.20	12.0	.	.	.	.	.	.	.	.	.	.	18
5143	PENNISETUM POLYSTACHYON	W	.	0.0	17.60	2.7	67.8	27.10	11.9	.	.	.	.	.	.	.	.	.	.	18
5144	PENNISETUM PURPUREUM	SH	16	93.7	2.60	0.2	2.1	1.10	1.4	11	71	2.0	.	.	.	.	.	.	.	1
5145	PENNISETUM PURPUREUM	SH	.	0.0	13.20	2.4	74.1	32.90	10.3	.	.	.	.	.	.	.	.	.	.	18
5146	PENNISETUM PURPUREUM	H	.	0.0	15.10	2.4	70.4	34.90	12.1	.	.	.	.	.	.	.	.	.	.	18
5147	PENNISETUM PURPUREUM	W	.	0.0	6.00	2.0	80.6	37.70	11.4	.	.	.	.	.	.	.	.	.	.	6
5148	PENNISETUM PURPUREUM	L	.	0.0	12.30	2.9	78.2	32.50	6.6	.	.	.	.	.	.	.	.	.	.	6
5149	PENNISETUM SP.	S	346	10.8	8.90	3.7	74.9	1.70	1.7	21	279	3.7	.	.	10	0.33	0.14	1.80	.	1
5150	PENNISETUM SQUAMULATUM	W	.	0.0	7.00	2.5	81.9	39.30	8.6	.	.	.	.	.	.	.	.	.	.	18
5151	PENNISETUM STRAMINEUM	W	.	0.0	11.90	3.2	74.3	35.80	10.6	.	.	.	.	.	.	.	.	.	.	18
5152	PENNISETUM TRACHYPHYLLUM	W	.	0.0	10.20	2.0	77.4	30.00	10.4	.	.	.	.	.	.	.	.	.	.	18
5153	PENNISETUM VILLOSUM	S	.	0.0	11.20	4.2	.	.	.	.	.	.	.	.	.	.	.	.	.	20
5154	PENSTEMON ALBIDUS	S	.	0.0	14.60	20.4	.	.	3.7	.	.	.	.	.	.	.	.	.	.	19
5155	PENSTEMON AUSTRALIS	S	.	0.0	18.80	23.0	.	.	.	.	.	.	.	.	.	.	.	.	.	20
5156	PENSTEMON CAMPANULATUS	S	.	0.0	13.10	20.1	.	.	.	.	.	.	.	.	.	.	.	.	.	20
5157	PENSTEMON CAMPANULATUS	S	.	0.0	14.60	23.3	.	.	.	.	.	.	.	.	.	.	.	.	.	21
5158	PENSTEMON CENTRANTHIFOLIUS	S	.	0.0	14.40	28.7	.	.	.	.	.	.	.	.	.	.	.	.	.	20
5159	PENSTEMON CF. KUNTHII	S	.	0.0	12.50	21.7	.	.	.	.	.	.	.	.	.	.	.	.	.	20
5160	PENSTEMON CF. KUNTHII	S	.	0.0	13.10	21.8	.	.	.	.	.	.	.	.	.	.	.	.	.	20
5161	PENSTEMON FENDLERI	S	.	0.0	17.00	20.3	.	.	.	.	.	.	.	.	.	.	.	.	.	20
5162	PENSTEMON GENTIANOIDES	S	.	0.0	13.20	29.4	.	.	.	.	.	.	.	.	.	.	.	.	.	21
5163	PENSTEMON GRANDIFLORUS	S	.	0.0	14.60	16.9	.	.	4.0	.	.	.	.	.	.	.	.	.	.	19
5164	PENSTEMON HAVARDII	S	.	0.0	12.50	21.5	.	.	.	.	.	.	.	.	.	.	.	.	.	20
5165	PENSTEMON KUNTHII	S	.	0.0	13.10	20.1	.	.	.	.	.	.	.	.	.	.	.	.	.	20
5166	PENSTEMON KUNTHII	S	.	0.0	12.30	24.7	.	.	.	.	.	.	.	.	.	.	.	.	.	21
5167	PENSTEMON MULTIFLORUS	S	.	0.0	14.80	29.2	.	.	.	.	.	.	.	.	.	.	.	.	.	21
5168	PENSTEMON MURRAYANUS X GRANDIFLORUS	S	.	0.0	19.70	25.8	.	.	3.8	.	.	.	.	.	.	.	.	.	.	19
5169	PENSTEMON PALMERI	S	.	0.0	11.80	23.4	.	.	.	.	.	.	.	.	.	.	.	.	.	21
5170	PENSTEMON PALMERI	S	.	0.0	11.60	27.5	.	.	.	.	.	.	.	.	.	.	.	.	.	21
5171	PENSTEMON SP.	S	.	0.0	12.50	27.3	.	.	.	.	.	.	.	.	.	.	.	.	.	20
5172	PENSTEMON SP.	S	.	0.0	11.80	18.7	.	.	.	.	.	.	.	.	.	.	.	.	.	21
5173	PENSTEMON SP.	S	.	0.0	14.40	18.0	.	.	.	.	.	.	.	.	.	.	.	.	.	21
5174	PENSTEMON SPECTABILIS	S	.	0.0	11.10	30.2	.	.	2.6	.	.	.	.	.	.	.	.	.	.	19
5175	PENSTEMON SPECTABILIS	S	.	0.0	10.00	26.6	.	.	.	.	.	.	.	.	.	.	.	.	.	20
5176	PENTACLETHRA MACROLOBA	S	597	0.0	24.20	49.5	24.2	2.70	2.5	203	184	17.1	.	.	.	0.07	0.34	0.96	.	10
5177	PENTACLETHRA MACROPHYLLA	S	558	6.2	22.60	46.3	22.6	2.50	2.3	190	172	16.0	.	.	.	0.07	0.32	0.90	.	1
5178	PENTADESMA BUTYRACEA	S	539	5.6	2.30	37.6	53.1	6.70	1.4	190	160	.	.	.	.	.	.	.	.	1

TABLE 2 - DATA DIRECTLY TRANSCRIBED FROM ORIGINAL SOURCES.

OBS	PLANA	PART	CALORIES	WATER	PROTEIN	FAT	TOTCARB	FIBER	ASH	CALCIUM	PHOSPHO	IRON	SODIUM	POTASS	CAROTE	THIAMINE	RIBOFLAV	NIACIN	ASCORBIC	REF
5179	PENTZIA INCANA	SH	.	0.0	13.90	2.6	70.6	25.50	12.9	.	.	.	.	.	.	.	.	.	.	18
5180	PENTZIA INCANA		.	0.0	8.27	.	79.1	33.28	5.8	620.0	200	82.0	.	.	.	.	.	.	.	8
5181	PENTZIA SP.	S	.	0.0	13.80	8.8	.	.	.	.	.	.	.	.	.	.	.	.	.	20
5182	PENTZIA SPHAEROCEPHALA	S	.	0.0	10.00	6.6	.	.	.	.	.	.	.	.	.	.	.	.	.	20
5183	PEPEROMIA PELLUCIDA	L	25	92.2	0.50	0.3	5.9	1.00	1.1	124.0	34	3.2	8	277	2500	0.03	0.07	0.60	10	2
5184	PEPEROMIA PERESKIIFOLIA	L	42	85.0	3.00	0.0	0.8	2.30	2.3	150.0	40	5.7	7	548	13800	0.15	0.25	0.70	5	2
5185	PERESKIA ACULEATA	F	32	91.4	1.00	0.7	6.3	0.70	0.6	174.0	26	0.0	.	.	1930	0.03	0.03	0.50	2	2
5186	PERESKIA ACULEATA	F	11	96.8	0.70	0.1	2.1	0.20	0.3	13.0	9	0.8	.	.	1780	0.02	0.03	1.00	4	3
5187	PERESKIA ACULEATA	F	.	91.4	1.00	0.7	7.0	0.70	0.6	174.0	26	0.0	.	.	1929	0.03	0.03	0.50	2	5
5188	PEREZIA PLATYPHYLLA	S	.	0.0	19.90	11.0	.	.	.	.	.	.	.	.	.	.	.	.	.	20
5189	PERIDERIDIA GAIRDNERI	R	.	64.1	.	.	.	.	.	32.3	.	7.9	.	.	.	0.05	0.08	.	23	9
5190	PERILLA FRUTESCENS	S	.	0.0	31.80	42.0	.	.	4.3	.	.	.	.	.	.	.	.	.	.	19
5191	PERILLA FRUTESCENS	S	42	88.2	2.60	1.3	6.9	3.70	1.0	140.0	63	1.3	.	.	780	0.08	0.09	0.50	20	2
5192	PERILLA FRUTESCENS	L	42	86.0	3.40	0.6	8.0	1.50	2.0	197.0	73	6.7	20	650	4380	0.07	0.32	0.80	46	2
5193	PERILLA FRUTESCENS	S	425	17.8	15.70	26.3	37.0	28.00	3.2	350.0	33	11.1	.	.	10	0.32	0.11	3.10	0	2
5194	PERILLA FRUTESCENS	S	.	0.0	20.20	35.3	.	.	.	.	.	.	.	.	.	.	.	.	.	21
5195	PERILLA FRUTESCENS	S	.	6.3	23.10	45.1	20.6	10.30	4.6	.	.	.	.	.	.	.	.	4.00	.	5
5196	PERIPLOCA GRAECA	S	.	0.0	29.60	18.6	.	.	.	.	.	.	.	.	.	.	.	.	.	21
5197	PERITYLE SP.	S	.	0.0	18.10	23.5	.	.	.	.	.	.	.	.	.	.	.	.	.	20
5198	PERSEA AMERICANA	F	121	80.3	1.40	11.3	6.1	1.80	0.9	19.0	46	1.4	.	.	530	0.05	0.15	2.00	18	1
5199	PERSEA AMERICANA	F	102	79.0	1.10	6.1	13.2	1.00	0.6	12.0	26	0.7	2	278	205	0.05	0.10	1.40	8	2
5200	PERSEA AMERICANA	F	112	81.2	1.50	9.9	6.6	2.00	0.8	14.0	42	0.7	.	.	120	0.06	0.12	1.70	17	3
5201	PERSEA AMERICANA	F	154	77.0	1.70	15.8	4.4	1.80	1.1	10.0	42	1.0	.	.	120	0.08	0.12	1.50	11	3
5202	PERSEA AMERICANA	F	94	83.9	1.30	8.2	5.7	1.30	0.9	6.0	31	0.8	.	.	30	0.06	0.10	1.50	16	3
5203	PERSEA AMERICANA	F	167	74.0	2.10	16.4	6.3	1.60	1.2	10.0	42	0.6	4	604	174	0.11	0.20	1.60	14	4
5204	PERSEA AMERICANA	F	.	73.6	1.70	22.8	0.8	.	1.1	10.0	80	0.7	.	.	.	.	.	.	13	5
5205	PERSEA SCHIEDEANA	F	131	75.9	1.50	9.9	11.9	2.00	0.8	14.0	44	0.5	.	.	60	0.06	0.09	1.10	12	3
5206	PETALOSTEMON CARNEUS	S	.	0.0	31.20	4.9	.	.	.	.	.	.	.	.	.	.	.	.	.	20
5207	PETALOSTEMON EMARGINATUS	S	.	0.0	36.90	4.6	.	.	.	.	.	.	.	.	.	.	.	.	.	20
5208	PETALOSTEMON FEAYI	S	.	0.0	28.00	6.9	.	.	.	.	.	.	.	.	.	.	.	.	.	20
5209	PETALOSTEMON PULCHERRIMUM	S	.	0.0	38.10	5.2	.	.	.	.	.	.	.	.	.	.	.	.	.	20
5210	PETALOSTEMON PURPUREUM	S	.	0.0	36.90	4.4	.	.	.	.	.	.	.	.	.	.	.	.	.	20
5211	PETALOSTEMON SP.	W	.	0.0	10.40	5.6	77.0	25.70	7.0	2820.0	430	.	.	.	.	.	.	.	.	6
5212	PETALOSTEMUM CANDIDUM	S	.	0.0	24.10	6.2	.	.	6.9	.	.	.	.	.	.	.	.	.	.	19
5213	PETALOSTEMUM COMPACTUM	S	.	0.0	34.90	8.5	.	.	3.8	.	.	.	.	.	.	.	.	.	.	19
5214	PETALOSTEMUM PURPUREUM	S	.	0.0	37.20	10.8	.	.	3.9	.	.	.	.	.	.	.	.	.	.	19
5215	PETALOSTEMUM PURPUREUM	S	.	0.0	14.90	3.1	.	.	9.0	.	.	.	.	.	.	.	.	.	.	19
5216	PETALOSTEMUM VILLOSUM	S	.	0.0	29.40	10.6	.	.	3.5	.	.	.	.	.	.	.	.	.	.	19
5217	PETALOSTEMUM VILLOSUM	S	.	0.0	14.30	1.5	.	.	7.6	.	.	.	.	.	.	.	.	.	.	19
5218	PETASITES JAPONICA	L	9	96.4	0.70	0.1	1.9	0.70	0.9	43.0	20	0.1	33	450	10	0.02	0.02	0.20	2	2
5219	PETROSELINUM CRISPUM	L	42	85.8	3.20	0.3	8.9	1.50	1.8	206.0	47	2.5	.	.	.	.	.	.	172	1
5220	PETROSELINUM CRISPUM	L	276	9.0	22.40	4.4	51.7	10.30	12.5	1468.0	351	97.9	452	3805	14004	0.17	1.23	7.93	122	16
5221	PETROSELINUM CRISPUM	L	50	83.8	3.70	0.8	9.5	1.40	2.2	228.0	54	4.6	28	900	1550	0.18	0.25	1.00	153	2
5222	PETROSELINUM CRISPUM	S	.	0.0	19.10	33.0	.	.	.	.	.	.	.	.	.	.	.	.	.	21
5223	PETROSELINUM CRISPUM	L	43	85.9	3.20	0.6	8.5	1.30	1.8	195.0	52	3.1	.	.	3640	0.12	0.24	1.00	146	3

TABLE 2 - DATA DIRECTLY TRANSCRIBED FROM ORIGINAL SOURCES.

OBS	PLANA	PART	CALORIES	WATER	PROTEIN	FAT	TOTCARB	FIBER	ASH	CALCIUM	PHOSPHO	IRON	SODIUM	POTASS	CAROT	THIAMINE	RIBOFLAV	NIACIN	ASCORBIC	REFCO
5224	PETROSELINUM CRISPUM	L	44	85.1	3.6	0.6	8.5	1.5	2.2	203	63	6.2	45	727	5100	0.12	0.26	1.2	172	4
5225	PETROSELINUM CRISPUM	L	.	68.4	5.9	1.0	21.5	1.8	3.2	390	200	17.9	.	.	1920	0.04	.	0.5	281	5
5226	PEUCEDANUM CAPENSE	S	.	0.0	13.1	20.3	.	.	.	.	.	.	.	.	.	.	.	.	.	21
5227	PEUCEDANUM FERULACEUM	S	.	0.0	14.0	22.3	.	.	.	.	.	.	.	.	.	.	.	.	.	21
5228	PHACELIA CONGESTA	S	.	0.0	15.6	2.6	.	.	.	.	.	.	.	.	.	.	.	.	.	20
5229	PHACELIA CONGESTA	S	.	0.0	13.1	6.5	.	.	.	.	.	.	.	.	.	.	.	.	.	20
5230	PHACELIA HETEROPHYLLA	S	.	0.0	16.1	16.5	.	.	.	.	.	.	.	.	.	.	.	.	.	21
5231	PHACELIA HETEROPHYLLA	I	.	0.0	12.8	2.3	70.3	15.3	14.6	.	.	.	.	.	.	.	.	.	.	6
5232	PHACELIA HETEROPHYLLA	W	.	0.0	15.1	2.0	65.3	12.2	17.6	.	.	.	.	.	.	.	.	.	.	6
5233	PHACELIA INTEGRIFOLIA	S	.	0.0	16.6	8.0	.	.	.	.	.	.	.	.	.	.	.	.	.	20
5234	PHACELIA PATULIFLORA	S	.	0.0	10.0	16.4	.	.	.	.	.	.	.	.	.	.	.	.	.	19
5235	PHACELIA ROBUSTA	S	.	0.0	16.9	12.7	.	.	.	.	.	.	.	.	.	.	.	.	.	20
5236	PHACELIA SP.	S	.	0.0	10.4	13.2	.	.	2.3	.	.	.	.	.	.	.	.	.	.	19
5237	PHACELIA SP.	W	.	0.0	16.4	2.1	65.4	22.9	16.1	3870	320	.	.	3010	.	.	.	.	.	6
5238	PHACELIA TANACETIFOLIA	S	.	0.0	18.2	10.4	.	.	5.4	.	.	.	.	.	.	.	.	.	.	19
5239	PHALARIS ANGUSTA(IMMATURE)	W	.	0.0	20.1	3.0	67.6	25.6	9.3	510	250	.	.	2700	.	.	.	.	.	6
5240	PHALARIS AQUATICA	W	.	0.0	12.8	5.1	71.3	23.7	10.8	.	.	.	.	.	.	.	.	.	.	18
5241	PHALARIS AQUATICA	W	.	0.0	15.1	3.6	68.1	28.1	13.2	.	.	.	.	.	.	.	.	.	.	18
5242	PHALARIS AQUATICA	L	.	0.0	5.2	1.4	84.7	33.2	8.6	.	300	.	.	.	.	.	.	.	.	5
5243	PHALARIS ARUNDINACEA	L	.	0.0	14.5	2.2	71.3	30.2	10.2	790	980	.	.	.	.	.	.	.	.	5
5244	PHALARIS ARUNDINACEA	H	.	0.0	8.8	2.2	81.8	34.3	7.2	340	250	15.0	.	2350	.	.	.	.	.	6
5245	PHALARIS ARUNDINACEA	W	.	0.0	13.2	3.7	74.7	26.8	8.4	400	300	.	.	3640	.	.	.	.	.	6
5246	PHALARIS CANARIENSIS	S	.	0.0	20.5	6.0	.	.	5.0	.	.	.	.	.	.	.	.	.	.	19
5247	PHALARIS CANARIENSIS	S	.	0.0	21.3	6.2	.	.	4.8	.	.	.	.	.	.	.	.	.	.	19
5248	PHALARIS CANARIENSIS	S	.	0.0	22.0	8.2	.	.	.	.	.	.	.	.	.	.	.	.	.	21
5249	PHALARIS CANARIENSIS	W	.	0.0	13.1	3.7	73.8	26.8	9.4	440	290	.	.	3170	.	.	.	.	.	6
5250	PHALARIS CAROLINIANA(IMMATURE)	W	.	0.0	10.3	2.6	71.5	28.8	15.6	490	270	.	.	.	.	.	.	.	.	6
5251	PHALARIS COERULESCENS	S	.	0.0	25.0	10.8	.	.	6.1	.	.	.	.	.	.	.	.	.	.	19
5252	PHALARIS MINOR	W	.	0.0	10.0	3.7	68.8	21.2	17.5	.	.	.	.	.	.	.	.	.	.	18
5253	PHALARIS MINOR	L	.	0.0	19.0	3.7	59.7	21.2	17.5	1300	890	.	.	.	.	.	.	.	.	5
5254	PHALARIS MINOR	L	.	0.0	13.9	5.2	61.2	29.9	19.7	720	550	.	.	.	.	.	.	.	.	5
5255	PHALARIS MINOR	W	.	0.0	10.3	2.6	67.9	26.3	19.2	.	.	.	.	.	.	.	.	.	.	6
5256	PHALARIS PARADOXA	S	.	0.0	26.5	9.4	.	.	.	.	.	.	.	.	.	.	.	.	.	21
5257	PHALARIS TUBEROSA STENOPTERA	H	.	0.0	5.4	2.0	86.2	34.8	6.4	.	.	.	.	.	.	.	.	.	.	6
5258	PHASEOLUS ACONITIFOLIUS	H	.	0.0	17.2	1.7	69.1	29.4	12.0	.	.	.	.	.	.	.	.	.	.	6
5259	PHASEOLUS ACUTIFOLIUS	S	353	8.6	19.3	1.2	67.8	4.8	3.1	112	310	.	.	.	.	0.33	0.12	2.8	0	1
5260	PHASEOLUS ACUTIFOLIUS	H	.	0.0	18.5	3.2	66.6	28.1	11.7	.	.	.	.	.	.	.	.	.	.	6
5261	PHASEOLUS ACUTIFOLIUS VAR. LATIFOLIUS	S	.	0.0	32.2	1.1	.	.	4.2	.	.	.	.	.	.	.	.	.	.	19
5262	PHASEOLUS ANGULARIS	S	.	0.0	22.9	0.4	.	.	3.7	.	.	.	.	.	.	.	.	.	.	19
5263	PHASEOLUS AUREUS	S	.	0.0	28.2	0.8	.	.	3.6	.	.	.	.	.	.	.	.	.	.	19
5264	PHASEOLUS AUREUS	S	.	0.0	29.0	1.0	.	.	3.8	.	.	.	.	.	.	.	.	.	.	19
5265	PHASEOLUS AUREUS	S	.	0.0	25.7	0.9	.	.	3.9	.	.	.	.	.	.	.	.	.	.	19
5266	PHASEOLUS AUREUS	S	.	0.0	23.4	1.5	.	.	4.7	.	.	.	.	.	.	.	.	.	.	19
5267	PHASEOLUS AUREUS	S	.	0.0	22.8	1.1	.	.	4.1	.	.	.	.	.	.	.	.	.	.	19
5268	PHASEOLUS AUREUS	S	.	0.0	25.5	0.9	.	.	3.9	.	.	.	.	.	.	.	.	.	.	19

TABLE 2 - DATA DIRECTLY TRANSCRIBED FROM ORIGINAL SOURCES.

OBS	PLNA	PART	CALORIES	WATER	PROTEIN	FAT	TOTLCARB	FIBER	ASH	CALCIUM	PHOSPHO	IRON	SODIUM	POTASS	CAROT	THIAMINE	RIBOFLAV	NIACIN	ASCORBIC	REFCO
5269	PHASEOLUS AUREUS	S	.	0.0	26.1	0.9	.	.	3.7	.	.	.	.	.	.	.	.	.	.	19
5270	PHASEOLUS AUREUS	S	.	0.0	28.4	1.1	.	.	4.6	.	.	.	.	.	.	.	.	.	.	19
5271	PHASEOLUS AUREUS	S	.	0.0	25.6	0.9	.	.	4.6	.	.	.	.	.	.	.	.	.	.	19
5272	PHASEOLUS AUREUS	S	.	0.0	23.6	0.7	.	.	.	.	.	.	.	.	.	.	.	.	.	21
5273	PHASEOLUS AUREUS	S	.	0.0	23.9	0.9	.	.	.	.	.	.	.	.	.	.	.	.	.	21
5274	PHASEOLUS AUREUS	W	.	0.0	17.6	2.3	67.0	25.6	13.1	.	.	.	.	.	.	.	.	.	.	6
5275	PHASEOLUS AUREUS	H	.	0.0	11.0	2.4	77.7	26.2	8.9	.	.	.	.	.	.	.	.	.	.	6
5276	PHASEOLUS COCCINEUS	S	.	0.0	26.1	1.8	.	.	5.7	.	.	.	.	.	.	.	.	.	.	19
5277	PHASEOLUS COCCINEUS	S	.	0.0	22.1	1.9	.	.	4.8	.	.	.	.	.	.	.	.	.	.	19
5278	PHASEOLUS COCCINEUS	S	.	0.0	27.2	1.8	.	.	6.1	.	.	.	.	.	.	.	.	.	.	19
5279	PHASEOLUS COCCINEUS	S	.	0.0	19.5	2.4	.	.	4.8	.	.	.	.	.	.	.	.	.	.	19
5280	PHASEOLUS COCCINEUS	IS	250	34.2	16.4	0.3	46.3	12.2	2.8	61	277	4.1	.	.	30	0.54	0.14	2.3	0	3
5281	PHASEOLUS COCCINEUS	S	328	13.6	22.8	1.5	58.1	5.0	4.0	128	328	5.4	.	.	10	0.30	0.11	2.1	.	3
5282	PHASEOLUS COCCINEUS	S	338	12.5	20.3	1.8	62.0	4.8	3.4	114	354	9.0	.	.	.	0.50	0.19	2.3	2	3
5283	PHASEOLUS COCCINEUS	G	.	58.3	7.4	1.0	31.7	1.9	1.6	50	160	2.6	.	.	34	0.34	0.19	.	27	5
5284	PHASEOLUS LUNATUS	S	335	12.7	21.4	1.4	61.1	4.3	3.4	116	387	4.9	.	.	.	0.33	0.16	2.1	1	1
5285	PHASEOLUS LUNATUS	F	.	0.0	18.8	0.6	76.6	17.5	4.0	.	.	.	.	.	.	.	.	.	.	18
5286	PHASEOLUS LUNATUS	S	.	0.0	27.2	0.9	66.4	5.2	5.5	.	.	.	.	.	.	.	.	.	.	18
5287	PHASEOLUS LUNATUS	S	.	0.0	23.8	0.8	.	.	5.6	.	.	.	.	.	.	.	.	.	.	19
5288	PHASEOLUS LUNATUS	S	.	0.0	23.4	1.0	.	.	5.3	.	.	.	.	.	.	.	.	.	.	19
5289	PHASEOLUS LUNATUS	S	.	0.0	26.5	0.6	.	.	4.2	.	.	.	.	.	.	.	.	.	.	19
5290	PHASEOLUS LUNATUS	S	.	0.0	20.4	1.4	.	.	4.8	.	.	.	.	.	.	.	.	.	.	19
5291	PHASEOLUS LUNATUS	S	.	0.0	24.2	1.2	.	.	4.2	.	.	.	.	.	.	.	.	.	.	19
5292	PHASEOLUS LUNATUS	S	346	10.5	19.8	1.3	65.4	.	3.0	90	240	5.6	18	295	0	0.46	0.21	1.4	0	2
5293	PHASEOLUS LUNATUS	IS	119	68.5	8.4	0.5	21.0	1.0	1.6	25	119	2.2	2	747	90	0.16	0.16	1.5	30	2
5294	PHASEOLUS LUNATUS	L	8	97.2	0.6	0.0	1.7	.	0.5	8	36	2.3	.	.	.	.	.	.	.	2
5295	PHASEOLUS LUNATUS	X	110	64.0	13.0	0.8	20.0	0.6	2.2	109	382	8.2	.	.	30	0.17	0.14	2.0	7	2
5296	PHASEOLUS LUNATUS	IS	146	61.3	9.2	0.1	27.9	1.9	1.5	36	113	3.2	.	.	380	0.25	0.11	1.5	3	3
5297	PHASEOLUS LUNATUS	IS	137	63.7	10.4	0.4	23.9	2.6	1.6	44	142	2.4	.	.	0	0.24	0.13	1.0	14	3
5298	PHASEOLUS LUNATUS	S	336	12.0	20.7	1.2	62.4	4.9	3.7	113	330	4.8	.	.	0	0.34	0.21	2.2	0	3
5299	PHASEOLUS LUNATUS	IS	123	67.5	8.4	0.5	22.1	1.8	1.5	52	142	2.8	2	650	174	0.24	0.12	1.4	29	4
5300	PHASEOLUS LUNATUS	H	.	0.0	12.5	1.6	74.3	29.3	11.6	.	.	.	.	.	.	.	.	.	.	6
5301	PHASEOLUS LUNATUS	L	.	0.0	17.3	4.8	67.4	11.1	10.5	.	.	.	.	.	.	.	.	.	.	6
5302	PHASEOLUS LUNATUS	Y	.	0.0	9.2	2.4	.	39.2	.	.	.	.	.	.	.	.	.	.	.	6
5303	PHASEOLUS TRILOBUS	S	323	11.2	24.5	0.6	57.2	7.7	6.5	.	.	.	.	.	.	.	.	.	.	1
5304	PHASEOLUS VULGARIS	S	336	12.3	21.7	1.5	60.9	4.4	3.6	120	323	8.2	.	.	10	0.37	0.16	2.4	1	1
5305	PHASEOLUS VULGARIS	L	36	86.8	3.6	0.4	6.6	2.8	2.6	274	75	9.2	.	.	3240	0.18	0.06	1.3	110	1
5306	PHASEOLUS VULGARIS	F	36	88.7	2.5	0.2	7.9	1.8	0.7	43	48	1.4	.	.	750	0.08	0.12	0.5	27	1
5307	PHASEOLUS VULGARIS	H	.	0.0	6.3	0.7	85.8	43.4	7.2	1150	90	.	.	.	.	.	.	.	.	18
5308	PHASEOLUS VULGARIS	S	.	0.0	24.2	2.4	68.7	6.7	4.7	260	430	.	.	.	.	.	.	.	.	18
5309	PHASEOLUS VULGARIS	S	.	0.0	24.2	1.2	.	.	4.7	.	.	.	.	.	.	.	.	.	.	19
5310	PHASEOLUS VULGARIS	S	.	0.0	27.6	1.0	.	.	4.2	.	.	.	.	.	.	.	.	.	.	19
5311	PHASEOLUS VULGARIS	S	.	0.0	28.8	1.0	.	.	4.4	.	.	.	.	.	.	.	.	.	.	19
5312	PHASEOLUS VULGARIS	S	.	0.0	29.4	0.9	.	.	4.2	.	.	.	.	.	.	.	.	.	.	19
5313	PHASEOLUS VULGARIS	S	.	0.0	29.2	1.0	.	.	5.3	.	.	.	.	.	.	.	.	.	.	19

TABLE 2 - DATA DIRECTLY TRANSCRIBED FROM ORIGINAL SOURCES.

OBS	PLNA	PART	CALORIES	WATER	PROTEIN	FAT	TOTLCARB	FIBER	ASH	CALCIUM	PHOSPHO	IRON	SODIUM	POTASS	CAROT	THIAMINE	RIBOFLAV	NIACIN	ASCORBIC	REFCO
5314	PHASEOLUS VULGARIS	S	.	0	31.9	1.1	.	.	5.2	.	.	.	.	.	.	.	.	.	.	19
5315	PHASEOLUS VULGARIS	S	.	0	27.8	0.9	.	.	4.9	.	.	.	.	.	.	.	.	.	.	19
5316	PHASEOLUS VULGARIS	S	.	0	28.9	1.3	.	.	4.1	.	.	.	.	.	.	.	.	.	.	19
5317	PHASEOLUS VULGARIS	S	.	0	21.1	1.1	.	.	4.2	.	.	.	.	.	.	.	.	.	.	19
5318	PHASEOLUS VULGARIS	S	.	0	26.4	1.1	.	.	3.8	.	.	.	.	.	.	.	.	.	.	19
5319	PHASEOLUS VULGARIS	S	.	0	28.9	0.9	.	.	5.6	.	.	.	.	.	.	.	.	.	.	19
5320	PHASEOLUS VULGARIS	S	.	0	28.2	1.1	.	.	6.2	.	.	.	.	.	.	.	.	.	.	19
5321	PHASEOLUS VULGARIS	S	.	0	26.6	1.0	.	.	5.2	.	.	.	.	.	.	.	.	.	.	19
5322	PHASEOLUS VULGARIS	S	.	0	24.9	1.0	.	.	4.8	.	.	.	.	.	.	.	.	.	.	19
5323	PHASEOLUS VULGARIS	S	.	0	26.2	0.8	.	.	4.9	.	.	.	.	.	.	.	.	.	.	19
5324	PHASEOLUS VULGARIS	S	.	0	.	1.0	.	.	4.4	.	.	.	.	.	.	.	.	.	.	19
5325	PHASEOLUS VULGARIS	S	.	0	25.8	0.9	.	.	5.6	.	.	.	.	.	.	.	.	.	.	19
5326	PHASEOLUS VULGARIS	S	.	0	26.1	1.0	.	.	4.5	.	.	.	.	.	.	.	.	.	.	19
5327	PHASEOLUS VULGARIS	S	.	0	25.9	1.2	.	.	5.0	.	.	.	.	.	.	.	.	.	.	19
5328	PHASEOLUS VULGARIS	S	.	0	26.9	1.0	.	.	4.8	.	.	.	.	.	.	.	.	.	.	19
5329	PHASEOLUS VULGARIS	S	.	0	26.6	1.0	.	.	4.9	.	.	.	.	.	.	.	.	.	.	19
5330	PHASEOLUS VULGARIS	S	.	0	25.0	1.2	.	.	5.2	.	.	.	.	.	.	.	.	.	.	19
5331	PHASEOLUS VULGARIS	S	.	0	24.5	1.0	.	.	4.7	.	.	.	.	.	.	.	.	.	.	19
5332	PHASEOLUS VULGARIS	S	.	0	25.6	1.2	.	.	4.5	.	.	.	.	.	.	.	.	.	.	19
5333	PHASEOLUS VULGARIS	S	.	0	25.5	1.1	.	.	4.1	.	.	.	.	.	.	.	.	.	.	19
5334	PHASEOLUS VULGARIS	S	.	0	22.9	1.3	.	.	4.0	.	.	.	.	.	.	.	.	.	.	19
5335	PHASEOLUS VULGARIS	S	.	0	25.9	1.0	.	.	4.9	.	.	.	.	.	.	.	.	.	.	19
5336	PHASEOLUS VULGARIS	S	.	0	26.8	1.0	.	.	4.8	.	.	.	.	.	.	.	.	.	.	19
5337	PHASEOLUS VULGARIS	S	.	0	23.6	1.0	.	.	4.4	.	.	.	.	.	.	.	.	.	.	19
5338	PHASEOLUS VULGARIS	S	.	0	23.0	1.1	.	.	4.1	.	.	.	.	.	.	.	.	.	.	19
5339	PHASEOLUS VULGARIS	S	.	0	25.2	1.1	.	.	4.4	.	.	.	.	.	.	.	.	.	.	19
5340	PHASEOLUS VULGARIS	S	.	0	25.4	2.1	.	.	4.6	.	.	.	.	.	.	.	.	.	.	19
5341	PHASEOLUS VULGARIS	S	.	0	29.4	1.0	.	.	5.4	.	.	.	.	.	.	.	.	.	.	19
5342	PHASEOLUS VULGARIS	S	.	0	23.7	1.1	.	.	4.5	.	.	.	.	.	.	.	.	.	.	19
5343	PHASEOLUS VULGARIS	S	.	0	22.4	1.1	.	.	4.2	.	.	.	.	.	.	.	.	.	.	19
5344	PHASEOLUS VULGARIS	S	.	0	23.2	1.1	.	.	4.7	.	.	.	.	.	.	.	.	.	.	19
5345	PHASEOLUS VULGARIS	S	.	0	24.2	1.4	.	.	5.0	.	.	.	.	.	.	.	.	.	.	19
5346	PHASEOLUS VULGARIS	S	.	0	28.5	1.0	.	.	5.2	.	.	.	.	.	.	.	.	.	.	19
5347	PHASEOLUS VULGARIS	S	.	0	26.6	0.9	.	.	4.9	.	.	.	.	.	.	.	.	.	.	19
5348	PHASEOLUS VULGARIS	S	.	0	27.2	0.8	.	.	4.6	.	.	.	.	.	.	.	.	.	.	19
5349	PHASEOLUS VULGARIS	S	.	0	26.0	1.2	.	.	5.4	.	.	.	.	.	.	.	.	.	.	19
5350	PHASEOLUS VULGARIS	S	.	0	25.8	1.3	.	.	4.5	.	.	.	.	.	.	.	.	.	.	19
5351	PHASEOLUS VULGARIS	S	.	0	30.6	0.8	.	.	5.3	.	.	.	.	.	.	.	.	.	.	19
5352	PHASEOLUS VULGARIS	S	.	0	29.9	0.9	.	.	4.5	.	.	.	.	.	.	.	.	.	.	19
5353	PHASEOLUS VULGARIS	S	.	0	26.0	1.0	.	.	4.4	.	.	.	.	.	.	.	.	.	.	19
5354	PHASEOLUS VULGARIS	S	.	0	30.8	1.3	.	.	5.3	.	.	.	.	.	.	.	.	.	.	19
5355	PHASEOLUS VULGARIS	S	.	0	28.6	1.1	.	.	4.6	.	.	.	.	.	.	.	.	.	.	19
5356	PHASEOLUS VULGARIS	S	.	0	31.0	1.1	.	.	4.5	.	.	.	.	.	.	.	.	.	.	19
5357	PHASEOLUS VULGARIS	S	.	0	27.7	1.1	.	.	4.4	.	.	.	.	.	.	.	.	.	.	19
5358	PHASEOLUS VULGARIS	S	.	0	27.2	1.2	.	.	5.3	.	.	.	.	.	.	.	.	.	.	19

TABLE 2 - DATA DIRECTLY TRANSCRIBED FROM ORIGINAL SOURCES.

OBS	PLNA	PART	CALORIES	WATER	PROTEIN	FAT	TOTLCARB	FIBER	ASH	CALCIUM	PHOSPHO	IRON	SODIUM	POTASS	CAROT	THIAMINE	RIBOFLAV	NIACIN	ASCORBIC	REFCO
5359	PHASEOLUS VULGARIS	S	.	0.0	28.1	1.0	.	.	4.7	.	.	.	.	.	.	.	.	.	.	19
5360	PHASEOLUS VULGARIS	S	.	0.0	26.9	1.0	.	.	5.6	.	.	.	.	.	.	.	.	.	.	19
5361	PHASEOLUS VULGARIS	S	.	0.0	24.7	1.0	.	.	4.7	.	.	.	.	.	.	.	.	.	.	19
5362	PHASEOLUS VULGARIS	S	.	0.0	27.9	1.2	.	.	4.8	.	.	.	.	.	.	.	.	.	.	19
5363	PHASEOLUS VULGARIS	S	.	0.0	30.2	1.2	.	.	5.6	.	.	.	.	.	.	.	.	.	.	19
5364	PHASEOLUS VULGARIS	S	.	0.0	33.8	1.0	.	.	4.7	.	.	.	.	.	.	.	.	.	.	19
5365	PHASEOLUS VULGARIS	S	.	0.0	33.9	0.9	.	.	4.9	.	.	.	.	.	.	.	.	.	.	19
5366	PHASEOLUS VULGARIS	S	.	0.0	33.5	0.9	.	.	4.9	.	.	.	.	.	.	.	.	.	.	19
5367	PHASEOLUS VULGARIS	S	.	0.0	30.2	1.1	.	.	4.5	.	.	.	.	.	.	.	.	.	.	19
5368	PHASEOLUS VULGARIS	S	.	0.0	34.9	1.0	.	.	4.8	.	.	.	.	.	.	.	.	.	.	19
5369	PHASEOLUS VULGARIS	S	.	0.0	30.9	1.0	.	.	4.9	.	.	.	.	.	.	.	.	.	.	19
5370	PHASEOLUS VULGARIS	S	.	0.0	27.8	1.1	.	.	4.0	.	.	.	.	.	.	.	.	.	.	19
5371	PHASEOLUS VULGARIS	S	.	0.0	29.5	1.2	.	.	4.4	.	.	.	.	.	.	.	.	.	.	19
5372	PHASEOLUS VULGARIS	S	.	0.0	31.1	1.1	.	.	4.2	.	.	.	.	.	.	.	.	.	.	19
5373	PHASEOLUS VULGARIS	S	.	0.0	29.9	1.3	.	.	4.6	.	.	.	.	.	.	.	.	.	.	19
5374	PHASEOLUS VULGARIS	S	.	0.0	31.5	1.4	.	.	4.3	.	.	.	.	.	.	.	.	.	.	19
5375	PHASEOLUS VULGARIS	S	.	0.0	30.3	1.1	.	.	4.1	.	.	.	.	.	.	.	.	.	.	19
5376	PHASEOLUS VULGARIS	S	.	0.0	30.4	1.0	.	.	4.2	.	.	.	.	.	.	.	.	.	.	19
5377	PHASEOLUS VULGARIS	S	.	0.0	31.4	1.1	.	.	4.6	.	.	.	.	.	.	.	.	.	.	19
5378	PHASEOLUS VULGARIS	S	.	0.0	39.4	0.8	.	.	5.1	.	.	.	.	.	.	.	.	.	.	19
5379	PHASEOLUS VULGARIS	S	.	0.0	27.9	1.1	.	.	4.7	.	.	.	.	.	.	.	.	.	.	19
5380	PHASEOLUS VULGARIS	S	.	0.0	27.1	1.1	.	.	4.6	.	.	.	.	.	.	.	.	.	.	19
5381	PHASEOLUS VULGARIS	S	.	0.0	30.9	0.9	.	.	4.8	.	.	.	.	.	.	.	.	.	.	19
5382	PHASEOLUS VULGARIS	S	.	0.0	29.9	0.9	.	.	4.6	.	.	.	.	.	.	.	.	.	.	19
5383	PHASEOLUS VULGARIS	S	.	0.0	27.2	0.8	.	.	5.3	.	.	.	.	.	.	.	.	.	.	19
5384	PHASEOLUS VULGARIS	S	.	0.0	30.9	1.0	.	.	5.4	.	.	.	.	.	.	.	.	.	.	19
5385	PHASEOLUS VULGARIS	S	.	0.0	34.1	0.8	.	.	4.5	.	.	.	.	.	.	.	.	.	.	19
5386	PHASEOLUS VULGARIS	S	.	0.0	30.5	1.0	.	.	4.2	.	.	.	.	.	.	.	.	.	.	19
5387	PHASEOLUS VULGARIS	S	.	0.0	29.9	0.9	.	.	5.5	.	.	.	.	.	.	.	.	.	.	19
5388	PHASEOLUS VULGARIS	S	.	0.0	28.9	0.8	.	.	5.2	.	.	.	.	.	.	.	.	.	.	19
5389	PHASEOLUS VULGARIS	S	.	0.0	28.7	1.0	.	.	5.2	.	.	.	.	.	.	.	.	.	.	19
5390	PHASEOLUS VULGARIS	S	.	0.0	30.3	0.9	.	.	4.7	.	.	.	.	.	.	.	.	.	.	19
5391	PHASEOLUS VULGARIS	S	.	0.0	29.2	0.9	.	.	3.9	.	.	.	.	.	.	.	.	.	.	19
5392	PHASEOLUS VULGARIS	S	.	0.0	27.6	1.0	.	.	5.1	.	.	.	.	.	.	.	.	.	.	19
5393	PHASEOLUS VULGARIS	S	.	0.0	30.9	0.9	.	.	4.8	.	.	.	.	.	.	.	.	.	.	19
5394	PHASEOLUS VULGARIS	S	.	0.0	29.1	0.9	.	.	4.7	.	.	.	.	.	.	.	.	.	.	19
5395	PHASEOLUS VULGARIS	S	.	0.0	26.2	1.9	.	.	4.2	.	.	.	.	.	.	.	.	.	.	19
5396	PHASEOLUS VULGARIS	F	28	91.5	1.9	0.4	5.6	1.0	0.6	75	49	1.2	2	196	323	0.05	0.10	0.6	18	2
5397	PHASEOLUS VULGARIS	S	337	12.0	22.0	1.6	60.8	4.3	3.6	86	247	7.6	.	.	10	0.54	0.19	2.1	3	3
5398	PHASEOLUS VULGARIS	F	36	90.5	2.0	0.2	6.6	1.2	0.7	55	45	1.7	.	.	220	0.08	0.11	0.6	18	3
5399	PHASEOLUS VULGARIS	IS	150	60.4	9.8	0.3	27.8	2.3	1.7	59	213	3.6	.	.	20	0.38	0.12	1.5	7	3
5400	PHASEOLUS VULGARIS	F	32	90.1	1.9	0.2	7.1	1.0	0.7	56	44	0.8	7	243	360	0.08	0.11	0.5	19	4
5401	PHASEOLUS VULGARIS	S	343	10.4	22.5	1.5	61.9	4.2	3.7	110	406	6.9	10	984	12	0.51	0.20	2.3	.	4
5402	PHASEOLUS VULGARIS	H	.	0.0	8.2	1.8	82.7	36.7	7.3	.	.	.	.	.	.	.	.	.	.	6
5403	PHASEOLUS VULGARIS	L	.	0.0	29.7	5.4	48.2	14.0	16.7	.	.	.	.	.	.	.	.	.	.	6

TABLE 2 - DATA DIRECTLY TRANSCRIBED FROM ORIGINAL SOURCES.

OBS	PLNA	PART	CALORIES	WATER	PROTEIN	FAT	TOTLCARB	FIBER	ASH	CALCIUM	PHOSPHO	IRON	SODIUM	POTASS	CAROT	THIAMINE	RIBOFLAV	NIACIN	ASCORBIC	REFCO
5404	PHELLODENDRON AMURENSE	S	.	0.0	35.8	51.7	.	.	.	.	.	.	.	.	.	.	.	.	.	21
5405	PHILADELPHUS GRANDIFLORUS	S	.	0.0	22.6	36.4	.	.	.	.	.	.	.	.	.	.	.	.	.	20
5406	PHILLYREA ANGUSTIFOLIA	S	.	0.0	20.2	21.0	.	.	.	.	.	.	.	.	.	.	.	.	.	21
5407	PHILLYREA LATIFOLIA	S	.	0.0	13.6	13.4	.	.	4.3	.	.	.	.	.	.	.	.	.	.	19
5408	PHILLYREA LATIFOLIA	S	.	0.0	3.3	1.6	.	.	3.8	.	.	.	.	.	.	.	.	.	.	19
5409	PHILLYREA LATIFOLIA	S	.	0.0	12.4	9.2	.	.	.	.	.	.	.	.	.	.	.	.	.	21
5410	PHILLYREA MEDIA	S	.	0.0	16.0	18.1	.	.	.	.	.	.	.	.	.	.	.	.	.	21
5411	PHLEUM PRATENSE	S	.	0.0	21.9	4.4	.	.	.	.	.	.	.	.	.	.	.	.	.	20
5412	PHLEUM PRATENSE	S	.	0.0	19.4	4.8	.	.	.	.	.	.	.	.	.	.	.	.	.	20
5413	PHLEUM PRATENSE	L	.	76.1	4.7	0.9	15.7	4.6	2.6	140	90	.	.	500	.	.	.	.	.	5
5414	PHLEUM PRATENSE	L	.	11.0	9.7	2.7	70.1	27.9	6.5	.	.	.	.	.	.	.	.	.	.	5
5415	PHLEUM PRATENSE	H	.	0.0	7.6	2.4	84.4	34.3	5.6	370	190	13.0	.	1660	1	0.18	1.01	2.95	.	6
5416	PHLEUM PRATENSE	W	.	0.0	11.3	3.4	78.3	28.1	7.0	470	340	23.0	.	1740	22	0.29	1.14	.	.	6
5417	PHLEUM SP.	S	.	0.0	27.1	22.1	.	.	.	.	.	.	.	.	.	.	.	.	.	21
5418	PHLOMIS ARMENIACA	S	.	0.0	35.9	26.3	.	.	.	.	.	.	.	.	.	.	.	.	.	21
5419	PHLOMIS HERBA-VENTI	S	.	0.0	34.9	31.1	.	.	.	.	.	.	.	.	.	.	.	.	.	21
5420	PHLOX PANICULATA	S	.	0.0	41.2	23.5	.	.	.	.	.	.	.	.	.	.	.	.	.	19
5421	PHLOX SP.	W	.	0.0	8.5	1.9	65.3	22.4	24.3	1600	210	.	.	3170	.	.	.	.	.	6
5422	PHOENIX DACTYLIFERA	F	142	60.2	1.6	0.6	36.6	3.3	1.0	34	350	6.0	.	.	145	.	.	5.40	30	1
5423	PHOENIX DACTYLIFERA	F	.	0.0	2.9	1.0	90.4	6.5	5.7	.	.	.	.	.	.	.	.	.	.	18
5424	PHOENIX DACTYLIFERA	S	.	0.0	5.8	7.1	85.3	30.0	1.8	.	.	.	.	.	.	.	.	.	.	18
5425	PHOENIX DACTYLIFERA	F	143	59.5	0.9	0.3	38.2	1.7	1.1	51	30	1.3	.	.	100	0.07	0.05	0.60	6	3
5426	PHOENIX DACTYLIFERA	F	224	36.8	1.4	0.4	59.9	2.7	1.5	60	29	0.7	.	.	40	0.09	0.08	0.10	1	3
5427	PHOENIX DACTYLIFERA	F	274	22.5	2.2	0.5	72.9	2.3	1.9	59	63	3.0	1	648	30	0.09	0.10	2.20	0	4
5428	PHOENIX DACTYLIFERA	S	.	7.9	5.2	6.8	79.1	13.6	0.9	.	.	.	.	.	.	.	.	.	.	5
5429	PHOENIX DACTYLIFERA	F	.	15.3	2.5	0.4	79.7	3.9	2.1	120	50	.	.	.	26	0.01	0.02	0.90	3	5
5430	PHOENIX FARINIFERA	S	.	0.0	7.5	9.3	.	.	1.4	.	.	.	.	.	.	.	.	.	.	19
5431	PHOENIX RECLINATA	S	227	42.6	1.8	0.2	53.8	2.9	1.6	32	56	4.7	.	.	.	0.00	.	.	.	1
5432	PHOENIX SYLVESTRIS	L	.	0.0	11.6	1.6	78.8	27.7	8.0	600	180	.	.	.	.	.	.	.	.	18
5433	PHORADENDRON SEROTINUM	L	.	0.0	22.0	5.6	66.5	19.8	5.9	.	.	.	.	.	.	.	.	.	.	6
5434	PHOTINIA GLABRA	S	.	0.0	29.4	21.8	.	.	.	.	.	.	.	.	.	.	.	.	.	21
5435	PHRAGMITES AUSTRALIS	L	.	0.0	11.4	2.3	74.2	31.0	10.8	940	390	.	.	.	.	.	.	.	.	5
5436	PHRAGMITES AUSTRALIS	SH	.	5.3	5.2	0.9	89.0	32.0	5.8	.	.	.	.	.	.	.	.	.	.	5
5437	PHRAGMITES AUSTRALIS	W	.	0.0	10.6	2.1	72.7	31.9	14.5	480	60	.	.	.	.	.	.	.	.	6
5438	PHRAGMITES AUSTRALIS	L	.	0.0	17.1	3.5	63.7	27.4	15.7	.	.	.	.	.	.	.	.	.	.	6
5439	PHRAGMITES AUSTRALIS	Y	.	0.0	4.8	0.8	90.0	41.2	4.4	.	.	.	.	.	.	.	.	.	.	6
5440	PHYLICA EXCELSA	S	.	0.0	19.4	19.7	.	.	.	.	.	.	.	.	.	.	.	.	.	21
5441	PHYLICA EXCELSA VAR. PAPILLOSA	S	.	0.0	24.6	9.1	.	.	.	.	.	.	.	.	.	.	.	.	.	21
5442	PHYLICA OLEAEFOLIA	S	.	0.0	17.1	15.1	.	.	.	.	.	.	.	.	.	.	.	.	.	21
5443	PHYLICA PINEA	S	.	0.0	22.0	24.3	.	.	.	.	.	.	.	.	.	.	.	.	.	21
5444	PHYLICA PLUMIGERA	S	.	0.0	20.1	19.6	.	.	.	.	.	.	.	.	.	.	.	.	.	21
5445	PHYLICA SPICATA	S	.	0.0	21.1	10.9	.	.	.	.	.	.	.	.	.	.	.	.	.	21
5446	PHYLLANTHUS ACIDUS	F	34	90.5	0.8	0.4	7.9	1.3	0.4	6	20	1.1	6	151	0	0.01	0.01	0.40	2	2
5447	PHYLLANTHUS ACIDUS	F	37	91.4	1.0	1.8	5.2	2.1	0.6	20	21	3.2	.	.	70	0.05	0.02	0.50	11	3
5448	PHYLLANTHUS EMBLICA	F	58	84.1	0.7	0.5	14.3	2.4	0.4	29	21	0.5	22	171	60	0.03	0.04	0.20	276	2

TABLE 2 - DATA DIRECTLY TRANSCRIBED FROM ORIGINAL SOURCES.

OBS	PLANA	PART	CALORIES	WATER	PROTEIN	FAT	TOTCARB	FIBER	ASH	CALCIUM	PHOSPHO	IRON	SODIUM	POTASS	CAROTENE	THIAMIN	RIBOFLAV	NIACIN	ASCORBIC	REFCO
5449	PHYLLANTHUS EMBLICA	F	81	77.1	0.4	0.2	21.8	1.9	0.5	12	26	0.5	.	.	0	0.03	0.05	0.20	625	3
5450	PHYLLANTHUS EMBLICA	F	.	81.2	0.5	0.1	14.1	3.4	0.7	50	20	1.2	.	.	.	.	.	0.20	600	5
5451	PHYLLANTHUS GUINEENSIS	SH	.	0.0	5.3	1.8	82.3	49.0	4.6	.	.	.	.	.	.	.	.	.	.	18
5452	PHYLLANTHUS SP.	S	.	0.0	13.6	30.4	.	.	1.8	.	.	.	.	.	.	.	.	.	.	19
5453	PHYLLOSTACHYS SP.	SH	22	92.5	2.5	0.2	4.1	1.2	0.7	4	51	0.4	2	.	0	0.01	0.08	0.80	10	2
5454	PHYMASPERMUM PARVIFOLIUM	C	.	0.0	10.8	5.9	72.7	33.3	10.6	.	.	.	.	.	.	.	.	.	.	18
5455	PHYSALIS ALKEKENGI	S	.	0.0	16.2	31.8	.	.	.	.	.	.	.	.	.	.	.	.	.	20
5456	PHYSALIS ANGULATA	F	27	89.9	3.2	0.4	4.5	1.4	2.0	81	135	.	.	.	.	.	.	.	.	1
5457	PHYSALIS IXOCARPA	F	.	91.7	0.7	0.6	6.4	0.6	0.6	7	40	1.4	0	243	48	0.05	0.02	2.10	2	5
5458	PHYSALIS NICANDROIDES	S	.	0.0	15.0	18.8	.	.	.	.	.	.	.	.	.	.	.	.	.	20
5459	PHYSALIS NICANDROIDES	S	.	0.0	16.6	20.2	.	.	.	.	.	.	.	.	.	.	.	.	.	21
5460	PHYSALIS PERUVIANA	F	73	78.9	0.3	0.2	19.6	4.9	1.0	8	55	1.2	.	.	1460	0.10	0.03	1.70	43	3
5461	PHYSALIS PERUVIANA	F	.	82.9	1.8	0.2	14.3	3.2	0.8	10	67	2.0	1	320	1428	0.05	0.02	0.30	49	5
5462	PHYSALIS SP.	S	.	0.0	11.9	17.6	.	.	.	.	.	.	.	.	.	.	.	.	.	19
5463	PHYSALIS SP.	S	.	0.0	16.2	19.1	.	.	.	.	.	.	.	.	.	.	.	.	.	19
5464	PHYSALIS SP.	F	48	83.0	1.8	0.1	11.3	.	3.8	7	55	0.9	.	.	.	0.15	0.05	1.60	4	2
5465	PHYSALIS SP.	F	40	88.3	1.6	0.5	8.8	1.7	0.8	10	34	0.9	.	.	50	0.09	0.04	2.40	6	3
5466	PHYSALIS SP.	F	53	85.4	1.9	0.7	11.2	2.8	0.8	9	40	1.0	.	.	432	0.11	0.04	2.80	11	4
5467	PHYSANTHYLLIS TETRAPHYLLA	S	.	0.0	23.5	3.7	.	.	.	.	.	.	.	.	.	.	.	.	.	21
5468	PHYSOCARPUS SP.	Y	.	0.0	3.8	2.8	90.1	43.2	3.3	.	.	.	.	.	.	.	.	.	.	6
5469	PHYSOSTEGIA VIRGINIANA	S	.	0.0	25.0	35.2	.	.	.	.	.	.	.	.	.	.	.	.	.	19
5470	PHYTELEPHAS MACROCARPA	S	.	0.0	5.3	1.6	91.6	9.3	1.5	.	.	.	.	.	.	.	.	.	.	18
5471	PHYTOLACCA AMERICANA	S	.	0.0	9.8	13.0	.	.	2.6	.	.	.	.	.	.	.	.	.	.	19
5472	PHYTOLACCA AMERICANA	S	.	0.0	11.7	10.7	.	.	.	.	.	.	.	.	.	.	.	.	.	21
5473	PHYTOLACCA AMERICANA	SH	23	91.6	2.6	0.4	3.7	.	1.7	53	44	1.7	.	.	5220	0.08	0.33	1.20	136	4
5474	PHYTOLACCA SP.	W	.	0.0	14.7	9.3	72.6	13.4	3.4	.	.	.	.	.	.	.	.	.	.	6
5475	PICEA ABIES	S	.	0.0	22.5	38.1	.	.	.	.	.	.	.	.	.	.	.	.	.	20
5476	PICRAMNIA PENTANDRA	S	.	0.0	12.5	54.7	.	.	2.1	.	.	.	.	.	.	.	.	.	.	19
5477	PICRAMNIA PENTANDRA	S	.	0.0	.	12.7	.	.	.	.	.	.	.	.	.	.	.	.	.	19
5478	PICRAMNIA PENTANDRA	S	.	0.0	13.1	56.8	.	.	.	.	.	.	.	.	.	.	.	.	.	20
5479	PICRAMNIA PENTANDRA	S	.	0.0	13.1	57.9	.	.	.	.	.	.	.	.	.	.	.	.	.	20
5480	PIERIS JAPONICA	S	.	0.0	10.4	12.0	.	.	.	.	.	.	.	.	.	.	.	.	.	21
5481	PILIOSTIGMA MALABARICUM VAR. ACIDUM	L	47	86.9	3.4	1.1	8.2	1.8	0.4	46	31	0.3	.	.	865	0.02	0.17	1.40	10	2
5482	PILIOSTIGMA RETICULATUM	L	64	78.3	4.8	0.1	14.4	6.8	2.4	435	80	.	.	.	.	.	.	.	.	1
5483	PILIOSTIGMA THONNINGII	L	.	0.0	11.7	3.3	80.0	26.9	5.0	910	130	.	.	.	.	.	.	.	.	18
5484	PILIOSTIGMA THONNINGII	F	.	0.0	7.0	2.0	86.7	26.2	4.3	.	.	.	.	.	.	.	.	.	.	18
5485	PIMENTA DIOICA	L	118	64.2	4.2	1.3	27.1	6.4	3.2	360	42	3.4	.	.	3680	0.04	0.10	0.50	74	1
5486	PIMENTA DIOICA	F	263	8.5	6.1	8.7	72.1	21.6	4.6	661	113	7.1	77	1044	324	0.10	0.06	2.86	39	16
5487	PIMENTA DIOICA	L	115	65.7	4.6	1.4	25.7	6.0	2.6	383	48	2.6	.	.	4130	0.06	0.12	0.60	69	3
5488	PIMPINELLA ANISUM	S	337	9.5	17.6	15.9	50.0	14.6	7.0	646	440	37.0	16	1441	.	.	.	.	.	16
5489	PIMPINELLA ANISUM	S	.	0.0	19.6	23.9	.	.	6.6	.	.	.	.	.	.	.	.	.	.	19
5490	PIMPINELLA ANISUM	S	.	0.0	21.2	30.4	.	.	.	.	.	.	.	.	.	.	.	.	.	19
5491	PIMPINELLA ANISUM	S	415	11.0	19.0	24.7	45.3	.	.	698	.	34.8	.	.	.	.	.	.	0	2
5492	PIMPINELLA ANISUM	S	.	0.0	18.8	24.3	.	.	.	.	.	.	.	.	.	.	.	.	.	20
5493	PIMPINELLA PEREGRINA	S	.	0.0	22.1	19.8	.	.	.	.	.	.	.	.	.	.	.	.	.	21

TABLE 2 - DATA DIRECTLY TRANSCRIBED FROM ORIGINAL SOURCES.

OBS	PLANA	PART	CALORIES	WATER	PROTEIN	FAT	TOTLCARB	FIBER	ASH	CALCIUM	PHOSPHO	IRON	SODIUM	POTASS	CAROTE	THIAMINE	RIBOFLAV	NIACIN	ASCORBIC	REFCO
5494	PINUS CEMBROIDES	S	634	3.7	15.3	61.3	16.8	2.6	2.9	14	515	4.4	.	.	20	0.76	0.23	9.80	1	3
5495	PINUS CEMBROIDES	S	635	3.1	13.0	60.5	20.5	1.1	2.9	12	604	5.2	.	.	18	1.28	0.23	4.50	0	4
5496	PINUS DENSIFLORA	S	.	0.0	31.2	35.1	.	.	.	.	.	.	.	.	.	.	.	.	.	21
5497	PINUS FLEXILIS	S	.	0.0	25.0	59.8	.	.	.	.	.	.	.	.	.	.	.	.	.	20
5498	PINUS PINEA	S	552	5.6	31.1	47.4	11.6	0.9	4.3	.	.	.	.	.	.	0.62	.	.	.	4
5499	PINUS SP.	S	640	3.5	13.8	62.0	18.0	2.0	2.7	15	392	6.4	4	620	0	0.16	0.40	5.50	2	2
5500	PINUS THUNBERGII	S	.	0.0	27.9	31.5	.	.	.	.	.	.	.	.	.	.	.	.	.	21
5501	PIPER AURITUM	L	57	82.2	3.9	1.4	10.1	2.2	2.4	257	52	5.6	.	.	3440	0.12	0.24	0.90	60	3
5502	PIPER NIGRUM	F	255	10.5	11.0	3.3	64.8	13.1	4.3	437	173	28.9	44	1259	114	0.11	0.24	1.14	.	16
5503	PIPER NIGRUM	S	325	12.0	12.2	8.2	63.6	.	4.0	.	.	.	.	.	.	.	.	.	.	2
5504	PIPER SARMENTOSUM	L	101	69.5	5.4	2.5	18.8	4.6	3.8	601	30	7.6	.	.	5290	0.13	0.11	16.20	10	2
5505	PIPTOCHAETIUM BICOLOR	S	.	0.0	22.5	4.4	.	.	.	.	.	.	.	.	.	.	.	.	.	21
5506	PIPTOCHAETIUM CONFUSUM	S	.	0.0	15.5	2.8	.	.	.	.	.	.	.	.	.	.	.	.	.	21
5507	PIPTOCHAETIUM VERRUCULOSUM	S	.	0.0	12.2	3.0	.	.	.	.	.	.	.	.	.	.	.	.	.	21
5508	PISTACIA LENTISCUS	S	.	0.0	7.2	13.0	.	.	.	.	.	.	.	.	.	.	.	.	.	21
5509	PISTACIA TEREBINTHUS	S	.	0.0	24.2	60.3	.	.	.	.	.	.	.	.	.	.	.	.	.	21
5510	PISTACIA TEREBINTHUS	S	.	0.0	7.4	2.6	.	.	.	.	.	.	.	.	.	.	.	.	.	21
5511	PISTACIA VERA	S	598	4.2	18.9	54.0	19.7	.	3.2	.	.	.	.	.	.	.	.	.	.	1
5512	PISTACIA VERA	S	594	5.3	19.3	53.7	19.0	1.9	2.7	131	500	7.3	.	972	138	0.67	.	1.40	0	4
5513	PISUM MACROCARPUM	F	317	0.0	24.4	1.6	69.1	13.8	4.9	398	439	8.9	73	1098	1991	1.38	0.81	7.32	293	10
5514	PISUM SATIVUM	S	339	11.0	22.3	1.1	62.0	5.7	3.6	90	382	17.9	.	.	15	0.88	0.17	3.00	0	1
5515	PISUM SATIVUM	F	82	74.9	6.7	0.5	17.2	2.5	0.7	29	101	1.9	.	.	330	0.32	0.13	2.00	24	1
5516	PISUM SATIVUM	W	.	0.0	17.2	3.0	67.1	26.9	12.7	1870	370	.	.	.	.	.	.	.	.	18
5517	PISUM SATIVUM	H	.	0.0	14.9	2.7	74.2	31.5	8.3	.	.	.	.	.	.	.	.	.	.	18
5518	PISUM SATIVUM	S	.	0.0	25.3	1.1	68.9	6.2	4.7	.	.	.	.	.	.	.	.	.	.	18
5519	PISUM SATIVUM	S	94	75.6	6.2	0.4	16.9	2.4	0.9	32	102	1.2	6	350	405	0.28	0.11	2.80	27	2
5520	PISUM SATIVUM	F	39	87.7	3.0	0.2	8.5	1.7	0.6	49	54	1.1	9	135	245	0.17	0.10	0.90	36	2
5521	PISUM SATIVUM	S	.	0.0	26.1	1.0	.	.	.	.	.	.	.	.	.	.	.	.	.	21
5522	PISUM SATIVUM	IS	97	70.2	7.6	0.4	21.0	3.0	0.8	24	124	2.0	.	.	250	0.38	0.14	2.20	26	3
5523	PISUM SATIVUM	S	343	12.0	22.5	2.0	61.0	4.7	2.5	80	290	5.8	.	.	50	0.57	0.17	3.00	1	3
5524	PISUM SATIVUM	F	53	83.3	3.4	0.2	12.0	1.2	1.1	62	90	0.7	.	170	408	0.28	0.12	.	21	4
5525	PISUM SATIVUM	S	84	78.0	6.3	0.4	14.4	2.0	0.9	26	116	1.9	2	316	384	0.35	0.14	2.90	27	4
5526	PISUM SATIVUM ARVENSE	W	.	0.0	19.7	3.3	68.0	25.9	9.0	1210	230	40.0	.	1500	.	.	.	.	.	6
5527	PISUM SATIVUM VAR. ARVENSE	S	.	0.0	25.9	0.9	.	.	2.7	.	.	.	.	.	.	.	.	.	.	19
5528	PISUM SATIVUM VAR. ARVENSE	S	.	0.0	19.6	1.5	.	.	3.6	.	.	.	.	.	.	.	.	.	.	19
5529	PISUM SATIVUM VAR. ARVENSE	S	.	0.0	21.1	1.4	.	.	3.1	.	.	.	.	.	.	.	.	.	.	19
5530	PISUM SATIVUM VAR. ARVENSE	S	.	0.0	28.1	1.0	.	.	.	.	.	.	.	.	.	.	.	.	.	20
5531	PISUM SYRIACUM	S	.	0.0	29.1	0.7	.	.	.	.	.	.	.	.	.	.	.	.	.	21
5532	PITHECELLOBIUM BREVIFOLIUM	S	.	0.0	31.2	13.3	.	.	3.9	.	.	.	.	.	.	.	.	.	.	19
5533	PITHECELLOBIUM DULCE	L	.	0.0	29.0	4.4	61.0	17.5	5.6	.	.	.	.	.	.	.	.	.	.	18
5534	PITHECELLOBIUM DULCE	SH	.	0.0	23.5	3.0	68.5	23.1	5.0	.	.	.	.	.	.	.	.	.	.	18
5535	PITHECELLOBIUM DULCE	S	.	0.0	24.4	13.2	.	.	3.3	.	.	.	.	.	.	.	.	.	.	19
5536	PITHECELLOBIUM DULCE	S	.	0.0	23.8	13.3	.	.	.	.	.	.	.	.	.	.	.	.	.	19
5537	PITHECELLOBIUM DULCE	S	.	0.0	20.0	12.5	.	.	.	.	.	.	.	.	.	.	.	.	.	19
5538	PITHECELLOBIUM DULCE	F	78	77.8	3.0	0.4	18.2	1.2	0.6	13	42	0.5	19	222	15	0.24	0.10	0.60	133	2

TABLE 2 - DATA DIRECTLY TRANSCRIBED FROM ORIGINAL SOURCES.

OBS	PLANA	PART	CALORIES	WATER	PROTEIN	FAT	TOTLCARB	FIBER	ASH	CALCIUM	PHOSPHON	IRON	SODIUM	POTASS	CAROTE	THIAMINE	RIBOFLAV	NIACIN	ASCORBIC	REFCO
5539	PITHECELLOBIUM JIRINGA	F	92	76.3	6.2	0.2	16.9	1.3	0.4	23	38	0.7	.	.	395	0.14	0.01	0.4	8	2
5540	PITHECELLOBIUM MEXICANUM	S	.	0.0	35.6	14.5	.	.	.	.	.	.	.	.	.	.	.	.	.	20
5541	PITHECELLOBIUM UNDULATUM	S	.	0.0	27.5	6.4	.	.	.	.	.	.	.	.	.	.	.	.	.	20
5542	PITTOSPORUM TOBIRA	S	.	0.0	13.8	13.0	.	.	.	.	.	.	.	.	.	.	.	.	.	19
5543	PITTOSPORUM TOBIRA	S	.	0.0	11.3	13.5	.	.	.	.	.	.	.	.	.	.	.	.	.	20
5544	PITURANTHOS APHYLLA	S	.	0.0	19.8	27.1	.	.	.	.	.	.	.	.	.	.	.	.	.	20
5545	PITURANTHOS APHYLLA	S	.	0.0	22.8	26.2	.	.	.	.	.	.	.	.	.	.	.	.	.	21
5546	PLAGIOBOTHRYS NOTHOFULVUS	W	.	0.0	14.2	.	.	22.0	14.7	1940	520	.	.	3940	.	.	.	.	.	6
5547	PLANTAGO ALBICANS	S	.	0.0	17.0	6.4	.	.	.	.	.	.	.	.	.	.	.	.	.	21
5548	PLANTAGO ARISTATA	S	.	0.0	16.2	6.2	.	.	.	.	.	.	.	.	.	.	.	.	.	20
5549	PLANTAGO LAGOPUS	S	.	0.0	20.2	7.1	.	.	.	.	.	.	.	.	.	.	.	.	.	21
5550	PLANTAGO LANCEOLATA	L	.	0.0	15.9	2.2	69.0	13.3	12.9	1540	290	.	.	.	.	.	.	.	.	18
5551	PLANTAGO LANCEOLATA	SH	.	0.0	22.8	3.8	58.7	16.1	14.7	2460	430	.	.	.	.	.	.	.	.	18
5552	PLANTAGO LANCEOLATA	S	.	0.0	18.8	7.3	.	.	.	.	.	.	.	.	.	.	.	.	.	19
5553	PLANTAGO LANCEOLATA	S	.	0.0	17.2	5.1	.	.	.	.	.	.	.	.	.	.	.	.	.	21
5554	PLANTAGO LANCEOLATA	S	.	0.0	16.4	7.8	.	.	.	.	.	.	.	.	.	.	.	.	.	21
5555	PLANTAGO MAJOR	L	61	81.4	2.5	0.3	14.6	.	1.2	184	52	1.2	16	277	2520	.	0.28	0.8	8	2
5556	PLANTAGO MAJOR	S	.	0.0	19.1	15.5	.	.	.	.	.	.	.	.	.	.	.	.	.	21
5557	PLANTAGO OVATA	F	.	0.0	20.9	2.4	68.7	31.9	8.0	650	570	.	.	.	.	.	.	.	.	18
5558	PLANTAGO OVATA	S	.	0.0	16.2	5.9	.	.	3.3	.	.	.	.	.	.	.	.	.	.	19
5559	PLANTAGO OVATA	S	.	0.0	14.6	5.4	.	.	2.4	.	.	.	.	.	.	.	.	.	.	19
5560	PLANTAGO PSYLLIUM	S	.	0.0	14.8	6.9	.	.	.	.	.	.	.	.	.	.	.	.	.	21
5561	PLANTAGO RHODOSPERMA	S	.	0.0	15.0	6.0	.	.	.	.	.	.	.	.	.	.	.	.	.	19
5562	PLANTAGO RHODOSPERMA	S	.	0.0	17.5	5.7	.	.	.	.	.	.	.	.	.	.	.	.	.	19
5563	PLANTAGO RHODOSPERMA	S	.	0.0	18.8	6.0	.	.	.	.	.	.	.	.	.	.	.	.	.	20
5564	PLANTAGO RHODOSPERMA	W	.	0.0	14.5	1.5	71.1	13.1	12.9	5240	190	.	.	2700	.	.	.	.	.	6
5565	PLANTAGO SP.	S	.	0.0	17.5	5.2	.	.	.	.	.	.	.	.	.	.	.	.	.	19
5566	PLANTAGO SP.	S	.	0.0	15.0	8.9	.	.	.	.	.	.	.	.	.	.	.	.	.	19
5567	PLANTAGO SP.	S	.	0.0	34.8	8.3	.	.	.	.	.	.	.	.	.	.	.	.	.	21
5568	PLANTAGO SP.	W	.	0.0	13.3	1.9	75.0	15.9	9.8	4160	190	.	.	2350	.	.	.	.	.	6
5569	PLANTAGO SQUARROSA	S	.	0.0	17.0	5.6	.	.	.	.	.	.	.	.	.	.	.	.	.	21
5570	PLANTAGO WRIGHTIANA	S	.	0.0	15.6	9.9	.	.	.	.	.	.	.	.	.	.	.	.	.	19
5571	PLANTAGO WRIGHTIANA	S	.	0.0	20.6	8.6	.	.	.	.	.	.	.	.	.	.	.	.	.	20
5572	PLATANUS OCCIDENTALIS	S	.	0.0	9.4	4.8	.	.	.	.	.	.	.	.	.	.	.	.	.	20
5573	PLATANUS OCCIDENTALIS	S	.	0.0	13.8	7.8	.	.	.	.	.	.	.	.	.	.	.	.	.	20
5574	PLATONIA ESCULENTA	F	113	70.0	1.8	2.0	25.0	5.3	1.2	.	.	.	.	.	.	.	.	.	18	3
5575	PLATYCODON GRANDIFLORUM	R	287	24.2	2.4	0.9	71.0	8.9	1.5	232	189	6.2	.	.	0	0.10	0.36	7.8	0	2
5576	PLECTRANTHUS ESCULENTUS	R	84	78.0	1.9	0.4	18.8	1.0	0.9	18	80	6.4	.	.	540	0.00	0.62	0.7	.	1
5577	PLEIOGYNIUM SOLANDRI	S	.	0.0	23.1	60.0	.	.	.	.	.	.	.	.	.	.	.	.	.	20
5578	PLUCHEA INDICA	L	42	86.0	1.8	0.5	9.4	.	2.3	256	49	5.6	.	.	2390	0.02	.	.	30	2
5579	PLUCHEA ODORATA	S	.	0.0	25.0	15.9	.	.	.	.	.	.	.	.	.	.	.	.	.	19
5580	PLUMERIA ACUMINATA	L	43	86.3	1.6	0.0	11.0	.	1.1	69	60	.	.	.	.	0.03	.	.	224	2
5581	POA ANNUA	L	.	0.0	11.0	2.8	60.3	26.3	8.8	440	390	.	.	.	.	.	.	.	.	5
5582	POA ANNUA	L	.	0.0	14.7	3.3	79.2	33.0	16.1	1270	780	.	.	.	.	.	.	.	.	5
5583	POA ARACHNIFERA	W	.	0.0	12.4	4.7	74.8	32.4	8.1	470	200	.	.	.	.	.	.	.	.	6

TABLE 2 - DATA DIRECTLY TRANSCRIBED FROM ORIGINAL SOURCES.

OBS	PLNA	PART	CALORIES	WATER	PROTEIN	FAT	TOTCARB	FIBER	ASH	CALCIUM	PHOSPHO	IRON	SODIUM	POTASS	CAROTE	THIAMINE	RIBOFLAV	NIACIN	ASCORBIC	REFCO
5584	POA BULBOSA	BU	.	0.0	8.8	3.7	.	.	.	.	.	.	.	.	.	.	.	.	.	20
5585	POA BULBOSA	L	.	0.0	8.4	3.2	79.8	26.8	7.4	620	610	.	.	.	.	.	.	.	.	5
5586	POA COMPRESSA	L	.	66.8	3.0	1.2	16.4	10.3	2.6	400	100	.	.	500	.	.	.	.	.	5
5587	POA COMPRESSA	H	.	0.0	11.6	2.7	77.8	28.9	7.9	300	290	.	.	1590	.	.	.	.	.	6
5588	POA COMPRESSA	W	.	0.0	17.0	3.7	70.4	26.4	8.9	390	390	.	2040	.	.	.	.	.	.	6
5589	POA EPILIS	H	.	0.0	3.4	1.6	88.2	38.7	6.8	.	.	.	.	.	.	.	.	.	.	6
5590	POA FENDLERIANA	W	.	0.0	7.0	2.4	79.5	33.5	11.1	410	340	.	.	.	.	.	.	.	.	6
5591	POA NEMORALIS	S	.	0.0	11.4	1.2	.	.	.	.	.	.	.	.	.	.	.	.	.	21
5592	POA PALUSTRIS	H	.	0.0	8.1	1.9	81.8	38.0	8.2	260	260	.	.	.	.	.	.	.	.	6
5593	POA PRATENSIS	S	.	0.0	14.4	2.9	.	.	8.5	.	.	.	.	.	.	.	.	.	.	19
5594	POA PRATENSIS	S	.	0.0	9.0	2.0	.	.	2.7	.	.	.	.	.	.	.	.	.	.	19
5595	POA PRATENSIS	L	.	0.0	7.7	1.6	69.6	30.6	8.0	160	130	6.0	70	590	8	0.26	.	.	.	5
5596	POA PRATENSIS	L	.	0.0	12.5	3.2	87.2	37.8	12.8	.	.	.	.	.	.	.	.	.	.	5
5597	POA PRATENSIS	H	.	0.0	12.0	3.1	76.7	29.9	8.2	400	260	26.0	.	1720	.	.	.	.	.	6
5598	POA PRATENSIS	W	.	0.0	15.8	3.6	72.7	26.5	7.9	450	390	24.0	.	2060	.	.	.	.	.	6
5599	POA REFLEXA	H	.	0.0	9.5	2.3	82.0	29.2	6.2	.	.	.	.	.	.	.	.	.	.	6
5600	POA SCABRELLA	W	.	0.0	9.9	.	.	35.7	5.3	360	250	.	.	1630	.	.	.	.	.	6
5601	POA SECUNDA	H	.	0.0	8.6	3.1	81.9	32.6	6.4	.	.	.	.	.	.	.	.	.	.	6
5602	POA SECUNDA	W	.	0.0	9.0	2.4	79.1	39.0	9.5	410	270	.	.	.	.	.	.	.	.	6
5603	POCHOTA GLABRA	S	362	10.7	6.6	1.9	79.5	1.9	1.3	.	.	.	.	.	.	.	.	.	.	1
5604	PODACHAENIUM EMINENS	S	.	0.0	18.8	28.0	.	.	.	.	.	.	.	.	.	.	.	.	.	20
5605	PODALYRIA CALYPTRATA	S	.	0.0	36.2	9.0	.	.	.	.	.	.	.	.	.	.	.	.	.	21
5606	PODALYRIA CUNEIFOLIA	S	.	0.0	30.0	6.2	.	.	.	.	.	.	.	.	.	.	.	.	.	21
5607	PODOCARPUS MACROPHYLLA	S	.	0.0	6.8	2.8	.	.	.	.	.	.	.	.	.	.	.	.	.	20
5608	PODOCARPUS NAGI	S	.	0.0	10.0	39.9	.	.	2.4	.	.	.	.	.	.	.	.	.	.	19
5609	PODOCARPUS NAGI	S	.	0.0	1.4	0.9	.	.	5.6	.	.	.	.	.	.	.	.	.	.	19
5610	POGOSTEMON PARVIFLORUS	S	.	0.0	25.0	35.8	.	.	.	.	.	.	.	.	.	.	.	.	.	21
5611	POLAKOWSKIA TACACO	F	.	.	1.9	.	.	.	0.7	9	53	0.9	.	.	40	0.03	0.13	.	23	3
5612	POLANISIA SP.	S	.	0.0	14.4	26.7	.	.	.	.	.	.	.	.	.	.	.	.	.	20
5613	POLANISIA TRACHYSPERMA	S	.	0.0	17.5	25.2	.	.	3.3	.	.	.	.	.	.	.	.	.	.	19
5614	POLANISIA TRACHYSPERMA	S	.	0.0	18.1	20.7	.	.	.	.	.	.	.	.	.	.	.	.	.	19
5615	POLANISIA TRACHYSPERMA	S	.	0.0	19.4	20.6	.	.	.	.	.	.	.	.	.	.	.	.	.	19
5616	POLANISIA TRACHYSPERMA	S	.	0.0	16.9	29.8	.	.	.	.	.	.	.	.	.	.	.	.	.	21
5617	POLANISIA VISCOSA	S	.	0.0	16.2	26.2	.	.	.	.	.	.	.	.	.	.	.	.	.	19
5618	POLEMONIUM ALBIFLORUM	W	.	0.0	10.6	2.3	.	.	11.7	1350	430	.	.	.	.	.	.	.	.	6
5619	POLEMONIUM CAERULEUM	S	.	0.0	23.9	26.1	.	.	4.6	.	.	.	.	.	.	.	.	.	.	19
5620	POLEMONIUM CAERULEUM	S	.	0.0	17.2	12.0	.	.	.	.	.	.	.	.	.	.	.	.	.	21
5621	POLYALTHIA LONGIFOLIA	L	36	87.0	4.6	0.1	6.7	.	1.6	5	96	.	.	.	.	.	.	.	.	2
5622	POLYGONUM CESPITOSUM VAR. LONGISETUM	S	.	0.0	13.8	3.2	.	.	.	.	.	.	.	.	.	.	.	.	.	19
5623	POLYGONUM CONVOLVULUS	S	.	0.0	14.4	1.9	.	.	1.8	.	.	.	.	.	.	.	.	.	.	19
5624	POLYGONUM DUMETORUM	S	.	0.0	12.1	2.1	.	.	.	.	.	.	.	.	.	.	.	.	.	21
5625	POLYGONUM KITAIBELIANUM	S	.	0.0	14.4	3.5	.	.	.	.	.	.	.	.	.	.	.	.	.	20
5626	POLYGONUM LAPATHIFOLIUM	S	.	0.0	8.8	4.0	.	.	.	.	.	.	.	.	.	.	.	.	.	19
5627	POLYGONUM ORIENTALE	S	.	0.0	11.6	3.7	.	.	1.5	.	.	.	.	.	.	.	.	.	.	19
5628	POLYGONUM PENSYLVANICUM	S	.	0.0	9.9	4.4	.	.	1.8	.	.	.	.	.	.	.	.	.	.	19

TABLE 2 - DATA DIRECTLY TRANSCRIBED FROM ORIGINAL SOURCES.

OBS	PLNA	PART	CALORIES	WATER	PROTEIN	FAT	TOTCARB	FIBER	ASH	CALCIUM	PHOSPHO	IRON	SODIUM	POTASS	CAROT	THIAMINE	RIBOFLAV	NIACIN	ASCORBIC	REFCO
5629	POLYGONUM PENSYLVANICUM	S	.	0.0	7.7	1.9	.	.	.	.	.	.	.	.	.	.	.	.	.	20
5630	POLYGONUM PUNCTATUM	S	.	0.0	8.1	1.7	.	.	.	.	.	.	.	.	.	.	.	.	.	20
5631	POLYGONUM RAMOSISSIMUM	S	.	0.0	11.0	2.6	.	.	1.4	.	.	.	.	.	.	.	.	.	.	19
5632	POLYGONUM RAMOSISSIMUM	S	.	0.0	5.4	1.6	.	.	6.0	.	.	.	.	.	.	.	.	.	.	19
5633	POLYGONUM SALICIFOLIUM	L	64	80.0	3.6	0.3	14.7	3.5	1.4	150	46	.	.	.	.	.	.	.	.	1
5634	POLYMNIA EDULIS	L	.	0.0	17.1	7.4	59.5	10.0	16.0	.	.	.	.	.	.	.	.	.	.	18
5635	POLYMNIA EDULIS	S	.	0.0	11.4	2.3	76.1	26.9	10.2	.	.	.	.	.	.	.	.	.	.	18
5636	POLYMNIA EDULIS	R	.	0.0	7.3	0.4	85.6	5.7	6.7	.	.	.	.	.	.	.	.	.	.	18
5637	POLYMNIA UVEDALIA	S	.	0.0	34.2	59.0	.	.	.	.	.	.	.	.	.	.	.	.	.	20
5638	POLYPOGON MONSPELIENSIS	S	.	0.0	28.8	5.8	.	.	.	.	.	.	.	.	.	.	.	.	.	21
5639	POLYPOGON MONSPELIENSIS	L	.	0.0	11.8	1.3	62.9	24.9	21.9	1200	900	.	.	.	.	.	.	.	.	5
5640	POLYPOGON MONSPELIENSIS	W	.	0.0	13.6	2.7	69.1	28.3	14.6	340	330	.	.	.	.	.	.	.	.	6
5641	POLYSCIAS SCUTILLARIA	L	54	82.0	3.7	0.3	11.8	.	2.2	474	49	4.0	.	.	3270	0.06	.	.	83	2
5642	POLYTAENIA NUTTALLII	S	.	0.0	14.4	26.6	.	.	.	.	.	.	.	.	.	.	.	.	.	20
5643	POLYTAENIA NUTTALLII	S	.	0.0	15.7	32.6	.	.	.	.	.	.	.	.	.	.	.	.	.	21
5644	PONGAMIA PINNATA	S	.	0.0	21.9	34.6	.	.	3.2	.	.	.	.	.	.	.	.	.	.	19
5645	PONGAMIA PINNATA	S	.	19.0	17.4	27.5	13.9	7.3	2.4	.	.	.	.	.	.	.	.	.	.	5
5646	PONGAMIA SP.	S	.	0.0	21.9	18.1	.	.	.	.	.	.	.	.	.	.	.	.	.	20
5647	POPULUS ANGUSTIFOLIA	L	.	0.0	24.8	6.9	61.3	13.9	7.0	.	.	.	.	.	.	.	.	.	.	6
5648	POPULUS TREMULOIDES(IMMATURE)	L	.	0.0	19.3	4.7	69.8	13.5	6.2	1350	510	.	.	.	.	.	.	.	.	6
5649	POPULUS TREMULOIDES(MATURE)	L	.	0.0	9.5	6.2	75.8	15.0	8.5	3020	300	.	.	.	.	.	.	.	.	6
5650	PORTULACA OLERACEA	L	23	90.6	2.4	0.3	4.2	0.8	2.5	104	31	1.4	.	.	.	0.03	0.15	.	58	1
5651	PORTULACA OLERACEA	L	255	0.0	34.5	5.3	35.5	10.5	24.7	2078	774	.	55	505	.	.	.	6.72	.	12
5652	PORTULACA OLERACEA	S	.	0.0	21.0	18.9	.	.	3.4	.	.	.	.	.	.	.	.	.	.	19
5653	PORTULACA OLERACEA	L	37	87.5	2.2	0.3	7.9	1.1	2.0	115	40	1.4	.	390	1320	0.06	0.14	0.80	21	2
5654	PORTULACA OLERACEA	L	26	91.2	2.0	0.4	5.0	0.9	1.4	79	32	3.6	.	.	1500	0.02	0.10	0.50	23	3
5655	PORTULACA OLERACEA	L	21	92.5	1.7	0.4	3.8	0.9	1.6	103	39	3.5	.	.	1500	0.03	0.10	0.50	25	4
5656	PORTULACA OLERACEA	L	.	90.5	2.4	0.6	4.2	1.3	2.3	111	45	1.0	67	716	2292	0.10	0.22	0.70	29	5
5657	PORTULACA OLERACEA	W	.	0.0	21.4	2.9	57.3	14.6	18.4	.	.	.	.	.	.	.	.	.	.	6
5658	PORTULACA QUADRIFIDA	L	38	85.8	1.6	0.8	7.8	2.0	4.0	150	70	.	.	.	.	.	.	.	.	1
5659	PORTULACARIA AFRA	L	.	0.0	11.8	3.9	72.5	14.6	11.8	2870	350	.	.	.	.	.	.	.	.	18
5660	PORTULACARIA AFRA	SH	.	0.0	10.7	3.8	66.4	11.9	19.1	1400	120	.	.	.	.	.	.	.	.	18
5661	PORTULACARIA AFRA	H	.	0.0	8.5	3.8	77.9	20.9	9.8	.	.	.	.	.	.	.	.	.	.	18
5662	PORTULACARIA AFRA	L	.	0.0	8.5	3.8	77.9	20.9	9.8	.	.	.	.	.	.	.	.	.	.	5
5663	PORTULACEA SP.	W	.	0.0	9.9	5.0	59.9	17.9	25.2	2840	320	.	.	5390	.	.	.	.	.	6
5664	POTAMOGETON PECTINATUS	L	.	.	7.6	0.6	64.6	19.1	16.2	.	.	.	.	.	10	.	.	.	.	5
5665	POTENTILLA ARGUTA SSP. CONVALLARIA	S	.	0.0	21.1	32.5	.	.	.	.	.	.	.	.	.	.	.	.	.	20
5666	POTENTILLA ARGUTA SUBSP. CONVALLARIA	S	.	0.0	.	31.1	.	.	.	.	.	.	.	.	.	.	.	.	.	21
5667	POTENTILLA GLANDULOSA	S	.	0.0	13.9	24.7	.	.	.	.	.	.	.	.	.	.	.	.	.	21
5668	POTENTILLA HIRTA	S	.	0.0	19.9	25.1	.	.	.	.	.	.	.	.	.	.	.	.	.	20
5669	POTENTILLA PULCHERRIMA	S	.	0.0	21.8	13.9	.	.	.	.	.	.	.	.	.	.	.	.	.	20
5670	POTENTILLA RECTA	S	.	0.0	17.5	20.7	.	.	.	.	.	.	.	.	.	.	.	.	.	20
5671	POTENTILLA SP.	W	.	0.0	22.1	3.9	65.9	11.3	8.1	.	.	.	.	.	.	.	.	.	.	6
5672	POTERIUM SPINOSUM	S	.	0.0	7.6	8.5	.	.	.	.	.	.	.	.	.	.	.	.	.	21
5673	POTERIUM SPINOSUM	S	.	0.0	8.5	7.0	.	.	.	.	.	.	.	.	.	.	.	.	.	21

TABLE 2 - DATA DIRECTLY TRANSCRIBED FROM ORIGINAL SOURCES.

OBS	PLANA	PART	CALORIES	WATER	PROTEIN	FAT	TOTLCARB	FIBER	ASH	CALCIUM	PHOSPHO	IRON	SODIUM	POTASS	CAROT	THIAMINE	RIBOFLAV	NIACIN	ASCORBIC	REFCO
5674	POTERIUM SPINOSUM	S	.	0.0	8.70	7.9	.	.	.	.	.	.	.	.	.	.	.	.	.	21
5675	POTHOMORPHE UMBELLATA	L	69	82.1	4.60	3.4	8.2	1.8	1.7	169	40	.	.	.	.	.	.	.	.	1
5676	POUTERIA CAIMITO	F	140	60.6	1.80	0.4	36.3	0.9	0.9	22	41	1.0	.	.	260	0.02	0.02	3.4	49	3
5677	POUTERIA CAMPECHIANA	F	154	57.2	2.50	0.6	39.1	7.5	0.6	30	30	1.1	.	.	1235	0.02	0.04	0.4	43	2
5678	POUTERIA CAMPECHIANA	F	145	59.7	1.30	0.8	37.3	1.0	0.9	36	36	0.7	.	.	210	0.18	0.00	3.3	40	3
5679	POUTERIA CAMPECHIANA	F	140	60.6	1.80	0.4	36.3	0.9	0.9	22	41	1.0	.	.	260	0.02	0.02	3.4	49	3
5680	POUTERIA VIRIDIS	F	110	68.4	1.60	0.2	28.6	1.8	1.2	23	28	0.7	.	.	50	0.01	0.03	1.9	43	3
5681	PRANGOS PABULARIA	S	.	0.0	33.10	24.0	.	.	.	.	.	.	.	.	.	.	.	.	.	21
5682	PREMNA ODORATA	L	64	80.0	3.50	0.4	14.7	2.6	1.4	64	76	3.0	5	574	.	0.04	0.09	0.9	5	2
5683	PRIONOSCIADIUM SP.	S	.	0.0	26.20	29.1	.	.	.	.	.	.	.	.	.	.	.	.	.	20
5684	PRISMATOCARPUS ROELLOIDES	S	.	0.0	17.60	25.9	.	.	.	.	.	.	.	.	.	.	.	.	.	21
5685	PROBOSCIDEA ALTHEAEFOLIA	S	.	0.0	26.20	36.2	.	.	.	.	.	.	.	.	.	.	.	.	.	20
5686	PROBOSCIDEA FRAGRANS	L	.	0.0	18.10	9.5	46.7	6.2	25.7	2190	250	.	.	3010	.	.	.	.	.	6
5687	PROBOSCIDEA SP.	S	.	0.0	24.40	36.5	.	.	.	.	.	.	.	.	.	.	.	.	.	20
5688	PROSOPIS AFRICANA	S	361	0.0	16.00	1.7	78.5	7.6	3.4	438	329	.	.	.	.	.	.	.	.	10
5689	PROSOPIS CHILENSIS	F	.	0.0	11.00	2.2	79.5	11.9	7.3	440	160	.	.	.	.	.	.	.	.	18
5690	PROSOPIS CHILENSIS	S	.	0.0	35.20	5.2	54.5	9.1	5.1	.	.	.	.	.	.	.	.	.	.	18
5691	PROSOPIS CHILENSIS	L	.	0.0	23.50	2.9	66.9	24.7	4.7	860	250	.	.	.	.	.	.	.	.	18
5692	PROSOPIS CHILENSIS	L	.	7.1	26.30	8.5	56.6	24.8	1.4	.	.	.	.	.	.	.	.	.	.	5
5693	PROSOPIS CHILENSIS	L	.	0.0	21.10	3.4	69.6	27.3	5.9	1940	190	.	.	1410	.	.	.	.	.	6
5694	PROSOPIS CHILENSIS	F	.	0.0	14.40	1.9	80.1	28.8	3.6	660	210	.	.	1420	.	.	.	.	.	6
5695	PROSOPIS JULIFLORA	I	.	0.0	21.00	3.2	65.8	15.5	10.0	1310	400	.	.	.	.	.	.	.	.	18
5696	PROSOPIS JULIFLORA	L	.	0.0	19.00	2.9	69.6	21.6	8.5	2080	220	.	.	.	.	.	.	.	.	18
5697	PROSOPIS JULIFLORA	F	.	0.0	13.90	3.0	78.3	27.7	4.8	.	.	.	.	.	.	.	.	.	.	18
5698	PROSOPIS JULIFLORA	S	.	0.0	65.20	7.8	21.8	2.8	5.2	.	.	.	.	.	.	.	.	.	.	18
5699	PROSOPIS JULIFLORA	S	.	0.0	33.90	6.9	.	.	3.0	.	.	.	.	.	.	.	.	.	.	19
5700	PROSOPIS JULIFLORA	S	.	0.0	7.00	1.8	.	.	4.8	.	.	.	.	.	.	.	.	.	.	19
5701	PROSOPIS JULIFLORA	S	.	0.0	39.90	6.6	.	.	.	.	.	.	.	.	.	.	.	.	.	20
5702	PROSOPIS JULIFLORA	S	.	0.0	38.20	3.4	.	.	.	.	.	.	.	.	.	.	.	.	.	21
5703	PROSOPIS JULIFLORA VAR. VELUTINA	S	.	0.0	36.70	7.6	.	.	3.5	.	.	.	.	.	.	.	.	.	.	19
5704	PROSOPIS JULIFLORA VAR. VELUTINA	S	.	0.0	34.90	6.7	.	.	3.3	.	.	.	.	.	.	.	.	.	.	19
5705	PROSOPIS SPICIGERA	L	.	0.0	15.30	3.2	71.5	17.5	10.0	2650	240	.	.	.	.	.	.	.	.	18
5706	PROSOPIS STEPHANIANA	F	.	0.0	14.50	2.9	78.8	15.7	3.8	.	.	.	.	.	.	.	.	.	.	18
5707	PROSOPIS STEPHANIANA	S	.	0.0	20.10	3.1	73.4	9.3	3.4	.	.	.	.	.	.	.	.	.	.	18
5708	PROSOPIS TAMARUGO	SH	.	0.0	10.70	2.7	81.6	32.1	5.0	3980	120	.	.	.	.	.	.	.	.	18
5709	PROSOPIS TAMARUGO	L	.	0.0	9.90	2.0	67.8	15.8	20.3	6210	110	.	.	.	.	.	.	.	.	18
5710	PROSOPIS TAMARUGO	F	.	0.0	11.70	1.6	82.1	41.5	4.6	330	130	.	.	.	.	.	.	.	.	18
5711	PROSOPIS TAMARUGO	S	.	0.0	10.10	0.7	81.7	38.4	7.5	360	1500	.	.	.	.	.	.	.	.	18
5712	PROSOPIS TAMARUGO	S	.	0.0	30.00	5.3	.	.	.	.	.	.	.	.	.	.	.	.	.	20
5713	PROSOPIS VELUTINA	L	.	0.0	8.30	5.4	76.7	32.1	9.6	.	.	.	.	.	.	.	.	.	.	18
5714	PROSOPIS VELUTINA	F	.	0.0	13.60	2.0	79.3	29.1	5.1	.	.	.	.	.	.	.	.	.	.	18
5715	PROSOPIS VELUTINA	S	.	0.0	55.30	8.9	33.2	4.5	4.4	.	.	.	.	.	.	.	.	.	.	18
5716	PROTEA BARBIGERA	S	.	0.0	68.60	31.4	.	.	.	.	.	.	.	.	.	.	.	.	.	21
5717	PROTEA CYNAROIDES		.	0.0	4.31	.	.	.	.	1120	34	3.7	400	120	.	.	.	.	.	8
5718	PROTEA GRANDIFLORA	S	.	0.0	2.10	0.7	.	.	.	.	.	.	.	.	.	.	.	.	.	21

TABLE 2 - DATA DIRECTLY TRANSCRIBED FROM ORIGINAL SOURCES.

OBS	PLNA	PART	CALORIES	WATER	PROTEIN	FAT	TOTCARB	FIBER	ASH	CALCIUM	PHOSPHO	IRON	SODIUM	POTASS	CAROT	THIAMINE	RIBOFLAV	NIACIN	ASCORBIC	REFCO
5719	PROTEA LATIFOLIA	S	.	0.0	90.3	18.2	.	.	.	.	.	.	.	.	.	.	.	.	.	21
5720	PROTEA MARGINATA	S	.	0.0	4.6	1.8	.	.	.	.	.	.	.	.	.	.	.	.	.	21
5721	PROTEA MELLIFERA	S	.	0.0	3.0	1.2	.	.	.	.	.	.	.	.	.	.	.	.	.	21
5722	PROTEA MELLIFERA	S	.	0.0	8.0	4.2	.	.	.	.	.	.	.	.	.	.	.	.	.	21
5723	PROTEA MELLIFERA	S	.	0.0	8.9	.	.	.	.	.	.	.	.	.	.	.	.	.	.	21
5724	PROTEA MELLIFERA	S	.	0.0	2.7	0.7	.	.	.	.	.	.	.	.	.	.	.	.	.	21
5725	PRUNELLA VULGARIS	S	.	0.0	19.0	19.2	.	.	.	.	.	.	.	.	.	.	.	.	.	21
5726	PRUNELLA VULGARIS VAR. LANCEOLATA	S	.	0.0	16.9	22.4	.	.	.	.	.	.	.	.	.	.	.	.	.	19
5727	PRUNUS ARMENIACA	F	58	83.4	1.2	0.4	14.1	1.1	0.9	13	24	0.7	.	.	3145	.	.	.	.	1
5728	PRUNUS ARMENIACA	F	38	88.6	0.8	0.4	9.0	1.5	1.2	21	34	0.9	1	218	1740	0.02	0.05	0.7	6	2
5729	PRUNUS ARMENIACA	S	549	8.0	29.0	47.3	12.9	3.0	2.8	140	276	4.4	.	716	0	0.14	0.49	1.6	.	2
5730	PRUNUS ARMENIACA	F	57	84.2	0.8	0.6	13.8	1.1	0.6	30	32	1.1	.	.	1340	0.04	0.06	0.4	10	3
5731	PRUNUS ARMENIACA	F	51	85.3	1.0	0.2	12.8	0.6	0.7	17	23	0.5	1	281	1620	0.03	0.04	0.6	10	4
5732	PRUNUS AVIUM	F	.	83.0	1.1	0.5	14.8	0.3	0.6	18	20	0.4	.	.	372	0.05	0.06	0.4	8	5
5733	PRUNUS CERASIFERA	F	39	89.0	0.7	0.2	9.7	0.6	0.4	12	15	0.4	.	.	100	0.02	0.03	0.4	8	3
5734	PRUNUS CERASUS	F	46	87.4	1.1	0.6	10.3	1.0	0.6	19	29	0.9	3	112	110	0.03	0.03	0.4	11	2
5735	PRUNUS CERASUS	F	63	82.6	1.8	0.4	14.8	1.0	0.4	34	32	0.3	.	.	60	0.05	0.01	0.2	15	3
5736	PRUNUS CERASUS	F	46	86.9	1.2	0.3	11.0	0.8	0.6	14	18	0.5	.	.	.	.	.	.	.	3
5737	PRUNUS CERASUS	F	.	83.4	1.1	0.5	14.2	0.4	0.8	24	25	1.3	.	320	.	0.08	0.08	0.3	7	5
5738	PRUNUS DEMISSA	F	.	.	.	.	.	.	.	.	.	.	.	.	.	0.02	0.30	.	11	9
5739	PRUNUS DOMESTICA	F	47	87.0	0.6	0.2	11.9	0.4	0.3	8	15	0.4	.	.	80	0.03	0.04	0.5	6	3
5740	PRUNUS DOMESTICA	F	.	86.3	0.7	0.4	12.1	0.4	0.5	10	20	1.4	.	190	100	.	.	0.1	0	5
5741	PRUNUS DOMESTICA	S	.	9.6	20.2	39.5	27.2	12.2	3.4	.	.	.	.	.	.	.	.	.	.	5
5742	PRUNUS DULCIS	S	605	4.8	16.8	54.9	21.5	.	2.0	.	.	.	.	.	.	.	.	.	.	1
5743	PRUNUS DULCIS	S	.	0.0	18.6	54.1	24.3	2.7	3.0	270	520	.	.	.	.	.	.	.	.	18
5744	PRUNUS DULCIS	S	603	4.8	21.0	54.9	17.3	3.0	2.0	282	485	5.2	14	432	5	0.25	0.67	6.0	0	2
5745	PRUNUS DULCIS	S	547	4.7	18.6	54.1	19.6	2.7	3.0	254	475	4.4	.	.	0	0.25	0.67	4.6	0	3
5746	PRUNUS DULCIS	S	598	4.7	18.6	54.2	19.5	2.6	3.0	234	504	4.7	4	773	0	0.24	0.92	3.5	0	4
5747	PRUNUS DULCIS	S	.	5.2	20.8	58.9	12.2	1.7	2.9	230	490	4.5	.	.	0	0.24	0.15	2.5	0	5
5748	PRUNUS ILICIFOLIA	S	.	0.0	10.0	2.7	.	.	.	.	.	.	.	.	.	.	.	.	.	20
5749	PRUNUS INSITITIA	S	.	0.0	33.1	30.9	.	.	.	.	.	.	.	.	.	.	.	.	.	21
5750	PRUNUS INSITITIA	S	.	0.0	30.1	47.2	.	.	.	.	.	.	.	.	.	.	.	.	.	21
5751	PRUNUS LAUROCERASUS	S	.	0.0	31.2	44.4	.	.	.	.	.	.	.	.	.	.	.	.	.	21
5752	PRUNUS MAHALEB	S	.	0.0	29.9	43.1	.	.	.	.	.	.	.	.	.	.	.	.	.	21
5753	PRUNUS MICROCARPA	S	.	0.0	21.0	52.2	.	.	.	.	.	.	.	.	.	.	.	.	.	21
5754	PRUNUS MINUTIFLORA	L	.	0.0	23.4	4.2	65.4	10.8	7.0	2010	270	.	.	1670	.	.	.	.	.	6
5755	PRUNUS MUME	F	52	86.6	1.3	1.8	9.1	1.4	1.2	11	36	1.8	8	312	20	0.05	0.05	0.5	10	2
5756	PRUNUS PERSICA	F	31	91.1	0.4	0.1	8.1	0.4	0.3	4	13	0.3	2	137	270	0.01	0.03	0.6	3	
5757	PRUNUS PERSICA	F	50	85.8	0.9	0.2	12.5	0.9	0.6	7	19	0.5	.	.	25	.	.	.	16	1
5758	PRUNUS PERSICA	F	39	88.7	0.6	0.1	10.0	0.8	0.6	10	17	0.4	.	.	.	.	.	.	.	1
5759	PRUNUS PERSICA	F	39	88.9	0.6	0.2	9.9	1.8	0.4	4	10	1.1	4	245	.	0.01	0.01	0.7	.	2
5760	PRUNUS PERSICA	S	.	0.0	23.9	45.6	.	.	.	.	.	.	.	.	.	.	.	.	.	20
5761	PRUNUS PERSICA	F	52	85.3	0.8	0.2	13.3	0.9	0.4	12	26	1.1	.	.	10	0.03	0.06	0.4	28	3
5762	PRUNUS SALICINA	F	53	85.1	0.7	0.2	13.5	0.7	0.5	16	32	0.4	3	145	10	0.02	0.02	0.5	2	2
5763	PRUNUS SALICINA	F	53	85.2	0.6	0.2	13.6	0.4	0.4	7	20	0.4	.	.	60	0.01	0.02	0.4	10	3

TABLE 2 - DATA DIRECTLY TRANSCRIBED FROM ORIGINAL SOURCES.

OBS	PLANA	PART	CALORIES	WATER	PROTEIN	FAT	TOTCARB	FIBER	ASH	CALCIUM	PHOSPHON	IRON	SODIUM	POTASS	CAROTENE	THIAMIN	RIBOFLAV	NIACIN	ASCORBIC	REFCO
5764	PRUNUS SEROTINA SUBSP. CAPULI	F	81	77.2	1.3	0.2	20.7	0.6	0.6	24	24	0.8	.	.	90	0.04	0.04	1.1	18	3
5765	PRUNUS SP.	F	49	86.1	0.6	0.2	12.5	1.1	0.6	10	12	0.6	.	.	770	.	.	.	.	1
5766	PRUNUS SP.	S	.	0.0	33.4	43.3	.	.	2.7	.	.	.	.	.	.	.	.	.	.	19
5767	PRUNUS VIRGINIANA	S	.	0.0	34.6	39.3	.	.	.	.	.	.	.	.	.	.	.	.	.	20
5768	PRUNUS VIRGINIANA	S	.	0.0	36.6	42.6	.	.	.	.	.	.	.	.	.	.	.	.	.	21
5769	PRUNUS VIRGINIANA(IMMATURE)	L	.	0.0	15.9	3.2	73.8	10.8	7.1	1800	410	.	.	.	.	.	.	.	.	6
5770	PRUNUS VIRGINIANA(MATURE)	L	.	0.0	8.1	6.5	75.2	11.5	10.2	3030	670	.	.	.	.	.	.	.	.	6
5771	PRUNUS WEBBII	S	.	0.0	24.2	52.2	.	.	.	.	.	.	.	.	.	.	.	.	.	21
5772	PSEUDERANTHEMUM RETICULATUM	L	46	85.0	3.4	0.2	10.1	1.7	1.3	.	67	.	.	.	2605	0.02	0.10	1.4	51	2
5773	PSEUDOPHOENIX SAONAE	S	.	0.0	6.2	19.2	.	.	.	.	.	.	.	.	.	.	.	.	.	19
5774	PSEUDOPHOENIX VINIFERA	S	.	0.0	6.4	21.4	.	.	1.3	.	.	.	.	.	.	.	.	.	.	19
5775	PSEUDOPHOENIX VINIFERA	S	.	0.0	1.6	1.6	.	.	C.8	.	.	.	.	.	.	.	.	.	.	19
5776	PSIDIUM CATTLEIANUM	F	56	84.3	0.4	0.3	14.5	5.2	0.5	29	17	0.2	.	.	145	0.03	0.03	0.6	33	2
5777	PSIDIUM CATTLEIANUM	F	59	83.4	0.4	0.3	15.3	5.5	0.6	31	18	0.2	.	.	0	0.03	0.03	0.4	21	2
5778	PSIDIUM CATTLEIANUM	F	65	81.8	1.0	0.6	15.8	6.4	0.8	23	42	0.9	4	289	54	0.03	0.03	0.6	37	4
5779	PSIDIUM CATTLEIANUM	F	.	85.3	0.1	0.2	13.8	4.8	0.6	50	20	1.2	.	.	0	0.02	0.02	0.3	15	5
5780	PSIDIUM GUAJAVA	F	64	82.2	1.1	0.4	15.7	5.3	0.6	24	31	1.3	.	.	290	0.06	0.04	1.3	326	1
5781	PSIDIUM GUAJAVA	L	.	0.0	11.7	8.7	71.9	16.1	7.7	1340	160	.	.	.	.	.	.	.	.	18
5782	PSIDIUM GUAJAVA	F	69	80.6	1.0	0.4	17.3	5.6	0.7	15	24	0.7	4	291	75	0.05	0.04	1.1	132	2
5783	PSIDIUM GUAJAVA	F	69	80.8	0.9	0.4	17.3	5.3	0.6	22	26	0.7	.	.	160	0.04	0.04	1.0	218	3
5784	PSIDIUM GUAJAVA	P	57	83.8	0.5	0.1	15.2	2.4	0.4	20	13	0.6	.	.	140	0.01	0.04	0.5	72	3
5785	PSIDIUM GUAJAVA	P	62	83.0	0.8	0.6	15.0	5.6	0.6	23	42	0.9	4	289	168	0.05	0.05	1.2	242	4
5786	PSIDIUM GUAJAVA	F	.	81.7	0.9	0.3	16.4	5.2	0.7	10	28	1.4	6	91	0	0.03	0.03	0.4	212	5
5787	PSIDIUM SARTORIANUM	S	.	0.0	3.1	3.2	.	.	.	.	.	.	.	.	.	.	.	.	.	19
5788	PSIDIUM SARTORIANUM	S	.	0.0	3.8	0.9	.	.	.	.	.	.	.	.	.	.	.	.	.	19
5789	PSILOSTROPHE CF. GNAPHALIOIDES	S	.	0.0	18.1	10.5	.	.	.	.	.	.	.	.	.	.	.	.	.	20
5790	PSOPHOCARPUS TETRAGONOLOBUS	S	379	14.9	29.7	15.1	36.8	9.4	3.5	.	.	.	.	.	.	.	.	.	.	1
5791	PSOPHOCARPUS TETRAGONOLOBUS	G	25	92.2	2.1	0.3	4.9	1.7	0.5	.	.	.	.	.	.	.	.	.	.	1
5792	PSOPHOCARPUS TETRAGONOLOBUS	R	91	75.4	2.8	0.6	20.3	1.5	0.9	.	.	.	.	.	.	.	.	.	.	1
5793	PSOPHOCARPUS TETRAGONOLOBUS	F	34	89.5	1.9	0.1	7.9	1.6	0.6	53	48	0.2	3	205	340	0.19	0.08	1.0	21	2
5794	PSOPHOCARPUS TETRAGONOLOBUS	L	47	85.0	5.0	0.5	8.5	.	1.0	134	81	6.2	.	.	3145	0.28	.	.	29	2
5795	PSOPHOCARPUS TETRAGONOLOBUS	F	.	90.4	2.9	0.2	7.1	1.3	0.7	63	37	0.3	3	205	357	0.24	0.09	1.2	19	5
5796	PSOPHOCARPUS TETRAGONOLOBUS	S	.	8.5	41.9	13.1	31.2	.	5.3	.	.	.	.	.	.	.	.	.	.	5
5797	PSOPHOCARPUS TETRAGONOLOBUS	R	.	9.0	24.6	1.0	61.5	5.4	3.9	.	.	.	.	.	.	.	.	.	.	5
5798	PSOPHOCARPUS TETRAGONOLOBUS	L	.	78.9	6.3	1.0	12.0	4.1	1.8	370	120	.	.	.	.	.	.	.	.	5
5799	PSORALEA BITUMINOSA	S	.	0.0	32.9	8.2	.	.	.	.	.	.	.	.	.	.	.	.	.	21
5800	PSORALEA BITUMINOSA	S	.	0.0	34.1	12.4	.	.	.	.	.	.	.	.	.	.	.	.	.	21
5801	PSORALEA CANDICANS	C	.	0.0	14.7	5.4	75.0	39.3	4.9	780	150	.	.	.	.	.	.	.	.	18
5802	PSORALEA SPICATA	S	.	0.0	31.5	10.8	.	.	.	.	.	.	.	.	.	.	.	.	.	21
5803	PTELEA ANGUSTIFOLIA	S	.	0.0	13.8	11.0	.	.	.	.	.	.	.	.	.	.	.	.	.	19
5804	PTELEA TRIFOLIATA	S	.	0.0	44.0	45.9	.	.	3.1	.	.	.	.	.	.	.	.	.	.	19
5805	PTELEA TRIFOLIATA	S	.	0.0	6.6	2.4	.	.	4.1	.	.	.	.	.	.	.	.	.	.	19
5806	PTELEA TRIFOLIATA VAR. MOLLIS	S	.	0.0	10.2	6.0	.	.	.	.	.	.	.	.	.	.	.	.	.	20
5807	PTEROCARPUS SP.	L	131	61.5	13.9	3.4	19.3	4.7	1.9	233	133	4.1	.	.	590	1.25	0.83	0.8	25	1
5808	PTEROCARYA STENOPTERA	S	.	0.0	8.9	11.5	.	.	3.1	.	.	.	.	.	.	.	.	.	.	19

TABLE 2 - DATA DIRECTLY TRANSCRIBED FROM ORIGINAL SOURCES.

OBS	PLNA	PART	CALORIES	WATER	PROTEIN	FAT	TOTLCARB	FIBER	ASH	CALCIUM	PHOSPHO	IRON	SODIUM	POTASS	CAROT	THIAMINE	RIBOFLAV	NIACIN	ASCORBIC	REFCO
5809	PTERONIA ADENOCARPA	S	.	0.0	28.2	11.4	.	.	.	.	.	.	.	.	.	.	.	.	.	21
5810	PUCCINELLIA DISTANS	S	.	0.0	16.4	3.2	.	.	.	.	.	.	.	.	.	.	.	.	.	21
5811	PUERARIA LOBATA	L	.	0.0	19.6	3.9	68.7	21.5	7.8	.	.	.	.	.	.	.	.	.	.	18
5812	PUERARIA LOBATA	C	.	0.0	5.7	1.0	87.7	46.2	5.6	.	.	.	.	.	.	.	.	.	.	18
5813	PUERARIA LOBATA	H	.	0.0	13.3	2.5	74.9	40.3	9.3	.	.	.	.	.	.	.	.	.	.	18
5814	PUERARIA LOBATA	L	36	89.0	0.4	0.1	9.7	7.7	0.8	35	20	4.9	.	.	0	0.03	0.91	0.8	0	2
5815	PUERARIA LOBATA	R	113	68.6	2.1	0.1	27.8	0.7	1.4	15	18	0.6	.	.	.	.	.	.	.	2
5816	PUERARIA LOBATA	R	.	68.6	2.1	0.1	27.1	0.7	1.4	15	18	0.6	.	.	.	.	.	.	.	5
5817	PUERARIA LOBATA	L	.	5.7	15.9	3.3	73.6	25.9	12.9	.	230	.	.	3400	4	.	0.80	.	.	5
5818	PUERARIA PHASEOLOIDES	W	.	0.0	18.0	2.4	73.5	42.9	6.1	710	180	.	.	.	.	.	.	.	.	18
5819	PUERARIA PHASEOLOIDES	S	.	0.0	32.7	3.2	.	.	4.1	.	.	.	.	.	.	.	.	.	.	19
5820	PUERARIA PHASEOLOIDES	L	.	80.9	3.8	0.4	13.4	5.5	1.5	140	30	.	.	.	.	.	.	.	.	5
5821	PUERARIA SP.	W	.	0.0	17.2	2.1	72.7	33.9	8.0	.	.	.	.	.	.	.	.	.	.	6
5822	PUERARIA SP.	H	.	0.0	13.8	2.5	77.1	33.6	6.6	1610	470	.	.	.	.	.	.	.	.	6
5823	PUERARIA THUNBERGIANA	S	.	0.0	36.9	5.7	.	.	5.7	.	.	.	.	.	.	.	.	.	.	19
5824	PUERARIA THUNBERGIANA	S	.	0.0	36.4	5.6	.	.	5.5	.	.	.	.	.	.	.	.	.	.	19
5825	PULICARIA UNDULATA	S	.	0.0	9.0	8.7	.	.	.	.	.	.	.	.	.	.	.	.	.	21
5826	PULSATILLA VULGARIS	S	.	0.0	23.8	24.4	.	.	.	.	.	.	.	.	.	.	.	.	.	20
5827	PUNICA GRANATUM	F	78	78.1	1.6	0.3	19.6	4.5	0.4	12	37	1.0	.	.	40	.	.	.	.	1
5828	PUNICA GRANATUM	S	.	0.0	2.5	4.0	.	.	2.7	.	.	.	.	.	.	.	.	.	.	19
5829	PUNICA GRANATUM	F	72	80.0	1.0	0.6	17.7	1.1	0.7	13	23	0.7	7	379	.	0.07	0.01	0.3	7	2
5830	PUNICA GRANATUM	F	67	81.8	0.8	0.7	16.2	2.0	0.5	10	34	0.6	.	.	0	0.07	0.03	0.9	8	3
5831	PUNICA GRANATUM	F	63	82.3	0.5	0.3	16.4	0.2	0.5	3	8	0.3	3	259	0	0.03	0.03	0.3	4	4
5832	PUNICA GRANATUM	S	.	35.0	9.4	6.9	35.0	22.4	1.5	.	.	.	.	.	.	.	.	.	.	5
5833	PUNICA GRANATUM	F	.	78.0	1.6	0.1	19.6	5.1	0.7	10	70	0.3	1	133	0	0.06	0.10	0.3	14	5
5834	PURSHIA TRIDENTATA	W	.	0.0	11.8	5.0	78.3	21.4	4.9	1840	250	.	.	.	.	.	.	.	.	6
5835	PYCNANTHEMUM MUTICUM	S	.	0.0	30.6	34.8	.	.	4.8	.	.	.	.	.	.	.	.	.	.	19
5836	PYRACANTHA COCCINEA	S	.	0.0	11.2	6.6	.	.	.	.	.	.	.	.	.	.	.	.	.	19
5837	PYRACANTHA COCCINEA	S	.	0.0	8.6	6.3	.	.	.	.	.	.	.	.	.	.	.	.	.	20
5838	PYRUS ARBUTIFOLIA	S	.	0.0	18.1	10.4	.	.	.	.	.	.	.	.	.	.	.	.	.	19
5839	PYRUS ARBUTIFOLIA	S	.	0.0	3.8	3.0	.	.	.	.	.	.	.	.	.	.	.	.	.	19
5840	PYRUS COMMUNIS	F	58	84.2	0.4	0.6	14.5	1.8	0.3	6	12	0.2	.	.	.	.	.	.	.	1
5841	PYRUS COMMUNIS	F	44	87.6	0.4	0.2	11.4	1.0	0.4	10	15	0.6	7	141	0	0.03	0.03	0.1	4	2
5842	PYRUS COMMUNIS	F	56	84.4	0.3	0.2	14.8	1.9	0.3	6	10	0.5	.	.	10	0.02	0.03	0.2	5	3
5843	PYRUS COMMUNIS	F	61	83.2	0.7	0.4	15.3	1.4	0.4	8	11	0.3	2	130	12	0.02	0.04	0.1	4	4
5844	QUAMOCLIT PINNATA	S	.	0.0	21.8	12.7	.	.	.	.	.	.	.	.	.	.	.	.	.	21
5845	QUERCUS ACUTISSIMA	S	.	0.0	31.6	5.1	.	.	2.9	.	.	.	.	.	.	.	.	.	.	19
5846	QUERCUS ACUTISSIMA	S	.	0.0	4.4	0.9	.	.	2.1	.	.	.	.	.	.	.	.	.	.	19
5847	QUERCUS ALBA	S	.	0.0	6.2	7.8	.	.	.	.	.	.	.	.	.	.	.	.	.	19
5848	QUERCUS COCCIFERA	S	.	0.0	3.7	0.9	.	.	.	.	.	.	.	.	.	.	.	.	.	21
5849	QUERCUS DILATATA	L	.	0.0	9.6	4.5	80.8	29.1	5.1	1610	300	.	.	.	.	.	.	.	.	18
5850	QUERCUS DURANDII	L	.	0.0	9.6	2.5	82.5	25.3	5.4	1440	110	.	.	760	.	.	.	.	.	6
5851	QUERCUS EMORYI	S	.	0.0	8.1	41.2	.	.	.	.	.	.	.	.	.	.	.	.	.	19
5852	QUERCUS GAMBELII	S	.	0.0	6.9	8.2	.	.	.	.	.	.	.	.	.	.	.	.	.	19
5853	QUERCUS GLAUCA	L	.	0.0	9.6	4.1	78.7	29.0	7.6	1870	230	.	.	.	.	.	.	.	.	18

TABLE 2 - DATA DIRECTLY TRANSCRIBED FRCM ORIGINAL SOURCES.

OBS	PLANA	PART	CALORIES	WATER	PROTEIN	FAT	TOTCARB	FIBER	ASH	CALCIUM	PHOSPHO	IRON	SODIUM	POTASS	CAROT	THIAMINE	RIBOFLAV	NIACIN	ASCORBIC	REFCO
5854	QUERCUS INCANA	L	.	0.0	10.2	4.8	79.8	31.3	5.2	990	150	.	.	.	.	.	.	.	.	18
5855	QUERCUS PALUSTRIS	S	.	0.0	6.9	26.1	.	.	.	.	.	.	.	.	.	.	.	.	.	19
5856	QUERCUS VIRGINIANA	L	.	0.0	9.9	2.8	80.2	31.9	7.1	710	230	.	.	.	.	.	.	.	.	18
5857	QUERCUS VIRGINIANA	L	.	0.0	10.6	2.4	81.3	28.6	5.7	.	.	.	.	.	.	.	.	.	.	6
5858	QUINCULA SP.	W	.	0.0	31.5	4.1	46.7	14.5	17.7	2000	340	.	.	5590	.	.	.	.	.	6
5859	QUISQUALIS INDICA	L	76	75.4	4.8	0.0	18.1	2.0	1.7	104	97	.	.	.	6665	0.04	.	.	70	2
5860	RADYERA URENS	S	.	0.0	19.0	18.5	.	.	.	.	.	.	.	.	.	.	.	.	.	21
5861	RANDIA SP.	S	.	0.0	14.4	10.4	.	.	.	.	.	.	.	.	.	.	.	.	.	19
5862	RANDIA SP.	S	.	0.0	13.8	11.6	.	.	.	.	.	.	.	.	.	.	.	.	.	20
5863	RANUNCULUS ARVENSIS	S	.	0.0	11.1	20.4	.	.	.	.	.	.	.	.	.	.	.	.	.	20
5864	RANUNCULUS CONSTANTINOPOLITANUS	S	.	0.0	17.0	22.4	.	.	.	.	.	.	.	.	.	.	.	.	.	20
5865	RANUNCULUS LAETUS	S	.	0.0	11.8	20.7	.	.	.	.	.	.	.	.	.	.	.	.	.	21
5866	RANUNCULUS SCELERATUS	S	.	0.0	18.1	25.9	.	.	.	.	.	.	.	.	.	.	.	.	.	20
5867	RANUNCULUS SP.	S	.	0.0	12.5	9.1	.	.	5.0	.	.	.	.	.	.	.	.	.	.	19
5868	RAPHANUS CAUDATUS	S	.	0.0	38.7	32.2	.	.	.	.	.	.	.	.	.	.	.	.	.	21
5869	RAPHANUS RAPHANISTRUM	S	.	0.0	31.9	30.9	.	.	.	.	.	.	.	.	.	.	.	.	.	20
5870	RAPHANUS SATIVUS	I	19	93.6	1.1	0.1	4.0	0.1	1.2	42	30	1.0	.	.	.	.	.	.	.	1
5871	RAPHANUS SATIVUS	S	.	0.0	33.2	29.8	.	.	4.2	.	.	.	.	.	.	.	.	.	.	19
5872	RAPHANUS SATIVUS	S	.	0.0	33.6	37.1	.	.	3.8	.	.	.	.	.	.	.	.	.	.	19
5873	RAPHANUS SATIVUS	S	.	0.0	32.2	33.1	.	.	4.0	.	.	.	.	.	.	.	.	.	.	19
5874	RAPHANUS SATIVUS	S	.	0.0	30.9	37.4	.	.	4.2	.	.	.	.	.	.	.	.	.	.	19
5875	RAPHANUS SATIVUS	S	.	0.0	23.6	35.6	.	.	4.5	.	.	.	.	.	.	.	.	.	.	19
5876	RAPHANUS SATIVUS	S	.	0.0	30.0	40.4	.	.	4.2	.	.	.	.	.	.	.	.	.	.	19
5877	RAPHANUS SATIVUS	F	34	89.9	2.6	0.5	6.6	1.4	0.4	76	26	0.4	5	207	265	0.08	0.07	0.6	79	2
5878	RAPHANUS SATIVUS	L	33	88.5	3.3	0.6	5.7	1.1	1.9	220	30	4.1	110	500	2470	0.08	0.28	4.0	81	2
5879	RAPHANUS SATIVUS	R	26	92.6	1.0	0.2	5.6	1.3	0.6	32	34	1.4	10	242	0	0.04	0.03	0.4	26	2
5880	RAPHANUS SATIVUS	X	19	93.8	1.8	0.1	3.8	1.1	0.5	140	63	1.2	.	.	3600	0.08	0.25	0.5	70	2
5881	RAPHANUS SATIVUS	S	.	0.0	30.6	41.7	.	.	.	.	.	.	.	.	.	.	.	.	.	20
5882	RAPHANUS SATIVUS	R	23	93.2	0.9	0.1	5.0	0.7	0.8	26	30	1.2	.	.	0	0.03	0.03	0.3	28	3
5883	RAPHANUS SATIVUS	L	52	85.6	2.8	0.5	9.9	1.3	1.2	238	44	2.8	.	.	3290	0.14	0.26	0.6	122	3
5884	RAPHANUS SATIVUS	R	25	92.6	1.1	0.0	5.5	0.7	0.8	28	26	0.6	.	.	0	0.02	0.06	0.5	46	3
5885	RAPHANUS SATIVUS	R	17	94.5	1.0	0.1	3.6	0.7	0.8	30	31	1.0	18	322	6	0.03	0.03	0.3	26	4
5886	RAPHANUS SATIVUS	F	.	92.3	1.3	0.3	5.4	1.1	0.7	78	24	.	.	.	.	.	.	.	.	5
5887	RAPHANUS SATIVUS	F	.	90.5	2.3	0.3	6.1	1.4	0.8	80	100	2.8	.	.	30	0.07	0.05	0.2	69	5
5888	RAPHIA HOOKERI	S	326	10.6	7.8	1.0	71.4	8.1	9.2	1519	277	.	.	.	.	.	.	.	.	1
5889	RAPHIONACME BROWNII	R	16	95.4	0.2	0.0	4.1	0.5	0.3	50	.	1.0	.	.	.	0.01	0.01	0.1	0	1
5890	RAPISTRUM RUGOSUM	S	.	0.0	34.4	38.3	.	.	.	.	.	.	.	.	.	.	.	.	.	20
5891	RAPISTRUM RUGOSUM	S	.	0.0	32.4	35.8	.	.	.	.	.	.	.	.	.	.	.	.	.	21
5892	RAPISTRUM RUGOSUM	S	.	0.0	41.6	24.6	.	.	.	.	.	.	.	.	.	.	.	.	.	21
5893	RAPISTRUM RUGOSUM	S	.	0.0	31.1	39.2	.	.	.	.	.	.	.	.	.	.	.	.	.	21
5894	RAPISTRUM RUGOSUM	S	.	0.0	34.5	36.6	.	.	.	.	.	.	.	.	.	.	.	.	.	21
5895	RAPISTRUM RUGOSUM	S	.	0.0	33.3	39.1	.	.	.	.	.	.	.	.	.	.	.	.	.	21
5896	RAPISTRUM RUGOSUM SUBSP. ORIENTALE	S	.	0.0	33.2	36.2	.	.	.	.	.	.	.	.	.	.	.	.	.	21
5897	RATIBIDA COLUMNARIS	S	.	0.0	26.9	26.1	.	.	4.3	.	.	.	.	.	.	.	.	.	.	19
5898	RATIBIDA COLUMNARIS	S	.	0.0	18.8	23.8	.	.	.	.	.	.	.	.	.	.	.	.	.	19

TABLE 2 - DATA DIRECTLY TRANSCRIBED FROM ORIGINAL SOURCES.

OBS	PLNA	PART	CALORIES	WATER	PROTEIN	FAT	TOTLCARB	FIBER	ASH	CALCIUM	PHOSPHO	IRON	SODIUM	POTASS	CAROT	THIAMINE	RIBOFLAV	NIACIN	ASCORBIC	REFCO
5899	RATIBIDA COLUMNIFERA	S	.	0.0	18.9	18.6	.	.	.	.	.	.	.	.	.	.	.	.	.	20
5900	RATIBIDA SP.	L	.	0.0	19.7	4.1	60.1	14.0	16.1	3280	260	.	.	4530	.	.	.	.	.	6
5901	RATIBIDA SP.	I	.	0.0	13.6	6.1	73.1	37.2	7.2	1830	310	.	.	1300	.	.	.	.	.	6
5902	RATIBIDA TAGETES	S	.	0.0	16.7	2.8	.	.	.	.	.	.	.	.	.	.	.	.	.	20
5903	REBOUDIA PINNATA	S	.	0.0	35.2	30.6	.	.	.	.	.	.	.	.	.	.	.	.	.	21
5904	REPTONIA BUXIFOLIA	S	.	0.0	6.8	24.2	.	.	.	.	.	.	.	.	.	.	.	.	.	21
5905	RESEDA ERECTA	S	.	0.0	21.9	35.7	.	.	.	.	.	.	.	.	.	.	.	.	.	21
5906	RESEDA LUTEA	S	.	0.0	19.7	35.0	.	.	.	.	.	.	.	.	.	.	.	.	.	20
5907	RESEDA LUTEOLA	S	.	0.0	23.7	33.8	.	.	.	.	.	.	.	.	.	.	.	.	.	21
5908	RESEDA LUTEOLA	S	.	0.0	21.0	22.4	.	.	.	.	.	.	.	.	.	.	.	.	.	21
5909	RESEDA ODORATA	S	.	0.0	18.1	32.3	.	.	.	.	.	.	.	.	.	.	.	.	.	20
5910	RETAMA RAETAM	S	.	0.0	25.8	2.8	.	.	.	.	.	.	.	.	.	.	.	.	.	21
5911	RETAMA SPHAEROCARPA	S	.	0.0	25.0	5.3	.	.	.	.	.	.	.	.	.	.	.	.	.	21
5912	REVERCHONIA ARENARIA	S	.	0.0	10.6	27.8	.	.	.	.	.	.	.	.	.	.	.	.	.	20
5913	RHAMNUS CATHARTICA	S	.	0.0	20.5	24.6	.	.	2.5	.	.	.	.	.	.	.	.	.	.	19
5914	RHAMNUS DAVURICA	S	.	0.0	20.4	27.3	.	.	2.5	.	.	.	.	.	.	.	.	.	.	19
5915	RHAMNUS PRINOIDES	L	305	9.6	8.2	2.9	73.1	28.0	6.2	1113	125	25.0	.	.	6900	0.00	0.08	2.6	.	1
5916	RHAMNUS PURSHIANA	S	.	0.0	25.4	56.9	.	.	2.3	.	.	.	.	.	.	.	.	.	.	19
5917	RHAMNUS PURSHIANA	S	.	0.0	6.7	13.4	.	.	1.3	.	.	.	.	.	.	.	.	.	.	19
5918	RHAZYA STRICTA	S	.	0.0	17.8	17.8	.	.	.	.	.	.	.	.	.	.	.	.	.	21
5919	RHEEDIA ACUMINATA	F	46	87.2	0.6	0.1	11.9	1.0	0.2	12	22	0.4	.	.	0	0.06	0.04	0.3	6	3
5920	RHEUM OFFICINALE	Z	16	94.9	0.5	0.1	3.8	0.7	0.7	51	25	0.5	.	.	20	0.01	0.02	0.1	10	3
5921	RHEUM RHABARBARUM	Z	20	93.8	0.4	0.2	4.8	0.8	0.8	60	20	0.5	2	299	20	0.04	0.03	0.2	16	2
5922	RHEUM RHABARBARUM	L	18	94.0	0.7	0.2	4.1	0.7	1.0	62	16	0.3	.	.	30	0.02	0.02	0.2	11	3
5923	RHEUM RHABARBARUM	L	20	92.8	1.3	0.1	4.4	0.9	1.4	47	25	1.8	.	.	.	.	.	.	.	3
5924	RHEUM RHABARBARUM	Z	16	94.8	0.6	0.1	3.7	0.7	0.8	96	18	0.8	2	251	60	0.03	0.07	0.3	9	4
5925	RHEUM RHAPONTICUM	L	.	0.0	27.6	6.5	.	8.0	.	.	.	.	.	.	.	.	.	.	.	6
5926	RHIGOZUM OBOVATUM	SH	.	0.0	11.3	4.1	78.6	15.5	6.0	920	130	.	.	.	.	.	.	.	.	18
5927	RHIGOZUM TRICHOTOMUM	S	.	0.0	26.8	20.1	.	.	.	.	.	.	.	.	.	.	.	.	.	21
5928	RHINANTHUS SP.	W	.	0.0	24.7	1.6	67.1	23.4	6.6	420	130	.	.	.	.	.	.	.	.	6
5929	RHIZOPHORA MANGLE	L	.	0.0	10.7	3.4	77.0	14.5	8.9	.	.	.	.	.	.	.	.	.	.	6
5930	RHODODENDRON PONTICUM	S	.	0.0	24.8	27.6	.	.	.	.	.	.	.	.	.	.	.	.	.	21
5931	RHODODENDRON SP.	L	.	0.0	6.3	6.0	82.9	21.3	4.8	1420	80	37.0	.	.	.	.	.	.	.	6
5932	RHODOTYPOS SCANDENS	S	.	0.0	50.2	37.5	.	.	.	.	.	.	.	.	.	.	.	.	.	21
5933	RHODOTYPOS TETRAPETALA	S	.	0.0	13.1	8.3	.	.	.	.	.	.	.	.	.	.	.	.	.	20
5934	RHODOTYPOS TETRAPETALA	S	.	0.0	48.8	39.7	.	.	.	.	.	.	.	.	.	.	.	.	.	20
5935	RHUS CANADENSIS	S	.	0.0	16.3	13.4	.	.	2.0	.	.	.	.	.	.	.	.	.	.	19
5936	RHUS COPALLINA	S	.	0.0	13.8	10.1	.	.	.	.	.	.	.	.	.	.	.	.	.	19
5937	RHUS COPALLINA	S	.	0.0	4.4	26.3	.	.	.	.	.	.	.	.	.	.	.	.	.	19
5938	RHUS COPALLINA	S	.	0.0	9.4	13.6	.	.	.	.	.	.	.	.	.	.	.	.	.	19
5939	RHUS COPALLINA	S	.	0.0	8.8	10.9	.	.	.	.	.	.	.	.	.	.	.	.	.	20
5940	RHUS COPALLINA	S	.	0.0	4.4	14.8	.	.	.	.	.	.	.	.	.	.	.	.	.	20
5941	RHUS CORIARIA	S	.	0.0	5.2	6.3	.	.	.	.	.	.	.	.	.	.	.	.	.	21
5942	RHUS GLABRA	S	.	0.0	10.0	8.7	.	.	.	.	.	.	.	.	.	.	.	.	.	19
5943	RHUS GLABRA	S	.	0.0	3.1	17.7	.	.	.	.	.	.	.	.	.	.	.	.	.	19

TABLE 2 - DATA DIRECTLY TRANSCRIBED FROM ORIGINAL SOURCES.

OBS	PLANA	PART	CALORIES	WATER	PROTEIN	FAT	TOTCARB	FIBER	ASH	CALCIUM	PHOSPHO	IRON	SODIUM	POTASS	CAROTE	THIAMINE	RIBOFLAV	NIACIN	ASCORBIC	REFCO
5944	RHUS GLABRA	S	.	0.0	10.0	12.0	.	.	.	.	.	.	.	.	.	.	.	.	.	19
5945	RHUS GLABRA	S	.	0.0	10.0	13.1	.	.	.	.	.	.	.	.	.	.	.	.	.	20
5946	RHUS GLABRA	S	.	0.0	6.9	13.5	.	.	.	.	.	.	.	.	.	.	.	.	.	20
5947	RHUS GLABRA	S	.	0.0	6.6	11.0	.	.	.	.	.	.	.	.	.	.	.	.	.	21
5948	RHUS LANCEA	SH	.	0.0	12.7	7.3	72.7	22.3	7.3	800	190	.	.	.	.	.	.	.	.	18
5949	RHUS MICROPHYLLA	S	.	0.0	10.0	7.8	.	.	.	.	.	.	.	.	.	.	.	.	.	19
5950	RHUS MICROPHYLLA	S	.	0.0	10.0	23.0	.	.	.	.	.	.	.	.	.	.	.	.	.	19
5951	RHUS MICROPHYLLA	L	.	0.0	14.9	4.7	72.8	11.4	7.6	1950	170	.	.	1880	.	.	.	.	.	6
5952	RHUS NATALENSIS	SH	.	0.0	9.0	3.5	78.7	19.6	8.8	1430	100	.	.	.	.	.	.	.	.	18
5953	RHUS NATALENSIS	F	.	0.0	11.1	5.8	79.1	25.3	4.0	270	190	.	.	.	.	.	.	.	.	18
5954	RHUS RADICANS	S	.	0.0	10.6	22.4	.	.	.	.	.	.	.	.	.	.	.	.	.	19
5955	RHUS SEMPERVIRENS	L	.	0.0	9.9	6.3	79.0	18.7	4.8	1670	50	.	.	730	.	.	.	.	.	6
5956	RHUS SP.	S	.	0.0	3.1	2.6	.	.	.	.	.	.	.	.	.	.	.	.	.	19
5957	RHUS SP.	S	.	0.0	2.5	0.2	.	.	.	.	.	.	.	.	.	.	.	.	.	19
5958	RHUS TRILOBATA	S	.	0.0	13.4	13.4	.	.	1.8	.	.	.	.	.	.	.	.	.	.	19
5959	RHUS TRILOBATA	S	.	0.0	12.5	6.6	.	.	.	.	.	.	.	.	.	.	.	.	.	19
5960	RHUS TRILOBATA	S	.	0.0	9.4	31.2	.	.	.	.	.	.	.	.	.	.	.	.	.	19
5961	RHUS TRILOBATA	L	.	0.0	8.0	4.4	82.1	13.7	5.5	1930	110	.	.	1690	.	.	.	.	.	6
5962	RHUS TRILOBATA VAR. PILOSSISSIMA	S	.	0.0	12.8	9.7	.	.	.	.	.	.	.	.	.	.	.	.	.	20
5963	RHUS TYPHINA	S	.	0.0	9.4	11.1	.	.	.	.	.	.	.	.	.	.	.	.	.	19
5964	RHUS TYPHINA	S	.	0.0	4.4	21.9	.	.	.	.	.	.	.	.	.	.	.	.	.	19
5965	RHUS VERNIX	S	.	0.0	16.2	9.2	.	.	.	.	.	.	.	.	.	.	.	.	.	19
5966	RHYNCHELYTRUM REPENS	W	.	0.0	11.3	2.3	76.6	31.8	9.8	.	.	.	.	.	.	.	.	.	.	18
5967	RHYNCHELYTRUM REPENS	H	.	0.0	5.6	1.4	85.0	41.6	8.0	.	.	.	.	.	.	.	.	.	.	18
5968	RHYNCHELYTRUM ROSEUM	H	.	0.0	3.9	1.5	89.4	42.8	5.2	530	350	.	.	.	.	.	.	.	.	6
5969	RHYNCHOSIA CAPITATA	S	.	0.0	23.1	2.4	.	.	.	.	.	.	.	.	.	.	.	.	.	21
5970	RHYNCHOSIA MINIMA	S	.	0.0	23.6	0.6	.	.	.	.	.	.	.	.	.	.	.	.	.	21
5971	RHYNCHOSIA PYRAMIDALIS	S	.	0.0	18.8	0.7	.	.	.	.	.	.	.	.	.	.	.	.	.	20
5972	RHYNCHOSIA PYRAMIDALIS	S	.	0.0	18.8	1.0	.	.	.	.	.	.	.	.	.	.	.	.	.	20
5973	RHYNCHOSIA TEXANA	W	.	0.0	18.8	3.3	71.1	24.9	6.8	1650	230	.	.	1370	.	.	.	.	.	6
5974	RHYNCHOSPORA INDIANOLENSIS	S	.	0.0	10.6	10.5	.	.	.	.	.	.	.	.	.	.	.	.	.	19
5975	RHYNCHOSPORA INDIANOLENSIS	S	.	0.0	11.2	9.6	.	.	.	.	.	.	.	.	.	.	.	.	.	19
5976	RIBES LACUSTRE	F	361	0.0	15.0	4.0	78.0	.	4.0	319	.	5.0	.	.	.	.	.	.	333	15
5977	RIBES PUNCTATUM	F	31	92.4	0.7	1.0	5.6	0.7	0.3	11	20	1.2	.	.	30	0.01	0.01	0.4	4	3
5978	RIBES SP.	F	73	80.1	2.2	1.3	15.3	2.9	1.1	14	39	1.1	.	.	.	.	.	.	.	1
5979	RIBES SP.		39	88.9	0.8	0.2	9.7	1.9	0.4	18	15	0.5	1	155	174	.	.	.	33	4
5980	RIBES SP.	F	54	84.2	1.4	0.1	13.1	2.4	0.9	60	40	1.1	3	372	138	0.05	0.05	0.3	200	4
5981	RIBES SP.	F	50	85.7	1.4	0.2	12.1	3.4	0.6	32	23	1.0	2	257	72	0.04	0.05	0.1	41	4
5982	RICINODENDRON HEUDELOTII	S	530	5.5	21.2	43.1	23.4	1.7	6.8	611	926	0.4	.	.	0	0.01	0.00	0.0	0	1
5983	RICINUS COMMUNIS	L	.	0.0	24.8	5.4	57.4	10.3	12.4	2670	460	.	.	.	.	.	.	.	.	18
5984	RICINUS COMMUNIS	S	.	0.0	26.2	65.9	.	.	.	.	.	.	.	.	.	.	.	.	.	19
5985	RICINUS COMMUNIS VAR. CAMBODGENSIS	S	.	0.0	20.6	46.4	.	.	.	.	.	.	.	.	.	.	.	.	.	20
5986	RICINUS COMMUNIS VAR. GIBSONII	S	.	0.0	21.9	49.0	.	.	.	.	.	.	.	.	.	.	.	.	.	20
5987	RICINUS COMMUNIS VAR. SANGUINEUS	S	.	0.0	25.0	67.3	.	.	.	.	.	.	.	.	.	.	.	.	.	20
5988	RICINUS COMMUNIS VAR. ZANZIBARENSIS	S	.	0.0	20.0	50.6	.	.	.	.	.	.	.	.	.	.	.	.	.	20

TABLE 2 - DATA DIRECTLY TRANSCRIBED FROM ORIGINAL SOURCES.

OBS	PLNA	PART	CALORIES	WATER	PROTEIN	FAT	TOTLCARB	FIBER	ASH	CALCIUM	PHOSPHO	IRON	SODIUM	POTASS	CAROT	THIAMINE	RIBOFLAV	NIACIN	ASCORBIC	REFCO
5989	RIDOLFIA SEGETUM	S	.	0.0	21.2	12.0	.	.	.	.	.	.	.	.	.	.	.	.	.	21
5990	RIVEA CORYMBOSA	S	.	0.0	18.1	9.2	.	.	.	.	.	.	.	.	.	.	.	.	.	20
5991	ROBINIA NEO-MEXICANA	S	.	0.0	41.2	12.9	.	.	.	.	.	.	.	.	.	.	.	.	.	19
5992	ROBINIA PSEUDO-ACACIA	S	.	0.0	43.2	10.8	.	.	4.4	.	.	.	.	.	.	.	.	.	.	19
5993	ROBINIA SP.	S	.	0.0	20.0	2.3	.	.	3.9	.	.	.	.	.	.	.	.	.	.	19
5994	ROEMERIA HYBRIDA	S	.	0.0	21.8	37.9	.	.	.	.	.	.	.	.	.	.	.	.	.	21
5995	ROGERIA LONGIFLORA	S	.	0.0	20.1	31.1	.	.	.	.	.	.	.	.	.	.	.	.	.	21
5996	ROMULEA SP.	S	.	0.0	15.1	8.5	.	.	.	.	.	.	.	.	.	.	.	.	.	21
5997	ROSA AGRESTIS	S	.	0.0	7.4	8.9	.	.	.	.	.	.	.	.	.	.	.	.	.	21
5998	ROSA ARKANSANA	S	.	0.0	4.6	4.2	.	.	.	.	.	.	.	.	.	.	.	.	.	21
5999	ROSA CANINA	S	.	0.0	9.6	8.5	.	.	.	.	.	.	.	.	.	.	.	.	.	21
6000	ROSA CANINA VAR. DUMETORUM	S	.	0.0	8.8	10.8	.	.	.	.	.	.	.	.	.	.	.	.	.	21
6001	ROSA CAROLINA	S	.	0.0	16.2	14.3	.	.	.	.	.	.	.	.	.	.	.	.	.	19
6002	ROSA SP.	S	.	0.0	8.1	8.2	.	.	.	.	.	.	.	.	.	.	.	.	.	20
6003	ROSA SP.	S	.	0.0	9.8	7.0	.	.	.	.	.	.	.	.	.	.	.	.	.	21
6004	ROSA SPALDINGII(IMMATURE)	L	.	0.0	18.6	3.7	70.1	11.7	7.6	1500	530	.	.	.	.	.	.	.	.	6
6005	ROSA SPALDINGII(MATURE)	L	.	0.0	11.6	6.2	73.9	11.4	8.3	2340	650	.	.	.	.	.	.	.	.	6
6006	ROSA SPINOSISSIMA	S	.	0.0	6.9	5.4	.	.	.	.	.	.	.	.	.	.	.	.	.	21
6007	ROSA WOODSII FENDLERI	C	.	0.0	6.0	3.0	87.8	32.7	3.2	.	.	.	.	.	.	.	.	.	.	6
6008	ROSMARINUS OFFICINALIS	H	331	9.3	4.9	15.2	64.1	17.6	6.5	1280	70	29.2	50	955	1877	0.51	.	1.0	61	16
6009	ROTTBOELLIA EXALTATA	W	.	0.0	11.1	2.3	75.7	32.9	10.9	.	.	.	.	.	.	.	.	.	.	18
6010	ROTTBOELLIA EXALTATA	H	.	0.0	5.3	0.8	83.1	31.1	10.8	.	.	.	.	.	.	.	.	.	.	18
6011	ROYENA LYCIOIDES	S	.	0.0	7.5	4.9	.	.	.	.	.	.	.	.	.	.	.	.	.	21
6012	ROYENA LYCIOIDES	S	.	0.0	6.8	4.3	.	.	.	.	.	.	.	.	.	.	.	.	.	21
6013	ROYENA SP.	S	.	0.0	8.1	3.4	.	.	.	.	.	.	.	.	.	.	.	.	.	21
6014	ROYSTONEA REGIA	F	.	0.0	6.1	26.6	63.0	23.6	4.3	.	.	.	.	.	.	.	.	.	.	18
6015	RUBIA PURPUREA	S	.	0.0	12.6	15.5	.	.	.	.	.	.	.	.	.	.	.	.	.	21
6016	RUBIA TINCTORUM	S	.	0.0	16.0	11.6	.	.	.	.	.	.	.	.	.	.	.	.	.	21
6017	RUBUS IDAEUS VAR. STRIGOSUS	S	.	0.0	7.4	13.8	.	.	.	.	.	.	.	.	.	.	.	.	.	20
6018	RUBUS OCCIDENTALIS X RUBUS IDAEUS	F	73	80.8	1.5	1.4	15.7	5.1	0.6	30	22	0.9	1	199	0	0.03	0.09	0.9	18	4
6019	RUBUS PARVIFLORUS	F	363	0.0	10.0	4.0	82.0	.	4.0	428	.	3.0	.	.	.	.	.	.	259	15
6020	RUBUS PARVIFLORUS	S	.	0.0	10.6	22.8	.	.	.	.	.	.	.	.	.	.	.	.	.	20
6021	RUBUS SP.	S	.	0.0	9.4	14.6	.	.	.	.	.	.	.	.	.	.	.	.	.	20
6022	RUBUS SP.	F	57	84.4	1.2	0.6	13.2	3.9	0.6	34	36	2.0	.	.	20	0.02	0.04	0.5	18	3
6023	RUBUS SP.	F	57	84.2	1.2	0.5	13.6	3.0	0.5	22	22	0.9	1	168	78	0.03	0.09	0.9	25	4
6024	RUBUS SP.	F	58	84.5	1.2	0.9	12.9	4.1	0.5	32	19	0.9	1	170	120	0.03	0.04	0.4	21	4
6025	RUBUS SPECTABILIS	F	373	0.0	35.0	5.0	59.0	.	1.0	225	.	5.0	.	.	.	.	.	.	371	15
6026	RUBUS URSINUS	F	374	0.0	18.0	6.0	72.0	.	4.0	353	.	4.0	.	.	.	.	.	.	201	15
6027	RUBUS URSINUS	F	48	86.8	1.2	0.3	11.4	2.7	0.3	25	24	1.6	1	153	102	0.02	0.13	1.0	13	4
6028	RUBUS URSINUS	F	62	83.0	1.0	0.6	14.9	3.0	0.5	35	17	1.2	1	170	120	0.03	0.04	0.4	24	4
6029	RUDBECKIA BICOLOR VAR. SUPERBA	S	.	0.0	29.1	32.0	.	.	4.3	.	.	.	.	.	.	.	.	.	.	19
6030	RUDBECKIA BICOLOR VAR. SUPERBA	S	.	0.0	28.8	28.1	.	.	4.0	.	.	.	.	.	.	.	.	.	.	19
6031	RUDBECKIA BICOLOR VAR. SUPERBA	S	.	0.0	30.6	27.6	.	.	.	.	.	.	.	.	.	.	.	.	.	19
6032	RUDBECKIA BICOLOR VAR. SUPERBA	S	.	0.0	22.5	22.8	.	.	.	.	.	.	.	.	.	.	.	.	.	19
6033	RUDBECKIA FULGIDA	S	.	0.0	30.6	27.1	.	.	.	.	.	.	.	.	.	.	.	.	.	21

TABLE 2 - DATA DIRECTLY TRANSCRIBED FROM ORIGINAL SOURCES.

OBS	PLNA	PART	CALORIES	WATER	PROTEIN	FAT	TOTCARB	FIBER	ASH	CALCIUM	PHOSPHO	IRON	SODIUM	POTASS	CAROT	THIAMINE	RIBOFLAV	NIACIN	ASCORBIC	REFCO
6034	RUDBECKIA GRANDIFLORA VAR. ALISMAEFOLIA	S	.	0.0	13.8	6.6	.	.	.	.	.	.	.	.	.	.	.	.	.	20
6035	RUDBECKIA HIRTA	S	.	0.0	30.0	29.5	.	.	.	.	.	.	.	.	.	.	.	.	.	19
6036	RUDBECKIA HIRTA	S	.	0.0	27.5	31.3	.	.	.	.	.	.	.	.	.	.	.	.	.	19
6037	RUDBECKIA HIRTA VAR. PULCHERRIMA	S	.	0.0	24.4	30.6	.	.	.	.	.	.	.	.	.	.	.	.	.	20
6038	RUDBECKIA HIRTA VAR. PULCHERRIMA	S	.	0.0	28.8	30.0	.	.	.	.	.	.	.	.	.	.	.	.	.	20
6039	RUDBECKIA HIRTA VAR. PULCHERRIMA	S	.	0.0	25.6	28.1	.	.	.	.	.	.	.	.	.	.	.	.	.	21
6040	RUDBECKIA LACINIATA	S	.	0.0	37.8	26.3	.	.	4.3	.	.	.	.	.	.	.	.	.	.	19
6041	RUDBECKIA MAXIMA	S	.	0.0	25.4	13.6	.	.	.	.	.	.	.	.	.	.	.	.	.	20
6042	RUDBECKIA OCCIDENTALIS	S	.	0.0	21.5	13.4	.	.	.	.	.	.	.	.	.	.	.	.	.	21
6043	RUDBECKIA OCCIDENTALIS	W	.	0.0	21.7	6.0	59.0	11.4	13.3	2930	440	.	.	.	.	.	.	.	.	6
6044	RUDBECKIA OCCIDENTALIS	L	.	0.0	34.4	2.2	51.4	8.7	12.0	.	.	.	.	.	.	.	.	.	.	6
6045	RUMEX ACETOSELLA	S	.	0.0	5.6	1.7	.	.	.	.	.	.	.	.	.	.	.	.	.	20
6046	RUMEX ACETOSELLA	S	.	0.0	9.6	3.6	.	.	.	.	.	.	.	.	.	.	.	.	.	21
6047	RUMEX ALTISSIMUS	S	.	0.0	13.4	3.7	.	.	2.0	.	.	.	.	.	.	.	.	.	.	19
6048	RUMEX ALTISSIMUS	S	.	0.0	12.5	2.4	.	.	.	.	.	.	.	.	.	.	.	.	.	20
6049	RUMEX CONGLOMERATUS	S	.	0.0	12.1	2.2	.	.	.	.	.	.	.	.	.	.	.	.	.	21
6050	RUMEX CRISPUS	S	.	0.0	13.9	3.0	.	.	2.1	.	.	.	.	.	.	.	.	.	.	19
6051	RUMEX CRISPUS	S	.	0.0	9.6	1.3	.	.	7.9	.	.	.	.	.	.	.	.	.	.	19
6052	RUMEX CRISPUS	S	.	0.0	16.2	3.0	.	.	2.4	.	.	.	.	.	.	.	.	.	.	19
6053	RUMEX CRISPUS	S	.	0.0	13.1	2.4	.	.	.	.	.	.	.	.	.	.	.	.	.	20
6054	RUMEX CRISPUS	S	.	0.0	13.4	3.2	.	.	.	.	.	.	.	.	.	.	.	.	.	20
6055	RUMEX CRISPUS	S	.	0.0	13.6	3.0	.	.	.	.	.	.	.	.	.	.	.	.	.	20
6056	RUMEX CRISPUS	S	.	0.0	14.7	2.6	.	.	.	.	.	.	.	.	.	.	.	.	.	21
6057	RUMEX CRISPUS	S	.	0.0	12.8	2.9	.	.	.	.	.	.	.	.	.	.	.	.	.	21
6058	RUMEX CRISPUS	S	.	0.0	21.9	4.0	.	.	.	.	.	.	.	.	.	.	.	.	.	21
6059	RUMEX CRISPUS	S	.	0.0	16.2	2.8	.	.	.	.	.	.	.	.	.	.	.	.	.	21
6060	RUMEX CRISPUS	L	21	92.6	1.5	0.3	4.1	0.9	1.5	74	56	5.6	.	.	2770	0.06	0.08	0.4	30	3
6061	RUMEX CRISPUS	W	.	0.0	13.0	3.2	80.2	23.7	3.6	.	.	.	.	.	.	.	.	.	.	6
6062	RUMEX DENTATUS SUBSP. KLOTZSCHIANUS	S	.	0.0	12.3	2.8	.	.	.	.	.	.	.	.	.	.	.	.	.	21
6063	RUMEX HYDROLAPATHUM	S	.	0.0	13.0	2.9	.	.	.	.	.	.	.	.	.	.	.	.	.	21
6064	RUMEX OBTUSIFOLIUS	S	.	0.0	12.7	2.8	.	.	.	.	.	.	.	.	.	.	.	.	.	20
6065	RUMEX PULCHER	L	.	0.0	26.8	.	.	.	12.9	697	.	.	1830	2400	.	.	.	.	.	17
6066	RUMEX PULCHER	S	.	0.0	11.9	2.2	.	.	.	.	.	.	.	.	.	.	.	.	.	19
6067	RUMEX PULCHER	S	.	0.0	23.5	2.3	.	.	.	.	.	.	.	.	.	.	.	.	.	21
6068	RUMEX PULCHER	S	.	0.0	13.3	2.6	.	.	.	.	.	.	.	.	.	.	.	.	.	21
6069	RUMEX SALICIFOLIUS	S	.	0.0	13.5	2.6	.	.	2.0	.	.	.	.	.	.	.	.	.	.	19
6070	RUMEX SALICIFOLIUS	S	.	0.0	10.2	1.2	.	.	7.7	.	.	.	.	.	.	.	.	.	.	19
6071	RUMEX SCUTATUS	S	.	0.0	24.5	7.4	.	.	3.3	.	.	.	.	.	.	.	.	.	.	19
6072	RUMEX SCUTATUS	S	.	0.0	17.2	4.5	.	.	.	.	.	.	.	.	.	.	.	.	.	21
6073	RUMEX SP.	S	.	0.0	12.8	2.1	.	.	.	.	.	.	.	.	.	.	.	.	.	20
6074	RUMEX SP.	L	28	90.9	2.1	0.3	5.6	0.8	1.1	66	41	1.6	5	338	7740	0.09	0.22	0.5	119	4
6075	RUMEX SP.	L	28	90.9	2.1	0.3	5.6	0.8	1.1	66	41	1.6	5	338	7740	0.09	0.22	0.5	119	4
6076	RUMEX VENOSUS	S	.	0.0	20.0	4.2	.	.	1.4	.	.	.	.	.	.	.	.	.	.	19
6077	RUPRECHTIA CORIACEA	S	.	0.0	11.2	2.3	.	.	2.0	.	.	.	.	.	.	.	.	.	.	19
6078	RUTA CHALEPENSIS	S	.	0.0	26.4	33.2	.	.	.	.	.	.	.	.	.	.	.	.	.	21

TABLE 2 - DATA DIRECTLY TRANSCRIBED FROM ORIGINAL SOURCES.

OBS	PLANA	PART	CALORIES	WATER	PROTEIN	FAT	TOTCARB	FIBER	ASH	CALCIUM	PHOSPHO	IRON	SODIUM	POTASS	CAROTE	THIAMINE	RIBOFLAV	NIACIN	ASCORBIC	REFCO
6079	SABA SENEGALENSIS	F	71	80.0	0.8	0.2	18.5	1.3	0.5	51	28	1.0	.	.	0	0.15	0.03	0.50	48	1
6080	SACCHARUM OFFICINARUM	L	75	77.5	1.8	0.8	17.9	3.0	2.0	.	.	.	.	.	.	.	.	.	.	1
6081	SACCHARUM OFFICINARUM	SH	62	82.5	0.6	0.1	16.5	3.1	0.3	8	6	1.4	.	.	0	0.02	0.01	0.10	3	1
6082	SACCHARUM OFFICINARUM	I	25	91.0	4.6	0.4	3.0	.	1.0	40	80	2.0	.	.	0	0.08	.	.	50	2
6083	SACCHARUM OFFICINARUM	J	82	78.8	0.3	0.1	20.5	0.4	0.3	13	12	0.7	.	.	0	0.02	0.01	0.10	2	3
6084	SACCHARUM OFFICINARUM	H	.	0.0	2.6	1.2	92.1	43.1	4.1	3600	.	.	.	.	.	.	.	.	.	6
6085	SACCHARUM OFFICINARUM	W	.	0.0	3.6	1.3	91.4	28.2	3.7	470	170	8.0	.	.	.	.	.	.	.	6
6086	SACCHARUM SINENSE	W	.	0.0	10.3	2.5	80.9	32.1	6.3	.	.	.	.	.	.	.	.	.	.	18
6087	SACCHARUM SINENSE	L	.	0.0	9.5	2.8	78.9	33.4	8.8	.	.	.	.	.	.	.	.	.	.	18
6088	SACCHARUM SINENSE	S	.	0.0	5.3	1.5	82.9	35.9	10.5	.	.	.	.	.	.	.	.	.	.	18
6089	SACCHARUM SPONTANEUM	W	.	0.0	5.4	1.2	88.2	40.2	7.2	.	.	.	.	.	.	.	.	.	.	18
6090	SAGITTARIA PAPILLOSA	S	.	0.0	19.4	23.0	.	.	.	.	.	.	.	.	.	.	.	.	.	19
6091	SAGITTARIA SAGITTIFOLIA	R	107	70.6	5.0	0.3	22.4	0.9	1.7	13	165	2.6	.	729	0	0.16	0.04	1.40	5	2
6092	SAGITTARIA SAGITTIFOLIA	S	.	0.0	21.2	17.1	.	.	.	.	.	.	.	.	.	.	.	.	.	21
6093	SALACCA EDULIS	F	76	78.0	0.4	0.0	20.9	.	0.7	28	18	4.2	.	.	0	0.04	.	.	2	2
6094	SALAMALIA MALABARICA	L	.	0.0	12.6	6.4	71.7	22.3	9.3	2700	190	.	.	.	.	.	.	.	.	18
6095	SALIX BABYLONICA	L	.	0.0	16.7	2.9	70.2	18.0	10.2	1600	260	.	.	.	.	.	.	.	.	18
6096	SALIX LUTEA	L	.	0.0	14.4	2.9	75.9	14.0	6.8	250	240	.	.	.	.	.	.	.	.	6
6097	SALSOLA APHYLLA	SH	.	0.0	20.0	2.0	49.4	14.9	28.6	1550	160	.	.	.	.	.	.	.	.	18
6098	SALSOLA GLABRESCENS	SH	.	0.0	16.5	2.1	68.2	25.2	13.2	1950	150	.	.	.	.	.	.	.	.	18
6099	SALSOLA KALI	S	.	0.0	40.5	27.0	.	.	.	.	.	.	.	.	.	.	.	.	.	21
6100	SALSOLA KALI	W	.	0.0	13.1	2.0	69.4	31.7	15.5	2730	130	.	.	4630	.	.	.	.	.	6
6101	SALSOLA PESTIFER	S	.	0.0	40.3	22.4	.	.	7.0	.	.	.	.	.	.	.	.	.	.	19
6102	SALSOLA RABIANA	C	.	0.0	11.9	1.2	76.6	37.1	10.3	1330	50	.	.	.	.	.	.	.	.	18
6103	SALSOLA SP.	S	.	0.0	16.9	1.3	.	.	17.2	.	.	.	.	.	.	.	.	.	.	19
6104	SALSOLA SP.	W	.	0.0	12.3	1.8	70.7	31.7	15.2	2470	170	.	.	6460	.	.	.	.	.	6
6105	SALSOLA VERMICULATA	S	.	0.0	47.4	15.2	.	.	.	.	.	.	.	.	.	.	.	.	.	21
6106	SALVADORA PERSICA	L	.	0.0	16.1	2.3	61.3	14.1	20.3	2440	150	.	.	.	.	.	.	.	.	18
6107	SALVIA AETHIOPIS	S	.	0.0	20.9	22.8	.	.	.	.	.	.	.	.	.	.	.	.	.	21
6108	SALVIA AETHIOPIS	S	.	0.0	23.0	22.4	.	.	.	.	.	.	.	.	.	.	.	.	.	21
6109	SALVIA APIANA	S	.	0.0	7.9	6.9	.	.	3.8	.	.	.	.	.	.	.	.	.	.	19
6110	SALVIA CARDUACEA	S	.	0.0	26.2	30.0	.	.	.	.	.	.	.	.	.	.	.	.	.	20
6111	SALVIA CERATOPHYLLA	S	.	0.0	16.9	19.3	.	.	.	.	.	.	.	.	.	.	.	.	.	20
6112	SALVIA COLUMBARIAE	S	.	0.0	20.2	34.4	.	.	5.6	.	.	.	.	.	.	.	.	.	.	19
6113	SALVIA CRYPTANTHA	S	.	0.0	16.9	11.4	.	.	.	.	.	.	.	.	.	.	.	.	.	21
6114	SALVIA FARINACEA	S	.	0.0	15.0	17.0	.	.	.	.	.	.	.	.	.	.	.	.	.	20
6115	SALVIA GLUTINOSA	S	.	0.0	21.8	34.3	.	.	.	.	.	.	.	.	.	.	.	.	.	21
6116	SALVIA GRANDIFLORA	S	.	0.0	16.9	15.9	.	.	.	.	.	.	.	.	.	.	.	.	.	21
6117	SALVIA GRANDIFLORA	S	.	0.0	17.4	13.7	.	.	.	.	.	.	.	.	.	.	.	.	.	21
6118	SALVIA HISPANICA	S	.	0.0	17.5	28.6	.	.	.	.	.	.	.	.	.	.	.	.	.	20
6119	SALVIA HORMINUM	S	.	0.0	22.1	24.0	.	.	.	.	.	.	.	.	.	.	.	.	.	21
6120	SALVIA HYDRANGEA	S	.	0.0	20.1	16.2	.	.	.	.	.	.	.	.	.	.	.	.	.	21
6121	SALVIA JUDAICA	S	.	0.0	19.0	20.2	.	.	.	.	.	.	.	.	.	.	.	.	.	21
6122	SALVIA LANIGERA	S	.	0.0	19.4	22.9	.	.	.	.	.	.	.	.	.	.	.	.	.	21
6123	SALVIA LAVANDULAEFOLIA	S	.	0.0	19.4	20.6	.	.	.	.	.	.	.	.	.	.	.	.	.	21

TABLE 2 - DATA DIRECTLY TRANSCRIBED FROM ORIGINAL SOURCES.

OBS	PLNA	PART	CALORIES	WATER	PROTEIN	FAT	TOTCARB	FIBER	ASH	CALCIUM	PHOSPHO	IRON	SODIUM	POTASS	CAROT	THIAMINE	RIBOFLAV	NIACIN	ASCORBIC	REFCO
6124	SALVIA LONGISPICATA	S	.	0.0	18.7	18.8	.	.	.	.	.	.	.	.	.	.	.	.	.	21
6125	SALVIA LYRATA	S	.	0.0	20.6	22.6	.	.	.	.	.	.	.	.	.	.	.	.	.	20
6126	SALVIA MOORCROFTIANA	S	.	0.0	23.3	27.1	.	.	.	.	.	.	.	.	.	.	.	.	.	21
6127	SALVIA OFFICINALIS	L	315	8.0	10.6	12.7	60.7	18.0	8.0	1652	91	28.1	11	1070	3540	0.75	0.34	5.72	32	16
6128	SALVIA PLEBEIA	S	.	0.0	18.4	21.3	.	.	.	.	.	.	.	.	.	.	.	.	.	21
6129	SALVIA POLYSTACHYA	S	.	0.0	23.1	32.7	.	.	.	.	.	.	.	.	.	.	.	.	.	21
6130	SALVIA REFLEXA	S	.	0.0	19.4	23.3	.	.	.	.	.	.	.	.	.	.	.	.	.	19
6131	SALVIA REFLEXA	S	.	0.0	15.6	26.6	.	.	.	.	.	.	.	.	.	.	.	.	.	20
6132	SALVIA RUGOSA	S	.	0.0	18.8	27.5	.	.	.	.	.	.	.	.	.	.	.	.	.	21
6133	SALVIA RUGOSA	S	.	0.0	24.1	27.4	.	.	.	.	.	.	.	.	.	.	.	.	.	21
6134	SALVIA SCLAREA	S	.	0.0	20.7	31.8	.	.	.	.	.	.	.	.	.	.	.	.	.	21
6135	SALVIA SIMILATA	S	.	0.0	18.9	10.8	.	.	.	.	.	.	.	.	.	.	.	.	.	21
6136	SALVIA SONOMENSIS	S	.	0.0	15.5	21.2	.	.	.	.	.	.	.	.	.	.	.	.	.	21
6137	SALVIA SP.	S	.	0.0	23.1	35.0	.	.	.	.	.	.	.	.	.	.	.	.	.	20
6138	SALVIA SP.	W	.	0.0	24.4	3.9	57.7	12.2	14.0	1490	360	.	.	4630	.	.	.	.	.	6
6139	SALVIA SUFFRUTICOSA	S	.	0.0	26.2	40.5	.	.	.	.	.	.	.	.	.	.	.	.	.	21
6140	SALVIA SYRIACA	S	.	0.0	15.6	12.4	.	.	.	.	.	.	.	.	.	.	.	.	.	20
6141	SALVIA TCHIHATCHEFFII	S	.	0.0	20.4	22.3	.	.	.	.	.	.	.	.	.	.	.	.	.	20
6142	SALVIA TEXANA	S	.	0.0	17.1	16.4	.	.	.	.	.	.	.	.	.	.	.	.	.	20
6143	SALVIA TRILOBA	S	.	0.0	19.4	23.2	.	.	.	.	.	.	.	.	.	.	.	.	.	21
6144	SAMANEA SAMAN	C	.	0.0	24.7	2.8	68.1	22.1	4.4	550	260	.	.	.	.	.	.	.	.	18
6145	SAMANEA SAMAN	L	.	0.0	22.1	7.0	64.9	29.4	6.0	1420	210	.	.	.	.	.	.	.	.	18
6146	SAMANEA SAMAN	F	.	0.0	12.8	0.7	84.1	14.5	2.4	290	320	.	.	.	.	.	.	.	.	18
6147	SAMANEA SAMAN	S	.	0.0	31.6	6.0	58.1	14.0	4.3	160	340	.	.	.	.	.	.	.	.	18
6148	SAMANEA SAMAN	F	.	15.3	12.7	2.1	66.7	11.4	3.2	.	.	.	.	.	.	.	.	.	.	5
6149	SAMANEA SAMAN	S	.	7.6	28.6	4.3	56.1	14.1	3.5	.	.	.	.	.	.	.	.	.	.	5
6150	SAMBUCUS CAERULEA	F	344	0.0	16.0	5.0	68.0	.	10.0	115	.	5.0	.	.	.	.	.	.	156	15
6151	SAMBUCUS CANADENSIS	L	.	47.8	10.2	2.1	37.9	15.7	2.0	.	.	.	.	.	.	.	.	.	.	
6152	SAMBUCUS CANADENSIS	S	.	0.0	10.6	20.1	.	.	.	.	.	.	.	.	.	.	.	.	.	19
6153	SAMBUCUS CANADENSIS	F	72	79.8	2.6	0.5	16.4	7.0	0.7	38	28	1.6	.	300	360	0.07	0.06	0.50	36	4
6154	SAMBUCUS EBULUS	S	.	0.0	14.2	14.3	.	.	.	.	.	.	.	.	.	.	.	.	.	21
6155	SAMBUCUS NIGRA	S	.	0.0	12.3	26.9	.	.	.	.	.	.	.	.	.	.	.	.	.	21
6156	SAMBUCUS PUBENS	S	.	0.0	10.6	28.4	.	.	.	.	.	.	.	.	.	.	.	.	.	20
6157	SAMBUCUS RACEMOSA	F	372	0.0	17.0	6.0	73.0	.	4.0	356	.	7.0	.	.	.	.	.	.	533	15
6158	SAMBUCUS WIGHTIANA	S	.	0.0	12.8	14.0	.	.	.	.	.	.	.	.	.	.	.	.	.	21
6159	SANDORICUM KOETJAPE	F	57	84.5	0.4	0.7	13.9	1.0	0.5	9	17	1.2	3	328	5	0.05	0.03	0.90	14	2
6160	SANDORICUM KOETJAPE	F	46	87.0	0.8	0.1	11.8	0.1	0.3	4	17	0.4	.	.	0	0.05	0.03	0.70	86	3
6161	SANGUISORBA MINOR	S	.	0.0	13.5	20.3	.	.	6.2	.	.	.	.	.	.	.	.	.	.	19
6162	SANGUISORBA MINOR	S	.	0.0	14.1	19.9	.	.	7.2	.	.	.	.	.	.	.	.	.	.	19
6163	SANGUISORBA MINOR	S	.	0.0	13.1	17.7	.	.	8.6	.	.	.	.	.	.	.	.	.	.	19
6164	SANGUISORBA MINOR	S	.	0.0	12.1	9.4	.	.	.	.	.	.	.	.	.	.	.	.	.	20
6165	SANGUISORBA MINOR	W	.	0.0	11.1	2.0	80.4	18.0	6.5	.	.	.	.	.	.	.	.	.	.	6
6166	SANGUISORBA OFFICINALIS	S	.	0.0	12.8	13.5	.	.	.	.	.	.	.	.	.	.	.	.	.	21
6167	SANTIRIA TRIMERA	P	72	80.8	1.3	1.2	16.0	.	0.7	46	.	5.0	.	.	.	.	.	.	22	1
6168	SANTOLINA ROSMARINIFOLIA	S	.	0.0	25.9	25.3	.	.	.	.	.	.	.	.	.	.	.	.	.	21

TABLE 2 - DATA DIRECTLY TRANSCRIBED FROM ORIGINAL SOURCES.

OBS	PLNA	PART	CALORIES	WATER	PROTEIN	FAT	TOTCARB	FIBER	ASH	CALCIUM	PHOSPHO	IRON	SODIUM	POTASS	CAROTE	THIAMINE	RIBOFLAV	NIACIN	ASCORBIC	REFCO
6169	SAPINDUS ABYSSINICUS	F	99	71.4	1.5	0.0	26.1	0.5	1.0	23	90	3.0	.	.	360	0.03	.	.	68	1
6170	SAPINDUS MUKOROSSI	S	.	0.0	31.2	35.4	.	.	5.2	.	.	.	.	.	.	.	.	.	.	19
6171	SAPINDUS MUKOROSSI	S	.	0.0	28.6	38.1	.	.	.	.	.	.	.	.	.	.	.	.	.	21
6172	SAPINDUS SAPONARIA	S	.	0.0	29.2	30.0	.	.	5.0	.	.	.	.	.	.	.	.	.	.	19
6173	SAPINDUS SAPONARIA	S	.	0.0	27.5	34.7	.	.	.	.	.	.	.	.	.	.	.	.	.	20
6174	SAPINDUS SP.	S	.	0.0	26.1	34.7	.	.	5.2	.	.	.	.	.	.	.	.	.	.	19
6175	SAPINDUS SP.	S	.	0.0	2.7	0.8	.	.	3.1	.	.	.	.	.	.	.	.	.	.	19
6176	SAPINDUS TRIFOLIATA	S	.	0.0	25.0	57.4	.	.	.	.	.	.	.	.	.	.	.	.	.	20
6177	SAPIUM SEBIFERUM	L	.	0.0	17.8	5.3	70.9	12.4	6.0	1360	220	.	.	.	.	.	.	.	.	18
6178	SAPIUM SEBIFERUM	S	.	0.0	8.1	50.7	.	.	.	.	.	.	.	.	.	.	.	.	.	21
6179	SAPIUM SEBIFERUM	S	.	0.0	9.2	40.5	.	.	.	.	.	.	.	.	.	.	.	.	.	21
6180	SAPONARIA OCYMOIDES	S	.	0.0	13.8	3.9	.	.	3.9	.	.	.	.	.	.	.	.	.	.	19
6181	SAPONARIA OFFICINALIS	S	.	0.0	16.2	5.6	.	.	2.9	.	.	.	.	.	.	.	.	.	.	19
6182	SAPONARIA OFFICINALIS	S	.	0.0	16.6	5.9	.	.	.	.	.	.	.	.	.	.	.	.	.	21
6183	SAPONARIA PROSTRATA	S	.	0.0	15.5	4.8	.	.	.	.	.	.	.	.	.	.	.	.	.	21
6184	SAPONARIA VISCOSA	S	.	0.0	20.0	4.7	.	.	.	.	.	.	.	.	.	.	.	.	.	20
6185	SARCOBATUS SP.	L	.	0.0	21.9	4.1	59.8	21.3	14.2	840	150	.	.	.	.	.	.	.	.	6
6186	SARCOCOCCA SALIGNA	S	.	0.0	20.1	39.4	.	.	.	.	.	.	.	.	.	.	.	.	.	21
6187	SARCOPHRYNIUM ARNOLDIANUM	SH	43	87.0	1.9	0.7	9.0	.	1.4	.	.	.	.	.	.	.	.	.	.	1
6188	SASSAFRAS ALBIDUM	S	.	0.0	23.8	60.4	.	.	2.4	.	.	.	.	.	.	.	.	.	.	19
6189	SASSAFRAS ALBIDUM	S	.	0.0	6.2	11.9	.	.	2.6	.	.	.	.	.	.	.	.	.	.	19
6190	SATUREJA HORTENSIS	H	272	9.0	6.7	5.9	68.7	15.3	9.6	2132	140	37.9	24	1051	3078	0.37	.	4.08	.	16
6191	SATUREJA HORTENSIS	S	.	0.0	24.1	41.5	.	.	5.7	.	.	.	.	.	.	.	.	.	.	19
6192	SATUREJA HORTENSIS	S	.	0.0	23.1	39.1	.	.	.	.	.	.	.	.	.	.	.	.	.	19
6193	SATUREJA THYMBRA	S	.	0.0	12.6	17.9	.	.	.	.	.	.	.	.	.	.	.	.	.	21
6194	SAURAUIA PAUCISERRATA	F	59	83.4	0.7	0.4	14.7	1.3	0.8	65	32	0.5	.	.	50	0.01	0.03	0.20	13	3
6195	SAUROMATUM GUTTATUM	S	.	0.0	12.3	2.9	.	.	.	.	.	.	.	.	.	.	.	.	.	21
6196	SAUROPUS ANDROGYNUS	L	58	81.0	4.8	0.9	10.9	.	2.4	51	86	2.7	28	503	6220	0.07	0.39	2.20	83	2
6197	SAVIGNYA PARVIFLORA	S	.	0.0	25.2	19.0	.	.	.	.	.	.	.	.	.	.	.	.	.	21
6198	SCABIOSA ATROPURPUREA	S	.	0.0	34.8	25.4	.	.	5.1	.	.	.	.	.	.	.	.	.	.	19
6199	SCABIOSA ATROPURPUREA	S	.	0.0	7.0	2.2	.	.	6.6	.	.	.	.	.	.	.	.	.	.	19
6200	SCABIOSA ATROPURPUREA	S	.	0.0	31.6	28.6	.	.	.	.	.	.	.	.	.	.	.	.	.	21
6201	SCABIOSA JAPONICA	S	.	0.0	31.9	29.1	.	.	.	.	.	.	.	.	.	.	.	.	.	19
6202	SCABIOSA JAPONICA	S	.	0.0	31.9	29.5	.	.	.	.	.	.	.	.	.	.	.	.	.	20
6203	SCABIOSA ROTATA	S	.	0.0	11.6	10.3	.	.	.	.	.	.	.	.	.	.	.	.	.	21
6204	SCABIOSA STELLATA	S	.	0.0	13.8	8.6	.	.	.	.	.	.	.	.	.	.	.	.	.	21
6205	SCABIOSA UCRANICA	S	.	0.0	12.2	13.5	.	.	.	.	.	.	.	.	.	.	.	.	.	21
6206	SCAEVOLA PLUMIERI	S	.	0.0	19.2	66.7	.	.	.	.	.	.	.	.	.	.	.	.	.	21
6207	SCALIGERIA MEIFOLIA	S	.	0.0	28.4	26.4	.	.	.	.	.	.	.	.	.	.	.	.	.	21
6208	SCANDIX IBERICA	S	.	0.0	12.0	2.6	.	.	.	.	.	.	.	.	.	.	.	.	.	21
6209	SCANDIX PECTEN-VENERIS	S	.	0.0	16.7	4.4	.	.	.	.	.	.	.	.	.	.	.	.	.	21
6210	SCHEDONNARDUS PANICULATUS	W	.	0.0	9.7	2.1	77.4	30.5	10.8	680	150	.	.	.	.	.	.	.	.	6
6211	SCHIMPERA ARABICA	S	.	0.0	36.6	38.6	.	.	.	.	.	.	.	.	.	.	.	.	.	21
6212	SCHIMPERA ARABICA	S	.	0.0	40.6	36.8	.	.	.	.	.	.	.	.	.	.	.	.	.	21
6213	SCHINUS MOLLE	S	.	0.0	7.5	12.4	.	.	.	.	.	.	.	.	.	.	.	.	.	20

TABLE 2 - DATA DIRECTLY TRANSCRIBED FROM ORIGINAL SOURCES.

OBS	PLNA	PART	CALORIES	WATER	PROTEIN	FAT	TOTLCARB	FIBER	ASH	CALCIUM	PHOSPHO	IRON	SODIUM	POTASS	CAROT	THIAMINE	RIBOFLAV	NIACIN	ASCORBIC	REFCO
6214	SCHINUS MOLLE	S	.	0.0	8.1	10.6	.	.	.	.	.	.	.	.	.	.	.	.	.	20
6215	SCHINUS MOLLE	S	.	0.0	8.5	7.7	.	.	.	.	.	.	.	.	.	.	.	.	.	21
6216	SCHINUS TEREBINTHIFOLIUS	S	.	0.0	10.8	32.2	.	.	.	.	.	.	.	.	.	.	.	.	.	21
6217	SCHIZANTHUS WISETONENSIS	S	.	0.0	20.6	21.8	.	.	.	.	.	.	.	.	.	.	.	.	.	20
6218	SCHKUHRIA WRIGHTII	S	.	0.0	20.6	22.2	.	.	.	.	.	.	.	.	.	.	.	.	.	20
6219	SCHOTIA AFRA	S	.	0.0	11.2	4.6	.	.	.	.	.	.	.	.	.	.	.	.	.	21
6220	SCHOTIA BRACHYPETALA	S	.	0.0	10.5	2.6	.	.	.	.	.	.	.	.	.	.	.	.	.	21
6221	SCHOTIA BRACHYPETALA	A	.	0.0	5.8	0.4	.	.	.	.	.	.	.	.	.	.	.	.	.	21
6222	SCHRANKIA ANGUSTISILIQUA		.	.	.	.	.	.	.	1040	230	.	.	1320	.	.	.	.	.	6
6223	SCHRANKIA ROEMERIANA	L	.	0.0	18.7	2.4	68.2	14.2	10.7	4370	200	.	.	1380	.	.	.	.	.	6
6224	SCHRANKIA UNCINATA	S	.	0.0	35.6	4.5	.	.	3.1	.	.	.	.	.	.	.	.	.	.	19
6225	SCIRPUS ARTICULATUS	S	.	0.0	8.2	3.0	.	.	.	.	.	.	.	.	.	.	.	.	.	21
6226	SCIRPUS MARITIMUS	S	.	0.0	6.0	2.4	.	.	.	.	.	.	.	.	.	.	.	.	.	21
6227	SCIRPUS RUBRICOSUS	S	.	0.0	18.2	26.0	.	.	.	.	.	.	.	.	.	.	.	.	.	21
6228	SCIRPUS SP.	H	.	0.0	8.8	2.4	75.1	28.4	13.7	760	160	.	.	.	.	.	.	.	.	6
6229	SCIRPUS SP.	I	.	0.0	7.4	2.5	84.7	23.4	5.4	.	.	.	.	.	.	.	.	.	.	6
6230	SCIRPUS SP.	L	.	0.0	11.7	1.5	70.1	23.0	16.7	.	.	.	.	.	.	.	.	.	.	6
6231	SCLERANTHUS ANNUUS	S	.	0.0	22.6	9.1	.	.	.	.	.	.	.	.	.	.	.	.	.	21
6232	SCLEROCARPUS AFRICANUS	S	.	0.0	39.2	41.7	.	.	.	.	.	.	.	.	.	.	.	.	.	21
6233	SCLEROCARPUS SPATHULATUS	S	.	0.0	12.5	11.1	.	.	.	.	.	.	.	.	.	.	.	.	.	20
6234	SCLEROCARYA BIERREA	F	.	87.0	.	.	.	.	.	.	.	.	.	.	.	.	.	.	179	1
6235	SCLEROCARYA CAFFRA	F	30	91.7	0.5	0.1	7.5	0.5	0.2	6	19	0.1	.	.	.	0.03	0.05	0.2	68	1
6236	SCLEROCARYA CAFFRA	S	604	3.9	24.6	57.5	9.2	2.7	4.8	143	1248	0.4	.	.	.	0.04	0.12	0.7	.	1
6237	SCLEROCHLOA DURA	S	.	0.0	17.9	2.9	.	.	.	.	.	.	.	.	.	.	.	.	.	21
6238	SCLEROPOGON SP.	W	.	0.0	7.1	.	.	31.4	9.4	510	110	.	.	.	.	.	.	.	.	6
6239	SCOHTIA AFRA	S	.	0.0	10.6	4.5	.	.	.	.	.	.	.	.	.	.	.	.	.	21
6240	SCOLYMUS HISPANICUS	S	.	0.0	18.2	20.6	.	.	.	.	.	.	.	.	.	.	.	.	.	21
6241	SCORPIURUS MURICATA	S	.	0.0	33.4	2.9	.	.	.	.	.	.	.	.	.	.	.	.	.	21
6242	SCORPIURUS SUBVILLOSA	S	.	0.0	30.9	2.7	.	.	.	.	.	.	.	.	.	.	.	.	.	21
6243	SCORPIURUS SULCATA	S	.	0.0	35.3	2.3	.	.	4.1	.	.	.	.	.	.	.	.	.	.	19
6244	SCORZONERA HISPANICA	S	.	0.0	21.9	17.1	.	.	.	.	.	.	.	.	.	.	.	.	.	20
6245	SCORZONERA HISPANICA	S	.	0.0	23.1	18.3	.	.	.	.	.	.	.	.	.	.	.	.	.	20
6246	SCROPHULARIA CANINA	S	.	0.0	16.6	31.5	.	.	.	.	.	.	.	.	.	.	.	.	.	21
6247	SCROPHULARIA STRIATA	S	.	0.0	18.9	35.4	.	.	.	.	.	.	.	.	.	.	.	.	.	21
6248	SCROPHULARIA XANTHOGLOSSA	S	.	0.0	19.4	31.3	.	.	.	.	.	.	.	.	.	.	.	.	.	21
6249	SCROPHULARIA XANTHOGLOSSA	S	.	0.0	17.8	29.3	.	.	.	.	.	.	.	.	.	.	.	.	.	21
6250	SCUTELLARIA DRUMMONDII	S	.	0.0	22.5	33.0	.	.	.	.	.	.	.	.	.	.	.	.	.	20
6251	SCUTELLARIA RESINOSA	W	.	0.0	4.2	10.2	74.6	19.5	11.0	1200	450	.	.	.	.	.	.	.	.	6
6252	SCUTELLARIA SP.	W	.	0.0	7.0	6.5	77.1	25.3	9.4	1530	170	.	.	1720	.	.	.	.	.	6
6253	SECALE CEREALE	S	336	11.1	10.5	1.8	75.0	2.0	1.6	40	351	4.1	.	.	.	0.25	.	.	.	1
6254	SECALE CEREALE	W	.	0.0	21.4	4.8	58.6	24.8	15.2	.	.	.	.	.	.	.	.	.	.	18
6255	SECALE CEREALE	S	.	0.0	13.7	1.9	82.2	2.6	2.2	40	410	.	.	.	.	.	.	.	.	18
6256	SECALE CEREALE	S	.	0.0	17.1	1.7	.	.	2.2	.	.	.	.	.	.	.	.	.	.	19
6257	SECALE CEREALE	S	334	12.5	12.8	2.6	70.4	1.9	1.7	38	330	3.0	6	.	0	0.47	0.20	1.7	.	2
6258	SECALE CEREALE	S	334	11.0	12.1	1.7	73.4	2.0	1.8	38	376	3.7	1	467	0	0.43	0.22	1.6	0	4

TABLE 2 - DATA DIRECTLY TRANSCRIBED FROM ORIGINAL SOURCES.

OBS	PLNA	PART	CALORIES	WATER	PROTEIN	FAT	TOTLCARB	FIBER	ASH	CALCIUM	PHOSPHO	IRON	SODIUM	POTASS	CAROT	THIAMINE	RIBOFLAV	NIACIN	ASCORBIC	REFCO
6259	SECALE CEREALE	W	.	0.0	23.2	4.2	61.6	24.1	11.0	550	450	.	.	3400	.	.	.	.	.	6
6260	SECALE CEREALE	H	.	0.0	9.4	2.8	82.0	38.7	5.8	350	370	.	.	1050	.	.	.	.	.	6
6261	SECALE MONTANUM	S	.	0.0	16.4	2.2	.	.	.	.	.	.	.	.	.	.	.	.	.	21
6262	SECHIUM EDULE	F	20	94.0	1.2	0.2	4.4	1.2	0.2	18	15	0.9	.	.	.	.	.	.	10	1
6263	SECHIUM EDULE	L	.	88.4	3.3	.	.	4.7	1.7	.	.	.	.	.	4560	.	.	.	.	1
6264	SECHIUM EDULE	F	19	94.3	0.7	0.1	4.6	0.6	0.3	17	14	0.4	2	108	15	0.01	0.02	0.40	14	2
6265	SECHIUM EDULE	L	25	91.2	4.0	0.3	3.5	0.8	1.0	62	58	1.4	1	310	1515	0.07	0.09	0.80	24	2
6266	SECHIUM EDULE	F	31	90.8	0.9	0.2	7.7	0.6	0.4	12	30	0.6	.	.	10	0.03	0.04	0.40	20	3
6267	SECHIUM EDULE	L	60	89.7	4.0	0.4	4.7	1.2	1.2	58	108	2.5	.	.	1230	0.08	0.18	1.10	16	3
6268	SECHIUM EDULE	R	79	79.0	2.0	0.2	17.8	0.4	1.0	7	34	0.8	.	.	0	0.05	0.03	0.90	19	3
6269	SECHIUM EDULE	F	28	91.8	0.6	0.1	7.1	0.7	0.4	13	26	0.5	5	102	12	0.03	0.03	0.40	19	4
6270	SECURIGERA SECURIDACA	S	.	0.0	44.8	6.6	.	.	.	.	.	.	.	.	.	.	.	.	.	21
6271	SECURIGERA SECURIDACA	S	.	0.0	31.4	3.4	.	.	.	.	.	.	.	.	.	.	.	.	.	21
6272	SEDUM MAKINOI	L	22	93.9	0.8	1.0	3.4	1.8	0.9	233	54	2.5	.	.	35	0.15	0.12	.	30	2
6273	SEDUM SP.	W	.	0.0	8.5	6.1	64.7	11.7	20.7	7510	400	.	.	2640	.	.	.	.	.	6
6274	SEHIMA NERVOSUM	W	.	0.0	7.0	1.3	82.9	38.7	8.8	.	.	.	.	.	.	.	.	.	.	18
6275	SELENIA GRANDIS	S	.	0.0	21.2	18.5	.	.	.	.	.	.	.	.	.	.	.	.	.	20
6276	SENECIO AMBROSIOIDES	S	.	0.0	19.1	26.0	.	.	.	.	.	.	.	.	.	.	.	.	.	21
6277	SENECIO HARTWEGII	S	.	0.0	35.6	24.3	.	.	.	.	.	.	.	.	.	.	.	.	.	20
6278	SENECIO HIERACIFOLIUS	S	.	0.0	21.2	36.6	.	.	.	.	.	.	.	.	.	.	.	.	.	20
6279	SENECIO ORIENTALIS	S	.	0.0	15.0	3.2	.	.	.	.	.	.	.	.	.	.	.	.	.	21
6280	SENECIO SCHAFFNERI	S	.	0.0	21.9	19.1	.	.	.	.	.	.	.	.	.	.	.	.	.	20
6281	SENECIO SERRA	W	.	0.0	20.9	3.6	61.9	13.3	13.6	3320	470	.	.	.	.	.	.	.	.	6
6282	SENECIO SP.	S	.	0.0	22.5	22.0	.	.	.	.	.	.	.	.	.	.	.	.	.	20
6283	SENECIO TRIANGULARIS		.	0.0	29.1	3.0	54.8	16.5	13.1	.	.	.	.	.	.	.	.	.	.	6
6284	SERICOCARPUS ASTEROIDES	S	.	0.0	30.0	17.2	.	.	.	.	.	.	.	.	.	.	.	.	.	20
6285	SERRURIA ACROCARPA	S	.	0.0	1.8	1.2	.	.	.	.	.	.	.	.	.	.	.	.	.	21
6286	SERRURIA ELONGATA	S	.	0.0	4.6	2.2	.	.	.	.	.	.	.	.	.	.	.	.	.	21
6287	SESAMUM ALATUM	S	414	7.9	10.8	18.1	55.2	30.5	8.0	432	221	.	.	.	.	.	.	.	.	1
6288	SESAMUM INDICUM	S	558	5.2	17.9	48.4	22.3	4.5	6.2	816	600	8.1	.	.	30	0.68	0.19	3.40	0	1
6289	SESAMUM INDICUM	S	588	4.8	26.4	54.8	9.4	3.0	4.6	131	776	7.8	40	407	40	0.72	0.08	4.68	.	16
6290	SESAMUM INDICUM	S	.	0.0	19.7	51.5	23.0	1.7	5.8	810	500	.	.	.	.	.	.	.	.	18
6291	SESAMUM INDICUM	S	582	5.8	17.2	52.8	19.8	5.4	4.4	750	614	12.0	49	508	15	0.72	0.17	5.10	0	2
6292	SESAMUM INDICUM	L	45	85.5	3.4	0.7	8.6	2.4	1.8	77	203	.	.	.	.	.	0.30	.	.	2
6293	SESAMUM INDICUM	S	.	0.0	14.1	32.8	.	.	.	.	.	.	.	.	.	.	.	.	.	21
6294	SESAMUM INDICUM	S	584	4.1	17.6	52.2	21.1	5.3	5.0	1212	620	10.4	.	.	10	0.98	0.25	5.00	0	3
6295	SESAMUM INDICUM	S	563	5.4	18.6	49.1	21.6	6.3	5.3	1160	616	10.5	60	725	18	0.98	0.24	5.40	0	4
6296	SESAMUM SP.	L	55	81.2	4.7	0.4	11.3	2.0	2.4	354	57	.	.	.	.	.	.	.	.	1
6297	SESBANIA BISPINOSA	S	.	0.0	32.7	2.9	59.4	10.7	5.0	370	590	.	.	.	.	.	.	.	.	18
6298	SESBANIA CANNABINA	S	.	0.0	29.2	6.4	.	.	.	.	.	.	.	.	.	.	.	.	.	21
6299	SESBANIA DRUMMONDII	S	.	0.0	31.6	5.0	.	.	3.0	.	.	.	.	.	.	.	.	.	.	19
6300	SESBANIA DRUMMONDII	S	.	0.0	31.2	2.9	.	.	.	.	.	.	.	.	.	.	.	.	.	20
6301	SESBANIA EXALTATA	S	.	0.0	40.3	4.4	.	.	3.6	.	.	.	.	.	.	.	.	.	.	19
6302	SESBANIA EXALTATA	S	.	0.0	37.4	4.0	.	.	3.6	.	.	.	.	.	.	.	.	.	.	19
6303	SESBANIA EXALTATA	S	.	0.0	38.2	4.6	.	.	3.2	.	.	.	.	.	.	.	.	.	.	19

TABLE 2 - DATA DIRECTLY TRANSCRIBED FROM ORIGINAL SOURCES.

OBS	PLANA	PART	CALORIES	WATER	PROTEIN	FAT	TOTCARB	FIBER	ASH	CALCIUM	PHOSPHO	IRON	SODIUM	POTASS	CAROT	THIAMINE	RIBOFLAV	NIACIN	ASCORBIC	REFCO
6304	SESBANIA EXALTATA	S	.	0.0	37.9	4.6	.	.	.	.	.	.	.	.	.	.	.	.	.	21
6305	SESBANIA EXALTATA	S	.	0.0	33.5	4.6	.	.	.	.	.	.	.	.	.	.	.	.	.	21
6306	SESBANIA GRANDIFLORA	I	.	0.0	15.1	1.8	76.6	13.2	6.5	.	.	.	.	.	.	.	.	.	.	18
6307	SESBANIA GRANDIFLORA	L	.	0.0	33.4	2.6	52.4	5.7	11.6	2330	340	.	.	.	.	.	.	.	.	18
6308	SESBANIA GRANDIFLORA	F	.	0.0	1.6	4.8	87.1	32.7	6.5	.	.	.	.	.	.	.	.	.	.	18
6309	SESBANIA GRANDIFLORA	L	77	76.0	8.7	1.8	11.3	2.2	2.2	404	62	.	5	481	6230	0.24	0.25	2.2	58	2
6310	SESBANIA GRANDIFLORA	F	38	89.0	1.6	0.4	8.5	1.2	0.5	16	32	0.6	32	154	70	0.10	0.08	1.6	52	2
6311	SESBANIA JAVANICA	L	45	86.0	3.6	0.4	9.1	3.9	0.9	51	56	8.2	.	.	2005	0.26	.	2.8	24	2
6312	SESBANIA MACROCARPA	S	.	0.0	38.2	6.4	.	.	3.6	.	.	.	.	.	.	.	.	.	.	19
6313	SESBANIA PUNICEA	S	.	0.0	33.0	3.8	.	.	.	.	.	.	.	.	.	.	.	.	.	21
6314	SESBANIA PUNICEA	S	.	0.0	34.4	3.9	.	.	.	.	.	.	.	.	.	.	.	.	.	21
6315	SESBANIA ROXBURGII	L	321	0.0	25.7	2.9	51.0	27.8	6.4	364	400	58.5	.	.	14316	1.85	.	20.0	171	10
6316	SESBANIA SESBAN	L	.	0.0	26.5	0.9	62.6	12.2	10.0	2780	430	.	.	.	.	.	.	.	.	18
6317	SESBANIA SESBAN	F	.	0.0	7.8	0.5	85.5	10.0	6.2	1370	370	.	.	.	.	.	.	.	.	18
6318	SESBANIA SESBAN	S	.	0.0	21.2	2.5	69.0	8.5	7.2	440	680	.	.	.	.	.	.	.	.	18
6319	SESBANIA SESBAN	S	.	0.0	27.4	3.6	.	.	.	.	.	.	.	.	.	.	.	.	.	21
6320	SESBANIA SESBAN	S	.	0.0	31.3	8.6	.	.	.	.	.	.	.	.	.	.	.	.	.	21
6321	SESBANIA SONORAE	S	.	0.0	36.1	6.0	.	.	4.4	.	.	.	.	.	.	.	.	.	.	19
6322	SESBANIA SP.	S	357	9.2	32.0	4.5	51.2	11.4	3.1	272	503	.	.	.	.	.	.	.	.	1
6323	SESBANIA SP.	S	.	0.0	37.9	6.4	.	.	3.7	.	.	.	.	.	.	.	.	.	.	19
6324	SESBANIA SP.	S	.	0.0	37.5	5.8	.	.	.	.	.	.	.	.	.	.	.	.	.	19
6325	SESBANIA VESICARIA	S	.	0.0	39.7	5.4	.	.	4.2	.	.	.	.	.	.	.	.	.	.	19
6326	SESBANIA VESICARIA	S	.	0.0	40.6	6.7	.	.	3.5	.	.	.	.	.	.	.	.	.	.	19
6327	SESUVIUM PORTULACASTRUM	L	25	89.0	2.1	0.4	4.7	0.8	3.8	68	24	6.3	.	322	1850	0.04	0.14	0.5	27	2
6328	SETARIA CHEVALIERI	W	.	0.0	8.3	1.8	77.8	33.5	12.1	.	.	.	.	.	.	.	.	.	.	18
6329	SETARIA CHEVALIERI	H	.	0.0	12.3	2.4	72.9	26.4	12.4	.	.	.	.	.	.	.	.	.	.	18
6330	SETARIA ITALICA	W	.	0.0	9.0	2.2	78.7	33.7	10.1	.	.	.	.	.	.	.	.	.	.	18
6331	SETARIA ITALICA	H	.	0.0	7.6	1.7	81.0	45.1	9.7	.	.	.	.	.	.	.	.	.	.	18
6332	SETARIA ITALICA	S	341	11.3	9.5	2.9	74.7	1.2	1.6	33	244	5.5	7	249	0	0.43	0.12	2.2	0	2
6333	SETARIA ITALICA	H	.	0.0	9.5	3.0	78.3	29.4	8.2	.	.	.	.	.	.	.	.	.	.	6
6334	SETARIA LUTESCENS	S	.	0.0	17.8	5.4	.	.	10.8	.	.	.	.	.	.	.	.	.	.	19
6335	SETARIA LUTESCENS	W	.	0.0	5.8	1.7	82.0	34.6	9.5	260	80	.	.	.	.	.	.	.	.	6
6336	SETARIA MACROSTACHYA	I	.	0.0	8.4	1.1	.	.	11.6	.	.	.	.	.	.	.	.	.	.	19
6337	SETARIA MACROSTACHYA	W	.	0.0	17.5	2.3	68.4	27.6	11.8	330	200	.	.	.	.	.	.	.	.	6
6338	SETARIA SP.	S	.	0.0	16.1	5.6	.	.	6.3	.	.	.	.	.	.	.	.	.	.	19
6339	SETARIA SPHACELATA	W	.	0.0	4.1	1.3	86.8	36.3	7.8	.	.	.	.	.	.	.	.	.	.	18
6340	SETARIA SPHACELATA	W	.	0.0	18.4	4.7	63.8	24.0	13.1	.	.	.	.	.	.	.	.	.	.	18
6341	SETARIA SPLENDIDA	W	.	0.0	11.4	3.0	73.5	27.8	12.1	.	.	.	.	.	.	.	.	.	.	18
6342	SETARIA VERTICILLATA	S	.	0.0	26.9	9.5	.	.	.	.	.	.	.	.	.	.	.	.	.	20
6343	SETARIA VERTICILLATA	S	.	0.0	18.2	6.8	.	.	.	.	.	.	.	.	.	.	.	.	.	21
6344	SETARIA VIRIDIS	S	.	0.0	17.0	8.0	.	.	6.8	.	.	.	.	.	.	.	.	.	.	19
6345	SETARIA VIRIDIS	S	.	0.0	15.4	3.6	.	.	.	.	.	.	.	.	.	.	.	.	.	21
6346	SETARIA VIRIDIS	S	.	0.0	16.9	8.6	.	.	.	.	.	.	.	.	.	.	.	.	.	21
6347	SETARIA VIRIDIS	W	.	0.0	9.6	2.5	77.1	31.3	10.8	300	210	.	.	.	.	.	.	.	.	6
6348	SEYMERIA GLANDULOSA	S	.	0.0	31.2	32.8	.	.	.	.	.	.	.	.	.	.	.	.	.	20

TABLE 2 - DATA DIRECTLY TRANSCRIBED FROM ORIGINAL SOURCES.

OBS	PLNA	PART	CALORIES	WATER	PROTEIN	FAT	TOTLCARB	FIBER	ASH	CALCIUM	PHOSPHO	IRON	SODIUM	POTASS	CAROT	THIAMINE	RIBOFLAV	NIACIN	ASCORBIC	REFCO
6349	SHOREA ROBUSTA	L	.	0.0	10.1	3.2	82.8	27.4	3.9	770	120	.	.	.	.	.	.	.	.	18
6350	SHOREA ROBUSTA	S	.	0.0	8.0	16.7	72.9	5.2	2.4	.	.	.	.	.	.	.	.	.	.	18
6351	SHOREA STENOPTERA	S	.	0.0	17.5	3.1	77.2	6.8	2.2	.	.	.	.	.	.	.	.	.	.	18
6352	SIBARA VIRGINICA	S	.	0.0	20.0	31.1	.	.	.	.	.	.	.	.	.	.	.	.	.	20
6353	SICANA ODORIFERA	F	43	87.4	0.8	0.1	11.1	0.8	0.6	31	26	0.6	.	.	100	0.05	0.04	0.80	16	3
6354	SICYOS ANGULATA	S	.	0.0	31.2	27.8	.	.	.	.	.	.	.	.	.	.	.	.	.	20
6355	SIDA ACUTA	S	.	0.0	23.2	16.8	.	.	.	.	.	.	.	.	.	.	.	.	.	21
6356	SIDA CORYMBOSA	S	.	0.0	19.4	12.7	.	.	.	.	.	.	.	.	.	.	.	.	.	20
6357	SIDA RHOMBIFOLIA	L	63	80.2	7.4	1.4	9.4	3.3	1.6	466	58	5.0	.	.	6050	0.22	0.47	2.10	90	3
6358	SIDA SCHIMPERIANA	SH	.	0.0	13.7	1.3	73.7	29.9	11.3	1000	240	.	.	.	.	.	.	.	.	18
6359	SIDA SP.	W	.	0.0	19.1	1.8	69.2	17.5	9.9	2630	260	.	.	1800	.	.	.	.	.	6
6360	SIDA SPINOSA	S	.	0.0	23.1	10.6	.	.	5.0	.	.	.	.	.	.	.	.	.	.	19
6361	SIDERITIS MONTANA	S	.	0.0	19.2	37.6	.	.	.	.	.	.	.	.	.	.	.	.	.	20
6362	SIDEROXYLON ANGUSTIFOLIUM	S	.	0.0	9.4	9.7	.	.	.	.	.	.	.	.	.	.	.	.	.	20
6363	SIDEROXYLON CAPIRI	L	132	64.6	2.2	1.7	30.7	1.7	0.8	74	24	.	.	.	220	0.08	0.06	0.50	149	3
6364	SILENE ANTIRRHINA	S	.	0.0	17.0	5.0	.	.	4.3	.	.	.	.	.	.	.	.	.	.	19
6365	SILENE COMPACTA	S	.	0.0	16.0	5.6	.	.	.	.	.	.	.	.	.	.	.	.	.	21
6366	SILENE CONOIDEA	S	.	0.0	14.3	4.8	.	.	.	.	.	.	.	.	.	.	.	.	.	21
6367	SILENE DENSIFLORA	S	.	0.0	24.2	7.0	.	.	.	.	.	.	.	.	.	.	.	.	.	20
6368	SILENE DICHOTOMA	S	.	0.0	14.9	4.8	.	.	.	.	.	.	.	.	.	.	.	.	.	21
6369	SILENE GALLICA	W	.	0.0	10.7	.	.	21.1	12.8	1530	480	.	.	3600	.	.	.	.	.	6
6370	SILENE NOCTIFLORA	S	.	0.0	21.6	4.3	.	.	2.9	.	.	.	.	.	.	.	.	.	.	19
6371	SILENE SUPINA	S	.	0.0	19.2	4.9	.	.	.	.	.	.	.	.	.	.	.	.	.	20
6372	SILENE VULGARIS	S	.	0.0	18.1	2.7	.	.	.	.	.	.	.	.	.	.	.	.	.	20
6373	SILYBUM MARIANUM	S	.	0.0	18.6	29.2	.	.	.	.	.	.	.	.	.	.	.	.	.	21
6374	SILYBUM MARIANUM	S	.	0.0	18.9	30.4	.	.	.	.	.	.	.	.	.	.	.	.	.	21
6375	SILYBUM MARIANUM	S	.	0.0	19.1	29.0	.	.	.	.	.	.	.	.	.	.	.	.	.	21
6376	SILYBUM MARIANUM	S	.	0.0	19.1	27.1	.	.	.	.	.	.	.	.	.	.	.	.	.	21
6377	SIMAROUBA GLAUCA	S	.	0.0	24.0	45.5	.	.	.	.	.	.	.	.	.	.	.	.	.	21
6378	SIMMONDSIA CHINENSIS	S	.	0.0	15.6	45.6	.	.	.	.	.	.	.	.	.	.	.	.	.	20
6379	SIMSIA AMPLEXICAULIS	S	.	0.0	26.2	29.7	.	.	.	.	.	.	.	.	.	.	.	.	.	20
6380	SIMSIA AMPLEXICAULIS	S	.	0.0	24.4	28.8	.	.	.	.	.	.	.	.	.	.	.	.	.	20
6381	SIMSIA CALVA	S	.	0.0	20.0	21.3	.	.	.	.	.	.	.	.	.	.	.	.	.	19
6382	SIMSIA CALVA	S	.	0.0	17.5	14.1	.	.	.	.	.	.	.	.	.	.	.	.	.	20
6383	SIMSIA EXAURICULATA(IMMATURE)	W	.	0.0	20.4	1.3	62.6	18.9	15.7	4420	230	.	.	3530	.	.	.	.	.	6
6384	SIMSIA EXAURICULATA(MATURE)	W	.	0.0	16.4	1.4	68.8	21.6	13.4	3740	210	.	.	3070	.	.	.	.	.	6
6385	SIMSIA FOETIDA	S	.	0.0	21.9	24.5	.	.	.	.	.	.	.	.	.	.	.	.	.	21
6386	SIMSIA LAGASCAEFORMIS	S	.	0.0	23.1	27.8	.	.	.	.	.	.	.	.	.	.	.	.	.	20
6387	SINAPIS ALBA	S	469	6.9	24.9	28.8	34.9	6.6	4.5	521	841	10.0	5	682	37	0.54	0.38	7.89	.	16
6388	SINAPIS ALBA	S	469	5.0	26.4	36.3	28.2	5.2	4.1	410	613	20.9	.	.	630	0.40	0.31	7.30	0	2
6389	SINAPIS ALBA	S	.	0.0	31.7	28.0	.	.	.	.	.	.	.	.	.	.	.	.	.	21
6390	SINAPIS ARVENSIS	S	.	0.0	29.5	31.0	.	.	.	.	.	.	.	.	.	.	.	.	.	20
6391	SINAPIS ARVENSIS	S	.	0.0	30.0	28.4	.	.	.	.	.	.	.	.	.	.	.	.	.	21
6392	SINAPIS ARVENSIS	S	.	0.0	26.8	28.1	.	.	.	.	.	.	.	.	.	.	.	.	.	21
6393	SISYMBRIUM ALTISSIMUM	S	.	0.0	31.3	35.2	.	.	.	.	.	.	.	.	.	.	.	.	.	21

TABLE 2 - DATA DIRECTLY TRANSCRIBED FROM ORIGINAL SOURCES.

OBS	PLNA	PART	CALORIES	WATER	PROTEIN	FAT	TOTCARB	FIBER	ASH	CALCIUM	PHOSPHO	IRON	SODIUM	POTASS	CAROT	THIAMINE	RIBOFLAV	NIACIN	ASCORBIC	REFCO
6394	SISYMBRIUM ALTISSIMUM	S	.	0.0	30.6	35.9	.	.	.	.	.	.	.	.	.	.	.	.	.	21
6395	SISYMBRIUM ERYSIMOIDES	S	.	0.0	34.9	30.0	.	.	.	.	.	.	.	.	.	.	.	.	.	21
6396	SISYMBRIUM GARIEPINUM	S	.	0.0	30.4	32.3	.	.	.	.	.	.	.	.	.	.	.	.	.	21
6397	SISYMBRIUM IRIO	S	.	0.0	33.8	31.0	.	.	.	.	.	.	.	.	.	.	.	.	.	20
6398	SISYMBRIUM IRIO	S	.	0.0	40.0	13.0	.	.	.	.	.	.	.	.	.	.	.	.	.	20
6399	SISYMBRIUM IRIO	S	.	0.0	36.2	30.2	.	.	.	.	.	.	.	.	.	.	.	.	.	21
6400	SISYMBRIUM IRIO	S	.	0.0	36.9	20.7	.	.	.	.	.	.	.	.	.	.	.	.	.	21
6401	SISYMBRIUM SP.	S	.	0.0	31.2	36.2	.	.	3.9	.	.	.	.	.	.	.	.	.	.	19
6402	SISYNDITE SPARTEA	S	.	0.0	27.0	15.7	.	.	.	.	.	.	.	.	.	.	.	.	.	21
6403	SISYRINCHIUM SP.	S	.	0.0	13.8	8.5	.	.	.	.	.	.	.	.	.	.	.	.	.	20
6404	SITANION HYSTRIX	W	.	0.0	6.9	2.5	76.0	36.5	14.6	1010	100	.	.	1940	.	.	.	.	.	6
6405	SKIMMIA JAPONICA	S	.	0.0	26.2	43.1	.	.	1.8	.	.	.	.	.	.	.	.	.	.	19
6406	SMILACINA RACEMOSA	F	364	0.0	10.0	.	86.0	3.0	2.0	161	.	3.0	.	.	.	.	.	.	508	15
6407	SMILACINA RACEMOSA	S	.	0.0	11.2	4.8	.	.	2.0	.	.	.	.	.	.	.	.	.	.	19
6408	SMILACINA RACEMOSA	S	.	0.0	6.2	4.8	.	.	.	.	.	.	.	.	.	.	.	.	.	19
6409	SMILAX EXCELSA	S	.	0.0	8.4	10.3	.	.	.	.	.	.	.	.	.	.	.	.	.	21
6410	SMILAX ROTUNDIFOLIA	S	.	0.0	10.6	10.5	.	.	.	.	.	.	.	.	.	.	.	.	.	19
6411	SMILEX ROTUNDIFOLIA	S	.	0.0	11.2	10.5	.	.	.	.	.	.	.	.	.	.	.	.	.	19
6412	SMYRNIUM PERFOLIATUM	S	.	0.0	24.6	16.2	.	.	.	.	.	.	.	.	.	.	.	.	.	21
6413	SOLANUM AETHIOPICUM	L	51	82.1	4.8	0.3	10.3	2.4	2.5	523	94	6.0	.	.	6400	0.23	0.44	1.8	67	1
6414	SOLANUM AETHIOPICUM	F	32	90.6	1.5	0.1	7.2	2.0	0.6	28	47	1.5	.	.	350	0.07	0.06	0.8	8	1
6415	SOLANUM CAROLINIENSE	S	.	0.0	16.9	21.4	.	.	.	.	.	.	.	.	.	.	.	.	.	19
6416	SOLANUM ELAEAGNIFOLIUM	S	.	0.0	16.2	14.2	.	.	.	.	.	.	.	.	.	.	.	.	.	20
6417	SOLANUM ELAEAGNIFOLIUM	W	.	0.0	21.8	3.2	66.0	25.6	9.0	1070	230	.	.	2140	.	.	.	.	.	6
6418	SOLANUM INCANUM	F	29	92.0	1.2	0.7	5.7	1.9	0.4	103	.	.	.	.	.	.	.	.	176	1
6419	SOLANUM INCANUM	L	.	0.0	30.3	2.0	54.9	21.0	12.8	1560	320	.	.	.	.	.	.	.	.	18
6420	SOLANUM INCANUM	S	.	0.0	17.2	11.3	.	.	.	.	.	.	.	.	.	.	.	.	.	21
6421	SOLANUM MACROCARPON	F	40	89.0	1.4	1.0	8.0	1.5	0.6	13	.	.	.	.	.	.	.	.	.	1
6422	SOLANUM MACROCARPON	L	42	85.6	4.6	1.0	6.4	1.6	2.4	391	49	.	.	.	.	.	.	.	.	1
6423	SOLANUM MELONGENA	L	38	86.4	4.6	0.4	6.6	.	2.0	254	38	15.5	.	.	.	.	.	.	79	1
6424	SOLANUM MELONGENA	F	32	90.5	1.0	0.2	7.7	1.3	0.6	14	26	1.3	.	.	34	0.05	0.05	0.5	9	1
6425	SOLANUM MELONGENA	F	26	92.0	1.6	0.3	5.6	1.0	0.5	22	37	0.9	7	221	50	0.08	0.07	0.7	6	2
6426	SOLANUM MELONGENA	F	27	91.8	1.0	0.3	6.3	1.2	0.6	23	31	0.8	.	.	0	0.04	0.04	0.8	5	3
6427	SOLANUM MELONGENA	F	25	92.4	1.2	0.2	5.6	0.9	0.6	12	26	0.7	2	214	6	0.05	0.05	0.6	5	4
6428	SOLANUM MURICATUM	F	32	92.0	0.4	1.0	6.3	0.4	0.3	18	14	0.8	.	.	120	0.08	0.04	0.5	32	3
6429	SOLANUM NIGRUM	S	.	0.0	17.5	27.2	.	.	.	.	.	.	.	.	.	.	.	.	.	19
6430	SOLANUM NIGRUM	L	44	85.0	4.6	0.4	8.4	1.1	1.6	216	88	4.2	.	.	1650	0.12	0.24	1.3	30	2
6431	SOLANUM NIGRUM	S	.	0.0	20.5	30.9	.	.	.	.	.	.	.	.	.	.	.	.	.	21
6432	SOLANUM NIGRUM	L	45	85.0	5.0	0.8	7.4	1.4	1.8	199	60	9.9	.	.	460	0.18	0.35	1.0	61	3
6433	SOLANUM NODIFLORUM	L	39	87.8	3.2	1.0	6.4	2.2	1.6	200	54	0.3	.	.	3660	.	.	.	24	1
6434	SOLANUM NODIFLORUM	S	.	0.0	14.4	25.3	.	.	.	.	.	.	.	.	.	.	.	.	.	19
6435	SOLANUM QUITOENSE	F	45	87.0	1.2	0.2	10.9	4.0	0.7	11	41	0.6	.	.	140	0.07	0.04	1.5	48	3
6436	SOLANUM QUITOENSE	P	28	91.8	0.7	0.1	6.8	0.4	0.6	8	14	0.4	.	.	100	0.06	0.04	1.5	65	3
6437	SOLANUM SURATTENSE	S	.	0.0	12.5	21.3	.	.	.	.	.	.	.	.	.	.	.	.	.	21
6438	SOLANUM TORREYI	W	.	0.0	16.8	2.7	71.4	19.9	9.1	1970	160	.	.	2740	.	.	.	.	.	6

TABLE 2 - DATA DIRECTLY TRANSCRIBED FROM ORIGINAL SOURCES.

OBS	PLNA	PART	CALORIES	WATER	PROTEIN	FAT	TOTCARB	FIBER	ASH	CALCIUM	PHOSPHO	IRON	SODIUM	POTASS	CAROTENE	THIAMINE	RIBOFLAV	NIACIN	ASCORBIC	REFCO
6439	SOLANUM TORVUM	F	47	85.4	2.4	0.4	10.7	6.1	1.1	104	70	4.6	.	.	390	0.12	0.09	2.6	4	2
6440	SOLANUM TRIQUETRUM	W	.	0.0	22.4	4.3	63.7	23.9	9.6	2770	260	.	.	2760	.	.	.	.	.	6
6441	SOLANUM TUBEROSUM	R	82	77.7	1.7	0.1	18.9	0.6	1.6	13	51	1.1	.	.	25	0.07	0.03	1.3	21	1
6442	SOLANUM TUBEROSUM	L	.	0.0	17.5	6.4	65.3	39.1	10.8	1390	140	.	.	.	.	.	.	.	.	18
6443	SOLANUM TUBEROSUM	H	.	0.0	10.9	4.3	71.3	27.0	13.5	.	.	.	.	.	.	.	.	.	.	18
6444	SOLANUM TUBEROSUM	R	.	0.0	12.1	0.9	82.0	3.9	5.0	100	200	.	.	.	.	.	.	.	.	18
6445	SOLANUM TUBEROSUM	R	82	78.3	2.0	0.1	18.7	0.4	0.9	9	52	0.8	7	396	0	0.10	0.04	1.6	18	2
6446	SOLANUM TUBEROSUM	R	79	77.9	2.8	0.2	18.2	0.6	0.9	10	50	1.0	.	.	0	0.11	0.04	1.5	20	3
6447	SOLANUM TUBEROSUM	R	76	79.8	2.1	0.1	17.1	0.5	0.9	7	53	0.6	3	407	0	0.10	0.04	1.5	20	4
6448	SOLANUM TUBEROSUM	W	.	0.0	12.4	2.8	66.7	25.9	18.1	2120	200	.	.	3950	.	.	.	.	.	6
6449	SOLANUM VERBASCIFOLIUM	W	.	0.0	26.5	3.9	56.7	17.7	12.9	.	.	.	.	.	.	.	.	.	.	18
6450	SOLENOSTEMON ROTUNDIFOLIUS	R	94	75.6	1.3	0.2	21.9	1.1	1.0	17	.	6.0	.	.	.	0.05	0.02	1.0	1	1
6451	SOLIDAGO CANADENSIS VAR. GILVOCANESCENS	S	.	0.0	25.1	30.4	.	.	5.7	.	.	.	.	.	.	.	.	.	.	19
6452	SOLIDAGO GRAMINEA	S	.	0.0	.	29.2	.	.	.	.	.	.	.	.	.	.	.	.	.	21
6453	SOLIDAGO SEROTINA	S	.	0.0	14.4	18.2	.	.	5.6	.	.	.	.	.	.	.	.	.	.	19
6454	SOLIDAGO SP.	W	.	0.0	8.7	14.2	70.0	12.6	7.1	1130	520	.	.	.	.	.	.	.	.	6
6455	SONCHUS CRASSIFOLIUS	S	.	0.0	20.1	20.5	.	.	.	.	.	.	.	.	.	.	.	.	.	21
6456	SONCHUS OLERACEUS	L	18	93.2	1.9	0.3	3.0	0.4	1.6	131	37	3.1	.	.	1425	0.13	0.80	.	0	1
6457	SONCHUS OLERACEUS	S	.	0.0	18.8	31.5	.	.	.	.	.	.	.	.	.	.	.	.	.	20
6458	SONCHUS OLERACEUS	L	20	92.0	2.4	.	3.9	.	1.7	93	35	.	.	.	4370	0.07	0.12	0.4	5	3
6459	SONCHUS SP.	L	18	93.2	1.9	0.3	3.0	0.4	1.6	131	37	3.1	.	.	1430	.	0.13	0.8	0	1
6460	SOPHIA OCHROLEUCA	S	.	0.0	26.9	35.9	.	.	.	.	.	.	.	.	.	.	.	.	.	19
6461	SOPHORA ALOPECUROIDES	S	.	0.0	27.4	5.4	.	.	.	.	.	.	.	.	.	.	.	.	.	21
6462	SOPHORA GRIFFITHII	S	.	0.0	18.5	16.1	.	.	.	.	.	.	.	.	.	.	.	.	.	21
6463	SOPHORA JAPONICA	S	.	0.0	21.4	6.9	.	.	3.7	.	.	.	.	.	.	.	.	.	.	19
6464	SOPHORA JAPONICA	S	414	4.5	16.4	11.6	63.8	9.5	3.7	253	260	.	.	1018	.	.	.	.	.	2
6465	SOPHORA JAPONICA	S	.	0.0	23.0	7.3	.	.	.	.	.	.	.	.	.	.	.	.	.	21
6466	SOPHORA RETICULATA	S	.	0.0	26.5	7.4	.	.	.	.	.	.	.	.	.	.	.	.	.	21
6467	SOPHORA SECUNDIFLORA	S	.	0.0	21.2	25.8	.	.	.	.	.	.	.	.	.	.	.	.	.	19
6468	SOPHORA SECUNDIFLORA	S	.	0.0	22.4	25.2	.	.	.	.	.	.	.	.	.	.	.	.	.	20
6469	SOPHORA SECUNDIFLORA	S	.	0.0	15.6	13.4	.	.	.	.	.	.	.	.	.	.	.	.	.	20
6470	SOPHORA SECUNDIFLORA	L	.	0.0	19.1	2.6	73.1	31.1	5.2	1990	140	.	.	890	.	.	.	.	.	6
6471	SOPHORA SECUNDIFLORA	F	.	0.0	18.2	0.6	78.0	15.1	3.2	500	140	.	.	1150	.	.	.	.	.	6
6472	SORBUS AMERICANA	S	.	0.0	31.9	23.9	.	.	.	.	.	.	.	.	.	.	.	.	.	19
6473	SORBUS AMERICANA	S	.	0.0	5.0	1.4	.	.	.	.	.	.	.	.	.	.	.	.	.	19
6474	SORBUS DOMESTICA	S	.	0.0	28.4	22.9	.	.	.	.	.	.	.	.	.	.	.	.	.	21
6475	SORBUS TORMINALIS	S	.	0.0	26.8	1.3	.	.	.	.	.	.	.	.	.	.	.	.	.	21
6476	SORGHASTRUM NUTANS	S	.	0.0	26.9	5.6	.	.	2.2	.	.	.	.	.	.	.	.	.	.	19
6477	SORGHASTRUM NUTANS	S	.	0.0	7.5	3.0	.	.	7.2	.	.	.	.	.	.	.	.	.	.	19
6478	SORGHASTRUM NUTANS	W	.	0.0	8.1	2.3	81.0	32.9	8.6	530	150	.	.	.	.	.	.	.	.	6
6479	SORGHUM ALMUM	W	.	0.0	11.3	3.8	75.5	29.4	9.4	.	.	.	.	.	.	.	.	.	.	18
6480	SORGHUM ALMUM	W	.	0.0	7.8	2.6	82.3	36.6	7.3	.	.	.	.	.	.	.	.	.	.	18
6481	SORGHUM ALMUM	S	.	0.0	8.1	4.0	.	.	5.9	.	.	.	.	.	.	.	.	.	.	19
6482	SORGHUM ALMUM	S	.	0.0	23.1	3.0	.	.	.	.	.	.	.	.	.	.	.	.	.	19
6483	SORGHUM ALMUM	S	.	0.0	18.8	5.3	.	.	.	.	.	.	.	.	.	.	.	.	.	20

TABLE 2 - DATA DIRECTLY TRANSCRIBED FROM ORIGINAL SOURCES.

OBS	PLNA	PART	CALORIES	WATER	PROTEIN	FAT	TOTCARB	FIBER	ASH	CALCIUM	PHOSPHO	IRON	SODIUM	POTASS	CAROTENE	THIAMINE	RIBOFLAV	NIACIN	ASCORBIC	REFCO
6484	SORGHUM BICOLOR	S	350	9.1	8.7	3.2	77.3	3.1	1.7	23	446	3.9	.	.	.	0.35	0.12	3.4	.	1
6485	SORGHUM BICOLOR	S	362	12.3	7.8	1.1	78.1	0.2	0.7	11	236	3.0	.	.	0	0.21	0.08	1.1	.	1
6486	SORGHUM BICOLOR	S	345	9.4	15.1	3.0	70.4	2.2	2.1	27	416	12.5	.	.	.	.	.	.	.	1
6487	SORGHUM BICOLOR	S	355	9.1	16.6	5.0	67.2	1.2	2.1	18	572	7.4	.	.	.	.	.	.	.	1
6488	SORGHUM BICOLOR	S	343	9.4	11.3	2.1	75.4	1.4	1.8	18	390	3.4	.	.	.	.	.	.	.	1
6489	SORGHUM BICOLOR	S	352	9.9	12.8	4.7	70.6	1.4	2.0	18	390	4.5	.	.	.	.	.	.	.	1
6490	SORGHUM BICOLOR	S	353	9.4	11.6	4.3	72.8	1.3	1.9	32	392	4.2	.	.	.	0.53	0.07	2.6	.	1
6491	SORGHUM BICOLOR	S	358	8.5	14.1	4.6	71.0	0.9	1.8	18	420	6.0	.	.	.	.	.	.	.	1
6492	SORGHUM BICOLOR	S	347	9.2	15.1	3.0	70.9	2.5	1.8	18	481	13.6	.	.	.	.	.	.	.	1
6493	SORGHUM BICOLOR	S	354	9.2	13.7	4.4	70.8	2.0	1.9	18	408	4.7	.	.	.	.	.	.	.	1
6494	SORGHUM BICOLOR	S	343	9.4	15.6	2.4	70.6	1.9	2.0	18	407	4.3	.	.	.	.	.	.	.	1
6495	SORGHUM BICOLOR	W	.	0.0	8.2	1.5	84.5	22.6	5.8	.	.	.	.	.	.	.	.	.	.	18
6496	SORGHUM BICOLOR	H	.	0.0	4.2	0.8	87.1	32.1	7.9	.	.	.	.	.	.	.	.	.	.	18
6497	SORGHUM BICOLOR	S	342	12.0	10.0	3.7	72.7	2.2	1.5	22	242	3.8	8	44	0	0.33	0.18	3.9	0	2
6498	SORGHUM BICOLOR	S	.	0.0	11.2	3.0	.	.	.	.	.	.	.	.	.	.	.	.	.	20
6499	SORGHUM BICOLOR	S	.	0.0	16.9	5.0	.	.	.	.	.	.	.	.	.	.	.	.	.	20
6500	SORGHUM BICOLOR	S	342	10.0	8.8	3.2	76.3	2.3	1.7	19	299	3.7	.	.	20	0.41	0.12	3.2	0	3
6501	SORGHUM BICOLOR	W	.	0.0	9.4	2.5	79.5	29.3	8.6	380	180	.	.	1580	.	.	.	.	.	6
6502	SORGHUM BICOLOR	H	.	0.0	7.9	2.5	81.1	26.1	8.5	400	170	5.0	.	1410	.	.	.	.	.	6
6503	SORGHUM HALAPENSE	S	.	0.0	17.7	3.8	.	.	.	.	.	.	.	.	.	.	.	.	.	21
6504	SORGHUM HALEPENSE	W	.	0.0	16.1	2.8	70.0	29.6	11.1	.	.	.	.	.	.	.	.	.	.	18
6505	SORGHUM HALEPENSE	H	.	0.0	6.6	1.9	85.6	34.6	5.9	.	.	.	.	.	.	.	.	.	.	18
6506	SORGHUM HALEPENSE	S	.	0.0	9.4	4.6	.	.	6.3	.	.	.	.	.	.	.	.	.	.	19
6507	SORGHUM HALEPENSE	H	.	0.0	7.7	2.1	81.4	33.3	8.8	810	310	.	.	1350	.	.	.	.	.	6
6508	SORGHUM HALEPENSE	W	.	0.0	11.4	2.7	75.4	30.9	10.5	940	240	.	.	3120	.	.	.	.	.	6
6509	SORGHUM LANCEOLATUM	W	.	0.0	17.3	1.4	70.4	37.2	10.9	.	.	.	.	.	.	.	.	.	.	18
6510	SORGHUM SP.	S	.	0.0	21.8	4.5	.	.	2.5	.	.	.	.	.	.	.	.	.	.	19
6511	SORGHUM SP.	S	.	0.0	8.2	2.2	.	.	10.4	.	.	.	.	.	.	.	.	.	.	19
6512	SORGHUM SP.	S	332	11.0	11.0	3.3	73.0	1.7	1.7	28	287	4.4	.	350	0	0.38	0.15	3.9	0	4
6513	SORGHUM SUDANENSE	W	.	0.0	15.4	3.7	71.0	23.4	9.9	.	.	.	.	.	.	.	.	.	.	18
6514	SORGHUM SUDANENSE	H	.	0.0	7.3	2.0	81.8	35.7	8.9	.	.	.	.	.	.	.	.	.	.	18
6515	SORGHUM SUDANENSE	S	.	0.0	11.9	4.0	.	.	4.6	.	.	.	.	.	.	.	.	.	.	19
6516	SORGHUM SUDANENSE	W	.	0.0	11.2	2.5	76.5	32.2	9.8	430	410	21.0	.	2140	.	.	.	.	.	6
6517	SORGHUM VERTICILLIFLORUM	W	.	0.0	8.3	1.5	79.2	35.0	11.0	.	.	.	.	.	.	.	.	.	.	18
6518	SORGHUM VULGARE	S	.	0.0	7.5	1.6	.	.	7.9	.	.	.	.	.	.	.	.	.	.	19
6519	SPARGANIUM RAMOSUM	S	.	0.0	6.5	0.8	.	.	.	.	.	.	.	.	.	.	.	.	.	21
6520	SPARGANIUM RAMOSUM	S	.	0.0	5.9	0.9	.	.	.	.	.	.	.	.	.	.	.	.	.	21
6521	SPARTINA PATENS(IMMATURE)	W	.	0.0	11.1	2.4	75.1	31.9	11.4	.	.	.	.	.	.	.	.	.	.	6
6522	SPARTINA PATENS(MATURE)	W	.	0.0	5.0	2.4	84.1	33.3	8.5	370	90	.	.	.	.	.	.	.	.	6
6523	SPARTINA PECTINATA	W	.	0.0	7.1	2.1	85.5	35.6	5.3	250	110	.	.	.	.	.	.	.	.	6
6524	SPARTINA SPARTINAE(IMMATURE)	W	.	0.0	11.1	2.5	76.4	32.1	10.0	560	160	.	.	.	.	.	.	.	.	6
6525	SPARTINA SPARTINAE(MATURE)	W	.	0.0	5.0	2.4	84.1	33.3	8.5	320	100	.	.	.	.	.	.	.	.	6
6526	SPARTIUM JUNCEUM	S	.	0.0	37.5	10.3	.	.	.	.	.	.	.	.	.	.	.	.	.	20
6527	SPARTIUM JUNCEUM	S	.	0.0	34.5	12.4	.	.	.	.	.	.	.	.	.	.	.	.	.	21
6528	SPARTIUM JUNCEUM	S	.	0.0	37.8	10.0	.	.	.	.	.	.	.	.	.	.	.	.	.	21

TABLE 2 - DATA DIRECTLY TRANSCRIBED FROM ORIGINAL SOURCES.

OBS	PLANA	PART	CALORIES	WATER	PROTEIN	FAT	TOTCARB	FIBER	ASH	CALCIUM	PHOSPHO	IRON	SODIUM	POTASS	CAROT	THIAMINE	RIBOFLAV	NIACIN	ASCORBIC	REFCO
6529	SPARTIUM JUNCEUM	S	.	0.0	31.0	10.3	.	.	.	.	.	.	.	.	.	.	.	.	.	21
6530	SPARTIUM JUNCEUM	Y	.	0.0	11.0	4.6	80.0	39.7	4.4	650	260	.	.	.	.	.	.	.	.	5
6531	SPATHIPHYLLUM PHRYNIIFOLIUM	I	55	82.4	6.3	1.2	8.4	1.5	1.7	403	136	1.4	.	.	20	0.04	0.49	1.8	13	3
6532	SPERGULA ARVENSIS	S	.	0.0	16.1	14.7	.	.	.	.	.	.	.	.	.	.	.	.	.	21
6533	SPERGULA ARVENSIS	S	.	0.0	14.1	14.0	.	.	.	.	.	.	.	.	.	.	.	.	.	21
6534	SPERGULA ARVENSIS	L	.	0.0	19.5	4.0	61.7	20.3	14.8	1960	1810	18.4	200	.	.	.	.	.	.	5
6535	SPERMOLEPIS INERMIS	S	.	0.0	17.5	5.1	.	.	.	.	.	.	.	.	.	.	.	.	.	20
6536	SPHAERALCEA ANGUSTIFOLIA	W	.	0.0	20.4	2.4	65.5	23.2	11.7	3340	310	.	.	2690	.	.	.	.	.	6
6537	SPHAERALCEA ANGUSTIFOLIA VAR. CUSPIDATA	S	.	0.0	25.7	17.0	.	.	.	.	.	.	.	.	.	.	.	.	.	21
6538	SPHAERALCEA COCCINEA	S	.	0.0	12.9	7.4	.	.	3.8	.	.	.	.	.	.	.	.	.	.	19
6539	SPHENOPHOLIS OBTUSATA	W	.	0.0	6.7	5.1	78.3	34.5	9.9	.	.	.	.	.	.	.	.	.	.	6
6540	SPHENOSTYLIS STENOCARPA	R	129	64.8	3.8	0.2	30.4	0.4	0.8	10	80	.	.	.	.	.	.	.	.	1
6541	SPHENOSTYLIS STENOCARPA	S	350	9.4	19.2	1.1	67.4	5.2	2.9	55	398	.	.	.	.	0.69	.	.	.	1
6542	SPHINTOSPERMUM CONSTRICTUM	S	.	0.0	30.0	2.9	.	.	.	.	.	.	.	.	.	.	.	.	.	19
6543	SPILANTHES ACMELLA	L	32	89.0	1.9	0.3	7.1	.	1.7	162	41	4.0	.	.	2350	0.03	.	.	20	2
6544	SPILANTHES OLERACEA	L	26	90.4	2.0	.	6.0	3.1	1.6	.	.	.	.	.	1560	.	.	.	.	1
6545	SPINACIA OLERACEA	L	26	90.6	2.1	0.2	5.3	0.8	1.8	61	46	1.7	.	.	.	0.03	0.27	.	46	1
6546	SPINACIA OLERACEA	S	.	0.0	15.2	4.2	.	.	2.6	.	.	.	.	.	.	.	.	.	.	19
6547	SPINACIA OLERACEA	S	.	0.0	14.8	3.9	.	.	6.9	.	.	.	.	.	.	.	.	.	.	19
6548	SPINACIA OLERACEA	S	.	0.0	24.1	7.0	.	.	4.0	.	.	.	.	.	.	.	.	.	.	19
6549	SPINACIA OLERACEA	S	.	0.0	10.9	1.9	.	.	4.7	.	.	.	.	.	.	.	.	.	.	19
6550	SPINACIA OLERACEA	L	19	93.0	2.4	0.4	2.8	0.7	1.4	62	39	3.9	38	461	3640	0.06	0.22	0.7	56	2
6551	SPINACIA OLERACEA	S	.	0.0	20.0	6.9	.	.	.	.	.	.	.	.	.	.	.	.	.	20
6552	SPINACIA OLERACEA	S	.	0.0	25.6	7.5	.	.	.	.	.	.	.	.	.	.	.	.	.	20
6553	SPINACIA OLERACEA	S	.	0.0	15.0	3.1	.	.	.	.	.	.	.	.	.	.	.	.	.	20
6554	SPINACIA OLERACEA	L	30	89.8	2.8	0.7	4.9	0.7	1.8	60	30	3.2	.	.	2340	0.06	0.17	0.6	46	3
6555	SPINACIA OLERACEA	L	26	90.7	3.2	0.3	4.3	0.6	1.5	93	51	3.1	71	470	4860	0.10	0.20	0.6	51	4
6556	SPINACIA OLERACEA	L	.	92.1	2.0	0.7	3.5	0.6	1.7	73	21	10.9	58	206	5580	0.03	0.07	0.5	28	5
6557	SPINACIA OLERACEA	L	.	0.0	31.0	4.8	33.1	8.5	31.1	.	.	.	.	.	.	.	.	.	.	6
6558	SPODIOPOGON BYRONIS	L	.	0.0	7.6	1.2	76.7	32.5	14.4	.	.	.	.	.	.	.	.	.	.	
6559	SPONDIAS DULCIS	F	46	86.9	0.2	0.1	12.4	1.1	0.4	56	67	0.3	1	95	205	0.05	0.02	1.4	36	2
6560	SPONDIAS DULCIS	F	.	59.6	0.8	1.8	11.6	3.6	0.6	.	.	.	.	.	260	.	.	.	42	5
6561	SPONDIAS MOMBIN	F	42	88.0	0.8	0.2	10.6	0.5	0.4	30	39	1.0	.	.	1400	0.04	0.03	1.4	16	1
6562	SPONDIAS MOMBIN	F	70	82.7	0.8	2.1	13.8	1.0	0.6	26	31	2.2	.	.	140	0.08	0.06	0.5	28	3
6563	SPONDIAS MOMBIN	P	.	.	1.4	0.6	10.6	1.2	0.7	.	.	.	.	.	.	.	.	.	.	5
6564	SPONDIAS PINNATA	P	.	90.3	0.7	3.0	5.5	1.0	0.5	36	11	3.9	.	.	270	0.02	0.02	0.3	21	5
6565	SPONDIAS PURPUREA	SH	35	88.6	4.3	0.8	5.1	1.6	1.2	12	73	2.8	.	.	.	0.08	0.12	1.5	23	2
6566	SPONDIAS PURPUREA	L	59	81.0	3.5	0.3	13.4	.	1.8	540	82	6.2	.	.	1740	0.06	.	.	29	2
6567	SPONDIAS PURPUREA	F	86	75.8	1.0	0.3	22.3	0.5	0.6	14	35	0.9	2	270	225	0.09	0.05	0.4	49	2
6568	SPONDIAS PURPUREA	F	83	76.2	0.9	0.1	22.0	0.4	0.8	22	40	0.6	.	.	60	0.07	0.03	1.0	45	3
6569	SPOROBOLUS AIROIDES	W	.	0.0	6.8	1.8	82.5	34.2	8.9	380	130	13.0	.	380	.	.	.	.	.	6
6570	SPOROBOLUS ASPER	W	.	0.0	4.2	1.9	90.7	37.9	3.2	290	110	.	.	.	.	.	.	.	.	6
6571	SPOROBOLUS CRYPTANDRUS	S	.	0.0	14.4	2.8	.	.	4.9	.	.	.	.	.	.	.	.	.	.	19
6572	SPOROBOLUS CRYPTANDRUS	W	.	0.0	7.4	1.5	82.5	36.6	8.6	380	110	.	.	1530	.	.	.	.	.	6
6573	SPOROBOLUS DIANDER	L	.	0.0	6.4	3.6	72.8	23.0	16.1	590	440	.	.	.	.	.	.	.	.	5

TABLE 2 - DATA DIRECTLY TRANSCRIBED FROM ORIGINAL SOURCES.

OBS	PLNA	PART	CALORIES	WATER	PROTEIN	FAT	TOTCARB	FIBER	ASH	CALCIUM	PHOSPHON	IRON	SODIUM	POTASS	CAROTE	THIAMINE	RIBOFLAV	NIACIN	ASCORBIC	REFCO
6574	SPOROBOLUS FLEXUOSUS	W	.	0.0	6.4	1.5	86.4	37.1	5.7	190.0	90	.	.	.	.	.	.	.	.	6
6575	SPOROBOLUS GLAUCIFOLIUS	L	.	0.0	4.2	1.0	86.2	26.5	8.7	340.0	420	.	.	.	.	.	.	.	.	5
6576	SPOROBOLUS HELVOLUS	W	.	0.0	16.8	4.0	69.1	28.0	10.1	.	.	.	.	.	.	.	.	.	.	18
6577	SPOROBOLUS HETEROLEPIS	W	.	0.0	5.7	1.4	87.8	33.5	5.1	.	.	.	.	.	.	.	.	.	.	6
6578	SPOROBOLUS LONGIBRACHIATUS	W	.	0.0	16.2	2.2	69.4	22.4	12.2	.	.	.	.	.	.	.	.	.	.	18
6579	SPOROBOLUS MARGINATUS	W	.	0.0	23.3	3.1	64.3	25.7	9.3	.	.	.	.	.	.	.	.	.	.	18
6580	SPOROBOLUS MARGINATUS	N	.	0.0	6.1	1.0	84.0	34.4	8.8	230.0	140	.	.	.	.	.	.	.	.	5
6581	SPOROBOLUS MARGINATUS	L	.	0.0	9.1	1.4	80.1	31.5	9.4	290.0	230	.	.	.	.	.	.	.	.	5
6582	SPOROBOLUS POIRETII	W	.	0.0	6.7	1.8	83.2	32.7	8.3	420.0	110	.	.	.	.	.	.	.	.	6
6583	SPOROBOLUS PYRAMIDALIS	W	.	0.0	7.7	1.8	84.3	35.9	6.2	.	.	.	.	.	.	.	.	.	.	18
6584	SPOROBOLUS PYRAMIDALIS	H	.	0.0	10.0	1.9	79.1	34.4	9.0	.	.	.	.	.	.	.	.	.	.	18
6585	SPOROBOLUS TREMULUS	L	.	0.0	6.1	0.9	81.6	24.6	11.4	.	.	.	.	.	.	.	.	.	.	5
6586	SPOROBOLUS WALLICHII	N	.	0.0	7.1	1.9	73.6	43.8	17.4	520.0	400	.	.	.	.	.	.	.	.	5
6587	STACHYS AFFINIS	R	.	78.0	4.3	0.2	14.6	.	1.0	.	.	.	.	.	.	.	.	.	.	5
6588	STACHYS CRETICA	S	.	0.0	16.0	29.2	.	.	.	.	.	.	.	.	.	.	.	.	.	21
6589	STACHYS OLYMPICA	S	.	0.0	20.4	31.9	.	.	7.0	.	.	.	.	.	.	.	.	.	.	19
6590	STACHYS VITICINA	S	.	0.0	13.8	30.9	.	.	.	.	.	.	.	.	.	.	.	.	.	21
6591	STAPHYLEA PINNATA	S	.	0.0	35.0	49.2	.	.	.	.	.	.	.	.	.	.	.	.	.	20
6592	STAPHYLEA PINNATA	S	.	0.0	33.6	42.9	.	.	.	.	.	.	.	.	.	.	.	.	.	21
6593	STELLARIA MEDIA	S	.	0.0	17.8	5.9	.	.	.	.	.	.	.	.	.	.	.	.	.	21
6594	STELLARIA MEDIA	L	.	91.7	1.2	0.2	5.3	1.7	1.6	.	.	.	.	.	600	.	.	.	.	5
6595	STELLARIA MEDIA	L	.	0.0	.	.	.	.	.	.	.	.	.	.	32500	0.02	0.14	0.51	375	5
6596	STENOTAPHRUM SECUNDATUM	H	.	0.0	6.7	2.7	86.9	32.5	3.7	.	.	.	.	.	.	.	.	.	.	18
6597	STENOTAPHRUM SECUNDATUM	W	.	0.0	9.2	1.8	80.3	29.2	8.7	510.0	210	20.0	.	630	.	.	.	.	.	6
6598	STERCULIA APETELA	S	.	0.0	19.4	25.9	.	.	.	.	.	.	.	.	.	.	.	.	.	20
6599	STERCULIA FOETIDA	S	.	0.0	20.5	51.5	.	.	3.8	.	.	.	.	.	.	.	.	.	.	19
6600	STERCULIA FOETIDA	S	.	35.6	11.4	35.5	.	.	2.4	33.0	415	1.7	.	517	.	0.06	0.08	1.10	5	5
6601	STERCULIA MONOSPERMA	F	146	58.5	4.5	0.9	34.3	1.2	1.8	6.0	180	1.1	14	770	870	0.09	0.18	1.20	31	2
6602	STEREOSPERMUM PERSONATUM	L	.	0.0	9.8	2.1	.	25.5	12.5	1870.0	415	.	225	855	.	.	.	.	.	5
6603	STEVIA RHOMBIFOLIA VAR. UNIARISTATA	S	.	0.0	28.8	16.4	.	.	.	.	.	.	.	.	.	.	.	.	.	20
6604	STEVIA SERRATA	S	.	0.0	24.4	18.3	.	.	.	.	.	.	.	.	.	.	.	.	.	20
6605	STEVIA SERRATA	S	.	0.0	26.2	16.4	.	.	.	.	.	.	.	.	.	.	.	.	.	20
6606	STEVIA SERRATA VAR. ARGUTA	S	.	0.0	28.6	16.2	.	.	.	.	.	.	.	.	.	.	.	.	.	20
6607	STEVIA SERRATA VAR. ARGUTA	S	.	0.0	29.4	20.7	.	.	.	.	.	.	.	.	.	.	.	.	.	20
6608	STEVIA SERRATA VAR. LINOIDES	S	.	0.0	33.8	25.1	.	.	.	.	.	.	.	.	.	.	.	.	.	20
6609	STEVIA SERRATA VAR. LINOIDES	S	.	0.0	22.5	11.9	.	.	.	.	.	.	.	.	.	.	.	.	.	20
6610	STEVIA SERRATA VAR. LINOIDES	S	.	0.0	21.0	14.3	.	.	.	.	.	.	.	.	.	.	.	.	.	21
6611	STEVIA SP.	S	.	0.0	27.5	24.0	.	.	.	.	.	.	.	.	.	.	.	.	.	20
6612	STEVIA VISCIDA	S	.	0.0	29.4	14.9	.	.	.	.	.	.	.	.	.	.	.	.	.	20
6613	STILLINGIA SP.	W	.	0.0	21.7	3.2	63.4	13.9	11.7	310.0	240	.	.	2590	.	.	.	.	.	6
6614	STIPA CAPILLATA	L	.	0.0	4.1	1.6	82.6	31.7	10.8	680.0	170	.	.	.	.	.	.	.	.	5
6615	STIPA CLARAZII	S	.	0.0	16.2	3.2	.	.	.	.	.	.	.	.	.	.	.	.	.	21
6616	STIPA CLARAZII	S	.	0.0	25.9	4.4	.	.	.	.	.	.	.	.	.	.	.	.	.	21
6617	STIPA COMATA	W	.	0.0	7.6	2.7	79.3	34.5	10.4	300.0	90	.	.	.	.	.	.	.	.	6
6618	STIPA LAGASCAE	S	.	0.0	21.5	2.6	.	.	.	.	.	.	.	.	.	.	.	.	.	21

TABLE 2 - DATA DIRECTLY TRANSCRIBED FROM ORIGINAL SOURCES.

OBS	PLANT	PART	CALORIES	WATER	PROTEIN	FAT	TOTCARB	FIBER	ASH	CALCIUM	PHOSPHO	IRON	SODIUM	POTASS	CAROTENE	THIAMINE	RIBOFLAV	NIACIN	ASCORBIC	REFCO
6619	STIPA LETTERMANII	H	.	0.0	10.9	3.2	78.8	31.3	7.1	84.5	210	.	.	.	.	.	.	.	.	6
6620	STIPA LEUCOTRICHA	W	.	0.0	9.2	2.7	76.2	31.2	11.9	560.0	150	.	.	1350	.	.	.	.	.	6
6621	STIPA MEGAPOTAMIA VAR. LATIFLORA	S	.	0.0	14.9	2.6	.	.	.	.	.	.	.	.	.	.	.	.	.	21
6622	STIPA ORIENTALIS	L	.	0.0	5.7	4.6	78.2	21.7	9.3	1630.0	530	.	.	.	.	.	.	.	.	5
6623	STIPA SIBIRICA	L	.	0.0	7.7	2.3	81.7	31.1	7.2	710.0	340	.	.	.	.	.	.	.	.	5
6624	STIPA SPARTEA	W	.	0.0	5.6	0.9	85.0	41.8	8.5	.	.	.	.	.	.	.	.	.	.	6
6625	STIPA VIRIDULA	S	.	0.0	30.6	3.1	.	.	1.3	.	.	.	.	.	.	.	.	.	.	19
6626	STIPA VIRIDULA	W	.	0.0	7.6	1.1	84.4	40.7	6.9	.	.	.	.	.	.	.	.	.	.	6
6627	STIZOLOBIUM DEERINGIANUM	S	.	0.0	29.2	5.6	.	.	3.4	.	.	.	.	.	.	.	.	.	.	19
6628	STIZOLOBIUM SP.	S	.	0.0	28.1	4.2	.	.	3.8	.	.	.	.	.	.	.	.	.	.	19
6629	STIZOLOBIUM SP.	S	.	0.0	28.8	3.2	.	.	3.5	.	.	.	.	.	.	.	.	.	.	19
6630	STIZOLOBIUM SP.	H	.	0.0	17.7	3.3	71.0	29.6	8.0	.	.	.	.	.	.	.	.	.	.	6
6631	STIZOLOBIUM SP.	W	.	0.0	19.6	3.9	65.3	28.5	11.2	.	.	.	.	.	.	.	.	.	.	6
6632	STOCKSIA BRAHUICA	S	.	0.0	11.9	19.4	.	.	.	.	.	.	.	.	.	.	.	.	.	21
6633	STRELITZIA REGINAE	R	26	91.9	0.4	0.1	6.7	1.6	0.9	45.0	22	1.0	.	.	.	0.01	0.03	0.1	.	2
6634	STREPTOPUS AMPLEXIFOLIUS	S	.	0.0	17.5	12.8	.	.	.	.	.	.	.	.	.	.	.	.	.	20
6635	STROPHANTHUS KOMBE	S	.	0.0	27.0	35.4	.	.	.	.	.	.	.	.	.	.	.	.	.	21
6636	STROPHOSTYLES HELVEOLA	S	.	0.0	30.0	0.5	.	.	3.7	.	.	.	.	.	.	.	.	.	.	19
6637	STRYCHNOS COCCULOIDES	F	.	79.7	.	.	.	.	.	.	.	.	.	.	.	.	.	.	1	1
6638	STRYCHNOS INNOCUA	F	.	70.6	.	.	.	.	.	.	.	.	.	.	.	.	.	.	14	1
6639	STRYCHNOS POTATORUM	S	.	0.0	6.5	2.8	.	.	.	.	.	.	.	.	.	.	.	.	.	21
6640	STRYCHNOS SPINOSA	F	72	79.7	1.6	0.6	17.1	0.6	1.0	28.0	42	0.7	.	.	.	0.11	0.17	1.9	18	1
6641	STRYCHNOS SPINOSA	S	.	0.0	10.6	4.0	.	.	1.4	.	.	.	.	.	.	.	.	.	.	19
6642	STYLOSANTHES BOJERI	W	.	0.0	10.1	1.2	80.1	38.4	8.6	1230.0	430	.	.	.	.	.	.	.	.	18
6643	STYLOSANTHES GRACILIS	S	.	0.0	40.0	17.1	.	.	4.1	.	.	.	.	.	.	.	.	.	.	19
6644	STYLOSANTHES GUIANENSIS	W	.	0.0	18.1	2.1	71.5	26.8	8.3	.	.	.	.	.	.	.	.	.	.	18
6645	STYLOSANTHES GUIANENSIS	H	.	0.0	16.7	1.7	71.6	31.7	10.0	1550.0	560	.	.	.	.	.	.	.	.	18
6646	STYLOSANTHES GUIANENSIS	L	.	76.0	4.0	0.4	17.2	7.6	2.4	520.0	310	.	.	.	.	.	.	.	.	5
6647	STYLOSANTHES HUMILIS	W	.	0.0	22.0	3.0	67.0	21.0	8.0	800.0	220	.	.	.	.	.	.	.	.	18
6648	STYLOSANTHES MUCRONATA	L	.	0.0	14.6	2.0	73.2	30.6	10.1	2430.0	120	.	.	.	.	.	.	.	.	5
6649	STYRAX AMERICANA	S	.	0.0	18.8	42.7	.	.	2.8	.	.	.	.	.	.	.	.	.	.	19
6650	STYRAX GRANDIFOLIA	S	.	0.0	12.4	36.2	.	.	.	.	.	.	.	.	.	.	.	.	.	21
6651	STYRAX JAPONICA	S	.	0.0	13.8	50.0	.	.	.	.	.	.	.	.	.	.	.	.	.	21
6652	STYRAX OBESSIA	S	.	0.0	17.5	46.6	.	.	.	.	.	.	.	.	.	.	.	.	.	20
6653	STYRAX OFFICINALIS	S	.	0.0	16.5	50.8	.	.	.	.	.	.	.	.	.	.	.	.	.	21
6654	SUAEDA MARITIMA	L	.	89.3	2.0	0.4	4.5	0.9	3.8	50.0	79	1.6	.	.	2100	0.01	0.05	0.2	13	5
6655	SUAEDA PROSTRATA	S	.	0.0	25.1	13.9	.	.	.	.	.	.	.	.	.	.	.	.	.	21
6656	SUAEDA SETIGERA	S	.	0.0	21.4	15.9	.	.	.	.	.	.	.	.	.	.	.	.	.	21
6657	SUAEDA SUFFRUTESCENS	S	.	0.0	.	36.4	.	.	.	.	.	.	.	.	.	.	.	.	.	21
6658	SUTHERLANDIA FRUTESCENS	S	.	0.0	29.2	2.8	.	.	.	.	.	.	.	.	.	.	.	.	.	21
6659	SYMPHORICARPOS ORBICULATUS	S	.	0.0	10.0	15.6	.	.	.	.	.	.	.	.	.	.	.	.	.	20
6660	SYMPHORICARPOS ROTUNDIFOLIUS	C	.	0.0	4.2	2.0	90.0	41.1	3.8	.	.	.	.	.	.	.	.	.	.	6
6661	SYMPHORICARPOS ROTUNDIFOLIUS(IMMATURE)	L	.	0.0	16.0	3.9	73.2	11.7	6.9	1220.0	380	.	.	.	.	.	.	.	.	6
6662	SYMPHORICARPOS ROTUNDIFOLIUS(MATURE)	L	.	0.0	8.7	4.6	77.0	13.6	9.7	2580.0	520	.	.	.	.	.	.	.	.	6
6663	SYMPHORICARPOS SP.	L	.	0.0	10.6	5.3	76.6	14.6	7.5	1680.0	390	.	.	.	.	.	.	.	.	6

TABLE 2 - DATA DIRECTLY TRANSCRIBED FROM ORIGINAL SOURCES.

OBS	PLNA	PART	CALORIES	WATER	PROTEIN	FAT	TOTCARB	FIBER	ASH	CALCIUM	PHOSPHO	IRON	SODIUM	POTASS	CAROT	THIAMINE	RIBOFLAV	NIACIN	ASCORBIC	REFCO
6664	SYMPHORICARPOS VACCINIOIDES	L	.	0.0	3.9	2.2	89.8	38.8	4.1	770	140	.	.	.	.	.	.	.	.	6
6665	SYMPHYTUM ASPERRINUM	W	.	0.0	19.8	2.5	59.2	13.6	18.5	.	.	.	.	.	.	.	.	.	.	6
6666	SYMPHYTUM ASPERUM	L	.	.	19.8	2.5	59.2	13.6	18.5	.	.	.	.	.	.	.	.	.	.	5
6667	SYMPHYTUM PEREGRINUM	W	.	0.0	19.0	2.9	44.4	14.0	33.7	.	.	.	.	.	.	.	.	.	.	18
6668	SYMPHYTUM PEREGRINUM	W	.	0.0	20.1	3.3	55.7	11.3	20.9	1460	410	.	.	.	.	.	.	.	.	18
6669	SYMPHYTUM X UPLANDICUM	L	.	.	22.7	5.4	50.6	4.2	21.2	2020	570	.	.	.	.	.	.	.	.	5
6670	SYMPLOCOS PANICULATA	S	.	0.0	21.2	51.7	.	.	2.5	.	.	.	.	.	.	.	.	.	.	19
6671	SYMPLOCOS PANICULATA	S	.	0.0	21.0	52.0	.	.	.	.	.	.	.	.	.	.	.	.	.	5
6672	SYMPLOCOS RAMOSISSIMA	L	.	0.0	11.9	.	.	.	11.2	2950	.	.	.	.	.	.	.	.	.	5
6673	SYNELCOSCIADIUM CARMELLI	S	.	0.0	15.8	8.7	.	.	.	.	.	.	.	.	.	.	.	.	.	21
6674	SYRINGA VULGARIS	S	.	0.0	20.0	18.8	.	.	.	.	.	.	.	.	.	.	.	.	.	20
6675	SYZYGIUM AQUEUM	F	46	87.0	0.6	0.2	11.8	0.9	0.4	8	9	1.1	.	.	0	0.00	0.03	0.60	5	2
6676	SYZYGIUM AROMATICUM	I	323	6.9	6.0	20.1	61.2	9.6	5.9	646	105	8.7	243	1102	318	0.12	0.27	1.46	81	16
6677	SYZYGIUM AROMATICUM	I	.	25.2	5.2	8.9	55.5	9.5	5.2	740	100	4.9	.	.	253	0.08	0.13	1.51	.	5
6678	SYZYGIUM CUMINI	L	.	0.0	7.9	2.6	82.5	20.7	7.0	1380	120	.	.	.	.	.	.	.	.	18
6679	SYZYGIUM CUMINI	F	60	82.7	0.7	0.1	15.8	0.3	0.7	8	13	0.2	9	116	0	0.00	0.01	0.20	23	2
6680	SYZYGIUM CUMINI	F	51	85.8	0.8	0.2	12.9	0.3	0.3	8	16	1.6	.	.	0	0.01	0.01	0.30	6	3
6681	SYZYGIUM CUMINI	F	.	83.7	0.7	0.3	14.9	0.9	0.4	15	15	1.2	26	55	48	0.03	0.01	0.20	18	5
6682	SYZYGIUM CUMINI	S	.	.	8.5	1.2	.	16.9	21.7	410	170	.	.	.	.	.	.	.	.	5
6683	SYZYGIUM CUMINI	L	.	.	9.1	4.3	.	17.0	6.0	1300	190	.	.	.	.	.	.	.	.	5
6684	SYZYGIUM GUINEENSE	F	.	88.1	.	.	.	.	.	.	.	.	.	.	.	.	.	.	0	1
6685	SYZYGIUM JAMBOS	F	63	82.4	0.6	0.2	16.4	1.4	0.4	36	15	0.4	.	.	150	0.02	0.03	0.60	22	3
6686	SYZYGIUM JAMBOS	F	.	89.1	0.7	0.2	10.9	1.2	0.3	10	30	0.5	34	50	141	0.01	0.05	0.40	3	5
6687	SYZYGIUM MALACCENSE	F	30	91.4	0.5	0.1	7.6	1.0	0.4	18	12	0.4	2	114	0	0.03	0.02	0.20	17	2
6688	SYZYGIUM MALACCENSE	F	32	90.9	0.6	0.1	8.0	0.7	0.4	6	16	0.4	.	.	0	0.03	0.03	0.30	13	3
6689	SYZYGIUM MALACCENSE	F	.	93.5	0.1	0.4	5.9	2.2	0.1	17	3	0.0	.	.	.	.	0.02	0.13	30	5
6690	SYZYGIUM SAMARANGENSE	F	30	91.5	0.4	0.1	7.8	0.8	0.2	17	7	0.3	2	105	0	0.03	0.01	0.30	13	2
6691	TABEBUIA PALMERI	S	.	0.0	23.1	35.4	.	.	.	.	.	.	.	.	.	.	.	.	.	20
6692	TABEBUIA PALMERI	S	.	0.0	23.1	36.3	.	.	.	.	.	.	.	.	.	.	.	.	.	20
6693	TACCA LEONTOPETALODES	R	121	68.8	1.5	0.1	28.8	0.4	0.8	27	64	.	.	.	.	.	.	.	13	1
6694	TAGETES ERECTA	S	.	0.0	23.8	20.5	.	.	.	.	.	.	.	.	.	.	.	.	.	19
6695	TAGETES ERECTA	S	.	0.0	24.0	20.0	.	.	.	.	.	.	.	.	.	.	.	.	.	5
6696	TAGETES LUCIDA	S	.	0.0	18.8	9.4	.	.	.	.	.	.	.	.	.	.	.	.	.	20
6697	TAGETES LUCIDA	S	.	0.0	21.2	11.4	.	.	.	.	.	.	.	.	.	.	.	.	.	20
6698	TAGETES SUBULATA	S	.	0.0	33.8	18.8	.	.	.	.	.	.	.	.	.	.	.	.	.	20
6699	TALINUM TRIANGULARE	L	25	90.8	2.4	0.4	4.4	1.0	2.0	121	67	5.0	.	.	.	0.08	0.18	0.30	31	1
6700	TALINUM TRIANGULARE	L	.	0.0	21.1	1.5	42.8	10.3	34.6	.	.	.	.	.	.	.	.	.	.	18
6701	TALINUM TRIANGULARE	L	23	92.4	1.9	0.4	4.1	0.6	1.2	90	21	4.8	10	392	2985	0.09	0.19	0.60	58	2
6702	TALINUM TRIANGULARE	L	26	90.3	2.7	0.6	4.1	0.8	2.3	120	27	5.5	.	.	.	0.10	0.17	0.40	54	3
6703	TALINUM TRIANGULARE	L	.	91.2	1.9	0.7	4.3	.	1.9	120	16	0.9	.	.	3720	0.13	0.24	0.46	58	5
6704	TAMARINDUS INDICA	F	68	80.2	2.0	0.2	16.6	8.4	1.0	60	97	.	.	.	.	.	.	.	8	1
6705	TAMARINDUS INDICA	L	73	78.0	3.1	0.8	16.5	4.1	1.6	506	22	.	.	.	.	.	.	.	.	1
6706	TAMARINDUS INDICA	L	.	0.0	15.8	9.6	66.7	23.7	7.9	2860	260	.	.	.	.	.	.	.	.	18
6707	TAMARINDUS INDICA	S	.	0.0	18.3	7.4	70.8	26.4	3.5	140	300	.	.	.	.	.	.	.	.	18
6708	TAMARINDUS INDICA	L	78	77.2	5.1	1.0	16.1	1.3	0.6	24	52	2.0	8	273	2510	0.10	0.11	1.50	6	2

TABLE 2 - DATA DIRECTLY TRANSCRIBED FROM ORIGINAL SOURCES.

OBS	PLNA	PART	CALORIES	WATER	PROTEIN	FAT	TOTLCARB	FIBER	ASH	CALCIUM	PHOSPHO	IRON	SODIUM	POTASS	CAROTE	THIAMINE	RIBOFLAV	NIACIN	ASCORBIC	REFCO
6709	TAMARINDUS INDICA	I	75	80.0	2.5	1.8	15.0	1.2	0.7	53	44	1.4	5	254	205	0.08	0.12	1.2	12	2
6710	TAMARINDUS INDICA	F	81	14.7	0.9	0.1	21.5	0.7	0.8	31	33	0.5	1	217	0	0.08	0.03	0.4	1	2
6711	TAMARINDUS INDICA	G	27	30.2	0.9	0.0	6.6	0.3	0.3	22	11	0.3	1	120	0	0.06	0.02	0.2	4	2
6712	TAMARINDUS INDICA	P	214	38.7	2.3	0.2	56.7	1.9	2.1	81	86	1.3	3	570	10	0.22	0.08	1.1	3	2
6713	TAMARINDUS INDICA	G	71	79.5	2.4	0.1	17.2	0.8	0.8	58	29	0.7	3	316	10	0.15	0.05	0.4	12	2
6714	TAMARINDUS INDICA	F	272	22.6	3.1	0.4	71.8	3.0	2.1	54	108	1.0	.	.	40	0.44	0.16	2.1	6	3
6715	TAMARINDUS INDICA	P	239	31.4	2.8	0.6	62.5	5.1	2.7	74	113	2.8	51	781	18	0.34	0.14	1.2	2	4
6716	TAMARIX GALLICA	S	.	0.0	26.2	4.5	.	.	.	.	.	.	.	.	.	.	.	.	.	20
6717	TAMARIX GALLICA	S	.	0.0	36.9	17.4	.	.	.	.	.	.	.	.	.	.	.	.	.	20
6718	TAMUS COMMUNIS	S	.	0.0	12.1	11.5	.	.	.	.	.	.	.	.	.	.	.	.	.	21
6719	TANACETUM CAPITATUM	W	.	0.0	4.1	7.6	77.9	26.8	10.4	.	.	.	.	.	.	.	.	.	.	6
6720	TANACETUM MICROPHYLLUM	S	.	0.0	21.9	25.5	.	.	.	.	.	.	.	.	.	.	.	.	.	21
6721	TANACETUM VULGARE	S	.	0.0	28.9	26.5	.	.	.	.	.	.	.	.	.	.	.	.	.	21
6722	TARAXACUM OFFICINALE	L	45	85.6	2.7	0.7	9.2	1.6	1.8	187	66	3.1	76	397	8400	0.19	0.26	.	35	2
6723	TARAXACUM OFFICINALE	S	.	0.0	30.0	26.7	.	.	.	.	.	.	.	.	.	.	.	.	.	20
6724	TARAXACUM OFFICINALE	L	44	85.7	2.7	0.7	8.9	1.8	2.0	150	.	3.1	.	.	5000	0.19	0.14	0.8	28	3
6725	TARAXACUM OFFICINALE	L	.	88.8	3.6	1.6	3.7	0.4	2.3	473	59	3.3	.	.	4020	0.19	0.14	0.8	73	5
6726	TARAXACUM OFFICINALE	W	.	0.0	23.0	4.8	59.3	11.1	12.9	.	440	62.0	.	.	.	.	.	.	.	6
6727	TARCHONANTHUS CAMPHORATUS	SH	.	0.0	11.4	6.7	76.1	26.8	5.8	370	100	.	.	.	.	.	.	.	.	18
6728	TARCHONANTHUS MINOR	SH	.	0.0	12.1	11.6	71.6	25.6	6.7	1420	130	.	.	.	.	.	.	.	.	18
6729	TAXODIUM DISTICHUM	S	.	0.0	40.6	37.4	.	.	.	.	.	.	.	.	.	.	.	.	.	20
6730	TELANTHERA VERSICOLOR	L	73	77.5	2.4	0.4	17.9	1.4	1.8	320	54	12.1	.	.	4111	0.01	0.15	1.0	28	2
6731	TELEPHIUM IMPERATI	S	.	0.0	14.9	3.7	.	.	.	.	.	.	.	.	.	.	.	.	.	21
6732	TELFAIRIA OCCIDENTALIS	L	47	86.4	2.9	1.8	7.0	1.7	1.9	.	.	.	.	.	.	.	.	.	.	1
6733	TELFAIRIA OCCIDENTALIS	S	543	6.2	20.5	45.0	23.5	2.2	4.8	84	572	.	.	.	.	.	.	.	.	1
6734	TELFAIRIA OCCIDENTALIS	L	.	0.0	21.2	12.9	52.0	12.8	13.9	.	.	.	.	.	.	.	.	.	.	18
6735	TELFAIRIA PEDATA	S	.	4.4	29.7	63.3	.	.	2.6	10	570	4.1	.	.	.	.	.	.	.	5
6736	TELOSMA MINOR	L	65	80.5	5.0	1.1	12.1	1.6	1.3	70	90	1.0	.	.	1890	0.10	0.12	1.5	45	2
6737	TEPHROSIA CANDIDA	S	.	0.0	40.8	10.4	.	.	.	.	.	.	.	.	.	.	.	.	.	21
6738	TEPHROSIA GRANDIFLORA	S	.	0.0	38.2	13.4	.	.	.	.	.	.	.	.	.	.	.	.	.	21
6739	TEPHROSIA LEIOCARPA	S	.	0.0	43.8	11.4	.	.	.	.	.	.	.	.	.	.	.	.	.	20
6740	TEPHROSIA LONGIPES	S	.	0.0	33.9	14.1	.	.	.	.	.	.	.	.	.	.	.	.	.	21
6741	TEPHROSIA NOCTIFLORA	L	.	0.0	22.0	8.3	61.7	22.1	7.9	1750	360	.	.	.	.	.	.	.	.	5
6742	TEPHROSIA NYIKENSIS	L	.	0.0	17.1	2.0	73.0	28.7	7.8	1520	400	.	.	.	.	.	.	.	.	5
6743	TEPHROSIA PURPUREA	S	.	0.0	35.8	9.1	.	.	.	.	.	.	.	.	.	.	.	.	.	21
6744	TEPHROSIA PURPUREA	L	.	0.0	24.4	2.4	65.4	28.0	7.7	1650	520	.	.	.	.	.	.	.	.	5
6745	TEPHROSIA PURPUREA VAR.PUMILA	L	.	0.0	14.7	2.7	74.8	27.9	7.9	2020	260	.	.	.	.	.	.	.	.	5
6746	TEPHROSIA SP.	S	.	0.0	44.4	10.4	.	.	3.2	.	.	.	.	.	.	.	.	.	.	19
6747	TEPHROSIA SP.	W	.	0.0	8.6	2.3	84.0	42.6	5.1	2420	370	.	.	.	.	.	.	.	.	6
6748	TEPHROSIA TENELLA	S	.	0.0	36.2	9.4	.	.	.	.	.	.	.	.	.	.	.	.	.	19
6749	TEPHROSIA VOGELII	S	.	0.0	43.0	10.9	.	.	.	.	.	.	.	.	.	.	.	.	.	21
6750	TERMINALIA BELLERICA	L	.	0.0	8.6	4.7	78.7	18.6	8.0	2080	270	.	.	.	.	.	.	.	.	18
6751	TERMINALIA BELLERICA	S	.	0.0	40.9	43.8	.	.	.	.	.	.	.	.	.	.	.	.	.	21
6752	TERMINALIA BROWNEI	SH	.	0.0	10.3	2.9	81.2	16.7	5.6	820	130	.	.	.	.	.	.	.	.	18
6753	TERMINALIA BUCERAS	S	.	0.0	5.0	1.0	.	.	.	.	.	.	.	.	.	.	.	.	.	20

TABLE 2 - DATA DIRECTLY TRANSCRIBED FROM ORIGINAL SOURCES.

OBS	PLANA	PART	CALORIES	WATER	PROTEIN	FAT	TOTLCARB	FIBER	ASH	CALCIUM	PHOSPHO	IRON	SODIUM	POTASS	CAROT	THIAMINE	RIBOFLAV	NIACIN	ASCORBIC	REFCO
6754	TERMINALIA CATAPPA	S	596	6.0	19.1	56.0	14.9	1.8	4.0	394	864	.	.	.	.	.	.	.	.	1
6755	TERMINALIA CATAPPA	S	594	4.0	20.8	54.0	17.2	2.3	4.0	32	789	9.2	70	784	.	0.32	0.08	0.60	0	2
6756	TERMINALIA CATAPPA	F	35	0.4	1.2	3.2	1.0	0.1	0.2	2	47	0.6	4	47	.	0.02	0.00	0.00	0	2
6757	TERMINALIA CATAPPA	S	607	2.7	23.1	54.6	17.2	9.2	2.4	497	957	2.4	.	.	0	0.71	0.28	0.70	.	3
6758	TERMINALIA CHEBULA	L	56	85.9	1.2	1.7	10.8	2.5	0.4	18	18	0.0	.	.	300	0.00	0.01	2.00	116	2
6759	TERMINALIA TOMENTOSA	L	.	0.0	8.9	4.9	75.9	21.8	10.3	3200	250	.	.	.	.	.	.	.	.	18
6760	TERMINALIA TOMENTOSA	L	.	0.0	12.7	1.4	.	16.9	10.3	3110	370	.	70	980	.	.	.	.	.	5
6761	TETRACARPIDIUM CONOPHORUM	S	419	8.1	28.7	18.3	40.9	2.8	4.0	.	.	.	.	.	.	.	.	.	.	1
6762	TETRACHNE DREGEI	S	.	0.0	30.6	4.2	.	.	2.9	.	.	.	.	.	.	.	.	.	.	19
6763	TETRACLEA COULTERI	S	.	0.0	16.9	8.4	.	.	.	.	.	.	.	.	.	.	.	.	.	19
6764	TETRADYMIA SPINOSA	W	.	0.0	8.7	6.3	79.6	36.8	5.4	940	250	.	.	.	.	.	.	.	.	6
6765	TETRAGONIA ARBUSCULA	S	.	0.0	12.2	6.0	.	.	3.2	.	.	.	.	.	.	.	.	.	.	19
6766	TETRAGONIA ARBUSCULA	S	.	0.0	13.5	2.6	.	.	7.8	.	.	.	.	.	.	.	.	.	.	19
6767	TETRAGONIA ARBUSCULA	S	.	0.0	12.8	6.1	.	.	4.0	.	.	.	.	.	.	.	.	.	.	19
6768	TETRAGONIA ARBUSCULA	S	.	0.0	15.1	1.5	.	.	9.2	.	.	.	.	.	.	.	.	.	.	19
6769	TETRAGONIA ARBUSCULA	S	.	0.0	13.1	5.8	.	.	3.5	.	.	.	.	.	.	.	.	.	.	19
6770	TETRAGONIA ARBUSCULA	S	.	0.0	15.0	3.2	.	.	7.8	.	.	.	.	.	.	.	.	.	.	19
6771	TETRAGONIA ARBUSCULA	S	.	0.0	13.1	5.3	.	.	3.4	.	.	.	.	.	.	.	.	.	.	19
6772	TETRAGONIA TETRAGONIOIDES	L	19	92.9	1.4	.	4.5	2.4	1.2	.	.	.	.	.	1620	.	.	.	.	1
6773	TETRAGONIA TETRAGONIOIDES	L	22	91.5	2.8	0.4	3.2	0.8	2.1	178	61	3.8	88	555	3540	0.05	0.24	1.20	27	2
6774	TETRAGONIA TETRAGONIOIDES	L	22	92.1	2.4	0.4	3.5	0.7	1.6	65	34	3.4	.	.	2160	0.03	0.16	0.50	31	3
6775	TETRAGONIA TETRAGONIOIDES	L	19	92.6	2.2	0.3	3.1	0.7	1.8	58	46	2.6	159	795	2580	0.04	0.17	0.60	30	4
6776	TETRAGONIA TETRAGONIOIDES	L	.	88.6	4.1	0.5	.	1.0	2.2	376	68	2.5	.	.	7640	0.03	0.19	1.49	36	5
6777	TETRAGONOTHECA TEXANA	S	.	0.0	13.8	8.2	.	.	.	.	.	.	.	.	.	.	.	.	.	19
6778	TETRAPANAX PAPYRIFERUS	Y	.	.	1.1	1.1	98.1	48.7	5.9	.	.	.	.	.	.	.	.	.	.	5
6779	TETRAPOGON SPATHACEUS	W	.	0.0	12.6	1.7	77.0	34.6	8.7	.	.	.	.	.	.	.	.	.	.	18
6780	TEUCRIUM CRETICUM	S	.	0.0	36.8	52.1	.	.	.	.	.	.	.	.	.	.	.	.	.	21
6781	TEUCRIUM POLIUM	S	.	0.0	14.8	14.6	.	.	.	.	.	.	.	.	.	.	.	.	.	21
6782	THALIA GENICULATA	S	.	0.0	14.1	2.3	.	.	.	.	.	.	.	.	.	.	.	.	.	21
6783	THALICTRUM DIPTEROCARPUM	S	.	0.0	28.1	36.7	.	.	.	.	.	.	.	.	.	.	.	.	.	20
6784	THALICTRUM FENDLERI	W	.	0.0	8.9	3.1	.	.	9.6	.	.	.	.	.	.	.	.	.	.	6
6785	THALICTRUM FENDLERI	L	.	0.0	16.2	5.0	.	.	12.1	.	.	.	.	.	.	.	.	.	.	6
6786	THALICTRUM FENDLERI	Y	.	0.0	4.4	1.4	.	.	7.4	.	.	.	.	.	.	.	.	.	.	6
6787	THALICTRUM LUCIDUM	S	.	0.0	32.5	37.6	.	.	.	.	.	.	.	.	.	.	.	.	.	21
6788	THALICTRUM MINUS	S	.	0.0	29.2	41.1	.	.	.	.	.	.	.	.	.	.	.	.	.	21
6789	THALICTRUM POLYCARPUM	S	.	0.0	24.5	33.1	.	.	3.4	.	.	.	.	.	.	.	.	.	.	19
6790	THALICTRUM POLYCARPUM	S	.	0.0	9.9	7.4	.	.	6.6	.	.	.	.	.	.	.	.	.	.	19
6791	THALICTRUM POLYCARPUM	S	.	0.0	26.9	36.8	.	.	.	.	.	.	.	.	.	.	.	.	.	21
6792	THALICTRUM REVOLUTUM	S	.	0.0	27.5	38.4	.	.	3.1	.	.	.	.	.	.	.	.	.	.	19
6793	THEA SINENSIS	L	293	8.0	24.5	2.8	58.8	8.7	5.9	327	313	24.3	50	.	2700	0.07	0.80	7.60	9	2
6794	THEA SINENSIS	L	300	7.7	28.3	4.8	53.6	9.6	5.6	245	415	18.9	60	.	8400	0.38	1.24	4.60	230	2
6795	THEA SINENSIS	L	299	8.1	24.1	3.5	59.0	9.7	5.3	320	185	31.6	.	.	8400	0.07	0.79	7.30	85	2
6796	THELEPOGON ELEGANS	L	.	0.0	5.6	1.3	88.0	28.2	5.1	.	.	.	.	.	.	.	.	.	.	5
6797	THELESPERMA AMBIGUUM	S	.	0.0	12.3	9.1	.	.	.	.	.	.	.	.	.	.	.	.	.	20
6798	THELESPERMA AMBIGUUM	S	.	0.0	10.0	10.1	.	.	.	.	.	.	.	.	.	.	.	.	.	20

TABLE 2 - DATA DIRECTLY TRANSCRIBED FROM ORIGINAL SOURCES.

OBS	PLANA	PART	CALORIES	WATER	PROTEIN	FAT	TOTLCARB	FIBER	ASH	CALCIUM	PHOSPHO	IRON	SODIUM	POTASS	CAROTE	THIAMINE	RIBOFLAV	NIACIN	ASCORBIC	REFCO
6799	THELESPERMA MEGAPOTAMICUM	S	.	0.0	15.6	13.7	.	.	.	.	.	.	.	.	.	.	.	.	.	20
6800	THELESPERMA SP.	L	.	0.0	10.0	1.6	83.2	33.6	5.2	1300	150	.	.	1600	.	.	.	.	.	6
6801	THELYPODIUM TEXANUM	S	.	0.0	27.5	35.6	.	.	.	.	.	.	.	.	.	.	.	.	.	20
6802	THEMEDA ANATHERA	L	.	.	7.3	1.9	80.2	34.4	10.7	480	100	.	.	.	.	.	.	.	.	5
6803	THEMEDA ARGUENS	N	.	0.0	4.0	.	.	31.2	3.6	720	210	490.0	.	.	.	.	.	.	.	5
6804	THEMEDA ARGUENS	L	.	0.0	5.5	1.2	80.7	34.2	12.5	.	.	.	.	.	.	.	.	.	.	5
6805	THEMEDA ARUNDINACEA	N	.	0.0	1.9	1.0	92.4	48.9	4.6	460	140	.	.	.	.	.	.	.	.	5
6806	THEMEDA CYMBARIA	L	.	0.0	4.6	2.1	87.1	37.8	9.1	320	200	.	.	.	.	.	.	.	.	5
6807	THEMEDA QUADRIVALVIS	L	.	0.0	3.0	3.0	.	26.2	12.5	330	140	.	.	330	.	.	.	.	.	5
6808	THEMEDA TREMULA	L	.	0.0	9.0	3.5	76.8	30.9	10.8	630	290	.	.	.	.	.	.	.	.	5
6809	THEMEDA TRIANDRA	W	.	0.0	6.9	3.0	78.1	32.3	12.0	.	.	.	.	.	.	.	.	.	.	18
6810	THEMEDA TRIANDRA	H	.	0.0	4.4	1.6	83.7	40.4	10.3	.	.	.	.	.	.	.	.	.	.	18
6811	THEMEDA TRIANDRA	S	.	0.0	23.8	5.5	.	.	.	.	.	.	.	.	.	.	.	.	.	20
6812	THEMEDA TRIANDRA	S	.	0.0	26.2	8.3	.	.	.	.	.	.	.	.	.	.	.	.	.	20
6813	THEMEDA TRIANDRA	S	.	0.0	21.2	6.7	.	.	.	.	.	.	.	.	.	.	.	.	.	20
6814	THEMEDA TRIANDRA	S	.	0.0	11.2	3.6	.	.	.	.	.	.	.	.	.	.	.	.	.	20
6815	THEMEDA TRIANDRA	S	.	0.0	11.2	2.6	.	.	.	.	.	.	.	.	.	.	.	.	.	20
6816	THEMEDA TRIANDRA	S	.	0.0	15.0	6.8	.	.	.	.	.	.	.	.	.	.	.	.	.	20
6817	THEMEDA TRIANDRA	S	.	0.0	11.9	4.5	.	.	.	.	.	.	.	.	.	.	.	.	.	20
6818	THEMEDA TRIANDRA	S	.	0.0	12.5	2.9	.	.	.	.	.	.	.	.	.	.	.	.	.	20
6819	THEMEDA TRIANDRA	S	.	0.0	11.2	3.7	.	.	.	.	.	.	.	.	.	.	.	.	.	20
6820	THEMEDA TRIANDRA	S	.	0.0	25.0	3.4	.	.	.	.	.	.	.	.	.	.	.	.	.	20
6821	THEMEDA TRIANDRA	S	.	0.0	13.8	4.0	.	.	.	.	.	.	.	.	.	.	.	.	.	20
6822	THEMEDA TRIANDRA	S	.	0.0	13.8	4.8	.	.	.	.	.	.	.	.	.	.	.	.	.	20
6823	THEMEDA TRIANDRA	S	.	0.0	13.1	10.6	.	.	.	.	.	.	.	.	.	.	.	.	.	20
6824	THEMEDA TRIANDRA	S	.	0.0	12.5	6.1	.	.	.	.	.	.	.	.	.	.	.	.	.	20
6825	THEMEDA TRIANDRA	S	.	0.0	15.0	5.7	.	.	.	.	.	.	.	.	.	.	.	.	.	20
6826	THEMEDA TRIANDRA	S	.	0.0	15.6	5.9	.	.	.	.	.	.	.	.	.	.	.	.	.	20
6827	THEMEDA TRIANDRA	S	.	0.0	11.9	9.6	.	.	.	.	.	.	.	.	.	.	.	.	.	20
6828	THEMEDA TRIANDRA	L	.	0.0	2.5	2.2	.	32.2	5.7	260	90	.	.	210	.	.	.	.	.	5
6829	THEOBROMA BICOLOR	P	71	79.2	2.8	0.3	16.5	1.1	1.2	6	41	0.7	.	.	190	1.80	0.15	3.20	21	3
6830	THEOBROMA BICOLOR	S	602	3.6	17.1	54.2	21.9	2.6	3.2	65	549	3.4	.	.	0	0.22	0.34	1.60	0	3
6831	THEOBROMA CACAO	S	.	0.0	15.1	37.1	44.5	5.9	3.3	80	360	.	.	.	.	.	.	.	.	18
6832	THEOBROMA CACAO	S	456	3.6	12.0	46.3	34.7	8.6	3.4	106	537	3.6	.	.	30	0.17	0.14	1.70	3	3
6833	THERMOPSIS DIVARICARPA	S	.	0.0	32.3	10.8	.	.	.	.	.	.	.	.	.	.	.	.	.	21
6834	THERMOPSIS RHOMBIFOLIA	S	.	0.0	34.9	11.6	.	.	3.1	.	.	.	.	.	.	.	.	.	.	19
6835	THEVETIA SP.	S	.	0.0	16.2	57.5	.	.	2.5	.	.	.	.	.	.	.	.	.	.	19
6836	THEVETIA THEVETIOIDES	S	.	0.0	17.5	62.2	.	.	.	.	.	.	.	.	.	.	.	.	.	20
6837	THLASPI ARVENSE	S	.	0.0	25.6	28.2	.	.	.	.	.	.	.	.	.	.	.	.	.	20
6838	THLASPI ARVENSE	S	.	0.0	23.8	32.0	.	.	.	.	.	.	.	.	.	.	.	.	.	20
6839	THLASPI ARVENSE	S	.	0.0	24.2	32.9	.	.	.	.	.	.	.	.	.	.	.	.	.	21
6840	THLASPI ARVENSE	S	.	0.0	26.0	28.9	.	.	.	.	.	.	.	.	.	.	.	.	.	21
6841	THLASPI ARVENSE	L	.	85.8	7.7	.	4.7	.	.	.	.	.	.	.	.	.	.	.	270	5
6842	THLASPI PERFOLIATUM	S	.	0.0	23.8	30.8	.	.	.	.	.	.	.	.	.	.	.	.	.	20
6843	THYMBRA SPICATA	S	.	0.0	10.4	6.4	.	.	.	.	.	.	.	.	.	.	.	.	.	21

TABLE 2 - DATA DIRECTLY TRANSCRIBED FROM ORIGINAL SOURCES.

OBS	PLNA	PART	CALORIES	WATER	PROTEIN	FAT	TOTLCARB	FIBER	ASH	CALCIUM	PHOSPHO	IRON	SODIUM	POTASS	CAROT	THIAMINE	RIBOFLAV	NIACIN	ASCORBIC	REFCO
6844	THYMUS CAPITATUS	S	.	0.0	8.9	5.3	.	.	.	.	.	.	.	.	.	.	.	.	.	21
6845	THYMUS CHAUBARDII	S	.	0.0	33.7	37.7	.	.	.	.	.	.	.	.	.	.	.	.	.	21
6846	THYMUS SP.	L	68	78.3	4.3	0.8	14.1	5.5	2.5	56	45	9.2	.	.	.	.	.	.	.	1
6847	THYMUS VULGARIS	H	276	7.8	9.1	7.4	63.9	18.6	11.7	1890	201	123.6	55	814	2280	0.51	0.40	4.94	.	16
6848	THYMUS VULGARIS	S	.	0.0	28.2	37.2	.	.	4.3	.	.	.	.	.	.	.	.	.	.	19
6849	THYMUS VULGARIS	S	.	0.0	25.6	38.9	.	.	.	.	.	.	.	.	.	.	.	.	.	19
6850	THYRALLIS GLAUCA	S	.	0.0	20.6	1.2	.	.	.	.	.	.	.	.	.	.	.	.	.	19
6851	THYSANOCARPUS RADIANS	S	.	0.0	18.4	10.3	.	.	.	.	.	.	.	.	.	.	.	.	.	21
6852	THYSANOLAENA MAXIMA	L	.	.	9.5	5.6	73.2	30.4	11.8	230	110	.	.	.	.	.	.	.	.	5
6853	TIDESTROMIA SP.	W	.	0.0	16.2	1.3	71.4	22.5	11.1	.	.	.	.	.	.	.	.	.	.	6
6854	TILIA X EUROPAEA	L	.	12.0	16.2	2.9	57.5	13.2	11.4	.	.	.	.	.	.	.	.	.	.	5
6855	TILIA X EUROPAEA	F	.	68.1	3.6	3.0	23.2	13.5	2.1	.	.	.	.	.	.	.	.	.	.	5
6856	TILLANDSIA USNEOIDES	W	.	0.0	5.2	2.4	84.5	30.2	7.9	.	.	.	.	.	.	.	.	.	.	6
6857	TINOSPORA CORDIFOLIA	L	.	.	11.2	2.5	78.5	17.5	7.8	1060	570	.	.	.	.	.	.	.	.	5
6858	TITHONIA CALVA	S	.	0.0	16.2	24.6	.	.	.	.	.	.	.	.	.	.	.	.	.	20
6859	TITHONIA CALVA	S	.	0.0	16.2	31.5	.	.	.	.	.	.	.	.	.	.	.	.	.	20
6860	TITHONIA SP.	S	.	0.0	21.2	18.8	.	.	.	.	.	.	.	.	.	.	.	.	.	20
6861	TITHONIA SPECIOSA	S	.	0.0	20.9	18.3	.	.	5.2	.	.	.	.	.	.	.	.	.	.	19
6862	TITHONIA THURBERI	S	.	0.0	22.5	22.2	.	.	.	.	.	.	.	.	.	.	.	.	.	20
6863	TITHONIA THURBERI	S	.	0.0	21.9	22.0	.	.	.	.	.	.	.	.	.	.	.	.	.	20
6864	TITHONIA TUBAEFORMIS	S	.	0.0	22.5	23.0	.	.	.	.	.	.	.	.	.	.	.	.	.	20
6865	TOFIELDIA OCCIDENTALIS	S	.	0.0	31.1	35.0	.	.	.	.	.	.	.	.	.	.	.	.	.	20
6866	TOONA SINENSIS	SH	51	83.6	6.0	1.0	7.9	1.3	1.5	30	102	3.2	.	576	.	.	.	.	.	2
6867	TORDYLIUM AEGYPTIACUM	S	.	0.0	13.6	10.6	.	.	.	.	.	.	.	.	.	.	.	.	.	21
6868	TORILIS ARVENSIS SUBSP. NEGLECTA	S	.	0.0	19.9	23.0	.	.	.	.	.	.	.	.	.	.	.	.	.	21
6869	TORILIS JAPONICA	S	.	0.0	21.2	16.9	.	.	.	.	.	.	.	.	.	.	.	.	.	20
6870	TORILIS LEPTOPHYLLA	S	.	0.0	16.6	10.6	.	.	.	.	.	.	.	.	.	.	.	.	.	21
6871	TORILIS LEPTOPHYLLA	S	.	0.0	16.7	14.4	.	.	.	.	.	.	.	.	.	.	.	.	.	21
6872	TORILIS NODOSA	S	.	0.0	20.6	11.2	.	.	.	.	.	.	.	.	.	.	.	.	.	19
6873	TORREYA NUCIFERA	S	582	6.6	11.1	51.2	28.2	5.7	2.9	40	275	3.6	.	850	25	0.04	0.11	1.90	7	2
6874	TORULARIA TORULOSA	S	.	0.0	37.4	34.3	.	.	.	.	.	.	.	.	.	.	.	.	.	21
6875	TRACHYPOGON MONTUFARII	W	.	0.0	3.5	2.3	88.2	38.0	6.0	240	110	.	.	.	.	.	.	.	.	6
6876	TRACHYPOGON SPICATUS	W	.	0.0	5.7	1.8	82.9	40.2	9.6	.	.	.	.	.	.	.	.	.	.	18
6877	TRACHYSPERMUM AMMI	F	327	9.8	11.3	6.3	57.8	.	14.8	.	.	.	.	.	480	0.43	0.75	.	2	1
6878	TRAGIA SP.	L	.	0.0	16.2	3.2	73.0	19.7	7.6	.	.	.	.	.	.	.	.	.	.	6
6879	TRAGOPOGON DUBIUS	S	.	0.0	16.3	12.1	.	.	.	.	.	.	.	.	.	.	.	.	.	21
6880	TRAGOPOGON DUBIUS	S	.	0.0	18.2	11.6	.	.	.	.	.	.	.	.	.	.	.	.	.	21
6881	TRAGOPOGON MAJOR	S	.	0.0	15.0	14.7	.	.	.	.	.	.	.	.	.	.	.	.	.	20
6882	TRAGOPOGON PORRIFOLIUS	S	.	0.0	17.7	22.0	.	.	.	.	.	.	.	.	.	.	.	.	.	21
6883	TRAGOPOGON PORRIFOLIUS	R	89	76.8	1.4	0.2	20.6	1.8	1.0	48	50	1.4	.	.	0	0.04	0.04	0.30	10	3
6884	TRAGOPOGON PORRIFOLIUS	R	48	77.6	2.9	0.6	18.0	1.8	0.9	47	66	1.5	.	380	6	0.04	0.04	0.30	11	4
6885	TRAGOPOGON PORRIFOLIUS	R	.	76.8	1.4	0.2	20.6	1.8	1.0	48	50	1.4	.	.	.	0.04	0.04	0.30	10	5
6886	TRAGOPOGON PRATENSIS	S	.	0.0	18.8	11.4	.	.	4.2	.	.	.	.	.	.	.	.	.	.	19
6887	TRAGOPOGON PRATENSIS	S	.	0.0	17.5	19.5	.	.	.	.	.	.	.	.	.	.	.	.	.	20
6888	TRAGOPOGON PRATENSIS	S	.	0.0	18.5	9.5	.	.	.	.	.	.	.	.	.	.	.	.	.	20

TABLE 2 - DATA DIRECTLY TRANSCRIBED FROM ORIGINAL SOURCES.

OBS	PLNA	PART	CALORIES	WATER	PROTEIN	FAT	TOTCARB	FIBER	ASH	CALCIUM	PHOSPHO	IRON	SODIUM	POTASS	CAROTE	THIAMINE	RIBOFLAV	NIACIN	ASCORBIC	REFCO
6889	TRAGOPOGON PRATENSIS	S	.	0.0	15.9	15.7	.	.	.	.	.	.	.	.	.	.	.	.	.	21
6890	TRAPA BISPINOSA	F	117	66.4	4.1	0.4	27.8	0.8	1.3	54	114	1.2	21	452	0	0.13	0.06	2.0	7	2
6891	TRAPA BISPINOSA	S	.	70.0	4.7	0.3	23.9	0.6	1.1	20	150	0.8	49	650	.	0.05	0.07	0.6	9	5
6892	TRAPA NATANS	S	.	0.0	11.9	1.0	.	.	.	.	.	.	.	.	.	.	.	.	.	21
6893	TRECULIA AFRICANA	S	377	9.2	12.6	5.6	70.4	1.6	2.2	127	317	.	.	.	.	.	.	.	.	1
6894	TREWIA NUDIFLORA	F	.	81.4	1.4	0.2	15.8	.	1.1	30	60	1.8	.	.	.	.	.	.	.	5
6895	TRIANTHEMA PENTANDRA	W	.	0.0	16.5	2.5	63.1	23.1	17.9	1040	270	.	.	.	.	.	.	.	.	18
6896	TRIANTHEMA PORTULACASTRUM	L	31	87.9	3.5	0.2	5.8	1.2	2.6	211	74	4.3	.	.	6160	0.04	0.12	1.0	30	1
6897	TRIANTHEMA PORTULACASTRUM	L	.	91.3	2.0	0.4	4.1	0.9	2.2	100	30	38.5	.	.	2300	.	.	.	70	5
6898	TRIBULUS TERRESTRIS	L	70	72.0	5.4	0.6	14.6	6.4	7.4	.	95	.	.	.	.	.	.	.	.	1
6899	TRIBULUS TERRESTRIS	S	.	0.0	46.2	29.2	.	.	.	.	.	.	.	.	.	.	.	.	.	21
6900	TRIBULUS TERRESTRIS	S	.	0.0	38.5	37.8	.	.	.	.	.	.	.	.	.	.	.	.	.	21
6901	TRIBULUS TERRESTRIS	N	.	.	12.1	2.6	68.6	27.8	16.7	4210	250	.	.	.	.	.	.	.	.	5
6902	TRIBULUS TERRESTRIS	L	.	79.1	7.2	0.5	.	.	4.6	1550	80	9.2	.	.	.	.	.	.	42	5
6903	TRIBULUS TERRESTRIS	W	.	0.0	18.9	2.7	66.6	25.7	11.8	3350	240	.	.	2840	.	.	.	.	.	6
6904	TRICHACHNE CALIFORNICA	W	.	0.0	8.9	1.9	77.7	36.1	11.5	.	.	.	.	.	.	.	.	.	.	6
6905	TRICHILIA SP.	S	518	6.0	10.4	36.3	43.9	18.8	3.4	272	294	1.3	.	.	.	0.05	0.47	.	13	1
6906	TRICHILIA SP.	S	.	0.0	14.4	53.8	.	.	4.0	.	.	.	.	.	.	.	.	.	.	19
6907	TRICHOSANTHES ANGUINA	F	.	94.6	0.5	0.3	4.1	0.8	0.5	50	20	1.1	25	34	96	0.04	0.06	0.3	0	5
6908	TRICHOSANTHES CUCUMERINA	F	18	94.3	0.9	0.0	4.5	0.5	0.3	19	34	1.0	.	123	810	0.04	0.05	0.7	6	2
6909	TRICHOSANTHES CUCUMERINA	S	.	0.0	33.8	56.3	.	.	.	.	.	.	.	.	.	.	.	.	.	21
6910	TRICHOSANTHES DIOICA	F	.	92.0	2.0	0.3	5.2	3.0	0.5	30	40	1.7	3	83	153	0.05	0.06	0.5	29	5
6911	TRICHOSANTHES DIOICA	L	.	80.5	5.4	1.1	10.0	4.2	3.0	531	73	.	.	.	.	.	.	.	.	5
6912	TRICOSCYPHA SP.	F	61	83.4	0.7	0.5	15.1	.	0.3	42	60	2.0	.	.	.	.	.	.	75	1
6913	TRIFOLIUM AGRARIUM	W	.	0.0	16.7	3.7	73.2	20.5	6.4	1170	310	25.0	.	1850	.	.	.	.	.	6
6914	TRIFOLIUM ALEXANDRINUM	W	.	0.0	26.7	3.0	54.7	14.9	15.6	1480	310	.	.	.	.	.	.	.	.	18
6915	TRIFOLIUM ALEXANDRINUM	H	.	0.0	16.6	2.4	69.9	26.2	11.1	1310	210	.	.	.	.	.	.	.	.	18
6916	TRIFOLIUM ALEXANDRINUM	S	.	0.0	34.2	5.3	.	.	.	.	.	.	.	.	.	.	.	.	.	21
6917	TRIFOLIUM ALEXANDRINUM	W	.	0.0	19.2	3.7	60.2	22.1	16.9	3560	320	.	.	2390	.	.	.	.	.	6
6918	TRIFOLIUM ANGUSTIFOLIUM	S	.	0.0	45.8	7.9	.	.	.	.	.	.	.	.	.	.	.	.	.	21
6919	TRIFOLIUM ARMENIUM	S	.	0.0	39.8	5.6	.	.	.	.	.	.	.	.	.	.	.	.	.	21
6920	TRIFOLIUM ARVENSE	S	.	0.0	38.2	6.1	.	.	.	.	.	.	.	.	.	.	.	.	.	21
6921	TRIFOLIUM CHERLERI	S	.	0.0	35.8	5.6	.	.	.	.	.	.	.	.	.	.	.	.	.	21
6922	TRIFOLIUM CLYPEATUM	S	.	0.0	29.8	7.9	.	.	.	.	.	.	.	.	.	.	.	.	.	21
6923	TRIFOLIUM DUBIUM	S	.	0.0	39.5	9.8	.	.	.	.	.	.	.	.	.	.	.	.	.	21
6924	TRIFOLIUM FRAGIFERUM	S	.	0.0	31.4	7.9	.	.	3.5	.	.	.	.	.	.	.	.	.	.	19
6925	TRIFOLIUM FRAGIFERUM	S	.	0.0	32.4	7.4	.	.	.	.	.	.	.	.	.	.	.	.	.	21
6926	TRIFOLIUM GLOMERATUM	S	.	0.0	35.4	5.4	.	.	3.5	.	.	.	.	.	.	.	.	.	.	19
6927	TRIFOLIUM GLOMERATUM	S	.	0.0	24.0	9.6	.	.	9.3	.	.	.	.	.	.	.	.	.	.	19
6928	TRIFOLIUM HIRTUM	S	.	0.0	34.0	4.2	.	.	3.6	.	.	.	.	.	.	.	.	.	.	19
6929	TRIFOLIUM HIRTUM	S	.	0.0	37.5	4.8	.	.	.	.	.	.	.	.	.	.	.	.	.	21
6930	TRIFOLIUM HYBRIDUM	S	.	0.0	36.6	5.2	.	.	4.1	.	.	.	.	.	.	.	.	.	.	19
6931	TRIFOLIUM HYBRIDUM	S	.	0.0	35.4	5.5	.	.	4.4	.	.	.	.	.	.	.	.	.	.	19
6932	TRIFOLIUM HYBRIDUM	W	.	0.0	17.5	3.1	69.4	25.3	10.0	1270	300	43.0	.	2700	.	.	.	.	.	6
6933	TRIFOLIUM INCARNATUM	W	.	0.0	18.9	3.2	66.4	14.8	11.5	1320	200	.	.	.	.	.	.	.	.	18

TABLE 2 - DATA DIRECTLY TRANSCRIBED FROM ORIGINAL SOURCES.

OBS	PLANA	PART	CALORIES	WATER	PROTEIN	FAT	TOTCARB	FIBER	ASH	CALCIUM	PHOSPHO	IRON	SODIUM	POTASS	CAROT	THIAMINE	RIBOFLAV	NIACIN	ASCORBIC	REFCO
6934	TRIFOLIUM INCARNATUM	S	.	0.0	43.5	3.9	.	.	4.1	.	.	.	.	.	.	.	.	.	.	19
6935	TRIFOLIUM INCARNATUM	S	.	0.0	38.4	4.7	.	.	4.1	.	.	.	.	.	.	.	.	.	.	19
6936	TRIFOLIUM INCARNATUM	L	.	0.0	4.2	0.3	14.9	5.0	1.6	160	70	.	.	.	.	.	.	.	.	5
6937	TRIFOLIUM INCARNATUM	W	.	0.0	16.7	3.1	70.9	28.3	9.3	1620	350	70.0	.	2410	.	.	.	.	.	6
6938	TRIFOLIUM LAPPACEUM	H	.	0.0	20.5	3.2	66.3	25.6	10.0	1340	230	63.0	.	.	.	.	.	.	.	6
6939	TRIFOLIUM LUPINASTER	S	.	0.0	32.3	6.0	.	.	4.3	.	.	.	.	.	.	.	.	.	.	19
6940	TRIFOLIUM LUPINASTER	S	.	0.0	30.8	6.3	.	.	4.0	.	.	.	.	.	.	.	.	.	.	19
6941	TRIFOLIUM MEDIUM	S	.	0.0	38.9	9.0	.	.	4.1	.	.	.	.	.	.	.	.	.	.	19
6942	TRIFOLIUM MEDIUM	S	.	0.0	39.1	9.1	.	.	4.0	.	.	.	.	.	.	.	.	.	.	19
6943	TRIFOLIUM MEDIUM	S	.	0.0	37.5	8.6	.	.	.	.	.	.	.	.	.	.	.	.	.	19
6944	TRIFOLIUM NIGRESCENS	S	.	0.0	31.6	6.8	.	.	11.3	.	.	.	.	.	.	.	.	.	.	19
6945	TRIFOLIUM NIGRESCENS SUBSP. PETRISAVII	S	.	0.0	34.3	5.0	.	.	.	.	.	.	.	.	.	.	.	.	.	21
6946	TRIFOLIUM PALAESTINUM	S	.	0.0	43.6	9.9	.	.	.	.	.	.	.	.	.	.	.	.	.	21
6947	TRIFOLIUM PILULARE	S	.	0.0	38.4	12.4	.	.	.	.	.	.	.	.	.	.	.	.	.	21
6948	TRIFOLIUM PRATENSE	W	.	0.0	20.7	5.3	63.0	22.5	11.0	1420	320	.	.	.	.	.	.	.	.	18
6949	TRIFOLIUM PRATENSE	H	.	0.0	9.8	2.9	81.7	33.8	5.6	980	130	.	.	.	.	.	.	.	.	18
6950	TRIFOLIUM PRATENSE	S	.	0.0	36.2	8.5	.	.	4.1	.	.	.	.	.	.	.	.	.	.	19
6951	TRIFOLIUM PRATENSE	S	.	0.0	38.1	8.6	.	.	.	.	.	.	.	.	.	.	.	.	.	21
6952	TRIFOLIUM PRATENSE	L	.	0.0	13.5	1.9	76.4	27.8	8.1	1270	220	.	.	.	.	.	.	.	.	5
6953	TRIFOLIUM PRATENSE	H	.	0.0	14.9	2.9	74.3	30.1	7.9	1610	220	13.0	.	1760	.	.	.	.	.	6
6954	TRIFOLIUM PROCUMBENS	S	.	0.0	39.1	7.5	.	.	3.6	.	.	.	.	.	.	.	.	.	.	19
6955	TRIFOLIUM PROCUMBENS	S	.	0.0	30.9	6.1	.	.	4.1	.	.	.	.	.	.	.	.	.	.	19
6956	TRIFOLIUM PURPUREUM	S	.	0.0	39.9	12.8	.	.	.	.	.	.	.	.	.	.	.	.	.	21
6957	TRIFOLIUM REPENS	W	.	0.0	21.0	1.4	68.1	23.8	9.5	510	350	.	.	.	.	.	.	.	.	18
6958	TRIFOLIUM REPENS	S	.	0.0	35.5	6.0	.	.	3.8	.	.	.	.	.	.	.	.	.	.	19
6959	TRIFOLIUM REPENS	S	.	0.0	33.9	6.2	.	.	4.5	.	.	.	.	.	.	.	.	.	.	19
6960	TRIFOLIUM REPENS	S	.	0.0	36.1	6.2	.	.	4.0	.	.	.	.	.	.	.	.	.	.	19
6961	TRIFOLIUM REPENS	S	.	0.0	34.8	6.9	.	.	.	.	.	.	.	.	.	.	.	.	.	21
6962	TRIFOLIUM REPENS	L	.	0.0	25.8	3.4	59.2	18.6	11.4	1850	340	.	.	.	.	.	.	.	.	5
6963	TRIFOLIUM REPENS	W	.	0.0	27.8	3.2	56.9	15.6	12.1	1480	460	34.0	.	2200	.	.	.	.	.	6
6964	TRIFOLIUM REPENS "LADINO"	H	.	0.0	23.1	3.4	64.0	19.2	9.5	1380	400	65.0	.	2170	.	.	.	.	.	6
6965	TRIFOLIUM REPENS "LADINO"	W	.	0.0	25.1	4.8	58.7	14.5	11.4	1330	370	37.0	.	2190	.	.	.	.	.	6
6966	TRIFOLIUM REPENS "WHITE"	H	.	0.0	22.0	2.6	65.3	23.3	10.1	1240	390	106.0	.	2030	.	.	.	.	.	6
6967	TRIFOLIUM RESUPINATUM	W	.	0.0	27.8	4.0	52.1	15.6	16.1	1830	430	.	.	.	.	.	.	.	.	18
6968	TRIFOLIUM RESUPINATUM	S	.	0.0	36.0	7.0	.	.	4.3	.	.	.	.	.	.	.	.	.	.	19
6969	TRIFOLIUM RESUPINATUM	S	.	0.0	36.8	7.1	.	.	.	.	.	.	.	.	.	.	.	.	.	21
6970	TRIFOLIUM RESUPINATUM	H	.	0.0	16.3	1.8	71.6	30.8	10.3	1660	230	.	.	.	.	.	.	.	.	6
6971	TRIFOLIUM SQUARROSUM	S	.	0.0	32.2	3.3	.	.	3.6	.	.	.	.	.	.	.	.	.	.	19
6972	TRIFOLIUM STELLATUM	S	.	0.0	47.7	10.5	.	.	.	.	.	.	.	.	.	.	.	.	.	21
6973	TRIFOLIUM STRIATUM	S	.	0.0	28.5	1.6	.	.	3.6	.	.	.	.	.	.	.	.	.	.	19
6974	TRIFOLIUM SUBTERRANEUM	W	.	0.0	12.5	4.4	73.3	28.7	9.8	1490	330	.	.	.	.	.	.	.	.	18
6975	TRIFOLIUM SUBTERRANEUM	H	.	0.0	8.0	2.2	81.5	36.9	8.3	730	230	.	.	.	.	.	.	.	.	18
6976	TRIFOLIUM SUBTERRANEUM	S	.	0.0	36.4	15.2	.	.	4.9	.	.	.	.	.	.	.	.	.	.	19
6977	TRIFOLIUM SUBTERRANEUM	S	.	0.0	38.6	16.2	.	.	.	.	.	.	.	.	.	.	.	.	.	21
6978	TRIFOLIUM TRIDENTATUM	W	.	0.0	14.3	.	.	24.8	10.2	.	.	.	.	.	.	.	.	.	.	6

TABLE 2 - DATA DIRECTLY TRANSCRIBED FROM ORIGINAL SOURCES.

OBS	PLNA	PART	CALORIES	WATER	PROTEIN	FAT	TOTLCARB	FIBER	ASH	CALCIUM	PHOSPHO	IRON	SODIUM	POTASS	CAROT	THIAMINE	RIBOFLAV	NIACIN	ASCORBIC	REFCO
6979	TRIGLOCHIN BULBOSUM	S	.	0.0	37.0	28.7	.	.	.	.	.	.	.	.	.	.	.	.	.	21
6980	TRIGONELLA ARABICA	S	.	0.0	28.3	3.3	.	.	.	.	.	.	.	.	.	.	.	.	.	21
6981	TRIGONELLA BALANSAE	S	.	0.0	35.2	6.3	.	.	4.7	.	.	.	.	.	.	.	.	.	.	19
6982	TRIGONELLA COERULEA	S	.	0.0	27.8	6.5	.	.	4.4	.	.	.	.	.	.	.	.	.	.	19
6983	TRIGONELLA FOENUM-GRAECUM	S	323	8.8	23.0	6.4	58.4	10.1	3.4	176	296	33.5	67	770	.	0.32	0.37	1.64	3	16
6984	TRIGONELLA FOENUM-GRAECUM	W	.	0.0	15.7	2.1	73.4	31.1	8.8	1470	220	.	.	.	.	.	.	.	.	18
6985	TRIGONELLA FOENUM-GRAECUM	S	.	0.0	27.2	7.8	61.0	1.0	4.0	880	300	.	.	.	.	.	.	.	.	18
6986	TRIGONELLA FOENUM-GRAECUM	F	.	0.0	11.3	2.9	82.8	4.2	3.0	660	140	.	.	.	.	.	.	.	.	18
6987	TRIGONELLA FOENUM-GRAECUM	H	.	0.0	10.0	2.2	77.0	35.1	10.8	.	.	.	.	.	.	.	.	.	.	18
6988	TRIGONELLA FOENUM-GRAECUM	S	.	0.0	28.9	5.0	.	.	3.7	.	.	.	.	.	.	.	.	.	.	19
6989	TRIGONELLA FOENUM-GRAECUM	S	.	0.0	32.0	5.4	.	.	4.1	.	.	.	.	.	.	.	.	.	.	19
6990	TRIGONELLA FOENUM-GRAECUM	S	.	0.0	27.9	5.4	.	.	3.8	.	.	.	.	.	.	.	.	.	.	19
6991	TRIGONELLA FOENUM-GRAECUM	S	.	0.0	29.4	5.3	.	.	3.8	.	.	.	.	.	.	.	.	.	.	19
6992	TRIGONELLA FOENUM-GRAECUM	S	.	0.0	29.4	5.3	.	.	4.5	.	.	.	.	.	.	.	.	.	.	19
6993	TRIGONELLA FOENUM-GRAECUM	S	.	0.0	30.3	6.0	.	.	3.5	.	.	.	.	.	.	.	.	.	.	19
6994	TRIGONELLA FOENUM-GRAECUM	S	.	0.0	31.8	5.9	.	.	4.4	.	.	.	.	.	.	.	.	.	.	19
6995	TRIGONELLA FOENUM-GRAECUM	S	.	0.0	32.5	6.5	.	.	3.9	.	.	.	.	.	.	.	.	.	.	19
6996	TRIGONELLA FOENUM-GRAECUM	S	.	0.0	29.5	5.2	.	.	4.3	.	.	.	.	.	.	.	.	.	.	19
6997	TRIGONELLA FOENUM-GRAECUM	S	.	0.0	27.2	5.0	.	.	3.4	.	.	.	.	.	.	.	.	.	.	19
6998	TRIGONELLA FOENUM-GRAECUM	S	.	0.0	28.7	5.5	.	.	3.2	.	.	.	.	.	.	.	.	.	.	19
6999	TRIGONELLA FOENUM-GRAECUM	S	.	0.0	28.3	5.5	.	.	3.2	.	.	.	.	.	.	.	.	.	.	19
7000	TRIGONELLA FOENUM-GRAECUM	S	.	0.0	31.0	5.5	.	.	3.2	.	.	.	.	.	.	.	.	.	.	19
7001	TRIGONELLA FOENUM-GRAECUM	S	.	0.0	30.4	5.4	.	.	5.0	.	.	.	.	.	.	.	.	.	.	19
7002	TRIGONELLA FOENUM-GRAECUM	S	.	0.0	28.8	7.3	.	.	2.4	.	.	.	.	.	.	.	.	.	.	19
7003	TRIGONELLA FOENUM-GRAECUM	S	.	0.0	30.3	6.8	.	.	3.0	.	.	.	.	.	.	.	.	.	.	19
7004	TRIGONELLA FOENUM-GRAECUM	S	.	0.0	27.4	6.9	.	.	3.2	.	.	.	.	.	.	.	.	.	.	19
7005	TRIGONELLA FOENUM-GRAECUM	S	.	0.0	27.6	6.2	.	.	3.6	.	.	.	.	.	.	.	.	.	.	19
7006	TRIGONELLA FOENUM-GRAECUM	S	.	0.0	39.8	6.5	.	.	4.1	.	.	.	.	.	.	.	.	.	.	19
7007	TRIGONELLA FOENUM-GRAECUM	S	.	0.0	34.1	6.0	.	.	3.5	.	.	.	.	.	.	.	.	.	.	19
7008	TRIGONELLA FOENUM-GRAECUM	S	.	0.0	32.8	6.3	.	.	4.2	.	.	.	.	.	.	.	.	.	.	19
7009	TRIGONELLA FOENUM-GRAECUM	S	.	0.0	26.1	6.3	.	.	3.5	.	.	.	.	.	.	.	.	.	.	19
7010	TRIGONELLA FOENUM-GRAECUM	S	.	0.0	26.2	4.9	.	.	3.6	.	.	.	.	.	.	.	.	.	.	19
7011	TRIGONELLA FOENUM-GRAECUM	S	.	0.0	26.2	6.8	.	.	.	.	.	.	.	.	.	.	.	.	.	20
7012	TRIGONELLA FOENUM-GRAECUM	S	.	13.7	26.2	5.8	51.3	7.2	3.0	160	370	14.1	19	530	96	0.34	0.29	1.10	.	5
7013	TRIGONELLA FOENUM-GRAECUM	N	.	0.0	26.2	5.6	57.3	13.6	11.0	1290	460	.	.	.	.	.	.	.	.	5
7014	TRIGONELLA FOENUM-GRAECUM	L	.	86.1	4.4	0.9	7.1	1.1	1.5	395	51	16.5	76	31	2340	0.04	0.31	0.80	52	5
7015	TRIGONELLA GLADIATA	S	.	0.0	26.4	3.3	.	.	.	.	.	.	.	.	.	.	.	.	.	21
7016	TRIGONELLA KOTSCHYI	S	.	0.0	27.6	5.1	.	.	7.4	.	.	.	.	.	.	.	.	.	.	19
7017	TRIODIA ALBESCENS	W	.	0.0	9.7	2.0	78.1	33.7	10.2	520	200	.	.	1610	.	.	.	.	.	6
7018	TRIODIA FLAVA	S	.	0.0	36.4	3.1	.	.	.	.	.	.	.	.	.	.	.	.	.	21
7019	TRIODIA GRANDIFLORA	W	.	.	.	.	.	.	.	910	100	.	.	.	.	.	.	.	.	6
7020	TRIODIA MUTICA	W	.	.	.	.	.	.	.	1010	80	.	.	.	.	.	.	.	.	6
7021	TRIODIA PULCHELLA	W	.	0.0	7.6	1.4	68.3	29.1	22.7	1190	60	.	.	.	.	.	.	.	.	6
7022	TRIODIA TEXANA	W	.	0.0	12.2	2.0	75.1	33.4	10.7	440	170	.	.	.	.	.	.	.	.	6
7023	TRIPSACUM DACTYLOIDES	W	.	0.0	8.5	2.0	79.3	30.2	10.2	620	310	.	.	.	.	.	.	.	.	6

TABLE 2 - DATA DIRECTLY TRANSCRIBED FROM ORIGINAL SOURCES.

OBS	PLNA	PART	CALORIES	WATER	PROTEIN	FAT	TOTLCARB	FIBER	ASH	CALCIUM	PHOSPHO	IRON	SODIUM	POTASS	CAROT	THIAMINE	RIBOFLAV	NIACIN	ASCORBIC	REFCO
7024	TRIPSACUM FASCICULATUM	W	.	0.0	15.9	2.8	71.7	31.4	9.6	.	.	.	.	.	.	.	.	.	.	18
7025	TRIPSACUM FASCICULATUM	L	.	0.0	6.1	3.4	85.2	40.4	5.3	.	.	.	.	.	.	.	.	.	.	18
7026	TRIPSACUM FASCICULATUM	S	.	0.0	4.6	1.9	87.1	36.1	6.4	.	.	.	.	.	.	.	.	.	.	18
7027	TRIPSACUM LATIFOLIUM	W	.	0.0	9.3	0.7	83.0	32.5	7.0	.	.	.	.	.	.	.	.	.	.	18
7028	TRIPSACUM LAXUM	W	.	0.0	6.4	1.1	80.1	32.8	12.4	.	.	.	.	.	.	.	.	.	.	6
7029	TRISETUM INTERRUPTUM	W	.	0.0	11.2	1.8	76.3	31.3	10.7	.	.	.	.	.	.	.	.	.	.	6
7030	TRITICUM AEGILOPOIDES	S	.	0.0	25.8	3.2	.	.	.	.	.	.	.	.	.	.	.	.	.	21
7031	TRITICUM AESTIVUM	W	.	0.0	18.1	3.7	70.6	15.5	7.6	.	.	.	.	.	.	.	.	.	.	18
7032	TRITICUM AESTIVUM	H	.	0.0	3.1	1.1	85.6	45.4	10.2	.	.	.	.	.	.	.	.	.	.	18
7033	TRITICUM AESTIVUM	S	.	0.0	13.5	2.1	81.1	3.3	3.3	30	480	.	.	.	.	.	.	.	.	18
7034	TRITICUM AESTIVUM	S	330	12.5	12.3	1.8	71.7	2.3	1.7	46	354	3.4	.	.	0	0.52	0.12	4.3	0	3
7035	TRITICUM AESTIVUM	S	330	13.0	14.0	2.2	69.1	2.3	1.7	36	383	3.1	.	.	0	0.57	0.12	4.3	0	3
7036	TRITICUM AESTIVUM	S	326	14.0	10.2	2.0	72.1	2.3	1.7	42	400	3.5	.	.	0	0.43	0.11	3.6	0	3
7037	TRITICUM AESTIVUM	S	330	13.0	14.0	2.2	69.1	2.3	1.7	36	383	3.1	3	370	0	0.57	0.12	4.3	0	4
7038	TRITICUM AESTIVUM	W	.	0.0	22.6	3.8	61.7	20.8	11.9	400	350	29.0	.	3370	.	.	.	.	.	6
7039	TRITICUM AESTIVUM	H	.	0.0	3.6	1.7	86.6	41.5	8.1	170	80	15.0	.	1110	.	.	.	.	.	6
7040	TRITICUM DICOCCON	S	333	12.3	12.5	2.4	71.0	2.7	1.8	38	360	4.7	.	.	0	0.39	0.13	4.7	.	1
7041	TRITICUM DICOCCUM	H	.	0.0	10.8	2.2	76.9	36.4	10.1	.	.	.	.	.	.	.	.	.	.	6
7042	TRITICUM DURUM	S	322	14.0	12.8	1.6	69.5	2.4	2.1	48	300	.	.	.	0	0.22	0.10	3.0	.	1
7043	TRITICUM DURUM	S	332	12.5	11.6	2.2	72.1	2.1	1.6	48	382	3.3	5	500	0	0.37	0.12	4.6	0	2
7044	TRITICUM DURUM	S	.	0.0	15.0	2.2	80.8	2.5	2.0	.	.	.	.	.	.	.	.	.	.	6
7045	TRITICUM SPELTA	S	.	0.0	13.3	2.2	80.6	9.1	3.9	.	.	.	.	.	.	.	.	.	.	6
7046	TRIUMFETTA BARTRAMIA	S	.	0.0	23.8	17.2	.	.	.	.	.	.	.	.	.	.	.	.	.	21
7047	TRIUMFETTA BREVIPES	S	.	0.0	30.6	22.8	.	.	.	.	.	.	.	.	.	.	.	.	.	20
7048	TRIUMFETTA POLYANDRA	S	.	0.0	16.8	9.8	.	.	.	.	.	.	.	.	.	.	.	.	.	21
7049	TRIUMFETTA SP.	L	68	78.4	4.2	0.4	15.2	3.4	1.8	392	76	29.2	.	.	.	.	.	.	.	1
7050	TRIXIS RADIALIS	S	.	0.0	23.8	6.4	.	.	.	.	.	.	.	.	.	.	.	.	.	20
7051	TROPAEOLUM SP.	S	.	0.0	26.2	10.2	.	.	.	.	.	.	.	.	.	.	.	.	.	20
7052	TROPAEOLUM TUBEROSUM	R	52	86.0	1.6	0.6	11.0	0.8	0.8	7	42	1.2	.	.	30	0.06	0.08	0.6	67	3
7053	TULIPA SP.	S	.	0.0	20.8	3.8	.	.	.	.	.	.	.	.	.	.	.	.	.	21
7054	TURGENIA LATIFOLIA	S	.	0.0	14.5	19.5	.	.	.	.	.	.	.	.	.	.	.	.	.	21
7055	TURGENIA LATIFOLIA	S	.	0.0	11.6	5.3	.	.	.	.	.	.	.	.	.	.	.	.	.	21
7056	TYPHONIUM TRILOBATUM	R	.	69.9	1.4	0.1	27.0	1.0	1.6	35	20	1.3	9	237	78	0.07	.	0.7	.	5
7057	UAPACA GUINEENSIS	F	92	73.4	1.8	0.1	23.8	.	0.9	125	.	.	.	.	.	.	.	.	.	1
7058	UAPACA KIRKIANA	F	.	76.0	.	.	.	.	.	.	.	.	.	.	.	.	.	.	2	1
7059	ULEX EUROPAEUS	S	.	0.0	30.6	11.7	.	.	4.3	.	.	.	.	.	.	.	.	.	.	19
7060	ULEX EUROPAEUS	S	.	0.0	37.3	12.5	.	.	.	.	.	.	.	.	.	.	.	.	.	21
7061	ULEX EUROPAEUS	L	.	52.7	4.6	0.9	40.3	14.3	1.6	.	.	.	.	.	.	.	.	.	.	5
7062	ULLUCUS TUBEROSUS	R	51	85.9	1.0	0.0	12.5	0.6	0.6	3	35	0.8	.	.	0	0.04	0.02	0.3	23	3
7063	ULMUS AMERICANA	S	.	0.0	42.9	34.5	.	.	4.7	.	.	.	.	.	.	.	.	.	.	19
7064	ULMUS AMERICANA	S	.	0.0	13.4	5.9	.	.	13.2	.	.	.	.	.	.	.	.	.	.	19
7065	ULMUS PUMILA	S	.	0.0	43.8	27.7	.	.	.	.	.	.	.	.	.	.	.	.	.	20
7066	UNGNADIA SPECIOSA	W	.	0.0	12.8	2.4	78.4	23.4	6.4	2420	170	.	.	1230	.	.	.	.	.	6
7067	URENA LOBATA	L	54	81.8	3.2	0.1	12.8	1.8	2.1	558	67	.	.	.	.	.	.	.	62	1
7068	URENA LOBATA	S	.	0.0	28.8	19.9	.	.	.	.	.	.	.	.	.	.	.	.	.	19

TABLE 2 - DATA DIRECTLY TRANSCRIBED FROM ORIGINAL SOURCES.

OBS	PLANA	PART	CALORIES	WATER	PROTEIN	FAT	TOTCARB	FIBER	ASH	CALCIUM	PHOSPHO	IRON	SODIUM	POTASS	CAROT	THIAMINE	RIBOFLAV	NIACIN	ASCORBIC	REFCO
7069	URERA OBOVATA	L	53	81.2	2.8	0.6	11.6	2.7	3.8	.	41	.	.	.	.	.	.	.	.	1
7070	UROCHLOA BOLBODES	W	.	0.0	13.7	2.8	72.9	30.0	10.6	.	.	.	.	.	.	.	.	.	.	18
7071	UROCHLOA BOLBODES	H	.	0.0	17.7	2.0	67.4	24.6	12.9	.	.	.	.	.	.	.	.	.	.	18
7072	UROCHLOA MOSAMBICENSIS	W	.	0.0	11.0	2.0	58.8	22.4	28.2	.	.	.	.	.	.	.	.	.	.	18
7073	UROCHLOA MOSAMBICENSIS	H	.	0.0	14.9	1.5	70.7	26.1	12.9	.	.	.	.	.	.	.	.	.	.	18
7074	UROCHLOA PANICOIDES	W	.	0.0	14.7	1.7	68.7	29.9	14.9	.	.	.	.	.	.	.	.	.	.	18
7075	UROCHLOA PANICOIDES	L	.	0.0	14.7	1.7	67.7	29.9	14.9	480	250	.	.	.	.	.	.	.	.	5
7076	UROCHLOA PULLULANS	L	.	0.0	5.8	1.4	79.7	33.4	13.2	510	310	.	.	.	.	.	.	.	.	5
7077	UROSPERMUM PICROIDES	S	.	0.0	13.8	16.9	.	.	.	.	.	.	.	.	.	.	.	.	.	21
7078	URSINIA ANETHEMOIDES	S	.	0.0	23.7	13.8	.	.	5.3	.	.	.	.	.	.	.	.	.	.	19
7079	URSINIA ANETHEMOIDES	S	.	0.0	13.4	7.5	.	.	5.6	.	.	.	.	.	.	.	.	.	.	19
7080	URSINIA ANETHEMOIDES	S	.	0.0	22.5	14.0	.	.	.	.	.	.	.	.	.	.	.	.	.	19
7081	URSINIA CAKILEFOLIA	S	.	0.0	24.8	13.6	.	.	.	.	.	.	.	.	.	.	.	.	.	21
7082	URTICA DIOICA	S	.	0.0	15.5	23.1	.	.	.	.	.	.	.	.	.	.	.	.	.	21
7083	URTICA DIOICA VAR. PROCERA	S	.	0.0	20.1	38.2	.	.	.	.	.	.	.	.	.	.	.	.	.	21
7084	UVARIA RUFA	F	35	90.2	0.7	0.4	8.1	1.5	0.6	23	15	0.6	.	.	0	0.07	0.03	0.7	12	2
7085	VACCARIA PYRAMIDATA	S	.	0.0	13.9	3.2	.	.	.	.	.	.	.	.	.	.	.	.	.	21
7086	VACCARIA PYRAMIDATA	S	.	0.0	13.8	1.6	.	.	.	.	.	.	.	.	.	.	.	.	.	21
7087	VACCARIA PYRAMIDATA	S	.	0.0	16.1	2.9	.	.	.	.	.	.	.	.	.	.	.	.	.	21
7088	VACCINIUM FLORIBUNDUM	F	76	79.6	0.8	0.8	18.5	2.6	0.3	20	16	0.9	.	.	70	0.01	0.01	0.7	8	3
7089	VACCINIUM MACROCARPON	F	46	87.9	0.4	0.7	10.8	1.4	0.2	14	10	0.5	2	82	25	0.03	0.02	0.1	11	2
7090	VACCINIUM OVATUM(DRY)	F	352	0.0	6.0	0.0	92.0	.	2.0	174	.	2.0	.	.	.	.	.	.	384	15
7091	VACCINIUM OVATUM(FRESH)	F	360	0.0	8.0	1.0	89.0	.	1.0	156	.	2.0	.	.	.	.	.	.	346	15
7092	VACCINIUM PARVIFOLIUM	F	394	0.0	16.0	9.0	72.0	.	2.0	216	.	4.0	.	.	.	.	.	.	354	15
7093	VACCINIUM RETICULATUM	F	36	90.1	0.4	0.2	9.1	1.3	0.3	7	10	0.1	.	.	500	0.02	0.04	0.3	6	2
7094	VACCINIUM SP.	F	62	83.2	0.7	0.5	15.3	1.5	0.3	15	13	1.0	1	81	60	0.03	0.06	0.5	14	4
7095	VACCINIUM SYMPLOCIFOLIUM	F	.	79.5	0.8	0.6	18.8	7.3	0.3	20	10	1.4	.	.	48	.	.	.	.	5
7096	VALERIANA OFFICINALIS	S	.	0.0	19.9	30.0	.	.	5.8	.	.	.	.	.	.	.	.	.	.	19
7097	VALERIANA OFFICINALIS	S	.	0.0	19.4	34.4	.	.	.	.	.	.	.	.	.	.	.	.	.	19
7098	VALERIANELLA LOCUSTA	L	21	92.8	2.0	0.4	3.6	0.8	1.2	.	.	.	.	.	.	.	.	.	.	4
7099	VALERIANELLA RADIATA	S	.	0.0	15.6	21.9	.	.	.	.	.	.	.	.	.	.	.	.	.	20
7100	VALERIANELLA VESICARIA	S	.	0.0	9.6	10.1	.	.	.	.	.	.	.	.	.	.	.	.	.	21
7101	VALERIANOPSIS POLYSTACHYA	S	.	0.0	18.4	30.5	.	.	.	.	.	.	.	.	.	.	.	.	.	21
7102	VALLISNERIA SPIRALIS	L	.	92.2	.	.	.	.	1.1	90	250	11.0	.	.	.	.	.	.	.	5
7103	VANGUERIA INFAUSTA	F	.	53.1	.	.	.	.	.	.	.	.	.	.	.	.	.	.	0	1
7104	VANILLA SP.	S	.	0.0	10.0	21.2	62.8	26.3	6.0	1900	70	.	.	.	.	.	.	.	.	6
7105	VERBASCUM COROMANDELICUM	S	.	0.0	21.2	28.7	.	.	.	.	.	.	.	.	.	.	.	.	.	21
7106	VERBASCUM PULVERULENTUM	S	.	0.0	22.5	35.5	.	.	.	.	.	.	.	.	.	.	.	.	.	21
7107	VERBASCUM SINUATUM	S	.	0.0	19.4	38.3	.	.	.	.	.	.	.	.	.	.	.	.	.	21
7108	VERBASCUM SINUATUM	S	.	0.0	21.0	34.2	.	.	.	.	.	.	.	.	.	.	.	.	.	21
7109	VERBASCUM SINUATUM	S	.	0.0	21.1	38.4	.	.	.	.	.	.	.	.	.	.	.	.	.	21
7110	VERBASCUM SP.	S	.	0.0	20.0	30.1	.	.	.	.	.	.	.	.	.	.	.	.	.	20
7111	VERBASCUM THAPSUS	S	.	0.0	18.3	27.1	.	.	.	.	.	.	.	.	.	.	.	.	.	20
7112	VERBASCUM THAPSUS	S	.	0.0	17.8	26.8	.	.	.	.	.	.	.	.	.	.	.	.	.	21
7113	VERBENA BIPINNATIFIDA	S	.	0.0	15.2	17.5	.	.	.	.	.	.	.	.	.	.	.	.	.	20

TABLE 2 - DATA DIRECTLY TRANSCRIBED FROM ORIGINAL SOURCES.

OBS	PLANA	PART	CALORIES	WATER	PROTEIN	FAT	TOTCARB	FIBER	ASH	CALCIUM	PHOSPHO	IRON	SODIUM	POTASS	CAROTENE	THIAMIN	RIBOFLAV	NIACIN	ASCORBIC	REFCO
7114	VERBENA BIPINNATIFIDA	L	.	0.0	11.5	1.8	78.3	18.1	8.4	1990	180	.	.	1600	.	.	.	.	.	6
7115	VERBENA BRACTEATA	S	.	0.0	9.8	16.9	.	.	.	.	.	.	.	.	.	.	.	.	.	21
7116	VERBENA BRASILIENSIS	S	.	0.0	12.9	19.9	.	.	.	.	.	.	.	.	.	.	.	.	.	21
7117	VERBENA CAROLINA	S	.	0.0	8.9	11.2	.	.	.	.	.	.	.	.	.	.	.	.	.	21
7118	VERBENA MACDOUGALII	S	.	0.0	11.6	20.0	.	.	.	.	.	.	.	.	.	.	.	.	.	21
7119	VERBENA RIGIDA	S	.	0.0	11.2	14.6	.	.	.	.	.	.	.	.	.	.	.	.	.	19
7120	VERBENA RIGIDA	S	.	0.0	10.6	16.2	.	.	.	.	.	.	.	.	.	.	.	.	.	20
7121	VERBENA SP.	L	.	0.0	11.5	3.0	75.7	15.6	9.8	2460	170	.	.	1590	.	.	.	.	.	6
7122	VERBENA STRICTA	S	.	0.0	8.9	13.8	.	.	5.1	.	.	.	.	.	.	.	.	.	.	19
7123	VERBESINA ENCELIOIDES	S	.	0.0	28.8	13.1	.	.	.	.	.	.	.	.	.	.	.	.	.	19
7124	VERBESINA ENCELIOIDES VAR. EXAURICULATA	S	.	0.0	30.0	24.5	.	.	.	.	.	.	.	.	.	.	.	.	.	20
7125	VERBESINA ENCELIOIDES VAR. EXAURICULATA	S	.	0.0	28.8	26.6	.	.	.	.	.	.	.	.	.	.	.	.	.	20
7126	VERBESINA ENCELIOIDES VAR. EXAURICULATA	S	.	0.0	27.9	29.3	.	.	.	.	.	.	.	.	.	.	.	.	.	21
7127	VERBESINA MYRIOCEPHALA	S	.	0.0	31.2	31.9	.	.	.	.	.	.	.	.	.	.	.	.	.	19
7128	VERBESINA ROTHROCKII	S	.	0.0	47.5	24.9	.	.	.	.	.	.	.	.	.	.	.	.	.	19
7129	VERBESINA SERRATA	S	.	0.0	33.8	14.4	.	.	.	.	.	.	.	.	.	.	.	.	.	20
7130	VERBESINA SERRATA VAR. PRINGLEI	S	.	0.0	32.5	15.9	.	.	.	.	.	.	.	.	.	.	.	.	.	20
7131	VERNONIA ALTISSIMA	S	.	0.0	23.7	22.7	.	.	.	.	.	.	.	.	.	.	.	.	.	21
7132	VERNONIA AMYGDALINA	L	52	82.6	5.3	0.4	10.0	1.5	1.7	145	67	5.0	.	.	.	.	.	.	51	1
7133	VERNONIA AMYGDALINA	L	194	27.4	32.5	1.5	28.6	.	10.0	105	536	5.5	.	.	.	.	.	.	20	1
7134	VERNONIA ANTHELMINTICA	S	.	0.0	18.1	26.5	.	.	5.9	.	.	.	.	.	.	.	.	.	.	19
7135	VERNONIA ANTHELMINTICA	S	.	0.0	16.9	26.1	.	.	.	.	.	.	.	.	.	.	.	.	.	20
7136	VERNONIA ANTHELMINTICA	S	.	0.0	18.8	25.6	.	.	.	.	.	.	.	.	.	.	.	.	.	20
7137	VERNONIA ANTHELMINTICA	S	.	0.0	15.8	24.3	.	.	.	.	.	.	.	.	.	.	.	.	.	21
7138	VERNONIA ANTHELMINTICA	S	.	0.0	14.6	24.0	.	.	.	.	.	.	.	.	.	.	.	.	.	21
7139	VERNONIA BALDWINI	S	.	0.0	23.1	29.2	.	.	6.2	.	.	.	.	.	.	.	.	.	.	19
7140	VERNONIA DEPPEANA	S	.	0.0	23.8	21.1	.	.	7.6	.	.	.	.	.	.	.	.	.	.	19
7141	VERNONIA DEPPEANA	S	.	0.0	20.0	10.0	.	.	.	.	.	.	.	.	.	.	.	.	.	20
7142	VERNONIA FASCICULATA	S	.	0.0	21.8	27.6	.	.	6.5	.	.	.	.	.	.	.	.	.	.	19
7143	VERNONIA LIATROIDES	S	.	0.0	14.4	11.8	.	.	.	.	.	.	.	.	.	.	.	.	.	20
7144	VERNONIA MISSURICA	S	.	0.0	23.4	21.5	.	.	4.8	.	.	.	.	.	.	.	.	.	.	19
7145	VERNONIA PALLENS	S	.	0.0	15.0	3.4	.	.	.	.	.	.	.	.	.	.	.	.	.	20
7146	VERNONIA SP.	S	.	0.0	28.1	18.5	.	.	5.7	.	.	.	.	.	.	.	.	.	.	19
7147	VERNONIA SP.	S	.	0.0	16.9	13.6	.	.	.	.	.	.	.	.	.	.	.	.	.	19
7148	VERONICA LONGIFOLIA	S	.	0.0	27.5	18.5	.	.	.	.	.	.	.	.	.	.	.	.	.	20
7149	VERONICA PEREGRINA	S	.	0.0	15.0	13.7	.	.	.	.	.	.	.	.	.	.	.	.	.	20
7150	VERONICA SALINA	S	.	0.0	21.4	15.8	.	.	.	.	.	.	.	.	.	.	.	.	.	21
7151	VERONICA SPICATA	S	.	0.0	26.2	26.6	.	.	6.0	.	.	.	.	.	.	.	.	.	.	19
7152	VERONICA SPICATA	S	.	0.0	26.2	35.4	.	.	.	.	.	.	.	.	.	.	.	.	.	20
7153	VETIVERIA ZIZANIOIDES	L	.	.	6.4	1.6	84.6	38.4	7.1	295	325	.	.	.	.	.	.	.	.	5
7154	VIBURNUM ACERIFOLIUM	S	.	0.0	18.1	38.9	.	.	3.4	.	.	.	.	.	.	.	.	.	.	19
7155	VIBURNUM ACERIFOLIUM	S	.	0.0	4.4	7.2	.	.	4.8	.	.	.	.	.	.	.	.	.	.	19
7156	VIBURNUM CASSINOIDES	S	.	0.0	19.4	30.4	.	.	.	.	.	.	.	.	.	.	.	.	.	19
7157	VIBURNUM DENTATUM	S	.	0.0	16.9	41.5	.	.	3.3	.	.	.	.	.	.	.	.	.	.	19
7158	VIBURNUM DENTATUM	S	.	0.0	5.1	23.3	.	.	3.0	.	.	.	.	.	.	.	.	.	.	19

TABLE 2 -- DATA DIRECTLY TRANSCRIBED FROM ORIGINAL SOURCES.

OBS	PLNA	PART	CALORIES	WATER	PROTEIN	FAT	TOTLCARB	FIBER	ASH	CALCIUM	PHOSPHO	IRON	SODIUM	POTASS	CAROT	THIAMINE	RIBOFLAV	NIACIN	ASCORBIC	REFCO
7159	VIBURNUM LANTANA	S	.	0.0	4.6	10.4	.	.	.	.	.	.	.	.	.	.	.	.	.	21
7160	VIBURNUM ORIENTALE	S	.	0.0	4.8	7.9	.	.	.	.	.	.	.	.	.	.	.	.	.	21
7161	VIBURNUM TINUS	S	.	0.0	12.2	32.0	.	.	.	.	.	.	.	.	.	.	.	.	.	21
7162	VICIA AMERICANA(IMMATURE)	H	.	0.0	27.3	2.3	61.9	24.2	8.5	1300	390	.	.	.	.	.	.	.	.	6
7163	VICIA ANGUSTIFOLIA	S	.	0.0	47.8	6.2	.	.	3.1	.	.	.	.	.	.	.	.	.	.	19
7164	VICIA ANGUSTIFOLIA	S	.	0.0	29.5	2.2	.	.	.	.	.	.	.	.	.	.	.	.	.	20
7165	VICIA ANGUSTIFOLIA	S	.	0.0	31.9	1.0	.	.	.	.	.	.	.	.	.	.	.	.	.	20
7166	VICIA ANGUSTIFOLIA	S	.	0.0	31.2	0.7	.	.	.	.	.	.	.	.	.	.	.	.	.	21
7167	VICIA ATROPURPUREA	S	.	0.0	26.7	0.5	.	.	3.2	.	.	.	.	.	.	.	.	.	.	19
7168	VICIA BENGHALENSIS	W	.	0.0	23.6	3.0	65.3	28.8	8.1	8500	350	.	.	.	.	.	.	.	.	18
7169	VICIA BENGHALENSIS	S	.	0.0	27.2	0.6	.	.	3.5	.	.	.	.	.	.	.	.	.	.	19
7170	VICIA CASSUBICA	S	.	0.0	43.8	1.0	.	.	.	.	.	.	.	.	.	.	.	.	.	21
7171	VICIA CRACCA	S	.	0.0	34.9	0.6	.	.	3.4	.	.	.	.	.	.	.	.	.	.	19
7172	VICIA CRACCA	S	.	0.0	31.2	0.7	.	.	3.4	.	.	.	.	.	.	.	.	.	.	19
7173	VICIA DASYCARPA	S	.	0.0	33.1	1.1	.	.	3.6	.	.	.	.	.	.	.	.	.	.	19
7174	VICIA DASYCARPA	S	.	0.0	35.3	1.0	.	.	3.5	.	.	.	.	.	.	.	.	.	.	19
7175	VICIA DASYCARPA	S	.	0.0	31.2	1.7	.	.	.	.	.	.	.	.	.	.	.	.	.	20
7176	VICIA ERYILIA	S	.	0.0	22.8	1.0	.	.	.	.	.	.	.	.	.	.	.	.	.	21
7177	VICIA FABA	X	104	65.3	13.3	0.6	18.6	0.7	2.2	3	316	6.2	88	778	75	0.30	0.15	1.0	110	
7178	VICIA FABA	S	344	10.1	26.2	1.3	59.4	6.8	3.0	104	301	6.7	.	.	130	0.38	0.24	2.1	16	1
7179	VICIA FABA	IS	75	76.3	7.1	0.4	15.3	3.2	0.9	38	127	.	.	.	.	0.10	0.22	.	140	1
7180	VICIA FABA	X	111	64.5	10.9	0.3	22.9	5.6	1.4	.	.	.	.	.	.	.	.	.	.	1
7181	VICIA FABA	S	.	0.0	25.2	1.2	.	.	3.2	.	.	.	.	.	.	.	.	.	.	19
7182	VICIA FABA	IS	92	75.6	8.2	0.5	14.7	1.7	1.0	26	148	1.8	4	438	135	0.34	0.17	2.1	26	2
7183	VICIA FABA	IS	118	69.0	9.3	0.4	20.3	3.8	1.0	31	140	2.3	.	.	120	0.28	0.17	1.7	28	3
7184	VICIA FABA	S	339	12.6	24.0	2.2	58.2	5.9	3.0	77	374	6.3	.	.	60	0.53	0.30	2.5	6	3
7185	VICIA FABA	F	.	85.4	4.5	0.1	9.2	2.0	0.8	50	74	.	44	39	9	0.08	.	0.8	12	5
7186	VICIA FABA	L	71	77.6	5.6	0.3	15.2	3.7	1.3	111	149	.	.	.	.	.	.	.	.	5
7187	VICIA FABA	W	.	0.0	20.5	2.7	67.3	22.7	9.5	.	.	.	.	.	.	.	.	.	.	6
7188	VICIA FABA	H	.	0.0	14.6	0.9	78.5	24.0	6.0	.	.	.	.	.	.	.	.	.	.	6
7189	VICIA GIGANTEA	S	.	0.0	38.1	1.2	.	.	3.0	.	.	.	.	.	.	.	.	.	.	19
7190	VICIA HIRSUTA	S	.	0.0	27.6	0.6	.	.	3.8	.	.	.	.	.	.	.	.	.	.	19
7191	VICIA LEAVENWORTHII	S	.	0.0	33.2	0.7	.	.	3.2	.	.	.	.	.	.	.	.	.	.	19
7192	VICIA LEAVENWORTHII	S	.	0.0	33.1	0.9	.	.	.	.	.	.	.	.	.	.	.	.	.	20
7193	VICIA LUTEA	S	.	0.0	31.6	0.8	.	.	.	.	.	.	.	.	.	.	.	.	.	21
7194	VICIA MONATHA	S	.	0.0	27.5	0.8	.	.	.	.	.	.	.	.	.	.	.	.	.	21
7195	VICIA MONATHA	S	.	0.0	28.6	1.0	.	.	.	.	.	.	.	.	.	.	.	.	.	21
7196	VICIA MONATHA	S	.	0.0	26.4	1.0	.	.	.	.	.	.	.	.	.	.	.	.	.	21
7197	VICIA NARBONENSIS	W	.	0.0	24.7	3.2	59.9	22.7	12.2	590	360	.	.	.	.	.	.	.	.	18
7198	VICIA NARBONENSIS	S	.	0.0	27.8	0.7	.	.	.	.	.	.	.	.	.	.	.	.	.	21
7199	VICIA NARBONENSIS	S	.	0.0	25.6	1.2	.	.	.	.	.	.	.	.	.	.	.	.	.	21
7200	VICIA NARBONENSIS	S	.	0.0	27.1	0.9	.	.	.	.	.	.	.	.	.	.	.	.	.	21
7201	VICIA NARBONENSIS	S	.	0.0	23.8	0.6	.	.	.	.	.	.	.	.	.	.	.	.	.	21
7202	VICIA NOEANA	S	.	0.0	28.2	2.4	.	.	.	.	.	.	.	.	.	.	.	.	.	20
7203	VICIA PANNONICA	S	.	0.0	27.1	1.7	.	.	3.2	.	.	.	.	.	.	.	.	.	.	19

TABLE 2 - DATA DIRECTLY TRANSCRIBED FROM ORIGINAL SOURCES.

OBS	PLNA	PART	CALORIES	WATER	PROTEIN	FAT	TOTCARB	FIBER	ASH	CALCIUM	PHOSPHO	IRON	SODIUM	POTASS	CAROT	THIAMINE	RIBOFLAV	NIACIN	ASCORBIC	REFCO
7204	VICIA PEREGRINA	S	.	0.0	29.5	1.0	.	.	.	.	.	.	.	.	.	.	.	.	.	21
7205	VICIA PEREGRINA	S	.	0.0	28.8	0.8	.	.	.	.	.	.	.	.	.	.	.	.	.	21
7206	VICIA SATIVA	W	.	0.0	35.9	2.8	52.1	21.3	9.2	490	610	.	.	.	.	.	.	.	.	18
7207	VICIA SATIVA	H	.	0.0	19.0	2.4	69.5	28.5	9.1	1110	300	.	.	.	.	.	.	.	.	18
7208	VICIA SATIVA	S	.	0.0	26.1	0.8	.	.	3.2	.	.	.	.	.	.	.	.	.	.	19
7209	VICIA SATIVA	S	.	0.0	29.9	0.8	.	.	3.3	.	.	.	.	.	.	.	.	.	.	19
7210	VICIA SATIVA	S	.	0.0	30.2	1.0	.	.	.	.	.	.	.	.	.	.	.	.	.	20
7211	VICÍA SATIVA	S	.	0.0	31.6	0.6	.	.	.	.	.	.	.	.	.	.	.	.	.	20
7212	VICIA SATIVA	S	.	0.0	31.6	1.0	.	.	.	.	.	.	.	.	.	.	.	.	.	21
7213	VICIA SATIVA	H	.	0.0	19.1	2.1	69.8	27.6	9.0	.	.	.	.	.	.	.	.	.	.	6
7214	VICIA SATIVA SUBSP. SATIVA	S	343	11.1	27.6	1.7	57.2	2.0	2.4	137	240	9.3	.	.	0	0.15	0.10	2.7	.	1
7215	VICIA SINENSIS	S	.	0.0	27.2	1.4	.	.	3.8	.	.	.	.	.	.	.	.	.	.	19
7216	VICIA SP.	S	.	0.0	30.0	1.2	.	.	.	.	.	.	.	.	.	.	.	.	.	19
7217	VICIA VILLOSA	W	.	0.0	30.5	2.4	53.8	24.2	13.3	640	460	.	.	.	.	.	.	.	.	18
7218	VICIA VILLOSA	S	.	0.0	25.6	0.9	.	.	3.3	.	.	.	.	.	.	.	.	.	.	19
7219	VICIA VILLOSA	S	.	0.0	30.2	0.7	.	.	.	.	.	.	.	.	.	.	.	.	.	20
7220	VICIA VILLOSA	W	.	0.0	23.1	2.7	62.1	27.5	12.1	1100	330	.	.	2250	.	.	.	.	.	6
7221	VICIA VILLOSA	H	.	0.0	22.6	3.2	64.3	28.7	9.9	.	.	.	.	.	.	.	.	.	.	6
7222	VIGNA ACONITIFOLIA	W	.	0.0	15.5	1.9	70.0	19.7	12.6	2370	540	.	.	.	.	.	.	.	.	18
7223	VIGNA ACONITIFOLIA	H	.	0.0	10.6	1.9	73.7	26.8	13.8	2340	180	.	.	.	.	.	.	.	.	18
7224	VIGNA ACONITIFOLIA	F	.	0.0	9.6	2.9	73.1	19.4	14.4	2010	250	.	.	.	.	.	.	.	.	18
7225	VIGNA ACONITIFOLIA	S	.	0.0	26.6	0.6	67.2	5.3	5.6	350	380	.	.	.	.	.	.	.	.	18
7226	VIGNA ACONITIFOLIA	N	.	.	10.6	1.9	73.8	26.8	13.8	2340	180	.	.	.	.	.	.	.	.	5
7227	VIGNA ANGULARIS	IS	135	65.0	9.6	0.1	24.9	0.6	0.4	24	80	2.1	.	.	0	0.01	0.01	0.2	0	2
7228	VIGNA ANGULARIS	S	325	13.1	25.3	0.6	57.1	5.7	3.9	252	317	7.6	.	.	10	0.57	0.18	3.2	.	3
7229	VIGNA ANGULARIS	S	.	13.0	25.3	0.6	57.1	5.7	3.9	252	.	7.6	.	.	9	0.57	0.18	3.2	.	5
7230	VIGNA CYLINDRICA	S	.	0.0	25.7	1.4	.	.	3.6	.	.	.	.	.	.	.	.	.	.	19
7231	VIGNA HOSEI	L	.	79.9	4.4	.	.	.	2.3	430	180	.	.	390	.	.	.	.	.	5
7232	VIGNA LUTEOLA	S	.	0.0	25.2	0.6	.	.	.	.	.	.	.	.	.	.	.	.	.	21
7233	VIGNA MUNGO	S	334	10.0	24.0	1.1	59.4	4.9	5.5	110	382	8.9	.	.	45	0.48	0.21	2.3	0	1
7234	VIGNA MUNGO	W	.	0.0	19.4	2.5	62.1	26.8	16.0	1970	240	.	.	.	.	.	.	.	.	18
7235	VIGNA MUNGO	H	.	0.0	8.9	2.8	75.7	28.6	12.6	1740	160	.	.	.	.	.	.	.	.	18
7236	VIGNA MUNGO	F	.	0.0	9.0	2.3	76.5	29.9	12.2	2710	190	.	.	.	.	.	.	.	.	18
7237	VIGNA MUNGO	S	.	0.0	26.8	0.9	66.7	5.3	5.6	220	390	.	.	.	.	.	.	.	.	18
7238	VIGNA MUNGO	S	344	10.6	21.0	1.6	63.4	4.4	3.4	110	348	8.4	.	171	20	0.58	0.20	2.3	2	2
7239	VIGNA MUNGO	S	347	10.9	24.0	1.4	60.5	0.9	3.2	154	385	9.1	40	800	38	0.42	0.20	2.0	.	5
7240	VIGNA RADIATA	S	341	9.8	23.5	0.9	61.8	4.2	4.0	102	438	.	.	.	125	0.52	0.20	2.4	.	1
7241	VIGNA RADIATA	W	.	0.0	13.0	3.7	71.9	21.0	11.4	2470	340	.	.	.	.	.	.	.	.	18
7242	VIGNA RADIATA	H	.	0.0	9.0	2.3	76.6	29.9	12.1	2710	200	.	.	.	.	.	.	.	.	18
7243	VIGNA RADIATA	F	.	0.0	8.9	2.8	75.2	28.1	13.1	2490	150	.	.	.	.	.	.	.	.	18
7244	VIGNA RADIATA	S	.	0.0	23.3	1.0	70.4	6.7	5.3	80	.	.	.	.	.	.	.	.	.	18
7245	VIGNA RADIATA	X	30	90.1	4.2	0.2	5.0	0.9	0.5	15	71	1.2	7	222	20	0.11	0.10	0.8	18	2
7246	VIGNA RADIATA	X	35	88.8	3.8	0.2	6.6	0.7	0.6	19	64	1.3	5	223	12	0.13	0.13	0.8	19	4
7247	VIGNA RADIATA	S	334	10.4	24.0	1.3	60.8	4.1	3.5	124	326	7.3	28	843	94	0.47	0.27	2.1	0	5
7248	VIGNA RADIATA	X	.	88.8	3.8	0.2	6.6	0.7	0.6	19	64	1.3	5	223	.	.	.	.	.	5

TABLE 2 - DATA DIRECTLY TRANSCRIBED FROM ORIGINAL SOURCES.

OBS	PLANA	PART	CALORIES	WATER	PROTEIN	FAT	TOTLCARB	FIBER	ASH	CALCIUM	PHOSPHO	IRON	SODIUM	POTASS	CAROTE	THIAMINE	RIBOFLAV	NIACIN	ASCORBIC	REFCO
7249	VIGNA SINENSIS	S	.	0.0	25.7	1.4	.	.	3.6	.	.	.	.	.	.	.	.	.	.	19
7250	VIGNA SINENSIS	S	.	0.0	27.0	1.3	.	.	3.5	.	.	.	.	.	.	.	.	.	.	19
7251	VIGNA TRILOBATA	N	.	10.7	11.4	1.3	63.5	22.1	13.4	2690	400	.	.	.	.	.	.	.	.	5
7252	VIGNA UMBELLATA	S	327	13.3	20.9	0.9	60.7	4.8	4.2	200	390	10.9	.	.	.	0.49	0.21	2.4	.	1
7253	VIGNA UMBELLATA	W	.	0.0	16.9	1.9	73.4	30.6	7.8	1030	260	.	.	.	.	.	.	.	.	18
7254	VIGNA UMBELLATA	S	.	0.0	23.3	0.6	71.5	6.0	4.6	670	390	.	.	.	.	.	.	.	.	18
7255	VIGNA UMBELLATA	S	335	14.0	18.5	1.0	64.5	.	2.0	80	400	5.0	.	.	0	0.30	0.21	2.4	9	2
7256	VIGNA UMBELLATA	S	338	10.5	19.3	0.6	65.4	5.8	4.2	218	313	7.2	.	.	20	0.58	0.08	2.2	.	3
7257	VIGNA UMBELLATA	S	.	12.8	19.1	0.6	.	5.5	4.3	360	700	10.9	.	.	.	0.59	0.21	2.4	.	5
7258	VIGNA UMBELLATA	F	.	9.1	15.8	5.6	65.6	7.1	3.9	320	240	8.0	.	.	.	0.47	0.21	2.2	.	5
7259	VIGNA UMBELLATA	SH	.	68.0	5.4	0.6	23.5	9.8	2.5	460	190	.	.	.	.	.	.	.	.	5
7260	VIGNA UNGUICULATA	W	.	0.0	17.6	2.2	73.1	21.4	7.1	550	290	.	.	.	.	.	.	.	.	18
7261	VIGNA UNGUICULATA	H	.	0.0	10.6	1.8	74.5	16.2	13.1	1810	1180	.	.	.	.	.	.	.	.	18
7262	VIGNA UNGUICULATA	F	.	0.0	7.2	1.2	80.1	37.7	11.5	1660	130	.	.	.	.	.	.	.	.	18
7263	VIGNA UNGUICULATA	S	.	0.0	23.8	1.1	68.4	6.4	6.7	400	290	.	.	.	.	.	.	.	.	18
7264	VIGNA UNGUICULATA	F	.	85.3	3.5	0.2	10.1	2.0	0.9	72	59	2.5	.	.	564	0.07	0.09	0.9	14	5
7265	VIGNA UNGUICULATA	L	38	89.0	3.4	0.7	5.3	1.2	1.6	290	58	20.1	.	.	6072	0.05	0.18	0.6	4	5
7266	VIGNA UNGUICULATA	W	.	0.0	15.1	4.8	67.4	24.8	12.7	.	.	.	.	.	.	.	.	.	.	6
7267	VIGNA UNGUICULATA	H	.	0.0	18.4	2.9	67.1	27.3	11.6	1340	320	90.0	.	1990	.	.	.	.	.	6
7268	VIGNA UNGUICULATA SUBSP. CYLINDRICA	IS	49	83.6	4.8	0.4	9.5	2.1	1.7	151	68	.	.	.	150	.	.	.	.	1
7269	VIGNA UNGUICULATA SUBSP. CYLINDRICA	L	44	85.0	4.7	0.3	8.3	2.0	1.7	256	63	5.7	.	.	7970	0.20	0.37	2.1	56	1
7270	VIGNA UNGUICULATA SUBSP. CYLINDRICA	S	342	10.8	23.1	1.4	61.4	4.8	3.3	101	383	7.6	.	.	70	0.75	0.18	2.5	1	1
7271	VIGNA UNGUICULATA SUBSP. SESQUIPEDALIS	F	37	88.3	3.0	0.2	7.9	1.6	0.6	44	45	0.7	6	233	225	0.12	0.11	1.0	22	2
7272	VIGNA UNGUICULATA SUBSP. SESQUIPEDALIS	L	34	88.4	4.2	0.4	5.8	1.7	1.2	108	106	4.7	25	366	2400	0.28	0.24	1.2	35	2
7273	VIGNA UNGUICULATA SUBSP. SESQUIPEDALIS	F	34	89.0	3.3	0.1	6.9	1.3	0.7	54	68	1.4	.	.	470	0.09	0.12	1.1	26	3
7274	VIGNA UNGUICULATA SUBSP. UNGUICULATA	S	.	11.8	22.0	0.5	62.6	5.3	3.1	280	390	7.6	.	.	71	.	.	1.5	.	
7275	VIGNA UNGUICULATA SUBSP. UNGUICULATA	S	338	11.4	22.5	1.4	61.0	5.4	3.7	104	416	.	.	.	.	0.08	0.09	4.0	2	1
7276	VIGNA UNGUICULATA SUBSP. UNGUICULATA	SH	30	89.0	4.8	0.3	4.4	.	1.8	73	106	2.2	.	.	.	0.35	0.18	1.1	36	2
7277	VIGNA UNGUICULATA SUBSP. UNGUICULATA	F	36	88.7	3.7	0.6	6.2	1.2	0.8	54	59	1.4	2	220	455	0.14	0.10	1.0	24	2
7278	VIGNA UNGUICULATA SUBSP. UNGUICULATA	F	47	85.3	3.6	0.3	10.0	1.8	0.8	45	52	1.2	.	.	340	0.13	0.10	1.0	22	3
7279	VIGNA UNGUICULATA SUBSP. UNGUICULATA	S	341	10.6	24.1	1.2	60.7	4.9	3.4	77	420	7.2	.	.	20	0.87	0.23	1.9	3	3
7280	VIGNA VEXILLATA	S	.	0.0	33.1	3.1	.	.	4.3	.	.	.	.	.	.	.	.	.	.	19
7281	VIGUIERA CILIATA	S	.	0.0	26.2	30.6	.	.	.	.	.	.	.	.	.	.	.	.	.	20
7282	VIGUIERA DECURRENS	S	.	0.0	19.4	11.2	.	.	.	.	.	.	.	.	.	.	.	.	.	20
7283	VIGUIERA DENTATA	S	.	0.0	25.0	24.9	.	.	.	.	.	.	.	.	.	.	.	.	.	20
7284	VIGUIERA EXCELSA	S	.	0.0	21.9	14.0	.	.	.	.	.	.	.	.	.	.	.	.	.	21
7285	VIGUIERA HYPARGYREA	S	.	0.0	29.4	19.8	.	.	.	.	.	.	.	.	.	.	.	.	.	20
7286	VIGUIERA LACINIATA	S	.	0.0	19.3	19.9	.	.	4.2	.	.	.	.	.	.	.	.	.	.	19
7287	VIGUIERA LINEARIS	S	.	0.0	27.5	24.8	.	.	.	.	.	.	.	.	.	.	.	.	.	20
7288	VIGUIERA MULTIFLORA	S	.	0.0	24.1	32.7	.	.	.	.	.	.	.	.	.	.	.	.	.	21
7289	VIGUIERA MULŢIFLORA	S	.	0.0	24.0	35.9	.	.	.	.	.	.	.	.	.	.	.	.	.	21
7290	VIGUIERA SP.	S	.	0.0	31.9	18.7	.	.	.	.	.	.	.	.	.	.	.	.	.	20
7291	VIGUIERA STENOLOBA	S	.	0.0	20.6	19.4	.	.	.	.	.	.	.	.	.	.	.	.	.	20
7292	VIGUIERA STENOLOBA	S	.	0.0	20.4	20.4	.	.	.	.	.	.	.	.	.	.	.	.	.	20
7293	VINCA ROSEA	S	.	0.0	19.4	34.4	.	.	.	.	.	.	.	.	.	.	.	.	.	20

TABLE 2 - DATA DIRECTLY TRANSCRIBED FROM ORIGINAL SOURCES.

OBS	PLNA	PART	CALORIES	WATER	PROTEIN	FAT	TOTCARB	FIBER	ASH	CALCIUM	PHOSPHO	IRON	SODIUM	POTASS	CAROT	THIAMINE	RIBOFLAV	NIACIN	ASCORBIC	REFCO
7294	VIROLA GUATEMALENSIS	S	.	0.0	11.20	70.70	.	.	.	.	.	.	.	.	.	.	.	.	.	20
7295	VITEX AGNUS-CASTUS	S	.	0.0	8.10	9.00	.	.	.	.	.	.	.	.	.	.	.	.	.	19
7296	VITEX AGNUS-CASTUS	S	.	0.0	6.10	6.50	.	.	.	.	.	.	.	.	.	.	.	.	.	21
7297	VITEX AGNUS-CASTUS	S	.	0.0	5.40	3.80	.	.	.	.	.	.	.	.	.	.	.	.	.	21
7298	VITEX DONIANA	F	104	70.6	0.70	0.40	27.40	1.30	0.9	34	47	2.7	.	.	.	0.02	.	.	9	1
7299	VITEX NEGUNDO	S	.	0.0	2.50	0.40	.	.	.	.	.	.	.	.	.	.	.	.	.	21
7300	VITIS AMURENSIS	F	69	80.5	1.00	0.60	16.90	3.50	1.0	73	10	1.7	.	.	0	0.05	0.03	0.5	8	2
7301	VITIS RIPARIA	S	.	0.0	10.60	17.40	.	.	.	.	.	.	.	.	.	.	.	.	.	19
7302	VITIS TRIFOLIA	S	.	0.0	11.30	15.80	.	.	.	.	.	.	.	.	.	.	.	.	.	21
7303	VITIS TRIFOLIA	F	57	83.9	2.00	0.40	12.90	3.50	0.8	62	41	2.2	.	.	20	0.04	0.04	1.2	6	3
7304	VITIS VINIFERA	F	62	82.7	0.60	0.40	15.80	2.00	0.5	21	19	0.8	.	.	50	.	.	.	.	1
7305	VITIS VINIFERA	L	.	0.0	15.80	13.40	62.20	9.20	8.6	2250	160	.	.	.	.	.	.	.	.	18
7306	VITIS VINIFERA	F	50	86.0	0.50	0.30	12.80	0.90	0.4	9	20	0.6	6	111	50	0.10	0.06	0.2	4	2
7307	VITIS VINIFERA	F	68	81.6	0.60	0.70	16.70	0.50	0.4	12	15	0.9	.	.	0	0.05	0.04	0.5	3	3
7308	VITIS VINIFERA	F	282	20.2	2.50	0.30	75.20	1.00	1.8	50	73	3.0	.	.	0	0.12	0.13	0.4	12	3
7309	VITIS VINIFERA	F	69	81.6	1.30	1.00	15.70	0.60	0.4	16	12	0.4	3	158	60	0.05	0.03	0.3	4	4
7310	VITIS VINIFERA	F	67	81.4	0.60	0.30	17.30	0.50	0.4	12	20	0.4	3	173	60	0.05	0.03	0.3	4	4
7311	VITIS VULPINA	S	.	0.0	10.00	15.90	.	.	3.8	.	.	.	.	.	.	.	.	.	.	19
7312	VOANDZEIA SUBTERRANEA	S	152	57.3	7.80	3.10	30.00	3.00	1.8	14	258	1.2	.	.	.	.	.	.	.	1
7313	VOANDZEIA SUBTERRANEA	S	367	10.3	18.80	6.20	61.30	4.80	3.4	62	276	12.2	.	.	10	0.47	0.14	1.8	0	1
7314	VOANDZEIA SUBTERRANEA	H	.	0.0	15.90	1.80	74.80	31.70	7.5	.	.	.	.	.	.	.	.	.	.	18
7315	VOANDZEIA SUBTERRANEA	F	.	0.0	18.20	5.50	70.90	14.20	5.4	.	.	.	.	.	.	.	.	.	.	18
7316	VOANDZEIA SUBTERRANEA	S	.	0.0	21.30	7.50	67.90	6.70	3.3	10	300	.	.	.	.	.	.	.	.	18
7317	VOANDZEIA SUBTERRANEA	S	370	10.1	16.00	6.00	65.00	.	3.0	85	264	4.2	.	.	0	0.18	.	.	0	2
7318	VOANDZEIA SUBTERRANEA	S	.	12.4	17.50	9.20	54.10	4.80	3.4	38	224	2.7	.	.	.	0.24	0.16	2.0	.	5
7319	WATSONIA MERIANA	S	.	0.0	14.60	2.70	.	.	.	.	.	.	.	.	.	.	.	.	.	21
7320	WATSONIA PYRAMIDATA	S	.	0.0	17.80	3.40	.	.	.	.	.	.	.	.	.	.	.	.	.	21
7321	WATSONIA SCHLECHTERI	S	.	0.0	17.10	9.70	.	.	.	.	.	.	.	.	.	.	.	.	.	21
7322	WATSONIA SP.	S	.	0.0	14.20	7.40	.	.	.	.	.	.	.	.	.	.	.	.	.	21
7323	WIEDEMANNIA ORIENTALIS	S	.	0.0	25.70	40.20	.	.	.	.	.	.	.	.	.	.	.	.	.	20
7324	WILDENOWIA STRIATA		.	0.0	3.92	1.06	92.42	34.41	2.6	150	60	6.8	.	.	.	.	.	.	.	8
7325	WILLARDIA MEXICANA	S	.	0.0	26.20	33.70	.	.	.	.	.	.	.	.	.	.	.	.	.	19
7326	WILLDENOWIA ARGENTEA	S	.	0.0	46.90	1.30	.	.	.	.	.	.	.	.	.	.	.	.	.	21
7327	WILLDENOWIA STRIATA	S	.	0.0	44.10	0.60	.	.	.	.	.	.	.	.	.	.	.	.	.	21
7328	WISLIZENIA REFRACTA	S	.	0.0	18.60	23.00	.	.	.	.	.	.	.	.	.	.	.	.	.	21
7329	WISTERIA FLORIBUNDA VAR. MULTIJUGA	S	.	0.0	36.20	11.00	.	.	3.4	.	.	.	.	.	.	.	.	.	.	19
7330	WITHANIA SOMNIFERA	S	.	0.0	18.40	17.50	.	.	.	.	.	.	.	.	.	.	.	.	.	21
7331	WITHANIA SOMNIFERA	S	.	0.0	18.50	19.60	.	.	.	.	.	.	.	.	.	.	.	.	.	21
7332	WYETHIA ANGUSTIFOLIA	S	.	0.0	24.10	5.40	.	.	5.6	.	.	.	.	.	.	.	.	.	.	19
7333	WYETHIA HELENIOIDES	S	.	0.0	21.20	15.20	.	.	5.6	.	.	.	.	.	.	.	.	.	.	19
7334	XANTHISMA TEXANUM	S	.	0.0	17.10	11.20	.	.	3.3	.	.	.	.	.	.	.	.	.	.	19
7335	XANTHIUM PENSYLVANICUM	S	.	0.0	41.40	37.80	.	.	.	.	.	.	.	.	.	.	.	.	.	21
7336	XANTHIUM SP.	S	.	0.0	43.80	34.30	.	.	.	.	.	.	.	.	.	.	.	.	.	19
7337	XANTHIUM SPECIOSUM	S	.	0.0	47.40	34.80	.	.	.	.	.	.	.	.	.	.	.	.	.	21
7338	XANTHIUM STRUMARIUM	S	.	0.0	49.10	35.70	.	.	.	.	.	.	.	.	.	.	.	.	.	21

TABLE 2 - DATA DIRECTLY TRANSCRIBED FROM ORIGINAL SOURCES.

OBS	PLNA	PART	CALORIES	WATER	PROTEIN	FAT	TOTLCARB	FIBER	ASH	CALCIUM	PHOSPHO	IRON	SODIUM	POTASS	CAROT	THIAMINE	RIBOFLAV	NIACIN	ASCORBIC	REFCO
7339	XANTHIUM STRUMARIUM	S	.	0.0	41.2	38.2	.	.	.	.	.	.	.	.	.	.	.	.	.	21
7340	XANTHIUM STRUMARIUM	L	.	.	17.6	5.3	63.3	.	13.8	3150	1130	.	470	2420	.	.	.	.	47	5
7341	XANTHOSOMA SAGITTIFOLIUM	R	137	64.4	2.2	0.2	32.0	1.0	1.2	16	47	0.9	.	.	.	.	.	.	8	1
7342	XANTHOSOMA SAGITTIFOLIUM	L	34	89.9	2.5	1.0	5.3	2.1	1.3	95	388	2.0	.	.	3300	.	.	.	37	1
7343	XANTHOSOMA SAGITTIFOLIUM	R	341	15.0	0.5	0.6	83.0	.	0.9	21	160	2.0	.	.	.	.	.	.	.	1
7344	XANTHOSOMA SAGITTIFOLIUM	SH	33	89.5	3.1	0.6	5.7	3.2	1.1	49	80	0.3	.	.	.	.	.	.	82	1
7345	XANTHOSOMA SAGITTIFOLIUM	L	.	0.0	22.9	7.2	55.3	10.0	14.6	1780	640	.	.	.	.	.	.	.	.	18
7346	XANTHOSOMA SAGITTIFOLIUM	C	.	0.0	13.0	2.4	63.2	15.6	21.4	.	.	.	.	.	.	.	.	.	.	18
7347	XANTHOSOMA SAGITTIFOLIUM	R	.	0.0	6.9	0.2	88.7	2.6	4.2	190	200	.	.	.	.	.	.	.	.	18
7348	XANTHOSOMA SAGITTIFOLIUM	R	132	65.9	1.7	0.3	30.9	0.6	1.2	14	56	0.8	.	.	20	0.13	0.03	0.7	5	3
7349	XANTHOXYLUM SP.	S	.	0.0	9.4	11.0	.	.	8.0	.	.	.	.	.	.	.	.	.	.	19
7350	XANTHOXYLUM SP.	S	.	0.0	12.5	2.3	.	.	5.5	.	.	.	.	.	.	.	.	.	.	19
7351	XERANTHEMUM ANNUUM	S	.	0.0	30.8	14.5	.	.	4.4	.	.	.	.	.	.	.	.	.	.	19
7352	XERANTHEMUM ANNUUM	S	.	0.0	24.4	9.3	.	.	7.3	.	.	.	.	.	.	.	.	.	.	19
7353	XERANTHEMUM ANNUUM	S	.	0.0	19.9	18.4	.	.	.	.	.	.	.	.	.	.	.	.	.	21
7354	XEROPHYLLUM SP.	L	.	0.0	7.3	2.3	83.0	42.0	7.4	.	.	.	.	.	.	.	.	.	.	6
7355	XIMENESIA ENCELIOIDES	S	.	0.0	35.6	30.8	.	.	4.6	.	.	.	.	.	.	.	.	.	.	19
7356	XIMENIA AMERICANA	S	.	0.0	20.0	62.4	.	.	.	.	.	.	.	.	.	.	.	.	.	19
7357	XIMENIA AMERICANA	S	667	5.1	17.5	60.6	13.0	.	3.8	.	.	.	.	.	.	.	.	.	9	5
7358	XIMENIA CAFFRA	F	.	66.4	.	.	.	.	.	.	.	.	.	.	.	.	.	.	49	1
7359	YUCCA ALOIFOLIA	I	33	90.0	3.1	0.2	6.0	.	0.7	47	73	0.5	.	.	20	0.14	0.09	0.6	.	3
7360	YUCCA ARIZONICA	S	.	0.0	14.4	28.4	.	.	.	.	.	.	.	.	.	.	.	.	.	19
7361	YUCCA BACCATA	S	.	0.0	14.4	27.2	.	.	.	.	.	.	.	.	.	.	.	.	.	19
7362	YUCCA CONSTRICTA	S	.	0.0	21.4	25.9	.	.	3.1	.	.	.	.	.	.	.	.	.	.	19
7363	YUCCA ELATA	S	.	0.0	20.6	29.1	.	.	3.0	.	.	.	.	.	.	.	.	.	.	19
7364	YUCCA ELATA	S	.	0.0	22.2	29.7	.	.	2.8	.	.	.	.	.	.	.	.	.	.	19
7365	YUCCA ELATA	S	.	0.0	21.2	31.7	.	.	.	.	.	.	.	.	.	.	.	.	.	19
7366	YUCCA ELATA	S	.	0.0	22.5	32.6	.	.	.	.	.	.	.	.	.	.	.	.	.	20
7367	YUCCA ELATA	Y	.	0.0	4.5	1.4	87.6	29.6	6.5	.	.	.	.	.	.	.	.	.	.	6
7368	YUCCA ELEPHANTIPES	L	30	90.2	1.2	0.3	6.4	1.1	1.9	342	24	0.9	.	.	0	0.20	0.04	0.3	26	3
7369	YUCCA ELEPHANTIPES	I	61	83.2	2.0	0.3	13.7	0.9	0.8	34	69	1.4	.	.	20	0.16	0.15	1.5	393	3
7370	YUCCA GLAUCA	S	.	0.0	20.9	26.6	.	.	3.4	.	.	.	.	.	.	.	.	.	.	19
7371	YUCCA GLAUCA	W	.	0.0	5.7	1.9	84.9	32.9	7.5	.	.	.	.	.	.	.	.	.	.	6
7372	YUCCA PENINSULARIS	S	.	0.0	15.0	27.5	.	.	.	.	.	.	.	.	.	.	.	.	.	19
7373	YUCCA SCHIDIGERA	S	.	0.0	11.9	34.5	.	.	.	.	.	.	.	.	.	.	.	.	.	20
7374	YUCCA SCHOTTII	S	.	0.0	12.5	29.4	.	.	.	.	.	.	.	.	.	.	.	.	.	20
7375	YUCCA SP.	S	.	0.0	13.8	28.3	.	.	.	.	.	.	.	.	.	.	.	.	.	19
7376	YUCCA TORREYI	L	.	0.0	12.0	1.9	75.6	36.1	10.5	860	140	12.0	.	860	.	.	.	.	.	6
7377	YUCCA TRECULEANA	L	.	0.0	4.2	2.1	86.8	30.0	6.9	3820	170	.	.	1780	.	.	.	.	.	6
7378	YUCCA WHIPPLEI	S	.	0.0	24.4	23.4	.	.	.	.	.	.	.	.	.	.	.	.	.	19
7379	ZALUZANIA DISCOIDEA	S	.	0.0	33.1	29.2	.	.	.	.	.	.	.	.	.	.	.	.	.	20
7380	ZALUZANIA SP.	S	.	0.0	32.5	26.5	.	.	.	.	.	.	.	.	.	.	.	.	.	20
7381	ZANTHOXYLUM PIPERITUM	S	.	0.0	15.8	31.8	.	.	.	.	.	.	.	.	.	.	.	.	.	21
7382	ZEA MAYS	W	.	0.0	10.3	2.4	77.0	28.6	10.3	.	.	.	.	.	.	.	.	.	.	18
7383	ZEA MAYS	C	.	0.0	7.7	0.8	83.0	46.2	8.5	.	.	.	.	.	.	.	.	.	.	18

TABLE 2 - DATA DIRECTLY TRANSCRIBED FROM ORIGINAL SOURCES.

OBS	PLNA	PART	CALORIES	WATER	PROTEIN	FAT	TOTCARB	FIBER	ASH	CALCIUM	PHOSPHO	IRON	SODIUM	POTASS	CAROT	THIAMINE	RIBOFLAV	NIACIN	ASCORBIC	REFCO
7384	ZEA MAYS	H	.	0.0	7.0	1.4	84.7	27.0	6.9	.	.	.	.	.	.	.	.	.	.	18
7385	ZEA MAYS	H	.	0.0	8.9	2.4	82.0	25.9	6.7	500	250	10.0	.	930	.	.	.	.	.	6
7386	ZEA MAYS	W	.	0.0	8.2	2.7	83.4	25.3	5.7	270	220	9.0	.	1260	.	.	.	.	.	6
7387	ZEA MAYS SUBSP MEXICANA	S	334	12.0	21.6	2.5	63.1	0.4	0.8	9	194	3.3	.	.	0	0.14	0.07	0.70	0	3
7388	ZEA MAYS SUBSP MEXICANA	L	.	68.0	1.4	0.4	26.7	10.3	3.5	340	120	.	.	720	.	.	.	.	.	5
7389	ZEA MAYS SUBSP MEXICANA	S	.	14.0	11.0	3.4	69.8	16.8	3.6	.	.	.	.	.	.	.	.	.	.	5
7390	ZEA MAYS SUBSP. MAYS	S	357	11.6	9.4	4.2	73.6	1.9	1.2	16	220	3.6	.	.	0	0.33	0.10	2.20	0	1
7391	ZEA MAYS SUBSP. MAYS	S	357	11.8	8.6	4.3	73.9	1.4	1.4	6	182	2.8	.	.	20	0.36	0.08	1.40	5	1
7392	ZEA MAYS SUBSP. MAYS	S	364	10.4	10.0	4.8	73.6	2.0	1.2	13	219	4.9	.	.	100	0.32	0.12	1.70	4	1
7393	ZEA MAYS SUBSP. MAYS	X	319	20.6	7.4	2.8	68.3	2.9	0.9	.	210	2.9	.	.	0	0.00	0.16	3.00	.	1
7394	ZEA MAYS SUBSP. MAYS	G	134	62.5	4.2	1.7	30.7	1.1	0.9	5	126	0.9	3	259	160	0.20	0.11	1.50	8	2
7395	ZEA MAYS SUBSP. MAYS	G	134	62.5	4.2	1.7	30.7	1.1	0.9	5	126	0.9	3	259	0	0.20	0.11	1.50	8	2
7396	ZEA MAYS SUBSP. MAYS	S	349	13.6	9.1	4.2	71.7	2.3	1.4	14	245	2.8	5	.	0	0.29	0.11	2.10	0	2
7397	ZEA MAYS SUBSP. MAYS	S	349	13.6	9.1	4.2	71.7	2.3	1.4	14	245	2.8	5	.	270	0.29	0.11	2.10	0	2
7398	ZEA MAYS SUBSP. MAYS	S	361	10.6	9.4	4.3	74.4	1.8	1.3	9	290	2.5	.	.	140	0.43	0.10	1.90	0	3
7399	ZEA MAYS SUBSP. MAYS	S	129	63.5	4.1	1.3	30.3	1.0	0.8	5	128	1.1	.	.	70	0.18	0.08	1.90	9	3
7400	ZEA MAYS SUBSP. MAYS	S	96	72.7	3.5	1.0	22.1	0.7	0.7	3	111	0.7	0	280	240	0.15	0.12	1.70	12	4
7401	ZEA MAYS SUBSP. MAYS	S	362	9.8	11.9	4.7	72.1	2.1	1.5	10	264	2.5	3	.	.	0.39	0.11	2.10	0	4
7402	ZEA MAYS SUSP. MEXICANA	W	.	0.0	9.5	2.9	79.6	23.0	8.0	.	.	.	.	.	.	.	.	.	.	18
7403	ZELKOVA SERRATA	S	.	0.0	18.2	21.3	.	.	8.3	.	.	.	.	.	.	.	.	.	.	19
7404	ZEPHYRANTHES LONGIFOLIA	S	.	0.0	5.0	9.6	.	.	.	.	.	.	.	.	.	.	.	.	.	19
7405	ZEXMANIA HISPIDA(BUDDING)	W	.	0.0	17.5	3.2	63.8	21.3	15.5	3510	200	.	.	2480	.	.	.	.	.	6
7406	ZEXMENIA HISPIDA(FLOWERING)	W	.	0.0	12.2	1.3	70.0	25.7	16.5	2930	180	.	.	3150	.	.	.	.	.	6
7407	ZEXMENIA SEEMANNII	S	.	0.0	17.0	6.6	.	.	.	.	.	.	.	.	.	.	.	.	.	21
7408	ZINGIBER MIOGA	I	17	94.2	1.2	0.1	3.8	1.7	0.7	33	16	0.6	5	100	15	0.07	0.06	0.50	5	2
7409	ZINGIBER OFFICINALE	R	301	10.2	7.6	2.9	72.4	.	6.9	180	.	.	.	.	120	0.16	0.27	8.40	0	1
7410	ZINGIBER OFFICINALE	R	347	9.4	9.1	6.0	70.8	5.9	4.8	116	148	11.5	32	1342	88	0.05	0.18	5.16	.	16
7411	ZINGIBER OFFICINALE	R	46	87.4	1.6	0.8	9.2	1.3	1.0	19	32	1.3	7	316	55	0.01	0.03	1.70	4	2
7412	ZINGIBER OFFICINALE	SH	22	93.0	0.8	0.6	4.2	1.2	1.4	42	55	0.7	.	475	.	.	.	.	.	2
7413	ZINGIBER OFFICINALE	R	47	87.6	1.6	0.8	9.0	0.9	1.0	44	66	1.8	.	.	0	0.02	0.06	0.70	2	3
7414	ZINGIBER OFFICINALE	R	49	87.0	1.4	1.0	9.5	1.1	1.1	23	36	2.1	6	264	6	0.02	0.04	0.70	4	4
7415	ZINGIBER OFFICINALE	R	.	10.9	12.4	12.5	60.2	7.2	6.6	.	.	.	.	.	.	.	.	.	.	5
7416	ZINGIBER SP.?	R	.	80.9	2.3	0.9	14.7	2.4	1.2	20	60	2.6	.	.	.	0.06	0.03	0.60	6	5
7417	ZINNIA ELEGANS	S	.	0.0	37.5	28.3	.	.	.	.	.	.	.	.	.	.	.	.	.	19
7418	ZIZANIA AQUATICA	S	353	8.5	14.1	0.7	75.3	1.0	1.4	19	339	4.2	7	220	0	0.45	0.63	6.20	0	4
7419	ZIZANIA AQUATICA	S	.	0.0	15.1	0.7	82.7	1.0	1.5	.	.	.	.	.	.	.	.	.	.	6
7420	ZIZANIA PALUSTRIS	SH	29	91.4	1.6	0.2	6.1	1.1	0.7	8	56	1.0	.	293	35	0.06	0.04	0.40	12	2
7421	ZIZIPHUS JUJUBA	L	.	0.0	11.8	4.3	75.3	14.3	8.6	1970	200	.	.	.	.	.	.	.	.	18
7422	ZIZIPHUS JUJUBA	SH	.	0.0	8.6	1.7	78.9	30.1	10.8	2160	230	.	.	.	.	.	.	.	.	18
7423	ZIZIPHUS JUJUBA	F	82	76.9	1.6	0.4	20.5	1.1	0.6	37	49	0.8	3	278	40	0.03	0.03	0.70	46	2
7424	ZIZIPHUS JUJUBA	S	.	0.0	39.4	27.6	.	.	.	.	.	.	.	.	.	.	.	.	.	20
7425	ZIZIPHUS JUJUBA	S	.	0.0	36.9	33.5	.	.	.	.	.	.	.	.	.	.	.	.	.	20
7426	ZIZIPHUS JUJUBA	F	59	83.0	1.8	0.1	14.5	0.6	0.6	25	27	0.8	.	.	20	0.02	0.04	0.90	66	3
7427	ZIZIPHUS JUJUBA	F	105	70.2	1.2	0.2	27.6	1.4	0.8	29	37	0.7	3	269	24	0.02	0.04	0.90	69	4
7428	ZIZIPHUS JUJUBA	F	.	65.4	1.2	0.3	25.4	1.3	0.9	.	.	.	.	.	70	.	.	.	.	5

TABLE 2 - DATA DIRECTLY TRANSCRIBED FROM ORIGINAL SOURCES.

OBS	PLNA	PART	CALORIES	WATER	PROTEIN	FAT	TOTLCARB	FIBER	ASH	CALCIUM	PHOSPHO	IRON	SODIUM	POTASS	CAROT	THIAMINE	RIBOFLAV	NIACIN	ASCORBIC	REFCO
7429	ZIZIPHUS JUJUBA	F	.	16.2	4.4	.	63.2	3.2	2.3	.	.	.	.	.	.	.	.	.	.	5
7430	ZIZIPHUS JUJUBA	S	.	21.3	2.3	0.4	74.8	2.2	1.3	.	.	.	.	.	.	.	.	.	.	5
7431	ZIZIPHUS MAURITIANA	F	97	71.5	1.9	0.0	25.2	2.1	1.4	51	20	.	.	.	.	.	.	.	58	1
7432	ZIZIPHUS MAURITIANA	F	240	30.8	3.7	0.1	63.0	2.8	2.4	170	64	3.1	.	.	0	0.02	.	2.1	35	1
7433	ZIZIPHUS MAURITIANA	F	47	86.1	1.9	0.1	11.1	0.9	0.8	32	33	0.6	.	221	60	0.02	0.02	0.5	62	2
7434	ZIZIPHUS MAURITIANA	F	.	81.6	0.8	0.3	17.0	.	0.3	4	9	1.8	.	.	21	0.02	0.02	0.7	76	5
7435	ZIZIPHUS MAURITIANA	L	.	.	14.9	2.1	71.3	15.3	10.9	2505	270	.	.	.	.	.	.	.	.	5
7436	ZIZIPHUS MUCRONATA	SH	.	0.0	14.3	2.6	73.7	8.4	9.4	4680	170	.	.	.	.	.	.	.	.	18
7437	ZIZIPHUS NUMMULARIA	L	.	0.0	11.5	1.6	80.7	33.8	6.2	1900	310	.	.	.	.	.	.	.	.	18
7438	ZIZIPHUS NUMMULARIA	L	.	.	14.2	3.2	72.4	20.6	.	1700	190	.	.	.	.	.	.	.	.	5
7439	ZIZIPHUS NUMMULARIA	L	.	.	14.6	5.2	69.5	11.7	.	2170	160	.	.	.	.	.	.	.	.	5
7440	ZIZIPHUS NUMMULARIA	L	.	.	14.7	5.9	68.9	8.2	.	2070	160	.	.	.	.	.	.	.	.	5
7441	ZIZIPHUS RUGOSA	F	.	55.3	3.2	1.3	38.2	4.9	2.0	270	94	.	.	.	.	.	.	.	.	5
7442	ZIZIPHUS SPINA-CHRISTI	F	314	9.3	4.8	0.9	80.6	.	4.4	140	.	3.0	.	.	0	0.04	0.13	3.7	30	1
7443	ZORNIA DIPHYLLA	L	.	0.0	15.3	3.8	.	23.4	10.0	1020	180	.	.	530	.	.	.	.	.	5
7444	ZOYSIA JAPONICA	S	.	0.0	8.3	2.2	.	.	3.4	.	.	.	.	.	.	.	.	.	.	19
7445	ZOYSIA MATRELLA	L	.	0.0	5.0	1.3	86.0	35.3	7.6	.	.	.	.	.	.	.	.	.	.	5
7446	ZYGOPHYLLUM FABAGO	S	.	0.0	22.5	7.7	.	.	.	.	.	.	.	.	.	.	.	.	.	20

TABLE 3 - DATA CONVERTED TO A ZERO-MOISTURE BASIS.

OBS	PLANA	PART	CALORIES	WATER	PROTEIN	FAT	TOTLCARB	FIBER	ASH	CALCIUM	PHOSPHO	IRON	SODIUMS	POTASS	CAROT	THIAMINE	RIBOFLAV	NIACIN	ASCORBIC	REFCO
1	ABELMOSCHUS ESCULENTUS	F	324	0	21.6	2.7	68.5	9.0	7.2	828.8	459.5	5.4	27	2243.2	2810.81	1.53	1.89	9.01	279.3	4
2	ABRUS PRECATORIUS	L	295	0	26.7	.	65.2	17.6	8.1	126.6	309.0	.	.	.	.	.	.	.	.	10
3	ABRUS PRECATORIUS	S	.	0	19.4	2.8	.	.	.	.	.	.	.	.	.	.	.	.	.	20
4	ABRUS PRECATORIUS	L	295	0	26.7	.	65.2	17.6	8.1	1266.7	309.5	.	.	.	.	.	.	.	.	1
5	ABUTILON AMERICANUM	S	.	0	23.1	20.2	.	.	.	.	.	.	.	.	.	.	.	.	.	20
6	ABUTILON BIDENTATUM	S	.	0	20.6	14.6	.	.	.	.	.	.	.	.	.	.	.	.	.	21
7	ABUTILON INCANUM	S	.	0	24.4	14.7	.	.	.	.	.	.	.	.	.	.	.	.	.	19
8	ABUTILON INCANUM	L	.	0	19.6	1.8	68.7	22.5	9.9	2320.0	280.0	.	.	2444.0	.	.	.	.	.	6
9	ABUTILON MOLLE	S	.	0	23.8	14.7	.	.	.	.	.	.	.	.	.	.	.	.	.	21
10	ABUTILON THEOPHRASTI	S	.	0	23.1	15.3	.	.	4.9	.	.	.	.	.	.	.	.	.	.	19
11	ABUTILON THEOPHRASTI	S	.	0	20.9	17.7	.	.	.	.	.	.	.	.	.	.	.	.	.	21
12	ABUTILON TRISULCATUM	S	.	0	16.8	12.7	.	.	.	.	.	.	.	.	.	.	.	.	.	21
13	ACACIA ALBIDA	L	.	0	17.1	2.3	72.2	12.4	8.4	.	.	.	.	.	.	.	.	.	.	18
14	ACACIA ALBIDA	F	.	0	14.3	1.5	77.9	24.7	6.3	1110.0	140.0	.	.	.	.	.	.	.	.	18
15	ACACIA ALBIDA	L	.	0	19.7	1.6	71.5	19.6	7.2	1000.0	230.0	.	.	.	.	.	.	.	.	18
16	ACACIA ANEURA	SH	.	0	9.2	4.6	80.6	30.1	5.6	870.0	70.0	.	.	.	.	.	.	.	.	18
17	ACACIA ANGUSTISSIMA	S	.	0	47.5	4.4	.	.	.	.	.	.	.	.	.	.	.	.	.	20
18	ACACIA ANGUSTISSIMA	L	.	0	26.6	3.8	64.7	16.9	4.9	970.0	290.0	.	.	1280.0	.	.	.	.	.	6
19	ACACIA AURICULAEFORMIS	S	.	0	33.6	7.4	.	.	.	.	.	.	.	.	.	.	.	.	.	21
20	ACACIA BAILEYANA	S	.	0	18.8	13.6	.	.	.	.	.	.	.	.	.	.	.	.	.	20
21	ACACIA BREVISPICA	L	.	0	18.8	5.7	69.0	18.0	6.5	1860.0	310.0	.	.	.	.	.	.	.	.	18
22	ACACIA BREVISPICA	F	.	0	17.5	1.1	76.8	24.0	4.6	460.0	270.0	.	.	.	.	.	.	.	.	18
23	ACACIA CATECHU	L	.	0	13.0	4.6	72.6	22.6	9.8	2740.0	170.0	.	.	.	.	.	.	.	.	18
24	ACACIA CAVEN	S	.	0	24.5	4.7	.	.	.	.	.	.	.	.	.	.	.	.	.	21
25	ACACIA CONSTRICTA	S	.	0	40.6	9.5	.	.	3.9	.	.	.	.	.	.	.	.	.	.	19
26	ACACIA CONSTRICTA	S	.	0	41.9	11.0	.	.	.	.	.	.	.	.	.	.	.	.	.	20
27	ACACIA CONSTRICTA VAR. PAUCISPINA	S	.	0	31.9	12.6	.	.	.	.	.	.	.	.	.	.	.	.	.	19
28	ACACIA CYANOPHYLLA	S	.	0	28.1	13.6	.	.	.	.	.	.	.	.	.	.	.	.	.	20
29	ACACIA CYANOPHYLLA	S	.	0	26.8	12.8	.	.	.	.	.	.	.	.	.	.	.	.	.	21
30	ACACIA DECURRENS	S	.	0	21.2	12.3	.	.	3.8	.	.	.	.	.	.	.	.	.	.	19
31	ACACIA ELATA	S	.	0	25.0	9.0	.	.	.	.	.	.	.	.	.	.	.	.	.	20
32	ACACIA FARNESIANA	SH	.	0	17.2	2.1	75.0	17.9	5.2	.	.	.	.	.	.	.	.	.	.	18
33	ACACIA FARNESIANA	F	.	0	17.2	1.6	77.3	19.4	4.1	550.0	200.0	.	.	.	.	.	.	.	.	18
34	ACACIA FARNESIANA	S	.	0	20.9	2.3	73.0	18.3	3.8	390.0	270.0	.	.	.	.	.	.	.	.	18
35	ACACIA FARNESIANA	S	.	0	55.0	7.8	.	.	4.3	.	.	.	.	.	.	.	.	.	.	19
36	ACACIA FARNESIANA	S	.	0	4.0	3.5	.	.	.	.	.	.	.	.	.	.	.	.	.	20
37	ACACIA FARNESIANA	S	.	0	22.4	3.0	.	.	.	.	.	.	.	.	.	.	.	.	.	21
38	ACACIA GALPINII	F	.	0	12.7	5.4	78.1	26.3	3.8	.	.	.	.	.	.	.	.	.	.	18
39	ACACIA GERRARDII	F	.	0	20.9	1.3	72.7	27.4	5.1	.	.	.	.	.	.	.	.	.	.	18
40	ACACIA GERRARDII	SH	.	0	17.7	1.6	74.7	24.2	6.0	630.0	250.0	.	.	.	.	.	.	.	.	18
41	ACACIA GIRAFFAE	F	.	0	13.8	1.7	80.9	32.9	3.6	710.0	110.0	.	.	.	.	.	.	.	.	18
42	ACACIA GIRAFFAE	S	.	0	26.3	4.7	.	.	.	.	.	.	.	.	.	.	.	.	.	21
43	ACACIA GREGGII	S	.	0	22.0	11.6	.	.	2.0	.	.	.	.	.	.	.	.	.	.	19
44	ACACIA GREGGII	S	.	0	13.2	2.2	.	.	3.5	.	.	.	.	.	.	.	.	.	.	19
45	ACACIA GREGGII	S	.	0	14.4	9.2	.	.	.	.	.	.	.	.	.	.	.	.	.	19

TABLE 3 - DATA CONVERTED TO A ZERO-MOISTURE BASIS.

OBS	PLANA	PART	CALORIES	WATER	PROTEIN	FAT	TOTCARB	FIBER	ASH	CALCIUM	PHOSPHO	IRON	SODIUM	POTASS	CAROTE	THIAMINE	RIBOFLAV	NIACIN	ASCORBIC	REFCO
46	ACACIA GREGGII	S	.	0	17.5	8.8	.	.	.	.	.	.	.	.	.	.	.	.	.	20
47	ACACIA GREGGII	S	.	0	18.1	8.4	.	.	.	.	.	.	.	.	.	.	.	.	.	20
48	ACACIA LAETA	L	.	0	26.4	3.0	61.4	18.9	9.2	.	.	.	.	.	.	.	.	.	.	18
49	ACACIA LINEARIS	S	.	0	23.1	9.4	.	.	.	.	.	.	.	.	.	.	.	.	.	20
50	ACACIA LONGIFOLIA	S	.	0	26.9	13.2	.	.	.	.	.	.	.	.	.	.	.	.	.	20
51	ACACIA MELANOXYLON	S	.	0	39.4	7.6	.	.	3.9	.	.	.	.	.	.	.	.	.	.	19
52	ACACIA MELLIFERA	L	.	0	42.8	2.6	48.4	16.2	6.2	510	580	.	.	.	.	.	.	.	.	18
53	ACACIA MELLIFERA	L	.	0	21.3	2.6	67.5	14.1	8.6	1140	180	.	.	.	.	.	.	.	.	18
54	ACACIA MILLEFOLIA	S	.	0	38.8	13.1	.	.	.	.	.	.	.	.	.	.	.	.	.	19
55	ACACIA MODESTA	S	.	0	38.9	8.4	.	.	.	.	.	.	.	.	.	.	.	.	.	21
56	ACACIA NILOTICA	L	.	0	12.9	12.6	68.1	11.3	6.4	1090	280	.	.	.	.	.	.	.	.	18
57	ACACIA NILOTICA	S	.	0	18.2	4.2	.	.	.	.	.	.	.	.	.	.	.	.	.	21
58	ACACIA NILOTICA	S	.	0	23.4	10.6	.	.	.	.	.	.	.	.	.	.	.	.	.	21
59	ACACIA NILOTICA SUBSP. SUBALATA	S	.	0	13.6	5.0	77.6	10.2	3.8	.	.	.	.	.	.	.	.	.	.	18
60	ACACIA NUBICA	L	.	0	32.4	1.7	57.1	15.8	8.8	1710	440	.	.	.	.	.	.	.	.	18
61	ACACIA NUBICA	F	.	0	15.2	1.9	75.4	37.4	7.5	1460	230	.	.	.	.	.	.	.	.	18
62	ACACIA OCCIDENTALIS	S	.	0	18.9	7.4	.	.	2.9	.	.	.	.	.	.	.	.	.	.	19
63	ACACIA OCCIDENTALIS	S	.	0	22.5	8.6	.	.	.	.	.	.	.	.	.	.	.	.	.	19
64	ACACIA PENNATULA	S	.	0	24.4	3.8	.	.	.	.	.	.	.	.	.	.	.	.	.	20
65	ACACIA PYCNANTHA	S	.	0	26.9	13.1	.	.	3.4	.	.	.	.	.	.	.	.	.	.	19
66	ACACIA RADDIANA	L	.	0	17.9	4.7	69.7	17.2	7.7	1550	210	.	.	.	.	.	.	.	.	18
67	ACACIA RADDIANA	F	.	0	17.5	1.8	74.5	20.4	6.2	900	260	.	.	.	.	.	.	.	.	18
68	ACACIA ROEMERIANA	L	.	0	20.5	3.6	70.6	20.2	5.3	1510	200	.	.	1050	.	.	.	.	.	6
69	ACACIA SALINGA	S	.	0	26.9	12.7	.	.	4.8	.	.	.	.	.	.	.	.	.	.	19
70	ACACIA SCHAFFNERI	S	.	0	27.5	3.5	.	.	.	.	.	.	.	.	.	.	.	.	.	20
71	ACACIA SENEGAL	L	.	0	18.2	6.7	66.7	11.2	8.4	1320	140	.	.	.	.	.	.	.	.	18
72	ACACIA SENEGAL	F	.	0	22.0	1.0	69.9	39.0	7.1	.	.	.	.	.	.	.	.	.	.	18
73	ACACIA SENEGAL	S	.	0	40.1	5.8	49.1	13.7	5.0	750	310	.	.	.	.	.	.	.	.	18
74	ACACIA SEYAL	L	.	0	22.6	4.0	64.4	8.4	9.0	3830	210	.	.	.	.	.	.	.	.	18
75	ACACIA SEYAL	F	.	0	20.9	1.9	67.9	20.2	9.3	1330	370	.	.	.	.	.	.	.	.	18
76	ACACIA SEYAL	E	.	0	10.6	0.9	79.7	22.2	8.8	2500	70	.	.	.	.	.	.	.	.	18
77	ACACIA SIEBERIANA	F	.	0	8.4	1.9	83.0	25.1	6.7	100	140	.	.	.	.	.	.	.	.	18
78	ACACIA SIEBERIANA	L	.	0	15.8	6.3	66.3	23.6	11.6	2500	100	.	.	.	.	.	.	.	.	18
79	ACACIA SP.	L	307	0	43.0	3.2	48.4	30.1	5.4	500	452	19.9	.	.	.	1.08	0.91	45.73	264	10
80	ACACIA SP.	S	.	0	44.8	4.7	.	.	4.4	.	.	.	.	.	.	.	.	.	.	19
81	ACACIA SUAVEOLENS	S	.	0	22.5	12.4	.	.	.	.	.	.	.	.	.	.	.	.	.	20
82	ACACIA TORTILIS	L	.	0	19.2	6.1	66.0	11.6	8.7	2270	170	.	.	.	.	.	.	.	.	18
83	ACACIA TORTILIS	F	.	0	17.3	3.1	73.9	24.8	5.7	790	340	.	.	.	.	.	.	.	.	18
84	ACACIA TORTILIS	S	.	0	37.8	6.0	50.6	10.9	5.9	560	730	.	.	.	.	.	.	.	.	18
85	ACACIA WILLARDIANA	S	.	0	35.2	21.3	.	.	4.9	.	.	.	.	.	.	.	.	.	.	19
86	ACALYPHA FRUTICOSA	SH	.	0	14.1	3.8	69.3	18.1	12.8	.	.	.	.	.	.	.	.	.	.	18
87	ACANTHOPANAX SPINOSUM	S	.	0	30.6	36.6	.	.	3.5	.	.	.	.	.	.	.	.	.	.	19
88	ACANTHOPANAX SPINOSUM	S	.	0	5.6	5.6	.	.	2.1	.	.	.	.	.	.	.	.	.	.	19
89	ACANTHUS HIRSUTUS	S	.	0	22.1	2.5	.	.	.	.	.	.	.	.	.	.	.	.	.	20
90	ACANTHUS MOLLIS	W	.	0	25.3	2.3	56.7	11.2	15.7	830	340	.	.	.	.	.	.	.	.	18

TABLE 3 - DATA CONVERTED TO A ZERO-MOISTURE BASIS.

OBS	PLANA	PART	CALORIES	WATER	PROTEIN	FAT	TOTLCARB	FIBER	ASH	CALCIUM	PHOSPHO	IRON	SODIUM	POTASS	CAROT	THIAMINE	RIBOFLAV	NIACIN	ASCORBIC	REFCO
91	ACANTHUS MOLLIS	S	.	0	26.1	1.4	69.7	2.8	2.8	.	.	.	.	.	.	.	.	.	.	18
92	ACANTHUS MOLLIS	S	.	0	27.5	1.2	.	.	.	.	.	.	.	.	.	.	.	.	.	20
93	ACANTHUS SYRIACUS	S	.	0	22.4	2.2	.	.	.	.	.	.	.	.	.	.	.	.	.	21
94	ACANTHUS SYRIACUS	S	.	0	24.9	2.4	.	.	.	.	.	.	.	.	.	.	.	.	.	21
95	ACER ARGUTUM	S	.	0	7.2	2.3	.	.	.	.	.	.	.	.	.	.	.	.	.	21
96	ACER BUERGERIANUM	S	.	0	15.6	9.5	.	.	.	.	.	.	.	.	.	.	.	.	.	20
97	ACER CAMPESTRE	S	.	0	7.2	1.3	.	.	.	.	.	.	.	.	.	.	.	.	.	21
98	ACER CISSIFOLIUM	S	.	0	23.8	17.4	.	.	.	.	.	.	.	.	.	.	.	.	.	20
99	ACER GLABRUM	S	.	0	8.3	5.8	.	.	.	.	.	.	.	.	.	.	.	.	.	21
100	ACER JAPONICUM	S	.	0	12.9	10.6	.	.	.	.	.	.	.	.	.	.	.	.	.	21
101	ACER NEGUNDO	S	.	0	48.8	28.2	.	.	.	.	.	.	.	.	.	.	.	.	.	20
102	ACER PALMATUM	S	.	0	12.9	5.3	.	.	.	.	.	.	.	.	.	.	.	.	.	21
103	ACER RUBRUM	S	.	0	31.8	3.9	.	.	4.3	.	.	.	.	.	.	.	.	.	.	19
104	ACER RUBRUM	L	.	0	.	.	.	.	.	850.0	220.0	29.0	.	.	.	.	.	.	.	6
105	ACER TATARICUM	S	.	0	17.2	8.9	.	.	.	.	.	.	.	.	.	.	.	.	.	21
106	ACER TRUNCATUM	S	.	0	17.5	33.9	.	.	.	.	.	.	.	.	.	.	.	.	.	20
107	ACER TRUNCATUM	S	.	0	15.6	36.6	.	.	.	.	.	.	.	.	.	.	.	.	.	20
108	ACHILLEA AEGYPTIACA	S	.	0	33.1	30.4	.	.	.	.	.	.	.	.	.	.	.	.	.	21
109	ACHILLEA FILIPENDULINA	S	.	0	33.8	27.6	.	.	.	.	.	.	.	.	.	.	.	.	.	20
110	ACHILLEA LANULOSA	W	.	0	12.8	3.9	71.8	23.6	10.4	1330.0	360.0	.	.	.	.	.	.	.	.	6
111	ACHILLEA MILLEFOLIUM	S	.	0	33.8	25.4	.	.	.	.	.	.	.	.	.	.	.	.	.	19
112	ACHILLEA MILLEFOLIUM	S	.	0	36.9	22.3	.	.	.	.	.	.	.	.	.	.	.	.	.	20
113	ACHILLEA MILLEFOLIUM	S	.	0	28.6	33.4	.	.	.	.	.	.	.	.	.	.	.	.	.	21
114	ACHILLEA MILLEFOLIUM	W	.	0	14.4	1.8	71.3	20.1	12.5	.	.	.	.	.	.	.	.	.	.	6
115	ACHILLEA SANTOLINA	S	.	0	31.9	24.6	.	.	.	.	.	.	.	.	.	.	.	.	.	21
116	ACHILLEA SANTOLINA	S	.	0	40.9	27.5	.	.	.	.	.	.	.	.	.	.	.	.	.	21
117	ACHRAS ZAPOTA	S	.	0	9.6	22.5	.	.	4.7	.	.	.	.	.	.	.	.	.	.	19
118	ACHRAS ZAPOTA	S	.	0	1.4	0.7	.	.	0.3	.	.	.	.	.	.	.	.	.	.	19
119	ACHRYANTHES ASPERA	S	.	0	20.5	3.5	.	.	.	.	.	.	.	.	.	.	.	.	.	21
120	ACHYRANTHES ASPERA	W	.	0	26.4	1.4	53.6	20.2	18.6	1790.0	310.0	.	.	.	.	.	.	.	.	18
121	ACHYRANTHES ASPERA	I	307	0	42.7	6.7	41.3	5.3	9.3	1053.3	393.3	0.0	.	.	43700.0	0.87	3.20	6.67	200.0	2
122	ACLEISANTHES SP.	L	.	0	24.4	2.0	62.4	24.7	11.2	2650.0	150.0	.	.	3370	.	.	.	.	.	6
123	ACNIDA CANNABINA	S	.	0	16.9	12.7	.	.	.	.	.	.	.	.	.	.	.	.	.	19
124	ACONITUM NAPELLUS	S	.	0	28.1	37.4	.	.	.	.	.	.	.	.	.	.	.	.	.	20
125	ACROCERAS AMPLECTENS	F	371	0	7.3	5.9	78.1	.	8.7	11.2	.	4.5	.	.	.	0.13	0.26	3.91	.	1
126	ACROCERAS MACRUM	H	.	0	8.5	1.8	83.6	31.8	6.1	.	.	.	.	.	.	.	.	.	.	18
127	ACROCERAS MACRUM	L	.	0	21.3	3.8	66.5	30.0	8.4	.	.	.	.	.	.	.	.	.	.	18
128	ACROCLINIUM ROSEUM	S	.	0	37.5	25.9	.	.	.	.	.	.	.	.	.	.	.	.	.	19
129	ACROCOMIA MEXICANA	SH	315	0	19.4	3.2	67.7	5.6	9.7	1266.1	475.8	8.1	.	.	0.0	0.32	0.65	10.48	161.3	3
130	ACROCOMIA MEXICANA	F	479	0	9.1	28.6	57.9	27.8	4.4	412.9	118.3	0.4	.	.	290.5	0.29	0.19	2.07	58.1	3
131	ACTAEA RUBRA	S	.	0	14.4	26.0	.	.	.	.	.	.	.	.	.	.	.	.	.	20
132	ACTINIDIA ARGUTA	S	.	0	16.2	22.4	.	.	3.0	.	.	.	.	.	.	.	.	.	.	19
133	ACTINIDIA CHINENSIS	S	.	0	15.0	34.1	.	.	.	.	.	.	.	.	.	.	.	.	.	19
134	ACTINOLEMA MACROLEMA	S	.	0	21.3	48.4	.	.	.	.	.	.	.	.	.	.	.	.	.	21
135	ACTINOMERIS ALTERNIFOLIA	S	.	0	43.1	38.8	.	.	.	.	.	.	.	.	.	.	.	.	.	19

TABLE 3 - DATA CONVERTED TO A ZERO-MOISTURE BASIS.

OBS	PLNA	PART	CALORIES	WATER	PROTEIN	FAT	TOTLCARB	FIBER	ASH	CALCIUM	PHOSPHO	IRON	SODIUM	POTASS	CAROT	THIAMINE	RIBOFLAV	NIACIN	ASCORBIC	REFCO
136	ACTINOPHLOEUS MACARTHURII	S	.	0	5.9	1.6	.	.	1.4	.	.	.	.	.	.	.	.	.	.	19
137	ADANSONIA DIGITATA	P	345	0	2.6	1.0	91.3	8.1	5.1	338.1	140.5	8.8	.	.	83.3	0.44	0.07	2.50	321.4	1
138	ADANSONIA DIGITATA	L	320	0	13.9	3.5	71.7	11.0	10.9	2540.8	311.8	27.2	.	.	11009.1	0.15	0.93	4.99	0.0	1
139	ADANSONIA DIGITATA	S	490	0	32.5	32.1	26.8	3.5	8.6	285.2	1620.4	15.1	.	.	0.0	1.50	0.14	1.52	.	1
140	ADENANTHERA MICROSPERMA	S	.	0	22.5	14.3	.	.	.	.	.	.	.	.	.	.	.	.	.	19
141	ADENANTHERA PAVONINA	SH	.	0	24.8	3.5	65.9	14.0	5.8	190.0	240.0	.	.	.	.	.	.	.	.	18
142	ADENANTHERA PAVONINA	S	.	0	40.0	26.1	.	.	4.2	.	.	.	.	.	.	.	.	.	.	19
143	ADENANTHERA PAVONINA	S	.	0	4.0	1.0	.	.	4.5	.	.	.	.	.	.	.	.	.	.	19
144	ADENANTHERA PAVONINA	S	.	0	21.2	13.5	.	.	.	.	.	.	.	.	.	.	.	.	.	20
145	ADENIA GUMMIFERA	L	311	0	18.9	5.8	60.5	10.0	14.7	978.9	252.6	22.1	.	.	.	.	.	.	.	1
146	ADENOCARPUS HISPANICUS	S	.	0	38.9	6.2	.	.	.	.	.	.	.	.	.	.	.	.	.	21
147	ADENOCARPUS SP.	S	.	0	39.2	7.4	.	.	.	.	.	.	.	.	.	.	.	.	.	21
148	ADENOPHORA LILIFOLIA	S	.	0	19.4	35.1	.	.	.	.	.	.	.	.	.	.	.	.	.	20
149	ADENOPHORA POTANINII	S	.	0	21.2	37.0	.	.	.	.	.	.	.	.	.	.	.	.	.	20
150	ADLUMIA FUNGOSA	S	.	0	14.4	31.0	.	.	.	.	.	.	.	.	.	.	.	.	.	20
151	ADONIDIA MERRILLII	S	.	0	4.1	1.3	.	.	1.5	.	.	.	.	.	.	.	.	.	.	19
152	ADONIS AESTIVALIS	S	.	0	10.0	17.8	.	.	.	.	.	.	.	.	.	.	.	.	.	20
153	ADONIS AESTIVALIS	S	.	0	11.9	20.4	.	.	.	.	.	.	.	.	.	.	.	.	.	20
154	ADONIS AUTUMNALIS	S	.	0	11.2	16.4	.	.	.	.	.	.	.	.	.	.	.	.	.	20
155	ADONIS FLAMMEA	S	.	0	14.4	20.5	.	.	.	.	.	.	.	.	.	.	.	.	.	20
156	AEGILOPS CYLINDRICA	S	.	0	21.6	2.1	.	.	.	.	.	.	.	.	.	.	.	.	.	21
157	AEGILOPS TRIUNCIALIS	S	.	0	23.1	2.7	.	.	.	.	.	.	.	.	.	.	.	.	.	21
158	AEGLE MARMELOS	F	345	0	4.7	0.8	90.1	7.5	4.4	220.8	129.9	1.6	.	1558.4	142.9	3.38	3.09	2.86	20.8	2
159	AELUROPUS LITTORALIS	S	.	0	15.5	3.1	.	.	.	.	.	.	.	.	.	.	.	.	.	21
160	AERVA LANATA	L	284	0	23.1	3.6	55.6	12.4	17.8	2917.2	508.9	.	.	.	.	.	.	.	.	1
161	AERVA PERSICA	W	.	0	10.3	0.8	73.0	41.7	8.4	.	.	.	.	.	.	.	.	.	.	18
162	AESCHYNOMENE AMERICANA	S	.	0	37.3	7.0	.	.	4.5	.	.	.	.	.	.	.	.	.	.	19
163	AESCHYNOMENE AMERICANA	S	.	0	35.8	6.8	.	.	4.7	.	.	.	.	.	.	.	.	.	.	19
164	AESCHYNOMENE AMORPHOIDES	S	.	0	44.4	10.8	.	.	.	.	.	.	.	.	.	.	.	.	.	21
165	AESCHYNOMENE INDICA	S	.	0	33.1	9.1	.	.	.	.	.	.	.	.	.	.	.	.	.	20
166	AESCHYNOMENE RUDIS	S	.	0	33.4	11.5	.	.	.	.	.	.	.	.	.	.	.	.	.	21
167	AESCHYNOMENE SCABRA	S	.	0	31.4	7.2	.	.	.	.	.	.	.	.	.	.	.	.	.	21
168	AESCULUS OCTANDRA	S	.	0	16.2	3.7	.	.	.	.	.	.	.	.	.	.	.	.	.	20
169	AESCULUS TURBINATA	S	.	0	6.9	6.6	.	.	.	.	.	.	.	.	.	.	.	.	.	20
170	AESCULUS TURBINATA	S	431	0	3.0	7.1	88.6	4.6	1.3	280.0	.	.	.	.	0.0	.	.	2.45	0.0	2
171	AFRAEGLE PANICULATA	L	315	0	27.2	3.0	62.6	14.9	7.2	2000.0	229.8	.	.	.	.	.	.	.	.	1
172	AFRAMOMUM MELEGUETA	F	.	0	9.5	.	35.2	.	5.3	949.4	0.0	.	.	.	.	.	.	.	.	1
173	AGASTACHE URTICIFOLIA	S	.	0	21.1	32.0	.	.	.	.	.	.	.	.	.	.	.	.	.	20
174	AGASTACHE URTICIFOLIA	S	.	0	17.7	31.1	.	.	.	.	.	.	.	.	.	.	.	.	.	21
175	AGASTACHE URTICIFOLIA	W	.	0	22.8	3.0	62.3	15.8	11.9	1490.0	420.0	.	.	.	.	.	.	.	.	6
176	AGAVE AMERICANA	H	.	0	6.1	1.3	82.2	15.6	12.4	.	.	.	.	.	.	.	.	.	.	18
177	AGAVE ATROVIRENS	L	.	0	2.3	2.8	86.0	32.4	8.9	.	.	.	.	.	.	.	.	.	.	18
178	AGAVE ATROVIRENS	I	333	0	10.0	2.2	81.1	.	6.7	1266.7	300.0	10.0	.	.	3555.6	1.22	0.56	2.22	655.6	3
179	AGAVE GEMINIFLORA	S	.	0	32.5	12.8	.	.	.	.	.	.	.	.	.	.	.	.	.	20
180	AGAVE LECHEGUILLA	S	.	0	24.9	17.4	.	.	.	.	.	.	.	.	.	.	.	.	.	21

TABLE 3 - DATA CONVERTED TO A ZERO-MOISTURE BASIS.

OBS	PLNA	PART	CALORIES	WATER	PROTEIN	FAT	TOTCARB	FIBER	ASH	CALCIUM	PHOSPHO	IRON	SODIUM	POTASS	CAROT	THIAMINE	RIBOFLAV	NIACIN	ASCORBIC	REFCO
181	AGAVE PERRYI	S	.	0	26.9	25.1	.	.	.	.	.	.	.	.	.	.	.	.	.	19
182	AGAVE PERRYI	S	.	0	8.1	0.8	.	.	.	.	.	.	.	.	.	.	.	.	.	19
183	AGAVE POTRERANA	S	.	0	27.5	18.1	.	.	.	.	.	.	.	.	.	.	.	.	.	20
184	AGAVE SCHOTTII	S	.	0	29.4	21.4	.	.	0.3	.	.	.	.	.	.	.	.	.	.	19
185	AGERATUM HOUSTONIANUM	S	.	0	25.2	30.8	.	.	7.4	.	.	.	.	.	.	.	.	.	.	19
186	AGERATUM HOUSTONIANUM	S	.	0	24.4	31.4	.	.	5.7	.	.	.	.	.	.	.	.	.	.	19
187	AGRIMONIA EUPATORIA	S	.	0	35.3	5.8	.	.	.	.	.	.	.	.	.	.	.	.	.	21
188	AGRIMONIA GRYPOSEPALA	S	.	0	16.8	10.1	.	.	.	.	.	.	.	.	.	.	.	.	.	21
189	AGROPYRON CRISTATUM	S	.	0	24.4	2.6	.	.	.	.	.	.	.	.	.	.	.	.	.	20
190	AGROPYRON CRISTATUM	S	.	0	18.6	2.3	.	.	.	.	.	.	.	.	.	.	.	.	.	21
191	AGROPYRON CRISTATUM	W	.	0	12.1	2.9	76.1	30.4	8.9	410	210	.	.	.	25	0.33	0.86	.	.	6
192	AGROPYRON DASYSTACHYUM	S	.	0	13.8	2.5	.	.	6.9	.	.	.	.	.	.	.	.	.	.	19
193	AGROPYRON INERME	W	.	0	11.3	3.7	77.3	29.7	7.7	580	240	.	.	.	32	.	.	.	.	6
194	AGROPYRON INTERMEDIUM	W	.	0	19.3	5.9	62.3	26.6	12.5	.	.	.	.	.	.	.	.	.	.	18
195	AGROPYRON INTERMEDIUM	H	.	0	8.7	2.3	79.8	34.8	9.2	360	.	.	.	.	.	.	.	.	.	6
196	AGROPYRON JUNCEUM	S	.	0	20.9	3.1	.	.	.	.	.	.	.	.	.	.	.	.	.	21
197	AGROPYRON REPENS	S	.	0	21.2	12.1	.	.	.	.	.	.	.	.	.	.	.	.	.	20
198	AGROPYRON REPENS	S	.	0	16.8	1.8	.	.	.	.	.	.	.	.	.	.	.	.	.	21
199	AGROPYRON REPENS	H	.	0	10.6	2.8	79.7	33.9	6.9	.	.	.	.	.	.	.	.	.	.	6
200	AGROPYRON REPENS	W	.	0	11.2	3.8	77.2	31.7	7.8	330	280	.	.	.	12	.	.	.	.	6
201	AGROPYRON SMITHII	H	.	0	7.5	3.0	80.9	33.5	8.6	300	140	11	.	.	13	.	.	.	.	6
202	AGROPYRON SPICATUM	W	.	0	10.2	3.1	78.8	31.4	7.9	420	240	.	.	.	23	.	.	.	.	6
203	AGROPYRON SUBSECUNDUM	H	.	0	3.4	4.1	80.9	35.0	11.6	.	.	.	.	.	.	.	.	.	.	6
204	AGROPYRON TRACHYCAULUM	S	.	0	14.5	2.0	.	.	5.1	.	.	.	.	.	.	.	.	.	.	19
205	AGROPYRON TRACHYCAULUM	H	.	0	9.9	2.6	78.6	34.2	8.9	360	240	.	.	2680	.	0.51	1.06	.	.	6
206	AGROPYRON X. TRITICUM	S	.	0	26.2	2.6	.	.	.	.	.	.	.	.	.	.	.	.	.	20
207	AGROPYRON X. TRITICUM	S	.	0	26.2	2.0	.	.	.	.	.	.	.	.	.	.	.	.	.	20
208	AGROSTEMMA GITHAGO	S	.	0	15.4	6.4	.	.	4.1	.	.	.	.	.	.	.	.	.	.	19
209	AGROSTEMMA GITHAGO	S	.	0	16.2	6.8	.	.	.	.	.	.	.	.	.	.	.	.	.	20
210	AGROSTEMMA GITHAGO	S	.	0	14.4	6.4	.	.	.	.	.	.	.	.	.	.	.	.	.	21
211	AGROSTIS ALBA	H	.	0	8.6	2.7	81.4	31.7	7.3	450	230	27	.	1690	0	.	.	.	.	6
212	AGROSTIS ALBA	W	.	0	13.3	3.8	74.5	25.0	8.4	620	370	25	.	2350	20	.	.	.	.	6
213	AGROSTIS CANINA	W	.	0	19.3	3.0	67.5	20.6	10.2	.	490	45	.	.	.	.	.	.	.	6
214	AGROSTIS HIEMALIS	W	.	0	9.6	2.8	78.8	34.4	8.8	390	260	.	.	.	.	.	.	.	.	6
215	AGROSTIS PALUSTRIS	H	.	0	4.3	1.4	88.5	30.8	5.8	.	.	.	.	.	.	.	.	.	.	6
216	AGROSTIS SCHIMPERANA	W	.	0	17.9	2.1	69.3	28.2	10.7	.	.	.	.	.	.	.	.	.	.	18
217	AGROSTIS TENUIS	H	.	0	7.1	2.3	83.5	31.4	7.1	.	.	.	.	.	.	.	.	.	.	6
218	AGROSTIS TENUIS	W	.	0	17.7	3.5	68.6	23.5	10.2	660	390	.	.	2270	13	.	.	.	.	6
219	AGROSTIS VERTICILLATA	W	.	0	9.3	2.4	73.8	29.7	14.5	530	180	.	.	.	.	.	.	.	.	6
220	AILANTHUS ALTISSIMA	S	.	0	27.6	55.5	.	.	4.1	.	.	.	.	.	.	.	.	.	.	19
221	AILANTHUS ALTISSIMA	S	.	0	27.5	59.1	.	.	.	.	.	.	.	.	.	.	.	.	.	19
222	AJUGA CHIA	S	.	0	11.8	20.8	.	.	.	.	.	.	.	.	.	.	.	.	.	20
223	AKEBIA QUINATA	S	.	0	15.6	38.3	.	.	.	.	.	.	.	.	.	.	.	.	.	20
224	AKEBIA TRIFOLIATA	S	.	0	15.6	38.9	.	.	.	.	.	.	.	.	.	.	.	.	.	20
225	ALBIZIA AMARA	F	.	0	26.8	1.7	64.5	21.4	7.0	1020	280	.	.	.	.	.	.	.	.	18

TABLE 3 - DATA CONVERTED TO A ZERO-MOISTURE BASIS.

OBS	PLANA	PART	CALORIES	WATER	PROTEIN	FAT	TOTCARB	FIBER	ASH	CALCIUM	PHOSPHO	IRON	SODIUM	POTASS	CAROT	THIAMINE	RIBOFLAV	NIACIN	ASCORBIC	REFCO
226	ALBIZIA AMARA	SH	.	0	26.8	1.9	65.2	26.8	6.1	470.0	250.0	.	.	.	.	.	.	.	.	18
227	ALBIZIA AMARA	F	.	0	13.7	7.5	74.5	35.2	4.3	.	.	.	.	.	.	.	.	.	.	18
228	ALBIZIA CHINENSIS	L	.	0	15.1	4.4	75.2	31.6	5.3	1170.0	140.0	.	.	.	.	.	.	.	.	18
229	ALBIZIA JULIBRISSIN	S	.	0	33.0	11.1	.	.	4.7	.	.	.	.	.	.	.	.	.	.	19
230	ALBIZIA JULIBRISSIN	S	.	0	32.1	8.5	.	.	.	.	.	.	.	.	.	.	.	.	.	21
231	ALBIZIA LEBBEK	L	.	0	18.1	4.7	69.2	26.5	8.0	2020.0	140.0	.	.	.	.	.	.	.	.	18
232	ALBIZIA LEBBEK	F	.	0	21.1	4.6	69.7	23.0	4.6	.	.	.	.	.	.	.	.	.	.	18
233	ALBIZIA LEBBEK	S	.	0	30.2	5.3	.	.	.	.	.	.	.	.	.	.	.	.	.	21
234	ALBIZIA LOPHANTHA	S	.	0	32.5	8.0	.	.	1.3	.	.	.	.	.	.	.	.	.	.	19
235	ALBIZIA SP.	L	310	0	35.0	1.0	61.0	16.0	3.0	2100.0	310.0	.	.	.	.	.	.	.	.	1
236	ALBIZZIA SP.	L	310	0	35.0	1.0	61.0	16.0	3.0	2100.0	310.0	.	.	.	.	.	.	.	.	10
237	ALCEA ACAULIS	S	.	0	12.2	6.6	.	.	.	.	.	.	.	.	.	.	.	.	.	21
238	ALCEA PALLIDA	S	.	0	16.4	12.8	.	.	.	.	.	.	.	.	.	.	.	.	.	21
239	ALCEA SETOSA	S	.	0	15.1	6.9	.	.	.	.	.	.	.	.	.	.	.	.	.	21
240	ALCEA SP.	S	.	0	16.4	7.0	.	.	.	.	.	.	.	.	.	.	.	.	.	21
241	ALCEA SP.	S	.	0	15.2	3.2	.	.	.	.	.	.	.	.	.	.	.	.	.	21
242	ALCEA STRIATA SUBSP. STRIATA	S	.	0	33.4	17.4	.	.	.	.	.	.	.	.	.	.	.	.	.	21
243	ALEURITES MOLUCCANA	S	.	0	23.1	64.0	.	.	3.6	.	.	.	.	.	.	.	.	.	.	19
244	ALEURITES MOLUCCANA	S	673	0	20.4	67.7	8.6	.	3.2	86.0	215.1	2.2	.	.	0.0	0.06	.	.	0.0	2
245	ALEURITES TRISPERMA	S	.	0	4.9	65.4	.	.	.	.	.	.	.	.	.	.	.	.	.	20
246	ALHAGI CAMELORUM	S	.	0	21.8	4.0	.	.	.	.	.	.	.	.	.	.	.	.	.	21
247	ALHAGI CAMELORUM	S	.	0	27.7	5.0	.	.	.	.	.	.	.	.	.	.	.	.	.	21
248	ALISMA PLANTAGO-AQUATICA	S	.	0	13.2	4.2	.	.	.	.	.	.	.	.	.	.	.	.	.	21
249	ALLAEANTHUS LUZONICUS	I	313	0	34.4	5.4	50.9	10.7	9.4	1258.9	553.6	30.4	22.3	3500.0	5223.2	0.67	1.56	7.14	107.1	2
250	ALLANBLACKIA FLORIBUNDA	S	689	0	3.8	68.5	25.3	3.3	2.3	129.8	179.8	.	.	.	.	.	.	.	.	1
251	ALLANBLACKIA FLORIBUNDA	L	314	0	28.1	5.2	57.1	11.4	9.5	1300.0	404.8	17.6	23.8	2281.0	14190.5	0.67	1.67	5.71	452.4	2
252	ALLANBLACKIA FLORIBUNDA	F	368	0	3.0	3.8	91.7	7.5	1.5	37.6	97.7	3.0	.	.	375.9	0.45	0.30	2.26	37.6	3
253	ALLIUM AMPELOPRASUM	R	328	0	16.1	2.2	75.2	.	6.6	379.6	365.0	8.0	.	.	182.5	0.80	.	.	124.1	2
254	ALLIUM AMPELOPRASUM	L	330	0	21.7	4.7	66.0	9.4	7.5	547.2	452.8	25.5	47.2	2981.1	22971.7	0.85	0.94	5.66	301.9	2
255	ALLIUM AMPELOPRASUM	I	325	0	32.5	3.0	62.1	6.5	2.4	136.1	224.9	5.3	.	.	15088.8	0.83	1.12	5.33	236.7	2
256	ALLIUM AMPELOPRASUM	L	356	0	15.1	2.1	76.7	8.9	6.2	356.2	342.5	7.5	34.2	2376.7	164.4	0.75	0.41	3.42	116.4	4
257	ALLIUM CANADENSE	BU	.	0	13.1	1.9	.	.	.	.	.	.	.	.	.	.	.	.	.	19
258	ALLIUM CEPA	R	358	0	14.2	2.2	77.6	6.0	6.0	194.0	320.9	5.2	59.7	2216.4	0.0	0.30	0.37	7.46	44.8	2
259	ALLIUM CEPA	R	333	0	14.0	1.8	78.9	6.1	5.3	263.2	386.0	8.8	78.9	1456.1	0.0	0.53	0.35	1.75	78.9	2
260	ALLIUM CEPA	Q	375	0	14.4	1.9	79.8	9.6	3.8	326.9	403.8	13.5	.	.	0.0	0.38	0.38	3.85	125.0	3
261	ALLIUM CEPA	L	333	0	23.1	7.7	60.3	14.1	9.0	538.5	551.3	43.6	.	.	15769.2	0.64	1.41	8.97	500.0	3
262	ALLIUM CEPA	R	349	0	13.8	0.9	79.8	5.5	5.5	247.7	330.3	4.6	91.7	1440.4	220.2	0.28	0.37	1.83	91.7	4
263	ALLIUM CF. DRUMMONDII	S	.	0	24.4	17.7	.	.	.	.	.	.	.	.	.	.	.	.	.	20
264	ALLIUM FISTULOSUM	R	349	0	18.6	3.5	70.9	9.3	7.0	639.5	476.7	12.8	174.4	2232.6	7325.6	0.70	0.93	5.81	220.9	2
265	ALLIUM ODORUM	L	293	0	25.9	3.4	58.6	6.9	12.1	508.6	569.0	22.4	51.7	2017.2	8793.1	0.69	1.29	7.76	310.3	2
266	ALLIUM PANICULATUM	S	.	0	30.5	11.3	.	.	.	.	.	.	.	.	.	.	.	.	.	21
267	ALLIUM PORRUM	S	.	0	25.9	15.9	.	.	4.0	.	.	.	.	.	.	.	.	.	.	19
268	ALLIUM PORRUM	S	.	0	28.2	13.9	.	.	4.4	.	.	.	.	.	.	.	.	.	.	19
269	ALLIUM ROTUNDUM	BU	.	0	15.8	2.2	.	.	.	.	.	.	.	.	.	.	.	.	.	21
270	ALLIUM ROTUNDUM	S	.	0	28.5	15.3	.	.	.	.	.	.	.	.	.	.	.	.	.	21

TABLE 3 - DATA CONVERTED TO A ZERO-MOISTURE BASIS.

OBS	PLNA	PART	CALORIES	WATER	PROTEIN	FAT	TOTLCARB	FIBER	ASH	CALCIUM	PHOSPHO	IRON	SODIUM	POTASS	CAROT	THIAMINE	RIBOFLAV	NIACIN	ASCORBIC	REFCO
271	ALLIUM SATIVUM	R	363	0	10.9	0.9	85.1	2.2	3.1	55.9	273.3	4.7	55.9	1158.4	0.0	0.75	0.16	1.24	31.1	2
272	ALLIUM SATIVUM	L	324	0	19.1	3.7	69.9	13.2	7.4	426.5	338.2	4.4	29.4	2397.1	6764.7	0.81	1.03	4.41	286.8	2
273	ALLIUM SATIVUM	SH	341	0	5.4	1.3	90.1	7.6	3.1	53.8	233.2	7.6	.	1224.2	896.9	0.63	0.27	2.24	188.3	2
274	ALLIUM SATIVUM	I	336	0	12.1	1.7	81.0	6.9	5.2	215.5	396.6	7.8	.	.	517.2	0.95	0.52	3.45	379.3	2
275	ALLIUM SATIVUM	R	354	0	16.0	0.5	79.6	3.9	3.9	74.9	522.0	3.9	49.1	1366.9	0.0	0.65	0.21	1.29	38.8	4
276	ALLIUM SCHOENOPRASUM	L	338	0	33.8	7.5	53.8	8.8	5.0	1037.5	512.5	10.0	.	.	.	1.25	0.75	6.25	400.0	2
277	ALLIUM SCHOENOPRASUM	L	322	0	20.7	3.4	66.7	12.6	9.2	793.1	505.7	19.5	.	2873.6	40000.0	0.92	1.49	5.75	643.7	4
278	ALLIUM SP.	BU	.	0	14.4	1.8	.	.	2.9	.	.	.	.	.	.	.	.	.	.	19
279	ALLIUM SP.	S	.	0	31.8	10.5	.	.	.	.	.	.	.	.	.	.	.	.	.	21
280	ALLIUM VINEALE	BU	.	0	27.8	2.8	.	.	3.9	.	.	.	.	.	.	.	.	.	.	19
281	ALLOTEROPSIS SEMIALATA	W	.	0	15.6	1.1	74.3	14.7	9.0	.	.	.	.	.	.	.	.	.	.	18
282	ALNUS CORDIFOLIA	S	.	0	24.4	23.2	.	.	.	.	.	.	.	.	.	.	.	.	.	20
283	ALNUS CORDIFOLIA	S	.	0	15.6	16.8	.	.	.	.	.	.	.	.	.	.	.	.	.	20
284	ALNUS GLUTINOSA	S	.	0	19.4	16.1	.	.	.	.	.	.	.	.	.	.	.	.	.	19
285	ALNUS GLUTINOSA	S	.	0	11.1	11.7	.	.	.	.	.	.	.	.	.	.	.	.	.	21
286	ALNUS INCANA	S	.	0	16.2	9.8	.	.	.	.	.	.	.	.	.	.	.	.	.	20
287	ALOCASIA MACRORRHIZA	L	.	0	24.0	6.3	52.0	12.5	17.7	1570.0	140.0	.	.	.	.	.	.	.	.	18
288	ALOCASIA MACRORRHIZA	R	381	0	3.8	1.9	92.5	.	1.9	187.5	312.5	6.3	.	.	0.0	0.31	.	.	31.3	2
289	ALONSOA WARSCEWICZII	S	.	0	20.0	26.2	.	.	4.3	.	.	.	.	.	.	.	.	.	.	19
290	ALOPECURUS ARUNDINACEUS	S	.	0	24.4	13.9	.	.	.	.	.	.	.	.	.	.	.	.	.	20
291	ALOPECURUS ARUNDINACEUS	S	.	0	21.0	14.0	.	.	.	.	.	.	.	.	.	.	.	.	.	21
292	ALOPECURUS PRATENSIS	S	.	0	20.6	15.8	.	.	.	.	.	.	.	.	.	.	.	.	.	20
293	ALOPECURUS PRATENSIS	W	.	0	17.2	4.6	67.5	21.5	10.7	570.0	460.0	.	.	.	.	.	.	.	.	6
294	ALPINIA GALANGA	R	362	0	7.1	2.8	83.0	22.0	7.1	219.9	177.3	14.9	.	.	10780.1	0.35	0.14	7.09	184.4	2
295	ALTERNANTHERA MARITIMA	L	256	0	23.3	4.2	46.5	10.2	26.0	660.5	162.8	.	.	.	.	.	.	.	.	1
296	ALTERNANTHERA PHILOXEROIDES	SH	.	0	7.3	1.6	77.5	8.5	13.6	140.0	.	.	.	.	.	.	.	.	.	18
297	ALTERNANTHERA SESSILIS	S	.	0	18.2	11.4	.	.	.	.	.	.	.	.	.	.	.	.	.	21
298	ALTERNANTHERA SESSILIS	L	303	0	23.7	4.0	59.6	10.6	12.6	737.4	227.3	.	.	.	.	.	.	.	.	1
299	ALTERNANTHERA SP.		.	0	21.1	1.6	55.9	14.8	21.4	3060.0	300.0	.	.	3750.0	.	.	.	.	.	6
300	ALTERNANTHERA SP.	L	301	0	12.6	1.4	72.0	.	14.0	1412.6	503.5	111.9	804.2	3272.7	5909.1	0.14	1.96	10.49	83.9	2
301	ALTHAEA OFFICINALIS	S	.	0	23.7	12.8	.	.	.	.	.	.	.	.	.	.	.	.	.	21
302	ALTHAEA OFFICINALIS	S	.	0	22.7	17.2	.	.	.	.	.	.	.	.	.	.	.	.	.	21
303	ALTHAEA ROSEA	S	.	0	25.6	12.8	.	.	.	.	.	.	.	.	.	.	.	.	.	19
304	ALVARADOA AMORPHOIDES	S	.	0	25.0	58.8	.	.	.	.	.	.	.	.	.	.	.	.	.	20
305	ALYSICARPUS RUGOSUS	SH	.	0	14.7	1.8	73.5	39.1	10.0	1260.0	230.0	.	.	.	.	.	.	.	.	18
306	ALYSICARPUS VAGINALIS	L	.	0	14.0	6.3	69.6	29.3	10.1	.	.	.	.	.	.	.	.	.	.	18
307	ALYSICARPUS VAGINALIS	SH	.	0	13.9	2.5	70.7	21.4	12.9	.	.	.	.	.	.	.	.	.	.	18
308	ALYSICARPUS VAGINALIS	S	.	0	12.0	1.4	75.6	48.9	11.0	.	.	.	.	.	.	.	.	.	.	18
309	ALYSICARPUS VAGINALIS	S	.	0	34.2	6.6	.	.	4.1	.	.	.	.	.	.	.	.	.	.	19
310	ALYSICARPUS VAGINALIS	S	.	0	36.6	6.2	.	.	4.0	.	.	.	.	.	.	.	.	.	.	19
311	ALYSICARPUS VAGINALIS	S	.	0	35.6	5.6	.	.	.	.	.	.	.	.	.	.	.	.	.	20
312	ALYSICARPUS VAGINALIS	H	.	0	12.6	1.7	79.2	39.4	6.5	.	.	.	.	.	.	.	.	.	.	6
313	ALYSSUM CAMPESTRE	S	.	0	25.8	18.1	.	.	.	.	.	.	.	.	.	.	.	.	.	20
314	ALYSSUM CONSTELLATUM	S	.	0	26.6	9.5	.	.	.	.	.	.	.	.	.	.	.	.	.	21
315	ALYSSUM DASYCARPUM	S	.	0	33.7	29.2	.	.	.	.	.	.	.	.	.	.	.	.	.	20

TABLE 3 - DATA CONVERTED TO A ZERO-MOISTURE BASIS.

OBS	PLNA	PART	CALORIES	WATER	PROTEIN	FAT	TOTLCARB	FIBER	ASH
316	ALYSSUM MINIMUM	S	.	0	25.0	20.9	.	.	.
317	ALYSSUM SAXATILE	S	.	0	26.9	18.3	.	.	.
318	ALYSSUM SAXATILE	S	.	0	27.5	16.8	.	.	.
319	ALYSSUM SAXATILE VAR. LUTEUM	S	.	0	28.8	15.7	.	.	.
320	ALYSSUM TORTUOSUM	S	.	0	31.8	25.4	.	.	.
321	AMARANTHUS ALBUS	S	.	0	16.3	7.6	.	.	.
322	AMARANTHUS BLITOIDES	S	.	0	16.1	6.8	.	.	.
323	AMARANTHUS BLITOIDES	W	.	0	27.0	1.4	50.4	11.7	21.2
324	AMARANTHUS BLITOIDES	W	.	0	18.5	1.3	53.9	13.1	26.3
325	AMARANTHUS CAUDATUS	L	265	0	26.7	2.8	49.5	8.7	21.0
326	AMARANTHUS CAUDATUS	S	.	0	17.5	6.5	.	.	2.8
327	AMARANTHUS CAUDATUS	S	.	0	16.9	6.9	.	.	.
328	AMARANTHUS CAUDATUS	S	408	0	14.7	8.2	74.2	7.6	2.9
329	AMARANTHUS DEFLEXUS	S	.	0	17.2	6.5	.	.	.
330	AMARANTHUS GRACILIS	L	256	0	31.9	3.0	43.0	9.2	22.2
331	AMARANTHUS GRAECIZANS	L	269	0	23.2	3.2	52.0	13.5	21.6
332	AMARANTHUS GRAECIZANS	SH	.	0	19.9	1.5	61.6	21.0	17.0
333	AMARANTHUS HYBRIDUS	S	.	0	16.2	6.6	.	.	.
334	AMARANTHUS HYBRIDUS	S	.	0	16.7	4.8	.	.	.
335	AMARANTHUS LIVIDUS	S	.	0	15.4	7.6	.	.	.
336	AMARANTHUS PALMERI	S	.	0	15.8	8.1	.	.	.
337	AMARANTHUS RETROFLEXUS	S	.	0	17.5	7.2	.	.	3.4
338	AMARANTHUS RETROFLEXUS	S	.	0	16.8	6.9	.	.	.
339	AMARANTHUS RETROFLEXUS	S	.	0	16.9	7.4	.	.	.

OBS	CALCIUM	PHOSPHO	IRON	SODIUM	POTASS	CAROT	THIAMINE	RIBOFLAV	NIACIN	ASCORBIC	REFCO
316	.	.	.	.	.	.	.	.	.	.	20
317	.	.	.	.	.	.	.	.	.	.	20
318	.	.	.	.	.	.	.	.	.	.	20
319	.	.	.	.	.	.	.	.	.	.	20
320	.	.	.	.	.	.	.	.	.	.	20
321	.	.	.	.	.	.	.	.	.	.	21
322	.	.	.	.	.	.	.	.	.	.	21
323	4190.0	490.0	.	.	5200.0	.	.	.	.	.	6
324	3200.0	180.0	.	.	4560.0	.	.	.	.	.	6
325	2341.0	304.0	.	11.0	263.0	.	.	.	4.74	.	12
326	.	.	.	.	.	.	.	.	.	.	19
327	.	.	.	.	.	.	.	.	.	.	20
328	281.6	570.1	3.9	.	.	0.0	0.16	0.36	1.14	3.4	3
329	.	.	.	.	.	.	.	.	.	.	21
330	2877.0	559.0	.	14.0	403.0	.	.	.	7.19	.	12
331	1827.0	455.0	.	11.0	226.0	.	.	.	6.23	.	12
332	1810.0	540.0	.	.	.	.	.	.	.	.	18
333	.	.	.	.	.	.	.	.	.	.	19
334	.	.	.	.	.	.	.	.	.	.	21
335	.	.	.	.	.	.	.	.	.	.	21
336	.	.	.	.	.	.	.	.	.	.	21
337	.	.	.	.	.	.	.	.	.	.	19
338	.	.	.	.	.	.	.	.	.	.	21
339	.	.	.	.	.	.	.	.	.	.	21

TABLE 3 - DATA CONVERTED TO A ZERO-MOISTURE BASIS.

OBS	PLNA	PART	CALORIES	WATER	PROTEIN	FAT	TOTLCARB	FIBER	ASH
340	AMARANTHUS RETROFLEXUS	S	.	0	16.9	6.6	.	.	.
341	AMARANTHUS RETROFLEXUS	S	.	0	17.5	7.4	.	.	.
342	AMARANTHUS RETROFLEXUS	W	.	0	25.7	1.5	54.0	10.8	18.8
343	AMARANTHUS SP.	L	263	0	28.8	1.3	51.9	11.3	18.1
344	AMARANTHUS SP.	S	408	0	16.0	6.9	72.3	10.8	4.8
345	AMARANTHUS SP.	L	275	0	26.7	3.8	49.6	9.9	19.8
346	AMARANTHUS SPINOSUS	L	267	0	28.4	4.5	45.0	10.4	22.1
347	AMARANTHUS SPINOSUS	L	276	0	34.4	3.7	45.4	9.8	16.6
348	AMARANTHUS SPINOSUS	L	.	0	20.0	2.0	54.0	.	24.0
349	AMARANTHUS TRICOLOR	L	276	0	25.9	4.3	49.1	11.2	20.7
350	AMARANTHUS TRICOLOR	L	270	0	27.0	2.5	50.8	13.1	19.7
351	AMARANTHUS TRICOLOR	L	288	0	27.1	3.4	53.4	12.7	16.1
352	AMARANTHUS VIRIDIS	S	.	0	16.2	7.0	.	.	.
353	AMARANTHUS VIRIDIS	S	.	0	14.1	4.7	.	.	.
354	AMARANTHUS VIRIDIS	L	283	0	34.2	5.3	44.1	6.6	16.4
355	AMBLYOLEPIS SETIGERA	S	.	0	16.9	13.6	.	.	.
356	AMBLYOPYRUM MUTICUM	S	.	0	24.5	3.1	.	.	.
357	AMBLYOPYRUM MUTICUM	S	.	0	26.3	3.4	.	.	.
358	AMBROSIA ARTEMISIAEFOLIA	W	.	0	18.7	9.8	66.4	34.4	5.1
359	AMBROSIA PSILOSTACHYA	W	.	0	23.8	5.5	54.9	12.0	15.8
360	AMBROSIA TRIFIDA	S	.	0	21.6	24.7	.	.	3.7
361	AMBROSIA TRIFIDA	S	.	0	15.2	19.9	.	.	.
362	AMBURANA SP.	S	.	0	25.6	35.2	.	.	.
363	AMELANCHIER ALNIFOLIA	S	.	0	18.1	13.0	.	.	.

OBS	CALCIUM	PHOSPHO	IRON	SODIUM	POTASS	CAROT	THIAMINE	RIBOFLAV	NIACIN	ASCORBIC	REFCO
340	.	.	.	.	.	.	.	.	.	.	21
341	.	.	.	.	.	.	.	.	.	.	21
342	3950.0	510.0	.	.	5150.0	10.0	.	.	.	.	6
343	2562.5	643.8	55.6	.	.	35725.0	0.31	2.63	7.50	400.0	1
344	561.3	521.2	.	.	.	.	.	.	.	.	1
345	2038.2	511.5	29.8	.	3137.4	27938.9	0.61	1.22	10.69	610.7	4
346	1795.0	430.0	.	13.0	337.0	.	.	.	7.73	.	12
347	2920.2	460.1	13.5	36.8	3527.6	40828.2	0.06	2.02	8.59	503.1	2
348	5333.3	333.3	152.7	.	.	.	.	.	.	.	5
349	2043.1	577.6	48.3	34.5	3586.2	26594.8	0.52	1.72	10.34	456.9	2
350	2360.7	1008.2	44.3	.	4475.4	37622.9	0.66	1.31	11.48	729.5	2
351	2440.7	678.0	51.7	.	.	550.8	0.68	2.37	2.54	228.8	2
352	.	.	.	.	.	.	.	.	.	.	20
353	.	.	.	.	.	.	.	.	.	.	21
354	2243.4	500.0	27.0	335.5	2907.9	50756.6	0.07	2.43	11.84	789.5	2
355	.	.	.	.	.	.	.	.	.	.	20
356	.	.	.	.	.	.	.	.	.	.	21
357	.	.	.	.	.	.	.	.	.	.	21
358	.	.	.	.	.	.	.	.	.	.	6
359	.	.	.	.	.	.	.	.	.	.	6
360	.	.	.	.	.	.	.	.	.	.	19
361	.	.	.	.	.	.	.	.	.	.	21
362	.	.	.	.	.	.	.	.	.	.	20
363	.	.	.	.	.	.	.	.	.	.	20

TABLE 3 - DATA CONVERTED TO A ZERO-MOISTURE BASIS.

OBS	PLNA	PART	CALORIES	WATER	PROTEIN	FAT	TOTLCARB	FIBER	ASH
364	AMELANCHIER ALNIFOLIA	S	.	0	15.0	10.2	.	.	.
365	AMELANCHIER ALNIFOLIA	S	.	0	14.2	10.3	.	.	.
366	AMELANCHIER ALNIFOLIA	L	.	0	10.9	6.2	74.4	14.8	8.5
367	AMELANCHIER ALNIFOLIA	L	.	0	18.0	3.9	71.4	13.4	6.7
368	AMELANCHIER ALNIFOLIA	C	.	0	5.5	3.4	88.3	34.8	2.8
369	AMELANCHIER OVALIS	S	.	0	11.8	7.8	.	.	.
370	AMICIA ZYGOMERIS	S	.	0	22.5	20.6	.	.	.
371	AMMANIA BACCIFERA	L	275	0	39.6	0.0	49.5	.	11.0
372	AMMI MAJUS	S	.	0	18.8	11.6	.	.	.
373	AMMI MAJUS	S	.	0	17.8	8.9	.	.	.
374	AMMI MAJUS	S	.	0	18.9	12.5	.	.	.
375	AMMI MAJUS	S	.	0	19.1	19.4	.	.	.
376	AMMI VISNAGA	S	.	0	18.1	19.2	.	.	8.9
377	AMMI VISNAGA	S	.	0	17.3	15.3	.	.	.
378	AMMI VISNAGA	S	.	0	19.9	16.9	.	.	.
379	AMMI VISNAGA	S	.	0	18.4	27.0	.	.	.
380	AMMI VISNAGA	S	.	0	10.9	16.9	.	.	.
381	AMMOBIUM ALATUM	S	.	0	28.1	27.0	.	.	.
382	AMMOBIUM ALATUM VAR. GRANDIFLORUM	S	.	0	30.0	25.0	.	.	.
383	AMMOPHILA ARENARIA	S	.	0	23.7	12.5	.	.	.
384	AMMOSELINUM POPEI	S	.	0	13.1	7.8	.	.	.
385	AMMOSELINUM POPEI	S	.	0	16.3	9.9	.	.	.
386	AMOREUXIA PALMATIFIDA	S	.	0	11.9	14.0	.	.	.
387	AMORPHA CANESCENS	S	.	0	17.6	3.4	.	.	4.9

OBS	CALCIUM	PHOSPHO	IRON	SODIUM	POTASS	CAROT	THIAMINE	RIBOFLAV	NIACIN	ASCORBIC	REFCO
364	.	.	.	.	.	.	.	.	.	.	20
365	.	.	.	.	.	.	.	.	.	.	20
366	2320.0	620.0	.	.	.	.	.	.	.	.	6
367	1600.0	480.0	.	.	.	.	.	.	.	.	6
368	.	.	.	.	.	.	.	.	.	.	6
369	.	.	.	.	.	.	.	.	.	.	21
370	.	.	.	.	.	.	.	.	.	.	20
371	659.3	747.3	70.3	.	.	42582.4	0.77	0.55	.	582.4	2
372	.	.	.	.	.	.	.	.	.	.	19
373	.	.	.	.	.	.	.	.	.	.	20
374	.	.	.	.	.	.	.	.	.	.	21
375	.	.	.	.	.	.	.	.	.	.	21
376	.	.	.	.	.	.	.	.	.	.	19
377	.	.	.	.	.	.	.	.	.	.	21
378	.	.	.	.	.	.	.	.	.	.	21
379	.	.	.	.	.	.	.	.	.	.	21
380	.	.	.	.	.	.	.	.	.	.	21
381	.	.	.	.	.	.	.	.	.	.	20
382	.	.	.	.	.	.	.	.	.	.	20
383	.	.	.	.	.	.	.	.	.	.	21
384	.	.	.	.	.	.	.	.	.	.	20
385	.	.	.	.	.	.	.	.	.	.	21
386	.	.	.	.	.	.	.	.	.	.	20
387	.	.	.	.	.	.	.	.	.	.	19

TABLE 3 - DATA CONVERTED TO A ZERO-MOISTURE BASIS.

OBS	PLNA	PART	CALORIES	WATER	PROTEIN	FAT	TOTLCARB	FIBER	ASH
388	AMORPHA CANESCENS	S	.	0	39.8	6.6	.	.	4.3
389	AMORPHA CANESCENS	S	.	0	12.4	7.0	.	.	5.3
390	AMORPHA CANESCENS	W	.	0	8.2	8.5	72.9	21.5	10.4
391	AMORPHA FRUTICOSA	S	.	0	32.3	11.0	.	.	4.8
392	AMORPHA FRUTICOSA	S	.	0	21.2	7.2	.	.	5.9
393	AMORPHA FRUTICOSA	S	.	0	28.4	10.0	.	.	4.5
394	AMORPHA FRUTICOSA	S	.	0	10.5	9.4	.	.	6.0
395	AMORPHA FRUTICOSA	S	.	0	30.7	8.6	.	.	4.1
396	AMORPHA FRUTICOSA	S	.	0	11.5	10.1	.	.	6.1
397	AMORPHOPHALLUS APHYLLUS	R	375	0	8.3	1.4	84.4	4.9	5.9
398	AMORPHOPHALLUS CAMPANULATUS	SH	.	0	15.3	3.3	65.1	14.2	16.3
399	AMORPHOPHALLUS CAMPANULATUS	R	.	0	19.8	1.6	72.0	2.3	6.6
400	AMORPHOPHALLUS CAMPANULATUS	R	377	0	9.3	0.9	85.6	3.3	4.2
401	AMORPHOPHALLUS CONJAC	L	308	0	3.8	0.0	88.5	3.8	7.7
402	AMSINCKIA DOUGLASIANA		.	0	13.9	.	.	23.0	16.2
403	ANACARDIUM OCCIDENTALE	S	587	0	18.8	47.0	31.6	1.5	2.6
404	ANACARDIUM OCCIDENTALE	F	368	0	6.9	4.9	85.4	4.2	2.8
405	ANACARDIUM OCCIDENTALE	L	332	0	17.3	2.0	76.7	.	4.0
406	ANACARDIUM OCCIDENTALE	S	548	0	15.6	38.0	43.2	1.4	3.2
407	ANACARDIUM OCCIDENTALE	S	592	0	18.1	48.2	30.9	1.5	2.7
408	ANACOLOSA LUZONIENSIS	P	393	0	12.5	1.6	82.9	3.1	3.1
409	ANACOLOSA LUZONIENSIS	S	416	0	6.5	4.7	86.9	3.3	1.8
410	ANACYCLUS CLAVATUS	S	.	0	21.3	19.8	.	.	.
411	ANAGALLIS ARVENSIS	S	.	0	14.4	17.7	.	.	.

OBS	CALCIUM	PHOSPHO	IRON	SODIUM	POTASS	CAROT	THIAMINE	RIBOFLAV	NIACIN	ASCORBIC	REFCO
388	.	.	.	.	.	.	.	.	.	.	19
389	.	.	.	.	.	.	.	.	.	.	19
390	2800.0	920.0	.	.	.	.	.	.	.	.	6
391	.	.	.	.	.	.	.	.	.	.	19
392	.	.	.	.	.	.	.	.	.	.	19
393	.	.	.	.	.	.	.	.	.	.	19
394	.	.	.	.	.	.	.	.	.	.	19
395	.	.	.	.	.	.	.	.	.	.	19
396	.	.	.	.	.	.	.	.	.	.	19
397	527.8	378.5	.	.	.	.	.	.	.	.	1
398	1520.0	810.0	.	.	.	.	.	.	.	.	18
399	.	.	.	.	.	.	.	.	.	.	18
400	176.7	176.7	11.2	23.3	1934.9	0.00	0.28	0.09	7.91	27.9	2
401	653.8	269.2	11.5	.	.	.	0.00	0.00	0.00	0.0	2
402	2280.0	520.0	.	.	4420.0	.	.	.	.	.	6
403	82.3	625.5	19.5	.	.	.	0.70	0.27	1.73	7.6	1
404	83.3	312.5	9.7	.	.	5277.78	0.21	1.67	2.08	1750.0	1
405	.	.	.	.	.	2043.19	0.03	0.03	4.65	295.7	2
406	24.7	596.1	1.8	.	.	0.00	0.87	0.33	2.16	.	3
407	40.1	393.5	4.0	15.8	489.5	63.29	0.45	0.26	1.90	.	4
408	105.1	194.6	5.1	.	.	2821.01	0.27	0.31	4.28	77.8	2
409	41.7	230.5	1.8	.	.	.	0.25	0.25	1.63	16.3	2
410	.	.	.	.	.	.	.	.	.	.	21
411	.	.	.	.	.	.	.	.	.	.	20

TABLE 3 - DATA CONVERTED TO A ZERO-MOISTURE BASIS.

OBS	PLANA	PART	CALORIES	WATER	PROTEIN	FAT	TOTCARB	FIBER	ASH	CALCIUM	PHOSPHO	IRON	SODIUM	POTASS	CAROT	THIAMINE	RIBOFLAV	NIACIN	ASCORBIC	REFCO
412	ANAGALLIS ARVENSIS	S	.	0	15.3	20.7	.	.	.	.	.	.	.	.	.	.	.	.	.	21
413	ANAGYRIS FOETIDA	S	.	0	21.8	10.3	.	.	.	.	.	.	.	.	.	.	.	.	.	21
414	ANAGYRIS FOETIDA	S	.	0	22.8	9.6	.	.	.	.	.	.	.	.	.	.	.	.	.	21
415	ANAGYRIS FOETIDA	S	.	0	24.8	10.1	.	.	.	.	.	.	.	.	.	.	.	.	.	21
416	ANAGYRIS FOETIDA	S	.	0	26.4	11.4	.	.	.	.	.	.	.	.	.	.	.	.	.	21
417	ANAGYRIS FOETIDA	S	.	0	24.7	11.2	.	.	.	.	.	.	.	.	.	.	.	.	.	21
418	ANANAS COMOSUS	W	.	0	11.9	3.4	75.9	20.3	8.8	.	.	.	.	.	.	.	.	.	.	6
419	ANANAS COMOSUS	F	362	0	5.4	2.3	89.2	3.8	3.1	130.8	92.3	3.8	15.4	961.5	269.23	0.46	0.23	2.31	169.2	2
420	ANANAS COMOSUS	F	356	0	2.7	1.4	93.8	2.7	2.1	123.3	54.8	3.4	.	.	219.18	0.55	0.27	1.37	417.8	3
421	ANANAS COMOSUS	F	354	0	2.7	1.4	93.2	2.7	2.7	115.6	54.4	3.4	6.8	993.2	285.71	0.61	0.20	1.36	115.6	4
422	ANCHOMANES DIFFORMIS	R	366	0	12.3	0.4	81.9	5.1	5.4	728.3	250.0	.	.	.	.	.	.	.	.	1
423	ANCHUSA AZUREA	S	.	0	16.4	21.4	.	.	16.8	.	.	.	.	.	.	.	.	.	.	19
424	ANCHUSA CAPENSIS	S	.	0	19.4	29.9	.	.	15.7	.	.	.	.	.	.	.	.	.	.	19
425	ANCHUSA HYBRIDA	S	.	0	20.4	20.5	.	.	.	.	.	.	.	.	.	.	.	.	.	20
426	ANDROPOGON BARBINODIS	W	.	0	7.2	2.1	82.2	33.0	8.5	690.0	390.0	.	.	.	.	.	.	.	.	6
427	ANDROPOGON CANALICULATUS	S	.	0	13.3	2.0	74.9	31.9	9.8	.	.	.	.	.	.	.	.	.	.	18
428	ANDROPOGON CIRRATUS	W	.	0	3.7	1.8	89.7	36.5	4.8	290.0	70.0	.	.	520.0	.	.	.	.	.	6
429	ANDROPOGON GAYANUS	W	.	0	6.5	3.1	80.5	29.6	9.9	.	.	.	.	.	.	.	.	.	.	18
430	ANDROPOGON GAYANUS	H	.	0	6.1	1.7	84.3	35.1	7.9	.	.	.	.	.	.	.	.	.	.	18
431	ANDROPOGON GERARDII	S	.	0	31.9	8.3	.	.	.	.	.	.	.	.	.	.	.	.	.	20
432	ANDROPOGON GERARDII	W	.	0	7.0	2.3	82.9	33.2	7.8	360.0	160.0	17.0	.	1220.0	13.00	.	.	.	.	6
433	ANDROPOGON GLOMERATUS(IMMATURE)	W	.	0	10.6	2.1	76.7	31.3	10.6	.	.	.	.	.	.	.	.	.	.	6
434	ANDROPOGON GLOMERATUS(MATURE)	W	.	0	4.6	1.5	86.5	34.1	7.4	.	.	.	.	.	.	.	.	.	.	6
435	ANDROPOGON HALLII	S	.	0	26.2	6.0	.	.	.	.	.	.	.	.	.	.	.	.	.	20
436	ANDROPOGON HALLII	W	.	0	5.8	2.3	84.3	33.5	7.6	.	.	.	.	.	.	.	.	.	.	6
437	ANDROPOGON ISCHAEMUM	S	.	0	30.0	9.8	.	.	.	.	.	.	.	.	.	.	.	.	.	19
438	ANDROPOGON ISCHAEMUM	W	.	0	8.0	2.0	81.5	33.5	8.5	.	.	.	.	.	.	.	.	.	.	6
439	ANDROPOGON PERTUSUS	S	.	0	26.2	8.6	.	.	.	.	.	.	.	.	.	.	.	.	.	19
440	ANDROPOGON SACCHAROIDES	W	.	0	7.1	1.8	80.3	33.8	10.8	.	.	.	.	.	.	.	.	.	.	6
441	ANDROPOGON SCOPARIUS	S	.	0	33.8	7.2	.	.	.	.	.	.	.	.	.	.	.	.	.	20
442	ANDROPOGON SCOPARIUS	W	.	0	6.2	2.2	83.9	32.8	7.7	350.0	120.0	.	.	.	7.00	.	.	.	.	6
443	ANDROPOGON TENER	W	.	0	6.3	2.4	82.3	33.8	9.0	490.0	100.0	.	.	.	.	.	.	.	.	6
444	ANDROPOGON TERNARIUS	S	.	0	37.5	9.2	.	.	.	.	.	.	.	.	.	.	.	.	.	20
445	ANDROPOGON VIRGINICUS	W	.	0	13.5	1.8	73.8	30.9	10.9	.	.	.	.	.	.	.	.	.	.	6
446	ANDROSACE MAXIMA	S	.	0	11.1	8.9	.	.	.	.	.	.	.	.	.	.	.	.	.	21
447	ANEILEMA PEDUNCULOSUM	L	276	0	24.1	4.6	49.4	14.9	21.8	1758.6	609.2	.	.	.	.	.	.	.	.	1
448	ANEMONE CORONARIA	S	.	0	25.6	26.6	.	.	4.5	.	.	.	.	.	.	.	.	.	.	19
449	ANEMONE CORONARIA (ST.BRIGID)	S	.	0	26.0	20.7	.	.	4.1	.	.	.	.	.	.	.	.	.	.	19
450	ANEMONE CYLINDRICA	S	.	0	20.6	27.9	.	.	.	.	.	.	.	.	.	.	.	.	.	21
451	ANEMONE DECAPETALA	S	.	0	22.5	32.9	.	.	.	.	.	.	.	.	.	.	.	.	.	20
452	ANETHUM GRAVEOLENS	S	.	0	21.3	17.7	.	.	7.6	.	.	.	.	.	.	.	.	.	.	19
453	ANETHUM GRAVEOLENS	S	.	0	22.6	9.9	.	.	8.4	.	.	.	.	.	.	.	.	.	.	19
454	ANETHUM GRAVEOLENS	S	.	0	18.6	6.8	.	.	7.3	.	.	.	.	.	.	.	.	.	.	19
455	ANETHUM GRAVEOLENS	S	330	0	17.3	15.7	59.8	22.9	7.2	1642.5	300.1	17.7	21.7	1284.9	34.67	0.46	0.30	3.04	.	16
456	ANETHUM GRAVEOLENS	H	273	0	21.6	4.7	60.2	12.8	13.6	1924.5	585.8	52.6	224.4	3568.5	.	0.45	0.30	3.03	.	16

TABLE 3 - DATA CONVERTED TO A ZERO-MOISTURE BASIS.

OBS	PLNA	PART	CALORIES	WATER	PROTEIN	FAT	TOTLCARB	FIBER	ASH
457	ANGELICA AMPLA	S	.	0	29.5	26.0	.	.	.
458	ANGELICA ARCHANGELICA	S	.	0	31.9	18.9	.	.	.
459	ANGELICA ARCHANGELICA	S	.	0	32.5	20.3	.	.	.
460	ANISOMELES INDICA	S	.	0	14.6	27.5	.	.	.
461	ANNONA CHERIMOLA	F	351	0	4.8	0.3	92.7	8.0	2.2
462	ANNONA CHERIMOLA	F	350	0	4.7	0.9	91.0	8.1	3.4
463	ANNONA CHERIMOLA	F	355	0	4.9	1.5	90.6	8.3	3.0
464	ANNONA DIVERSIFOLIA	F	344	0	9.8	0.7	84.6	4.6	4.9
465	ANNONA GLABRA	F	394	0	3.0	12.1	78.0	28.8	6.8
466	ANNONA MURICATA	F	351	0	6.0	1.2	89.9	3.6	3.0
467	ANNONA MURICATA	F	355	0	5.9	2.4	88.2	6.5	3.6
468	ANNONA MURICATA	F	355	0	5.5	1.6	89.1	6.0	3.8
469	ANNONA PURPUREA	F	673	0	4.7	70.0	19.3	8.0	6.0
470	ANNONA RETICULATA	F	350	0	6.9	1.4	87.6	6.5	4.1
471	ANNONA RETICULATA	F	350	0	8.9	0.9	86.4	10.3	3.7
472	ANNONA RETICULATA	F	354	0	6.0	2.1	88.4	11.9	3.5
473	ANNONA SQUAMOSA	F	347	0	6.2	0.9	88.9	7.1	4.0
474	ANNONA SQUAMOSA	F	353	0	5.9	0.7	90.4	5.9	2.9
475	ANNONA SQUAMOSA	F	352	0	6.7	1.1	88.8	6.4	3.4
476	ANODA CRISTATA	S	.	0	19.9	18.1	.	.	.
477	ANTHEMIS SANCTI-JOHANNIS	S	.	0	17.7	24.4	.	.	6.8
478	ANTHEMIS TINCTORIA	S	.	0	18.6	19.2	.	.	.
479	ANTHEMIS TINCTORIA	S	.	0	33.7	38.1	.	.	.
480	ANTHEMIS TINCTORIA VAR. KELWAYI	S	.	0	16.2	22.6	.	.	.

OBS	CALCIUM	PHOSPHO	IRON	SODIUM	POTASS	CAROT	THIAMINE	RIBOFLAV	NIACIN	ASCORBIC	REFCO
457	.	.	.	.	.	.	.	.	.	.	20
458	.	.	.	.	.	.	.	.	.	.	20
459	.	.	.	.	.	.	.	.	.	.	20
460	.	.	.	.	.	.	.	.	.	.	21
461	28.8	76.7	0.6	.	.	0.0	0.35	0.35	3.19	38.3	2
462	145.3	149.6	2.6	.	.	0.0	0.38	0.56	3.85	72.6	3
463	86.8	150.9	1.9	.	.	22.6	0.38	0.42	4.91	34.0	4
464	112.3	182.5	2.5	.	.	0.0	0.07	0.11	7.72	49.1	3
465	393.9	181.8	17.4	.	.	2348.5	0.30	0.53	4.55	159.1	3
466	83.3	125.0	3.0	47.6	1744.0	0.0	0.48	0.60	7.74	142.9	2
467	142.0	165.7	3.0	.	.	59.2	0.41	0.30	5.33	153.8	3
468	76.5	147.5	3.3	76.5	1448.1	32.8	0.38	0.27	4.92	109.3	4
469	126.7	146.7	6.7	4800.0	.	.	0.33	0.47	5.33	186.7	3
470	124.4	147.5	2.3	27.6	2281.1	92.2	0.51	0.32	2.76	96.8	2
471	112.1	121.5	4.7	.	.	0.0	0.33	0.56	3.27	140.2	3
472	94.7	70.2	2.8	.	.	0.0	0.28	0.35	1.75	77.2	4
473	133.3	160.0	2.7	22.2	1328.9	22.2	0.49	0.44	3.56	160.0	2
474	102.9	132.4	6.6	.	.	0.0	0.40	0.55	3.31	128.7	3
475	82.4	153.6	2.2	41.2	1030.0	22.5	0.37	0.52	3.75	127.3	4
476	.	.	.	.	.	.	.	.	.	.	20
477	.	.	.	.	.	.	.	.	.	.	19
478	.	.	.	.	.	.	.	.	.	.	20
479	.	.	.	.	.	.	.	.	.	.	21
480	.	.	.	.	.	.	.	.	.	.	19

TABLE 3 - DATA CONVERTED TO A ZERO-MOISTURE BASIS.

OBS	PLNA	PART	CALORIES	WATER	PROTEIN	FAT	TOTLCARB	FIBER	ASH
481	ANTHEMIS TINCTORIA VAR. KELWAYI	S	.	0	16.2	22.3	.	.	.
482	ANTHEMIS TINCTORIA VAR. KELWAYI	S	.	0	15.0	23.0	.	.	.
483	ANTHEPHORA NIGRITANA	W	.	0	14.1	1.8	79.4	38.0	8.4
484	ANTHOXANTHUM ODORATUM	W	.	0	20.0	4.6	65.4	17.8	10.0
485	ANTHRISCUS CEREFOLIUM	S	.	0	18.8	15.6	.	.	.
486	ANTHRISCUS CEREFOLIUM	S	.	0	19.4	10.5	.	.	.
487	ANTHRISCUS CEREFOLIUM	L	255	0	25.0	4.2	52.9	12.2	17.9
488	ANTHYLLIS CORNICINA	S	.	0	45.3	9.0	.	.	.
489	ANTHYLLIS LOTOIDES	S	.	0	43.8	7.3	.	.	.
490	ANTIDESMA BUNIUS	F	356	0	5.9	5.9	80.2	7.9	7.9
491	ANTIDESMA PLATYPHYLLUM	F	347	0	2.1	0.5	93.3	0.5	4.1
492	ANTIGONON LEPTOPUS	S	.	0	10.6	1.7	.	.	.
493	ANTIRRHINUM MAJUS	S	.	0	16.2	42.9	.	.	5.0
494	ANTIRRHINUM MAJUS	S	.	0	.	36.2	.	.	.
495	ANTIRRHINUM ORONTIUM	S	.	0	18.6	39.1	.	.	.
496	ANTIRRHINUM SICULUM	S	.	0	15.1	37.6	.	.	.
497	ANTROCARYON MICRASTER	S	709	0	11.2	72.7	15.2	2.6	0.9
498	APHANOSTEPHUS ARIZONICUS	S	.	0	12.5	11.8	.	.	.
499	APHANOSTEPHUS SP.	W	.	0	14.4	3.1	73.9	22.6	8.6
500	APIUM GRAVEOLENS	L	.	0	25.6	.	.	.	16.0
501	APIUM GRAVEOLENS	L	.	0	27.2	6.9	46.9	3.5	19.0
502	APIUM GRAVEOLENS	S	.	0	14.8	32.9	.	.	6.2
503	APIUM GRAVEOLENS	S	.	0	23.3	30.4	.	.	7.3
504	APIUM GRAVEOLENS	S	.	0	20.5	36.2	.	.	.

OBS	CALCIUM	PHOSPHO	IRON	SODIUM	POTASS	CAROT	THIAMINE	RIBOFLAV	NIACIN	ASCORBIC	REFCO
481	.	.	.	.	.	.	.	.	.	.	20
482	.	.	.	.	.	.	.	.	.	.	20
483	.	.	.	.	.	.	.	.	.	.	18
484	635.0	400.0	61.0	.	.	.	.	.	.	.	6
485	.	.	.	.	.	.	.	.	.	.	19
486	.	.	.	.	.	.	.	.	.	.	20
487	1450.4	484.9	34.5	89.4	5107.8	.	.	.	.	.	16
488	.	.	.	.	.	.	.	.	.	.	21
489	.	.	.	.	.	.	.	.	.	.	21
490	297.0	237.6	7.9	.	.	247.5	0.10	0.40	3.96	59.4	2
491	82.9	217.6	9.3	.	.	673.6	0.05	0.16	2.59	88.1	3
492	.	.	.	.	.	.	.	.	.	.	20
493	.	.	.	.	.	.	.	.	.	.	19
494	.	.	.	.	.	.	.	.	.	.	21
495	.	.	.	.	.	.	.	.	.	.	21
496	.	.	.	.	.	.	.	.	.	.	21
497	69.6	579.4	.	.	.	.	.	.	.	.	1
498	.	.	.	.	.	.	.	.	.	.	20
499	1050.0	200.0	.	.	2540.0	.	.	.	.	.	6
500	1310.0	.	.	2340.0	2050.0	.	.	.	.	.	17
501	2700.0	1700.0	.	.	.	.	.	.	.	.	18
502	.	.	.	.	.	.	.	.	.	.	19
503	.	.	.	.	.	.	.	.	.	.	19
504	.	.	.	.	.	.	.	.	.	.	21

TABLE 3 - DATA CONVERTED TO A ZERO-MOISTURE BASIS.

OBS	PLANA	PART	CALORIES	WATER	PROTEIN	FAT	TOTLCARB	FIBER	ASH	CALCIUM	PHOSPHO	IRON	SODIUM	POTASS	CAROT	THIAMINE	RIBOFLAV	NIACIN	ASCORBIC	REFCO
505	APIUM GRAVEOLENS	L	.	0	28.2	6.2	.	7.9	.	.	.	.	.	.	.	.	.	.	.	6
506	APIUM GRAVEOLENS	S	417	0	19.3	26.9	44.0	12.6	9.9	1879.8	581.9	47.8	170.2	1489.4	33.0	.	.	.	18.1	16
507	APIUM GRAVEOLENS VAR. DULCE	L	274	0	15.1	1.4	61.6	13.7	21.9	1191.8	684.9	12.3	.	.	.	0.41	3.42	.	232.9	1
508	APIUM GRAVEOLENS VAR. DULCE	L	292	0	19.4	4.2	58.3	13.9	18.1	861.1	513.9	34.7	1333.3	4527.8	14444.4	0.83	0.97	5.56	277.8	2
509	APIUM GRAVEOLENS VAR. DULCE	L	288	0	15.3	1.7	66.1	10.2	16.9	661.0	474.6	5.1	2135.6	5779.7	2440.7	0.51	0.51	5.08	152.5	4
510	APIUM SELLOWIANUM	S	.	0	20.2	25.6	.	.	.	.	.	.	.	.	.	.	.	.	.	21
511	APOCYNUM CANNABINUM	S	.	0	28.7	22.6	.	.	4.7	.	.	.	.	.	.	.	.	.	.	19
512	APOCYNUM CANNABINUM	S	.	0	33.1	23.8	.	.	.	.	.	.	.	.	.	.	.	.	.	19
513	APODANTHERA UNDULATA	S	.	0	41.9	47.7	.	.	.	.	.	.	.	.	.	.	.	.	.	19
514	AQUILEGIA ALPINA	S	.	0	21.2	27.6	.	.	.	.	.	.	.	.	.	.	.	.	.	20
515	AQUILEGIA CAERULEA	S	.	0	23.8	30.2	.	.	.	.	.	.	.	.	.	.	.	.	.	20
516	AQUILEGIA SP.	S	.	0	20.6	26.4	.	.	.	.	.	.	.	.	.	.	.	.	.	20
517	AQUILEGIA VULGARIS	S	.	0	19.6	26.9	.	.	.	.	.	.	.	.	.	.	.	.	.	21
518	ARABIS ALPINA	S	.	0	24.4	30.6	.	.	.	.	.	.	.	.	.	.	.	.	.	20
519	ARABIS ALPINA	S	.	0	21.2	32.6	.	.	.	.	.	.	.	.	.	.	.	.	.	20
520	ARABIS ALPINA	S	.	0	20.6	32.5	.	.	.	.	.	.	.	.	.	.	.	.	.	20
521	ARABIS GLABRA	S	.	0	24.9	37.7	.	.	.	.	.	.	.	.	.	.	.	.	.	20
522	ARABIS LAEVIGATA	S	.	0	23.8	38.7	.	.	.	.	.	.	.	.	.	.	.	.	.	20
523	ARABIS SP.	S	.	0	28.6	38.0	.	.	.	.	.	.	.	.	.	.	.	.	.	20
524	ARACHIS HYPOGAEA	SH	.	0	17.5	2.2	71.7	20.1	8.6	930.0	200.0	.	.	.	.	.	.	.	.	18
525	ARACHIS HYPOGAEA	H	.	0	26.2	2.7	59.5	27.9	11.6	3150.0	810.0	.	.	.	.	.	.	.	.	18
526	ARACHIS HYPOGAEA	W	.	0	16.4	9.5	66.8	25.1	7.3	1710.0	290.0	.	.	970.0	.	.	.	.	.	6
527	ARACHIS HYPOGAEA	L	321	0	20.5	2.8	69.3	21.4	7.4	1218.6	381.4	19.5	.	.	35976.7	1.07	2.70	7.44	455.8	1
528	ARACHIS HYPOGAEA	S	524	0	26.0	33.6	37.7	1.9	2.8	96.9	423.9	3.6	6.9	728.4	34.6	1.68	0.31	16.78	19.0	2
529	ARACHIS HYPOGAEA	S	597	0	27.5	50.3	19.7	2.5	2.4	73.1	424.8	2.2	5.3	714.0	.	1.21	0.14	18.22	0.0	4
530	ARALIA CORDATA	L	298	0	21.3	4.3	61.7	10.6	12.8	276.6	531.9	6.4	42.6	6595.7	0.0	1.28	0.43	17.02	106.4	2
531	ARALIA ELATA	L	300	0	34.4	5.6	47.5	13.8	12.5	506.3	868.8	46.3	.	.	.	0.63	0.31	6.25	62.5	2
532	ARALIA SPINOSA	S	.	0	17.5	46.3	.	.	3.7	.	.	.	.	.	.	.	.	.	.	19
533	ARALIA SPINOSA	S	.	0	5.8	4.2	.	.	5.7	.	.	.	.	.	.	.	.	.	.	19
534	ARBUTUS ANDRACHNE	S	.	0	10.6	1.6	.	.	.	.	.	.	.	.	.	.	.	.	.	21
535	ARBUTUS ANDRACHNE	S	.	0	11.1	0.7	.	.	.	.	.	.	.	.	.	.	.	.	.	21
536	ARBUTUS MENZIESII	S	.	0	15.6	22.2	.	.	.	.	.	.	.	.	.	.	.	.	.	20
537	ARBUTUS UNEDO	S	.	0	33.1	36.8	.	.	.	.	.	.	.	.	.	.	.	.	.	20
538	ARBUTUS UNEDO	S	.	0	3.1	3.4	.	.	.	.	.	.	.	.	.	.	.	.	.	20
539	ARBUTUS UNEDO	S	.	0	23.5	16.5	.	.	.	.	.	.	.	.	.	.	.	.	.	21
540	ARCTIUM LAPPA	S	.	0	16.8	21.4	.	.	.	.	.	.	.	.	.	.	.	.	.	21
541	ARCTIUM LAPPA	R	379	0	10.6	0.4	85.5	7.2	3.4	212.8	246.8	5.1	127.7	766.0	0.0	1.06	0.34	1.28	8.5	2
542	ARCTIUM MINUS	S	.	0	16.9	20.8	.	.	4.2	.	.	.	.	.	.	.	.	.	.	19
543	ARCTIUM MINUS	S	.	0	18.8	16.6	.	.	.	.	.	.	.	.	.	.	.	.	.	20
544	ARCTIUM MINUS	S	.	0	21.3	16.1	.	.	.	.	.	.	.	.	.	.	.	.	.	21
545	ARCTIUM MINUS	S	.	0	17.6	16.5	.	.	.	.	.	.	.	.	.	.	.	.	.	21
546	ARCTIUM MINUS	S	.	0	18.6	13.8	.	.	.	.	.	.	.	.	.	.	.	.	.	21
547	ARCTIUM MINUS	S	.	0	21.6	21.4	.	.	.	.	.	.	.	.	.	.	.	.	.	21
548	ARCTIUM MINUS	S	.	0	19.9	15.4	.	.	.	.	.	.	.	.	.	.	.	.	.	21
549	ARCTOSTAPHYLOS GLAUCA	S	.	0	24.9	55.7	.	.	3.2	.	.	.	.	.	.	.	.	.	.	19

TABLE 3 - DATA CONVERTED TO A ZERO-MOISTURE BASIS.

OBS	PLNA	PART	CALORIES	WATER	PROTEIN	FAT	TOTLCARB	FIBER	ASH
550	ARCTOSTAPHYLOS PATULA	S	.	0	3.1	1.8	.	.	.
551	ARCTOSTAPHYLOS PUNGENS	S	.	0	2.5	3.6	.	.	.
552	ARCTOSTAPHYLOS PUNGENS	S	.	0	2.5	3.7	.	.	.
553	ARCTOSTAPHYLOS TOMENTOSA	F	366	0	4.0	3.6	89.4	4.6	3.0
554	ARCTOSTAPHYLOS UVA-URSI	S	.	0	25.6	50.0	.	.	3.3
555	ARCTOSTAPHYLOS UVA-URSI	S	.	0	2.5	2.0	.	.	0.6
556	ARCTOTIS GRANDIS	S	.	0	36.2	47.7	.	.	6.3
557	ARCTOTIS GRANDIS	S	.	0	12.5	10.2	.	.	7.1
558	ARDISIA CRENULATA	S	.	0	7.5	0.7	.	.	2.7
559	ARDISIA ELLIPTCA	F	358	0	4.7	3.2	87.8	11.1	4.3
560	ARECA CATECHU	SH	316	0	24.3	2.2	66.2	.	7.4
561	ARECA CATECHU	S	449	0	6.8	12.3	79.1	18.1	1.7
562	ARECASTRUM ROMANZOFFIANUM	S	.	0	12.8	64.7	.	.	.
563	ARENGA PINNATA	SH	358	0	1.9	3.8	92.5	9.4	1.9
564	ARGANIA SPINOSA	S	.	0	9.0	71.7	17.2	6.8	2.1
565	ARGEMONE ALBIFLORA	S	.	0	17.5	33.0	.	.	.
566	ARGEMONE ALBIFLORA	S	.	0	16.9	33.0	.	.	.
567	ARGEMONE HISPIDA	S	.	0	17.0	31.5	.	.	7.3
568	ARGEMONE INTERMEDIA	S	.	0	18.2	34.0	.	.	6.8
569	ARGEMONE INTERMEDIA	S	.	0	19.4	40.1	.	.	5.4
570	ARGEMONE INTERMEDIA	S	.	0	18.1	38.8	.	.	.
571	ARGEMONE MEXICANA	S	.	0	18.1	36.4	.	.	.
572	ARGEMONE MEXICANA	S	.	0	17.6	39.3	.	.	.
573	ARGEMONE MUNITA SUBSP. ROTUNDATA	S	.	0	15.4	34.4	.	.	.

OBS	CALCIUM	PHOSPHO	IRON	SODIUM	POTASS	CAROT	THIAMINE	RIBOFLAV	NIACIN	ASCORBIC	REFCO
550	.	.	.	.	.	.	.	.	.	.	21
551	.	.	.	.	.	.	.	.	.	.	19
552	.	.	.	.	.	.	.	.	.	.	19
553	.	108.9	.	.	.	3366.34	0.10	0.00	1.32	257.4	3
554	.	.	.	.	.	.	.	.	.	.	19
555	.	.	.	.	.	.	.	.	.	.	19
556	.	.	.	.	.	.	.	.	.	.	19
557	.	.	.	.	.	.	.	.	.	.	19
558	.	.	.	.	.	.	.	.	.	.	19
559	465.9	136.2	2.9	.	.	0.00	0.00	0.04	5.73	35.8	3
560	44.1	654.4	14.7	.	.	.	.	.	.	.	2
561	618.0	71.8	6.5	86.7	508.6	0.00	0.19	0.79	0.68	0.0	2
562	.	.	.	.	.	.	.	.	.	.	19
563	396.2	56.6	9.4	37.7	132.1	.	0.00	0.19	1.89	.	2
564	.	.	.	.	.	.	.	.	.	.	18
565	.	.	.	.	.	.	.	.	.	.	20
566	.	.	.	.	.	.	.	.	.	.	20
567	.	.	.	.	.	.	.	.	.	.	19
568	.	.	.	.	.	.	.	.	.	.	19
569	.	.	.	.	.	.	.	.	.	.	19
570	.	.	.	.	.	.	.	.	.	.	20
571	.	.	.	.	.	.	.	.	.	.	19
572	.	.	.	.	.	.	.	.	.	.	21
573	.	.	.	.	.	.	.	.	.	.	21

TABLE 3 - DATA CONVERTED TO A ZERO-MOISTURE BASIS.

OBS	PLNA	PART	CALORIES	WATER	PROTEIN	FAT	TOTLCARB	FIBER	ASH	CALCIUM	PHOSPHO	IRON	SODIUM	POTASS	CAROT	THIAMINE	RIBOFLAV	NIACIN	ASCORBIC	REFCO
574	ARGEMONE PLATYCERAS	S	.	0	24.4	33.7	.	.	6.6	.	.	.	.	.	.	.	.	.	.	19
575	ARGEMONE POLYANTHEMOS	S	.	0	18.6	36.2	.	.	.	.	.	.	.	.	.	.	.	.	.	21
576	ARGEMONE SP.	S	.	0	19.4	36.8	.	.	.	.	.	.	.	.	.	.	.	.	.	19
577	ARGYROLOBIUM UNIFLORUM	S	.	0	30.8	4.6	.	.	.	.	.	.	.	.	.	.	.	.	.	21
578	ARISAEMA JACQUEMONTII	S	.	0	14.0	1.8	.	.	.	.	.	.	.	.	.	.	.	.	.	21
579	ARISAEMA TRIPHYLLUM	S	.	0	15.0	2.8	.	.	.	.	.	.	.	.	.	.	.	.	.	19
580	ARISTIDA ADSCENSIONIS	W	.	0	10.5	1.6	78.4	33.6	9.5	.	.	.	.	.	.	.	.	.	.	18
581	ARISTIDA ADSCENSIONIS	H	.	0	2.0	0.9	89.2	45.7	7.9	.	.	.	.	.	.	.	.	.	.	18
582	ARISTIDA LONGISETA	W	.	0	5.6	1.4	81.9	36.1	11.1	760.0	90.0	.	.	.	.	.	.	.	.	6
583	ARISTIDA LONGSPICA	W	.	0	6.8	2.0	80.6	31.9	10.6	.	.	.	.	.	.	.	.	.	.	6
584	ARISTIDA OLIGANTHA	W	.	0	5.2	2.2	82.8	31.4	9.8	.	.	.	.	.	.	.	.	.	.	6
585	ARISTIDA PURPURASCENS	W	.	0	5.2	1.2	82.0	35.9	11.6	.	.	.	.	.	.	.	.	.	.	6
586	ARISTIDA PURPUREA	W	.	0	6.8	1.4	80.0	33.9	11.8	.	.	.	.	.	.	.	.	.	.	6
587	ARISTIDA WRIGHTII	W	.	0	7.8	1.7	79.1	32.4	11.4	.	.	.	.	.	.	.	.	.	.	6
588	ARISTOLOCHIA MAURORUM	S	.	0	21.1	23.4	.	.	.	.	.	.	.	.	.	.	.	.	.	20
589	ARMERIA PSEUD-ARMERIA	S	.	0	37.5	6.5	.	.	.	.	.	.	.	.	.	.	.	.	.	20
590	ARMERIA PSEUD-ARMERIA	S	.	0	32.5	7.4	.	.	.	.	.	.	.	.	.	.	.	.	.	20
591	ARMORACIA RUSTICANA	R	348	0	13.6	1.0	78.3	7.1	7.1	121.2	328.3	10.6	.	.	25.3	0.15	0.35	2.02	111.1	2
592	ARMORACIA RUSTICANA	R	343	0	12.6	1.2	77.6	9.4	8.7	551.2	252.0	5.5	31.5	2220.5	.	0.28	.	.	318.9	4
593	ARRABIDAEA SP.	S	.	0	20.0	22.8	.	.	.	.	.	.	.	.	.	.	.	.	.	20
594	ARRACACIA XANTHORRHIZA	L	383	0	3.0	0.8	91.7	3.8	4.5	97.7	195.5	3.4	.	.	0.0	0.26	0.23	10.53	86.5	3
595	ARRHENATHERUM ELATIUS	S	.	0	25.0	8.3	.	.	.	.	.	.	.	.	.	.	.	.	.	20
596	ARRHENATHERUM ELATIUS	S	.	0	23.8	8.7	.	.	.	.	.	.	.	.	.	.	.	.	.	20
597	ARRHENATHERUM ELATIUS	W	.	0	9.5	3.1	80.4	31.2	7.0	.	.	.	.	.	47.0	.	.	.	.	6
598	ARTEDIA SQUAMATA	S	.	0	11.2	9.5	.	.	.	.	.	.	.	.	.	.	.	.	.	21
599	ARTEDIA SQUAMATA	S	.	0	8.3	4.8	.	.	.	.	.	.	.	.	.	.	.	.	.	21
600	ARTEMESIA ABSINTHIUM	S	.	0	25.8	33.4	.	.	6.6	.	.	.	.	.	.	.	.	.	.	19
601	ARTEMISIA ABSINTHIUM	S	.	0	28.1	38.1	.	.	.	.	.	.	.	.	.	.	.	.	.	20
602	ARTEMISIA ARBUSCULA	W	.	0	8.6	7.3	77.3	20.2	6.8	.	.	.	.	.	.	.	.	.	.	6
603	ARTEMISIA CAMPESTRIS	S	.	0	33.1	35.6	.	.	.	.	.	.	.	.	.	.	.	.	.	21
604	ARTEMISIA DRACUNCULUS	S	.	0	31.5	38.1	.	.	.	.	.	.	.	.	.	.	.	.	.	21
605	ARTEMISIA DRACUNCULUS	H	320	0	24.7	7.8	54.4	8.0	13.0	1234.0	339.1	35.0	67.2	3271.9	2730.2	0.27	1.45	9.70	.	16
606	ARTEMISIA FILIFOLIA	W	.	0	11.2	8.2	75.4	23.5	5.2	750.0	230.0	.	.	.	.	.	.	.	.	6
607	ARTEMISIA FRIGIDA	S	.	0	30.4	31.4	.	.	.	.	.	.	.	.	.	.	.	.	.	21
608	ARTEMISIA FRIGIDA	W	.	0	9.4	4.0	72.3	30.8	14.3	600.0	290.0	.	.	.	.	.	.	.	.	6
609	ARTEMISIA GNAPHALODES	W	.	0	11.1	3.1	77.7	26.9	8.1	.	.	.	.	.	.	.	.	.	.	6
610	ARTEMISIA HERBA-ALBA	S	.	0	30.1	30.6	.	.	.	.	.	.	.	.	.	.	.	.	.	21
611	ARTEMISIA LACTIFLORA	L	336	0	23.5	6.7	61.3	11.8	8.4	378.2	403.4	18.5	.	.	22689.1	1.01	2.18	3.36	302.5	2
612	ARTEMISIA MARITIMA	S	.	0	43.8	22.0	.	.	.	.	.	.	.	.	.	.	.	.	.	21
613	ARTEMISIA NOVA	W	508	0	8.6	10.0	75.3	27.5	6.1	650.0	200.0	.	.	.	2.0	.	.	.	.	6
614	ARTEMISIA PEDATIFIDA	W	.	0	7.0	2.4	62.8	23.7	27.8	.	.	.	.	.	.	.	.	.	.	6
615	ARTEMISIA RIGIDA	W	.	0	21.1	3.7	68.6	22.0	6.6	.	.	.	.	.	.	.	.	.	.	6
616	ARTEMISIA SPINESCENS	W	424	0	18.5	3.3	61.8	20.1	16.4	1570.0	420.0	.	.	.	2.0	.	.	.	.	6
617	ARTEMISIA TRIDENTATA	W	.	0	9.9	13.0	71.1	25.3	6.0	720.0	180.0	.	.	.	2.0	.	.	.	.	6
618	ARTEMISIA TRIPARTITA	W	501	0	13.1	.	.	.	.	790.0	210.0	.	.	.	.	.	.	.	.	6

TABLE 3 - DATA CONVERTED TO A ZERO-MOISTURE BASIS.

OBS	PLANA	PART	CALORIES	WATER	PROTEIN	FAT	TOTCARB	FIBER	ASH	CALCIUM	PHOSPHO	IRON	SODIUM	POTASS	CAROT	THIAMINE	RIBOFLAV	NIACIN	ASCORBIC	REFCO
619	ARTEMISIA VARIABILIS	S	.	0	24.8	38.7	.	.	.	.	.	.	.	.	.	.	.	.	.	21
620	ARTEMISIA VULGARIS	W	.	0	11.4	6.0	76.3	33.5	6.3	860.0	190.0	.	.	2200.0	.	.	.	.	.	6
621	ARTEMISIA VULGARIS	L	276	0	40.9	6.3	35.4	26.8	17.3	645.7	315.0	11.8	.	.	16850.4	1.18	1.26	23.62	566.9	2
622	ARTOCARPUS ALTILIS	F	354	0	4.8	1.1	91.1	4.8	3.0	107.0	147.6	2.6	48.0	1461.3	36.9	0.30	0.22	4.43	44.3	2
623	ARTOCARPUS ALTILIS	S	544	0	18.9	36.3	42.6	3.1	2.1	82.7	401.0	8.4	51.4	476.2	350.9	1.10	0.69	1.00	15.0	2
624	ARTOCARPUS ALTILIS	F	353	0	5.8	1.0	89.7	4.1	3.4	113.0	109.6	4.1	51.4	1503.4	82.2	0.38	0.10	3.08	99.3	4
625	ARTOCARPUS HETEROPHYLLUS	L	.	0	18.5	5.0	66.3	26.2	10.2	2000.0	110.0	.	.	.	.	.	.	.	.	18
626	ARTOCARPUS HETEROPHYLLUS	F	347	0	6.3	1.1	87.5	3.3	5.2	99.6	140.2	2.2	7.4	1501.8	867.2	0.33	0.41	2.58	33.2	2
627	ARTOCARPUS HETEROPHYLLUS	G	346	0	13.7	2.6	77.8	18.3	5.9	313.7	163.4	2.0	19.6	2111.1	65.4	0.65	0.26	1.96	104.6	2
628	ARTOCARPUS ODORATISSIMA	S	386	0	11.0	1.0	83.4	3.8	4.6	89.5	322.3	3.1	56.3	2150.9	63.9	0.46	0.13	1.28	43.5	2
629	ARTOCARPUS ODORATISSIMA	L	306	0	20.4	0.0	71.4	.	8.2	693.9	244.9	71.4	.	.	.	0.41	.	.	285.7	2
630	ARTOCARPUS ODORATISSIMA	F	354	0	1.7	0.7	94.9	2.7	2.7	40.8	112.2	11.9	.	.	.	0.20	0.17	2.04	102.0	2
631	ARTOCARPUS ODORATISSIMA	U	450	0	5.0	11.8	81.9	1.4	1.2	52.4	227.2	1.2	.	.	.	0.50	0.08	1.17	.	2
632	ARUM ORIENTALE	S	.	0	14.6	2.0	.	.	.	.	.	.	.	.	.	.	.	.	.	21
633	ARUNDINARIA CANNAVIEIRA	L	.	0	7.8	2.4	86.5	45.0	3.3	.	.	.	.	.	.	.	.	.	.	18
634	ARUNDINARIA JAPONICA	H	.	0	2.4	2.3	91.6	33.9	3.7	370.0	150.0	.	.	680.0	.	.	.	.	.	6
635	ARUNDINARIA JAPONICA	H	.	0	2.2	2.3	92.2	34.1	3.3	.	.	.	.	.	.	.	.	.	.	6
636	ARUNDINARIA JAPONICA	C	.	0	2.7	1.1	92.6	28.0	3.6	.	.	.	.	.	.	.	.	.	.	6
637	ARUNDINARIA TECTA	W	.	0	15.1	3.5	70.8	29.4	10.6	210.0	190.0	20.0	.	.	.	.	.	.	.	6
638	ARUNDINARIA TECTA	L	.	0	12.1	.	.	.	.	240.0	60.0	.	.	.	.	.	.	.	.	6
639	ARUNDO DONAX	W	.	0	13.2	1.9	69.9	28.5	15.0	.	.	.	.	.	.	.	.	.	.	18
640	ARUNDO DONAX	H	.	0	8.8	1.1	77.7	33.0	12.4	.	.	.	.	.	.	.	.	.	.	18
641	ARUNDO DONAX	W	.	0	6.9	1.3	82.6	37.5	9.2	490.0	110.0	.	.	2550.0	.	.	.	.	.	6
642	ASCLEPIAS CAPRICORNUS	S	.	0	33.8	22.3	.	.	.	.	.	.	.	.	.	.	.	.	.	19
643	ASCLEPIAS CURASSAVICA	S	.	0	22.4	25.9	.	.	.	.	.	.	.	.	.	.	.	.	.	21
644	ASCLEPIAS ENGELMANNIANA	S	.	0	41.5	38.4	.	.	6.2	.	.	.	.	.	.	.	.	.	.	19
645	ASCLEPIAS ENGELMANNIANA	S	.	0	32.6	18.1	.	.	3.7	.	.	.	.	.	.	.	.	.	.	19
646	ASCLEPIAS INCARNATA	S	.	0	30.9	26.8	.	.	3.9	.	.	.	.	.	.	.	.	.	.	19
647	ASCLEPIAS LINARIA	S	.	0	26.2	24.4	.	.	.	.	.	.	.	.	.	.	.	.	.	19
648	ASCLEPIAS SPECIOSA	S	.	0	31.8	25.2	.	.	.	.	.	.	.	.	.	.	.	.	.	21
649	ASCLEPIAS SUBVERTICILLATA	S	.	0	26.1	20.1	.	.	.	.	.	.	.	.	.	.	.	.	.	20
650	ASCLEPIAS SYRIACA	S	.	0	33.8	31.8	.	.	3.7	.	.	.	.	.	.	.	.	.	.	19
651	ASCLEPIAS SYRIACA	S	.	0	37.2	28.2	.	.	4.0	.	.	.	.	.	.	.	.	.	.	19
652	ASCLEPIAS SYRIACA	S	.	0	17.6	9.1	.	.	4.2	.	.	.	.	.	.	.	.	.	.	19
653	ASCLEPIAS TUBEROSA	S	.	0	32.5	24.1	.	.	.	.	.	.	.	.	.	.	.	.	.	20
654	ASCOPHYLLUM NODOSUM	W	.	0	8.4	3.2	.	.	20.9	1160.0	110.0	.	.	.	.	.	.	.	.	6
655	ASIMINA TRILOBA	F	363	0	22.2	3.8	71.8	.	2.1	.	.	.	.	.	.	.	.	.	.	4
656	ASPALATHUS CONTAMINATUS	L	348	0	6.2	1.7	89.3	.	2.8	147.9	244.7	51.5	.	.	.	.	.	.	12.2	1
657	ASPALATHUS SP.	S	.	0	41.0	4.5	.	.	.	.	.	.	.	.	.	.	.	.	.	21
658	ASPARAGUS ACUTIFOLIUS	S	.	0	18.5	19.8	.	.	.	.	.	.	.	.	.	.	.	.	.	21
659	ASPARAGUS OFFICINALIS	S	.	0	16.2	15.4	.	.	2.4	.	.	.	.	.	.	.	.	.	.	19
660	ASPARAGUS OFFICINALIS	SH	288	0	34.2	4.1	49.3	12.3	12.3	219.2	808.2	19.2	68.5	3027.4	5205.5	2.05	2.47	24.66	274.0	2
661	ASPARAGUS OFFICINALIS	SH	280	0	29.3	3.7	50.0	9.8	17.1	304.9	1024.4	11.0	36.6	.	365.8	1.95	3.66	36.59	122.0	2
662	ASPARAGUS OFFICINALIS	SH	313	0	30.1	2.4	60.2	8.4	7.2	265.1	747.0	12.0	24.1	3349.4	6506.0	2.17	2.41	18.07	397.6	4
663	ASPARAGUS OFFICINALIS	SH	.	0	31.4	2.9	55.7	10.0	10.0	357.1	557.1	14.3	.	.	12000.0	2.57	1.86	.	571.4	5

TABLE 3 - DATA CONVERTED TO A ZERO-MOISTURE BASIS.

OBS	PLNA	PART	CALORIES	WATER	PROTEIN	FAT	TOTLCARB	FIBER
664	ASPARAGUS THUNBERGIANUS	S	.	0	13.5	8.9	.	.
665	ASPHODELINE ISTHMOCARPA	S	.	0	28.4	24.6	.	.
666	ASPHODELUS FISTULOSUS	S	.	0	21.4	23.8	.	.
667	ASPHODELUS MICROCARPUS	S	.	0	38.1	13.6	.	.
668	ASPHODELUS TENUIFOLIUS	S	.	0	18.6	16.8	.	.
669	ASPHODELUS TENUIFOLIUS	S	.	0	15.2	28.8	.	.
670	ASTER ALPINUS	S	.	0	19.3	22.7	.	.
671	ASTER AMELLUS	L	305	0	32.8	5.5	50	8.6
672	ASTER ERICOIDES	S	.	0	28.1	30.4	.	.
673	ASTER ERICOIDES	W	.	0	7.6	5.0	81	21.1
674	ASTER FREMONTII	C	.	0	5.3	1.5	.	.
675	ASTER FREMONTII	W	.	0	9.1	3.1	.	.
676	ASTER FREMONTII	L	.	0	12.2	4.8	.	.
677	ASTER NOVAE-ANGLIAE	S	.	0	32.5	28.7	.	.
678	ASTER NOVI-BELGII	S	.	0	27.5	24.3	.	.
679	ASTER SQUAMATUS	S	.	0	27.9	29.8	.	.
680	ASTER TANACETIFOLIUS	S	.	0	15.8	8.4	.	.
681	ASTER TONGOLENSIS	S	.	0	20.0	19.3	.	.
682	ASTER WASATCHENSIS	S	.	0	24.9	25.2	.	.
683	ASTRAGALUS ADSURGENS	S	.	0	42.1	5.3	.	.
684	ASTRAGALUS ANGUSTIFOLIUS	S	.	0	43.5	5.7	.	.
685	ASTRAGALUS ANGUSTIFOLIUS	S	.	0	43.1	5.4	.	.
686	ASTRAGALUS BOETICUS	S	.	0	29.5	1.2	.	.
687	ASTRAGALUS BRAZOENSIS	S	.	0	43.8	1.6	.	.

OBS	ASH	CALCIUM	PHOSPHO	IRON	SODIUM	POTASS	CAROT	THIAMINE	RIBOFLAV	NIACIN	ASCORBIC	REFCO
664	.	.	.	.	.	.	.	.	.	.	.	21
665	.	.	.	.	.	.	.	.	.	.	.	21
666	.	.	.	.	.	.	.	.	.	.	.	21
667	.	.	.	.	.	.	.	.	.	.	.	21
668	.	.	.	.	.	.	.	.	.	.	.	21
669	.	.	.	.	.	.	.	.	.	.	.	21
670	5.6	.	.	.	.	.	.	.	.	.	.	19
671	11.7	328.1	593.8	31.3	.	4164.1	26132.8	1.41	2.81	8.59	687.5	2
672	5.5	.	.	.	.	.	.	.	.	.	.	19
673	6.4	2470.0	100.0	.	.	940.0	.	.	.	.	.	6
674	5.8	910.0	730.0	.	.	.	.	.	.	.	.	6
675	8.8	1260.0	350.0	.	.	.	.	.	.	.	.	6
676	11.7	1600.0	510.0	.	.	.	.	.	.	.	.	6
677	.	.	.	.	.	.	.	.	.	.	.	20
678	.	.	.	.	.	.	.	.	.	.	.	20
679	.	.	.	.	.	.	.	.	.	.	.	21
680	.	.	.	.	.	.	.	.	.	.	.	20
681	.	.	.	.	.	.	.	.	.	.	.	20
682	.	.	.	.	.	.	.	.	.	.	.	21
683	3.7	.	.	.	.	.	.	.	.	.	.	19
684	.	.	.	.	.	.	.	.	.	.	.	21
685	.	.	.	.	.	.	.	.	.	.	.	21
686	.	.	.	.	.	.	.	.	.	.	.	21
687	.	.	.	.	.	.	.	.	.	.	.	19

TABLE 3 - DATA CONVERTED TO A ZERO-MOISTURE BASIS.

OBS	PLNA	PART	CALORIES	WATER	PROTEIN	FAT	TOTLCARB	FIBER
688	ASTRAGALUS CANADENSIS	S	.	0	40.2	2.8	.	.
689	ASTRAGALUS CICER	S	.	0	39.9	3.2	.	.
690	ASTRAGALUS CICER	S	.	0	37.0	3.6	.	.
691	ASTRAGALUS CICER	S	.	0	40.1	3.5	.	.
692	ASTRAGALUS CICER	S	.	0	38.8	4.2	.	.
693	ASTRAGALUS CRASSICARPUS	S	.	0	47.5	5.7	.	.
694	ASTRAGALUS FALCATUS	S	.	0	41.0	3.2	.	.
695	ASTRAGALUS FLEXUOSUS	S	.	0	43.8	4.0	.	.
696	ASTRAGALUS FLEXUOSUS	S	.	0	42.9	4.3	.	.
697	ASTRAGALUS GALEGIFORMIS	S	.	0	43.7	5.2	.	.
698	ASTRAGALUS GALEGIFORMIS	S	.	0	43.9	5.4	.	.
699	ASTRAGALUS GALEGIFORMIS	S	.	0	44.5	6.0	.	.
700	ASTRAGALUS GLYCYPHYLLOS	S	.	0	36.1	2.2	.	.
701	ASTRAGALUS GLYCYPHYLLOS	S	.	0	40.0	3.0	.	.
702	ASTRAGALUS HAMOSUS	S	.	0	34.2	1.4	.	.
703	ASTRAGALUS HAMOSUS	S	.	0	26.4	1.2	.	.
704	ASTRAGALUS HAMOSUS	S	.	0	34.8	1.7	.	.
705	ASTRAGALUS LONGIDENTATUS	S	.	0	30.4	2.0	.	.
706	ASTRAGALUS MACROCEPHALUS	S	.	0	50.9	5.5	.	.
707	ASTRAGALUS MACROCEPHALUS	S	.	0	31.2	2.7	.	.
708	ASTRAGALUS MENZIESII	S	.	0	36.5	4.9	.	.
709	ASTRAGALUS MEXICANUS	S	.	0	38.9	4.6	.	.
710	ASTRAGALUS MOLLISSIMUS	S	.	0	37.6	6.0	.	.
711	ASTRAGALUS NERVULOSUS	S	.	0	52.6	9.1	.	.

OBS	ASH	CALCIUM	PHOSPHO	IRON	SODIUM	POTASS	CAROT	THIAMINE	RIBOFLAV	NIACIN	ASCORBIC	REFCO
688	3.0	.	.	.	.	.	.	.	.	.	.	19
689	4.6	.	.	.	.	.	.	.	.	.	.	19
690	6.2	.	.	.	.	.	.	.	.	.	.	19
691	4.3	.	.	.	.	.	.	.	.	.	.	19
692	4.2	.	.	.	.	.	.	.	.	.	.	19
693	.	.	.	.	.	.	.	.	.	.	.	20
694	4.7	.	.	.	.	.	.	.	.	.	.	19
695	5.6	.	.	.	.	.	.	.	.	.	.	19
696	4.9	.	.	.	.	.	.	.	.	.	.	19
697	4.5	.	.	.	.	.	.	.	.	.	.	19
698	3.8	.	.	.	.	.	.	.	.	.	.	19
699	.	.	.	.	.	.	.	.	.	.	.	21
700	4.3	.	.	.	.	.	.	.	.	.	.	19
701	.	.	.	.	.	.	.	.	.	.	.	21
702	.	.	.	.	.	.	.	.	.	.	.	20
703	.	.	.	.	.	.	.	.	.	.	.	21
704	.	.	.	.	.	.	.	.	.	.	.	21
705	.	.	.	.	.	.	.	.	.	.	.	21
706	.	.	.	.	.	.	.	.	.	.	.	21
707	.	.	.	.	.	.	.	.	.	.	.	21
708	2.0	.	.	.	.	.	.	.	.	.	.	19
709	4.4	.	.	.	.	.	.	.	.	.	.	19
710	4.0	.	.	.	.	.	.	.	.	.	.	19
711	.	.	.	.	.	.	.	.	.	.	.	21

TABLE 3 - DATA CONVERTED TO A ZERO-MOISTURE BASIS.

OBS	PLNA	PART	CALORIFS	WATER	PROTEIN	FAT	TOTLCARB	FIBER
712	ASTRAGALUS NUTTALLIANUS	S	.	0	35.0	2.6	.	.
713	ASTRAGALUS NUTTALLIANUS VAR. TRICHOCARPUS	S	.	0	35.0	1.8	.	.
714	ASTRAGALUS ODORATUS	S	.	0	41.0	3.2	.	.
715	ASTRAGALUS OVALIS	S	.	0	49.8	6.8	.	.
716	ASTRAGALUS PANDURATUS	S	.	0	52.4	8.5	.	.
717	ASTRAGALUS PANDURATUS	S	.	0	50.9	7.1	.	.
718	ASTRAGALUS PARISHII	S	.	0	38.1	5.3	.	.
719	ASTRAGALUS PODOCARPUS	S	.	0	37.0	2.6	.	.
720	ASTRAGALUS PONTICUS	S	.	0	46.3	5.7	.	.
721	ASTRAGALUS PONTICUS	S	.	0	40.5	6.3	.	.
722	ASTRAGALUS PYRRHOTRICHUS	S	.	0	41.7	7.4	.	.
723	ASTRAGALUS RACEMOSUS	S	.	0	40.9	6.5	.	.
724	ASTRAGALUS RUBYI	S	.	0	35.4	5.1	.	.
725	ASTRAGALUS SEMIBILOCULARIS	S	.	0	41.9	5.3	.	.
726	ASTRAGALUS SETULOSUS	S	.	0	43.4	4.8	.	.
727	ASTRAGALUS SINICUS	S	.	0	36.6	5.3	.	.
728	ASTRAGALUS SP.	S	.	0	33.8	4.7	.	.
729	ASTRAGALUS SP.	S	.	0	44.5	4.0	.	.
730	ASTRAGALUS SP.	S	.	0	36.7	4.2	.	.
731	ASTRAGALUS SP.	S	.	0	61.1	7.6	.	.
732	ASTRAGALUS SP.	W	.	0	21.2	2.2	63.5	20.9
733	ASTRAGALUS SP.	W	.	0	23.2	2.7	65.4	19.1
734	ASTRAGALUS STIPULATUS	S	.	0	33.3	5.2	.	.
735	ASTRAGALUS TENELLUS	S	.	0	32.1	5.0	.	.

OBS	ASH	CALCIUM	PHOSPHO	IRON	SODIUM	POTASS	CAROT	THIAMINE	RIBOFLAV	NIACIN	ASCORBIC	REFCO
712	.	.	.	.	.	.	.	.	.	.	.	20
713	.	.	.	.	.	.	.	.	.	.	.	19
714	.	.	.	.	.	.	.	.	.	.	.	21
715	.	.	.	.	.	.	.	.	.	.	.	21
716	.	.	.	.	.	.	.	.	.	.	.	20
717	.	.	.	.	.	.	.	.	.	.	.	21
718	.	.	.	.	.	.	.	.	.	.	.	21
719	4.4	.	.	.	.	.	.	.	.	.	.	19
720	.	.	.	.	.	.	.	.	.	.	.	21
721	.	.	.	.	.	.	.	.	.	.	.	21
722	.	.	.	.	.	.	.	.	.	.	.	21
723	.	.	.	.	.	.	.	.	.	.	.	20
724	4.5	.	.	.	.	.	.	.	.	.	.	19
725	4.7	.	.	.	.	.	.	.	.	.	.	19
726	.	.	.	.	.	.	.	.	.	.	.	21
727	3.3	.	.	.	.	.	.	.	.	.	.	19
728	4.0	.	.	.	.	.	.	.	.	.	.	19
729	3.9	.	.	.	.	.	.	.	.	.	.	19
730	4.8	.	.	.	.	.	.	.	.	.	.	19
731	.	.	.	.	.	.	.	.	.	.	.	21
732	13.1	1180.0	300.0	.	.	2800	.	.	.	.	.	6
733	8.7	1650.0	390.0	.	.	2800	.	.	.	.	.	6
734	4.8	.	.	.	.	.	.	.	.	.	.	19
735	3.7	.	.	.	.	.	.	.	.	.	.	19

TABLE 3 - DATA CONVERTED TO A ZERO-MOISTURE BASIS.

OBS	PLANA	PART	CALORIES	WATER	PROTEIN	FAT	TOTLCARB	FIBER	ASH	CALCIUM	PHOSPHO	IRON	SODIUM	POTASS	CAROT	THIAMINE	RIBOFLAV	NIACIN	ASCORBIC	REFCO
736	ASTRAGALUS VULNERARIAE	S	.	0	45.6	5.1	.	.	.	.	.	.	.	.	.	.	.	.	.	21
737	ASTRANTIA MAXIMA	S	.	0	28.9	39.2	.	.	.	.	.	.	.	.	.	.	.	.	.	21
738	ASTROCARYUM STANDLEYANUM	F	352	0	6.0	2.5	86.5	20.3	5.0	220.6	145.9	5.3	.	.	43736.6	0.21	0.25	1.78	14.2	3
739	ASTRODAUCUS ORIENTALIS	S	.	0	17.1	11.8	.	.	.	.	.	.	.	.	.	.	.	.	.	21
740	ASYSTASIA GANGETICA	L	322	0	21.3	6.9	59.8	6.9	12.1	1298.9	172.4	27.0	.	.	35919.5	1.09	1.21	5.75	241.4	2
741	ATHANASIA TOMENTOSA	S	.	0	21.7	5.7	.	.	.	.	.	.	.	.	.	.	.	.	.	21
742	ATRAPHAXIS BILLARDIERI	S	.	0	13.6	2.0	.	.	.	.	.	.	.	.	.	.	.	.	.	20
743	ATRIPLEX ARGENTEA	W	.	0	10.3	1.5	67.8	28.9	20.4	.	.	.	.	.	.	.	.	.	.	6
744	ATRIPLEX CANESCENS	S	.	0	5.4	1.0	.	.	6.5	.	.	.	.	.	.	.	.	.	.	19
745	ATRIPLEX CANESCENS	S	.	0	5.6	1.5	.	.	.	.	.	.	.	.	.	.	.	.	.	19
746	ATRIPLEX CANESCENS	S	.	0	23.6	10.6	.	.	.	.	.	.	.	.	.	.	.	.	.	21
747	ATRIPLEX CANESCENS	S	.	0	22.1	11.5	.	.	.	.	.	.	.	.	.	.	.	.	.	21
748	ATRIPLEX CANESCENS	W	.	0	14.5	2.1	71.6	16.2	11.8	1190.0	150.0	10.0	.	810.0	.	.	.	.	.	6
749	ATRIPLEX CONFERTIFOLIA	W	389	0	7.9	2.2	68.2	30.7	21.7	2610.0	200.0	.	.	.	2.0	.	.	.	.	6
750	ATRIPLEX CONFERTIFOLIA	W	.	0	8.0	2.2	67.9	31.1	21.9	2270.0	140.0	.	.	.	.	.	.	.	.	6
751	ATRIPLEX CONFERTIFOLIA	L	.	0	14.9	1.9	52.1	10.8	31.1	.	.	.	.	.	.	.	.	.	.	6
752	ATRIPLEX GARDNERI	W	.	0	8.4	3.0	73.0	30.6	15.6	.	.	.	.	.	.	.	.	.	.	6
753	ATRIPLEX GARDNERI	L	.	0	16.9	2.7	56.3	11.3	24.1	.	.	.	.	.	.	.	.	.	.	6
754	ATRIPLEX GARDNERI	C	.	0	7.7	5.0	79.6	35.7	7.7	.	.	.	.	.	.	.	.	.	.	6
755	ATRIPLEX HASTATA	S	.	0	21.8	9.0	.	.	.	.	.	.	.	.	.	.	.	.	.	21
756	ATRIPLEX HORTENSIS	S	.	0	30.3	5.9	.	.	.	.	.	.	.	.	.	.	.	.	.	21
757	ATRIPLEX HORTENSIS	S	.	0	31.2	5.6	.	.	.	.	.	.	.	.	.	.	.	.	.	21
758	ATRIPLEX LAEVIS	S	.	0	20.8	8.4	.	.	.	.	.	.	.	.	.	.	.	.	.	21
759	ATRIPLEX LENTIFORMIS	S	.	0	12.5	2.1	.	.	13.6	.	.	.	.	.	.	.	.	.	.	19
760	ATRIPLEX NUMMULARIA	SH	.	0	21.3	2.6	53.9	10.3	22.2	1220.0	170.0	.	.	.	.	.	.	.	.	18
761	ATRIPLEX ROSEA	L	.	0	15.5	1.8	59.7	14.5	23.0	.	.	.	.	.	.	.	.	.	.	6
762	ATRIPLEX ROSEA	L	.	0	23.7	2.1	42.5	8.1	31.7	.	.	.	.	.	.	.	.	.	.	6
763	ATRIPLEX SEMIBACCATA	W	.	0	16.6	1.6	58.2	18.7	23.6	1810.0	140.0	.	.	.	.	.	.	.	.	18
764	ATRIPLEX SEMIBACCATA	SH	.	0	21.4	1.4	57.9	17.0	19.3	.	.	.	.	.	.	.	.	.	.	18
765	ATRIPLEX SEMIBACCATA	W	.	0	17.5	1.6	59.2	18.8	21.7	.	.	.	.	.	.	.	.	.	.	6
766	ATRIPLEX SP.	S	.	0	23.8	1.9	.	.	.	.	.	.	.	.	.	.	.	.	.	19
767	ATRIPLEX TATARICA	S	.	0	20.6	11.1	.	.	.	.	.	.	.	.	.	.	.	.	.	21
768	ATRIPLEX TATARICA	S	.	0	19.9	6.7	.	.	.	.	.	.	.	.	.	.	.	.	.	21
769	ATROPA ACUMINATA	S	.	0	18.4	41.3	.	.	.	.	.	.	.	.	.	.	.	.	.	21
770	ATROPA BELLADONA	S	.	0	21.0	27.6	.	.	.	.	.	.	.	.	.	.	.	.	.	21
771	AUCUBA JAPONICA	S	.	0	8.8	0.8	.	.	3.8	.	.	.	.	.	.	.	.	.	.	19
772	AULAX PALLASIA	S	.	0	71.1	23.7	.	.	.	.	.	.	.	.	.	.	.	.	.	21
773	AVENA BARBATA	S	.	0	16.2	11.3	.	.	2.0	.	.	.	.	.	.	.	.	.	.	19
774	AVENA BARBATA	S	.	0	17.5	9.7	.	.	2.6	.	.	.	.	.	.	.	.	.	.	19
775	AVENA BARBATA	W	.	0	8.6	.	.	34.1	8.5	320.0	310.0	.	.	2620.0	.	.	.	.	.	6
776	AVENA BYZANTINA	W	.	0	26.1	5.3	56.9	18.1	11.7	.	.	.	.	.	.	.	.	.	.	18
777	AVENA BYZANTINA	W	.	0	8.5	3.0	81.7	18.4	6.8	.	.	.	.	.	.	.	.	.	.	18
778	AVENA BYZANTINA	S	.	0	12.8	5.6	77.6	13.9	4.0	.	.	.	.	.	.	.	.	.	.	6
779	AVENA FATUA	S	.	0	14.8	11.5	.	.	2.8	.	.	.	.	.	.	.	.	.	.	19
780	AVENA FATUA	H	.	0	7.5	2.9	82.0	34.1	7.6	.	.	.	.	.	.	.	.	.	.	6

TABLE 3 - DATA CONVERTED TO A ZERO-MOISTURE BASIS.

OBS	PLANA	PART	CALORIES	WATER	PROTEIN	FAT	TOTALCARB	FIBER	ASH	CALCIUM	PHOSPHO	IRON	SODIUM	POTASS	CAROT	THIAMINE	RIBOFLAV	NIACIN	ASCORBIC	REFCO
781	AVENA SATIVA		.	0	34.1	6.5	49.4	21.4	10.0	.	.	.	.	.	.	.	.	.	.	18
782	AVENA SATIVA	H	.	0	7.1	1.9	86.1	30.9	4.9	.	.	.	.	.	.	.	.	.	.	18
783	AVENA SATIVA	W	.	0	18.0	3.9	67.4	26.3	10.7	400.0	370.0	57.0	.	3230.0	43.0	0.13	1.10	4.94	.	6
784	AVENA SATIVA	H	454	0	9.2	3.1	80.2	31.0	7.5	260.0	240.0	53.0	.	970.0	10.0	0.35	1.01	.	.	6
785	AVENA SATIVA	S	471	0	13.3	5.1	77.5	12.0	4.1	110.0	390.0	8.0	.	420.0	0.0	0.70	0.17	1.79	.	6
786	AVENA SATIVA	S	420	0	14.7	6.9	75.7	6.5	2.7	66.3	477.5	5.2	11.2	.	0.0	0.39	0.10	2.47	0.0	2
787	AVERRHOA BILIMBI	F	360	0	8.0	4.0	84.0	8.0	4.0	66.7	173.3	8.0	53.3	1733.3	1400.0	0.27	0.53	2.67	466.7	2
788	AVERRHOA BILIMBI	F	345	0	13.8	1.7	79.3	10.3	5.2	51.7	189.7	17.2	.	.	689.7	0.17	0.52	5.17	275.9	3
789	AVERRHOA CARAMBOLA	F	357	0	3.2	2.4	91.3	6.3	3.2	47.6	127.0	7.1	.	.	1666.7	0.24	0.16	3.17	309.5	1
790	AVERRHOA CARAMBOLA	F	364	0	3.9	5.2	87.0	13.0	3.9	103.9	194.8	11.7	26.0	2350.6	2077.9	0.65	0.52	5.19	493.5	2
791	AVERRHOA CARAMBOLA	F	383	0	5.3	3.2	93.6	6.4	4.3	53.2	191.5	4.3	.	.	1914.9	0.43	0.21	3.19	372.3	3
792	AVERRHOA CARAMBOLA	F	365	0	7.3	5.2	83.3	9.4	4.2	41.7	177.1	15.6	20.8	2000.0	7500.0	0.42	0.21	3.13	364.6	4
793	AVICENNIA GERMINANS	S	392	0	6.2	0.6	90.1	4.4	3.1	229.5	129.7	.	.	.	.	.	.	.	.	1
794	AVICENNIA NITIDA	L	.	0	10.7	4.4	69.2	23.9	15.7	.	.	.	.	.	.	.	.	.	.	6
795	AXONOPUS AFFINIS	W	.	0	8.3	1.7	84.1	30.4	5.9	.	.	.	.	.	.	.	.	.	.	18
796	AXONOPUS COMPRESSUS	W	.	0	11.4	1.8	76.4	42.4	10.4	.	.	.	.	.	.	.	.	.	.	18
797	AXONOPUS SP.	W	.	0	9.3	1.7	77.9	27.7	11.1	460.0	190.0	44.0	.	850.0	14.0	.	.	.	.	6
798	AZADIRACHTA INDICA	L	.	0	15.4	4.2	69.2	12.7	11.2	2650.0	240.0	.	.	.	.	.	.	.	.	18
799	AZADIRACHTA INDICA	F	370	0	7.2	5.5	83.4	.	3.9	138.1	226.5	.	.	.	.	.	.	.	.	2
800	BACCAUREA MOTLEYANA	F	310	0	8.1	1.0	76.7	.	14.3	61.9	95.2	3.8	.	.	0.0	0.00	.	.	23.8	2
801	BACCHARIS HALIMIFOLIA	S	.	0	28.8	35.3	.	.	.	.	.	.	.	.	.	.	.	.	.	20
802	BACCHARIS SAROTHROIDES	S	.	0	23.8	31.3	.	.	.	.	.	.	.	.	.	.	.	.	.	20
803	BACCHARIS VACCINIOIDES	S	.	0	25.0	22.5	.	.	.	.	.	.	.	.	.	.	.	.	.	20
804	BACTRIS MINOR	F	343	0	5.9	1.0	87.3	10.3	5.9	68.6	137.3	5.9	.	.	833.3	0.15	0.29	1.96	39.2	3
805	BAHIA GLANDULOSA	S	.	0	19.4	22.7	.	.	.	.	.	.	.	.	.	.	.	.	.	20
806	BAILEYA PLENIRADIATA	S	.	0	20.0	13.0	.	.	.	.	.	.	.	.	.	.	.	.	.	21
807	BALANITES AEGYPTIACA	L	.	0	11.6	4.2	71.5	13.6	12.7	.	.	.	.	.	.	.	.	.	.	18
808	BALANITES AEGYPTIACA	SH	.	0	27.5	1.5	64.4	23.3	6.6	480.0	380.0	.	.	.	.	.	.	.	.	18
809	BALANITES AEGYPTIACA	F	.	0	11.1	1.7	79.1	10.2	8.1	380.0	400.0	.	.	.	.	.	.	.	.	18
810	BALANITES AEGYPTIACA	S	.	0	21.9	45.7	.	.	3.3	.	.	.	.	.	.	.	.	.	.	19
811	BALANITES AEGYPTIACA	F	339	0	6.1	0.0	88.6	.	5.3	130.6	.	.	.	.	.	.	.	.	38.9	1
812	BALSAMORHIZA SAGITTATA		.	0	15.1	3.4	.	20.2	.	1790.0	260.0	.	.	.	.	.	.	.	.	6
813	BAMBUSA ARUNDINACEA	L	.	0	18.6	4.1	65.5	24.1	11.8	.	.	.	.	.	.	.	.	.	.	18
814	BAMBUSA ARUNDINACEA	SH	312	0	24.7	2.2	65.6	5.4	7.5	354.8	440.9	4.3	.	.	215.0	1.61	0.75	6.45	43.0	3
815	BAMBUSA ARUNDINACEA	S	.	0	13.3	0.7	84.7	1.9	1.3	.	.	.	.	.	.	.	.	.	.	5
816	BAMBUSA SP.	SH	308	0	26.0	2.9	61.5	8.7	9.6	96.2	557.7	5.8	86.5	4673.1	144.2	1.25	0.96	6.73	96.2	2
817	BAMBUSA SP.	SH	303	0	26.3	2.6	60.5	25.0	10.5	92.1	618.4	13.2	.	5236.8	0.0	.	.	.	.	2
818	BAMBUSA SP.	SH	310	0	28.3	4.4	56.6	6.2	10.6	292.0	424.8	1.8	.	6141.6	442.5	0.62	0.80	5.31	79.6	2
819	BAMBUSA SP.	SH	300	0	28.9	3.3	57.8	7.8	10.0	144.4	655.6	5.6	.	5922.2	133.3	1.67	0.78	6.67	44.4	4
820	BAMBUSA VULGARIS	L	.	0	19.0	0.5	63.6	28.8	16.9	.	.	.	.	.	.	.	.	.	.	18
821	BAPTISIA AUSTRALIS	S	.	0	34.8	10.4	.	.	3.4	.	.	.	.	.	.	.	.	.	.	19
822	BAPTISIA LEUCANTHA	S	.	0	37.7	10.2	.	.	.	.	.	.	.	.	.	.	.	.	.	20
823	BAPTISIA VIRIDIS	S	.	0	38.2	11.4	.	.	.	.	.	.	.	.	.	.	.	.	.	20
824	BARBAREA VULGARIS	S	.	0	15.6	31.0	.	.	.	.	.	.	.	.	.	.	.	.	.	20
825	BARLERIA OPACA	L	289	0	18.9	2.6	61.6	14.2	16.8	4600.0	200.0	.	.	.	.	.	.	.	.	1

TABLE 3 - DATA CONVERTED TO A ZERO-MOISTURE BASIS.

OBS	PLNA	PART	CALORIES	WATER	PROTEIN	FAT	TOTLCARB	FIBER	ASH
826	BAROSMA BETULINA	S	.	0	27.1	7.8	.	.	.
827	BAROSMA BETULINA	S	.	0	29.9	33.8	.	.	.
828	BARRINGTONIA ACUTANGULA	L	318	0	16.7	1.5	73.5	.	8.3
829	BASELLA ALBA	L	288	0	24.2	4.5	53.0	9.1	18.2
830	BASELLA ALBA	L	271	0	17.1	2.9	58.6	8.6	21.4
831	BASELLA ALBA	L	275	0	26.1	4.3	49.3	10.1	20.3
832	BASSIA HYSSOPIFOLIA	S	.	0	26.8	14.0	.	.	.
833	BASSIA HYSSOPIFOLIA	S	.	0	24.3	17.6	.	.	.
834	BAUHINIA ESCULENTA	L	296	0	15.5	0.4	71.8	24.5	8.7
835	BAUHINIA ESCULENTA	S	560	0	29.5	42.9	24.4	20.5	3.2
836	BAUHINIA GALPINII	F	.	0	17.2	3.1	76.1	20.1	3.6
837	BAUHINIA HERRERAE	S	.	0	19.4	2.4	.	.	4.0
838	BAUHINIA PURPUREA	S	.	0	27.5	20.0	.	.	.
839	BAUHINIA RETICULATA	L	295	0	22.1	0.5	66.4	31.3	11.1
840	BAUHINIA SP.	S	.	0	26.9	7.5	.	.	.
841	BAUHINIA VARIEGATA	L	.	0	18.5	9.0	62.5	10.5	10.0
842	BAUHINIA VARIEGATA	S	.	0	32.5	19.3	.	.	.
843	BECKMANNIA ERUCAEFORMIS	S	.	0	17.7	14.8	.	.	.
844	BECKMANNIA SYZIGACHNE		.	0	5.0	1.3	82.1	37.9	11.6
845	BECKMANNIA SYZIGACHNE		.	0	3.3	1.1	83.2	39.2	12.4
846	BEILSCHMIEDIA MANNII	V	390	0	6.9	0.6	88.8	1.9	3.7
847	BELLARDIA TRIXAGO	S	.	0	30.1	33.8	.	.	.
848	BELLARDIA TRIXAGO	S	.	0	30.0	31.6	.	.	.
849	BENINCASA HISPIDA	F	316	0	13.2	2.6	76.3	15.8	7.9

OBS	CALCIUM	PHOSPHO	IRON	SODIUM	POTASS	CAROT	THIAMINE	RIBOFLAV	NIACIN	ASCORBIC	REFCO
826	.	.	.	.	.	.	.	.	.	.	21
827	.	.	.	.	.	.	.	.	.	.	21
828	303.0	469.7	20.5	90.9	2833.3	9848.5	0.15	.	.	.	2
829	1606.1	590.9	24.2	348.5	7393.9	52878.8	0.91	2.58	9.09	1303.0	2
830	4942.9	157.1	55.7	.	.	24571.4	0.43	1.14	5.71	985.7	3
831	1579.7	753.6	17.4	.	.	69565.2	0.72	.	7.25	1478.3	4
832	.	.	.	.	.	.	.	.	.	.	21
833	.	.	.	.	.	.	.	.	.	.	21
834	1574.0	310.5	.	.	.	.	.	.	.	245.5	1
835	196.2	479.3	6.6	.	.	.	0.08	0.98	1.92	.	1
836	200.0	200.0	.	.	.	.	.	.	.	.	18
837	.	.	.	.	.	.	.	.	.	.	19
838	.	.	.	.	.	.	.	.	.	.	19
839	2005.0	369.0	.	.	.	.	.	.	.	.	10
840	.	.	.	.	.	.	.	.	.	.	20
841	1730.0	160.0	.	.	.	.	.	.	.	.	18
842	.	.	.	.	.	.	.	.	.	.	21
843	.	.	.	.	.	.	.	.	.	.	21
844	.	.	.	.	.	.	.	.	.	.	6
845	.	.	.	.	.	.	.	.	.	.	6
846	257.6	117.1	.	.	.	.	.	.	.	.	1
847	.	.	.	.	.	.	.	.	.	.	21
848	.	.	.	.	.	.	.	.	.	.	21
849	447.4	500.0	10.5	131.6	2921.1	131.6	0.79	0.79	5.26	526.3	2

TABLE 3 - DATA CONVERTED TO A ZERO-MOISTURE BASIS.

OBS	PLNA	PART	CALORIES	WATER	PROTEIN	FAT	TOTLCARB	FIBER	ASH
850	BENINCASA HISPIDA	F	333	0	11.9	2.4	78.6	19.0	7.1
851	BENINCASA HISPIDA	F	333	0	10.3	5.1	76.9	12.8	7.7
852	BEQUAERTIODENDRON MAGALISMONTANUM	S	.	0	5.9	5.8	.	.	.
853	BERBERIS CRATAEGINA	S	.	0	17.2	17.9	.	.	.
854	BERBERIS NERVOSA	F	379	0	18.0	8.0	71.0	.	4.0
855	BERTHOLLETIA EXCELSA	S	684	0	15.6	70.5	9.7	.	4.1
856	BERTHOLLETIA EXCELSA	S	676	0	18.3	68.2	10.1	4.1	3.5
857	BERTHOLLETIA EXCELSA	S	657	0	13.6	61.9	21.0	1.2	3.5
858	BERTHOLLETIA EXCELSA	S	686	0	15.0	70.1	11.4	3.2	3.5
859	BETA LOMATOGONA	S	.	0	11.7	2.6	.	.	.
860	BETA VULGARIS	S	.	0	15.4	6.3	.	.	5.9
861	BETA VULGARIS	S	.	0	14.3	6.4	.	.	5.1
862	BETA VULGARIS	S	.	0	15.4	7.1	.	.	5.6
863	BETA VULGARIS	S	.	0	12.6	4.0	.	.	5.5
864	BETA VULGARIS	S	.	0	14.6	3.4	.	.	6.8
865	BETA VULGARIS	S	.	0	13.2	5.8	.	.	6.5
866	BETA VULGARIS	S	.	0	11.1	2.8	.	.	6.6
867	BETA VULGARIS	S	.	0	9.3	4.6	.	.	.
868	BETA VULGARIS	W	.	0	16.0	2.7	59.8	10.9	21.5
869	BETA VULGARIS	R	346	0	14.3	0.8	78.9	9.0	6.0
870	BETA VULGARIS	L	276	0	26.3	2.6	47.4	5.3	23.7
871	BETA VULGARIS	R	349	0	12.7	1.6	79.4	7.1	6.3
872	BETA VULGARIS	L	269	0	23.1	5.8	46.2	10.6	25.0
873	BETA VULGARIS	L	284	0	24.3	5.4	50.0	12.2	20.3

OBS	CALCIUM	PHOSPHO	IRON	SODIUM	POTASS	CAROT	THIAMINE	RIBOFLAV	NIACIN	ASCORBIC	REFCO
850	500.0	166.7	11.9	.	.	238.1	0.00	0.24	7.14	642.9	3
851	487.2	487.2	10.3	153.8	2846.2	0.0	1.03	2.82	10.26	333.3	4
852	.	.	.	.	.	.	.	.	.	.	21
853	.	.	.	.	.	.	.	.	.	.	21
854	191.0	.	3.0	.	.	.	.	.	.	145.0	15
855	.	.	.	.	.	.	.	.	.	.	1
856	177.3	650.6	3.8	2.1	.	5.3	0.21	0.72	0.21	2.1	2
857	170.4	677.6	3.3	.	.	41.1	1.12	0.12	7.91	10.3	3
858	195.0	726.4	3.6	1.0	749.5	0.0	1.01	0.13	1.68	.	4
859	.	.	.	.	.	.	.	.	.	.	21
860	.	.	.	.	.	.	.	.	.	.	19
861	.	.	.	.	.	.	.	.	.	.	19
862	.	.	.	.	.	.	.	.	.	.	19
863	.	.	.	.	.	.	.	.	.	.	19
864	.	.	.	.	.	.	.	.	.	.	19
865	.	.	.	.	.	.	.	.	.	.	19
866	.	.	.	.	.	.	.	.	.	.	19
867	.	.	.	.	.	.	.	.	.	.	21
868	1030.0	230.0	.	.	5700.0	.	0.59	0.61	4.73	.	6
869	180.5	323.3	6.0	.	.	.	.	.	.	22.6	1
870	1736.8	447.4	9.2	.	.	.	0.00	2.63	.	.	1
871	182.5	277.8	8.7	285.7	2619.0	0.0	0.16	0.32	2.38	47.6	2
872	923.1	365.4	8.7	.	4211.5	25625.0	0.67	2.31	5.77	115.4	2
873	945.9	594.6	39.2	.	.	23243.2	1.35	2.57	6.76	297.3	2

TABLE 3 - DATA CONVERTED TO A ZERO-MOISTURE BASIS.

OBS	PLNA	PART	CALORIES	WATER	PROTEIN	FAT	TOTLCARB	FIBER	ASH	CALCIUM	PHOSPHO	IRON	SODIUM	POTASS	CAROT	THIAMINE	RIBOFLAV	NIACIN	ASCORBIC	REFCO
874	BETA VULGARIS	R	361	0	13.9	0.8	77.9	8.2	7.4	114.8	311.5	6.6	.	.	0.0	0.08	0.33	1.64	41.0	3
875	BETA VULGARIS	L	331	0	23.5	2.9	59.6	27.9	14.0	838.2	250.0	22.8	.	.	23161.8	0.51	1.62	4.41	367.6	3
876	BETA VULGARIS	L	293	0	17.4	4.3	60.9	10.9	17.4	1195.7	315.2	39.1	.	.	19021.7	0.33	0.98	4.35	369.6	3
877	BETA VULGARIS	R	339	0	12.6	0.8	78.0	6.3	8.7	126.0	259.8	5.5	472.4	2637.8	94.5	0.24	0.39	3.15	78.7	4
878	BETA VULGARIS	L	264	0	24.2	3.3	50.5	14.3	22.0	1307.7	439.6	36.3	1428.6	5494.5	40219.8	1.10	2.42	4.40	329.7	4
879	BETA VULGARIS	L	281	0	27.0	3.4	51.7	9.0	18.0	988.8	438.2	36.0	1651.7	6179.8	43820.2	0.67	1.91	5.62	359.6	4
880	BETULA LENTA	L	.	0	28.1	8.6	55.6	16.9	7.7	.	.	.	.	.	.	.	.	.	.	6
881	BETULA PLATYPHYLLA	S	.	0	12.4	19.0	.	.	.	.	.	.	.	.	.	.	.	.	.	20
882	BETULA POPULIFOLIA	S	.	0	6.2	5.8	.	.	.	.	.	.	.	.	.	.	.	.	.	21
883	BIDENS BITERNATA	S	.	0	23.1	14.5	.	.	.	.	.	.	.	.	.	.	.	.	.	21
884	BIDENS FRONDOSA	S	.	0	33.8	34.3	.	.	.	.	.	.	.	.	.	.	.	.	.	19
885	BIDENS GENTRYI	S	.	0	22.9	14.4	.	.	.	.	.	.	.	.	.	.	.	.	.	21
886	BIDENS PILOSA	L	295	0	24.5	4.0	56.4	12.1	15.1	1721.0	273.0	.	11.0	267.0	.	.	.	.	.	12
887	BIDENS PILOSA	S	.	0	17.5	17.1	.	.	.	.	.	.	.	.	.	.	.	.	.	20
888	BIDENS PILOSA	L	289	0	25.5	3.4	56.4	26.2	14.8	2281.9	449.7	.	.	.	12080.5	.	.	.	.	1
889	BIDENS PILOSA F. RADIATA	S	.	0	20.6	18.9	.	.	.	.	.	.	.	.	.	.	.	.	.	20
890	BIDENS SP.	S	.	0	35.6	33.1	.	.	7.1	.	.	.	.	.	.	.	.	.	.	19
891	BIDENS SP.	S	.	0	26.4	27.0	.	.	.	.	.	.	.	.	.	.	.	.	.	21
892	BIDENS TRIPARTITA	S	.	0	20.6	6.7	.	.	.	.	.	.	.	.	.	.	.	.	.	21
893	BIFORA AMERICANA	S	.	0	9.1	17.9	.	.	.	.	.	.	.	.	.	.	.	.	.	20
894	BIGNONIA UNGUIS-CATI	S	.	0	20.0	27.9	.	.	4.2	.	.	.	.	.	.	.	.	.	.	19
895	BIGNONIA UNGUIS-CATI	S	.	0	26.3	38.9	.	.	.	.	.	.	.	.	.	.	.	.	.	21
896	BISERRULA PELECINUS	S	.	0	32.8	1.9	.	.	.	.	.	.	.	.	.	.	.	.	.	21
897	BIXA ORELLANA	S	.	0	13.1	5.0	.	.	5.4	.	.	.	.	.	.	.	.	.	.	19
898	BIXA ORELLANA	S	.	0	13.2	3.8	.	.	.	.	.	.	.	.	.	.	.	.	.	20
899	BIXA ORELLANA	P	346	0	0.0	1.9	91.7	3.2	6.4	44.9	64.1	5.1	.	.	576.9	0.00	0.32	1.92	12.8	3
900	BLEPHAROCALYX TWEEDIEI	S	.	0	7.4	24.3	.	.	.	.	.	.	.	.	.	.	.	.	.	21
901	BLIGHIA SAPIDA	A	636	0	16.2	64.9	14.9	5.2	3.9	129.9	51.9	8.8	.	.	1801.9	0.42	0.45	4.55	84.4	1
902	BLIGHIA SAPIDA	S	317	0	26.7	5.9	56.9	15.3	10.4	99.0	.	17.3	.	.	.	0.74	0.45	8.42	.	1
903	BLIGHIA SAPIDA	S	632	0	14.7	61.1	19.3	1.1	4.9	129.8	245.6	6.0	.	.	.	0.25	0.60	5.61	273.7	3
904	BOCCONIA ARBOREA	S	.	0	14.4	35.1	.	.	.	.	.	.	.	.	.	.	.	.	.	20
905	BOEHMERIA CYLINDRICA	S	.	0	20.6	25.9	.	.	.	.	.	.	.	.	.	.	.	.	.	19
906	BOEHMERIA NIVEA	W	.	0	30.1	4.4	47.3	13.1	18.2	.	.	.	.	.	.	.	.	.	.	18
907	BOEHMERIA NIVEA	H	.	0	22.4	3.2	56.7	11.9	17.7	450.0	140.0	.	.	.	.	.	.	.	.	18
908	BOEHMERIA NIVEA	W	.	0	21.9	4.1	60.6	25.9	13.4	4600.0	240.0	.	.	.	30.0	0.73	1.65	6.70	.	6
909	BOEHMERIA NIVEA	L	292	0	25.4	3.8	54.6	18.5	16.2	3530.8	346.2	12.3	.	.	40307.7	0.23	2.15	5.38	561.5	3
910	BOERHAAVIA DIFFUSA	K	363	0	9.2	5.9	78.6	22.4	6.2	812.9	267.3	.	.	.	.	.	.	.	.	1
911	BOERHAAVIA DIFFUSA	L	289	0	25.0	2.8	57.2	12.2	15.0	2400.0	400.0	.	.	.	.	.	.	.	.	1
912	BOISSIERA SQUARROSA	S	.	0	29.6	2.2	.	.	.	.	.	.	.	.	.	.	.	.	.	21
913	BOLTONIA ASTEROIDES	S	.	0	19.1	21.1	.	.	6.2	.	.	.	.	.	.	.	.	.	.	19
914	BOMBAX BUONOPOZENSE	L	310	0	17.0	3.0	72.5	17.5	7.5	1380.0	160.0	.	.	.	.	.	.	.	.	1
915	BOMBAX BUONOPOZENSE	I	356	0	6.7	4.4	82.2	13.3	6.7	966.7	150.0	.	.	.	.	.	.	.	.	1
916	BOMBAX BUONOPOZENSE	S	495	0	20.5	29.3	43.9	13.0	6.3	370.2	719.1	.	.	.	.	.	.	.	.	1
917	BORAGO OFFICINALIS	S	.	0	20.9	38.3	.	.	11.0	.	.	.	.	.	.	.	.	.	.	19
918	BORASSUS FLABELLIFER	L	.	0	13.3	4.6	74.7	38.0	7.4	.	.	.	.	.	.	.	.	.	.	18

TABLE 3 - DATA CONVERTED TO A ZERO-MOISTURE BASIS.

OBS	PLNA	PART	CALORIES	WATER	PROTEIN	FAT	TOTCARB	FIBER	ASH	CALCIUM	PHOSPHO	IRON	SODIUM	POTASS	CAROT	THIAMINE	RIBOFLAV	NIACIN	ASCORBIC	REFCO
919	BORASSUS FLABELLIFER	F	.	0	6.5	0.8	87.9	16.1	4.8	210.0	240.0	.	.	.	.	.	.	.	.	18
920	BORASSUS FLABELLIFER	F	347	0	6.5	0.8	87.9	16.1	4.8	217.7	241.9	8.1	.	.	0.00	0.32	0.16	2.42	40.3	1
921	BORASSUS FLABELLIFER	IS	401	0	5.1	0.6	92.7	7.9	1.7	169.5	180.8	.	.	.	.	.	.	.	.	1
922	BORASSUS FLABELLIFER	SH	338	0	8.9	0.7	87.2	7.2	3.3	59.0	459.0	.	.	.	.	0.16	0.59	2.95	26.2	2
923	BORASSUS FLABELLIFER	F	.	0	3.1	0.9	93.4	.	3.1	.	.	.	.	.	.	.	.	.	105.3	5
924	BORASSUS FLABELLIFER	S	.	0	8.1	1.4	85.1	.	3.5	.	.	.	.	.	.	.	.	.	175.7	5
925	BOREAVA ORIENTALIS	S	.	0	36.2	41.2	.	.	.	.	.	.	.	.	.	.	.	.	.	20
926	BORONIA MEGASTIGMA	S	.	0	20.0	32.3	.	.	.	.	.	.	.	.	.	.	.	.	.	20
927	BOSCIA ALBITRUNCA	L	.	0	15.8	.	.	30.3	7.4	1440.0	120.0	.	.	.	.	.	.	.	.	18
928	BOSCIA ANGUSTIFOLIA	SH	.	0	19.3	1.3	72.3	39.2	7.1	610.0	180.0	.	.	.	.	.	.	.	.	18
929	BOSCIA FOETIDA	S	.	0	25.2	23.3	.	.	.	.	.	.	.	.	.	.	.	.	.	21
930	BOSCIA SALICIFOLIA	L	344	0	17.2	8.9	63.3	20.6	10.6	1111.1	72.2	17.8	.	.	.	.	.	.	.	1
931	BOSCIA SENEGALENSIS	S	385	0	23.9	1.8	70.4	1.9	3.8	138.8	364.6	7.7	.	.	186.23	0.02	0.03	9.93	5.6	1
932	BOTHRIOCHLOA CAUCASICA	S	.	0	21.9	7.0	.	.	.	.	.	.	.	.	.	.	.	.	.	20
933	BOTHRIOCHLOA INSCULPTA	W	.	0	9.7	2.6	75.0	32.5	12.7	.	.	.	.	.	.	.	.	.	.	18
934	BOTHRIOCHLOA INSCULPTA	H	.	0	4.6	1.8	83.7	35.2	9.9	.	.	.	.	.	.	.	.	.	.	18
935	BOTHRIOCHLOA INTERMEDIA	W	.	0	3.9	1.3	85.3	38.9	9.5	.	.	.	.	.	.	.	.	.	.	18
936	BOTHRIOCHLOA INTERMEDIA	H	.	0	4.2	1.5	85.1	43.2	9.2	.	.	.	.	.	.	.	.	.	.	18
937	BOTHRIOCHLOA ISCHAEMUM	S	.	0	28.8	9.6	.	.	.	.	.	.	.	.	.	.	.	.	.	20
938	BOTHRIOCHLOA PERTUSA	H	.	0	3.0	1.2	85.1	39.2	10.7	.	.	.	.	.	.	.	.	.	.	18
939	BOTHRIOCHLOA PERTUSA	W	.	0	5.7	1.0	81.6	34.9	11.7	.	.	.	.	.	.	.	.	.	.	18
940	BOTHRIOCHLOA RADICANS	W	.	0	10.2	3.7	77.5	28.8	8.6	.	.	.	.	.	.	.	.	.	.	18
941	BOTHRIOCHLOA SACCHAROIDES	S	.	0	27.5	10.8	.	.	.	.	.	.	.	.	.	.	.	.	.	20
942	BOUEA MACROPHYLLA	F	351	0	4.7	0.0	93.9	4.1	1.4	40.5	74.3	2.0	.	.	270.27	0.20	0.20	2.03	506.8	3
943	BOUEA SP.	F	353	0	4.7	0.6	92.4	.	2.4	70.6	129.4	5.3	.	870.6	1411.76	0.24	0.24	2.35	629.4	2
944	BOUEA SP.	L	316	0	16.3	1.6	73.7	.	8.4	210.5	236.8	24.7	.	.	1894.74	0.00	.	.	321.1	2
945	BOURGAEA TOURNEFORTII	S	.	0	42.5	22.6	.	.	.	.	.	.	.	.	.	.	.	.	.	21
946	BOUTELOUA CURTIPENDULA		.	0	5.8	1.7	79.9	32.1	12.6	390.0	90.0	.	.	710.0	3.00	.	.	.	.	6
947	BOUTELOUA ERIOPODA		.	0	8.2	1.7	82.7	31.9	7.4	240.0	120.0	.	.	.	4.00	.	.	.	.	6
948	BOUTELOUA FILIFORMIS		.	0	9.0	2.0	73.1	30.4	14.9	.	.	.	.	.	.	.	.	.	.	6
949	BOUTELOUA GRACILIS		.	0	8.1	1.9	77.8	29.8	12.2	570.0	240.0	.	.	.	10.00	.	.	.	.	6
950	BOUTELOUA HIRSUTA		.	0	7.2	1.6	73.6	29.0	17.6	580.0	90.0	.	.	870.0	.	.	.	.	.	6
951	BOUTELOUA RIGIDISETA		.	0	8.0	1.9	75.7	30.9	14.4	620.0	100.0	.	.	.	.	.	.	.	.	6
952	BOUTELOUA TRIFIDA		.	0	7.6	2.3	81.4	33.0	8.7	610.0	100.0	.	.	600.0	.	.	.	.	.	6
953	BRACHIARIA BRIZANTHA	W	.	0	4.0	1.3	83.6	31.1	11.1	.	.	.	.	.	.	.	.	.	.	18
954	BRACHIARIA BRIZANTHA	H	.	0	8.6	1.9	78.9	31.3	10.6	.	.	.	.	.	.	.	.	.	.	18
955	BRACHIARIA DECUMBENS	W	.	0	11.2	2.8	76.1	28.0	9.9	.	.	.	.	.	.	.	.	.	.	18
956	BRACHIARIA DISTACHYA	W	.	0	10.1	3.0	73.4	26.6	13.5	.	.	.	.	.	.	.	.	.	.	18
957	BRACHIARIA LEERSIOIDES	W	.	0	11.8	1.9	72.8	31.5	13.5	.	.	.	.	.	.	.	.	.	.	18
958	BRACHIARIA MUTICA	L	.	0	14.0	2.2	75.7	31.0	8.1	.	.	.	.	.	.	.	.	.	.	18
959	BRACHIARIA MUTICA	Y	.	0	5.9	0.8	87.4	41.5	6.3	.	.	.	.	.	.	.	.	.	.	18
960	BRACHIARIA MUTICA	H	.	0	12.4	2.0	74.9	30.3	10.7	.	.	.	.	.	.	.	.	.	.	18
961	BRACHIARIA NIGROPEDATA	W	.	0	8.5	1.6	80.9	30.7	9.0	.	.	.	.	.	.	.	.	.	.	18
962	BRACHIARIA RAMOSA	S	.	0	15.8	4.6	.	.	.	.	.	.	.	.	.	.	.	.	.	21
963	BRACHIARIA RUZIZIENSIS	W	.	0	13.9	2.3	73.8	27.2	9.0	.	.	.	.	.	.	.	.	.	.	18

TABLE 3 - DATA CONVERTED TO A ZERO-MOISTURE BASIS.

OBS	PLNA	PART	CALORIES	WATER	PROTEIN	FAT	TOTLCARB	FIBER	ASH
964	BRACHYCHITON ACERIFOLIUM	S	.	0	36.2	29.9	.	.	5.7
965	BRACHYCHITON ACERIFOLIUM	S	.	0	3.4	1.5	.	.	2.6
966	BRACHYCOME IBERIDIFOLIA	S	.	0	26.2	27.2	.	.	7.4
967	BRACHYPODIUM DISTACHYON	S	.	0	10.2	1.2	.	.	.
968	BRACHYPODIUM PHOENICOIDES	S	.	0	21.6	2.5	.	.	.
969	BRACHYPODIUM PINNATUM	S	.	0	16.3	2.3	.	.	.
970	BRASENIA SCHREBERI	L	135	0	9.5	2.7	24.3	1.4	63.5
971	BRASSICA CAMPESTRIS	S	.	0	21.6	43.7	28.7	6.3	6.0
972	BRASSICA CAMPESTRIS	S	.	0	22.6	42.9	.	.	4.3
973	BRASSICA CAMPESTRIS	S	.	0	28.8	38.8	.	.	.
974	BRASSICA CAMPESTRIS	S	.	0	22.5	44.2	.	.	.
975	BRASSICA CAMPESTRIS	S	.	0	21.2	44.8	.	.	.
976	BRASSICA CAMPESTRIS	S	.	0	26.8	39.2	.	.	.
977	BRASSICA CAMPESTRIS	S	.	0	27.4	29.1	.	.	.
978	BRASSICA CARINATA	S	.	0	34.4	26.4	.	.	5.5
979	BRASSICA CARINATA	S	.	0	32.6	29.7	.	.	5.6
980	BRASSICA JAPONICA	L	288	0	31.3	5.0	46.3	17.5	17.5
981	BRASSICA JUNCEA	S	.	0	28.8	34.2	.	.	.
982	BRASSICA JUNCEA	L	293	0	29.3	4.9	52.4	12.2	13.4
983	BRASSICA JUNCEA	R	257	0	12.8	2.0	59.5	13.5	25.7
984	BRASSICA JUNCEA	I	296	0	24.1	5.6	51.9	9.3	18.5
985	BRASSICA JUNCEA	SH	323	0	27.7	7.7	52.3	12.3	12.3
986	BRASSICA JUNCEA	L	287	0	27.7	4.3	51.1	10.6	17.0
987	BRASSICA NAPUS	S	.	0	19.3	43.4	33.8	27.2	3.5

OBS	CALCIUM	PHOSPHO	IRON	SODIUM	POTASS	CAROT	THIAMINE	RIBOFLAV	NIACIN	ASCORBIC	REFCO
964	.	.	.	.	.	.	.	.	.	.	19
965	.	.	.	.	.	.	.	.	.	.	19
966	.	.	.	.	.	.	.	.	.	.	19
967	.	.	.	.	.	.	.	.	.	.	21
968	.	.	.	.	.	.	.	.	.	.	21
969	.	.	.	.	.	.	.	.	.	.	21
970	121.6	310.8	27.0	.	.	135.1	0.41	0.41	4.05	0.0	2
971	490.0	710.0	.	.	.	.	.	.	.	.	18
972	.	.	.	.	.	.	.	.	.	.	19
973	.	.	.	.	.	.	.	.	.	.	20
974	.	.	.	.	.	.	.	.	.	.	20
975	.	.	.	.	.	.	.	.	.	.	20
976	.	.	.	.	.	.	.	.	.	.	21
977	.	.	.	.	.	.	.	.	.	.	21
978	.	.	.	.	.	.	.	.	.	.	19
979	.	.	.	.	.	.	.	.	.	.	19
980	1750.0	525.0	33.8	1500.0	4250.0	6250.0	1.00	1.63	12.50	875.0	2
981	.	.	.	.	.	.	.	.	.	.	20
982	1951.2	585.4	32.9	292.7	3622.0	22256.1	0.73	1.71	9.76	890.2	2
983	750.0	439.2	10.8	.	3020.3	304.0	0.34	0.81	4.73	141.9	2
984	2666.7	648.1	18.5	.	4759.3	33611.1	0.56	1.48	.	518.5	2
985	1092.3	276.9	33.8	.	3415.4	.	.	.	.	.	2
986	851.1	425.5	42.6	.	.	12978.7	0.74	2.23	65.96	659.6	3
987	320.0	650.0	.	.	.	.	.	.	.	.	18

TABLE 3 - DATA CONVERTED TO A ZERO-MOISTURE BASIS.

OBS	PLNA	PART	CALORIES	WATER	PROTEIN	FAT	TOTLCARB	FIBER	ASH
988	BRASSICA NAPUS	S	.	0	23.8	45.9	.	.	.
989	BRASSICA NAPUS	S	.	0	26.9	41.6	.	.	.
990	BRASSICA NAPUS	S	.	0	23.1	43.9	.	.	.
991	BRASSICA NAPUS	S	.	0	21.2	47.4	.	.	.
992	BRASSICA NAPUS	S	.	0	20.0	47.8	.	.	.
993	BRASSICA NAPUS	S	.	0	21.9	44.6	.	.	.
994	BRASSICA NAPUS	S	.	0	29.9	36.8	.	.	.
995	BRASSICA NAPUS	S	.	0	29.8	34.5	.	.	.
996	BRASSICA NAPUS	S	.	0	30.1	36.4	.	.	.
997	BRASSICA NAPUS	W	.	0	16.0	3.6	67.2	17.2	13.2
998	BRASSICA NAPUS	I	294	0	34.1	3.2	50.8	.	11.9
999	BRASSICA NAPUS	L	365	0	17.4	10.2	67.1	10.8	5.4
1000	BRASSICA NAPUS	R	324	0	25.0	1.5	61.8	11.8	11.8
1001	BRASSICA NAPUS	R	354	0	8.5	0.8	84.6	8.5	6.2
1002	BRASSICA NAPUS NAPOBRASSICA GROUP	W	.	0	16.4	2.6	55.4	12.3	25.6
1003	BRASSICA NAPUS NAPOBRASSICA GROUP	Y	.	0	18.5	.	.	14.9	.
1004	BRASSICA NAPUS NAPOBRASSICA GROUP	R	354	0	8.5	0.8	84.6	8.5	6.2
1005	BRASSICA NARINOSA	L	267	0	32.6	3.5	46.5	10.5	18.6
1006	BRASSICA NIGRA	S	.	0	36.2	31.7	.	.	.
1007	BRASSICA NIGRA	S	.	0	31.9	32.0	.	.	.
1008	BRASSICA NIGRA	S	.	0	29.8	30.1	.	.	.
1009	BRASSICA NIGRA	L	295	0	28.6	4.8	53.3	10.5	13.3
1010	BRASSICA OLERACEA ACEPHALA GROUP	Y	.	0	17.6	3.7	.	11.2	.
1011	BRASSICA OLERACEA ACEPHALA GROUP	L	279	0	22.7	5.2	50.0	10.4	22.1

OBS	CALCIUM	PHOSPHO	IRON	SODIUM	POTASS	CAROT	THIAMINE	RIBOFLAV	NIACIN	ASCORBIC	REFCO
988	.	.	.	.	.	.	.	.	.	.	20
989	.	.	.	.	.	.	.	.	.	.	20
990	.	.	.	.	.	.	.	.	.	.	20
991	.	.	.	.	.	.	.	.	.	.	20
992	.	.	.	.	.	.	.	.	.	.	20
993	.	.	.	.	.	.	.	.	.	.	20
994	.	.	.	.	.	.	.	.	.	.	21
995	.	.	.	.	.	.	.	.	.	.	21
996	.	.	.	.	.	.	.	.	.	.	21
997	1190.0	340.0	18.0	.	2580.0	.	.	.	.	.	6
998	928.6	769.8	27.0	.	.	25396.8	0.32	0.16	9.52	333.3	3
999	814.4	227.5	27.5	.	.	16047.9	0.48	0.90	2.99	718.6	3
1000	294.1	294.1	22.1	.	.	0.0	0.59	0.59	10.29	382.4	3
1001	507.7	300.0	3.1	38.5	1838.5	2676.9	0.54	0.54	8.46	330.8	4
1002	.	.	.	.	.	.	.	.	.	.	6
1003	.	.	.	.	.	1.0	.	0.86	.	.	6
1004	507.7	300.0	3.1	38.5	1838.5	2692.3	0.54	0.54	8.46	330.8	2
1005	2325.6	674.4	44.2	.	6965.1	23720.9	0.93	1.74	6.98	674.4	2
1006	.	.	.	.	.	.	.	.	.	.	20
1007	.	.	.	.	.	.	.	.	.	.	21
1008	.	.	.	.	.	.	.	.	.	.	21
1009	1742.9	476.2	28.6	304.8	3590.5	40000.0	1.05	2.10	7.62	923.8	4
1010	.	.	.	.	.	.	.	.	.	.	6
1011	857.1	500.0	8.4	.	.	5844.2	.	.	.	714.3	1

TABLE 3 - DATA CONVERTED TO A ZERO-MOISTURE BASIS.

OBS	PLNA	PART	CALORIES	WATER	PROTEIN	FAT	TOTLCARB	FIBER	ASH
1012	BRASSICA OLERACEA ACEPHALA GROUP	L	318	0	27.3	3.6	61.8	10.9	7.3
1013	BRASSICA OLERACEA ACEPHALA GROUP	L	301	0	30.8	4.8	51.4	8.9	13.0
1014	BRASSICA OLERACEA ACEPHALA GROUP	L	306	0	32.7	5.4	51.0	8.2	10.9
1015	BRASSICA OLERACEA BOTRYTIS GROUP	Y	.	0	21.1	3.3	.	16.6	.
1016	BRASSICA OLERACEA BOTRYTIS GROUP	Y	.	0	17.1	.	.	17.3	.
1017	BRASSICA OLERACEA BOTRYTIS GROUP	L	305	0	29.5	2.1	60.0	9.5	8.4
1018	BRASSICA OLERACEA BOTRYTIS GROUP	L	307	0	35.4	4.7	50.4	12.6	9.4
1019	BRASSICA OLERACEA BOTRYTIS GROUP	L	300	0	30.0	2.2	57.8	11.1	10.0
1020	BRASSICA OLERACEA BOTRYTIS GROUP	L	294	0	33.0	2.8	54.1	13.8	10.1
1021	BRASSICA OLERACEA CAPITATA GROUP	L	.	0	24.0	4.0	62.5	12.6	9.5
1022	BRASSICA OLERACEA CAPITATA GROUP	L	302	0	19.8	1.2	69.8	14.0	9.3
1023	BRASSICA OLERACEA CAPITATA GROUP	L	316	0	22.1	2.2	67.6	11.8	8.1
1024	BRASSICA OLERACEA CAPITATA GROUP	L	314	0	22.9	4.3	62.9	11.4	10.0
1025	BRASSICA OLERACEA CAPITATA GROUP	L	326	0	19.8	2.3	70.9	11.6	7.0
1026	BRASSICA OLERACEA CAPITATA GROUP	L	316	0	20.4	2.0	70.4	10.2	7.1
1027	BRASSICA OLERACEA CAPITATA GROUP	L	316	0	17.1	2.6	71.1	10.5	9.2
1028	BRASSICA OLERACEA GEMMIFERA GROUP	SH	318	0	26.4	2.7	64.2	10.8	7.4
1029	BRASSICA OLERACEA GEMMIFERA GROUP	SH	299	0	31.1	1.8	59.3	11.4	7.8
1030	BRASSICA OLERACEA GEMMIFERA GROUP	SH	304	0	33.1	2.7	56.1	10.8	8.1
1031	BRASSICA OLERACEA GONGYLODES GROUP	L	299	0	26.4	2.3	58.6	16.1	12.6
1032	BRASSICA OLERACEA GONGYLODES GROUP	L	307	0	25.7	4.0	59.4	9.9	10.9
1033	BRASSICA OLERACEA GONGYLODES GROUP	L	344	0	22.2	1.1	67.8	11.1	8.9
1034	BRASSICA OLERACEA GONGYLODES GROUP	L	299	0	20.6	1.0	68.0	10.3	10.3
1035	BRASSICA OLERACEA VAR. CAPITATA	S	.	0	33.8	25.9	.	.	5.8

OBS	CALCIUM	PHOSPHO	IRON	SODIUM	POTASS	CAROT	THIAMINE	RIBOFLAV	NIACIN	ASCORBIC	REFCO
1012	2090.9	509.1	18.2	.	.	2454.5	0.91	1.18	3.64	845.5	2
1013	1726.0	452.1	15.1	.	.	27602.7	1.10	1.64	8.22	856.2	3
1014	1700.7	557.8	10.2	.	3061.2	37959.2	1.09	2.11	11.56	1034.0	4
1015	.	.	.	.	.	.	.	.	.	.	6
1016	.	.	.	.	.	.	.	.	.	.	6
1017	315.8	610.5	10.5	210.5	3673.7	578.9	0.74	0.84	7.37	757.9	2
1018	913.4	637.8	10.2	.	.	8818.9	0.94	1.42	8.66	740.2	3
1019	277.8	622.2	12.2	144.4	3277.8	400.0	1.22	1.11	7.78	866.7	4
1020	945.0	715.6	10.1	137.6	3504.6	13761.5	0.92	2.11	8.26	1036.7	4
1021	630.0	210.0	.	.	.	.	.	0.89	.	.	6
1022	546.5	465.1	8.1	.	.	1162.8	0.47	0.47	3.49	627.9	1
1023	522.1	441.2	11.0	29.4	1955.9	257.3	0.88	0.59	4.41	625.0	2
1024	785.7	442.9	11.4	142.9	3400.0	4000.0	0.86	0.86	4.29	657.1	2
1025	500.0	418.6	8.1	.	.	697.7	0.70	0.47	3.49	500.0	3
1026	428.6	357.1	8.2	265.3	2734.7	244.9	0.92	0.61	4.08	622.4	4
1027	644.7	381.6	5.3	263.2	3065.8	1026.3	0.66	0.66	3.95	618.4	4
1028	270.3	500.0	9.5	94.6	3243.2	337.8	1.08	0.88	3.38	479.7	2
1029	281.4	550.9	10.2	.	.	1736.5	1.02	0.96	6.59	491.0	3
1030	243.2	540.5	10.1	94.6	2635.1	2229.7	0.68	1.08	6.08	689.2	4
1031	413.8	413.8	18.4	.	.	.	.	.	.	.	1
1032	514.9	475.2	12.9	79.2	2465.3	19802.0	0.59	0.99	5.94	673.3	2
1033	355.6	533.3	3.3	.	.	0.0	0.56	0.33	4.44	666.7	3
1034	422.7	525.8	5.2	82.5	3835.1	123.7	0.62	0.41	3.09	680.4	4
1035	.	.	.	.	.	.	.	.	.	.	19

TABLE 3 - DATA CONVERTED TO A ZERO-MOISTURE BASIS.

OBS	PLNA	PART	CALORIES	WATER	PROTEIN	FAT	TOTLCARB	FIBER
1036	BRASSICA RAPA	S	.	0	26.2	30.7	.	.
1037	BRASSICA RAPA	R	323	0	15.4	0.0	72.3	13.8
1038	BRASSICA RAPA	L	302	0	30.2	4.2	55.2	4.2
1039	BRASSICA RAPA	L	294	0	29.4	3.5	52.9	9.4
1040	BRASSICA RAPA	R	292	0	13.9	2.8	61.1	11.1
1041	BRASSICA RAPA	L	315	0	26.0	2.7	63.0	13.7
1042	BRASSICA RAPA	R	360	0	10.7	2.7	76.0	10.7
1043	BRASSICA RAPA	L	365	0	17.4	10.2	67.1	10.8
1044	BRASSICA RAPA	R	353	0	11.8	2.4	77.6	10.6
1045	BRASSICA RAPA	L	289	0	30.9	3.1	51.5	8.2
1046	BRASSICA RAPA CHINENSIS GROUP	L	275	0	22.0	3.3	54.9	15.4
1047	BRASSICA RAPA CHINENSIS GROUP	S	450	0	24.0	19.8	49.3	17.2
1048	BRASSICA RAPA CHINENSIS GROUP	L	293	0	29.3	3.4	53.4	12.1
1049	BRASSICA RAPA CHINENSIS GROUP	L	281	0	28.1	3.5	50.9	10.5
1050	BRASSICA RAPA CHINENSIS GROUP	L	280	0	24.0	2.0	60.0	12.0
1051	BRASSICA RAPA PEKINENSIS GROUP	L	236	0	23.6	3.6	41.8	16.4
1052	BRASSICA RAPA RAPIFERA GROUP	Y	.	0	18.0	.	.	10.3
1053	BRASSICA RAPA RAPIFERA GROUP	W	.	0	17.5	2.4	61.9	11.2
1054	BRASSICA SP.	S	.	0	26.3	33.1	.	.
1055	BRASSICA SP.	S	.	0	26.2	40.3	.	.
1056	BRASSICA SP.	S	.	0	26.2	39.3	.	.
1057	BRASSICA SP.	S	.	0	31.2	28.1	.	.
1058	BRASSICA SP.	S	.	0	35.6	32.9	.	.
1059	BRASSICA SP.	S	.	0	17.5	51.4	.	.

OBS	ASH	CALCIUM	PHOSPHO	IRON	SODIUM	POTASS	CAROT	THIAMINE	RIBOFLAV	NIACIN	ASCORBIC	REFCO
1036	.	.	.	.	.	.	.	.	.	.	.	20
1037	12.3	584.6	646.2	7.7	.	.	.	0.46	0.77	.	1153.8	1
1038	10.4	1645.8	1364.6	19.8	.	.	.	0.21	2.60	.	.	1
1039	14.1	1494.1	623.5	14.1	.	4564.7	30588.2	1.65	1.06	8.24	529.4	2
1040	11.1	597.2	458.3	12.5	833.3	3902.8	277.8	0.56	0.56	6.94	305.6	2
1041	8.2	2301.4	712.3	35.6	1068.5	5753.4	18219.2	1.37	2.47	9.59	643.8	2
1042	10.7	426.7	346.7	5.3	.	.	0.0	0.40	0.40	6.67	373.3	3
1043	5.4	814.4	227.5	27.5	.	.	16047.9	0.48	0.90	2.99	718.6	3
1044	8.2	458.8	352.9	5.9	576.5	3152.9	0.0	0.47	0.82	7.06	423.5	4
1045	14.4	2536.1	597.9	18.6	.	.	47010.3	2.16	4.02	8.25	1433.0	4
1046	19.8	637.4	131.9	0.0	.	.	25054.9	0.00	1.98	71.43	.	1
1047	6.9	.	.	.	.	.	.	0.58	0.43	.	.	1
1048	13.8	1758.6	793.1	44.8	379.3	4810.3	39741.4	1.21	2.24	13.79	913.8	2
1049	17.5	2894.7	771.9	14.0	456.1	5368.4	32631.6	0.88	1.75	14.04	438.6	4
1050	14.0	860.0	800.0	12.0	460.0	5060.0	1800.0	1.00	0.80	12.00	500.0	4
1051	30.9	636.4	709.1	9.1	.	.	.	.	.	.	.	1
1052	.	.	.	.	.	.	.	.	.	.	.	6
1053	18.2	2870.0	580.0	38.0	.	3790.0	5.0	0.14	0.24	.	.	6
1054	5.2	.	.	.	.	.	.	.	.	.	.	19
1055	.	.	.	.	.	.	.	.	.	.	.	19
1056	.	.	.	.	.	.	.	.	.	.	.	19
1057	.	.	.	.	.	.	.	.	.	.	.	19
1058	.	.	.	.	.	.	.	.	.	.	.	20
1059	.	.	.	.	.	.	.	.	.	.	.	20

TABLE 3 - DATA CONVERTED TO A ZERO-MOISTURE BASIS.

OBS	PLNA	PART	CALORIES	WATER	PROTEIN	FAT	TOTLCARB	FIBER
1060	BRASSICA SP.	L	328	0	22.1	5.7	63.9	9.8
1061	BRASSICA SP. X SINAPIS SP.	S	.	0	30.4	28.2	.	.
1062	BRASSICA TOURNEFORTII	S	.	0	26.6	28.1	.	.
1063	BRASSICA TOURNEFORTII	S	.	0	25.4	29.3	.	.
1064	BRASSICA TOURNEFORTII	S	.	0	29.0	22.6	.	.
1065	BRASSICA TOURNEFORTII	S	.	0	22.1	29.6	.	.
1066	BRASSICA TOURNEFORTII	S	.	0	21.6	32.2	.	.
1067	BRASSICA TOURNEFORTII	S	.	0	27.2	29.8	.	.
1068	BRASSICA TOURNEFORTII	S	.	0	27.6	30.2	.	.
1069	BRAZORIA SCUTELLARIOIDES	S	.	0	18.5	36.1	.	.
1070	BREYNIA RHAMNOIDES	S	.	0	16.4	27.2	.	.
1071	BRICKELLIA CALIFORNICA	S	.	0	15.4	5.6	.	.
1072	BRICKELLIA SCOPARIA VAR. SUBAURICULATA	S	.	0	32.5	21.1	.	.
1073	BRIDELIA STIPULARIA	S	.	0	15.2	18.4	.	.
1074	BRIZA FUSCA	S	.	0	31.8	17.2	.	.
1075	BRIZA MAXIMA	S	.	0	21.2	5.5	.	.
1076	BRIZA SPICATA	S	.	0	15.4	18.3	.	.
1077	BRIZA SPICATA	S	.	0	15.9	21.0	.	.
1078	BROMELIA PINGUIN	L	313	0	11.3	3.8	71.3	7.5
1079	BROMELIA PLUMIERI	L	333	0	9.3	1.6	82.9	3.1
1080	BROMUS ARENARIUS		.	0	8.0	.	.	31.4
1081	BROMUS ARVENSIS	S	.	0	11.2	1.3	.	.
1082	BROMUS ARVENSIS	S	.	0	15.1	1.3	.	.
1083	BROMUS BREVIS	S	.	0	14.8	1.4	.	.

OBS	ASH	CALCIUM	PHOSPHO	IRON	SODIUM	POTASS	CAROT	THIAMINE	RIBOFLAV	NIACIN	ASCORBIC	REFCO
1060	8.2	1573.8	475.4	19.7	.	2975.4	9836.07	0.33	1.15	4.10	327.9	2
1061	.	.	.	.	.	.	.	.	.	.	.	21
1062	.	.	.	.	.	.	.	.	.	.	.	21
1063	.	.	.	.	.	.	.	.	.	.	.	21
1064	.	.	.	.	.	.	.	.	.	.	.	21
1065	.	.	.	.	.	.	.	.	.	.	.	21
1066	.	.	.	.	.	.	.	.	.	.	.	21
1067	.	.	.	.	.	.	.	.	.	.	.	21
1068	.	.	.	.	.	.	.	.	.	.	.	21
1069	.	.	.	.	.	.	.	.	.	.	.	20
1070	.	.	.	.	.	.	.	.	.	.	.	21
1071	.	.	.	.	.	.	.	.	.	.	.	21
1072	.	.	.	.	.	.	.	.	.	.	.	20
1073	.	.	.	.	.	.	.	.	.	.	.	21
1074	.	.	.	.	.	.	.	.	.	.	.	21
1075	.	.	.	.	.	.	.	.	.	.	.	21
1076	.	.	.	.	.	.	.	.	.	.	.	20
1077	.	.	.	.	.	.	.	.	.	.	.	21
1078	13.8	1450.0	300.0	7.5	.	.	0.00	0.38	0.50	5.00	262.5	3
1079	6.2	372.1	46.5	4.7	.	.	77.52	0.47	0.16	2.33	364.3	3
1080	5.5	360.0	270.0	.	.	1410.0	.	.	.	.	.	6
1081	3.0	.	.	.	.	.	.	.	.	.	.	19
1082	.	.	.	.	.	.	.	.	.	.	.	21
1083	2.1	.	.	.	.	.	.	.	.	.	.	19

TABLE 3 - DATA CONVERTED TO A ZERO-MOISTURE BASIS.

OBS	PLNA	PART	CALORIES	WATER	PROTEIN	FAT	TOTLCARB	FIBER	ASH	CALCIUM	PHOSPHO	IRON	SODIUM	POTASS	CAROT	THIAMINE	RIBOFLAV	NIACIN	ASCORBIC	REFCO
1084	BROMUS BREVIS	S	.	0	6.9	2.1	.	.	6.7	.	.	.	.	.	.	.	.	.	.	19
1085	BROMUS CAPPADOCICUS	S	.	0	13.9	1.2	.	.	.	.	.	.	.	.	.	.	.	.	.	21
1086	BROMUS CARINATUS		.	0	7.6	1.9	83.2	37.6	7.3	390.0	240.0	.	.	.	.	.	.	.	.	6
1087	BROMUS CATHARTICUS	S	.	0	17.5	0.8	.	.	.	.	.	.	.	.	.	.	.	.	.	19
1088	BROMUS CATHARTICUS	S	.	0	18.8	0.8	.	.	.	.	.	.	.	.	.	.	.	.	.	19
1089	BROMUS CATHARTICUS		.	0	16.3	3.2	67.0	24.1	13.5	560.0	290.0	.	.	3850	.	.	.	.	.	6
1090	BROMUS INERMIS	S	.	0	19.4	1.5	.	.	3.0	.	.	.	.	.	.	.	.	.	.	19
1091	BROMUS INERMUS		.	0	11.4	3.0	77.0	28.4	8.6	440.0	250.0	.	.	2700	32.00	0.31	0.77	.	.	6
1092	BROMUS JAPONICUS		.	0	16.1	3.1	70.7	28.7	10.1	400.0	260.0	.	.	.	.	.	.	.	.	6
1093	BROMUS MOLLIS		.	0	11.2	2.4	77.4	29.3	9.0	506.0	366.0	.	.	2480	.	.	.	.	.	6
1094	BROMUS RIGIDUS		.	0	14.0	.	.	26.6	8.5	480.0	390.0	.	.	.	.	.	.	.	.	6
1095	BROMUS RUBENS	S	.	0	17.9	2.0	.	.	2.8	.	.	.	.	.	.	.	.	.	.	19
1096	BROMUS RUBENS		.	0	8.5	.	.	31.6	7.2	280.0	230.0	.	.	1450	.	.	.	.	.	6
1097	BROMUS SCOPARIUS	S	.	0	16.1	1.4	.	.	.	.	.	.	.	.	.	.	.	.	.	21
1098	BROMUS SP.	S	.	0	16.9	1.6	.	.	.	.	.	.	.	.	.	.	.	.	.	20
1099	BROMUS TECTORUM	S	.	0	14.1	1.2	.	.	.	.	.	.	.	.	.	.	.	.	.	21
1100	BROMUS TECTORUM		.	0	9.7	2.0	78.9	24.4	9.4	440.0	260.0	.	.	.	.	.	.	.	.	6
1101	BROMUS VALDIVIANUS	S	.	0	8.1	2.5	.	.	.	.	.	.	.	.	.	.	.	.	.	19
1102	BROMUS WILLDENOWII	S	.	0	14.5	1.0	.	.	.	.	.	.	.	.	.	.	.	.	.	21
1103	BRONGNIARTIA ALAMOSSANA	S	.	0	26.9	33.9	.	.	.	.	.	.	.	.	.	.	.	.	.	20
1104	BROSIMUM ALICASTRUM	L	.	0	13.9	3.4	74.8	26.5	7.9	.	.	.	.	.	.	.	.	.	.	18
1105	BROSIMUM ALICASTRUM	S	.	0	9.9	1.1	.	.	4.1	.	.	.	.	.	.	.	.	.	.	19
1106	BROSIMUM ALICASTRUM	L	334	0	8.4	3.2	80.5	23.4	7.9	1394.7	178.9	14.2	.	.	2157.89	0.63	1.34	3.68	144.7	3
1107	BROSIMUM ALICASTRUM	F	350	0	15.6	3.1	75.6	7.5	5.6	281.3	225.0	5.0	.	.	3500.00	0.31	9.50	5.00	175.0	3
1108	BROSIMUM ALICASTRUM	S	388	0	12.2	1.7	81.4	6.6	4.7	225.7	151.9	4.9	.	.	139.04	0.03	0.15	2.25	.	3
1109	BROUSSONETIA PAPYRIFERA	S	.	0	18.5	30.1	.	.	7.7	.	.	.	.	.	.	.	.	.	.	19
1110	BROWALLIA AMERICANA	S	.	0	25.0	20.8	.	.	.	.	.	.	.	.	.	.	.	.	.	20
1111	BRYONIA SP.	S	.	0	19.1	23.3	.	.	.	.	.	.	.	.	.	.	.	.	.	21
1112	BUCHLOE DACTYLOIDES	H	.	0	9.2	1.8	78.6	27.7	12.4	520.0	160.0	.	.	710	9.00	.	.	.	.	6
1113	BUDDLEIA DAVIDII	S	.	0	18.8	27.7	.	.	4.2	.	.	.	.	.	.	.	.	.	.	19
1114	BUMELIA TENAX	L	.	0	13.0	10.3	70.1	25.3	6.6	3040.0	120.0	.	.	570	.	.	.	.	.	6
1115	BUNCHOSIA PALMERI	S	.	0	19.3	1.8	.	.	.	.	.	.	.	.	.	.	.	.	.	21
1116	BUNIUM PERSICUM	S	.	0	19.9	13.6	.	.	.	.	.	.	.	.	.	.	.	.	.	21
1117	BUPLEURUM CROECEUM	S	.	0	14.9	9.5	.	.	.	.	.	.	.	.	.	.	.	.	.	21
1118	BUPLEURUM HELDREICHII	S	.	0	15.1	8.2	.	.	.	.	.	.	.	.	.	.	.	.	.	21
1119	BUPLEURUM SUBOVATUM	S	.	0	10.3	6.8	.	.	.	.	.	.	.	.	.	.	.	.	.	21
1120	BURSERA CONFUSA	S	.	0	11.9	16.7	.	.	.	.	.	.	.	.	.	.	.	.	.	19
1121	BURSERA CONFUSA	S	.	0	6.2	16.9	.	.	.	.	.	.	.	.	.	.	.	.	.	19
1122	BURSERA CONFUSA	S	.	0	3.8	6.7	.	.	.	.	.	.	.	.	.	.	.	.	.	20
1123	BURSERA GRANDIFOLIA	S	.	0	1.2	0.4	.	.	.	.	.	.	.	.	.	.	.	.	.	19
1124	BURSERA GRANDIFOLIA	S	.	0	3.2	2.7	.	.	.	.	.	.	.	.	.	.	.	.	.	20
1125	BURSERA INOPINATA	S	.	0	7.5	7.6	.	.	.	.	.	.	.	.	.	.	.	.	.	20
1126	BUTIA CAPITATA	S	.	0	15.7	56.5	.	.	.	.	.	.	.	.	.	.	.	.	.	21
1127	BUTIA CAPITATA X EREOSPATHA	S	.	0	12.9	44.1	.	.	1.8	.	.	.	.	.	.	.	.	.	.	19
1128	BUTYROSPERMUM PARADOXUM	S	622	0	7.3	52.6	38.2	5.6	1.8	107.4	43.0	3.2	.	.	.	0.56	.	.	.	1

TABLE 3 - DATA CONVERTED TO A ZERO-MCISTURE BASIS.

OBS	PLANA	PART	CALORIES	WATER	PROTEIN	FAT	TOTCARB	FIBER	ASH	CALCIUM	PHOSPHO	IRON	SODIUM	POTASS	CAROT	THIAMINE	RIBOFLAV	NIACIN	ASCORBIC	REFCO
1129	BUXUS SEMPERVIRENS	S	.	0	21.4	41.6	.	.	1.6	.	.	.	.	.	.	.	.	.	.	19
1130	BUXUS SEMPERVIRENS	S	.	0	21.8	44.3	.	.	.	.	.	.	.	.	.	.	.	.	.	21
1131	BYRSONIMA CRASSIFOLIA	F	384	0	5.2	7.6	83.7	12.8	3.5	191.9	98.8	11.6	.	.	232.6	0.12	0.23	2.33	488.4	3
1132	CACALIA DECOMPOSITA	S	.	0	33.8	34.2	.	.	.	.	.	.	.	.	.	.	.	.	.	20
1133	CACALIA PLANTAGINEA	S	.	0	26.2	36.5	.	.	.	.	.	.	.	.	.	.	.	.	.	19
1134	CACALIA TUBEROSA	S	.	0	27.9	28.3	.	.	.	.	.	.	.	.	.	.	.	.	.	20
1135	CACHRYS GONIOCARPA	S	.	0	37.5	21.5	.	.	.	.	.	.	.	.	.	.	.	.	.	21
1136	CADABA FARINOSA	L	.	0	30.6	1.8	52.1	9.3	15.4	920.0	160.0	.	.	.	.	.	.	.	.	18
1137	CADABA FARINOSA	SH	.	0	18.2	1.8	73.0	27.3	7.0	390.0	170.0	.	.	.	.	.	.	.	.	18
1138	CADABA FARINOSA	L	283	0	30.7	4.3	48.7	6.3	16.3	2460.0	120.0	12.7	.	.	.	.	.	.	.	1
1139	CAESALPINIA PLATYLOBA	S	.	0	19.4	12.8	.	.	.	.	.	.	.	.	.	.	.	.	.	19
1140	CAESALPINIA SPINOSA	F	.	0	8.8	1.6	81.4	9.6	8.2	220.0	200.0	.	.	.	.	.	.	.	.	18
1141	CAJANUS CAJAN	W	.	0	21.4	6.0	66.8	30.8	5.8	890.0	240.0	.	.	.	.	.	.	.	.	18
1142	CAJANUS CAJAN	L	.	0	11.0	6.9	63.6	18.3	18.5	.	.	.	.	.	.	.	.	.	.	18
1143	CAJANUS CAJAN	F	.	0	20.3	1.7	74.7	35.2	3.3	.	.	.	.	.	.	.	.	.	.	18
1144	CAJANUS CAJAN	S	.	0	23.4	0.9	71.4	10.6	4.3	140.0	450.0	.	.	.	.	.	.	.	.	18
1145	CAJANUS CAJAN	S	.	0	21.8	1.4	.	.	4.0	.	.	.	.	.	.	.	.	.	.	19
1146	CAJANUS CAJAN	S	.	0	22.5	.	.	.	.	.	.	.	.	.	.	.	.	.	.	20
1147	CAJANUS CAJAN	S	.	0	21.5	.	.	.	.	.	.	.	.	.	.	.	.	.	.	21
1148	CAJANUS CAJAN	W	.	0	14.5	4.4	76.1	37.8	5.0	.	.	.	.	.	.	.	.	.	.	6
1149	CAJANUS CAJAN	V	383	0	21.6	1.4	72.7	8.1	4.2	178.7	316.3	16.6	.	.	61.0	0.80	0.16	3.22	0.0	1
1150	CAJANUS CAJAN	V	383	0	23.1	1.4	71.6	5.0	4.0	116.4	253.1	5.5	29.4	739.0	180.8	0.55	0.24	2.49	0.0	2
1151	CAJANUS CAJAN	G	320	0	24.4	1.7	68.8	10.1	5.1	202.2	488.8	5.6	14.0	1747.2	407.3	1.24	0.45	5.06	89.9	2
1152	CAJANUS CAJAN	IS	383	0	24.1	1.9	69.5	.	4.5	93.2	434.1	4.2	16.1	1810.3	466.2	1.29	0.80	7.72	83.6	2
1153	CAJANUS CAJAN	V	384	0	21.9	1.7	72.1	9.2	4.3	156.0	366.7	5.7	.	.	45.6	0.82	0.19	2.96	0.0	3
1154	CAJANUS CAJAN	F	386	0	22.9	2.0	70.9	11.4	4.2	114.4	405.2	5.6	.	.	130.7	1.11	0.69	7.52	160.1	3
1155	CAJANUS CAJAN	IS	384	0	23.6	2.0	69.8	10.8	4.6	137.7	416.4	5.2	16.4	1809.8	275.4	1.31	0.56	7.21	127.9	4
1156	CAKILE EDENTULA	S	.	0	31.2	48.7	.	.	.	.	.	.	.	.	.	.	.	.	.	19
1157	CAKILE MARITIMA	S	.	0	33.4	43.8	.	.	.	.	.	.	.	.	.	.	.	.	.	21
1158	CAKILE MARITIMA	S	.	0	32.7	45.6	.	.	.	.	.	.	.	.	.	.	.	.	.	21
1159	CALAMOGROSTIS CANADENSIS	W	.	0	9.2	2.7	78.9	34.1	9.2	.	.	.	.	.	.	.	.	.	.	6
1160	CALAMOVILFA LONGIFOLIA	W	.	0	5.9	1.9	86.8	37.8	5.4	.	.	.	.	.	.	.	.	.	.	6
1161	CALAMUS ORNATUS	F	376	0	2.9	5.7	88.6	2.4	2.9	90.5	47.6	8.1	.	.	.	0.29	0.05	4.29	23.8	2
1162	CALANDRINIA MICRANTHA	L	271	0	31.4	4.3	44.3	12.9	20.0	685.7	442.9	45.7	.	.	13714.3	1.14	1.29	5.71	242.9	3
1163	CALATHEA MACROSEPALA		286	0	23.4	2.6	57.1	19.5	16.9	259.7	635.4	13.0	.	.	0.0	0.52	1.30	9.09	90.9	3
1164	CALEA URTICAEFOLIA	S	.	0	18.8	21.0	.	.	.	.	.	.	.	.	.	.	.	.	.	20
1165	CALEA URTICAEFOLIA	S	.	0	12.9	7.5	.	.	.	.	.	.	.	.	.	.	.	.	.	21
1166	CALENDULA ARVENSIS	S	.	0	41.1	40.6	.	.	.	.	.	.	.	.	.	.	.	.	.	21
1167	CALENDULA OFFICINALIS	S	.	0	33.1	42.2	.	.	.	.	.	.	.	.	.	.	.	.	.	19
1168	CALENDULA OFFICINALIS	S	.	0	8.1	3.3	.	.	.	.	.	.	.	.	.	.	.	.	.	19
1169	CALENDULA OFFICINALIS	S	.	0	30.0	45.3	.	.	.	.	.	.	.	.	.	.	.	.	.	20
1170	CALENDULA OFFICINALIS	S	.	0	33.8	42.1	.	.	.	.	.	.	.	.	.	.	.	.	.	20
1171	CALENDULA OFFICINALIS	S	.	0	31.2	43.6	.	.	.	.	.	.	.	.	.	.	.	.	.	20
1172	CALENDULA OFFICINALIS	S	.	0	36.9	40.8	.	.	.	.	.	.	.	.	.	.	.	.	.	20
1173	CALENDULA OFFICINALIS	S	.	0	33.8	44.2	.	.	.	.	.	.	.	.	.	.	.	.	.	20

TABLE 3 - DATA CONVERTED TO A ZERO-MOISTURE BASIS.

OBS	PLNA	PART	CALORIES	WATER	PROTEIN	FAT	TOTLCARB	FIBER	ASH	CALCIUM	PHOSPHO	IRON	SODIUM	POTASS	CAROT	THIAMINE	RIBOFLAV	NIACIN	ASCORBIC	REFCO
1174	CALENDULA OFFICINALIS	S	.	0	31.9	43.6	.	.	.	.	.	.	.	.	.	.	.	.	.	20
1175	CALENDULA OFFICINALIS	S	.	0	30.6	45.1	.	.	.	.	.	.	.	.	.	.	.	.	.	20
1176	CALENDULA OFFICINALIS	S	.	0	30.6	45.8	.	.	.	.	.	.	.	.	.	.	.	.	.	20
1177	CALENDULA OFFICINALIS	S	.	0	32.5	44.7	.	.	.	.	.	.	.	.	.	.	.	.	.	20
1178	CALENDULA OFFICINALIS	S	.	0	30.6	45.2	.	.	.	.	.	.	.	.	.	.	.	.	.	20
1179	CALENDULA OFFICINALIS	S	.	0	32.5	43.2	.	.	.	.	.	.	.	.	.	.	.	.	.	20
1180	CALENDULA OFFICINALIS	S	.	0	32.5	44.3	.	.	.	.	.	.	.	.	.	.	.	.	.	20
1181	CALENDULA OFFICINALIS	S	.	0	32.5	44.0	.	.	.	.	.	.	.	.	.	.	.	.	.	20
1182	CALENDULA OFFICINALIS	S	.	0	33.1	44.1	.	.	.	.	.	.	.	.	.	.	.	.	.	20
1183	CALENDULA OFFICINALIS	S	.	0	33.8	43.1	.	.	.	.	.	.	.	.	.	.	.	.	.	20
1184	CALENDULA OFFICINALIS	S	.	0	31.9	43.6	.	.	.	.	.	.	.	.	.	.	.	.	.	20
1185	CALENDULA OFFICINALIS	S	.	0	33.8	43.5	.	.	.	.	.	.	.	.	.	.	.	.	.	20
1186	CALENDULA OFFICINALIS	S	.	0	32.5	43.7	.	.	.	.	.	.	.	.	.	.	.	.	.	20
1187	CALENDULA OFFICINALIS	S	.	0	.	43.1	.	.	.	.	.	.	.	.	.	.	.	.	.	21
1188	CALLIANDRA ERIOPHYLLA	S	.	0	38.6	15.9	.	.	4.1	.	.	.	.	.	.	.	.	.	.	19
1189	CALLIANDRA ERIOPHYLLA	S	.	0	38.9	16.4	.	.	4.4	.	.	.	.	.	.	.	.	.	.	19
1190	CALLIANDRA GRACILIS	S	.	0	26.7	2.2	.	.	.	.	.	.	.	.	.	.	.	.	.	21
1191	CALLICARPA AMERICANA	S	.	0	6.2	9.8	.	.	.	.	.	.	.	.	.	.	.	.	.	19
1192	CALLIOPSIS DRUMMONDII	S	.	0	18.1	22.4	.	.	.	.	.	.	.	.	.	.	.	.	.	19
1193	CALLIRHOE PEDATA	W	.	0	10.4	2.7	77.6	30.2	9.3	2461.0	450.0	223.0	.	.	.	.	.	.	.	6
1194	CALOCHORTUS SP.	S	.	0	14.6	6.0	.	.	.	.	.	.	.	.	.	.	.	.	.	20
1195	CALOPOGONIUM MUCUNOIDES	W	.	0	15.6	2.3	75.9	31.5	6.2	1420.0	190.0	.	.	.	.	.	.	.	.	18
1196	CALOTROPIS GIGANTEA	S	.	0	28.8	36.1	.	.	.	.	.	.	.	.	.	.	.	.	.	20
1197	CALOTROPIS PROCERA	S	.	0	27.5	36.6	.	.	.	.	.	.	.	.	.	.	.	.	.	21
1198	CALOTROPIS PROCERA	S	.	0	30.1	34.8	.	.	.	.	.	.	.	.	.	.	.	.	.	21
1199	CALYCANTHUS FLORIDUS	S	.	0	24.4	48.3	.	.	2.6	.	.	.	.	.	.	.	.	.	.	19
1200	CALYCOTOME VILLOSA	S	.	0	31.4	3.0	.	.	.	.	.	.	.	.	.	.	.	.	.	21
1201	CAMASSIA QUAMASH	TU	.	0	2.3	0.8	90.3	4.3	6.7	.	.	.	.	.	.	0.23	0.17	.	13.3	9
1202	CAMELINA MICROCARPA	S	.	0	25.6	34.1	.	.	.	.	.	.	.	.	.	.	.	.	.	20
1203	CAMELINA RUMELICA	S	.	0	29.1	35.2	.	.	.	.	.	.	.	.	.	.	.	.	.	20
1204	CAMELINA SATIVA	S	.	0	33.1	33.2	.	.	.	.	.	.	.	.	.	.	.	.	.	19
1205	CAMELLIA JAPONICA	S	.	0	8.8	66.4	.	.	.	.	.	.	.	.	.	.	.	.	.	20
1206	CAMELLIA OLEIFERA	S	.	0	9.4	41.6	.	.	.	.	.	.	.	.	.	.	.	.	.	19
1207	CAMPANULA CARPATICA	S	.	0	16.9	38.4	.	.	.	.	.	.	.	.	.	.	.	.	.	20
1208	CAMPANULA PERSICIFOLIA	S	.	0	21.2	37.8	.	.	.	.	.	.	.	.	.	.	.	.	.	20
1209	CAMPSIS RADICANS	S	.	0	22.5	24.4	.	.	.	.	.	.	.	.	.	.	.	.	.	19
1210	CAMPSIS RADICANS	S	.	0	22.5	20.2	.	.	.	.	.	.	.	.	.	.	.	.	.	19
1211	CAMPSIS RADICANS	S	.	0	24.4	28.8	.	.	.	.	.	.	.	.	.	.	.	.	.	19
1212	CAMPTOSEMA RUBICUNDUM	S	.	0	21.9	4.1	.	.	.	.	.	.	.	.	.	.	.	.	.	21
1213	CANARIUM ALBUM	F	364	0	6.1	6.1	81.8	17.2	6.1	90.9	146.5	10.6	237.4	2101.0	1666.67	0.10	0.56	2.02	101.0	2
1214	CANARIUM OVATUM	S	714	0	12.2	75.9	9.0	2.9	3.0	149.4	591.2	3.6	3.2	521.9	25.61	0.94	0.10	0.53	0.0	
1215	CANARIUM OVATUM	S	708	0	15.6	75.2	6.0	3.5	3.2	130.6	557.6	2.9	3.3	536.8	27.44	1.04	0.13	0.44	0.0	2
1216	CANARIUM OVATUM	S	699	0	12.4	73.2	10.8	2.3	3.6	179.5	71.1	4.8	.	.	34.97	0.75	0.07	0.58	24.5	3
1217	CANARIUM SCHWEINFURTHII	S	581	0	19.5	48.7	26.0	3.6	5.9	170.2	1340.4	.	.	.	.	.	.	.	.	1
1218	CANARIUM SCHWEINFURTHII	F	471	0	3.0	26.6	66.1	.	4.3	167.7	1242.6	9.9	.	.	.	.	.	.	.	1

TABLE 3 - DATA CONVERTED TO A ZERO-MOISTURE BASIS.

OBS	PLANT NAME	PART	CALORIES	WATER	PROTEIN	FAT	TOTCARB	FIBER	ASH	CALCIUM	PHOSPHO	IRON	SODIUM	POTASS	CAROT	THIAMINE	RIBOFLAV	NIACIN	ASCORBIC	REFCO
1219	CANAVALIA ENSIFORMIS	W	.	0	22.5	2.1	63.8	27.4	11.6	.	.	.	.	.	.	.	.	.	.	18
1220	CANAVALIA ENSIFORMIS	S	.	0	36.3	2.5	58.3	9.9	2.9	190.0	270.0	.	.	.	.	.	.	.	.	18
1221	CANAVALIA ENSIFORMIS	F	.	0	4.5	1.5	90.2	48.1	3.8	300.0	10.0	.	.	.	.	.	.	.	.	18
1222	CANAVALIA ENSIFORMIS	S	.	0	28.1	2.3	.	.	.	.	.	.	.	.	.	.	.	.	.	19
1223	CANAVALIA ENSIFORMIS	S	.	0	29.4	2.8	.	.	.	.	.	.	.	.	.	.	.	.	.	19
1224	CANAVALIA ENSIFORMIS	S	.	0	27.5	3.2	.	.	.	.	.	.	.	.	.	.	.	.	.	19
1225	CANAVALIA ENSIFORMIS	S	.	0	30.0	2.4	.	.	.	.	.	.	.	.	.	.	.	.	.	19
1226	CANAVALIA ENSIFORMIS	S	.	0	27.2	2.8	.	.	.	.	.	.	.	.	.	.	.	.	.	21
1227	CANAVALIA ENSIFORMIS	S	.	0	28.7	2.8	.	.	.	.	.	.	.	.	.	.	.	.	.	21
1228	CANAVALIA ENSIFORMIS	G	325	0	21.1	2.6	71.1	15.8	5.3	526.3	350.9	17.5	.	.	.	.	.	.	.	1
1229	CANAVALIA ENSIFORMIS	S	389	0	27.4	2.9	66.1	8.3	3.6	176.9	333.7	7.8	.	.	.	0.86	0.17	2.02	.	1
1230	CANAVALIA ENSIFORMIS	G	377	0	23.7	1.8	69.3	15.8	5.3	526.3	350.9	17.5	.	.	219.3	0.88	.	.	0.0	2
1231	CANAVALIA ENSIFORMIS	S	392	0	23.6	3.6	68.7	8.6	4.1	150.9	337.8	9.7	39.4	844.6	.	0.73	0.15	3.49	2.3	2
1232	CANAVALIA ENSIFORMIS	G	381	0	32.1	2.3	61.9	15.3	3.7	153.5	307.0	5.6	.	.	139.5	1.02	0.47	9.30	148.8	3
1233	CANAVALIA ENSIFORMIS	S	376	0	28.9	1.5	64.9	5.6	4.8	109.1	389.8	5.6	.	.	22.7	0.52	0.17	2.39	.	3
1234	CANAVALIA GLADIATA	S	.	0	33.2	2.6	60.6	8.0	3.6	.	.	.	.	.	.	.	.	.	.	18
1235	CANAVALIA GLADIATA	S	374	0	31.8	0.7	63.2	13.6	4.2	.	.	.	.	.	.	.	.	.	.	1
1236	CANAVALIA GLADIATA	G	315	0	25.9	1.9	67.6	13.9	4.6	.	.	.	.	.	.	.	.	.	.	1
1237	CANAVALIA GLADIATA	G	.	0	23.7	1.8	69.3	13.2	5.3	.	.	.	.	.	210.5	.	.	.	.	5
1238	CANAVALIA SP.	S	.	0	33.1	1.4	.	.	.	.	.	.	.	.	.	.	.	.	.	20
1239	CANAVALIA SP.	W	.	0	22.4	2.2	63.8	27.6	11.6	.	.	.	.	.	.	.	.	.	.	6
1240	CANNA EDULIS	L	.	0	14.1	2.5	71.1	24.6	12.3	.	.	.	.	.	.	.	.	.	.	18
1241	CANNA EDULIS	SH	.	0	10.2	5.1	68.2	19.6	16.5	.	.	.	.	.	.	.	.	.	.	18
1242	CANNA EDULIS	R	.	0	3.6	0.8	88.2	3.4	7.4	.	.	.	.	.	.	.	.	.	.	18
1243	CANNA EDULIS	R	392	0	2.7	0.3	94.3	1.5	2.7	45.2	189.8	4.2	.	.	0.0	0.09	0.03	1.20	21.1	3
1244	CANNA FLACCIDA	S	.	0	9.4	0.8	.	.	.	.	.	.	.	.	.	.	.	.	.	20
1245	CANNA INDICA	R	.	0	3.6	0.7	90.2	4.0	5.3	130.0	220.0	.	.	.	.	.	.	.	.	18
1246	CANNA INDICA	S	.	0	10.0	0.8	.	.	.	.	.	.	.	.	.	.	.	.	.	20
1247	CANNA INDICA	S	.	0	8.7	0.5	.	.	.	.	.	.	.	.	.	.	.	.	.	20
1248	CANNABIS SATIVA	S	.	0	20.0	35.8	39.6	16.5	4.6	.	.	.	.	.	.	.	.	.	.	18
1249	CANNABIS SATIVA	S	487	0	31.4	29.6	31.9	23.5	7.1	138.9	1122.7	13.9	.	.	5.8	0.37	0.20	2.43	0.0	2
1250	CANNABIS SATIVA	S	.	0	23.6	33.3	38.0	20.6	5.0	.	.	.	.	.	.	.	.	.	.	5
1251	CANNOMOIS VIRGATA	S	.	0	7.6	0.8	.	.	.	.	.	.	.	.	.	.	.	.	.	21
1252	CAPPARIS CORYMBOSA	L	300	0	30.0	4.3	53.6	14.6	12.1	3307.1	107.1	20.4	.	.	.	.	.	.	.	1
1253	CAPPARIS OVATA	S	.	0	22.4	30.6	.	.	.	.	.	.	.	.	.	.	.	.	.	21
1254	CAPPARIS SPINOSA	S	.	0	22.0	34.6	.	.	4.8	.	.	.	.	.	.	.	.	.	.	19
1255	CAPPARIS SPINOSA	S	.	0	19.0	31.6	.	.	.	.	.	.	.	.	.	.	.	.	.	21
1256	CAPSELLA BURSA-PASTORIS	S	.	0	32.5	25.9	.	.	.	.	.	.	.	.	.	.	.	.	.	20
1257	CAPSELLA BURSA-PASTORIS	S	.	0	24.4	39.7	.	.	.	.	.	.	.	.	.	.	.	.	.	20
1258	CAPSELLA BURSA-PASTORIS	S	.	0	27.8	26.0	.	.	.	.	.	.	.	.	.	.	.	.	.	21
1259	CAPSELLA BURSA-PASTORIS	L	280	0	35.6	4.2	44.1	10.2	16.1	1762.7	728.8	40.7	.	3339.0	21949.1	2.12	1.44	3.39	305.1	2
1260	CAPSICUM ABYSSINICUM	F	368	0	11.0	8.8	74.3	33.8	5.9	69.9	404.4	20.6	.	.	5183.8	.	1.10	5.88	540.4	1
1261	CAPSICUM ANNUUM	F	319	0	16.4	14.4	61.5	23.1	7.7	195.6	381.2	26.1	37.6	2590.1	40179.0	0.71	1.92	16.93	78.5	16
1262	CAPSICUM ANNUUM VAR. ANNUUM	F	344	0	15.3	6.1	72.5	13.0	6.1	84.0	358.8	6.9	.	2855.0	36412.2	0.69	0.92	3.05	656.5	2
1263	CAPSICUM ANNUUM VAR. ANNUUM	F	325	0	16.3	2.5	75.0	17.5	6.3	150.0	425.0	11.3	62.5	3425.0	21875.0	0.88	1.00	10.00	1287.5	2

TABLE 3 - DATA CONVERTED TO A ZERO-MOISTURE BASIS.

OBS	PLANA	PART	CALORIES	WATER	PROTEIN	FAT	TOTLCARB	FIBER	ASH	CALCIUM	PHOSPHO	IRON	SODIUM	POTASS	CAROT	THIAMINE	RIBOFLAV	NIACIN	ASCORBIC	REFCO
1264	CAPSICUM ANNUUM VAR. ANNUUM	F	341	0	15.2	5.9	71.3	12.0	7.6	172.8	237.3	8.9	.	.	7442.4	0.25	1.20	17.28	91.0	3
1265	CAPSICUM ANNUUM VAR. ANNUUM	F	368	0	18.4	11.2	65.1	26.3	5.3	190.8	322.4	9.2	.	.	6447.4	0.92	0.59	10.53	677.6	3
1266	CAPSICUM ANNUUM VAR. ANNUUM	F	360	0	14.5	8.8	70.6	35.1	6.1	114.0	302.6	4.8	.	.	4122.8	0.83	0.70	11.40	381.6	3
1267	CAPSICUM ANNUUM VAR. ANNUUM	F	333	0	14.0	2.6	78.1	15.8	5.3	140.4	298.2	7.0	.	.	6666.7	0.61	0.53	11.40	1587.7	3
1268	CAPSICUM ANNUUM VAR. ANNUUM	F	337	0	13.0	3.3	77.2	14.1	6.5	87.0	293.5	6.5	.	.	3152.2	0.65	0.65	10.87	1239.1	3
1269	CAPSICUM ANNUUM VAR. ANNUUM	S	417	0	18.4	20.0	57.4	19.5	4.1	95.0	491.3	6.1	.	.	414.8	0.60	0.73	14.85	28.4	3
1270	CAPSICUM ANNUUM VAR. ANNUUM	S	334	0	17.4	1.9	77.0	37.8	3.7	61.6	503.2	7.6	.	.	324.0	0.69	0.31	12.74	31.3	3
1271	CAPSICUM ANNUUM VAR. ANNUUM	S	422	0	17.9	20.8	57.4	37.7	3.9	40.7	493.0	8.4	.	.	278.7	0.13	0.23	9.00	24.7	3
1272	CAPSICUM ANNUUM VAR. ANNUUM	S	445	0	17.8	24.8	54.2	26.3	3.2	54.5	479.1	7.7	.	.	2365.8	0.28	0.26	10.80	33.2	3
1273	CAPSICUM ANNUUM VAR. ANNUUM	F	362	0	14.4	8.9	70.4	35.0	6.2	112.8	303.5	4.7	.	.	50428.0	0.86	1.40	17.12	1435.8	4
1274	CAPSICUM ANNUUM VAR. ANNUUM	F	330	0	11.6	1.8	81.3	16.1	5.4	89.3	223.2	6.3	.	.	4125.0	0.80	0.54	15.18	2098.2	4
1275	CAPSICUM FRUTESCENS	S	.	0	16.2	18.2	.	.	.	.	.	.	.	.	.	.	.	.	.	19
1276	CAPSICUM FRUTESCENS	S	.	0	20.0	21.0	.	.	.	.	.	.	.	.	.	.	.	.	.	19
1277	CAPSICUM FRUTESCENS	S	.	0	16.2	17.5	.	.	.	.	.	.	.	.	.	.	.	.	.	19
1278	CAPSICUM FRUTESCENS	S	.	0	20.0	24.8	.	.	4.6	.	.	.	.	.	.	.	.	.	.	19
1279	CAPSICUM FRUTESCENS	S	.	0	20.0	26.1	.	.	4.6	.	.	.	.	.	.	.	.	.	.	19
1280	CAPSICUM FRUTESCENS	F	333	0	8.7	0.8	84.9	11.1	5.6	39.7	142.9	9.5	.	.	2619.0	0.32	0.40	15.87	317.5	
1281	CAPSICUM FRUTESCENS	F	364	0	15.9	8.9	69.8	23.3	5.4	224.8	391.5	11.2	.	.	27674.4	0.97	0.78	9.30	469.0	1
1282	CAPSICUM FRUTESCENS	P	348	0	11.3	5.0	78.7	10.6	5.0	156.0	227.0	13.5	.	.	5106.4	0.50	0.64	7.80	886.5	3
1283	CAPSICUM PUBESCENS	F	336	0	11.2	1.9	81.3	9.3	5.6	56.1	186.9	6.5	.	.	2803.7	0.56	0.75	19.63	327.1	3
1284	CAPSICUM SP.	F	386	0	15.4	15.6	61.1	18.3	7.9	144.8	211.6	.	.	.	1180.4	0.70	0.30	9.24	200.4	1
1285	CAPSICUM SP.	L	296	0	32.4	5.6	47.5	10.6	14.5	1374.3	312.8	7.8	27.9	3988.8	34692.7	2.23	1.84	10.61	379.9	2
1286	CAPSICUM SP.	F	339	0	17.0	5.4	71.4	19.6	6.3	178.6	250.0	15.2	.	.	8392.9	0.80	1.16	13.39	812.5	3
1287	CARAGANA ARBORESCENS	S	.	0	35.5	13.6	.	.	3.4	.	.	.	.	.	.	.	.	.	.	19
1288	CARAGANA ARBORESCENS	S	.	0	36.4	13.2	.	.	3.6	.	.	.	.	.	.	.	.	.	.	19
1289	CARDARIA DRABA	S	.	0	26.4	19.3	.	.	.	.	.	.	.	.	.	.	.	.	.	21
1290	CARDIOSPERMUM HALICACABUM	S	.	0	16.8	18.6	.	.	.	.	.	.	.	.	.	.	.	.	.	21
1291	CARDIOSPERMUM MICROCARPUM	S	.	0	16.6	25.9	.	.	.	.	.	.	.	.	.	.	.	.	.	21
1292	CARDOPATIUM CORYMBOSUM	S	.	0	43.4	34.2	.	.	.	.	.	.	.	.	.	.	.	.	.	21
1293	CARDUUS NUTANS	S	.	0	17.3	29.1	.	.	.	.	.	.	.	.	.	.	.	.	.	20
1294	CARDUUS SP.	W	.	0	9.3	4.2	75.7	30.4	10.8	.	.	.	.	.	.	.	.	.	.	6
1295	CAREX AQUATILIS	W	.	0	13.2	4.0	74.4	30.2	8.4	400.0	310.0	.	.	.	.	.	.	.	.	6
1296	CAREX ATHROSTACHYA	W	.	0	8.0	3.1	82.6	32.8	6.3	.	.	.	.	.	.	.	.	.	.	6
1297	CAREX BICKNELLII	S	.	0	7.7	3.5	.	.	2.2	.	.	.	.	.	.	.	.	.	.	19
1298	CAREX CRINATA	S	.	0	14.8	5.7	.	.	.	.	.	.	.	.	.	.	.	.	.	20
1299	CAREX CRUS-CORVI	S	.	0	10.6	8.4	.	.	.	.	.	.	.	.	.	.	.	.	.	19
1300	CAREX DOUGLASII	W	.	0	10.1	4.0	79.1	29.7	6.8	.	.	.	.	.	.	.	.	.	.	6
1301	CAREX EBENEA	S	.	0	23.8	16.7	.	.	.	.	.	.	.	.	.	.	.	.	.	21
1302	CAREX FESTIVELLA	W	.	0	12.9	2.5	79.7	29.6	4.9	.	.	.	.	.	.	.	.	.	.	6
1303	CAREX FILIFOLIA	W	.	0	7.5	3.2	78.5	32.9	10.8	.	.	.	.	.	.	.	.	.	.	6
1304	CAREX HELIOPHILA	W	.	0	15.9	2.4	73.9	22.5	7.8	.	.	.	.	.	.	.	.	.	.	6
1305	CAREX LANUGINOSA	W	.	0	9.4	2.3	81.0	32.3	7.3	.	.	.	.	.	.	.	.	.	.	6
1306	CAREX NEBRASKENSIS	S	.	0	6.7	4.6	.	.	2.0	.	.	.	.	.	.	.	.	.	.	19
1307	CAREX NEBRASKENSIS	W	.	0	9.6	3.2	79.5	30.2	7.7	.	.	.	.	.	.	.	.	.	.	6
1308	CAREX PRESLII	W	.	0	22.8	2.9	65.6	20.9	8.7	.	.	.	.	.	.	.	.	.	.	6

TABLE 3 - DATA CONVERTED TO A ZERO-MOISTURE BASIS.

OBS	PLNA	PART	CALORIES	WATER	PROTEIN	FAT	TOTLCARB	FIBER	ASH	CALCIUM	PHOSPHO	IRON	SODIUM	POTASS	CAROT	THIAMINE	RIBOFLAV	NIACIN	ASCORBIC	REFCO
1309	CAREX ROSTRATA	S	.	0	11.3	4.8	.	.	.	.	.	.	.	.	.	.	.	.	.	20
1310	CAREX ROSTRATA	W	.	0	11.2	3.4	76.6	28.9	8.8	380.0	280.0	.	.	.	.	.	.	.	.	6
1311	CAREX ROSTRATA	W	.	0	11.7	2.8	76.9	29.1	8.6	570.0	310.0	.	.	.	.	.	.	.	.	6
1312	CAREX SICCATA	W	.	0	8.9	2.8	81.4	30.6	6.9	.	.	.	.	.	.	.	.	.	.	6
1313	CAREX SP.	S	.	0	13.1	6.1	.	.	.	.	.	.	.	.	.	.	.	.	.	20
1314	CAREX ULTRA	S	.	0	13.0	12.9	.	.	6.6	.	.	.	.	.	.	.	.	.	.	19
1315	CAREX VESICARIA	S	.	0	12.9	8.5	.	.	.	.	.	.	.	.	.	.	.	.	.	21
1316	CARICA CANDAMARCENSIS	F	343	0	11.4	2.9	77.1	17.1	8.6	171.4	200.0	7.1	.	.	1000.0	0.29	0.43	14.29	671.4	3
1317	CARICA PAPAYA	L	.	0	32.6	0.8	55.6	7.3	11.0	.	.	.	.	.	.	.	.	.	.	18
1318	CARICA PAPAYA	L	.	0	20.9	13.6	50.1	14.5	15.4	2380.0	220.0	.	.	.	.	.	.	.	.	18
1319	CARICA PAPAYA	F	.	0	11.1	1.2	78.3	11.7	9.4	230.0	160.0	.	.	.	.	.	.	.	.	18
1320	CARICA PAPAYA	F	348	0	4.3	1.1	90.2	9.8	4.3	228.3	163.0	6.5	.	.	10326.1	0.33	0.33	4.35	565.2	1
1321	CARICA PAPAYA	F	349	0	3.9	0.8	91.5	3.9	3.9	186.0	170.5	5.4	31.0	1713.2	5503.9	0.23	0.39	3.10	565.9	2
1322	CARICA PAPAYA	L	329	0	31.1	8.9	50.2	8.0	9.8	1528.9	631.1	3.6	71.1	2897.8	51400.0	0.40	2.13	9.33	622.2	2
1323	CARICA PAPAYA	G	329	0	12.7	1.3	78.5	11.4	7.6	481.0	253.2	3.8	88.6	2721.5	189.9	0.25	0.38	3.80	506.3	2
1324	CARICA PAPAYA	F	345	0	5.3	0.9	88.5	8.0	5.3	177.0	141.6	2.7	26.5	2070.8	9292.0	0.35	0.35	2.65	495.6	4
1325	CARICA PENTAGONA	F	338	0	13.2	1.5	79.4	7.4	5.9	161.8	205.9	5.9	.	.	882.3	0.29	0.88	14.71	426.5	3
1326	CARISSA MACROCARPA	F	376	0	2.2	5.0	90.6	4.4	2.2	60.8	38.7	7.2	.	.	138.1	0.22	0.33	1.10	309.4	2
1327	CARISSA MACROCARPA	F	365	0	2.6	6.8	83.3	4.7	2.1	.	.	.	.	.	125.0	0.21	0.31	1.04	197.9	4
1328	CARPHOCHAETE WISLIZENII	S	.	0	20.0	13.3	.	.	.	.	.	.	.	.	.	.	.	.	.	20
1329	CARRICHTERA ANNUA	S	.	0	31.4	18.6	.	.	.	.	.	.	.	.	.	.	.	.	.	21
1330	CARRICHTERA ANNUA	S	.	0	35.1	13.4	.	.	.	.	.	.	.	.	.	.	.	.	.	21
1331	CARTHAMUS ARBORESCENS	S	.	0	8.5	12.7	.	.	.	.	.	.	.	.	.	.	.	.	.	21
1332	CARTHAMUS FLAVESCENS	S	.	0	14.9	19.9	.	.	.	.	.	.	.	.	.	.	.	.	.	21
1333	CARTHAMUS GLAUCUS	S	.	0	16.9	20.4	.	.	.	.	.	.	.	.	.	.	.	.	.	21
1334	CARTHAMUS LANATUS	S	.	0	12.4	16.9	.	.	.	.	.	.	.	.	.	.	.	.	.	21
1335	CARTHAMUS LANATUS	S	.	0	13.8	21.3	.	.	.	.	.	.	.	.	.	.	.	.	.	21
1336	CARTHAMUS NITIDUS	S	.	0	16.8	14.0	.	.	.	.	.	.	.	.	.	.	.	.	.	21
1337	CARTHAMUS OXYACANTHA	S	.	0	15.1	26.9	.	.	.	.	.	.	.	.	.	.	.	.	.	21
1338	CARTHAMUS TINCTORIUS	W	.	0	17.0	1.6	69.2	10.9	12.2	1700.0	340.0	.	.	.	.	.	.	.	.	18
1339	CARTHAMUS TINCTORIUS	H	.	0	12.2	2.4	76.9	31.1	8.5	.	.	.	.	.	.	.	.	.	.	18
1340	CARTHAMUS TINCTORIUS	H	.	0	4.2	5.1	89.2	58.2	1.5	.	.	.	.	.	.	.	.	.	.	6
1341	CARTHAMUS TINCTORIUS	S	506	0	13.2	29.2	53.0	26.4	4.5	132.4	325.6	10.2	.	.	0.0	0.62	0.15	0.53	0.0	1
1342	CARTHAMUS TINCTORIUS	S	647	0	20.1	62.6	13.1	.	4.2	.	.	.	.	.	.	.	.	.	.	4
1343	CARUM CARVI	S	370	0	22.0	16.2	55.4	14.0	6.5	764.7	630.4	18.0	18.9	1499.4	241.9	0.42	0.42	4.01	.	16
1344	CARYA ILLINOENSIS	S	714	0	9.6	74.4	14.4	2.3	1.7	75.4	299.6	2.5	3.1	624.0	82.6	0.89	0.13	0.93	2.1	2
1345	CARYA ILLINOENSIS	S	718	0	9.7	75.3	13.4	2.3	1.6	76.3	334.0	2.5	.	.	20.6	0.74	0.11	0.93	2.1	3
1346	CARYA ILLINOENSIS	S	711	0	9.5	73.7	15.1	2.4	1.7	75.6	299.2	2.5	0.0	624.2	80.8	0.89	0.13	0.93	2.1	4
1347	CASIMIROA EDULIS	S	.	0	16.3	0.8	.	.	5.8	.	.	.	.	.	.	.	.	.	.	19
1348	CASIMIROA EDULIS	S	.	0	2.8	0.4	.	.	1.1	.	.	.	.	.	.	.	.	.	.	19
1349	CASIMIROA EDULIS	F	361	0	7.8	2.2	87.2	9.4	2.8	44.4	100.0	1.1	.	.	166.7	0.22	0.39	2.78	127.8	3
1350	CASIMIROA TETRAMERIA	S	.	0	12.5	0.9	.	.	5.1	.	.	.	.	.	.	.	.	.	.	19
1351	CASIMIROA TETRAMERIA	S	.	0	5.6	0.7	.	.	1.8	.	.	.	.	.	.	.	.	.	.	19
1352	CASSIA ALATA	S	.	0	18.1	2.9	.	.	.	.	.	.	.	.	.	.	.	.	.	20
1353	CASSIA ALATA	S	.	0	16.1	3.5	.	.	.	.	.	.	.	.	.	.	.	.	.	21

TABLE 3 - DATA CONVERTED TO A ZERO-MOISTURE BASIS.

OBS	PLNA	PART	CALORIES	WATER	PROTEIN	FAT	TOTCARB	FIBER	ASH	CALCIUM	PHOSPHO	IRON	SODIUM	POTASS	CAROT	THIAMINE	RIBOFLAV	NIACIN	ASCORBIC	REFCO
1354	CASSIA ARMATA	S	.	0	23.6	12.4	.	.	.	.	.	.	.	.	.	.	.	.	.	21
1355	CASSIA BAUHINIOIDES	S	.	0	23.1	3.4	.	.	.	.	.	.	.	.	.	.	.	.	.	19
1356	CASSIA BIFLORA	S	.	0	23.1	4.4	.	.	.	.	.	.	.	.	.	.	.	.	.	20
1357	CASSIA BONARIENSIS	S	.	0	20.9	3.3	.	.	.	.	.	.	.	.	.	.	.	.	.	21
1358	CASSIA BONARIENSIS	S	.	0	20.0	3.1	.	.	.	.	.	.	.	.	.	.	.	.	.	21
1359	CASSIA CF. LINDHEIMERIANA	S	.	0	19.4	5.1	.	.	.	.	.	.	.	.	.	.	.	.	.	20
1360	CASSIA CF. UNIFLORA	S	.	0	24.4	5.3	.	.	.	.	.	.	.	.	.	.	.	.	.	20
1361	CASSIA CORYMBOSA	S	.	0	20.0	4.3	.	.	.	.	.	.	.	.	.	.	.	.	.	20
1362	CASSIA COVESII	S	.	0	20.6	3.6	.	.	.	.	.	.	.	.	.	.	.	.	.	20
1363	CASSIA DENSIFLORA	S	.	0	19.7	6.6	.	.	.	.	.	.	.	.	.	.	.	.	.	21
1364	CASSIA DIDYMOBOTRYA	S	.	0	23.9	4.0	.	.	6.0	.	.	.	.	.	.	.	.	.	.	19
1365	CASSIA DURANGENSIS	S	.	0	18.7	3.6	.	.	.	.	.	.	.	.	.	.	.	.	.	20
1366	CASSIA EMARGINATA	S	.	0	25.0	2.5	.	.	.	.	.	.	.	.	.	.	.	.	.	19
1367	CASSIA FASCICULATA	S	.	0	37.2	4.3	.	.	4.0	.	.	.	.	.	.	.	.	.	.	19
1368	CASSIA FASCICULATA	S	.	0	37.4	5.2	.	.	4.4	.	.	.	.	.	.	.	.	.	.	19
1369	CASSIA FASCICULATA	S	.	0	33.8	4.6	.	.	3.7	.	.	.	.	.	.	.	.	.	.	19
1370	CASSIA FASCICULATA	S	.	0	32.2	4.4	.	.	3.9	.	.	.	.	.	.	.	.	.	.	19
1371	CASSIA FASCICULATA	S	.	0	35.1	4.2	.	.	.	.	.	.	.	.	.	.	.	.	.	20
1372	CASSIA FISTULA	L	.	0	17.6	7.8	66.8	30.2	7.8	3270.0	330.0	.	.	.	.	.	.	.	.	18
1373	CASSIA FISTULA	S	.	0	19.8	3.6	.	.	.	.	.	.	.	.	.	.	.	.	.	21
1374	CASSIA HIRSUTA	S	.	0	17.5	2.4	.	.	.	.	.	.	.	.	.	.	.	.	.	20
1375	CASSIA JAVANICA	S	.	0	11.9	1.4	.	.	.	.	.	.	.	.	.	.	.	.	.	20
1376	CASSIA LAEVIGATA	SH	.	0	17.9	6.2	65.8	24.7	7.4	.	.	.	.	.	.	.	.	.	.	18
1377	CASSIA LEPTADENIA	S	.	0	31.3	3.5	.	.	.	.	.	.	.	.	.	.	.	.	.	21
1378	CASSIA LEPTOCARPA	S	.	0	20.3	3.6	.	.	4.0	.	.	.	.	.	.	.	.	.	.	19
1379	CASSIA LEPTOCARPA	S	.	0	21.5	4.4	.	.	4.1	.	.	.	.	.	.	.	.	.	.	19
1380	CASSIA LEPTOCARPA	S	.	0	21.5	4.4	.	.	4.1	.	.	.	.	.	.	.	.	.	.	19
1381	CASSIA LEPTOCARPA	S	.	0	18.8	4.8	.	.	.	.	.	.	.	.	.	.	.	.	.	19
1382	CASSIA LEPTOCARPA	S	.	0	22.5	3.7	.	.	.	.	.	.	.	.	.	.	.	.	.	20
1383	CASSIA LEPTOCARPA	S	.	0	18.9	4.7	.	.	.	.	.	.	.	.	.	.	.	.	.	21
1384	CASSIA LINDHEIMERIANA	S	.	0	18.8	4.6	.	.	.	.	.	.	.	.	.	.	.	.	.	20
1385	CASSIA MARILANDICA	S	.	0	16.9	3.6	.	.	3.4	.	.	.	.	.	.	.	.	.	.	19
1386	CASSIA MARILANDICA	S	.	0	21.9	3.2	.	.	.	.	.	.	.	.	.	.	.	.	.	21
1387	CASSIA OBTUSIFOLIA	S	.	0	19.0	6.5	.	.	.	.	.	.	.	.	.	.	.	.	.	21
1388	CASSIA OBTUSIFOLIA	L	296	0	27.6	1.0	61.6	11.3	9.9	2901.5	472.9	29.1	.	.	39113.3	1.13	3.50	7.39	556.7	1
1389	CASSIA OCCIDENTALIS	S	.	0	22.5	3.4	.	.	4.4	.	.	.	.	.	.	.	.	.	.	19
1390	CASSIA OCCIDENTALIS	S	.	0	18.6	3.2	.	.	.	.	.	.	.	.	.	.	.	.	.	21
1391	CASSIA OCCIDENTALIS	S	.	0	18.1	.	.	.	.	.	.	.	.	.	.	.	.	.	.	21
1392	CASSIA OCCIDENTALIS	S	.	0	18.2	3.6	.	.	.	.	.	.	.	.	.	.	.	.	.	21
1393	CASSIA OCCIDENTALIS	S	.	0	24.9	3.6	.	.	.	.	.	.	.	.	.	.	.	.	.	21
1394	CASSIA OCCIDENTALIS	S	.	0	19.1	3.9	.	.	.	.	.	.	.	.	.	.	.	.	.	21
1395	CASSIA OCCIDENTALIS	S	.	0	20.6	2.7	.	.	.	.	.	.	.	.	.	.	.	.	.	21
1396	CASSIA PURPUREA	S	.	0	19.1	5.5	.	.	.	.	.	.	.	.	.	.	.	.	.	21
1397	CASSIA ROEMERIANA	S	.	0	21.5	3.7	.	.	.	.	.	.	.	.	.	.	.	.	.	20
1398	CASSIA SIAMEA	C	.	0	20.0	5.6	69.1	16.5	5.3	1140.0	140.0	.	.	.	.	.	.	.	.	18

TABLE 3 - DATA CONVERTED TO A ZERO-MOISTURE BASIS.

OBS	PLNA	PART	CALORIES	WATER	PROTEIN	FAT	TOTLCARB	FIBER
1399	CASSIA SIAMEA	L	.	0	16.8	11.2	66.6	19.8
1400	CASSIA SIAMEA	L	331	0	21.5	2.9	70.9	10.8
1401	CASSIA SKINNERI	S	.	0	23.2	7.6	.	.
1402	CASSIA SP.	S	.	0	17.5	3.9	.	.
1403	CASSIA SP.	S	.	0	30.0	6.2	.	.
1404	CASSIA SP.	S	.	0	31.2	5.3	.	.
1405	CASSIA SP.	S	.	0	16.3	5.0	.	.
1406	CASSIA SP.	S	386	0	20.1	1.4	74.6	7.2
1407	CASSIA SPECTABILIS	S	.	0	22.9	2.8	.	.
1408	CASSIA TOMENTOSA	S	.	0	17.4	3.5	.	.
1409	CASSIA TORA	S	.	0	18.2	7.4	65.3	4.6
1410	CASSIA TORA	SH	.	0	13.2	4.4	72.6	19.9
1411	CASSIA TORA	S	.	0	21.9	3.3	.	.
1412	CASSIA TORA	S	.	0	18.8	5.7	.	.
1413	CASSIA TORA	S	.	0	36.0	3.8	.	.
1414	CASSIA UNIFLORA	S	.	0	21.2	6.3	.	.
1415	CASSIA UNIFLORA	S	.	0	25.0	4.6	.	.
1416	CASSIA WISLIZENII	S	.	0	33.1	5.3	.	.
1417	CASSINE EUCLEAEFORMIS	S	.	0	14.5	47.9	.	.
1418	CASSINE EUCLEAEFORMIS	S	.	0	5.8	65.0	.	.
1419	CASSINE EUCLEAEFORMIS	S	.	0	11.3	51.0	.	.
1420	CASSINE EUCLEAEFORMIS	S	.	0	5.4	70.4	.	.
1421	CASSINE PAPILLOSA	S	.	0	15.8	60.3	.	.
1422	CASTALIS NUDICAULIS VAR. GRAMINIFOLIA	S	.	0	.	7.2	.	.

OBS	ASH	CALCIUM	PHOSPHO	IRON	SODIUM	POTASS	CAROT	THIAMINE	RIBOFLAV	NIACIN	ASCORBIC	REFCO
1399	5.4	.	.	.	.	.	.	.	.	.	.	18
1400	4.7	290.7	450.6	15.4	.	.	19302.3	0.12	2.01	3.78	162.8	2
1401	.	.	.	.	.	.	.	.	.	.	.	21
1402	.	.	.	.	.	.	.	.	.	.	.	19
1403	.	.	.	.	.	.	.	.	.	.	.	20
1404	.	.	.	.	.	.	.	.	.	.	.	20
1405	.	.	.	.	.	.	.	.	.	.	.	21
1406	3.9	228.5	205.1	.	.	.	.	1.16	.	.	.	1
1407	5.3	.	.	.	.	.	.	.	.	.	.	19
1408	.	.	.	.	.	.	.	.	.	.	.	21
1409	9.1	720.0	200.0	.	.	.	.	.	.	.	.	18
1410	9.9	2810.0	570.0	.	.	.	.	.	.	.	.	18
1411	.	.	.	.	.	.	.	.	.	.	.	19
1412	.	.	.	.	.	.	.	.	.	.	.	20
1413	.	.	.	.	.	.	.	.	.	.	.	21
1414	.	.	.	.	.	.	.	.	.	.	.	20
1415	.	.	.	.	.	.	.	.	.	.	.	20
1416	5.4	.	.	.	.	.	.	.	.	.	.	19
1417	.	.	.	.	.	.	.	.	.	.	.	21
1418	.	.	.	.	.	.	.	.	.	.	.	21
1419	.	.	.	.	.	.	.	.	.	.	.	21
1420	.	.	.	.	.	.	.	.	.	.	.	21
1421	.	.	.	.	.	.	.	.	.	.	.	21
1422	.	.	.	.	.	.	.	.	.	.	.	21

TABLE 3 - DATA CONVERTED TO A ZERO-MOISTURE BASIS.

OBS	PLNA	PART	CALORIES	WATER	PROTEIN	FAT	TOTLCARB	FIBER
1423	CASTALIS NUDICAULIS VAR. NUDICAULIS	S	.	0	.	10.8	.	.
1424	CASTANEA CRENATA	S	399	0	7.0	1.4	89.2	2.3
1425	CASTANEA MOLLISSIMA	S	403	0	11.9	2.7	83.2	.
1426	CASTANEA SATIVA	S	406	0	7.5	2.8	87.7	2.4
1427	CASTANEA SATIVA	S	408	0	6.1	3.2	88.6	2.3
1428	CASTILLEJA SP.	W	.	0	9.9	2.6	75.0	14.4
1429	CASUARINA TORULOSA	S	.	0	48.8	41.9	.	.
1430	CATALPA BIGNONIOIDES	S	.	0	36.3	33.0	.	.
1431	CATANANCHE CAERULEA	S	.	0	33.8	27.1	.	.
1432	CAUCALIS DAUCOIDES	S	.	0	14.7	11.6	.	.
1433	CAULANTHUS INFLATUS	S	.	0	28.8	30.5	.	.
1434	CAYAPONIA MAXIMOWICZII	S	.	0	30.2	40.3	.	.
1435	CAYAPONIA SP.	S	.	0	16.2	17.9	.	.
1436	CEDRELA ODORATA	S	.	0	25.0	21.4	.	.
1437	CEDRELA SINENSIS	SH	311	0	36.6	6.1	48.2	7.9
1438	CEDRUS ATLANTICA	S	.	0	14.4	43.0	.	.
1439	CEDRUS DEODARA	S	.	0	13.1	44.7	.	.
1440	CEIBA ACUMINATA	S	.	0	31.2	32.0	.	.
1441	CEIBA ACUMINATA	S	.	0	39.4	42.7	.	.
1442	CEIBA PENTANDRA	S	.	0	30.4	23.1	38.3	19.6
1443	CEIBA PENTANDRA	S	530	0	33.2	39.2	21.6	1.6
1444	CEIBA SP.	S	.	0	35.6	43.5	.	.
1445	CELASTRUS ORBICULATUS	S	.	0	21.9	46.1	.	.
1446	CELOSIA ANTHELMINTICA	SH	.	0	15.0	2.1	76.4	32.9

OBS	ASH	CALCIUM	PHOSPHO	IRON	SODIUM	POTASS	CAROT	THIAMINE	RIBOFLAV	NIACIN	ASCORBIC	REFCO
1423	.	.	.	.	.	.	.	.	.	.	.	21
1424	2.3	79.8	187.8	3.8	37.6	.	.	0.89	0.42	3.76	68.1	2
1425	2.2	36.1	167.9	3.8	.	.	216.6	0.29	0.32	1.44	65.0	2
1426	2.0	30.3	183.8	3.4	32.3	1705.1	.	0.46	0.46	1.21	.	2
1427	2.1	56.8	185.3	3.6	12.6	955.8	.	0.46	0.46	1.26	.	4
1428	12.5	910.0	490.0	.	.	.	2.0	.	.	.	.	6
1429	5.2	.	.	.	.	.	.	.	.	.	.	19
1430	4.7	.	.	.	.	.	.	.	.	.	.	19
1431	.	.	.	.	.	.	.	.	.	.	.	19
1432	.	.	.	.	.	.	.	.	.	.	.	20
1433	.	.	.	.	.	.	.	.	.	.	.	21
1434	.	.	.	.	.	.	.	.	.	.	.	21
1435	2.9	.	.	.	.	.	.	.	.	.	.	19
1436	8.3	.	.	.	.	.	.	.	.	.	.	19
1437	9.1	182.9	622.0	19.5	.	3512.2	.	.	.	.	.	2
1438	.	.	.	.	.	.	.	.	.	.	.	20
1439	.	.	.	.	.	.	.	.	.	.	.	20
1440	3.2	.	.	.	.	.	.	.	.	.	.	19
1441	.	.	.	.	.	.	.	.	.	.	.	19
1442	8.2	470.0	970.0	.	.	.	.	.	.	.	.	18
1443	6.1	229.6	1269.3	.	.	.	.	.	.	.	.	1
1444	.	.	.	.	.	.	.	.	.	.	.	19
1445	3.6	.	.	.	.	.	.	.	.	.	.	19
1446	6.5	560.0	180.0	.	.	.	.	.	.	.	.	18

TABLE 3 - DATA CONVERTED TO A ZERO-MOISTURE BASIS.

OBS	PLNA	PART	CALORIES	WATER	PROTEIN	FAT	TOTCARB	FIBER	ASH	CALCIUM	PHOSPHO	IRON	SODIUM	POTASS	CAROT	THIAMINE	RIBOFLAV	NIACIN	ASCORBIC	REFCO
1447	CELOSIA ARGENTEA	S	.	0	14.9	11.2	.	.	3.3	.	.	.	.	.	.	.	.	.	.	19
1448	CELOSIA ARGENTEA	S	.	0	13.0	9.3	.	.	.	.	.	.	.	.	.	.	.	.	.	21
1449	CELOSIA ARGENTEA	L	272	0	29.0	4.3	46.9	11.1	19.8	1604.9	265.4	48.1	.	.	.	.	.	.	.	1
1450	CELOSIA CRISTATA	S	.	0	18.1	13.4	.	.	.	.	.	.	.	.	.	.	.	.	.	20
1451	CELOSIA LAXA	L	297	0	24.3	3.6	57.7	12.6	14.4	1387.4	288.3	45.0	.	.	17342.3	.	0.90	.	90.1	1
1452	CELOSIA TRIGYNA	L	297	0	24.3	3.6	57.7	12.6	14.4	1387.4	288.3	45.0	.	.	17342.3	.	0.90	.	90.1	1
1453	CELTIS AUSTRALIS	L	.	0	15.5	4.6	69.8	22.0	10.1	2830.0	110.0	.	.	.	.	.	.	.	.	18
1454	CELTIS AUSTRALIS	S	.	0	11.9	14.5	.	.	.	.	.	.	.	.	.	.	.	.	.	20
1455	CELTIS MISSISSIPPIENSIS	L	284	0	9.0	5.4	60.7	.	25.0	.	186.1	22.1	.	.	.	.	.	.	.	1
1456	CELTIS OCCIDENTALIS	S	.	0	11.2	9.5	.	.	.	.	.	.	.	.	.	.	.	.	.	20
1457	CELTIS PALLIDA	F	351	0	4.9	1.5	89.3	4.4	4.4	307.3	107.3	13.7	.	.	146.3	0.0	0.00	1.46	248.8	3
1458	CELTIS RETICULATA	L	.	0	15.8	1.8	66.6	18.5	15.8	4910.0	180.0	.	.	1380.0	.	.	.	.	.	6
1459	CENCHRUS BIFLORUS	W	.	0	11.6	2.8	77.9	33.9	7.7	.	.	.	.	.	.	.	.	.	.	18
1460	CENCHRUS BIFLORUS	H	.	0	9.0	1.0	78.4	32.9	11.6	.	.	.	.	.	.	.	.	.	.	18
1461	CENCHRUS BIFLORUS	S	410	0	19.7	9.4	69.1	0.7	1.8	159.6	299.3	24.4	.	.	.	.	.	.	.	1
1462	CENCHRUS CILIARIS	W	.	0	11.0	2.6	73.2	31.9	13.2	.	.	.	.	.	.	.	.	.	.	18
1463	CENCHRUS CILIARIS	H	.	0	7.4	1.7	79.2	35.2	11.7	.	.	.	.	.	.	.	.	.	.	18
1464	CENCHRUS SETIGERAS	W	.	0	18.6	1.9	67.6	28.3	11.9	.	.	.	.	.	.	.	.	.	.	18
1465	CENCHRUS SP.	W	.	0	10.7	2.3	80.0	36.5	7.0	.	.	.	.	.	.	.	.	.	.	6
1466	CENCHRUS TRIBULOIDES	S	.	0	25.6	8.8	.	.	2.2	.	.	.	.	.	.	.	.	.	.	19
1467	CENTAUREA AMERICANA	S	.	0	20.2	14.3	.	.	3.2	.	.	.	.	.	.	.	.	.	.	19
1468	CENTAUREA AMERICANA	S	.	0	15.1	10.6	.	.	.	.	.	.	.	.	.	.	.	.	.	20
1469	CENTAUREA AMERICANA	S	.	0	19.6	16.9	.	.	.	.	.	.	.	.	.	.	.	.	.	20
1470	CENTAUREA ASPERA	S	.	0	15.4	20.5	.	.	.	.	.	.	.	.	.	.	.	.	.	21
1471	CENTAUREA CALCITRAPA	S	.	0	24.6	21.2	.	.	.	.	.	.	.	.	.	.	.	.	.	21
1472	CENTAUREA CALCITRAPA	S	.	0	20.0	24.1	.	.	.	.	.	.	.	.	.	.	.	.	.	21
1473	CENTAUREA CYANUS	S	.	0	15.8	23.6	.	.	3.4	.	.	.	.	.	.	.	.	.	.	19
1474	CENTAUREA CYANUS	S	.	0	15.8	25.4	.	.	.	.	.	.	.	.	.	.	.	.	.	20
1475	CENTAUREA CYANUS	S	.	0	18.1	25.0	.	.	.	.	.	.	.	.	.	.	.	.	.	20
1476	CENTAUREA DEALBATA	S	.	0	23.8	19.4	.	.	.	.	.	.	.	.	.	.	.	.	.	20
1477	CENTAUREA DEPRESSA	S	.	0	17.4	24.8	.	.	.	.	.	.	.	.	.	.	.	.	.	21
1478	CENTAUREA GYMNOCARPA	S	.	0	18.8	23.5	.	.	.	.	.	.	.	.	.	.	.	.	.	20
1479	CENTAUREA IBERICA	S	.	0	21.6	24.8	.	.	.	.	.	.	.	.	.	.	.	.	.	21
1480	CENTAUREA INTYBACEA	S	.	0	13.2	18.5	.	.	.	.	.	.	.	.	.	.	.	.	.	21
1481	CENTAUREA MACROCEPHALA	S	.	0	35.0	19.9	.	.	.	.	.	.	.	.	.	.	.	.	.	20
1482	CENTAUREA MACROCEPHALA	S	.	0	33.2	18.3	.	.	.	.	.	.	.	.	.	.	.	.	.	21
1483	CENTAUREA MONTANA	S	.	0	21.2	22.0	.	.	.	.	.	.	.	.	.	.	.	.	.	20
1484	CENTAUREA MOSCHATA	S	.	0	22.5	28.4	.	.	.	.	.	.	.	.	.	.	.	.	.	20
1485	CENTAUREA ORNATA	S	.	0	17.4	8.5	.	.	.	.	.	.	.	.	.	.	.	.	.	21
1486	CENTAUREA REPENS	S	.	0	12.4	9.0	.	.	.	.	.	.	.	.	.	.	.	.	.	21
1487	CENTAUREA REPENS	S	.	0	16.7	9.1	.	.	.	.	.	.	.	.	.	.	.	.	.	21
1488	CENTAUREA RIGIDA	S	.	0	25.9	20.7	.	.	.	.	.	.	.	.	.	.	.	.	.	21
1489	CENTAUREA SQUARROSA	S	.	0	25.3	11.6	.	.	.	.	.	.	.	.	.	.	.	.	.	21
1490	CENTAUREA TRIUMFETTII	S	.	0	13.6	23.2	.	.	.	.	.	.	.	.	.	.	.	.	.	21
1491	CENTAUREA VERUTUM	S	.	0	18.9	20.4	.	.	.	.	.	.	.	.	.	.	.	.	.	21

TABLE 3 - DATA CONVERTED TO A ZERO-MOISTURE BASIS.

OBS	PLNA	PART	CALORIES	WATER	PROTEIN	FAT	TOTLCARB	FIBER
1492	CENTELLA ASIATICA	L	318	0	15.0	5.6	64.5	18.7
1493	CENTROSEMA PUBESCENS	W	.	0	25.3	3.2	63.7	31.8
1494	CEPHALANTHUS GLABRATUS	S	.	0	8.9	8.5	.	.
1495	CEPHALANTHUS OCCIDENTALIS	S	.	0	9.8	9.4	.	.
1496	CEPHALARIA JOPPICA	S	.	0	16.4	23.9	.	.
1497	CEPHALARIA LEUCANTHA	S	.	0	17.0	13.7	.	.
1498	CEPHALARIA MICROCEPHALA	S	.	0	31.8	19.3	.	.
1499	CEPHALARIA SETOSA	S	.	0	17.1	26.0	.	.
1500	CEPHALARIA TRANSYLVANICA	S	.	0	21.4	24.1	.	.
1501	CEPHALOTAXUS HARRINGTONIA VAR. DRUPACEAE	S	.	0	12.5	65.2	.	.
1502	CERASTIUM DICHOTOMUM	S	.	0	15.7	4.1	.	.
1503	CERASTIUM PERFOLIATUM	S	.	0	17.4	3.8	.	.
1504	CERASTIUM TOMENTOSUM	S	.	0	18.8	5.2	.	.
1505	CERASTIUM VISCOSUM	S	.	0	16.9	5.9	.	.
1506	CERATONIA SILIQUA	F	.	0	6.9	1.2	88.2	12.0
1507	CERATONIA SILIQUA	S	.	0	18.5	2.0	76.0	8.0
1508	CERATONIA SILIQUA	S	.	0	17.5	1.5	.	.
1509	CERATONIA SILIQUA		203	0	5.1	1.6	90.9	8.7
1510	CERATOPETALUM GUMMIFERUM	S	.	0	8.8	8.6	.	.
1511	CERATOTHECA SESAMOIDES	L	284	0	22.1	2.6	57.9	12.1
1512	CERATOTHECA SESAMOIDES	S	591	0	15.3	50.0	29.6	18.6
1513	CERCIDIUM FLORIDUM	S	.	0	54.4	15.7	.	.
1514	CERCIDIUM FLORIDUM	S	.	0	6.6	1.1	.	.
1515	CERCIDIUM FLORIDUM	S	.	0	27.5	5.0	.	.

OBS	ASH	CALCIUM	PHOSPHO	IRON	SODIUM	POTASS	CAROT	THIAMINE	RIBOFLAV	NIACIN	ASCORBIC	REFCO
1492	15.0	1588.8	280.4	29.0	.	3869.2	61495.3	1.40	1.31	11.21	37.4	2
1493	7.8	1040.0	320.0	.	.	.	.	.	.	.	.	18
1494	.	.	.	.	.	.	.	.	.	.	.	21
1495	3.2	.	.	.	.	.	.	.	.	.	.	19
1496	.	.	.	.	.	.	.	.	.	.	.	21
1497	.	.	.	.	.	.	.	.	.	.	.	21
1498	.	.	.	.	.	.	.	.	.	.	.	21
1499	.	.	.	.	.	.	.	.	.	.	.	21
1500	.	.	.	.	.	.	.	.	.	.	.	21
1501	.	.	.	.	.	.	.	.	.	.	.	20
1502	.	.	.	.	.	.	.	.	.	.	.	21
1503	.	.	.	.	.	.	.	.	.	.	.	20
1504	3.8	.	.	.	.	.	.	.	.	.	.	19
1505	.	.	.	.	.	.	.	.	.	.	.	20
1506	3.7	560.0	120.0	.	.	.	.	.	.	.	.	18
1507	3.5	.	.	.	.	.	.	.	.	.	.	18
1508	.	.	.	.	.	.	.	.	.	.	.	19
1509	2.5	396.4	91.2	.	.	.	.	.	.	.	.	4
1510	.	.	.	.	.	.	.	.	.	.	.	20
1511	17.4	1578.9	452.6	16.8	.	.	.	.	.	.	147.4	1
1512	5.2	953.8	.	40.9	.	.	.	0.81	0.32	4.73	.	1
1513	5.1	.	.	.	.	.	.	.	.	.	.	19
1514	2.6	.	.	.	.	.	.	.	.	.	.	19
1515	.	.	.	.	.	.	.	.	.	.	.	20

TABLE 3 - DATA CONVERTED TO A ZERO-MOISTURE BASIS.

OBS	PLANA	PART	CALORIES	WATER	PROTEIN	FAT	TOTLCARB	FIBER	ASH	CALCIUM	PHOSPHO	IRON	SODIUM	POTASS	CAROT	THIAMINE	RIBOFLAV	NIACIN	ASCORBIC	REFCO
1516	CERCIDIUM TORREYANUM	S	.	0	27.5	4.3	.	.	.	.	.	.	.	.	.	.	.	.	.	20
1517	CERCIS CANADENSIS	S	.	0	22.9	8.2	.	.	3.1	.	.	.	.	.	.	.	.	.	.	19
1518	CERCIS CANADENSIS	S	.	0	23.1	8.8	.	.	2.9	.	.	.	.	.	.	.	.	.	.	19
1519	CERCIS CANADENSIS	S	.	0	27.3	8.3	.	.	3.1	.	.	.	.	.	.	.	.	.	.	19
1520	CERCIS CANADENSIS	S	.	0	27.5	7.7	.	.	.	.	.	.	.	.	.	.	.	.	.	19
1521	CERCIS CHINENSIS	S	.	0	16.6	9.2	.	.	2.8	.	.	.	.	.	.	.	.	.	.	19
1522	CERCIS OCCIDENTALIS	S	.	0	18.6	8.8	.	.	2.6	.	.	.	.	.	.	.	.	.	.	19
1523	CERCIS SILIQUASTRUM	S	.	0	20.4	8.9	.	.	3.0	.	.	.	.	.	.	.	.	.	.	19
1524	CERCIS SILIQUASTRUM	S	.	0	23.1	8.0	.	.	.	.	.	.	.	.	.	.	.	.	.	21
1525	CERCOCARPUS MONTANUS	S	.	0	16.9	8.3	.	.	.	.	.	.	.	.	.	.	.	.	.	20
1526	CEREUS HEXAGONUS	I	322	0	.	.	91.1	.	8.9	1022.2	300.0	48.9	.	.	2888.9	0.33	0.11	5.56	444.4	3
1527	CEREUS HEXAGONUS	F	363	0	10.3	4.9	79.9	.	4.9	215.7	112.7	.	.	.	343.1	0.20	0.15	2.45	156.9	3
1528	CERINTHE MAJOR	S	.	0	13.1	25.3	.	.	.	.	.	.	.	.	.	.	.	.	.	21
1529	CERINTHE MINOR	S	.	0	14.1	10.4	.	.	.	.	.	.	.	.	.	.	.	.	.	20
1530	CEVALLIA SINUATA	S	.	0	20.9	21.7	.	.	.	.	.	.	.	.	.	.	.	.	.	21
1531	CHAENACTIS ALPINA	S	.	0	21.8	18.8	.	.	.	.	.	.	.	.	.	.	.	.	.	20
1532	CHAENACTIS ALPINA	S	.	0	21.2	21.0	.	.	.	.	.	.	.	.	.	.	.	.	.	20
1533	CHAENACTIS SP.	S	.	0	18.1	27.8	.	.	.	.	.	.	.	.	.	.	.	.	.	20
1534	CHAEROPHYLLUM AUREUM	S	.	0	24.1	12.8	.	.	.	.	.	.	.	.	.	.	.	.	.	21
1535	CHAEROPHYLLUM TAINTURIERI	S	.	0	20.6	10.5	.	.	.	.	.	.	.	.	.	.	.	.	.	20
1536	CHAEROPHYLLUM TAINTURIERI	S	.	0	14.9	9.1	.	.	.	.	.	.	.	.	.	.	.	.	.	20
1537	CHAEROPHYLLUM TAINTURIERI	S	.	0	13.4	10.6	.	.	.	.	.	.	.	.	.	.	.	.	.	21
1538	CHAMAECYPARIS LAWSONIANA	S	.	0	11.9	24.7	.	.	.	.	.	.	.	.	.	.	.	.	.	20
1539	CHAMAECYTISUS PYGMAEUS	S	.	0	34.5	8.9	.	.	.	.	.	.	.	.	.	.	.	.	.	21
1540	CHAMAEDOREA SP.		300	0	26.7	4.7	55.3	8.0	13.3	2460.0	706.7	9.3	.	.	66.7	0.53	0.67	6.00	93.3	3
1541	CHAMAEPEUCE HISPANICA	S	.	0	19.9	24.1	.	.	.	.	.	.	.	.	.	.	.	.	.	21
1542	CHAMAEROPS HUMILIS	S	.	0	5.0	8.7	.	.	.	.	.	.	.	.	.	.	.	.	.	20
1543	CHAMAESARACHA CORONOPUS	S	.	0	20.6	20.3	.	.	.	.	.	.	.	.	.	.	.	.	.	19
1544	CHARDINIA ORIENTALIS	S	.	0	23.2	26.1	.	.	.	.	.	.	.	.	.	.	.	.	.	21
1545	CHASCANUM CARIPENSE	F	.	0	14.3	2.4	74.5	27.3	8.8	1910.0	110.0	.	.	.	.	.	.	.	.	18
1546	CHEIRANTHUS CHEIRI	S	.	0	27.5	29.5	.	.	.	.	.	.	.	.	.	.	.	.	.	20
1547	CHELONE BARBATA	S	.	0	18.9	24.2	.	.	3.3	.	.	.	.	.	.	.	.	.	.	19
1548	CHENOPODIUM ALBUM	L	.	0	21.7	3.7	41.8	8.3	32.8	.	.	.	.	.	.	.	.	.	.	18
1549	CHENOPODIUM ALBUM	S	.	0	19.4	9.3	.	.	.	.	.	.	.	.	.	.	.	.	.	19
1550	CHENOPODIUM ALBUM	S	.	0	16.1	7.0	69.0	20.0	5.9	.	.	.	.	.	.	.	.	.	.	5
1551	CHENOPODIUM ALBUM	W	.	0	26.9	1.7	43.8	11.2	27.6	3330.0	510.0	.	.	8710	.	.	.	.	.	6
1552	CHENOPODIUM ALBUM	L	253	0	30.7	4.1	40.7	6.6	24.5	1539.4	332.0	.	.	.	.	.	.	.	.	1
1553	CHENOPODIUM ALBUM	L	267	0	17.5	5.8	50.0	21.7	26.7	1333.3	3683.3	25.0	.	.	31583.3	0.67	1.58	2.50	.	3
1554	CHENOPODIUM AMBROSIOIDES	S	.	0	15.6	9.2	.	.	.	.	.	.	.	.	.	.	.	.	.	19
1555	CHENOPODIUM AMBROSIOIDES	S	.	0	17.5	2.5	.	.	.	.	.	.	.	.	.	.	.	.	.	19
1556	CHENOPODIUM AMBROSIOIDES	S	.	0	12.1	8.5	.	.	.	.	.	.	.	.	.	.	.	.	.	21
1557	CHENOPODIUM AMBROSIOIDES	L	290	0	26.2	4.8	52.4	9.0	16.6	2096.6	358.6	35.9	.	.	16689.7	0.41	1.93	4.14	75.9	3
1558	CHENOPODIUM BERLANDIERI	L	259	0	42.4	5.3	31.8	14.7	20.6	917.6	400.0	17.6	.	.	32529.4	1.00	2.76	7.06	641.2	3
1559	CHENOPODIUM BOTRYS	S	.	0	15.2	5.2	.	.	.	.	.	.	.	.	.	.	.	.	.	21
1560	CHENOPODIUM FOLIOSUM	S	.	0	13.5	2.0	.	.	.	.	.	.	.	.	.	.	.	.	.	21

TABLE 3 - DATA CONVERTED TO A ZERO-MOISTURE BASIS.

OBS	PLNA	PART	CALORIES	WATER	PROTEIN	FAT	TOTLCARB	FIBER	ASH	CALCIUM	PHOSPHO	IRON	SODIUM	POTAS	CAROT	THIAMINE	RIBOFLAV	NIACIN	ASCORBIC	REFCO
1561	CHENOPODIUM LEPTOPHYLLUM	S	.	0	16.7	6.9	.	.	.	.	.	.	.	.	.	.	.	.	.	20
1562	CHENOPODIUM MURALE	S	.	0	15.4	8.6	.	.	.	.	.	.	.	.	.	.	.	.	.	21
1563	CHENOPODIUM OPULIFOLIUM	S	.	0	17.2	9.3	.	.	.	.	.	.	.	.	.	.	.	.	.	21
1564	CHENOPODIUM OPULIFOLIUM	S	.	0	18.6	9.3	.	.	.	.	.	.	.	.	.	.	.	.	.	21
1565	CHENOPODIUM PALLIDICAULE	S	.	0	16.9	8.8	71.5	3.9	2.8	.	.	.	.	.	.	.	.	.	.	18
1566	CHENOPODIUM PALLIDICAULE	S	374	0	16.2	3.1	75.4	12.6	5.3	153.1	441.1	12.0	.	.	.	0.98	0.95	1.26	2.3	3
1567	CHENOPODIUM PALLIDICAULE	S	381	0	15.9	5.1	73.3	11.2	5.7	125.6	428.1	15.2	.	.	.	0.54	0.74	1.26	1.1	3
1568	CHENOPODIUM PALLIDICAULE	S	370	0	15.9	1.7	77.4	13.6	5.0	151.3	419.2	11.5	.	.	.	0.99	0.35	1.50	1.2	3
1569	CHENOPODIUM QUINOA	S	.	0	12.1	7.3	.	.	3.1	.	.	.	.	.	.	.	.	.	.	19
1570	CHENOPODIUM QUINOA	S	.	0	12.5	7.2	.	.	.	.	.	.	.	.	.	.	.	.	.	19
1571	CHENOPODIUM QUINOA	S	394	0	13.8	6.9	76.1	5.2	3.3	125.8	321.3	8.4	.	.	0.0	0.40	0.47	1.57	3.4	3
1572	CHENOPODIUM RUBRUM	S	.	0	14.8	9.1	.	.	.	.	.	.	.	.	.	.	.	.	.	21
1573	CHENOPODIUM SP.	S	.	0	14.3	2.7	.	.	3.1	.	.	.	.	.	.	.	.	.	.	19
1574	CHENOPODIUM SP.	S	.	0	12.7	3.7	.	.	1.6	.	.	.	.	.	.	.	.	.	.	19
1575	CHENOPODIUM SP.	L	274	0	26.8	5.1	46.5	13.4	21.7	1968.2	458.6	7.6	.	.	44331.2	1.02	2.80	7.64	509.6	4
1576	CHILOPSIS LINEARIS	S	.	0	17.3	29.0	.	.	3.3	.	.	.	.	.	.	.	.	.	.	19
1577	CHILOPSIS LINEARIS	S	.	0	22.1	35.0	.	.	4.0	.	.	.	.	.	.	.	.	.	.	19
1578	CHILOPSIS LINEARIS	S	.	0	22.5	31.9	.	.	.	.	.	.	.	.	.	.	.	.	.	19
1579	CHIMONANTHUS PRAECOX	S	.	0	34.4	36.8	.	.	2.7	.	.	.	.	.	.	.	.	.	.	19
1580	CHIMONANTHUS PRAECOX	S	.	0	1.6	5.9	.	.	1.0	.	.	.	.	.	.	.	.	.	.	19
1581	CHLORIS CUCULLATA	W	.	0	9.9	2.2	71.7	28.6	16.2	4800.0	160.0	.	.	.	.	.	.	.	.	6
1582	CHLORIS GAYANA	W	.	0	8.0	2.0	76.9	37.2	13.1	.	.	.	.	.	.	.	.	.	.	18
1583	CHLORIS GAYANA	H	.	0	9.9	27.0	78.5	33.4	8.9	.	.	.	.	.	.	.	.	.	.	18
1584	CHLORIS GAYANA	S	.	0	21.2	4.7	.	.	.	.	.	.	.	.	.	.	.	.	.	19
1585	CHLORIS GAYANA	S	.	0	21.9	4.6	.	.	.	.	.	.	.	.	.	.	.	.	.	19
1586	CHLORIS GAYANA	S	.	0	23.1	5.5	.	.	.	.	.	.	.	.	.	.	.	.	.	20
1587	CHLORIS GAYANA	W	.	0	7.4	1.6	80.5	37.5	10.5	.	.	.	.	.	.	.	.	.	.	6
1588	CHLORIS PRIEURII	W	.	0	9.2	1.8	79.8	37.1	9.2	.	.	.	.	.	.	.	.	.	.	18
1589	CHLORIS ROXBURGHIANA	W	.	0	16.1	0.9	74.2	30.8	8.8	.	.	.	.	.	.	.	.	.	.	18
1590	CHLORIS VERTICILLATA	W	.	0	8.2	2.0	74.0	29.2	15.8	640.0	180.0	.	.	.	.	.	.	.	.	6
1591	CHLORIS VIRGATA	W	.	0	12.9	1.8	71.5	31.1	13.8	.	.	.	.	.	.	.	.	.	.	18
1592	CHLORIS VIRGATA	H	.	0	10.3	1.8	75.2	34.2	12.7	.	.	.	.	.	.	.	.	.	.	18
1593	CHLORIS VIRGATA	S	.	0	28.1	10.7	.	.	.	.	.	.	.	.	.	.	.	.	.	20
1594	CHONDROPETALUM MACROCARPUM	S	.	0	34.4	3.1	.	.	.	.	.	.	.	.	.	.	.	.	.	21
1595	CHONDROPETALUM MICROCARPUM	S	.	0	38.4	1.0	.	.	.	.	.	.	.	.	.	.	.	.	.	21
1596	CHROZOPHORA HIEROSOLYMITANA	S	.	0	22.2	28.1	.	.	.	.	.	.	.	.	.	.	.	.	.	21
1597	CHROZOPHORA TINCTORIA	S	.	0	18.3	25.3	.	.	.	.	.	.	.	.	.	.	.	.	.	21
1598	CHROZOPHORA TINCTORIA	S	.	0	37.2	48.6	.	.	.	.	.	.	.	.	.	.	.	.	.	21
1599	CHROZOPHORA TINCTORIA	S	.	0	34.7	56.5	.	.	.	.	.	.	.	.	.	.	.	.	.	21
1600	CHRYSALIDOCARPUS LUCUBENSIS	S	.	0	2.9	8.2	.	.	.	.	.	.	.	.	.	.	.	.	.	20
1601	CHRYSALIDOCARPUS LUTESCENS	S	.	0	6.9	7.2	.	.	.	.	.	.	.	.	.	.	.	.	.	20
1602	CHRYSANTHEMOIDES INCANA	S	.	0	.	34.9	.	.	.	.	.	.	.	.	.	.	.	.	.	21
1603	CHRYSANTHEMOIDES MONILIFERA	S	.	0	.	50.4	.	.	.	.	.	.	.	.	.	.	.	.	.	20
1604	CHRYSANTHEMOIDES MONILIFERA	S	.	0	.	49.0	.	.	.	.	.	.	.	.	.	.	.	.	.	21
1605	CHRYSANTHEMOIDES MONILIFERA	S	.	0	.	50.0	.	.	.	.	.	.	.	.	.	.	.	.	.	21

TABLE 3 - DATA CONVERTED TO A ZERO-MOISTURE BASIS.

OBS	PLNA	PART	CALORIES	WATER	PROTEIN	FAT	TOTLCARB	FIBER
1606	CHRYSANTHEMUM BALSAMITA	S	.	0	24.6	20.0	.	.
1607	CHRYSANTHEMUM CARINATUM	S	.	0	10.2	13.6	.	.
1608	CHRYSANTHEMUM CINERARIIFOLIUM	SH	.	0	13.0	0.5	79.4	23.6
1609	CHRYSANTHEMUM CORONARIUM	S	.	0	13.2	21.9	.	.
1610	CHRYSANTHEMUM CORONARIUM	S	.	0	17.6	19.3	.	.
1611	CHRYSANTHEMUM CORONARIUM	S	.	0	15.0	19.5	.	.
1612	CHRYSANTHEMUM CORONARIUM	S	.	0	13.8	12.8	.	.
1613	CHRYSANTHEMUM CORONARIUM	S	.	0	11.4	4.4	.	.
1614	CHRYSANTHEMUM CORONARIUM	S	.	0	6.9	4.4	.	.
1615	CHRYSANTHEMUM CORONARIUM	S	.	0	33.6	44.6	.	.
1616	CHRYSANTHEMUM CORONARIUM	S	.	0	12.8	7.1	.	.
1617	CHRYSANTHEMUM CORONARIUM	S	.	0	9.2	8.7	.	.
1618	CHRYSANTHEMUM CORONARIUM	L	292	0	27.7	4.6	50.8	13.8
1619	CHRYSANTHEMUM LEUCANTHEMUM	S	.	0	24.3	23.4	.	.
1620	CHRYSANTHEMUM LEUCANTHEMUM	S	.	0	19.4	22.9	.	.
1621	CHRYSANTHEMUM LEUCANTHEMUM	S	.	0	16.4	18.7	.	.
1622	CHRYSANTHEMUM LEUCANTHEMUM CV. ALASKA	S	.	0	23.8	27.9	.	.
1623	CHRYSANTHEMUM PARTHENIUM	S	.	0	22.2	31.2	.	.
1624	CHRYSANTHEMUM SEGETUM	L	279	0	17.6	4.4	57.4	10.3
1625	CHRYSOBALANUS ICACO	F	361	0	3.9	1.9	91.6	7.1
1626	CHRYSOBALANUS ICACO	F	343	0	2.9	0.7	90.5	7.3
1627	CHRYSOBALANUS ICACO	S	.	0	11.0	.	65.4	.
1628	CHRYSOPHYLLUM CAINITO	F	368	0	5.4	3.8	88.6	10.8
1629	CHRYSOPHYLLUM CAINITO	F	395	0	4.7	9.3	84.3	5.8

OBS	ASH	CALCIUM	PHOSPHO	IRON	SODIUM	POTASS	CAROT	THIAMINE	RIBOFLAV	NIACIN	ASCORBIC	REFCO
1606	.	.	.	.	.	.	.	.	.	.	.	21
1607	4.5	.	.	.	.	.	.	.	.	.	.	19
1608	7.1	530.0	240.0	.	.	.	.	.	.	.	.	18
1609	4.6	.	.	.	.	.	.	.	.	.	.	19
1610	5.7	.	.	.	.	.	.	.	.	.	.	19
1611	.	.	.	.	.	.	.	.	.	.	.	19
1612	.	.	.	.	.	.	.	.	.	.	.	20
1613	.	.	.	.	.	.	.	.	.	.	.	21
1614	.	.	.	.	.	.	.	.	.	.	.	21
1615	.	.	.	.	.	.	.	.	.	.	.	21
1616	.	.	.	.	.	.	.	.	.	.	.	21
1617	.	.	.	.	.	.	.	.	.	.	.	21
1618	16.9	969.2	523.1	38.5	1630.8	3938.5	48615.4	1.38	2.92	9.23	415.4	2
1619	5.3	.	.	.	.	.	.	.	.	.	.	19
1620	.	.	.	.	.	.	.	.	.	.	.	20
1621	.	.	.	.	.	.	.	.	.	.	.	21
1622	.	.	.	.	.	.	.	.	.	.	.	19
1623	.	.	.	.	.	.	.	.	.	.	.	21
1624	20.6	1823.5	735.3	44.1	.	.	17058.8	1.91	1.47	10.29	132.4	3
1625	2.6	187.1	103.2	7.7	.	.	503.2	0.13	0.19	5.16	141.9	
1626	5.8	277.4	124.1	4.4	.	.	0.0	0.29	0.22	2.19	65.7	3
1627	1.7	.	.	.	.	.	.	.	.	.	.	5
1628	2.2	91.9	86.5	2.2	27.0	756.8	0.0	0.05	0.11	4.86	43.2	2
1629	1.7	122.1	98.8	4.7	.	.	58.1	0.23	0.17	5.81	64.0	3

TABLE 3 - DATA CONVERTED TO A ZERO-MOISTURE BASIS.

OBS	PLNA	PART	CALORIES	WATER	PROTEIN	FAT	TOTLCARB	FIBER
1630	CHRYSOPHYLLUM CAINITO	G	422	0	7.0	16.1	73.4	16.6
1631	CHRYSOPOGON FULVUS	W	.	0	9.0	2.1	78.0	32.6
1632	CHRYSOPOGON FULVUS	H	.	0	4.7	1.1	85.5	36.6
1633	CHRYSOPOGON GRYLLUS	S	.	0	40.5	6.0	.	.
1634	CHRYSOPOGON LANCEARIUS	L	.	0	4.7	.	.	27.6
1635	CHRYSOPOGON MONTANUS	L	.	0	4.5	1.8	84.1	26.5
1636	CHRYSOPOGON ORIENTALIS	L	.	0	4.8	1.2	84.8	38.6
1637	CHRYSOPSIS FOLIOSA	S	.	0	24.8	21.2	.	.
1638	CHRYSOTHAMNUS LANCEOLATUS	W	.	0	6.1	5.0	82.6	28.6
1639	CHRYSOTHAMNUS NAUSEOSUS	W	.	0	12.8	3.2	75.2	22.6
1640	CHRYSOTHAMNUS PARRYI SUBSP. NEVADENSIS	S	.	0	32.2	28.0	.	.
1641	CHRYSOTHAMNUS STENOPHYLLUS	W	.	0	6.6	8.9	76.6	29.8
1642	CHRYSOTHAMNUS VISCIDIFLORUS	W	.	0	13.4	5.4	73.2	23.8
1643	CHUSQUEA BACULIFERA	L	.	0	8.8	3.1	77.6	34.4
1644	CHUSQUEA PINIFOLIA	L	.	0	7.6	5.0	80.1	41.2
1645	CHUSQUEA QUITA	L	.	0	10.0	3.6	75.0	34.5
1646	CICER ANATOLICUM	S	.	0	24.1	3.4	.	.
1647	CICER ARIETINUM	SH	.	0	11.3	2.2	75.1	27.2
1648	CICER ARIETINUM	H	.	0	12.9	1.5	74.4	35.3
1649	CICER ARIETINUM	F	.	0	17.0	2.9	70.5	25.0
1650	CICER ARIETINUM	S	.	0	18.1	4.9	73.5	9.8
1651	CICER ARIETINUM	S	.	0	19.4	5.5	.	.
1652	CICER ARIETINUM	S	397	0	21.8	4.1	70.6	7.4
1653	CICER ARIETINUM	S	411	0	20.6	7.0	69.0	3.8

OBS	ASH	CALCIUM	PHOSPHO	IRON	SODIUM	POTASS	CAROT	THIAMINE	RIBOFLAV	NIACIN	ASCORBIC	REFCO
1630	3.5	110.6	105.5	3.0	.	.	201.0	0.10	0.05	4.52	15.1	3
1631	11.0	.	.	.	.	.	.	.	.	.	.	18
1632	8.7	.	.	.	.	.	.	.	.	.	.	18
1633	.	.	.	.	.	.	.	.	.	.	.	21
1634	16.3	1860.0	620.0	.	2230.0	.	.	.	.	.	.	5
1635	9.6	.	.	.	.	.	.	.	.	.	.	5
1636	9.1	.	.	.	.	.	.	.	.	.	.	5
1637	.	.	.	.	.	.	.	.	.	.	.	21
1638	6.3	.	.	.	.	.	.	.	.	.	.	6
1639	8.8	950.0	380.0	.	.	.	.	.	.	.	.	6
1640	.	.	.	.	.	.	.	.	.	.	.	21
1641	7.9	2210.0	110.0	.	.	.	.	.	.	.	.	6
1642	8.0	1400.0	340.0	.	.	.	.	.	.	.	.	6
1643	10.5	.	.	.	.	.	.	.	.	.	.	18
1644	7.3	.	.	.	.	.	.	.	.	.	.	18
1645	11.3	.	.	.	.	.	.	.	.	.	.	18
1646	.	.	.	.	.	.	.	.	.	.	.	21
1647	11.4	1410.0	250.0	.	.	.	.	.	.	.	.	18
1648	11.2	1530.0	200.0	.	.	.	.	.	.	.	.	18
1649	9.6	1390.0	240.0	.	.	.	.	.	.	.	.	18
1650	3.5	260.0	410.0	.	.	.	.	.	.	.	.	18
1651	2.7	.	.	.	.	.	.	.	.	.	.	19
1652	3.6	280.0	301.1	12.3	.	.	66.7	0.53	0.18	2.00	.	1
1653	3.4	151.4	366.1	8.2	.	.	33.9	0.52	0.18	1.92	1.1	3

TABLE 3 - DATA CONVERTED TO A ZERO-MOISTURE BASIS.

OBS	PLNA	PART	CALORIES	WATER	PROTEIN	FAT	TOTLCARB	FIBER
1654	CICER ARIETINUM	S	403	0	23.0	5.4	68.3	5.6
1655	CICHORIUM ENDIVIA	L	292	0	20.8	2.8	59.7	15.3
1656	CICHORIUM ENDIVIA	L	290	0	24.6	2.9	59.4	13.0
1657	CICHORIUM ENDIVIA	L	290	0	24.6	1.4	59.4	13.0
1658	CICHORIUM ENDIVIA	SH	306	0	20.4	2.0	65.3	.
1659	CICHORIUM ENDIVIA	L	278	0	25.0	4.2	52.8	11.1
1660	CICHORIUM INTYBUS	S	.	0	18.9	17.4	.	.
1661	CICHORIUM INTYBUS	S	.	0	26.8	21.4	.	.
1662	CICHORIUM INTYBUS	S	.	0	23.2	23.2	.	.
1663	CICHORIUM INTYBUS	S	.	0	21.8	28.1	.	.
1664	CICHORIUM INTYBUS	W	.	0	14.4	2.2	65.4	15.3
1665	CICHORIUM INTYBUS	L	290	0	24.6	2.9	59.4	13.0
1666	CICUTA DOUGLASII	S	.	0	25.4	16.8	.	.
1667	CICUTA DOUGLASII	S	.	0	26.2	16.5	.	.
1668	CICUTA MACULATA	S	.	0	23.8	15.9	.	.
1669	CICUTA MEXICANA	S	.	0	16.2	32.6	.	.
1670	CIMICIFUGA RACEMOSA	S	.	0	23.8	45.1	.	.
1671	CINCHONA SP.	S	.	0	18.4	16.0	.	.
1672	CINNAMOMUM CASSIA & CINNAMOMUM VERUM	E	288	0	4.3	3.5	88.2	27.0
1673	CINNAMOMUM VERUM	E	346	0	3.5	1.4	91.4	.
1674	CIRCAEA QUADRISULCATA	S	.	0	17.5	18.5	.	.
1675	CIRSIUM ACARNA	S	.	0	18.5	29.2	.	.
1676	CIRSIUM ALTISSIMUM	S	.	0	19.1	22.2	.	.
1677	CIRSIUM ARVENSE	S	.	0	17.6	15.7	.	.

OBS	ASH	CALCIUM	PHOSPHO	IRON	SODIUM	POTASS	CAROT	THIAMINE	RIBOFLAV	NIACIN	ASCORBIC	REFCO
1654	3.4	168.0	370.7	7.7	29.1	892.5	33.6	0.35	0.17	2.24	.	4
1655	16.7	833.3	500.0	16.7	.	.	.	.	.	.	.	1
1656	13.0	1144.9	.	24.6	.	.	28695.6	1.01	1.74	5.80	159.4	3
1657	14.5	1173.9	782.6	24.6	202.9	4260.9	28695.6	1.01	2.03	7.25	144.9	4
1658	12.2	367.3	428.6	10.2	142.9	3714.3	0.0	.	.	.	.	4
1659	18.1	1194.4	555.6	12.5	.	5833.3	33333.3	0.83	1.39	6.94	305.6	4
1660	.	.	.	.	.	.	.	.	.	.	.	20
1661	.	.	.	.	.	.	.	.	.	.	.	20
1662	.	.	.	.	.	.	.	.	.	.	.	21
1663	.	.	.	.	.	.	.	.	.	.	.	21
1664	18.0	1890.0	220.0	.	.	.	.	.	.	.	.	6
1665	13.0	1144.9	.	24.6	.	.	22898.5	1.01	1.74	5.80	159.4	3
1666	.	.	.	.	.	.	.	.	.	.	.	20
1667	.	.	.	.	.	.	.	.	.	.	.	20
1668	.	.	.	.	.	.	.	.	.	.	.	20
1669	.	.	.	.	.	.	.	.	.	.	.	20
1670	.	.	.	.	.	.	.	.	.	.	.	19
1671	6.1	.	.	.	.	.	.	.	.	.	.	19
1672	4.0	1356.9	67.4	42.1	28.7	552.5	172.4	0.09	0.15	1.44	30.9	16
1673	3.7	532.9	.	.	.	.	.	0.02	0.08	0.79	.	1
1674	.	.	.	.	.	.	.	.	.	.	.	19
1675	.	.	.	.	.	.	.	.	.	.	.	21
1676	3.4	.	.	.	.	.	.	.	.	.	.	19
1677	.	.	.	.	.	.	.	.	.	.	.	21

TABLE 3 - DATA CONVERTED TO A ZERO-MOISTURE BASIS.

OBS	PLNA	PART	CALORIES	WATER	PROTEIN	FAT	TOTLCARB	FIBER	ASH	CALCIUM	PHOSPHO	IRON	SODIUM	POTASS	CAROT	THIAMINE	RIBOFLAV	NIACIN	ASCORBIC	REFCO
1678	CIRSIUM CAROLINIANUM	S	.	0	15.9	17.4	.	.	.	.	.	.	.	.	.	.	.	.	.	20
1679	CIRSIUM DIPSACOLEPSII	L	330	0	15.7	0.5	80.1	7.9	3.7	104.7	413.6	44.5	.	.	0.0	0.10	0.31	1.05	0.0	2
1680	CIRSIUM MEGACEPHALUM	S	.	0	16.8	12.3	.	.	.	.	.	.	.	.	.	.	.	.	.	20
1681	CIRSIUM SP.	W	.	0	24.9	2.7	57.6	14.4	14.8	4260.0	210.0	.	.	3180.0	2.0	.	.	.	.	6
1682	CISSUS POPULNEA	G	360	0	6.5	9.0	76.0	16.0	8.5	1960.0	.	.	.	.	.	.	.	.	.	1
1683	CISSUS STRIATA	SH	.	0	20.8	2.5	68.6	14.2	8.1	2500.0	350.0	.	.	.	.	.	.	.	.	18
1684	CISTUS LAURIFOLIUS	S	.	0	20.6	10.3	.	.	.	.	.	.	.	.	.	.	.	.	.	21
1685	CISTUS PARVIFLORUS	S	.	0	17.7	7.2	.	.	.	.	.	.	.	.	.	.	.	.	.	21
1686	CISTUS PARVIFLORUS	S	.	0	21.6	7.4	.	.	.	.	.	.	.	.	.	.	.	.	.	21
1687	CISTUS SALVIFOLIUS	S	.	0	13.3	7.2	.	.	.	.	.	.	.	.	.	.	.	.	.	21
1688	CISTUS VILLOSUS	S	.	0	18.9	7.4	.	.	.	.	.	.	.	.	.	.	.	.	.	21
1689	CITRULLUS COLOCYNTHIS	S	.	0	12.2	18.5	.	.	.	.	.	.	.	.	.	.	.	.	.	21
1690	CITRULLUS COLOCYNTHIS	S	.	0	16.6	22.7	.	.	.	.	.	.	.	.	.	.	.	.	.	21
1691	CITRULLUS COLOCYNTHIS	S	596	0	25.3	50.6	20.9	1.6	3.2	49.3	621.7	.	.	.	.	0.12	.	.	.	1
1692	CITRULLUS LANATUS	SH	.	0	20.0	2.2	59.4	17.1	18.4	1960.0	460.0	.	.	.	.	.	.	.	.	18
1693	CITRULLUS LANATUS	F	.	0	10.0	8.9	73.3	25.7	7.8	340.0	290.0	.	.	.	.	.	.	.	.	18
1694	CITRULLUS LANATUS	S	.	0	24.4	35.4	36.0	31.6	4.2	.	.	.	.	.	.	.	.	.	.	18
1695	CITRULLUS LANATUS	F	344	0	7.8	1.6	85.9	6.3	4.7	125.0	140.6	4.7	.	.	3906.3	0.63	0.78	1.56	125.0	1
1696	CITRULLUS LANATUS	S	559	0	19.8	41.7	34.5	24.2	3.9	129.4	830.6	.	.	.	.	.	.	.	.	1
1697	CITRULLUS LANATUS	K	601	0	27.4	52.7	16.0	4.2	3.9	56.2	.	7.8	.	.	.	0.11	0.13	1.48	.	1
1698	CITRULLUS LANATUS	F	351	0	6.8	2.7	86.5	4.1	4.1	94.6	135.1	6.8	13.5	1351.4	4783.8	0.41	0.41	2.70	94.6	4
1699	CITRULLUS VULGARIS	S	.	0	37.5	51.2	.	.	3.7	.	.	.	.	.	.	.	.	.	.	19
1700	CITRULLUS VULGARIS	S	.	0	37.5	51.4	.	.	3.8	.	.	.	.	.	.	.	.	.	.	19
1701	CITRULLUS VULGARIS	S	.	0	38.1	51.8	.	.	3.6	.	.	.	.	.	.	.	.	.	.	19
1702	CITRULLUS VULGARIS	S	.	0	37.5	52.0	.	.	3.6	.	.	.	.	.	.	.	.	.	.	19
1703	CITRULLUS VULGARIS	S	.	0	12.3	17.5	.	.	1.4	.	.	.	.	.	.	.	.	.	.	19
1704	CITRULLUS VULGARIS	S	.	0	11.5	17.0	.	.	.	.	.	.	.	.	.	.	.	.	.	21
1705	CITRUS AURANTIFOLIA	F	305	0	5.7	7.6	82.9	6.7	3.8	181.0	200.0	6.7	.	.	95.2	0.29	0.19	2.86	428.6	1
1706	CITRUS AURANTIFOLIA	F	400	0	5.6	26.7	65.6	3.3	2.2	144.4	122.2	0.0	22.2	911.1	111.1	0.33	0.22	1.11	511.1	2
1707	CITRUS AURANTIFOLIA	F	356	0	4.4	15.6	77.8	3.3	2.2	266.7	155.6	4.4	.	.	111.1	0.33	0.22	2.22	444.4	3
1708	CITRUS AURANTIFOLIA	F	262	0	6.5	1.9	88.8	4.7	2.8	308.4	168.2	5.6	18.7	953.3	56.1	0.28	0.19	1.87	345.8	4
1709	CITRUS AURANTIUM	F	343	0	5.6	0.0	89.8	3.7	4.6	166.7	111.1	.	.	.	2685.2	.	.	.	444.4	1
1710	CITRUS AURANTIUM	F	352	0	5.6	0.8	89.6	16.0	4.0	336.0	160.0	3.2	.	.	560.0	0.56	0.24	2.40	344.0	2
1711	CITRUS AURANTIUM	F	350	0	4.9	0.7	90.9	6.3	3.5	300.7	118.9	4.2	.	.	279.7	0.42	0.28	2.10	293.7	3
1712	CITRUS GRANDIS	F	329	0	6.6	1.3	82.9	6.6	9.2	342.1	289.5	3.9	.	.	.	0.92	0.26	.	394.7	1
1713	CITRUS GRANDIS	F	351	0	6.3	2.7	85.6	3.6	5.4	243.2	198.2	4.5	9.0	2117.1	270.3	0.45	0.18	2.70	477.5	2
1714	CITRUS GRANDIS	F	351	0	6.2	2.1	87.6	8.2	4.1	268.0	268.0	5.2	.	.	0.0	0.41	0.21	2.06	360.8	3
1715	CITRUS HYSTRIX	F	413	0	6.5	14.1	76.1	4.3	3.3	250.0	87.0	6.5	.	.	.	0.43	0.22	1.09	467.4	2
1716	CITRUS IYO	F	351	0	5.3	2.6	87.7	3.5	4.4	184.2	149.1	2.6	.	.	263.2	0.79	0.53	7.02	263.2	2
1717	CITRUS JAPONICA	F	431	0	3.1	18.5	73.8	0.0	4.6	246.2	184.6	10.8	46.2	12307.7	0.0	0.15	0.15	3.08	661.5	2
1718	CITRUS LIMETTA	F	297	0	6.9	5.9	83.2	9.9	4.0	277.2	198.0	5.0	.	.	0.0	0.50	0.30	1.98	475.2	3
1719	CITRUS LIMON	F	284	0	6.9	3.9	86.3	8.8	2.9	245.1	176.5	4.9	.	.	.	.	0.20	.	.	1
1720	CITRUS LIMON	F	292	0	7.3	8.3	79.2	7.3	5.2	333.3	197.9	3.1	41.7	1427.1	156.3	0.42	0.21	1.04	520.8	2
1721	CITRUS LIMON	F	299	0	6.2	6.2	83.5	6.2	4.1	422.7	154.6	7.2	.	.	103.1	0.62	0.21	1.03	525.8	3
1722	CITRUS LIMON	F	273	0	11.1	3.0	82.8	4.0	3.0	262.6	161.6	6.1	20.2	1393.9	121.2	0.40	0.20	1.01	535.4	4

TABLE 3 - DATA CONVERTED TO A ZERO-MOISTURE BASIS.

OBS	PLNA	PART	CALORIES	WATER	PROTEIN	FAT	TOTLCARB	FIBER	ASH
1723	CITRUS MEDICA	L	308	0	17.6	0.0	74.2	.	8.2
1724	CITRUS MEDICA	F	354	0	5.3	0.9	90.3	12.4	3.5
1725	CITRUS PARADISI	F	340	0	8.0	1.0	86.0	6.0	5.0
1726	CITRUS PARADISI	F	342	0	5.3	0.9	88.6	4.4	5.3
1727	CITRUS PARADISI	F	352	0	5.6	1.9	88.9	1.9	3.7
1728	CITRUS PARADISI	F	353	0	4.3	0.9	91.4	1.7	3.4
1729	CITRUS POONENSIS	F	353	0	6.9	1.7	87.9	3.4	3.4
1730	CITRUS RETICULATA	F	353	0	5.8	2.2	88.5	4.3	3.6
1731	CITRUS RETICULATA	F	360	0	6.1	1.8	89.5	2.6	2.6
1732	CITRUS RETICULATA	F	352	0	5.7	1.6	89.3	3.3	3.3
1733	CITRUS RETICULATA	F	354	0	6.2	1.5	89.2	3.8	3.1
1734	CITRUS RETICULATA X C. SINENSIS	F	355	0	6.5	1.6	88.7	1.6	3.2
1735	CITRUS SINENSIS	F	358	0	5.0	3.3	87.5	5.0	4.2
1736	CITRUS SINENSIS	F	351	0	7.0	1.8	86.8	3.5	4.4
1737	CITRUS SINENSIS	F	341	0	6.5	1.6	85.4	3.3	6.5
1738	CITRUS SINENSIS	F	350	0	7.1	1.4	87.1	3.6	4.3
1739	CITRUS TANKAN	F	358	0	5.7	3.8	85.8	5.7	4.7
1740	CLADIUM JAMAICENSE	S	.	0	7.2	5.4	.	.	.
1741	CLARKIA AMOENA	S	.	0	31.2	34.6	.	.	.
1742	CLARKIA ELEGANS	S	.	0	28.7	35.5	.	.	7.0
1743	CLARKIA ELEGANS	S	.	0	28.8	36.7	.	.	7.9
1744	CLARKIA ELEGANS	S	.	0	29.4	37.1	.	.	.
1745	CLAUSENA LANSIUM	F	344	0	5.6	0.6	88.1	5.0	5.6
1746	CLEISTOPHOLIS SP	F	356	0	18.8	3.1	75.0	.	3.1

OBS	CALCIUM	PHOSPHO	IRON	SODIUM	POTASS	CAROT	THIAMINE	RIBOFLAV	NIACIN	ASCORBIC	REFCO
1723	186.8	423.1	24.2	.	.	65576.9	0.22	.	.	.	2
1724	371.7	177.0	3.5	.	.	88.5	0.53	0.35	1.77	283.2	3
1725	210.0	180.0	6.0	.	.	250.0	0.50	0.30	2.00	440.0	1
1726	105.3	193.0	8.8	17.5	1184.2	526.3	0.61	0.53	4.39	333.3	2
1727	166.7	194.4	4.6	.	.	185.2	0.46	0.19	1.85	398.1	3
1728	137.9	137.9	3.4	8.6	1163.8	413.8	0.34	0.17	1.72	327.6	4
1729	181.0	163.8	2.6	.	1439.7	2931.0	0.86	0.34	5.17	465.5	2
1730	273.4	122.3	7.9	.	.	1654.7	0.58	0.36	1.44	201.4	1
1731	228.1	122.8	1.8	8.8	1307.0	4078.9	0.79	0.35	3.51	368.4	2
1732	245.9	131.1	3.3	.	.	655.7	0.66	0.25	2.46	270.5	3
1733	307.7	138.5	3.1	15.4	969.2	1938.5	0.46	0.15	0.77	238.5	4
1734	298.4	129.0	1.6	.	.	2661.3	0.81	0.24	1.61	290.3	3
1735	233.3	141.7	0.8	.	.	625.0	0.17	0.25	1.67	383.3	1
1736	184.2	175.4	2.6	17.5	1421.1	1315.8	0.61	0.35	3.51	377.2	2
1737	276.4	162.6	5.7	.	.	650.4	0.73	0.24	1.63	479.7	3
1738	292.9	142.9	2.9	7.1	1428.6	857.1	0.71	0.29	2.86	357.1	4
1739	311.3	188.7	2.8	.	3047.2	7924.5	0.75	0.47	4.72	537.7	2
1740	.	.	.	.	.	.	.	.	.	.	20
1741	.	.	.	.	.	.	.	.	.	.	20
1742	.	.	.	.	.	.	.	.	.	.	19
1743	.	.	.	.	.	.	.	.	.	.	19
1744	.	.	.	.	.	.	.	.	.	.	20
1745	93.8	118.8	0.0	100.0	1756.3	0.0	0.13	0.69	20.63	925.0	2
1746	143.8	250.0	13.8	.	.	.	.	.	.	62.5	1

TABLE 3 - DATA CONVERTED TO A ZERO-MOISTURE BASIS.

OBS	PLNA	PART	CALORIES	WATER	PROTEIN	FAT	TOTCARB	FIBER	ASH	CALCIUM	PHOSPHO	IRON	SODIUM	POTASS	CAROTE	THIAMINE	RIBOFLAV	NIACIN	ASCORBIC	REFCO
1747	CLEMATIS DRUMMONDI	W	.	0	15.8	2.0	76.8	27.6	5.4	1470.0	210.0	.	.	1550.0	.	.	.	.	.	6
1748	CLEMATIS FLAMMULA	S	.	0	13.1	15.4	.	.	.	.	.	.	.	.	.	.	.	.	.	20
1749	CLEMATIS FLAMMULA	S	.	0	13.1	13.8	.	.	.	.	.	.	.	.	.	.	.	.	.	20
1750	CLEMATIS ORIENTALIS	S	.	0	22.8	25.3	.	.	.	.	.	.	.	.	.	.	.	.	.	21
1751	CLEMATIS ORIENTALIS	S	.	0	16.3	18.7	.	.	.	.	.	.	.	.	.	.	.	.	.	21
1752	CLEMATIS PANICULATA	S	.	0	19.4	17.9	.	.	.	.	.	.	.	.	.	.	.	.	.	20
1753	CLEMATIS RECTA	S	.	0	13.8	18.1	.	.	.	.	.	.	.	.	.	.	.	.	.	20
1754	CLEMATIS VIRGINIANA	S	.	0	20.6	14.0	.	.	4.6	.	.	.	.	.	.	.	.	.	.	19
1755	CLEMATIS VITALBA	S	.	0	15.1	14.0	.	.	.	.	.	.	.	.	.	.	.	.	.	21
1756	CLEMATIS VITICELLA	S	.	0	15.0	19.6	.	.	.	.	.	.	.	.	.	.	.	.	.	20
1757	CLEMATIS VITICELLA	S	.	0	15.0	16.7	.	.	.	.	.	.	.	.	.	.	.	.	.	20
1758	CLEMATIS VITICELLA	S	.	0	15.1	19.2	.	.	.	.	.	.	.	.	.	.	.	.	.	21
1759	CLEOME ARABICA	S	.	0	17.0	21.4	.	.	.	.	.	.	.	.	.	.	.	.	.	21
1760	CLEOME DIANDRA	S	.	0	20.8	27.8	.	.	.	.	.	.	.	.	.	.	.	.	.	21
1761	CLEOME PILOSA	S	.	0	18.8	30.1	.	.	.	.	.	.	.	.	.	.	.	.	.	21
1762	CLEOME SERRULATA	S	.	0	23.1	28.6	.	.	3.1	.	.	.	.	.	.	.	.	.	.	19
1763	CLEOME SERRULATA	S	.	0	20.8	33.6	.	.	2.7	.	.	.	.	.	.	.	.	.	.	19
1764	CLEOME SERRULATA	S	.	0	18.1	23.7	.	.	.	.	.	.	.	.	.	.	.	.	.	20
1765	CLEOME SONORAE	S	.	0	22.9	28.9	.	.	.	.	.	.	.	.	.	.	.	.	.	21
1766	CLEOME SP.	L	300	0	28.9	4.7	53.2	8.9	13.2	2389.5	310.5	14.2	.	.	.	.	.	.	.	1
1767	CLEOME SPINOSA	S	.	0	19.8	33.0	.	.	2.6	.	.	.	.	.	.	.	.	.	.	19
1768	CLEOME SPINOSA	S	.	0	19.4	35.7	.	.	2.6	.	.	.	.	.	.	.	.	.	.	19
1769	CLEOME SPINOSA	S	.	0	18.8	36.6	.	.	2.6	.	.	.	.	.	.	.	.	.	.	19
1770	CLEOME TENUIS	S	.	0	18.8	30.0	.	.	.	.	.	.	.	.	.	.	.	.	.	19
1771	CLEOME VISCOSA	S	.	0	14.8	23.4	.	.	.	.	.	.	.	.	.	.	.	.	.	21
1772	CLERODENDRUM INDICUM	L	305	0	21.2	0.0	70.6	.	8.2	197.0	316.0	.	.	.	29390	.	.	.	66.9	2
1773	CLERODENDRUM TRICHOTOMUM	S	.	0	25.0	60.7	.	.	.	.	.	.	.	.	.	.	.	.	.	20
1774	CLITORIA TERNATEA	SH	.	0	19.5	4.5	67.4	11.8	8.6	3290.0	170.0	.	.	.	.	.	.	.	.	18
1775	CLITORIA TERNATEA	S	.	0	46.9	12.1	.	.	4.3	.	.	.	.	.	.	.	.	.	.	19
1776	CLITORIA TERNATEA	S	.	0	46.3	12.3	.	.	4.7	.	.	.	.	.	.	.	.	.	.	19
1777	CLITORIA TERNATEA	S	.	0	46.9	.	.	.	.	.	.	.	.	.	.	.	.	.	.	20
1778	CLITORIA TERNATEA	F	330	0	19.0	2.0	75.0	24.0	4.0	200.0	120.0	2.0	20.0	1545.0	3350	0.20	0.90	7.00	1240.0	2
1779	CLUTIA AFFINIS	S	.	0	23.2	41.7	.	.	.	.	.	.	.	.	.	.	.	.	.	21
1780	CNICUS BENEDICTUS	S	.	0	17.5	27.1	.	.	4.1	.	.	.	.	.	.	.	.	.	.	19
1781	CNICUS BENEDICTUS	S	.	0	13.8	25.8	.	.	.	.	.	.	.	.	.	.	.	.	.	21
1782	CNIDOSCOLUS ACONITIFOLIUM	L	317	0	30.7	6.4	53.0	9.9	9.9	1158.4	376.2	13.9	.	.	23465	0.99	1.98	7.92	960.4	3
1783	CNIDOSCOLUS ANGUSTIDENS	S	.	0	42.5	43.7	.	.	.	.	.	.	.	.	.	.	.	.	.	19
1784	CNIDOSCOLUS ELASTICUS	S	.	0	12.7	26.2	.	.	.	.	.	.	.	.	.	.	.	.	.	21
1785	CNIDOSCOLUS SP.	S	.	0	19.4	30.8	.	.	.	.	.	.	.	.	.	.	.	.	.	19
1786	CNIDOSCOLUS TEPIQUENSIS	S	.	0	13.4	19.8	.	.	.	.	.	.	.	.	.	.	.	.	.	21
1787	COBAEA SCANDENS	S	.	0	18.2	22.1	.	.	3.6	.	.	.	.	.	.	.	.	.	.	19
1788	COCCINEA CORDIFOLIA	L	298	0	43.6	4.3	44.7	10.6	7.4	1340.4	319.1	48.9	.	.	115372	1.81	1.38	40.43	510.6	2
1789	COCCINEA CORDIFOLIA	G	.	0	17.4	1.4	73.9	23.2	7.2	579.7	434.8	20.3	.	.	2261	.	.	.	405.8	5
1790	COCCOLOBA BARBADENSIS	S	.	0	4.8	1.2	.	.	.	.	.	.	.	.	.	.	.	.	.	21
1791	COCCOLOBA UVIFERA	F	349	0	6.7	1.4	87.1	10.0	4.8	258.4	157.9	5.7	.	.	0	0.14	0.14	2.87	81.3	3

TABLE 3 - DATA CONVERTED TO A ZERO-MOISTURE BASIS.

OBS	PLNA	PART	CALORIES	WATER	PROTEIN	FAT	TOTCARB	FIBER	ASH	CALCIUM	PHOSPHO	IRON	SODIUM	POTASS	CAROT	THIAMINE	RIBOFLAV	NIACIN	ASCORBIC	REFCO
1792	COCCULUS CAROLINUS	S	.	0	16.9	18.4	.	.	.	.	.	.	.	.	.	.	.	.	.	19
1793	COCHLOSPERMUM RELIGIOSUM	S	.	0	23.0	15.6	55.5	16.1	5.7	.	.	.	.	.	.	.	.	.	.	5
1794	COCHLOSPERMUM VITIFOLIUM	S	.	0	14.6	15.6	.	.	5.2	.	.	.	.	.	.	.	.	.	.	19
1795	COCOS NUCIFERA	S	676	0	6.3	67.9	24.0	11.5	1.7	36.6	179.4	4.4	.	.	43.5	0.05	0.05	1.05	3.5	1
1796	COCOS NUCIFERA	IS	625	0	6.3	54.4	36.6	11.6	2.8	.	.	.	.	.	.	.	.	.	.	1
1797	COCOS NUCIFERA	S	646	0	6.6	58.4	33.1	6.2	1.9	47.6	231.9	5.2	14.5	1149.1	0.0	0.10	0.06	1.24	6.2	2
1798	COCOS NUCIFERA	IS	481	0	8.8	22.5	64.4	2.5	4.4	262.5	350.0	6.3	318.8	1606.3	0.0	0.25	0.19	5.00	37.5	2
1799	COCOS NUCIFERA	SH	353	0	13.2	9.6	66.9	7.4	10.3	198.5	441.2	3.7	169.1	4889.7	.	0.22	0.29	8.82	66.2	2
1800	COCOS NUCIFERA	S	652	0	7.7	59.9	30.2	8.4	2.2	28.6	182.8	4.0	.	.	0.0	0.09	0.07	1.32	8.8	3
1801	COCOS NUCIFERA	IS	656	0	10.2	64.0	21.5	3.8	4.3	59.1	225.8	5.9	.	.	0.0	0.27	0.16	4.30	37.6	3
1802	COCOS NUCIFERA	S	705	0	7.1	71.9	19.1	8.1	1.8	26.5	193.5	3.5	46.8	521.4	0.0	0.10	0.04	1.02	6.1	4
1803	COCOS NUCIFERA	Z	.	0	.	.	.	.	5.9	304.4	191.7	.	.	293.1	.	.	.	.	.	5
1804	CODON ROYENII	S	.	0	14.2	8.4	.	.	.	.	.	.	.	.	.	.	.	.	.	21
1805	CODONOPSIS LANCEOLATA	L	416	0	12.9	19.7	61.2	36.0	6.2	505.6	679.8	11.8	.	.	0.0	0.67	1.24	4.49	0.0	2
1806	COFFEA ARABICA	L	.	0	9.9	5.9	71.2	18.7	13.0	.	.	.	.	.	.	.	.	.	.	18
1807	COFFEA ARABICA	L	327	0	9.9	5.9	71.2	18.7	13.0	2040.6	181.6	103.2	.	.	2521.4	0.00	0.22	5.56	.	1
1808	COFFEA ARABICA	S	217	0	12.5	11.5	72.8	24.4	3.2	128.1	190.0	3.1	.	.	21.3	0.23	0.06	1.39	.	3
1809	COIX LACRYMA-JOBI	W	.	0	8.5	2.8	79.8	27.9	8.9	.	.	.	.	.	.	.	.	.	.	18
1810	COIX LACRYMA-JOBI	S	.	0	8.1	3.6	82.4	17.0	5.9	.	.	.	.	.	.	.	.	.	.	18
1811	COIX LACRYMA-JOBI	S	.	0	21.9	7.9	.	.	1.8	.	.	.	.	.	.	.	.	.	.	19
1812	COIX LACRYMA-JOBI	S	.	0	2.0	6.4	.	.	5.8	.	.	.	.	.	.	.	.	.	.	19
1813	COIX LACRYMA-JOBI	S	.	0	26.2	5.0	.	.	.	.	.	.	.	.	.	.	.	.	.	21
1814	COIX LACRYMA-JOBI	S	366	0	11.4	5.8	73.0	11.5	9.8	.	.	.	.	.	.	.	.	.	.	1
1815	COIX LACRYMA-JOBI	S	428	0	17.3	7.0	73.5	0.9	2.1	28.2	489.9	5.6	.	.	0.0	0.32	0.21	4.84	0.0	3
1816	COLA ACUMINATA	F	399	0	5.9	1.1	90.8	3.8	2.2	156.3	231.8	5.4	.	.	67.4	0.08	0.08	1.62	145.6	1
1817	COLA ACUMINATA	A	371	0	9.0	3.6	86.2	4.8	1.2	18.0	101.8	8.4	.	.	179.6	0.06	0.30	4.19	59.9	3
1818	COLA CORDIFOLIA	P	347	0	4.6	0.0	92.4	1.5	3.1	209.9	179.4	9.5	.	.	1755.7	0.04	.	1.53	38.2	1
1819	COLCHICUM BORNMULLERI	S	.	0	8.9	1.1	.	.	.	.	.	.	.	.	.	.	.	.	.	21
1820	COLEUS AMBOINICUS	L	304	0	16.1	7.1	55.4	8.9	21.4	4142.9	142.9	69.6	.	.	17500.0	0.54	1.25	3.57	214.3	3
1821	COLLADONIA CRENATA	S	.	0	22.9	6.0	.	.	.	.	.	.	.	.	.	.	.	.	.	21
1822	COLLADONIA CRENATA	S	.	0	30.7	13.1	.	.	.	.	.	.	.	.	.	.	.	.	.	21
1823	COLLINSIA HETEROPHYLLA	S	.	0	16.9	8.2	.	.	3.9	.	.	.	.	.	.	.	.	.	.	19
1824	COLLINSIA HETEROPHYLLA	S	.	0	13.2	9.5	.	.	.	.	.	.	.	.	.	.	.	.	.	21
1825	COLLINSIA PARVIFLORA	S	.	0	14.7	13.0	.	.	.	.	.	.	.	.	.	.	.	.	.	21
1826	COLLINSIA TINCTORIA	S	.	0	15.7	5.6	.	.	.	.	.	.	.	.	.	.	.	.	.	21
1827	COLOCASIA ANTIQUORUM	L	281	0	26.8	8.0	41.6	14.6	15.5	993.0	401.0	.	11.0	213.0	.	.	.	7.12	.	12
1828	COLOCASIA ESCULENTA	L	.	0	25.0	10.7	52.9	12.1	12.4	1740.0	580.0	.	.	.	.	.	.	.	.	18
1829	COLOCASIA ESCULENTA	R	.	0	9.5	0.1	86.3	1.7	4.1	.	.	.	.	.	.	.	.	.	.	18
1830	COLOCASIA ESCULENTA	R	379	0	6.7	0.4	88.5	3.7	4.5	189.6	327.1	4.5	.	.	0.0	0.37	0.11	2.97	29.7	1
1831	COLOCASIA ESCULENTA	L	310	0	24.0	6.0	57.0	15.0	13.0	980.0	490.0	20.0	.	.	18000.0	1.70	3.50	8.00	110.0	1
1832	COLOCASIA ESCULENTA	L	328	0	22.0	5.4	64.0	6.5	8.6	871.0	371.0	5.4	48.4	5177.4	29758.1	0.70	1.83	8.06	338.7	2
1833	COLOCASIA ESCULENTA	Z	329	0	6.8	2.7	79.5	12.3	11.0	671.2	342.5	12.3	54.8	4575.3	2465.8	0.27	0.55	5.48	178.1	2
1834	COLOCASIA ESCULENTA	R	382	0	8.9	1.6	85.4	3.3	4.1	138.2	252.0	4.9	40.7	1821.1	0.0	0.49	0.16	4.07	32.5	2
1835	COLOCASIA ESCULENTA	R	362	0	6.3	0.8	88.2	3.1	4.7	378.0	346.5	4.7	.	.	39.4	0.31	0.16	2.76	27.6	3
1836	COLOCASIA ESCULENTA	R	363	0	7.0	0.7	87.8	3.0	4.4	103.7	225.9	3.7	25.9	1903.7	74.1	0.48	0.15	4.07	14.8	4

TABLE 3 - DATA CONVERTED TO A ZERO-MOISTURE BASIS.

OBS	PLNA	PART	CALORIES	WATER	PROTEIN	FAT	TOTLCARB	FIBER	ASH
1837	COLOCASIA ESCULENTA	L	313	0	23.4	6.3	57.8	10.9	12.5
1838	COLOCASIA ESCULENTA	R	.	0	11.2	0.4	82.2	.	6.3
1839	COLOCASIA ESCULENTA	Z	.	0	4.5	4.5	71.2	9.1	18.2
1840	COLUBRINA TEXENSIS	S	.	0	20.0	17.9	.	.	.
1841	COLUBRINA TEXENSIS	W	.	0	19.5	4.5	67.3	11.8	8.7
1842	COLUTEA ARBORESCENS	S	.	0	40.6	3.0	.	.	4.7
1843	COLUTEA ARBORESCENS	S	.	0	40.5	4.1	.	.	.
1844	COLUTEA CILICICA	S	.	0	45.4	2.6	.	.	.
1845	COLUTEA ISTRIA	S	.	0	39.2	2.3	.	.	.
1846	COMANDRA PALLIDA	S	.	0	8.1	24.5	.	.	.
1847	COMBRETUM APICULATUM	L	.	0	12.2	3.8	74.0	21.6	10.0
1848	COMMELINA BENGHALENSIS	L	312	0	16.7	2.9	68.8	17.4	11.6
1849	CONDALIA LYCIOIDES VAR. CANESCENS	S	.	0	38.1	17.8	.	.	.
1850	CONDALIA OBTUSIFOLIA	W	.	0	22.5	2.4	67.4	10.1	7.7
1851	CONDALIA SP.	S	.	0	11.2	68.6	.	.	.
1852	CONIUM MACULATUM	S	.	0	29.2	16.4	.	.	.
1853	CONIUM MACULATUM	S	.	0	23.5	15.5	.	.	.
1854	CONIUM MACULATUM	S	.	0	22.5	17.7	.	.	.
1855	CONIUM MACULATUM	S	.	0	21.8	14.3	.	.	.
1856	CONIUM MACULATUM	S	.	0	20.6	17.5	.	.	.
1857	CONIUM MACULATUM	S	.	0	24.9	11.6	.	.	.
1858	CONIUM MACULATUM	S	.	0	23.7	17.9	.	.	.
1859	CONRINGIA ORIENTALIS	S	.	0	30.6	27.3	.	.	.
1860	CONRINGIA PLANISILIQUA	S	.	0	29.9	34.4	.	.	.

OBS	CALCIUM	PHOSPHO	IRON	SODIUM	POTASS	CAROT	THIAMINE	RIBOFLAV	NIACIN	ASCORBIC	REFCO
1837	593.8	460.9	7.8	.	.	.	.	.	.	242.2	4
1838	148.7	520.4	7.8	.	.	89.2	.	.	.	0.0	5
1839	909.1	303.0	7.6	.	.	.	.	.	.	.	5
1840	.	.	.	.	.	.	.	.	.	.	19
1841	3290.0	170.0	.	.	1280.0	.	.	.	.	.	6
1842	.	.	.	.	.	.	.	.	.	.	19
1843	.	.	.	.	.	.	.	.	.	.	21
1844	.	.	.	.	.	.	.	.	.	.	21
1845	.	.	.	.	.	.	.	.	.	.	21
1846	.	.	.	.	.	.	.	.	.	.	20
1847	2840.0	190.0	.	.	.	.	.	.	.	.	18
1848	1521.7	376.8	.	.	.	.	.	.	.	.	1
1849	.	.	.	.	.	.	.	.	.	.	19
1850	1500.0	220.0	.	.	2390.0	.	.	.	.	.	6
1851	.	.	.	.	.	.	.	.	.	.	20
1852	.	.	.	.	.	.	.	.	.	.	20
1853	.	.	.	.	.	.	.	.	.	.	21
1854	.	.	.	.	.	.	.	.	.	.	21
1855	.	.	.	.	.	.	.	.	.	.	21
1856	.	.	.	.	.	.	.	.	.	.	21
1857	.	.	.	.	.	.	.	.	.	.	21
1858	.	.	.	.	.	.	.	.	.	.	21
1859	.	.	.	.	.	.	.	.	.	.	20
1860	.	.	.	.	.	.	.	.	.	.	20

TABLE 3 - DATA CONVERTED TO A ZERO-MOISTURE BASIS.

OBS	PLNA	PART	CALORIES	WATER	PROTEIN	FAT	TOTLCARB	FIBER	ASH
1861	CONSOLIDA ORIENTALIS	S	.	0	21.1	45.7	.	.	.
1862	CONVOLVULUS ARVENSIS	S	.	0	22.0	8.8	.	.	.
1863	CONVOLVULUS INCANUS	S	.	0	20.1	6.8	.	.	.
1864	CONVOLVULUS SECUNDUS	S	.	0	24.2	8.0	.	.	.
1865	CONVOLVULUS TRICOLOR	S	.	0	22.2	7.7	.	.	5.7
1866	CONVOLVULUS TRICOLOR	S	.	0	16.9	10.7	.	.	.
1867	CONYZA BONARIENSIS	S	.	0	26.4	24.3	.	.	.
1868	CORCHORUS ACUTANGULUS	S	.	0	22.6	12.8	.	.	.
1869	CORCHORUS CAPSULARIS	S	.	0	21.8	11.7	.	.	7.1
1870	CORCHORUS CLITORIUS	S	.	0	21.0	14.1	.	.	7.8
1871	CORCHORUS OLITORIUS	L	296	0	23.0	1.5	63.3	10.2	12.2
1872	CORCHORUS OLITORIUS	L	270	0	35.2	1.9	47.8	10.7	15.1
1873	CORCHORUS TRIDENS	L	308	0	23.6	3.4	62.3	8.4	10.8
1874	CORCHORUS TRILOCULARIS	L	318	0	30.4	4.8	57.2	7.9	7.7
1875	CORDEAUXIA EDULIS	S	448	0	12.1	13.5	71.9	1.6	2.5
1876	CORDIA ALBA	F	338	0	9.0	3.0	78.9	0.8	9.0
1877	CORDIA MYXA	L	.	0	15.8	6.8	64.3	14.7	13.1
1878	CORDIA MYXA	S	533	0	36.9	39.2	19.0	2.1	5.0
1879	CORDIA MYXA	L	181	0	14.2	0.8	38.9	10.0	46.1
1880	CORDYLA PINNATA	G	351	0	12.0	1.0	83.8	7.9	3.1
1881	CORDYLA PINNATA	F	350	0	7.0	0.5	89.0	5.5	3.5
1882	CORDYLA PINNATA	F	351	0	12.0	1.0	83.8	7.9	3.1
1883	CORDYLINE AUSTRALIS	S	.	0	19.9	48.3	.	.	2.8
1884	CORDYLINE INDIVISA	S	.	0	15.6	29.1	.	.	.

OBS	CALCIUM	PHOSPHO	IRON	SODIUM	POTASS	CAROT	THIAMINE	RIBOFLAV	NIACIN	ASCORBIC	REFCO
1861	.	.	.	.	.	.	.	.	.	.	21
1862	.	.	.	.	.	.	.	.	.	.	21
1863	.	.	.	.	.	.	.	.	.	.	20
1864	.	.	.	.	.	.	.	.	.	.	21
1865	.	.	.	.	.	.	.	.	.	.	19
1866	.	.	.	.	.	.	.	.	.	.	20
1867	.	.	.	.	.	.	.	.	.	.	21
1868	.	.	.	.	.	.	.	.	.	.	21
1869	.	.	.	.	.	.	.	.	.	.	19
1870	.	.	.	.	.	.	.	.	.	.	19
1871	1836.7	622.4	36.7	.	.	32704.1	0.77	2.70	6.12	408.2	1
1872	1673.0	610.1	48.4	75.5	2792.5	49371.1	0.82	1.64	6.92	333.3	2
1873	1912.0	308.0	.	14.0	198.0	.	.	.	5.31	.	12
1874	1428.0	292.0	.	135.0	270.0	.	.	.	7.00	.	12
1875	36.0	208.1	7.2	.	.	.	.	.	.	.	1
1876	52.6	308.3	9.8	.	.	75.2	0.00	0.30	3.76	157.9	3
1877	2560.0	220.0	.	.	.	.	.	.	.	.	18
1878	.	.	.	.	.	.	.	.	.	.	1
1879	225.0	186.1	20.0	.	.	916.7	0.28	0.81	0.56	0.0	2
1880	120.4	659.7	.	.	.	.	.	.	.	486.9	1
1881	145.0	710.0	9.0	.	.	1550.0	0.10	.	4.00	370.0	1
1882	120.4	659.7	.	.	.	.	.	.	.	486.9	1
1883	.	.	.	.	.	.	.	.	.	.	19
1884	.	.	.	.	.	.	.	.	.	.	20

TABLE 3 - DATA CONVERTED TO A ZERO-MOISTURE BASIS.

OBS	PLNA	PART	CALORIES	WATER	PROTEIN	FAT	TOTLCARB	FIBER	ASH
1885	CORDYLINE TERMINALIS	L	.	0	13.9	3.1	73.9	30.9	9.1
1886	COREOPSIS BASILIS	S	.	0	14.4	21.6	.	.	.
1887	COREOPSIS GRANDIFLORA	S	.	0	12.5	13.2	.	.	.
1888	COREOPSIS LANCEOLATA	S	.	0	14.6	14.6	.	.	4.1
1889	COREOPSIS TINCTORIA	S	.	0	20.6	24.2	.	.	4.7
1890	CORIANDRUM SATIVUM	S	.	0	13.8	22.2	.	.	.
1891	CORIANDRUM SATIVUM	S	.	0	14.4	18.1	.	.	.
1892	CORIANDRUM SATIVUM	S	.	0	11.1	11.5	.	.	.
1893	CORIANDRUM SATIVUM	L	301	0	23.5	5.2	56.2	11.2	15.1
1894	CORIANDRUM SATIVUM	S	327	0	13.6	19.5	60.4	31.9	6.6
1895	CORIANDRUM SATIVUM	L	298	0	21.0	4.8	58.9	12.9	15.3
1896	CORIANDRUM SATIVUM	L	300	0	23.6	5.0	57.1	12.1	14.3
1897	CORIANDRUM SATIVUM	F	.	0	15.9	18.1	61.0	36.7	5.0
1898	CORNUS FLORIDA	S	.	0	34.9	49.4	.	.	3.7
1899	CORNUS FLORIDA	S	.	0	2.3	2.5	.	.	3.6
1900	CORNUS FLORIDA	S	.	0	38.8	43.8	.	.	3.6
1901	CORNUS FLORIDA	S	.	0	2.6	2.4	.	.	2.8
1902	CORNUS FLORIDA	S	.	0	35.3	43.3	.	.	3.6
1903	CORNUS FLORIDA	L	.	0	12.1	5.4	73.3	16.0	9.2
1904	CORNUS MAS	S	.	0	31.1	49.5	.	.	3.1
1905	CORNUS MAS	S	.	0	1.8	2.3	.	.	1.0
1906	CORNUS MAS	S	.	0	4.1	7.3	.	.	.
1907	CORNUS MAS	S	.	0	3.8	4.5	.	.	.
1908	CORNUS SANGUINEA SUBSP. AUSTRALIS	S	.	0	6.9	22.2	.	.	.

OBS	CALCIUM	PHOSPHO	IRON	SODIUM	POTASS	CAROT	THIAMINE	RIBOFLAV	NIACIN	ASCORBIC	REFCO
1885	.	.	.	.	.	.	.	.	.	.	18
1886	.	.	.	.	.	.	.	.	.	.	20
1887	.	.	.	.	.	.	.	.	.	.	20
1888	.	.	.	.	.	.	.	.	.	.	19
1889	.	.	.	.	.	.	.	.	.	.	19
1890	.	.	.	.	.	.	.	.	.	.	20
1891	.	.	.	.	.	.	.	.	.	.	20
1892	.	.	.	.	.	.	.	.	.	.	21
1893	1344.1	518.9	45.8	227.6	4817.7	.	1.35	1.62	11.54	611.7	16
1894	778.3	449.0	17.9	38.4	1390.8	0.0	0.26	0.32	2.34	.	16
1895	1072.6	645.2	36.3	758.1	4516.1	23064.5	0.89	1.21	10.48	629.0	2
1896	1342.9	514.3	21.4	.	.	22857.1	1.07	2.00	11.43	535.7	3
1897	709.5	416.7	20.2	.	.	5855.9	.	.	.	281.5	5
1898	.	.	.	.	.	.	.	.	.	.	19
1899	.	.	.	.	.	.	.	.	.	.	19
1900	.	.	.	.	.	.	.	.	.	.	19
1901	.	.	.	.	.	.	.	.	.	.	19
1902	.	.	.	.	.	.	.	.	.	.	19
1903	2590.0	150.0	.	.	.	.	.	.	.	.	6
1904	.	.	.	.	.	.	.	.	.	.	19
1905	.	.	.	.	.	.	.	.	.	.	19
1906	.	.	.	.	.	.	.	.	.	.	21
1907	.	.	.	.	.	.	.	.	.	.	21
1908	.	.	.	.	.	.	.	.	.	.	21

TABLE 3 - DATA CONVERTED TO A ZERO-MOISTURE BASIS.

OBS	PLNA	PART	CALORIES	WATER	PROTEIN	FAT	TOTLCARB	FIBER	ASH	CALCIUM	PHOSPHO	IRON	SODIUM	POTASS	CAROT	THIAMINE	RIBOFLAV	NIACIN	ASCORBIC	REFCO
1909	CORONILLA CRETICA	S	.	0	28.8	4.0	.	.	3.5	.	.	.	.	.	.	.	.	.	.	19
1910	CORONILLA CRETICA	S	.	0	24.9	3.9	.	.	3.8	.	.	.	.	.	.	.	.	.	.	19
1911	CORONILLA CRETICA	S	.	0	28.4	4.8	.	.	3.7	.	.	.	.	.	.	.	.	.	.	19
1912	CORONILLA EMEROIDES	S	.	0	36.9	7.2	.	.	.	.	.	.	.	.	.	.	.	.	.	21
1913	CORONILLA JUNCEA	S	.	0	17.4	6.3	.	.	.	.	.	.	.	.	.	.	.	.	.	21
1914	CORONILLA VARIA	S	.	0	31.5	8.3	.	.	3.5	.	.	.	.	.	.	.	.	.	.	19
1915	CORONILLA VARIA	S	.	0	33.9	6.7	.	.	3.8	.	.	.	.	.	.	.	.	.	.	19
1916	CORONILLA VARIA	S	.	0	35.8	6.8	.	.	3.5	.	.	.	.	.	.	.	.	.	.	19
1917	CORONILLA VARIA	S	.	0	31.7	8.2	.	.	3.5	.	.	.	.	.	.	.	.	.	.	19
1918	CORONOPUS DIDYMUS	S	.	0	27.4	40.9	.	.	.	.	.	.	.	.	.	.	.	.	.	21
1919	CORYDALIS AUREA	L	.	0	38.9	3.4	43.2	9.3	14.5	2500.0	720.0	.	.	4250.0	.	.	.	.	.	6
1920	CORYLUS AVELLANA	S	620	0	21.0	54.3	22.9	5.9	1.8	.	.	.	.	.	.	.	.	.	.	1
1921	CORYLUS AVELLANA	S	634	0	16.4	58.5	21.4	3.3	3.7	200.9	462.2	4.5	.	1044.3	10.80	0.17	0.44	5.40	2.2	2
1922	CORYLUS SP.	S	670	0	11.2	65.4	20.5	2.4	2.9	262.9	330.2	3.7	.	.	132.51	0.48	0.57	5.18	7.2	3
1923	CORYLUS SP.	S	673	0	13.4	66.2	17.7	3.2	2.7	221.9	357.7	3.6	2.1	747.3	.	.	0.49	.	9.6	4
1924	CORYPHA UTAN	F	326	0	3.7	0.5	93.7	6.8	2.1	73.7	89.5	1.1	.	.	.	0.05	0.11	3.16	57.9	2
1925	COSMOS BIPINNATUS	S	.	0	19.0	20.2	.	.	4.1	.	.	.	.	.	.	.	.	.	.	19
1926	COSMOS BIPINNATUS	S	.	0	20.0	17.4	.	.	.	.	.	.	.	.	.	.	.	.	.	20
1927	COSMOS BIPINNATUS	S	.	0	15.6	21.4	.	.	.	.	.	.	.	.	.	.	.	.	.	20
1928	COSMOS SEEMANNII	S	.	0	21.2	11.6	.	.	.	.	.	.	.	.	.	.	.	.	.	20
1929	COSMOS SULPHUREUS	S	.	0	18.6	19.2	.	.	.	.	.	.	.	.	.	.	.	.	.	21
1930	COTONEASTER ACUMINATA	S	.	0	8.8	5.7	.	.	.	.	.	.	.	.	.	.	.	.	.	20
1931	COTONEASTER BACILLARIS	S	.	0	5.4	5.0	.	.	.	.	.	.	.	.	.	.	.	.	.	21
1932	COTONEASTER NUMMULARIA	S	.	0	6.3	5.3	.	.	.	.	.	.	.	.	.	.	.	.	.	21
1933	COUEPIA SP.	S	.	0	12.5	73.9	.	.	.	.	.	.	.	.	.	.	.	.	.	20
1934	COULA EDULIS	S	505	0	7.9	25.7	64.3	2.4	2.1	179.9	269.4	.	.	.	.	.	.	.	.	1
1935	COUMEROUNA ALATA	S	.	0	15.1	45.3	.	.	2.6	.	.	.	.	.	.	.	.	.	.	19
1936	COUMEROUNA ALATA	S	.	0	1.2	9.3	.	.	0.4	.	.	.	.	.	.	.	.	.	.	19
1937	COURSETIA GLANDULOSA	S	.	0	55.6	17.1	.	.	.	.	.	.	.	.	.	.	.	.	.	19
1938	COUSINIA BRACHYPTERA	S	.	0	25.4	17.5	.	.	.	.	.	.	.	.	.	.	.	.	.	21
1939	COWANIA STANSBURIANA	S	.	0	14.4	7.6	.	.	.	.	.	.	.	.	.	.	.	.	.	20
1940	CRAMBE ABYSSINICA	S	.	0	15.5	55.0	26.5	10.8	3.0	.	.	.	.	.	.	.	.	.	.	18
1941	CRAMBE ABYSSINICA	S	.	0	31.5	35.5	.	.	12.6	.	.	.	.	.	.	.	.	.	.	19
1942	CRAMBE ABYSSINICA	S	.	0	22.6	17.9	.	.	8.5	.	.	.	.	.	.	.	.	.	.	19
1943	CRAMBE ABYSSINICA	S	.	0	25.0	48.1	.	.	4.2	.	.	.	.	.	.	.	.	.	.	19
1944	CRAMBE ABYSSINICA	S	.	0	9.8	12.7	.	.	8.7	.	.	.	.	.	.	.	.	.	.	19
1945	CRAMBE ABYSSINICA	S	.	0	25.0	33.8	.	.	.	.	.	.	.	.	.	.	.	.	.	20
1946	CRAMBE ABYSSINICA	S	.	0	29.9	40.5	.	.	.	.	.	.	.	.	.	.	.	.	.	21
1947	CRAMBE ABYSSINICA	S	.	0	27.9	46.3	.	.	.	.	.	.	.	.	.	.	.	.	.	21
1948	CRAMBE ABYSSINICA	S	.	0	23.0	36.8	.	.	.	.	.	.	.	.	.	.	.	.	.	21
1949	CRAMBE ABYSSINICA	S	.	0	30.4	43.3	.	.	.	.	.	.	.	.	.	.	.	.	.	21
1950	CRAMBE ABYSSINICA	S	.	0	22.0	31.1	.	.	.	.	.	.	.	.	.	.	.	.	.	21
1951	CRAMBE ABYSSINICA	S	.	0	30.2	43.1	.	.	.	.	.	.	.	.	.	.	.	.	.	21
1952	CRAMBE ABYSSINICA	S	.	0	22.2	30.7	.	.	.	.	.	.	.	.	.	.	.	.	.	21
1953	CRAMBE ABYSSINICA	S	.	0	30.7	43.3	.	.	.	.	.	.	.	.	.	.	.	.	.	21

TABLE 3 - DATA CONVERTED TO A ZERO-MOISTURE BASIS.

OBS	PLNA	PART	CALORIES	WATER	PROTEIN	FAT	TOTLCARB	FIBER
1954	CRAMBE ABYSSINICA	S	.	0	22.0	30.8	.	.
1955	CRAMBE ABYSSINICA	S	.	0	30.4	43.9	.	.
1956	CRAMBE ABYSSINICA	S	.	0	23.3	30.7	.	.
1957	CRAMBE ABYSSINICA	S	.	0	31.1	42.5	.	.
1958	CRAMBE ABYSSINICA	S	.	0	22.6	30.3	.	.
1959	CRAMBE ABYSSINICA	L	291	0	38.2	5.5	41.8	14.5
1960	CRAMBE CORDIFOLIA	S	.	0	45.2	25.7	.	.
1961	CRAMBE CORDIFOLIA	S	.	0	29.4	16.4	.	.
1962	CRAMBE HISPANICA	S	.	0	29.2	44.8	.	.
1963	CRAMBE ORIENTALIS	S	.	0	31.0	43.1	.	.
1964	CRAMBE TATARICA	S	.	0	40.0	33.4	.	.
1965	CRAMBE TATARICA	S	.	0	40.8	33.3	.	.
1966	CRASSOCEPHALUM RUBENS	L	318	0	15.9	3.5	69.7	9.5
1967	CRATAEGUS AZAROLUS	S	.	0	3.2	0.6	.	.
1968	CRATAEGUS AZAROLUS	S	.	0	4.1	0.7	.	.
1969	CRATAEGUS CHRYSOCARPA	S	.	0	4.4	2.0	.	.
1970	CRATAEGUS CRUS-GALLI	S	.	0	41.9	35.4	.	.
1971	CRATAEGUS MOLLIS	S	.	0	46.9	31.3	.	.
1972	CRATAEGUS MOLLIS	S	.	0	1.9	0.8	.	.
1973	CRATAEGUS MONOGYNA	S	.	0	4.6	0.9	.	.
1974	CRATAEGUS MONOGYNA	S	.	0	5.1	2.4	.	.
1975	CRATAEGUS ORIENTALIS	S	.	0	5.1	3.7	.	.
1976	CRATAEGUS PINNATIFIDA	F	352	0	1.5	3.8	91.6	6.9
1977	CRATAEGUS PUBESCENS	F	353	0	2.8	1.2	93.3	9.5

OBS	ASH	CALCIUM	PHOSPHO	IRON	SODIUM	POTASS	CAROT	THIAMINE	RIBOFLAV	NIACIN	ASCORBIC	REFCO
1954	.	.	.	.	.	.	.	.	.	.	.	21
1955	.	.	.	.	.	.	.	.	.	.	.	21
1956	.	.	.	.	.	.	.	.	.	.	.	21
1957	.	.	.	.	.	.	.	.	.	.	.	21
1958	.	.	.	.	.	.	.	.	.	.	.	21
1959	14.5	1600.0	636.4	26.4	.	.	772.7	0.00	0.82	6.36	.	1
1960	.	.	.	.	.	.	.	.	.	.	.	21
1961	.	.	.	.	.	.	.	.	.	.	.	21
1962	.	.	.	.	.	.	.	.	.	.	.	21
1963	.	.	.	.	.	.	.	.	.	.	.	20
1964	.	.	.	.	.	.	.	.	.	.	.	20
1965	.	.	.	.	.	.	.	.	.	.	.	21
1966	10.9	1293.5	258.7	.	.	.	.	.	.	.	.	1
1967	.	.	.	.	.	.	.	.	.	.	.	21
1968	.	.	.	.	.	.	.	.	.	.	.	21
1969	.	.	.	.	.	.	.	.	.	.	.	21
1970	.	.	.	.	.	.	.	.	.	.	.	20
1971	5.1	.	.	.	.	.	.	.	.	.	.	19
1972	0.9	.	.	.	.	.	.	.	.	.	.	19
1973	.	.	.	.	.	.	.	.	.	.	.	21
1974	.	.	.	.	.	.	.	.	.	.	.	21
1975	.	.	.	.	.	.	.	.	.	.	.	21
1976	3.1	325.7	95.8	8.0	.	1252.9	.	.	.	.	.	2
1977	2.8	325.4	111.1	6.3	.	.	6627.0	0.12	0.16	1.59	313.5	3

TABLE 3 - DATA CONVERTED TO A ZERO-MOISTURE BASIS.

OBS	PLNA	PART	CALORIES	WATER	PROTEIN	FAT	TOTLCARB	FIBER
1978	CRATAEGUS SP.	S	.	0	44.2	39.2	.	.
1979	CRATEVA RELIGIOSA	L	292	0	24.6	0.4	64.2	15.4
1980	CRATEVA RELIGIOSA	L	350	0	12.8	4.9	77.4	18.4
1981	CREPIS ACUMINATA	W	.	0	14.1	.	.	.
1982	CREPIS FOETIDA	S	.	0	24.3	23.9	.	.
1983	CREPIS OCCIDENTALIS	S	.	0	.	22.4	.	.
1984	CREPIS RUBRA	S	.	0	.	22.2	.	.
1985	CRESCENTIA ALATA	S	549	0	31.3	41.1	23.7	2.5
1986	CROCUS SATIVUS	I	352	0	12.9	6.6	74.2	4.4
1987	CROTALARIA ALATA	S	.	0	28.2	2.8	.	.
1988	CROTALARIA ANAGYROIDES	SH	.	0	23.0	2.2	67.6	28.0
1989	CROTALARIA ANAGYROIDES	S	.	0	35.0	4.3	.	.
1990	CROTALARIA ANAGYROIDES	S	.	0	34.9	6.1	.	.
1991	CROTALARIA ARENARIA	SH	.	0	14.3	1.7	71.9	37.5
1992	CROTALARIA AUSTRALIS	S	.	0	23.5	2.0	.	.
1993	CROTALARIA BREVIDENS	SH	.	0	28.9	3.3	57.9	22.1
1994	CROTALARIA BREVIDENS VAR. INTERMEDIA	L	.	0	34.5	.	.	.
1995	CROTALARIA CAPENSIS	S	.	0	27.4	8.0	.	.
1996	CROTALARIA CF. INCANA	S	.	0	31.9	4.5	.	.
1997	CROTALARIA CF. LONGIROSTRATA	S	.	0	31.2	1.7	.	.
1998	CROTALARIA ERIOCARPA	S	.	0	33.8	4.4	.	.
1999	CROTALARIA FALCATA	S	.	0	27.8	2.2	.	.
2000	CROTALARIA GRANTIANA	S	.	0	30.3	1.9	.	.
2001	CROTALARIA GRANTIANA	S	.	0	30.4	.	.	.

OBS	ASH	CALCIUM	PHOSPHO	IRON	SODIUM	POTASS	CAROT	THIAMINE	RIBOFLAV	NIACIN	ASCORBIC	REFCO
1978	.	.	.	.	.	.	.	.	.	.	.	21
1979	10.8	1687.5	645.8	35.0	.	.	20666.7	1.63	2.96	5.83	800.0	1
1980	4.9	466.2	75.2	19.9	.	.	13721.8	0.30	0.94	5.64	18.8	2
1981	.	.	.	.	.	.	.	.	.	.	.	6
1982	.	.	.	.	.	.	.	.	.	.	.	21
1983	.	.	.	.	.	.	.	.	.	.	.	21
1984	.	.	.	.	.	.	.	.	.	.	.	21
1985	3.9	51.8	1002.1	9.7	.	.	20.7	0.76	0.12	0.93	.	3
1986	6.1	126.0	286.0	12.6	168	1956.9	.	.	.	.	.	16
1987	.	.	.	.	.	.	.	.	.	.	.	21
1988	7.2	570.0	280.0	.	.	.	.	.	.	.	.	18
1989	.	.	.	.	.	.	.	.	.	.	.	20
1990	.	.	.	.	.	.	.	.	.	.	.	21
1991	12.1	1230.0	220.0	.	.	.	.	.	.	.	.	18
1992	.	.	.	.	.	.	.	.	.	.	.	21
1993	9.9	440.0	270.0	.	.	.	.	.	.	.	.	18
1994	6.3	870.6	.	3.1	.	.	.	.	.	.	.	1
1995	.	.	.	.	.	.	.	.	.	.	.	21
1996	.	.	.	.	.	.	.	.	.	.	.	20
1997	.	.	.	.	.	.	.	.	.	.	.	20
1998	.	.	.	.	.	.	.	.	.	.	.	20
1999	.	.	.	.	.	.	.	.	.	.	.	21
2000	.	.	.	.	.	.	.	.	.	.	.	21
2001	.	.	.	.	.	.	.	.	.	.	.	21

TABLE 3 - DATA CONVERTED TO A ZERO-MOISTURE BASIS.

OBS	PLANA	PART	CALORIES	WATER	PROTEIN	FAT	TOTCARB	FIBER	ASH	CALCIUM	PHOSPHO	IRON	SODIUM	POTASS	CAROT	THIAMINE	RIBOFLAV	NIACIN	ASCORBIC	REFCO
2002	CROTALARIA INCANA	SH	.	0	23.8	2.9	63.4	27.1	9.9	810.0	260.0	.	.	.	.	.	.	.	.	18
2003	CROTALARIA INCANA	S	.	0	30.2	3.0	.	.	3.3	.	.	.	.	.	.	.	.	.	.	19
2004	CROTALARIA INCANA	S	.	0	29.5	2.7	.	.	3.4	.	.	.	.	.	.	.	.	.	.	19
2005	CROTALARIA INCANA	S	.	0	33.5	3.0	.	.	3.7	.	.	.	.	.	.	.	.	.	.	19
2006	CROTALARIA INCANA	S	.	0	27.9	.	.	.	.	.	.	.	.	.	.	.	.	.	.	21
2007	CROTALARIA INCANA	S	.	0	33.1	.	.	.	.	.	.	.	.	.	.	.	.	.	.	21
2008	CROTALARIA INCANA	S	.	0	28.6	.	.	.	.	.	.	.	.	.	.	.	.	.	.	21
2009	CROTALARIA INCANA	S	.	0	29.1	.	.	.	.	.	.	.	.	.	.	.	.	.	.	21
2010	CROTALARIA INTERMEDIA	S	.	0	30.5	0.4	.	.	3.8	.	.	.	.	.	.	.	.	.	.	19
2011	CROTALARIA INTERMEDIA	S	.	0	34.8	3.2	.	.	3.5	.	.	.	.	.	.	.	.	.	.	19
2012	CROTALARIA INTERMEDIA	S	.	0	30.6	2.7	.	.	.	.	.	.	.	.	.	.	.	.	.	19
2013	CROTALARIA INTERMEDIA	S	.	0	31.8	2.6	.	.	.	.	.	.	.	.	.	.	.	.	.	20
2014	CROTALARIA INTERMEDIA	S	.	0	33.5	3.2	.	.	.	.	.	.	.	.	.	.	.	.	.	20
2015	CROTALARIA INTERMEDIA	S	.	0	31.4	.	.	.	.	.	.	.	.	.	.	.	.	.	.	21
2016	CROTALARIA INTERMEDIA	S	.	0	31.8	.	.	.	.	.	.	.	.	.	.	.	.	.	.	21
2017	CROTALARIA INTERMEDIA	S	.	0	31.6	.	.	.	.	.	.	.	.	.	.	.	.	.	.	21
2018	CROTALARIA INTERMEDIA	W	.	0	14.5	2.1	78.3	40.3	5.1	1820.0	590.0	.	.	.	.	.	.	.	.	6
2019	CROTALARIA INTERMEDIA(BUD)	W	.	0	13.7	2.2	79.2	40.1	4.9	2930.0	810.0	.	.	.	.	.	.	.	.	6
2020	CROTALARIA INTERMEDIA(POD)	W	.	0	11.9	1.5	82.7	47.9	3.9	620.0	190.0	.	.	.	.	.	.	.	.	6
2021	CROTALARIA JUNCEA	W	.	0	15.0	5.8	71.0	30.0	8.2	2080.0	510.0	.	.	.	.	.	.	.	.	18
2022	CROTALARIA JUNCEA	H	.	0	12.5	1.0	81.0	45.7	5.5	.	.	.	.	.	.	.	.	.	.	18
2023	CROTALARIA JUNCEA	S	.	0	40.3	1.4	52.7	10.0	5.6	.	.	.	.	.	.	.	.	.	.	18
2024	CROTALARIA JUNCEA	S	.	0	41.1	3.5	.	.	4.6	.	.	.	.	.	.	.	.	.	.	19
2025	CROTALARIA JUNCEA	S	.	0	39.0	4.1	.	.	4.6	.	.	.	.	.	.	.	.	.	.	19
2026	CROTALARIA JUNCEA	S	.	0	39.8	3.7	.	.	4.7	.	.	.	.	.	.	.	.	.	.	19
2027	CROTALARIA JUNCEA	S	.	0	38.7	2.7	.	.	1.2	.	.	.	.	.	.	.	.	.	.	19
2028	CROTALARIA JUNCEA	S	.	0	37.8	3.8	.	.	4.5	.	.	.	.	.	.	.	.	.	.	19
2029	CROTALARIA JUNCEA	S	.	0	40.0	3.2	.	.	.	.	.	.	.	.	.	.	.	.	.	20
2030	CROTALARIA JUNCEA	S	.	0	24.7	.	.	.	.	.	.	.	.	.	.	.	.	.	.	21
2031	CROTALARIA JUNCEA	S	.	0	27.8	.	.	.	.	.	.	.	.	.	.	.	.	.	.	21
2032	CROTALARIA LABURNIFOLIA	S	.	0	31.4	3.8	.	.	.	.	.	.	.	.	.	.	.	.	.	21
2033	CROTALARIA LANCEOLATA	S	.	0	31.4	2.5	.	.	3.5	.	.	.	.	.	.	.	.	.	.	19
2034	CROTALARIA LANCEOLATA	S	.	0	30.9	.	.	.	.	.	.	.	.	.	.	.	.	.	.	21
2035	CROTALARIA LEIOLOBA	S	.	0	26.5	2.0	.	.	.	.	.	.	.	.	.	.	.	.	.	21
2036	CROTALARIA LONGIROSTRATA	S	.	0	33.9	3.4	.	.	.	.	.	.	.	.	.	.	.	.	.	21
2037	CROTALARIA LONGIROSTRATA	S	.	0	33.0	4.4	.	.	.	.	.	.	.	.	.	.	.	.	.	21
2038	CROTALARIA LONGIROSTRATA	L	304	0	38.0	4.3	49.5	10.9	8.2	1559.8	391.3	25.5	.	.	33315.2	7.23	2.66	10.87	543.5	3
2039	CROTALARIA MACROCALYX	W	.	0	15.1	2.4	76.3	37.7	6.2	560.0	120.0	.	.	.	.	.	.	.	.	18
2040	CROTALARIA MEDICAGINEA	S	.	0	35.2	3.1	.	.	.	.	.	.	.	.	.	.	.	.	.	21
2041	CROTALARIA MUCRONATA	S	.	0	30.5	2.7	.	.	3.5	.	.	.	.	.	.	.	.	.	.	19
2042	CROTALARIA MUCRONATA	S	.	0	25.0	2.1	.	.	.	.	.	.	.	.	.	.	.	.	.	20
2043	CROTALARIA MUCRONATA	S	.	0	33.5	1.9	.	.	.	.	.	.	.	.	.	.	.	.	.	20
2044	CROTALARIA MUCRONATA	S	.	0	28.0	.	.	.	.	.	.	.	.	.	.	.	.	.	.	21
2045	CROTALARIA MUCRONATA	S	.	0	27.1	.	.	.	.	.	.	.	.	.	.	.	.	.	.	21
2046	CROTALARIA MUCRONATA	S	.	0	24.1	.	.	.	.	.	.	.	.	.	.	.	.	.	.	21

TABLE 3 - DATA CONVERTED TO A ZERO-MOISTURE BASIS.

OBS	PLNA	PART	CALORIES	WATER	PROTEIN	FAT	TOTCARB	FIBER	ASH	CALCIUM	PHOSPHO	IRON	SODIUM	POTASS	CAROT	THIAMINE	RIBOFLAV	NIACIN	ASCORBIC	REFCO
2047	CROTALARIA MUCRONATA	S	.	0	24.4	.	.	.	.	.	.	.	.	.	.	.	.	.	.	21
2048	CROTALARIA MUCRONATA	S	.	0	23.8	.	.	.	.	.	.	.	.	.	.	.	.	.	.	21
2049	CROTALARIA MUCRONATA	S	.	0	28.0	.	.	.	.	.	.	.	.	.	.	.	.	.	.	21
2050	CROTALARIA MUCRONATA	S	.	0	29.5	.	.	.	.	.	.	.	.	.	.	.	.	.	.	21
2051	CROTALARIA MUCRONATA	S	.	0	24.6	.	.	.	.	.	.	.	.	.	.	.	.	.	.	21
2052	CROTALARIA MUCRONATA	S	.	0	31.5	2.9	.	.	.	.	.	.	.	.	.	.	.	.	.	21
2053	CROTALARIA OCHROLEUCA	S	.	0	31.0	2.6	.	.	.	.	.	.	.	.	.	.	.	.	.	21
2054	CROTALARIA PUMILA	S	.	0	24.4	3.2	.	.	4.1	.	.	.	.	.	.	.	.	.	.	19
2055	CROTALARIA PUMILA	S	.	0	31.8	4.4	.	.	3.3	.	.	.	.	.	.	.	.	.	.	19
2056	CROTALARIA PUMILA	S	.	0	36.2	3.0	.	.	.	.	.	.	.	.	.	.	.	.	.	20
2057	CROTALARIA PUMILA	S	.	0	31.9	2.9	.	.	.	.	.	.	.	.	.	.	.	.	.	20
2058	CROTALARIA PUMILA	S	.	0	31.8	3.4	.	.	.	.	.	.	.	.	.	.	.	.	.	21
2059	CROTALARIA PURPUREA	S	.	0	31.4	3.8	.	.	.	.	.	.	.	.	.	.	.	.	.	21
2060	CROTALARIA QUINQUEFOLIA	S	.	0	24.2	2.6	.	.	.	.	.	.	.	.	.	.	.	.	.	20
2061	CROTALARIA RETUSA	S	.	0	27.5	.	.	.	.	.	.	.	.	.	.	.	.	.	.	20
2062	CROTALARIA RETUSA	S	.	0	21.6	.	.	.	.	.	.	.	.	.	.	.	.	.	.	21
2063	CROTALARIA RETUSA	S	.	0	24.3	.	.	.	.	.	.	.	.	.	.	.	.	.	.	21
2064	CROTALARIA RETUSA	S	.	0	22.2	.	.	.	.	.	.	.	.	.	.	.	.	.	.	21
2065	CROTALARIA RHODESIAE	S	.	0	29.4	2.2	.	.	.	.	.	.	.	.	.	.	.	.	.	21
2066	CROTALARIA SAGITTALIS	S	.	0	33.8	4.5	.	.	.	.	.	.	.	.	.	.	.	.	.	19
2067	CROTALARIA SALTIANA	S	.	0	29.0	2.1	.	.	.	.	.	.	.	.	.	.	.	.	.	21
2068	CROTALARIA SERICAE	S	.	0	29.2	2.3	.	.	.	.	.	.	.	.	.	.	.	.	.	21
2069	CROTALARIA SP.	S	.	0	36.8	2.0	.	.	3.6	.	.	.	.	.	.	.	.	.	.	19
2070	CROTALARIA SP.	S	.	0	26.0	.	.	.	.	.	.	.	.	.	.	.	.	.	.	21
2071	CROTALARIA SP.	S	.	0	33.2	.	.	.	.	.	.	.	.	.	.	.	.	.	.	21
2072	CROTALARIA SP.	S	.	0	32.1	.	.	.	.	.	.	.	.	.	.	.	.	.	.	21
2073	CROTALARIA SP.	S	.	0	28.5	3.2	.	.	.	.	.	.	.	.	.	.	.	.	.	21
2074	CROTALARIA SPECTABILIS	S	.	0	30.8	2.6	.	.	3.9	.	.	.	.	.	.	.	.	.	.	19
2075	CROTALARIA SPECTABILIS	S	.	0	29.4	.	.	.	.	.	.	.	.	.	.	.	.	.	.	20
2076	CROTALARIA SPECTABILIS	S	.	0	32.5	2.3	.	.	.	.	.	.	.	.	.	.	.	.	.	20
2077	CROTALARIA SPECTABILIS	S	.	0	28.1	2.6	.	.	.	.	.	.	.	.	.	.	.	.	.	20
2078	CROTALARIA SPECTABILIS	S	.	0	30.2	3.4	.	.	.	.	.	.	.	.	.	.	.	.	.	20
2079	CROTALARIA SPECTABILIS	S	.	0	28.6	2.7	.	.	.	.	.	.	.	.	.	.	.	.	.	20
2080	CROTALARIA USARAMOENSIS	S	.	0	30.2	2.2	.	.	.	.	.	.	.	.	.	.	.	.	.	21
2081	CROTALARIA USARAMOENSIS	S	.	0	32.9	.	.	.	.	.	.	.	.	.	.	.	.	.	.	21
2082	CROTALARIA USARAMOENSIS	S	.	0	29.8	.	.	.	.	.	.	.	.	.	.	.	.	.	.	21
2083	CROTALARIA USARAMOENSIS	S	.	0	32.3	2.8	.	.	.	.	.	.	.	.	.	.	.	.	.	21
2084	CROTALARIA VERRUCOSA	S	.	0	33.3	3.0	.	.	.	.	.	.	.	.	.	.	.	.	.	21
2085	CROTALARIA VERRUCOSA	S	.	0	25.4	.	.	.	.	.	.	.	.	.	.	.	.	.	.	21
2086	CROTALARIA VITELLINA	S	.	0	32.5	3.7	.	.	.	.	.	.	.	.	.	.	.	.	.	21
2087	CROTALARIA VITELLINA	S	.	0	30.6	5.0	.	.	.	.	.	.	.	.	.	.	.	.	.	21
2088	CROTON CAPITATUS	S	.	0	20.9	14.8	.	.	2.9	.	.	.	.	.	.	.	.	.	.	19
2089	CROTON CORYMBULOSUS	S	.	0	22.9	23.6	.	.	.	.	.	.	.	.	.	.	.	.	.	21
2090	CROTON DICHOGAMUS	W	.	0	25.1	2.0	62.4	20.1	10.5	1860	490	.	.	.	.	.	.	.	.	18
2091	CROTON FRAGILIS	S	.	0	26.9	25.4	.	.	.	.	.	.	.	.	.	.	.	.	.	20

TABLE 3 - DATA CONVERTED TO A ZERO-MOISTURE BASIS.

OBS	PLNA	PART	CALORIES	WATER	PROTEIN	FAT	TOTLCARB	FIBER	ASH	CALCIUM	PHOSPHO	IRON	SODIUM	POTASS	CAROT	THIAMINE	RIBOFLAV	NIACIN	ASCORBIC	REFCO
2092	CROTON SP.	W	.	0	16.0	4.4	72.2	23.0	7.4	2120.0	160.0	.	.	1510.0	.	.	.	.	.	6
2093	CROTON SP.(FRUITING)	W	.	0	16.3	4.2	72.8	23.8	6.7	2290.0	180.0	.	.	1370.0	.	.	.	.	.	6
2094	CROTON SP.(IMMATURE)	W	.	0	19.0	3.5	69.6	20.5	7.9	1890.0	160.0	.	.	1750.0	.	.	.	.	.	6
2095	CROTON TEXENSIS	S	.	0	24.9	26.4	.	.	3.2	.	.	.	.	.	.	.	.	.	.	19
2096	CRUCIANELLA SP.	S	.	0	19.7	10.0	.	.	.	.	.	.	.	.	.	.	.	.	.	21
2097	CRYPTANTHA BRADBURIANA	S	.	0	21.1	27.2	.	.	.	.	.	.	.	.	.	.	.	.	.	21
2098	CRYPTANTHA SP.	W	.	0	7.5	.	.	27.7	13.3	2060.0	300.0	.	.	2520.0	.	.	.	.	.	6
2099	CRYPTOMERIA JAPONICA	S	.	0	10.6	10.6	.	.	2.3	.	.	.	.	.	.	.	.	.	.	19
2100	CRYPTOSTEMMA CALENDULACEUM	S	.	0	15.3	24.0	.	.	.	.	.	.	.	.	.	.	.	.	.	21
2101	CRYPTOTAENIA JAPONICA	L	277	0	30.8	1.5	52.3	20.0	15.4	1246.2	692.3	27.7	107.7	7538.5	12000.0	2.31	3.08	7.69	923.1	2
2102	CUCUMEROPSIS MANNII	S	555	0	16.9	40.2	39.4	16.8	3.5	149.7	719.7	.	.	.	.	.	.	.	.	1
2103	CUCUMIS ANGURIA	S	.	0	25.8	25.3	.	.	.	.	.	.	.	.	.	.	.	.	.	21
2104	CUCUMIS DIPSACEUS	L	347	0	9.9	6.9	75.2	0.0	7.9	138.6	247.5	9.9	.	.	.	0.30	0.50	.	207.9	1
2105	CUCUMIS MELO	SH	.	0	27.4	2.8	65.3	16.9	4.5	2980.0	680.0	.	.	.	.	.	.	.	.	18
2106	CUCUMIS MELO	F	.	0	21.2	3.0	66.8	21.2	9.0	450.0	300.0	.	.	.	.	.	.	.	.	18
2107	CUCUMIS MELO	S	.	0	35.8	44.6	.	.	5.6	.	.	.	.	.	.	.	.	.	.	5
2108	CUCUMIS MELO	F	346	0	9.6	1.9	80.8	9.6	7.7	384.6	538.5	32.7	.	.	576.9	0.77	0.38	11.54	230.8	1
2109	CUCUMIS MELO	L	273	0	28.0	1.3	54.0	.	16.7	.	.	.	.	.	.	.	.	.	.	1
2110	CUCUMIS MELO	F	295	0	22.7	2.3	63.6	13.6	11.4	681.8	318.2	9.1	454.5	.	0.0	0.45	0.45	4.55	318.2	2
2111	CUCUMIS MELO	F	353	0	4.4	2.9	85.3	5.9	7.4	117.6	279.4	4.4	.	441.2	25073.5	0.59	0.29	10.29	558.8	2
2112	CUCUMIS MELO	F	342	0	13.2	3.9	73.7	6.6	9.2	236.8	184.2	6.6	184.2	2723.7	0.0	0.66	0.26	5.26	328.9	2
2113	CUCUMIS MELO	F	342	0	13.2	3.9	73.7	6.6	9.2	236.8	184.2	6.6	184.2	2723.7	1184.2	0.66	0.26	5.26	328.9	2
2114	CUCUMIS MELO	F	347	0	6.9	1.4	86.1	6.9	5.6	208.3	208.3	16.7	.	.	9722.2	0.56	0.42	8.33	402.8	3
2115	CUCUMIS MELO	F	349	0	4.8	2.4	88.1	3.2	4.8	254.0	254.0	5.6	.	.	0.0	0.24	0.16	4.76	182.5	3
2116	CUCUMIS MELO	F	341	0	8.0	1.1	85.2	3.4	5.7	159.1	181.8	4.5	136.4	2852.3	23181.8	0.45	0.34	6.82	375.0	4
2117	CUCUMIS MELO	F	318	0	14.1	0.0	76.5	5.9	9.4	164.7	188.2	4.7	141.2	2952.9	211.8	0.47	0.35	7.06	152.9	4
2118	CUCUMIS MELO	F	351	0	8.5	3.2	81.9	6.4	6.4	148.9	170.2	4.3	127.7	2670.2	255.3	0.43	0.32	6.38	244.7	4
2119	CUCUMIS SATIVUS	S	.	0	27.5	33.0	.	.	.	.	.	.	.	.	.	.	.	.	.	20
2120	CUCUMIS SATIVUS	S	.	0	26.9	34.0	.	.	.	.	.	.	.	.	.	.	.	.	.	20
2121	CUCUMIS SATIVUS	S	.	0	26.2	27.2	.	.	.	.	.	.	.	.	.	.	.	.	.	20
2122	CUCUMIS SATIVUS	S	.	0	26.9	33.5	.	.	.	.	.	.	.	.	.	.	.	.	.	20
2123	CUCUMIS SATIVUS	S	.	0	31.2	33.0	.	.	.	.	.	.	.	.	.	.	.	.	.	20
2124	CUCUMIS SATIVUS	S	.	0	42.0	42.5	.	.	.	.	620.0	.	.	.	.	.	.	.	.	5
2125	CUCUMIS SATIVUS	F	306	0	16.3	2.0	69.4	16.3	12.2	265.3	612.2	10.2	.	.	0.0	0.41	0.20	6.12	285.7	1
2126	CUCUMIS SATIVUS	F	316	0	15.8	2.6	71.1	13.2	10.5	552.6	631.6	10.5	342.1	4052.6	2236.8	0.79	1.05	5.26	289.5	2
2127	CUCUMIS SATIVUS	F	326	0	15.2	2.2	73.9	8.7	8.7	347.8	521.7	13.0	.	.	217.4	0.65	0.87	4.35	304.3	3
2128	CUCUMIS SATIVUS	F	306	0	18.4	2.0	69.4	12.2	10.2	510.2	551.0	22.4	122.4	3265.3	3061.2	0.61	0.82	4.08	224.5	4
2129	CUCURBITA DIGITATA	S	.	0	27.5	22.8	.	.	4.3	.	.	.	.	.	.	.	.	.	.	19
2130	CUCURBITA DIGITATA	S	.	0	36.9	48.2	.	.	.	.	.	.	.	.	.	.	.	.	.	19
2131	CUCURBITA DIGITATA	S	.	0	22.5	27.5	.	.	.	.	.	.	.	.	.	.	.	.	.	19
2132	CUCURBITA FICIFOLIA	S	.	0	29.4	19.5	.	.	.	.	.	.	.	.	.	.	.	.	.	19
2133	CUCURBITA FICIFOLIA	G	329	0	11.0	1.4	82.2	4.1	5.5	164.4	561.6	2.7	.	.	0.0	0.82	0.55	9.59	630.1	3
2134	CUCURBITA FICIFOLIA	F	328	0	12.5	1.6	79.7	6.3	6.3	234.4	296.9	6.3	.	.	312.5	0.63	0.47	4.69	171.9	3
2135	CUCURBITA FOETIDISSIMA	S	.	0	32.5	33.9	.	.	.	.	.	.	.	.	.	.	.	.	.	19
2136	CUCURBITA FOETIDISSIMA	S	.	0	33.8	31.1	.	.	.	.	.	.	.	.	.	.	.	.	.	20

TABLE 3 - DATA CONVERTED TO A ZERO-MOISTURE BASIS.

OBS	PLNA	PART	CALORIES	WATER	PROTEIN	FAT	TOTCARB	FIBER	ASH	CALCIUM	PHOSPHO	IRON	SODIUM	POTASS	CAROT	THIAMINE	RIBOFLAV	NIACIN	ASCORBIC	REFCO
2137	CUCURBITA FOETIDISSIMA	L	.	0	25.3	4.0	49.2	10.2	21.5	7760.0	260.0	.	.	2530.0	.	.	.	.	.	6
2138	CUCURBITA LUNDELLIANA	S	.	0	28.8	23.0	.	.	.	.	.	.	.	.	.	.	.	.	.	20
2139	CUCURBITA MAXIMA	S	.	0	35.6	51.4	.	.	.	.	.	.	.	.	.	.	.	.	.	19
2140	CUCURBITA MAXIMA	S	.	0	39.4	50.7	.	.	.	.	.	.	.	.	.	.	.	.	.	19
2141	CUCURBITA MAXIMA	S	.	0	41.9	43.6	.	.	.	.	.	.	.	.	.	.	.	.	.	19
2142	CUCURBITA MAXIMA	S	.	0	38.8	51.0	.	.	.	.	.	.	.	.	.	.	.	.	.	19
2143	CUCURBITA MAXIMA	S	.	0	36.2	51.5	.	.	.	.	.	.	.	.	.	.	.	.	.	19
2144	CUCURBITA MAXIMA	S	.	0	38.1	51.9	.	.	.	.	.	.	.	.	.	.	.	.	.	19
2145	CUCURBITA MAXIMA	S	.	0	40.0	50.3	.	.	.	.	.	.	.	.	.	.	.	.	.	19
2146	CUCURBITA MAXIMA	L	250	0	37.0	1.9	40.7	22.2	20.4	4416.7	1259.3	7.4	.	.	33333.3	0.56	2.96	.	740.7	1
2147	CUCURBITA MAXIMA	L	318	0	25.5	8.2	52.7	12.7	13.6	872.7	745.5	15.5	54.5	4127.3	13272.7	1.18	1.64	10.00	154.5	2
2148	CUCURBITA MAXIMA	F	338	0	9.5	1.4	84.5	5.4	4.7	182.4	290.5	4.1	47.3	2371.6	9966.2	0.61	0.41	6.76	94.6	2
2149	CUCURBITA MAXIMA	L	288	0	25.0	3.8	55.8	11.5	15.4	1692.3	769.2	59.6	96.2	4615.4	2692.3	0.96	1.54	15.38	519.2	2
2150	CUCURBITA MAXIMA	G	333	0	13.9	2.8	76.4	5.6	6.9	263.9	444.4	8.3	.	.	416.7	0.69	0.56	6.94	263.9	3
2151	CUCURBITA MAXIMA	F	321	0	15.6	1.8	74.3	8.3	8.3	293.6	220.2	21.1	.	.	21009.2	0.64	0.46	7.34	100.9	3
2152	CUCURBITA MAXIMA	S	575	0	31.9	48.2	15.5	2.3	4.8	40.0	1118.8	9.7	.	.	31.5	0.24	0.17	3.05	0.0	3
2153	CUCURBITA MAXIMA	F	336	0	9.4	2.0	83.2	9.4	5.4	147.7	255.0	4.0	6.7	2476.5	14765.1	0.34	0.74	4.03	87.2	4
2154	CUCURBITA MAXIMA	F	.	0	18.9	1.4	71.6	.	8.1	135.1	405.4	9.5	.	.	675.7	.	.	.	27.0	5
2155	CUCURBITA MOSCHATA	S	.	0	40.0	47.7	.	.	.	.	.	.	.	.	.	.	.	.	.	19
2156	CUCURBITA MOSCHATA	S	.	0	40.0	48.7	.	.	.	.	.	.	.	.	.	.	.	.	.	19
2157	CUCURBITA MOSCHATA	S	.	0	41.2	48.1	.	.	.	.	.	.	.	.	.	.	.	.	.	19
2158	CUCURBITA MOSCHATA	S	.	0	36.9	49.4	.	.	.	.	.	.	.	.	.	.	.	.	.	19
2159	CUCURBITA MOSCHATA	S	.	0	38.1	52.9	.	.	.	.	.	.	.	.	.	.	.	.	.	19
2160	CUCURBITA MOSCHATA	S	.	0	36.9	51.1	.	.	.	.	.	.	.	.	.	.	.	.	.	19
2161	CUCURBITA MOSCHATA	F	326	0	12.8	2.3	76.7	14.0	9.3	383.7	558.1	19.8	.	.	3604.6	1.28	1.05	8.14	279.1	2
2162	CUCURBITA MOSCHATA	L	291	0	35.0	5.8	43.7	.	15.5	1339.8	961.2	35.9	.	.	16019.4	1.36	.	.	349.5	2
2163	CUCURBITA MOSCHATA	F	342	0	10.3	2.6	83.8	5.1	3.4	102.6	230.8	6.0	.	.	18034.2	0.43	0.34	5.13	359.0	3
2164	CUCURBITA MOSCHATA	S	504	0	33.0	30.2	33.4	5.3	3.4	37.9	785.5	1.2	.	.	0.0	0.14	0.13	2.38	0.0	3
2165	CUCURBITA OKEECHOBEENSIS	S	.	0	38.1	52.6	.	.	.	.	.	.	.	.	.	.	.	.	.	19
2166	CUCURBITA PALMERI	S	.	0	32.4	39.4	.	.	.	.	.	.	.	.	.	.	.	.	.	21
2167	CUCURBITA PALMERI	S	.	0	30.6	43.5	.	.	.	.	.	.	.	.	.	.	.	.	.	21
2168	CUCURBITA PEPO	S	.	0	38.1	48.1	.	.	.	.	.	.	.	.	.	.	.	.	.	19
2169	CUCURBITA PEPO	S	.	0	40.6	48.3	.	.	.	.	.	.	.	.	.	.	.	.	.	19
2170	CUCURBITA PEPO	S	.	0	39.4	49.8	.	.	.	.	.	.	.	.	.	.	.	.	.	19
2171	CUCURBITA PEPO	S	.	0	38.8	50.7	.	.	.	.	.	.	.	.	.	.	.	.	.	19
2172	CUCURBITA PEPO	S	.	0	36.9	48.7	.	.	.	.	.	.	.	.	.	.	.	.	.	19
2173	CUCURBITA PEPO	S	.	0	39.3	46.2	.	.	4.8	.	.	.	.	.	.	.	.	.	.	19
2174	CUCURBITA PEPO	S	.	0	38.1	50.4	.	.	.	.	.	.	.	.	.	.	.	.	.	19
2175	CUCURBITA PEPO	S	.	0	40.0	48.7	.	.	5.2	.	.	.	.	.	.	.	.	.	.	19
2176	CUCURBITA PEPO	S	.	0	27.5	5.1	.	.	1.6	.	.	.	.	.	.	.	.	.	.	19
2177	CUCURBITA PEPO	G	289	0	19.7	1.3	65.8	13.2	13.2	315.8	500.0	5.3	.	.	.	0.00	2.24	10.53	.	1
2178	CUCURBITA PEPO	F	311	0	13.5	1.4	74.3	10.8	10.8	337.8	432.4	18.9	.	.	48175.7	0.68	0.27	6.76	108.1	1
2179	CUCURBITA PEPO	L	250	0	37.0	1.9	40.7	22.2	20.4	4416.7	1259.3	7.4	.	.	33333.3	.	0.56	2.96	740.7	1
2180	CUCURBITA PEPO	L	284	0	40.5	5.4	40.5	.	13.5	500.0	1337.8	28.4	81.1	6378.4	26216.2	1.22	1.76	12.16	148.6	2
2181	CUCURBITA PEPO	S	563	0	30.5	42.0	26.1	2.1	1.5	34.3	741.4	10.3	461.1	304.3	244.0	0.42	0.15	3.32	0.0	2

TABLE 3 - DATA CONVERTED TO A ZERO-MOISTURE BASIS.

OBS	PLANA	PART	CALORIES	WATER	PROTEIN	FAT	TOTCARB	FIBER	ASH	CALCIUM	PHOSPHO	IRON	SODIUM	POTASS	CAROT	THIAMINE	RIBOFLAV	NIACIN	ASCORBIC	REFCO
2182	CUCURBITA PEPO	F	333	0	8.6	2.5	81.5	9.9	7.4	296.3	407.4	8.6	98.8	4321.0	9691.4	0.37	0.49	6.17	172.8	2
2183	CUCURBITA PEPO	I	308	0	26.9	5.8	51.9	11.5	15.4	903.8	1653.8	19.2	.	.	7692.3	0.38	2.12	11.54	346.2	3
2184	CUCURBITA PEPO	L	271	0	43.8	4.2	35.4	15.6	16.7	1322.9	1000.0	60.4	.	.	16979.2	1.46	1.77	18.75	604.2	3
2185	CUCURBITA PEPO	G	333	0	13.9	2.8	76.4	5.6	6.9	263.9	444.4	8.3	.	.	416.7	0.69	0.56	6.94	263.9	3
2186	CUCURBITA PEPO	F	333	0	6.7	2.2	84.4	7.8	6.7	211.1	244.4	5.6	.	.	20444.4	0.44	0.44	5.56	166.7	3
2187	CUCURBITA PEPO	F	308	0	26.9	5.8	51.9	11.5	15.4	903.8	1653.8	19.2	.	.	7692.3	0.38	2.12	11.54	346.2	3
2188	CUCURBITA PEPO	L	271	0	43.8	4.2	35.4	15.6	16.7	1322.9	1000.0	60.4	.	.	16979.2	1.46	1.77	18.75	604.2	3
2189	CUCURBITA PEPO	F	313	0	16.7	2.1	72.9	8.3	8.3	375.0	291.7	47.9	.	.	0.0	0.63	0.63	8.33	354.2	3
2190	CUCURBITA PEPO	F	317	0	18.3	1.7	70.0	10.0	10.0	466.7	483.3	6.7	16.7	3366.7	4100.0	0.83	1.50	16.67	366.7	4
2191	CUCURBITA PEPO	F	310	0	11.9	1.2	77.4	13.1	9.5	250.0	523.8	9.5	11.9	4047.6	11428.6	0.60	1.31	7.14	107.1	4
2192	CUCURBITA PEPO	S	578	0	30.3	48.8	15.7	2.0	5.1	53.3	1196.7	11.7	.	.	43.9	0.25	0.20	2.51	.	4
2193	CUMINUM CYMINUM	S	408	0	19.4	24.3	48.1	11.4	8.3	1013.1	543.0	72.3	182.8	1945.6	829.2	0.69	0.36	4.98	7.6	16
2194	CUMINUM CYMINUM	S	377	0	18.9	15.2	57.0	.	8.9	1551.1	.	27.6	.	.	590.9	.	.	.	0.0	2
2195	CUMINUM CYMINUM	S	.	0	21.2	17.0	55.2	13.6	6.6	1225.9	556.2	35.2	.	.	592.5	.	.	.	3.4	5
2196	CUNNINGHAMIA LANCEOLATA	S	.	0	4.4	1.6	.	.	.	.	.	.	.	.	.	.	.	.	.	20
2197	CUPHEA CARTHAGENENSIS	S	.	0	16.2	32.7	.	.	.	.	.	.	.	.	.	.	.	.	.	21
2198	CUPHEA HOOKERIANA	S	.	0	18.6	15.8	.	.	.	.	.	.	.	.	.	.	.	.	.	21
2199	CUPHEA IGNEA	S	.	0	18.1	33.6	.	.	.	.	.	.	.	.	.	.	.	.	.	20
2200	CUPHEA LLAVEA	S	.	0	21.2	30.0	.	.	.	.	.	.	.	.	.	.	.	.	.	20
2201	CUPHEA LLAVEA	S	.	0	15.2	17.0	.	.	.	.	.	.	.	.	.	.	.	.	.	21
2202	CUPHEA LLAVEA	S	.	0	15.2	22.8	.	.	.	.	.	.	.	.	.	.	.	.	.	21
2203	CUPHEA LLAVEA VAR. MINIATA	S	.	0	16.9	20.9	.	.	6.8	.	.	.	.	.	.	.	.	.	.	19
2204	CUPHEA PAINTERI	S	.	0	23.4	35.8	.	.	.	.	.	.	.	.	.	.	.	.	.	21
2205	CUPRESSUS ARIZONICA	S	.	0	4.4	4.8	.	.	.	.	.	.	.	.	.	.	.	.	.	20
2206	CUPRESSUS SEMPERVIRENS	S	.	0	9.8	9.4	.	.	.	.	.	.	.	.	.	.	.	.	.	21
2207	CURCUMA DOMESTICA	R	400	0	8.8	11.2	73.3	7.6	6.8	205.4	302.5	46.7	42.9	2849.9	0.0	0.17	0.26	5.80	29.3	16
2208	CURCUMA DOMESTICA	R	390	0	10.2	15.3	66.9	7.6	7.6	228.8	542.4	19.5	25.4	4127.1	0.0	0.17	0.25	4.24	0.0	2
2209	CURCUMA DOMESTICA	R	301	0	23.3	0.0	61.9	6.8	14.8	289.8	630.7	.	.	.	0.0	0.85	1.19	6.25	28.4	3
2210	CURCUMA DOMESTICA	R	.	0	7.2	5.9	82.9	3.0	4.0	.	.	.	.	34.5	.	.	.	.	.	5
2211	CUSCUTA GRONOVII	S	.	0	13.8	4.7	.	.	.	.	.	.	.	.	.	.	.	.	.	19
2212	CUSCUTA SP.	W	.	0	12.0	3.4	82.1	19.6	2.5	.	.	.	.	.	.	.	.	.	.	6
2213	CUSCUTA SP.	Y	.	0	7.6	1.7	87.8	24.4	2.9	110.0	130.0	.	.	1240.0	.	.	.	.	.	6
2214	CYAMOPSIS TETRAGONOLOBA	C	.	0	6.8	1.0	87.7	48.0	4.5	750.0	120.0	.	.	.	.	.	.	.	.	18
2215	CYAMOPSIS TETRAGONOLOBA	L	.	0	22.5	3.5	59.6	9.7	14.4	.	.	.	.	.	.	.	.	.	.	18
2216	CYAMOPSIS TETRAGONOLOBA	H	.	0	25.2	0.9	57.4	13.8	16.5	2420.0	270.0	.	.	.	.	.	.	.	.	18
2217	CYAMOPSIS TETRAGONOLOBA	SH	.	0	16.0	1.9	65.1	22.7	17.0	1800.0	100.0	.	.	.	.	.	.	.	.	18
2218	CYAMOPSIS TETRAGONOLOBA	S	.	0	28.0	3.7	64.3	5.0	4.0	.	420.0	.	.	.	.	.	.	.	.	18
2219	CYAMOPSIS TETRAGONOLOBA	S	.	0	32.1	3.1	.	.	4.2	.	.	.	.	.	.	.	.	.	.	19
2220	CYAMOPSIS TETRAGONOLOBA	S	.	0	36.2	3.3	.	.	4.3	.	.	.	.	.	.	.	.	.	.	19
2221	CYAMOPSIS TETRAGONOLOBA	S	.	0	35.4	2.9	.	.	3.8	.	.	.	.	.	.	.	.	.	.	19
2222	CYAMOPSIS TETRAGONOLOBA	S	.	0	36.1	2.8	.	.	3.8	.	.	.	.	.	.	.	.	.	.	19
2223	CYAMOPSIS TETRAGONOLOBA	S	.	0	35.5	3.0	.	.	4.0	.	.	.	.	.	.	.	.	.	.	19
2224	CYAMOPSIS TETRAGONOLOBA	S	.	0	35.1	2.7	.	.	3.8	.	.	.	.	.	.	.	.	.	.	19
2225	CYAMOPSIS TETRAGONOLOBA	S	.	0	34.9	2.8	.	.	3.9	.	.	.	.	.	.	.	.	.	.	19
2226	CYAMOPSIS TETRAGONOLOBA	S	.	0	28.4	2.8	.	.	.	.	.	.	.	.	.	.	.	.	.	19

TABLE 3 - DATA CONVERTED TO A ZERO-MCISTURE BASIS.

OBS	PLNA	PART	CALORIES	WATER	PROTEIN	FAT	TOTLCARB	FIBER	ASH
2227	CYAMOPSIS TETRAGONOLOBA	S	.	0	29.4	2.7	.	.	3.6
2228	CYAMOPSIS TETRAGONOLOBA	S	.	0	25.1	2.9	.	.	3.9
2229	CYAMOPSIS TETRAGONOLOBA	S	.	0	24.2	3.7	.	.	3.3
2230	CYAMOPSIS TETRAGONOLOBA	S	.	0	31.2	3.5	.	.	3.4
2231	CYAMOPSIS TETRAGONOLOBA	S	.	0	34.4	2.9	.	.	3.5
2232	CYAMOPSIS TETRAGONOLOBA	S	.	0	33.8	3.2	.	.	.
2233	CYAMOPSIS TETRAGONOLOBA	W	.	0	14.6	2.5	70.7	22.8	12.2
2234	CYAMOPSIS TETRAGONOLOBA	G	.	0	21.1	1.1	69.7	13.1	8.0
2235	CYAMOPSIS TETRAGONOLOBA (FRUITING)	W	.	0	11.9	2.4	78.2	25.8	7.5
2236	CYCAS REVOLUTA	S	.	0	15.4	0.9	.	.	.
2237	CYCAS REVOLUTA	S	.	0	13.9	1.0	.	.	.
2238	CYCLANTHERA PEDATA	F	309	0	10.9	1.8	72.7	12.7	14.5
2239	CYCLOPIA SUBTERNATA	S	.	0	24.8	4.6	.	.	.
2240	CYDONIA OBLONGA	F	356	0	2.3	1.7	93.1	17.8	2.9
2241	CYDONIA OBLONGA	F	358	0	3.4	1.7	92.6	12.5	2.3
2242	CYDONIA OBLONGA	F	352	0	2.5	0.6	94.4	10.5	2.5
2243	CYDONIA OBLONGA	F	.	0	2.1	0.7	95.1	11.9	2.1
2244	CYMBOPOGON CITRATUS	L	358	0	3.9	5.4	85.2	16.3	5.4
2245	CYMBOPOGON GIGANTEUS	W	.	0	5.0	4.1	84.4	33.2	6.5
2246	CYNANCHUM ACUTUM	S	.	0	28.0	24.1	.	.	.
2247	CYNANCHUM ACUTUM	S	.	0	16.0	11.9	.	.	.
2248	CYNARA ALBA	S	.	0	27.4	17.1	.	.	.
2249	CYNARA CARDUNCULUS	SH	.	0	16.9	6.2	62.3	26.9	14.6
2250	CYNARA CARDUNCULUS	S	.	0	27.1	42.2	.	.	5.3

OBS	CALCIUM	PHOSPHO	IRON	SODIUM	POTASS	CAROT	THIAMINE	RIBOFLAV	NIACIN	ASCORBIC	REFCO
2227	.	.	.	.	.	.	.	.	.	.	19
2228	.	.	.	.	.	.	.	.	.	.	19
2229	.	.	.	.	.	.	.	.	.	.	19
2230	.	.	.	.	.	.	.	.	.	.	19
2231	.	.	.	.	.	.	.	.	.	.	19
2232	.	.	.	.	.	.	.	.	.	.	20
2233	.	.	.	.	.	.	.	.	.	.	6
2234	742.9	285.7	33.1	.	.	1131.43	.	.	.	280.0	5
2235	1980.0	160.0	.	.	1910.0	.	.	.	.	.	6
2236	.	.	.	.	.	.	.	.	.	.	21
2237	.	.	.	.	.	.	.	.	.	.	21
2238	236.4	472.7	10.9	.	.	545.45	0.73	0.73	5.45	254.5	3
2239	.	.	.	.	.	.	.	.	.	.	21
2240	80.5	86.2	5.2	.	.	.	.	.	.	.	1
2241	34.1	85.2	3.4	.	.	113.64	0.17	0.17	2.27	96.6	3
2242	67.9	104.9	4.3	24.7	1216.0	148.15	0.12	0.19	1.23	92.6	4
2243	.	.	.	.	.	.	.	.	.	69.9	5
2244	124.5	116.7	7.0	.	.	1653.70	0.19	0.08	8.56	3.9	2
2245	.	.	.	.	.	.	.	.	.	.	18
2246	.	.	.	.	.	.	.	.	.	.	21
2247	.	.	.	.	.	.	.	.	.	.	21
2248	.	.	.	.	.	.	.	.	.	.	21
2249	1450.0	330.0	.	.	.	.	.	.	.	.	18
2250	.	.	.	.	.	.	.	.	.	.	19

TABLE 3 - DATA CONVERTED TO A ZERO-MOISTURE BASIS.

OBS	PLNA	PART	CALORIES	WATER	PROTEIN	FAT	TOTLCARB	FIBER	ASH	CALCIUM	PHOSPHO	IRON	SODIUM	POTASS	CAROT	THIAMINE	RIBOFLAV	NIACIN	ASCORBIC	REFCO
2251	CYNARA CARDUNCULUS	S	.	0	7.4	4.6	.	.	2.5	.	.	.	.	.	.	.	.	.	.	19
2252	CYNARA CARDUNCULUS	S	.	0	27.5	43.0	.	.	.	.	.	.	.	.	.	.	.	.	.	20
2253	CYNARA CARDUNCULUS	S	.	0	21.4	21.5	.	.	.	.	.	.	.	.	.	.	.	.	.	21
2254	CYNARA CARDUNCULUS	SH	290	0	8.1	3.2	66.1	16.1	22.6	1838.7	161.3	24.2	.	.	0.00	0.16	0.48	3.23	16.1	3
2255	CYNARA HUMILIS	S	.	0	21.9	21.0	.	.	.	.	.	.	.	.	.	.	.	.	.	21
2256	CYNARA SCOLYMUS	S	.	0	21.9	17.1	.	.	.	.	.	.	.	.	.	.	.	.	.	20
2257	CYNARA SCOLYMUS	W	.	0	5.1	1.1	86.1	18.0	7.7	.	.	.	.	.	.	.	.	.	.	6
2258	CYNARA SCOLYMUS	I	298	0	24.1	1.4	63.8	5.7	10.6	453.9	560.3	1.4	.	.	.	.	.	.	.	1
2259	CYNARA SCOLYMUS	I	323	0	19.9	1.3	73.5	16.8	5.3	354.0	610.6	8.8	296.5	2977.9	663.72	0.53	0.35	7.08	0.0	2
2260	CYNARA SCOLYMUS	I	296	0	27.6	2.0	60.2	22.4	10.2	449.0	591.8	8.2	.	.	1938.78	0.61	0.71	8.16	51.0	3
2261	CYNARA SCOLYMUS	I	193	0	20.0	1.4	73.1	16.6	5.5	351.7	606.9	9.0	296.6	2965.5	662.07	0.55	0.34	6.90	82.8	4
2262	CYNARA SCOLYMUS	I	.	0	15.9	0.4	75.8	5.3	7.9	528.6	440.5	10.1	.	.	158.59	.	.	.	0.0	5
2263	CYNODON AETHIOPICUS	W	.	0	12.2	1.7	74.2	26.6	11.9	.	.	.	.	.	.	.	.	.	.	18
2264	CYNODON AETHIOPICUS	H	.	0	6.4	0.6	86.2	35.6	6.8	.	.	.	.	.	.	.	.	.	.	18
2265	CYNODON DACTYLON	W	.	0	14.2	1.9	71.5	26.6	12.4	.	.	.	.	.	.	.	.	.	.	18
2266	CYNODON DACTYLON	H	.	0	11.1	1.4	75.0	18.4	12.5	.	.	.	.	.	.	.	.	.	.	18
2267	CYNODON DACTYLON	L	.	0	10.5	1.8	76.0	28.2	11.8	.	.	.	.	.	.	.	.	.	.	5
2268	CYNODON DACTYLON	W	.	0	11.6	2.1	75.9	25.9	10.4	530.0	220.0	112.0	.	1630.0	28.00	.	.	.	.	6
2269	CYNODON PLECTOSTACHYUS	W	.	0	19.1	2.9	65.0	22.6	13.0	.	.	.	.	.	.	.	.	.	.	18
2270	CYNODON PLECTOSTACHYUS	H	.	0	5.2	0.9	85.8	39.7	8.1	.	.	.	.	.	.	.	.	.	.	18
2271	CYNOGLOSSUM AMABILE	S	.	0	17.6	22.6	.	.	11.6	.	.	.	.	.	.	.	.	.	.	19
2272	CYNOGLOSSUM CRETICUM	S	.	0	24.8	42.2	.	.	.	.	.	.	.	.	.	.	.	.	.	21
2273	CYNOGLOSSUM OFFICINALE	S	.	0	20.4	21.2	.	.	.	.	.	.	.	.	.	.	.	.	.	21
2274	CYNOGLOSSUM PICTUM	S	.	0	16.7	22.3	.	.	.	.	.	.	.	.	.	.	.	.	.	21
2275	CYNOSURUS CRISTATUS	S	.	0	15.1	4.7	.	.	.	.	.	.	.	.	.	.	.	.	.	21
2276	CYPERUS ESCULENTUS	R	.	0	6.1	27.4	.	.	2.4	.	.	.	.	.	.	.	.	.	.	19
2277	CYPERUS ESCULENTUS	R	476	0	5.5	20.0	72.6	11.7	1.9	39.4	321.3	12.6	.	.	.	0.44	0.14	2.05	4.7	1
2278	CYPERUS ESCULENTUS	R	461	0	6.5	25.5	65.1	10.5	2.8	87.5	230.0	3.6	.	.	.	0.13	.	.	.	3
2279	CYPERUS KALLI	S	.	0	17.4	20.0	.	.	.	.	.	.	.	.	.	.	.	.	.	21
2280	CYPERUS VEGETUS	S	.	0	10.0	8.7	.	.	.	.	.	.	.	.	.	.	.	.	.	21
2281	CYPERUS VIRENS	S	.	0	11.5	8.2	.	.	6.9	.	.	.	.	.	.	.	.	.	.	19
2282	CYPERUS VIRENS	S	.	0	8.8	2.6	.	.	8.3	.	.	.	.	.	.	.	.	.	.	19
2283	CYPHOLEPSIS YEMINICA	W	.	0	15.5	1.8	74.9	37.9	7.8	.	.	.	.	.	.	.	.	.	.	18
2284	CYPHOMANDRA BETACEA	F	329	0	8.7	1.2	83.8	24.3	6.4	410.4	173.4	4.0	.	.	.	.	.	.	.	1
2285	CYPHOMANDRA BETACEA	F	340	0	10.6	2.1	80.1	15.6	7.1	92.2	170.2	5.7	.	.	3262.41	0.28	0.28	7.09	120.6	2
2286	CYPHOMANDRA BETACEA	F	355	0	15.6	6.4	73.0	11.3	5.0	63.8	340.4	5.7	.	.	3262.41	0.71	0.28	8.51	205.7	3
2287	CYPHOMANDRA BETACEA	F	.	0	8.7	1.2	83.8	24.3	6.4	578.0	173.4	4.0	.	.	1872.83	.	.	.	.	5
2288	CYTISUS ALBUS	S	.	0	33.8	6.9	.	.	.	.	.	.	.	.	.	.	.	.	.	20
2289	CYTISUS CANTABRICUS	S	.	0	30.1	7.7	.	.	.	.	.	.	.	.	.	.	.	.	.	21
2290	CYTISUS MULTIFLORUS	S	.	0	34.0	7.0	.	.	3.2	.	.	.	.	.	.	.	.	.	.	19
2291	CYTISUS MULTIFLORUS	S	.	0	32.9	7.5	.	.	.	.	.	.	.	.	.	.	.	.	.	21
2292	CYTISUS NIGRICANS	S	.	0	27.6	7.7	.	.	3.0	.	.	.	.	.	.	.	.	.	.	19
2293	CYTISUS PROLIFERUS	W	.	0	25.0	4.0	66.0	22.0	5.0	570.0	190.0	.	.	.	.	.	.	.	.	18
2294	CYTISUS SCOPARIUS	S	.	0	32.5	6.0	.	.	3.8	.	.	.	.	.	.	.	.	.	.	19
2295	DACRYODES EDULIS	F	600	0	10.5	53.0	34.0	18.7	2.5	98.2	137.0	1.8	.	.	.	.	.	.	43.4	2

TABLE 3 - DATA CONVERTED TO A ZERO-MOISTURE BASIS.

OBS	PLNA	PART	CALORIES	WATER	PROTEIN	FAT	TOTCARB	FIBER	ASH	CALCIUM	PHOSPHO	IRON	SODIUM	POTASS	CAROT	THIAMINE	RIBOFLAV	NIACIN	ASCORBIC	REFCO
2296	DACTYLIS GLOMERATA	W	.	0	12.6	3.5	74.9	33.1	9.0											18
2297	DACTYLIS GLOMERATA	H	.	0	16.1	3.3	68.4	37.7	12.2	.	.	.	.	.	.	.	.	.	.	18
2298	DACTYLIS GLOMERATA	S	.	0	29.6	11.6	.	.	.	.	.	.	.	.	.	.	.	.	.	19
2299	DACTYLIS GLOMERATA	S	.	0	6.9	2.3	.	.	10.3	.	.	.	.	.	.	.	.	.	.	19
2300	DACTYLIS GLOMERATA	S	.	0	26.2	10.8	.	.	2.7	.	.	.	.	.	.	.	.	.	.	19
2301	DACTYLIS GLOMERATA	S	.	0	27.5	10.9	.	.	.	.	.	.	.	.	.	.	.	.	.	20
2302	DACTYLIS GLOMERATA	S	.	0	25.3	10.9	.	.	.	.	.	.	.	.	.	.	.	.	.	21
2303	DACTYLIS GLOMERATA	W	.	0	13.8	4.3	72.7	27.9	9.2	530.0	510.0	23	.	3440	32	0.17	.	.	.	6
2304	DACTYLOCTENIUM AEGYPTIUM	W	.	0	15.8	1.8	72.4	26.8	10.6	.	.	.	.	.	.	.	.	.	.	18
2305	DACTYLOCTENIUM AEGYPTIUM	H	.	0	8.3	0.8	84.4	36.9	6.5	.	.	.	.	.	.	.	.	.	.	18
2306	DACTYLOCTENIUM AEGYPTIUM	S	388	0	17.7	2.1	76.9	3.2	3.3	925.5	519.2	.	.	.	.	.	.	.	.	1
2307	DAHLIA PINNATA	S	.	0	31.9	17.8	.	.	.	.	.	.	.	.	.	.	.	.	.	20
2308	DALBERGIA CULTRATA	L	309	0	21.5	0.0	71.9	.	6.6	109.4	304.7	.	.	.	.	.	.	.	101.6	1
2309	DALBERGIA SISSOO	L	.	0	24.1	2.0	67.3	12.5	6.6	840.0	420.0	.	.	.	.	.	.	.	.	18
2310	DALEA ALBIFLORA	S	.	0	33.8	5.8	.	.	.	.	.	.	.	.	.	.	.	.	.	19
2311	DALEA ALBIFLORA	S	.	0	35.0	7.0	.	.	.	.	.	.	.	.	.	.	.	.	.	20
2312	DALEA ALOPECUROIDES	S	.	0	36.6	5.7	.	.	3.5	.	.	.	.	.	.	.	.	.	.	19
2313	DALEA ALOPECUROIDES	S	.	0	30.2	6.5	.	.	4.7	.	.	.	.	.	.	.	.	.	.	19
2314	DALEA ALOPECUROIDES	S	.	0	31.2	4.7	.	.	.	.	.	.	.	.	.	.	.	.	.	21
2315	DALEA CITRIODORA	S	.	0	31.2	7.1	.	.	.	.	.	.	.	.	.	.	.	.	.	20
2316	DALEA DIFFUSA	S	.	0	32.5	6.4	.	.	.	.	.	.	.	.	.	.	.	.	.	19
2317	DALEA DIFFUSA & DALEA NUTANS	S	.	0	28.8	11.9	.	.	.	.	.	.	.	.	.	.	.	.	.	21
2318	DALEA GRAYI	S	.	0	34.0	11.8	.	.	3.2	.	.	.	.	.	.	.	.	.	.	19
2319	DALEA LACHNOSTACHYS	S	.	0	45.0	11.9	.	.	.	.	.	.	.	.	.	.	.	.	.	19
2320	DALEA LAGOPINA	S	.	0	33.1	6.3	.	.	.	.	.	.	.	.	.	.	.	.	.	20
2321	DALEA LUMHOLTZII	S	.	0	33.4	6.9	.	.	4.5	.	.	.	.	.	.	.	.	.	.	19
2322	DALEA MUCRONATA	S	.	0	37.5	4.9	.	.	.	.	.	.	.	.	.	.	.	.	.	20
2323	DALEA NEGLECTA	S	.	0	41.2	13.7	.	.	.	.	.	.	.	.	.	.	.	.	.	20
2324	DALEA NUTANS	S	.	0	26.9	10.9	.	.	.	.	.	.	.	.	.	.	.	.	.	20
2325	DALEA NUTANS	S	.	0	31.5	9.7	.	.	.	.	.	.	.	.	.	.	.	.	.	21
2326	DALEA OAXACANA	S	.	0	36.6	6.8	.	.	6.6	.	.	.	.	.	.	.	.	.	.	19
2327	DALEA PSORALEOIDES	S	.	0	27.5	2.7	.	.	.	.	.	.	.	.	.	.	.	.	.	20
2328	DALEA SP.	S	.	0	28.0	3.4	.	.	.	.	.	.	.	.	.	.	.	.	.	21
2329	DALEA SP.	S	.	0	33.9	3.3	.	.	.	.	.	.	.	.	.	.	.	.	.	21
2330	DALEA SP.	W	.	0	19.0	6.5	67.8	14.5	6.7	2270.0	140.0	.	.	1210	.	.	.	.	.	6
2331	DALEA TOMENTOSA	S	.	0	26.4	3.4	.	.	.	.	.	.	.	.	.	.	.	.	.	20
2332	DALEA TUBERCULINA	S	.	0	36.9	8.2	.	.	.	.	.	.	.	.	.	.	.	.	.	20
2333	DALEA VIRIDIFLORA	S	.	0	36.2	2.7	.	.	.	.	.	.	.	.	.	.	.	.	.	20
2334	DANAE RACEMOSA	S	.	0	13.8	0.4	.	.	.	.	.	.	.	.	.	.	.	.	.	20
2335	DAPHNE MEZEREUM	S	.	0	24.4	65.4	.	.	.	.	.	.	.	.	.	.	.	.	.	19
2336	DAPHNE MEZEREUM	S	.	0	7.5	6.0	.	.	.	.	.	.	.	.	.	.	.	.	.	19
2337	DAPHNE PONTICA	S	.	0	7.8	28.0	.	.	.	.	.	.	.	.	.	.	.	.	.	21
2338	DAPHNIPHYLLUM HUMILE	S	.	0	15.0	34.8	.	.	2.3	.	.	.	.	.	.	.	.	.	.	19
2339	DAPHNIPHYLLUM MACROPODUM	S	.	0	18.1	37.4	.	.	.	.	.	.	.	.	.	.	.	.	.	20
2340	DASYLIRION TEXANUM	S	.	0	27.5	23.0	.	.	.	.	.	.	.	.	.	.	.	.	.	19

TABLE 3 - DATA CONVERTED TO A ZERO-MOISTURE BASIS.

OBS	PLNA	PART	CALORIES	WATER	PROTEIN	FAT	TOTCARB	FIBER	ASH	CALCIUM	PHOSPHO	IRON	SODIUM	POTASS	CAROT	THIAMINE	RIBOFLAV	NIACIN	ASCORBIC	REFCO
2341	DASYLIRION TEXANUM	W	.	0	7.1	2.2	87.7	39.0	3.0	690.0	80.0	.	.	760.0	.	.	.	.	.	6
2342	DASYLIRION WHEELERI	S	.	0	26.3	22.1	.	.	2.1	.	.	.	.	.	.	.	.	.	.	19
2343	DASYLIRION WHEELERI	S	.	0	7.1	1.5	.	.	6.0	.	.	.	.	.	.	.	.	.	.	19
2344	DASYLIRION WHEELERI	S	.	0	26.2	23.8	.	.	.	.	.	.	.	.	.	.	.	.	.	19
2345	DASYLIRION WHEELERI	S	.	0	42.5	28.3	.	.	.	.	.	.	.	.	.	.	.	.	.	19
2346	DASYLIRION WHEELERI	S	.	0	50.0	21.7	.	.	.	.	.	.	.	.	.	.	.	.	.	19
2347	DASYLIRION WHEELERI	Y	.	0	3.7	.	.	.	11.6	890.0	210.0	19.0	2100.0	.	.	.	.	.	.	6
2348	DASYLIRION WHEELERI	L	.	0	4.1	.	.	.	7.7	1440.0	260.0	13.0	.	1630.0	.	.	.	.	.	6
2349	DATISCA CANNABINA	S	.	0	25.8	40.9	.	.	.	.	.	.	.	.	.	.	.	.	.	21
2350	DATURA CF. INOXIA	S	.	0	17.5	22.7	.	.	.	.	.	.	.	.	.	.	.	.	.	20
2351	DATURA DISCOLOR	S	.	0	21.2	20.8	.	.	.	.	.	.	.	.	.	.	.	.	.	20
2352	DATURA FEROX	S	.	0	18.1	17.6	.	.	.	.	.	.	.	.	.	.	.	.	.	21
2353	DATURA INOXIA	S	.	0	13.8	19.6	.	.	.	.	.	.	.	.	.	.	.	.	.	20
2354	DATURA INOXIA	S	.	0	15.1	17.3	.	.	.	.	.	.	.	.	.	.	.	.	.	21
2355	DATURA INOXIA	S	.	0	20.0	12.0	.	.	.	.	.	.	.	.	.	.	.	.	.	21
2356	DATURA METAL	S	.	0	13.4	18.8	.	.	4.5	.	.	.	.	.	.	.	.	.	.	19
2357	DATURA METAL	S	.	0	12.5	17.2	.	.	5.6	.	.	.	.	.	.	.	.	.	.	19
2358	DATURA METALOIDES	S	.	0	17.5	21.8	.	.	.	.	.	.	.	.	.	.	.	.	.	19
2359	DATURA QUERCIFOLIA	S	.	0	21.2	16.1	.	.	.	.	.	.	.	.	.	.	.	.	.	19
2360	DATURA REBURRA	S	.	0	12.3	19.5	.	.	.	.	.	.	.	.	.	.	.	.	.	21
2361	DATURA SP.	S	.	0	14.4	22.3	.	.	.	.	.	.	.	.	.	.	.	.	.	19
2362	DATURA STRAMONIUM	S	.	0	18.3	25.8	.	.	2.7	.	.	.	.	.	.	.	.	.	.	19
2363	DATURA STRAMONIUM	S	.	0	18.2	22.8	.	.	2.7	.	.	.	.	.	.	.	.	.	.	19
2364	DATURA STRAMONIUM	S	.	0	18.1	27.1	.	.	.	.	.	.	.	.	.	.	.	.	.	20
2365	DATURA STRAMONIUM	S	.	0	19.3	24.9	.	.	.	.	.	.	.	.	.	.	.	.	.	21
2366	DATURA STRAMONIUM	S	.	0	14.0	18.4	.	.	.	.	.	.	.	.	.	.	.	.	.	21
2367	DATURA STRAMONIUM	S	.	0	17.2	19.6	.	.	.	.	.	.	.	.	.	.	.	.	.	21
2368	DATURA STRAMONIUM	S	.	0	19.4	28.5	.	.	.	.	.	.	.	.	.	.	.	.	.	21
2369	DATURA WRIGHTII	S	.	0	13.4	24.2	.	.	.	.	.	.	.	.	.	.	.	.	.	21
2370	DAUCUS AUREUS	S	.	0	15.6	15.0	.	.	.	.	.	.	.	.	.	.	.	.	.	21
2371	DAUCUS BROTERI	S	.	0	18.9	18.6	.	.	.	.	.	.	.	.	.	.	.	.	.	21
2372	DAUCUS CAROTA	SH	.	0	12.1	2.3	67.4	16.7	18.2	.	.	.	.	.	.	.	.	.	.	18
2373	DAUCUS CAROTA	R	.	0	12.7	1.5	77.6	10.4	8.2	.	.	.	.	.	.	.	.	.	.	18
2374	DAUCUS CAROTA	S	.	0	28.4	20.1	.	.	6.6	.	.	.	.	.	.	.	.	.	.	19
2375	DAUCUS CAROTA	S	.	0	28.9	28.3	.	.	7.4	.	.	.	.	.	.	.	.	.	.	19
2376	DAUCUS CAROTA	S	.	0	26.9	28.2	.	.	6.7	.	.	.	.	.	.	.	.	.	.	19
2377	DAUCUS CAROTA	S	.	0	28.5	19.8	.	.	7.4	.	.	.	.	.	.	.	.	.	.	19
2378	DAUCUS CAROTA	S	.	0	23.9	16.3	.	.	7.6	.	.	.	.	.	.	.	.	.	.	19
2379	DAUCUS CAROTA	S	.	0	35.8	16.4	.	.	4.1	.	.	.	.	.	.	.	.	.	.	19
2380	DAUCUS CAROTA	S	.	0	22.5	18.2	.	.	6.5	.	.	.	.	.	.	.	.	.	.	19
2381	DAUCUS CAROTA	S	.	0	25.4	26.5	.	.	8.4	.	.	.	.	.	.	.	.	.	.	19
2382	DAUCUS CAROTA	S	.	0	29.8	30.2	.	.	6.7	.	.	.	.	.	.	.	.	.	.	19
2383	DAUCUS CAROTA	S	.	0	13.4	8.7	.	.	11.8	.	.	.	.	.	.	.	.	.	.	19
2384	DAUCUS CAROTA	S	.	0	26.2	19.6	.	.	.	.	.	.	.	.	.	.	.	.	.	19
2385	DAUCUS CAROTA	S	.	0	20.0	19.2	.	.	.	.	.	.	.	.	.	.	.	.	.	19

TABLE 3 - DATA CONVERTED TO A ZERO-MOISTURE BASIS.

OBS	PLNA	PART	CALORIES	WATER	PROTEIN	FAT	TOTLCARB	FIBER
2386	DAUCUS CAROTA	S	.	0	20.0	12.9	.	.
2387	DAUCUS CAROTA	S	.	0	25.6	15.6	.	.
2388	DAUCUS CAROTA	S	.	0	22.4	27.2	.	.
2389	DAUCUS CAROTA	L	.	0	16.4	4.6	64.0	11.1
2390	DAUCUS CAROTA	Y	.	0	10.4	5.5	.	18.1
2391	DAUCUS CAROTA	R	351	0	7.9	0.9	84.2	12.3
2392	DAUCUS CAROTA	R	356	0	10.6	2.9	79.8	8.7
2393	DAUCUS CAROTA	R	376	0	7.3	3.7	81.7	7.3
2394	DAUCUS CAROTA	R	356	0	9.3	1.7	82.2	8.5
2395	DAUCUS CAROTA	R	.	0	6.4	0.7	85.0	8.6
2396	DAUCUS CAROTA VAR. MAXIMUS	S	.	0	19.1	18.9	.	.
2397	DAUCUS GUTTATUS	S	.	0	20.9	23.3	.	.
2398	DAUCUS PUSILLUS	S	.	0	20.0	27.1	.	.
2399	DAUCUS PUSILLUS	S	.	0	21.4	22.7	.	.
2400	DAUCUS PUSILLUS	W	.	0	11.7	6.4	72.6	18.9
2401	DELONIX REGIA	F	.	0	11.1	.	.	49.7
2402	DELONIX REGIA	S	.	0	17.5	3.3	.	.
2403	DELONIX REGIA	S	.	0	16.9	3.0	.	.
2404	DELONIX REGIA	S	378	0	35.5	5.2	51.7	.
2405	DELONIX REGIA	S	.	0	64.4	10.4	17.3	.
2406	DELPHINIUM GRANDIFLORUM	S	.	0	29.4	27.2	.	.
2407	DELPHINIUM HANSENII	S	.	0	25.1	27.3	.	.
2408	DELPHINIUM HYBRIDUM	S	.	0	31.2	28.2	.	.
2409	DELPHINIUM OCCIDENTALE SSP. CUCULLATUM	S	.	0	27.6	42.3	.	.

OBS	ASH	CALCIUM	PHOSPHO	IRON	SODIUM	POTASS	CAROT	THIAMINE	RIBOFLAV	NIACIN	ASCORBIC	REFCO
2386	.	.	.	.	.	.	.	.	.	.	.	20
2387	.	.	.	.	.	.	.	.	.	.	.	20
2388	.	.	.	.	.	.	.	.	.	.	.	21
2389	15.0	.	.	.	.	.	.	.	.	.	.	6
2390	.	.	.	.	.	.	.	.	.	.	.	6
2391	7.0	307.0	333.3	6.1	.	.	48070.2	0.35	0.35	5.26	70.2	1
2392	6.7	346.2	365.4	11.5	673.1	2355.8	67307.7	0.58	0.48	6.73	76.9	2
2393	7.3	311.9	238.5	8.3	.	.	64770.6	0.55	0.37	5.50	45.9	3
2394	6.8	313.6	305.1	5.9	398.3	2889.8	55932.2	0.51	0.42	5.08	67.8	4
2395	7.9	571.4	214.3	10.7	.	.	13285.7	.	.	.	21.4	5
2396	.	.	.	.	.	.	.	.	.	.	.	21
2397	.	.	.	.	.	.	.	.	.	.	.	21
2398	6.7	.	.	.	.	.	.	.	.	.	.	19
2399	.	.	.	.	.	.	.	.	.	.	.	20
2400	9.3	2350.0	350.0	.	.	2590.0	.	.	.	.	.	6
2401	4.7	230.0	140.0	.	.	.	.	.	.	.	.	18
2402	.	.	.	.	.	.	.	.	.	.	.	19
2403	.	.	.	.	.	.	.	.	.	.	.	19
2404	7.6	674.4	383.7	6.4	5.8	2296.5	1744.2	.	1.16	4.65	.	2
2405	7.9	.	.	.	.	.	.	.	.	.	.	5
2406	.	.	.	.	.	.	.	.	.	.	.	20
2407	.	.	.	.	.	.	.	.	.	.	.	21
2408	.	.	.	.	.	.	.	.	.	.	.	20
2409	.	.	.	.	.	.	.	.	.	.	.	20

TABLE 3 - DATA CONVERTED TO A ZERO-MOISTURE BASIS.

OBS	PLNA	PART	CALORIES	WATER	PROTEIN	FAT	TOTLCARB	FIBER	ASH	CALCIUM	PHOSPHO	IRON	SODIUM	POTASS	CAROT	THIAMINE	RIBOFLAV	NIACIN	ASCORBIC	REFCO
2410	DELPHINIUM SP.	S	.	0	22.5	6.6	.	.	5.5	.	.	.	.	.	.	.	.	.	.	19
2411	DELPHINIUM TENUISECTUM	S	.	0	28.8	33.6	.	.	.	.	.	.	.	.	.	.	.	.	.	20
2412	DENDROCALAMUS STRICTUS	L	.	0	20.5	7.5	57.2	26.0	14.8	.	.	.	.	.	.	.	.	.	.	18
2413	DESCURAINIA PINNATA	S	.	0	24.4	38.4	.	.	.	.	.	.	.	.	.	.	.	.	.	20
2414	DESCURAINIA SOPHIA	S	.	0	29.5	35.3	.	.	3.9	.	.	.	.	.	.	.	.	.	.	19
2415	DESCURAINIA SOPHIA	S	.	0	25.5	39.7	.	.	3.6	.	.	.	.	.	.	.	.	.	.	19
2416	DESCURAINIA SOPHIA	S	.	0	28.1	38.3	.	.	.	.	.	.	.	.	.	.	.	.	.	20
2417	DESCURAINIA SOPHIA	S	.	0	29.2	39.4	.	.	.	.	.	.	.	.	.	.	.	.	.	20
2418	DESCURAINIA SOPHIA	S	.	0	28.8	38.9	.	.	.	.	.	.	.	.	.	.	.	.	.	20
2419	DESCURAINIA SOPHIA	S	.	0	27.6	37.6	.	.	.	.	.	.	.	.	.	.	.	.	.	20
2420	DESCURAINIA SOPHIA	S	.	0	29.9	26.9	.	.	.	.	.	.	.	.	.	.	.	.	.	21
2421	DESCURAINIA SOPHIA	S	.	0	29.3	38.1	.	.	.	.	.	.	.	.	.	.	.	.	.	21
2422	DESCURAINIA SP.	S	.	0	36.9	33.6	.	.	.	.	.	.	.	.	.	.	.	.	.	21
2423	DESCURAINIA SP.	W	.	0	7.6	2.4	.	.	5.6	1490.0	490.0	.	.	4450	.	.	.	.	.	6
2424	DESMANTHUS ACUMINATUS	S	.	0	33.8	2.6	.	.	.	.	.	.	.	.	.	.	.	.	.	20
2425	DESMANTHUS ILLINOENSIS	S	.	0	34.9	2.3	.	.	3.7	.	.	.	.	.	.	.	.	.	.	19
2426	DESMANTHUS ILLINOENSIS	S	.	0	34.4	2.2	.	.	3.8	.	.	.	.	.	.	.	.	.	.	19
2427	DESMANTHUS ILLINOENSIS	S	.	0	34.4	2.3	.	.	3.6	.	.	.	.	.	.	.	.	.	.	19
2428	DESMANTHUS SUBULATUS	S	.	0	32.6	3.3	.	.	.	.	.	.	.	.	.	.	.	.	.	21
2429	DESMANTHUS VELUTINUS	S	.	0	31.2	3.9	.	.	.	.	.	.	.	.	.	.	.	.	.	19
2430	DESMANTHUS VIRGATUS	SH	.	0	11.5	2.4	80.3	39.3	5.8	.	.	.	.	.	.	.	.	.	.	18
2431	DESMANTHUS VIRGATUS	S	.	0	31.9	4.4	.	.	.	.	.	.	.	.	.	.	.	.	.	20
2432	DESMANTHUS VIRGATUS	S	.	0	35.0	4.3	.	.	.	.	.	.	.	.	.	.	.	.	.	20
2433	DESMANTHUS VIRGATUS	W	.	0	12.6	2.2	79.5	38.8	5.7	1750.0	350.0	.	.	1810	.	.	.	.	.	6
2434	DESMODIUM AFFINE	S	.	0	27.5	14.0	.	.	.	.	.	.	.	.	.	.	.	.	.	21
2435	DESMODIUM BARBATUM	SH	.	0	10.4	4.0	81.7	34.7	3.9	.	.	.	.	.	.	.	.	.	.	18
2436	DESMODIUM BELLUM	S	.	0	34.4	8.5	.	.	.	.	.	.	.	.	.	.	.	.	.	20
2437	DESMODIUM CF. DISTORTUM	S	.	0	28.5	8.2	.	.	.	.	.	.	.	.	.	.	.	.	.	21
2438	DESMODIUM CINERASCENS	S	.	0	37.7	12.0	.	.	4.2	.	.	.	.	.	.	.	.	.	.	19
2439	DESMODIUM CINEREUM	L	340	0	22.4	6.9	63.7	23.6	6.9	876.4	417.0	14.7	.	.	13745.2	0.97	0.42	5.79	633.2	3
2440	DESMODIUM DILLENII	S	.	0	39.4	14.1	.	.	.	.	.	.	.	.	.	.	.	.	.	20
2441	DESMODIUM DISCOLOR	SH	.	0	18.7	0.3	71.0	33.4	10.0	.	.	.	.	.	.	.	.	.	.	18
2442	DESMODIUM DISCOLOR	SH	.	0	12.1	2.8	78.4	35.1	6.7	1250.0	190.0	.	.	.	.	.	.	.	.	18
2443	DESMODIUM GYROIDES	SH	.	0	11.5	0.7	79.3	45.2	8.5	970.0	400.0	.	.	.	.	.	.	.	.	18
2444	DESMODIUM HETEROPHYLLUM	SH	.	0	8.9	1.4	77.2	27.1	12.5	740.0	100.0	.	.	.	.	.	.	.	.	18
2445	DESMODIUM INTORTUM	SH	.	0	20.7	3.3	66.6	25.4	9.4	1410.0	520.0	.	.	.	.	.	.	.	.	18
2446	DESMODIUM NICARAQUENSE	S	.	0	31.9	9.0	.	.	4.3	.	.	.	.	.	.	.	.	.	.	19
2447	DESMODIUM OVALIFOLIUM	S	.	0	32.7	3.7	.	.	.	.	.	.	.	.	.	.	.	.	.	21
2448	DESMODIUM PABULARIS	SH	.	0	23.8	2.3	63.7	23.4	10.2	.	.	.	.	.	.	.	.	.	.	18
2449	DESMODIUM SCORPIARUS	SH	.	0	15.5	3.0	71.8	29.3	9.7	.	.	.	.	.	.	.	.	.	.	18
2450	DESMODIUM SP.	W	.	0	15.0	2.5	71.7	26.1	10.8	1240.0	280.0	47.0	.	.	18.0	0.37	0.77	5.25	.	6
2451	DESMODIUM TILIAEFOLIUM	S	.	0	44.2	13.1	.	.	.	.	.	.	.	.	.	.	.	.	.	21
2452	DESMODIUM TRIFLORUM	SH	.	0	8.8	2.8	79.6	29.2	8.8	740.0	120.0	.	.	.	.	.	.	.	.	18
2453	DESMODIUM UNCINATUM	SH	.	0	12.8	3.4	74.7	29.7	9.1	1060.0	410.0	.	.	.	.	.	.	.	.	18
2454	DESMOSTACHYA BIPINNATA	L	.	0	6.8	1.6	82.5	40.3	9.1	.	.	.	.	.	.	.	.	.	.	5

TABLE 3 - DATA CONVERTED TO A ZERO-MOISTURE BASIS.

OBS	PLNA	PART	CALORIES	WATER	PROTEIN	FAT	TOTLCARB	FIBER
2455	DETARIUM MICROCARPUM	F	350	0	5.5	0.4	91.1	7.4
2456	DEUTZIA SCABRA	S	.	0	26.2	33.8	.	.
2457	DIALIUM GUINEENSE	F	355	0	4.1	0.1	94.6	2.7
2458	DIANTHUS BARBATUS	S	.	0	18.8	6.1	.	.
2459	DIANTHUS CHINENSIS	S	.	0	18.1	6.0	.	.
2460	DICENTRA OCHROLEUCA	S	.	0	13.6	30.9	.	.
2461	DICHANTHIUM ANNULATUM	W	.	0	10.4	1.7	75.8	34.9
2462	DICHANTHIUM ANNULATUM	H	.	0	2.7	1.2	84.6	39.1
2463	DICHANTHIUM ANNULATUM	L	.	0	7.0	5.3	75.1	30.6
2464	DICHANTHIUM ANNULATUM	L	.	0	6.5	5.0	76.8	34.0
2465	DICHANTHIUM ANNULATUM	L	.	0	5.8	4.9	78.3	41.3
2466	DICHANTHIUM CARICOSUM	W	.	0	7.0	1.5	79.8	36.7
2467	DICHANTHIUM CARICOSUM	H	.	0	6.1	1.7	80.4	40.3
2468	DICHROSTACHYS CINEREA	F	.	0	16.6	5.4	72.4	22.1
2469	DICHROSTACHYS GLOMERATA	F	.	0	10.8	1.4	82.4	26.6
2470	DICLIPTERA ROXBURGHIANA	S	.	0	23.4	21.2	.	.
2471	DICTAMNUS ALBUS	S	.	0	24.4	36.2	.	.
2472	DIDISCUS CAERULEUS	S	.	0	22.5	37.4	.	.
2473	DIGITALIS FERRUGINEA	S	.	0	19.1	28.2	.	.
2474	DIGITALIS LANATA	S	.	0	17.1	34.7	.	.
2475	DIGITALIS ORIENTALIS	S	.	0	17.9	35.8	.	.
2476	DIGITALIS PURPUREA	S	.	0	16.8	36.1	.	.
2477	DIGITALIS PURPUREA	S	.	0	17.5	36.7	.	.
2478	DIGITALIS PURPUREA	S	.	0	14.8	38.9	.	.

OBS	ASH	CALCIUM	PHOSPHO	IRON	SODIUM	POTASS	CAROT	THIAMINE	RIBOFLAV	NIACIN	ASCORBIC	REFCO
2455	2.9	92.2	94.5	2.0	.	.	.	0.03	.	.	36.0	1
2456	.	.	.	.	.	.	.	.	.	.	.	19
2457	1.2	236.4	92.9	4.5	.	.	.	.	.	.	.	1
2458	.	.	.	.	.	.	.	.	.	.	.	20
2459	.	.	.	.	.	.	.	.	.	.	.	20
2460	2.8	.	.	.	.	.	.	.	.	.	.	19
2461	12.1	.	.	.	.	.	.	.	.	.	.	18
2462	11.5	.	.	.	.	.	.	.	.	.	.	18
2463	12.6	.	.	.	.	.	.	.	.	.	.	5
2464	10.9	.	.	.	.	.	.	.	.	.	.	5
2465	11.0	.	.	.	.	.	.	.	.	.	.	5
2466	11.7	.	.	.	.	.	.	.	.	.	.	18
2467	11.8	.	.	.	.	.	.	.	.	.	.	18
2468	5.6	920.0	120.0	.	.	.	.	.	.	.	.	18
2469	5.4	510.0	200.0	.	.	.	.	.	.	.	.	18
2470	.	.	.	.	.	.	.	.	.	.	.	21
2471	.	.	.	.	.	.	.	.	.	.	.	20
2472	3.0	.	.	.	.	.	.	.	.	.	.	19
2473	.	.	.	.	.	.	.	.	.	.	.	21
2474	.	.	.	.	.	.	.	.	.	.	.	21
2475	.	.	.	.	.	.	.	.	.	.	.	21
2476	3.6	.	.	.	.	.	.	.	.	.	.	19
2477	.	.	.	.	.	.	.	.	.	.	.	20
2478	.	.	.	.	.	.	.	.	.	.	.	21

TABLE 3 - DATA CONVERTED TO A ZERO-MOISTURE BASIS.

OBS	PLNA	PART	CALORIES	WATER	PROTEIN	FAT	TOTLCARB	FIBER
2479	DIGITALIS PURPUREA VAR. GLOXINIAEFLORA	S	.	0	16.2	40.4	.	.
2480	DIGITALIS THASPI	S	.	0	14.9	35.0	.	.
2481	DIGITARIA DECUMBENS	W	.	0	14.9	3.0	70.7	31.0
2482	DIGITARIA DECUMBENS	H	.	0	3.2	1.3	89.6	33.5
2483	DIGITARIA DECUMBENS	W	.	0	10.8	2.0	77.4	29.8
2484	DIGITARIA EXILIS	H	.	0	5.4	1.7	85.9	30.1
2485	DIGITARIA EXILIS	S	374	0	8.0	3.4	83.8	8.3
2486	DIGITARIA GAZENSIS	W	.	0	13.7	3.7	74.5	28.2
2487	DIGITARIA GAZENSIS	H	.	0	11.5	2.4	72.8	30.1
2488	DIGITARIA IBURUA	S	382	0	9.9	3.3	84.3	6.9
2489	DIGITARIA MACROBLEPHERA	W	.	0	12.3	1.8	74.7	32.2
2490	DIGITARIA MILANJIANA	W	.	0	8.1	2.3	79.9	36.8
2491	DIGITARIA MILANJIANA	H	.	0	12.5	2.3	76.0	29.2
2492	DIGITARIA PENTZII	W	.	0	14.0	3.0	68.7	28.4
2493	DIGITARIA PENTZII	H	.	0	10.1	2.0	80.1	32.9
2494	DIGITARIA SANGUINALIS	W	.	0	13.6	2.9	71.7	30.9
2495	DIGITARIA SCALARUM	W	.	0	14.7	3.8	72.0	29.0
2496	DIGITARIA SETIVALVA	H	.	0	13.2	2.6	72.8	27.7
2497	DIGITARIA SETIVALVA	W	.	0	10.4	2.4	75.8	29.8
2498	DIGITARIA SWAZILANDENSIS	H	.	0	13.2	1.3	75.8	34.6
2499	DIGITARIA SWAZILANDENSIS	W	.	0	7.5	1.9	79.0	30.5
2500	DILLENIA BURBIDGEI	F	349	0	3.6	2.4	89.2	10.8
2501	DILLENIA PHILIPPINENSIS	F	349	0	3.6	2.4	89.2	10.8
2502	DIMOCARPUS LONGAN	S	.	0	6.9	4.4	.	.

OBS	ASH	CALCIUM	PHOSPHO	IRON	SODIUM	POTASS	CAROT	THIAMINE	RIBOFLAV	NIACIN	ASCORBIC	REFCO
2479	.	.	.	.	.	.	.	.	.	.	.	20
2480	.	.	.	.	.	.	.	.	.	.	.	21
2481	11.4	.	.	.	.	.	.	.	.	.	.	18
2482	5.9	.	.	.	.	.	.	.	.	.	.	18
2483	9.8	450.0	350.0	.	.	.	.	.	.	.	.	6
2484	7.0	.	.	.	.	.	.	.	.	.	.	18
2485	4.8	46.2	215.1	9.6	.	.	.	0.27	0.11	2.14	.	1
2486	8.1	.	.	.	.	.	.	.	.	.	.	18
2487	13.3	.	.	.	.	.	.	.	.	.	.	18
2488	2.5	.	260.9	11.1	.	.	.	.	.	.	.	1
2489	11.2	.	.	.	.	.	.	.	.	.	.	18
2490	9.7	.	.	.	.	.	.	.	.	.	.	18
2491	9.2	.	.	.	.	.	.	.	.	.	.	18
2492	14.3	.	.	.	.	.	.	.	.	.	.	18
2493	7.8	.	.	.	.	.	.	.	.	.	.	18
2494	11.8	680.0	250.0	.	.	4220.0	.	.	.	.	.	6
2495	9.5	.	.	.	.	.	.	.	.	.	.	18
2496	11.4	.	.	.	.	.	.	.	.	.	.	18
2497	11.3	.	.	.	.	.	.	.	.	.	.	18
2498	9.7	.	.	.	.	.	.	.	.	.	.	18
2499	11.6	.	.	.	.	.	.	.	.	.	.	18
2500	4.8	337.3	60.2	1.2	120.5	1385.5	301.2	0.24	0.48	2.41	24.1	2
2501	4.8	337.3	60.2	1.2	120.5	1385.5	301.2	0.24	0.48	2.41	24.1	2
2502	1.5	.	.	.	.	.	.	.	.	.	.	19

TABLE 3 - DATA CONVERTED TO A ZERO-MOISTURE BASIS.

OBS	PLNA	PART	CALORIES	WATER	PROTEIN	FAT	TOTCARB	FIBER	ASH	CALCIUM	PHOSPHO	IRON	SODIUM	POTASS	CAROT	THIAMINE	RIBOFLAV	NIACIN	ASCORBIC	REFCO
2503	DIMOCARPUS LONGAN	F	374	0	5.3	7.4	82.1	1.6	5.3	121.1	189.5	2.1	.	.	.	0.16	0.74	1.58	294.7	2
2504	DIMOCARPUS LONGAN	F	347	0	5.7	0.6	89.8	2.3	4.0	56.8	238.6	6.8	.	.	.	.	.	.	34.1	4
2505	DIMOCARPUS LONGAN	F	.	0	8.0	2.3	51.1	3.4	.	.	.	.	.	.	.	.	.	.	.	5
2506	DIMORPHOTHECA ANNUA VAR. RINGENS	S	.	0	35.6	40.9	.	.	.	.	.	.	.	.	.	.	.	.	.	19
2507	DIMORPHOTHECA AURANTIACA	S	.	0	37.8	34.0	.	.	5.4	.	.	.	.	.	.	.	.	.	.	19
2508	DIMORPHOTHECA AURANTIACA	S	.	0	14.4	12.5	.	.	5.5	.	.	.	.	.	.	.	.	.	.	19
2509	DIMORPHOTHECA AURANTIACA	S	.	0	40.0	33.6	.	.	.	.	.	.	.	.	.	.	.	.	.	19
2510	DIMORPHOTHECA AURANTIACA	S	.	0	12.5	4.9	.	.	.	.	.	.	.	.	.	.	.	.	.	19
2511	DIMORPHOTHECA AURANTIACA	S	.	0	36.2	39.2	.	.	.	.	.	.	.	.	.	.	.	.	.	19
2512	DIMORPHOTHECA AURANTIACA	S	.	0	38.8	37.9	.	.	.	.	.	.	.	.	.	.	.	.	.	19
2513	DIMORPHOTHECA AURANTIACA	S	.	0	37.5	35.0	.	.	.	.	.	.	.	.	.	.	.	.	.	19
2514	DIMORPHOTHECA CALENDULACEA	S	.	0	38.1	35.5	.	.	.	.	.	.	.	.	.	.	.	.	.	19
2515	DIMORPHOTHECA CHRYSANTHEMIFOLIA	S	.	0	28.2	48.0	.	.	.	.	.	.	.	.	.	.	.	.	.	20
2516	DIMORPHOTHECA CUNEATA	S	.	0	38.2	32.2	.	.	.	.	.	.	.	.	.	.	.	.	.	20
2517	DIMORPHOTHECA PLUVIALIS	S	.	0	38.1	36.2	.	.	.	.	.	.	.	.	.	.	.	.	.	20
2518	DIMORPHOTHECA PLUVIALIS	S	.	0	27.1	44.6	.	.	.	.	.	.	.	.	.	.	.	.	.	20
2519	DIMORPHOTHECA PLUVIALIS	S	.	0	32.4	41.5	.	.	.	.	.	.	.	.	.	.	.	.	.	20
2520	DIMORPHOTHECA PLUVIALIS	S	.	0	25.4	32.9	.	.	.	.	.	.	.	.	.	.	.	.	.	20
2521	DIMORPHOTHECA PLUVIALIS	S	.	0	35.3	41.7	.	.	.	.	.	.	.	.	.	.	.	.	.	20
2522	DIMORPHOTHECA SINUATA	S	.	0	41.9	34.4	.	.	.	.	.	.	.	.	.	.	.	.	.	19
2523	DIMORPHOTHECA SINUATA	S	.	0	32.5	43.7	.	.	.	.	.	.	.	.	.	.	.	.	.	19
2524	DIMORPHOTHECA SINUATA	S	.	0	22.5	20.3	.	.	.	.	.	.	.	.	.	.	.	.	.	20
2525	DIMORPHOTHECA SINUATA	S	.	0	39.4	42.5	.	.	.	.	.	.	.	.	.	.	.	.	.	20
2526	DIMORPHOTHECA SINUATA	S	.	0	35.6	39.2	.	.	.	.	.	.	.	.	.	.	.	.	.	20
2527	DIMORPHOTHECA SINUATA	S	.	0	33.0	42.1	.	.	.	.	.	.	.	.	.	.	.	.	.	20
2528	DIMORPHOTHECA SINUATA	S	.	0	32.3	42.4	.	.	.	.	.	.	.	.	.	.	.	.	.	20
2529	DIMORPHOTHECA SINUATA	S	.	0	38.8	38.2	.	.	.	.	.	.	.	.	.	.	.	.	.	20
2530	DIMORPHOTHECA SINUATA	S	.	0	.	33.4	.	.	.	.	.	.	.	.	.	.	.	.	.	21
2531	DIMORPHOTHECA SINUATA	S	.	0	.	32.0	.	.	.	.	.	.	.	.	.	.	.	.	.	21
2532	DIMORPHOTHECA SINUATA	S	.	0	.	30.7	.	.	.	.	.	.	.	.	.	.	.	.	.	21
2533	DIMORPHOTHECA SP.	S	.	0	36.2	41.2	.	.	.	.	.	.	.	.	.	.	.	.	.	20
2534	DIMORPHOTHECA SP.	S	.	0	36.9	41.4	.	.	.	.	.	.	.	.	.	.	.	.	.	20
2535	DIMORPHOTHECA ZEYHERI	S	.	0	42.4	36.2	.	.	.	.	.	.	.	.	.	.	.	.	.	20
2536	DINEBRA RETROFLEXA	L	.	0	3.6	2.9	86.3	22.7	7.6	.	.	.	.	.	.	.	.	.	.	5
2537	DINEBRA RETROFLEXA	L	.	0	5.3	3.0	84.4	20.9	7.6	.	.	.	.	.	.	.	.	.	.	5
2538	DINEBRA RETROFLEXA	L	.	0	3.3	3.9	81.6	37.2	11.5	.	.	.	.	.	.	.	.	.	.	5
2539	DIODIA TERES VAR. SETIFERA	S	.	0	13.6	9.1	.	.	.	.	.	.	.	.	.	.	.	.	.	21
2540	DIOSCOREA ALATA	W	.	0	12.0	2.3	77.8	25.3	7.9	950.0	160.0	.	.	.	.	.	.	.	.	18
2541	DIOSCOREA ALATA	R	.	0	8.1	0.8	85.9	2.6	5.2	120.0	180.0	.	.	.	.	.	.	.	.	18
2542	DIOSCOREA ALATA	R	387	0	6.6	0.3	91.1	4.3	2.0	80.2	149.0	4.6	.	.	28.65	0.14	0.09	1.43	34.4	1
2543	DIOSCOREA BULBIFERA	R	386	0	5.2	0.3	91.4	3.1	3.1	237.9	100.0	.	.	.	.	.	.	.	.	1
2544	DIOSCOREA CAYENENSIS	R	.	0	6.2	0.4	90.2	2.4	3.2	.	.	.	.	.	.	.	.	.	.	18
2545	DIOSCOREA CAYENENSIS	R	370	0	7.8	0.5	85.4	3.1	6.3	187.5	88.5	27.1	.	.	.	.	.	.	.	1
2546	DIOSCOREA DUMETORUM	R	379	0	9.8	0.3	86.5	2.4	3.4	159.0	137.6	.	.	.	.	.	.	.	.	1
2547	DIOSCOREA ESCULENTA	R	.	0	8.7	0.5	86.6	3.5	4.2	.	.	.	.	.	.	.	.	.	.	18

TABLE 3 - DATA CONVERTED TO A ZERO-MOISTURE BASIS.

OBS	PLNA	PART	CALORIES	WATER	PROTEIN	FAT	TOTLCARB	FIBER	ASH
2548	DIOSCOREA ESCULENTA	R	.	0	10.8	1.7	77.8	6.5	9.6
2549	DIOSCOREA ESCULENTA	R	386	0	5.7	0.8	90.5	2.3	3.0
2550	DIOSCOREA ESCULENTA	R	376	0	11.7	0.3	84.6	1.7	3.4
2551	DIOSCOREA ESCULENTA	R	380	0	6.0	0.4	89.6	.	4.0
2552	DIOSCOREA LUZONENSIS	R	383	0	9.9	0.4	87.4	1.6	2.4
2553	DIOSCOREA MINUTIFLORA	R	373	0	9.5	0.3	85.4	31.6	4.7
2554	DIOSCOREA MINUTIFLORA	R	373	0	9.5	0.3	85.4	31.6	4.7
2555	DIOSCOREA PREUSII	R	372	0	9.6	0.3	84.6	4.2	5.4
2556	DIOSCOREA ROTUNDATA	R	.	0	5.9	0.5	89.3	2.4	4.3
2557	DIOSCOREA SP.	R	383	0	7.1	0.4	88.7	1,8	3.5
2558	DIOSCOREA SP.	L	.	0	2.9	0.0	.	.	.
2559	DIOSCOREA SP.	R	365	0	7.3	0.7	88.7	2.2	3.3
2560	DIOSCOREA SP.	R	381	0	7.9	0.8	87.5	3.4	3.8
2561	DIOSPYROS ANISANDRA	S	.	0	7.5	3.0	.	.	.
2562	DIOSPYROS DISCOLOR	F	350	0	4.3	1.4	90.0	11.4	4.3
2563	DIOSPYROS DISCOLOR	F	.	0	16.5	1.2	79.4	10.6	2.4
2564	DIOSPYROS DISCOLOR	F	.	0	3.2	0.7	91.8	6.1	4.3
2565	DIOSPYROS EBENUM	F	509	0	3.2	33.8	59.9	0.9	3.2
2566	DIOSPYROS EBENUM	F	367	0	3.9	6.7	83.3	8.9	6.1
2567	DIOSPYROS KAKI	F	.	0	.	.	.	.	2.6
2568	DIOSPYROS KAKI	F	344	0	3.7	1.2	89.6	7.4	5.5
2569	DIOSPYROS KAKI	F	358	0	3.7	1.8	91.7	8.7	2.8
2570	DIOSPYROS KAKI	F	360	0	3.3	1.9	92.1	7.5	2.8
2571	DIOSPYROS KAKI	F	.	0	3.9	1.0	93.1	.	2.0

OBS	CALCIUM	PHOSPHO	IRON	SODIUM	POTASS	CAROT	THIAMINE	RIBOFLAV	NIACIN	ASCORBIC	REFCO
2548	.	940.0	.	.	.	.	.	.	.	.	5
2549	45.5	132.6	3.0	30.3	1386.4	.	0.38	0.04	3.03	56.8	
2550	208.1	177.9	.	.	.	.	.	.	.	.	1
2551	56.0	196.0	3.2	.	.	0.0	0.20	.	.	16.0	2
2552	237.2	162.1	2.8	31.6	1205.5	0.0	0.43	0.08	2.77	71.1	2
2553	218.4	129.7	.	.	.	.	.	.	.	.	1
2554	218.4	129.7	.	.	.	.	.	.	.	.	1
2555	192.3	160.3	.	.	.	.	.	.	.	.	1
2556	.	.	.	.	.	.	.	.	.	.	18
2557	78.0	138.3	3.5	35.5	1042.6	0.0	0.35	0.14	2.48	28.4	2
2558	1435.7	.	7.9	.	.	46571.4	1.07	.	.	221.4	2
2559	51.1	156.9	4.7	.	.	0.0	0.47	0.07	1.46	10.9	3
2560	75.5	260.4	2.3	.	2264.2	0.0	0.38	0.15	1.89	34.0	4
2561	.	.	.	.	.	.	.	.	.	.	19
2562	271.4	135.7	5.7	21.4	2528.6	107.1	0.00	0.29	2.14	200.0	2
2563	.	.	.	.	.	.	.	.	.	.	5
2564	.	.	.	.	.	.	.	.	.	.	5
2565	67.6	112.6	1.4	.	.	90.1	0.00	0.23	1.35	18.0	2
2566	100.0	144.4	6.7	.	.	444.4	0.11	0.17	1.11	161.1	3
2567	125.7	.	.	.	.	7434.5	.	.	.	26.2	1
2568	55.2	147.2	2.5	12.3	1079.8	3681.0	0.18	0.25	1.84	122.7	2
2569	27.5	119.3	1.4	.	.	6880.7	0.23	0.23	0.00	50.5	3
2570	28.0	121.5	1.4	28.0	813.1	7598.1	0.14	0.09	0.47	51.4	4
2571	49.0	49.0	14.7	.	.	5029.4	0.15	0.15	0.25	49.0	5

TABLE 3 - DATA CONVERTED TO A ZERO-MOISTURE BASIS.

OBS	PLNA	PART	CALORIES	WATER	PROTEIN	FAT	TOTLCARB	FIBER	ASH
2572	DIOSPYROS LOTUS	S	.	0	9.3	2.6	.	.	.
2573	DIOSPYROS LYCIOIDES SUBSP. SERICEA	F	.	0	.	.	.	.	.
2574	DIOSPYROS MESPILIFORMIS	F	352	0	2.8	0.0	95.5	1.7	1.7
2575	DIOSPYROS TEXANA	S	.	0	8.1	2.8	.	.	.
2576	DIOSPYROS TEXANA	L	.	0	13.0	7.4	69.2	21.0	10.4
2577	DIOSPYROS VIRGINIANA	S	.	0	10.0	2.6	.	.	.
2578	DIOSPYROS VIRGINIANA	F	357	0	2.2	1.1	94.1	4.2	2.5
2579	DIOTIS MARITIMA	S	.	0	8.9	7.0	.	.	.
2580	DIPCADI GLAUCUM	S	.	0	27.1	10.3	.	.	.
2581	DIPLOKNEMA BUTYRACEA	S	.	0	5.5	58.8	31.6	.	4.0
2582	DIPLOTAXIS CATHOLICA	S	.	0	25.4	31.4	.	.	.
2583	DIPLOTAXIS ERUCOIDES	S	.	0	32.1	32.1	.	.	.
2584	DIPLOTAXIS ERUCOIDES	S	.	0	26.7	35.9	.	.	.
2585	DIPLOTAXIS SIIFOLIA	S	.	0	27.8	33.0	.	.	.
2586	DIPSACUS SYLVESTRIS	S	.	0	23.6	27.7	.	.	.
2587	DIPSACUS SYLVESTRIS	S	.	0	22.6	28.3	.	.	.
2588	DISANTHUS CERCIDIFOLIUS	S	.	0	12.5	16.2	.	.	.
2589	DISTICHLIS SPICATA	W	.	0	6.9	1.7	82.1	29.8	9.3
2590	DISTICHLIS STRICTA	W	.	0	5.9	1.7	85.6	29.7	6.8
2591	DITHYREA WISLIZENII	S	.	0	16.9	3.9	.	.	.
2592	DODONAEA VISCOSA	S	.	0	21.4	19.0	.	.	.
2593	DODONAEA VISCOSA	S	.	0	20.4	7.6	.	.	.
2594	DOLICHOS AXILLARIS	S	.	0	28.8	0.6	.	.	.
2595	DOLICHOS BIFLORUS	S	.	0	33.1	0.6	.	.	6.3

OBS	CALCIUM	PHOSPHO	IRON	SODIUM	POTASS	CAROT	THIAMINE	RIBOFLAV	NIACIN	ASCORBIC	REFCO
2572	.	.	.	.	.	.	.	.	.	.	21
2573	.	.	.	.	.	.	.	.	.	87.7	1
2574	118.3	129.6	5.6	.	.	.	0.03	.	0.56	36.6	1
2575	.	.	.	.	.	.	.	.	.	.	19
2576	4110.0	120.0	.	.	530.0	.	.	.	.	.	6
2577	.	.	.	.	.	.	.	.	.	.	19
2578	75.8	73.0	7.0	2.8	870.8	.	.	.	.	185.4	4
2579	.	.	.	.	.	.	.	.	.	.	21
2580	.	.	.	.	.	.	.	.	.	.	21
2581	.	.	.	.	.	.	.	.	.	.	5
2582	.	.	.	.	.	.	.	.	.	.	21
2583	.	.	.	.	.	.	.	.	.	.	21
2584	.	.	.	.	.	.	.	.	.	.	21
2585	.	.	.	.	.	.	.	.	.	.	21
2586	.	.	.	.	.	.	.	.	.	.	21
2587	.	.	.	.	.	.	.	.	.	.	21
2588	.	.	.	.	.	.	.	.	.	.	20
2589	330.0	120.0	19.0	.	240.0	.	.	.	.	.	6
2590	160.0	90.0	.	.	.	.	.	.	.	.	6
2591	.	.	.	.	.	.	.	.	.	.	19
2592	.	.	.	.	.	.	.	.	.	.	21
2593	.	.	.	.	.	.	.	.	.	.	21
2594	.	.	.	.	.	.	.	.	.	.	19
2595	.	.	.	.	.	.	.	.	.	.	19

TABLE 3 - DATA CONVERTED TO A ZERO-MOISTURE BASIS.

OBS	PLNA	PART	CALORIES	WATER	PROTEIN	FAT	TOTLCARB	FIBER
2596	DOLICHOS BIFLORUS	W	.	0	16.4	0.2	72.3	30.2
2597	DOLICHOS LABLAB	SH	.	0	14.2	3.5	67.5	28.1
2598	DOLICHOS LABLAB	H	.	0	16.6	2.9	71.2	37.1
2599	DOLICHOS LABLAB	F	.	0	10.1	0.8	82.9	36.2
2600	DOLICHOS LABLAB	S	.	0	28.0	1.2	66.6	8.6
2601	DOLICHOS LABLAB	S	.	0	24.9	1.0	.	.
2602	DOLICHOS LABLAB	S	.	0	21.8	1.2	.	.
2603	DOLICHOS LABLAB	S	382	0	25.6	1.1	69.7	9.7
2604	DOLICHOS LABLAB	S	374	0	27.1	0.9	67.2	.
2605	DOLICHOS LABLAB	S	380	0	24.5	1.4	69.9	7.7
2606	DOLICHOS LABLAB	S	384	0	25.4	1.2	70.4	9.1
2607	DOLICHOS LABLAB	S	383	0	25.2	1.7	69.2	7.8
2608	DOLICHOS LABLAB	F	313	0	25.0	2.7	65.2	16.1
2609	DOLICHOS LABLAB	F	.	0	25.6	0.6	68.2	11.4
2610	DOLICHOS LIGNOSUS	S	374	0	27.1	0.9	67.0	.
2611	DOLICHOS SP.	S	.	0	28.8	0.5	.	.
2612	DORONICUM CAUCASICUM	S	.	0	20.6	38.4	.	.
2613	DORYCNIUM ANATOLICUM	S	.	0	42.2	5.7	.	.
2614	DORYCNIUM INTERMEDIUM	S	.	0	31.2	5.3	.	.
2615	DOVYALIS HEBECARPA	F	366	0	7.0	4.7	84.9	10.5
2616	DRABA SP.	W	.	0	15.8	2.1	67.7	17.6
2617	DRACOCEPHALUM MOLDAVICA	S	.	0	21.2	25.3	.	.
2618	DRACOPIS AMPLEXICAULIS	S	.	0	19.3	29.9	.	.
2619	DUCROSIA ANETHIFOLIA	S	.	0	13.4	11.6	.	.

OBS	ASH	CALCIUM	PHOSPHO	IRON	SODIUM	POTASS	CAROT	THIAMINE	RIBOFLAV	NIACIN	ASCORBIC	REFCO
2596	11.1	.	.	.	.	.	.	.	.	.	.	6
2597	14.8	1980.0	260.0	.	.	.	.	.	.	.	.	18
2598	9.3	.	.	.	.	.	.	.	.	.	.	18
2599	6.2	900.0	120.0	.	.	.	.	.	.	.	.	18
2600	4.2	490.0	360.0	.	.	.	.	.	.	.	.	18
2601	4.2	.	.	.	.	.	.	.	.	.	.	19
2602	3.7	.	.	.	.	.	.	.	.	.	.	19
2603	3.6	101.0	368.1	10.1	.	.	.	0.61	0.16	2.58	0.0	1
2604	4.8	196.0	427.3	4.3	.	.	.	0.19	.	.	.	2
2605	4.3	111.5	392.5	4.4	.	.	.	0.46	0.14	2.05	0.0	2
2606	3.0	52.9	431.5	8.4	.	.	.	0.51	0.14	1.73	.	3
2607	4.0	82.8	473.9	5.8	.	.	.	0.70	0.20	2.38	.	4
2608	7.1	508.9	473.2	8.9	17.9	2544.6	3107.14	0.80	0.98	8.04	178.6	4
2609	5.7	284.1	340.9	9.7	.	.	.	.	.	4.55	51.1	5
2610	4.8	196.0	196.0	.	.	.	.	.	.	.	.	10
2611	.	.	.	.	.	.	.	.	.	.	.	19
2612	.	.	.	.	.	.	.	.	.	.	.	20
2613	.	.	.	.	.	.	.	.	.	.	.	21
2614	.	.	.	.	.	.	.	.	.	.	.	21
2615	3.5	75.6	151.2	7.0	.	.	1220.93	0.12	0.23	1.74	569.8	3
2616	14.4	3310.0	330.0	.	.	1780.0	.	.	.	.	.	6
2617	.	.	.	.	.	.	.	.	.	.	.	20
2618	.	.	.	.	.	.	.	.	.	.	.	20
2619	.	.	.	.	.	.	.	.	.	.	.	21

TABLE 3 - DATA CONVERTED TO A ZERO-MOISTURE BASIS.

OBS	PLNA	PART	CALORIES	WATER	PROTEIN	FAT	TOTLCARB	FIBER
2620	DURANTA REPENS	S	.	0	13.1	1.4	.	.
2621	DURIO ZIBETHINUS	F	373	0	7.5	4.8	85.2	4.2
2622	DURIO ZIBETHINUS	S	392	0	5.4	0.8	89.9	.
2623	DURIO ZIBETHINUS	F	354	0	11.6	4.2	78.3	8.5
2624	DURIO ZIBETHINUS	P	.	0	6.7	9.3	81.2	.
2625	DYSSODIA CANCELLATA	S	.	0	20.6	15.2	.	.
2626	DYSSODIA CANCELLATA	S	.	0	23.8	19.4	.	.
2627	DYSSODIA HARTWEGI & DYSSODIA PENTACHAETA	S	.	0	21.2	18.0	.	.
2628	DYSSODIA PINNATA	S	.	0	18.1	11.8	.	.
2629	DYSSODIA ROSEATA	S	.	0	18.8	20.3	.	.
2630	DYSSODIA TAGETIFLORA	S	.	0	23.8	21.4	.	.
2631	EBENUS HIRSUTA	S	.	0	53.7	10.7	.	.
2632	ECBALLIUM ELATERIUM	S	.	0	28.2	57.0	.	.
2633	ECBALLIUM ELATERIUM	S	.	0	29.7	57.6	.	.
2634	ECHINACEA ANGUSTIFOLIA	S	.	0	49.4	37.1	.	.
2635	ECHINACEA ANGUSTIFOLIA	S	.	0	11.0	6.7	.	.
2636	ECHINACEA ANGUSTIFOLIA	S	.	0	23.2	9.4	.	.
2637	ECHINACEA SP.	S	.	0	30.6	16.0	.	.
2638	ECHINARIA CAPITATA	S	.	0	17.2	1.6	.	.
2639	ECHINOCHLOA COLONUM	SH	.	0	12.5	2.4	72.7	28.6
2640	ECHINOCHLOA COLONUM	S	.	0	12.5	7.0	.	.
2641	ECHINOCHLOA COLONUM	W	.	0	17.4	2.3	66.7	29.6
2642	ECHINOCHLOA CRUS-GALLI	SH	.	0	7.4	2.9	81.1	31.1
2643	ECHINOCHLOA CRUS-GALLI	H	.	0	13.5	2.5	73.6	22.7

OBS	ASH	CALCIUM	PHOSPHO	IRON	SODIUM	POTASS	CAROT	THIAMINE	RIBOFLAV	NIACIN	ASCORBIC	REFCO
2620	.	.	.	.	.	.	.	.	.	.	.	21
2621	2.4	60.2	189.8	2.7	3.0	1810.2	30.12	0.81	0.87	3.61	111.4	2
2622	3.9	35.1	140.2	2.1	6.2	1983.5	515.46	.	0.10	1.86	.	2
2623	5.8	42.3	201.1	3.7	.	.	105.82	1.85	1.06	3.70	127.0	3
2624	2.9	23.8	119.0	2.4	.	.	28.57	.	.	.	59.5	5
2625	.	.	.	.	.	.	.	.	.	.	.	20
2626	.	.	.	.	.	.	.	.	.	.	.	20
2627	.	.	.	.	.	.	.	.	.	.	.	20
2628	.	.	.	.	.	.	.	.	.	.	.	20
2629	.	.	.	.	.	.	.	.	.	.	.	20
2630	.	.	.	.	.	.	.	.	.	.	.	21
2631	.	.	.	.	.	.	.	.	.	.	.	21
2632	.	.	.	.	.	.	.	.	.	.	.	21
2633	.	.	.	.	.	.	.	.	.	.	.	21
2634	2.7	.	.	.	.	.	.	.	.	.	.	19
2635	4.3	.	.	.	.	.	.	.	.	.	.	19
2636	.	.	.	.	.	.	.	.	.	.	.	20
2637	.	.	.	.	.	.	.	.	.	.	.	20
2638	.	.	.	.	.	.	.	.	.	.	.	21
2639	12.4	.	.	.	.	.	.	.	.	.	.	18
2640	.	.	.	.	.	.	.	.	.	.	.	19
2641	13.6	830.0	240.0	.	.	3670.0	.	.	.	.	.	6
2642	8.6	.	.	.	.	.	.	.	.	.	.	18
2643	10.4	.	.	.	.	.	.	.	.	.	.	18

TABLE 3 - DATA CONVERTED TO A ZERO-MOISTURE BASIS.

OBS	PLNA	PART	CALORIES	WATER	PROTEIN	FAT	TOTLCARB	FIBER
2644	ECHINOCHLOA CRUS-GALLI	S	.	0	15.6	4.3	.	.
2645	ECHINOCHLOA CRUS-GALLI	S	.	0	12.1	7.1	.	.
2646	ECHINOCHLOA CRUS-GALLI	W	.	0	13.7	2.9	72.2	22.0
2647	ECHINOCHLOA CRUS-GALLI	W	.	0	9.0	2.7	79.9	29.9
2648	ECHINOCHLOA CRUS-GALLI	S	385	0	11.0	4.4	81.7	8.1
2649	ECHINOCHLOA CRUS-GALLI	S	.	0	7.0	2.5	85.5	11.1
2650	ECHINOCHLOA CRUSGALLI VAR. FRUMENTACEA	S	.	0	11.9	5.9	.	.
2651	ECHINOCHLOA HAPLOCLADA	SH	.	0	13.7	3.0	72.0	30.7
2652	ECHINOCHLOA POLYSTACHYA	SH	.	0	13.8	4.2	69.3	32.0
2653	ECHINOCHLOA POLYSTACHYA	H	.	0	7.9	2.1	79.5	35.5
2654	ECHINOCHLOA PYRAMIDALIS	SH	.	0	7.0	1.1	83.3	31.4
2655	ECHINOCHLOA PYRAMIDALIS	H	.	0	15.6	2.4	71.8	33.5
2656	ECHINOCHLOA SP.	S	.	0	18.8	6.2	.	.
2657	ECHINOCHLOA SP.	S	.	0	10.0	3.1	.	.
2658	ECHINOCHLOA SP.	S	.	0	18.8	6.4	.	.
2659	ECHINOCHLOA STAGNINA	SH	.	0	11.3	2.2	76.6	32.5
2660	ECHINOCYSTIS FABACEA	S	.	0	31.2	51.1	.	.
2661	ECHINOCYSTIS OREGANA	S	.	0	33.5	47.0	.	.
2662	ECHINOPEPON HORRIDUS	S	.	0	20.2	21.4	.	.
2663	ECHINOPS EXALTATUS	S	.	0	37.4	34.7	.	.
2664	ECHINOPS STRIGOSUS	S	.	0	36.9	38.8	.	.
2665	ECHINOPS VISCOSUS	S	.	0	37.6	27.8	.	.
2666	ECHIUM ITALICUM	S	.	0	16.1	17.4	.	.
2667	ECHIUM PLANTAGINEUM	S	.	0	19.4	29.9	.	.

OBS	ASH	CALCIUM	PHOSPHO	IRON	SODIUM	POTASS	CAROT	THIAMINE	RIBOFLAV	NIACIN	ASCORBIC	REFCO
2644	4.2	.	.	.	.	.	.	.	.	.	.	19
2645	.	.	.	.	.	.	.	.	.	.	.	21
2646	11.2	.	.	.	.	.	.	.	.	.	.	6
2647	8.4	.	.	.	.	.	.	.	.	.	.	6
2648	2.9	46.1	371.2	4.8	.	.	0.0	0.37	0.11	4.50	0.0	2
2649	5.0	22.7	31.8	3.3	.	.	0.0	.	.	.	.	5
2650	3.2	.	.	.	.	.	.	.	.	.	.	19
2651	11.3	.	.	.	.	.	.	.	.	.	.	18
2652	12.7	.	.	.	.	.	.	.	.	.	.	18
2653	10.5	.	.	.	.	.	.	.	.	.	.	18
2654	8.6	.	.	.	.	.	.	.	.	.	.	18
2655	10.2	.	.	.	.	.	.	.	.	.	.	18
2656	.	.	.	.	.	.	.	.	.	.	.	19
2657	.	.	.	.	.	.	.	.	.	.	.	19
2658	.	.	.	.	.	.	.	.	.	.	.	20
2659	9.9	.	.	.	.	.	.	.	.	.	.	18
2660	3.2	.	.	.	.	.	.	.	.	.	.	19
2661	3.7	.	.	.	.	.	.	.	.	.	.	19
2662	.	.	.	.	.	.	.	.	.	.	.	21
2663	5.1	.	.	.	.	.	.	.	.	.	.	19
2664	.	.	.	.	.	.	.	.	.	.	.	21
2665	.	.	.	.	.	.	.	.	.	.	.	21
2666	.	.	.	.	.	.	.	.	.	.	.	21
2667	.	.	.	.	.	.	.	.	.	.	.	20

TABLE 3 - DATA CONVERTED TO A ZERO-MOISTURE BASIS.

OBS	PLANA	PART	CALORIES	WATER	PROTEIN	FAT	TOTCARB	FIBER	ASH	CALCIUM	PHOSPHO	IRON	SODIUM	POTASS	CAROT	THIAMINE	RIBOFLAV	NIACIN	ASCORBIC	REFCO
2668	ECLIPTA ALBA	S	.	0	15.6	13.1	.	.	.	.	.	.	.	.	.	.	.	.	.	21
2669	EHRHARTA CALYCINA	S	.	0	13.1	2.2	.	.	2.0	.	.	.	.	.	.	.	.	.	.	19
2670	EHRHARTA CALYCINA	S	.	0	8.1	2.1	.	.	6.3	.	.	.	.	.	.	.	.	.	.	19
2671	EICHHORNIA CRASSIPES	L	.	0	22.3	2.0	60.6	19.5	15.1	745.0	890.0	158.0	1860.0	2780.0	245.0	0.59	3.07	7.94	.	13
2672	EICHHORNIA CRASSIPES	L	.	0	23.4	2.2	54.0	17.1	20.4	560.0	850.0	256.0	1630.0	3450.0	.	.	.	.	.	13
2673	EICHHORNIA CRASSIPES	L	.	0	17.1	1.6	70.2	18.6	11.1	1720.0	450.0	212.0	1660.0	1640.0	0.0	.	.	.	.	13
2674	EICHHORNIA CRASSIPES	L	.	0	9.7	1.7	68.7	19.2	19.9	792.0	310.0	615.0	2020.0	2660.0	.	.	.	.	.	13
2675	EICHHORNIA CRASSIPES	W	.	0	13.1	1.3	70.3	18.2	15.3	2160.0	410.0	.	.	.	.	.	.	.	.	18
2676	EICHHORNIA CRASSIPES	H	.	0	11.6	0.7	69.9	24.2	17.8	2190.0	640.0	.	.	.	.	.	.	.	.	18
2677	EICHHORNIA CRASSIPES	R	.	0	5.8	0.9	89.6	20.5	3.7	.	.	.	.	.	.	.	.	.	.	18
2678	EICHHORNIA CRASSIPES	L	294	0	4.9	1.0	73.5	23.5	20.6	.	.	.	.	.	.	.	.	.	.	2
2679	ELAEAGNUS ANGUSTIFOLIA	S	.	0	41.9	26.3	.	.	2.4	.	.	.	.	.	.	.	.	.	.	19
2680	ELAEAGNUS ANGUSTIFOLIA	S	.	0	9.2	1.8	.	.	.	.	.	.	.	.	.	.	.	.	.	21
2681	ELAEAGNUS ANGUSTIFOLIA	S	.	0	9.5	5.3	.	.	.	.	.	.	.	.	.	.	.	.	.	21
2682	ELAEAGNUS COMMUTATA	F	370	0	9.4	6.5	79.0	3.6	5.1	50.7	144.9	2.9	.	.	144.9	0.22	0.36	2.90	72.5	2
2683	ELAEAGNUS PUNGENS	S	.	0	42.2	23.1	.	.	.	.	.	.	.	.	.	.	.	.	.	21
2684	ELAEAGNUS UMBELLATA	S	.	0	26.0	14.5	.	.	.	.	.	.	.	.	.	.	.	.	.	21
2685	ELAEIS GUINEENSIS	F	.	0	7.9	54.0	36.4	3.9	1.7	90.0	310.0	.	.	.	.	.	.	.	.	18
2686	ELAEIS GUINEENSIS	S	.	0	9.9	54.4	.	.	1.6	.	.	.	.	.	.	.	.	.	.	19
2687	ELAEIS GUINEENSIS	F	732	0	2.6	79.1	16.9	4.3	1.4	111.1	63.7	6.1	.	.	57479.7	0.27	0.14	1.90	16.3	1
2688	ELAEIS GUINEENSIS	F	746	0	2.2	81.9	14.6	3.8	1.3	136.1	61.1	5.6	.	.	50680.6	0.35	0.17	1.81	12.5	3
2689	ELAEIS GUINEENSIS	O	882	0	0.0	99.6	0.4	0.0	0.0	7.0	8.0	5.5	.	.	27417.1	0.00	0.03	0.00	.	3
2690	ELAEOCARPUS SERRATUS	F	353	0	4.3	0.4	92.6	12.0	2.7	170.5	139.5	6.2	.	.	620.2	0.08	0.04	1.55	391.5	3
2691	ELEGIA GALPINII	S	.	0	3.1	0.3	.	.	.	.	.	.	.	.	.	.	.	.	.	21
2692	ELEGIA PARVIFLORA	S	.	0	17.6	0.7	.	.	.	.	.	.	.	.	.	.	.	.	.	21
2693	ELEOCHARIS DULCIS	R	360	0	7.4	1.6	85.2	3.2	5.8	26.5	407.4	3.7	52.9	2545.0	0.0	0.16	0.11	5.29	31.7	2
2694	ELEOCHARIS DULCIS	R	364	0	6.5	0.9	87.6	3.7	5.1	18.4	299.5	2.8	92.2	2304.1	0.0	0.65	0.92	4.61	18.4	4
2695	ELEOCHARIS DULCIS	R	.	0	8.5	0.7	84.8	3.9	6.0	.	.	.	.	.	.	.	.	.	.	5
2696	ELEOCHARIS PALUSTRIS	W	.	0	10.3	.	.	29.0	13.9	450.0	210.0	.	.	2720.0	.	.	.	.	.	6
2697	ELEPHANTOPUS SCABER	S	.	0	26.9	21.5	.	.	.	.	.	.	.	.	.	.	.	.	.	20
2698	ELEPHANTORRHIZA ELEPHANTINA	S	.	0	22.2	39.0	.	.	.	.	.	.	.	.	.	.	.	.	.	21
2699	ELETTARIA CARDAMOMUM	S	339	0	11.8	7.3	74.7	12.3	6.3	417.7	194.1	15.3	19.6	1285.7	.	0.22	0.20	1.20	.	16
2700	ELETTARIA CARDAMOMUM	F	.	0	12.8	2.8	77.8	25.1	6.8	162.5	200.0	6.3	.	.	.	.	.	.	.	5
2701	ELEUSINE CORACANA	W	.	0	7.6	1.1	76.2	33.6	15.1	.	.	.	.	.	.	.	.	.	.	18
2702	ELEUSINE CORACANA	H	.	0	3.2	1.3	87.6	34.2	7.9	.	.	.	.	.	.	.	.	.	.	18
2703	ELEUSINE CORACANA	S	369	0	8.3	1.5	87.2	4.8	3.0	445.6	273.8	19.2	.	.	0.0	0.20	0.12	0.90	.	1
2704	ELEUSINE CORACANA	S	376	0	7.0	1.6	88.6	1.4	2.8	372.6	287.7	6.0	.	.	0.0	0.37	0.11	1.47	0.0	2
2705	ELEUSINE CORACANA	S	.	0	8.2	1.5	87.8	.	2.5	379.7	310.7	6.2	.	.	48.3	0.46	.	1.27	.	5
2706	ELEUSINE CORACANA	L	.	0	9.9	2.6	68.8	28.1	18.2	.	.	.	.	.	.	.	.	.	.	5
2707	ELEUSINE INDICA	W	.	0	13.3	2.8	71.9	28.5	12.0	.	.	.	.	.	.	.	.	.	.	18
2708	ELEUSINE INDICA	H	.	0	3.5	1.1	85.5	35.8	9.9	.	.	.	.	.	.	.	.	.	.	18
2709	ELEUSINE INDICA	H	.	0	3.5	1.1	85.5	35.8	9.9	.	.	.	.	.	.	.	.	.	.	6
2710	ELIONURUS TRIPSACOIDES	L	.	0	5.3	1.7	87.6	39.1	5.4	250.0	100.0	.	.	.	.	.	.	.	.	6
2711	ELYMUS AMBIGUUS	H	.	0	7.9	1.8	84.5	36.5	5.8	.	.	.	.	.	.	.	.	.	.	6
2712	ELYMUS CANADENSIS	S	.	0	26.9	2.2	.	.	2.9	.	.	.	.	.	.	.	.	.	.	19

TABLE 3 - DATA CONVERTED TO A ZERO-MOISTURE BASIS.

OBS	PLNA	PART	CALORIES	WATER	PROTEIN	FAT	TOTCARB	FIBER	ASH	CALCIUM	PHOSPHO	IRON	SODIUM	POTASS	CAROT	THIAMINE	RIBOFLAV	NIACIN	ASCORBIC	REFCO
2713	ELYMUS CANADENSIS	S	.	0	16.2	3.2	.	.	.	.	.	.	.	.	.	.	.	.	.	19
2714	ELYMUS CANADENSIS	W	.	0	8.4	2.3	77.8	30.6	11.5	760.0	210.0	.	.	.	.	.	.	.	.	6
2715	ELYMUS CAPUT-MEDUSAE	S	.	0	14.2	1.9	.	.	.	.	.	.	.	.	.	.	.	.	.	21
2716	ELYMUS CONDENSATUS	W	.	0	9.3	3.2	77.4	35.9	10.1	390.0	170.0	.	.	.	.	.	.	.	.	6
2717	ELYMUS GIGANTEUS	S	.	0	16.3	2.1	.	.	3.0	.	.	.	.	.	.	.	.	.	.	19
2718	ELYMUS GLAUCUS	W	.	0	7.3	2.3	.	.	5.5	.	.	.	.	.	.	.	.	.	.	6
2719	ELYMUS JUNCEUS	S	.	0	17.1	1.9	.	.	6.2	.	.	.	.	.	.	.	.	.	.	19
2720	ELYMUS JUNCEUS	W	.	0	14.1	2.6	75.3	22.4	8.0	.	.	.	.	.	.	.	.	.	.	6
2721	ELYMUS VILLOSUS	W	.	0	21.5	3.0	59.9	21.7	15.6	530.0	350.0	.	.	.	.	.	.	.	.	6
2722	ELYMUS VIRGINICUS	W	.	0	5.9	2.3	85.5	34.4	6.3	400.0	210.0	.	.	.	.	.	.	.	.	6
2723	ELYONURUS ARGENTEUS	W	.	0	8.0	1.9	82.6	37.8	7.5	.	.	.	.	.	.	.	.	.	.	18
2724	EMILIA COCCINEA	L	318	0	15.9	3.5	69.7	9.5	10.9	1293.5	258.7	.	.	.	.	.	.	.	.	1
2725	EMILIA SONCHIFOLIA	L	308	0	22.0	3.3	64.3	11.0	10.4	2186.8	648.4	.	.	.	.	.	.	.	.	1
2726	ENARTHROCARPUS STRANGULATUS	S	.	0	35.8	39.6	.	.	.	.	.	.	.	.	.	.	.	.	.	21
2727	ENCELIA FARINOSA	S	.	0	23.1	23.9	.	.	4.5	.	.	.	.	.	.	.	.	.	.	19
2728	ENDOPLEURA UCHI	F	480	0	0.6	26.2	71.8	5.6	1.4	40.0	34.0	4.0	.	.	660.00	0.02	0.02	5.40	12.0	3
2729	ENGELMANNIA PINNATIFIDA	S	.	0	26.6	25.5	.	.	.	.	.	.	.	.	.	.	.	.	.	20
2730	ENGELMANNIA PINNATIFIDA	S	.	0	30.6	26.9	.	.	.	.	.	.	.	.	.	.	.	.	.	21
2731	ENGELMANNIA PINNATIFIDA	L	.	0	23.2	1.2	58.2	9.9	17.4	2630.0	370.0	.	.	4730	.	.	.	.	.	6
2732	ENKIANTHUS PERULATUS	S	.	0	14.2	18.6	.	.	.	.	.	.	.	.	.	.	.	.	.	21
2733	ENSETE VENTRICOSUM	HE	391	0	2.7	0.5	94.5	3.7	2.3	274.6	.	12.1	.	.	.	0.07	0.09	0.23	.	1
2734	ENSETE VENTRICOSUM	HE	400	0	0.4	0.2	98.4	0.2	1.1	78.2	.	13.7	.	.	.	.	0.04	0.00	.	1
2735	ENSETE VENTRICOSUM	S	402	0	15.2	1.6	82.4	0.6	0.8	.	.	.	.	.	.	.	.	.	.	1
2736	ENTEROLOBIUM CONTORTISILIQUA	S	.	0	25.6	2.3	.	.	.	.	.	.	.	.	.	.	.	.	.	21
2737	ENTEROLOBIUM CYCLOCARPUM	C	.	0	16.8	5.3	72.9	12.2	5.0	670.0	140.0	.	.	.	.	.	.	.	.	18
2738	ENTEROLOBIUM CYCLOCARPUM	S	.	0	21.8	1.7	72.5	13.5	4.1	.	.	.	.	.	.	.	.	.	.	18
2739	ENTEROLOBIUM CYCLOCARPUM	S	.	0	43.4	2.5	.	.	3.3	.	.	.	.	.	.	.	.	.	.	19
2740	ENTEROLOBIUM CYCLOCARPUM	S	.	0	3.9	0.3	.	.	5.0	.	.	.	.	.	.	.	.	.	.	19
2741	ENTEROLOBIUM CYCLOCARPUM	S	.	0	26.0	1.6	.	.	.	.	.	.	.	.	.	.	.	.	.	20
2742	ENTEROLOBIUM CYCLOCARPUM	S	391	0	31.7	4.0	61.1	2.0	3.2	30.4	554.8	4.7	.	.	30.43	2.78	0.23	1.83	7.1	3
2743	ENTEROPOGON MONOSTACHYOS	L	.	0	4.7	1.5	84.9	36.6	8.7	.	.	.	.	.	.	.	.	.	.	5
2744	ENTOLASIA IMBRICATA	W	.	0	19.5	1.1	72.9	14.8	8.5	.	.	.	.	.	.	.	.	.	.	18
2745	EPHEDRA ANTISYPHILITICA	Y	.	0	7.0	1.5	85.2	33.7	6.3	3060.0	80.0	.	.	620	.	.	.	.	.	6
2746	EPHEDRA NEVADENSIS	L	.	0	5.0	.	.	.	.	5810.0	500.0	.	.	.	.	.	.	.	.	6
2747	EPHEDRA TRIFURCA	L	.	0	6.0	.	.	.	7.7	3310.0	80.0	13.0	.	420	.	.	.	.	.	6
2748	EPHEDRA TWEEDIANA	S	.	0	15.1	14.0	.	.	.	.	.	.	.	.	.	.	.	.	.	21
2749	EPILOBIUM ANGUSTIFOLIUM	L	.	0	2.8	7.1	82.5	12.7	7.6	.	.	.	.	.	.	.	.	.	.	6
2750	EPILOBIUM HIRSUTUM	S	.	0	25.2	31.7	.	.	.	.	.	.	.	.	.	.	.	.	.	21
2751	EQUISETUM ARVENSE	W	.	0	5.3	2.4	73.8	23.5	18.5	.	.	.	.	.	.	.	.	.	.	6
2752	EQUISETUM ARVENSE	L	317	0	15.9	3.2	69.8	17.5	11.1	920.6	1476.2	69.8	.	.	4761.90	0.00	1.11	88.89	793.7	2
2753	ERAGROSTIS BEYRICHII	W	.	0	7.2	1.6	79.5	32.7	11.7	.	.	.	.	.	.	.	.	.	.	6
2754	ERAGROSTIS CAESPITOSA	W	.	0	8.7	1.8	81.4	37.5	8.1	.	.	.	.	.	.	.	.	.	.	18
2755	ERAGROSTIS CHLOROMELAS	W	.	0	10.4	3.0	79.8	34.5	7.8	.	.	.	.	.	.	.	.	.	.	6
2756	ERAGROSTIS CILIANENSIS	W	.	0	15.3	2.4	71.7	29.0	10.6	.	.	.	.	.	.	.	.	.	.	18
2757	ERAGROSTIS CILIANENSIS	W	.	0	12.2	1.6	75.3	27.8	10.9	950.0	240.0	.	.	1610	.	.	.	.	.	6

TABLE 3 - DATA CONVERTED TO A ZERO-MOISTURE BASIS.

OBS	PLANA	PART	CALORIES	WATER	PROTEIN	FAT	TOTCARB	FIBER	ASH	CALCIUM	PHOSPHO	IRON	SODIUM	POTASS	CAROT	THIAMINE	RIBOFLAV	NIACIN	ASCORBIC	REFCO
2758	ERAGROSTIS CURVULA	W	.	0	13.1	2.1	78.8	31.5	6.0	.	.	.	.	.	.	.	.	.	.	18
2759	ERAGROSTIS CURVULA	W	.	0	9.3	3.0	82.1	32.4	5.6	310.0	180.0	.	.	.	28.00	.	.	.	.	6
2760	ERAGROSTIS DIFFUSA	W	.	0	11.8	3.7	76.1	30.8	8.4	380.0	200.0	.	.	.	.	.	.	.	.	6
2761	ERAGROSTIS INTERMEDIA	W	.	0	9.9	2.0	76.6	32.8	11.5	410.0	170.0	.	.	.	.	.	.	.	.	6
2762	ERAGROSTIS LEHMANNIANA	H	.	0	10.4	1.7	79.2	31.9	8.7	.	.	.	.	.	.	.	.	.	.	18
2763	ERAGROSTIS LEHMANNIANA	W	.	0	6.7	1.7	84.0	35.0	7.6	.	.	.	.	.	.	.	.	.	.	6
2764	ERAGROSTIS LUGENS	W	.	0	13.2	2.1	73.9	29.4	10.8	370.0	160.0	.	.	.	.	.	.	.	.	6
2765	ERAGROSTIS PLUMOSA	L	.	0	9.9	2.2	79.7	33.6	8.2	550.0	870.0	.	.	.	.	.	.	.	.	5
2766	ERAGROSTIS SECUNDIFLORA	W	.	0	6.5	2.0	81.4	37.1	10.1	340.0	120.0	.	.	.	.	.	.	.	.	6
2767	ERAGROSTIS SUPERBA	W	.	0	11.3	1.9	80.2	32.3	6.6	.	.	.	.	.	.	.	.	.	.	18
2768	ERAGROSTIS SUPERBA	S	.	0	21.2	5.0	.	.	3.4	.	.	.	.	.	.	.	.	.	.	19
2769	ERAGROSTIS TEF	H	.	0	10.5	1.1	83.1	34.2	5.3	.	.	.	.	.	.	.	.	.	.	18
2770	ERAGROSTIS TEF	S	375	0	11.0	3.0	82.3	3.1	3.8	196.6	357.7	86.3	.	.	5.71	0.15	0.14	1.60	100.6	1
2771	ERAGROSTIS TEF	S	374	0	9.8	2.1	85.1	2.6	2.9	164.6	428.4	23.6	.	.	0.00	0.43	0.16	2.14	.	1
2772	ERAGROSTIS TEF	S	.	0	11.4	2.4	.	4.7	2.9	.	.	.	.	.	.	.	.	.	.	5
2773	ERAGROSTIS TEF	L	.	0	9.0	2.0	79.2	36.0	10.3	.	.	.	.	.	.	.	.	.	.	5
2774	ERAGROSTIS TENELLA	L	.	0	.	.	.	.	4.0	493.0	613.0	0.0	162.0	1441.0	.	.	.	.	.	5
2775	ERAGROSTIS TREMULA	L	.	0	3.3	1.8	89.7	41.0	5.1	.	.	.	.	.	.	.	.	.	.	5
2776	ERAGROSTIS TRICHODES	W	.	0	9.8	2.5	79.3	27.8	8.4	490.0	180.0	.	.	.	.	.	.	.	.	6
2777	ERAGROSTIS TRICHODES	W	.	0	4.8	2.6	89.6	34.3	3.0	380.0	180.0	.	.	.	.	.	.	.	.	6
2778	ERAGROSTIS TRICHODES	W	.	0	3.1	2.1	89.5	36.3	5.3	370.0	120.0	.	.	.	.	.	.	.	.	6
2779	EREMIA TOTTA	S	.	0	2.1	3.1	.	.	.	.	.	.	.	.	.	.	.	.	.	21
2780	EREMOCARPUS SETIGERUS	S	.	0	28.1	28.0	.	.	.	.	.	.	.	.	.	.	.	.	.	20
2781	EREMOCHLOA OPHIUROIDES	W	.	0	11.3	2.8	80.3	31.6	5.6	580.0	260.0	13.0	.	.	.	.	.	.	.	6
2782	EREMOPOGON FOVEOLATUS	L	.	0	2.5	1.4	88.1	40.7	8.0	.	.	.	.	.	.	.	.	.	.	5
2783	EREMOSTACHYS LACINIATA	S	.	0	26.4	10.4	.	.	.	.	.	.	.	.	.	.	.	.	.	21
2784	EREMOSTACHYS LACINIATA	S	.	0	21.1	12.2	.	.	.	.	.	.	.	.	.	.	.	.	.	21
2785	EREMOSTACHYS SPECIOSA	S	.	0	24.2	21.1	.	.	.	.	.	.	.	.	.	.	.	.	.	21
2786	EREMOSTACHYS VICARYI	S	.	0	25.0	19.0	.	.	.	.	.	.	.	.	.	.	.	.	.	21
2787	EREMURUS AUCHERIANUS	S	.	0	30.6	29.5	.	.	.	.	.	.	.	.	.	.	.	.	.	21
2788	EREMURUS AURANTIACUS	S	.	0	34.1	27.3	.	.	.	.	.	.	.	.	.	.	.	.	.	21
2789	ERIANTHUS LONGESETOSUS	L	.	0	8.0	2.2	80.6	35.0	9.2	210.0	160.0	.	.	.	.	.	.	.	.	5
2790	ERIANTHUS RAVENNAE	H	.	0	1.9	1.6	89.6	43.9	6.9	.	.	.	.	.	.	.	.	.	.	18
2791	ERIANTHUS SP.	W	.	0	6.1	3.5	88.5	48.9	1.9	610.0	100.0	.	.	.	1.00	.	.	.	.	6
2792	ERIGERON CANADENSIS	W	.	0	14.9	1.8	75.1	26.1	8.2	1010.0	280.0	.	.	2610.0	.	.	.	.	.	6
2793	ERIGERON SPECIOSUS	S	.	0	26.1	21.8	.	.	.	.	.	.	.	.	.	.	.	.	.	21
2794	ERIOBOTRYA JAPONICA	F	351	0	4.4	1.8	89.5	4.4	4.4	157.9	219.3	1.8	35.1	2763.2	6798.25	0.18	0.35	1.75	35.1	2
2795	ERIOBOTRYA JAPONICA	F	370	0	1.7	5.0	89.9	6.7	3.4	151.3	117.6	6.7	.	.	2184.87	0.17	0.42	2.52	84.0	3
2796	ERIOBOTRYA JAPONICA	F	356	0	3.0	1.5	91.9	3.7	3.7	148.1	266.7	3.0	.	2577.8	2977.78	.	.	.	7.4	4
2797	ERIOBOTRYA JAPONICA	F	.	0	5.6	2.4	88.1	7.1	4.0	238.1	158.7	5.6	.	.	.	.	.	.	.	5
2798	ERIOCHLOA GRACILIS	W	.	0	13.3	2.4	73.5	29.5	10.8	680.0	220.0	.	.	3040.0	.	.	.	.	.	6
2799	ERIOCHLOA NUBICA	W	.	0	10.2	1.4	72.2	33.5	16.2	.	.	.	.	.	.	.	.	.	.	18
2800	ERIOCHLOA POLYSTACHYA	SH	.	0	10.3	1.9	77.2	32.9	10.6	.	.	.	.	.	.	.	.	.	.	18
2801	ERIOCHLOA POLYSTACHYA	C	.	0	7.1	1.3	81.0	37.1	10.6	.	.	.	.	.	.	.	.	.	.	18
2802	ERIOCHLOA POLYSTACHYA	L	.	0	14.4	2.8	72.1	27.5	10.7	.	.	.	.	.	.	.	.	.	.	18

TABLE 3 - DATA CONVERTED TO A ZERO-MOISTURE BASIS.

OBS	PLNA	PART	CALORIES	WATER	PROTEIN	FAT	TOTLCARB	FIBER	ASH
2803	ERIOGONUM ALATUM	S	.	0	18.8	2.6	.	.	.
2804	ERIOGONUM ALATUM	S	.	0	7.5	1.9	.	.	.
2805	ERIOGONUM ALATUM	S	.	0	18.8	2.9	.	.	3.2
2806	ERIOGONUM ALATUM	S	.	0	6.2	1.7	.	.	5.0
2807	ERIOGONUM ANNUUM	S	.	0	19.2	5.9	.	.	.
2808	ERIOGONUM SP.	W	.	0	6.6	.	.	.	13.7
2809	ERIOGONUM SP.	I	.	0	17.0	2.2	74.3	13.5	6.5
2810	ERIOSEMA GRANDIFLORUM	S	.	0	23.2	1.9	.	.	.
2811	ERODIUM BOTRYS	W	.	0	17.0	4.8	65.7	22.0	12.5
2812	ERODIUM BOTRYS	W	.	0	13.5	2.8	81.1	21.2	12.6
2813	ERODIUM CICONIUM	S	.	0	21.0	12.0	.	.	.
2814	ERODIUM CICUTARIUM	W	.	0	16.5	.	.	18.2	13.1
2815	ERODIUM GRUINUM	S	.	0	20.6	10.7	.	.	.
2816	ERODIUM MOSCHATUM	W	.	0	16.9	5.1	65.2	22.1	12.8
2817	ERODIUM MOSCHATUM	W	.	0	16.2	.	.	18.1	15.0
2818	ERODIUM TEXANUM	S	.	0	25.6	15.8	.	.	.
2819	ERODIUM TEXANUM	W	.	0	24.4	2.4	62.0	8.1	11.2
2820	ERUCA LONGIROSTRIS	S	.	0	31.6	25.5	.	.	.
2821	ERUCA SATIVA	S	.	0	37.4	22.5	.	.	4.8
2822	ERUCA SATIVA	S	.	0	30.6	34.6	.	.	.
2823	ERUCA SATIVA	S	.	0	29.6	35.1	.	.	.
2824	ERUCA SATIVA	S	.	0	35.1	24.2	.	.	.
2825	ERUCA VESICARIA	S	.	0	33.8	27.2	.	.	.
2826	ERUCARIA HISPANICA	S	.	0	35.1	31.9	.	.	.

OBS	CALCIUM	PHOSPHO	IRON	SODIUM	POTASS	CAROT	THIAMINE	RIBOFLAV	NIACIN	ASCORBIC	REFCO
2803	.	.	.	.	.	.	.	.	.	.	19
2804	.	.	.	.	.	.	.	.	.	.	19
2805	.	.	.	.	.	.	.	.	.	.	19
2806	.	.	.	.	.	.	.	.	.	.	19
2807	.	.	.	.	.	.	.	.	.	.	21
2808	1030.0	90.0	.	.	.	1.0	.	.	.	.	6
2809	.	.	.	.	.	.	.	.	.	.	6
2810	.	.	.	.	.	.	.	.	.	.	21
2811	2400.0	540.0	.	.	.	.	.	.	.	.	18
2812	1950.0	330.0	.	.	2950.0	.	.	.	.	.	6
2813	.	.	.	.	.	.	.	.	.	.	21
2814	2510.0	470.0	.	.	3370.0	.	.	.	.	.	6
2815	.	.	.	.	.	.	.	.	.	.	20
2816	2490.0	530.0	.	.	.	.	.	.	.	.	18
2817	2210.0	440.0	.	.	3450.0	.	.	.	.	.	6
2818	.	.	.	.	.	.	.	.	.	.	19
2819	1960.0	260.0	.	.	2280.0	.	.	.	.	.	6
2820	.	.	.	.	.	.	.	.	.	.	21
2821	.	.	.	.	.	.	.	.	.	.	19
2822	.	.	.	.	.	.	.	.	.	.	20
2823	.	.	.	.	.	.	.	.	.	.	21
2824	.	.	.	.	.	.	.	.	.	.	21
2825	.	.	.	.	.	.	.	.	.	.	21
2826	.	.	.	.	.	.	.	.	.	.	21

TABLE 3 - DATA CONVERTED TO A ZERO-MOISTURE BASIS.

OBS	PLNA	PART	CALORIES	WATER	PROTEIN	FAT	TOTLCARB	FIBER
2827	ERUCASTRUM STRIGOSUM	S	.	0	25.6	28.2	.	.
2828	ERUCASTRUM STRIGOSUM	S	.	0	28.8	25.2	.	.
2829	ERYNGIUM AQUATICUM VAR. RAVENELII	L	308	0	16.1	4.9	64.3	14.0
2830	ERYNGIUM CRETICUM	S	.	0	15.9	9.1	.	.
2831	ERYNGIUM GIGANTEUM	S	.	0	20.9	26.7	.	.
2832	ERYNGIUM MARITIMUM	S	.	0	13.8	23.2	.	.
2833	ERYNGIUM PLANUM	S	.	0	30.0	27.9	.	.
2834	ERYNGIUM SP.	S	.	0	27.6	31.6	.	.
2835	ERYSIMUM CUSPIDATUM	S	.	0	35.8	32.8	.	.
2836	ERYSIMUM PEROFSKIANUM	S	.	0	28.0	35.3	.	.
2837	ERYTHEA SP.	S	.	0	5.8	6.6	.	.
2838	ERYTHEA SP.	S	.	0	5.6	9.2	.	.
2839	ERYTHRINA BERTEROANA	SH	304	0	27.8	1.3	63.3	15.2
2840	ERYTHRINA CRISTA-GALLI	S	.	0	30.3	16.7	.	.
2841	ERYTHRINA FLABELLIFORMIS	S	.	0	36.2	26.1	.	.
2842	ERYTHRINA FLABELLIFORMIS	S	.	0	27.5	20.1	.	.
2843	ERYTHRINA FLABELLIFORMIS	S	.	0	25.6	19.6	.	.
2844	ERYTHRINA FUSCA	L	324	0	24.9	4.3	63.2	22.2
2845	ERYTHRINA HERBACEA	S	.	0	26.0	21.3	.	.
2846	ERYTHRINA POEPPIGIANA	S	.	0	36.8	12.4	.	.
2847	ERYTHRINA SP.	S	.	0	42.5	19.0	.	.
2848	ERYTHRINA SP.	S	.	0	5.4	1.4	.	.
2849	ERYTHROCOCCA BONGESIS	L	331	0	22.5	8.4	57.8	13.4
2850	ERYTHROXYLUM COCA	L	326	0	20.2	5.3	64.8	15.4

OBS	ASH	CALCIUM	PHOSPHO	IRON	SODIUM	POTASS	CAROT	THIAMINE	RIBOFLAV	NIACIN	ASCORBIC	REFCO
2827	.	.	.	.	.	.	.	.	.	.	.	21
2828	.	.	.	.	.	.	.	.	.	.	.	21
2829	14.7	2062.9	321.7	39.9	.	.	23216.8	0.63	1.54	6.99	195.8	3
2830	.	.	.	.	.	.	.	.	.	.	.	21
2831	7.1	.	.	.	.	.	.	.	.	.	.	19
2832	.	.	.	.	.	.	.	.	.	.	.	21
2833	.	.	.	.	.	.	.	.	.	.	.	20
2834	.	.	.	.	.	.	.	.	.	.	.	21
2835	.	.	.	.	.	.	.	.	.	.	.	21
2836	5.4	.	.	.	.	.	.	.	.	.	.	19
2837	.	.	.	.	.	.	.	.	.	.	.	19
2838	.	.	.	.	.	.	.	.	.	.	.	21
2839	7.6	683.5	506.3	13.9	.	.	1392.4	1.20	1.20	7.59	234.2	3
2840	.	.	.	.	.	.	.	.	.	.	.	21
2841	.	.	.	.	.	.	.	.	.	.	.	19
2842	.	.	.	.	.	.	.	.	.	.	.	20
2843	.	.	.	.	.	.	.	.	.	.	.	20
2844	7.6	308.1	216.2	9.7	.	.	12432.4	1.30	0.92	25.41	16.2	2
2845	.	.	.	.	.	.	.	.	.	.	.	21
2846	5.3	.	.	.	.	.	.	.	.	.	.	19
2847	6.2	.	.	.	.	.	.	.	.	.	.	19
2848	2.4	.	.	.	.	.	.	.	.	.	.	19
2849	11.3	2118.8	334.4	.	.	.	.	.	.	.	.	1
2850	9.6	1647.1	974.3	49.0	43.9	2160.4	11764.7	0.37	2.04	1.38	1.1	14

TABLE 3 - DATA CONVERTED TO A ZERO-MOISTURE BASIS.

OBS	PLNA	PART	CALORIES	WATER	PROTEIN	FAT	TOTLCARB	FIBER
2851	ERYTHROXYLUM MEXICANUM	S	.	0	15.6	6.2	.	.
2852	ESCHSCHOLZIA CALIFORNICA	S	.	0	25.0	46.8	.	.
2853	ESCOBEDIA CF. CRASSIPES	S	.	0	20.6	15.0	.	.
2854	ESCOBEDIA PEDUNCULARIS	S	.	0	21.2	12.5	.	.
2855	EUADENIA SP	L	308	0	21.6	3.6	62.8	12.4
2856	EUCHLAENA MEXICANA	S	.	0	10.4	3.8	.	.
2857	EUCHLAENA SP.	W	.	0	8.4	2.3	79.5	31.1
2858	EUCLEA ACUTIFOLIA	S	.	0	7.0	5.1	.	.
2859	EUCLEA POLYANDRA	S	.	0	5.8	3.8	.	.
2860	EUGENIA DOMBEYI	S	.	0	5.3	0.8	.	.
2861	EUGENIA DOMBEYI	F	361	0	4.1	2.0	91.2	4.1
2862	EUGENIA JAMBOLANA	S	.	0	6.6	1.4	.	.
2863	EUGENIA JAMBOS	S	.	0	5.8	3.8	.	.
2864	EUGENIA SP.	F	356	0	5.7	1.1	90.8	6.9
2865	EUGENIA SP.	F	359	0	5.6	2.8	88.0	4.2
2866	EUGENIA UNIFLORA	S	.	0	5.6	0.5	.	.
2867	EUGENIA UNIFLORA	F	345	0	4.5	0.9	92.7	2.7
2868	EUGENIA UNIFLORA	F	359	0	5.6	2.8	88.0	4.2
2869	EUONYMUS ALATUS	S	.	0	20.5	44.4	.	.
2870	EUONYMUS EUROPAEUS	S	.	0	20.7	51.5	.	.
2871	EUONYMUS HAMILTONIANUS	S	.	0	16.1	51.8	.	.
2872	EUONYMUS LATIFOLIUS	S	.	0	20.8	47.9	.	.
2873	EUONYMUS VERRUCOSUS	S	.	0	27.2	36.2	.	.
2874	EUONYMUS VERRUCOSUS	S	.	0	19.2	52.5	.	.

OBS	ASH	CALCIUM	PHOSPHO	IRON	SODIUM	POTASS	CAROT	THIAMINE	RIBOFLAV	NIACIN	ASCORBIC	REFCO
2851	.	.	.	.	.	.	.	.	.	.	.	20
2852	.	.	.	.	.	.	.	.	.	.	.	20
2853	.	.	.	.	.	.	.	.	.	.	.	20
2854	.	.	.	.	.	.	.	.	.	.	.	20
2855	12.0	1680.0	160.0	.	.	.	.	.	.	.	.	1
2856	5.3	.	.	.	.	.	.	.	.	.	.	19
2857	9.8	.	.	.	.	.	.	.	.	.	.	6
2858	.	.	.	.	.	.	.	.	.	.	.	21
2859	.	.	.	.	.	.	.	.	.	.	.	21
2860	2.7	.	.	.	.	.	.	.	.	.	.	19
2861	2.7	272.1	95.2	3.4	.	.	272.1	0.27	0.20	2.04	129.3	3
2862	.	.	.	.	.	.	.	.	.	.	.	21
2863	2.0	.	.	.	.	.	.	.	.	.	.	19
2864	2.3	126.4	126.4	4.6	92	689.7	0.0	0.23	0.23	4.60	149.4	2
2865	3.5	63.4	77.5	1.4	.	.	8943.7	0.21	0.42	2.11	98.6	3
2866	2.9	.	.	.	.	.	.	.	.	.	.	19
2867	1.8	63.6	81.8	0.9	.	.	10181.8	0.18	0.45	1.82	172.7	2
2868	3.5	63.4	77.5	1.4	.	.	6338.0	0.21	0.28	2.11	211.3	4
2869	3.9	.	.	.	.	.	.	.	.	.	.	19
2870	.	.	.	.	.	.	.	.	.	.	.	21
2871	.	.	.	.	.	.	.	.	.	.	.	21
2872	.	.	.	.	.	.	.	.	.	.	.	21
2873	.	.	.	.	.	.	.	.	.	.	.	21
2874	.	.	.	.	.	.	.	.	.	.	.	21

TABLE 3 - DATA CONVERTED TO A ZERO-MOISTURE BASIS.

OBS	PLNA	PART	CALORIES	WATER	PROTEIN	FAT	TOTLCARB	FIBER
2875	EUPATORIUM HYSSOPINUM & EUPATORIUM SPP.	S	.	0	30.6	21.7	.	.
2876	EUPATORIUM LIEBMANNII	S	.	0	29.4	30.8	.	.
2877	EUPATORIUM ODORATUM	S	.	0	20.6	15.7	.	.
2878	EUPATORIUM ODORATUM	S	.	0	21.2	27.5	.	.
2879	EUPATORIUM RUGOSUM	S	.	0	25.6	39.6	.	.
2880	EUPATORIUM SEROTINUM	S	.	0	23.3	34.2	.	.
2881	EUPATORIUM SP.	S	.	0	15.6	26.5	.	.
2882	EUPATORIUM SP.	S	.	0	23.8	24.9	.	.
2883	EUPHORBIA AKENOCARPA	S	.	0	23.8	42.0	.	.
2884	EUPHORBIA AMYGDALOIDES	S	.	0	21.6	40.0	.	.
2885	EUPHORBIA BICOLOR	S	.	0	23.1	26.8	.	.
2886	EUPHORBIA COERULESCENS	SH	.	0	15.1	22.6	50.5	15.3
2887	EUPHORBIA CUPHOSPERMA	S	.	0	7.5	7.1	.	.
2888	EUPHORBIA HETEROPHYLLA	S	.	0	25.2	37.4	.	.
2889	EUPHORBIA HETEROPHYLLA	S	.	0	26.2	37.9	.	.
2890	EUPHORBIA HETEROPHYLLA	S	.	0	26.9	29.6	.	.
2891	EUPHORBIA HETEROPHYLLA	S	.	0	23.1	40.1	.	.
2892	EUPHORBIA KOTSCHYANA	S	.	0	39.4	36.0	.	.
2893	EUPHORBIA LATHYRUS	S	.	0	15.8	44.3	.	.
2894	EUPHORBIA LATHYRUS	S	.	0	8.4	6.5	.	.
2895	EUPHORBIA LATHYRUS	S	.	0	15.0	47.5	.	.
2896	EUPHORBIA MACROCLADA	S	.	0	32.2	24.5	.	.
2897	EUPHORBIA MARGINATA	S	.	0	21.4	32.0	.	.
2898	EUPHORBIA MARGINATA	S	.	0	20.9	31.7	.	.

OBS	ASH	CALCIUM	PHOSPHO	IRON	SODIUM	POTASS	CAROT	THIAMINE	RIBOFLAV	NIACIN	ASCORBIC	REFCO
2875	.	.	.	.	.	.	.	.	.	.	.	20
2876	.	.	.	.	.	.	.	.	.	.	.	20
2877	.	.	.	.	.	.	.	.	.	.	.	19
2878	.	.	.	.	.	.	.	.	.	.	.	20
2879	3.8	.	.	.	.	.	.	.	.	.	.	19
2880	.	.	.	.	.	.	.	.	.	.	.	21
2881	.	.	.	.	.	.	.	.	.	.	.	20
2882	.	.	.	.	.	.	.	.	.	.	.	20
2883	.	.	.	.	.	.	.	.	.	.	.	21
2884	.	.	.	.	.	.	.	.	.	.	.	21
2885	.	.	.	.	.	.	.	.	.	.	.	21
2886	11.8	170.0	240.0	.	.	.	.	.	.	.	.	18
2887	.	.	.	.	.	.	.	.	.	.	.	20
2888	5.7	.	.	.	.	.	.	.	.	.	.	19
2889	6.5	.	.	.	.	.	.	.	.	.	.	19
2890	.	.	.	.	.	.	.	.	.	.	.	19
2891	.	.	.	.	.	.	.	.	.	.	.	20
2892	.	.	.	.	.	.	.	.	.	.	.	21
2893	2.6	.	.	.	.	.	.	.	.	.	.	19
2894	9.3	.	.	.	.	.	.	.	.	.	.	19
2895	.	.	.	.	.	.	.	.	.	.	.	20
2896	.	.	.	.	.	.	.	.	.	.	.	21
2897	3.2	.	.	.	.	.	.	.	.	.	.	19
2898	3.2	.	.	.	.	.	.	.	.	.	.	19

TABLE 3 - DATA CONVERTED TO A ZERO-MOISTURE BASIS.

OBS	PLANA	PART	CALORIES	WATER	PROTEIN	FAT	TOTCARB	FIBER	ASH	CALCIUM	PHOSPHO	IRON	SODIUM	POTASS	CAROT	THIAMINE	RIBOFLAV	NIACIN	ASCORBIC	REFCO
2899	EUPHORBIA MARGINATA	S	.	0	16.4	17.9	.	.	3.2	.	.	.	.	.	.	.	.	.	.	19
2900	EUPHORBIA MAURITANICA	S	.	0	17.6	45.0	.	.	.	.	.	.	.	.	.	.	.	.	.	21
2901	EUPHORBIA MYRSINITES	S	.	0	27.2	32.3	.	.	.	.	.	.	.	.	.	.	.	.	.	20
2902	EUPHORBIA PARALIAS	S	.	0	21.7	41.0	.	.	.	.	.	.	.	.	.	.	.	.	.	21
2903	EUPHORBIA PARALIAS	S	.	0	20.4	49.4	.	.	.	.	.	.	.	.	.	.	.	.	.	21
2904	EUPHORBIA PARRYI	S	.	0	21.9	44.4	.	.	.	.	.	.	.	.	.	.	.	.	.	20
2905	EUPHORBIA SALICIFOLIA	S	.	0	24.2	30.6	.	.	.	.	.	.	.	.	.	.	.	.	.	21
2906	EUPHORBIA SEGETALIS	S	.	0	22.6	39.8	.	.	.	.	.	.	.	.	.	.	.	.	.	21
2907	EUPHORBIA SERRATA	S	.	0	21.3	42.1	.	.	.	.	.	.	.	.	.	.	.	.	.	21
2908	EUPHORBIA SP.	S	.	0	20.0	24.3	.	.	.	.	.	.	.	.	.	.	.	.	.	20
2909	EUPHORBIA SP.	S	.	0	9.3	13.8	.	.	.	.	.	.	.	.	.	.	.	.	.	21
2910	EUPHORBIA SP.	W	.	0	14.4	4.1	70.5	13.0	11.0	1190.0	260.0	.	.	1940.0	.	.	.	.	.	6
2911	EUPHORBIA TERRACINA	S	.	0	17.6	39.8	.	.	.	.	.	.	.	.	.	.	.	.	.	21
2912	EUPHORBIA TERRACINA	S	.	0	17.8	32.5	.	.	.	.	.	.	.	.	.	.	.	.	.	21
2913	EUPHORBIA THAMNOIDES	S	.	0	25.6	34.5	.	.	.	.	.	.	.	.	.	.	.	.	.	21
2914	EUPTELEA POLYANDRA	S	.	0	7.5	11.0	.	.	.	.	.	.	.	.	.	.	.	.	.	19
2915	EUROTIA LANATA	W	.	0	11.3	2.3	72.5	31.5	13.9	1900.0	120.0	.	.	.	2.00	.	.	.	.	6
2916	EURYALE FEROX	S	400	0	9.6	0.3	89.4	0.2	0.7	45.1	99.5	1.4	.	66.0	.	.	.	.	.	2
2917	EURYALE FEROX	S	.	0	11.1	0.1	88.2	.	0.6	22.9	103.2	1.6	.	.	0.00	.	.	.	.	5
2918	EURYOPS ATHANASIAE	S	.	0	22.8	40.4	.	.	.	.	.	.	.	.	.	.	.	.	.	21
2919	EUSTACHYS PASPALOIDES	W	.	0	9.7	2.6	77.7	33.8	10.0	.	.	.	.	.	.	.	.	.	.	18
2920	EUTERPE LONGEPETIOLATA	SH	289	0	24.4	2.2	57.8	6.7	15.6	955.6	877.8	8.9	.	.	0.00	0.44	1.00	7.78	188.9	3
2921	EUTERPE OLERACEA	F	449	0	5.8	20.7	71.5	30.5	2.0	.	.	.	.	.	.	0.61	.	.	15.3	3
2922	EUTREMA WASABI	R	318	0	21.9	0.9	71.7	6.0	5.6	399.1	309.0	3.4	.	.	128.76	0.64	0.43	2.15	343.3	3
2923	EXOCHORDA RACEMOSA	S	.	0	48.1	27.7	.	.	4.1	.	.	.	.	.	.	.	.	.	.	19
2924	EXOCHORDA RACEMOSA	S	.	0	10.6	5.2	.	.	2.2	.	.	.	.	.	.	.	.	.	.	19
2925	FAGARA CHALYBEA	L	.	0	10.3	3.3	79.6	19.3	6.8	1700.0	470.0	.	.	.	.	.	.	.	.	18
2926	FAGOPYRUM ESCULENTUM	S	.	0	12.9	2.6	.	.	1.9	.	.	.	.	.	.	.	.	.	.	19
2927	FAGOPYRUM ESCULENTUM	S	379	0	11.6	2.7	83.2	7.4	2.5	49.6	360.8	3.4	.	502.8	0.00	0.41	0.20	2.25	0.0	2
2928	FAGOPYRUM ESCULENTUM	S	376	0	13.1	2.7	81.9	11.1	2.2	128.1	316.9	3.5	.	503.4	0.00	0.67	.	4.94	0.0	4
2929	FAGOPYRUM ESCULENTUM	S	.	0	11.6	2.7	83.0	9.7	2.7	78.9	338.2	14.9	.	.	.	1.01	.	4.96	.	5
2930	FAGOPYRUM SP.	W	.	0	12.6	2.5	75.1	21.9	9.8	.	.	.	.	.	.	.	.	.	.	6
2931	FAGOPYRUM SP.	H	.	0	4.8	1.1	84.3	40.2	9.8	1420.0	200.0	.	.	3220.0	.	.	.	.	.	6
2932	FAGOPYRUM TARTARICUM	S	.	0	11.7	2.7	.	.	2.3	.	.	.	.	.	.	.	.	.	.	19
2933	FAGOPYRUM TARTARICUM	S	.	0	11.2	2.4	.	.	5.3	.	.	.	.	.	.	.	.	.	.	19
2934	FAGUS GRANDIFOLIA	S	608	0	20.8	53.5	21.7	4.0	4.0	.	.	.	.	.	.	.	.	.	.	4
2935	FAGUS ORIENTALIS	S	.	0	31.8	44.6	.	.	.	.	.	.	.	.	.	.	.	.	.	21
2936	FALCARIA VULGARIS	S	.	0	24.8	11.4	.	.	.	.	.	.	.	.	.	.	.	.	.	20
2937	FALLUGIA PARADOXA	S	.	0	30.6	36.8	.	.	.	.	.	.	.	.	.	.	.	.	.	20
2938	FEIJOA SELLOWIANA	F	.	0	.	.	.	.	2.4	173.1	.	.	.	.	.	.	.	.	163.5	1
2939	FELICIA AMELLOIDES	S	.	0	18.8	28.1	.	.	.	.	.	.	.	.	.	.	.	.	.	20
2940	FERNALDIA PANDURATA	L	296	0	24.1	1.9	63.0	13.0	11.1	537.0	425.9	10.2	.	.	1018.52	5.93	1.02	21.30	111.1	3
2941	FEROCACTUS SP.	S	.	0	18.8	17.2	.	.	.	.	.	.	.	.	.	.	.	.	.	19
2942	FEROCACTUS SP.	S	.	0	17.5	17.4	.	.	.	.	.	.	.	.	.	.	.	.	.	19
2943	FERONIA LIMONIA	F	.	0	23.9	2.0	67.9	17.0	6.2	426.2	360.7	2.0	.	.	.	.	0.56	.	6.6	5

TABLE 3 - DATA CONVERTED TO A ZERO-MOISTURE BASIS.

OBS	PLNA	PART	CALORIES	WATER	PROTEIN	FAT	TOTLCARB	FIBER	ASH
2944	FERULA COMMUNIS	S	.	0	28.3	9.2	.	.	.
2945	FERULA COPODA	S	.	0	25.4	19.1	.	.	.
2946	FERULA GALVANIFLUA	S	.	0	13.1	46.9	.	.	.
2947	FERULA OVINA	S	.	0	21.4	19.1	.	.	.
2948	FERULAGO SP.	S	.	0	18.9	6.9	.	.	.
2949	FESTUCA ARUNDINACEA	S	.	0	11.9	1.7	.	.	6.1
2950	FESTUCA ARUNDINACEA	S	.	0	22.5	1.4	.	.	.
2951	FESTUCA ARUNDINACEA	W	.	0	11.6	3.3	76.8	29.6	8.3
2952	FESTUCA ELATIOR	S	.	0	21.9	1.6	.	.	2.4
2953	FESTUCA ELATIOR	W	.	0	15.1	4.1	71.8	27.1	9.0
2954	FESTUCA IDAHOENSIS	W	.	0	7.8	3.8	80.9	35.9	7.5
2955	FESTUCA MEGALURA	W	.	0	17.2	4.6	67.5	21.5	10.7
2956	FESTUCA OCTOFLORA	W	.	0	13.0	2.5	74.5	27.4	10.0
2957	FESTUCA OVINA	S	.	0	26.1	2.2	.	.	.
2958	FESTUCA OVINA	W	.	0	16.3	3.2	73.2	27.1	8.3
2959	FESTUCA RUBRA	W	.	0	13.9	3.1	73.5	27.8	9.5
2960	FESTUCA RUBRA SUBSP. COMMUTATA	H	.	0	13.9	.	.	.	.
2961	FIBIGIA CLYPEATA	S	.	0	22.8	11.9	.	.	.
2962	FIBIGIA CLYPEATA	S	.	0	27.8	12.4	.	.	.
2963	FICUS BENGHALENSIS	L	.	0	9.7	2.9	73.0	22.6	14.4
2964	FICUS BENGHALENSIS	F	351	0	5.2	0.0	92.8	0.0	2.1
2965	FICUS CARICA	L	.	0	14.2	5.9	63.2	17.1	16.7
2966	FICUS CARICA	F	351	0	5.4	1.4	89.2	8.8	4.1
2967	FICUS CARICA	F	360	0	6.1	2.4	88.4	9.1	3.0

OBS	CALCIUM	PHOSPHO	IRON	SODIUM	POTASS	CAROT	THIAMINE	RIBOFLAV	NIACIN	ASCORBIC	REFCO
2944	.	.	.	.	.	.	.	.	.	.	21
2945	.	.	.	.	.	.	.	.	.	.	21
2946	.	.	.	.	.	.	.	.	.	.	20
2947	.	.	.	.	.	.	.	.	.	.	21
2948	.	.	.	.	.	.	.	.	.	.	21
2949	.	.	.	.	.	.	.	.	.	.	19
2950	.	.	.	.	.	.	.	.	.	.	20
2951	.	.	.	.	.	.	.	.	.	.	6
2952	.	.	.	.	.	.	.	.	.	.	19
2953	510.0	380.0	.	.	2000.0	9.00	1.52	0.55	.	.	6
2954	390.0	220.0	.	.	.	.	.	.	.	.	6
2955	570.0	460.0	.	.	.	.	.	.	.	.	6
2956	440.0	220.0	.	.	1720.0	.	.	.	.	.	6
2957	.	.	.	.	.	.	.	.	.	.	21
2958	490.0	360.0	43.0	.	1850.0	.	.	.	.	.	6
2959	540.0	440.0	.	.	.	.	.	.	.	.	6
2960	500.0	380.0	.	.	1580.0	.	.	.	.	.	6
2961	.	.	.	.	.	.	.	.	.	.	20
2962	.	.	.	.	.	.	.	.	.	.	21
2963	2560.0	190.0	.	.	.	.	.	.	.	.	18
2964	113.4	113.4	4.1	.	.	0.00	0.00	0.00	3.09	154.6	3
2965	3160.0	160.0	.	.	.	.	.	.	.	.	18
2966	297.3	135.1	3.4	.	.	202.70	.	.	.	101.4	1
2967	250.0	176.8	3.0	36.6	1079.3	853.66	0.18	0.18	2.44	12.2	2

TABLE 3 - DATA CONVERTED TO A ZERO-MOISTURE BASIS.

OBS	PLNA	PART	CALORIES	WATER	PROTEIN	FAT	TOTLCARB	FIBER	ASH
2968	FICUS CARICA	F	348	0	6.7	1.1	87.6	9.0	4.5
2969	FICUS CARICA	G	341	0	13.0	1.6	79.7	15.4	5.7
2970	FICUS CARICA	F	356	0	5.3	1.3	90.2	5.3	3.1
2971	FICUS CUNIA	F	.	0	10.1	6.6	70.3	20.5	13.1
2972	FICUS ELASTICA	L	.	0	13.0	4.4	75.3	32.2	7.3
2973	FICUS GLABRATA	F	340	0	5.2	1.0	86.6	17.5	7.2
2974	FICUS GLOMERATA	L	.	0	11.2	2.4	71.3	12.3	15.1
2975	FICUS GLOMERATA	F	.	0	8.6	6.5	77.4	20.7	7.5
2976	FICUS INFECTORIA	L	.	0	12.5	3.5	70.9	20.0	13.1
2977	FICUS ITEOPHYLLA	F	332	0	13.1	0.9	78.2	25.3	7.9
2978	FICUS PLATYPHYLLA	F	329	0	11.2	0.6	80.6	21.2	7.6
2979	FICUS RELIGIOSA	L	.	0	9.0	2.7	68.3	15.9	20.0
2980	FICUS RELIGIOSA	SH	.	0	11.7	2.9	70.1	26.1	15.3
2981	FICUS SYCOMORUS	F	358	0	7.4	5.6	79.5	23.7	7.4
2982	FICUS SYCOMORUS	L	292	0	31.3	1.0	57.9	12.3	9.7
2983	FILAGO GALLICA	W	.	0	10.5	.	.	28.1	9.8
2984	FINGERHUTHIA SESLERIAEFORMIS	S	.	0	33.1	4.0	.	.	3.6
2985	FIRMIANA SIMPLEX	S	.	0	25.6	36.9	.	.	5.0
2986	FIRMIANA SIMPLEX	S	.	0	24.2	37.8	.	.	.
2987	FIRMIANA SIMPLEX	S	547	0	20.5	39.8	35.1	2.3	4.6
2988	FLACOURTIA INDICA	S	.	0	6.8	12.0	.	.	.
2989	FLACOURTIA INDICA	F	.	0	.	.	.	.	.
2990	FLACOURTIA INERMIS	F	364	0	1.9	2.3	93.8	4.7	1.9
2991	FLACOURTIA INERMIS	F	.	0	3.0	1.5	93.3	8.2	3.0

OBS	CALCIUM	PHOSPHO	IRON	SODIUM	POTASS	CAROT	THIAMINE	RIBOFLAV	NIACIN	ASCORBIC	REFCO
2968	280.9	168.5	2.8	.	.	337.1	0.22	0.28	2.25	22.5	3
2969	422.8	276.4	5.7	.	.	1626.0	0.33	0.57	3.25	162.6	3
2970	155.6	97.8	2.7	8.9	862.2	355.6	0.27	0.22	1.78	8.9	4
2971	.	1294.8	.	.	.	.	.	.	.	.	5
2972	1110.0	130.0	.	.	.	.	.	.	.	.	18
2973	639.2	72.2	23.7	.	.	515.5	0.31	0.21	2.06	0.0	3
2974	2680.0	310.0	.	.	.	.	.	.	.	.	18
2975	.	1053.2	.	.	.	.	.	.	.	.	5
2976	2210.0	290.0	.	.	.	.	.	.	.	.	18
2977	1266.4	327.5	43.7	.	.	1026.2	0.17	.	.	109.2	1
2978	1205.9	388.2	23.5	.	.	.	0.24	0.24	1.76	129.4	1
2979	2970.0	210.0	.	.	.	.	.	.	.	.	18
2980	3690.0	270.0	.	.	.	.	.	.	.	.	18
2981	1162.8	302.3	27.9	.	.	1023.3	0.14	0.28	2.33	69.8	1
2982	379.5	835.9	81.0	.	.	.	.	.	.	179.5	1
2983	1050.0	330.0	.	.	2310.0	.	.	.	.	.	6
2984	.	.	.	.	.	.	.	.	.	.	19
2985	.	.	.	.	.	.	.	.	.	.	19
2986	.	.	.	.	.	.	.	.	.	.	21
2987	27.7	217.4	4.9	.	1236.9	.	.	.	.	.	2
2988	.	.	.	.	.	.	.	.	.	.	21
2989	.	.	.	.	.	.	.	.	.	0.0	1
2990	127.9	65.9	2.7	3.9	662.8	116.3	0.04	0.08	1.55	19.4	2
2991	.	.	.	.	.	.	.	.	.	.	5

TABLE 3 - DATA CONVERTED TO A ZERO-MOISTURE BASIS.

OBS	PLNA	PART	CALORIES	WATER	PROTEIN	FAT	TOTCARB	FIBER	ASH	CALCIUM	PHOSPHO	IRON	SODIUM	POTASS	CAROT	THIAMINE	RIBOFLAV	NIACIN	ASCORBIC	REFCO
2992	FLACOURTIA JANGOMAS	F	350	0	2.2	0.4	93.7	4.5	3.6	192.8	112.1	.	.	.	.	.	.	.	.	2
2993	FLACOURTIA JANGOMAS	F	354	0	1.6	0.0	96.7	1.6	1.6	65.6	144.3	3.9	.	.	98.4	0.07	0.10	1.97	45.9	3
2994	FLACOURTIA MONTANA	S	.	0	10.8	1.1	.	.	.	.	.	.	.	.	.	.	.	.	.	21
2995	FLAVERIA TRINERVIA	S	.	0	26.9	27.7	.	.	.	.	.	.	.	.	.	.	.	.	.	20
2996	FLAVERIA TRINERVIA	S	.	0	26.9	27.5	.	.	.	.	.	.	.	.	.	.	.	.	.	20
2997	FLEMINGIA CONGESTA	S	.	0	24.2	0.7	.	.	.	.	.	.	.	.	.	.	.	.	.	21
2998	FLEMINGIA SP.	S	.	0	22.5	0.5	.	.	.	.	.	.	.	.	.	.	.	.	.	20
2999	FLOURENSIA CERNUS	S	.	0	16.2	6.6	.	.	.	.	.	.	.	.	.	.	.	.	.	20
3000	FLOURENSIA PRINGLEI	S	.	0	46.2	30.6	.	.	.	.	.	.	.	.	.	.	.	.	.	20
3001	FOENICULUM VULGARE	S	.	0	20.1	26.6	.	.	7.1	.	.	.	.	.	.	.	.	.	.	19
3002	FOENICULUM VULGARE	S	.	0	20.0	22.9	.	.	.	.	.	.	.	.	.	.	.	.	.	19
3003	FOENICULUM VULGARE	S	.	0	25.0	21.5	.	.	.	.	.	.	.	.	.	.	.	.	.	21
3004	FOENICULUM VULGARE	L	287	0	26.9	4.6	51.9	4.6	16.7	1055.6	500.0	26.9	.	3129.6	24166.7	1.11	1.39	6.48	314.8	1
3005	FOENICULUM VULGARE	S	378	0	17.3	16.3	57.3	17.2	9.0	1311.4	534.0	20.3	9.6	1857.5	88.8	0.45	0.38	6.63	.	16
3006	FOENICULUM VULGARE	L	280	0	28.0	4.0	51.0	5.0	17.0	1000.0	510.0	27.0	.	3970.0	21000.0	.	.	.	310.0	4
3007	FOENICULUM VULGARE VAR. DULCE	S	.	0	19.7	24.0	.	.	8.2	.	.	.	.	.	.	.	.	.	.	19
3008	FORESTIERA NEOMEXICANA	L	.	0	19.7	2.9	71.0	9.4	6.4	1000.0	250.0	.	.	2550.0	.	.	.	.	.	6
3009	FORESTIERA PUBESCENS	S	.	0	12.5	18.9	.	.	.	.	.	.	.	.	.	.	.	.	.	20
3010	FORTUNELLA MARGARITA	F	361	0	3.0	3.8	89.5	11.3	3.8	120.3	488.7	6.0	834.6	1172.9	188.0	0.68	0.45	3.76	300.8	2
3011	FORTUNELLA MARGARITA	F	352	0	7.4	1.1	88.6	8.5	2.8	352.3	125.0	4.5	.	.	1818.2	0.80	0.34	4.55	312.5	3
3012	FORTUNELLA SP.	F	348	0	4.8	0.5	91.4	19.8	3.2	336.9	123.0	2.1	37.4	1262.0	1925.1	0.43	0.53	.	192.5	4
3013	FOUQUIERIA SPLENDENS	S	.	0	28.8	18.6	.	.	.	.	.	.	.	.	.	.	.	.	.	19
3014	FOUQUIERIA SPLENDENS	W	.	0	3.7	.	.	.	3.5	630.0	70.0	17.0	.	500.0	.	.	.	.	.	6
3015	FRAGARIA SP.	F	336	0	7.3	2.7	80.9	10.9	9.1	163.6	163.6	6.4	.	.	727.3	.	.	17.27	481.8	1
3016	FRAGARIA SP.	F	362	0	8.5	5.3	80.9	18.1	5.3	266.0	319.1	9.6	10.6	2053.2	106.4	0.43	0.32	4.26	563.8	1
3017	FRAGARIA SP.	F	360	0	8.0	3.0	85.0	13.0	4.0	290.0	290.0	10.0	.	.	200.0	0.30	0.40	4.00	700.0	3
3018	FRAGARIA SP.	F	366	0	6.9	5.0	83.2	12.9	5.0	207.9	207.9	9.9	9.9	1623.8	356.4	0.30	0.69	5.94	584.2	4
3019	FRANGULA ALNUS	S	.	0	12.9	15.8	.	.	.	.	.	.	.	.	.	.	.	.	.	21
3020	FRANGULA ALNUS	S	.	0	15.7	10.7	.	.	.	.	.	.	.	.	.	.	.	.	.	21
3021	FRANSERIA ACANTHICARPA	S	.	0	14.9	20.7	.	.	.	.	.	.	.	.	.	.	.	.	.	21
3022	FRASERA PARRYI	S	.	0	15.6	27.0	.	.	.	.	.	.	.	.	.	.	.	.	.	20
3023	FRAXINUS AMERICANA	S	.	0	25.2	33.4	.	.	3.5	.	.	.	.	.	.	.	.	.	.	19
3024	FRAXINUS AMERICANA	S	.	0	3.9	2.7	.	.	5.6	.	.	.	.	.	.	.	.	.	.	19
3025	FRAXINUS ORNUS	S	.	0	15.2	26.5	.	.	.	.	.	.	.	.	.	.	.	.	.	21
3026	FRAXINUS OXYCARPA	S	.	0	18.1	29.9	.	.	.	.	.	.	.	.	.	.	.	.	.	21
3027	FRAXINUS XANTHOXYLOIDES	S	.	0	16.6	26.7	.	.	.	.	.	.	.	.	.	.	.	.	.	21
3028	FROELICHIA FLORIDANA	S	.	0	21.8	9.0	.	.	.	.	.	.	.	.	.	.	.	.	.	20
3029	FUMARIA DENSIFLORA	S	.	0	20.0	32.7	.	.	.	.	.	.	.	.	.	.	.	.	.	21
3030	GAILLARDIA ARISTATA	S	.	0	45.9	27.6	.	.	5.3	.	.	.	.	.	.	.	.	.	.	19
3031	GAILLARDIA ARISTATA	S	.	0	15.0	8.5	.	.	10.3	.	.	.	.	.	.	.	.	.	.	19
3032	GAILLARDIA PINNATIFIDA	S	.	0	29.1	19.2	.	.	.	.	.	.	.	.	.	.	.	.	.	20
3033	GAILLARDIA PULCHELLA	S	.	0	31.5	18.3	.	.	7.2	.	.	.	.	.	.	.	.	.	.	19
3034	GAILLARDIA PULCHELLA	S	.	0	22.6	17.2	.	.	.	.	.	.	.	.	.	.	.	.	.	20
3035	GAILLARDIA PULCHELLA	W	.	0	13.0	4.7	71.0	28.6	11.3	2680.0	280.0	.	.	3270.0	.	.	.	.	.	6
3036	GALACTIA WRIGHTII	S	.	0	25.0	1.6	.	.	3.6	.	.	.	.	.	.	.	.	.	.	19

TABLE 3 - DATA CONVERTED TO A ZERO-MOISTURE BASIS.

OBS	PLNA	PART	CALORIES	WATER	PROTEIN	FAT	TOTLCARB	FIBER	ASH
3037	GALACTITES TOMENTOSA	S	.	0	20.2	23.6	.	.	.
3038	GALEGA CORONILLOIDES	S	.	0	40.7	3.7	.	.	.
3039	GALEGA OFFICINALIS	S	.	0	41.8	2.8	.	.	.
3040	GALENIA AFRICANA	C	.	0	10.2	4.9	75.0	34.2	9.9
3041	GALINSOGA PARVIFLORA	L	302	0	23.0	3.6	59.7	9.4	13.7
3042	GALIUM APARINE	S	.	0	12.7	3.1	.	.	.
3043	GALIUM ARTICULATUM	S	.	0	17.0	10.7	.	.	.
3044	GALIUM CORONATUM	S	.	0	13.3	9.3	.	.	.
3045	GALIUM PARISIENSE	S	.	0	17.8	9.6	.	.	.
3046	GAMOLEPIS CHRYSANTHEMOIDES	S	.	0	14.6	25.6	.	.	.
3047	GARCINIA HUILLENSIS	F	.	0	.	.	.	.	.
3048	GARCINIA MANGOSTANA	F	363	0	3.2	1.9	93.6	31.8	1.3
3049	GARCINIA MANGOSTANA	P	.	0	3.3	0.7	94.7	.	1.3
3050	GARCINIA TINCTORIA	S	.	0	4.2	26.1	.	.	1.8
3051	GARCINIA TINCTORIA	S	.	0	5.1	7.7	.	.	0.2
3052	GARDENIA JASMINOIDES	S	.	0	14.4	20.0	.	.	2.5
3053	GAULTHERIA SHALLON	F	371	0	13.0	5.0	79.0	.	3.0
3054	GAULTHERIA SHALLON	F	351	0	6.0	1.0	88.0	.	4.0
3055	GAURA NEOMEXICANA	S	.	0	31.2	48.3	.	.	.
3056	GAURA PARVIFLORA	S	.	0	11.4	13.7	.	.	4.6
3057	GAURA PARVIFLORA	S	.	0	30.6	41.2	.	.	.
3058	GAURA SP.	W	.	0	25.4	2.0	59.8	15.7	12.8
3059	GAURA SUFFULTA	S	.	0	9.9	8.9	.	.	.
3060	GAZANIA KREBSIANA SUBSP. KREBSIANA	S	.	0	45.1	27.5	.	.	.

OBS	CALCIUM	PHOSPHO	IRON	SODIUM	POTASS	CAROT	THIAMINE	RIBOFLAV	NIACIN	ASCORBIC	REFCO
3037	.	.	.	.	.	.	.	.	.	.	21
3038	.	.	.	.	.	.	.	.	.	.	21
3039	.	.	.	.	.	.	.	.	.	.	21
3040	1540.0	60.0	.	.	.	.	.	.	.	.	18
3041	1762.6	323.7	51.1	.	.	16115.1	0.79	1.94	15.11	215.8	3
3042	.	.	.	.	.	.	.	.	.	.	20
3043	.	.	.	.	.	.	.	.	.	.	20
3044	.	.	.	.	.	.	.	.	.	.	21
3045	.	.	.	.	.	.	.	.	.	.	21
3046	.	.	.	.	.	.	.	.	.	.	21
3047	.	.	.	.	.	.	.	.	.	33.2	1
3048	63.7	63.7	3.2	6.4	859.9	0.0	0.19	0.13	3.82	25.5	2
3049	66.2	132.5	1.3	.	.	.	.	.	.	.	5
3050	.	.	.	.	.	.	.	.	.	.	19
3051	.	.	.	.	.	.	.	.	.	.	19
3052	.	.	.	.	.	.	.	.	.	.	19
3053	377.0	.	4.0	.	.	.	.	.	.	487.0	15
3054	344.0	.	4.0	.	.	.	.	.	.	710.0	15
3055	.	.	.	.	.	.	.	.	.	.	21
3056	.	.	.	.	.	.	.	.	.	.	19
3057	.	.	.	.	.	.	.	.	.	.	19
3058	.	.	.	.	.	.	.	.	.	.	6
3059	.	.	.	.	.	.	.	.	.	.	20
3060	.	.	.	.	.	.	.	.	.	.	21

TABLE 3 - DATA CONVERTED TO A ZERO-MOISTURE BASIS.

OBS	PLNA	PART	CALORIES	WATER	PROTEIN	FAT	TOTCARB	FIBER	ASH	CALCIUM	PHOSPHO	IRON	SODIUM	POTASS	CAROT	THIAMINE	RIBOFLAV	NIACIN	ASCORBIC	REFCO
3061	GAZANIA PINNATA	S	.	0	45.8	27.2	.	.	.	.	.	.	.	.	.	.	.	.	.	21
3062	GAZANIA PINNATA	S	.	0	42.7	26.6	.	.	.	.	.	.	.	.	.	.	.	.	.	21
3063	GENISTA AETHNENSIS	S	.	0	36.9	8.5	.	.	3.0	.	.	.	.	.	.	.	.	.	.	19
3064	GENISTA CARINALIS	S	.	0	36.2	4.8	.	.	.	.	.	.	.	.	.	.	.	.	.	21
3065	GENISTA JAUBERTII	S	.	0	43.9	11.0	.	.	.	.	.	.	.	.	.	.	.	.	.	21
3066	GENISTA LYDIA	S	.	0	40.9	11.3	.	.	.	.	.	.	.	.	.	.	.	.	.	21
3067	GENISTA MONOSPERMA	S	.	0	28.1	5.2	.	.	.	.	.	.	.	.	.	.	.	.	.	19
3068	GENISTA TINCTORIA	S	.	0	41.2	9.7	.	.	.	.	.	.	.	.	.	.	.	.	.	21
3069	GENTIANA LUTEA	S	.	0	31.0	26.0	.	.	.	.	.	.	.	.	.	.	.	.	.	21
3070	GEONOMA EDULIS	SH	297	0	27.1	2.5	59.3	12.7	11.0	347.5	694.9	15.3	.	.	0.00	0.59	1.10	10.17	152.5	3
3071	GEOPRUMNON CRASSICARPUM	S	.	0	45.0	5.4	.	.	3.9	.	.	.	.	.	.	.	.	.	.	19
3072	GERANIUM CAROLINIANUM	S	.	0	22.9	17.0	.	.	.	.	.	.	.	.	.	.	.	.	.	20
3073	GERANIUM CAROLINIANUM	S	.	0	22.3	19.5	.	.	.	.	.	.	.	.	.	.	.	.	.	21
3074	GERANIUM CAROLINIANUM	W	.	0	13.3	2.0	71.7	16.6	13.0	1340.0	570.0	.	.	2160.0	.	.	.	.	.	6
3075	GERANIUM FREMONTII	W	.	0	18.8	3.3	70.7	8.4	7.2	1250.0	480.0	.	.	.	.	.	.	.	.	6
3076	GERANIUM VISCOSISSIMUM	W	.	0	28.9	2.4	60.4	8.2	8.3	.	.	.	.	.	.	.	.	.	.	6
3077	GEROPOGON GLABER	S	.	0	19.4	16.5	.	.	.	.	.	.	.	.	.	.	.	.	.	21
3078	GEUM CHILOENSE	S	.	0	29.6	24.4	.	.	4.7	.	.	.	.	.	.	.	.	.	.	19
3079	GEUM CHILOENSE	S	.	0	13.1	7.4	.	.	8.4	.	.	.	.	.	.	.	.	.	.	19
3080	GEUM MACROPHYLLUM	S	.	0	15.4	21.7	.	.	.	.	.	.	.	.	.	.	.	.	.	21
3081	GEUM SP.	S	.	0	14.4	18.8	.	.	.	.	.	.	.	.	.	.	.	.	.	19
3082	GEUM SP.	H	.	0	17.4	17.9	60.8	21.2	3.9	.	.	.	.	.	.	.	.	.	.	6
3083	GEUM URBANUM	S	.	0	13.9	21.0	.	.	.	.	.	.	.	.	.	.	.	.	.	21
3084	GILIA CANDIDA	S	.	0	16.8	16.1	.	.	.	.	.	.	.	.	.	.	.	.	.	21
3085	GILIA CAPITATA	S	.	0	15.6	22.7	.	.	.	.	.	.	.	.	.	.	.	.	.	21
3086	GILIA PINNATA	S	.	0	18.1	21.0	.	.	.	.	.	.	.	.	.	.	.	.	.	20
3087	GILIA RIGIDULA	S	.	0	14.1	17.5	.	.	.	.	.	.	.	.	.	.	.	.	.	20
3088	GILIA SP.	S	.	0	26.9	30.7	.	.	.	.	.	.	.	.	.	.	.	.	.	20
3089	GILIA TRICOLOR		.	0	12.0	.	.	23.1	8.8	1440.0	450.0	.	.	2600.0	.	.	.	.	.	6
3090	GILIA VISCIDULA	W	.	0	8.0	.	.	24.9	8.6	1550.0	180.0	.	.	1190.0	.	.	.	.	.	6
3091	GINGKO BILOBA	S	.	0	10.2	3.1	.	.	3.8	.	.	.	.	.	.	.	.	.	.	19
3092	GINKGO BILOBA	S	403	0	10.5	3.5	83.0	1.3	3.1	10.9	326.8	2.6	15.3	1139.4	392.16	0.52	0.26	6.10	54.5	2
3093	GLADIOLUS UNGUICULATUS	R	339	0	5.0	0.8	89.7	7.8	4.5	540.2	.	47.7	.	.	0.00	0.25	0.20	2.76	7.5	1
3094	GLAUCIUM ARABICUM	S	.	0	23.4	34.7	.	.	.	.	.	.	.	.	.	.	.	.	.	21
3095	GLAUCIUM CORNICULATUM	S	.	0	20.6	33.5	.	.	.	.	.	.	.	.	.	.	.	.	.	21
3096	GLAUCIUM CORNICULATUM	S	.	0	22.8	31.3	.	.	.	.	.	.	.	.	.	.	.	.	.	21
3097	GLEDITSIA JAPONICA	S	.	0	22.4	2.8	.	.	.	.	.	.	.	.	.	.	.	.	.	21
3098	GLEDITSIA TRIACANTHOS	F	.	0	23.1	4.6	66.9	12.7	5.4	.	.	.	.	.	.	.	.	.	.	18
3099	GLEDITSIA TRIACANTHOS	S	.	0	10.6	0.8	84.7	21.1	3.9	280.0	320.0	.	.	.	.	.	.	.	.	18
3100	GLEDITSIA TRIACANTHOS	S	.	0	20.6	2.6	.	.	.	.	.	.	.	.	.	.	.	.	.	19
3101	GLEDITSIA TRIACANTHOS	S	.	0	21.8	2.3	.	.	3.8	.	.	.	.	.	.	.	.	.	.	19
3102	GLEDITSIA TRIACANTHOS	S	.	0	22.1	1.7	.	.	4.0	.	.	.	.	.	.	.	.	.	.	19
3103	GLEDITSIA TRIACANTHOS	S	.	0	24.1	4.3	.	.	.	.	.	.	.	.	.	.	.	.	.	21
3104	GLIRICIDIA MACULATA	C	.	0	18.8	3.7	71.2	15.5	6.3	660.0	110.0	.	.	.	.	.	.	.	.	18
3105	GLIRICIDIA MACULATA	L	.	0	30.0	4.3	57.7	14.1	8.0	.	.	.	.	.	.	.	.	.	.	18

TABLE 3 - DATA CONVERTED TO A ZERO-MOISTURE BASIS.

OBS	PLNA	PART	CALORIES	WATER	PROTEIN	FAT	TOTCARB	FIBER	ASH	CALCIUM	PHOSPHO	IRON	SODIUM	POTASS	CAROT	THIAMINE	RIBOFLAV	NIACIN	ASCORBIC	REFCO
3106	GLIRICIDIA SEPIUM	L	340	0	15.7	3.3	77.8	8.5	3.3	111.1	222.2	5.2	.	.	.	0.92	0.52	6.54	78.4	2
3107	GLIRICIDIA SEPIUM	I	329	0	17.1	1.4	77.1	11.4	4.3	157.1	264.3	12.9	.	.	500.0	0.79	0.57	4.29	314.3	3
3108	GLIRICIDIA SEPIUM	I	.	0	25.5	10.3	57.9	16.6	6.9	.	.	.	.	.	.	.	.	.	.	5
3109	GLIRICIDIA SP.	S	.	0	47.5	20.4	.	.	4.6	.	.	.	.	.	.	.	.	.	.	19
3110	GLOBULARIA ORIENTALIS	S	.	0	27.5	39.0	.	.	.	.	.	.	.	.	.	.	.	.	.	21
3111	GLOBULARIA TRICHOSANTHA	S	.	0	36.1	35.2	.	.	.	.	.	.	.	.	.	.	.	.	.	21
3112	GLYCERIA GRANDIS	W	.	0	12.5	2.4	76.7	32.8	8.4	260.0	550.0	.	.	.	14.0	.	.	.	.	6
3113	GLYCERIA OBTUSA	S	.	0	23.1	3.4	.	.	.	.	.	.	.	.	.	.	.	.	.	20
3114	GLYCERIA STRIATA	S	.	0	22.5	2.1	.	.	.	.	.	.	.	.	.	.	.	.	.	20
3115	GLYCINE GRACILIS	S	.	0	44.0	14.3	.	.	5.9	.	.	.	.	.	.	.	.	.	.	19
3116	GLYCINE MAX	W	.	0	16.1	6.2	65.8	29.6	11.9	1220.0	240.0	.	.	.	.	.	.	.	.	18
3117	GLYCINE MAX	H	.	0	13.9	2.5	78.1	35.5	5.5	940.0	180.0	.	.	.	.	.	.	.	.	18
3118	GLYCINE MAX	S	.	0	31.2	21.8	.	.	5.5	.	.	.	.	.	.	.	.	.	.	19
3119	GLYCINE MAX	W	.	0	16.6	4.0	68.8	28.1	10.6	1300.0	530.0	21.0	.	1060.0	29.0	.	.	.	.	6
3120	GLYCINE MAX	S	448	0	37.2	19.8	37.5	5.2	5.5	202.2	597.8	6.7	.	.	60.8	0.78	0.28	2.21	.	1
3121	GLYCINE MAX	S	439	0	38.0	17.1	40.4	4.9	4.6	242.9	580.4	10.8	.	467.5	11.4	0.74	0.26	3.19	0.0	2
3122	GLYCINE MAX	S	437	0	40.9	17.9	35.8	6.0	5.3	245.3	496.9	11.9	47.2	1908.8	1132.1	1.26	0.53	4.72	84.9	2
3123	GLYCINE MAX	X	335	0	41.6	9.7	43.2	3.8	5.4	281.1	313.5	5.9	162.2	1508.1	135.1	1.03	0.81	4.32	54.1	2
3124	GLYCINE MAX	S	438	0	36.8	18.1	39.1	6.3	6.1	244.5	804.0	12.7	.	.	0.0	0.97	0.30	2.42	.	3
3125	GLYCINE MAX	G	435	0	35.4	16.6	42.9	4.5	5.2	217.5	730.5	9.1	.	.	1344.2	1.43	0.52	4.55	94.2	4
3126	GLYCINE MAX	X	336	0	45.3	10.2	38.7	5.8	5.8	350.4	489.1	7.3	.	.	350.4	1.68	1.46	5.84	94.9	4
3127	GLYCINE SP.	S	.	0	32.4	4.5	.	.	5.0	.	.	.	.	.	.	.	.	.	.	19
3128	GLYCINE WIGHTII	W	.	0	10.1	0.7	80.6	32.7	8.6	530.0	330.0	.	.	.	.	.	.	.	.	18
3129	GLYCINE WIGHTII	H	.	0	14.4	2.6	74.5	34.2	8.5	990.0	310.0	.	.	.	.	.	.	.	.	18
3130	GLYCYRRHIZA ECHINATA	S	.	0	42.2	3.1	.	.	.	.	.	.	.	.	.	.	.	.	.	21
3131	GLYCYRRHIZA GLABRA	S	.	0	26.8	1.3	.	.	.	.	.	.	.	.	.	.	.	.	.	21
3132	GLYCYRRHIZA GLABRA	S	.	0	31.3	9.0	.	.	.	.	.	.	.	.	.	.	.	.	.	21
3133	GLYCYRRHIZA GLABRA	S	.	0	21.1	11.2	.	.	.	.	.	.	.	.	.	.	.	.	.	21
3134	GLYCYRRHIZA LEPIDOTA	S	.	0	38.5	6.4	.	.	3.6	.	.	.	.	.	.	.	.	.	.	19
3135	GLYCYRRHIZA LEPIDOTA	S	.	0	27.6	2.1	.	.	.	.	.	.	.	.	.	.	.	.	.	21
3136	GNETUM BUCHHOLZIANUM	L	341	0	19.9	4.3	71.5	18.2	4.3	430.5	178.8	18.5	.	.	.	.	.	.	331.1	1
3137	GNETUM GNEMON	L	333	0	15.4	3.3	74.8	10.6	6.5	520.3	406.5	11.0	4.1	337.4	28902.4	0.69	0.98	6.50	459.3	2
3138	GNETUM GNEMON	F	330	0	25.0	3.5	66.5	.	5.0	815.0	375.0	14.0	.	.	3000.0	0.50	.	.	500.0	2
3139	GNETUM GNEMON	S	397	0	13.8	1.7	82.2	.	2.3	114.9	459.8	5.7	.	.	0.0	0.23	.	.	0.0	2
3140	GNETUM GNEMON	L	352	0	17.2	6.4	71.5	34.3	4.9	976.7	212.2	7.8	.	.	26133.7	0.32	0.96	4.36	625.0	3
3141	GNETUM GNEMON	L	.	0	.	.	.	.	7.2	607.7	1326.0	55.2	.	.	.	.	.	.	.	5
3142	GNETUM GNEMON	S	.	0	15.6	2.3	78.4	1.3	2.4	.	.	.	.	.	.	.	.	.	.	5
3143	GODETIA AMOENA	W	.	0	6.1	.	.	19.1	8.0	1865.0	265.0	.	.	1445.0	.	.	.	.	.	6
3144	GODETIA GRANDIFLORA	S	.	0	32.5	33.1	.	.	.	.	.	.	.	.	.	.	.	.	.	19
3145	GODETIA QUADRIVULNERA	W	.	0	4.0	.	.	32.6	6.9	1760.0	240.0	.	.	1140.0	.	.	.	.	.	6
3146	GODETIA QUADRIVULNERA	W	.	0	6.8	.	.	23.8	7.5	1840.0	380.0	.	.	1460.0	.	.	.	.	.	6
3147	GOLDMANIA FOETIDA	S	.	0	28.9	7.1	.	.	.	.	.	.	.	.	.	.	.	.	.	21
3148	GONOLOBUS SALVINII	F	372	0	14.1	1.3	79.5	10.3	5.1	153.8	217.9	5.1	.	.	128.2	1.28	0.26	2.56	1538.5	3
3149	GOSSYPIUM HIRSUTUM	S	.	0	37.1	38.4	.	.	5.2	.	.	.	.	.	.	.	.	.	.	19
3150	GOSSYPIUM HIRSUTUM	S	433	0	35.8	17.5	39.9	5.2	6.7	162.1	1112.1	11.1	.	.	0.0	0.15	0.03	3.92	.	3

TABLE 3 - DATA CONVERTED TO A ZERO-MOISTURE BASIS.

OBS	PLNA	PART	CALORIES	WATER	PROTEIN	FAT	TOTCARB	FIBER	ASH	CALCIUM	PHOSPHO	IRON	SODIUM	POTASS	CAROT	THIAMINE	RIBOFLAV	NIACIN	ASCORBIC	REFCO
3151	GOSSYPIUM SP.	L	.	0	16.8	7.8	57.8	11.2	17.6	3510.0	290.0	.	.	1170	.	.	.	.	.	6
3152	GOSSYPIUM SP.	Y	.	0	6.2	0.9	86.1	47.2	6.8	.	.	.	.	.	.	.	.	.	.	6
3153	GOSSYPIUM SP.	S	475	0	22.0	23.0	50.7	20.1	4.3	171.7	895.7	.	.	.	.	.	.	.	.	1
3154	GRAYIA SPINOSA	W	.	0	8.9	2.0	84.0	38.3	5.1	1100.0	190.0	.	.	.	.	.	.	.	.	6
3155	GRAYIA SPINOSA	L	.	0	10.2	1.7	57.5	9.2	30.6	.	.	.	.	.	.	.	.	.	.	6
3156	GREWIA ASIATICA	S	.	0	9.3	6.1	.	.	.	.	.	.	.	.	.	.	.	.	.	21
3157	GREWIA BICOLOR	SH	.	0	12.7	3.5	77.8	23.0	6.0	1600.0	300.0	.	.	.	.	.	.	.	.	18
3158	GREWIA BICOLOR	L	.	0	5.8	7.4	73.3	17.5	13.5	3020.0	140.0	.	.	.	.	.	.	.	.	18
3159	GREWIA KAKOTHAMNOS	SH	.	0	8.7	2.5	81.1	32.6	7.7	1530.0	360.0	.	.	.	.	.	.	.	.	18
3160	GREWIA OPPOSITIFOLIA	L	.	0	16.4	8.4	70.3	16.6	14.9	.	.	.	.	.	.	.	.	.	.	18
3161	GREWIA OPTIVA	S	.	0	33.4	20.6	.	.	.	.	.	.	.	.	.	.	.	.	.	21
3162	GREWIA OPTIVA	L	.	0	10.1	6.8	68.9	14.1	14.2	4180.0	250.0	.	.	.	.	.	.	.	.	5
3163	GREWIA TENAX	S	.	0	29.4	23.0	.	.	.	.	.	.	.	.	.	.	.	.	.	21
3164	GRIELUM SP.	S	.	0	18.6	2.5	.	.	.	.	.	.	.	.	.	.	.	.	.	21
3165	GRINDELIA LANCEOLATA	S	.	0	16.0	12.1	.	.	.	.	.	.	.	.	.	.	.	.	.	21
3166	GRINDELIA OXYLEPIS	S	.	0	13.1	9.0	.	.	.	.	.	.	.	.	.	.	.	.	.	20
3167	GRINDELIA ROBUSTA	S	.	0	17.4	14.5	.	.	.	.	.	.	.	.	.	.	.	.	.	21
3168	GRINDELIA SQUARROSA	S	.	0	13.8	19.6	.	.	3.4	.	.	.	.	.	.	.	.	.	.	19
3169	GRINDELIA SUBALPINA VAR. ERECTA	S	.	0	16.5	19.1	.	.	.	.	.	.	.	.	.	.	.	.	.	21
3170	GUAIACUM COULTERI	S	.	0	14.4	3.4	.	.	.	.	.	.	.	.	.	.	.	.	.	19
3171	GUARA VILLOSA	S	.	0	37.2	43.9	.	.	.	.	.	.	.	.	.	.	.	.	.	21
3172	GUETTARDA URUGUENSIS	S	.	0	6.2	3.6	.	.	.	.	.	.	.	.	.	.	.	.	.	21
3173	GUIZOTIA ABYSSINICA	H	.	0	12.8	10.9	64.8	24.0	11.5	.	.	.	.	.	.	.	.	.	.	18
3174	GUIZOTIA ABYSSINICA	S	.	0	24.2	40.0	30.5	16.8	5.3	.	.	.	.	.	.	.	.	.	.	18
3175	GUIZOTIA ABYSSINICA	S	.	0	21.1	39.0	.	.	10.0	.	.	.	.	.	.	.	.	.	.	19
3176	GUIZOTIA ABYSSINICA	S	515	0	18.4	36.1	36.5	14.4	9.0	501.1	852.9	.	.	.	0	0.46	0.59	3.2	0	1
3177	GUIZOTIA ABYSSINICA	S	.	0	21.0	33.9	43.1	.	2.0	54.2	195.2	.	.	.	.	.	.	.	.	5
3178	GUNDELIA TOURNEFORTII	S	.	0	44.2	39.3	.	.	.	.	.	.	.	.	.	.	.	.	.	21
3179	GUTIERREZIA MICROCEPHALA	W	.	0	10.4	15.8	69.6	22.5	4.2	740.0	310.0	.	.	1350	.	.	.	.	.	6
3180	GUTIERREZIA SAROTHRAE	S	.	0	32.3	31.8	.	.	.	.	.	.	.	.	.	.	.	.	.	21
3181	GUTIERREZIA SAROTHRAE	W	.	0	6.7	11.2	76.8	29.3	5.3	1320.0	200.0	.	.	.	.	.	.	.	.	6
3182	GUTIERREZIA TEXANA	W	.	0	14.7	7.2	72.1	19.0	6.0	650.0	170.0	.	.	2980	.	.	.	.	.	6
3183	GYMNOCLADUS DIOICA	S	.	0	29.6	24.2	.	.	4.0	.	.	.	.	.	.	.	.	.	.	19
3184	GYMNOCLADUS DIOICA	S	.	0	4.2	0.5	.	.	5.4	.	.	.	.	.	.	.	.	.	.	19
3185	GYMNOCLADUS DIOICA	S	.	0	30.7	21.7	.	.	3.8	.	.	.	.	.	.	.	.	.	.	19
3186	GYNANDROPSIS GYNANDRA	SH	.	0	25.7	2.7	52.5	18.6	19.1	1690.0	590.0	.	.	.	.	.	.	.	.	18
3187	GYNANDROPSIS GYNANDRA	S	.	0	18.5	25.5	.	.	.	.	.	.	.	.	.	.	.	.	.	21
3188	GYNANDROPSIS GYNANDRA	L	254	0	35.8	3.0	38.8	9.0	22.4	2149.3	828.4	44.8	.	.	.	.	.	.	97	1
3189	GYPSOPHILA ANATOLICA	S	.	0	19.3	7.1	.	.	.	.	.	.	.	.	.	.	.	.	.	21
3190	GYPSOPHILA ELEGANS	S	.	0	16.9	6.6	.	.	5.0	.	.	.	.	.	.	.	.	.	.	19
3191	GYPSOPHILA ELEGANS	S	.	0	16.3	5.7	.	.	4.6	.	.	.	.	.	.	.	.	.	.	19
3192	GYPSOPHILA ELEGANS	S	.	0	16.9	6.9	.	.	.	.	.	.	.	.	.	.	.	.	.	20
3193	GYPSOPHILA PORRIGENS	S	.	0	19.1	8.4	.	.	.	.	.	.	.	.	.	.	.	.	.	20
3194	GYPSOPHILA RUSCIFOLIA	S	.	0	19.8	7.6	.	.	.	.	.	.	.	.	.	.	.	.	.	21
3195	HALOCNEMUM STROBILACEUM	S	.	0	27.1	12.6	.	.	.	.	.	.	.	.	.	.	.	.	.	21

TABLE 3 - DATA CONVERTED TO A ZERO-MOISTURE BASIS.

OBS	PLNA	PART	CALORIES	WATER	PROTEIN	FAT	TOTLCARB	FIBER	ASH	CALCIUM	PHOSPHO	IRON	SODIUM	POTASS	CAROT	THIAMINE	RIBOFLAV	NIACIN	ASCORBIC	REFCO
3196	HAMAMELIS JAPONICA	S	.	0	14.9	16.4	.	.	.	.	.	.	.	.	.	.	.	.	.	21
3197	HAPLOPAPPUS CILIATUS	S	.	0	12.8	11.8	.	.	.	.	.	.	.	.	.	.	.	.	.	21
3198	HAPLOPAPPUS CROCEUS	S	.	0	23.9	13.6	.	.	.	.	.	.	.	.	.	.	.	.	.	21
3199	HAYNALDIA VILLOSA	S	.	0	26.2	2.7	.	.	.	.	.	.	.	.	.	.	.	.	.	21
3200	HEDEOMA DRUMMONDII	S	.	0	8.8	6.9	.	.	.	.	.	.	.	.	.	.	.	.	.	20
3201	HEDEOMA DRUMMONDII	W	.	0	10.8	3.8	77.3	30.1	8.1	1990	190	.	.	1730	.	.	.	.	.	6
3202	HEDERA HELIX	S	.	0	16.2	35.1	.	.	.	.	.	.	.	.	.	.	.	.	.	19
3203	HEDYPNOIS CRETICA	S	.	0	15.0	12.1	.	.	.	.	.	.	.	.	.	.	.	.	.	21
3204	HEDYSARUM BOREALE	S	.	0	43.0	8.8	.	.	5.1	.	.	.	.	.	.	.	.	.	.	19
3205	HEDYSARUM CORONARIUM	W	.	0	19.6	2.4	64.0	24.0	14.0	1120	340	.	.	.	.	.	.	.	.	18
3206	HEDYSARUM CORONARIUM	H	.	0	5.8	0.3	86.4	51.8	7.5	.	.	.	.	.	.	.	.	.	.	18
3207	HEDYSARUM CORONARIUM	F	.	0	13.8	2.1	68.0	23.6	15.5	.	.	.	.	.	.	.	.	.	.	18
3208	HEDYSARUM CORONARIUM	L	.	0	13.5	2.4	74.1	30.6	10.6	.	.	.	.	.	.	.	.	.	.	5
3209	HEDYSARUM HUMILE	S	.	0	55.8	10.0	.	.	.	.	.	.	.	.	.	.	.	.	.	21
3210	HEDYSARUM SPINOSISSIMUM	S	.	0	61.6	11.1	.	.	.	.	.	.	.	.	.	.	.	.	.	21
3211	HEDYSARUM VARIUM	S	.	0	50.1	8.3	.	.	.	.	.	.	.	.	.	.	.	.	.	20
3212	HEIMIA SALICIFOLIA	S	.	0	21.6	28.7	.	.	.	.	.	.	.	.	.	.	.	.	.	21
3213	HELENIUM AUTUMNALE	S	.	0	25.2	22.8	.	.	7.8	.	.	.	.	.	.	.	.	.	.	19
3214	HELENIUM AUTUMNALE	S	.	0	27.5	26.3	.	.	.	.	.	.	.	.	.	.	.	.	.	20
3215	HELENIUM LACINIATUM	S	.	0	23.1	28.6	.	.	.	.	.	.	.	.	.	.	.	.	.	20
3216	HELENIUM OOCLINIUM	S	.	0	30.3	26.3	.	.	.	.	.	.	.	.	.	.	.	.	.	20
3217	HELENIUM SP.	W	.	0	18.7	4.8	64.2	11.0	12.3	2440	610	.	.	3950	.	.	.	.	.	6
3218	HELIANTHELLA CALIFORNICA	S	.	0	30.8	21.3	.	.	.	.	.	.	.	.	.	.	.	.	.	21
3219	HELIANTHELLA UNIFLORA	W	.	0	7.4	4.3	.	.	9.4	2030	310	.	.	.	.	.	.	.	.	6
3220	HELIANTHEMUM LEDIFOLIUM	S	.	0	19.8	5.0	.	.	.	.	.	.	.	.	.	.	.	.	.	21
3221	HELIANTHEMUM NUMMULARIUM VAR. MUTABILE	S	.	0	17.5	4.8	.	.	.	.	.	.	.	.	.	.	.	.	.	20
3222	HELIANTHEMUM STIPULATUM	S	.	0	22.8	8.9	.	.	.	.	.	.	.	.	.	.	.	.	.	21
3223	HELIANTHUS ANNUUS	SH	.	0	13.0	1.9	70.3	20.4	14.8	1670	370	.	.	.	.	.	.	.	.	18
3224	HELIANTHUS ANNUUS	S	.	0	12.3	42.0	42.6	27.9	3.1	210	590	.	.	.	.	.	.	.	.	18
3225	HELIANTHUS ANNUUS	I	.	0	12.7	13.7	64.3	23.9	9.3	630	80	.	.	.	.	.	.	.	.	18
3226	HELIANTHUS ANNUUS	S	.	0	34.4	48.8	.	.	4.4	.	.	.	.	.	.	.	.	.	.	19
3227	HELIANTHUS ANNUUS	S	.	0	26.9	54.4	.	.	4.0	.	.	.	.	.	.	.	.	.	.	19
3228	HELIANTHUS ANNUUS	S	.	0	41.2	42.4	.	.	4.6	.	.	.	.	.	.	.	.	.	.	19
3229	HELIANTHUS ANNUUS	S	.	0	35.0	45.3	.	.	5.0	.	.	.	.	.	.	.	.	.	.	19
3230	HELIANTHUS ANNUUS	S	.	0	36.9	42.2	.	.	4.6	.	.	.	.	.	.	.	.	.	.	19
3231	HELIANTHUS ANNUUS	S	.	0	30.0	49.0	.	.	4.8	.	.	.	.	.	.	.	.	.	.	19
3232	HELIANTHUS ANNUUS	S	.	0	29.4	48.8	.	.	5.0	.	.	.	.	.	.	.	.	.	.	19
3233	HELIANTHUS ANNUUS	S	.	0	32.5	44.1	.	.	4.9	.	.	.	.	.	.	.	.	.	.	19
3234	HELIANTHUS ANNUUS	S	.	0	36.2	43.5	.	.	5.4	.	.	.	.	.	.	.	.	.	.	19
3235	HELIANTHUS ANNUUS	S	.	0	31.2	46.0	.	.	4.6	.	.	.	.	.	.	.	.	.	.	19
3236	HELIANTHUS ANNUUS	S	.	0	33.8	45.4	.	.	4.6	.	.	.	.	.	.	.	.	.	.	19
3237	HELIANTHUS ANNUUS	S	.	0	27.5	50.8	.	.	4.2	.	.	.	.	.	.	.	.	.	.	19
3238	HELIANTHUS ANNUUS	S	.	0	26.2	51.6	.	.	4.9	.	.	.	.	.	.	.	.	.	.	19
3239	HELIANTHUS ANNUUS	S	.	0	31.2	44.3	.	.	5.0	.	.	.	.	.	.	.	.	.	.	19
3240	HELIANTHUS ANNUUS	S	.	0	26.2	54.5	.	.	4.3	.	.	.	.	.	.	.	.	.	.	19

TABLE 3 - DATA CONVERTED TO A ZERO-MOISTURE BASIS.

OBS	PLANA	PART	CALORIES	WATER	PROTEIN	FAT	TOTLCARB	FIBER	ASH
3241	HELIANTHUS ANNUUS	S	.	0	34.4	42.1	.	.	5.6
3242	HELIANTHUS ANNUUS	S	.	0	33.8	49.5	.	.	5.1
3243	HELIANTHUS ANNUUS	S	.	0	40.0	43.8	.	.	5.0
3244	HELIANTHUS ANNUUS	S	.	0	40.6	43.8	.	.	5.1
3245	HELIANTHUS ANNUUS	S	.	0	41.9	39.9	.	.	5.2
3246	HELIANTHUS ANNUUS	S	.	0	43.8	40.2	.	.	5.2
3247	HELIANTHUS ANNUUS	S	.	0	40.0	40.0	.	.	4.6
3248	HELIANTHUS ANNUUS	S	.	0	40.0	44.5	.	.	4.6
3249	HELIANTHUS ANNUUS	S	.	0	37.5	45.1	.	.	4.5
3250	HELIANTHUS ANNUUS	S	.	0	19.3	28.2	.	.	3.1
3251	HELIANTHUS ANNUUS	S	.	0	18.8	28.6	.	.	.
3252	HELIANTHUS ANNUUS	S	.	0	15.1	24.4	.	.	.
3253	HELIANTHUS ANNUUS	S	.	0	18.1	32.4	.	.	.
3254	HELIANTHUS ANNUUS	S	.	0	16.8	33.1	.	.	.
3255	HELIANTHUS ANNUUS	S	.	0	16.5	34.4	.	.	.
3256	HELIANTHUS ANNUUS	S	.	0	16.7	34.3	.	.	.
3257	HELIANTHUS ANNUUS	S	.	0	17.3	45.7	.	.	.
3258	HELIANTHUS ANNUUS	S	.	0	18.9	46.7	.	.	.
3259	HELIANTHUS ANNUUS	S	.	0	18.2	43.7	.	.	.
3260	HELIANTHUS ANNUUS	S	.	0	19.4	40.4	.	.	.
3261	HELIANTHUS ANNUUS	S	.	0	18.7	43.6	.	.	.
3262	HELIANTHUS ANNUUS	S	.	0	19.9	45.7	.	.	.
3263	HELIANTHUS ANNUUS	S	.	0	17.2	43.6	.	.	.
3264	HELIANTHUS ANNUUS	S	.	0	19.1	43.9	.	.	.
3265	HELIANTHUS ANNUUS	S	.	0	19.6	44.2	.	.	.
3266	HELIANTHUS ANNUUS	S	.	0	20.2	41.2	.	.	.
3267	HELIANTHUS ANNUUS	S	.	0	16.3	39.0	.	.	.
3268	HELIANTHUS ANNUUS	S	.	0	18.6	37.9	.	.	.
3269	HELIANTHUS ANNUUS	W	.	0	8.8	2.9	77.2	30.3	11.1
3270	HELIANTHUS ANNUUS	S	514	0	13.8	29.3	54.3	27.4	2.6
3271	HELIANTHUS ANNUUS	S	536	0	18.3	35.8	42.2	4.0	3.7
3272	HELIANTHUS ANNUUS	S	609	0	23.7	54.4	17.5	5.3	4.3
3273	HELIANTHUS ANNUUS	S	588	0	25.2	49.7	20.9	4.0	4.2
3274	HELIANTHUS ANNUUS(IMMATURE)	W	.	0	20.5	6.9	55.7	10.7	16.9
3275	HELIANTHUS CILIARIA	S	.	0	15.7	15.6	.	.	4.4
3276	HELIANTHUS CILIARIS	S	.	0	15.6	23.6	.	.	.
3277	HELIANTHUS MAXIMILIANI	S	.	0	30.8	30.4	.	.	4.3
3278	HELIANTHUS MAXIMILIANI	S	.	0	30.6	29.9	.	.	.
3279	HELIANTHUS MOLLIS	S	.	0	25.6	28.6	.	.	.
3280	HELIANTHUS NUTTALLII	S	.	0	30.6	23.4	.	.	.
3281	HELIANTHUS PETIOLARIS	S	.	0	19.3	29.9	.	.	2.8
3282	HELIANTHUS PETIOLARIS	S	.	0	16.9	28.4	.	.	.
3283	HELIANTHUS PETIOLARIS	S	.	0	21.2	24.1	.	.	.
3284	HELIANTHUS PETIOLARIS	W	.	0	10.2	6.9	71.4	35.2	11.5
3285	HELIANTHUS PETIOLARIS SUBSP. FALLAX	S	.	0	22.6	32.5	.	.	.

OBS	CALCIUM	PHOSPHO	IRON	SODIUM	POTASS	CAROT	THIAMINE	RIBOFLAV	NIACIN	ASCORBIC	REFCO
3241	.	.	.	.	.	.	.	.	.	.	19
3242	.	.	.	.	.	.	.	.	.	.	19
3243	.	.	.	.	.	.	.	.	.	.	19
3244	.	.	.	.	.	.	.	.	.	.	19
3245	.	.	.	.	.	.	.	.	.	.	19
3246	.	.	.	.	.	.	.	.	.	.	19
3247	.	.	.	.	.	.	.	.	.	.	19
3248	.	.	.	.	.	.	.	.	.	.	19
3249	.	.	.	.	.	.	.	.	.	.	19
3250	.	.	.	.	.	.	.	.	.	.	19
3251	.	.	.	.	.	.	.	.	.	.	19
3252	.	.	.	.	.	.	.	.	.	.	20
3253	.	.	.	.	.	.	.	.	.	.	21
3254	.	.	.	.	.	.	.	.	.	.	21
3255	.	.	.	.	.	.	.	.	.	.	21
3256	.	.	.	.	.	.	.	.	.	.	21
3257	.	.	.	.	.	.	.	.	.	.	21
3258	.	.	.	.	.	.	.	.	.	.	21
3259	.	.	.	.	.	.	.	.	.	.	21
3260	.	.	.	.	.	.	.	.	.	.	21
3261	.	.	.	.	.	.	.	.	.	.	21
3262	.	.	.	.	.	.	.	.	.	.	21
3263	.	.	.	.	.	.	.	.	.	.	21
3264	.	.	.	.	.	.	.	.	.	.	21
3265	.	.	.	.	.	.	.	.	.	.	21
3266	.	.	.	.	.	.	.	.	.	.	21
3267	.	.	.	.	.	.	.	.	.	.	21
3268	.	.	.	.	.	.	.	.	.	.	21
3269	.	.	.	.	.	.	.	.	.	.	6
3270	.	.	.	.	.	.	.	.	.	.	1
3271	100.5	690.7	6.3	.	890.7	0.00	.	0.08	2.62	0	2
3272	111.2	919.5	8.2	.	.	10.59	2.12	0.20	8.05	0	3
3273	126.1	879.2	7.5	31.5	966.4	31.51	2.06	0.24	5.67	.	4
3274	2130.0	220.0	.	.	4470.0	.	.	.	.	.	6
3275	.	.	.	.	.	.	.	.	.	.	19
3276	.	.	.	.	.	.	.	.	.	.	20
3277	.	.	.	.	.	.	.	.	.	.	19
3278	.	.	.	.	.	.	.	.	.	.	20
3279	.	.	.	.	.	.	.	.	.	.	19
3280	.	.	.	.	.	.	.	.	.	.	21
3281	.	.	.	.	.	.	.	.	.	.	19
3282	.	.	.	.	.	.	.	.	.	.	19
3283	.	.	.	.	.	.	.	.	.	.	21
3284	.	.	.	.	.	.	.	.	.	.	6
3285	.	.	.	.	.	.	.	.	.	.	21

TABLE 3 - DATA CONVERTED TO A ZERO-MOISTURE BASIS.

OBS	PLNA	PART	CALORIES	WATER	PROTEIN	FAT	TOTLCARB	FIBER
3286	HELIANTHUS PUMILIS	S	.	0	24.2	27.7	.	.
3287	HELIANTHUS RYDBERGII	S	.	0	26.1	30.3	.	.
3288	HELIANTHUS SCABERRIMUS	S	.	0	24.3	29.5	.	.
3289	HELIANTHUS SP.	S	.	0	32.2	32.2	.	.
3290	HELIANTHUS SP.	S	.	0	22.5	37.8	.	.
3291	HELIANTHUS TUBEROSUS	L	.	0	20.7	2.3	60.9	12.4
3292	HELIANTHUS TUBEROSUS	SH	.	0	10.5	3.4	70.6	16.7
3293	HELIANTHUS TUBEROSUS	R	.	0	8.3	0.6	85.1	4.8
3294	HELIANTHUS TUBEROSUS	S	.	0	30.6	26.5	.	.
3295	HELIANTHUS TUBEROSUS	R	332	0	9.9	0.5	84.7	5.9
3296	HELIANTHUS TUBEROSUS	R	352	0	9.2	0.5	84.2	4.1
3297	HELIANTHUS TUBEROSUS	R	325	0	11.0	0.5	82.5	4.0
3298	HELIANTHUS TUBEROSUS	R	203	0	11.4	0.5	82.7	4.0
3299	HELIANTHUS TUBEROSUS	SH	.	0	5.1	1.1	86.0	18.0
3300	HELICHRYSUM BRACTEATUM	S	.	0	21.9	24.2	.	.
3301	HELICHRYSUM BRACTEATUM	S	.	0	21.9	27.8	.	.
3302	HELICHRYSUM FULGIDUM	S	.	0	22.2	25.6	.	.
3303	HELICHRYSUM MONSTROSUM	S	.	0	21.9	29.0	.	.
3304	HELICONIA BREVISPATHA	L	321	0	4.9	1.2	82.7	19.8
3305	HELICTERES ISORA	S	.	0	19.5	9.4	.	.
3306	HELIOPSIS HELIANTHOIDES VAR. PITCHERIANA	S	.	0	18.9	22.6	.	.
3307	HELIOPSIS HELIANTHOIDES VAR. PITCHERIANA	S	.	0	22.5	23.0	.	.
3308	HELIOPSIS LAEVIS	S	.	0	18.1	11.3	.	.
3309	HELIOPSIS SCABRA VAR. ZINNIAEFLORA	S	.	0	25.0	21.3	.	.

OBS	ASH	CALCIUM	PHOSPHO	IRON	SODIUM	POTASS	CAROT	THIAMINE	RIBOFLAV	NIACIN	ASCORBIC	REFCO
3286	.	.	.	.	.	.	.	.	.	.	.	21
3287	3.8	.	.	.	.	.	.	.	.	.	.	19
3288	3.6	.	.	.	.	.	.	.	.	.	.	19
3289	4.3	.	.	.	.	.	.	.	.	.	.	19
3290	.	.	.	.	.	.	.	.	.	.	.	20
3291	16.1	2040.0	360.0	.	.	.	.	.	.	.	.	18
3292	15.5	.	.	.	.	.	.	.	.	.	.	18
3293	6.0	210.0	350.0	.	.	.	.	.	.	.	.	18
3294	4.8	.	.	.	.	.	.	.	.	.	.	19
3295	5.0	158.4	435.6	2.0	.	.	.	.	.	.	.	1
3296	6.1	148.0	336.7	8.2	.	1949.0	0.0	1.02	0.26	1.02	30.6	2
3297	6.0	160.0	.	3.0	.	.	50.0	1.00	0.25	0.50	30.0	3
3298	5.4	69.3	386.1	16.8	.	.	59.4	0.99	0.30	6.44	19.8	4
3299	7.7	1617.6	110.3	.	.	1360.3	.	.	.	.	.	5
3300	5.7	.	.	.	.	.	.	.	.	.	.	19
3301	.	.	.	.	.	.	.	.	.	.	.	20
3302	.	.	.	.	.	.	.	.	.	.	.	21
3303	.	.	.	.	.	.	.	.	.	.	.	20
3304	11.1	555.6	271.6	12.3	1049.4	1963.0	.	0.12	0.37	1.23	.	2
3305	.	.	.	.	.	.	.	.	.	.	.	21
3306	3.5	.	.	.	.	.	.	.	.	.	.	19
3307	.	.	.	.	.	.	.	.	.	.	.	20
3308	.	.	.	.	.	.	.	.	.	.	.	20
3309	.	.	.	.	.	.	.	.	.	.	.	20

TABLE 3 - DATA CONVERTED TO A ZERO-MOISTURE BASIS.

OBS	PLNA	PART	CALORIES	WATER	PROTEIN	FAT	TOTLCARB	FIBER
3310	HELIOTROPIUM EUROPAEUM	S	.	0	20.9	23.8	.	.
3311	HELIOTROPIUM INDICUM	S	.	0	15.0	11.6	.	.
3312	HELIPTERUM MANGLESII	S	.	0	35.6	25.5	.	.
3313	HELIPTERUM MANGLESII VAR. MACULATUM	S	.	0	33.8	25.9	.	.
3314	HELIPTERUM ROSEUM	S	.	0	37.5	26.5	.	.
3315	HELLEBORUS NIGER	S	.	0	21.2	35.2	.	.
3316	HEMEROCALLIS FULVA	S	.	0	38.8	27.2	.	.
3317	HEMEROCALLIS LILIO-ASPHODELUS	I	328	0	15.6	3.1	75.0	9.4
3318	HEMEROCALLIS LILIO-ASPHODELUS	I	.	0	11.7	2.8	80.7	8.3
3319	HERACLEUM LANATUM	S	.	0	28.2	15.0	.	.
3320	HERACLEUM LANATUM	S	.	0	10.9	1.0	.	.
3321	HERACLEUM LANATUM	S	.	0	21.5	17.8	.	.
3322	HERACLEUM LANATUM	S	.	0	20.0	14.0	.	.
3323	HERACLEUM LANATUM	I	.	0	25.3	5.2	57.9	16.4
3324	HERACLEUM LANATUM	L	.	0	20.2	3.6	61.7	12.4
3325	HERACLEUM LANATUM	H	.	0	24.5	3.1	56.6	12.8
3326	HERACLEUM MAXIMUM	S	.	0	29.3	21.0	.	.
3327	HERACLEUM SPHONDYLIUM	S	.	0	15.8	15.2	.	.
3328	HERNIARIA GLABRA	S	.	0	16.9	8.6	.	.
3329	HERNIARIA INCANA	S	.	0	8.4	2.7	.	.
3330	HESPERIS MATRONALIS	S	.	0	24.9	31.8	.	.
3331	HESPEROCHLOA KINGII	W	.	0	11.5	4.2	76.9	33.9
3332	HESPEROCHLOA KINGII(IMMATURE)	W	.	0	18.8	4.4	69.2	29.2
3333	HESPEROCHLOA KINGII(MATURE)	W	.	0	6.6	3.8	82.1	37.2

OBS	ASH	CALCIUM	PHOSPHO	IRON	SODIUM	POTASS	CAROT	THIAMINE	RIBOFLAV	NIACIN	ASCORBIC	REFCO
3310	.	.	.	.	.	.	.	.	.	.	.	21
3311	.	.	.	.	.	.	.	.	.	.	.	19
3312	.	.	.	.	.	.	.	.	.	.	.	20
3313	.	.	.	.	.	.	.	.	.	.	.	20
3314	.	.	.	.	.	.	.	.	.	.	.	20
3315	.	.	.	.	.	.	.	.	.	.	.	20
3316	3.5	.	.	.	.	.	.	.	.	.	.	19
3317	6.3	679.7	1367.2	9.4	187.5	1328.1	23789.1	1.25	1.64	6.25	687.5	2
3318	5.5	.	.	.	.	.	.	.	.	.	.	5
3319	5.2	.	.	.	.	.	.	.	.	.	.	19
3320	7.6	.	.	.	.	.	.	.	.	.	.	19
3321	8.1	.	.	.	.	.	.	.	.	.	.	19
3322	.	.	.	.	.	.	.	.	.	.	.	20
3323	11.6	.	.	.	.	.	.	.	.	.	.	6
3324	14.5	.	.	.	.	.	.	.	.	.	.	6
3325	15.8	.	.	.	.	.	.	.	.	.	.	6
3326	.	.	.	.	.	.	.	.	.	.	.	20
3327	.	.	.	.	.	.	.	.	.	.	.	21
3328	.	.	.	.	.	.	.	.	.	.	.	21
3329	.	.	.	.	.	.	.	.	.	.	.	21
3330	4.6	.	.	.	.	.	.	.	.	.	.	19
3331	7.4	440.0	300.0	.	.	.	.	.	.	.	.	6
3332	7.6	490.0	380.0	.	.	.	.	.	.	.	.	6
3333	7.5	440.0	200.0	.	.	.	.	.	.	.	.	6

TABLE 3 - DATA CONVERTED TO A ZERO-MOISTURE BASIS.

OBS	PLNA	PART	CALORIES	WATER	PROTEIN	FAT	TOTCARB	FIBER	ASH	CALCIUM	PHOSPHO	IRON	SODIUM	POTASS	CAROT	THIAMINE	RIBOFLAV	NIACIN	ASCORBIC	REFCO
3334	HETERANTHERA RENIFORMIS	L	290	0	14.5	3.2	64.5	12.9	17.7	1016.1	596.8	298.4	.	.	23548.4	0.48	1.29	4.84	258.1	3
3335	HETEROPHRAGMA ADENOPHYLLUM	S	.	0	62.5	20.2	.	.	.	.	.	.	.	.	.	.	.	.	.	20
3336	HETEROPOGON CONTORTUS	W	.	0	9.4	1.0	83.3	28.9	6.3	.	.	.	.	.	.	.	.	.	.	18
3337	HETEROPOGON CONTORTUS	H	.	0	3.5	1.0	87.3	40.1	8.2	.	.	.	.	.	.	.	.	.	.	18
3338	HETEROPOGON CONTORTUS	L	.	0	7.5	3.2	78.4	28.9	.	950.0	400.0	.	.	.	.	.	.	.	.	5
3339	HETEROPOGON CONTORTUS	L	.	0	4.4	3.2	83.7	26.9	.	460.0	320.0	.	.	.	.	.	.	.	.	5
3340	HETEROPOGON CONTORTUS	L	.	0	2.3	2.5	87.5	28.9	.	340.0	370.0	.	.	.	.	.	.	.	.	5
3341	HETEROPOGON CONTORTUS(IMMATURE)	W	.	0	9.6	2.0	78.4	31.0	10.0	470.0	160.0	.	.	.	.	.	.	.	.	6
3342	HETEROPOGON CONTORTUS(MATURE)	W	.	0	3.4	1.9	88.1	34.4	6.6	220.0	50.0	.	.	.	.	.	.	.	.	6
3343	HETEROTHECA LATIFOLIA VAR. LATIFOLIA	S	.	0	17.2	11.8	.	.	.	.	.	.	.	.	.	.	.	.	.	21
3344	HETEROTHECA SUBAXILLARIS	S	.	0	13.9	13.4	.	.	2.8	.	.	.	.	.	.	.	.	.	.	19
3345	HEVEA BRASILIENSIS	F	.	0	18.0	52.3	27.1	1.9	2.6	.	.	.	.	.	.	.	.	.	.	18
3346	HEVEA BRASILIENSIS	S	.	0	19.2	53.0	25.0	.	2.7	131.1	469.9	.	.	.	.	.	.	.	.	5
3347	HIBISCUS ABELMOSCHUS	S	.	0	27.8	21.5	.	.	.	.	.	.	.	.	.	.	.	.	.	20
3348	HIBISCUS ASPER	L	.	0	31.1	3.5	55.6	13.5	9.8	1040.0	90.0	.	.	.	.	.	.	.	.	18
3349	HIBISCUS CANNABINUS	C	.	0	15.0	5.8	71.0	19.9	8.2	2080.0	500.0	.	.	.	.	.	.	.	.	18
3350	HIBISCUS CANNABINUS	L	.	0	13.1	2.1	73.0	11.6	11.8	3310.0	350.0	.	.	.	.	.	.	.	.	18
3351	HIBISCUS CANNABINUS	S	.	0	27.2	15.4	51.3	25.1	6.1	600.0	630.0	.	.	.	.	.	.	.	.	18
3352	HIBISCUS CANNABINUS	S	.	0	27.9	18.5	.	.	5.6	.	.	.	.	.	.	.	.	.	.	19
3353	HIBISCUS CANNABINUS	S	.	0	28.8	22.9	.	.	.	.	.	.	.	.	.	.	.	.	.	19
3354	HIBISCUS CANNABINUS	S	.	0	32.5	17.5	.	.	.	.	.	.	.	.	.	.	.	.	.	19
3355	HIBISCUS CANNABINUS	S	.	0	29.8	19.8	.	.	5.5	.	.	.	.	.	.	.	.	.	.	19
3356	HIBISCUS CANNABINUS	S	.	0	31.4	19.1	.	.	6.0	.	.	.	.	.	.	.	.	.	.	19
3357	HIBISCUS CANNABINUS	S	.	0	27.4	20.7	.	.	4.9	.	.	.	.	.	.	.	.	.	.	19
3358	HIBISCUS CANNABINUS	S	.	0	37.0	21.7	.	.	4.3	.	.	.	.	.	.	.	.	.	.	19
3359	HIBISCUS CANNABINUS	S	.	0	37.0	22.9	.	.	4.6	.	.	.	.	.	.	.	.	.	.	19
3360	HIBISCUS CANNABINUS	S	.	0	28.1	18.8	.	.	6.1	.	.	.	.	.	.	.	.	.	.	19
3361	HIBISCUS CANNABINUS	S	.	0	30.1	17.9	.	.	6.0	.	.	.	.	.	.	.	.	.	.	19
3362	HIBISCUS CANNABINUS	S	.	0	32.7	18.4	.	.	6.0	.	.	.	.	.	.	.	.	.	.	19
3363	HIBISCUS CANNABINUS	S	.	0	30.6	18.3	.	.	5.8	.	.	.	.	.	.	.	.	.	.	19
3364	HIBISCUS CANNABINUS	S	.	0	28.8	18.7	.	.	5.8	.	.	.	.	.	.	.	.	.	.	19
3365	HIBISCUS CANNABINUS	S	.	0	28.9	19.0	.	.	5.6	.	.	.	.	.	.	.	.	.	.	19
3366	HIBISCUS CANNABINUS	S	.	0	26.6	14.2	.	.	5.5	.	.	.	.	.	.	.	.	.	.	19
3367	HIBISCUS CANNABINUS	S	.	0	28.9	19.0	.	.	6.3	.	.	.	.	.	.	.	.	.	.	19
3368	HIBISCUS CANNABINUS	S	.	0	32.1	20.7	.	.	5.6	.	.	.	.	.	.	.	.	.	.	19
3369	HIBISCUS CANNABINUS	S	.	0	27.5	20.9	.	.	4.6	.	.	.	.	.	.	.	.	.	.	19
3370	HIBISCUS CANNABINUS	S	.	0	26.2	20.7	.	.	4.6	.	.	.	.	.	.	.	.	.	.	19
3371	HIBISCUS CANNABINUS	S	.	0	25.6	20.4	.	.	4.5	.	.	.	.	.	.	.	.	.	.	19
3372	HIBISCUS CANNABINUS	S	.	0	27.5	21.5	.	.	4.3	.	.	.	.	.	.	.	.	.	.	19
3373	HIBISCUS CANNABINUS	S	.	0	26.2	20.9	.	.	4.6	.	.	.	.	.	.	.	.	.	.	19
3374	HIBISCUS CANNABINUS	S	.	0	26.2	20.6	.	.	4.8	.	.	.	.	.	.	.	.	.	.	19
3375	HIBISCUS CANNABINUS	S	.	0	28.8	20.7	.	.	4.8	.	.	.	.	.	.	.	.	.	.	19
3376	HIBISCUS CANNABINUS	S	.	0	25.6	20.5	.	.	4.3	.	.	.	.	.	.	.	.	.	.	19
3377	HIBISCUS CANNABINUS	S	.	0	26.9	21.3	.	.	4.4	.	.	.	.	.	.	.	.	.	.	19
3378	HIBISCUS CANNABINUS	S	.	0	28.1	21.5	.	.	4.4	.	.	.	.	.	.	.	.	.	.	19

TABLE 3 - DATA CONVERTED TO A ZERO-MOISTURE BASIS.

OBS	PLNA	PART	CALORIES	WATER	PROTEIN	FAT	TOTCARB	FIBER	ASH	CALCIUM	PHOSPHO	IRON	SODIUM	POTASS	CAROT	THIAMINE	RIBOFLAV	NIACIN	ASCORBIC	REFCO
3379	HIBISCUS CANNABINUS	S	.	0	26.2	21.2	.	.	4.2	.	.	.	.	.	.	.	.	.	.	19
3380	HIBISCUS CANNABINUS	S	.	0	25.6	21.5	.	.	4.6	.	.	.	.	.	.	.	.	.	.	19
3381	HIBISCUS CANNABINUS	S	.	0	28.8	20.2	.	.	4.8	.	.	.	.	.	.	.	.	.	.	19
3382	HIBISCUS CANNABINUS	S	.	0	28.1	21.1	.	.	4.9	.	.	.	.	.	.	.	.	.	.	19
3383	HIBISCUS CANNABINUS	S	.	0	27.5	19.6	.	.	5.4	.	.	.	.	.	.	.	.	.	.	19
3384	HIBISCUS CANNABINUS	S	.	0	28.1	20.2	.	.	4.8	.	.	.	.	.	.	.	.	.	.	19
3385	HIBISCUS CANNABINUS	S	.	0	26.2	21.1	.	.	4.3	.	.	.	.	.	.	.	.	.	.	19
3386	HIBISCUS CANNABINUS	S	.	0	24.2	21.7	.	.	.	.	.	.	.	.	.	.	.	.	.	20
3387	HIBISCUS CANNABINUS	S	.	0	22.5	15.1	.	.	.	.	.	.	.	.	.	.	.	.	.	20
3388	HIBISCUS CANNABINUS	L	319	0	26.2	5.7	58.1	11.0	10.0	2304.8	85.7	57.6	.	.	.	.	.	.	357.1	1
3389	HIBISCUS CANNABINUS	L	311	0	23.0	0.0	71.1	.	5.9	459.3	281.5	9.6	.	.	.	0.52	0.44	.	.	2
3390	HIBISCUS ESCULENTUS	S	.	0	23.8	14.7	.	.	5.0	.	.	.	.	.	.	.	.	.	.	19
3391	HIBISCUS ESCULENTUS	S	.	0	25.6	15.4	.	.	.	.	.	.	.	.	.	.	.	.	.	20
3392	HIBISCUS ESCULENTUS	S	.	0	22.5	15.1	.	.	.	.	.	.	.	.	.	.	.	.	.	20
3393	HIBISCUS GIBSONII	S	.	0	23.2	15.1	.	.	.	.	.	.	.	.	.	.	.	.	.	21
3394	HIBISCUS GRANDIFLORUS	S	.	0	24.4	11.7	.	.	.	.	.	.	.	.	.	.	.	.	.	20
3395	HIBISCUS LASIOCARPUS	S	.	0	21.2	12.1	.	.	.	.	.	.	.	.	.	.	.	.	.	20
3396	HIBISCUS LUNARIFOLIUS	W	.	0	17.8	1.6	71.4	37.4	9.2	1930.0	330.0	.	.	.	.	.	.	.	.	18
3397	HIBISCUS MOSCHEUTOS	S	.	0	22.5	11.8	.	.	3.4	.	.	.	.	.	.	.	.	.	.	19
3398	HIBISCUS MOSCHEUTOS	S	.	0	21.9	12.7	.	.	.	.	.	.	.	.	.	.	.	.	.	19
3399	HIBISCUS MOSCHEUTOS	S	.	0	23.8	11.6	.	.	.	.	.	.	.	.	.	.	.	.	.	20
3400	HIBISCUS PALUSTRIS	S	.	0	21.9	12.4	.	.	.	.	.	.	.	.	.	.	.	.	.	19
3401	HIBISCUS PLATANIFOLIUS	S	.	0	13.2	8.9	.	.	2.1	.	.	.	.	.	.	.	.	.	.	19
3402	HIBISCUS ROSA-SINENSIS	W	.	0	15.4	3.5	69.7	15.5	11.4	1670.0	520.0	.	.	.	.	.	.	.	.	18
3403	HIBISCUS ROSA-SINENSIS	F	353	0	3.9	3.9	86.3	15.7	5.9	39.2	264.7	16.7	.	.	.	0.29	0.49	5.88	39.2	2
3404	HIBISCUS ROSA-SINENSIS	I	.	0	3.9	3.9	.	15.7	.	39.2	264.7	16.7	.	.	.	.	.	.	.	5
3405	HIBISCUS SABDARIFFA	S	.	0	21.4	17.5	55.7	12.0	5.4	610.0	320.0	.	.	.	.	.	.	.	.	18
3406	HIBISCUS SABDARIFFA	L	.	0	12.3	7.8	72.5	7.8	7.4	1310.0	280.0	.	.	.	.	.	.	.	.	18
3407	HIBISCUS SABDARIFFA	SH	.	0	30.1	1.9	58.9	10.9	9.1	.	.	.	.	.	.	.	.	.	.	18
3408	HIBISCUS SABDARIFFA	L	.	0	25.7	3.3	60.4	11.4	10.6	.	.	.	.	.	.	.	.	.	.	18
3409	HIBISCUS SABDARIFFA	F	316	0	12.3	0.6	79.4	14.8	7.7	1109.7	367.7	18.7	.	.	1935.5	.	.	.	90.3	1
3410	HIBISCUS SABDARIFFA	L	299	0	22.9	2.1	63.9	11.1	11.1	1479.2	645.8	33.3	.	.	28715.3	1.18	3.13	8.33	375.0	1
3411	HIBISCUS SABDARIFFA	I	319	0	11.6	0.7	80.4	18.1	7.2	1159.4	434.8	27.5	.	.	2065.2	0.29	0.43	3.62	101.4	1
3412	HIBISCUS SABDARIFFA	F	390	0	7.0	11.0	76.0	14.0	6.0	1740.0	180.0	1.0	.	.	1100.0	0.10	0.20	4.00	100.0	2
3413	HIBISCUS SABDARIFFA	L	324	0	14.0	2.2	76.5	9.6	7.4	852.9	352.9	11.0	.	.	55514.7	0.15	1.10	13.24	250.0	2
3414	HIBISCUS SABDARIFFA	I	355	0	11.0	6.5	77.4	6.5	5.2	709.7	251.6	14.2	.	.	129.0	0.26	0.39	2.58	116.1	3
3415	HIBISCUS SCHIZOPETALUS	L	.	0	17.4	8.5	62.8	14.9	11.3	.	.	.	.	.	.	.	.	.	.	18
3416	HIBISCUS SYRIACUS	S	.	0	28.8	29.4	.	.	.	.	.	.	.	.	.	.	.	.	.	19
3417	HIBISCUS SYRIACUS	S	.	0	30.6	28.0	.	.	.	.	.	.	.	.	.	.	.	.	.	19
3418	HIBISCUS SYRIACUS	S	.	0	28.8	15.2	.	.	.	.	.	.	.	.	.	.	.	.	.	19
3419	HIBISCUS SYRIACUS	S	.	0	28.8	27.7	.	.	.	.	.	.	.	.	.	.	.	.	.	20
3420	HIBISCUS SYRIACUS	S	.	0	27.5	24.4	.	.	.	.	.	.	.	.	.	.	.	.	.	20
3421	HIBISCUS TILIACEUS	S	.	0	18.2	6.3	.	.	.	.	.	.	.	.	.	.	.	.	.	21
3422	HIBISCUS TRIONUM	S	.	0	18.6	14.0	.	.	5.7	.	.	.	.	.	.	.	.	.	.	19
3423	HILARIA BELANGERI(ALL)	W	.	0	7.5	2.2	74.6	28.4	15.7	560.0	120.0	.	.	670	.	.	.	.	.	6

TABLE 3 - DATA CONVERTED TO A ZERO-MOISTURE BASIS.

OBS	PLNA	PART	CALORIES	WATER	PROTEIN	FAT	TOTCARB	FIBER	ASH	CALCIUM	PHOSPHO	IRON	SODIUM	POTASS	CAROT	THIAMINE	RIBOFLAV	NIACIN	ASCORBIC	REFCO
3424	HILARIA BELANGERI(IMMATURE)	W	.	0	17.2	2.5	64.6	25.6	15.7	1040.0	260.0	.	.	790.0	.	.	.	.	.	6
3425	HILARIA BELANGERI(MATURE)	W	.	0	6.0	2.2	74.7	28.8	17.1	550.0	90.0	.	.	390.0	.	.	.	.	.	6
3426	HILARIA BELANGERI(VERY MATURE)	W	.	0	4.9	2.0	79.6	29.3	13.5	330.0	60.0	.	.	420.0	.	.	.	.	.	6
3427	HILARIA JAMESII	W	.	0	4.9	1.5	79.3	32.9	14.3	670.0	110.0	.	.	.	.	.	.	.	.	6
3428	HILARIA JAMESII(MATURE)	W	.	0	6.2	2.6	.	.	19.1	1160.0	80.0	.	.	.	.	.	.	.	.	6
3429	HILARIA JAMESII(VERY MATURE)	W	.	0	3.8	1.2	81.0	33.0	14.0	590.0	80.0	.	.	.	.	.	.	.	.	6
3430	HILARIA MUTICA(ALL)	W	.	0	7.0	1.5	80.6	33.4	10.9	250.0	130.0	.	.	.	.	.	.	.	.	6
3431	HILLERIA LATIFOLIA	L	280	0	26.1	5.1	49.7	.	19.1	2222.9	.	26.1	.	.	.	.	.	.	140.1	1
3432	HIPPOCREPIS UNISILIQUOSA	S	.	0	22.1	1.8	.	.	.	.	.	.	.	.	.	.	.	.	.	21
3433	HIPPOMARATHRUM BOISSIERI	S	.	0	15.4	20.2	.	.	.	.	.	.	.	.	.	.	.	.	.	21
3434	HIPPOMARATHRUM PAUCIRADIATUM	S	.	0	36.9	31.3	.	.	.	.	.	.	.	.	.	.	.	.	.	21
3435	HIPPOPHAE RHAMNOIDES	S	.	0	28.1	9.5	.	.	.	.	.	.	.	.	.	.	.	.	.	20
3436	HIPPOPHAE RHAMNOIDES	S	.	0	22.6	9.2	.	.	.	.	.	.	.	.	.	.	.	.	.	21
3437	HIRPICIUM INTEGRIFOLIUM	S	.	0	46.3	22.6	.	.	.	.	.	.	.	.	.	.	.	.	.	21
3438	HIRSCHFELDIA INCANA	S	.	0	28.1	30.6	.	.	.	.	.	.	.	.	.	.	.	.	.	21
3439	HIRSCHFELDIA INCANA	S	.	0	32.8	32.7	.	.	.	.	.	.	.	.	.	.	.	.	.	21
3440	HOFFMANNSEGGIA SP.	W	.	0	14.7	2.5	74.9	17.1	7.9	3180.0	180.0	.	.	1210.0	.	.	.	.	.	6
3441	HOLCUS LANATUS	S	.	0	20.6	17.9	.	.	5.4	.	.	.	.	.	.	.	.	.	.	19
3442	HOLCUS LANATUS	S	.	0	23.4	20.5	.	.	.	.	.	.	.	.	.	.	.	.	.	21
3443	HORDEUM BULBOSUM	S	.	0	27.5	2.6	.	.	4.3	.	.	.	.	.	.	.	.	.	.	19
3444	HORDEUM BULBOSUM	S	.	0	23.2	2.0	.	.	.	.	.	.	.	.	.	.	.	.	.	21
3445	HORDEUM BULBOSUM	S	.	0	21.6	3.1	.	.	.	.	.	.	.	.	.	.	.	.	.	21
3446	HORDEUM MURINUM	S	.	0	23.5	1.4	.	.	.	.	.	.	.	.	.	.	.	.	.	21
3447	HORDEUM VULGARE	F	.	0	10.3	2.4	83.8	6.2	3.5	80.0	290.0	.	.	.	.	.	.	.	.	18
3448	HORDEUM VULGARE	W	.	0	22.7	5.1	59.8	20.2	12.4	.	.	.	.	.	.	.	.	.	.	18
3449	HORDEUM VULGARE	H	.	0	2.5	1.4	78.5	44.6	17.6	.	.	.	.	.	.	.	.	.	.	18
3450	HORDEUM VULGARE	S	374	0	11.1	1.8	84.6	6.7	2.6	67.7	278.6	19.9	.	.	.	0.40	.	.	.	1
3451	HORDEUM VULGARE	S	379	0	12.2	2.4	83.2	2.9	2.2	57.9	438.0	7.0	4.6	651.2	0	0.36	0.12	6.03	0.0	2
3452	HORDEUM VULGARE	S	389	0	10.8	2.1	84.2	7.3	2.8	61.5	381.0	5.0	.	.	11	0.42	0.22	8.04	0.0	3
3453	HORDEUM VULGARE	S	393	0	9.2	1.1	88.6	0.6	1.0	18.0	212.6	2.2	3.4	180.0	0	0.13	0.06	3.49	0.0	4
3454	HORDEUM VULGARE(ALL)	W	.	0	21.2	3.9	62.0	20.8	12.9	480.0	370.0	.	.	3060.0	.	.	.	.	.	6
3455	HOSTA VENTRICOSA	S	.	0	25.6	36.4	.	.	.	.	.	.	.	.	.	.	.	.	.	19
3456	HOVENIA DULCIS	S	.	0	15.0	7.8	.	.	.	.	.	.	.	.	.	.	.	.	.	20
3457	HULSEA HETEROCHROMA	S	.	0	20.0	21.9	.	.	4.6	.	.	.	.	.	.	.	.	.	.	19
3458	HUMULUS SCANDENS	S	.	0	23.9	31.0	.	.	10.3	.	.	.	.	.	.	.	.	.	.	19
3459	HYACINTHUS SP.	W	.	0	8.0	3.3	76.2	18.0	12.5	.	.	.	.	.	.	.	.	.	.	6
3460	HYDRANGEA PANICULATA	S	.	0	25.0	31.3	.	.	.	.	.	.	.	.	.	.	.	.	.	20
3461	HYDRANGEA PETIOLARIS	S	.	0	29.1	38.0	.	.	.	.	.	.	.	.	.	.	.	.	.	21
3462	HYDROCOTYLE BONARIENSIS	S	.	0	10.0	13.4	.	.	.	.	.	.	.	.	.	.	.	.	.	20
3463	HYGROPHILA SPINOSA	L	315	0	23.0	.	71.9	.	5.1	309.0	410.1	.	.	.	144213	.	.	.	561.8	2
3464	HYLOCEREUS UNDATUS	F	346	0	9.0	2.6	84.6	9.0	3.8	64.1	166.7	8.3	.	.	0	0.26	0.26	1.92	51.3	3
3465	HYMENACHNE AMPLEXICAULIS	S	.	0	8.9	1.0	78.6	36.7	11.5	.	.	.	.	.	.	.	.	.	.	18
3466	HYMENACHNE AMPLEXICAULIS	L	.	0	22.6	2.8	67.4	32.4	7.2	.	.	.	.	.	.	.	.	.	.	18
3467	HYMENACHNE AMPLEXICAULIS	W	.	0	9.4	2.3	76.1	22.1	12.2	.	.	.	.	.	.	.	.	.	.	18
3468	HYMENACHNE AMPLEXICAULIS	H	.	0	7.5	1.4	78.2	29.2	12.9	.	.	.	.	.	.	.	.	.	.	18

TABLE 3 - DATA CONVERTED TO A ZERO-MOISTURE BASIS.

OBS	PLANA	PART	CALORIES	WATER	PROTEIN	FAT	TOTCARB	FIBER	ASH	CALCIUM	PHOSPHO	IRON	SODIUM	POTASS	CAROT	THIAMINE	RIBOFLAV	NIACIN	ASCORBIC	REFCO
3469	HYMENACHNE AMPLEXICAULIS	L	.	0	9.4	2.3	76.1	22.1	12.2	130.0	206.0	.	.	.	.	.	.	.	.	5
3470	HYMENACHNE AMPLEXICAULIS	L	.	0	7.5	1.4	79.1	29.2	12.9	141.0	186.0	.	.	.	.	.	.	.	.	5
3471	HYMENAEA COURBARIL	F	361	0	6.9	2.5	88.1	15.7	2.3	32.0	167.0	3.7	.	.	1	0.27	0.2	4.80	13	10
3472	HYMENOCARPUS CIRCINNATA	S	.	0	46.0	12.8	.	.	.	.	.	.	.	.	.	.	.	.	.	21
3473	HYMENOCARPUS CIRCINNATA	S	.	0	39.4	17.4	.	.	.	.	.	.	.	.	.	.	.	.	.	21
3474	HYMENOCLEA MONOGYRA	S	.	0	38.8	37.1	.	.	.	.	.	.	.	.	.	.	.	.	.	19
3475	HYMENOPAPPUS ARTEMISIAEFOLIUS	S	.	0	16.9	19.6	.	.	.	.	.	.	.	.	.	.	.	.	.	19
3476	HYMENOPAPPUS ARTEMISIAEFOLIUS	S	.	0	15.6	22.5	.	.	.	.	.	.	.	.	.	.	.	.	.	20
3477	HYMENOXYS ODORATA(IMMATURE)	W	.	0	33.3	3.3	51.0	11.0	12.4	1840.0	460.0	.	.	4590	.	.	.	.	.	6
3478	HYMENOXYS SP.(PAST FLOWERING)	W	.	0	12.3	6.1	73.4	23.9	8.2	1040.0	180.0	.	.	2540	.	.	.	.	.	6
3479	HYOSCYAMUS AUREUS	S	.	0	19.4	31.7	.	.	.	.	.	.	.	.	.	.	.	.	.	21
3480	HYOSCYAMUS NIGER	S	.	0	21.2	38.9	.	.	.	.	.	.	.	.	.	.	.	.	.	21
3481	HYOSCYAMUS RETICULATUS	S	.	0	20.2	36.1	.	.	.	.	.	.	.	.	.	.	.	.	.	20
3482	HYOSCYAMUS RETICULATUS	S	.	0	22.0	34.8	.	.	.	.	.	.	.	.	.	.	.	.	.	21
3483	HYPARRHENIA CYMBARIA	W	.	0	7.3	2.6	75.0	30.3	15.1	.	.	.	.	.	.	.	.	.	.	18
3484	HYPARRHENIA DISSOLUTA	W	.	0	12.9	3.0	75.3	33.7	8.8	.	.	.	.	.	.	.	.	.	.	18
3485	HYPARRHENIA DISSOLUTA	H	.	0	6.8	1.5	85.9	36.9	5.8	.	.	.	.	.	.	.	.	.	.	18
3486	HYPARRHENIA FILIPENDULA	W	.	0	6.6	1.8	85.8	36.3	5.8	.	.	.	.	.	.	.	.	.	.	18
3487	HYPARRHENIA LINTONII	W	.	0	11.4	5.0	75.0	31.8	8.6	.	.	.	.	.	.	.	.	.	.	18
3488	HYPARRHENIA RUFA	W	.	0	9.2	2.6	73.3	28.9	14.9	.	.	.	.	.	.	.	.	.	.	18
3489	HYPARRHENIA RUFA	H	.	0	6.5	2.3	73.3	35.0	17.9	.	.	.	.	.	.	.	.	.	.	18
3490	HYPARRHENIA RUFA	S	.	0	11.2	2.0	.	.	9.4	.	.	.	.	.	.	.	.	.	.	19
3491	HYPERICUM ANDROSAEMUM	S	.	0	14.7	28.9	.	.	.	.	.	.	.	.	.	.	.	.	.	21
3492	HYPERICUM CALYCINUM	S	.	0	.	31.6	.	.	.	.	.	.	.	.	.	.	.	.	.	21
3493	HYPERICUM CALYCINUM	S	.	0	14.9	44.9	.	.	.	.	.	.	.	.	.	.	.	.	.	21
3494	HYPERICUM CISTIFOLIUM	S	.	0	23.1	38.6	.	.	.	.	.	.	.	.	.	.	.	.	.	20
3495	HYPERICUM PERFORATUM	S	.	0	20.7	33.2	.	.	.	.	.	.	.	.	.	.	.	.	.	21
3496	HYPERICUM PERFORATUM	S	.	0	18.1	32.8	.	.	.	.	.	.	.	.	.	.	.	.	.	21
3497	HYPERICUM TETRAPTERUM	S	.	0	24.2	34.9	.	.	.	.	.	.	.	.	.	.	.	.	.	20
3498	HYPERICUM TRIQUETRIFOLIUM	S	.	0	16.2	32.6	.	.	.	.	.	.	.	.	.	.	.	.	.	21
3499	HYPHAENE SP.	S	.	0	9.0	7.0	81.2	7.3	2.8	.	.	.	.	.	.	.	.	.	.	18
3500	HYPHAENE SP.	F	.	0	4.5	2.6	89.3	24.7	3.6	.	.	.	.	.	.	.	.	.	.	18
3501	HYPHAENE THEBAICA	S	420	0	4.1	6.8	85.7	10.0	3.3	153.2	240.4	.	.	.	.	.	.	.	.	1
3502	HYPHAENE TURBINATA	S	.	0	8.1	13.4	.	.	2.3	.	.	.	.	.	.	.	.	.	.	19
3503	HYPOCALYPTUS OBCORDATUS	S	.	0	24.1	4.1	.	.	.	.	.	.	.	.	.	.	.	.	.	21
3504	HYPTIS CF. AMERICANA	S	.	0	13.1	14.4	.	.	.	.	.	.	.	.	.	.	.	.	.	20
3505	HYPTIS DECURRENS	S	.	0	15.0	19.5	.	.	.	.	.	.	.	.	.	.	.	.	.	20
3506	HYPTIS SPICIGERA	S	.	0	14.4	18.8	.	.	.	.	.	.	.	.	.	.	.	.	.	19
3507	HYPTIS SPICIGERA	S	501	0	17.8	30.4	45.4	18.0	6.4	223.8	667.4	.	.	.	.	.	.	.	.	1
3508	HYPTIS STELLULATA	S	.	0	19.7	22.3	.	.	.	.	.	.	.	.	.	.	.	.	.	21
3509	HYPTIS SUAVEOLENS	S	.	0	21.9	24.4	.	.	.	.	.	.	.	.	.	.	.	.	.	19
3510	HYPTIS SUAVEOLENS	S	.	0	18.8	14.3	.	.	.	.	.	.	.	.	.	.	.	.	.	20
3511	IBERIS AMARA	S	.	0	28.1	26.8	.	.	.	.	.	.	.	.	.	.	.	.	.	20
3512	IBERIS UMBELLATA	S	.	0	34.4	25.6	.	.	.	.	.	.	.	.	.	.	.	.	.	20
3513	ICACINA SENEGALENSIS	R	379	0	11.7	0.8	84.4	.	3.2	169.9	.	7.9	.	.	.	0.05	0.2	1.59	.	1

TABLE 3 - DATA CONVERTED TO A ZERO-MOISTURE BASIS.

OBS	PLANA	PART	CALORIES	WATER	PROTEIN	FAT	TOTCARB	FIBER	ASH	CALCIUM	PHOSPHO	IRON	SODIUM	POTASS	CAROT	THIAMINE	RIBOFLAV	NIACIN	ASCORBIC	REFCO
3514	IDESIA POLYCARPA	S	.	0	11.9	29.4	.	.	.	.	.	.	.	.	.	.	.	.	.	20
3515	ILEX AQUIFOLIUM	S	.	0	6.6	1.9	.	.	.	.	.	.	.	.	.	.	.	.	.	21
3516	ILEX CRENATA	S	.	0	24.2	48.0	.	.	.	.	.	.	.	.	.	.	.	.	.	21
3517	ILEX LATIFOLIA	S	.	0	7.0	9.9	.	.	.	.	.	.	.	.	.	.	.	.	.	21
3518	ILEX PARAGUARIENSIS	S	.	0	6.9	17.3	.	56.0	1.4	.	.	.	.	.	.	.	.	.	.	5
3519	ILEX PURPUREA	S	.	0	13.0	8.5	.	.	2.2	.	.	.	.	.	.	.	.	.	.	19
3520	ILEX SERRATA	S	.	0	21.0	30.4	.	.	.	.	.	.	.	.	.	.	.	.	.	20
3521	ILEX VERTICILLATA	S	.	0	12.5	15.9	.	.	.	.	.	.	.	.	.	.	.	.	.	19
3522	IMPATIENS BALSAMINA	S	.	0	16.2	21.8	.	.	3.3	.	.	.	.	.	.	.	.	.	.	19
3523	IMPATIENS GLANDULIFERA	S	.	0	22.5	53.5	.	.	.	.	.	.	.	.	.	.	.	.	.	5
3524	IMPERATA CYLINDRICA	W	.	0	11.8	1.9	79.2	32.1	7.1	.	.	.	.	.	.	.	.	.	.	18
3525	IMPERATA CYLINDRICA	H	.	0	3.8	0.7	87.7	39.7	7.8	.	.	.	.	.	.	.	.	.	.	18
3526	IMPERATA CYLINDRICA	SH	.	0	6.6	3.3	82.2	34.6	7.9	390.0	220.0	.	.	.	.	.	.	.	.	5
3527	INDIGOFERA ARRECTA	W	.	0	26.2	2.2	62.7	25.5	8.9	1360.0	350.0	.	.	.	.	.	.	.	.	18
3528	INDIGOFERA BRACHYSTACHYA	S	.	0	34.1	3.1	.	.	.	.	.	.	.	.	.	.	.	.	.	21
3529	INDIGOFERA DENSIFLORA	S	.	0	35.0	3.0	.	.	.	.	.	.	.	.	.	.	.	.	.	20
3530	INDIGOFERA DIPHYLLA	W	.	0	15.8	5.4	72.9	38.8	5.9	570.0	150.0	.	.	.	.	.	.	.	.	18
3531	INDIGOFERA ECHINATA	S	.	0	36.2	4.0	.	.	2.9	.	.	.	.	.	.	.	.	.	.	19
3532	INDIGOFERA ENDECAPHYLLA	S	.	0	28.9	3.5	.	.	4.0	.	.	.	.	.	.	.	.	.	.	19
3533	INDIGOFERA ENDECAPHYLLA	W	.	0	13.0	2.2	79.4	33.5	5.4	.	.	.	.	.	.	.	.	.	.	6
3534	INDIGOFERA FRUTESCENS	S	.	0	32.1	4.8	.	.	.	.	.	.	.	.	.	.	.	.	.	21
3535	INDIGOFERA GLANDULOSA	S	.	0	34.7	2.4	59.4	8.5	3.5	.	.	.	.	.	.	.	.	.	.	5
3536	INDIGOFERA GLANDULOSA	S	.	0	40.4	5.3	49.3	10.1	4.5	.	.	.	.	.	.	.	.	.	.	5
3537	INDIGOFERA HIRSUTA	W	.	0	19.5	2.9	67.3	22.7	10.3	1590.0	200.0	.	.	.	.	.	.	.	.	18
3538	INDIGOFERA HIRSUTA	S	.	0	27.1	1.9	.	.	4.0	.	.	.	.	.	.	.	.	.	.	19
3539	INDIGOFERA HIRSUTA	S	.	0	30.1	1.5	.	.	3.8	.	.	.	.	.	.	.	.	.	.	19
3540	INDIGOFERA HIRSUTA	S	.	0	30.2	1.6	.	.	.	.	.	.	.	.	.	.	.	.	.	21
3541	INDIGOFERA JALISCENSIS	S	.	0	15.6	1.4	.	.	.	.	.	.	.	.	.	.	.	.	.	20
3542	INDIGOFERA LEPTOSEPALA	S	.	0	39.9	5.0	.	.	3.7	.	.	.	.	.	.	.	.	.	.	19
3543	INDIGOFERA LINIFOLIA	S	.	0	37.9	3.1	.	.	.	.	.	.	.	.	.	.	.	.	.	21
3544	INDIGOFERA PILOSA	S	.	0	28.9	3.3	.	.	3.8	.	.	.	.	.	.	.	.	.	.	19
3545	INDIGOFERA RETROFLEXA	W	.	0	16.8	1.1	72.6	31.8	9.5	.	.	.	.	.	.	.	.	.	.	18
3546	INDIGOFERA SECUNDIFLORA	W	.	0	16.9	1.9	64.0	19.1	17.2	4260.0	230.0	.	.	.	.	.	.	.	.	18
3547	INDIGOFERA SP.	H	.	0	12.0	2.5	79.8	24.9	5.7	1500.0	210.0	.	.	.	.	.	.	.	.	6
3548	INDIGOFERA SPHAEROCARPA	S	.	0	31.2	3.0	.	.	.	.	.	.	.	.	.	.	.	.	.	19
3549	INDIGOFERA SPICATA	W	.	0	12.3	0.8	75.2	38.4	11.7	1230.0	400.0	.	.	.	.	.	.	.	.	18
3550	INDIGOFERA SPICATA	L	.	0	21.0	3.1	64.6	21.5	11.3	.	.	.	.	.	.	.	.	.	.	5
3551	INDIGOFERA SUFFRUTICOSA	S	.	0	27.8	1.6	.	.	3.2	.	.	.	.	.	.	.	.	.	.	19
3552	INDIGOFERA SULCATA	S	.	0	27.2	4.7	.	.	.	.	.	.	.	.	.	.	.	.	.	21
3553	INDIGOFERA ZOLLINGERIANA	S	.	0	28.8	1.3	.	.	.	.	.	.	.	.	.	.	.	.	.	20
3554	INGA EDULIS	S	322	0	29.2	1.9	65.4	4.4	3.5	523.2	348.8	8.4	.	.	0	0.52	0.35	3.27	35.4	3
3555	INGA SP.	P	353	0	5.9	0.6	91.2	7.1	2.4	123.5	117.6	5.3	.	.	0	0.24	0.35	2.35	52.9	3
3556	INGA SP.	S	388	0	21.6	2.4	72.0	3.9	4.0	145.3	292.9	5.5	.	.	0	0.37	0.03	1.60	3.4	3
3557	INOCARPUS FAGIFERUS	S	426	0	6.7	7.9	82.8	4.4	2.6	.	.	.	.	.	0	0.46	.	.	3.5	2
3558	INULA GRAVEOLENS	S	.	0	27.8	29.8	.	.	.	.	.	.	.	.	.	.	.	.	.	21

TABLE 3 - DATA CONVERTED TO A ZERO-MOISTURE BASIS.

OBS	PLANA	PART	CALORIES	WATER	PROTEIN	FAT	TOTCARB	FIBER	ASH	CALCIUM	PHOSPHO	IRON	SODIUM	POTASS	CAROT	THIAMINE	RIBOFLAV	NIACIN	ASCORBIC	REFCO
3559	INULA VISCOSA	S	.	0	20.8	23.9	.	.	.	.	.	.	.	.	.	.	.	.	.	21
3560	IOSTEPHANE HETEROPHYLLA	S	.	0	26.2	14.9	.	.	.	.	.	.	.	.	.	.	.	.	.	20
3561	IPOMOEA AQUATICA	W	.	0	18.8	2.1	60.9	20.9	18.2	710.0	320.0	.	.	.	.	.	.	.	.	18
3562	IPOMOEA AQUATICA	L	.	0	24.0	2.7	60.0	12.7	13.3	1200.0	280.0	.	.	.	.	.	.	.	.	18
3563	IPOMOEA AQUATICA	SH	.	0	34.3	3.9	48.9	10.2	12.9	.	.	.	.	.	.	.	.	.	.	18
3564	IPOMOEA AQUATICA	L	293	0	24.0	2.7	60.0	12.7	13.3	1200.0	280.0	36.0	.	.	.	.	.	.	.	1
3565	IPOMOEA AQUATICA	L	300	0	27.0	4.0	56.0	11.0	13.0	600.0	420.0	25.0	440.0	4690.0	28650.0	0.90	1.60	11.00	470.0	2
3566	IPOMOEA AQUATICA	L	282	0	29.1	2.9	52.4	10.7	15.5	708.7	495.1	24.3	.	1456.3	36699.0	0.68	1.17	6.80	310.7	4
3567	IPOMOEA BATATAS	L	.	0	19.4	3.7	51.0	10.2	25.9	.	.	.	.	.	.	.	.	.	.	18
3568	IPOMOEA BATATAS	W	.	0	21.9	3.4	56.7	15.0	18.0	1790.0	240.0	.	.	.	.	.	.	.	.	18
3569	IPOMOEA BATATAS	H	.	0	16.4	5.2	65.8	27.4	12.6	.	.	.	.	.	.	.	.	.	.	18
3570	IPOMOEA BATATAS	R	.	0	8.8	0.6	88.5	2.4	2.1	230.0	200.0	.	.	.	.	.	.	.	.	18
3571	IPOMOEA BATATAS		.	0	.	.	.	.	.	.	.	.	.	.	.	.	.	.	.	6
3572	IPOMOEA BATATAS	R	388	0	5.1	0.6	91.3	3.2	2.9	105.8	121.8	6.4	.	.	240.4	0.29	0.13	2.24	118.6	1
3573	IPOMOEA BATATAS	R	388	0	5.1	0.6	91.3	3.2	2.9	105.8	121.8	6.4	.	.	4022.4	0.29	0.13	2.24	118.6	1
3574	IPOMOEA BATATAS	L	288	0	27.1	1.2	60.0	14.1	11.8	929.4	494.1	36.5	.	.	34529.4	0.59	1.65	5.29	411.8	1
3575	IPOMOÊA BATATAS	R	390	0	3.6	1.1	92.4	2.9	2.5	75.8	180.5	3.2	111.9	758.1	126.3	0.51	0.18	2.53	75.8	2
3576	IPOMOEA BATATAS	R	392	0	4.1	1.0	92.5	2.7	2.4	122.9	191.1	3.1	122.9	1037.5	5733.8	0.41	0.17	2.05	102.4	2
3577	IPOMOEA BATATAS	L	316	0	24.1	5.3	60.2	12.0	10.5	646.6	609.0	33.8	37.6	4225.6	20300.8	0.98	1.95	6.77	157.9	2
3578	IPOMOEA BATATAS	L	350	0	23.4	5.8	62.0	11.7	8.8	810.2	219.0	21.2	.	.	21313.9	0.58	1.90	5.11	423.4	3
3579	IPOMOEA BATATAS	R	388	0	5.8	1.4	89.5	2.4	3.4	108.8	159.9	2.4	34.0	826.5	17959.2	0.34	0.20	2.04	71.4	4
3580	IPOMOEA CARDIOPHYLLA	S	.	0	31.4	10.1	.	.	.	.	.	.	.	.	.	.	.	.	.	21
3581	IPOMOEA CARDIOPHYLLA	S	.	0	23.5	10.0	.	.	.	.	.	.	.	.	.	.	.	.	.	21
3582	IPOMOEA CARDIOPHYLLA	S	.	0	24.4	10.7	.	.	.	.	.	.	.	.	.	.	.	.	.	21
3583	IPOMOEA CARDIOPHYLLA	S	.	0	26.6	12.7	.	.	.	.	.	.	.	.	.	.	.	.	.	21
3584	IPOMOEA CARNEA	S	.	0	19.9	9.8	.	.	.	.	.	.	.	.	.	.	.	.	.	21
3585	IPOMOEA CISSOIDES	S	.	0	17.4	6.3	.	.	.	.	.	.	.	.	.	.	.	.	.	21
3586	IPOMOEA HEDERACEA	S	.	0	25.0	11.8	.	.	.	.	.	.	.	.	.	.	.	.	.	20
3587	IPOMOEA HISPIDA	S	.	0	24.8	9.5	.	.	.	.	.	.	.	.	.	.	.	.	.	21
3588	IPOMOEA LEPTOPHYLLA	S	.	0	31.2	11.1	.	.	.	.	.	.	.	.	.	.	.	.	.	19
3589	IPOMOEA MURICATA	S	.	0	15.0	18.8	.	.	.	.	.	.	.	.	.	.	.	.	.	21
3590	IPOMOEA NIL	S	.	0	25.4	11.3	.	.	.	.	.	.	.	.	.	.	.	.	.	21
3591	IPOMOEA PARASITICA	S	.	0	25.6	18.6	.	.	.	.	.	.	.	.	.	.	.	.	.	20
3592	IPOMOEA PARASITICA	S	.	0	24.3	20.1	.	.	.	.	.	.	.	.	.	.	.	.	.	21
3593	IPOMOEA PENDURATA	S	.	0	16.2	7.1	.	.	3.9	.	.	.	.	.	.	.	.	.	.	19
3594	IPOMOEA PENTAPHYLLA	S	.	0	19.9	8.2	.	.	.	.	.	.	.	.	.	.	.	.	.	21
3595	IPOMOEA PES-TIGRIDIS	S	.	0	26.7	9.7	.	.	.	.	.	.	.	.	.	.	.	.	.	21
3596	IPOMOEA PES-TIGRIDIS	L	.	0	12.3	3.0	74.4	26.2	10.3	917.0	479.0	.	.	.	.	.	.	.	.	5
3597	IPOMOEA PES-TIGRIDIS	L	.	0	13.6	3.5	71.0	27.0	11.9	1480.0	344.0	.	.	.	.	.	.	.	.	5
3598	IPOMOEA PILOSA	S	.	0	31.6	9.3	.	.	.	.	.	.	.	.	.	.	.	.	.	21
3599	IPOMOEA PURPUREA	S	.	0	22.5	11.3	.	.	3.4	.	.	.	.	.	.	.	.	.	.	19
3600	IPOMOEA SP.	S	.	0	25.6	17.6	.	.	.	.	.	.	.	.	.	.	.	.	.	19
3601	IPOMOEA SP.	S	.	0	18.1	5.8	.	.	.	.	.	.	.	.	.	.	.	.	.	20
3602	IPOMOEA SP.	S	.	0	23.8	15.6	.	.	.	.	.	.	.	.	.	.	.	.	.	21
3603	IPOMOEA SP.	S	.	0	19.9	8.6	.	.	.	.	.	.	.	.	.	.	.	.	.	21

TABLE 3 - DATA CONVERTED TO A ZERO-MOISTURE BASIS.

OBS	PLNA	PART	CALORIES	WATER	PROTEIN	FAT	TOTLCARB	FIBER	ASH	CALCIUM	PHOSPHO	IRON	SODIUM	POTASS	CAROT	THIAMINE	RIBOFLAV	NIACIN	ASCORBIC	REFCO
3604	IPOMOEA SP.	S	.	0	25.6	17.7	.	.	.	.	.	.	.	.	.	.	.	.	.	21
3605	IPOMOEA TRILOBA	S	.	0	20.4	11.8	.	.	.	.	.	.	.	.	.	.	.	.	.	21
3606	IRIS GERMANICA	S	.	0	14.8	17.3	.	.	5.1	.	.	.	.	.	.	.	.	.	.	19
3607	IRIS GERMANICA	S	.	0	15.8	19.2	.	.	.	.	.	.	.	.	.	.	.	.	.	21
3608	IRIS KERNERIANA	S	.	0	21.9	11.6	.	.	.	.	.	.	.	.	.	.	.	.	.	21
3609	IRIS MISSOURIENSIS	S	.	0	13.1	13.9	.	.	.	.	.	.	.	.	.	.	.	.	.	20
3610	IRIS MISSOURIENSIS	S	.	0	16.9	14.3	.	.	.	.	.	.	.	.	.	.	.	.	.	20
3611	IRIS MISSOURIENSIS	S	.	0	16.2	14.7	.	.	.	.	.	.	.	.	.	.	.	.	.	20
3612	IRIS MISSOURIENSIS	S	.	0	11.2	13.9	.	.	.	.	.	.	.	.	.	.	.	.	.	21
3613	IRIS OCHROLEUCA	S	.	0	14.4	11.6	.	.	.	.	.	.	.	.	.	.	.	.	.	21
3614	IRIS SP.	S	.	0	14.2	16.3	.	.	.	.	.	.	.	.	.	.	.	.	.	21
3615	IRIS SPURIA	S	.	0	12.6	14.9	.	.	.	.	.	.	.	.	.	.	.	.	.	21
3616	IRVINGIA GABONENSIS	S	706	0	7.9	72.6	17.5	2.3	2.0	132.8	212.9	3.6	.	.	.	0.17	0.08	0.74	.	1
3617	ISACHNE GLOBOSA	L	.	0	10.5	1.8	72.0	32.0	15.7	.	.	.	.	.	.	.	.	.	.	5
3618	ISACHNE MILIACEA	L	.	0	10.4	1.5	70.8	29.0	17.3	.	.	.	.	.	.	.	.	.	.	5
3619	ISATIS AUCHERI	S	.	0	36.1	33.2	.	.	.	.	.	.	.	.	.	.	.	.	.	20
3620	ISATIS LUSITANICA	S	.	0	11.6	8.8	.	.	.	.	.	.	.	.	.	.	.	.	.	21
3621	ISATIS TINCTORIA	S	.	0	32.9	38.4	.	.	.	.	.	.	.	.	.	.	.	.	.	21
3622	ISATIS TINCTORIS	S	.	0	12.5	12.6	.	.	.	.	.	.	.	.	.	.	.	.	.	20
3623	ISATIS TINCTORIS	S	.	0	34.0	37.4	.	.	.	.	.	.	.	.	.	.	.	.	.	20
3624	ISCHAEMUM ARISTATUM	W	.	0	14.2	2.5	72.7	31.6	10.6	.	.	.	.	.	.	.	.	.	.	18
3625	ISCHAEMUM ARISTATUM	W	.	0	8.2	1.3	81.6	35.0	8.9	.	.	.	.	.	.	.	.	.	.	18
3626	ISCHAEMUM ARISTATUM	L	.	0	4.9	1.6	88.4	43.9	5.3	.	.	.	.	.	5.29	.	.	.	.	5
3627	ISCHAEMUM MUTICUM	L	.	0	9.2	1.5	81.9	24.6	7.3	192.3	115.4	.	.	.	.	.	.	.	.	5
3628	ISCHAEMUM PILOSUM	L	.	0	7.3	3.3	78.7	30.0	10.3	.	.	.	.	.	.	.	.	.	.	5
3629	ISCHAEMUM TIMORENSE	L	.	0	15.3	1.6	73.4	30.0	9.7	.	.	.	.	.	.	.	.	.	.	18
3630	ISCHAEMUM TIMORENSE	C	.	0	6.2	0.8	84.6	38.3	8.4	.	.	.	.	.	.	.	.	.	.	18
3631	ISCHAEMUM TIMORENSE	W	.	0	10.7	2.0	77.0	32.7	10.3	.	.	.	.	.	.	.	.	.	.	18
3632	ISEILEMA ANTHEPHOROIDES	L	.	0	4.6	2.2	81.3	34.7	12.0	.	.	.	.	.	.	.	.	.	.	5
3633	ISEILEMA LAXUM	H	.	0	6.4	0.6	86.2	35.6	6.8	.	.	.	.	.	.	.	.	.	.	18
3634	ISEILEMA LAXUM	W	.	0	5.1	1.4	81.9	34.2	11.6	.	.	.	.	.	.	.	.	.	.	18
3635	ISEILEMA LAXUM	L	.	0	5.7	2.7	79.8	28.0	11.8	1300.0	630.0	.	300	1450	.	.	.	.	.	5
3636	ISOMERIS ARBOREA	S	.	0	37.2	44.8	.	.	3.5	.	.	.	.	.	.	.	.	.	.	19
3637	ISOMERIS ARBOREA	S	.	0	6.9	2.8	.	.	1.2	.	.	.	.	.	.	.	.	.	.	19
3638	ISOMERIS ARBOREA	S	.	0	40.6	42.0	.	.	.	.	.	.	.	.	.	.	.	.	.	20
3639	IVA AXILLARIS	S	.	0	3.6	8.5	.	.	.	.	.	.	.	.	.	.	.	.	.	21
3640	IVA FRUTESCENS	S	.	0	18.8	11.9	.	.	.	.	.	.	.	.	.	.	.	.	.	19
3641	IVA XANTHIFOLIA	S	.	0	19.4	22.8	.	.	5.2	.	.	.	.	.	.	.	.	.	.	19
3642	IVA XANTHIFOLIA	S	.	0	19.6	31.8	.	.	.	.	.	.	.	.	.	.	.	.	.	21
3643	IXOPHORUS UNISETUS	W	.	0	14.4	3.0	67.9	30.0	14.8	.	.	.	.	.	.	.	.	.	.	18
3644	IXOPHORUS UNISETUS	H	.	0	5.4	0.9	80.4	35.3	13.3	.	.	.	.	.	.	.	.	.	.	18
3645	JACARANDA MIMOSIFOLIA	S	.	0	33.1	32.7	.	.	4.3	.	.	.	.	.	.	.	.	.	.	19
3646	JACQUINIA PUNGENS	S	.	0	11.2	11.5	.	.	.	.	.	.	.	.	.	.	.	.	.	19
3647	JACQUINIA PUNGENS	S	.	0	11.2	11.6	.	.	.	.	.	.	.	.	.	.	.	.	.	20
3648	JACQUINIA PUNGENS	S	.	0	10.6	11.1	.	.	.	.	.	.	.	.	.	.	.	.	.	20

TABLE 3 - DATA CONVERTED TO A ZERO-MOISTURE BASIS.

OBS	PLANA	PART	CALORIES	WATER	PROTEIN	FAT	TOTCARB	FIBER	ASH	CALCIUM	PHOSPHO	IRON	SODIUM	POTASS	CAROT	THIAMINE	RIBOFLAV	NIACIN	ASCORBIC	REFCO
3649	JASMINUM FRUTICANS	S	.	0	10.8	22.5	.	.	.	.	.	.	.	.	.	.	.	.	.	20
3650	JATROPHA CORDATA	S	.	0	15.6	39.1	.	.	.	.	.	.	.	.	.	.	.	.	.	20
3651	JATROPHA CURCAS	S	.	0	18.6	31.1	.	.	.	.	.	.	.	.	.	.	.	.	.	21
3652	JATROPHA CURCAS	S	.	0	19.5	40.7	35.9	16.6	4.8	.	.	.	.	.	.	.	.	.	.	5
3653	JATROPHA MACRORHIZA	S	.	0	36.2	52.8	.	.	.	.	.	.	.	.	.	.	.	.	.	20
3654	JATROPHA SPATHULATA	S	.	0	26.2	58.2	.	.	.	.	.	.	.	.	.	.	.	.	.	19
3655	JUBAEA CHILENSIS	S	714	0	8.2	75.3	15.5	6.8	1.0	36.7	202.1	6.1	.	.	0.0	0.30	0.06	0.46	0.0	3
3656	JUGLANS CINEREA	S	654	0	24.6	63.6	8.7	.	3.0	.	.	7.1	.	.	.	.	.	.	.	4
3657	JUGLANS NIGRA	S	648	0	21.2	61.2	15.3	1.8	2.4	0.0	588.2	6.2	3.1	474.7	185.8	0.23	0.11	0.72	.	4
3658	JUGLANS REGIA	S	.	0	15.1	58.2	24.8	9.2	1.9	.	280.0	.	.	.	.	.	.	.	.	18
3659	JUGLANS REGIA	S	674	0	16.0	66.1	16.2	.	1.7	.	.	.	.	.	.	.	.	.	.	1
3660	JUGLANS REGIA	S	671	0	18.9	66.0	13.1	1.7	2.1	110.0	338.2	3.1	3.1	556.0	0.0	0.52	0.08	3.11	5.2	2
3661	JUGLANS REGIA	S	693	0	14.3	70.1	13.8	2.1	1.8	96.0	395.6	3.4	.	.	52.2	0.28	0.53	3.13	0.0	3
3662	JUGLANS REGIA	S	675	0	15.3	66.3	16.4	2.2	2.0	102.6	393.8	3.2	2.1	466.3	18.6	0.34	0.13	0.93	2.1	4
3663	JUNCUS ACUTUS	S	.	0	20.5	5.8	.	.	.	.	.	.	.	.	.	.	.	.	.	21
3664	JUNCUS ACUTUS	S	.	0	20.9	6.8	.	.	.	.	.	.	.	.	.	.	.	.	.	21
3665	JUNCUS BALTICUS	W	.	0	10.0	1.8	82.5	33.0	5.7	.	.	.	.	.	.	.	.	.	.	6
3666	JUNCUS BUFONIUS	W	.	0	10.0	.	.	27.7	5.7	280.0	390.0	.	.	1910.0	.	.	.	.	.	6
3667	JUNCUS EFFUSUS	S	.	0	13.1	5.0	.	.	.	.	.	.	.	.	.	.	.	.	.	20
3668	JUNCUS GERARDII	H	.	0	8.4	2.8	80.7	28.0	8.1	.	.	.	.	.	.	.	.	.	.	6
3669	JUNCUS INFLEXUS	S	.	0	15.7	6.4	.	.	.	.	.	.	.	.	.	.	.	.	.	21
3670	JUNCUS INFLEXUS	S	.	0	18.6	6.0	.	.	.	.	.	.	.	.	.	.	.	.	.	21
3671	JUNCUS LONGISTYLIS	H	.	0	10.8	1.4	81.5	33.8	6.3	.	.	.	.	.	.	.	.	.	.	6
3672	JUNCUS MARGINATUS	S	.	0	12.8	2.7	.	.	.	.	.	.	.	.	.	.	.	.	.	21
3673	JUNCUS OXYMERIS		.	0	10.8	.	.	31.3	8.1	290.0	210.0	.	.	2790.0	.	.	.	.	.	6
3674	JUNCUS SP.	W	.	0	10.0	1.9	81.5	31.8	6.6	300.0	250.0	.	.	2360.0	.	.	.	.	.	6
3675	JUNIPERUS COMMUNIS SUBSP. NANA	S	.	0	30.9	53.9	.	.	.	.	.	.	.	.	.	.	.	.	.	21
3676	JUNIPERUS EXCELSA	S	.	0	4.7	3.4	.	.	.	.	.	.	.	.	.	.	.	.	.	21
3677	JUNIPERUS OSTEOSPERMA	W	.	0	6.4	16.3	72.8	22.0	4.5	1590.0	170.0	.	.	.	.	.	.	.	.	6
3678	JUNIPERUS OXYCEDRUS	S	.	0	3.1	5.1	.	.	.	.	.	.	.	.	.	.	.	.	.	21
3679	JUNIPERUS OXYCEDRUS	S	.	0	4.4	10.0	.	.	.	.	.	.	.	.	.	.	.	.	.	21
3680	JUNIPERUS PINCHOTII	W	.	0	8.6	10.3	76.3	19.9	4.8	.	.	.	.	.	.	.	.	.	.	6
3681	JUNIPERUS SP.	S	.	0	6.4	14.1	.	.	.	.	.	.	.	.	.	.	.	.	.	21
3682	JUNIPERUS THURIFERA	S	.	0	3.8	2.3	.	.	.	.	.	.	.	.	.	.	.	.	.	21
3683	JUNIPERUS VIRGINIANA	S	.	0	13.8	16.4	.	.	.	.	.	.	.	.	.	.	.	.	.	20
3684	JURINEA ANATOLICA	S	.	0	20.9	24.0	.	.	.	.	.	.	.	.	.	.	.	.	.	21
3685	JUSSIAEA LEPTOCARPA	S	.	0	13.8	12.5	.	.	.	.	.	.	.	.	.	.	.	.	.	20
3686	JUSTICIA EXIGUA	SH	.	0	14.0	3.2	63.9	21.9	18.9	1670.0	450.0	.	.	.	.	.	.	.	.	18
3687	JUSTICIA PECTORALIS	L	293	0	26.0	4.0	54.7	18.7	15.3	4420.0	233.3	49.3	.	.	17800.0	0.27	1.33	16.67	186.7	3
3688	JUSTICIA SP.	L	250	0	25.0	3.0	46.2	12.9	25.8	3863.6	530.3	.	.	.	.	.	.	.	.	1
3689	KAEMPFERIA CANDIDA	R	362	0	10.5	7.6	69.5	.	12.4	266.7	381.0	19.0	.	.	28571.4	0.76	0.19	.	.	2
3690	KALIMERIS YOMENA	L	296	0	22.4	4.1	58.2	5.1	15.3	816.3	673.5	40.8	.	.	18979.6	1.53	2.04	20.41	510.2	2
3691	KALMIA LATIFOLIA	W	.	0	8.7	9.5	78.2	14.7	3.6	960.0	110.0	5.0	.	.	.	.	.	.	.	6
3692	KALOPANAX SEPTEMLOBUS	S	.	0	3.2	21.7	.	.	.	.	.	.	.	.	.	.	.	.	.	21
3693	KERRIA JAPONICA	S	.	0	44.9	45.3	.	.	.	.	.	.	.	.	.	.	.	.	.	21

TABLE 3 - DATA CONVERTED TO A ZERO-MOISTURE BASIS.

OBS	PLNA	PART	CALORIES	WATER	PROTEIN	FAT	TOTLCARB	FIBER	ASH
3694	KERSTINGIELLA GEOCARPA	S	386	0	21.5	1.2	73.9	6.1	3.6
3695	KIGELIA AFRICANA	F	370	0	5.5	6.2	83.6	29.5	4.8
3696	KIGELIA PINNATA	F	.	0	5.8	6.0	83.7	29.4	4.5
3697	KLAINEDOXA GABONESIS	S	664	0	9.1	64.9	22.0	1.8	4.0
3698	KLEINHOVIA HOSPITA	S	.	0	21.3	8.5	.	.	.
3699	KOCHIA PROSTRATA	L	.	0	9.8	3.0	74.5	35.3	.
3700	KOCHIA SCOPARIA	S	.	0	23.1	11.2	.	.	4.5
3701	KOCHIA SCOPARIA	S	.	0	20.4	8.8	.	.	9.4
3702	KOCHIA SCOPARIA	S	.	0	27.5	16.0	.	.	3.4
3703	KOCHIA SCOPARIA	H	.	0	13.8	1.5	71.3	24.9	13.4
3704	KOCHIA SCOPARIA	L	.	0	21.5	2.4	56.8	19.7	19.2
3705	KOCHIA SCOPARIA(FLOWERING)	H	.	0	13.1	1.6	71.4	27.6	13.9
3706	KOCHIA SCOPARIA(IMMATURE)	H	.	0	18.8	1.4	66.1	19.7	13.7
3707	KOCHIA SCOPARIA(MATURE)	H	.	0	11.2	1.5	75.1	35.6	12.2
3708	KOCHIA SP.	H	.	0	13.7	1.6	71.2	25.7	13.5
3709	KOCHIA VESTITA	W	.	0	9.0	3.7	62.9	15.5	24.4
3710	KOELERIA CRISTATA	L	.	0	8.9	2.6	79.2	38.5	8.3
3711	KOELERIA CRISTATA	W	.	0	7.3	2.1	82.5	37.8	8.1
3712	KOELERIA CRISTATA(MATURE)	W	.	0	5.8	2.5	83.6	38.6	8.1
3713	KOELERIA CRISTATA(VERY MATURE)	W	.	0	3.8	1.8	86.7	41.0	7.7
3714	KOELERIA GLAUCOVIRENS	S	.	0	.	5.6	.	.	.
3715	KOELREUTERIA BIPINNATA	S	.	0	28.5	54.0	.	.	3.1
3716	KOELREUTERIA FORMOSANA	S	.	0	28.9	52.7	.	.	5.9
3717	KOELREUTERIA PANICULATA	S	.	0	30.6	40.8	.	.	.

OBS	CALCIUM	PHOSPHO	IRON	SODIUM	POTASS	CAROT	THIAMINE	RIBOFLAV	NIACIN	ASCORBIC	REFCO
3694	114.0	435.0	16.6	.	.	.	0.84	0.21	2.55	0.0	10
3695	.	.	.	.	.	.	.	.	.	.	1
3696	.	.	.	.	.	.	.	.	.	.	18
3697	339.5	459.9	.	.	.	.	.	.	.	.	1
3698	.	.	.	.	.	.	.	.	.	.	21
3699	.	.	.	.	.	.	.	.	.	.	5
3700	.	.	.	.	.	.	.	.	.	.	19
3701	.	.	.	.	.	.	.	.	.	.	19
3702	.	.	.	.	.	.	.	.	.	.	19
3703	.	.	.	.	.	.	.	.	.	.	6
3704	.	.	.	.	.	4.6	.	.	.	.	5
3705	.	.	.	.	.	.	.	.	.	.	6
3706	.	.	.	.	.	.	.	.	.	.	6
3707	.	.	.	.	.	.	.	.	.	.	6
3708	.	.	.	.	.	.	.	.	.	.	6
3709	2370.0	120.0	.	.	.	.	.	.	.	.	6
3710	500.0	600.0	.	.	.	.	.	.	.	.	5
3711	320.0	240.0	.	.	.	8.0	.	.	.	.	6
3712	330.0	310.0	.	.	.	.	.	.	.	.	6
3713	290.0	140.0	.	.	.	.	.	.	.	.	6
3714	.	.	.	.	.	.	.	.	.	.	21
3715	.	.	.	.	.	.	.	.	.	.	19
3716	.	.	.	.	.	.	.	.	.	.	19
3717	.	.	.	.	.	.	.	.	.	.	19

TABLE 3 - DATA CONVERTED TO A ZERO-MOISTURE BASIS.

OBS	PLNA	PART	CALORIES	WATER	PROTEIN	FAT	TOTCARB	FIBER	ASH	CALCIUM	PHOSPHO	IRON	SODIUM	POTASS	CAROT	THIAMINE	RIBOFLAV	NIACIN	ASCORBIC	REFCO
3718	KOELREUTERIA PANICULATA	S	.	0	26.9	41.9	.	.	.	.	.	.	.	.	.	.	.	.	.	19
3719	KOELREUTERIA PANICULATA	S	.	0	27.5	45.8	.	.	.	.	.	.	.	.	.	.	.	.	.	19
3720	KOSTELETZKYA SP.	H	.	0	12.8	2.2	79.3	36.3	5.7	260.0	210.0	.	.	900.0	.	.	.	.	.	6
3721	KOSTELETZKYA VIRGINICA	S	.	0	26.9	18.9	.	.	4.0	.	.	.	.	.	.	.	.	.	.	19
3722	KUHNIA GLUTINOSA	S	.	0	33.1	21.5	.	.	4.3	.	.	.	.	.	.	.	.	.	.	19
3723	KUNDMANNIA SICULA	L	.	0	19.8	.	.	.	20.3	2270.0	.	.	1910.0	3130.0	.	.	.	.	.	17
3724	LACTUCA CRETICA	S	.	0	30.2	31.4	.	.	.	.	.	.	.	.	.	.	.	.	.	21
3725	LACTUCA SATIVA	S	.	0	26.8	41.6	.	.	3.8	.	.	.	.	.	.	.	.	.	.	19
3726	LACTUCA SATIVA	S	.	0	31.8	33.8	.	.	4.7	.	.	.	.	.	.	.	.	.	.	19
3727	LACTUCA SATIVA	L	317	0	19.0	3.2	68.3	9.5	9.5	412.7	634.9	11.1	.	.	.	.	.	.	158.7	1
3728	LACTUCA SATIVA	L	.	0	.	.	.	.	6.5	.	.	8.1	.	.	.	.	.	.	.	1
3729	LACTUCA SATIVA	L	302	0	22.6	3.8	60.4	7.5	13.2	1056.6	1000.0	20.8	188.7	.	13584.9	1.13	1.13	11.32	283.0	1
3730	LACTUCA SATIVA	L	313	0	21.9	4.7	60.9	9.4	12.5	875.0	531.2	32.8	218.8	3968.7	31796.9	0.94	1.88	7.81	265.6	2
3731	LACTUCA SATIVA	L	333	0	23.8	9.5	57.1	9.5	9.5	428.6	523.8	9.5	.	.	21071.4	0.95	0.95	4.76	95.2	2
3732	LACTUCA SATIVA	L	294	0	25.5	3.9	56.9	13.7	13.7	843.1	666.7	25.5	.	.	10196.1	1.57	1.57	7.84	235.3	3
3733	LACTUCA SATIVA	L	310	0	23.8	2.4	64.3	11.9	9.5	381.0	547.6	9.5	.	.	.	1.19	0.71	7.14	166.7	3
3734	LACTUCA SATIVA	L	300	0	21.7	5.0	58.3	11.7	15.0	1133.3	416.7	23.3	150.0	4400.0	19000.0	0.83	1.33	6.67	300.0	4
3735	LACTUCA SATIVA	L	286	0	24.5	4.1	51.0	10.2	20.4	714.3	530.6	40.8	183.7	5387.8	11877.5	1.22	1.22	6.12	163.3	4
3736	LACTUCA SCARIOLA	S	.	0	29.4	40.8	.	.	4.4	.	.	.	.	.	.	.	.	.	.	19
3737	LACTUCA SCARIOLA	S	.	0	25.9	39.5	.	.	.	.	.	.	.	.	.	.	.	.	.	21
3738	LACTUCA SP.	W	.	0	12.8	6.1	60.3	19.1	20.8	.	.	.	.	.	.	.	.	.	.	6
3739	LACTUCA TARAXAEIFOLIA	L	280	0	20.4	5.1	52.9	12.7	21.7	2076.4	369.4	.	.	.	.	.	.	.	.	1
3740	LAGENARIA SICERARIA	F	333	0	9.0	2.6	80.8	19.2	7.7	230.8	269.2	6.4	.	.	0.0	0.51	0.38	7.69	243.6	3
3741	LAGENARIA SICERARIA	S	.	0	31.5	53.8	10.2	1.6	4.5	112.8	2523.1	.	.	.	.	.	.	.	.	5
3742	LAGENARIA SICERARIA	F	.	0	5.4	2.7	78.4	.	13.5	540.5	.	18.9	297.3	2324.3	.	1.08	0.54	8.92	351.4	5
3743	LAGERSTROEMIA INDICA	S	.	0	21.2	8.3	.	.	.	.	.	.	.	.	.	.	.	.	.	20
3744	LAGERSTROEMIA SPECIOSA	S	.	0	4.4	1.5	.	.	.	.	.	.	.	.	.	.	.	.	.	20
3745	LAGERSTROEMIA TOMENTOSA	S	.	0	12.5	3.0	.	.	.	.	.	.	.	.	.	.	.	.	.	20
3746	LALLEMANTIA ROYLEANA	S	.	0	17.5	19.7	.	.	.	.	.	.	.	.	.	.	.	.	.	21
3747	LANARIA LANATA	S	.	0	16.7	17.0	.	.	.	.	.	.	.	.	.	.	.	.	.	21
3748	LANSIUM DOMESTICUM	F	348	0	5.7	0.6	89.9	5.1	3.8	75.9	151.9	5.7	6.3	1468.4	0.0	0.51	0.32	6.33	12.7	2
3749	LANSIUM DOMESTICUM	F	.	0	5.9	2.2	87.4	17.0	4.4	148.1	222.2	3.7	.	.	59.3	0.67	0.89	0.00	7.4	5
3750	LANTANA CAMARA	S	.	0	35.1	48.0	.	.	.	.	.	.	.	.	.	.	.	.	.	21
3751	LANTANA SP.	S	.	0	13.8	5.1	.	.	.	.	.	.	.	.	.	.	.	.	.	20
3752	LAPORTEA AESTUANS	L	265	0	29.0	2.0	50.0	15.0	19.0	2200.0	570.0	7.5	.	.	.	.	.	.	.	1
3753	LAPPULA REDOWSKII	S	.	0	17.8	18.8	.	.	.	.	.	.	.	.	.	.	.	.	.	20
3754	LAPPULA SP.	L	.	0	22.8	3.3	62.3	10.6	11.6	.	.	.	.	.	.	.	.	.	.	6
3755	LARIX SP	S	701	0	16.7	72.2	9.5	1.0	1.6	13.8	174.6	5.0	.	.	31.8	0.35	0.11	7.41	0.0	2
3756	LARREA DIVARICATA	S	.	0	29.0	12.9	.	.	.	.	.	.	.	.	.	.	.	.	.	20
3757	LARREA TRIDENTATA	S	.	0	30.0	11.7	.	.	5.8	.	.	.	.	.	.	.	.	.	.	19
3758	LARREA TRIDENTATA(IMMATURE)	W	.	0	15.8	11.4	62.5	14.5	10.3	2520.0	260.0	.	.	1390.0	.	.	.	.	.	6
3759	LASIURUS HIRSUTUS	W	.	0	10.4	8.2	69.5	34.1	11.9	.	.	.	.	.	.	.	.	.	.	18
3760	LASIURUS HIRSUTUS	H	.	0	5.9	0.3	82.7	38.0	11.0	.	.	.	.	.	.	.	.	.	.	18
3761	LATHYRUS CICERA	S	.	0	33.3	0.8	.	.	3.4	.	.	.	.	.	.	.	.	.	.	19
3762	LATHYRUS CICERA	S	.	0	34.7	0.7	.	.	3.3	.	.	.	.	.	.	.	.	.	.	19

TABLE 3 - DATA CONVERTED TO A ZERO-MOISTURE BASIS.

OBS	PLNA	PART	CALORIES	WATER	PROTEIN	FAT	TOTLCARB	FIBER	ASH	CALCIUM	PHOSPHO	IRON	SODIUM	POTASS	CAROT	THIAMINE	RIBOFLAV	NIACIN	ASCORBIC	REFCO
3763	LATHYRUS CICERA	S	.	0	30.3	0.7	.	.	3.6	.	.	.	.	.	.	.	.	.	.	19
3764	LATHYRUS CICERA	S	.	0	33.7	0.7	.	.	3.9	.	.	.	.	.	.	.	.	.	.	19
3765	LATHYRUS CICERA	S	.	0	34.4	0.7	.	.	3.6	.	.	.	.	.	.	.	.	.	.	19
3766	LATHYRUS CICERA	S	.	0	31.9	0.8	.	.	3.2	.	.	.	.	.	.	.	.	.	.	19
3767	LATHYRUS CICERA	S	.	0	32.0	0.7	.	.	3.0	.	.	.	.	.	.	.	.	.	.	19
3768	LATHYRUS CICERA	S	.	0	29.8	0.6	.	.	2.5	.	.	.	.	.	.	.	.	.	.	19
3769	LATHYRUS CICERA	S	.	0	32.0	0.8	.	.	3.5	.	.	.	.	.	.	.	.	.	.	19
3770	LATHYRUS CICERA	S	.	0	29.3	0.6	.	.	3.5	.	.	.	.	.	.	.	.	.	.	19
3771	LATHYRUS CICERA	S	.	0	27.8	1.0	.	.	3.1	.	.	.	.	.	.	.	.	.	.	19
3772	LATHYRUS CICERA	S	.	0	28.5	0.7	.	.	3.3	.	.	.	.	.	.	.	.	.	.	19
3773	LATHYRUS CICERA	S	.	0	27.8	0.9	.	.	3.3	.	.	.	.	.	.	.	.	.	.	19
3774	LATHYRUS CICERA	S	.	0	27.7	0.7	.	.	3.3	.	.	.	.	.	.	.	.	.	.	19
3775	LATHYRUS CICERA	S	.	0	27.8	0.8	.	.	3.3	.	.	.	.	.	.	.	.	.	.	19
3776	LATHYRUS CICERA	S	.	0	29.6	0.8	.	.	3.3	.	.	.	.	.	.	.	.	.	.	19
3777	LATHYRUS CICERA	S	.	0	30.1	0.9	.	.	3.6	.	.	.	.	.	.	.	.	.	.	19
3778	LATHYRUS CICERA	S	.	0	30.0	0.7	.	.	3.2	.	.	.	.	.	.	.	.	.	.	19
3779	LATHYRUS DIGITATUS	S	.	0	27.8	0.8	.	.	.	.	.	.	.	.	.	.	.	.	.	21
3780	LATHYRUS GORGONII	S	.	0	24.4	0.5	.	.	.	.	.	.	.	.	.	.	.	.	.	21
3781	LATHYRUS HIEROSOLYMITANUS	S	.	0	27.5	0.8	.	.	.	.	.	.	.	.	.	.	.	.	.	21
3782	LATHYRUS HIRSUTUS	S	.	0	28.6	0.7	.	.	3.5	.	.	.	.	.	.	.	.	.	.	19
3783	LATHYRUS HIRSUTUS	S	.	0	28.0	0.6	.	.	3.8	.	.	.	.	.	.	.	.	.	.	19
3784	LATHYRUS HIRSUTUS	S	.	0	26.9	0.6	.	.	4.4	.	.	.	.	.	.	.	.	.	.	19
3785	LATHYRUS INCURVUS	S	.	0	27.6	1.1	.	.	.	.	.	.	.	.	.	.	.	.	.	21
3786	LATHYRUS LATIFOLIUS	S	.	0	32.1	0.8	.	.	.	.	.	.	.	.	.	.	.	.	.	21
3787	LATHYRUS LEUCANTHUS	H	.	0	16.1	2.8	.	.	9.1	1350.0	310.0	.	.	.	.	.	.	.	.	6
3788	LATHYRUS ORNATUS	S	.	0	32.9	1.4	.	.	2.8	.	.	.	.	.	.	.	.	.	.	19
3789	LATHYRUS PALLESCENS	S	.	0	27.2	0.9	.	.	.	.	.	.	.	.	.	.	.	.	.	21
3790	LATHYRUS PSEUDO-CICERA	S	.	0	28.8	0.8	.	.	.	.	.	.	.	.	.	.	.	.	.	21
3791	LATHYRUS SATIVUS	W	.	0	20.7	3.4	65.0	23.0	10.9	1320.0	340.0	.	.	.	.	.	.	.	.	18
3792	LATHYRUS SATIVUS	H	.	0	18.2	3.3	67.4	30.5	11.1	.	.	.	.	.	.	.	.	.	.	18
3793	LATHYRUS SATIVUS	S	.	0	20.9	2.0	74.5	5.9	2.6	650.0	300.0	.	.	.	.	.	.	.	.	18
3794	LATHYRUS SATIVUS	S	.	0	28.1	0.9	.	.	3.4	.	.	.	.	.	.	.	.	.	.	19
3795	LATHYRUS SATIVUS	S	.	0	33.4	0.9	.	.	3.5	.	.	.	.	.	.	.	.	.	.	19
3796	LATHYRUS SATIVUS	S	.	0	33.1	0.8	.	.	3.2	.	.	.	.	.	.	.	.	.	.	19
3797	LATHYRUS SATIVUS	S	.	0	30.1	0.8	.	.	4.0	.	.	.	.	.	.	.	.	.	.	19
3798	LATHYRUS SATIVUS	S	.	0	33.6	0.8	.	.	3.0	.	.	.	.	.	.	.	.	.	.	19
3799	LATHYRUS SATIVUS	S	.	0	30.8	0.7	.	.	3.4	.	.	.	.	.	.	.	.	.	.	19
3800	LATHYRUS SATIVUS	S	.	0	.	0.7	.	.	3.6	.	.	.	.	.	.	.	.	.	.	19
3801	LATHYRUS SATIVUS	S	.	0	.	0.7	.	.	3.5	.	.	.	.	.	.	.	.	.	.	19
3802	LATHYRUS SATIVUS	S	.	0	.	0.5	.	.	2.8	.	.	.	.	.	.	.	.	.	.	19
3803	LATHYRUS SATIVUS	L	.	0	38.6	6.3	48.1	.	7.0	1012.7	632.9	46.2	.	.	22784.8	.	.	.	.	5
3804	LATHYRUS SP.	S	.	0	29.6	0.9	.	.	4.0	.	.	.	.	.	.	.	.	.	.	19
3805	LATHYRUS SP.	W	.	0	17.2	3.4	54.5	26.2	24.9	.	.	.	.	.	.	.	.	.	.	6
3806	LATHYRUS SP.	H	.	0	16.7	2.7	72.3	29.0	8.3	1390.0	310.0	.	.	.	.	.	.	.	.	6
3807	LATHYRUS SP.(LEAFMEAL)	W	.	0	21.3	6.6	61.9	14.8	10.2	1870.0	270.0	.	.	.	.	.	.	.	.	6

TABLE 3 - DATA CONVERTED TO A ZERO-MOISTURE BASIS.

OBS	PLNA	PART	CALORIES	WATER	PROTEIN	FAT	TOTLCARB	FIBER	ASH
3808	LATHYRUS SP.(STEMMEAL)	W	.	0	11.5	2.4	78.3	29.4	7.8
3809	LATHYRUS SYLVESTRIS	S	.	0	34.6	0.8	.	.	3.2
3810	LATHYRUS SYLVESTRIS	S	.	0	32.4	0.8	.	.	3.0
3811	LATHYRUS SYLVESTRIS(ALL)	H	.	0	28.1	3.9	60.9	28.6	7.1
3812	LATHYRUS SYLVESTRIS(IMMATURE)	H	.	0	32.4	3.7	56.9	26.6	7.0
3813	LATHYRUS SYLVESTRIS(PAST BLOOMING)	H	.	0	24.4	4.6	64.1	30.4	6.9
3814	LATHYRUS TINGITANUS	S	.	0	42.9	1.2	.	.	3.7
3815	LATHYRUS TUBEROSUS	S	.	0	27.5	0.9	.	.	3.6
3816	LATHYRUS TUBEROSUS	S	.	0	23.3	1.1	.	.	3.8
3817	LATHYRUS TUBEROSUS	S	.	0	25.0	1.0	.	.	.
3818	LATIPES SENEGALENSIS	W	.	0	9.3	1.3	81.9	38.9	7.5
3819	LATUS CORNICULATUS	W	.	0	21.6	2.5	67.8	22.2	8.1
3820	LAURUS NOBILIS	S	.	0	10.2	26.9	.	.	.
3821	LAURUS NOBILIS	S	.	0	8.2	25.2	.	.	.
3822	LAURUS NOBILIS	L	331	0	8.0	8.9	79.3	27.8	3.8
3823	LAURUS NOBILIS	L	343	0	7.7	2.2	85.9	8.4	4.2
3824	LAVANDULA LATIFOLIA	S	.	0	23.0	28.4	.	.	.
3825	LAVANDULA PEDUNCULATA	S	.	0	27.0	20.0	.	.	.
3826	LAVANDULA SP.	S	.	0	29.4	21.8	.	.	.
3827	LAVANDULA STOECHAS	S	.	0	23.3	20.7	.	.	.
3828	LAVATERA ASSURGENTIFLORA	S	.	0	18.8	11.2	.	.	.
3829	LAVATERA PUNCTATA	S	.	0	18.2	12.9	.	.	.
3830	LAVATERA PUNCTATA	S	.	0	19.8	14.4	.	.	.
3831	LAVATERA PUNCTATA	S	.	0	18.7	13.6	.	.	.

OBS	CALCIUM	PHOSPHO	IRON	SODIUM	POTASS	CAROT	THIAMINE	RIBOFLAV	NIACIN	ASCORBIC	REFCO
3808	600.0	180.0	.	.	.	.	.	.	.	.	6
3809	.	.	.	.	.	.	.	.	.	.	19
3810	.	.	.	.	.	.	.	.	.	.	19
3811	.	.	.	.	.	.	.	.	.	.	6
3812	520.0	490.0	.	.	.	.	.	.	.	.	6
3813	.	300.0	.	.	.	.	.	.	.	.	6
3814	.	.	.	.	.	.	.	.	.	.	19
3815	.	.	.	.	.	.	.	.	.	.	19
3816	.	.	.	.	.	.	.	.	.	.	19
3817	.	.	.	.	.	.	.	.	.	.	19
3818	.	.	.	.	.	.	.	.	.	.	18
3819	1760.0	200.0	.	.	1840.0	.	.	.	.	.	6
3820	.	.	.	.	.	.	.	.	.	.	21
3821	.	.	.	.	.	.	.	.	.	.	21
3822	881.6	119.5	45.5	24.3	559.2	3922.8	0.01	0.44	2.11	48.6	16
3823	341.2	127.7	9.7	.	.	1916.1	0.07	0.38	3.10	98.5	3
3824	.	.	.	.	.	.	.	.	.	.	21
3825	.	.	.	.	.	.	.	.	.	.	21
3826	.	.	.	.	.	.	.	.	.	.	20
3827	.	.	.	.	.	.	.	.	.	.	21
3828	.	.	.	.	.	.	.	.	.	.	20
3829	.	.	.	.	.	.	.	.	.	.	21
3830	.	.	.	.	.	.	.	.	.	.	21
3831	.	.	.	.	.	.	.	.	.	.	21

TABLE 3 - DATA CONVERTED TO A ZERO-MOISTURE BASIS.

OBS	PLNA	PART	CALORIES	WATER	PROTEIN	FAT	TOTLCARB	FIBER
3832	LAVATERA SP.	S	.	0	17.5	14.0	.	.
3833	LAVATERA TRILOBIA SUBSP. TRILOBA	S	.	0	22.0	14.8	.	.
3834	LAVATERA TRIMESTRIS	S	.	0	25.4	16.4	.	.
3835	LAVATERA TRIMESTRIS	S	.	0	25.0	17.1	.	.
3836	LAVATERA TRIMESTRIS	S	.	0	26.2	17.0	.	.
3837	LAVATERA TRIMESTRIS	S	.	0	23.1	17.2	.	.
3838	LAVATERA TRIMESTRIS	S	.	0	23.2	19.0	.	.
3839	LAVATERA TRIMESTRIS	S	.	0	26.2	16.9	.	.
3840	LAVATERA TRIMESTRIS	S	.	0	20.6	16.4	.	.
3841	LAVATERA TRIMESTRIS	S	.	0	22.5	16.4	.	.
3842	LAYIA SP.	W	.	0	10.0	.	.	24.0
3843	LEBECKIA SP.	S	.	0	44.3	8.6	.	.
3844	LECHEA TENUIFOLIA	S	.	0	22.9	3.6	.	.
3845	LECHEA VILLOSA	S	.	0	20.7	2.0	.	.
3846	LEERSIA HEXANDRA	W	.	0	10.1	1.8	77.7	25.6
3847	LEERSIA HEXANDRA	H	.	0	6.3	1.5	77.3	31.4
3848	LEERSIA HEXANDRA	L	.	0	5.8	2.1	75.4	28.4
3849	LEMNA MINOR	L	265	0	30.9	4.4	41.2	.
3850	LENS CULINARIS	S	.	0	26.7	1.2	60.1	0.5
3851	LENS CULINARIS	S	.	0	27.1	0.8	.	.
3852	LENS CULINARIS	S	383	0	27.6	1.3	67.8	4.3
3853	LENS CULINARIS	S	386	0	23.0	0.7	73.9	.
3854	LENS CULINARIS	S	387	0	27.0	1.5	69.1	3.6
3855	LENS CULINARIS	S	382	0	27.8	1.2	67.6	4.4

OBS	ASH	CALCIUM	PHOSPHO	IRON	SODIUM	POTASS	CAROT	THIAMINE	RIBOFLAV	NIACIN	ASCORBIC	REFCO
3832	.	.	.	.	.	.	.	.	.	.	.	20
3833	.	.	.	.	.	.	.	.	.	.	.	21
3834	5.3	.	.	.	.	.	.	.	.	.	.	19
3835	.	.	.	.	.	.	.	.	.	.	.	20
3836	.	.	.	.	.	.	.	.	.	.	.	20
3837	.	.	.	.	.	.	.	.	.	.	.	20
3838	.	.	.	.	.	.	.	.	.	.	.	20
3839	.	.	.	.	.	.	.	.	.	.	.	20
3840	.	.	.	.	.	.	.	.	.	.	.	20
3841	.	.	.	.	.	.	.	.	.	.	.	20
3842	10.8	1360.0	410.0	.	.	2900.0	.	.	.	.	.	6
3843	.	.	.	.	.	.	.	.	.	.	.	21
3844	.	.	.	.	.	.	.	.	.	.	.	20
3845	.	.	.	.	.	.	.	.	.	.	.	20
3846	10.4	.	.	.	.	.	.	.	.	.	.	18
3847	14.9	.	.	.	.	.	.	.	.	.	.	18
3848	16.7	190.0	140.0	.	.	.	.	.	.	.	.	5
3849	23.5	2088.2	58.8	.	.	.	16470.6	0.88	1.91	8.82	73.5	3
3850	12.0	.	.	.	.	.	.	.	.	.	.	18
3851	2.6	.	.	.	.	.	.	.	.	.	.	19
3852	3.2	71.0	330.7	7.8	.	.	66.6	0.46	0.21	2.44	0.0	1
3853	2.4	77.3	369.3	8.0	33.0	886.4	.	0.52	0.38	1.48	0.0	2
3854	2.4	77.4	402.1	8.0	.	.	22.8	0.52	0.38	2.73	5.7	3
3855	3.4	88.9	424.1	7.6	33.7	888.6	40.5	0.42	0.25	2.25	.	4

TABLE 3 - DATA CONVERTED TO A ZERO-MOISTURE BASIS.

OBS	PLNA	PART	CALORIES	WATER	PROTEIN	FAT	TOTLCARB	FIBER
3856	LEONOTIS NEPETAEFOLIA	S	.	0	26.6	38.0	.	.
3857	LEONOTIS NEPETAEFOLIA	S	.	0	25.9	30.7	36.8	15.7
3858	LEONOTIS NEPETIFOLIA	S	.	0	28.8	35.9	.	.
3859	LEONTICE LEONTOPETALUM	S	.	0	23.3	15.0	.	.
3860	LEONTICE LEONTOPETALUM	S	.	0	18.7	12.0	.	.
3861	LEONURUS CARDIACA	S	.	0	22.3	30.2	.	.
3862	LEONURUS SIBIRICUS	S	.	0	10.7	39.1	42.3	2.0
3863	LEPECHINIA SPICATA	S	.	0	14.0	15.2	.	.
3864	LEPECHINIA SPICATA	S	.	0	12.1	16.3	.	.
3865	LEPEDIUM SP.(FLOWERING)	W	.	0	24.4	2.0	65.0	22.3
3866	LEPIDAGATHIS CRISTATA	L	.	0	9.2	3.9	73.4	32.6
3867	LEPIDIUM DENSIFLORUM	S	.	0	27.0	23.6	.	.
3868	LEPIDIUM DRABA	S	.	0	24.1	15.9	.	.
3869	LEPIDIUM LASIOCARPUM	S	.	0	26.3	23.5	.	.
3870	LEPIDIUM LATIFOLIUM	S	.	0	25.6	18.7	.	.
3871	LEPIDIUM MONTANUM VAR. ANGUSTIFOLIUM	S	.	0	28.1	27.7	.	.
3872	LEPIDIUM PERFOLIATUM	S	.	0	29.4	19.2	.	.
3873	LEPIDIUM PERFOLIATUM	S	.	0	22.9	21.9	.	.
3874	LEPIDIUM SATIVUM	S	.	0	31.2	21.9	.	.
3875	LEPIDIUM SATIVUM	L	302	0	24.5	6.6	51.9	10.4
3876	LEPIDIUM SATIVUM	L	.	0	32.8	5.6	49.2	.
3877	LEPIDIUM SATIVUM	S	.	0	24.9	16.9	.	.
3878	LEPIDIUM SP.(IMMATURE)	W	.	0	38.9	1.9	45.7	11.9
3879	LEPIDIUM VIRGINICUM	S	.	0	26.2	19.7	.	.

OBS	ASH	CALCIUM	PHOSPHO	IRON	SODIUM	POTASS	CAROT	THIAMINE	RIBOFLAV	NIACIN	ASCORBIC	REFCO
3856	.	.	.	.	.	.	.	.	.	.	.	21
3857	6.5	.	.	.	.	.	.	.	.	.	.	5
3858	.	.	.	.	.	.	.	.	.	.	.	20
3859	.	.	.	.	.	.	.	.	.	.	.	21
3860	.	.	.	.	.	.	.	.	.	.	.	21
3861	6.4	.	.	.	.	.	.	.	.	.	.	19
3862	7.9	.	.	.	.	.	.	.	.	.	.	5
3863	.	.	.	.	.	.	.	.	.	.	.	21
3864	.	.	.	.	.	.	.	.	.	.	.	21
3865	8.6	1550.0	380.0	.	.	2050	.	.	.	.	.	6
3866	13.4	1970.0	310.0	1.3	.	.	.	.	.	.	.	5
3867	.	.	.	.	.	.	.	.	.	.	.	21
3868	.	.	.	.	.	.	.	.	.	.	.	20
3869	.	.	.	.	.	.	.	.	.	.	.	20
3870	.	.	.	.	.	.	.	.	.	.	.	21
3871	.	.	.	.	.	.	.	.	.	.	.	20
3872	.	.	.	.	.	.	.	.	.	.	.	20
3873	.	.	.	.	.	.	.	.	.	.	.	21
3874	5.7	.	.	.	.	.	.	.	.	.	.	19
3875	17.0	764.2	717.0	12.3	132.1	5717	52641.5	0.75	2.45	9.43	650.9	2
3876	12.4	2033.9	621.5	161.6	.	.	10067.8	0.62	0.96	5.65	491.5	5
3877	6.0	328.7	1749.7	.	.	.	.	.	.	.	.	5
3878	13.5	1950.0	600.0	.	.	3180	.	.	.	.	.	6
3879	.	.	.	.	.	.	.	.	.	.	.	19

TABLE 3 - DATA CONVERTED TO A ZERO-MOISTURE BASIS.

OBS	PLNA	PART	CALORIES	WATER	PROTEIN	FAT	TOTLCARB	FIBER	ASH	CALCIUM	PHOSPHO	IRON	SODIUM	POTASS	CAROT	THIAMINE	RIBOFLAV	NIACIN	ASCORBIC	REFCO
3880	LEPIDIUM VIRGINICUM	S	.	0	28.1	19.2	.	.	.	.	.	.	.	.	.	.	.	.	.	20
3881	LEPTADENIA HASTATA	L	284	0	25.8	1.1	59.5	24.7	13.7	2194.7	494.7	28.4	.	.	25868.4	1.32	1.84	10	410.5	1
3882	LEPTOCHLOA CHINENSIS	L	.	0	9.6	1.5	75.1	31.6	13.8	.	.	.	.	.	.	.	.	.	.	5
3883	LEPTOCHLOA DUBIA	W	.	0	11.6	2.1	76.8	32.4	9.5	400.0	230.0	.	.	1850	.	.	.	.	.	6
3884	LEPTOCHLOA OBTUSIFLORA	W	.	0	18.4	1.7	70.4	28.0	9.5	.	.	.	.	.	.	.	.	.	.	18
3885	LEPTOCHLOA OBTUSIFOLIA	I	.	0	4.3	1.3	86.4	39.2	8.2	561.8	851.8	.	.	.	.	.	.	.	.	5
3886	LEPTOCHLOA PANICEA	L	.	0	6.7	0.9	74.2	30.8	18.2	.	.	.	.	.	.	.	.	.	.	5
3887	LEPTOLOMA COGNATUM	W	.	0	9.7	2.3	79.9	37.1	8.1	540.0	150.0	.	.	2490	.	.	.	.	.	6
3888	LEPTOLOMA COGNATUM(FULL BLOOM)	W	.	0	12.3	1.9	75.9	36.6	9.9	580.0	190.0	.	.	1750	.	.	.	.	.	6
3889	LESPEDEZA BICOLOR	S	.	0	36.6	11.9	.	.	4.7	.	.	.	.	.	.	.	.	.	.	19
3890	LESPEDEZA BICOLOR	S	.	0	39.6	10.6	.	.	5.0	.	.	.	.	.	.	.	.	.	.	19
3891	LESPEDEZA BICOLOR	S	.	0	37.9	11.2	.	.	4.9	.	.	.	.	.	.	.	.	.	.	19
3892	LESPEDEZA BICOLOR	S	.	0	37.2	10.9	.	.	5.1	.	.	.	.	.	.	.	.	.	.	19
3893	LESPEDEZA BICOLOR	S	.	0	36.3	11.0	.	.	5.0	.	.	.	.	.	.	.	.	.	.	19
3894	LESPEDEZA BICOLOR	S	.	0	33.4	12.6	.	.	5.1	.	.	.	.	.	.	.	.	.	.	19
3895	LESPEDEZA CAPITATA	S	.	0	38.3	8.5	.	.	3.4	.	.	.	.	.	.	.	.	.	.	19
3896	LESPEDEZA CAPITATA	S	.	0	5.6	1.6	.	.	1.3	.	.	.	.	.	.	.	.	.	.	19
3897	LESPEDEZA CAPITATA	S	.	0	37.5	10.2	.	.	.	.	.	.	.	.	.	.	.	.	.	19
3898	LESPEDEZA CAPITATA	S	.	0	35.1	5.1	.	.	.	.	.	.	.	.	.	.	.	.	.	21
3899	LESPEDEZA CUNEATA	W	.	0	16.4	2.0	68.8	32.0	12.8	.	.	.	.	.	.	.	.	.	.	18
3900	LESPEDEZA CUNEATA	H	.	0	11.3	2.3	83.0	42.6	3.4	.	.	.	.	.	.	.	.	.	.	18
3901	LESPEDEZA CUNEATA	S	.	0	40.5	5.9	.	.	3.4	.	.	.	.	.	.	.	.	.	.	19
3902	LESPEDEZA CUNEATA	S	.	0	39.9	7.1	.	.	3.7	.	.	.	.	.	.	.	.	.	.	19
3903	LESPEDEZA CUNEATA	S	.	0	38.1	7.5	.	.	4.6	.	.	.	.	.	.	.	.	.	.	19
3904	LESPEDEZA CUNEATA	H	.	0	15.2	3.0	75.7	27.5	6.1	1460.0	230.0	29.0	.	1030	.	.	.	.	.	6
3905	LESPEDEZA CYRTOBOTRYA	S	.	0	33.2	14.2	.	.	5.5	.	.	.	.	.	.	.	.	.	.	19
3906	LESPEDEZA HEDYSARIOIDES	S	.	0	37.4	8.2	.	.	4.5	.	.	.	.	.	.	.	.	.	.	19
3907	LESPEDEZA HEDYSARIOIDES	S	.	0	35.8	7.4	.	.	3.7	.	.	.	.	.	.	.	.	.	.	19
3908	LESPEDEZA HEDYSARIOIDES	S	.	0	35.1	7.8	.	.	5.6	.	.	.	.	.	.	.	.	.	.	19
3909	LESPEDEZA JAPONICA VAR. INTERMEDIA	S	.	0	41.2	9.9	.	.	4.6	.	.	.	.	.	.	.	.	.	.	19
3910	LESPEDEZA LATISSIMA	S	.	0	36.1	8.2	.	.	5.0	.	.	.	.	.	.	.	.	.	.	19
3911	LESPEDEZA SP.	W	.	0	16.2	2.7	72.7	30.8	8.4	1140.0	310.0	31.0	.	1280	.	.	.	.	.	6
3912	LESPEDEZA STIPULACEA	S	.	0	52.0	7.1	.	.	4.4	.	.	.	.	.	.	.	.	.	.	19
3913	LESPEDEZA STIPULACEA	H	.	0	14.7	3.5	75.2	30.6	6.6	1000.0	220.0	21.0	.	1040	4.0	0.96	.	.	.	6
3914	LESPEDEZA STRIATA	S	.	0	51.9	7.3	.	.	4.5	.	.	.	.	.	.	.	.	.	.	19
3915	LESPEDEZA STRIATA	S	.	0	26.4	4.3	.	.	4.6	.	.	.	.	.	.	.	.	.	.	19
3916	LESPEDEZA STRIATA	H	.	0	13.7	2.7	78.1	32.0	5.5	1050.0	200.0	31.0	.	1050	5.0	1.00	.	.	.	6
3917	LESPEDEZA STRIATA	W	.	0	15.3	2.1	71.4	36.1	11.2	1130.0	270.0	32.0	.	1160	.	.	.	.	.	6
3918	LESQUERELLA ANGUSTIFOLIA	S	.	0	25.0	26.2	.	.	.	.	.	.	.	.	.	.	.	.	.	20
3919	LESQUERELLA ARGYRAEA	S	.	0	23.8	26.4	.	.	.	.	.	.	.	.	.	.	.	.	.	20
3920	LESQUERELLA ARGYRAEA	S	.	0	20.0	27.4	.	.	.	.	.	.	.	.	.	.	.	.	.	20
3921	LESQUERELLA DENSIPILA	S	.	0	20.6	24.4	.	.	.	.	.	.	.	.	.	.	.	.	.	20
3922	LESQUERELLA DENSIPILA	S	.	0	.	27.4	.	.	.	.	.	.	.	.	.	.	.	.	.	21
3923	LESQUERELLA ENGELMANNII	S	.	0	21.2	11.1	.	.	.	.	.	.	.	.	.	.	.	.	.	20
3924	LESQUERELLA ENGELMANNII	S	.	0	21.2	21.2	.	.	.	.	.	.	.	.	.	.	.	.	.	20

TABLE 3 - DATA CONVERTED TO A ZERO-MOISTURE BASIS.

OBS	PLNA	PART	CALORIES	WATER	PROTEIN	FAT	TOTLCARB	FIBER	ASH	CALCIUM	PHOSPHO	IRON	SODIUM	POTASS	CAROT	THIAMINE	RIBOFLAV	NIACIN	ASCORBIC	REFCO
3925	LESQUERELLA ENGELMANNII	S	.	0	21.2	19.8	.	.	.	.	.	.	.	.	.	.	.	.	.	20
3926	LESQUERELLA FENDLERI	S	.	0	21.2	19.8	.	.	.	.	.	.	.	.	.	.	.	.	.	20
3927	LESQUERELLA FENDLERI	S	.	0	22.5	28.1	.	.	.	.	.	.	.	.	.	.	.	.	.	20
3928	LESQUERELLA FENDLERI	S	.	0	20.6	22.8	.	.	.	.	.	.	.	.	.	.	.	.	.	20
3929	LESQUERELLA FENDLERI	S	.	0	21.0	24.6	.	.	.	.	.	.	.	.	.	.	.	.	.	21
3930	LESQUERELLA FENDLERI	S	.	0	24.1	21.6	.	.	.	.	.	.	.	.	.	.	.	.	.	21
3931	LESQUERELLA FENDLERI	S	.	0	20.9	27.6	.	.	.	.	.	.	.	.	.	.	.	.	.	21
3932	LESQUERELLA FENDLERI	S	.	0	25.8	26.2	.	.	.	.	.	.	.	.	.	.	.	.	.	21
3933	LESQUERELLA FENDLERI	S	.	0	26.1	25.8	.	.	.	.	.	.	.	.	.	.	.	.	.	21
3934	LESQUERELLA GLOBOSA	S	.	0	24.4	39.4	.	.	.	.	.	.	.	.	.	.	.	.	.	20
3935	LESQUERELLA GORDONII	S	.	0	23.1	28.8	.	.	.	.	.	.	.	.	.	.	.	.	.	20
3936	LESQUERELLA GORDONII	S	.	0	23.1	26.0	.	.	.	.	.	.	.	.	.	.	.	.	.	20
3937	LESQUERELLA GRANDIFLORA	S	.	0	19.4	37.2	.	.	.	.	.	.	.	.	.	.	.	.	.	20
3938	LESQUERELLA GRANDIFLORA	S	.	0	21.2	26.8	.	.	.	.	.	.	.	.	.	.	.	.	.	20
3939	LESQUERELLA LASIOCARPA	S	.	0	19.4	28.6	.	.	.	.	.	.	.	.	.	.	.	.	.	19
3940	LESQUERELLA LASIOCARPA	S	.	0	22.5	30.4	.	.	.	.	.	.	.	.	.	.	.	.	.	20
3941	LESQUERELLA LESCURII	S	.	0	21.2	28.1	.	.	.	.	.	.	.	.	.	.	.	.	.	20
3942	LESQUERELLA LESCURII	S	.	0	18.9	28.7	.	.	.	.	.	.	.	.	.	.	.	.	.	21
3943	LESQUERELLA LINDHEIMERI	S	.	0	21.2	25.5	.	.	.	.	.	.	.	.	.	.	.	.	.	19
3944	LESQUERELLA LINDHEIMERI	S	.	0	.	21.4	.	.	.	.	.	.	.	.	.	.	.	.	.	20
3945	LESQUERELLA LYRATA	S	.	0	18.4	25.9	.	.	.	.	.	.	.	.	.	.	.	.	.	21
3946	LESQUERELLA OVALIFOLIA	S	.	0	24.4	24.0	.	.	.	.	.	.	.	.	.	.	.	.	.	20
3947	LESQUERELLA PERFORATA	S	.	0	18.0	23.4	.	.	.	.	.	.	.	.	.	.	.	.	.	21
3948	LESQUERELLA PINETORUM	S	.	0	22.5	26.6	.	.	.	.	.	.	.	.	.	.	.	.	.	20
3949	LESQUERELLA SP.	W	.	0	21.4	2.2	65.8	17.4	10.6	4640	250	.	.	1680	.	.	.	.	.	6
3950	LESQUERELLA SP.	W	.	0	27.0	1.6	56.5	15.3	14.9	5020	350	.	.	2060	.	.	.	.	.	6
3951	LESQUERELLA STONENSIS	S	.	0	17.2	23.5	.	.	.	.	.	.	.	.	.	.	.	.	.	21
3952	LESSERTIA BRACHYSTACHYA	S	.	0	32.8	2.0	.	.	.	.	.	.	.	.	.	.	.	.	.	21
3953	LESSINGIA GERMANORUM	W	.	0	7.6	.	.	25.5	7.9	990	280	.	.	2180	.	.	.	.	.	6
3954	LEUCADENDRON AURANTIACUM	S	.	0	24.3	13.3	.	.	.	.	.	.	.	.	.	.	.	.	.	21
3955	LEUCADENDRON CORYMBOSUM	S	.	0	35.6	16.5	.	.	.	.	.	.	.	.	.	.	.	.	.	21
3956	LEUCADENDRON DREGEI	S	.	0	20.5	13.0	.	.	.	.	.	.	.	.	.	.	.	.	.	21
3957	LEUCADENDRON GALPINII	S	.	0	25.1	12.0	.	.	.	.	.	.	.	.	.	.	.	.	.	21
3958	LEUCADENDRON PLUMOSUM	S	.	0	56.5	21.2	.	.	.	.	.	.	.	.	.	.	.	.	.	21
3959	LEUCADENDRON PSEUDOSPATHULATUM	S	.	0	50.5	41.4	.	.	.	.	.	.	.	.	.	.	.	.	.	21
3960	LEUCADENDRON PSEUDOSPATHULATUM	S	.	0	53.8	42.0	.	.	.	.	.	.	.	.	.	.	.	.	.	21
3961	LEUCADENDRON TORTUM	S	.	0	69.3	28.5	.	.	.	.	.	.	.	.	.	.	.	.	.	21
3962	LEUCADENDRON TRUNCATUM	S	.	0	59.1	34.7	.	.	.	.	.	.	.	.	.	.	.	.	.	21
3963	LEUCADENDRON VENOSUM	S	.	0	56.3	39.9	.	.	.	.	.	.	.	.	.	.	.	.	.	21
3964	LEUCAENA GLAUCA	S	.	0	30.8	6.8	.	.	3.8	.	.	.	.	.	.	.	.	.	.	19
3965	LEUCAENA GLAUCA	S	.	0	32.5	6.4	.	.	.	.	.	.	.	.	.	.	.	.	.	19
3966	LEUCAENA LANCEOLATA	S	.	0	38.1	7.6	.	.	.	.	.	.	.	.	.	.	.	.	.	20
3967	LEUCAENA LANCEOLATA	S	.	0	36.9	4.7	.	.	.	.	.	.	.	.	.	.	.	.	.	20
3968	LEUCAENA LANCEOLATA	S	.	0	36.2	5.2	.	.	.	.	.	.	.	.	.	.	.	.	.	21
3969	LEUCAENA LEUCOCEPHALA	L	.	0	21.0	6.5	64.1	18.1	8.4	.	.	.	.	.	.	.	.	.	.	18

TABLE 3 - DATA CONVERTED TO A ZERO-MOISTURE BASIS.

OBS	PLNA	PART	CALORIES	WATER	PROTEIN	FAT	TOTLCARB	FIBER	ASH
3970	LEUCAENA LEUCOCEPHALA	C	.	0	27.8	3.2	65.5	10.4	3.5
3971	LEUCAENA LEUCOCEPHALA	SH	.	0	24.2	2.7	64.2	24.2	8.9
3972	LEUCAENA LEUCOCEPHALA	F	.	0	21.7	1.4	71.1	25.6	5.8
3973	LEUCAENA LEUCOCEPHALA	S	.	0	35.8	7.5	52.3	11.4	4.4
3974	LEUCAENA LEUCOCEPHALA	S	.	0	28.1	10.0	.	.	.
3975	LEUCAENA LEUCOCEPHALA	H	.	0	14.3	2.1	77.9	33.6	5.7
3976	LEUCAENA LEUCOCEPHALA	L	.	0	18.0	2.0	74.5	33.0	6.1
3977	LEUCAENA LEUCOCEPHALA	L	332	0	14.1	3.9	74.6	8.8	7.3
3978	LEUCAENA LEUCOCEPHALA	SH	306	0	43.5	4.7	45.6	19.7	6.2
3979	LEUCAENA LEUCOCEPHALA(IMMATURE)	L	.	0	18.3	2.1	72.9	37.7	6.7
3980	LEUCAENA LEUCOCEPHALA(MATURE)	L	.	0	12.8	1.2	78.4	43.4	7.6
3981	LEUCAENA RETUSA	S	.	0	29.4	7.5	.	.	.
3982	LEUCAENA RETUSA	S	.	0	32.5	6.5	.	.	.
3983	LEUCAENA RETUSA	S	.	0	12.3	9.2	.	.	.
3984	LEUCAENA RETUSA(IMMATURE)	L	.	0	17.6	3.9	72.1	24.4	6.4
3985	LEUCAS MICROPHYLLA	SH	.	0	11.4	3.3	76.5	34.1	8.8
3986	LEUCOPHYLUM SP.	W	.	0	7.0	5.8	83.5	25.8	3.7
3987	LEUCOSPAERA BAINESII	L	.	0	11.7	2.0	76.1	29.0	10.2
3988	LEUCOSPERMUM ATTENUATUM	S	.	0	49.8	44.1	.	.	.
3989	LEUCOSPERMUM CATHERINAE	S	.	0	47.6	46.8	.	.	.
3990	LEUCOSPERMUM CONOCARPUM	S	.	0	56.9	40.8	.	.	.
3991	LEUCOSPERMUM NUTANS	S	.	0	52.4	44.7	.	.	.
3992	LEVISTICUM OFFICINALE	S	.	0	20.0	14.7	.	.	.
3993	LEWISIA REDIVIVA	R	.	0	3.0	0.4	92.3	6.4	4.3

OBS	CALCIUM	PHOSPHO	IRON	SODIUM	POTASS	CAROT	THIAMINE	RIBOFLAV	NIACIN	ASCORBIC	REFCO
3970	.	.	.	.	.	.	.	.	.	.	18
3971	.	.	.	.	.	.	.	.	.	.	18
3972	.	.	.	.	.	.	.	.	.	.	18
3973	.	.	.	.	.	.	.	.	.	.	18
3974	.	.	.	.	.	.	.	.	.	.	20
3975	.	.	.	.	.	.	.	.	.	.	6
3976	.	.	.	.	.	.	.	.	.	.	
3977	2697.6	248.8	.	.	.	.	.	.	.	.	1
3978	709.8	57.0	47.7	.	.	24507.8	.	0.47	27.98	41.5	2
3979	.	.	.	.	.	.	.	.	.	.	6
3980	.	.	.	.	.	.	.	.	.	.	6
3981	.	.	.	.	.	.	.	.	.	.	19
3982	.	.	.	.	.	.	.	.	.	.	20
3983	.	.	.	.	.	.	.	.	.	.	20
3984	2810.0	120.0	.	.	1300.0	.	.	.	.	.	6
3985	910.0	620.0	.	.	.	.	.	.	.	.	18
3986	1140.0	100.0	.	.	700.0	.	.	.	.	.	6
3987	1750.0	70.0	.	.	.	.	.	.	.	.	18
3988	.	.	.	.	.	.	.	.	.	.	21
3989	.	.	.	.	.	.	.	.	.	.	21
3990	.	.	.	.	.	.	.	.	.	.	21
3991	.	.	.	.	.	.	.	.	.	.	21
3992	.	.	.	.	.	.	.	.	.	.	20
3993	79.9	.	6.0	.	.	.	0.38	0.09	.	115.4	9

TABLE 3 - DATA CONVERTED TO A ZERO-MOISTURE BASIS.

OBS	PLNA	PART	CALORIES	WATER	PROTEIN	FAT	TOTLCARB	FIBER
3994	LIATRIS PUNCTATA	S	.	0	47.2	29.4	.	.
3995	LIATRIS PUNCTATA	S	.	0	15.1	9.2	.	.
3996	LIATRIS PUNCTATA		.	0	7.5	8.8	73.7	29.4
3997	LIATRIS PYCNOSTACHYA	S	.	0	34.2	29.6	.	.
3998	LIATRIS PYCNOSTACHYA	S	.	0	15.0	3.7	.	.
3999	LIATRIS SP.	S	.	0	46.6	26.4	.	.
4000	LIATRIS SPICATA	S	.	0	35.6	35.9	.	.
4001	LIATRIS SPICATA	S	.	0	11.1	10.5	.	.
4002	LIBOCEDRUS DECURRENS	S	.	0	11.9	45.3	.	.
4003	LICANIA PLATYPUS	F	356	0	4.3	1.5	91.1	2.8
4004	LICANIA RIGIDA	S	.	0	8.3	49.4	.	.
4005	LIGUSTICUM PORTERI	S	.	0	27.5	20.5	.	.
4006	LIGUSTICUM PORTERI	S	.	0	26.4	23.9	.	.
4007	LIGUSTICUM PORTERI	I	.	0	18.9	2.7	67.2	15.9
4008	LIGUSTICUM PORTERI	L	.	0	16.1	3.0	68.1	14.4
4009	LIGUSTICUM SP.	W	.	0	28.5	3.0	57.0	12.8
4010	LIGUSTICUM SP.	L	.	0	18.0	3.2	66.1	14.0
4011	LIGUSTRUM IBOTA	S	.	0	14.1	22.0	.	.
4012	LIGUSTRUM JAPONICUM	S	.	0	14.4	25.7	.	.
4013	LIGUSTRUM JAPONICUM	S	.	0	12.2	20.1	.	.
4014	LILIUM LANCIFOLIUM	R	332	0	14.1	1.8	80.0	3.2
4015	LILIUM LANCIFOLIUM	R	368	0	9.2	2.1	84.8	1.6
4016	LILIUM PONTICUM	S	.	0	14.2	11.8	.	.
4017	LIMNANTHES ALBA VAR. ALBA	S	.	0	19.9	26.8	.	.

OBS	ASH	CALCIUM	PHOSPHO	IRON	SODIUM	POTASS	CAROT	THIAMINE	RIBOFLAV	NIACIN	ASCORBIC	REFCO
3994	3.5	.	.	.	.	.	.	.	.	.	.	19
3995	5.4	.	.	.	.	.	.	.	.	.	.	19
3996	10.0	.	.	.	.	.	.	.	.	.	.	6
3997	3.4	.	.	.	.	.	.	.	.	.	.	19
3998	.	.	.	.	.	.	.	.	.	.	.	20
3999	4.3	.	.	.	.	.	.	.	.	.	.	19
4000	4.5	.	.	.	.	.	.	.	.	.	.	19
4001	5.2	.	.	.	.	.	.	.	.	.	.	19
4002	.	.	.	.	.	.	.	.	.	.	.	20
4003	3.1	33.7	89.0	5.2	.	.	766.9	0.06	0.03	4.60	33.7	3
4004	.	.	.	.	.	.	.	.	.	.	.	21
4005	.	.	.	.	.	.	.	.	.	.	.	20
4006	.	.	.	.	.	.	.	.	.	.	.	20
4007	11.2	.	.	.	.	.	.	.	.	.	.	6
4008	12.8	.	.	.	.	.	.	.	.	.	.	6
4009	11.5	.	.	.	.	.	.	.	.	.	.	6
4010	12.7	.	.	.	.	.	.	.	.	.	.	6
4011	.	.	.	.	.	.	.	.	.	.	.	21
4012	.	.	.	.	.	.	.	.	.	.	.	20
4013	.	.	.	.	.	.	.	.	.	.	.	20
4014	4.1	14.7	705.9	1.5	29.4	.	0.0	0.15	0.12	0.59	58.8	2
4015	3.9	35.1	142.7	5.8	102.9	801.2	0.0	0.09	0.09	2.81	19.9	2
4016	.	.	.	.	.	.	.	.	.	.	.	21
4017	.	.	.	.	.	.	.	.	.	.	.	21

TABLE 3 - DATA CONVERTED TO A ZERO-MOISTURE BASIS.

OBS	PLNA	PART	CALORIES	WATER	PROTEIN	FAT	TOTLCARB	FIBER
4018	LIMNANTHES ALBA VAR. VERSICOLOR	S	.	0	19.2	30.9	.	.
4019	LIMNANTHES BAKERI	S	.	0	15.9	25.6	.	.
4020	LIMNANTHES DOUGLASII	S	.	0	24.7	24.5	.	.
4021	LIMNANTHES DOUGLASII	S	.	0	25.6	25.2	.	.
4022	LIMNANTHES DOUGLASII	S	.	0	22.5	27.4	.	.
4023	LIMNANTHES DOUGLASII	S	.	0	20.6	29.1	.	.
4024	LIMNANTHES DOUGLASII	S	.	0	23.1	25.6	.	.
4025	LIMNANTHES DOUGLASII	S	.	0	23.8	24.0	.	.
4026	LIMNANTHES DOUGLASII VAR. DOUGLASII	S	.	0	15.8	29.8	.	.
4027	LIMNANTHES DOUGLASII VAR. DOUGLASII	S	.	0	17.0	24.5	.	.
4028	LIMNANTHES DOUGLASII VAR. DOUGLASII	S	.	0	20.2	28.4	.	.
4029	LIMNANTHES DOUGLASII VAR. NIVEA	S	.	0	17.4	26.1	.	.
4030	LIMNANTHES DOUGLASII VAR. NIVEA	S	.	0	19.2	29.8	.	.
4031	LIMNANTHES DOUGLASII VAR. NIVEA	S	.	0	22.2	30.4	.	.
4032	LIMNANTHES DOUGLASII VAR. NIVEA	S	.	0	14.9	33.2	.	.
4033	LIMNANTHES DOUGLASII VAR. ROSEA	S	.	0	16.4	20.2	.	.
4034	LIMNANTHES DOUGLASII VAR. ROSEA	S	.	0	16.8	21.7	.	.
4035	LIMNANTHES FLOCCOSSA	S	.	0	24.6	28.3	.	.
4036	LIMNANTHES GRACILIS VAR. GRACILIS	S	.	0	16.6	28.8	.	.
4037	LIMNANTHES GRACILIS VAR. PARISHII	S	.	0	18.2	33.3	.	.
4038	LIMNANTHES MONTANA	S	.	0	20.7	25.8	.	.
4039	LIMNANTHES STRIATA	S	.	0	20.3	29.0	.	.
4040	LIMNOCHARIS FLAVA	L	330	0	17.0	2.0	77.0	.
4041	LIMNODEA ARKANSANA	W	.	0	11.4	2.4	79.6	29.2

OBS	ASH	CALCIUM	PHOSPHO	IRON	SODIUM	POTASS	CAROT	THIAMINE	RIBOFLAV	NIACIN	ASCORBIC	REFCO
4018	.	.	.	.	.	.	.	.	.	.	.	21
4019	.	.	.	.	.	.	.	.	.	.	.	21
4020	7.3	.	.	.	.	.	.	.	.	.	.	19
4021	.	.	.	.	.	.	.	.	.	.	.	19
4022	.	.	.	.	.	.	.	.	.	.	.	19
4023	.	.	.	.	.	.	.	.	.	.	.	19
4024	.	.	.	.	.	.	.	.	.	.	.	20
4025	.	.	.	.	.	.	.	.	.	.	.	20
4026	.	.	.	.	.	.	.	.	.	.	.	21
4027	.	.	.	.	.	.	.	.	.	.	.	21
4028	.	.	.	.	.	.	.	.	.	.	.	21
4029	.	.	.	.	.	.	.	.	.	.	.	21
4030	.	.	.	.	.	.	.	.	.	.	.	21
4031	.	.	.	.	.	.	.	.	.	.	.	21
4032	.	.	.	.	.	.	.	.	.	.	.	21
4033	.	.	.	.	.	.	.	.	.	.	.	21
4034	.	.	.	.	.	.	.	.	.	.	.	21
4035	.	.	.	.	.	.	.	.	.	.	.	21
4036	.	.	.	.	.	.	.	.	.	.	.	21
4037	.	.	.	.	.	.	.	.	.	.	.	21
4038	.	.	.	.	.	.	.	.	.	.	.	21
4039	.	.	.	.	.	.	.	.	.	.	.	21
4040	4.0	620.0	330.0	21.0	.	.	22800.0	0.70	.	.	540.0	2
4041	6.6	340.0	190.0	.	.	1770.0	.	.	.	.	.	6

TABLE 3 - DATA CONVERTED TO A ZERO-MOISTURE BASIS.

OBS	PLNA	PART	CALORIES	WATER	PROTEIN	FAT	TOTLCARB	FIBER	ASH	CALCIUM	PHOSPHO	IRON	SODIUM	POTASS	CAROT	THIAMINE	RIBOFLAV	NIACIN	ASCORBIC	REFCO
4042	LIMNOPHILA SP.	L	308	0	17.9	3.8	66.7	.	11.5	974.4	307.7	44.9	1576.9	1756.4	25641.0	0.38	1.79	7.69	12.8	2
4043	LIMNOSCIADIUM PUMILUM	S	.	0	12.5	16.2	.	.	.	.	.	.	.	.	.	.	.	.	.	20
4044	LIMONIUM GMELINII	S	.	0	10.3	2.3	.	.	.	.	.	.	.	.	.	.	.	.	.	21
4045	LIMONIUM HUMILE	S	.	0	33.5	7.8	.	.	.	.	.	.	.	.	.	.	.	.	.	21
4046	LIMONIUM SINUATUM	S	.	0	36.6	9.7	.	.	5.9	.	.	.	.	.	.	.	.	.	.	19
4047	LIMONIUM SINUATUM	S	.	0	7.8	1.6	.	.	5.0	.	.	.	.	.	.	.	.	.	.	19
4048	LIMONIUM TETARICUM	S	.	0	33.8	5.3	.	.	4.4	.	.	.	.	.	.	.	.	.	.	19
4049	LIMONIUM TETARICUM	S	.	0	12.3	2.3	.	.	5.3	.	.	.	.	.	.	.	.	.	.	19
4050	LINARIA LANIGERA	S	.	0	20.6	43.0	.	.	.	.	.	.	.	.	.	.	.	.	.	21
4051	LINARIA MAROCCANA	S	.	0	21.1	43.1	.	.	4.7	.	.	.	.	.	.	.	.	.	.	19
4052	LINARIA NIVEA	S	.	0	15.8	50.7	.	.	.	.	.	.	.	.	.	.	.	.	.	21
4053	LINARIA PYRAMIDATA	S	.	0	22.4	44.4	.	.	.	.	.	.	.	.	.	.	.	.	.	21
4054	LINDERA BENZOIN	S	.	0	21.2	61.3	.	.	.	.	.	.	.	.	.	.	.	.	.	19
4055	LINDERA PRAECOX	S	.	0	12.2	29.7	.	.	.	.	.	.	.	.	.	.	.	.	.	21
4056	LINUM AUSTRALE	S	.	0	22.2	28.8	.	.	.	.	.	.	.	.	.	.	.	.	.	20
4057	LINUM BIENNE	S	.	0	24.2	29.9	.	.	.	.	.	.	.	.	.	.	.	.	.	21
4058	LINUM GRANDIFLORUM	S	.	0	23.1	28.2	.	.	.	.	.	.	.	.	.	.	.	.	.	20
4059	LINUM HIRSUTUM	S	.	0	24.2	34.1	.	.	.	.	.	.	.	.	.	.	.	.	.	21
4060	LINUM PERENNE	S	.	0	23.1	28.8	.	.	.	.	.	.	.	.	.	.	.	.	.	20
4061	LINUM SP.	S	.	0	16.2	19.4	.	.	.	.	.	.	.	.	.	.	.	.	.	20
4062	LINUM STRICTUM	S	.	0	23.2	32.6	.	.	.	.	.	.	.	.	.	.	.	.	.	21
4063	LINUM USITATISSIMUM	H	.	0	7.6	6.1	80.1	36.5	6.2	.	30.0	.	.	.	.	.	.	.	.	18
4064	LINUM USITATISSIMUM	S	.	0	29.2	36.0	29.9	4.8	4.9	200.0	600.0	.	.	.	.	.	.	.	.	18
4065	LINUM USITATISSIMUM	S	.	0	25.6	40.4	.	.	3.7	.	.	.	.	.	.	.	.	.	.	19
4066	LINUM USITATISSIMUM	S	.	0	26.6	4.0	.	.	.	.	.	.	.	.	.	.	.	.	.	21
4067	LINUM USITATISSIMUM	H	.	0	7.8	3.3	81.7	46.2	7.2	720.0	110.0	.	.	1740.0	.	.	.	.	.	6
4068	LINUM USITATISSIMUM	S	531	0	19.2	36.3	39.7	9.4	4.8	289.2	493.1	46.7	.	.	0.0	0.18	0.17	1.49	.	1
4069	LINUM USITATISSIMUM	S	.	0	21.7	39.7	36.0	5.1	2.6	182.0	396.1	2.9	.	.	32.1	.	.	.	.	5
4070	LIPPIA DURANGENSE	S	.	0	18.1	23.8	.	.	.	.	.	.	.	.	.	.	.	.	.	20
4071	LIQUIDAMBAR STYRACIFLUA	S	.	0	33.1	29.7	.	.	5.6	.	.	.	.	.	.	.	.	.	.	19
4072	LISAEA PAPYRACEA	S	.	0	12.4	17.2	.	.	.	.	.	.	.	.	.	.	.	.	.	21
4073	LITCHI CHINENSIS	F	354	0	5.1	1.3	90.5	3.2	3.2	50.6	139.2	2.5	.	.	.	.	.	.	.	1
4074	LITCHI CHINENSIS	F	363	0	4.5	2.2	91.1	1.1	2.2	55.9	162.0	1.7	16.8	949.7	0.0	0.28	0.34	3.35	279.3	2
4075	LITCHI CHINENSIS	F	355	0	4.7	0.6	92.9	0.6	1.8	29.6	183.4	6.5	.	.	0.0	0.12	0.18	5.33	426.0	3
4076	LITCHI CHINENSIS	F	354	0	5.0	1.7	90.6	1.7	2.8	44.2	232.0	2.2	16.6	939.2	.	.	0.28	.	232.0	4
4077	LITHOSPERMUM OFFICINALE	S	.	0	34.6	54.3	7.7	2.8	3.7	.	.	.	.	.	.	.	.	.	.	5
4078	LITHOSPERMUM RUDERALE	W	.	0	27.5	3.9	54.3	12.6	14.3	.	.	.	.	.	.	.	.	.	.	6
4079	LITHOSPERMUM RUDERALE	L	.	0	17.2	3.0	64.5	15.8	15.3	2520.0	360.0	.	.	.	.	.	.	.	.	6
4080	LITHOSPERMUM TENUIFLORUM	S	.	0	20.1	16.4	.	.	.	.	.	.	.	.	.	.	.	.	.	21
4081	LITHRAEA BRASILIENSIS	S	.	0	7.1	11.0	.	.	.	.	.	.	.	.	.	.	.	.	.	21
4082	LITSEA GLAUCESCENS	S	.	0	15.0	10.9	.	.	.	.	.	.	.	.	.	.	.	.	.	20
4083	LITSEA GUATEMALENSIS	L	358	0	14.9	7.6	72.2	25.8	5.3	872.8	123.9	16.3	.	.	9021.7	0.11	0.71	2.72	.	3
4084	LOBELIA ERINUS	S	.	0	21.9	46.7	.	.	2.9	.	.	.	.	.	.	.	.	.	.	19
4085	LOBULARIA MARITIMA	S	.	0	26.2	35.6	.	.	.	.	.	.	.	.	.	.	.	.	.	20
4086	LOBULARIA MARITIMA	S	.	0	26.9	33.1	.	.	.	.	.	.	.	.	.	.	.	.	.	20

TABLE 3 - DATA CONVERTED TO A ZERO-MOISTURE BASIS.

OBS	PLNA	PART	CALORIES	WATER	PROTEIN	FAT	TOTCARB	FIBER	ASH	CALCIUM	PHOSPHO	IRON	SODIUM	POTASS	CAROTE	THIAMINE	RIBOFLAV	NIACIN	ASCORBIC	REFCO
4087	LOLIUM MULTIFLORUM	S	.	0	15.0	1.6	.	.	3.3	.	.	.	.	.	.	.	.	.	.	19
4088	LOLIUM MULTIFLORUM	S	.	0	2.1	2.0	.	.	4.4	.	.	.	.	.	.	.	.	.	.	19
4089	LOLIUM MULTIFLORUM	S	.	0	11.4	1.4	.	.	3.9	.	.	.	.	.	.	.	.	.	.	19
4090	LOLIUM MULTIFLORUM	W	.	0	18.5	4.2	58.1	20.0	19.2	640.0	410.0	.	.	2000.0	.	.	.	.	.	6
4091	LOLIUM MULTIFLORUM	H	.	0	8.0	1.8	81.9	32.6	8.3	620.0	410.0	.	.	1990.0	.	.	.	.	.	6
4092	LOLIUM MULTIFLORUM	L	.	0	11.4	4.8	74.5	25.1	9.2	479.7	295.2	.	.	1476.0	.	.	.	.	.	5
4093	LOLIUM MULTIFLORUM	L	.	0	9.1	2.1	80.2	31.4	8.5	.	.	.	.	.	.	.	.	.	.	5
4094	LOLIUM PERENNE	S	.	0	18.8	1.8	.	.	.	.	.	.	.	.	.	.	.	.	.	20
4095	LOLIUM PERENNE	W	.	0	24.7	3.9	55.4	18.7	16.0	600.0	480.0	.	.	.	.	.	.	.	.	6
4096	LOLIUM PERENNE	H	.	0	11.1	3.4	76.1	29.1	9.4	400.0	260.0	.	.	1910.0	.	.	.	.	.	6
4097	LOLIUM PERENNE	L	.	0	11.3	4.9	74.8	25.2	9.0	451.1	263.2	.	.	1917.3	.	.	.	.	.	5
4098	LOLIUM PERENNE	L	.	0	10.5	3.5	76.8	27.5	9.2	.	.	.	.	.	.	.	.	.	.	5
4099	LOLIUM SP.	W	.	0	16.2	3.9	66.7	23.2	13.2	550.0	450.0	.	.	3400.0	40	.	.	.	.	6
4100	LOLIUM TEMULENTUM	S	.	0	16.2	1.4	.	.	.	.	.	.	.	.	.	.	.	.	.	20
4101	LOMATIUM CANBYI	R	.	0	2.8	0.4	92.2	7.8	4.6	169.8	.	5.3	.	.	.	0.39	0.21	.	71.2	9
4102	LOMATIUM COUS	R	.	0	3.0	1.2	91.2	7.9	4.6	240.1	.	10.9	.	.	.	0.18	0.21	.	51.7	9
4103	LOMATIUM DAUCIFOLIUM	S	.	0	21.0	15.2	.	.	.	.	.	.	.	.	.	.	.	.	.	20
4104	LOMATIUM DAUCIFOLIUM	S	.	0	22.1	17.6	.	.	.	.	.	.	.	.	.	.	.	.	.	20
4105	LOMATIUM MACROCARPUM	S	.	0	23.9	23.5	.	.	7.0	.	.	.	.	.	.	.	.	.	.	19
4106	LOMATIUM MACROCARPUM	S	.	0	8.2	8.4	.	.	11.9	.	.	.	.	.	.	.	.	.	.	19
4107	LOMATIUM NUDICAULE	S	.	0	18.1	18.4	.	.	.	.	.	.	.	.	.	.	.	.	.	20
4108	LOMATIUM NUDICAULE	L	.	0	.	.	.	.	.	300.0	.	5.7	.	.	.	0.16	0.66	.	541.0	9
4109	LOMATIUM SP.	W	.	0	23.0	5.0	61.5	14.6	10.5	.	.	.	.	.	.	.	.	.	.	6
4110	LONAS ANNUS	S	.	0	25.6	27.8	.	.	.	.	.	.	.	.	.	.	.	.	.	20
4111	LONICERA INVOLUCRATA(IMMATURE)	L	.	0	18.2	2.4	72.6	10.5	6.8	2730.0	620.0	.	.	.	.	.	.	.	.	6
4112	LONICERA INVOLUCRATA(MATURE)	L	.	0	9.6	5.3	74.1	11.6	11.0	1050.0	420.0	.	.	.	.	.	.	.	.	6
4113	LONICERA JAPONICA	S	.	0	21.0	26.8	.	.	.	.	.	.	.	.	.	.	.	.	.	21
4114	LONICERA ORIENTALIS	S	.	0	7.2	16.5	.	.	.	.	.	.	.	.	.	.	.	.	.	21
4115	LONICERA SP.	L	.	0	12.0	3.5	76.1	17.0	8.4	1420.0	300.0	.	.	2390.0	.	.	.	.	.	6
4116	LONICERA TARTARICA	S	.	0	20.9	32.4	.	.	3.4	.	.	.	.	.	.	.	.	.	.	19
4117	LOPHIRA LANCEOLATA	S	595	0	15.0	47.9	34.8	1.3	2.3	109.8	169.6	.	.	.	.	.	.	.	.	1
4118	LOPHOPOGON TRIDENTATUS	L	.	0	2.6	1.6	74.6	31.7	21.3	.	.	.	.	.	.	.	.	.	.	5
4119	LOTONONIS BAINESII	W	.	0	19.3	4.0	68.6	27.0	8.1	650.0	320.0	.	.	.	.	.	.	.	.	18
4120	LOTUS AMERICANUS	S	.	0	48.3	6.5	.	.	3.2	.	.	.	.	.	.	.	.	.	.	19
4121	LOTUS AMERICANUS	W	.	0	15.0	1.9	75.5	22.5	7.6	1650.0	210.0	.	.	1680.0	.	.	.	.	.	6
4122	LOTUS ARABICUS	S	.	0	36.0	5.3	.	.	6.0	.	.	.	.	.	.	.	.	.	.	19
4123	LOTUS CORNICULATUS	W	.	0	13.4	4.3	73.3	26.4	8.8	1130.0	230.0	.	.	.	.	.	.	.	.	18
4124	LOTUS CORNICULATUS	S	.	0	35.5	5.8	.	.	7.8	.	.	.	.	.	.	.	.	.	.	19
4125	LOTUS CORNICULATUS	S	.	0	37.4	6.2	.	.	6.9	.	.	.	.	.	.	.	.	.	.	19
4126	LOTUS CORNICULATUS	S	.	0	35.4	6.1	.	.	9.9	.	.	.	.	.	.	.	.	.	.	19
4127	LOTUS CORNICULATUS	S	.	0	37.3	6.7	.	.	4.1	.	.	.	.	.	.	.	.	.	.	19
4128	LOTUS CORNICULATUS	S	.	0	34.9	6.2	.	.	4.3	.	.	.	.	.	.	.	.	.	.	19
4129	LOTUS CORNICULATUS	S	.	0	36.0	6.0	.	.	5.1	.	.	.	.	.	.	.	.	.	.	19
4130	LOTUS CORNICULATUS	S	.	0	33.8	6.7	.	.	.	.	.	.	.	.	.	.	.	.	.	19
4131	LOTUS CORNICULATUS	S	.	0	38.6	7.5	.	.	4.3	.	.	.	.	.	.	.	.	.	.	19

TABLE 3 - DATA CONVERTED TO A ZERO-MOISTURE BASIS.

OBS	PLNA	PART	CALORIES	WATER	PROTEIN	FAT	TOTLCARB	FIBER	ASH
4132	LOTUS CORNICULATUS	S	.	0	36.8	7.8	.	.	4.2
4133	LOTUS CORNICULATUS	S	.	0	36.1	6.3	.	.	4.5
4134	LOTUS CORNICULATUS	L	.	0	28.0	5.0	59.5	13.0	7.5
4135	LOTUS CORNICULATUS(FLOWERING)	W	.	0	19.9	2.0	70.5	21.6	7.6
4136	LOTUS CORNICULATUS(IMMATURE)	W	.	0	26.7	2.4	60.9	17.2	10.0
4137	LOTUS CORNICULATUS(PAST FLOWERING)	W	.	0	18.7	1.7	72.2	27.1	7.4
4138	LOTUS CRASSIFOLIUS	H	.	0	13.4	3.0	75.4	24.5	8.2
4139	LOTUS CRETICUS	S	.	0	31.6	7.6	.	.	.
4140	LOTUS CRETICUS	S	.	0	31.8	6.8	.	.	.
4141	LOTUS GEBELIA VAR. VILLOSUS	S	.	0	36.2	7.3	.	.	.
4142	LOTUS RIGIDUS	S	.	0	38.0	6.8	.	.	4.0
4143	LOTUS SCOPARIUS	S	.	0	35.3	8.4	.	.	3.7
4144	LOTUS SCOPARIUS	S	.	0	38.1	7.6	.	.	.
4145	LOTUS SCOPARIUS	S	.	0	40.0	7.8	.	.	.
4146	LOTUS SUBPINNATUS(IMMATURE)	W	.	0	23.4	.	.	16.7	13.1
4147	LOTUS TENUIS	S	.	0	34.2	7.0	.	.	.
4148	LOTUS ULIGINOSUS	S	.	0	35.2	6.4	.	.	8.1
4149	LOTUS ULIGINOSUS	S	.	0	33.9	5.1	.	.	4.9
4150	LOTUS ULIGINOSUS	S	.	0	34.5	5.6	.	.	3.8
4151	LUDWIGIA PEPLOIDES	L	315	0	25.4	3.1	63.8	25.4	7.7
4152	LUDWIGIA PEPLOIDES	L	344	0	20.0	8.0	63.2	6.4	8.8
4153	LUDWIGIA PILOSA	S	.	0	13.8	28.7	.	.	.
4154	LUFFA ACUTANGULA	S	.	0	27.5	23.5	.	.	3.2
4155	LUFFA ACUTANGULA	S	.	0	41.4	44.6	.	.	4.7

OBS	CALCIUM	PHOSPHO	IRON	SODIUM	POTASS	CAROT	THIAMINE	RIBOFLAV	NIACIN	ASCORBIC	REFCO
4132	.	.	.	.	.	.	.	.	.	.	19
4133	.	.	.	.	.	.	.	.	.	.	19
4134	.	.	.	.	.	.	.	.	.	.	5
4135	.	.	.	.	.	.	.	.	.	.	6
4136	.	.	.	.	.	.	.	.	.	.	6
4137	.	.	.	.	.	.	.	.	.	.	6
4138	.	.	.	.	.	.	.	.	.	.	6
4139	.	.	.	.	.	.	.	.	.	.	21
4140	.	.	.	.	.	.	.	.	.	.	21
4141	.	.	.	.	.	.	.	.	.	.	21
4142	.	.	.	.	.	.	.	.	.	.	19
4143	.	.	.	.	.	.	.	.	.	.	19
4144	.	.	.	.	.	.	.	.	.	.	20
4145	.	.	.	.	.	.	.	.	.	.	20
4146	2120.0	340.0	.	.	2350.0	.	.	.	.	.	6
4147	.	.	.	.	.	.	.	.	.	.	21
4148	.	.	.	.	.	.	.	.	.	.	19
4149	.	.	.	.	.	.	.	.	.	.	19
4150	.	.	.	.	.	.	.	.	.	.	19
4151	438.5	2307.7	97.7	.	.	45576.9	0.00	0.08	21.54	23.1	2
4152	1152.0	520.0	64.0	.	.	27600.0	0.32	0.96	6.40	696.0	3
4153	.	.	.	.	.	.	.	.	.	.	21
4154	.	.	.	.	.	.	.	.	.	.	19
4155	.	.	.	.	.	.	.	.	.	.	19

TABLE 3 - DATA CONVERTED TO A ZERO-MOISTURE BASIS.

OBS	PLNA	PART	CALORIES	WATER	PROTEIN	FAT	TOTLCARB	FIBER	ASH
4156	LUFFA ACUTANGULA	S	.	0	39.9	48.4	7.7	1.9	4.8
4157	LUFFA ACUTANGULA	F	315	0	13.0	1.9	75.9	11.1	9.3
4158	LUFFA ACUTANGULA	L	273	0	51.5	0.0	41.4	15.2	7.1
4159	LUFFA ACUTANGULA	F	327	0	14.5	3.6	74.5	9.1	7.3
4160	LUFFA ACUTANGULA	F	.	0	10.9	2.2	80.4	.	6.5
4161	LUFFA AEGYPTIACA	L	233	0	26.7	1.7	45.0	.	26.7
4162	LUFFA AEGYPTIACA	L	273	0	51.5	0.0	41.4	15.2	7.1
4163	LUFFA AEGYPTIACA	F	333	0	19.3	3.5	70.2	17.5	7.0
4164	LUFFA AEGYPTIACA	F	.	0	17.6	2.9	72.1	29.4	7.4
4165	LUFFA OPERCULATA	S	.	0	24.4	28.6	.	.	.
4166	LUNARIA ANNUA	S	.	0	24.6	32.9	.	.	4.2
4167	LUNARIA ANNUA	S	.	0	27.5	47.5	.	.	2.6
4168	LUNARIA ANNUA	S	.	0	28.1	30.3	.	.	.
4169	LUPINUS ALBUS	S	.	0	37.0	9.7	.	.	3.1
4170	LUPINUS ALBUS	S	.	0	36.0	9.0	.	.	3.4
4171	LUPINUS ALBUS	S	.	0	38.8	9.1	.	.	3.2
4172	LUPINUS ALBUS	S	.	0	35.7	9.7	.	.	3.8
4173	LUPINUS ANGUSTIFOLIUS	S	.	0	34.9	5.3	.	.	2.7
4174	LUPINUS ANGUSTIFOLIUS	S	.	0	35.3	3.0	.	.	.
4175	LUPINUS ANGUSTIFOLIUS	S	.	0	35.4	5.4	.	.	2.9
4176	LUPINUS ANGUSTIFOLIUS	S	.	0	36.6	5.4	.	.	3.0
4177	LUPINUS ANGUSTIFOLIUS	S	.	0	31.6	5.5	.	.	2.9
4178	LUPINUS ANGUSTIFOLIUS	S	.	0	36.8	5.2	.	.	3.4
4179	LUPINUS ANGUSTIFOLIUS	H	.	0	32.5	5.2	58.6	20.1	3.7

OBS	CALCIUM	PHOSPHO	IRON	SODIUM	POTASS	CAROT	THIAMINE	RIBOFLAV	NIACIN	ASCORBIC	REFCO
4156	.	.	.	.	.	.	.	.	.	.	5
4157	463.0	722.2	9.3	18.5	2629.6	555.6	0.56	0.56	3.70	111.1	2
4158	565.7	1414.1	116.2	.	.	93333.3	0.51	0.61	.	959.6	2
4159	345.5	600.0	16.4	.	.	4145.4	0.55	0.73	7.27	145.5	4
4160	869.6	869.6	34.8	.	.	739.1	.	.	.	.	5
4161	5500.0	550.0	.	.	.	.	.	.	.	.	1
4162	565.7	1414.1	116.2	.	.	93333.3	0.51	0.61	.	959.6	2
4163	35.1	526.3	12.3	52.6	2701.8	2982.5	0.53	0.70	5.26	175.4	2
4164	529.4	279.4	16.2	.	.	1764.7	0.29	0.88	5.88	0.0	5
4165	.	.	.	.	.	.	.	.	.	.	21
4166	.	.	.	.	.	.	.	.	.	.	19
4167	.	.	.	.	.	.	.	.	.	.	19
4168	.	.	.	.	.	.	.	.	.	.	20
4169	.	.	.	.	.	.	.	.	.	.	19
4170	.	.	.	.	.	.	.	.	.	.	19
4171	.	.	.	.	.	.	.	.	.	.	19
4172	.	.	.	.	.	.	.	.	.	.	19
4173	.	.	.	.	.	.	.	.	.	.	19
4174	.	.	.	.	.	.	.	.	.	.	19
4175	.	.	.	.	.	.	.	.	.	.	19
4176	.	.	.	.	.	.	.	.	.	.	19
4177	.	.	.	.	.	.	.	.	.	.	19
4178	.	.	.	.	.	.	.	.	.	.	19
4179	590.0	470.0	.	.	.	.	.	.	.	.	6

TABLE 3 - DATA CONVERTED TO A ZERO-MOISTURE BASIS.

OBS	PLNA	PART	CALORIES	WATER	PROTEIN	FAT	TOTLCARB	FIBER	ASH
4180	LUPINUS ANGUSTIFOLIUS	S	.	0	36.0	7.4	53.8	13.3	2.8
4181	LUPINUS BENTHAMI	W	.	0	18.3	.	.	23.9	10.4
4182	LUPINUS BICOLOR	W	.	0	17.1	.	.	24.9	9.1
4183	LUPINUS CAUDATUS	L	.	0	19.9	3.1	69.7	30.2	7.3
4184	LUPINUS HARTWEGII	S	.	0	38.0	9.1	.	.	4.8
4185	LUPINUS HIRSUTISSIMUS	S	.	0	41.3	10.3	.	.	.
4186	LUPINUS HISPANICUS	S	.	0	44.4	3.2	.	.	.
4187	LUPINUS HISPANICUS	S	.	0	39.1	3.7	.	.	.
4188	LUPINUS LATIFOLIUS	S	.	0	37.4	4.8	.	.	4.5
4189	LUPINUS LATIFOLIUS	S	.	0	38.8	5.9	.	.	.
4190	LUPINUS LUTEOLUS	S	.	0	44.1	9.5	.	.	.
4191	LUPINUS LUTEOLUS	S	.	0	40.7	10.5	.	.	.
4192	LUPINUS LUTEOLUS	S	.	0	54.0	10.0	.	.	.
4193	LUPINUS LUTEUS	S	.	0	44.1	5.4	.	.	5.3
4194	LUPINUS LUTEUS	S	.	0	42.2	5.5	.	.	4.2
4195	LUPINUS LUTEUS	H	.	0	35.6	5.2	54.2	19.6	5.0
4196	LUPINUS LUTEUS	S	.	0	44.8	5.5	44.7	15.7	5.1
4197	LUPINUS MICRANTHUS	S	.	0	46.7	5.3	.	.	5.4
4198	LUPINUS MUTABILIS	S	441	0	48.0	17.9	30.6	7.7	3.6
4199	LUPINUS MUTABILIS	F	441	0	47.9	17.8	31.5	2.1	2.8
4200	LUPINUS PALAESTINUS	S	.	0	27.7	5.3	.	.	.
4201	LUPINUS PARVIFLORUS	W	.	0	19.4	.	.	27.0	6.2
4202	LUPINUS PLATTENSIS	S	.	0	39.9	8.4	.	.	4.1
4203	LUPINUS POLYPHYLLUS VAR. CARMINEUS	S	.	0	41.5	9.2	.	.	5.0

OBS	CALCIUM	PHOSPHO	IRON	SODIUM	POTASS	CAROT	THIAMINE	RIBOFLAV	NIACIN	ASCORBIC	REFCO
4180	.	.	.	.	.	.	.	.	.	.	5
4181	1430.0	310.0	.	.	3070.0	.	.	.	.	.	6
4182	1410.0	290.0	.	.	2790.0	.	.	.	.	.	6
4183	1890.0	230.0	.	.	.	.	.	.	.	.	6
4184	.	.	.	.	.	.	.	.	.	.	19
4185	.	.	.	.	.	.	.	.	.	.	21
4186	.	.	.	.	.	.	.	.	.	.	21
4187	.	.	.	.	.	.	.	.	.	.	21
4188	.	.	.	.	.	.	.	.	.	.	19
4189	.	.	.	.	.	.	.	.	.	.	20
4190	.	.	.	.	.	.	.	.	.	.	21
4191	.	.	.	.	.	.	.	.	.	.	21
4192	.	.	.	.	.	.	.	.	.	.	21
4193	.	.	.	.	.	.	.	.	.	.	19
4194	.	.	.	.	.	.	.	.	.	.	19
4195	640.0	820.0	.	.	.	.	.	.	.	.	6
4196	258.7	438.7	.	.	911.1	.	.	.	.	.	5
4197	.	.	.	.	.	.	.	.	.	.	19
4198	97.5	590.5	6.8	.	.	0.0	0.30	0.54	2.82	.	3
4199	97.9	587.4	6.6	.	.	104.9	0.31	0.52	2.80	.	3
4200	.	.	.	.	.	.	.	.	.	.	21
4201	.	.	.	.	.	.	.	.	.	.	6
4202	.	.	.	.	.	.	.	.	.	.	19
4203	.	.	.	.	.	.	.	.	.	.	19

TABLE 3 - DATA CONVERTED TO A ZERO-MOISTURE BASIS.

OBS	PLNA	PART	CALORIES	WATER	PROTEIN	FAT	TOTLCARB	FIBER	ASH
4204	LUPINUS RIVULARIS	S	.	0	45.9	7.5	.	.	3.6
4205	LUPINUS SP.	W	.	0	26.6	2.6	56.9	19.1	13.9
4206	LUPINUS SP.	S	.	0	45.0	5.0	45.2	16.2	4.8
4207	LUPINUS SP.	S	.	0	42.6	10.4	.	.	4.3
4208	LUPINUS SP.	S	.	0	36.2	6.3	.	.	4.3
4209	LUPINUS SP.	S	.	0	41.2	8.7	.	.	3.9
4210	LUPINUS SP.	S	.	0	43.1	10.0	.	.	.
4211	LUPINUS SP.	S	.	0	41.2	8.3	.	.	.
4212	LUPINUS STIPULATUS	S	.	0	41.9	9.1	.	.	.
4213	LUPINUS SUBCARNOSUS	L	.	0	16.2	2.3	72.2	22.5	9.3
4214	LUPINUS SUCCULENTUS	S	.	0	38.2	5.3	.	.	4.9
4215	LUPINUS TEXENSIS	S	.	0	34.4	2.4	.	.	.
4216	LUPINUS TEXENSIS	S	.	0	30.6	2.7	.	.	.
4217	LUPINUS VARIUS	S	.	0	27.2	3.5	.	.	.
4218	LUPINUS VARIUS SUBSP. ORIENTALIS	S	.	0	26.9	3.6	.	.	.
4219	LYCHNIS CHALCEDONICA	S	.	0	16.9	6.5	.	.	.
4220	LYCHNIS VISCARIA	S	.	0	20.2	6.7	.	.	4.0
4221	LYCIUM CHINENSE	S	.	0	16.2	19.1	.	.	.
4222	LYCIUM CHINENSE	L	279	0	39.4	5.8	38.5	12.5	16.3
4223	LYCOPERSICON ESCULENTUM	W	.	0	26.4	1.7	45.3	15.4	26.6
4224	LYCOPERSICON ESCULENTUM	L	303	0	21.2	4.5	59.8	0.8	14.4
4225	LYCOPERSICON ESCULENTUM	F	324	0	14.7	2.9	72.1	5.9	10.3
4226	LYCOPERSICON ESCULENTUM	F	323	0	19.4	4.8	67.7	11.3	8.1
4227	LYCOPERSICON ESCULENTUM	F	333	0	16.7	4.2	70.8	12.5	8.3

OBS	CALCIUM	PHOSPHO	IRON	SODIUM	POTASS	CAROT	THIAMINE	RIBOFLAV	NIACIN	ASCORBIC	REFCO
4204	.	.	.	.	.	.	.	.	.	.	19
4205	1280.0	250.0	.	.	.	.	.	.	.	.	18
4206	370.0	200.0	.	.	.	.	.	.	.	.	18
4207	.	.	.	.	.	.	.	.	.	.	19
4208	.	.	.	.	.	.	.	.	.	.	19
4209	.	.	.	.	.	.	.	.	.	.	19
4210	.	.	.	.	.	.	.	.	.	.	20
4211	.	.	.	.	.	.	.	.	.	.	21
4212	.	.	.	.	.	.	.	.	.	.	20
4213	2570.0	240.0	.	.	1710.0	.	.	.	.	.	6
4214	.	.	.	.	.	.	.	.	.	.	19
4215	.	.	.	.	.	.	.	.	.	.	19
4216	.	.	.	.	.	.	.	.	.	.	20
4217	.	.	.	.	.	.	.	.	.	.	21
4218	.	.	.	.	.	.	.	.	.	.	21
4219	.	.	.	.	.	.	.	.	.	.	20
4220	.	.	.	.	.	.	.	.	.	.	19
4221	.	.	.	.	.	.	.	.	.	.	20
4222	1423.1	413.5	51.9	1836.5	4980.8	42788.5	0.77	2.98	7.69	76.9	2
4223	.	.	.	.	.	.	.	.	.	.	6
4224	3280.3	651.5	9.8	.	.	.	0.61	3.56	.	1325.8	1
4225	426.5	911.8	25.0	.	.	15294.1	0.74	0.59	10.29	735.3	1
4226	112.9	483.9	9.7	64.5	3790.3	8145.2	0.97	0.65	9.68	371.0	2
4227	156.3	364.6	6.3	.	.	6875.0	1.35	0.83	8.33	375.0	3

TABLE 3 - DATA CONVERTED TO A ZERO-MOISTURE BASIS.

OBS	PLANA	PART	CALORIES	WATER	PROTEIN	FAT	TOTCARB	FIBER	ASH	CALCIUM	PHOSPHO	IRON	SODIUM	POTASS	CAROT	THIAMINE	RIBOFLAV	NIACIN	ASCORBIC	REFCO
4228	LYCOPERSICON ESCULENTUM	F	339	0	12.9	4.8	74.2	9.7	8.1	112.9	387.1	9.7	.	.	5806.45	0.97	0.81	11.29	371.0	3
4229	LYCOPERSICON ESCULENTUM	F	328	0	18.8	3.1	71.9	9.4	6.3	93.8	312.5	9.4	.	.	1406.25	0.94	0.63	6.25	281.2	3
4230	LYCOPERSICON ESCULENTUM	G	343	0	17.1	2.9	72.9	7.1	7.1	185.7	385.7	7.1	42.9	3485.7	2314.29	0.86	0.57	7.14	285.7	4
4231	LYCOPERSICON ESCULENTUM	F	338	0	16.9	3.1	72.3	7.7	7.7	200.0	415.4	7.7	46.2	3753.8	8307.69	0.92	0.62	10.77	353.8	4
4232	LYCOPERSICON ESCULENTUM	S	.	0	30.3	26.8	38.5	14.9	4.4	.	.	.	.	.	.	.	.	.	.	5
4233	LYCOPUS ASPER	S	.	0	16.6	25.4	.	.	6.7	.	.	.	.	.	.	.	.	.	.	19
4234	LYCOPUS EUROPAEUS	S	.	0	19.3	33.0	.	.	.	.	.	.	.	.	.	.	.	.	.	21
4235	LYCOPUS EUROPAEUS	S	.	0	15.9	26.2	.	.	.	.	.	.	.	.	.	.	.	.	.	21
4236	LYSILOMA ACAPULCENSIS	S	.	0	40.6	9.1	.	.	.	.	.	.	.	.	.	.	.	.	.	20
4237	LYSILOMA BAHAMENSIS	S	.	0	25.0	4.9	.	.	.	.	.	.	.	.	.	.	.	.	.	19
4238	LYSILOMA DIVARICATA	S	.	0	33.1	12.9	.	.	4.7	.	.	.	.	.	.	.	.	.	.	19
4239	LYTHRUM SALICARIA	S	.	0	16.3	10.9	.	.	.	.	.	.	.	.	.	.	.	.	.	21
4240	MACADAMIA INTEGRIFOLIA	F	.	0	9.9	70.1	18.6	9.2	1.4	100.0	250.0	.	.	.	.	.	.	.	.	18
4241	MACADAMIA TENIFOLIA	S	712	0	8.0	73.8	16.4	2.6	1.8	49.5	166.0	2.1	.	272.2	0.00	0.35	0.11	1.34	0.0	4
4242	MACADAMIA TENIFOLIA	S	.	0	9.0	73.7	15.6	2.6	1.8	.	.	.	.	.	.	.	.	.	.	5
4243	MACHAERANTHERA TANACETIFOLIA	S	.	0	17.8	19.0	.	.	.	.	.	.	.	.	.	.	.	.	.	21
4244	MACHAERIUM ROSESCENS	S	.	0	41.9	23.1	.	.	6.2	.	.	.	.	.	.	.	.	.	.	19
4245	MACLEYA CORDATA VAR. JAPONICA	S	.	0	17.6	40.7	.	.	5.2	.	.	.	.	.	.	.	.	.	.	19
4246	MACLURA POMIFERA	S	.	0	38.2	44.5	.	.	3.0	.	.	.	.	.	.	.	.	.	.	19
4247	MACLURA POMIFERA	S	.	0	6.3	8.6	.	.	4.5	.	.	.	.	.	.	.	.	.	.	19
4248	MACLURA POMIFERA	S	.	0	46.9	35.8	.	.	3.4	.	.	.	.	.	.	.	.	.	.	19
4249	MACLURA POMIFERA	L	.	0	23.4	1.9	.	.	10.0	.	.	.	.	.	.	.	.	.	.	5
4250	MACROPTILIUM ATROPURPUREUM	W	.	0	23.0	3.1	60.1	30.4	13.7	1420.0	210.0	.	.	.	.	.	.	.	.	18
4251	MACROTYLOMA UNIFLORUM	S	369	0	24.9	1.1	67.0	5.2	7.0	.	.	.	.	.	.	.	.	.	.	1
4252	MACROTYLOMA UNIFLORUM	S	389	0	31.8	4.5	59.1	.	4.6	323.1	.	9.0	.	.	.	.	.	.	.	1
4253	MACROTYLOMA UNIFLORUM	S	.	0	24.9	0.6	71.0	6.0	3.5	317.5	442.2	8.6	.	.	80.50	.	.	1.70	.	5
4254	MADHUCA BUTYRACEA	S	.	0	5.5	58.8	31.6	.	4.0	.	.	.	.	.	.	.	.	.	.	5
4255	MADHUCA INDICA	L	.	0	9.1	4.1	79.0	18.7	7.8	1530.0	220.0	.	.	.	.	.	.	.	.	18
4256	MADHUCA INDICA	I	.	0	5.4	0.6	89.6	2.1	3.3	172.0	172.0	18.4	.	.	28.26	0.04	1.08	6.39	8.6	5
4257	MADHUCA INDICA	F	.	0	5.3	6.1	86.0	.	2.7	170.5	83.3	4.2	.	.	1162.88	.	.	.	151.5	5
4258	MADHUCA LONGIFOLIA	I	.	0	5.0	1.8	89.0	1.6	4.2	130.0	120.0	.	.	.	.	.	.	.	.	18
4259	MADHUCA LONGIFOLIA	L	.	0	9.1	3.9	79.4	19.0	7.6	1460.0	210.0	.	.	.	.	.	.	.	.	18
4260	MADIA SP.	W	.	0	13.0	.	.	18.0	11.1	1540.0	340.0	.	.	3270.0	.	.	.	.	.	6
4261	MADIA SP.(FLOWERING)	W	.	0	6.7	.	.	19.3	6.9	1640.0	470.0	.	.	4930.0	.	.	.	.	.	6
4262	MADIA SP.(IMMATURE)	W	.	0	22.9	.	.	12.8	15.7	2050.0	520.0	.	.	5230.0	.	.	.	.	.	6
4263	MADIA SP.(MATURE)	W	.	0	5.5	.	.	22.2	7.0	1040.0	110.0	.	.	1090.0	.	.	.	.	.	6
4264	MAERUA ANGOLENSIS	L	300	0	27.0	3.0	58.5	8.5	11.5	2100.0	420.0	.	.	.	.	.	.	.	.	1
4265	MAERUA CRASSIFOLIA	L	.	0	28.8	2.7	55.1	10.4	13.4	2060.0	170.0	.	.	.	.	.	.	.	.	18
4266	MAERUA CRASSIFOLIA	SH	.	0	11.3	1.5	77.1	33.7	10.1	2090.0	130.0	.	.	.	.	.	.	.	.	18
4267	MAGNOLIA GRANDIFLORA	S	.	0	8.8	56.4	.	.	.	.	.	.	.	.	.	.	.	.	.	20
4268	MAGNOLIA GRANDIFLORA	S	.	0	9.8	49.6	38.5	20.5	2.1	.	.	.	.	.	.	.	.	.	.	5
4269	MAGNOLIA MACROPHYLLA	S	.	0	15.6	46.1	.	.	.	.	.	.	.	.	.	.	.	.	.	20
4270	MAHONIA TRIFOLIATA	L	.	0	11.0	2.6	83.2	32.6	3.2	670.0	160.0	.	.	800.0	.	.	.	.	.	6
4271	MAJORANA HORTENSIS	S	.	0	31.5	39.7	.	.	5.2	.	.	.	.	.	.	.	.	.	.	19
4272	MAJORANA HORTENSIS	H	293	0	13.7	7.6	65.6	19.6	13.1	2153.7	331.2	89.5	83.3	1647.2	5239.18	0.31	0.35	4.46	55.2	16

TABLE 3 - DATA CONVERTED TO A ZERO-MOISTURE BASIS.

OBS	PLANA	PART	CALORIES	WATER	PROTEIN	FAT	TOTCARB	FIBER	ASH	CALCIUM	PHOSPHO	IRON	SODIUM	POTASS	CAROT	THIAMINE	RIBOFLAV	NIACIN	ASCORBIC	REFCO
4273	MALABALIA SEKAKUL	S	.	0	16.8	16.0	.	.	.	.	.	.	.	.	.	.	.	.	.	21
4274	MALACHRA ALCEAEFOLIA	S	.	0	23.4	17.7	.	.	.	.	.	.	.	.	.	.	.	.	.	21
4275	MALACHRA CAPITATA	S	.	0	22.5	15.5	.	.	.	.	.	.	.	.	.	.	.	.	.	20
4276	MALCOLMIA AFRICANA	S	.	0	32.4	30.6	.	.	.	.	.	.	.	.	.	.	.	.	.	21
4277	MALCOLMIA AFRICANA	S	.	0	37.0	29.8	.	.	.	.	.	.	.	.	.	.	.	.	.	21
4278	MALCOLMIA MARITIMA	S	.	0	23.8	20.4	.	.	.	.	.	.	.	.	.	.	.	.	.	20
4279	MALLOTUS PHILIPPINENSIS	S	.	0	18.5	25.8	.	.	.	.	.	.	.	.	.	.	.	.	.	21
4280	MALLOTUS PHILIPPINENSIS	S	.	0	20.0	23.7	.	.	.	.	.	.	.	.	.	.	.	.	.	21
4281	MALOPE TRIFIDA	S	.	0	23.1	16.2	.	.	.	.	.	.	.	.	.	.	.	.	.	20
4282	MALOPE TRIFIDA	S	.	0	26.9	17.5	.	.	.	.	.	.	.	.	.	.	.	.	.	20
4283	MALPIGHIA GLABRA	F	232	0	10.7	6.0	40.5	7.1	42.9	160.7	160.7	7.1	.	.	119.0	0.18	0.24	2.98	125.0	3
4284	MALPIGHIA UMBELLATA	S	.	0	41.9	45.8	.	.	.	.	.	.	.	.	.	.	.	.	.	19
4285	MALPIGHIA UMBELLATA	S	.	0	10.6	1.4	.	.	.	.	.	.	.	.	.	.	.	.	.	19
4286	MALUS SPECTABILIS	F	356	0	0.8	0.8	96.4	6.8	2.0	264.0	24.0	5.2	.	.	1840.0	0.04	0.08	0.80	8.0	2
4287	MALUS SYLVESTRIS	F	363	0	1.3	2.5	94.4	5.0	1.9	18.8	43.8	1.3	.	.	500.0	.	.	.	6.3	1
4288	MALUS SYLVESTRIS	F	359	0	2.8	2.1	93.0	4.2	2.1	70.4	70.4	3.5	14.1	915.5	140.8	0.14	0.21	1.41	28.2	2
4289	MALUS SYLVESTRIS	F	363	0	1.9	1.9	95.0	4.4	1.3	37.5	62.5	2.5	.	.	125.0	0.19	0.31	1.25	37.5	3
4290	MALUS SYLVESTRIS	F	372	0	1.3	3.8	92.9	6.4	1.9	44.9	64.1	1.9	6.4	705.1	346.1	0.19	0.13	0.64	25.6	4
4291	MALVA AEGYPTIA	S	.	0	25.4	17.3	.	.	.	.	.	.	.	.	.	.	.	.	.	21
4292	MALVA MONTANA	S	.	0	26.4	18.4	.	.	.	.	.	.	.	.	.	.	.	.	.	21
4293	MALVA NEGLECTA	S	.	0	21.0	15.2	.	.	.	.	.	.	.	.	.	.	.	.	.	21
4294	MALVA NICAEENSIS	S	.	0	20.8	26.1	.	.	.	.	.	.	.	.	.	.	.	.	.	21
4295	MALVA PARVIFLORA	S	.	0	21.2	13.0	.	.	.	.	.	.	.	.	.	.	.	.	.	19
4296	MALVA PARVIFLORA	S	.	0	19.0	14.4	.	.	.	.	.	.	.	.	.	.	.	.	.	21
4297	MALVA PARVIFLORA	S	.	0	30.4	14.7	.	.	.	.	.	.	.	.	.	.	.	.	.	21
4298	MALVA PARVIFLORA	L	263	0	35.0	1.5	46.7	10.9	16.8	2365.0	489.1	.	.	.	.	.	.	.	474.5	1
4299	MALVA PUSILLA	L	318	0	25.7	7.0	55.1	41.6	12.1	.	.	.	.	.	.	.	.	.	.	1
4300	MALVA SP.	S	.	0	15.6	11.0	.	.	.	.	.	.	.	.	.	.	.	.	.	20
4301	MALVA SP.	S	.	0	17.5	11.3	.	.	.	.	.	.	.	.	.	.	.	.	.	20
4302	MALVA SP.	S	.	0	23.6	16.2	.	.	.	.	.	.	.	.	.	.	.	.	.	21
4303	MALVA SP.	L	278	0	33.1	4.5	44.4	12.0	18.0	1872.2	518.8	95.5	.	.	32932.3	0.98	1.50	7.52	263.2	3
4304	MALVA SYLVESTRIS	S	.	0	19.0	14.2	.	.	.	.	.	.	.	.	.	.	.	.	.	21
4305	MALVA SYLVESTRIS	S	.	0	19.8	9.7	.	.	.	.	.	.	.	.	.	.	.	.	.	21
4306	MALVA SYLVESTRIS	L	333	0	42.9	16.7	25.0	10.7	15.5	1071.4	500.0	44.0	.	.	39464.9	2.02	3.45	5.95	285.7	2
4307	MALVA SYLVESTRIS	L	.	0	26.3	.	.	.	21.3	2762.5	450.0	121.3	.	.	62.5	1.25	3.00	5.50	.	5
4308	MALYASTRUM COROMANDELIANUM	S	.	0	26.0	13.7	.	.	.	.	.	.	.	.	.	.	.	.	.	21
4309	MALYASTRUM SPICATUM	S	.	0	26.2	15.2	.	.	.	.	.	.	.	.	.	.	.	.	.	21
4310	MAMMEA AFRICANA	S	441	0	2.1	9.6	86.8	3.3	1.5	359.9	179.4	.	.	.	.	.	.	.	.	1
4311	MAMMEA AMERICANA	F	356	0	4.5	1.5	91.7	7.6	2.3	98.5	90.9	3.0	.	.	1363.6	0.23	0.38	3.03	121.2	3
4312	MAMMEA AMERICANA	F	370	0	3.6	3.6	90.6	7.2	2.2	79.7	79.7	5.1	108.7	340.6	1000.0	0.14	0.29	2.90	101.4	4
4313	MAMMEA AMERICANA	F	.	0	4.4	2.2	91.1	11.1	2.2	96.3	.	3.0	.	.	1244.4	0.15	0.30	3.70	111.1	5
4314	MANDRAGORA OFFICINARUM	S	.	0	22.1	22.6	.	.	.	.	.	.	.	.	.	.	.	.	.	21
4315	MANFREDA SP.	S	.	0	20.0	18.4	.	.	.	.	.	.	.	.	.	.	.	.	.	20
4316	MANFREDA SP.	S	.	0	20.2	22.4	.	.	.	.	.	.	.	.	.	.	.	.	.	21
4317	MANGIFERA CAESIA	F	356	0	7.4	1.5	88.1	.	3.0	74.1	177.8	0.0	.	.	37.0	0.59	.	.	429.6	2

TABLE 3 - DATA CONVERTED TO A ZERO-MOISTURE BASIS.

OBS	PLNA	PART	CALORIES	WATER	PROTEIN	FAT	TOTLCARB	FIBER
4318	MANGIFERA INDICA	L	.	0	9.5	4.8	72.6	22.6
4319	MANGIFERA INDICA	F	351	0	3.5	1.2	92.4	5.3
4320	MANGIFERA INDICA	G	348	0	7.2	0.7	88.4	12.3
4321	MANGIFERA INDICA	F	356	0	3.4	1.7	91.4	2.9
4322	MANGIFERA INDICA	G	351	0	3.5	2.3	89.5	2.3
4323	MANGIFERA INDICA	L	302	0	21.8	1.1	67.6	8.9
4324	MANGIFERA INDICA	F	358	0	3.0	1.2	93.3	4.8
4325	MANGIFERA INDICA	G	355	0	3.2	1.6	92.7	6.5
4326	MANGIFERA INDICA	L	361	0	3.8	2.2	91.8	4.9
4327	MANGIFERA INDICA	L	.	0	13.8	1.8	75.7	7.3
4328	MANGIFERA ODORATA	F	348	0	4.5	0.5	92.0	4.0
4329	MANIHOT ESCULENTA	L	.	0	24.8	5.2	61.5	18.3
4330	MANIHOT ESCULENTA	R	.	0	2.4	0.7	94.0	2.0
4331	MANIHOT ESCULENTA	L	.	0	23.7	3.0	62.5	9.6
4332	MANIHOT ESCULENTA	L	322	0	24.7	3.5	64.7	14.1
4333	MANIHOT ESCULENTA	R	391	0	1.4	0.5	94.9	2.0
4334	MANIHOT ESCULENTA	R	392	0	3.2	0.5	93.9	2.9
4335	MANIHOT ESCULENTA	R	391	0	2.9	0.6	93.9	2.9
4336	MANIHOT ESCULENTA	L	316	0	36.3	6.8	48.4	11.1
4337	MANIHOT ESCULENTA	R	376	0	2.0	0.8	94.9	2.5
4338	MANIHOT ESCULENTA	R	379	0	2.9	1.1	94.3	2.9
4339	MANIHOT ESCULENTA	L	351	0	29.8	6.1	56.1	10.5
4340	MANIHOT ISOLOBA	S	.	0	21.2	32.0	.	.
4341	MANIHOT RUBRICAULIS	S	.	0	25.6	28.2	.	.

OBS	ASH	CALCIUM	PHOSPHO	IRON	SODIUM	POTASS	CAROT	THIAMINE	RIBOFLAV	NIACIN	ASCORBIC	REFCO
4318	13.1	3140.0	160.0	.	.	.	.	.	.	.	.	18
4319	2.9	140.4	128.7	7.0	.	.	18713	0.18	0.29	2.34	245.6	1
4320	3.6	413.0	253.6	9.4	.	.	7246	0.22	0.43	2.17	1956.5	1
4321	3.4	57.5	86.2	1.7	17.2	1229.9	10805	0.34	0.29	3.45	206.9	2
4322	4.7	58.5	87.7	1.2	35.1	812.9	643	0.35	0.29	3.51	362.6	2
4323	9.5	106.1	430.2	15.6	.	.	3408	0.34	0.50	8.38	150.8	2
4324	2.4	72.7	72.7	4.8	.	.	7636	0.30	0.36	2.42	321.2	3
4325	2.4	112.9	80.6	3.2	.	.	645	0.16	0.24	1.61	1032.3	3
4326	2.2	54.6	71.0	2.2	38.3	1032.8	15738	0.27	0.27	6.01	191.3	4
4327	8.7	133.0	330.3	28.4	.	.	4101	0.18	0.28	10.09	243.1	5
4328	3.0	19.9	89.6	1.0	.	.	1791	0.20	0.30	3.48	64.7	3
4329	8.5	980.0	520.0	.	.	.	.	.	.	.	.	18
4330	2.9	140.0	20.0	.	.	.	.	.	.	.	.	18
4331	10.8	380.0	380.0	12.0	.	.	.	.	.	.	.	6
4332	7.1	1070.7	420.5	26.9	.	.	41608	0.88	2.12	8.48	1098.9	1
4333	3.2	132.5	129.2	.	.	.	.	.	.	.	.	1
4334	2.4	178.9	110.5	5.0	.	.	79	0.11	0.13	1.58	81.6	1
4335	2.6	75.4	92.8	2.6	5.8	1142.0	0	0.14	0.12	1.74	98.6	2
4336	8.4	757.9	357.9	14.7	21.1	2152.6	43579	0.84	1.68	9.47	431.6	2
4337	2.3	91.4	121.8	2.8	.	.	25	0.15	0.10	1.78	101.5	3
4338	1.7	114.9	97.7	4.0	.	.	0	0.14	0.11	1.72	54.6	3
4339	7.9	903.5	377.2	8.8	.	.	263	0.53	1.18	7.46	1271.9	3
4340	.	.	.	.	.	.	.	.	.	.	.	20
4341	.	.	.	.	.	.	.	.	.	.	.	20

TABLE 3 - DATA CONVERTED TO A ZERO-MOISTURE BASIS.

OBS	PLNA	PART	CALORIES	WATER	PROTEIN	FAT	TOTLCARB	FIBER
4342	MANILKARA ZAPOTA	F	367	0	1.9	3.4	92.3	10.6
4343	MANILKARA ZAPOTA	F	376	0	2.0	4.4	92.0	6.4
4344	MARAH GILENSIS	S	.	0	27.8	57.1	.	.
4345	MARAH GILENSIS	S	.	0	3.8	2.8	.	.
4346	MARAH GILENSIS	S	.	0	30.0	56.7	.	.
4347	MARAH MACROCARPA	S	.	0	29.4	54.4	.	.
4348	MARANTA ARUNDINACEA	R	383	0	5.2	0.6	90.5	6.1
4349	MARANTA ARUNDINACEA	R	367	0	5.6	0.2	91.1	4.4
4350	MARRUBIUM PESTALOZZAE	S	.	0	24.4	34.0	.	.
4351	MARRUBIUM VULGARE	S	.	0	25.0	35.1	.	.
4352	MARRUBIUM VULGARE	S	.	0	24.3	34.0	.	.
4353	MARRUBIUM VULGARE	S	.	0	23.3	37.2	.	.
4354	MARRUBIUM VULGARE	S	.	0	18.5	26.4	.	.
4355	MARRUBIUM VULGARE	S	.	0	25.6	36.9	.	.
4356	MARSDENIA EDULIS	S	.	0	27.4	47.0	.	.
4357	MARSHALLIA CAESPITOSE VAR. SIGNATA	S	.	0	31.1	25.6	.	.
4358	MARSILEA CRENATA	L	386	0	14.3	17.1	57.1	47.1
4359	MARTINEZIA LINDENIANA	S	.	0	7.8	55.3	.	.
4360	MARTINEZIA LINDENIANA	S	.	0	2.2	0.6	.	.
4361	MARTINEZIA LINDENIANA	S	.	0	7.1	56.5	.	.
4362	MARTYNIA LOUISIANICA	S	.	0	25.0	59.8	.	.
4363	MARTYNIA PARVIFLORA	S	.	0	30.0	47.9	.	.
4364	MARTYNIA PARVIFLORA	S	.	0	15.0	3.5	.	.
4365	MARTYNIA SP.	S	.	0	31.9	56.5	.	.

OBS	ASH	CALCIUM	PHOSPHO	IRON	SODIUM	POTASS	CAROT	THIAMINE	RIBOFLAV	NIACIN	ASCORBIC	REFCO
4342	2.4	130.4	53.1	2.9	14.5	874.4	121	0.00	0.00	0.97	62.8	2
4343	1.6	96.0	40.0	4.0	.	.	80	0.04	0.04	0.80	60.0	3
4344	2.6	.	.	.	.	.	.	.	.	.	.	19
4345	1.1	.	.	.	.	.	.	.	.	.	.	19
4346	.	.	.	.	.	.	.	.	.	.	.	20
4347	3.0	.	.	.	.	.	.	.	.	.	.	19
4348	3.7	46.0	55.2	5.8	.	.	.	0.40	0.06	1.53	21.5	2
4349	3.0	46.7	56.1	7.5	.	.	0	0.19	0.07	1.64	21.0	3
4350	.	.	.	.	.	.	.	.	.	.	.	21
4351	.	.	.	.	.	.	.	.	.	.	.	19
4352	.	.	.	.	.	.	.	.	.	.	.	20
4353	.	.	.	.	.	.	.	.	.	.	.	20
4354	.	.	.	.	.	.	.	.	.	.	.	21
4355	.	.	.	.	.	.	.	.	.	.	.	21
4356	3.4	.	.	.	.	.	.	.	.	.	.	19
4357	.	.	.	.	.	.	.	.	.	.	.	20
4358	11.4	685.7	0.0	.	.	.	104286	1.43	1.00	48.57	42.9	2
4359	2.0	.	.	.	.	.	.	.	.	.	.	19
4360	1.0	.	.	.	.	.	.	.	.	.	.	19
4361	2.5	.	.	.	.	.	.	.	.	.	.	19
4362	.	.	.	.	.	.	.	.	.	.	.	19
4363	.	.	.	.	.	.	.	.	.	.	.	19
4364	.	.	.	.	.	.	.	.	.	.	.	19
4365	.	.	.	.	.	.	.	.	.	.	.	19

TABLE 3 - DATA CONVERTED TO A ZERO-MOISTURE BASIS.

OBS	PLNA	PART	CALORIES	WATER	PROTEIN	FAT	TOTLCARB	FIBER
4366	MASTICHODENDRON CAPIRI	L	373	0	6.2	4.8	86.7	4.8
4367	MATRICARIA CAPENSIS	S	.	0	26.2	26.6	.	.
4368	MATTHIOLA BICORNIS	S	.	0	35.3	29.4	.	.
4369	MATTHIOLA BICORNIS	S	.	0	33.1	28.8	.	.
4370	MATTHIOLA FRUTICULOSA	S	.	0	33.6	33.4	.	.
4371	MATTHIOLA LONGIPETALA SUBSP. BICORNIS	S	.	0	30.0	33.1	.	.
4372	MAURITIA VINIFERA	F	526	0	11.0	38.6	46.0	41.9
4373	MAXIMOWICZIA SONORAE	S	.	0	26.9	28.5	.	.
4374	MEDICAGO ARABICA	S	.	0	43.7	8.4	.	.
4375	MEDICAGO ARABICA	S	.	0	38.2	8.2	.	.
4376	MEDICAGO ARABICA	S	.	0	11.4	1.6	.	.
4377	MEDICAGO CILIARIS	S	.	0	38.6	10.3	.	.
4378	MEDICAGO DENTICULATA	L	284	0	35.8	2.2	50.0	6.7
4379	MEDICAGO FALCATA	S	.	0	42.7	9.9	.	.
4380	MEDICAGO FALCATA	S	.	0	36.8	10.8	.	.
4381	MEDICAGO FALCATA	S	.	0	42.5	9.8	.	.
4382	MEDICAGO HISPIDA	S	.	0	35.9	6.6	.	.
4383	MEDICAGO HISPIDA	S	.	0	36.5	7.3	.	.
4384	MEDICAGO HISPIDA	S	.	0	9.1	1.0	.	.
4385	MEDICAGO HISPIDA	S	.	0	37.9	7.3	.	.
4386	MEDICAGO LITTORALIS	S	.	0	46.3	12.5	.	.
4387	MEDICAGO LUPULINA	W	.	0	25.2	3.3	58.8	21.3
4388	MEDICAGO LUPULINA	S	.	0	41.7	6.6	.	.
4389	MEDICAGO LUPULINA	S	.	0	37.9	5.7	.	.

OBS	ASH	CALCIUM	PHOSPHO	IRON	SODIUM	POTASS	CAROT	THIAMINE	RIBOFLAV	NIACIN	ASCORBIC	REFCO
4366	2.3	209.0	67.8	.	.	.	621.5	0.23	0.17	1.41	420.9	3
4367	.	.	.	.	.	.	.	.	.	.	.	19
4368	4.5	.	.	.	.	.	.	.	.	.	.	19
4369	.	.	.	.	.	.	.	.	.	.	.	19
4370	.	.	.	.	.	.	.	.	.	.	.	21
4371	.	.	.	.	.	.	.	.	.	.	.	21
4372	4.4	415.4	69.9	12.9	.	.	90992.6	0.11	0.85	2.57	95.6	3
4373	.	.	.	.	.	.	.	.	.	.	.	20
4374	4.5	.	.	.	.	.	.	.	.	.	.	19
4375	3.6	.	.	.	.	.	.	.	.	.	.	19
4376	6.4	.	.	.	.	.	.	.	.	.	.	19
4377	.	.	.	.	.	.	.	.	.	.	.	21
4378	11.9	992.0	462.0	380.0	.	2342.0	18015.0	1.34	2.09	5.22	731.0	10
4379	3.9	.	.	.	.	.	.	.	.	.	.	19
4380	3.9	.	.	.	.	.	.	.	.	.	.	19
4381	4.3	.	.	.	.	.	.	.	.	.	.	19
4382	8.5	.	.	.	.	.	.	.	.	.	.	19
4383	3.0	.	.	.	.	.	.	.	.	.	.	19
4384	2.5	.	.	.	.	.	.	.	.	.	.	19
4385	3.8	.	.	.	.	.	.	.	.	.	.	19
4386	.	.	.	.	.	.	.	.	.	.	.	21
4387	12.7	1610.0	410.0	.	.	.	.	.	.	.	.	18
4388	3.5	.	.	.	.	.	.	.	.	.	.	19
4389	3.2	.	.	.	.	.	.	.	.	.	.	19

TABLE 3 - DATA CONVERTED TO A ZERO-MOISTURE BASIS.

OBS	PLNA	PART	CALORIES	WATER	PROTEIN	FAT	TOTLCARB	FIBER	ASH	CALCIUM	PHOSPHO	IRON	SODIUM	POTASS	CAROT	THIAMINE	RIBOFLAV	NIACIN	ASCORBIC	REFCO
4390	MEDICAGO LUPULINA	S	.	0	33.4	5.0	.	.	6.2	.	.	.	.	.	.	.	.	.	.	19
4391	MEDICAGO LUPULINA	S	.	0	29.9	4.5	.	.	.	.	.	.	.	.	.	.	.	.	.	21
4392	MEDICAGO LUPULINA	W	.	0	23.3	3.3	63.2	24.7	10.2	960.0	320.0	.	.	2280.0	.	.	.	.	.	6
4393	MEDICAGO LUPULINA	L	.	0	25.7	2.6	61.8	15.8	10.3	.	.	.	.	.	.	.	.	.	.	5
4394	MEDICAGO MARINA	S	.	0	38.6	14.9	.	.	.	.	.	.	.	.	.	.	.	.	.	21
4395	MEDICAGO MARINA	S	.	0	36.8	13.4	.	.	.	.	.	.	.	.	.	.	.	.	.	21
4396	MEDICAGO MINIMA	S	.	0	48.6	7.8	.	.	.	.	.	.	.	.	.	.	.	.	.	21
4397	MEDICAGO MINIMA	S	.	0	46.2	8.6	.	.	.	.	.	.	.	.	.	.	.	.	.	21
4398	MEDICAGO ORBICULARIS	S	.	0	31.4	5.2	.	.	5.7	.	.	.	.	.	.	.	.	.	.	19
4399	MEDICAGO ORBICULARIS	S	.	0	29.1	3.3	.	.	4.6	.	.	.	.	.	.	.	.	.	.	19
4400	MEDICAGO ORBICULARIS	S	.	0	36.6	3.4	.	.	.	.	.	.	.	.	.	.	.	.	.	21
4401	MEDICAGO ORBICULARIS	S	.	0	29.9	3.8	.	.	.	.	.	.	.	.	.	.	.	.	.	21
4402	MEDICAGO POLYMORPHA	W	.	0	26.7	3.6	55.2	21.4	14.5	1620.0	470.0	.	.	.	.	.	.	.	.	18
4403	MEDICAGO POLYMORPHA	S	.	0	17.4	2.5	72.7	38.6	7.4	1830.0	190.0	.	.	.	.	.	.	.	.	18
4404	MEDICAGO POLYMORPHA	H	.	0	19.0	2.8	68.5	25.1	9.7	.	.	.	.	.	.	.	.	.	.	6
4405	MEDICAGO POLYMORPHA	L	284	0	35.8	2.2	50.0	6.7	11.9	992.5	462.7	38.1	.	2343.3	18022.4	1.34	2.09	5.22	731.3	2
4406	MEDICAGO POLYMORPHA	L	.	0	24.5	8.2	56.3	18.8	11.1	.	.	.	.	.	.	.	.	.	4086.5	5
4407	MEDICAGO POLYMORPHA(IMMATURE)	H	.	0	26.0	2.3	58.3	23.1	13.4	.	.	.	.	.	.	.	.	.	.	6
4408	MEDICAGO POLYMORPHA(MATURE)	H	.	0	17.7	2.7	71.0	25.2	8.6	.	.	.	.	.	.	.	.	.	.	6
4409	MEDICAGO RIGIDULA	S	.	0	47.4	9.4	.	.	.	.	.	.	.	.	.	.	.	.	.	21
4410	MEDICAGO SATIVA	W	.	0	25.3	2.9	60.0	23.5	11.8	2410.0	350.0	.	.	.	.	.	.	.	.	18
4411	MEDICAGO SATIVA	L	.	0	21.7	2.4	58.9	20.1	17.0	1440.0	200.0	.	.	.	.	.	.	.	.	18
4412	MEDICAGO SATIVA	H	.	0	18.7	2.6	68.9	34.4	9.8	1260.0	180.0	.	.	.	.	.	.	.	.	18
4413	MEDICAGO SATIVA	S	.	0	37.0	12.3	.	.	3.7	.	.	.	.	.	.	.	.	.	.	19
4414	MEDICAGO SATIVA	S	.	0	36.9	11.3	.	.	3.7	.	.	.	.	.	.	.	.	.	.	19
4415	MEDICAGO SATIVA	S	.	0	38.5	10.1	.	.	.	.	.	.	.	.	.	.	.	.	.	21
4416	MEDICAGO SATIVA	S	.	0	38.0	10.6	.	.	.	.	.	.	.	.	.	.	.	.	.	21
4417	MEDICAGO SATIVA	H	.	0	17.3	2.1	71.7	31.4	8.9	1640.0	260.0	24.0	.	1770.0	6.0	0.35	1.47	4.00	.	6
4418	MEDICAGO SATIVA	W	.	0	19.3	3.0	68.7	27.4	9.0	1720.0	310.0	33.0	.	2030.0	20.0	0.64	1.61	4.42	.	6
4419	MEDICAGO SATIVA	L	301	0	34.7	2.3	54.9	17.9	8.1	69.4	294.8	31.2	.	.	39422.0	0.75	0.81	2.89	936.4	3
4420	MEDICAGO SCUTELLATA	S	.	0	33.5	11.5	.	.	4.9	.	.	.	.	.	.	.	.	.	.	19
4421	MEDICAGO SCUTELLATA	S	.	0	34.5	12.4	.	.	4.4	.	.	.	.	.	.	.	.	.	.	19
4422	MEDICAGO SCUTELLATA	S	.	0	39.2	9.8	.	.	.	.	.	.	.	.	.	.	.	.	.	21
4423	MEDICAGO SP.	S	.	0	34.9	10.3	.	.	5.1	.	.	.	.	.	.	.	.	.	.	19
4424	MEDICAGO TRIBULOIDES	S	.	0	43.8	15.9	.	.	3.2	.	.	.	.	.	.	.	.	.	.	19
4425	MEDICAGO TRUNCATULA	W	.	0	25.8	3.1	55.9	24.0	15.2	2130.0	460.0	.	.	.	.	.	.	.	.	18
4426	MEDICAGO TURBINATA	S	.	0	44.6	10.6	.	.	.	.	.	.	.	.	.	.	.	.	.	20
4427	MEDICAGO TURBINATA	S	.	0	40.8	15.5	.	.	.	.	.	.	.	.	.	.	.	.	.	21
4428	MEDICAGO TURBINATA	S	.	0	47.9	15.1	.	.	.	.	.	.	.	.	.	.	.	.	.	21
4429	MELAMPODIUM PERFOLIATUM	S	.	0	11.2	15.1	.	.	.	.	.	.	.	.	.	.	.	.	.	20
4430	MELAMPODIUM PERFOLIATUM	S	.	0	8.8	3.8	.	.	.	.	.	.	.	.	.	.	.	.	.	20
4431	MELAPODIUM LEUCANTHUM	W	.	0	13.5	10.0	66.6	15.9	9.9	1730.0	180.0	.	.	2900.0	.	.	.	.	.	6
4432	MELHANIA FUTTEYPORENSIS	S	.	0	19.3	4.8	.	.	.	.	.	.	.	.	.	.	.	.	.	21
4433	MELIA AZEDARACH	L	.	0	21.8	8.8	58.6	11.6	10.8	2310.0	220.0	.	.	.	.	.	.	.	.	18
4434	MELIA AZEDARACH	S	.	0	30.4	44.7	.	.	3.4	.	.	.	.	.	.	.	.	.	.	19

TABLE 3 - DATA CONVERTED TO A ZERO-MOISTURE BASIS.

OBS	PLNA	PART	CALORIES	WATER	PROTEIN	FAT	TOTLCARB	FIBER	ASH
4435	MELIA AZEDARACH	S	.	0	28.7	44.0	.	.	.
4436	MELIA AZEDERACH	S	.	0	2.5	2.4	.	.	1.9
4437	MELICA CILIATA VAR. NEBRODENSIS	S	.	0	25.2	3.5	.	.	.
4438	MELICA IMPERFECTA	W	.	0	11.0	2.6	79.9	30.3	6.5
4439	MELICOCCUS BIJUGATUS	F	.	0	10.1	1.3	85.9	10.8	2.7
4440	MELICOCCUS BIJUGATUS	S	.	0	11.6	1.2	84.5	11.2	2.7
4441	MELICOCCUS BIJUGATUS	F	356	0	5.1	0.9	92.1	6.5	1.9
4442	MELICOCCUS BIJUGATUS	F	273	0	5.1	0.9	92.1	6.5	1.9
4443	MELILOTUS ALBA	S	.	0	37.3	6.9	.	.	3.9
4444	MELILOTUS ALBA	S	.	0	40.6	5.6	.	.	4.0
4445	MELILOTUS ALBA	S	.	0	42.8	6.2	.	.	4.2
4446	MELILOTUS ALBA	S	.	0	35.7	7.5	.	.	4.5
4447	MELILOTUS ALBA	S	.	0	38.1	6.5	.	.	3.8
4448	MELILOTUS ALBA	S	.	0	39.1	5.9	.	.	.
4449	MELILOTUS ALBA	S	.	0	36.4	8.3	.	.	.
4450	MELILOTUS ALBA	S	.	0	39.4	7.6	26.7	12.9	4.1
4451	MELILOTUS ALBA VAR. ANNUA	S	.	0	39.5	6.7	.	.	4.3
4452	MELILOTUS INDICA	W	.	0	15.3	1.7	69.9	29.4	13.1
4453	MELILOTUS INDICA	F	.	0	25.3	1.9	64.3	14.9	8.5
4454	MELILOTUS INDICA	S	.	0	29.0	3.2	73.3	16.9	4.5
4455	MELILOTUS INDICA	S	.	0	32.4	4.6	.	.	6.9
4456	MELILOTUS INDICA	S	.	0	32.5	5.4	.	.	3.7
4457	MELILOTUS INDICUS	S	.	0	29.4	4.6	.	.	.
4458	MELILOTUS ITALICA	S	.	0	36.8	6.9	.	.	3.8

OBS	CALCIUM	PHOSPHO	IRON	SODIUM	POTASS	CAROT	THIAMINE	RIBOFLAV	NIACIN	ASCORBIC	REFCO
4435	.	.	.	.	.	.	.	.	.	.	21
4436	.	.	.	.	.	.	.	.	.	.	19
4437	.	.	.	.	.	.	.	.	.	.	21
4438	390.0	260.0	.	.	1940.0	.	.	.	.	.	6
4439	.	.	.	.	.	.	.	.	.	.	18
4440	.	.	.	.	.	.	.	.	.	.	18
4441	55.6	92.6	2.8	.	.	115.7	0.23	0.09	3.24	23.1	2
4442	55.6	92.6	2.8	.	.	231.5	0.23	0.09	3.24	23.1	3
4443	.	.	.	.	.	.	.	.	.	.	19
4444	.	.	.	.	.	.	.	.	.	.	19
4445	.	.	.	.	.	.	.	.	.	.	19
4446	.	.	.	.	.	.	.	.	.	.	19
4447	.	.	.	.	.	.	.	.	.	.	19
4448	.	.	.	.	.	.	.	.	.	.	21
4449	.	.	.	.	.	.	.	.	.	.	21
4450	.	.	.	.	.	.	.	.	.	.	5
4451	.	.	.	.	.	.	.	.	.	.	19
4452	.	.	.	.	.	.	.	.	.	.	18
4453	1340.0	350.0	.	.	.	.	.	.	.	.	18
4454	.	.	.	.	.	.	.	.	.	.	18
4455	.	.	.	.	.	.	.	.	.	.	19
4456	.	.	.	.	.	.	.	.	.	.	19
4457	.	.	.	.	.	.	.	.	.	.	20
4458	.	.	.	.	.	.	.	.	.	.	19

TABLE 3 - DATA CONVERTED TO A ZERO-MOISTURE BASIS.

OBS	PLNA	PART	CALORIES	WATER	PROTEIN	FAT	TOTLCARB	FIBER	ASH
4459	MELILOTUS NEAPOLITANA	S	.	0	35.40	1.9	.	.	.
4460	MELILOTUS OFFICINALIS	S	.	0	41.20	6.5	.	.	4.0
4461	MELILOTUS OFFICINALIS	S	.	0	41.60	7.6	.	.	4.7
4462	MELILOTUS OFFICINALIS	S	.	0	37.60	7.3	.	.	4.6
4463	MELILOTUS OFFICINALIS	S	.	0	38.40	4.8	.	.	4.3
4464	MELILOTUS OFFICINALIS	S	.	0	33.90	7.0	.	.	4.4
4465	MELILOTUS OFFICINALIS	S	.	0	12.80	1.9	.	.	14.3
4466	MELILOTUS OFFICINALIS	S	.	0	34.90	7.5	.	.	.
4467	MELILOTUS SP.	S	.	0	37.10	4.6	.	.	3.8
4468	MELILOTUS SP.	H	.	0	16.30	2.2	72.80	32.20	8.7
4469	MELILOTUS SP.	W	.	0	17.80	2.8	71.20	29.40	8.2
4470	MELILOTUS TAURICA	S	.	0	41.20	7.4	.	.	6.1
4471	MELINIS MINUTIFLORA	W	.	0	9.00	3.0	81.20	36.50	7.8
4472	MELINIS MINUTIFLORA	H	.	0	13.80	4.1	74.70	32.30	7.4
4473	MELINUS MINUTIFLORA	W	.	0	4.60	2.5	85.20	38.60	7.7
4474	MELIOSMA MYRIANTHA	S	.	0	8.10	9.6	.	.	.
4475	MELISSA OFFICINALIS	S	.	0	29.30	11.5	.	.	.
4476	MELOCHIA HIRSUTA	S	.	0	19.40	7.3	.	.	.
4477	MENISPERMUM CANADENSE	S	.	0	13.10	16.0	.	.	2.6
4478	MENTHA ARVENSIS	S	.	0	17.80	23.0	.	.	8.9
4479	MENTHA ARVENSIS		.	0	.	.	.	.	.
4480	MENTHA PULEGIUM	L	.	0	10.00	.	.	.	11.1
4481	MENTHA PULEGIUM	S	.	0	24.60	26.6	.	.	.
4482	MENTHA ROTUNDIFOLIA	L	.	0	14.40	.	.	.	9.7

OBS	CALCIUM	PHOSPHO	IRON	SODIUM	POTASS	CAROT	THIAMINE	RIBOFLAV	NIACIN	ASCORBIC	REFCO
4459	.	.	.	.	.	.	.	.	.	.	21
4460	.	.	.	.	.	.	.	.	.	.	19
4461	.	.	.	.	.	.	.	.	.	.	19
4462	.	.	.	.	.	.	.	.	.	.	19
4463	.	.	.	.	.	.	.	.	.	.	19
4464	.	.	.	.	.	.	.	.	.	.	19
4465	.	.	.	.	.	.	.	.	.	.	19
4466	.	.	.	.	.	.	.	.	.	.	20
4467	.	.	.	.	.	.	.	.	.	.	19
4468	1770.0	260.0	15.0	.	1340.0	12.0	.	.	.	.	6
4469	1320.0	270.0	14.0	.	1650.0	27.0	0.53	8.38	3.61	.	6
4470	.	.	.	.	.	.	.	.	.	.	19
4471	.	.	.	.	.	.	.	.	.	.	18
4472	.	.	.	.	.	.	.	.	.	.	18
4473	.	.	.	.	.	.	.	.	.	.	6
4474	.	.	.	.	.	.	.	.	.	.	20
4475	.	.	.	.	.	.	.	.	.	.	21
4476	.	.	.	.	.	.	.	.	.	.	20
4477	.	.	.	.	.	.	.	.	.	.	19
4478	.	.	.	.	.	.	.	.	.	.	19
4479	369.8	.	39.6	.	.	.	0.22	0.66	.	82.4	9
4480	944.0	.	.	541.0	1531.0	.	.	.	.	.	17
4481	.	.	.	.	.	.	.	.	.	.	21
4482	1630.0	.	.	252.0	1710.0	.	.	.	.	.	17

TABLE 3 - DATA CONVERTED TO A ZERO-MOISTURE BASIS.

OBS	PLNA	PART	CALORIES	WATER	PROTEIN	FAT	TOTLCARB	FIBER	ASH
4483	MENTHA ROTUNDIFOLIA	S	.	0	19.60	28.2	.	.	.
4484	MENTHA SP.	S	.	0	25.10	44.4	.	.	3.1
4485	MENTHA SP.	H	.	0	14.80	2.5	73.80	23.70	8.9
4486	MENTHA SP.	L	299	0	28.00	6.5	50.50	.	15.0
4487	MENTHA SP.	L	333	0	20.00	9.6	57.00	8.10	13.3
4488	MENTHA SPICATA	L	.	0	28.20	3.5	58.80	11.80	9.4
4489	MENTHA TOMENTOSA	S	.	0	19.10	28.4	.	.	.
4490	MENTZELIA ALBICAULIS	S	.	0	19.40	34.5	.	.	.
4491	MENTZELIA DECAPETALA	S	.	0	21.30	37.0	.	.	11.1
4492	MENTZELIA DISPERSA	S	.	0	22.50	46.5	.	.	.
4493	MENTZELIA NUDA	S	.	0	19.00	39.1	.	.	.
4494	MENTZELIA NUDA	S	.	0	21.20	39.3	.	.	.
4495	MENTZELLA DECAPETALA	S	.	0	21.40	42.4	.	.	12.5
4496	MERCURIALIS ANNUA	S	.	0	18.90	37.1	.	.	11.0
4497	MERREMIA PALMERI	S	.	0	18.20	9.0	.	.	.
4498	MERTENSIA SP.	W	.	0	25.30	3.3	69.80	17.10	11.6
4499	MESEMBRYANTHEMUM CRYSTALLINUM	S	.	0	31.20	7.8	.	.	5.0
4500	MESEMBRYANTHEMUM SP.		.	0	7.63	.	63.73	19.22	22.5
4501	MESPILUS GERMANICA	S	.	0	1.20	0.7	.	.	.
4502	MESPILUS GERMANICA	S	.	0	45.40	40.7	.	.	.
4503	MESPILUS GERMANICA	S	.	0	3.30	2.3	.	.	.
4504	METROXYLON SP.	HE	411	0	1.60	0.2	98.80	0.20	0.5
4505	METROXYLON SP.	HE	.	0	0.20	0.2	99.20	.	0.3
4506	MICROLONCHUS DURIAEI	S	.	0	20.90	23.5	.	.	.

OBS	CALCIUM	PHOSPHO	IRON	SODIUM	POTASS	CAROT	THIAMINE	RIBOFLAV	NIACIN	ASCORBIC	REFCO
4483	.	.	.	.	.	.	.	.	.	.	21
4484	.	.	.	.	.	.	.	.	.	.	19
4485	1710.0	220.0	.	.	.	.	.	.	.	.	6
4486	1813.1	448.6	35.5	18.7	1672.9	20186.9	1.21	1.50	6.54	598.1	2
4487	1022.2	303.7	49.6	.	.	37629.6	0.96	1.93	6.67	474.1	3
4488	1176.5	470.6	91.8	.	.	9529.4	0.29	0.47	2.35	.	5
4489	.	.	.	.	.	.	.	.	.	.	21
4490	.	.	.	.	.	.	.	.	.	.	19
4491	.	.	.	.	.	.	.	.	.	.	19
4492	.	.	.	.	.	.	.	.	.	.	21
4493	.	.	.	.	.	.	.	.	.	.	21
4494	.	.	.	.	.	.	.	.	.	.	21
4495	.	.	.	.	.	.	.	.	.	.	19
4496	.	.	.	.	.	.	.	.	.	.	19
4497	.	.	.	.	.	.	.	.	.	.	21
4498	.	.	.	.	.	.	.	.	.	.	6
4499	.	.	.	.	.	.	.	.	.	.	19
4500	1350.0	100.0	79.3	.	.	.	.	.	.	.	8
4501	.	.	.	.	.	.	.	.	.	.	20
4502	.	.	.	.	.	.	.	.	.	.	21
4503	.	.	.	.	.	.	.	.	.	.	21
4504	17.3	11.5	1.6	9.2	41.4	0.0	0.01	.	.	0.0	2
4505	11.4	11.4	1.5	.	.	.	.	.	0.23	.	5
4506	.	.	.	.	.	.	.	.	.	.	21

TABLE 3 - DATA CONVERTED TO A ZERO-MOISTURE BASIS.

OBS	PLANA	PART	CALORIES	WATER	PROTEIN	FAT	TOTCARB	FIBER	ASH	CALCIUM	PHOSPHO	IRON	SODIUM	POTASS	CAROT	THIAMINE	RIBOFLAV	NIACIN	ASCORBIC	REFCO
4507	MICROMERIA SERPYLLIFOLIA	S	.	0	20.4	27.9	.	.	.	.	.	.	.	.	.	.	.	.	.	21
4508	MIKANIA CORDATA	L	.	0	16.5	3.3	68.1	23.1	12.1	1318.7	879.1	.	.	.	.	.	.	.	.	5
4509	MILLERIA QUINQUEFLORA	S	.	0	12.1	13.4	.	.	.	.	.	.	.	.	.	.	.	.	.	21
4510	MILLETTIA AURICULATA	L	.	0	22.7	4.6	63.4	32.5	9.3	1920.0	290.0	.	.	.	.	.	.	.	.	18
4511	MILLETTIA OVALIFOLIA	S	.	0	26.9	37.0	.	.	4.7	.	.	.	.	.	.	.	.	.	.	19
4512	MILLETTIA OVALIFOLIA	S	.	0	32.1	7.8	.	.	.	.	.	.	.	.	.	.	.	.	.	21
4513	MIMOSA ACULEATICARPA	S	.	0	34.4	6.2	.	.	.	.	.	.	.	.	.	.	.	.	.	20
4514	MIMOSA HAMATA	L	.	0	11.9	7.9	70.4	12.1	9.8	.	.	.	.	.	.	.	.	.	.	5
4515	MIMOSA HIMALAYANA	S	.	0	35.9	2.2	.	.	.	.	.	.	.	.	.	.	.	.	.	21
4516	MIMOSA INVISA	W	.	0	16.5	2.5	73.7	21.8	7.3	1100.0	320.0	.	.	.	.	.	.	.	.	18
4517	MIMOSA PALMERI	S	.	0	36.9	6.1	.	.	.	.	.	.	.	.	.	.	.	.	.	19
4518	MIMOSA PUDICA	S	.	0	30.4	2.8	.	.	3.6	.	.	.	.	.	.	.	.	.	.	19
4519	MIMOSA PUDICA	S	.	0	11.6	2.4	.	.	5.4	.	.	.	.	.	.	.	.	.	.	19
4520	MIMOSA PUDICA	P	351	0	4.0	2.2	88.7	9.5	5.1	468.0	66.2	.	.	.	.	.	.	.	.	
4521	MIMOSA PUDICA	L	314	0	20.0	0.0	72.9	.	7.1	500.0	757.1	20.0	.	.	93000.0	1.71	0.71	.	257.1	2
4522	MIMOSA PUDICA	L	.	0	21.9	3.2	68.5	32.7	6.4	.	.	.	.	.	.	.	.	.	.	5
4523	MIMOSA SP.	L	.	0	20.6	3.4	68.9	21.1	7.1	2380.0	170.0	.	.	1140.0	.	.	.	.	.	6
4524	MIMOSA URUGUENSIS	S	.	0	31.8	5.2	.	.	.	.	.	.	.	.	.	.	.	.	.	21
4525	MIMULUS GUTTATUS	S	.	0	20.2	40.2	.	.	.	.	.	.	.	.	.	.	.	.	.	21
4526	MIMULUS LEWISII	W	.	0	15.6	5.9	64.6	17.1	13.9	.	.	.	.	.	.	.	.	.	.	6
4527	MIRABILIS JALAPA	S	.	0	17.1	3.9	.	.	3.2	.	.	.	.	.	.	.	.	.	.	19
4528	MIRABILIS JALAPA	S	.	0	18.1	4.6	.	.	.	.	.	.	.	.	.	.	.	.	.	21
4529	MIRABILIS NYCTAGINEA	S	.	0	21.2	7.0	.	.	3.5	.	.	.	.	.	.	.	.	.	.	19
4530	MNESITHEA LAEVIS	L	.	0	3.9	0.8	86.6	44.2	8.7	.	.	.	.	.	.	.	.	.	.	5
4531	MODIOLA CAROLINIANA	S	.	0	19.3	18.6	.	.	.	.	.	.	.	.	.	.	.	.	.	20
4532	MODIOLA CAROLINIANA	S	.	0	18.6	16.1	.	.	.	.	.	.	.	.	.	.	.	.	.	21
4533	MODIOLA CAROLINIANA	L	176	0	25.9	.	31.8	.	42.4	3094.1	458.8	70.6	.	.	21882.3	0.94	1.88	4.71	.	3
4534	MOLDAVICA PARVIFLORA	S	.	0	15.1	18.3	.	.	.	.	.	.	.	.	.	.	.	.	.	21
4535	MOLLUGO VERTICILLATA	S	.	0	13.1	15.6	.	.	.	.	.	.	.	.	.	.	.	.	.	20
4536	MOLTKIA CAERULEA	S	.	0	8.9	10.3	.	.	.	.	.	.	.	.	.	.	.	.	.	21
4537	MOLUCCELLA LAEVIS	S	.	0	27.5	33.4	.	.	.	.	.	.	.	.	.	.	.	.	.	20
4538	MOLUCCELLA LAEVIS	S	.	0	24.3	34.4	.	.	.	.	.	.	.	.	.	.	.	.	.	21
4539	MOLUCCELLA LAEVIS	S	.	0	25.1	37.2	.	.	.	.	.	.	.	.	.	.	.	.	.	21
4540	MOLUCCELLA SPINOSA	S	.	0	19.5	38.5	.	.	.	.	.	.	.	.	.	.	.	.	.	21
4541	MOMORDICA BALSAMINA	L	259	0	26.4	2.9	47.6	14.0	23.1	2157.0	270.0	.	19.0	193.0	.	.	.	10.30	.	12
4542	MOMORDICA BALSAMINA	F	246	0	20.8	2.1	49.8	17.6	9.8	1199.0	348.0	.	11.0	352.0	.	.	.	7.90	.	12
4543	MOMORDICA BALSAMINA	W	.	0	24.8	1.0	58.5	21.1	15.7	1440.0	290.0	.	.	.	.	.	.	.	.	18
4544	MOMORDICA BALSAMINA	S	.	0	19.2	22.8	.	.	2.4	.	.	.	.	.	.	.	.	.	.	19
4545	MOMORDICA BALSAMINA	S	.	0	28.1	42.7	.	.	3.2	.	.	.	.	.	.	.	.	.	.	19
4546	MOMORDICA BALSAMINA	S	.	0	28.1	39.7	.	.	3.3	.	.	.	.	.	.	.	.	.	.	19
4547	MOMORDICA BALSAMINA	S	.	0	30.6	38.8	.	.	3.2	.	.	.	.	.	.	.	.	.	.	19
4548	MOMORDICA BALSAMINA	S	.	0	20.6	20.4	.	.	.	.	.	.	.	.	.	.	.	.	.	19
4549	MOMORDICA BALSAMINA	S	.	0	32.5	37.6	.	.	.	.	.	.	.	.	.	.	.	.	.	20
4550	MOMORDICA CHARANTIA	S	.	0	32.8	46.2	.	.	3.4	.	.	.	.	.	.	.	.	.	.	19
4551	MOMORDICA CHARANTIA	S	.	0	5.3	1.0	.	.	1.1	.	.	.	.	.	.	.	.	.	.	19

TABLE 3 - DATA CONVERTED TO A ZERO-MOISTURE BASIS.

OBS	PLNA	PART	CALORIES	WATER	PROTEIN	FAT	TOTLCARB	FIBER
4552	MOMORDICA CHARANTIA	S	.	0	35.6	49.6	.	.
4553	MOMORDICA CHARANTIA	S	.	0	36.2	48.6	.	.
4554	MOMORDICA CHARANTIA	S	.	0	21.3	31.6	.	.
4555	MOMORDICA CHARANTIA	L	302	0	25.6	2.0	61.8	2.5
4556	MOMORDICA CHARANTIA	F	329	0	15.8	2.6	76.3	13.2
4557	MOMORDICA CHARANTIA	F	317	0	13.3	1.7	75.0	16.7
4558	MOMORDICA CHARANTIA	L	286	0	36.4	3.9	45.5	10.4
4559	MOMORDICA CHARANTIA	F	318	0	13.6	3.0	71.2	18.2
4560	MOMORDICA CHARANTIA	L	.	0	37.1	2.9	.	10.3
4561	MOMORDICA CHARANTIA	F	.	0	17.3	6.0	68.5	10.1
4562	MOMORDICA COCHINCHINENSIS	F	.	0	16.4	4.4	73.0	37.1
4563	MOMORDICA DIOICA	F	.	0	19.5	6.3	67.3	18.9
4564	MONANTHOCHLOE LITTORALIS		.	0	6.6	1.5	79.1	29.1
4565	MONARDA FISTULOSA	S	.	0	26.7	31.8	.	.
4566	MONARDA FISTULOSA VAR. MENTHAEFOLIA	S	.	0	27.5	36.5	.	.
4567	MONARDA PUNCTATA	S	.	0	21.3	31.3	.	.
4568	MONARDELLA SP.	S	.	0	19.4	26.8	.	.
4569	MONNINA WRIGHTII	S	.	0	10.0	13.7	.	.
4570	MONOCHORIA VAGINALIS	L	257	0	14.3	2.9	54.3	.
4571	MONODORA MYRSTICA	S	532	0	9.8	32.4	55.8	26.5
4572	MONTANOA ARBORESCENS	S	.	0	13.8	7.3	.	.
4573	MORINDA CITRIFOLIA	L	478	0	20.1	34.8	38.4	17.9
4574	MORINGA OLEIFERA	L	.	0	29.0	5.2	56.7	19.1
4575	MORINGA OLEIFERA	F	.	0	20.7	1.0	69.4	27.0

OBS	ASH	CALCIUM	PHOSPHO	IRON	SODIUM	POTASS	CAROT	THIAMINE	RIBOFLAV	NIACIN	ASCORBIC	REFCO
4552	.	.	.	.	.	.	.	.	.	.	.	19
4553	.	.	.	.	.	.	.	.	.	.	.	20
4554	.	.	.	.	.	.	.	.	.	.	.	21
4555	10.6	1326.6	3346.7	35.7	.	.	.	0.25	2.36	.	1241.2	1
4556	5.3	171.1	421.1	2.6	.	.	.	0.26	0.92	.	3644.7	1
4557	10.0	433.3	533.3	38.3	33.3	4500.0	1833	1.00	0.67	5.00	950.0	2
4558	14.3	1870.1	350.6	32.5	123.4	3311.7	33019	0.84	2.99	9.74	1103.9	2
4559	12.1	393.9	515.2	25.8	.	.	758	1.21	0.91	4.55	772.7	3
4560	14.9	1697.1	302.9	18.9	.	.	34	0.80	3.14	10.29	1200.0	5
4561	8.3	297.6	833.3	56.0	.	.	750	0.42	0.54	2.98	523.8	5
4562	6.3	132.1	930.8	16.4	.	.	.	.	.	.	1553.5	5
4563	6.9	207.5	264.2	28.9	.	.	10189	0.31	1.07	3.71	1729.6	5
4564	12.8	310.0	180.0	.	.	.	.	.	.	.	.	6
4565	6.7	.	.	.	.	.	.	.	.	.	.	19
4566	.	.	.	.	.	.	.	.	.	.	.	21
4567	.	.	.	.	.	.	.	.	.	.	.	20
4568	.	.	.	.	.	.	.	.	.	.	.	19
4569	.	.	.	.	.	.	.	.	.	.	.	20
4570	28.6	1142.9	642.9	52.9	.	.	8571	1.14	.	.	714.3	2
4571	2.0	115.0	.	10.4	.	.	.	0.10	0.11	0.56	.	1
4572	.	.	.	.	.	.	.	.	.	.	.	20
4573	6.7	160.7	67.0	10.3	.	.	161897	0.71	0.40	32.14	232.1	2
4574	9.1	2060.0	240.0	.	.	.	.	.	.	.	.	18
4575	8.9	.	.	.	.	.	.	.	.	.	.	18

TABLE 3 - DATA CONVERTED TO A ZERO-MOISTURE BASIS.

OBS	PLNA	PART	CALORIES	WATER	PROTEIN	FAT	TOTLCARB	FIBER	ASH
4576	MORINGA OLEIFERA	L	299	0	31.4	2.3	56.3	8.0	10.0
4577	MORINGA OLEIFERA	I	310	0	20.9	0.6	70.9	15.2	7.6
4578	MORINGA OLEIFERA	L	321	0	33.0	6.7	51.8	5.4	8.5
4579	MORINGA OLEIFERA	F	316	0	18.8	1.5	71.4	.	8.3
4580	MORINGA OLEIFERA	F	.	0	19.1	0.8	64.9	36.6	15.3
4581	MORINGA OLEIFERA	L	.	0	26.8	6.8	57.2	3.6	9.2
4582	MORTONIA SCABRELLA	S	.	0	8.8	10.5	.	.	3.6
4583	MORUS ALBA	L	.	0	15.0	7.4	63.3	15.3	14.3
4584	MORUS ALBA	L	.	0	14.0	6.8	49.7	.	13.8
4585	MORUS ALBA	F	353	0	11.3	2.7	81.3	6.0	4.7
4586	MORUS ALBA	F	.	0	12.0	3.2	77.6	.	7.2
4587	MORUS NIGRA	L	.	0	17.6	11.5	50.5	7.4	20.4
4588	MORUS NIGRA	F	347	0	11.6	2.5	81.0	5.8	5.0
4589	MORUS NIGRA	F	.	0	4.8	2.8	84.1	5.5	2.8
4590	MORUS RUBRA	S	.	0	25.6	37.6	.	.	.
4591	MORUS SP.	L	.	0	18.0	5.7	61.3	11.2	15.0
4592	MORUS SP.	F	.	0	.	.	.	.	.
4593	MUCUNA DEERINGIANA	W	.	0	15.5	4.3	74.6	34.4	5.6
4594	MUCUNA DEERINGIANA	D	.	0	21.0	2.6	71.9	15.6	4.5
4595	MUCUNA DEERINGIANA	S	.	0	28.6	0.7	66.7	9.5	4.0
4596	MUCUNA DEERINGIANA	L	.	0	15.1	2.1	67.8	19.3	14.9
4597	MUCUNA DEERINGIANA	S	397	0	23.9	5.0	67.2	5.2	4.0
4598	MUCUNA DEERINGIANA	F	.	0	20.1	4.9	70.3	14.4	4.7
4599	MUCUNA DEERINGIANA	S	.	0	26.0	6.3	64.3	7.1	3.3

OBS	CALCIUM	PHOSPHO	IRON	SODIUM	POTASS	CAROT	THIAMINE	RIBOFLAV	NIACIN	ASCORBIC	REFCO
4576	2187.7	570.9	43.7	.	.	45670.5	0.84	2.95	9.96	697.3	1
4577	139.2	500.0	32.9	.	.	2911.4	0.38	.	8.86	1569.6	1
4578	1325.9	401.8	16.1	17.9	2111.6	39531.3	0.89	3.26	15.18	745.5	2
4579	436.1	466.2	6.0	.	.	563.9	0.38	0.45	4.51	1195.5	2
4580	229.0	839.7	40.5	.	.	839.7	.	.	1.53	916.0	5
4581	1760.0	280.0	28.0	.	.	27120.0	0.84	.	3.20	880.0	5
4582	.	.	.	.	.	.	.	.	.	.	19
4583	2420.0	240.0	.	.	.	.	.	.	.	.	18
4584	2740.0	450.0	.	.	.	.	.	.	.	.	5
4585	200.0	213.3	24.7	246.7	1013.3	66.7	0.20	0.40	4.67	33.3	2
4586	640.0	320.0	15.2	.	.	832.0	0.08	1.44	6.40	104.0	5
4587	2150.0	130.0	.	.	.	.	.	.	.	.	18
4588	198.3	214.9	24.8	247.9	1016.5	124.0	0.33	0.66	5.79	322.3	2
4589	413.8	137.9	17.9	.	.	69.0	0.41	0.62	1.38	69.0	5
4590	.	.	.	.	.	.	.	.	.	.	20
4591	4640.0	170.0	.	.	1980.0	.	.	.	.	.	6
4592	327.6	284.5	12.9	.	.	.	0.69	.	.	525.9	1
4593	1210.0	130.0	.	.	.	.	.	.	.	.	18
4594	.	.	.	.	.	.	.	.	.	.	18
4595	.	.	.	.	.	.	.	.	.	.	18
4596	.	.	.	.	.	.	.	.	.	.	5
4597	149.0	361.2	.	.	.	.	0.14	0.11	3.39	2.3	1
4598	.	.	.	.	.	.	.	.	.	.	5
4599	.	.	.	.	.	.	.	.	.	.	5

TABLE 3 - DATA CONVERTED TO A ZERO-MCISTURE BASIS.

OBS	PLNA	PART	CALORIES	WATER	PROTEIN	FAT	TOTLCARB	FIBER	ASH
4600	MUCUNA HOLTONII	S	.	0	20.5	4.1	.	.	3.2
4601	MUCUNA PRURIENS	S	386	0	33.1	3.6	59.4	8.8	3.9
4602	MUCUNA PRURIENS	S	.	0	27.5	3.3	.	7.5	4.4
4603	MUCUNA SP.	S	403	0	27.9	6.3	62.2	4.9	3.7
4604	MUCUNA UTILIS	S	391	0	28.2	3.5	64.7	.	3.5
4605	MUHLENBERGIA ARENACEA	W	.	0	13.3	2.0	70.3	30.5	14.4
4606	MUHLENBERGIA CAPILLARIS(IMMATURE)	W	.	0	9.9	1.9	78.4	35.0	9.8
4607	MUHLENBERGIA CAPILLARIS(MATURE)	W	.	0	3.9	1.7	87.0	37.1	7.4
4608	MUHLENBERGIA CUSPIDATA	W	.	0	7.4	1.9	80.2	33.5	10.5
4609	MUHLENBERGIA EMERSLEYI	H	.	0	7.5	1.5	82.1	34.1	8.9
4610	MUHLENBERGIA PORTERI	W	.	0	7.3	1.8	85.0	36.9	5.9
4611	MUHLENBERGIA PUNGENS	W	.	0	18.3	1.7	68.9	26.9	11.1
4612	MUHLENBERGIA SP.(ALL)	W	.	0	8.8	1.8	81.4	35.2	8.0
4613	MUNDIA SPINOSA	S	.	0	29.6	52.1	.	.	.
4614	MUNDIA SPINOSA	S	.	0	30.7	44.2	.	.	.
4615	MUNROA SQUARROSA	W	.	0	12.5	1.7	71.9	17.3	13.9
4616	MUNTINGIA CALABURA	F	349	0	8.0	1.6	85.5	8.0	4.8
4617	MUNTINGIA CALABURA	F	384	0	8.9	9.7	75.5	25.3	5.9
4618	MURRAYA KOENIGII	L	326	0	35.9	6.3	52.2	.	5.6
4619	MURRAYA KOENIGII	L	.	0	18.1	3.0	66.5	19.0	12.5
4620	MUSA ACUMINATA	F	346	0	8.7	1.0	86.5	1.0	3.8
4621	MUSA ACUMINATA	F	352	0	6.6	0.7	89.8	3.6	3.0
4622	MUSA NANA	G	391	0	4.0	1.1	91.7	3.6	3.3
4623	MUSA NANA	F	394	0	4.3	1.4	91.5	3.2	2.8

OBS	CALCIUM	PHOSPHO	IRON	SODIUM	POTASS	CAROT	THIAMINE	RIBOFLAV	NIACIN	ASCORBIC	REFCO
4600	.	.	.	.	.	.	.	.	.	.	19
4601	.	.	.	.	.	.	0.57	0.27	1.91	.	1
4602	176.0	517.1	22.0	.	.	.	.	.	.	.	5
4603	.	.	.	.	.	.	.	.	.	.	1
4604	152.9	235.3	2.4	.	.	47.1	0.35	0.20	2.82	0.0	2
4605	1040.0	330.0	.	.	.	.	.	.	.	.	6
4606	430.0	280.0	.	.	.	.	.	.	.	.	6
4607	260.0	170.0	.	.	.	.	.	.	.	.	6
4608	.	.	.	.	.	.	.	.	.	.	6
4609	.	120.0	.	.	.	.	.	.	.	.	6
4610	270.0	90.0	.	.	.	.	.	.	.	.	6
4611	980.0	450.0	.	.	.	.	.	.	.	.	6
4612	420.0	170.0	.	.	.	.	.	.	.	.	6
4613	.	.	.	.	.	.	.	.	.	.	21
4614	.	.	.	.	.	.	.	.	.	.	21
4615	480.0	160.0	.	.	.	.	.	.	.	.	6
4616	257.0	289.2	4.8	8.0	1566.3	0.0	0.12	0.16	1.61	345.4	2
4617	527.4	396.6	5.1	.	.	126.6	0.25	0.21	2.11	379.7	3
4618	3003.7	.	11.5	.	.	22222.2	0.30	0.78	8.52	44.4	2
4619	2403.6	1780.4	9.2	.	.	22433.2	0.00	0.00	6.82	11.9	5
4620	48.1	115.4	6.3	86.5	2091.3	384.6	0.14	0.19	2.88	38.5	2
4621	26.3	95.4	1.6	.	.	.	0.20	0.20	2.30	42.8	3
4622	39.9	101.4	3.3	.	.	.	.	.	.	.	1
4623	35.5	92.2	2.8	.	.	1010.6	.	.	.	3.5	1

TABLE 3 - DATA CONVERTED TO A ZERO-MOISTURE BASIS.

OBS	PLNA	PART	CALORIES	WATER	PROTEIN	FAT	TOTCARB	FIBER	ASH	CALCIUM	PHOSPHO	IRON	SODIUM	POTASS	CAROT	THIAMINE	RIBOFLAV	NIACIN	ASCORBIC	REFCO
4624	MUSA NANA	F	350	0	5.1	0.7	91.0	2.2	3.2	36.1	115.5	3.2	.	.	577.6	0.18	0.32	2.17	28.9	3
4625	MUSA PARADISIACA SAPIENTUM	L	.	0	16.6	4.3	66.6	19.2	12.5	590.0	220.0	17.0	.	.	.	.	.	.	.	6
4626	MUSA TROGLODYTARUM	F	350	0	3.2	0.6	93.2	3.5	2.9	19.3	128.6	1.3	.	.	321.5	0.10	0.10	1.61	64.3	3
4627	MUSA URANOSCOPOS	F	351	0	3.9	0.7	91.8	3.2	3.5	53.2	88.7	1.4	.	.	1985.8	0.14	0.14	1.42	28.4	3
4628	MUSA X PARADISIACA	L	.	0	9.9	11.8	69.5	24.0	8.8	750.0	240.0	.	.	.	.	.	.	.	.	18
4629	MUSA X PARADISIACA	SH	.	0	2.4	2.3	81.0	20.5	14.3	1160.0	.	.	.	.	.	.	.	.	.	18
4630	MUSA X PARADISIACA	F	.	0	5.4	0.9	91.4	2.2	3.3	.	.	.	.	.	.	.	.	.	.	18
4631	MUSA X PARADISIACA	G	355	0	4.5	0.6	92.6	1.6	2.3	25.8	112.9	2.9	.	.	935.5	0.13	0.06	1.94	100.0	
4632	MUSA X PARADISIACA	F	390	0	3.5	0.9	92.8	1.4	2.9	23.1	109.8	3.8	.	.	2254.3	0.23	0.12	1.73	57.8	1
4633	MUSA X PARADISIACA	F	383	0	6.5	0.4	89.6	3.9	3.5	39.1	91.3	6.1	.	.	521.7	0.13	0.13	2.61	39.1	1
4634	MUSA X PARADISIACA	F	352	0	4.2	1.1	91.9	2.1	2.8	42.3	112.7	2.8	14.1	1412.0	792.3	0.11	0.14	2.11	49.3	2
4635	MUSA X PARADISIACA	F	351	0	5.0	0.8	90.7	2.3	3.5	34.7	42.5	2.3	.	.	270.3	0.69	0.19	1.93	30.9	3
4636	MUSA X PARADISIACA	F	355	0	2.9	0.9	93.9	1.5	2.3	23.3	98.8	2.3	.	.	1017.4	0.17	0.12	1.74	58.1	3
4637	MUSA X PARADISIACA	F	348	0	4.5	0.8	91.3	1.5	3.4	37.9	75.8	1.9	.	.	1250.0	0.15	0.15	2.27	30.3	3
4638	MUSA X PARADISIACA	G	353	0	3.2	0.3	94.4	1.1	2.1	21.4	107.0	2.1	.	.	2032.1	0.19	0.11	1.34	74.9	3
4639	MUSA X PARADISIACA	F	354	0	3.3	1.2	92.9	1.2	2.7	20.8	89.3	2.1	14.9	1145.8	1080.4	0.18	0.12	1.79	41.7	4
4640	MUSA X PARADISIACA	F	350	0	4.5	0.8	91.4	2.1	3.3	32.9	107.0	2.9	4.1	1522.6	469.1	0.21	0.25	2.88	41.2	4
4641	MUSCARI LONGIPES	S	.	0	15.4	10.4	.	.	.	.	.	.	.	.	.	.	.	.	.	21
4642	MUSCARI RACEMOSUM	S	.	0	14.9	10.0	.	.	.	.	.	.	.	.	.	.	.	.	.	21
4643	MUSCARI SP.	S	.	0	19.5	11.4	.	.	.	.	.	.	.	.	.	.	.	.	.	21
4644	MYOSOTIS SYLVATICA	S	.	0	16.9	42.8	.	.	.	.	.	.	.	.	.	.	.	.	.	20
4645	MYRCIARIA CAULIFLORA	F	400	0	3.0	10.9	83.6	6.1	2.4	369.7	103.0	1.2	.	.	0.0	0.06	0.12	1.82	103.0	2
4646	MYRCIARIA FLORIBUNDA	F	357	0	0.8	0.0	97.7	0.8	1.6	46.5	69.8	3.9	.	.	0.0	0.16	0.16	1.55	178.3	3
4647	MYRIANTHUS SP.	F	338	0	13.1	.	81.4	.	5.5	303.4	482.8	7.6	.	.	.	.	.	.	.	1
4648	MYRIANTHUS SP.	S	545	0	27.3	38.6	31.2	4.0	2.9	152.6	428.9	7.6	.	.	.	.	.	.	.	1
4649	MYRICA CERIFERA	S	.	0	4.9	13.3	.	.	.	.	.	.	.	.	.	.	.	.	.	20
4650	MYRICA CERIFERA	L	.	0	10.8	2.2	74.3	29.8	12.7	.	.	.	.	.	.	.	.	.	.	6
4651	MYRICA CORDIFOLIA	S	.	0	5.5	8.9	.	.	.	.	.	.	.	.	.	.	.	.	.	21
4652	MYRICA HETEROPHYLLA	S	.	0	5.6	31.2	.	.	.	.	.	.	.	.	.	.	.	.	.	19
4653	MYRICA PENSYLVANICA	S	.	0	6.2	35.3	.	.	.	.	.	.	.	.	.	.	.	.	.	20
4654	MYRICA PENSYLVANICA	S	.	0	9.4	4.3	.	.	.	.	.	.	.	.	.	.	.	.	.	20
4655	MYRISTICA CANARICA	S	.	0	7.9	53.2	37.5	21.7	1.3	.	.	.	.	.	.	.	.	.	.	5
4656	MYRISTICA FRAGRANS	A	517	0	7.3	35.3	55.0	5.2	2.4	274.5	119.8	15.1	87.1	504.4	522.9	0.34	0.49	1.47	.	16
4657	MYRISTICA FRAGRANS	S	560	0	6.2	38.7	52.6	4.3	2.5	196.2	227.1	3.2	17.1	373.1	65.0	0.37	0.06	1.39	.	16
4658	MYRISTICA FRAGRANS	S	.	0	8.8	42.5	46.8	13.5	2.0	140.0	280.0	5.4	.	.	.	.	.	.	.	5
4659	MYRISTICA FRAGRANS	A	.	0	7.7	29.0	61.4	4.5	1.9	214.0	118.9	15.0	.	.	.	.	.	.	.	5
4660	MYRISTICA FRAGRANS	F	.	0	7.6	3.0	84.8	.	4.5	303.0	75.8	15.2	.	.	37.9	.	.	.	.	5
4661	MYRISTICA MALABARICA	S	.	0	7.0	.	47.8	2.5	1.4	.	.	.	.	.	.	.	.	.	.	5
4662	MYRISTICA MALABARICA	A	.	0	7.6	.	24.8	3.2	1.6	.	.	.	.	.	.	.	.	.	.	5
4663	MYRSINE AFRICANA	S	.	0	10.3	19.7	.	.	.	.	.	.	.	.	.	.	.	.	.	21
4664	MYRTUS COMMUNIS	S	.	0	6.6	0.6	.	.	.	.	.	.	.	.	.	.	.	.	.	21
4665	MYRTUS COMMUNIS	S	.	0	6.9	6.3	.	.	.	.	.	.	.	.	.	.	.	.	.	21
4666	NAMA HAVARDII	S	.	0	22.3	14.5	.	.	.	.	.	.	.	.	.	.	.	.	.	20
4667	NAMA SP.	L	.	0	14.1	2.4	65.2	19.2	18.3	5020.0	120.0	.	.	2570.0	.	.	.	.	.	6
4668	NANDINA DOMESTICA	S	.	0	20.0	14.2	.	.	2.7	.	.	.	.	.	.	.	.	.	.	19

TABLE 3 - DATA CONVERTED TO A ZERO-MOISTURE BASIS.

OBS	PLNA	PART	CALORIES	WATER	PROTEIN	FAT	TOTLCARB	FIBER	ASH	CALCIUM	PHOSPHO	IRON	SODIUM	POTASS	CAROT	THIAMINE	RIBOFLAV	NIACIN	ASCORBIC	REFCO
4669	NASTURTIOPSIS ARABICA	S	.	0	30.6	33.4	.	.	.	.	.	.	.	.	.	.	.	.	.	21
4670	NASTURTIUM OFFICINALE	L	.	0	28.3	.	.	.	16.4	2060.0	.	.	335.0	3520.0	.	.	.	.	.	17
4671	NASTURTIUM OFFICINALE	S	.	0	29.4	27.6	.	.	.	.	.	.	.	.	.	.	.	.	.	20
4672	NASTURTIUM OFFICINALE	S	.	0	30.0	27.2	.	.	.	.	.	.	.	.	.	.	.	.	.	20
4673	NASTURTIUM OFFICINALE	L	281	0	29.7	3.1	53.1	12.5	14.1	1000.0	718.7	17.2	.	.	15000.0	0.47	4.22	.	.	1
4674	NASTURTIUM OFFICINALE	L	290	0	27.4	6.5	48.4	17.7	17.7	1838.7	677.4	30.6	661.3	3612.9	16532.3	1.13	2.10	8.06	741.9	2
4675	NASTURTIUM OFFICINALE	L	282	0	35.9	5.1	42.3	14.1	16.7	1500.0	974.4	24.4	.	.	28333.3	1.54	1.28	12.82	564.1	3
4676	NASTURTIUM OFFICINALE	L	284	0	32.8	4.5	44.8	10.4	17.9	2253.7	806.0	25.4	776.1	4209.0	43880.6	1.19	2.39	13.43	1179.1	4
4677	NASTURTIUM OFFICINALE	L	.	0	26.9	1.9	50.9	.	20.4	2685.2	1296.3	42.6	.	.	.	.	.	.	.	5
4678	NELUMBO NUCIFERA	S	318	0	24.2	1.0	70.2	13.1	4.5	298.0	298.0	.	.	.	.	.	.	.	.	1
4679	NELUMBO NUCIFERA	S	383	0	22.6	2.2	70.0	4.8	5.2	330.4	713.0	6.1	17.4	1665.2	0.0	0.65	0.26	7.83	43.5	2
4680	NELUMBO NUCIFERA	S	390	0	16.6	2.7	76.2	2.5	4.6	138.9	543.8	7.1	49.0	941.7	35.0	0.75	0.18	1.87	0.0	2
4681	NELUMBO NUCIFERA	R	.	0	16.7	0.6	74.1	4.9	6.8	370.4	.	.	.	.	.	1.36	0.37	12.96	92.6	5
4682	NEMESIA SUTTONII	S	.	0	28.8	47.7	.	.	4.7	.	.	.	.	.	.	.	.	.	.	19
4683	NEMOPHILA MENZIESII	S	.	0	11.5	3.3	.	.	3.1	.	.	.	.	.	.	.	.	.	.	19
4684	NEMOPHILA MENZIESII	S	.	0	13.1	4.6	.	.	3.4	.	.	.	.	.	.	.	.	.	.	19
4685	NENAX ACEROSA	S	.	0	33.2	6.1	.	.	.	.	.	.	.	.	.	.	.	.	.	21
4686	NEPETA CATARIA	S	.	0	18.4	21.2	.	.	3.2	.	.	.	.	.	.	.	.	.	.	19
4687	NEPETA CONGESTA	S	.	0	18.2	16.9	.	.	.	.	.	.	.	.	.	.	.	.	.	20
4688	NEPETA ITALICA	S	.	0	16.9	22.1	.	.	.	.	.	.	.	.	.	.	.	.	.	21
4689	NEPETA MUSSINI	S	.	0	21.7	25.3	.	.	8.0	.	.	.	.	.	.	.	.	.	.	19
4690	NEPETA NEPETELLA	S	.	0	18.6	26.8	.	.	.	.	.	.	.	.	.	.	.	.	.	21
4691	NEPETA TMOLEA	S	.	0	22.2	29.4	.	.	.	.	.	.	.	.	.	.	.	.	.	21
4692	NEPHELIUM LAPPACEUM	F	356	0	5.6	0.6	91.7	6.1	2.2	111.1	83.3	10.6	5.6	355.6	0.0	0.06	0.33	2.22	294.4	2
4693	NEPHELIUM LAPPACEUM	A	.	0	2.8	0.6	90.4	1.7	5.1	62.1	73.4	.	.	.	.	.	.	.	169.5	5
4694	NEPHELIUM MUTABILE	F	333	0	4.4	0.0	88.3	1.7	7.2	77.8	188.9	2.2	.	.	.	0.28	0.72	1.67	400.0	2
4695	NEPHELIUM MUTABILE	F	324	0	4.7	0.0	85.3	5.9	10.0	147.1	82.4	17.6	.	.	0.0	0.29	0.35	2.35	282.4	2
4696	NEPTUNIA OLERACEA	L	283	0	43.3	1.7	45.0	15.0	10.0	466.7	300.0	32.5	891.7	2308.3	29625.0	1.00	1.17	9.17	175.0	2
4697	NEPTUNIA PUBESCENS	S	.	0	26.2	3.9	.	.	.	.	.	.	.	.	.	.	.	.	.	20
4698	NERISYRENIA CAMPORUM	S	.	0	27.5	37.6	.	.	.	.	.	.	.	.	.	.	.	.	.	20
4699	NERIUM OLEANDER	S	.	0	30.0	29.4	.	.	.	.	.	.	.	.	.	.	.	.	.	20
4700	NESLIA PANICULATA	S	.	0	31.0	46.8	.	.	.	.	.	.	.	.	.	.	.	.	.	21
4701	NEUROLAENA LOBATA	S	.	0	21.9	34.8	.	.	.	.	.	.	.	.	.	.	.	.	.	20
4702	NICANDRA PHYSALODES	S	.	0	16.6	26.2	.	.	.	.	.	.	.	.	.	.	.	.	.	21
4703	NICOTIANA AFFINIS	S	.	0	20.0	39.2	.	.	4.0	.	.	.	.	.	.	.	.	.	.	19
4704	NICOTIANA ALATA	S	.	0	20.0	38.4	.	.	.	.	.	.	.	.	.	.	.	.	.	20
4705	NICOTIANA TABACUM	S	.	0	22.4	45.4	.	.	.	.	.	.	.	.	.	.	.	.	.	21
4706	NIGELLA CILIARIS	S	.	0	21.1	30.4	.	.	.	.	.	.	.	.	.	.	.	.	.	21
4707	NIGELLA DAMASCENA	S	.	0	25.7	42.7	.	.	5.3	.	.	.	.	.	.	.	.	.	.	19
4708	NIGELLA DAMASCENA	S	.	0	25.0	40.5	.	.	.	.	.	.	.	.	.	.	.	.	.	20
4709	NIGELLA HISPANICA	S	.	0	23.8	46.0	.	.	5.0	.	.	.	.	.	.	.	.	.	.	19
4710	NOLINA DURANGENSIS	S	.	0	14.4	23.3	.	.	.	.	.	.	.	.	.	.	.	.	.	20
4711	NOLINA MATAPENSIS	S	.	0	20.4	17.8	.	.	.	.	.	.	.	.	.	.	.	.	.	21
4712	NOLINA MICROCARPA	L	.	0	6.3	3.0	86.8	41.9	3.9	.	.	.	.	.	.	.	.	.	.	6
4713	NOLINA SP.	L	.	0	6.3	3.1	86.5	42.7	4.1	920.0	90.0	.	.	700.0	.	.	.	.	.	6

TABLE 3 - DATA CONVERTED TO A ZERO-MOISTURE BASIS.

OBS	PLNA	PART	CALORIES	WATER	PROTEIN	FAT	TOTLCARB	FIBER	ASH
4714	NOLINA TEXANA	S	.	0	15.6	20.3	.	.	.
4715	NOLINA TEXANA	S	.	0	11.2	10.2	.	.	.
4716	NOPALEA COCHENILLIFERA	F	361	0	7.8	1.8	88.6	16.9	1.8
4717	NOPALEA COCHENILLIFERA	SH	315	0	14.1	1.1	75.0	73.9	9.8
4718	NOTOBASIS SYRIACA	S	.	0	17.6	24.7	.	.	.
4719	NYMPHAEA ELEGANS	R	361	0	17.9	0.7	79.4	0.9	2.0
4720	NYMPHAEA LOTUS	R	333	0	14.4	0.6	81.9	2.8	3.1
4721	NYMPHAEA LOTUS	I	304	0	8.9	1.3	75.3	21.5	14.6
4722	NYMPHAEA LOTUS	S	405	0	9.0	2.4	87.2	6.5	1.5
4723	NYSSA OGECHE	S	.	0	5.6	7.4	.	.	2.0
4724	NYSSA OGECHE	S	.	0	3.8	1.6	.	.	2.7
4725	NYSSA OGECHE	S	.	0	5.6	8.9	.	.	.
4726	NYSSA OGECHE	S	.	0	2.2	1.2	.	.	.
4727	OCIMUM AMERICANUM	S	.	0	17.5	18.5	.	.	.
4728	OCIMUM BASILICUM	S	.	0	21.4	24.2	.	.	7.2
4729	OCIMUM BASILICUM	L	328	0	24.0	6.1	60.7	.	9.2
4730	OCIMUM BASILICUM	H	.	0	15.4	4.3	65.2	19.0	15.3
4731	OCIMUM BASILICUM	L	319	0	24.4	8.9	51.9	14.8	14.8
4732	OCIMUM CANUM	L	317	0	27.0	6.3	55.6	20.6	11.1
4733	OCIMUM KILIMANDSCHARICUM	S	.	0	20.1	18.6	54.3	28.8	6.9
4734	OCIMUM SANCTUM	S	.	0	13.8	15.5	.	.	.
4735	OCIMUM SP.	L	328	0	32.8	12.1	40.5	.	14.7
4736	OCOTEA ACUTIFOLIA	S	.	0	8.7	24.2	.	.	.
4737	OENANTHE GLOBULOSA	S	.	0	9.6	8.8	.	.	.

OBS	CALCIUM	PHOSPHO	IRON	SODIUM	POTASS	CAROT	THIAMINE	RIBOFLAV	NIACIN	ASCORBIC	REFCO
4714	.	.	.	.	.	.	.	.	.	.	19
4715	.	.	.	.	.	.	.	.	.	.	19
4716	84.3	126.5	1.8	.	.	0.0	0.06	0.18	3.01	102.4	3
4717	.	184.8	29.3	.	.	4782.6	0.33	0.43	4.35	173.9	3
4718	.	.	.	.	.	.	.	.	.	.	21
4719	124.7	346.9	6.3	.	.	0.0	0.39	0.09	14.74	2.3	3
4720	258.3	419.4	.	.	.	.	.	.	.	.	1
4721	278.5	101.3	.	.	.	.	.	.	.	.	1
4722	.	.	14.4	.	.	.	.	.	.	.	1
4723	.	.	.	.	.	.	.	.	.	.	19
4724	.	.	.	.	.	.	.	.	.	.	19
4725	.	.	.	.	.	.	.	.	.	.	21
4726	.	.	.	.	.	.	.	.	.	.	21
4727	.	.	.	.	.	.	.	.	.	.	21
4728	.	.	.	.	.	.	.	.	.	.	19
4729	2126.6	.	22.7	.	.	.	.	.	.	161.6	1
4730	2257.5	523.5	44.9	36.3	3667.7	6009.6	0.16	0.34	7.43	65.2	16
4731	2370.4	281.5	35.6	88.9	3177.8	33333.3	0.59	2.59	5.93	200.0	2
4732	730.2	460.3	15.9	.	.	63095.2	0.79	2.22	4.76	206.3	2
4733	.	.	.	.	.	.	.	.	.	.	5
4734	.	.	.	.	.	.	.	.	.	.	20
4735	5431.0	.	.	.	.	.	.	.	.	448.3	1
4736	.	.	.	.	.	.	.	.	.	.	21
4737	.	.	.	.	.	.	.	.	.	.	21

TABLE 3 - DATA CONVERTED TO A ZERO-MOISTURE BASIS.

OBS	PLNA	PART	CALORIES	WATER	PROTEIN	FAT	TOTLCARB	FIBER	ASH
4738	OENANTHE JAVANICA	L	298	0	19.1	3.2	62.8	12.8	14.9
4739	OENANTHE LACHENALII	S	.	0	17.6	22.1	.	.	.
4740	OENANTHE PIMPINELLOIDES	S	.	0	13.8	12.6	.	.	.
4741	OENOTHERA BIENNIS	S	.	0	16.4	24.6	.	.	6.3
4742	OENOTHERA BIENNIS	S	.	0	16.9	27.5	.	.	.
4743	OENOTHERA DEPRESSA	S	.	0	16.1	21.8	.	.	.
4744	OENOTHERA DEPRESSA	S	.	0	17.2	27.7	.	.	.
4745	OENOTHERA DEPRESSA	S	.	0	13.5	26.4	.	.	.
4746	OENOTHERA DRUMMONDII	S	.	0	16.9	25.0	.	.	.
4747	OENOTHERA DRUMMONDII	S	.	0	19.7	29.0	.	.	.
4748	OENOTHERA LACINIATA	S	.	0	11.9	17.3	.	.	.
4749	OENOTHERA LAMARCKIANA	S	.	0	17.2	26.1	.	.	6.8
4750	OENOTHERA LAMARCKIANA	S	.	0	20.0	28.4	.	.	8.7
4751	OENOTHERA LAMARCKIANA	S	.	0	20.0	28.3	.	.	.
4752	OENOTHERA MISSOURIENSIS VAR. INCANA	S	.	0	22.9	20.8	.	.	.
4753	OENOTHERA RHOMBIPETALA	S	.	0	15.6	30.7	.	.	9.3
4754	OENOTHERA SERRULATA	S	.	0	24.9	22.0	.	.	.
4755	OENOTHERA SP.	S	.	0	19.4	30.8	.	.	.
4756	OENOTHERA SP.	W	.	0	14.4	3.0	74.7	20.0	7.9
4757	OLEA EUROPAEA	L	.	0	13.1	7.3	73.5	17.7	6.1
4758	OLEA EUROPAEA	C	.	0	8.9	6.1	76.5	28.9	8.5
4759	OLEA EUROPAEA	F	532	0	6.4	58.3	6.0	6.0	29.4
4760	ONCOBA SPINOSA	S	.	0	12.5	29.1	.	.	.
4761	ONOBRYCHIS ARENARIA	S	.	0	.	6.1	.	.	.

OBS	CALCIUM	PHOSPHO	IRON	SODIUM	POTASS	CAROT	THIAMINE	RIBOFLAV	NIACIN	ASCORBIC	REFCO
4738	1202.1	585.1	31.9	191.5	4712.8	23297.9	0.64	2.34	10.64	148.9	2
4739	.	.	.	.	.	.	.	.	.	.	21
4740	.	.	.	.	.	.	.	.	.	.	21
4741	.	.	.	.	.	.	.	.	.	.	19
4742	.	.	.	.	.	.	.	.	.	.	19
4743	.	.	.	.	.	.	.	.	.	.	21
4744	.	.	.	.	.	.	.	.	.	.	21
4745	.	.	.	.	.	.	.	.	.	.	21
4746	.	.	.	.	.	.	.	.	.	.	20
4747	.	.	.	.	.	.	.	.	.	.	21
4748	.	.	.	.	.	.	.	.	.	.	20
4749	.	.	.	.	.	.	.	.	.	.	19
4750	.	.	.	.	.	.	.	.	.	.	19
4751	.	.	.	.	.	.	.	.	.	.	20
4752	.	.	.	.	.	.	.	.	.	.	20
4753	.	.	.	.	.	.	.	.	.	.	19
4754	.	.	.	.	.	.	.	.	.	.	20
4755	.	.	.	.	.	.	.	.	.	.	20
4756	1930.0	340.0	.	.	2940.0	.	.	.	.	.	6
4757	1180.0	90.0	.	.	.	.	.	.	.	.	18
4758	.	.	.	.	.	.	.	.	.	.	18
4759	279.8	78.0	7.3	11009.2	252.3	825.7	.	.	.	.	4
4760	.	.	.	.	.	.	.	.	.	.	20
4761	.	.	.	.	.	.	.	.	.	.	20

TABLE 3 - DATA CONVERTED TO A ZERO-MOISTURE BASIS.

OBS	PLANA	PART	CALORIES	WATER	PROTEIN	FAT	TOTLCARB	FIBER	ASH	CALCIUM	PHOSPHO	IRON	SODIUM	POTASS	CAROT	THIAMINE	RIBOFLAV	NIACIN	ASCORBIC	REFCO
4762	ONOBRYCHIS ARGYREA	S	.	0	49.2	16.8	.	.	.	.	.	.	.	.	.	.	.	.	.	20
4763	ONOBRYCHIS AURANTIACA	S	.	0	57.2	13.4	.	.	.	.	.	.	.	.	.	.	.	.	.	21
4764	ONOBRYCHIS CADMEA	S	.	0	41.5	5.7	.	.	.	.	.	.	.	.	.	.	.	.	.	21
4765	ONOBRYCHIS HYPARGYREA	S	.	0	47.6	13.8	.	.	.	.	.	.	.	.	.	.	.	.	.	21
4766	ONOBRYCHIS MATRITENSIS	S	.	0	47.0	7.3	.	.	.	.	.	.	.	.	.	.	.	.	.	21
4767	ONOBRYCHIS SP.	H	.	0	22.6	3.8	67.2	28.2	6.4	.	.	.	.	.	.	.	.	.	.	6
4768	ONOBRYCHIS SP.	W	.	0	17.4	3.3	70.9	25.5	8.4	.	.	.	.	.	.	.	.	.	.	6
4769	ONOBRYCHIS VICIAEFOLIA	S	.	0	40.9	7.6	.	.	3.4	.	.	.	.	.	.	.	.	.	.	19
4770	ONOBRYCHIS VICIAEFOLIA	S	.	0	33.3	7.9	.	.	4.8	.	.	.	.	.	.	.	.	.	.	19
4771	ONOBRYCHIS VULGARIS	S	.	0	40.5	7.0	.	.	3.9	.	.	.	.	.	.	.	.	.	.	19
4772	ONOBRYCHIS VULGARIS	S	.	0	40.9	6.8	.	.	3.9	.	.	.	.	.	.	.	.	.	.	19
4773	ONOBRYCHIS VULGARIS	S	.	0	39.8	6.7	.	.	4.2	.	.	.	.	.	.	.	.	.	.	19
4774	ONONIS ALOPECUROIDES	S	.	0	41.4	6.3	.	.	.	.	.	.	.	.	.	.	.	.	.	21
4775	ONONIS LEIOSPERMA	S	.	0	46.2	8.2	.	.	.	.	.	.	.	.	.	.	.	.	.	21
4776	ONONIS PUBESCENS	S	.	0	35.7	12.0	.	.	.	.	.	.	.	.	.	.	.	.	.	21
4777	ONOPORDUM ACANTHIUM	S	.	0	18.8	20.1	.	.	.	.	.	.	.	.	.	.	.	.	.	20
4778	ONOPORDUM ACANTHIUM	S	.	0	16.8	10.1	.	.	.	.	.	.	.	.	.	.	.	.	.	21
4779	ONOPORDUM ACANTHIUM	S	.	0	18.6	17.4	.	.	.	.	.	.	.	.	.	.	.	.	.	21
4780	ONOPORDUM ARABICUM	S	.	0	16.9	19.2	.	.	.	.	.	.	.	.	.	.	.	.	.	20
4781	ONOPORDUM ARABICUM	S	.	0	17.9	12.5	.	.	.	.	.	.	.	.	.	.	.	.	.	21
4782	ONOPORDUM CYNAROCEPHALUM	S	.	0	17.8	16.6	.	.	.	.	.	.	.	.	.	.	.	.	.	21
4783	ONOSMA SERICEUM	S	.	0	19.4	20.3	.	.	.	.	.	.	.	.	.	.	.	.	.	21
4784	ONOSMA STELLULATUM	S	.	0	24.4	22.9	.	.	.	.	.	.	.	.	.	.	.	.	.	21
4785	ONOSMODIUM MOLLE	S	.	0	11.9	17.2	.	.	.	.	.	.	.	.	.	.	.	.	.	21
4786	ONOSMODIUM MOLLE	S	.	0	33.0	53.6	.	.	.	.	.	.	.	.	.	.	.	.	.	21
4787	ONOSMODIUM OCCIDENTALE	S	.	0	35.7	56.4	.	.	3.6	.	.	.	.	.	.	.	.	.	.	19
4788	ONOSMODIUM OCCIDENTALE	S	.	0	6.8	7.1	.	.	59.0	.	.	.	.	.	.	.	.	.	.	19
4789	OOTHRINAX ANOMALA	S	.	0	4.9	1.8	.	.	1.6	.	.	.	.	.	.	.	.	.	.	19
4790	OOTHRINAX ANOMALA	S	.	0	2.8	0.6	.	.	0.6	.	.	.	.	.	.	.	.	.	.	19
4791	OPERCULINA TURPETHUM	L	320	0	26.0	4.7	60.0	9.3	9.3	800.0	546.7	42.7	.	.	15633.3	1.20	1.67	10.00	320.0	2
4792	OPOPANAX HISPIDUM	S	.	0	23.8	15.6	.	.	.	.	.	.	.	.	.	.	.	.	.	21
4793	OPUNTIA AMYCLAEA	SH	.	0	4.0	2.7	68.9	12.0	24.4	.	.	.	.	.	.	.	.	.	.	18
4794	OPUNTIA BRASILIENSIS	SH	.	0	6.5	3.3	.	10.9	17.6	.	.	.	.	.	.	.	.	.	.	18
4795	OPUNTIA CHOLLA	SH	.	0	5.5	4.1	76.4	13.8	14.0	.	.	.	.	.	.	.	.	.	.	18
4796	OPUNTIA ELATIOR	SH	.	0	6.3	3.3	71.6	13.5	18.8	.	.	.	.	.	.	.	.	.	.	18
4797	OPUNTIA FICUS-INDICA	SH	.	0	9.6	1.3	63.2	13.4	25.9	2810.0	280.0	.	.	.	.	.	.	.	.	18
4798	OPUNTIA FICUS-INDICA	F	360	0	5.9	2.2	89.2	5.9	2.7	306.5	172.0	6.5	.	.	.	0.05	0.11	1.61	96.8	3
4799	OPUNTIA FICUS-INDICA	S	.	0	14.0	6.3	55.8	12.8	24.1	.	.	.	.	.	.	.	.	.	.	5
4800	OPUNTIA MAXIMA	SH	.	0	6.5	1.6	74.2	11.3	17.7	.	.	.	.	.	.	.	.	.	.	18
4801	OPUNTIA MEGACANTHA	F	342	0	5.2	0.6	89.0	12.3	5.2	277.4	58.1	1.9	.	.	0.0	0.13	0.19	2.58	161.3	2
4802	OPUNTIA SP.	F	363	0	5.8	2.3	90.1	22.8	1.8	181.3	157.9	1.8	.	.	.	.	.	.	117.0	1
4803	OPUNTIA SP.	F	330	0	9.8	3.6	78.6	23.2	8.0	982.1	178.6	4.5	.	.	892.9	0.36	0.36	1.79	169.6	3
4804	OPUNTIA SP.	F	390	0	9.1	9.1	78.6	42.9	3.2	298.7	207.8	7.8	.	.	129.9	0.13	0.19	2.60	142.9	3
4805	OPUNTIA SP.	F	350	0	4.2	0.8	90.8	13.3	4.2	166.7	233.3	2.5	16.7	1383.3	300.0	0.08	0.25	3.33	183.3	4
4806	OPUNTIA SP.(ALL)	W	.	0	5.0	2.0	73.0	13.3	20.0	6290.0	80.0	90.0	.	1210.0	1.0	.	.	.	.	6

TABLE 3 - DATA CONVERTED TO A ZERO-MOISTURE BASIS.

OBS	PLNA	PART	CALORIES	WATER	PROTEIN	FAT	TOTLCARB	FIBER
4807	ORBIGNYA COHUNE	S	.	0	6.9	52.2	.	.
4808	ORBIGNYA COHUNE	S	.	0	1.2	0.5	.	.
4809	ORBIGNYA SPECIOSA	S	.	0	9.4	62.9	.	.
4810	ORBIGNYA SPECIOSA	S	.	0	16.2	0.2	.	.
4811	ORCHIS LAXIFLORA	S	.	0	10.9	0.7	.	.
4812	ORIGANUM VULGARE	H	330	0	11.9	11.0	69.4	16.2
4813	ORMOSIA JAMAICENSIS	S	.	0	12.5	9.3	.	.
4814	ORMOSIA PANAMENSIS	S	.	0	15.0	7.7	.	.
4815	ORNITHOGALUM NUTANS	S	.	0	17.5	6.3	.	.
4816	ORNITHOPUS COMPRESSUS	S	.	0	41.4	15.5	.	.
4817	ORNITHOPUS SATIVUS	W	.	0	24.4	3.7	60.4	18.7
4818	ORNITHOPUS SATIVUS	W	.	0	14.8	3.7	66.2	23.0
4819	ORTHOCARPUS LUTEUS	W	.	0	6.7	2.1	.	.
4820	ORYZA GLABERRIMA	S	386	0	7.4	2.1	83.2	.
4821	ORYZA SATIVA	W	.	0	9.0	1.8	73.3	28.3
4822	ORYZA SATIVA	H	.	0	8.2	1.8	74.3	32.0
4823	ORYZA SATIVA	H	.	0	4.2	1.4	77.5	35.1
4824	ORYZA SATIVA	S	394	0	6.9	2.2	85.2	11.5
4825	ORYZA SATIVA	S	395	0	6.7	2.7	85.1	12.1
4826	ORYZA SATIVA	S	410	0	8.3	1.7	89.2	0.9
4827	ORYZA SATIVA	S	409	0	8.5	2.2	88.0	1.0
4828	ORYZOPSIS HYMENOIDES	S	.	0	11.9	2.2	.	.
4829	ORYZOPSIS HYMENOIDES	W	.	0	4.0	1.9	85.4	38.0
4830	ORYZOPSIS MILIACEA	S	.	0	14.4	2.1	.	.

OBS	ASH	CALCIUM	PHOSPHO	IRON	SODIUM	POTASS	CAROT	THIAMINE	RIBOFLAV	NIACIN	ASCORBIC	REFCO
4807	.	.	.	.	.	.	.	.	.	.	.	19
4808	.	.	.	.	.	.	.	.	.	.	.	19
4809	.	.	.	.	.	.	.	.	.	.	.	19
4810	.	.	.	.	.	.	.	.	.	.	.	19
4811	.	.	.	.	.	.	.	.	.	.	.	21
4812	7.8	1698.3	215.5	47.4	16.2	1798.5	4463.36	0.37	.	6.70	.	16
4813	2.0	.	.	.	.	.	.	.	.	.	.	19
4814	1.8	.	.	.	.	.	.	.	.	.	.	19
4815	.	.	.	.	.	.	.	.	.	.	.	20
4816	.	.	.	.	.	.	.	.	.	.	.	21
4817	11.5	1210.0	240.0	.	.	.	.	.	.	.	.	18
4818	15.3	.	.	.	.	.	.	.	.	.	.	6
4819	9.2	1030.0	490.0	.	.	.	.	.	.	.	.	6
4820	7.3	28.1	264.9	2.2	.	.	.	0.35	.	5.61	.	1
4821	15.9	.	.	.	.	.	.	.	.	.	.	18
4822	15.7	.	.	.	.	.	.	.	.	.	.	18
4823	16.9	240.0	90.0	.	.	1320.0	.	.	.	.	.	6
4824	5.7	30.1	347.8	8.5	.	.	.	0.38	0.06	6.47	.	1
4825	5.6	27.8	273.5	1.6	93.9	180.8	.	0.38	0.07	6.49	.	2
4826	0.8	16.1	265.5	3.0	.	.	0.00	0.25	0.06	4.60	0	3
4827	1.4	36.4	251.1	1.8	10.2	243.2	0.00	0.39	0.06	5.34	0	4
4828	.	.	.	.	.	.	.	.	.	.	.	19
4829	8.7	580.0	50.0	.	.	.	.	.	.	.	.	6
4830	3.1	.	.	.	.	.	.	.	.	.	.	19

TABLE 3 - DATA CONVERTED TO A ZERO-MOISTURE BASIS.

OBS	PLNA	PART	CALORIES	WATER	PROTEIN	FAT	TOTLCARB	FIBER
4831	OSMANTHUS HETEROPHYLLUS	S	.	0	14.20	25.9	.	.
4832	OSMORHIZA CHILENSIS	S	.	0	24.70	8.5	.	.
4833	OSMORHIZA OCCIDENTALIS	S	.	0	20.80	10.4	.	.
4834	OSMORHIZA OCCIDENTALIS	W	.	0	10.50	2.9	.	.
4835	OSTEOSPERMUM AMPLECTANS	S	.	0	13.10	17.1	.	.
4836	OSTEOSPERMUM ASPERULUM	S	.	0	.	9.3	.	.
4837	OSTEOSPERMUM CALENDULACEUM	S	.	0	.	52.0	.	.
4838	OSTEOSPERMUM CAULESCENS	S	.	0	13.40	14.2	.	.
4839	OSTEOSPERMUM CLANDESTINUM	S	.	0	12.50	6.6	.	.
4840	OSTEOSPERMUM CLANDESTINUM	S	.	0	38.10	42.0	.	.
4841	OSTEOSPERMUM CLANDESTINUM	S	.	0	38.00	42.6	.	.
4842	OSTEOSPERMUM CORYMBOSUM	S	.	0	.	8.7	.	.
4843	OSTEOSPERMUM DREGEI	S	.	0	9.40	14.6	.	.
4844	OSTEOSPERMUM ECKLONIS	S	.	0	23.80	48.3	.	.
4845	OSTEOSPERMUM ECKLONIS	S	.	0	30.60	50.4	.	.
4846	OSTEOSPERMUM FRUTICOSUM	S	.	0	.	12.9	.	.
4847	OSTEOSPERMUM HYOSEROIDES	S	.	0	35.10	49.7	.	.
4848	OSTEOSPERMUM IMBRICATUM	S	.	0	.	4.7	.	.
4849	OSTEOSPERMUM IMBRICATUM	S	.	0	.	6.4	.	.
4850	OSTEOSPERMUM IMBRICATUM SUBSP. NERVATUM	S	.	0	.	10.3	.	.
4851	OSTEOSPERMUM JUCUNDUM	S	.	0	.	16.6	.	.
4852	OSTEOSPERMUM JUNCEUM	S	.	0	25.80	61.2	.	.
4853	OSTEOSPERMUM MURICATUM	S	.	0	11.90	9.9	.	.
4854	OSTEOSPERMUM MURICATUM	S	.	0	11.80	8.8	.	.

OBS	ASH	CALCIUM	PHOSPHO	IRON	SODIUM	POTASS	CAROT	THIAMINE	RIBOFLAV	NIACIN	ASCORBIC	REFCO
4831	.	.	.	.	.	.	.	.	.	.	.	21
4832	.	.	.	.	.	.	.	.	.	.	.	21
4833	.	.	.	.	.	.	.	.	.	.	.	20
4834	16.50	2940.0	450.0	.	.	.	.	.	.	.	.	6
4835	.	.	.	.	.	.	.	.	.	.	.	20
4836	.	.	.	.	.	.	.	.	.	.	.	21
4837	.	.	.	.	.	.	.	.	.	.	.	20
4838	.	.	.	.	.	.	.	.	.	.	.	20
4839	.	.	.	.	.	.	.	.	.	.	.	20
4840	.	.	.	.	.	.	.	.	.	.	.	20
4841	.	.	.	.	.	.	.	.	.	.	.	20
4842	.	.	.	.	.	.	.	.	.	.	.	21
4843	.	.	.	.	.	.	.	.	.	.	.	20
4844	.	.	.	.	.	.	.	.	.	.	.	20
4845	.	.	.	.	.	.	.	.	.	.	.	20
4846	.	.	.	.	.	.	.	.	.	.	.	21
4847	.	.	.	.	.	.	.	.	.	.	.	20
4848	.	.	.	.	.	.	.	.	.	.	.	21
4849	.	.	.	.	.	.	.	.	.	.	.	21
4850	.	.	.	.	.	.	.	.	.	.	.	21
4851	.	.	.	.	.	.	.	.	.	.	.	21
4852	.	.	.	.	.	.	.	.	.	.	.	20
4853	.	.	.	.	.	.	.	.	.	.	.	20
4854	.	.	.	.	.	.	.	.	.	.	.	20

TABLE 3 - DATA CONVERTED TO A ZERO-MOISTURE BASIS.

OBS	PLNA	PART	CALORIES	WATER	PROTEIN	FAT	TOTLCARB	FIBER	ASH
4855	OSTEOSPERMUM MURICATUM	S	.	0	11.90	10.6	.	.	.
4856	OSTEOSPERMUM MURICATUM	S	.	0	11.40	8.2	.	.	.
4857	OSTEOSPERMUM MURICATUM	S	.	0	.	5.6	.	.	.
4858	OSTEOSPERMUM SCARIOSUM	S	.	0	37.10	40.8	.	.	.
4859	OSTEOSPERMUM SCARIOSUM	S	.	0	.	39.5	.	.	.
4860	OSTEOSPERMUM SCARIOSUM	S	.	0	37.40	42.6	.	.	.
4861	OSTEOSPERMUM SINUATUM	S	.	0	41.20	43.9	.	.	.
4862	OSTEOSPERMUM SINUATUM	S	.	0	41.90	42.6	.	.	.
4863	OSTEOSPERMUM SINUATUM	S	.	0	39.40	43.3	.	.	.
4864	OSTEOSPERMUM SINUATUM	S	.	0	47.00	31.8	.	.	.
4865	OSTEOSPERMUM SINUATUM	S	.	0	48.20	30.8	.	.	.
4866	OSTEOSPERMUM SINUATUM	S	.	0	45.40	39.4	.	.	.
4867	OSTEOSPERMUM SINUATUM		.	0	14.01	.	60.58	20.22	22.04
4868	OSTEOSPERMUM SPINESCENS	C	.	0	9.20	3.2	72.30	26.40	15.30
4869	OSTEOSPERMUM SPINESCENS	S	.	0	40.00	43.3	.	.	.
4870	OSTEOSPERMUM SPINESCENS	S	.	0	.	41.2	.	.	.
4871	OSTEOSPERMUM SPINESCENS	S	.	0	.	36.8	.	.	.
4872	OSTEOSPERMUM SPINESCENS	S	.	0	.	4.5	.	.	.
4873	OSTEOSPERMUM SPINESCENS	S	.	0	.	7.2	.	.	.
4874	OSTEOSPERMUM SPINESCENS	S	.	0	41.90	42.6	.	.	.
4875	OSTEOSPERMUM SPINOSUM	S	.	0	.	6.6	.	.	.
4876	OSTEOSPERMUM SPINOSUM	S	.	0	.	6.9	.	.	.
4877	OSTRYA CARPINIFOLIA	S	.	0	13.00	12.7	.	.	.
4878	OSYRIS ALBA	S	.	0	10.60	36.1	.	.	.

OBS	CALCIUM	PHOSPHO	IRON	SODIUM	POTASS	CAROT	THIAMINE	RIBOFLAV	NIACIN	ASCORBIC	REFCO
4855	.	.	.	.	.	.	.	.	.	.	20
4856	.	.	.	.	.	.	.	.	.	.	20
4857	.	.	.	.	.	.	.	.	.	.	20
4858	.	.	.	.	.	.	.	.	.	.	20
4859	.	.	.	.	.	.	.	.	.	.	20
4860	.	.	.	.	.	.	.	.	.	.	20
4861	.	.	.	.	.	.	.	.	.	.	20
4862	.	.	.	.	.	.	.	.	.	.	20
4863	.	.	.	.	.	.	.	.	.	.	20
4864	.	.	.	.	.	.	.	.	.	.	20
4865	.	.	.	.	.	.	.	.	.	.	20
4866	.	.	.	.	.	.	.	.	.	.	21
4867	980.0	200.0	18.6	650.0	1780.0	.	.	.	.	.	8
4868	1040.0	240.0	.	.	.	.	.	.	.	.	18
4869	.	.	.	.	.	.	.	.	.	.	20
4870	.	.	.	.	.	.	.	.	.	.	20
4871	.	.	.	.	.	.	.	.	.	.	20
4872	.	.	.	.	.	.	.	.	.	.	21
4873	.	.	.	.	.	.	.	.	.	.	21
4874	.	.	.	.	.	.	.	.	.	.	21
4875	.	.	.	.	.	.	.	.	.	.	21
4876	.	.	.	.	.	.	.	.	.	.	21
4877	.	.	.	.	.	.	.	.	.	.	21
4878	.	.	.	.	.	.	.	.	.	.	21

TABLE 3 - DATA CONVERTED TO A ZERO-MOISTURE BASIS.

OBS	PLANA	PART	CALORIES	WATER	PROTEIN	FAT	TOTCARB	FIBER	ASH	CALCIUM	PHOSPHO	IRON	SODIUM	POTASS	CAROT	THIAMINE	RIBOFLAV	NIACIN	ASCORBIC	REFCO
4879	OTHONNOPSIS INTERMEDIA	S	.	0	31.8	44.1	.	.	.	.	.	.	.	.	.	.	.	.	.	21
4880	OXALIS EUROPAEA	S	.	0	21.9	47.6	.	.	.	.	.	.	.	.	.	.	.	.	.	20
4881	OXALIS TUBEROSA	R	389	0	6.2	3.7	85.2	4.9	4.9	24.7	209.9	4.9	.	.	0.00	0.31	0.43	2.47	228.4	3
4882	OXYTROPIS FLORIBUNDA	S	.	0	34.4	4.0	.	.	5.1	.	.	.	.	.	.	.	.	.	.	19
4883	OXYTROPIS LAGOPUS	W	.	0	17.1	2.0	74.3	29.3	6.6	.	.	.	.	.	.	.	.	.	.	6
4884	OXYTROPIS LAMBERTII	S	.	0	36.5	7.2	.	.	.	.	.	.	.	.	.	.	.	.	.	20
4885	OXYTROPIS SERICEA	S	.	0	41.2	6.0	.	.	.	.	.	.	.	.	.	.	.	.	.	20
4886	OXYTROPIS URALENSIS	S	.	0	27.4	4.3	.	.	3.6	.	.	.	.	.	.	.	.	.	.	19
4887	PACHIRA MACROCARPA	S	560	0	16.9	41.4	37.9	13.1	3.7	87.7	302.3	4.0	76.1	700.8	1300.21	0.03	0.06	4.02	25.4	2
4888	PACHYCEREUS PECTEN-ABORIGINUM	S	.	0	29.4	32.0	.	.	.	.	.	.	.	.	.	.	.	.	.	19
4889	PACHYRRHIZUS ANGULATUS	R	357	0	11.0	0.6	84.4	9.1	3.9	97.4	266.2	5.2	.	.	0.00	0.26	0.26	1.95	64.9	2
4890	PACHYRRHIZUS EROSUS	R	359	0	10.7	0.8	84.0	4.6	4.6	.	.	.	.	.	.	.	.	.	.	1
4891	PACHYRRHIZUS EROSUS	R	365	0	12.7	1.6	81.7	10.3	4.0	142.9	134.9	6.3	15.9	698.4	0.00	0.63	0.48	4.76	119.0	2
4892	PACHYRRHIZUS EROSUS	R	369	0	9.8	0.8	86.9	5.7	2.5	147.5	131.1	6.6	.	.	0.00	0.25	0.25	2.46	172.1	3
4893	PACHYRRHIZUS SP.	R	369	0	9.4	1.3	85.9	4.7	3.4	100.7	120.8	4.0	.	.	0.00	0.27	0.20	2.01	134.2	4
4894	PACHYRRHIZUS TUBEROSUS	R	.	0	8.5	0.6	88.6	3.4	2.8	90.9	.	6.3	.	.	.	0.28	0.11	1.14	79.5	5
4895	PACHYRRHIZUS VERNALIS	S	.	0	28.1	24.5	.	.	.	.	.	.	.	.	.	.	.	.	.	20
4896	PAEONIA BROWNII	S	.	0	14.1	18.7	.	.	3.0	.	.	.	.	.	.	.	.	.	.	19
4897	PAEONIA PEREGRINA	S	.	0	18.8	38.3	.	.	.	.	.	.	.	.	.	.	.	.	.	20
4898	PALIURIS SPINA-CHRISTI	S	.	0	31.5	19.2	.	.	.	.	.	.	.	.	.	.	.	.	.	21
4899	PALIURUS SPINA-CHRISTI	S	.	0	32.5	21.0	.	.	.	.	.	.	.	.	.	.	.	.	.	20
4900	PALLENSIS SPINOSA	S	.	0	28.8	25.2	.	.	.	.	.	.	.	.	.	.	.	.	.	21
4901	PANAX QUINQUEFOLIUS	R	.	0	8.7	0.6	86.1	6.4	4.6	250.0	270.0	36.6	52.0	660.0	.	.	.	.	.	7
4902	PANCRATIUM MARITIMUM	S	.	0	22.0	37.2	.	.	.	.	.	.	.	.	.	.	.	.	.	21
4903	PANDA OLEOSA	S	680	0	20.9	70.4	4.5	.	4.2	116.1	237.7	.	.	.	.	.	.	.	.	1
4904	PANDA OLEOSA	S	581	0	24.6	47.5	24.1	6.3	3.9	389.7	549.4	.	.	.	.	.	.	.	.	1
4905	PANDANUS SP.	F	476	0	15.6	26.0	56.8	21.6	1.6	.	.	.	.	.	31.75	0.29	.	4.44	28.6	2
4906	PANICUM ANTIDOTALE	W	.	0	18.8	3.0	68.1	28.0	10.1	.	.	.	.	.	.	.	.	.	.	18
4907	PANICUM ANTIDOTALE	S	.	0	15.0	5.8	.	.	4.0	.	.	.	.	.	.	.	.	.	.	19
4908	PANICUM ANTIDOTALE	S	.	0	9.2	5.4	.	.	.	.	.	.	.	.	.	.	.	.	.	21
4909	PANICUM ANTIDOTALE	W	.	0	18.9	2.3	65.7	23.5	13.1	610.0	260.0	.	.	.	.	.	.	.	.	6
4910	PANICUM ARIZONICUM	W	.	0	8.0	1.6	78.0	36.9	12.4	540.0	130.0	.	.	.	.	.	.	.	.	6
4911	PANICUM BULBOSUM	W	.	0	20.8	3.4	62.7	24.5	13.1	.	.	.	.	.	.	.	.	.	.	6
4912	PANICUM CAPILLARE	W	.	0	16.6	1.9	68.3	27.8	13.2	630.0	270.0	.	.	.	.	.	.	.	.	6
4913	PANICUM CAPILLARE OCCIDENTALE	W	.	0	9.6	2.2	77.6	33.1	10.6	610.0	220.0	.	.	.	.	.	.	.	.	6
4914	PANICUM COLORATUM	W	.	0	18.9	2.6	67.5	28.6	11.0	.	.	.	.	.	.	.	.	.	.	18
4915	PANICUM COLORATUM	H	.	0	9.0	2.8	80.3	36.7	7.9	.	.	.	.	.	.	.	.	.	.	18
4916	PANICUM COLORATUM	S	.	0	16.2	4.2	.	.	.	.	.	.	.	.	.	.	.	.	.	20
4917	PANICUM FASCICULATUM	W	.	0	18.3	2.5	65.5	27.6	13.7	.	.	.	.	.	.	.	.	.	.	6
4918	PANICUM GEMINATUM	W	.	0	9.9	2.1	73.0	31.1	14.9	490.0	230.0	.	.	.	.	.	.	.	.	6
4919	PANICUM HALLII	W	.	0	11.0	2.0	77.3	29.3	9.7	520.0	170.0	.	.	1520.0	.	.	.	.	.	6
4920	PANICUM HELLERI	W	.	0	10.3	2.5	72.6	28.1	14.6	320.0	100.0	.	.	.	.	.	.	.	.	6
4921	PANICUM HEMITOMON	W	.	0	16.5	1.9	71.3	29.9	10.3	250.0	380.0	11.0	.	1590.0	.	.	.	.	.	6
4922	PANICUM HIRTICAULE	W	.	0	11.9	1.8	75.3	40.8	11.0	480.0	180.0	.	.	1690.0	.	.	.	.	.	6
4923	PANICUM LAEVIFOLIUM	W	.	0	5.4	1.5	82.6	40.2	10.5	.	.	.	.	.	.	.	.	.	.	18

TABLE 3 - DATA CONVERTED TO A ZERO-MOISTURE BASIS.

OBS	PLANA	PART	CALORIES	WATER	PROTEIN	FAT	TOTCARB	FIBER	ASH	CALCIUM	PHOSPHO	IRON	SODIUM	POTASS	CAROT	THIAMINE	RIBOFLAV	NIACIN	ASCORBIC	REFCO
4924	PANICUM LAEVIFOLIUM	H	.	0	8.7	1.9	82.0	39.9	7.4	.	.	.	.	.	.	.	.	.	.	18
4925	PANICUM MASAIENSE	W	.	0	19.7	2.5	67.1	24.0	10.7	.	.	.	.	.	.	.	.	.	.	18
4926	PANICUM MAXIMUM	W	.	0	20.5	0.9	67.2	24.1	11.4	.	.	.	.	.	.	.	.	.	.	18
4927	PANICUM MAXIMUM	W	.	0	5.3	1.4	82.7	39.6	10.6	.	.	.	.	.	.	.	.	.	.	18
4928	PANICUM MAXIMUM	H	.	0	11.9	3.2	72.9	31.7	12.0	.	.	.	.	.	.	.	.	.	.	18
4929	PANICUM MAXIMUM	H	.	0	5.5	1.4	82.7	40.1	10.4	.	.	.	.	.	.	.	.	.	.	18
4930	PANICUM MAXIMUM	W	.	0	5.9	1.6	81.9	35.7	10.6	2090.0	590.0	.	.	.	.	.	.	.	.	6
4931	PANICUM MILIACEUM	H	.	0	12.5	2.5	78.4	33.9	6.6	.	.	.	.	.	.	.	.	.	.	18
4932	PANICUM MILIACEUM	W	.	0	8.4	2.6	81.5	29.0	7.5	.	.	.	.	.	.	.	.	.	.	6
4933	PANICUM MILIACEUM	S	377	0	14.7	4.0	76.9	10.5	4.4	23.1	312.1	4.0	.	.	0.00	0.46	0.12	4.62	0.0	2
4934	PANICUM MILIACEUM	S	371	0	11.2	3.3	82.7	3.6	2.8	22.7	352.6	7.7	.	487.5	0.00	0.83	0.43	2.61	0.0	4
4935	PANICUM MILIACEUM	S	.	0	14.2	1.2	80.7	2.5	3.9	11.4	374.6	6.5	.	.	0.00	0.89	0.11	1.14	0.0	5
4936	PANICUM MILIACEUM	L	.	0	8.1	2.4	82.2	30.0	7.3	.	.	.	.	.	.	.	.	.	.	5
4937	PANICUM MILIACEUM	L	.	0	10.3	2.4	79.2	26.5	8.1	.	.	.	.	.	.	.	.	.	.	5
4938	PANICUM OBTUSUM	W	.	0	5.4	1.8	82.3	35.8	10.5	400.0	200.0	.	.	2480.0	.	.	.	.	.	6
4939	PANICUM PRIONITIS	S	.	0	22.2	6.2	.	.	.	.	.	.	.	.	.	.	.	.	.	21
4940	PANICUM PURPURASCENS	W	.	0	7.7	1.7	80.2	32.6	10.4	600.0	340.0	15.0	.	1550.0	.	.	.	.	.	6
4941	PANICUM REPENS	W	.	0	14.0	2.1	70.5	32.6	13.4	.	.	.	.	.	.	.	.	.	.	18
4942	PANICUM REPENS	L	.	0	8.5	3.2	75.6	27.4	12.8	670.0	730.0	.	.	.	.	.	.	.	.	5
4943	PANICUM REPENS	L	.	0	9.0	1.7	83.4	24.4	5.8	.	.	.	.	.	.	.	.	.	.	5
4944	PANICUM SUMATRENSIS	S	.	0	8.1	5.5	82.5	10.8	3.9	27.0	359.6	7.9	.	.	.	0.38	.	.	.	5
4945	PANICUM TEXANUM	W	.	0	13.4	2.7	66.3	28.3	17.6	230.0	230.0	.	.	.	.	.	.	.	.	6
4946	PANICUM TRICHOCLADUM	W	.	0	9.6	2.5	79.4	35.0	8.5	.	.	.	.	.	.	.	.	.	.	18
4947	PANICUM TURGIDUM	S	.	0	12.4	4.2	.	.	.	.	.	.	.	.	.	.	.	.	.	21
4948	PANICUM VIRGATUM	W	.	0	6.0	2.3	85.5	35.0	6.2	330.0	120.0	.	.	.	.	.	.	.	.	6
4949	PAPAVER CLAVATUM	S	.	0	25.4	34.9	.	.	.	.	.	.	.	.	.	.	.	.	.	21
4950	PAPAVER MONATHUM	S	.	0	25.5	47.7	.	.	4.6	.	.	.	.	.	.	.	.	.	.	19
4951	PAPAVER RHOEAS	S	.	0	21.8	47.6	.	.	8.0	.	.	.	.	.	.	.	.	.	.	19
4952	PAPAVER RHOEAS	S	.	0	25.0	43.0	.	.	.	.	.	.	.	.	.	.	.	.	.	20
4953	PAPAVER RHOEAS	S	.	0	21.7	42.3	.	.	.	.	.	.	.	.	.	.	.	.	.	20
4954	PAPAVER RHOEAS	S	.	0	26.9	45.1	.	.	.	.	.	.	.	.	.	.	.	.	.	21
4955	PAPAVER RHOEAS	S	.	0	27.3	40.1	.	.	.	.	.	.	.	.	.	.	.	.	.	21
4956	PAPAVER SOMNIFERUM	S	.	0	22.5	35.0	34.4	9.3	8.1	.	.	.	.	.	.	.	.	.	.	18
4957	PAPAVER SOMNIFERUM	S	572	0	19.3	48.0	25.4	6.8	7.3	1553.6	909.9	10.1	22.5	751.1	0.00	0.91	0.18	1.05	.	16
4958	PARACARYUM ANGUSTIFOLIUM	S	.	0	29.4	39.0	.	.	.	.	.	.	.	.	.	.	.	.	.	21
4959	PARACARYUM CAELESTINUM	S	.	0	18.8	20.7	.	.	.	.	.	.	.	.	.	.	.	.	.	20
4960	PARINARI EXCELSA	F	352	0	3.9	0.6	93.1	4.5	2.4	114.5	81.3	5.1	.	.	873.49	.	.	.	93.4	1
4961	PARINARI GLABERRIMUM	F	288	0	10.0	2.5	65.0	.	22.5	375.0	625.0	57.5	.	.	0.00	.	.	.	475.0	2
4962	PARINARI MACROPHYLLA	F	351	0	3.5	0.2	93.8	5.9	2.5	104.0	133.7	4.2	.	.	99.01	0.07	0.20	1.73	235.1	1
4963	PARINARI MACROPHYLLA	S	672	0	15.9	66.5	14.8	5.2	2.8	94.8	512.4	6.5	.	.	216.49	0.59	0.07	0.52	4.1	1
4964	PARINARIUM SP.	S	.	0	9.4	73.6	.	.	.	.	.	.	.	.	.	.	.	.	.	20
4965	PARKIA FILICOIDEA	F	.	0	13.7	7.3	72.3	19.4	6.7	.	.	.	.	.	.	.	.	.	.	18
4966	PARKIA FILICOIDEA	S	.	0	31.8	17.4	46.4	9.4	4.4	.	.	.	.	.	.	.	.	.	.	18
4967	PARKIA SP.	S	458	0	34.7	21.0	39.9	4.4	4.4	312.9	412.9	35.7	.	.	.	0.32	0.22	3.23	6.5	1
4968	PARKIA SP.	P	349	0	3.9	0.5	92.3	14.5	3.3	142.9	184.3	4.1	.	.	2799.54	1.21	0.82	1.15	278.8	1

TABLE 3 - DATA CONVERTED TO A ZERO-MOISTURE BASIS.

OBS	PLNA	PART	CALORIES	WATER	PROTEIN	FAT	TOTLCARB	FIBER
4969	PARKIA SPECIOSA	F	444	0	27.3	27.6	40.6	1.7
4970	PARKINSONIA ACULEATA	S	.	0	24.3	6.8	.	.
4971	PARKINSONIA ACULEATA	S	.	0	5.9	0.8	.	.
4972	PARKINSONIA ACULEATA	S	.	0	20.5	3.1	.	.
4973	PARKINSONIA ACULEATA	S	.	0	21.2	3.8	.	.
4974	PARKINSONIA ACULEATA	L	.	0	7.5	1.8	73.8	29.0
4975	PARKINSONIA AFRICANA	S	.	0	23.8	4.2	.	.
4976	PARMENTIERA ACULEATA	F	383	0	11.5	12.5	66.6	53.3
4977	PARMENTIERA ACULEATA	F	355	0	11.3	0.8	85.5	2.4
4978	PARONYCHIA KURDICA	S	.	0	22.5	9.5	.	.
4979	PARROTIA JACQUEMONTIANA	S	.	0	14.8	16.8	.	.
4980	PARTHENIUM ARGENTATUM	S	.	0	34.9	24.0	.	.
4981	PARTHENIUM INCANUM	L	.	0	21.2	6.8	61.4	19.4
4982	PARTHENOCISSUS QUINQUEFOLIA	S	.	0	15.0	33.1	.	.
4983	PARTHENOCISSUS QUINQUEFOLIA	S	.	0	16.2	27.5	.	.
4984	PARTHENOCISSUS QUINQUEFOLIA	S	.	0	15.6	25.8	.	.
4985	PARTHENOCISSUS TRICUSPIDATA VAR. VEITCHII	S	.	0	14.4	31.3	.	.
4986	PASPALIDIUM DESERTORUM	W	.	0	13.8	1.2	72.0	29.4
4987	PASPALUM ALMUM		.	0	6.8	2.2	82.0	30.7
4988	PASPALUM CILIATIFOLIUM	W	.	0	7.3	1.8	76.2	33.3
4989	PASPALUM COMMERSONII	S	371	0	6.0	2.2	87.4	10.5
4990	PASPALUM CONJUGATUM	W	.	0	13.6	1.9	70.9	26.3
4991	PASPALUM CONJUGATUM	C	.	0	6.8	2.0	87.4	40.6
4992	PASPALUM CONJUGATUM	L	.	0	14.9	4.9	75.3	28.0

OBS	ASH	CALCIUM	PHOSPHO	IRON	SODIUM	POTASS	CAROT	THIAMINE	RIBOFLAV	NIACIN	ASCORBIC	REFCO
4969	4.4	259.4	283.3	2.4	.	.	.	0.38	0.03	3.41	20.5	2
4970	3.8	.	.	.	.	.	.	.	.	.	.	19
4971	7.0	.	.	.	.	.	.	.	.	.	.	19
4972	3.4	.	.	.	.	.	.	.	.	.	.	19
4973	.	.	.	.	.	.	.	.	.	.	.	20
4974	16.9	4160.0	170.0	.	.	.	.	.	.	.	.	5
4975	.	.	.	.	.	.	.	.	.	.	.	21
4976	9.4	140.3	372.4	.	.	.	306.12	0.51	0.26	3.06	66.3	3
4977	2.4	112.9	233.9	25.8	.	.	161.29	0.48	0.48	16.13	40.3	3
4978	.	.	.	.	.	.	.	.	.	.	.	21
4979	.	.	.	.	.	.	.	.	.	.	.	21
4980	6.6	.	.	.	.	.	.	.	.	.	.	19
4981	10.6	.	.	.	.	.	.	.	.	.	.	6
4982	.	.	.	.	.	.	.	.	.	.	.	19
4983	.	.	.	.	.	.	.	.	.	.	.	19
4984	.	.	.	.	.	.	.	.	.	.	.	20
4985	.	.	.	.	.	.	.	.	.	.	.	20
4986	13.0	.	.	.	.	.	.	.	.	.	.	18
4987	9.0	800.0	120.0	.	.	.	.	.	.	.	.	6
4988	14.7	640.0	150.0	.	.	.	.	.	.	.	.	6
4989	4.4	188.1	177.0	.	.	.	.	.	.	.	.	1
4990	13.6	.	.	.	.	.	.	.	.	.	.	18
4991	3.8	.	.	.	.	.	.	.	.	.	.	18
4992	4.9	.	.	.	.	.	.	.	.	.	.	18

TABLE 3 - DATA CONVERTED TO A ZERO-MOISTURE BASIS.

OBS	PLNA	PART	CALORIES	WATER	PROTEIN	FAT	TOTLCARB	FIBER	ASH
4993	PASPALUM CONJUGATUM	L	.	0	13.6	1.9	70.9	26.3	13.6
4994	PASPALUM DILATATUM	W	.	0	9.9	1.8	80.1	35.7	8.2
4995	PASPALUM DILATATUM	S	.	0	11.2	5.0	.	.	5.8
4996	PASPALUM DILATATUM	H	.	0	12.8	1.9	77.9	34.1	7.4
4997	PASPALUM DILATATUM	W	.	0	12.1	2.5	74.1	29.4	11.3
4998	PASPALUM DILATATUM	L	.	0	12.0	2.4	72.8	28.8	12.8
4999	PASPALUM DILATATUM(ALL)	W	.	0	10.1	2.1	79.1	33.9	8.7
5000	PASPALUM DISTICHUM	W	.	0	9.6	2.3	75.6	28.8	12.5
5001	PASPALUM FASCICULATUM	W	.	0	19.3	2.0	65.4	25.3	13.3
5002	PASPALUM FLORIDANUM	W	.	0	5.6	1.6	83.3	34.0	9.5
5003	PASPALUM HARTWEGIANUM(IMMATURE)	W	.	0	12.5	2.6	72.7	26.7	12.2
5004	PASPALUM HARTWEGIANUM(MATURE)	W	.	0	3.7	1.4	86.3	35.3	8.6
5005	PASPALUM LIVIDUM	W	.	0	5.4	1.6	83.0	30.2	10.0
5006	PASPALUM NOTATUM	W	.	0	13.0	4.0	76.2	34.5	6.8
5007	PASPALUM NOTATUM	H	.	0	12.3	1.4	74.9	33.5	11.4
5008	PASPALUM NOTATUM	S	.	0	7.4	3.1	.	.	3.1
5009	PASPALUM NOTATUM	W	.	0	8.5	1.6	79.7	31.6	10.2
5010	PASPALUM PLICATULUM	W	.	0	6.4	2.0	81.6	31.4	10.0
5011	PASPALUM PUBESCENS	W	.	0	4.9	1.4	84.5	35.6	9.2
5012	PASPALUM PUBIFLORUM	W	.	0	9.0	1.5	75.0	31.5	14.5
5013	PASPALUM SCORBICULATUM	H	.	0	3.5	1.5	82.7	34.3	12.3
5014	PASPALUM SCROBICULATUM	W	.	0	11.4	1.4	72.9	28.8	14.3
5015	PASPALUM SCROBICULATUM	S	.	0	12.0	4.8	78.3	11.3	5.0
5016	PASPALUM SCROBICULATUM	L	.	0	2.4	3.5	81.5	32.7	12.1

OBS	CALCIUM	PHOSPHO	IRON	SODIUM	POTASS	CAROT	THIAMINE	RIBOFLAV	NIACIN	ASCORBIC	REFCO
4993	431.9	201.9	.	.	.	.	.	.	.	.	5
4994	.	.	.	.	.	.	.	.	.	.	18
4995	.	.	.	.	.	.	.	.	.	.	19
4996	.	.	.	.	.	.	.	.	.	.	6
4997	550.0	270.0	17.0	.	1480.0	.	.	.	.	.	6
4998	560.0	200.0	16.0	.	1720.0	.	.	.	.	.	5
4999	510.0	200.0	12.0	.	.	.	.	.	.	.	6
5000	670.0	380.0	.	.	2390.0	30.00	.	.	.	.	6
5001	.	.	.	.	.	.	.	.	.	.	18
5002	470.0	90.0	.	.	.	.	.	.	.	.	6
5003	570.0	170.0	.	.	.	.	.	.	.	.	6
5004	450.0	60.0	.	.	.	.	.	.	.	.	6
5005	440.0	140.0	.	.	.	.	.	.	.	.	6
5006	.	.	.	.	.	.	.	.	.	.	18
5007	.	.	.	.	.	.	.	.	.	.	18
5008	.	.	.	.	.	.	.	.	.	.	19
5009	480.0	300.0	.	.	1450.0	.	.	.	.	.	6
5010	530.0	90.0	.	.	.	.	.	.	.	.	6
5011	320.0	80.0	.	.	.	.	.	.	.	.	6
5012	490.0	100.0	.	.	.	.	.	.	.	.	6
5013	.	.	.	.	.	.	.	.	.	.	18
5014	.	.	.	.	.	.	.	.	.	.	18
5015	56.6	321.3	6.8	.	.	.	0.45	.	.	.	5
5016	.	.	.	.	.	.	.	.	.	.	5

TABLE 3 - DATA CONVERTED TO A ZERO-MOISTURE BASIS.

OBS	PLNA	PART	CALORIES	WATER	PROTEIN	FAT	TOTCARB	FIBER	ASH	CALCIUM	PHOSPHO	IRON	SODIUM	POTASS	CAROT	THIAMINE	RIBOFLAV	NIACIN	ASCORBIC	REFCO
5017	PASPALUM SETACEUM	W	.	0	8.3	1.9	75.4	30.6	14.4	680.0	160.0	.	.	.	.	.	.	.	.	6
5018	PASPALUM STRAMINEUM	W	.	0	8.3	2.3	78.1	34.3	11.3	600.0	150.0	.	.	.	.	.	.	.	.	6
5019	PASPALUM UNISPICATUM	W	.	0	14.4	1.8	67.9	26.5	15.9	710.0	180.0	.	.	.	.	.	.	.	.	6
5020	PASPALUM URVILLEI	W	.	0	8.4	2.2	78.7	36.3	10.7	540.0	130.0	.	.	.	.	.	.	.	.	6
5021	PASSIFLORA EDULIS	F	.	0	.	.	.	.	.	119.0	228.6	.	.	.	.	0.14	0.48	.	.	1
5022	PASSIFLORA EDULIS	F	361	0	8.8	2.8	85.1	.	3.2	52.2	257.0	6.4	112.4	1397.6	1686.75	0.00	0.52	6.02	120.5	4
5023	PASSIFLORA EDULIS	S	.	0	11.7	25.2	62.2	56.8	1.9	84.6	676.5	19.0	.	.	.	.	.	.	.	5
5024	PASSIFLORA FOETIDA	S	.	0	10.8	24.8	.	.	.	.	.	.	.	.	.	.	.	.	.	21
5025	PASSIFLORA FOETIDA	F	356	0	10.0	2.2	84.4	.	3.3	111.1	266.7	3.9	16.7	1894.4	0.00	0.06	0.33	7.78	83.3	2
5026	PASSIFLORA INCARNATA	F	404	0	8.4	12.0	76.4	26.5	3.3	50.9	156.4	5.8	.	.	.	.	.	.	.	1
5027	PASSIFLORA LIGULARIS	F	397	0	10.1	11.8	73.0	17.7	5.1	42.2	270.0	3.8	.	.	42.19	0.00	0.46	6.75	84.4	3
5028	PASSIFLORA MOLLISSIMA	F	338	0	6.8	1.4	83.8	8.1	8.1	108.1	243.2	5.4	.	.	540.54	0.00	0.54	20.27	702.7	3
5029	PASSIFLORA QUADRANGULARIS	F	385	0	13.0	9.5	72.5	24.5	5.0	45.0	180.0	3.0	.	.	.	.	.	.	100.0	1
5030	PASSIFLORA QUADRANGULARIS	F	357	0	12.5	3.6	76.8	12.5	7.1	250.0	303.6	14.3	.	.	0.00	0.00	0.54	67.86	267.9	3
5031	PASSIFLORA QUADRANGULARIS	S	375	0	8.8	6.0	81.5	25.9	3.7	41.7	180.6	13.4	.	.	92.59	0.00	0.56	8.80	69.4	3
5032	PASTINACA SATIVA	S	.	0	15.9	27.3	.	.	4.9	.	.	.	.	.	.	.	.	.	.	19
5033	PASTINACA SATIVA	S	.	0	15.0	22.0	.	.	.	.	.	.	.	.	.	.	.	.	.	20
5034	PASTINACA SATIVA	S	.	0	15.5	28.8	.	.	.	.	.	.	.	.	.	.	.	.	.	21
5035	PASTINACA SATIVA	L	.	0	22.9	5.0	.	8.0	.	.	.	.	.	.	.	.	.	.	.	6
5036	PASTINACA SATIVA	Y	.	0	6.0	.	.	17.2	.	.	.	.	.	.	.	.	.	.	.	6
5037	PASTINACA SATIVA	R	346	0	8.5	1.6	80.9	8.0	9.0	255.3	345.7	11.2	.	.	.	.	.	.	.	1
5038	PASTINACA SATIVA	R	364	0	8.1	2.4	83.7	9.6	5.7	239.2	368.4	3.3	57.4	2588.5	95.69	0.38	0.43	0.96	76.6	2
5039	PASTINACA SATIVA	R	.	0	4.7	1.1	90.2	6.2	4.0	181.2	144.9	1.4	.	.	65.22	1.12	0.33	1.45	58.0	5
5040	PAULLINIA PINNATA	A	406	0	4.1	3.5	89.4	1.8	2.9	23.5	1005.9	.	.	.	.	.	.	.	.	1
5041	PAULOWNIA TOMENTOSA	S	.	0	11.9	23.6	.	.	3.8	.	.	.	.	.	.	.	.	.	.	19
5042	PAULOWNIA TOMENTOSA	S	.	0	7.8	13.5	.	.	5.2	.	.	.	.	.	.	.	.	.	.	19
5043	PAVETTA CRASSIPES	L	332	0	15.6	2.4	77.1	15.1	4.9	829.3	341.5	.	.	.	.	.	.	.	.	1
5044	PEDICULARIS GROENLANDICA	S	.	0	30.5	13.9	.	.	.	.	.	.	.	.	.	.	.	.	.	20
5045	PEDICULARIS RACEMOSA	W	.	0	15.0	1.8	71.6	17.7	11.6	.	.	.	.	.	.	.	.	.	.	6
5046	PEDILANTHUS MACROCARPUS	S	.	0	15.9	61.9	.	.	.	.	.	.	.	.	.	.	.	.	.	21
5047	PEGANUM HARMALA	S	.	0	29.4	15.4	.	.	.	.	.	.	.	.	.	.	.	.	.	21
5048	PEGANUM HARMALA	S	.	0	33.3	17.5	.	.	.	.	.	.	.	.	.	.	.	.	.	21
5049	PEGANUM HARMALA	S	.	0	35.2	13.5	.	.	.	.	.	.	.	.	.	.	.	.	.	21
5050	PEGANUM HARMALA	S	.	0	34.3	16.8	.	.	.	.	.	.	.	.	.	.	.	.	.	21
5051	PELTARIA ANGUSTIFOLIA	S	.	0	24.6	32.4	.	.	.	.	.	.	.	.	.	.	.	.	.	21
5052	PELTOPHORUM INERME	S	.	0	14.7	4.7	.	.	.	.	.	.	.	.	.	.	.	.	.	21
5053	PENNISETUM AMERICANUM	W	.	0	17.7	3.2	65.7	21.1	13.4	.	.	.	.	.	.	.	.	.	.	18
5054	PENNISETUM AMERICANUM	H	.	0	6.6	1.3	82.1	41.2	10.0	.	.	.	.	.	.	.	.	.	.	18
5055	PENNISETUM AMERICANUM	S	.	0	13.4	5.1	79.3	2.5	2.2	60.0	430.0	1.0	.	.	.	.	.	.	.	6
5056	PENNISETUM AMERICANUM	H	.	0	9.1	2.1	78.7	35.9	10.1	.	.	.	.	.	.	.	.	.	.	6
5057	PENNISETUM AMERICANUM	W	.	0	10.0	2.8	78.1	31.2	9.1	.	.	.	.	.	.	.	.	.	.	6
5058	PENNISETUM AMERICANUM	X	.	0	.	.	.	.	.	.	.	.	.	.	.	0.22	0.34	3.16	.	
5059	PENNISETUM AMERICANUM	S	388	0	11.8	4.5	81.4	2.2	2.3	25.0	325.0	23.5	.	.	0.00	0.34	0.25	1.93	3.4	1
5060	PENNISETUM AMERICANUM	S	390	0	10.7	4.5	83.0	2.3	1.8	25.7	342.6	8.4	.	.	.	0.33	0.16	1.79	.	1
5061	PENNISETUM AMERICANUM	S	412	0	10.6	6.2	81.1	2.6	2.1	18.6	385.8	18.9	.	.	.	0.35	0.21	1.75	.	1

TABLE 3 - DATA CONVERTED TO A ZERO-MOISTURE BASIS.

OBS	PLANA	PART	CALORIES	WATER	PROTEIN	FAT	TOTLCARB	FIBER	ASH	CALCIUM	PHOSPHO	IRON	SODIUM	POTASS	CAROT	THIAMINE	RIBOFLAV	NIACIN	ASCORBIC	REFCO
5062	PENNISETUM AMERICANUM	S	376	0	12.8	5.6	78.9	1.7	2.7	56.2	402.2	10.1	.	.	.	0.35	0.16	2.02	0	2
5063	PENNISETUM CILIARE	S	.	0	31.2	8.5	.	.	.	.	.	.	.	.	.	.	.	.	.	19
5064	PENNISETUM CILIARE	S	.	0	24.4	9.0	.	.	3.0	.	.	.	.	.	.	.	.	.	.	19
5065	PENNISETUM CILIARE	S	.	0	30.6	10.0	.	.	.	.	.	.	.	.	.	.	.	.	.	20
5066	PENNISETUM CILIARE	S	.	0	24.4	10.9	.	.	.	.	.	.	.	.	.	.	.	.	.	20
5067	PENNISETUM CILIARE	S	.	0	28.8	13.6	.	.	.	.	.	.	.	.	.	.	.	.	.	20
5068	PENNISETUM CILIARE	S	.	0	30.6	10.6	.	.	.	.	.	.	.	.	.	.	.	.	.	20
5069	PENNISETUM CILIARE	S	.	0	33.8	10.0	.	.	.	.	.	.	.	.	.	.	.	.	.	20
5070	PENNISETUM CLANDESTINUM	W	.	0	24.8	4.0	61.7	20.9	9.5	.	.	.	.	.	.	.	.	.	.	18
5071	PENNISETUM CLANDESTINUM	H	.	0	12.1	0.8	76.6	35.2	10.5	.	.	.	.	.	.	.	.	.	.	18
5072	PENNISETUM CLANDESTINUM	W	.	0	17.7	1.9	69.5	22.9	10.9	430.0	250.0	23.0	.	4590	.	.	.	.	.	6
5073	PENNISETUM GLAUCUM	S	.	0	11.5	6.1	.	.	1.8	.	.	.	.	.	.	.	.	.	.	19
5074	PENNISETUM GLAUCUM	S	.	0	17.5	6.0	.	.	2.1	.	.	.	.	.	.	.	.	.	.	19
5075	PENNISETUM GLAUCUM	S	.	0	15.6	5.3	.	.	1.8	.	.	.	.	.	.	.	.	.	.	19
5076	PENNISETUM GLAUCUM	S	.	0	16.2	5.7	.	.	2.0	.	.	.	.	.	.	.	.	.	.	19
5077	PENNISETUM ORIENTALE	W	.	0	11.0	7.7	74.0	37.3	7.3	.	.	.	.	.	.	.	.	.	.	18
5078	PENNISETUM ORIENTALE	S	.	0	13.9	8.1	.	.	.	.	.	.	.	.	.	.	.	.	.	21
5079	PENNISETUM PEDICELLATUM	W	.	0	12.5	1.9	73.6	35.2	12.0	.	.	.	.	.	.	.	.	.	.	18
5080	PENNISETUM POLYSTACHYON	W	.	0	17.6	2.7	67.8	27.1	11.9	.	.	.	.	.	.	.	.	.	.	18
5081	PENNISETUM PURPUREUM	SH	.	0	13.2	2.4	74.1	32.9	10.3	.	.	.	.	.	.	.	.	.	.	18
5082	PENNISETUM PURPUREUM	H	.	0	15.1	2.4	70.4	34.9	12.1	.	.	.	.	.	.	.	.	.	.	18
5083	PENNISETUM PURPUREUM	W	.	0	6.0	2.0	80.6	37.7	11.4	.	.	.	.	.	.	.	.	.	.	6
5084	PENNISETUM PURPUREUM	L	.	0	12.3	2.9	78.2	32.5	6.6	.	.	.	.	.	.	.	.	.	.	6
5085	PENNISETUM PURPUREUM	SH	254	0	41.3	3.2	33.3	17.5	22.2	174.6	1127.0	31.7	.	.	.	.	.	.	.	1
5086	PENNISETUM SP.	S	388	0	10.0	4.1	84.0	1.9	1.9	23.5	312.8	4.1	.	.	11.21	0.37	0.16	2.02	.	1
5087	PENNISETUM SQUAMULATUM	W	.	0	7.0	2.5	81.9	39.3	8.6	.	.	.	.	.	.	.	.	.	.	18
5088	PENNISETUM STRAMINEUM	W	.	0	11.9	3.2	74.3	35.8	10.6	.	.	.	.	.	.	.	.	.	.	18
5089	PENNISETUM TRACHYPHYLLUM	W	.	0	10.2	2.0	77.4	30.0	10.4	.	.	.	.	.	.	.	.	.	.	18
5090	PENNISETUM VILLOSUM	S	.	0	11.2	4.2	.	.	.	.	.	.	.	.	.	.	.	.	.	20
5091	PENSTEMON ALBIDUS	S	.	0	14.6	20.4	.	.	3.7	.	.	.	.	.	.	.	.	.	.	19
5092	PENSTEMON AUSTRALIS	S	.	0	18.8	23.0	.	.	.	.	.	.	.	.	.	.	.	.	.	20
5093	PENSTEMON CAMPANULATUS	S	.	0	13.1	20.1	.	.	.	.	.	.	.	.	.	.	.	.	.	20
5094	PENSTEMON CAMPANULATUS	S	.	0	14.6	23.3	.	.	.	.	.	.	.	.	.	.	.	.	.	21
5095	PENSTEMON CENTRANTHIFOLIUS	S	.	0	14.4	28.7	.	.	.	.	.	.	.	.	.	.	.	.	.	20
5096	PENSTEMON CF. KUNTHII	S	.	0	12.5	21.7	.	.	.	.	.	.	.	.	.	.	.	.	.	20
5097	PENSTEMON CF. KUNTHII	S	.	0	13.1	21.8	.	.	.	.	.	.	.	.	.	.	.	.	.	20
5098	PENSTEMON FENDLERI	S	.	0	17.0	20.3	.	.	.	.	.	.	.	.	.	.	.	.	.	20
5099	PENSTEMON GENTIANOIDES	S	.	0	13.2	29.4	.	.	.	.	.	.	.	.	.	.	.	.	.	21
5100	PENSTEMON GRANDIFLORUS	S	.	0	14.6	16.9	.	.	4.0	.	.	.	.	.	.	.	.	.	.	19
5101	PENSTEMON HAVARDII	S	.	0	12.5	21.5	.	.	.	.	.	.	.	.	.	.	.	.	.	20
5102	PENSTEMON KUNTHII	S	.	0	13.1	20.1	.	.	.	.	.	.	.	.	.	.	.	.	.	20
5103	PENSTEMON KUNTHII	S	.	0	12.3	24.7	.	.	.	.	.	.	.	.	.	.	.	.	.	21
5104	PENSTEMON MULTIFLORUS	S	.	0	14.8	29.2	.	.	.	.	.	.	.	.	.	.	.	.	.	21
5105	PENSTEMON MURRAYANUS X GRANDIFLORUS	S	.	0	19.7	25.8	.	.	3.8	.	.	.	.	.	.	.	.	.	.	19
5106	PENSTEMON PALMERI	S	.	0	11.8	23.4	.	.	.	.	.	.	.	.	.	.	.	.	.	21

TABLE 3 - DATA CONVERTED TO A ZERO-MOISTURE BASIS.

OBS	PLNA	PART	CALORIES	WATER	PROTEIN	FAT	TOTLCARB	FIBER	ASH
5107	PENSTEMON PALMERI	S	.	0	11.60	27.5	.	.	.
5108	PENSTEMON SP.	S	.	0	12.50	27.3	.	.	.
5109	PENSTEMON SP.	S	.	0	11.80	18.7	.	.	.
5110	PENSTEMON SP.	S	.	0	14.40	18.0	.	.	.
5111	PENSTEMON SPECTABILIS	S	.	0	11.10	30.2	.	.	2.6
5112	PENSTEMON SPECTABILIS	S	.	0	10.00	26.6	.	.	.
5113	PENTACLETHRA MACROLOBA	S	597	0	24.20	49.5	24.2	2.70	2.5
5114	PENTACLETHRA MACROPHYLLA	S	595	0	24.10	49.4	24.1	2.70	2.5
5115	PENTADESMA BUTYRACEA	S	571	0	2.40	39.8	56.3	7.10	1.5
5116	PENTZIA INCANA	SH	.	0	13.90	2.6	70.6	25.50	12.9
5117	PENTZIA INCANA		.	0	8.27	.	79.1	33.28	5.8
5118	PENTZIA SP.	S	.	0	13.80	8.8	.	.	.
5119	PENTZIA SPHAEROCEPHALA	S	.	0	10.00	6.6	.	.	.
5120	PEPEROMIA PELLUCIDA	L	321	0	6.40	3.8	75.6	12.80	14.1
5121	PERESKIA ACULEATA	F	372	0	11.60	8.1	73.3	8.10	7.0
5122	PERESKIA ACULEATA	F	344	0	21.90	3.1	65.6	6.30	9.4
5123	PERESKIA ACULEATA	F	.	0	11.60	8.1	81.4	8.10	7.0
5124	PEREZIA PLATYPHYLLA	S	.	0	19.90	11.0	.	.	.
5125	PERIDERIDIA GAIRDNERI	R	.	0	.	.	.	.	.
5126	PERILLA FRUTESCENS	S	.	0	31.80	42.0	.	.	4.3
5127	PERILLA FRUTESCENS	S	.	0	20.20	35.3	.	.	.
5128	PERILLA FRUTESCENS	S	356	0	22.00	11.0	58.5	31.40	8.5
5129	PERILLA FRUTESCENS	L	300	0	24.30	4.3	57.1	10.70	14.3
5130	PERILLA FRUTESCENS	S	517	0	19.10	32.0	45.0	34.10	3.9

OBS	CALCIUM	PHOSPHO	IRON	SODIUM	POTASS	CAROT	THIAMINE	RIBOFLAV	NIACIN	ASCORBIC	REFCO
5107	.	.	.	.	.	.	.	.	.	.	21
5108	.	.	.	.	.	.	.	.	.	.	20
5109	.	.	.	.	.	.	.	.	.	.	21
5110	.	.	.	.	.	.	.	.	.	.	21
5111	.	.	.	.	.	.	.	.	.	.	19
5112	.	.	.	.	.	.	.	.	.	.	20
5113	203.0	184.0	17.1	.	.	.	0.07	0.34	0.96	.	10
5114	202.6	183.4	17.1	.	.	.	0.07	0.34	0.96	.	1
5115	201.3	169.5	.	.	.	.	.	.	.	.	1
5116	.	.	.	.	.	.	.	.	.	.	18
5117	620.0	200.0	82.0	.	.	.	.	.	.	.	8
5118	.	.	.	.	.	.	.	.	.	.	20
5119	.	.	.	.	.	.	.	.	.	.	20
5120	1589.7	435.9	41.0	102.6	3551.3	32051.3	0.38	0.90	7.69	128.2	2
5121	2023.3	302.3	0.0	.	.	22441.9	0.35	0.35	5.81	23.3	2
5122	406.2	281.2	25.0	.	.	55625.0	0.63	0.94	31.25	125.0	3
5123	2023.3	302.3	0.0	.	.	22430.2	0.35	0.35	5.81	23.3	5
5124	.	.	.	.	.	.	.	.	.	.	20
5125	90.0	.	22.0	.	.	.	0.14	0.22	.	64.1	9
5126	.	.	.	.	.	.	.	.	.	.	19
5127	.	.	.	.	.	.	.	.	.	.	21
5128	1186.4	533.9	11.0	.	.	6610.2	0.68	0.76	4.24	169.5	2
5129	1407.1	521.4	47.9	142.9	4642.9	31285.7	0.50	2.29	5.71	328.6	2
5130	425.8	40.1	13.5	.	.	12.2	0.39	0.13	3.77	0.0	2

TABLE 3 - DATA CONVERTED TO A ZERO-MOISTURE BASIS.

OBS	PLNA	PART	CALORIES	WATER	PROTEIN	FAT	TOTCARB	FIBER	ASH	CALCIUM	PHOSPHO	IRON	SODIUM	POTASS	CAROT	THIAMINE	RIBOFLAV	NIACIN	ASCORBIC	REFCO
5131	PERILLA FRUTESCENS	S	.	0	24.7	48.1	22.0	11.0	4.9	.	.	.	.	.	.	.	.	4.27	.	5
5132	PERIPLOCA GRAECA	S	.	0	29.6	18.6	.	.	.	.	.	.	.	.	.	.	.	.	.	21
5133	PERITYLE SP.	S	.	0	18.1	23.5	.	.	.	.	.	.	.	.	.	.	.	.	.	20
5134	PERSEA AMERICANA	F	614	0	7.1	57.4	31.0	9.1	4.6	96.4	233.5	7.1	.	.	2690.4	0.25	0.76	10.15	91.4	1
5135	PERSEA AMERICANA	F	486	0	5.2	29.0	62.9	4.8	2.9	57.1	123.8	3.3	9.5	1323.8	976.2	0.24	0.48	6.67	38.1	2
5136	PERSEA AMERICANA	F	596	0	8.0	52.7	35.1	10.6	4.3	74.5	223.4	3.7	.	.	638.3	0.32	0.64	9.04	90.4	3
5137	PERSEA AMERICANA	F	670	0	7.4	68.7	19.1	7.8	4.8	43.5	182.6	4.3	.	.	521.7	0.35	0.52	6.52	47.8	3
5138	PERSEA AMERICANA	F	584	0	8.1	50.9	35.4	8.1	5.6	37.3	192.5	5.0	.	.	186.3	0.37	0.62	9.32	99.4	3
5139	PERSEA AMERICANA	F	642	0	8.1	63.1	24.2	6.2	4.6	38.5	161.5	2.3	15.4	2323.1	669.2	0.42	0.77	6.15	53.8	4
5140	PERSEA AMERICANA	F	.	0	6.4	86.4	3.0	.	4.2	37.9	303.0	2.7	.	.	.	.	.	.	49.2	5
5141	PERSEA SCHIEDEANA	F	544	0	6.2	41.1	49.4	8.3	3.3	58.1	182.6	2.1	.	.	249.0	0.25	0.37	4.56	49.8	3
5142	PETALOSTEMON CARNEUS	S	.	0	31.2	4.9	.	.	.	.	.	.	.	.	.	.	.	.	.	20
5143	PETALOSTEMON EMARGINATUS	S	.	0	36.9	4.6	.	.	.	.	.	.	.	.	.	.	.	.	.	20
5144	PETALOSTEMON FEAYI	S	.	0	28.0	6.9	.	.	.	.	.	.	.	.	.	.	.	.	.	20
5145	PETALOSTEMON PULCHERRIMUM	S	.	0	38.1	5.2	.	.	.	.	.	.	.	.	.	.	.	.	.	20
5146	PETALOSTEMON PURPUREUM	S	.	0	36.9	4.4	.	.	.	.	.	.	.	.	.	.	.	.	.	20
5147	PETALOSTEMON SP.	W	.	0	10.4	5.6	77.0	25.7	7.0	2820.0	430.0	.	.	.	.	.	.	.	.	6
5148	PETALOSTEMUM CANDIDUM	S	.	0	24.1	6.2	.	.	6.9	.	.	.	.	.	.	.	.	.	.	19
5149	PETALOSTEMUM COMPACTUM	S	.	0	34.9	8.5	.	.	3.8	.	.	.	.	.	.	.	.	.	.	19
5150	PETALOSTEMUM PURPUREUM	S	.	0	37.2	10.8	.	.	3.9	.	.	.	.	.	.	.	.	.	.	19
5151	PETALOSTEMUM PURPUREUM	S	.	0	14.9	3.1	.	.	9.0	.	.	.	.	.	.	.	.	.	.	19
5152	PETALOSTEMUM VILLOSUM	S	.	0	29.4	10.6	.	.	3.5	.	.	.	.	.	.	.	.	.	.	19
5153	PETALOSTEMUM VILLOSUM	S	.	0	14.3	1.5	.	.	7.6	.	.	.	.	.	.	.	.	.	.	19
5154	PETASITES JAPONICA	L	250	0	19.4	2.8	52.8	19.4	25.0	1194.4	555.6	2.8	916.7	12500.0	277.8	0.56	0.56	5.56	55.6	2
5155	PETROSELINUM CRISPUM	S	.	0	19.1	33.0	.	.	.	.	.	.	.	.	.	.	.	.	.	21
5156	PETROSELINUM CRISPUM	L	296	0	22.5	2.1	62.7	10.6	12.7	1450.7	331.0	17.6	.	.	.	.	.	.	1211.3	1
5157	PETROSELINUM CRISPUM	L	303	0	24.6	4.8	56.8	11.3	13.7	1613.2	385.7	107.6	496.7	4181.3	15389.0	0.19	1.35	8.71	134.1	16
5158	PETROSELINUM CRISPUM	L	309	0	22.8	4.9	58.6	8.6	13.6	1407.4	333.3	28.4	172.8	5555.6	9567.9	1.11	1.54	6.17	944.4	2
5159	PETROSELINUM CRISPUM	L	305	0	22.7	4.3	60.3	9.2	12.8	1383.0	368.8	22.0	.	.	25815.6	0.85	1.70	7.09	1035.5	3
5160	PETROSELINUM CRISPUM	L	295	0	24.2	4.0	57.0	10.1	14.8	1362.4	422.8	41.6	302.0	4879.2	34228.2	0.81	1.74	8.05	1154.4	4
5161	PETROSELINUM CRISPUM	L	.	0	18.7	3.2	68.0	5.7	10.1	1234.2	632.9	56.6	.	.	6075.9	0.13	.	1.58	889.2	5
5162	PEUCEDANUM CAPENSE	S	.	0	13.1	20.3	.	.	.	.	.	.	.	.	.	.	.	.	.	21
5163	PEUCEDANUM FERULACEUM	S	.	0	14.0	22.3	.	.	.	.	.	.	.	.	.	.	.	.	.	21
5164	PHACELIA CONGESTA	S	.	0	15.6	2.6	.	.	.	.	.	.	.	.	.	.	.	.	.	20
5165	PHACELIA CONGESTA	S	.	0	13.1	6.5	.	.	.	.	.	.	.	.	.	.	.	.	.	20
5166	PHACELIA HETEROPHYLLA	S	.	0	16.1	16.5	.	.	.	.	.	.	.	.	.	.	.	.	.	21
5167	PHACELIA HETEROPHYLLA	I	.	0	12.8	2.3	70.3	15.3	14.6	.	.	.	.	.	.	.	.	.	.	6
5168	PHACELIA HETEROPHYLLA	W	.	0	15.1	2.0	65.3	12.2	17.6	.	.	.	.	.	.	.	.	.	.	6
5169	PHACELIA INTEGRIFOLIA	S	.	0	16.6	8.0	.	.	.	.	.	.	.	.	.	.	.	.	.	20
5170	PHACELIA PATULIFLORA	S	.	0	10.0	16.4	.	.	.	.	.	.	.	.	.	.	.	.	.	19
5171	PHACELIA ROBUSTA	S	.	0	16.9	12.7	.	.	.	.	.	.	.	.	.	.	.	.	.	20
5172	PHACELIA SP.	S	.	0	10.4	13.2	.	.	2.3	.	.	.	.	.	.	.	.	.	.	19
5173	PHACELIA SP.	W	.	0	16.4	2.1	65.4	22.9	16.1	3870.0	320.0	.	.	3010.0	.	.	.	.	.	6
5174	PHACELIA TANACETIFOLIA	S	.	0	18.2	10.4	.	.	5.4	.	.	.	.	.	.	.	.	.	.	19
5175	PHALARIS ANGUSTA(IMMATURE)	W	.	0	20.1	3.0	67.6	25.6	9.3	510.0	250.0	.	.	2700.0	.	.	.	.	.	6

TABLE 3 - DATA CONVERTED TO A ZERO-MOISTURE BASIS.

OBS	PLANA	PART	CALORIES	WATER	PROTEIN	FAT	TOTCARB	FIBER	ASH	CALCIUM	PHOSPHO	IRON	SODIUM	POTASS	CAROT	THIAMINE	RIBOFLAV	NIACIN	ASCORBIC	REFCO
5176	PHALARIS AQUATICA	W	.	0	12.8	5.1	71.3	23.7	10.8	.	.	.	.	.	.	.	.	.	.	18
5177	PHALARIS AQUATICA	W	.	0	15.1	3.6	68.1	28.1	13.2	.	.	.	.	.	.	.	.	.	.	18
5178	PHALARIS AQUATICA	L	.	0	5.2	1.4	84.7	33.2	8.6	.	300.0	.	.	.	.	.	.	.	.	5
5179	PHALARIS ARUNDINACEA	L	.	0	14.5	2.2	71.3	30.2	10.2	790.0	980.0	.	.	.	.	.	.	.	.	5
5180	PHALARIS ARUNDINACEA	H	.	0	8.8	2.2	81.8	34.3	7.2	340.0	250.0	15.0	.	2350	.	.	.	.	.	6
5181	PHALARIS ARUNDINACEA	W	.	0	13.2	3.7	74.7	26.8	8.4	400.0	300.0	.	.	3640	.	.	.	.	.	6
5182	PHALARIS CANARIENSIS	S	.	0	20.5	6.0	.	.	5.0	.	.	.	.	.	.	.	.	.	.	19
5183	PHALARIS CANARIENSIS	S	.	0	21.3	6.2	.	.	4.8	.	.	.	.	.	.	.	.	.	.	19
5184	PHALARIS CANARIENSIS	S	.	0	22.0	8.2	.	.	.	.	.	.	.	.	.	.	.	.	.	21
5185	PHALARIS CANARIENSIS	W	.	0	13.1	3.7	73.8	26.8	9.4	440.0	290.0	.	.	3170	.	.	.	.	.	6
5186	PHALARIS CAROLINIANA(IMMATURE)	W	.	0	10.3	2.6	71.5	28.8	15.6	490.0	270.0	.	.	.	.	.	.	.	.	6
5187	PHALARIS COERULESCENS	S	.	0	25.0	10.8	.	.	6.1	.	.	.	.	.	.	.	.	.	.	19
5188	PHALARIS MINOR	W	.	0	10.0	3.7	68.8	21.2	17.5	.	.	.	.	.	.	.	.	.	.	18
5189	PHALARIS MINOR	L	.	0	19.0	3.7	59.7	21.2	17.5	1300.0	890.0	.	.	.	.	.	.	.	.	5
5190	PHALARIS MINOR	L	.	0	13.9	5.2	61.2	29.9	19.7	720.0	550.0	.	.	.	.	.	.	.	.	5
5191	PHALARIS MINOR	W	.	0	10.3	2.6	67.9	26.3	19.2	.	.	.	.	.	.	.	.	.	.	6
5192	PHALARIS PARADOXA	S	.	0	26.5	9.4	.	.	.	.	.	.	.	.	.	.	.	.	.	21
5193	PHALARIS TUBEROSA STENOPTERA	H	.	0	5.4	2.0	86.2	34.8	6.4	.	.	.	.	.	.	.	.	.	.	6
5194	PHASEOLUS ACONITIFOLIUS	H	.	0	17.2	1.7	69.1	29.4	12.0	.	.	.	.	.	.	.	.	.	.	6
5195	PHASEOLUS ACUTIFOLIUS	H	.	0	18.5	3.2	66.6	28.1	11.7	.	.	.	.	.	.	.	.	.	.	6
5196	PHASEOLUS ACUTIFOLIUS	S	386	0	21.1	1.3	74.2	5.3	3.4	122.5	339.2	.	.	.	.	0.36	0.13	3.06	0.0	1
5197	PHASEOLUS ACUTIFOLIUS VAR. LATIFOLIUS	S	.	0	32.2	1.1	.	.	4.2	.	.	.	.	.	.	.	.	.	.	19
5198	PHASEOLUS ANGULARIS	S	.	0	22.9	0.4	.	.	3.7	.	.	.	.	.	.	.	.	.	.	19
5199	PHASEOLUS AUREUS	S	.	0	28.2	0.8	.	.	3.6	.	.	.	.	.	.	.	.	.	.	19
5200	PHASEOLUS AUREUS	S	.	0	29.0	1.0	.	.	3.8	.	.	.	.	.	.	.	.	.	.	19
5201	PHASEOLUS AUREUS	S	.	0	25.7	0.9	.	.	3.9	.	.	.	.	.	.	.	.	.	.	19
5202	PHASEOLUS AUREUS	S	.	0	23.4	1.5	.	.	4.7	.	.	.	.	.	.	.	.	.	.	19
5203	PHASEOLUS AUREUS	S	.	0	22.8	1.1	.	.	4.1	.	.	.	.	.	.	.	.	.	.	19
5204	PHASEOLUS AUREUS	S	.	0	25.5	0.9	.	.	3.9	.	.	.	.	.	.	.	.	.	.	19
5205	PHASEOLUS AUREUS	S	.	0	26.1	0.9	.	.	3.7	.	.	.	.	.	.	.	.	.	.	19
5206	PHASEOLUS AUREUS	S	.	0	28.4	1.1	.	.	4.6	.	.	.	.	.	.	.	.	.	.	19
5207	PHASEOLUS AUREUS	S	.	0	25.6	0.9	.	.	4.6	.	.	.	.	.	.	.	.	.	.	19
5208	PHASEOLUS AUREUS	S	.	0	23.6	0.7	.	.	.	.	.	.	.	.	.	.	.	.	.	21
5209	PHASEOLUS AUREUS	S	.	0	23.9	0.9	.	.	.	.	.	.	.	.	.	.	.	.	.	21
5210	PHASEOLUS AUREUS	W	.	0	17.6	2.3	67.0	25.6	13.1	.	.	.	.	.	.	.	.	.	.	6
5211	PHASEOLUS AUREUS	H	.	0	11.0	2.4	77.7	26.2	8.9	.	.	.	.	.	.	.	.	.	.	6
5212	PHASEOLUS COCCINEUS	S	.	0	26.1	1.8	.	.	5.7	.	.	.	.	.	.	.	.	.	.	19
5213	PHASEOLUS COCCINEUS	S	.	0	22.1	1.9	.	.	4.8	.	.	.	.	.	.	.	.	.	.	19
5214	PHASEOLUS COCCINEUS	S	.	0	27.2	1.8	.	.	6.1	.	.	.	.	.	.	.	.	.	.	19
5215	PHASEOLUS COCCINEUS	S	.	0	19.5	2.4	.	.	4.8	.	.	.	.	.	.	.	.	.	.	19
5216	PHASEOLUS COCCINEUS	IS	380	0	24.9	0.5	70.4	18.5	4.3	92.7	421.0	6.2	.	.	45.59	0.82	0.21	3.50	0.0	3
5217	PHASEOLUS COCCINEUS	S	380	0	26.4	1.7	67.2	5.8	4.6	148.1	379.6	6.3	.	.	11.57	0.35	0.13	2.43	.	3
5218	PHASEOLUS COCCINEUS	S	386	0	23.2	2.1	70.9	5.5	3.9	130.3	404.6	10.3	.	.	.	0.57	0.22	2.63	2.3	3
5219	PHASEOLUS COCCINEUS	G	.	0	17.7	2.4	76.0	4.6	3.8	119.9	383.7	6.2	.	.	81.53	0.82	0.46	.	64.7	5
5220	PHASEOLUS LUNATUS	F	.	0	18.8	0.6	76.6	17.5	4.0	.	.	.	.	.	.	.	.	.	.	18

TABLE 3 - DATA CONVERTED TO A ZERO-MOISTURE BASIS.

OBS	PLNA	PART	CALORIES	WATER	PROTEIN	FAT	TOTLCARB	FIBER	ASH	CALCIUM	PHOSPHO	IRON	SODIUM	POTASS	CAROT	THIAMINE	RIBOFLAV	NIACIN	ASCORBIC	REFCO
5221	PHASEOLUS LUNATUS	S	.	0	27.2	0.9	66.4	5.2	5.5	.	.	.	.	.	.	.	.	.	.	18
5222	PHASEOLUS LUNATUS	S	.	0	23.8	0.8	.	.	5.6	.	.	.	.	.	.	.	.	.	.	19
5223	PHASEOLUS LUNATUS	S	.	0	23.4	1.0	.	.	5.3	.	.	.	.	.	.	.	.	.	.	19
5224	PHASEOLUS LUNATUS	S	.	0	26.5	0.6	.	.	4.2	.	.	.	.	.	.	.	.	.	.	19
5225	PHASEOLUS LUNATUS	S	.	0	20.4	1.4	.	.	4.8	.	.	.	.	.	.	.	.	.	.	19
5226	PHASEOLUS LUNATUS	S	.	0	24.2	1.2	.	.	4.2	.	.	.	.	.	.	.	.	.	.	19
5227	PHASEOLUS LUNATUS	H	.	0	12.5	1.6	74.3	29.3	11.6	.	.	.	.	.	.	.	.	.	.	6
5228	PHASEOLUS LUNATUS	L	.	0	17.3	4.8	67.4	11.1	10.5	.	.	.	.	.	.	.	.	.	.	6
5229	PHASEOLUS LUNATUS	Y	.	0	9.2	2.4	.	39.2	.	.	.	.	.	.	.	.	.	.	.	6
5230	PHASEOLUS LUNATUS	S	384	0	24.5	1.6	70.0	4.9	3.9	132.9	443.3	5.6	.	.	.	0.38	0.18	2.41	1.1	1
5231	PHASEOLUS LUNATUS	S	387	0	22.1	1.5	73.1	.	3.4	100.6	268.2	6.3	20.1	329.6	0.00	0.51	0.23	1.56	0.0	2
5232	PHASEOLUS LUNATUS	IS	378	0	26.7	1.6	66.7	3.2	5.1	79.4	377.8	7.0	6.3	2371.4	285.71	0.51	0.51	4.76	95.2	2
5233	PHASEOLUS LUNATUS	L	286	0	21.4	0.0	60.7	.	17.9	285.7	1285.7	82.1	.	.	.	.	.	.	.	2
5234	PHASEOLUS LUNATUS	X	306	0	36.1	2.2	55.6	1.7	6.1	302.8	1061.1	22.8	.	.	83.33	0.47	0.39	5.56	19.4	2
5235	PHASEOLUS LUNATUS	IS	377	0	23.8	0.3	72.1	4.9	3.9	93.0	292.0	8.3	.	.	981.91	0.65	0.28	3.88	7.8	3
5236	PHASEOLUS LUNATUS	IS	377	0	28.7	1.1	65.8	7.2	4.4	121.2	391.2	6.6	.	.	0.00	0.66	0.36	2.75	38.6	3
5237	PHASEOLUS LUNATUS	S	382	0	23.5	1.4	70.9	5.6	4.2	128.4	375.0	5.5	.	.	0.00	0.39	0.24	2.50	0.0	3
5238	PHASEOLUS LUNATUS	IS	378	0	25.8	1.5	68.0	5.5	4.6	160.0	436.9	8.6	6.2	2000.0	535.38	0.74	0.37	4.31	89.2	4
5239	PHASEOLUS TRILOBUS	S	364	0	27.6	0.7	64.4	8.7	7.3	.	.	.	.	.	.	.	.	.	.	1
5240	PHASEOLUS VULGARIS	H	.	0	6.3	0.7	85.8	43.4	7.2	1150.0	90.0	.	.	.	.	.	.	.	.	18
5241	PHASEOLUS VULGARIS	S	.	0	24.2	2.4	68.7	6.7	4.7	260.0	430.0	.	.	.	.	.	.	.	.	18
5242	PHASEOLUS VULGARIS	S	.	0	24.2	1.2	.	.	4.7	.	.	.	.	.	.	.	.	.	.	19
5243	PHASEOLUS VULGARIS	S	.	0	27.6	1.0	.	.	4.2	.	.	.	.	.	.	.	.	.	.	19
5244	PHASEOLUS VULGARIS	S	.	0	28.8	1.0	.	.	4.4	.	.	.	.	.	.	.	.	.	.	19
5245	PHASEOLUS VULGARIS	S	.	0	29.4	0.9	.	.	4.2	.	.	.	.	.	.	.	.	.	.	19
5246	PHASEOLUS VULGARIS	S	.	0	29.2	1.0	.	.	5.3	.	.	.	.	.	.	.	.	.	.	19
5247	PHASEOLUS VULGARIS	S	.	0	31.9	1.1	.	.	5.2	.	.	.	.	.	.	.	.	.	.	19
5248	PHASEOLUS VULGARIS	S	.	0	27.8	0.9	.	.	4.9	.	.	.	.	.	.	.	.	.	.	19
5249	PHASEOLUS VULGARIS	S	.	0	28.9	1.3	.	.	4.1	.	.	.	.	.	.	.	.	.	.	19
5250	PHASEOLUS VULGARIS	S	.	0	21.1	1.1	.	.	4.2	.	.	.	.	.	.	.	.	.	.	19
5251	PHASEOLUS VULGARIS	S	.	0	26.4	1.1	.	.	3.8	.	.	.	.	.	.	.	.	.	.	19
5252	PHASEOLUS VULGARIS	S	.	0	28.9	0.9	.	.	5.6	.	.	.	.	.	.	.	.	.	.	19
5253	PHASEOLUS VULGARIS	S	.	0	28.2	1.1	.	.	6.2	.	.	.	.	.	.	.	.	.	.	19
5254	PHASEOLUS VULGARIS	S	.	0	26.6	1.0	.	.	5.2	.	.	.	.	.	.	.	.	.	.	19
5255	PHASEOLUS VULGARIS	S	.	0	24.9	1.0	.	.	4.8	.	.	.	.	.	.	.	.	.	.	19
5256	PHASEOLUS VULGARIS	S	.	0	26.2	0.8	.	.	4.9	.	.	.	.	.	.	.	.	.	.	19
5257	PHASEOLUS VULGARIS	S	.	0	.	1.0	.	.	4.4	.	.	.	.	.	.	.	.	.	.	19
5258	PHASEOLUS VULGARIS	S	.	0	25.8	0.9	.	.	5.6	.	.	.	.	.	.	.	.	.	.	19
5259	PHASEOLUS VULGARIS	S	.	0	26.1	1.0	.	.	4.5	.	.	.	.	.	.	.	.	.	.	19
5260	PHASEOLUS VULGARIS	S	.	0	25.9	1.2	.	.	5.0	.	.	.	.	.	.	.	.	.	.	19
5261	PHASEOLUS VULGARIS	S	.	0	26.9	1.0	.	.	4.8	.	.	.	.	.	.	.	.	.	.	19
5262	PHASEOLUS VULGARIS	S	.	0	26.6	1.0	.	.	4.9	.	.	.	.	.	.	.	.	.	.	19
5263	PHASEOLUS VULGARIS	S	.	0	25.0	1.2	.	.	5.2	.	.	.	.	.	.	.	.	.	.	19
5264	PHASEOLUS VULGARIS	S	.	0	24.5	1.0	.	.	4.7	.	.	.	.	.	.	.	.	.	.	19
5265	PHASEOLUS VULGARIS	S	.	0	25.6	1.2	.	.	4.5	.	.	.	.	.	.	.	.	.	.	19

TABLE 3 - DATA CONVERTED TO A ZERO-MOISTURE BASIS.

OBS	PLNA	PART	CALORIES	WATER	PROTEIN	FAT	TOTCARB	FIBER	ASH	CALCIUM	PHOSPHO	IRON	SODIUM	POTASS	CAROT	THIAMINE	RIBOFLAV	NIACIN	ASCORBIC	REFCO
5266	PHASEOLUS VULGARIS	S	.	0	25.5	1.1	.	.	4.1	.	.	.	.	.	.	.	.	.	.	19
5267	PHASEOLUS VULGARIS	S	.	0	22.9	1.3	.	.	4.0	.	.	.	.	.	.	.	.	.	.	19
5268	PHASEOLUS VULGARIS	S	.	0	25.9	1.0	.	.	4.9	.	.	.	.	.	.	.	.	.	.	19
5269	PHASEOLUS VULGARIS	S	.	0	26.8	1.0	.	.	4.8	.	.	.	.	.	.	.	.	.	.	19
5270	PHASEOLUS VULGARIS	S	.	0	23.6	1.0	.	.	4.4	.	.	.	.	.	.	.	.	.	.	19
5271	PHASEOLUS VULGARIS	S	.	0	23.0	1.1	.	.	4.1	.	.	.	.	.	.	.	.	.	.	19
5272	PHASEOLUS VULGARIS	S	.	0	25.2	1.1	.	.	4.4	.	.	.	.	.	.	.	.	.	.	19
5273	PHASEOLUS VULGARIS	S	.	0	25.4	2.1	.	.	4.6	.	.	.	.	.	.	.	.	.	.	19
5274	PHASEOLUS VULGARIS	S	.	0	29.4	1.0	.	.	5.4	.	.	.	.	.	.	.	.	.	.	19
5275	PHASEOLUS VULGARIS	S	.	0	23.7	1.1	.	.	4.5	.	.	.	.	.	.	.	.	.	.	19
5276	PHASEOLUS VULGARIS	S	.	0	22.4	1.1	.	.	4.2	.	.	.	.	.	.	.	.	.	.	19
5277	PHASEOLUS VULGARIS	S	.	0	23.2	1.1	.	.	4.7	.	.	.	.	.	.	.	.	.	.	19
5278	PHASEOLUS VULGARIS	S	.	0	24.2	1.4	.	.	5.0	.	.	.	.	.	.	.	.	.	.	19
5279	PHASEOLUS VULGARIS	S	.	0	28.5	1.0	.	.	5.2	.	.	.	.	.	.	.	.	.	.	19
5280	PHASEOLUS VULGARIS	S	.	0	26.6	0.9	.	.	4.9	.	.	.	.	.	.	.	.	.	.	19
5281	PHASEOLUS VULGARIS	S	.	0	27.2	0.8	.	.	4.6	.	.	.	.	.	.	.	.	.	.	19
5282	PHASEOLUS VULGARIS	S	.	0	26.0	1.2	.	.	5.4	.	.	.	.	.	.	.	.	.	.	19
5283	PHASEOLUS VULGARIS	S	.	0	25.8	1.3	.	.	4.5	.	.	.	.	.	.	.	.	.	.	19
5284	PHASEOLUS VULGARIS	S	.	0	30.6	0.8	.	.	5.3	.	.	.	.	.	.	.	.	.	.	19
5285	PHASEOLUS VULGARIS	S	.	0	29.9	0.9	.	.	4.5	.	.	.	.	.	.	.	.	.	.	19
5286	PHASEOLUS VULGARIS	S	.	0	26.0	1.0	.	.	4.4	.	.	.	.	.	.	.	.	.	.	19
5287	PHASEOLUS VULGARIS	S	.	0	30.8	1.3	.	.	5.3	.	.	.	.	.	.	.	.	.	.	19
5288	PHASEOLUS VULGARIS	S	.	0	28.6	1.1	.	.	4.6	.	.	.	.	.	.	.	.	.	.	19
5289	PHASEOLUS VULGARIS	S	.	0	31.0	1.1	.	.	4.5	.	.	.	.	.	.	.	.	.	.	19
5290	PHASEOLUS VULGARIS	S	.	0	27.7	1.1	.	.	4.4	.	.	.	.	.	.	.	.	.	.	19
5291	PHASEOLUS VULGARIS	S	.	0	27.2	1.2	.	.	5.3	.	.	.	.	.	.	.	.	.	.	19
5292	PHASEOLUS VULGARIS	S	.	0	28.1	1.0	.	.	4.7	.	.	.	.	.	.	.	.	.	.	19
5293	PHASEOLUS VULGARIS	S	.	0	26.9	1.0	.	.	5.6	.	.	.	.	.	.	.	.	.	.	19
5294	PHASEOLUS VULGARIS	S	.	0	24.7	1.0	.	.	4.7	.	.	.	.	.	.	.	.	.	.	19
5295	PHASEOLUS VULGARIS	S	.	0	27.9	1.2	.	.	4.8	.	.	.	.	.	.	.	.	.	.	19
5296	PHASEOLUS VULGARIS	S	.	0	30.2	1.2	.	.	5.6	.	.	.	.	.	.	.	.	.	.	19
5297	PHASEOLUS VULGARIS	S	.	0	33.8	1.0	.	.	4.7	.	.	.	.	.	.	.	.	.	.	19
5298	PHASEOLUS VULGARIS	S	.	0	33.9	0.9	.	.	4.9	.	.	.	.	.	.	.	.	.	.	19
5299	PHASEOLUS VULGARIS	S	.	0	33.5	0.9	.	.	4.9	.	.	.	.	.	.	.	.	.	.	19
5300	PHASEOLUS VULGARIS	S	.	0	30.2	1.1	.	.	4.5	.	.	.	.	.	.	.	.	.	.	19
5301	PHASEOLUS VULGARIS	S	.	0	34.9	1.0	.	.	4.8	.	.	.	.	.	.	.	.	.	.	19
5302	PHASEOLUS VULGARIS	S	.	0	30.9	1.0	.	.	4.9	.	.	.	.	.	.	.	.	.	.	19
5303	PHASEOLUS VULGARIS	S	.	0	27.8	1.1	.	.	4.0	.	.	.	.	.	.	.	.	.	.	19
5304	PHASEOLUS VULGARIS	S	.	0	29.5	1.2	.	.	4.4	.	.	.	.	.	.	.	.	.	.	19
5305	PHASEOLUS VULGARIS	S	.	0	31.1	1.1	.	.	4.2	.	.	.	.	.	.	.	.	.	.	19
5306	PHASEOLUS VULGARIS	S	.	0	29.9	1.3	.	.	4.6	.	.	.	.	.	.	.	.	.	.	19
5307	PHASEOLUS VULGARIS	S	.	0	31.5	1.4	.	.	4.3	.	.	.	.	.	.	.	.	.	.	19
5308	PHASEOLUS VULGARIS	S	.	0	30.3	1.1	.	.	4.1	.	.	.	.	.	.	.	.	.	.	19
5309	PHASEOLUS VULGARIS	S	.	0	30.4	1.0	.	.	4.2	.	.	.	.	.	.	.	.	.	.	19
5310	PHASEOLUS VULGARIS	S	.	0	31.4	1.1	.	.	4.6	.	.	.	.	.	.	.	.	.	.	19

TABLE 3 - DATA CONVERTED TO A ZERO-MOISTURE BASIS.

OBS	PLNA	PART	CALORIES	WATER	PROTEIN	FAT	TOTLCARB	FIBER	ASH	CALCIUM	PHOSPHO	IRON	SODIUM	POTASS	CAROT	THIAMINE	RIBOFLAV	NIACIN	ASCORBIC	REFCO
5311	PHASEOLUS VULGARIS	S	.	0	39.4	0.8	.	.	5.1	.	.	.	.	.	.	.	.	.	.	19
5312	PHASEOLUS VULGARIS	S	.	0	27.9	1.1	.	.	4.7	.	.	.	.	.	.	.	.	.	.	19
5313	PHASEOLUS VULGARIS	S	.	0	27.1	1.1	.	.	4.6	.	.	.	.	.	.	.	.	.	.	19
5314	PHASEOLUS VULGARIS	S	.	0	30.9	0.9	.	.	4.8	.	.	.	.	.	.	.	.	.	.	19
5315	PHASEOLUS VULGARIS	S	.	0	29.9	0.9	.	.	4.6	.	.	.	.	.	.	.	.	.	.	19
5316	PHASEOLUS VULGARIS	S	.	0	27.2	0.8	.	.	5.3	.	.	.	.	.	.	.	.	.	.	19
5317	PHASEOLUS VULGARIS	S	.	0	30.9	1.0	.	.	5.4	.	.	.	.	.	.	.	.	.	.	19
5318	PHASEOLUS VULGARIS	S	.	0	34.1	0.8	.	.	4.5	.	.	.	.	.	.	.	.	.	.	19
5319	PHASEOLUS VULGARIS	S	.	0	30.5	1.0	.	.	4.2	.	.	.	.	.	.	.	.	.	.	19
5320	PHASEOLUS VULGARIS	S	.	0	29.9	0.9	.	.	5.5	.	.	.	.	.	.	.	.	.	.	19
5321	PHASEOLUS VULGARIS	S	.	0	28.9	0.8	.	.	5.2	.	.	.	.	.	.	.	.	.	.	19
5322	PHASEOLUS VULGARIS	S	.	0	28.7	1.0	.	.	5.2	.	.	.	.	.	.	.	.	.	.	19
5323	PHASEOLUS VULGARIS	S	.	0	30.3	0.9	.	.	4.7	.	.	.	.	.	.	.	.	.	.	19
5324	PHASEOLUS VULGARIS	S	.	0	29.2	0.9	.	.	3.9	.	.	.	.	.	.	.	.	.	.	19
5325	PHASEOLUS VULGARIS	S	.	0	27.6	1.0	.	.	5.1	.	.	.	.	.	.	.	.	.	.	19
5326	PHASEOLUS VULGARIS	S	.	0	30.9	0.9	.	.	4.8	.	.	.	.	.	.	.	.	.	.	19
5327	PHASEOLUS VULGARIS	S	.	0	29.1	0.9	.	.	4.7	.	.	.	.	.	.	.	.	.	.	19
5328	PHASEOLUS VULGARIS	S	.	0	26.2	1.9	.	.	4.2	.	.	.	.	.	.	.	.	.	.	19
5329	PHASEOLUS VULGARIS	H	.	0	8.2	1.8	82.7	36.7	7.3	.	.	.	.	.	.	.	.	.	.	6
5330	PHASEOLUS VULGARIS	L	.	0	29.7	5.4	48.2	14.0	16.7	.	.	.	.	.	.	.	.	.	.	6
5331	PHASEOLUS VULGARIS	S	383	0	24.7	1.7	69.4	5.0	4.1	136.8	368.3	9.4	.	.	11.4	0.42	0.18	2.74	1.1	1
5332	PHASEOLUS VULGARIS	L	273	0	27.3	3.0	50.0	21.2	19.7	2075.8	568.2	69.7	.	.	24545.4	1.36	0.45	9.85	833.3	1
5333	PHASEOLUS VULGARIS	F	319	0	22.1	1.8	69.9	15.9	6.2	380.5	424.8	12.4	.	.	6637.2	0.71	1.06	4.42	238.9	1
5334	PHASEOLUS VULGARIS	F	329	0	22.4	4.7	65.9	11.8	7.1	882.4	576.5	14.1	23.5	2305.9	3800.0	0.59	1.18	7.06	211.8	2
5335	PHASEOLUS VULGARIS	S	383	0	25.0	1.8	69.1	4.9	4.1	97.7	280.7	8.6	.	.	11.4	0.61	0.22	2.39	3.4	3
5336	PHASEOLUS VULGARIS	F	379	0	21.1	2.1	69.5	12.6	7.4	578.9	473.7	17.9	.	.	2315.8	0.84	1.16	6.32	189.5	3
5337	PHASEOLUS VULGARIS	IS	379	0	24.7	0.8	70.2	5.8	4.3	149.0	537.9	9.1	.	.	50.5	0.96	0.30	3.79	17.7	3
5338	PHASEOLUS VULGARIS	F	323	0	19.2	2.0	71.7	10.1	7.1	565.7	444.4	8.1	70.7	2454.5	3636.4	0.81	1.11	5.05	191.9	4
5339	PHASEOLUS VULGARIS	S	383	0	25.1	1.7	69.1	4.7	4.1	122.8	453.1	7.7	11.2	1098.2	13.4	0.57	0.22	2.57	.	4
5340	PHELLODENDRON AMURENSE	S	.	0	35.8	51.7	.	.	.	.	.	.	.	.	.	.	.	.	.	21
5341	PHILADELPHUS GRANDIFLORUS	S	.	0	22.6	36.4	.	.	.	.	.	.	.	.	.	.	.	.	.	20
5342	PHILLYREA ANGUSTIFOLIA	S	.	0	20.2	21.0	.	.	.	.	.	.	.	.	.	.	.	.	.	21
5343	PHILLYREA LATIFOLIA	S	.	0	13.6	13.4	.	.	4.3	.	.	.	.	.	.	.	.	.	.	19
5344	PHILLYREA LATIFOLIA	S	.	0	3.3	1.6	.	.	3.8	.	.	.	.	.	.	.	.	.	.	19
5345	PHILLYREA LATIFOLIA	S	.	0	12.4	9.2	.	.	.	.	.	.	.	.	.	.	.	.	.	21
5346	PHILLYREA MEDIA	S	.	0	16.0	18.1	.	.	.	.	.	.	.	.	.	.	.	.	.	21
5347	PHLEUM PRATENSE	S	.	0	21.9	4.4	.	.	.	.	.	.	.	.	.	.	.	.	.	20
5348	PHLEUM PRATENSE	S	.	0	19.4	4.8	.	.	.	.	.	.	.	.	.	.	.	.	.	20
5349	PHLEUM PRATENSE	H	.	0	7.6	2.4	84.4	34.3	5.6	370.0	190.0	13.0	.	1660.0	1.0	0.18	1.01	2.95	.	6
5350	PHLEUM PRATENSE	W	.	0	11.3	3.4	78.3	28.1	7.0	470.0	340.0	23.0	.	1740.0	22.0	0.29	1.14	.	.	6
5351	PHLEUM PRATENSE	L	.	0	19.7	3.8	65.7	19.2	10.9	585.8	376.6	.	.	2092.1	.	.	.	.	.	5
5352	PHLEUM PRATENSE	L	.	0	10.9	3.0	78.8	31.3	7.3	.	.	.	.	.	.	.	.	.	.	5
5353	PHLEUM SP.	S	.	0	27.1	22.1	.	.	.	.	.	.	.	.	.	.	.	.	.	21
5354	PHLOMIS ARMENIACA	S	.	0	35.9	26.3	.	.	.	.	.	.	.	.	.	.	.	.	.	21
5355	PHLOMIS HERBA-VENTI	S	.	0	34.9	31.1	.	.	.	.	.	.	.	.	.	.	.	.	.	21

TABLE 3 - DATA CONVERTED TO A ZERO-MOISTURE BASIS.

OBS	PLNA	PART	CALORIES	WATER	PROTEIN	FAT	TOTLCARB	FIBER	ASH
5356	PHLOX PANICULATA	S	.	0	41.2	23.5	.	.	.
5357	PHLOX SP.	W	.	0	8.5	1.9	65.3	22.4	24.3
5358	PHOENIX DACTYLIFERA	F	.	0	2.9	1.0	90.4	6.5	5.7
5359	PHOENIX DACTYLIFERA	S	.	0	5.8	7.1	85.3	30.0	1.8
5360	PHOENIX DACTYLIFERA	F	357	0	4.0	1.5	92.0	8.3	2.5
5361	PHOENIX DACTYLIFERA	F	353	0	2.2	0.7	94.3	4.2	2.7
5362	PHOENIX DACTYLIFERA	F	354	0	2.2	0.6	94.8	4.3	2.4
5363	PHOENIX DACTYLIFERA	F	354	0	2.8	0.6	94.1	3.0	2.5
5364	PHOENIX DACTYLIFERA	S	.	0	5.6	7.4	85.9	14.8	1.0
5365	PHOENIX DACTYLIFERA	F	.	0	3.0	0.5	94.1	4.6	2.5
5366	PHOENIX FARINIFERA	S	.	0	7.5	9.3	.	.	1.4
5367	PHOENIX RECLINATA	S	395	0	3.1	0.3	93.7	5.1	2.8
5368	PHOENIX SYLVESTRIS	L	.	0	11.6	1.6	78.8	27.7	8.0
5369	PHORADENDRON SEROTINUM	L	.	0	22.0	5.6	66.5	19.8	5.9
5370	PHOTINIA GLABRA	S	.	0	29.4	21.8	.	.	.
5371	PHRAGMITES AUSTRALIS	L	.	0	11.4	2.3	74.2	31.0	10.8
5372	PHRAGMITES AUSTRALIS	W	.	0	10.6	2.1	72.7	31.9	14.6
5373	PHRAGMITES AUSTRALIS	L	.	0	17.1	3.5	63.7	27.4	15.7
5374	PHRAGMITES AUSTRALIS	Y	.	0	4.8	0.8	90.0	41.2	4.4
5375	PHYLICA EXCELSA	S	.	0	19.4	19.7	.	.	.
5376	PHYLICA EXCELSA VAR. PAPILLOSA	S	.	0	24.6	9.1	.	.	.
5377	PHYLICA OLEAEFOLIA	S	.	0	17.1	15.1	.	.	.
5378	PHYLICA PINEA	S	.	0	22.0	24.3	.	.	.
5379	PHYLICA PLUMIGERA	S	.	0	20.1	19.6	.	.	.

OBS	CALCIUM	PHOSPHO	IRON	SODIUM	POTASS	CAROT	THIAMINE	RIBOFLAV	NIACIN	ASCORBIC	REFCO
5356	.	.	.	.	.	.	.	.	.	.	19
5357	1600.0	210.0	.	.	3170.0	.	.	.	.	.	6
5358	.	.	.	.	.	.	.	.	.	.	18
5359	.	.	.	.	.	.	.	.	.	.	18
5360	85.4	879.4	15.1	.	.	364.3	.	.	13.57	75.4	1
5361	125.9	74.1	3.2	.	.	246.9	0.17	0.12	1.48	14.8	3
5362	94.9	45.9	1.1	.	.	63.3	0.14	0.13	0.16	1.6	3
5363	76.1	81.3	3.9	1.3	836.1	38.7	0.12	0.13	2.84	0.0	4
5364	.	.	.	.	.	.	.	.	.	.	5
5365	141.7	59.0	.	.	.	30.7	0.01	0.02	1.06	3.5	5
5366	.	.	.	.	.	.	.	.	.	.	19
5367	55.7	97.6	8.2	.	.	.	0.00	.	.	.	1
5368	600.0	180.0	.	.	.	.	.	.	.	.	18
5369	.	.	.	.	.	.	.	.	.	.	6
5370	.	.	.	.	.	.	.	.	.	.	21
5371	940.0	390.0	.	.	.	.	.	.	.	.	5
5372	480.0	60.0	.	.	.	.	.	.	.	.	6
5373	.	.	.	.	.	.	.	.	.	.	6
5374	.	.	.	.	.	.	.	.	.	.	6
5375	.	.	.	.	.	.	.	.	.	.	21
5376	.	.	.	.	.	.	.	.	.	.	21
5377	.	.	.	.	.	.	.	.	.	.	21
5378	.	.	.	.	.	.	.	.	.	.	21
5379	.	.	.	.	.	.	.	.	.	.	21

TABLE 3 - DATA CONVERTED TO A ZERO-MOISTURE BASIS.

OBS	PLNA	PART	CALORIES	WATER	PROTEIN	FAT	TOTLCARB	FIBER
5380	PHYLICA SPICATA	S	.	0	21.1	10.9	.	.
5381	PHYLLANTHUS ACIDUS	F	358	0	8.4	4.2	83.2	13.7
5382	PHYLLANTHUS ACIDUS	F	430	0	11.6	20.9	60.5	24.4
5383	PHYLLANTHUS EMBLICA	F	365	0	4.4	3.1	89.9	15.1
5384	PHYLLANTHUS EMBLICA	F	354	0	1.7	0.9	95.2	8.3
5385	PHYLLANTHUS GUINEENSIS	SH	.	0	5.3	1.8	82.3	49.0
5386	PHYLLANTHUS SP.	S	.	0	13.6	30.4	.	.
5387	PHYLLOSTACHYS SP.	SH	293	0	33.3	2.7	54.7	16.0
5388	PHYMASPERMUM PARVIFOLIUM	C	.	0	10.8	5.9	72.7	33.3
5389	PHYSALIS ALKEKENGI	S	.	0	16.2	31.8	.	.
5390	PHYSALIS ANGULATA	F	267	0	31.7	4.0	44.6	13.9
5391	PHYSALIS IXOCARPA	F	.	0	8.4	7.2	77.1	7.2
5392	PHYSALIS NICANDROIDES	S	.	0	15.0	18.8	.	.
5393	PHYSALIS NICANDROIDES	S	.	0	16.6	20.2	.	.
5394	PHYSALIS PERUVIANA	F	346	0	1.4	0.9	92.9	23.2
5395	PHYSALIS PERUVIANA	F	.	0	10.5	1.2	83.6	18.7
5396	PHYSALIS SP.	S	.	0	11.9	17.6	.	.
5397	PHYSALIS SP.	S	.	0	16.2	19.1	.	.
5398	PHYSALIS SP.	F	282	0	10.6	0.6	66.5	.
5399	PHYSALIS SP.	F	342	0	13.7	4.3	75.2	14.5
5400	PHYSALIS SP.	F	363	0	13.0	4.8	76.7	19.2
5401	PHYSANTHYLLIS TETRAPHYLLA	S	.	0	23.5	3.7	.	.
5402	PHYSOCARPUS SP.	Y	.	0	3.8	2.8	90.1	43.2
5403	PHYSOSTEGIA VIRGINIANA	S	.	0	25.0	35.2	.	.

OBS	ASH	CALCIUM	PHOSPHO	IRON	SODIUM	POTASS	CAROT	THIAMINE	RIBOFLAV	NIACIN	ASCORBIC	REFCO
5380	.	.	.	.	.	.	.	.	.	.	.	21
5381	4.2	63.2	210.5	11.6	63.2	1589.5	0.0	0.11	0.11	4.21	21.1	2
5382	7.0	232.6	244.2	37.2	.	.	813.9	0.58	0.23	5.81	127.9	3
5383	2.5	182.4	132.1	3.1	138.4	1075.5	377.4	0.19	0.25	1.26	1735.8	2
5384	2.2	52.4	113.5	2.2	.	.	0.0	0.13	0.22	0.87	2729.3	3
5385	4.6	.	.	.	.	.	.	.	.	.	.	18
5386	1.8	.	.	.	.	.	.	.	.	.	.	19
5387	9.3	53.3	680.0	5.3	26.7	.	0.0	0.13	1.07	10.67	133.3	2
5388	10.6	.	.	.	.	.	.	.	.	.	.	18
5389	.	.	.	.	.	.	.	.	.	.	.	20
5390	19.8	802.0	1336.6	.	.	.	.	.	.	.	.	1
5391	7.2	84.3	481.9	16.9	0.0	2927.7	578.3	0.60	0.24	25.30	24.1	5
5392	.	.	.	.	.	.	.	.	.	.	.	20
5393	.	.	.	.	.	.	.	.	.	.	.	21
5394	4.7	37.9	260.7	5.7	.	.	6919.4	0.47	0.14	8.06	203.8	3
5395	4.7	58.5	391.8	11.7	5.8	1871.3	8350.9	0.29	0.12	1.75	286.5	5
5396	.	.	.	.	.	.	.	.	.	.	.	19
5397	.	.	.	.	.	.	.	.	.	.	.	19
5398	22.4	41.2	323.5	5.3	.	.	.	0.88	0.29	9.41	23.5	2
5399	6.8	85.5	290.6	7.7	.	.	427.3	0.77	0.34	20.51	51.3	3
5400	5.5	61.6	274.0	6.8	.	.	2958.9	0.75	0.27	19.18	75.3	4
5401	.	.	.	.	.	.	.	.	.	.	.	21
5402	3.3	.	.	.	.	.	.	.	.	.	.	6
5403	.	.	.	.	.	.	.	.	.	.	.	19

TABLE 3 - DATA CONVERTED TO A ZERO-MOISTURE BASIS.

OBS	PLNA	PART	CALORIES	WATER	PROTEIN	FAT	TOTLCARB	FIBER
5404	PHYTELEPHAS MACROCARPA	S	.	0	5.3	1.6	91.6	9.3
5405	PHYTOLACCA AMERICANA	S	.	0	9.8	13.0	.	.
5406	PHYTOLACCA AMERICANA	S	.	0	11.7	10.7	.	.
5407	PHYTOLACCA AMERICANA	SH	274	0	31.0	4.8	44.0	.
5408	PHYTOLACCA SP.	W	.	0	14.7	9.3	72.6	13.4
5409	PICEA ABIES	S	.	0	22.5	38.1	.	.
5410	PICRAMNIA PENTANDRA	S	.	0	12.5	54.7	.	.
5411	PICRAMNIA PENTANDRA	S	.	0	.	12.7	.	.
5412	PICRAMNIA PENTANDRA	S	.	0	13.1	56.8	.	.
5413	PICRAMNIA PENTANDRA	S	.	0	13.1	57.9	.	.
5414	PIERIS JAPONICA	S	.	0	10.4	12.0	.	.
5415	PILIOSTIGMA MALABARICUM VAR. ACIDUM	L	359	0	26.0	8.4	62.6	13.7
5416	PILIOSTIGMA RETICULATUM	L	295	0	22.1	0.5	66.4	31.3
5417	PILIOSTIGMA THONNINGII	L	.	0	11.7	3.3	80.0	26.9
5418	PILIOSTIGMA THONNINGII	F	.	0	7.0	2.0	86.7	26.2
5419	PIMENTA DIOICA	L	330	0	11.7	3.6	75.7	17.9
5420	PIMENTA DIOICA	F	287	0	6.7	9.5	78.8	23.6
5421	PIMENTA DIOICA	L	335	0	13.4	4.1	74.9	17.5
5422	PIMPINELLA ANISUM	S	.	0	19.6	23.9	.	.
5423	PIMPINELLA ANISUM	S	.	0	21.2	30.4	.	.
5424	PIMPINELLA ANISUM	S	.	0	18.8	24.3	.	.
5425	PIMPINELLA ANISUM	S	372	0	19.4	17.6	55.2	16.1
5426	PIMPINELLA ANISUM	S	466	0	21.3	27.8	50.9	.
5427	PIMPINELLA PEREGRINA	S	.	0	22.1	19.8	.	.

OBS	ASH	CALCIUM	PHOSPHO	IRON	SODIUM	POTASS	CAROT	THIAMINE	RIBOFLAV	NIACIN	ASCORBIC	REFCO
5404	1.5	.	.	.	.	.	.	.	.	.	.	18
5405	2.6	.	.	.	.	.	.	.	.	.	.	19
5406	.	.	.	.	.	.	.	.	.	.	.	21
5407	20.2	631.0	523.8	20.2	.	.	62142.9	0.95	3.93	14.29	1619.0	4
5408	3.4	.	.	.	.	.	.	.	.	.	.	6
5409	.	.	.	.	.	.	.	.	.	.	.	20
5410	2.1	.	.	.	.	.	.	.	.	.	.	19
5411	.	.	.	.	.	.	.	.	.	.	.	19
5412	.	.	.	.	.	.	.	.	.	.	.	20
5413	.	.	.	.	.	.	.	.	.	.	.	20
5414	.	.	.	.	.	.	.	.	.	.	.	21
5415	3.1	351.1	236.6	2.3	.	.	6603.0	0.15	1.30	10.69	76.3	2
5416	11.1	2004.6	368.7	.	.	.	.	.	.	.	.	1
5417	5.0	910.0	130.0	.	.	.	.	.	.	.	.	18
5418	4.3	.	.	.	.	.	.	.	.	.	.	18
5419	8.9	1005.6	117.3	9.5	.	.	10279.3	0.11	0.28	1.40	206.7	1
5420	5.0	722.4	123.5	7.8	84.2	1141.0	354.1	0.11	0.07	3.13	42.6	16
5421	7.6	1116.6	139.9	7.6	.	.	12040.8	0.17	0.35	1.75	201.2	3
5422	6.6	.	.	.	.	.	.	.	.	.	.	19
5423	.	.	.	.	.	.	.	.	.	.	.	19
5424	.	.	.	.	.	.	.	.	.	.	.	20
5425	7.7	713.8	486.2	40.9	17.7	1592.3	.	.	.	.	.	16
5426	.	784.3	.	39.1	.	.	.	.	.	.	0.0	2
5427	.	.	.	.	.	.	.	.	.	.	.	21

TABLE 3 - DATA CONVERTED TO A ZERO-MOISTURE BASIS.

OBS	PLANA	PART	CALORIES	WATER	PROTEIN	FAT	TOTCARB	FIBER	ASH	CALCIUM	PHOSPHO	IRON	SODIUM	POTASS	CAROT	THIAMINE	RIBOFLAV	NIACIN	ASCORBIC	REFCO
5428	PINUS CEMBROIDES	S	658	0	15.9	63.7	17.4	2.7	3.0	14.5	534.8	4.6	.	.	20.8	0.79	0.24	10.18	1.0	3
5429	PINUS CEMBROIDES	S	655	0	13.4	62.4	21.2	1.1	3.0	12.4	623.3	5.4	.	.	18.6	1.32	0.24	4.64	0.0	4
5430	PINUS DENSIFLORA	S	.	0	31.2	35.1	.	.	.	.	.	.	.	.	.	.	.	.	.	21
5431	PINUS FLEXILIS	S	.	0	25.0	59.8	.	.	.	.	.	.	.	.	.	.	.	.	.	20
5432	PINUS PINEA	S	585	0	32.9	50.2	12.3	1.0	4.6	.	.	.	.	.	.	0.66	.	.	.	4
5433	PINUS SP.	S	663	0	14.3	64.2	18.7	2.1	2.8	15.5	406.2	6.6	4.1	642.5	0.0	0.17	0.41	5.70	2.1	2
5434	PINUS THUNBERGII	S	.	0	27.9	31.5	.	.	.	.	.	.	.	.	.	.	.	.	.	21
5435	PIPER AURITUM	L	320	0	21.9	7.9	56.7	12.4	13.5	1443.8	292.1	31.5	.	.	19325.8	0.67	1.35	5.06	337.1	3
5436	PIPER NIGRUM	S	369	0	13.9	9.3	72.3	.	4.5	.	.	.	.	.	.	.	.	.	.	2
5437	PIPER SARMENTOSUM	L	331	0	17.7	8.2	61.6	15.1	12.5	1970.5	98.4	24.9	.	.	17344.3	0.43	0.36	53.11	32.8	2
5438	PIPTOCHAETIUM BICOLOR	S	.	0	22.5	4.4	.	.	.	.	.	.	.	.	.	.	.	.	.	21
5439	PIPTOCHAETIUM CONFUSUM	S	.	0	15.5	2.8	.	.	.	.	.	.	.	.	.	.	.	.	.	21
5440	PIPTOCHAETIUM VERRUCULOSUM	S	.	0	12.2	3.0	.	.	.	.	.	.	.	.	.	.	.	.	.	21
5441	PISTACIA LENTISCUS	S	.	0	7.2	13.0	.	.	.	.	.	.	.	.	.	.	.	.	.	21
5442	PISTACIA TEREBINTHUS	S	.	0	24.2	60.3	.	.	.	.	.	.	.	.	.	.	.	.	.	21
5443	PISTACIA TEREBINTHUS	S	.	0	7.4	2.6	.	.	.	.	.	.	.	.	.	.	.	.	.	21
5444	PISTACIA VERA	S	624	0	19.7	56.4	20.6	.	3.3	.	.	.	.	.	.	.	.	.	.	1
5445	PISTACIA VERA	S	627	0	20.4	56.7	20.1	2.0	2.9	138.3	528.0	7.7	.	1026.4	145.7	0.71	.	1.48	0.0	4
5446	PISUM MACROCARPUM	F	317	0	24.4	1.6	69.1	13.8	4.9	398.0	439.0	8.9	73.0	1098.0	1991.0	1.38	0.81	7.32	293.0	10
5447	PISUM SATIVUM	W	.	0	17.2	3.0	67.1	26.9	12.7	1870.0	370.0	.	.	.	.	.	.	.	.	18
5448	PISUM SATIVUM	H	.	0	14.9	2.7	74.2	31.5	8.3	.	.	.	.	.	.	.	.	.	.	18
5449	PISUM SATIVUM	S	.	0	25.3	1.1	68.9	6.2	4.7	.	.	.	.	.	.	.	.	.	.	18
5450	PISUM SATIVUM	S	.	0	26.1	1.0	.	.	.	.	.	.	.	.	.	.	.	.	.	21
5451	PISUM SATIVUM	S	381	0	25.1	1.2	69.7	6.4	4.0	101.1	429.2	20.1	.	.	16.8	0.99	0.19	3.37	0.0	1
5452	PISUM SATIVUM	F	327	0	26.7	2.0	68.5	10.0	2.8	115.5	402.4	7.6	.	.	1314.7	1.27	0.52	7.97	95.6	1
5453	PISUM SATIVUM	S	385	0	25.4	1.6	69.3	9.8	3.7	131.1	418.0	4.9	24.6	1434.4	1659.8	1.15	0.45	11.48	110.7	2
5454	PISUM SATIVUM	F	317	0	24.4	1.6	69.1	13.8	4.9	398.4	439.0	8.9	73.2	1097.6	1991.9	1.38	0.81	7.32	292.7	2
5455	PISUM SATIVUM	IS	326	0	25.5	1.3	70.5	10.1	2.7	80.5	416.1	6.7	.	.	838.9	1.28	0.47	7.38	87.2	3
5456	PISUM SATIVUM	S	390	0	25.6	2.3	69.3	5.3	2.8	90.9	329.5	6.6	.	.	56.8	0.65	0.19	3.41	1.1	3
5457	PISUM SATIVUM	F	317	0	20.4	1.2	71.9	7.2	6.6	371.3	538.9	4.2	.	1018.0	2443.1	1.68	0.72	.	125.7	4
5458	PISUM SATIVUM	S	382	0	28.6	1.8	65.5	9.1	4.1	118.2	527.3	8.6	9.1	1436.4	1745.4	1.59	0.64	13.18	122.7	4
5459	PISUM SATIVUM ARVENSE	W	.	0	19.7	3.3	68.0	25.9	9.0	1210.0	230.0	40.0	.	1500.0	.	.	.	.	.	6
5460	PISUM SATIVUM VAR. ARVENSE	S	.	0	25.9	0.9	.	.	2.7	.	.	.	.	.	.	.	.	.	.	19
5461	PISUM SATIVUM VAR. ARVENSE	S	.	0	19.6	1.5	.	.	3.6	.	.	.	.	.	.	.	.	.	.	19
5462	PISUM SATIVUM VAR. ARVENSE	S	.	0	21.1	1.4	.	.	3.1	.	.	.	.	.	.	.	.	.	.	19
5463	PISUM SATIVUM VAR. ARVENSE	S	.	0	28.1	1.0	.	.	.	.	.	.	.	.	.	.	.	.	.	20
5464	PISUM SYRIACUM	S	.	0	29.1	0.7	.	.	.	.	.	.	.	.	.	.	.	.	.	21
5465	PITHECELLOBIUM BREVIFOLIUM	S	.	0	31.2	13.3	.	.	3.9	.	.	.	.	.	.	.	.	.	.	19
5466	PITHECELLOBIUM DULCE	L	.	0	29.0	4.4	61.0	17.5	5.6	.	.	.	.	.	.	.	.	.	.	18
5467	PITHECELLOBIUM DULCE	SH	.	0	23.5	3.0	68.5	23.1	5.0	.	.	.	.	.	.	.	.	.	.	18
5468	PITHECELLOBIUM DULCE	S	.	0	24.4	13.2	.	.	3.3	.	.	.	.	.	.	.	.	.	.	19
5469	PITHECELLOBIUM DULCE	S	.	0	23.8	13.3	.	.	.	.	.	.	.	.	.	.	.	.	.	19
5470	PITHECELLOBIUM DULCE	S	.	0	20.0	12.5	.	.	.	.	.	.	.	.	.	.	.	.	.	19
5471	PITHECELLOBIUM DULCE	F	351	0	13.5	1.8	82.0	5.4	2.7	58.6	189.2	2.3	85.6	1000.0	67.6	1.08	0.45	2.70	599.1	2
5472	PITHECELLOBIUM JIRINGA	F	388	0	26.2	0.8	71.3	5.5	1.7	97.0	160.3	3.0	.	.	1666.7	0.59	0.04	1.69	33.8	2

TABLE 3 - DATA CONVERTED TO A ZERO-MOISTURE BASIS.

OBS	PLANA	PART	CALORIES	WATER	PROTEIN	FAT	TOTCARB	FIBER	ASH	CALCIUM	PHOSPHO	IRON	SODIUM	POTASS	CAROT	THIAMINE	RIBOFLAV	NIACIN	ASCORBIC	REFCO
5473	PITHECELLOBIUM MEXICANUM	S	.	0	35.6	14.5	.	.	.	.	.	.	.	.	.	.	.	.	.	20
5474	PITHECELLOBIUM UNDULATUM	S	.	0	27.5	6.4	.	.	.	.	.	.	.	.	.	.	.	.	.	20
5475	PITTOSPORUM TOBIRA	S	.	0	13.8	13.0	.	.	.	.	.	.	.	.	.	.	.	.	.	19
5476	PITTOSPORUM TOBIRA	S	.	0	11.3	13.5	.	.	.	.	.	.	.	.	.	.	.	.	.	20
5477	PITURANTHOS APHYLLA	S	.	0	19.8	27.1	.	.	.	.	.	.	.	.	.	.	.	.	.	20
5478	PITURANTHOS APHYLLA	S	.	0	22.8	26.2	.	.	.	.	.	.	.	.	.	.	.	.	.	21
5479	PLAGIOBOTHRYS NOTHOFULVUS	W	.	0	14.2	.	.	22.0	14.7	1940.0	520.0	.	.	3940.0	.	.	.	.	.	6
5480	PLANTAGO ALBICANS	S	.	0	17.0	6.4	.	.	.	.	.	.	.	.	.	.	.	.	.	21
5481	PLANTAGO ARISTATA	S	.	0	16.2	6.2	.	.	.	.	.	.	.	.	.	.	.	.	.	20
5482	PLANTAGO LAGOPUS	S	.	0	20.2	7.1	.	.	.	.	.	.	.	.	.	.	.	.	.	21
5483	PLANTAGO LANCEOLATA	L	.	0	15.9	2.2	69.0	13.3	12.9	1540.0	290.0	.	.	.	.	.	.	.	.	18
5484	PLANTAGO LANCEOLATA	SH	.	0	22.8	3.8	58.7	16.1	14.7	2460.0	430.0	.	.	.	.	.	.	.	.	18
5485	PLANTAGO LANCEOLATA	S	.	0	18.8	7.3	.	.	.	.	.	.	.	.	.	.	.	.	.	19
5486	PLANTAGO LANCEOLATA	S	.	0	17.2	5.1	.	.	.	.	.	.	.	.	.	.	.	.	.	21
5487	PLANTAGO LANCEOLATA	S	.	0	16.4	7.8	.	.	.	.	.	.	.	.	.	.	.	.	.	21
5488	PLANTAGO MAJOR	S	.	0	19.1	15.5	.	.	.	.	.	.	.	.	.	.	.	.	.	21
5489	PLANTAGO MAJOR	L	328	0	13.4	1.6	78.5	.	6.5	989.2	279.6	6.5	86	1489.2	13548.4	.	1.51	4.30	43.0	2
5490	PLANTAGO OVATA	F	.	0	20.9	2.4	68.7	31.9	8.0	650.0	570.0	.	.	.	.	.	.	.	.	18
5491	PLANTAGO OVATA	S	.	0	16.2	5.9	.	.	3.3	.	.	.	.	.	.	.	.	.	.	19
5492	PLANTAGO OVATA	S	.	0	14.6	5.4	.	.	2.4	.	.	.	.	.	.	.	.	.	.	19
5493	PLANTAGO PSYLLIUM	S	.	0	14.8	6.9	.	.	.	.	.	.	.	.	.	.	.	.	.	21
5494	PLANTAGO RHODOSPERMA	S	.	0	15.0	6.0	.	.	.	.	.	.	.	.	.	.	.	.	.	19
5495	PLANTAGO RHODOSPERMA	S	.	0	17.5	5.7	.	.	.	.	.	.	.	.	.	.	.	.	.	19
5496	PLANTAGO RHODOSPERMA	S	.	0	18.8	6.0	.	.	.	.	.	.	.	.	.	.	.	.	.	20
5497	PLANTAGO RHODOSPERMA	W	.	0	14.5	1.5	71.1	13.1	12.9	5240.0	190.0	.	.	2700.0	.	.	.	.	.	6
5498	PLANTAGO SP.	S	.	0	17.5	5.2	.	.	.	.	.	.	.	.	.	.	.	.	.	19
5499	PLANTAGO SP.	S	.	0	15.0	8.9	.	.	.	.	.	.	.	.	.	.	.	.	.	19
5500	PLANTAGO SP.	S	.	0	34.8	8.3	.	.	.	.	.	.	.	.	.	.	.	.	.	21
5501	PLANTAGO SP.	W	.	0	13.3	1.9	75.0	15.9	9.8	4160.0	190.0	.	.	2350.0	.	.	.	.	.	6
5502	PLANTAGO SQUARROSA	S	.	0	17.0	5.6	.	.	.	.	.	.	.	.	.	.	.	.	.	21
5503	PLANTAGO WRIGHTIANA	S	.	0	15.6	9.9	.	.	.	.	.	.	.	.	.	.	.	.	.	19
5504	PLANTAGO WRIGHTIANA	S	.	0	20.6	8.6	.	.	.	.	.	.	.	.	.	.	.	.	.	20
5505	PLATANUS OCCIDENTALIS	S	.	0	9.4	4.8	.	.	.	.	.	.	.	.	.	.	.	.	.	20
5506	PLATANUS OCCIDENTALIS	S	.	0	13.8	7.8	.	.	.	.	.	.	.	.	.	.	.	.	.	20
5507	PLATONIA ESCULENTA	F	377	0	6.0	6.7	83.3	17.7	4.0	.	.	.	.	.	.	.	.	.	60.0	3
5508	PLATYCODON GRANDIFLORUM	R	379	0	3.2	1.2	93.7	11.7	2.0	306.1	249.3	8.2	.	.	0.0	0.13	0.47	10.29	0.0	2
5509	PLECTRANTHUS ESCULENTUS	R	382	0	8.6	1.8	85.5	4.5	4.1	81.8	363.6	29.1	.	.	2454.5	0.00	2.82	3.18	.	1
5510	PLEIOGYNIUM SOLANDRI	S	.	0	23.1	60.0	.	.	.	.	.	.	.	.	.	.	.	.	.	20
5511	PLUCHEA INDICA	L	300	0	12.9	3.6	67.1	.	16.4	1828.6	350.0	40.0	.	.	17071.4	0.14	.	.	214.3	2
5512	PLUCHEA ODORATA	S	.	0	25.0	15.9	.	.	.	.	.	.	.	.	.	.	.	.	.	19
5513	PLUMERIA ACUMINATA	L	314	0	11.7	0.0	80.3	.	8.0	503.6	438.0	.	.	.	.	0.22	.	.	1635.0	2
5514	POA ANNUA	L	.	0	11.0	2.8	60.3	26.3	8.8	440.0	390.0	.	.	.	.	.	.	.	.	5
5515	POA ANNUA	L	.	0	14.7	3.3	79.2	33.0	16.1	1270.0	780.0	.	.	.	.	.	.	.	.	5
5516	POA ARACHNIFERA	W	.	0	12.4	4.7	74.8	32.4	8.1	470.0	200.0	.	.	.	.	.	.	.	.	6
5517	POA BULBOSA	BU	.	0	8.8	3.7	.	.	.	.	.	.	.	.	.	.	.	.	.	20

TABLE 3 - DATA CONVERTED TO A ZERO-MOISTURE BASIS.

OBS	PLNA	PART	CALORIES	WATER	PROTEIN	FAT	TOTCARB	FIBER	ASH	CALCIUM	PHOSPHO	IRON	SODIUM	POTASS	CAROTE	THIAMINE	RIBOFLAV	NIACIN	ASCORBIC	REFCO
5518	POA BULBOSA	L	.	0	8.4	3.2	79.8	26.8	7.4	620.0	610.0	.	.	.	.	.	.	.	.	5
5519	POA COMPRESSA	H	.	0	11.6	2.7	77.8	28.9	7.9	300.0	290.0	.	.	1590	.	.	.	.	.	6
5520	POA COMPRESSA	W	.	0	17.0	3.7	70.4	26.4	8.9	390.0	390.0	.	2040	.	.	.	.	.	.	6
5521	POA EPILIS	H	.	0	3.4	1.6	88.2	38.7	6.8	.	.	.	.	.	.	.	.	.	.	6
5522	POA FENDLERIANA	W	.	0	7.0	2.4	79.5	33.5	11.1	410.0	340.0	.	.	.	.	.	.	.	.	6
5523	POA NEMORALIS	S	.	0	11.4	1.2	.	.	.	.	.	.	.	.	.	.	.	.	.	21
5524	POA PALUSTRIS	H	.	0	8.1	1.9	81.8	38.0	8.2	260.0	260.0	.	.	.	.	.	.	.	.	6
5525	POA PRATENSIS	S	.	0	14.4	2.9	.	.	8.5	.	.	.	.	.	.	.	.	.	.	19
5526	POA PRATENSIS	S	.	0	9.0	2.0	.	.	2.7	.	.	.	.	.	.	.	.	.	.	19
5527	POA PRATENSIS	L	.	0	7.7	1.6	69.6	30.6	8.0	160.0	130.0	6	70	590	8	0.26	.	.	.	5
5528	POA PRATENSIS	L	.	0	12.5	3.2	87.2	37.8	12.8	.	.	.	.	.	.	.	.	.	.	5
5529	POA PRATENSIS	H	.	0	12.0	3.1	76.7	29.9	8.2	400.0	260.0	26	.	1720	.	.	.	.	.	6
5530	POA PRATENSIS	W	.	0	15.8	3.6	72.7	26.5	7.9	450.0	390.0	24	.	2060	.	.	.	.	.	6
5531	POA REFLEXA	H	.	0	9.5	2.3	82.0	29.2	6.2	.	.	.	.	.	.	.	.	.	.	6
5532	POA SCABRELLA	W	.	0	9.9	.	.	35.7	5.3	360.0	250.0	.	.	1630	.	.	.	.	.	6
5533	POA SECUNDA	H	.	0	8.6	3.1	81.9	32.6	6.4	.	.	.	.	.	.	.	.	.	.	6
5534	POA SECUNDA	W	.	0	9.0	2.4	79.1	39.0	9.5	410.0	270.0	.	.	.	.	.	.	.	.	6
5535	POCHOTA GLABRA	S	405	0	7.4	2.1	89.0	2.1	1.5	.	.	.	.	.	.	.	.	.	.	1
5536	PODACHAENIUM EMINENS	S	.	0	18.8	28.0	.	.	.	.	.	.	.	.	.	.	.	.	.	20
5537	PODALYRIA CALYPTRATA	S	.	0	36.2	9.0	.	.	.	.	.	.	.	.	.	.	.	.	.	21
5538	PODALYRIA CUNEIFOLIA	S	.	0	30.0	6.2	.	.	.	.	.	.	.	.	.	.	.	.	.	21
5539	PODOCARPUS MACROPHYLLA	S	.	0	6.8	2.8	.	.	.	.	.	.	.	.	.	.	.	.	.	20
5540	PODOCARPUS NAGI	S	.	0	10.0	39.9	.	.	2.4	.	.	.	.	.	.	.	.	.	.	19
5541	PODOCARPUS NAGI	S	.	0	1.4	0.9	.	.	5.6	.	.	.	.	.	.	.	.	.	.	19
5542	POGOSTEMON PARVIFLORUS	S	.	0	25.0	35.8	.	.	.	.	.	.	.	.	.	.	.	.	.	21
5543	POLANISIA SP.	S	.	0	14.4	26.7	.	.	.	.	.	.	.	.	.	.	.	.	.	20
5544	POLANISIA TRACHYSPERMA	S	.	0	17.5	25.2	.	.	3.3	.	.	.	.	.	.	.	.	.	.	19
5545	POLANISIA TRACHYSPERMA	S	.	0	18.1	20.7	.	.	.	.	.	.	.	.	.	.	.	.	.	19
5546	POLANISIA TRACHYSPERMA	S	.	0	19.4	20.6	.	.	.	.	.	.	.	.	.	.	.	.	.	19
5547	POLANISIA TRACHYSPERMA	S	.	0	16.9	29.8	.	.	.	.	.	.	.	.	.	.	.	.	.	21
5548	POLANISIA VISCOSA	S	.	0	16.2	26.2	.	.	.	.	.	.	.	.	.	.	.	.	.	19
5549	POLEMONIUM ALBIFLORUM	W	.	0	10.6	2.3	.	.	11.7	1350.0	430.0	.	.	.	.	.	.	.	.	6
5550	POLEMONIUM CAERULEUM	S	.	0	23.9	26.1	.	.	4.6	.	.	.	.	.	.	.	.	.	.	19
5551	POLEMONIUM CAERULEUM	S	.	0	17.2	12.0	.	.	.	.	.	.	.	.	.	.	.	.	.	21
5552	POLYALTHIA LONGIFOLIA	L	277	0	35.4	0.8	51.5	.	12.3	38.5	738.5	.	.	.	.	.	.	.	.	2
5553	POLYGONUM CESPITOSUM VAR. LONGISETUM	S	.	0	13.8	3.2	.	.	.	.	.	.	.	.	.	.	.	.	.	19
5554	POLYGONUM CONVOLVULUS	S	.	0	14.4	1.9	.	.	1.8	.	.	.	.	.	.	.	.	.	.	19
5555	POLYGONUM DUMETORUM	S	.	0	12.1	2.1	.	.	.	.	.	.	.	.	.	.	.	.	.	21
5556	POLYGONUM KITAIBELIANUM	S	.	0	14.4	3.5	.	.	.	.	.	.	.	.	.	.	.	.	.	20
5557	POLYGONUM LAPATHIFOLIUM	S	.	0	8.8	4.0	.	.	.	.	.	.	.	.	.	.	.	.	.	19
5558	POLYGONUM ORIENTALE	S	.	0	11.6	3.7	.	.	1.5	.	.	.	.	.	.	.	.	.	.	19
5559	POLYGONUM PENSYLVANICUM	S	.	0	9.9	4.4	.	.	1.8	.	.	.	.	.	.	.	.	.	.	19
5560	POLYGONUM PENSYLVANICUM	S	.	0	7.7	1.9	.	.	.	.	.	.	.	.	.	.	.	.	.	20
5561	POLYGONUM PUNCTATUM	S	.	0	8.1	1.7	.	.	.	.	.	.	.	.	.	.	.	.	.	20
5562	POLYGONUM RAMOSISSIMUM	S	.	0	11.0	2.6	.	.	1.4	.	.	.	.	.	.	.	.	.	.	19

TABLE 3 - DATA CONVERTED TO A ZERO-MOISTURE BASIS.

OBS	PLNA	PART	CALORIES	WATER	PROTEIN	FAT	TOTLCARB	FIBER
5563	POLYGONUM RAMOSISSIMUM	S	.	0	5.4	1.6	.	.
5564	POLYGONUM SALICIFOLIUM	L	320	0	18.0	1.5	73.5	17.5
5565	POLYMNIA EDULIS	L	.	0	17.1	7.4	59.5	10.0
5566	POLYMNIA EDULIS	S	.	0	11.4	2.3	76.1	26.9
5567	POLYMNIA EDULIS	R	.	0	7.3	0.4	85.6	5.7
5568	POLYMNIA UVEDALIA	S	.	0	34.2	59.0	.	.
5569	POLYPOGON MONSPELIENSIS	S	.	0	28.8	5.8	.	.
5570	POLYPOGON MONSPELIENSIS	L	.	0	11.8	1.3	62.9	24.9
5571	POLYPOGON MONSPELIENSIS	W	.	0	13.6	2.7	69.1	28.3
5572	POLYSCIAS SCUTILLARIA	L	300	0	20.6	1.7	65.6	.
5573	POLYTAENIA NUTTALLII	S	.	0	14.4	26.6	.	.
5574	POLYTAENIA NUTTALLII	S	.	0	15.7	32.6	.	.
5575	PONGAMIA PINNATA	S	.	0	21.9	34.6	.	.
5576	PONGAMIA SP.	S	.	0	21.9	18.1	.	.
5577	POPULUS ANGUSTIFOLIA	L	.	0	24.8	6.9	61.3	13.9
5578	POPULUS TREMULOIDES(IMMATURE)	L	.	0	19.3	4.7	69.8	13.5
5579	POPULUS TREMULOIDES(MATURE)	L	.	0	9.5	6.2	75.8	15.0
5580	PORTULACA OLERACEA	L	255	0	34.5	5.3	35.5	10.5
5581	PORTULACA OLERACEA	S	.	0	21.0	18.9	.	.
5582	PORTULACA OLERACEA	W	.	0	21.4	2.9	57.3	14.6
5583	PORTULACA OLERACEA	L	245	0	25.5	3.2	44.7	8.5
5584	PORTULACA OLERACEA	L	296	0	17.6	2.4	63.2	8.8
5585	PORTULACA OLERACEA	L	295	0	22.7	4.5	56.8	10.2
5586	PORTULACA OLERACEA	L	280	0	22.7	5.3	50.7	12.0

OBS	ASH	CALCIUM	PHOSPHO	IRON	SODIUM	POTASS	CAROT	THIAMINE	RIBOFLAV	NIACIN	ASCORBIC	REFCO
5563	6.0	.	.	.	.	.	.	.	.	.	.	19
5564	7.0	750.0	230.0	.	.	.	.	.	.	.	.	1
5565	16.0	.	.	.	.	.	.	.	.	.	.	18
5566	10.2	.	.	.	.	.	.	.	.	.	.	18
5567	6.7	.	.	.	.	.	.	.	.	.	.	18
5568	.	.	.	.	.	.	.	.	.	.	.	20
5569	.	.	.	.	.	.	.	.	.	.	.	21
5570	21.9	1200.0	900.0	.	.	.	.	.	.	.	.	5
5571	14.6	340.0	330.0	.	.	.	.	.	.	.	.	6
5572	12.2	2633.3	272.2	22.2	.	.	18166.7	0.33	.	.	461.1	2
5573	.	.	.	.	.	.	.	.	.	.	.	20
5574	.	.	.	.	.	.	.	.	.	.	.	21
5575	3.2	.	.	.	.	.	.	.	.	.	.	19
5576	.	.	.	.	.	.	.	.	.	.	.	20
5577	7.0	.	.	.	.	.	.	.	.	.	.	6
5578	6.2	1350.0	510.0	.	.	.	.	.	.	.	.	6
5579	8.5	3020.0	300.0	.	.	.	.	.	.	.	.	6
5580	24.7	2078.0	774.0	.	55.0	505.0	.	.	.	6.72	.	12
5581	3.4	.	.	.	.	.	.	.	.	.	.	19
5582	18.4	.	.	.	.	.	.	.	.	.	.	6
5583	26.6	1106.4	329.8	14.9	.	.	.	0.32	1.60	.	617.0	1
5584	16.0	920.0	320.0	11.2	.	3120.0	10560.0	0.48	1.12	6.40	168.0	2
5585	15.9	897.7	363.6	40.9	.	.	17045.4	0.23	1.14	5.68	261.4	3
5586	21.3	1373.3	520.0	46.7	.	.	20000.0	0.40	1.33	6.67	333.3	4

TABLE 3 - DATA CONVERTED TO A ZERO-MOISTURE BASIS.

OBS	PLNA	PART	CALORIES	WATER	PROTEIN	FAT	TOTLCARB	FIBER
5587	PORTULACA OLERACEA	L	.	0	25.3	6.3	44.2	13.7
5588	PORTULACA QUADRIFIDA	L	268	0	11.3	5.6	54.9	14.1
5589	PORTULACARIA AFRA	L	.	0	11.8	3.9	72.5	14.6
5590	PORTULACARIA AFRA	SH	.	0	10.7	3.8	66.4	11.9
5591	PORTULACARIA AFRA	H	.	0	8.5	3.8	77.9	20.9
5592	PORTULACARIA AFRA	L	.	0	8.5	3.8	77.9	20.9
5593	PORTULACEA SP.	W	.	0	9.9	5.0	59.9	17.9
5594	POTENTILLA ARGUTA SSP. CONVALLARIA	S	.	0	21.1	32.5	.	.
5595	POTENTILLA ARGUTA SUBSP. CONVALLARIA	S	.	0	.	31.1	.	.
5596	POTENTILLA GLANDULOSA	S	.	0	13.9	24.7	.	.
5597	POTENTILLA HIRTA	S	.	0	19.9	25.1	.	.
5598	POTENTILLA PULCHERRIMA	S	.	0	21.8	13.9	.	.
5599	POTENTILLA RECTA	S	.	0	17.5	20.7	.	.
5600	POTENTILLA SP.	W	.	0	22.1	3.9	65.9	11.3
5601	POTERIUM SPINOSUM	S	.	0	7.6	8.5	.	.
5602	POTERIUM SPINOSUM	S	.	0	8.5	7.0	.	.
5603	POTERIUM SPINOSUM	S	.	0	8.7	7.9	.	.
5604	POTHOMORPHE UMBELLATA	L	385	0	25.7	19.0	45.8	10.1
5605	POUTERIA CAIMITO	F	355	0	4.6	1.0	92.1	2.3
5606	POUTERIA CAMPECHIANA	F	360	0	5.8	1.4	91.4	17.5
5607	POUTERIA CAMPECHIANA	F	360	0	3.2	2.0	92.6	2.5
5608	POUTERIA CAMPECHIANA	F	355	0	4.6	1.0	92.1	2.3
5609	POUTERIA VIRIDIS	F	348	0	5.1	0.6	90.5	5.7
5610	PRANGOS PABULARIA	S	.	0	33.1	24.0	.	.

OBS	ASH	CALCIUM	PHOSPHO	IRON	SODIUM	POTASS	CAROT	THIAMINE	RIBOFLAV	NIACIN	ASCORBIC	REFCO
5587	24.2	1168.4	473.7	10.5	705.3	7536.8	24126.3	1.05	2.32	7.37	305.3	5
5588	28.2	1056.3	493.0	.	.	.	.	.	.	.	.	1
5589	11.8	2870.0	350.0	.	.	.	.	.	.	.	.	18
5590	19.1	1400.0	120.0	.	.	.	.	.	.	.	.	18
5591	9.8	.	.	.	.	.	.	.	.	.	.	18
5592	9.8	.	.	.	.	.	.	.	.	.	.	5
5593	25.2	2840.0	320.0	.	.	5390.0	.	.	.	.	.	6
5594	.	.	.	.	.	.	.	.	.	.	.	20
5595	.	.	.	.	.	.	.	.	.	.	.	21
5596	.	.	.	.	.	.	.	.	.	.	.	21
5597	.	.	.	.	.	.	.	.	.	.	.	20
5598	.	.	.	.	.	.	.	.	.	.	.	20
5599	.	.	.	.	.	.	.	.	.	.	.	20
5600	8.1	.	.	.	.	.	.	.	.	.	.	6
5601	.	.	.	.	.	.	.	.	.	.	.	21
5602	.	.	.	.	.	.	.	.	.	.	.	21
5603	.	.	.	.	.	.	.	.	.	.	.	21
5604	9.5	944.1	223.5	.	.	.	.	.	.	.	.	1
5605	2.3	55.8	104.1	2.5	.	.	659.9	0.05	0.05	8.63	124.4	3
5606	1.4	70.1	70.1	2.6	.	.	2885.5	0.05	0.09	0.93	100.5	2
5607	2.2	89.3	89.3	1.7	.	.	521.1	0.45	0.00	8.19	99.3	3
5608	2.3	55.8	104.1	2.5	.	.	659.9	0.05	0.05	8.63	124.4	3
5609	3.8	72.8	88.6	2.2	.	.	158.2	0.03	0.09	6.01	136.1	3
5610	.	.	.	.	.	.	.	.	.	.	.	21

TABLE 3 - DATA CONVERTED TO A ZERO-MOISTURE BASIS.

OBS	PLNA	PART	CALORIES	WATER	PROTEIN	FAT	TOTLCARB	FIBER	ASH
5611	PREMNA ODORATA	L	320	0	17.50	2.0	73.5	13.0	7.0
5612	PRIONOSCIADIUM SP.	S	.	0	26.20	29.1	.	.	.
5613	PRISMATOCARPUS ROELLOIDES	S	.	0	17.60	25.9	.	.	.
5614	PROBOSCIDEA ALTHEAEFOLIA	S	.	0	26.20	36.2	.	.	.
5615	PROBOSCIDEA FRAGRANS	L	.	0	18.10	9.5	46.7	6.2	25.7
5616	PROBOSCIDEA SP.	S	.	0	24.40	36.5	.	.	.
5617	PROSOPIS AFRICANA	S	361	0	16.00	1.7	78.5	7.6	3.4
5618	PROSOPIS CHILENSIS	F	.	0	11.00	2.2	79.5	11.9	7.3
5619	PROSOPIS CHILENSIS	S	.	0	35.20	5.2	54.5	9.1	5.1
5620	PROSOPIS CHILENSIS	L	.	0	23.50	2.9	66.9	24.7	4.7
5621	PROSOPIS CHILENSIS	L	.	0	21.10	3.4	69.6	27.3	5.9
5622	PROSOPIS CHILENSIS	F	.	0	14.40	1.9	80.1	28.8	3.6
5623	PROSOPIS CHILENSIS	L	.	0	28.30	9.1	60.9	26.7	1.5
5624	PROSOPIS JULIFLORA	I	.	0	21.00	3.2	65.8	15.5	10.0
5625	PROSOPIS JULIFLORA	L	.	0	19.00	2.9	69.6	21.6	8.5
5626	PROSOPIS JULIFLORA	F	.	0	13.90	3.0	78.3	27.7	4.8
5627	PROSOPIS JULIFLORA	S	.	0	65.20	7.8	21.8	2.8	5.2
5628	PROSOPIS JULIFLORA	S	.	0	33.90	6.9	.	.	3.0
5629	PROSOPIS JULIFLORA	S	.	0	7.00	1.8	.	.	4.8
5630	PROSOPIS JULIFLORA	S	.	0	39.90	6.6	.	.	.
5631	PROSOPIS JULIFLORA	S	.	0	38.20	3.4	.	.	.
5632	PROSOPIS JULIFLORA VAR. VELUTINA	S	.	0	36.70	7.6	.	.	3.5
5633	PROSOPIS JULIFLORA VAR. VELUTINA	S	.	0	34.90	6.7	.	.	3.3
5634	PROSOPIS SPICIGERA	L	.	0	15.30	3.2	71.5	17.5	10.0

OBS	CALCIUM	PHOSPHO	IRON	SODIUM	POTASS	CAROT	THIAMINE	RIBOFLAV	NIACIN	ASCORBIC	REFCO
5611	320.0	380.0	15.0	25.0	2870.0	.	0.20	0.45	4.50	25.0	2
5612	.	.	.	.	.	.	.	.	.	.	20
5613	.	.	.	.	.	.	.	.	.	.	21
5614	.	.	.	.	.	.	.	.	.	.	20
5615	2190.0	250.0	.	.	3010.0	.	.	.	.	.	6
5616	.	.	.	.	.	.	.	.	.	.	20
5617	438.0	329.0	.	.	.	.	.	.	.	.	10
5618	440.0	160.0	.	.	.	.	.	.	.	.	18
5619	.	.	.	.	.	.	.	.	.	.	18
5620	860.0	250.0	.	.	.	.	.	.	.	.	18
5621	1940.0	190.0	.	.	1410.0	.	.	.	.	.	6
5622	660.0	210.0	.	.	1420.0	.	.	.	.	.	6
5623	.	.	.	.	.	.	.	.	.	.	5
5624	1310.0	400.0	.	.	.	.	.	.	.	.	18
5625	2080.0	220.0	.	.	.	.	.	.	.	.	18
5626	.	.	.	.	.	.	.	.	.	.	18
5627	.	.	.	.	.	.	.	.	.	.	18
5628	.	.	.	.	.	.	.	.	.	.	19
5629	.	.	.	.	.	.	.	.	.	.	19
5630	.	.	.	.	.	.	.	.	.	.	20
5631	.	.	.	.	.	.	.	.	.	.	21
5632	.	.	.	.	.	.	.	.	.	.	19
5633	.	.	.	.	.	.	.	.	.	.	19
5634	2650.0	240.0	.	.	.	.	.	.	.	.	18

TABLE 3 - DATA CONVERTED TO A ZERO-MOISTURE BASIS.

OBS	PLNA	PART	CALORIES	WATER	PROTEIN	FAT	TOTLCARB	FIBER	ASH
5635	PROSOPIS STEPHANIANA	F	.	0	14.50	2.9	78.8	15.7	3.8
5636	PROSOPIS STEPHANIANA	S	.	0	20.10	3.1	73.4	9.3	3.4
5637	PROSOPIS TAMARUGO	SH	.	0	10.70	2.7	81.6	32.1	5.0
5638	PROSOPIS TAMARUGO	L	.	0	9.90	2.0	67.8	15.8	20.3
5639	PROSOPIS TAMARUGO	F	.	0	11.70	1.6	82.1	41.5	4.6
5640	PROSOPIS TAMARUGO	S	.	0	10.10	0.7	81.7	38.4	7.5
5641	PROSOPIS TAMARUGO	S	.	0	30.00	5.3	.	.	.
5642	PROSOPIS VELUTINA	L	.	0	8.30	5.4	76.7	32.1	9.6
5643	PROSOPIS VELUTINA	F	.	0	13.60	2.0	79.3	29.1	5.1
5644	PROSOPIS VELUTINA	S	.	0	55.30	8.9	33.2	4.5	4.4
5645	PROTEA BARBIGERA	S	.	0	68.60	31.4	.	.	.
5646	PROTEA CYNAROIDES		.	0	4.31	.	.	.	.
5647	PROTEA GRANDIFLORA	S	.	0	2.10	0.7	.	.	.
5648	PROTEA LATIFOLIA	S	.	0	90.30	18.2	.	.	.
5649	PROTEA MARGINATA	S	.	0	4.60	1.8	.	.	.
5650	PROTEA MELLIFERA	S	.	0	3.00	1.2	.	.	.
5651	PROTEA MELLIFERA	S	.	0	8.00	4.2	.	.	.
5652	PROTEA MELLIFERA	S	.	0	8.90	.	.	.	.
5653	PROTEA MELLIFERA	S	.	0	2.70	0.7	.	.	.
5654	PRUNELLA VULGARIS	S	.	0	19.00	19.2	.	.	.
5655	PRUNELLA VULGARIS VAR. LANCEOLATA	S	.	0	16.90	22.4	.	.	.
5656	PRUNUS ARMENIACA	F	349	0	7.20	2.4	84.9	6.6	5.4
5657	PRUNUS ARMENIACA	F	333	0	7.00	3.5	78.9	13.2	10.5
5658	PRUNUS ARMENIACA	S	597	0	31.50	51.4	14.0	3.3	3.0

OBS	CALCIUM	PHOSPHO	IRON	SODIUM	POTASS	CAROT	THIAMINE	RIBOFLAV	NIACIN	ASCORBIC	REFCO
5635	.	.	.	.	.	.	.	.	.	.	18
5636	.	.	.	.	.	.	.	.	.	.	18
5637	3980.0	120.0	.	.	.	.	.	.	.	.	18
5638	6210.0	110.0	.	.	.	.	.	.	.	.	18
5639	330.0	130.0	.	.	.	.	.	.	.	.	18
5640	360.0	1500.0	.	.	.	.	.	.	.	.	18
5641	.	.	.	.	.	.	.	.	.	.	20
5642	.	.	.	.	.	.	.	.	.	.	18
5643	.	.	.	.	.	.	.	.	.	.	18
5644	.	.	.	.	.	.	.	.	.	.	18
5645	.	.	.	.	.	.	.	.	.	.	21
5646	1120.0	34.0	3.7	400.0	120.0	.	.	.	.	.	8
5647	.	.	.	.	.	.	.	.	.	.	21
5648	.	.	.	.	.	.	.	.	.	.	21
5649	.	.	.	.	.	.	.	.	.	.	21
5650	.	.	.	.	.	.	.	.	.	.	21
5651	.	.	.	.	.	.	.	.	.	.	21
5652	.	.	.	.	.	.	.	.	.	.	21
5653	.	.	.	.	.	.	.	.	.	.	21
5654	.	.	.	.	.	.	.	.	.	.	21
5655	.	.	.	.	.	.	.	.	.	.	19
5656	78.3	144.6	4.2	.	.	18945.8	.	.	.	.	1
5657	184.2	298.2	7.9	8.8	1912.3	15263.2	0.18	0.44	6.14	52.6	2
5658	152.2	300.0	4.8	.	778.3	0.0	0.15	0.53	1.74	.	2

TABLE 3 - DATA CONVERTED TO A ZERO-MOISTURE BASIS.

OBS	PLNA	PART	CALORIES	WATER	PROTEIN	FAT	TOTCARB	FIBER	ASH	CALCIUM	PHOSPHO	IRON	SODIUM	POTASS	CAROT	THIAMINE	RIBOFLAV	NIACIN	ASCORBIC	REFCO
5659	PRUNUS ARMENIACA	F	361	0	5.1	3.8	87.3	7.0	3.8	189.9	202.5	7.0	.	.	8481.0	0.25	0.38	2.53	63.3	3
5660	PRUNUS ARMENIACA	F	347	0	6.8	1.4	87.1	4.1	4.8	115.6	156.5	3.4	6.8	1911.6	11020.4	0.20	0.27	4.08	68.0	4
5661	PRUNUS AVIUM	F	.	0	6.5	2.9	87.1	1.8	3.5	105.9	117.6	2.4	.	.	2188.2	0.29	0.35	2.35	47.1	5
5662	PRUNUS CERASIFERA	F	355	0	6.4	1.8	88.2	5.5	3.6	109.1	136.4	3.6	.	.	909.1	0.18	0.27	3.64	72.7	3
5663	PRUNUS CERASUS	F	365	0	8.7	4.8	81.7	7.9	4.8	150.8	230.2	7.1	23.8	888.9	873.0	0.24	0.24	3.17	87.3	2
5664	PRUNUS CERASUS	F	362	0	10.3	2.3	85.1	5.7	2.3	195.4	183.9	1.7	.	.	344.8	0.29	0.06	1.15	86.2	3
5665	PRUNUS CERASUS	F	351	0	9.2	2.3	84.0	6.1	4.6	106.9	137.4	3.8	.	.	.	.	.	.	.	3
5666	PRUNUS CERASUS	F	.	0	6.6	3.0	85.5	2.4	4.8	144.6	150.6	7.8	.	1927.7	.	0.48	0.48	1.81	42.2	5
5667	PRUNUS DOMESTICA	F	362	0	4.6	1.5	91.5	3.1	2.3	61.5	115.4	3.1	.	.	615.4	0.23	0.31	3.85	46.2	3
5668	PRUNUS DOMESTICA	F	.	0	5.1	2.9	88.3	2.9	3.6	73.0	146.0	10.2	.	1386.9	729.9	.	.	0.73	0.0	5
5669	PRUNUS DOMESTICA	S	.	0	22.3	43.7	30.1	13.5	3.8	.	.	.	.	.	.	.	.	.	.	5
5670	PRUNUS DULCIS	S	.	0	18.6	54.1	24.3	2.7	3.0	270.0	520.0	.	.	.	.	.	.	.	.	18
5671	PRUNUS DULCIS	S	636	0	17.6	57.7	22.6	.	2.1	.	.	.	.	.	.	.	.	.	.	1
5672	PRUNUS DULCIS	S	633	0	22.1	57.7	18.2	3.2	2.1	296.2	509.5	5.5	14.7	453.8	5.3	0.26	0.70	6.30	0.0	2
5673	PRUNUS DULCIS	S	574	0	19.5	56.8	20.6	2.8	3.1	266.5	498.4	4.6	.	.	0.0	0.26	0.70	4.83	0.0	3
5674	PRUNUS DULCIS	S	627	0	19.5	56.9	20.5	2.7	3.1	245.5	528.9	4.9	4.2	811.1	0.0	0.25	0.97	3.67	0.0	4
5675	PRUNUS DULCIS	S	.	0	21.9	62.1	12.9	1.8	3.1	242.6	516.9	4.7	.	.	0.0	0.25	0.16	2.64	0.0	5
5676	PRUNUS ILICIFOLIA	S	.	0	10.0	2.7	.	.	.	.	.	.	.	.	.	.	.	.	.	20
5677	PRUNUS INSITITIA	S	.	0	33.1	30.9	.	.	.	.	.	.	.	.	.	.	.	.	.	21
5678	PRUNUS INSITITIA	S	.	0	30.1	47.2	.	.	.	.	.	.	.	.	.	.	.	.	.	21
5679	PRUNUS LAUROCERASUS	S	.	0	31.2	44.4	.	.	.	.	.	.	.	.	.	.	.	.	.	21
5680	PRUNUS MAHALEB	S	.	0	29.9	43.1	.	.	.	.	.	.	.	.	.	.	.	.	.	21
5681	PRUNUS MICROCARPA	S	.	0	21.0	52.2	.	.	.	.	.	.	.	.	.	.	.	.	.	21
5682	PRUNUS MINUTIFLORA	L	.	0	23.4	4.2	65.4	10.8	7.0	2010.0	270.0	.	.	1670.0	.	.	.	.	.	6
5683	PRUNUS MUME	F	388	0	9.7	13.4	67.9	10.4	9.0	82.1	268.7	13.4	59.7	2328.4	149.3	0.37	0.37	3.73	74.6	2
5684	PRUNUS PERSICA	S	.	0	23.9	45.6	.	.	.	.	.	.	.	.	.	.	.	.	.	20
5685	PRUNUS PERSICA	F	348	0	4.5	1.1	91.0	4.5	3.4	44.9	146.1	3.4	22.5	1539.3	3033.7	0.11	0.34	6.74	33.7	
5686	PRUNUS PERSICA	F	352	0	6.3	1.4	88.0	6.3	4.2	49.3	133.8	3.5	.	.	176.1	.	.	.	112.7	1
5687	PRUNUS PERSICA	F	345	0	5.3	0.9	88.5	7.1	5.3	88.5	150.4	3.5	.	.	.	.	.	.	.	1
5688	PRUNUS PERSICA	F	351	0	5.4	1.8	89.2	16.2	3.6	36.0	90.1	9.9	36.0	2207.2	.	0.09	0.09	6.31	.	2
5689	PRUNUS PERSICA	F	354	0	5.4	1.4	90.5	6.1	2.7	81.6	176.9	7.5	.	.	68.0	0.20	0.41	2.72	190.5	3
5690	PRUNUS SALICINA	F	356	0	4.7	1.3	90.6	4.7	3.4	107.4	214.8	2.7	20.1	973.2	67.1	0.13	0.13	3.36	13.4	2
5691	PRUNUS SALICINA	F	358	0	4.1	1.4	91.9	2.7	2.7	47.3	135.1	2.7	.	.	405.4	0.07	0.14	2.70	67.6	3
5692	PRUNUS SEROTINA SUBSP. CAPULI	F	355	0	5.7	0.9	90.8	2.6	2.6	105.3	105.3	3.5	.	.	394.7	0.18	0.18	4.82	78.9	3
5693	PRUNUS SP.	S	.	0	33.4	43.3	.	.	2.7	.	.	.	.	.	.	.	.	.	.	19
5694	PRUNUS SP.	F	353	0	4.3	1.4	89.9	7.9	4.3	71.9	86.3	4.3	.	.	5539.6	.	.	.	.	1
5695	PRUNUS VIRGINIANA	S	.	0	34.6	39.3	.	.	.	.	.	.	.	.	.	.	.	.	.	20
5696	PRUNUS VIRGINIANA	S	.	0	36.6	42.6	.	.	.	.	.	.	.	.	.	.	.	.	.	21
5697	PRUNUS VIRGINIANA(IMMATURE)	L	.	0	15.9	3.2	73.8	10.8	7.1	1800.0	410.0	.	.	.	.	.	.	.	.	6
5698	PRUNUS VIRGINIANA(MATURE)	L	.	0	8.1	6.5	75.2	11.5	10.2	3030.0	670.0	.	.	.	.	.	.	.	.	6
5699	PRUNUS WEBBII	S	.	0	24.2	52.2	.	.	.	.	.	.	.	.	.	.	.	.	.	21
5700	PSEUDERANTHEMUM RETICULATUM	L	307	0	22.7	1.3	67.3	11.3	8.7	.	446.7	.	.	.	17366.7	0.13	0.67	9.33	340.0	2
5701	PSEUDOPHOENIX SAONAE	S	.	0	6.2	19.2	.	.	.	.	.	.	.	.	.	.	.	.	.	19
5702	PSEUDOPHOENIX VINIFERA	S	.	0	6.4	21.4	.	.	1.3	.	.	.	.	.	.	.	.	.	.	19
5703	PSEUDOPHOENIX VINIFERA	S	.	0	1.6	1.6	.	.	0.8	.	.	.	.	.	.	.	.	.	.	19

TABLE 3 - DATA CONVERTED TO A ZERO-MOISTURE BASIS.

OBS	PLNA	PART	CALORIES	WATER	PROTEIN	FAT	TOTCARB	FIBER	ASH	CALCIUM	PHOSPHO	IRON	SODIUM	POTASS	CAROT	THIAMINE	RIBOFLAV	NIACIN	ASCORBIC	REFCO
5704	PSIDIUM CATTLEIANUM	F	357	0	2.5	1.9	92.4	33.1	3.2	184.7	108.3	1.3	.	.	923.6	0.19	0.19	3.82	210.2	2
5705	PSIDIUM CATTLEIANUM	F	355	0	2.4	1.8	92.2	33.1	3.6	186.7	108.4	1.2	.	.	0.0	0.18	0.18	2.41	126.5	2
5706	PSIDIUM CATTLEIANUM	F	357	0	5.5	3.3	86.8	35.2	4.4	126.4	230.8	4.9	22.0	1587.9	296.7	0.16	0.16	3.30	203.3	4
5707	PSIDIUM CATTLEIANUM	F	.	0	0.7	1.4	93.9	32.7	4.1	340.1	136.1	8.2	.	.	0.0	0.14	0.14	2.04	102.0	5
5708	PSIDIUM GUAJAVA	L	.	0	11.7	8.7	71.9	16.1	7.7	1340.0	160.0	.	.	.	.	.	.	.	.	18
5709	PSIDIUM GUAJAVA	F	360	0	6.2	2.2	88.2	29.8	3.4	134.8	174.2	7.3	.	.	1629.2	0.34	0.22	7.30	1831.5	1
5710	PSIDIUM GUAJAVA	F	356	0	5.2	2.1	89.2	28.9	3.6	77.3	123.7	3.6	20.6	1500.0	386.6	0.26	0.21	5.67	680.4	2
5711	PSIDIUM GUAJAVA	F	359	0	4.7	2.1	90.1	27.6	3.1	114.6	135.4	3.6	.	.	833.3	0.21	0.21	5.21	1135.4	3
5712	PSIDIUM GUAJAVA	P	352	0	3.1	0.6	93.8	14.8	2.5	123.5	80.2	3.7	.	.	864.2	0.06	0.25	3.09	444.4	3
5713	PSIDIUM GUAJAVA	P	365	0	4.7	3.5	88.2	32.9	3.5	135.3	247.1	5.3	23.5	1700.0	988.2	0.29	0.29	7.06	1423.5	4
5714	PSIDIUM GUAJAVA	F	.	0	4.9	1.6	89.6	28.4	3.8	54.6	153.0	7.7	32.8	497.3	0.0	0.16	0.16	2.19	1158.5	5
5715	PSIDIUM SARTORIANUM	S	.	0	3.1	3.2	.	.	.	.	.	.	.	.	.	.	.	.	.	19
5716	PSIDIUM SARTORIANUM	S	.	0	3.8	0.9	.	.	.	.	.	.	.	.	.	.	.	.	.	19
5717	PSILOSTROPHE CF. GNAPHALIOIDES	S	.	0	18.1	10.5	.	.	.	.	.	.	.	.	.	.	.	.	.	20
5718	PSOPHOCARPUS TETRAGONOLOBUS	S	445	0	34.9	17.7	43.2	11.0	4.1	.	.	.	.	.	.	.	.	.	.	1
5719	PSOPHOCARPUS TETRAGONOLOBUS	G	321	0	26.9	3.8	62.8	21.8	6.4	.	.	.	.	.	.	.	.	.	.	1
5720	PSOPHOCARPUS TETRAGONOLOBUS	R	370	0	11.4	2.4	82.5	6.1	3.7	.	.	.	.	.	.	.	.	.	.	1
5721	PSOPHOCARPUS TETRAGONOLOBUS	F	324	0	18.1	1.0	75.2	15.2	5.7	504.8	457.1	1.9	28.6	1952.4	3238.1	1.81	0.76	9.52	200.0	2
5722	PSOPHOCARPUS TETRAGONOLOBUS	L	313	0	33.3	3.3	56.7	.	6.7	893.3	540.0	41.3	.	.	20966.7	1.87	.	.	193.3	2
5723	PSOPHOCARPUS TETRAGONOLOBUS	S	.	0	45.8	14.3	34.1	.	5.8	.	.	.	.	.	.	.	.	.	.	5
5724	PSOPHOCARPUS TETRAGONOLOBUS	R	.	0	27.0	1.1	67.6	5.9	4.3	.	.	.	.	.	.	.	.	.	.	5
5725	PSOPHOCARPUS TETRAGONOLOBUS	L	.	0	29.9	4.7	56.9	19.4	8.5	1753.6	568.7	.	.	.	.	.	.	.	.	5
5726	PSORALEA BITUMINOSA	S	.	0	32.9	8.2	.	.	.	.	.	.	.	.	.	.	.	.	.	21
5727	PSORALEA BITUMINOSA	S	.	0	34.1	12.4	.	.	.	.	.	.	.	.	.	.	.	.	.	21
5728	PSORALEA CANDICANS	C	.	0	14.7	5.4	75.0	39.3	4.9	780.0	150.0	.	.	.	.	.	.	.	.	18
5729	PSORALEA SPICATA	S	.	0	31.5	10.8	.	.	.	.	.	.	.	.	.	.	.	.	.	21
5730	PTELEA ANGUSTIFOLIA	S	.	0	13.8	11.0	.	.	.	.	.	.	.	.	.	.	.	.	.	19
5731	PTELEA TRIFOLIATA	S	.	0	44.0	45.9	.	.	3.1	.	.	.	.	.	.	.	.	.	.	19
5732	PTELEA TRIFOLIATA	S	.	0	6.6	2.4	.	.	4.1	.	.	.	.	.	.	.	.	.	.	19
5733	PTELEA TRIFOLIATA VAR. MOLLIS	S	.	0	10.2	6.0	.	.	.	.	.	.	.	.	.	.	.	.	.	20
5734	PTEROCARPUS SP.	L	340	0	36.1	8.8	50.1	12.2	4.9	605.2	345.5	10.6	.	.	1532.5	3.25	2.16	2.08	64.9	1
5735	PTEROCARYA STENOPTERA	S	.	0	8.9	11.5	.	.	3.1	.	.	.	.	.	.	.	.	.	.	19
5736	PTERONIA ADENOCARPA	S	.	0	28.2	11.4	.	.	.	.	.	.	.	.	.	.	.	.	.	21
5737	PUCCINELLIA DISTANS	S	.	0	16.4	3.2	.	.	.	.	.	.	.	.	.	.	.	.	.	21
5738	PUERARIA LOBATA	L	.	0	19.6	3.9	68.7	21.5	7.8	.	.	.	.	.	.	.	.	.	.	18
5739	PUERARIA LOBATA	C	.	0	5.7	1.0	87.7	46.2	5.6	.	.	.	.	.	.	.	.	.	.	18
5740	PUERARIA LOBATA	H	.	0	13.3	2.5	74.9	40.3	9.3	.	.	.	.	.	.	.	.	.	.	18
5741	PUERARIA LOBATA	L	327	0	3.6	0.9	88.2	70.0	7.3	318.2	181.8	44.5	.	.	0.0	0.27	8.27	7.27	0.0	2
5742	PUERARIA LOBATA	R	360	0	6.7	0.3	88.5	2.2	4.5	47.8	57.3	1.9	.	.	.	.	.	.	.	2
5743	PUERARIA LOBATA	R	.	0	6.7	0.3	86.3	2.2	4.5	47.8	57.3	1.9	.	.	.	.	.	.	.	5
5744	PUERARIA PHASEOLOIDES	W	.	0	18.0	2.4	73.5	42.9	6.1	710.0	180.0	.	.	.	.	.	.	.	.	18
5745	PUERARIA PHASEOLOIDES	S	.	0	32.7	3.2	.	.	4.1	.	.	.	.	.	.	.	.	.	.	19
5746	PUERARIA PHASEOLOIDES	L	.	0	19.9	2.1	70.2	28.8	7.9	733.0	157.1	.	.	.	.	.	.	.	.	5
5747	PUERARIA SP.	W	.	0	17.2	2.1	72.7	33.9	8.0	.	.	.	.	.	.	.	.	.	.	6
5748	PUERARIA SP.	H	.	0	13.8	2.5	77.1	33.6	6.6	1610.0	470.0	.	.	.	.	.	.	.	.	6

TABLE 3 - DATA CONVERTED TO A ZERO-MOISTURE BASIS.

OBS	PLNA	PART	CALORIES	WATER	PROTEIN	FAT	TOTLCARB	FIBER	ASH	CALCIUM	PHOSPHO	IRON	SODIUM	POTASS	CAROT	THIAMINE	RIBOFLAV	NIACIN	ASCORBIC	REFCO
5749	PUERARIA THUNBERGIANA	S	.	0	36.9	5.7	.	.	5.7	.	.	.	.	.	.	.	.	.	.	19
5750	PUERARIA THUNBERGIANA	S	.	0	36.4	5.6	.	.	5.5	.	.	.	.	.	.	.	.	.	.	19
5751	PULICARIA UNDULATA	S	.	0	9.0	8.7	.	.	.	.	.	.	.	.	.	.	.	.	.	21
5752	PULSATILLA VULGARIS	S	.	0	23.8	24.4	.	.	.	.	.	.	.	.	.	.	.	.	.	20
5753	PUNICA GRANATUM	S	.	0	2.5	4.0	.	.	2.7	.	.	.	.	.	.	.	.	.	.	19
5754	PUNICA GRANATUM	F	356	0	7.3	1.4	89.5	20.5	1.8	54.8	168.9	4.6	.	.	182.6	.	.	.	.	1
5755	PUNICA GRANATUM	F	360	0	5.0	3.0	88.5	5.5	3.5	65.0	115.0	3.5	35.0	1895.0	.	0.35	0.05	1.50	35.0	2
5756	PUNICA GRANATUM	F	368	0	4.4	3.8	89.0	11.0	2.7	54.9	186.8	3.3	.	.	0.0	0.38	0.16	4.95	44.0	3
5757	PUNICA GRANATUM	F	356	0	2.8	1.7	92.7	1.1	2.8	16.9	45.2	1.7	16.9	1463.3	0.0	0.17	0.17	1.69	22.6	4
5758	PUNICA GRANATUM	F	.	0	7.3	0.5	89.1	23.2	3.2	45.5	318.2	1.4	4.5	604.5	0.0	0.27	0.45	1.36	63.6	5
5759	PURSHIA TRIDENTATA	W	.	0	11.8	5.0	78.3	21.4	4.9	1840.0	250.0	.	.	.	.	.	.	.	.	6
5760	PYCNANTHEMUM MUTICUM	S	.	0	30.6	34.8	.	.	4.8	.	.	.	.	.	.	.	.	.	.	19
5761	PYRACANTHA COCCINEA	S	.	0	11.2	6.6	.	.	.	.	.	.	.	.	.	.	.	.	.	19
5762	PYRACANTHA COCCINEA	S	.	0	8.6	6.3	.	.	.	.	.	.	.	.	.	.	.	.	.	20
5763	PYRUS ARBUTIFOLIA	S	.	0	18.1	10.4	.	.	.	.	.	.	.	.	.	.	.	.	.	19
5764	PYRUS ARBUTIFOLIA	S	.	0	3.8	3.0	.	.	.	.	.	.	.	.	.	.	.	.	.	19
5765	PYRUS COMMUNIS	F	367	0	2.5	3.8	91.8	11.4	1.9	38.0	75.9	1.3	.	.	.	.	.	.	.	1
5766	PYRUS COMMUNIS	F	355	0	3.2	1.6	91.9	8.1	3.2	80.6	121.0	4.8	56.5	1137.1	0.0	0.24	0.24	0.81	32.3	2
5767	PYRUS COMMUNIS	F	359	0	1.9	1.3	94.9	12.2	1.9	38.5	64.1	3.2	.	.	64.1	0.13	0.19	1.28	32.1	3
5768	PYRUS COMMUNIS	F	363	0	4.2	2.4	91.1	8.3	2.4	47.6	65.5	1.8	11.9	773.8	71.4	0.12	0.24	0.60	23.8	4
5769	QUAMOCLIT PINNATA	S	.	0	21.8	12.7	.	.	.	.	.	.	.	.	.	.	.	.	.	21
5770	QUERCUS ACUTISSIMA	S	.	0	31.6	5.1	.	.	2.9	.	.	.	.	.	.	.	.	.	.	19
5771	QUERCUS ACUTISSIMA	S	.	0	4.4	0.9	.	.	2.1	.	.	.	.	.	.	.	.	.	.	19
5772	QUERCUS ALBA	S	.	0	6.2	7.8	.	.	.	.	.	.	.	.	.	.	.	.	.	19
5773	QUERCUS COCCIFERA	S	.	0	3.7	0.9	.	.	.	.	.	.	.	.	.	.	.	.	.	21
5774	QUERCUS DILATATA	L	.	0	9.6	4.5	80.8	29.1	5.1	1610.0	300.0	.	.	.	.	.	.	.	.	18
5775	QUERCUS DURANDII	L	.	0	9.6	2.5	82.5	25.3	5.4	1440.0	110.0	.	.	760.0	.	.	.	.	.	6
5776	QUERCUS EMORYI	S	.	0	8.1	41.2	.	.	.	.	.	.	.	.	.	.	.	.	.	19
5777	QUERCUS GAMBELII	S	.	0	6.9	8.2	.	.	.	.	.	.	.	.	.	.	.	.	.	19
5778	QUERCUS GLAUCA	L	.	0	9.6	4.1	78.7	29.0	7.6	1870.0	230.0	.	.	.	.	.	.	.	.	18
5779	QUERCUS INCANA	L	.	0	10.2	4.8	79.8	31.3	5.2	990.0	150.0	.	.	.	.	.	.	.	.	18
5780	QUERCUS PALUSTRIS	S	.	0	6.9	26.1	.	.	.	.	.	.	.	.	.	.	.	.	.	19
5781	QUERCUS VIRGINIANA	L	.	0	9.9	2.8	80.2	31.9	7.1	710.0	230.0	.	.	.	.	.	.	.	.	18
5782	QUERCUS VIRGINIANA	L	.	0	10.6	2.4	81.3	28.6	5.7	.	.	.	.	.	.	.	.	.	.	6
5783	QUINCULA SP.	W	.	0	31.5	4.1	46.7	14.5	17.7	2000.0	340.0	.	.	5590.0	.	.	.	.	.	6
5784	QUISQUALIS INDICA	L	309	0	19.5	0.0	73.6	8.1	6.9	422.8	394.3	.	.	.	27093.5	0.16	.	.	284.6	2
5785	RADYERA URENS	S	.	0	19.0	18.5	.	.	.	.	.	.	.	.	.	.	.	.	.	21
5786	RANDIA SP.	S	.	0	14.4	10.4	.	.	.	.	.	.	.	.	.	.	.	.	.	19
5787	RANDIA SP.	S	.	0	13.8	11.6	.	.	.	.	.	.	.	.	.	.	.	.	.	20
5788	RANUNCULUS ARVENSIS	S	.	0	11.1	20.4	.	.	.	.	.	.	.	.	.	.	.	.	.	20
5789	RANUNCULUS CONSTANTINOPOLITANUS	S	.	0	17.0	22.4	.	.	.	.	.	.	.	.	.	.	.	.	.	20
5790	RANUNCULUS LAETUS	S	.	0	11.8	20.7	.	.	.	.	.	.	.	.	.	.	.	.	.	21
5791	RANUNCULUS SCELERATUS	S	.	0	18.1	25.9	.	.	.	.	.	.	.	.	.	.	.	.	.	20
5792	RANUNCULUS SP.	S	.	0	12.5	9.1	.	.	5.0	.	.	.	.	.	.	.	.	.	.	19
5793	RAPHANUS CAUDATUS	S	.	0	38.7	32.2	.	.	.	.	.	.	.	.	.	.	.	.	.	21

TABLE 3 - DATA CONVERTED TO A ZERO-MOISTURE BASIS.

OBS	PLNA	PART	CALORIES	WATER	PROTEIN	FAT	TOTLCARB	FIBER	ASH
5794	RAPHANUS RAPHANISTRUM	S	.	0	31.9	30.9	.	.	.
5795	RAPHANUS SATIVUS	S	.	0	33.2	29.8	.	.	4.2
5796	RAPHANUS SATIVUS	S	.	0	33.6	37.1	.	.	3.8
5797	RAPHANUS SATIVUS	S	.	0	32.2	33.1	.	.	4.0
5798	RAPHANUS SATIVUS	S	.	0	30.9	37.4	.	.	4.2
5799	RAPHANUS SATIVUS	S	.	0	23.6	35.6	.	.	4.5
5800	RAPHANUS SATIVUS	S	.	0	30.0	40.4	.	.	4.2
5801	RAPHANUS SATIVUS	S	.	0	30.6	41.7	.	.	.
5802	RAPHANUS SATIVUS	I	297	0	17.2	1.6	62.5	1.6	18.8
5803	RAPHANUS SATIVUS	F	337	0	25.7	5.0	65.3	13.9	4.0
5804	RAPHANUS SATIVUS	L	287	0	28.7	5.2	49.6	9.6	16.5
5805	RAPHANUS SATIVUS	R	351	0	13.5	2.7	75.7	17.6	8.1
5806	RAPHANUS SATIVUS	X	306	0	29.0	1.6	61.3	17.7	8.1
5807	RAPHANUS SATIVUS	R	338	0	13.2	1.5	73.5	10.3	11.8
5808	RAPHANUS SATIVUS	L	361	0	19.4	3.5	68.8	9.0	8.3
5809	RAPHANUS SATIVUS	R	338	0	14.9	0.0	74.3	9.5	10.8
5810	RAPHANUS SATIVUS	R	309	0	18.2	1.8	65.5	12.7	14.5
5811	RAPHANUS SATIVUS	F	.	0	16.9	3.9	70.1	14.3	9.1
5812	RAPHANUS SATIVUS	F	.	0	24.2	3.2	64.2	14.7	8.4
5813	RAPHIA HOOKERI	S	365	0	8.7	1.1	79.9	9.1	10.3
5814	RAPHIONACME BROWNII	R	348	0	4.3	0.0	89.1	10.9	6.5
5815	RAPISTRUM RUGOSUM	S	.	0	34.4	38.3	.	.	.
5816	RAPISTRUM RUGOSUM	S	.	0	32.4	35.8	.	.	.
5817	RAPISTRUM RUGOSUM	S	.	0	41.6	24.6	.	.	.

OBS	CALCIUM	PHOSPHO	IRON	SODIUM	POTASS	CAROT	THIAMINE	RIBOFLAV	NIACIN	ASCORBIC	REFCO
5794	.	.	.	.	.	.	.	.	.	.	20
5795	.	.	.	.	.	.	.	.	.	.	19
5796	.	.	.	.	.	.	.	.	.	.	19
5797	.	.	.	.	.	.	.	.	.	.	19
5798	.	.	.	.	.	.	.	.	.	.	19
5799	.	.	.	.	.	.	.	.	.	.	19
5800	.	.	.	.	.	.	.	.	.	.	19
5801	.	.	.	.	.	.	.	.	.	.	20
5802	656.2	468.7	15.6	.	.	.	.	.	.	.	1
5803	752.5	257.4	4.0	49.5	2049.5	2623.8	0.79	0.69	5.94	782.2	2
5804	1913.0	260.9	35.7	956.5	4347.8	21478.3	0.70	2.43	34.78	704.3	2
5805	432.4	459.5	18.9	135.1	3270.3	0.0	0.54	0.41	5.41	351.4	2
5806	2258.1	1016.1	19.4	.	.	58064.5	1.29	4.03	8.06	1129.0	2
5807	382.4	441.2	17.6	.	.	0.0	0.44	0.44	4.41	411.8	3
5808	1652.8	305.6	19.4	.	.	22847.2	0.97	1.81	4.17	847.2	3
5809	378.4	351.4	8.1	.	.	0.0	0.27	0.81	6.76	621.6	3
5810	545.5	563.6	18.2	327.3	5854.5	109.1	0.55	0.55	5.45	472.7	4
5811	1013.0	311.7	.	.	.	.	.	.	.	.	5
5812	842.1	1052.6	29.5	.	.	315.8	0.74	0.53	2.11	726.3	5
5813	1699.1	309.8	.	.	.	.	.	.	.	.	1
5814	1087.0	.	21.7	.	.	.	0.22	0.22	2.17	0.0	1
5815	.	.	.	.	.	.	.	.	.	.	20
5816	.	.	.	.	.	.	.	.	.	.	21
5817	.	.	.	.	.	.	.	.	.	.	21

TABLE 3 - DATA CONVERTED TO A ZERO-MOISTURE BASIS.

OBS	PLNA	PART	CALORIES	WATER	PROTEIN	FAT	TOTLCARB	FIBER	ASH
5818	RAPISTRUM RUGOSUM	S	.	0	31.1	39.2	.	.	.
5819	RAPISTRUM RUGOSUM	S	.	0	34.5	36.6	.	.	.
5820	RAPISTRUM RUGOSUM	S	.	0	33.3	39.1	.	.	.
5821	RAPISTRUM RUGOSUM SUBSP. ORIENTALE	S	.	0	33.2	36.2	.	.	.
5822	RATIBIDA COLUMNARIS	S	.	0	26.9	26.1	.	.	4.3
5823	RATIBIDA COLUMNARIS	S	.	0	18.8	23.8	.	.	.
5824	RATIBIDA COLUMNIFERA	S	.	0	18.9	18.6	.	.	.
5825	RATIBIDA SP.	L	.	0	19.7	4.1	60.1	14.0	16.1
5826	RATIBIDA SP.	I	.	0	13.6	6.1	73.1	37.2	7.2
5827	RATIBIDA TAGETES	S	.	0	16.7	2.8	.	.	.
5828	REBOUDIA PINNATA	S	.	0	35.2	30.6	.	.	.
5829	REPTONIA BUXIFOLIA	S	.	0	6.8	24.2	.	.	.
5830	RESEDA ERECTA	S	.	0	21.9	35.7	.	.	.
5831	RESEDA LUTEA	S	.	0	19.7	35.0	.	.	.
5832	RESEDA LUTEOLA	S	.	0	23.7	33.8	.	.	.
5833	RESEDA LUTEOLA	S	.	0	21.0	22.4	.	.	.
5834	RESEDA ODORATA	S	.	0	18.1	32.3	.	.	.
5835	RETAMA RAETAM	S	.	0	25.8	2.8	.	.	.
5836	RETAMA SPHAEROCARPA	S	.	0	25.0	5.3	.	.	.
5837	REVERCHONIA ARENARIA	S	.	0	10.6	27.8	.	.	.
5838	RHAMNUS CATHARTICA	S	.	0	20.5	24.6	.	.	2.5
5839	RHAMNUS DAVURICA	S	.	0	20.4	27.3	.	.	2.5
5840	RHAMNUS PRINOIDES	L	337	0	9.1	3.2	80.9	31.0	6.9
5841	RHAMNUS PURSHIANA	S	.	0	25.4	56.9	.	.	2.3

OBS	CALCIUM	PHOSPHO	IRON	SODIUM	POTASS	CAROT	THIAMINE	RIBOFLAV	NIACIN	ASCORBIC	REFCO
5818	.	.	.	.	.	.	.	.	.	.	21
5819	.	.	.	.	.	.	.	.	.	.	21
5820	.	.	.	.	.	.	.	.	.	.	21
5821	.	.	.	.	.	.	.	.	.	.	21
5822	.	.	.	.	.	.	.	.	.	.	19
5823	.	.	.	.	.	.	.	.	.	.	19
5824	.	.	.	.	.	.	.	.	.	.	20
5825	3280.0	260.0	.	.	4530.0	.	.	.	.	.	6
5826	1830.0	310.0	.	.	1300.0	.	.	.	.	.	6
5827	.	.	.	.	.	.	.	.	.	.	20
5828	.	.	.	.	.	.	.	.	.	.	21
5829	.	.	.	.	.	.	.	.	.	.	21
5830	.	.	.	.	.	.	.	.	.	.	21
5831	.	.	.	.	.	.	.	.	.	.	20
5832	.	.	.	.	.	.	.	.	.	.	21
5833	.	.	.	.	.	.	.	.	.	.	21
5834	.	.	.	.	.	.	.	.	.	.	20
5835	.	.	.	.	.	.	.	.	.	.	21
5836	.	.	.	.	.	.	.	.	.	.	21
5837	.	.	.	.	.	.	.	.	.	.	20
5838	.	.	.	.	.	.	.	.	.	.	19
5839	.	.	.	.	.	.	.	.	.	.	19
5840	1231.2	138.3	27.7	.	.	7632.74	0.00	0.09	2.88	.	1
5841	.	.	.	.	.	.	.	.	.	.	19

TABLE 3 - DATA CONVERTED TO A ZERO-MOISTURE BASIS.

OBS	PLNA	PART	CALORIES	WATER	PROTEIN	FAT	TOTLCARB	FIBER	ASH
5842	RHAMNUS PURSHIANA	S	.	0	6.7	13.4	.	.	1.3
5843	RHAZYA STRICTA	S	.	0	17.8	17.8	.	.	.
5844	RHEEDIA ACUMINATA	F	359	0	4.7	0.8	93.0	7.8	1.6
5845	RHEUM OFFICINALE	Z	314	0	9.8	2.0	74.5	13.7	13.7
5846	RHEUM RHABARBARUM	Z	323	0	6.5	3.2	77.4	12.9	12.9
5847	RHEUM RHABARBARUM	L	300	0	11.7	3.3	68.3	11.7	16.7
5848	RHEUM RHABARBARUM	L	278	0	18.1	1.4	61.1	12.5	19.4
5849	RHEUM RHABARBARUM	Z	308	0	11.5	1.9	71.2	13.5	15.4
5850	RHEUM RHAPONTICUM	L	.	0	27.6	6.5	.	8.0	.
5851	RHIGOZUM OBOVATUM	SH	.	0	11.3	4.1	78.6	15.5	6.0
5852	RHIGOZUM TRICHOTOMUM	S	.	0	26.8	20.1	.	.	.
5853	RHINANTHUS SP.	W	.	0	24.7	1.6	67.1	23.4	6.6
5854	RHIZOPHORA MANGLE	L	.	0	10.7	3.4	77.0	14.5	8.9
5855	RHODODENDRON PONTICUM	S	.	0	24.8	27.6	.	.	.
5856	RHODODENDRON SP.	L	.	0	6.3	6.0	82.9	21.3	4.8
5857	RHODOTYPOS SCANDENS	S	.	0	50.2	37.5	.	.	.
5858	RHODOTYPOS TETRAPETALA	S	.	0	13.1	8.3	.	.	.
5859	RHODOTYPOS TETRAPETALA	S	.	0	48.8	39.7	.	.	.
5860	RHUS CANADENSIS	S	.	0	16.3	13.4	.	.	2.0
5861	RHUS COPALLINA	S	.	0	13.8	10.1	.	.	.
5862	RHUS COPALLINA	S	.	0	4.4	26.3	.	.	.
5863	RHUS COPALLINA	S	.	0	9.4	13.6	.	.	.
5864	RHUS COPALLINA	S	.	0	8.8	10.9	.	.	.
5865	RHUS COPALLINA	S	.	0	4.4	14.8	.	.	.

OBS	CALCIUM	PHOSPHO	IRON	SODIUM	POTASS	CAROT	THIAMINE	RIBOFLAV	NIACIN	ASCORBIC	REFCO
5842	.	.	.	.	.	.	.	.	.	.	19
5843	.	.	.	.	.	.	.	.	.	.	21
5844	93.8	171.9	3.1	.	.	0.00	0.47	0.31	2.34	46.9	3
5845	1000.0	490.2	9.8	.	.	392.16	0.20	0.39	1.96	196.1	3
5846	967.7	322.6	8.1	32.3	4822.6	322.58	0.65	0.48	3.23	258.1	2
5847	1033.3	266.7	5.0	.	.	500.00	0.33	0.33	3.33	183.3	3
5848	652.8	347.2	25.0	.	.	.	.	.	.	.	3
5849	1846.2	346.2	15.4	38.5	4826.9	1153.85	0.58	1.35	5.77	173.1	4
5850	.	.	.	.	.	.	.	.	.	.	6
5851	920.0	130.0	.	.	.	.	.	.	.	.	18
5852	.	.	.	.	.	.	.	.	.	.	21
5853	420.0	130.0	.	.	.	.	.	.	.	.	6
5854	.	.	.	.	.	.	.	.	.	.	6
5855	.	.	.	.	.	.	.	.	.	.	21
5856	1420.0	80.0	37.0	.	.	.	.	.	.	.	6
5857	.	.	.	.	.	.	.	.	.	.	21
5858	.	.	.	.	.	.	.	.	.	.	20
5859	.	.	.	.	.	.	.	.	.	.	20
5860	.	.	.	.	.	.	.	.	.	.	19
5861	.	.	.	.	.	.	.	.	.	.	19
5862	.	.	.	.	.	.	.	.	.	.	19
5863	.	.	.	.	.	.	.	.	.	.	19
5864	.	.	.	.	.	.	.	.	.	.	20
5865	.	.	.	.	.	.	.	.	.	.	20

TABLE 3 - DATA CONVERTED TO A ZERO-MOISTURE BASIS.

OBS	PLNA	PART	CALORIES	WATER	PROTEIN	FAT	TOTLCARB	FIBER
5866	RHUS CORIARIA	S	.	0	5.2	6.3	.	.
5867	RHUS GLABRA	S	.	0	10.0	8.7	.	.
5868	RHUS GLABRA	S	.	0	3.1	17.7	.	.
5869	RHUS GLABRA	S	.	0	10.0	12.0	.	.
5870	RHUS GLABRA	S	.	0	10.0	13.1	.	.
5871	RHUS GLABRA	S	.	0	6.9	13.5	.	.
5872	RHUS GLABRA	S	.	0	6.6	11.0	.	.
5873	RHUS LANCEA	SH	.	0	12.7	7.3	72.7	22.3
5874	RHUS MICROPHYLLA	S	.	0	10.0	7.8	.	.
5875	RHUS MICROPHYLLA	S	.	0	10.0	23.0	.	.
5876	RHUS MICROPHYLLA	L	.	0	14.9	4.7	72.8	11.4
5877	RHUS NATALENSIS	SH	.	0	9.0	3.5	78.7	19.6
5878	RHUS NATALENSIS	F	.	0	11.1	5.8	79.1	25.3
5879	RHUS RADICANS	S	.	0	10.6	22.4	.	.
5880	RHUS SEMPERVIRENS	L	.	0	9.9	6.3	79.0	18.7
5881	RHUS SP.	S	.	0	3.1	2.6	.	.
5882	RHUS SP.	S	.	0	2.5	0.2	.	.
5883	RHUS TRILOBATA	S	.	0	13.4	13.4	.	.
5884	RHUS TRILOBATA	S	.	0	12.5	6.6	.	.
5885	RHUS TRILOBATA	S	.	0	9.4	31.2	.	.
5886	RHUS TRILOBATA	L	.	0	8.0	4.4	82.1	13.7
5887	RHUS TRILOBATA VAR. PILOSSISSIMA	S	.	0	12.8	9.7	.	.
5888	RHUS TYPHINA	S	.	0	9.4	11.1	.	.
5889	RHUS TYPHINA	S	.	0	4.4	21.9	.	.

OBS	ASH	CALCIUM	PHOSPHO	IRON	SODIUM	POTASS	CAROT	THIAMINE	RIBOFLAV	NIACIN	ASCORBIC	REFCO
5866	.	.	.	.	.	.	.	.	.	.	.	21
5867	.	.	.	.	.	.	.	.	.	.	.	19
5868	.	.	.	.	.	.	.	.	.	.	.	19
5869	.	.	.	.	.	.	.	.	.	.	.	19
5870	.	.	.	.	.	.	.	.	.	.	.	20
5871	.	.	.	.	.	.	.	.	.	.	.	20
5872	.	.	.	.	.	.	.	.	.	.	.	21
5873	7.3	800.0	190.0	.	.	.	.	.	.	.	.	18
5874	.	.	.	.	.	.	.	.	.	.	.	19
5875	.	.	.	.	.	.	.	.	.	.	.	19
5876	7.6	1950.0	170.0	.	.	1880.0	.	.	.	.	.	6
5877	8.8	1430.0	100.0	.	.	.	.	.	.	.	.	18
5878	4.0	270.0	190.0	.	.	.	.	.	.	.	.	18
5879	.	.	.	.	.	.	.	.	.	.	.	19
5880	4.8	1670.0	50.0	.	.	730.0	.	.	.	.	.	6
5881	.	.	.	.	.	.	.	.	.	.	.	19
5882	.	.	.	.	.	.	.	.	.	.	.	19
5883	1.8	.	.	.	.	.	.	.	.	.	.	19
5884	.	.	.	.	.	.	.	.	.	.	.	19
5885	.	.	.	.	.	.	.	.	.	.	.	19
5886	5.5	1930.0	110.0	.	.	1690.0	.	.	.	.	.	6
5887	.	.	.	.	.	.	.	.	.	.	.	20
5888	.	.	.	.	.	.	.	.	.	.	.	19
5889	.	.	.	.	.	.	.	.	.	.	.	19

TABLE 3 - DATA CONVERTED TO A ZERO-MOISTURE BASIS.

OBS	PLNA	PART	CALORIES	WATER	PROTEIN	FAT	TOTLCARB	FIBER
5890	RHUS VERNIX	S	.	0	16.2	9.2	.	.
5891	RHYNCHELYTRUM REPENS	W	.	0	11.3	2.3	76.6	31.8
5892	RHYNCHELYTRUM REPENS	H	.	0	5.6	1.4	85.0	41.6
5893	RHYNCHELYTRUM ROSEUM	H	.	0	3.9	1.5	89.4	42.8
5894	RHYNCHOSIA CAPITATA	S	.	0	23.1	2.4	.	.
5895	RHYNCHOSIA MINIMA	S	.	0	23.6	0.6	.	.
5896	RHYNCHOSIA PYRAMIDALIS	S	.	0	18.8	0.7	.	.
5897	RHYNCHOSIA PYRAMIDALIS	S	.	0	18.8	1.0	.	.
5898	RHYNCHOSIA TEXANA	W	.	0	18.8	3.3	71.1	24.9
5899	RHYNCHOSPORA INDIANOLENSIS	S	.	0	10.6	10.5	.	.
5900	RHYNCHOSPORA INDIANOLENSIS	S	.	0	11.2	9.6	.	.
5901	RIBES LACUSTRE	F	361	0	15.0	4.0	78.0	.
5902	RIBES PUNCTATUM	F	408	0	9.2	13.2	73.7	9.2
5903	RIBES SP.	F	367	0	11.1	6.5	76.9	14.6
5904	RIBES SP.		351	0	7.2	1.8	87.4	17.1
5905	RIBES SP.	F	342	0	8.9	0.6	82.9	15.2
5906	RIBES SP.	F	350	0	9.8	1.4	84.6	23.8
5907	RICINODENDRON HEUDELOTII	S	561	0	22.4	45.6	24.8	1.8
5908	RICINUS COMMUNIS	L	.	0	24.8	5.4	57.4	10.3
5909	RICINUS COMMUNIS	S	.	0	26.2	65.9	.	.
5910	RICINUS COMMUNIS VAR. CAMBODGENSIS	S	.	0	20.6	46.4	.	.
5911	RICINUS COMMUNIS VAR. GIBSONII	S	.	0	21.9	49.0	.	.
5912	RICINUS COMMUNIS VAR. SANGUINEUS	S	.	0	25.0	67.3	.	.
5913	RICINUS COMMUNIS VAR. ZANZIBARENSIS	S	.	0	20.0	50.6	.	.

OBS	ASH	CALCIUM	PHOSPHO	IRON	SODIUM	POTASS	CAROT	THIAMINE	RIBOFLAV	NIACIN	ASCORBIC	REFCO
5890	.	.	.	.	.	.	.	.	.	.	.	19
5891	9.8	.	.	.	.	.	.	.	.	.	.	18
5892	8.0	.	.	.	.	.	.	.	.	.	.	18
5893	5.2	530.0	350.0	.	.	.	.	.	.	.	.	6
5894	.	.	.	.	.	.	.	.	.	.	.	21
5895	.	.	.	.	.	.	.	.	.	.	.	21
5896	.	.	.	.	.	.	.	.	.	.	.	20
5897	.	.	.	.	.	.	.	.	.	.	.	20
5898	6.8	1650.0	230.0	.	.	1370.0	.	.	.	.	.	6
5899	.	.	.	.	.	.	.	.	.	.	.	19
5900	.	.	.	.	.	.	.	.	.	.	.	19
5901	4.0	319.0	.	5.0	.	.	.	.	.	.	333.0	15
5902	3.9	144.7	263.2	15.8	.	.	394.74	0.13	0.13	5.26	52.6	3
5903	5.5	70.4	196.0	5.5	.	.	.	.	.	.	.	1
5904	3.6	162.2	135.1	4.5	9.0	1396.4	1567.57	.	.	.	297.3	4
5905	5.7	379.7	253.2	7.0	19.0	2354.4	873.42	0.32	0.32	1.90	1265.8	4
5906	4.2	223.8	160.8	7.0	14.0	1797.2	503.50	0.28	0.35	0.70	286.7	4
5907	7.2	646.6	979.9	0.4	.	.	0.00	0.01	0.00	0.00	0.0	1
5908	12.4	2670.0	460.0	.	.	.	.	.	.	.	.	18
5909	.	.	.	.	.	.	.	.	.	.	.	19
5910	.	.	.	.	.	.	.	.	.	.	.	20
5911	.	.	.	.	.	.	.	.	.	.	.	20
5912	.	.	.	.	.	.	.	.	.	.	.	20
5913	.	.	.	.	.	.	.	.	.	.	.	20

TABLE 3 - DATA CONVERTED TO A ZERO-MOISTURE BASIS.

OBS	PLNA	PART	CALORIES	WATER	PROTEIN	FAT	TOTLCARB	FIBER
5914	RIDOLFIA SEGETUM	S	.	0	21.2	12.0	.	.
5915	RIVEA CORYMBOSA	S	.	0	18.1	9.2	.	.
5916	ROBINIA NEO-MEXICANA	S	.	0	41.2	12.9	.	.
5917	ROBINIA PSEUDO-ACACIA	S	.	0	43.2	10.8	.	.
5918	ROBINIA SP.	S	.	0	20.0	2.3	.	.
5919	ROEMERIA HYBRIDA	S	.	0	21.8	37.9	.	.
5920	ROGERIA LONGIFLORA	S	.	0	20.1	31.1	.	.
5921	ROMULEA SP.	S	.	0	15.1	8.5	.	.
5922	ROSA AGRESTIS	S	.	0	7.4	8.9	.	.
5923	ROSA ARKANSANA	S	.	0	4.6	4.2	.	.
5924	ROSA CANINA	S	.	0	9.6	8.5	.	.
5925	ROSA CANINA VAR. DUMETORUM	S	.	0	8.8	10.8	.	.
5926	ROSA CAROLINA	S	.	0	16.2	14.3	.	.
5927	ROSA SP.	S	.	0	8.1	8.2	.	.
5928	ROSA SP.	S	.	0	9.8	7.0	.	.
5929	ROSA SPALDINGII(IMMATURE)	L	.	0	18.6	3.7	70.1	11.7
5930	ROSA SPALDINGII(MATURE)	L	.	0	11.6	6.2	73.9	11.4
5931	ROSA SPINOSISSIMA	S	.	0	6.9	5.4	.	.
5932	ROSA WOODSII FENDLERI	C	.	0	6.0	3.0	87.8	32.7
5933	ROSMARINUS OFFICINALIS	H	365	0	5.4	16.8	70.7	19.4
5934	ROTTBOELLIA EXALTATA	W	.	0	11.1	2.3	75.7	32.9
5935	ROTTBOELLIA EXALTATA	H	.	0	5.3	0.8	83.1	31.1
5936	ROYENA LYCIOIDES	S	.	0	7.5	4.9	.	.
5937	ROYENA LYCIOIDES	S	.	0	6.8	4.3	.	.

OBS	ASH	CALCIUM	PHOSPHO	IRON	SODIUM	POTASS	CAROT	THIAMINE	RIBOFLAV	NIACIN	ASCORBIC	REFCO
5914	.	.	.	.	.	.	.	.	.	.	.	21
5915	.	.	.	.	.	.	.	.	.	.	.	20
5916	.	.	.	.	.	.	.	.	.	.	.	19
5917	4.4	.	.	.	.	.	.	.	.	.	.	19
5918	3.9	.	.	.	.	.	.	.	.	.	.	19
5919	.	.	.	.	.	.	.	.	.	.	.	21
5920	.	.	.	.	.	.	.	.	.	.	.	21
5921	.	.	.	.	.	.	.	.	.	.	.	21
5922	.	.	.	.	.	.	.	.	.	.	.	21
5923	.	.	.	.	.	.	.	.	.	.	.	21
5924	.	.	.	.	.	.	.	.	.	.	.	21
5925	.	.	.	.	.	.	.	.	.	.	.	21
5926	.	.	.	.	.	.	.	.	.	.	.	19
5927	.	.	.	.	.	.	.	.	.	.	.	20
5928	.	.	.	.	.	.	.	.	.	.	.	21
5929	7.6	1500.0	530.0	.	.	.	.	.	.	.	.	6
5930	8.3	2340.0	650.0	.	.	.	.	.	.	.	.	6
5931	.	.	.	.	.	.	.	.	.	.	.	21
5932	3.2	.	.	.	.	.	.	.	.	.	.	6
5933	7.2	1411.2	77.2	32.2	55.1	1052.9	2069.46	0.56	.	1.10	67.3	16
5934	10.9	.	.	.	.	.	.	.	.	.	.	18
5935	10.8	.	.	.	.	.	.	.	.	.	.	18
5936	.	.	.	.	.	.	.	.	.	.	.	21
5937	.	.	.	.	.	.	.	.	.	.	.	21

TABLE 3 - DATA CONVERTED TO A ZERO-MOISTURE BASIS.

OBS	PLNA	PART	CALORIES	WATER	PROTEIN	FAT	TOTLCARB	FIBER
5938	ROYENA SP.	S	.	0	8.1	3.4	.	.
5939	ROYSTONEA REGIA	F	.	0	6.1	26.6	63.0	23.6
5940	RUBIA PURPUREA	S	.	0	12.6	15.5	.	.
5941	RUBIA TINCTORUM	S	.	0	16.0	11.6	.	.
5942	RUBUS IDAEUS VAR. STRIGOSUS	S	.	0	7.4	13.8	.	.
5943	RUBUS OCCIDENTALIS X RUBUS IDAEUS	F	380	0	7.8	7.3	81.8	26.6
5944	RUBUS PARVIFLORUS	F	363	0	10.0	4.0	82.0	.
5945	RUBUS PARVIFLORUS	S	.	0	10.6	22.8	.	.
5946	RUBUS SP.	S	.	0	9.4	14.6	.	.
5947	RUBUS SP.	F	365	0	7.7	3.8	84.6	25.0
5948	RUBUS SP.	F	361	0	7.6	3.2	86.1	19.0
5949	RUBUS SP.	F	374	0	7.7	5.8	83.2	26.5
5950	RUBUS SPECTABILIS	F	373	0	35.0	5.0	59.0	.
5951	RUBUS URSINUS	F	374	0	18.0	6.0	72.0	.
5952	RUBUS URSINUS	F	364	0	9.1	2.3	86.4	20.5
5953	RUBUS URSINUS	F	365	0	5.9	3.5	87.6	17.6
5954	RUDBECKIA BICOLOR VAR. SUPERBA	S	.	0	29.1	32.0	.	.
5955	RUDBECKIA BICOLOR VAR. SUPERBA	S	.	0	28.8	28.1	.	.
5956	RUDBECKIA BICOLOR VAR. SUPERBA	S	.	0	30.6	27.6	.	.
5957	RUDBECKIA BICOLOR VAR. SUPERBA	S	.	0	22.5	22.8	.	.
5958	RUDBECKIA FULGIDA	S	.	0	30.6	27.1	.	.
5959	RUDBECKIA GRANDIFLORA VAR. ALISMAEFOLIA	S	.	0	13.8	6.6	.	.
5960	RUDBECKIA HIRTA	S	.	0	30.0	29.5	.	.
5961	RUDBECKIA HIRTA	S	.	0	27.5	31.3	.	.

OBS	ASH	CALCIUM	PHOSPHO	IRON	SODIUM	POTASS	CAROT	THIAMINE	RIBOFLAV	NIACIN	ASCORBIC	REFCO
5938	.	.	.	.	.	.	.	.	.	.	.	21
5939	4.3	.	.	.	.	.	.	.	.	.	.	18
5940	.	.	.	.	.	.	.	.	.	.	.	21
5941	.	.	.	.	.	.	.	.	.	.	.	21
5942	.	.	.	.	.	.	.	.	.	.	.	20
5943	3.1	156.3	114.6	4.7	5.2	1036.5	0.0	0.16	0.47	4.69	93.8	4
5944	4.0	428.0	.	3.0	.	.	.	.	.	.	259.0	15
5945	.	.	.	.	.	.	.	.	.	.	.	20
5946	.	.	.	.	.	.	.	.	.	.	.	20
5947	3.8	217.9	230.8	12.8	.	.	128.2	0.13	0.26	3.21	115.4	3
5948	3.2	139.2	139.2	5.7	6.3	1063.3	493.7	0.19	0.57	5.70	158.2	4
5949	3.2	206.5	122.6	5.8	6.5	1096.8	774.2	0.19	0.26	2.58	135.5	4
5950	1.0	225.0	.	5.0	.	.	.	.	.	.	371.0	15
5951	4.0	353.0	.	4.0	.	.	.	.	.	.	201.0	15
5952	2.3	189.4	181.8	12.1	7.6	1159.1	772.7	0.15	0.98	7.58	98.5	4
5953	2.9	205.9	100.0	7.1	5.9	1000.0	705.9	0.18	0.24	2.35	141.2	4
5954	4.3	.	.	.	.	.	.	.	.	.	.	19
5955	4.0	.	.	.	.	.	.	.	.	.	.	19
5956	.	.	.	.	.	.	.	.	.	.	.	19
5957	.	.	.	.	.	.	.	.	.	.	.	19
5958	.	.	.	.	.	.	.	.	.	.	.	21
5959	.	.	.	.	.	.	.	.	.	.	.	20
5960	.	.	.	.	.	.	.	.	.	.	.	19
5961	.	.	.	.	.	.	.	.	.	.	.	19

TABLE 3 - DATA CONVERTED TO A ZERO-MOISTURE BASIS.

OBS	PLNA	PART	CALORIES	WATER	PROTEIN	FAT	TOTLCARB	FIBER
5962	RUDBECKIA HIRTA VAR. PULCHERRIMA	S	.	0	24.4	30.6	.	.
5963	RUDBECKIA HIRTA VAR. PULCHERRIMA	S	.	0	28.8	30.0	.	.
5964	RUDBECKIA HIRTA VAR. PULCHERRIMA	S	.	0	25.6	28.1	.	.
5965	RUDBECKIA LACINIATA	S	.	0	37.8	26.3	.	.
5966	RUDBECKIA MAXIMA	S	.	0	25.4	13.6	.	.
5967	RUDBECKIA OCCIDENTALIS	S	.	0	21.5	13.4	.	.
5968	RUDBECKIA OCCIDENTALIS	W	.	0	21.7	6.0	59.0	11.4
5969	RUDBECKIA OCCIDENTALIS	L	.	0	34.4	2.2	51.4	8.7
5970	RUMEX ACETOSELLA	S	.	0	5.6	1.7	.	.
5971	RUMEX ACETOSELLA	S	.	0	9.6	3.6	.	.
5972	RUMEX ALTISSIMUS	S	.	0	13.4	3.7	.	.
5973	RUMEX ALTISSIMUS	S	.	0	12.5	2.4	.	.
5974	RUMEX CONGLOMERATUS	S	.	0	12.1	2.2	.	.
5975	RUMEX CRISPUS	S	.	0	13.9	3.0	.	.
5976	RUMEX CRISPUS	S	.	0	9.6	1.3	.	.
5977	RUMEX CRISPUS	S	.	0	16.2	3.0	.	.
5978	RUMEX CRISPUS	S	.	0	13.1	2.4	.	.
5979	RUMEX CRISPUS	S	.	0	13.4	3.2	.	.
5980	RUMEX CRISPUS	S	.	0	13.6	3.0	.	.
5981	RUMEX CRISPUS	S	.	0	14.7	2.6	.	.
5982	RUMEX CRISPUS	S	.	0	12.8	2.9	.	.
5983	RUMEX CRISPUS	S	.	0	21.9	4.0	.	.
5984	RUMEX CRISPUS	S	.	0	16.2	2.8	.	.
5985	RUMEX CRISPUS	W	.	0	13.0	3.2	80.2	23.7

OBS	ASH	CALCIUM	PHOSPHO	IRON	SODIUM	POTASS	CAROT	THIAMINE	RIBOFLAV	NIACIN	ASCORBIC	REFCO
5962	.	.	.	.	.	.	.	.	.	.	.	20
5963	.	.	.	.	.	.	.	.	.	.	.	20
5964	.	.	.	.	.	.	.	.	.	.	.	21
5965	4.3	.	.	.	.	.	.	.	.	.	.	19
5966	.	.	.	.	.	.	.	.	.	.	.	20
5967	.	.	.	.	.	.	.	.	.	.	.	21
5968	13.3	2930.0	440.0	.	.	.	.	.	.	.	.	6
5969	12.0	.	.	.	.	.	.	.	.	.	.	6
5970	.	.	.	.	.	.	.	.	.	.	.	20
5971	.	.	.	.	.	.	.	.	.	.	.	21
5972	2.0	.	.	.	.	.	.	.	.	.	.	19
5973	.	.	.	.	.	.	.	.	.	.	.	20
5974	.	.	.	.	.	.	.	.	.	.	.	21
5975	2.1	.	.	.	.	.	.	.	.	.	.	19
5976	7.9	.	.	.	.	.	.	.	.	.	.	19
5977	2.4	.	.	.	.	.	.	.	.	.	.	19
5978	.	.	.	.	.	.	.	.	.	.	.	20
5979	.	.	.	.	.	.	.	.	.	.	.	20
5980	.	.	.	.	.	.	.	.	.	.	.	20
5981	.	.	.	.	.	.	.	.	.	.	.	21
5982	.	.	.	.	.	.	.	.	.	.	.	21
5983	.	.	.	.	.	.	.	.	.	.	.	21
5984	.	.	.	.	.	.	.	.	.	.	.	21
5985	3.6	.	.	.	.	.	.	.	.	.	.	6

TABLE 3 - DATA CONVERTED TO A ZERO-MOISTURE BASIS.

OBS	PLNA	PART	CALORIES	WATER	PROTEIN	FAT	TOTLCARB	FIBER
5986	RUMEX CRISPUS	L	284	0	20.3	4.1	55.4	12.2
5987	RUMEX DENTATUS SUBSP. KLOTZSCHIANUS	S	.	0	12.3	2.8	.	.
5988	RUMEX HYDROLAPATHUM	S	.	0	13.0	2.9	.	.
5989	RUMEX OBTUSIFOLIUS	S	.	0	12.7	2.8	.	.
5990	RUMEX PULCHER	L	.	0	26.8	.	.	.
5991	RUMEX PULCHER	S	.	0	11.9	2.2	.	.
5992	RUMEX PULCHER	S	.	0	23.5	2.3	.	.
5993	RUMEX PULCHER	S	.	0	13.3	2.6	.	.
5994	RUMEX SALICIFOLIUS	S	.	0	13.5	2.6	.	.
5995	RUMEX SALICIFOLIUS	S	.	0	10.2	1.2	.	.
5996	RUMEX SCUTATUS	S	.	0	24.5	7.4	.	.
5997	RUMEX SCUTATUS	S	.	0	17.2	4.5	.	.
5998	RUMEX SP.	S	.	0	12.8	2.1	.	.
5999	RUMEX SP.	L	308	0	23.1	3.3	61.5	8.8
6000	RUMEX SP.	L	308	0	23.1	3.3	61.5	8.8
6001	RUMEX VENOSUS	S	.	0	20.0	4.2	.	.
6002	RUPRECHTIA CORIACEA	S	.	0	11.2	2.3	.	.
6003	RUTA CHALEPENSIS	S	.	0	26.4	33.2	.	.
6004	SABA SENEGALENSIS	F	355	0	4.0	1.0	92.5	6.5
6005	SACCHARUM OFFICINARUM	H	.	0	2.6	1.2	92.1	43.1
6006	SACCHARUM OFFICINARUM	W	.	0	3.6	1.3	91.4	28.2
6007	SACCHARUM OFFICINARUM	L	333	0	8.0	3.6	79.6	13.3
6008	SACCHARUM OFFICINARUM	SH	354	0	3.4	0.6	94.3	17.7
6009	SACCHARUM OFFICINARUM	I	278	0	51.1	4.4	33.3	.

OBS	ASH	CALCIUM	PHOSPHO	IRON	SODIUM	POTASS	CAROT	THIAMINE	RIBOFLAV	NIACIN	ASCORBIC	REFCO
5986	20.3	1000.0	756.8	75.7	.	.	37432.4	0.81	1.08	5.41	405.4	3
5987	.	.	.	.	.	.	.	.	.	.	.	21
5988	.	.	.	.	.	.	.	.	.	.	.	21
5989	.	.	.	.	.	.	.	.	.	.	.	20
5990	12.9	697.0	.	.	1830.0	2400.0	.	.	.	.	.	17
5991	.	.	.	.	.	.	.	.	.	.	.	19
5992	.	.	.	.	.	.	.	.	.	.	.	21
5993	.	.	.	.	.	.	.	.	.	.	.	21
5994	2.0	.	.	.	.	.	.	.	.	.	.	19
5995	7.7	.	.	.	.	.	.	.	.	.	.	19
5996	3.3	.	.	.	.	.	.	.	.	.	.	19
5997	.	.	.	.	.	.	.	.	.	.	.	21
5998	.	.	.	.	.	.	.	.	.	.	.	20
5999	12.1	725.3	450.5	17.6	54.9	3714.3	85054.9	0.99	2.42	5.49	1307.7	4
6000	12.1	725.3	450.5	17.6	54.9	3714.3	85054.9	0.99	2.42	5.49	1307.7	4
6001	1.4	.	.	.	.	.	.	.	.	.	.	19
6002	2.0	.	.	.	.	.	.	.	.	.	.	19
6003	.	.	.	.	.	.	.	.	.	.	.	21
6004	2.5	255.0	140.0	5.0	.	.	0.0	0.75	0.15	2.50	240.0	1
6005	4.1	3600.0	.	.	.	.	.	.	.	.	.	6
6006	3.7	470.0	170.0	8.0	.	.	.	.	.	.	.	6
6007	8.9	.	.	.	.	.	.	.	.	.	.	1
6008	1.7	45.7	34.3	8.0	.	.	0.0	0.11	0.06	0.57	17.1	1
6009	11.1	444.4	888.9	22.2	.	.	0.0	0.89	.	.	555.6	2

TABLE 3 - DATA CONVERTED TO A ZERO-MOISTURE BASIS.

OBS	PLNA	PART	CALORIES	WATER	PROTEIN	FAT	TOTLCARB	FIBER	ASH	CALCIUM	PHOSPHO	IRON	SODIUM	POTASS	CAROT	THIAMINE	RIBOFLAV	NIACIN	ASCORBIC	REFCO
6010	SACCHARUM OFFICINARUM	J	387	0	1.4	0.5	96.7	1.9	1.4	61.3	56.6	3.3	.	.	0.00	0.09	0.05	0.47	9.4	3
6011	SACCHARUM SINENSE	W	.	0	10.3	2.5	80.9	32.1	6.3	.	.	.	.	.	.	.	.	.	.	18
6012	SACCHARUM SINENSE	L	.	0	9.5	2.8	78.9	33.4	8.8	.	.	.	.	.	.	.	.	.	.	18
6013	SACCHARUM SINENSE	S	.	0	5.3	1.5	82.9	35.9	10.5	.	.	.	.	.	.	.	.	.	.	18
6014	SACCHARUM SPONTANEUM	W	.	0	5.4	1.2	88.2	40.2	7.2	.	.	.	.	.	.	.	.	.	.	18
6015	SAGITTARIA PAPILLOSA	S	.	0	19.4	23.0	.	.	.	.	.	.	.	.	.	.	.	.	.	19
6016	SAGITTARIA SAGITTIFOLIA	S	.	0	21.2	17.1	.	.	.	.	.	.	.	.	.	.	.	.	.	21
6017	SAGITTARIA SAGITTIFOLIA	R	364	0	17.0	1.0	76.2	3.1	5.8	44.2	561.2	8.8	.	2479.6	0.00	0.54	0.14	4.76	17.0	2
6018	SALACCA EDULIS	F	345	0	1.8	0.0	95.0	.	3.2	127.3	81.8	19.1	.	.	0.00	0.18	.	.	9.1	2
6019	SALAMALIA MALABARICA	L	.	0	12.6	6.4	71.7	22.3	9.3	2700.0	190.0	.	.	.	.	.	.	.	.	18
6020	SALIX BABYLONICA	L	.	0	16.7	2.9	70.2	18.0	10.2	1600.0	260.0	.	.	.	.	.	.	.	.	18
6021	SALIX LUTEA	L	.	0	14.4	2.9	75.9	14.0	6.8	250.0	240.0	.	.	.	.	.	.	.	.	6
6022	SALSOLA APHYLLA	SH	.	0	20.0	2.0	49.4	14.9	28.6	1550.0	160.0	.	.	.	.	.	.	.	.	18
6023	SALSOLA GLABRESCENS	SH	.	0	16.5	2.1	68.2	25.2	13.2	1950.0	150.0	.	.	.	.	.	.	.	.	18
6024	SALSOLA KALI	S	.	0	40.5	27.0	.	.	.	.	.	.	.	.	.	.	.	.	.	21
6025	SALSOLA KALI	W	.	0	13.1	2.0	69.4	31.7	15.5	2730.0	130.0	.	.	4630.0	.	.	.	.	.	6
6026	SALSOLA PESTIFER	S	.	0	40.3	22.4	.	.	7.0	.	.	.	.	.	.	.	.	.	.	19
6027	SALSOLA RABIANA	C	.	0	11.9	1.2	76.6	37.1	10.3	1330.0	50.0	.	.	.	.	.	.	.	.	18
6028	SALSOLA SP.	S	.	0	16.9	1.3	.	.	17.2	.	.	.	.	.	.	.	.	.	.	19
6029	SALSOLA SP.	W	.	0	12.3	1.8	70.7	31.7	15.2	2470.0	170.0	.	.	6460.0	.	.	.	.	.	6
6030	SALSOLA VERMICULATA	S	.	0	47.4	15.2	.	.	.	.	.	.	.	.	.	.	.	.	.	21
6031	SALVADORA PERSICA	L	.	0	16.1	2.3	61.3	14.1	20.3	2440.0	150.0	.	.	.	.	.	.	.	.	18
6032	SALVIA AETHIOPIS	S	.	0	20.9	22.8	.	.	.	.	.	.	.	.	.	.	.	.	.	21
6033	SALVIA AETHIOPIS	S	.	0	23.0	22.4	.	.	.	.	.	.	.	.	.	.	.	.	.	21
6034	SALVIA APIANA	S	.	0	7.9	6.9	.	.	3.8	.	.	.	.	.	.	.	.	.	.	19
6035	SALVIA CARDUACEA	S	.	0	26.2	30.0	.	.	.	.	.	.	.	.	.	.	.	.	.	20
6036	SALVIA CERATOPHYLLA	S	.	0	16.9	19.3	.	.	.	.	.	.	.	.	.	.	.	.	.	20
6037	SALVIA COLUMBARIAE	S	.	0	20.2	34.4	.	.	5.6	.	.	.	.	.	.	.	.	.	.	19
6038	SALVIA CRYPTANTHA	S	.	0	16.9	11.4	.	.	.	.	.	.	.	.	.	.	.	.	.	21
6039	SALVIA FARINACEA	S	.	0	15.0	17.0	.	.	.	.	.	.	.	.	.	.	.	.	.	20
6040	SALVIA GLUTINOSA	S	.	0	21.8	34.3	.	.	.	.	.	.	.	.	.	.	.	.	.	21
6041	SALVIA GRANDIFLORA	S	.	0	16.9	15.9	.	.	.	.	.	.	.	.	.	.	.	.	.	21
6042	SALVIA GRANDIFLORA	S	.	0	17.4	13.7	.	.	.	.	.	.	.	.	.	.	.	.	.	21
6043	SALVIA HISPANICA	S	.	0	17.5	28.6	.	.	.	.	.	.	.	.	.	.	.	.	.	20
6044	SALVIA HORMINUM	S	.	0	22.1	24.0	.	.	.	.	.	.	.	.	.	.	.	.	.	21
6045	SALVIA HYDRANGEA	S	.	0	20.1	16.2	.	.	.	.	.	.	.	.	.	.	.	.	.	21
6046	SALVIA JUDAICA	S	.	0	19.0	20.2	.	.	.	.	.	.	.	.	.	.	.	.	.	21
6047	SALVIA LANIGERA	S	.	0	19.4	22.9	.	.	.	.	.	.	.	.	.	.	.	.	.	21
6048	SALVIA LAVANDULAEFOLIA	S	.	0	19.4	20.6	.	.	.	.	.	.	.	.	.	.	.	.	.	21
6049	SALVIA LONGISPICATA	S	.	0	18.7	18.8	.	.	.	.	.	.	.	.	.	.	.	.	.	21
6050	SALVIA LYRATA	S	.	0	20.6	22.6	.	.	.	.	.	.	.	.	.	.	.	.	.	20
6051	SALVIA MOORCROFTIANA	S	.	0	23.3	27.1	.	.	.	.	.	.	.	.	.	.	.	.	.	21
6052	SALVIA OFFICINALIS	L	342	0	11.5	13.8	66.0	19.6	8.7	1795.7	98.9	30.5	12	1163.0	3847.83	0.82	0.37	6.22	34.8	16
6053	SALVIA PLEBEIA	S	.	0	18.4	21.3	.	.	.	.	.	.	.	.	.	.	.	.	.	21
6054	SALVIA POLYSTACHYA	S	.	0	23.1	32.7	[illegible]	.	.	.	.	.	.	.	.	.	.	.	.	21

TABLE 3 - DATA CONVERTED TO A ZERO-MOISTURE BASIS.

OBS	PLANT	PART	CALORIES	WATER	PROTEIN	FAT	TOTCARB	FIBER	ASH	CALCIUM	PHOSPHO	IRON	SODIUM	POTASS	CAROT	THIAMINE	RIBOFLAV	NIACIN	ASCORBIC	REFCO
6055	SALVIA REFLEXA	S	.	0	19.4	23.3	.	.	.	.	.	.	.	.	.	.	.	.	.	19
6056	SALVIA REFLEXA	S	.	0	15.6	26.6	.	.	.	.	.	.	.	.	.	.	.	.	.	20
6057	SALVIA RUGOSA	S	.	0	18.8	27.5	.	.	.	.	.	.	.	.	.	.	.	.	.	21
6058	SALVIA RUGOSA	S	.	0	24.1	27.4	.	.	.	.	.	.	.	.	.	.	.	.	.	21
6059	SALVIA SCLAREA	S	.	0	20.7	31.8	.	.	.	.	.	.	.	.	.	.	.	.	.	21
6060	SALVIA SIMILATA	S	.	0	18.9	10.8	.	.	.	.	.	.	.	.	.	.	.	.	.	21
6061	SALVIA SONOMENSIS	S	.	0	15.5	21.2	.	.	.	.	.	.	.	.	.	.	.	.	.	21
6062	SALVIA SP.	S	.	0	23.1	35.0	.	.	.	.	.	.	.	.	.	.	.	.	.	20
6063	SALVIA SP.	W	.	0	24.4	3.9	57.7	12.2	14.0	1490.0	360.0	.	.	4630.0	.	.	.	.	.	6
6064	SALVIA SUFFRUTICOSA	S	.	0	26.2	40.5	.	.	.	.	.	.	.	.	.	.	.	.	.	21
6065	SALVIA SYRIACA	S	.	0	15.6	12.4	.	.	.	.	.	.	.	.	.	.	.	.	.	20
6066	SALVIA TCHIHATCHEFFII	S	.	0	20.4	22.3	.	.	.	.	.	.	.	.	.	.	.	.	.	20
6067	SALVIA TEXANA	S	.	0	17.1	16.4	.	.	.	.	.	.	.	.	.	.	.	.	.	20
6068	SALVIA TRILOBA	S	.	0	19.4	23.2	.	.	.	.	.	.	.	.	.	.	.	.	.	21
6069	SAMANEA SAMAN	C	.	0	24.7	2.8	68.1	22.1	4.4	550.0	260.0	.	.	.	.	.	.	.	.	18
6070	SAMANEA SAMAN	L	.	0	22.1	7.0	64.9	29.4	6.0	1420.0	210.0	.	.	.	.	.	.	.	.	18
6071	SAMANEA SAMAN	F	.	0	12.8	0.7	84.1	14.5	2.4	290.0	320.0	.	.	.	.	.	.	.	.	18
6072	SAMANEA SAMAN	S	.	0	31.6	6.0	58.1	14.0	4.3	160.0	340.0	.	.	.	.	.	.	.	.	18
6073	SAMANEA SAMAN	F	.	0	15.0	2.5	78.7	13.5	3.8	.	.	.	.	.	.	.	.	.	.	5
6074	SAMANEA SAMAN	S	.	0	31.0	4.7	60.7	15.3	3.8	.	.	.	.	.	.	.	.	.	.	5
6075	SAMBUCUS CAERULEA	F	344	0	16.0	5.0	68.0	.	10.0	115.0	.	5.0	.	.	.	.	.	.	156.0	15
6076	SAMBUCUS CANADENSIS	S	.	0	10.6	20.1	.	.	.	.	.	.	.	.	.	.	.	.	.	19
6077	SAMBUCUS CANADENSIS	L	.	0	19.5	4.0	72.6	30.1	3.8	.	.	.	.	.	.	.	.	.	.	.
6078	SAMBUCUS CANADENSIS	F	356	0	12.9	2.5	81.2	34.7	3.5	188.1	138.6	7.9	.	1485.1	1782.18	0.35	0.30	2.48	178.2	4
6079	SAMBUCUS EBULUS	S	.	0	14.2	14.3	.	.	.	.	.	.	.	.	.	.	.	.	.	21
6080	SAMBUCUS NIGRA	S	.	0	12.3	26.9	.	.	.	.	.	.	.	.	.	.	.	.	.	21
6081	SAMBUCUS PUBENS	S	.	0	10.6	28.4	.	.	.	.	.	.	.	.	.	.	.	.	.	20
6082	SAMBUCUS RACEMOSA	F	372	0	17.0	6.0	73.0	.	4.0	356.0	.	7.0	.	.	.	.	.	.	533.0	15
6083	SAMBUCUS WIGHTIANA	S	.	0	12.8	14.0	.	.	.	.	.	.	.	.	.	.	.	.	.	21
6084	SANDORICUM KOETJAPE	F	368	0	2.6	4.5	89.7	6.5	3.2	58.1	109.7	7.7	19.4	2116.1	32.26	0.32	0.19	5.81	90.3	2
6085	SANDORICUM KOETJAPE	F	354	0	6.2	0.8	90.8	0.8	2.3	30.8	130.8	3.1	.	.	0.00	0.38	0.23	5.38	661.5	3
6086	SANGUISORBA MINOR	S	.	0	13.5	20.3	.	.	6.2	.	.	.	.	.	.	.	.	.	.	19
6087	SANGUISORBA MINOR	S	.	0	14.1	19.9	.	.	7.2	.	.	.	.	.	.	.	.	.	.	19
6088	SANGUISORBA MINOR	S	.	0	13.1	17.7	.	.	8.6	.	.	.	.	.	.	.	.	.	.	19
6089	SANGUISORBA MINOR	S	.	0	12.1	9.4	.	.	.	.	.	.	.	.	.	.	.	.	.	20
6090	SANGUISORBA MINOR	W	.	0	11.1	2.0	80.4	18.0	6.5	.	.	.	.	.	.	.	.	.	.	6
6091	SANGUISORBA OFFICINALIS	S	.	0	12.8	13.5	.	.	.	.	.	.	.	.	.	.	.	.	.	21
6092	SANTIRIA TRIMERA	P	375	0	6.8	6.3	83.3	.	3.6	239.6	.	26.0	.	.	.	.	.	.	114.6	1
6093	SANTOLINA ROSMARINIFOLIA	S	.	0	25.9	25.3	.	.	.	.	.	.	.	.	.	.	.	.	.	21
6094	SAPINDUS ABYSSINICUS	F	346	0	5.2	0.0	91.3	1.7	3.5	80.4	314.7	10.5	.	.	1258.74	0.10	.	.	237.8	1
6095	SAPINDUS MUKOROSSI	S	.	0	31.2	35.4	.	.	5.2	.	.	.	.	.	.	.	.	.	.	19
6096	SAPINDUS MUKOROSSI	S	.	0	28.6	38.1	.	.	.	.	.	.	.	.	.	.	.	.	.	21
6097	SAPINDUS SAPONARIA	S	.	0	29.2	30.0	.	.	5.0	.	.	.	.	.	.	.	.	.	.	19
6098	SAPINDUS SAPONARIA	S	.	0	27.5	34.7	.	.	.	.	.	.	.	.	.	.	.	.	.	20
6099	SAPINDUS SP.	S	.	0	26.1	34.7	.	.	5.2	.	.	.	.	.	.	.	.	.	.	19

TABLE 3 - DATA CONVERTED TO A ZERO-MOISTURE BASIS.

OBS	PLANA	PART	CALORIES	WATER	PROTEIN	FAT	TOTLCARB	FIBER	ASH	CALCIUM	PHOSPHO	IRON	SODIUM	POTASS	CAROT	THIAMINE	RIBOFLAV	NIACIN	ASCORBIC	REFCO
6100	SAPINDUS SP.	S	.	0	2.7	0.8	.	.	3.1	.	.	.	.	.	.	.	.	.	.	19
6101	SAPINDUS TRIFOLIATA	S	.	0	25.0	57.4	.	.	.	.	.	.	.	.	.	.	.	.	.	20
6102	SAPIUM SEBIFERUM	L	.	0	17.8	5.3	70.9	12.4	6.0	1360.0	220.0	.	.	.	.	.	.	.	.	18
6103	SAPIUM SEBIFERUM	S	.	0	8.1	50.7	.	.	.	.	.	.	.	.	.	.	.	.	.	21
6104	SAPIUM SEBIFERUM	S	.	0	9.2	40.5	.	.	.	.	.	.	.	.	.	.	.	.	.	21
6105	SAPONARIA OCYMOIDES	S	.	0	13.8	3.9	.	.	3.9	.	.	.	.	.	.	.	.	.	.	19
6106	SAPONARIA OFFICINALIS	S	.	0	16.2	5.6	.	.	2.9	.	.	.	.	.	.	.	.	.	.	19
6107	SAPONARIA OFFICINALIS	S	.	0	16.6	5.9	.	.	.	.	.	.	.	.	.	.	.	.	.	21
6108	SAPONARIA PROSTRATA	S	.	0	15.5	4.8	.	.	.	.	.	.	.	.	.	.	.	.	.	21
6109	SAPONARIA VISCOSA	S	.	0	20.0	4.7	.	.	.	.	.	.	.	.	.	.	.	.	.	20
6110	SARCOBATUS SP.	L	.	0	21.9	4.1	59.8	21.3	14.2	840.0	150.0	.	.	.	.	.	.	.	.	6
6111	SARCOCOCCA SALIGNA	S	.	0	20.1	39.4	.	.	.	.	.	.	.	.	.	.	.	.	.	21
6112	SARCOPHRYNIUM ARNOLDIANUM	SH	331	0	14.6	5.4	69.2	.	10.8	.	.	.	.	.	.	.	.	.	.	1
6113	SASSAFRAS ALBIDUM	S	.	0	23.8	60.4	.	.	2.4	.	.	.	.	.	.	.	.	.	.	19
6114	SASSAFRAS ALBIDUM	S	.	0	6.2	11.9	.	.	2.6	.	.	.	.	.	.	.	.	.	.	19
6115	SATUREJA HORTENSIS	S	.	0	24.1	41.5	.	.	5.7	.	.	.	.	.	.	.	.	.	.	19
6116	SATUREJA HORTENSIS	S	.	0	23.1	39.1	.	.	.	.	.	.	.	.	.	.	.	.	.	19
6117	SATUREJA HORTENSIS	H	299	0	7.4	6.5	75.5	16.8	10.5	2342.9	153.8	41.6	26.4	1154.9	3382.4	0.41	.	4.48	.	16
6118	SATUREJA THYMBRA	S	.	0	12.6	17.9	.	.	.	.	.	.	.	.	.	.	.	.	.	21
6119	SAURAUIA PAUCISERRATA	F	355	0	4.2	2.4	88.6	7.8	4.8	391.6	192.8	3.0	.	.	301.2	0.06	0.18	1.20	78.3	3
6120	SAUROMATUM GUTTATUM	S	.	0	12.3	2.9	.	.	.	.	.	.	.	.	.	.	.	.	.	21
6121	SAUROPUS ANDROGYNUS	L	305	0	25.3	4.7	57.4	.	12.6	268.4	452.6	14.2	147.4	2647.4	32736.8	0.37	2.05	11.58	436.8	2
6122	SAVIGNYA PARVIFLORA	S	.	0	25.2	19.0	.	.	.	.	.	.	.	.	.	.	.	.	.	21
6123	SCABIOSA ATROPURPUREA	S	.	0	34.8	25.4	.	.	5.1	.	.	.	.	.	.	.	.	.	.	19
6124	SCABIOSA ATROPURPUREA	S	.	0	7.0	2.2	.	.	6.6	.	.	.	.	.	.	.	.	.	.	19
6125	SCABIOSA ATROPURPUREA	S	.	0	31.6	28.6	.	.	.	.	.	.	.	.	.	.	.	.	.	21
6126	SCABIOSA JAPONICA	S	.	0	31.9	29.1	.	.	.	.	.	.	.	.	.	.	.	.	.	19
6127	SCABIOSA JAPONICA	S	.	0	31.9	29.5	.	.	.	.	.	.	.	.	.	.	.	.	.	20
6128	SCABIOSA ROTATA	S	.	0	11.6	10.3	.	.	.	.	.	.	.	.	.	.	.	.	.	21
6129	SCABIOSA STELLATA	S	.	0	13.8	8.6	.	.	.	.	.	.	.	.	.	.	.	.	.	21
6130	SCABIOSA UCRANICA	S	.	0	12.2	13.5	.	.	.	.	.	.	.	.	.	.	.	.	.	21
6131	SCAEVOLA PLUMIERI	S	.	0	19.2	66.7	.	.	.	.	.	.	.	.	.	.	.	.	.	21
6132	SCALIGERIA MEIFOLIA	S	.	0	28.4	26.4	.	.	.	.	.	.	.	.	.	.	.	.	.	21
6133	SCANDIX IBERICA	S	.	0	12.0	2.6	.	.	.	.	.	.	.	.	.	.	.	.	.	21
6134	SCANDIX PECTEN-VENERIS	S	.	0	16.7	4.4	.	.	.	.	.	.	.	.	.	.	.	.	.	21
6135	SCHEDONNARDUS PANICULATUS	W	.	0	9.7	2.1	77.4	30.5	10.8	680.0	150.0	.	.	.	.	.	.	.	.	6
6136	SCHIMPERA ARABICA	S	.	0	36.6	38.6	.	.	.	.	.	.	.	.	.	.	.	.	.	21
6137	SCHIMPERA ARABICA	S	.	0	40.6	36.8	.	.	.	.	.	.	.	.	.	.	.	.	.	21
6138	SCHINUS MOLLE	S	.	0	7.5	12.4	.	.	.	.	.	.	.	.	.	.	.	.	.	20
6139	SCHINUS MOLLE	S	.	0	8.1	10.6	.	.	.	.	.	.	.	.	.	.	.	.	.	20
6140	SCHINUS MOLLE	S	.	0	8.5	7.7	.	.	.	.	.	.	.	.	.	.	.	.	.	21
6141	SCHINUS TEREBINTHIFOLIUS	S	.	0	10.8	32.2	.	.	.	.	.	.	.	.	.	.	.	.	.	21
6142	SCHIZANTHUS WISETONENSIS	S	.	0	20.6	21.8	.	.	.	.	.	.	.	.	.	.	.	.	.	20
6143	SCHKUHRIA WRIGHTII	S	.	0	20.6	22.2	.	.	.	.	.	.	.	.	.	.	.	.	.	20
6144	SCHOTIA AFRA	S	.	0	11.2	4.6	.	.	.	.	.	.	.	.	.	.	.	.	.	21

TABLE 3 - DATA CONVERTED TO A ZERO-MOISTURE BASIS.

OBS	PLNA	PART	CALORIES	WATER	PROTEIN	FAT	TOTCARB	FIBER	ASH	CALCIUM	PHOSPHO	IRON	SODIUM	POTASS	CAROT	THIAMINE	RIBOFLAV	NIACIN	ASCORBIC	REFCO
6145	SCHOTIA BRACHYPETALA	S	.	0	10.5	2.6	.	.	.	.	.	.	.	.	.	.	.	.	.	21
6146	SCHOTIA BRACHYPETALA	A	.	0	5.8	0.4	.	.	.	.	.	.	.	.	.	.	.	.	.	21
6147	SCHRANKIA ROEMERIANA	L	.	0	18.7	2.4	68.2	14.2	10.7	4370.0	200.0	.	.	1380.0	.	.	.	.	.	6
6148	SCHRANKIA UNCINATA	S	.	0	35.6	4.5	.	.	3.1	.	.	.	.	.	.	.	.	.	.	19
6149	SCIRPUS ARTICULATUS	S	.	0	8.2	3.0	.	.	.	.	.	.	.	.	.	.	.	.	.	21
6150	SCIRPUS MARITIMUS	S	.	0	6.0	2.4	.	.	.	.	.	.	.	.	.	.	.	.	.	21
6151	SCIRPUS RUBRICOSUS	S	.	0	18.2	26.0	.	.	.	.	.	.	.	.	.	.	.	.	.	21
6152	SCIRPUS SP.	H	.	0	8.8	2.4	75.1	28.4	13.7	760.0	160.0	.	.	.	.	.	.	.	.	6
6153	SCIRPUS SP.	I	.	0	7.4	2.5	84.7	23.4	5.4	.	.	.	.	.	.	.	.	.	.	6
6154	SCIRPUS SP.	L	.	0	11.7	1.5	70.1	23.0	16.7	.	.	.	.	.	.	.	.	.	.	6
6155	SCLERANTHUS ANNUUS	S	.	0	22.6	9.1	.	.	.	.	.	.	.	.	.	.	.	.	.	21
6156	SCLEROCARPUS AFRICANUS	S	.	0	39.2	41.7	.	.	.	.	.	.	.	.	.	.	.	.	.	21
6157	SCLEROCARPUS SPATHULATUS	S	.	0	12.5	11.1	.	.	.	.	.	.	.	.	.	.	.	.	.	20
6158	SCLEROCARYA BIERREA	F	.	0	.	.	.	.	.	.	.	.	.	.	.	.	.	.	1376.9	1
6159	SCLEROCARYA CAFFRA	F	361	0	6.0	1.2	90.4	6.0	2.4	72.3	228.9	1.2	.	.	.	0.36	0.60	2.41	819.3	1
6160	SCLEROCARYA CAFFRA	S	629	0	25.6	59.8	9.6	2.8	5.0	148.8	1298.6	0.4	.	.	.	0.04	0.12	0.73	.	1
6161	SCLEROCHLOA DURA	S	.	0	17.9	2.9	.	.	.	.	.	.	.	.	.	.	.	.	.	21
6162	SCLEROPOGON SP.	W	.	0	7.1	.	.	31.4	9.4	510.0	110.0	.	.	.	.	.	.	.	.	6
6163	SCOHTIA AFRA	S	.	0	10.6	4.5	.	.	.	.	.	.	.	.	.	.	.	.	.	21
6164	SCOLYMUS HISPANICUS	S	.	0	18.2	20.6	.	.	.	.	.	.	.	.	.	.	.	.	.	21
6165	SCORPIURUS MURICATA	S	.	0	33.4	2.9	.	.	.	.	.	.	.	.	.	.	.	.	.	21
6166	SCORPIURUS SUBVILLOSA	S	.	0	30.9	2.7	.	.	.	.	.	.	.	.	.	.	.	.	.	21
6167	SCORPIURUS SULCATA	S	.	0	35.3	2.3	.	.	4.1	.	.	.	.	.	.	.	.	.	.	19
6168	SCORZONERA HISPANICA	S	.	0	21.9	17.1	.	.	.	.	.	.	.	.	.	.	.	.	.	20
6169	SCORZONERA HISPANICA	S	.	0	23.1	18.3	.	.	.	.	.	.	.	.	.	.	.	.	.	20
6170	SCROPHULARIA CANINA	S	.	0	16.6	31.5	.	.	.	.	.	.	.	.	.	.	.	.	.	21
6171	SCROPHULARIA STRIATA	S	.	0	18.9	35.4	.	.	.	.	.	.	.	.	.	.	.	.	.	21
6172	SCROPHULARIA XANTHOGLOSSA	S	.	0	19.4	31.3	.	.	.	.	.	.	.	.	.	.	.	.	.	21
6173	SCROPHULARIA XANTHOGLOSSA	S	.	0	17.8	29.3	.	.	.	.	.	.	.	.	.	.	.	.	.	21
6174	SCUTELLARIA DRUMMONDII	S	.	0	22.5	33.0	.	.	.	.	.	.	.	.	.	.	.	.	.	20
6175	SCUTELLARIA RESINOSA	W	.	0	4.2	10.2	74.6	19.5	11.0	1200.0	450.0	.	.	.	.	.	.	.	.	6
6176	SCUTELLARIA SP.	W	.	0	7.0	6.5	77.1	25.3	9.4	1530.0	170.0	.	.	1720.0	.	.	.	.	.	6
6177	SECALE CEREALE	W	.	0	21.4	4.8	58.6	24.8	15.2	.	.	.	.	.	.	.	.	.	.	18
6178	SECALE CEREALE	S	.	0	13.7	1.9	82.2	2.6	2.2	40.0	410.0	.	.	.	.	.	.	.	.	18
6179	SECALE CEREALE	S	.	0	17.1	1.7	.	.	2.2	.	.	.	.	.	.	.	.	.	.	19
6180	SECALE CEREALE	W	.	0	23.2	4.2	61.6	24.1	11.0	550.0	450.0	.	.	3400.0	.	.	.	.	.	6
6181	SECALE CEREALE	H	.	0	9.4	2.8	82.0	38.7	5.8	350.0	370.0	.	.	1050.0	.	.	.	.	.	6
6182	SECALE CEREALE	S	378	0	11.8	2.0	84.4	2.2	1.8	45.0	394.8	4.6	.	.	.	0.28	.	.	.	1
6183	SECALE CEREALE	S	382	0	14.6	3.0	80.5	2.2	1.9	43.4	377.1	3.4	6.9	.	0.0	0.54	0.23	1.94	.	2
6184	SECALE CEREALE	S	375	0	13.6	1.9	82.5	2.2	2.0	42.7	422.5	4.2	1.1	524.7	0.0	0.48	0.25	1.80	0.0	4
6185	SECALE MONTANUM	S	.	0	16.4	2.2	.	.	.	.	.	.	.	.	.	.	.	.	.	21
6186	SECHIUM EDULE	F	333	0	20.0	3.3	73.3	20.0	3.3	300.0	250.0	15.0	.	.	.	.	.	.	166.7	1
6187	SECHIUM EDULE	L	.	0	28.4	.	.	40.5	14.7	.	.	.	.	.	39310.3	.	.	.	.	1
6188	SECHIUM EDULE	F	333	0	12.3	1.8	80.7	10.5	5.3	298.2	245.6	7.0	35.1	1894.7	263.2	0.18	0.35	7.02	245.6	2
6189	SECHIUM EDULE	L	284	0	45.5	3.4	39.8	9.1	11.4	704.5	659.1	15.9	11.4	3522.7	17215.9	0.80	1.02	9.09	272.7	2

TABLE 3 - DATA CONVERTED TO A ZERO-MOISTURE BASIS.

OBS	PLNA	PART	CALORIES	WATER	PROTEIN	FAT	TOTCARB	FIBER	ASH	CALCIUM	PHOSPHO	IRON	SODIUM	POTASS	CAROT	THIAMINE	RIBOFLAV	NIACIN	ASCORBIC	REFCO
6190	SECHIUM EDULE	F	337	0	9.8	2.2	83.7	6.5	4.3	130.4	326.1	6.5	.	.	108.7	0.33	0.43	4.35	217.4	3
6191	SECHIUM EDULE	L	583	0	38.8	3.9	45.6	11.7	11.7	563.1	1048.5	24.3	.	.	11941.8	0.78	1.75	10.68	155.3	3
6192	SECHIUM EDULE	R	376	0	9.5	1.0	84.8	1.9	4.8	33.3	161.9	3.8	.	.	0.0	0.24	0.14	4.29	90.5	3
6193	SECHIUM EDULE	F	341	0	7.3	1.2	86.6	8.5	4.9	158.5	317.1	6.1	61.0	1243.9	146.3	0.37	0.37	4.88	231.7	4
6194	SECURIGERA SECURIDACA	S	.	0	44.8	6.6	.	.	.	.	.	.	.	.	.	.	.	.	.	21
6195	SECURIGERA SECURIDACA	S	.	0	31.4	3.4	.	.	.	.	.	.	.	.	.	.	.	.	.	21
6196	SEDUM MAKINOI	L	361	0	13.1	16.4	55.7	29.5	14.8	3819.7	885.2	41.0	.	.	573.8	2.46	1.97	.	491.8	2
6197	SEDUM SP.	W	.	0	8.5	6.1	64.7	11.7	20.7	7510.0	400.0	.	.	2640.0	.	.	.	.	.	6
6198	SEHIMA NERVOSUM	W	.	0	7.0	1.3	82.9	38.7	8.8	.	.	.	.	.	.	.	.	.	.	18
6199	SELENIA GRANDIS	S	.	0	21.2	18.5	.	.	.	.	.	.	.	.	.	.	.	.	.	20
6200	SENECIO AMBROSIOIDES	S	.	0	19.1	26.0	.	.	.	.	.	.	.	.	.	.	.	.	.	21
6201	SENECIO HARTWEGII	S	.	0	35.6	24.3	.	.	.	.	.	.	.	.	.	.	.	.	.	20
6202	SENECIO HIERACIFOLIUS	S	.	0	21.2	36.6	.	.	.	.	.	.	.	.	.	.	.	.	.	20
6203	SENECIO ORIENTALIS	S	.	0	15.0	3.2	.	.	.	.	.	.	.	.	.	.	.	.	.	21
6204	SENECIO SCHAFFNERI	S	.	0	21.9	19.1	.	.	.	.	.	.	.	.	.	.	.	.	.	20
6205	SENECIO SERRA	W	.	0	20.9	3.6	61.9	13.3	13.6	3320.0	470.0	.	.	.	.	.	.	.	.	6
6206	SENECIO SP.	S	.	0	22.5	22.0	.	.	.	.	.	.	.	.	.	.	.	.	.	20
6207	SENECIO TRIANGULARIS		.	0	29.1	3.0	54.8	16.5	13.1	.	.	.	.	.	.	.	.	.	.	6
6208	SERICOCARPUS ASTEROIDES	S	.	0	30.0	17.2	.	.	.	.	.	.	.	.	.	.	.	.	.	20
6209	SERRURIA ACROCARPA	S	.	0	1.8	1.2	.	.	.	.	.	.	.	.	.	.	.	.	.	21
6210	SERRURIA ELONGATA	S	.	0	4.6	2.2	.	.	.	.	.	.	.	.	.	.	.	.	.	21
6211	SESAMUM ALATUM	S	450	0	11.7	19.7	59.9	33.1	8.7	469.1	240.0	.	.	.	.	.	.	.	.	1
6212	SESAMUM INDICUM	S	.	0	19.7	51.5	23.0	1.7	5.8	810.0	500.0	.	.	.	.	.	.	.	.	18
6213	SESAMUM INDICUM	S	.	0	14.1	32.8	.	.	.	.	.	.	.	.	.	.	.	.	.	21
6214	SESAMUM INDICUM	S	589	0	18.9	51.1	23.5	4.7	6.5	860.8	632.9	8.5	.	.	31.6	0.72	0.20	3.59	0.0	1
6215	SESAMUM INDICUM	S	618	0	27.7	57.6	9.9	3.2	4.8	137.6	815.1	8.2	42.0	427.5	42.0	0.76	0.08	4.92	.	16
6216	SESAMUM INDICUM	S	618	0	18.3	56.1	21.0	5.7	4.7	796.2	651.8	12.7	52.0	539.3	15.9	0.76	0.18	5.41	0.0	2
6217	SESAMUM INDICUM	L	310	0	23.4	4.8	59.3	16.6	12.4	531.0	1400.0	.	.	.	.	.	2.07	.	.	2
6218	SESAMUM INDICUM	S	609	0	18.4	54.4	22.0	5.5	5.2	1263.8	646.5	10.8	.	.	10.4	1.02	0.26	5.21	0.0	3
6219	SESAMUM INDICUM	S	595	0	19.7	51.9	22.8	6.7	5.6	1226.2	651.2	11.1	63.4	766.4	19.0	1.04	0.25	5.71	0.0	4
6220	SESAMUM SP.	L	293	0	25.0	2.1	60.1	10.6	12.8	1883.0	303.2	.	.	.	.	.	.	.	.	1
6221	SESBANIA BISPINOSA	S	.	0	32.7	2.9	59.4	10.7	5.0	370.0	590.0	.	.	.	.	.	.	.	.	18
6222	SESBANIA CANNABINA	S	.	0	29.2	6.4	.	.	.	.	.	.	.	.	.	.	.	.	.	21
6223	SESBANIA DRUMMONDII	S	.	0	31.6	5.0	.	.	3.0	.	.	.	.	.	.	.	.	.	.	19
6224	SESBANIA DRUMMONDII	S	.	0	31.2	2.9	.	.	.	.	.	.	.	.	.	.	.	.	.	20
6225	SESBANIA EXALTATA	S	.	0	40.3	4.4	.	.	3.6	.	.	.	.	.	.	.	.	.	.	19
6226	SESBANIA EXALTATA	S	.	0	37.4	4.0	.	.	3.6	.	.	.	.	.	.	.	.	.	.	19
6227	SESBANIA EXALTATA	S	.	0	38.2	4.6	.	.	3.2	.	.	.	.	.	.	.	.	.	.	19
6228	SESBANIA EXALTATA	S	.	0	37.9	4.6	.	.	.	.	.	.	.	.	.	.	.	.	.	21
6229	SESBANIA EXALTATA	S	.	0	33.5	4.6	.	.	.	.	.	.	.	.	.	.	.	.	.	21
6230	SESBANIA GRANDIFLORA	I	.	0	15.1	1.8	76.6	13.2	6.5	.	.	.	.	.	.	.	.	.	.	18
6231	SESBANIA GRANDIFLORA	L	.	0	33.4	2.6	52.4	5.7	11.6	2330.0	340.0	.	.	.	.	.	.	.	.	18
6232	SESBANIA GRANDIFLORA	F	.	0	1.6	4.8	87.1	32.7	6.5	.	.	.	.	.	.	.	.	.	.	18
6233	SESBANIA GRANDIFLORA	L	321	0	36.3	7.5	47.1	9.2	9.2	1683.3	258.3	.	20.8	2004.2	25958.3	1.00	1.04	9.17	241.7	2
6234	SESBANIA GRANDIFLORA	F	345	0	14.5	3.6	77.3	10.9	4.5	145.5	290.9	5.5	290.9	1400.0	636.4	0.91	0.73	14.55	472.7	2

TABLE 3 - DATA CONVERTED TO A ZERO-MOISTURE BASIS.

OBS	PLNA	PART	CALORIES	WATER	PROTEIN	FAT	TOTCARB	FIBER	ASH	CALCIUM	PHOSPHO	IRON	SODIUM	POTASS	CAROT	THIAMINE	RIBOFLAV	NIACIN	ASCORBIC	REFCO
6235	SESBANIA JAVANICA	L	321	0	25.7	2.9	65.0	27.9	6.4	364.3	400.0	58.6	.	.	14321.4	1.86	.	20.00	171.4	2
6236	SESBANIA MACROCARPA	S	.	0	38.2	6.4	.	.	3.6	.	.	.	.	.	.	.	.	.	.	19
6237	SESBANIA PUNICEA	S	.	0	33.0	3.8	.	.	.	.	.	.	.	.	.	.	.	.	.	21
6238	SESBANIA PUNICEA	S	.	0	34.4	3.9	.	.	.	.	.	.	.	.	.	.	.	.	.	21
6239	SESBANIA ROXBURGII	L	321	0	25.7	2.9	51.0	27.8	6.4	364.0	400.0	58.5	.	.	14316.0	1.85	.	20.00	171.0	10
6240	SESBANIA SESBAN	L	.	0	26.5	0.9	62.6	12.2	10.0	2780.0	430.0	.	.	.	.	.	.	.	.	18
6241	SESBANIA SESBAN	F	.	0	7.8	0.5	85.5	10.0	6.2	1370.0	370.0	.	.	.	.	.	.	.	.	18
6242	SESBANIA SESBAN	S	.	0	21.2	2.5	69.0	8.5	7.2	440.0	680.0	.	.	.	.	.	.	.	.	18
6243	SESBANIA SESBAN	S	.	0	27.4	3.6	.	.	.	.	.	.	.	.	.	.	.	.	.	21
6244	SESBANIA SESBAN	S	.	0	31.3	8.6	.	.	.	.	.	.	.	.	.	.	.	.	.	21
6245	SESBANIA SONORAE	S	.	0	36.1	6.0	.	.	4.4	.	.	.	.	.	.	.	.	.	.	19
6246	SESBANIA SP.	S	.	0	37.9	6.4	.	.	3.7	.	.	.	.	.	.	.	.	.	.	19
6247	SESBANIA SP.	S	.	0	37.5	5.8	.	.	.	.	.	.	.	.	.	.	.	.	.	19
6248	SESBANIA SP.	S	393	0	35.2	5.0	56.4	12.6	3.4	299.6	554.0	.	.	.	.	.	.	.	.	1
6249	SESBANIA VESICARIA	S	.	0	39.7	5.4	.	.	4.2	.	.	.	.	.	.	.	.	.	.	19
6250	SESBANIA VESICARIA	S	.	0	40.6	6.7	.	.	3.5	.	.	.	.	.	.	.	.	.	.	19
6251	SESUVIUM PORTULACASTRUM	L	227	0	19.1	3.6	42.7	7.3	34.5	618.2	218.2	57.3	.	2927.3	16818.2	0.36	1.27	4.55	245.5	2
6252	SETARIA CHEVALIERI	W	.	0	8.3	1.8	77.8	33.5	12.1	.	.	.	.	.	.	.	.	.	.	18
6253	SETARIA CHEVALIERI	H	.	0	12.3	2.4	72.9	26.4	12.4	.	.	.	.	.	.	.	.	.	.	18
6254	SETARIA ITALICA	W	.	0	9.0	2.2	78.7	33.7	10.1	.	.	.	.	.	.	.	.	.	.	18
6255	SETARIA ITALICA	H	.	0	7.6	1.7	81.0	45.1	9.7	.	.	.	.	.	.	.	.	.	.	18
6256	SETARIA ITALICA	H	.	0	9.5	3.0	78.3	29.4	8.2	.	.	.	.	.	.	.	.	.	.	6
6257	SETARIA ITALICA	S	384	0	10.7	3.3	84.2	1.4	1.8	37.2	275.1	6.2	7.9	280.7	0.0	0.48	0.14	2.48	0.0	2
6258	SETARIA LUTESCENS	S	.	0	17.8	5.4	.	.	10.8	.	.	.	.	.	.	.	.	.	.	19
6259	SETARIA LUTESCENS	W	.	0	5.8	1.7	82.0	34.6	9.5	260.0	80.0	.	.	.	.	.	.	.	.	6
6260	SETARIA MACROSTACHYA	I	.	0	8.4	1.1	.	.	11.6	.	.	.	.	.	.	.	.	.	.	19
6261	SETARIA MACROSTACHYA	W	.	0	17.5	2.3	68.4	27.6	11.8	330.0	200.0	.	.	.	.	.	.	.	.	6
6262	SETARIA SP.	S	.	0	16.1	5.6	.	.	6.3	.	.	.	.	.	.	.	.	.	.	19
6263	SETARIA SPHACELATA	W	.	0	4.1	1.3	86.8	36.3	7.8	.	.	.	.	.	.	.	.	.	.	18
6264	SETARIA SPHACELATA	W	.	0	18.4	4.7	63.8	24.0	13.1	.	.	.	.	.	.	.	.	.	.	18
6265	SETARIA SPLENDIDA	W	.	0	11.4	3.0	73.5	27.8	12.1	.	.	.	.	.	.	.	.	.	.	18
6266	SETARIA VERTICILLATA	S	.	0	26.9	9.5	.	.	.	.	.	.	.	.	.	.	.	.	.	20
6267	SETARIA VERTICILLATA	S	.	0	18.2	6.8	.	.	.	.	.	.	.	.	.	.	.	.	.	21
6268	SETARIA VIRIDIS	S	.	0	17.0	8.0	.	.	6.8	.	.	.	.	.	.	.	.	.	.	19
6269	SETARIA VIRIDIS	S	.	0	15.4	3.6	.	.	.	.	.	.	.	.	.	.	.	.	.	21
6270	SETARIA VIRIDIS	S	.	0	16.9	8.6	.	.	.	.	.	.	.	.	.	.	.	.	.	21
6271	SETARIA VIRIDIS	W	.	0	9.6	2.5	77.1	31.3	10.8	300.0	210.0	.	.	.	.	.	.	.	.	6
6272	SEYMERIA GLANDULOSA	S	.	0	31.2	32.8	.	.	.	.	.	.	.	.	.	.	.	.	.	20
6273	SHOREA ROBUSTA	L	.	0	10.1	3.2	82.8	27.4	3.9	770.0	120.0	.	.	.	.	.	.	.	.	18
6274	SHOREA ROBUSTA	S	.	0	8.0	16.7	72.9	5.2	2.4	.	.	.	.	.	.	.	.	.	.	18
6275	SHOREA STENOPTERA	S	.	0	17.5	3.1	77.2	6.8	2.2	.	.	.	.	.	.	.	.	.	.	18
6276	SIBARA VIRGINICA	S	.	0	20.0	31.1	.	.	.	.	.	.	.	.	.	.	.	.	.	20
6277	SICANA ODORIFERA	F	341	0	6.3	0.8	88.1	6.3	4.8	246.0	206.3	4.8	.	.	793.6	0.40	0.32	6.35	127.0	3
6278	SICYOS ANGULATA	S	.	0	31.2	27.8	.	.	.	.	.	.	.	.	.	.	.	.	.	20
6279	SIDA ACUTA	S	.	0	23.2	16.8	.	.	.	.	.	.	.	.	.	.	.	.	.	21

TABLE 3 - DATA CONVERTED TO A ZERO-MOISTURE BASIS.

OBS	PLNA	PART	CALORIES	WATER	PROTEIN	FAT	TOTCARB	FIBER	ASH	CALCIUM	PHOSPHO	IRON	SODIUM	POTASS	CAROT	THIAMINE	RIBOFLAV	NIACIN	ASCORBIC	REFCO
6280	SIDA CORYMBOSA	S	.	0	19.4	12.7	.	.	.	.	.	.	.	.	.	.	.	.	.	20
6281	SIDA RHOMBIFOLIA	L	318	0	37.4	7.1	47.5	16.7	8.1	2353.5	292.9	25.3	.	.	30555.6	1.11	2.37	10.61	454.5	3
6282	SIDA SCHIMPERIANA	SH	.	0	13.7	1.3	73.7	29.9	11.3	1000.0	240.0	.	.	.	.	.	.	.	.	18
6283	SIDA SP.	W	.	0	19.1	1.8	69.2	17.5	9.9	2630.0	260.0	.	.	1800.0	.	.	.	.	.	6
6284	SIDA SPINOSA	S	.	0	23.1	10.6	.	.	5.0	.	.	.	.	.	.	.	.	.	.	19
6285	SIDERITIS MONTANA	S	.	0	19.2	37.6	.	.	.	.	.	.	.	.	.	.	.	.	.	20
6286	SIDEROXYLON ANGUSTIFOLIUM	S	.	0	9.4	9.7	.	.	.	.	.	.	.	.	.	.	.	.	.	20
6287	SIDEROXYLON CAPIRI	L	373	0	6.2	4.8	86.7	4.8	2.3	209.0	67.8	.	.	.	621.5	0.23	0.17	1.41	420.9	3
6288	SILENE ANTIRRHINA	S	.	0	17.0	5.0	.	.	4.3	.	.	.	.	.	.	.	.	.	.	19
6289	SILENE COMPACTA	S	.	0	16.0	5.6	.	.	.	.	.	.	.	.	.	.	.	.	.	21
6290	SILENE CONOIDEA	S	.	0	14.3	4.8	.	.	.	.	.	.	.	.	.	.	.	.	.	21
6291	SILENE DENSIFLORA	S	.	0	24.2	7.0	.	.	.	.	.	.	.	.	.	.	.	.	.	20
6292	SILENE DICHOTOMA	S	.	0	14.9	4.8	.	.	.	.	.	.	.	.	.	.	.	.	.	21
6293	SILENE GALLICA	W	.	0	10.7	.	.	21.1	12.8	1530.0	480.0	.	.	3600.0	.	.	.	.	.	6
6294	SILENE NOCTIFLORA	S	.	0	21.6	4.3	.	.	2.9	.	.	.	.	.	.	.	.	.	.	19
6295	SILENE SUPINA	S	.	0	19.2	4.9	.	.	.	.	.	.	.	.	.	.	.	.	.	20
6296	SILENE VULGARIS	S	.	0	18.1	2.7	.	.	.	.	.	.	.	.	.	.	.	.	.	20
6297	SILYBUM MARIANUM	S	.	0	18.6	29.2	.	.	.	.	.	.	.	.	.	.	.	.	.	21
6298	SILYBUM MARIANUM	S	.	0	18.9	30.4	.	.	.	.	.	.	.	.	.	.	.	.	.	21
6299	SILYBUM MARIANUM	S	.	0	19.1	29.0	.	.	.	.	.	.	.	.	.	.	.	.	.	21
6300	SILYBUM MARIANUM	S	.	0	19.1	27.1	.	.	.	.	.	.	.	.	.	.	.	.	.	21
6301	SIMAROUBA GLAUCA	S	.	0	24.0	45.5	.	.	.	.	.	.	.	.	.	.	.	.	.	21
6302	SIMMONDSIA CHINENSIS	S	.	0	15.6	45.6	.	.	.	.	.	.	.	.	.	.	.	.	.	20
6303	SIMSIA AMPLEXICAULIS	S	.	0	26.2	29.7	.	.	.	.	.	.	.	.	.	.	.	.	.	20
6304	SIMSIA AMPLEXICAULIS	S	.	0	24.4	28.8	.	.	.	.	.	.	.	.	.	.	.	.	.	20
6305	SIMSIA CALVA	S	.	0	20.0	21.3	.	.	.	.	.	.	.	.	.	.	.	.	.	19
6306	SIMSIA CALVA	S	.	0	17.5	14.1	.	.	.	.	.	.	.	.	.	.	.	.	.	20
6307	SIMSIA EXAURICULATA(IMMATURE)	W	.	0	20.4	1.3	62.6	18.9	15.7	4420.0	230.0	.	.	3530.0	.	.	.	.	.	6
6308	SIMSIA EXAURICULATA(MATURE)	W	.	0	16.4	1.4	68.8	21.6	13.4	3740.0	210.0	.	.	3070.0	.	.	.	.	.	6
6309	SIMSIA FOETIDA	S	.	0	21.9	24.5	.	.	.	.	.	.	.	.	.	.	.	.	.	21
6310	SIMSIA LAGASCAEFORMIS	S	.	0	23.1	27.8	.	.	.	.	.	.	.	.	.	.	.	.	.	20
6311	SINAPIS ALBA	S	.	0	31.7	28.0	.	.	.	.	.	.	.	.	.	.	.	.	.	21
6312	SINAPIS ALBA	S	504	0	26.7	30.9	37.5	7.1	4.8	559.6	903.3	10.7	5.4	732.5	39.7	0.58	0.41	8.47	.	16
6313	SINAPIS ALBA	S	494	0	27.8	38.2	29.7	5.5	4.3	431.6	645.3	22.0	.	.	663.2	0.42	0.33	7.68	0.0	2
6314	SINAPIS ARVENSIS	S	.	0	29.5	31.0	.	.	.	.	.	.	.	.	.	.	.	.	.	20
6315	SINAPIS ARVENSIS	S	.	0	30.0	28.4	.	.	.	.	.	.	.	.	.	.	.	.	.	21
6316	SINAPIS ARVENSIS	S	.	0	26.8	28.1	.	.	.	.	.	.	.	.	.	.	.	.	.	21
6317	SISYMBRIUM ALTISSIMUM	S	.	0	31.3	35.2	.	.	.	.	.	.	.	.	.	.	.	.	.	21
6318	SISYMBRIUM ALTISSIMUM	S	.	0	30.6	35.9	.	.	.	.	.	.	.	.	.	.	.	.	.	21
6319	SISYMBRIUM ERYSIMOIDES	S	.	0	34.9	30.0	.	.	.	.	.	.	.	.	.	.	.	.	.	21
6320	SISYMBRIUM GARIEPINUM	S	.	0	30.4	32.3	.	.	.	.	.	.	.	.	.	.	.	.	.	21
6321	SISYMBRIUM IRIO	S	.	0	33.8	31.0	.	.	.	.	.	.	.	.	.	.	.	.	.	20
6322	SISYMBRIUM IRIO	S	.	0	40.0	13.0	.	.	.	.	.	.	.	.	.	.	.	.	.	20
6323	SISYMBRIUM IRIO	S	.	0	36.2	30.2	.	.	.	.	.	.	.	.	.	.	.	.	.	21
6324	SISYMBRIUM IRIO	S	.	0	36.9	20.7	.	.	.	.	.	.	.	.	.	.	.	.	.	21

TABLE 3 - DATA CONVERTED TO A ZERO-MOISTURE BASIS.

OBS	PLNA	PART	CALORIES	WATER	PROTEIN	FAT	TOTLCARB	FIBER
6325	SISYMBRIUM SP.	S	.	0	31.2	36.2	.	.
6326	SISYNDITE SPARTEA	S	.	0	27.0	15.7	.	.
6327	SISYRINCHIUM SP.	S	.	0	13.8	8.5	.	.
6328	SITANION HYSTRIX	W	.	0	6.9	2.5	76.0	36.5
6329	SKIMMIA JAPONICA	S	.	0	26.2	43.1	.	.
6330	SMILACINA RACEMOSA	F	364	0	10.0	.	86.0	3.0
6331	SMILACINA RACEMOSA	S	.	0	11.2	4.8	.	.
6332	SMILACINA RACEMOSA	S	.	0	6.2	4.8	.	.
6333	SMILAX EXCELSA	S	.	0	8.4	10.3	.	.
6334	SMILAX ROTUNDIFOLIA	S	.	0	10.6	10.5	.	.
6335	SMILEX ROTUNDIFOLIA	S	.	0	11.2	10.5	.	.
6336	SMYRNIUM PERFOLIATUM	S	.	0	24.6	16.2	.	.
6337	SOLANUM AETHIOPICUM	L	285	0	26.8	1.7	57.5	13.4
6338	SOLANUM AETHIOPICUM	F	340	0	16.0	1.1	76.6	21.3
6339	SOLANUM CAROLINIENSE	S	.	0	16.9	21.4	.	.
6340	SOLANUM ELAEAGNIFOLIUM	S	.	0	16.2	14.2	.	.
6341	SOLANUM ELAEAGNIFOLIUM	W	.	0	21.8	3.2	66.0	25.6
6342	SOLANUM INCANUM	L	.	0	30.3	2.0	54.9	21.0
6343	SOLANUM INCANUM	S	.	0	17.2	11.3	.	.
6344	SOLANUM INCANUM	F	363	0	15.0	8.8	71.3	23.8
6345	SOLANUM MACROCARPON	F	364	0	12.7	9.1	72.7	13.6
6346	SOLANUM MACROCARPON	L	292	0	31.9	6.9	44.4	11.1
6347	SOLANUM MELONGENA	L	279	0	33.8	2.9	48.5	.
6348	SOLANUM MELONGENA	F	337	0	10.5	2.1	81.1	13.7

OBS	ASH	CALCIUM	PHOSPHO	IRON	SODIUM	POTASS	CAROT	THIAMINE	RIBOFLAV	NIACIN	ASCORBIC	REFCO
6325	3.9	.	.	.	.	.	.	.	.	.	.	19
6326	.	.	.	.	.	.	.	.	.	.	.	21
6327	.	.	.	.	.	.	.	.	.	.	.	20
6328	14.6	1010.0	100.0	.	.	1940.0	.	.	.	.	.	6
6329	1.8	.	.	.	.	.	.	.	.	.	.	19
6330	2.0	161.0	.	3.0	.	.	.	.	.	.	508.0	15
6331	2.0	.	.	.	.	.	.	.	.	.	.	19
6332	.	.	.	.	.	.	.	.	.	.	.	19
6333	.	.	.	.	.	.	.	.	.	.	.	21
6334	.	.	.	.	.	.	.	.	.	.	.	19
6335	.	.	.	.	.	.	.	.	.	.	.	19
6336	.	.	.	.	.	.	.	.	.	.	.	21
6337	14.0	2921.8	525.1	33.5	.	.	35754.2	1.28	2.46	10.06	374.3	1
6338	6.4	297.9	500.0	16.0	.	.	3723.4	0.74	0.64	8.51	85.1	1
6339	.	.	.	.	.	.	.	.	.	.	.	19
6340	.	.	.	.	.	.	.	.	.	.	.	20
6341	9.0	1070.0	230.0	.	.	2140.0	.	.	.	.	.	6
6342	12.8	1560.0	320.0	.	.	.	.	.	.	.	.	18
6343	.	.	.	.	.	.	.	.	.	.	.	21
6344	5.0	1287.5	.	.	.	.	.	.	.	.	2200.0	1
6345	5.5	118.2	.	.	.	.	.	.	.	.	.	1
6346	16.7	2715.3	340.3	.	.	.	.	.	.	.	.	1
6347	14.7	1867.6	279.4	114.0	.	.	.	.	.	.	580.9	1
6348	6.3	147.4	273.7	13.7	.	.	357.9	0.53	0.53	5.26	94.7	1

TABLE 3 - DATA CONVERTED TO A ZERO-MOISTURE BASIS.

OBS	PLNA	PART	CALORIES	WATER	PROTEIN	FAT	TOTLCARB	FIBER
6349	SOLANUM MELONGENA	F	325	0	20.0	3.8	70.0	12.5
6350	SOLANUM MELONGENA	F	329	0	12.2	3.7	76.8	14.6
6351	SOLANUM MELONGENA	F	329	0	15.8	2.6	73.7	11.8
6352	SOLANUM MURICATUM	F	400	0	5.0	12.5	78.8	5.0
6353	SOLANUM NIGRUM	S	.	0	17.5	27.2	.	.
6354	SOLANUM NIGRUM	S	.	0	20.5	30.9	.	.
6355	SOLANUM NIGRUM	L	293	0	30.7	2.7	56.0	7.3
6356	SOLANUM NIGRUM	L	300	0	33.3	5.3	49.3	9.3
6357	SOLANUM NODIFLORUM	S	.	0	14.4	25.3	.	.
6358	SOLANUM NODIFLORUM	L	320	0	26.2	8.2	52.5	18.0
6359	SOLANUM QUITOENSE	F	346	0	9.2	1.5	83.8	30.8
6360	SOLANUM QUITOENSE	P	341	0	8.5	1.2	82.9	4.9
6361	SOLANUM SURATTENSE	S	.	0	12.5	21.3	.	.
6362	SOLANUM TORREYI	W	.	0	16.8	2.7	71.4	19.9
6363	SOLANUM TORVUM	F	322	0	16.4	2.7	73.3	41.8
6364	SOLANUM TRIQUETRUM	W	.	0	22.4	4.3	63.7	23.9
6365	SOLANUM TUBEROSUM	L	.	0	17.5	6.4	65.3	39.1
6366	SOLANUM TUBEROSUM	H	.	0	10.9	4.3	71.3	27.0
6367	SOLANUM TUBEROSUM	R	.	0	12.1	0.9	82.0	3.9
6368	SOLANUM TUBEROSUM	W	.	0	12.4	2.8	66.7	25.9
6369	SOLANUM TUBEROSUM	R	368	0	7.6	0.4	84.8	2.7
6370	SOLANUM TUBEROSUM	R	378	0	9.2	0.5	86.2	1.8
6371	SOLANUM TUBEROSUM	R	357	0	12.7	0.9	82.4	2.7
6372	SOLANUM TUBEROSUM	R	376	0	10.4	0.5	84.7	2.5

OBS	ASH	CALCIUM	PHOSPHO	IRON	SODIUM	POTASS	CAROT	THIAMINE	RIBOFLAV	NIACIN	ASCORBIC	REFCO
6349	6.3	275.0	462.5	11.3	87.5	2762.5	625.0	1.00	0.88	8.75	75.0	2
6350	7.3	280.5	378.0	9.8	.	.	0.0	0.49	0.49	9.76	61.0	3
6351	7.9	157.9	342.1	9.2	26.3	2815.8	78.9	0.66	0.66	7.89	65.8	4
6352	3.8	225.0	175.0	10.0	.	.	1500.0	1.00	0.50	6.25	400.0	3
6353	.	.	.	.	.	.	.	.	.	.	.	19
6354	.	.	.	.	.	.	.	.	.	.	.	21
6355	10.7	1440.0	586.7	28.0	.	.	11000.0	0.80	1.60	8.67	200.0	2
6356	12.0	1326.7	400.0	66.0	.	.	3066.7	1.20	2.33	6.67	406.7	3
6357	.	.	.	.	.	.	.	.	.	.	.	19
6358	13.1	1639.3	442.6	2.5	.	.	30000.0	.	.	.	196.7	1
6359	5.4	84.6	315.4	4.6	.	.	1076.9	0.54	0.31	11.54	369.2	3
6360	7.3	97.6	170.7	4.9	.	.	1219.5	0.73	0.49	18.29	792.7	3
6361	.	.	.	.	.	.	.	.	.	.	.	21
6362	9.1	1970.0	160.0	.	.	2740.0	.	.	.	.	.	6
6363	7.5	712.3	479.5	31.5	.	.	2671.2	0.82	0.62	17.81	27.4	2
6364	9.6	2770.0	260.0	.	.	2760.0	.	.	.	.	.	6
6365	10.8	1390.0	140.0	.	.	.	.	.	.	.	.	18
6366	13.5	.	.	.	.	.	.	.	.	.	.	18
6367	5.0	100.0	200.0	.	.	.	.	.	.	.	.	18
6368	18.1	2120.0	200.0	.	.	3950.0	.	.	.	.	.	6
6369	7.2	58.3	228.7	4.9	.	.	112.1	0.31	0.13	5.83	94.2	1
6370	4.1	41.5	239.6	3.7	32.3	1824.9	0.0	0.46	0.18	7.37	82.9	2
6371	4.1	45.2	226.2	4.5	.	.	0.0	0.50	0.18	6.79	90.5	3
6372	4.5	34.7	262.4	3.0	14.9	2014.9	0.0	0.50	0.20	7.43	99.0	4

TABLE 3 - DATA CONVERTED TO A ZERO-MOISTURE BASIS.

OBS	PLNA	PART	CALORIES	WATER	PROTEIN	FAT	TOTLCARB	FIBER
6373	SOLANUM VERBASCIFOLIUM	W	.	0	26.5	3.9	56.7	17.7
6374	SOLENOSTEMON ROTUNDIFOLIUS	R	385	0	5.3	0.8	89.8	4.5
6375	SOLIDAGO CANADENSIS VAR. GILVOCANESCENS	S	.	0	25.1	30.4	.	.
6376	SOLIDAGO GRAMINEA	S	.	0	.	29.2	.	.
6377	SOLIDAGO SEROTINA	S	.	0	14.4	18.2	.	.
6378	SOLIDAGO SP.	W	.	0	8.7	14.2	70.0	12.6
6379	SONCHUS CRASSIFOLIUS	S	.	0	20.1	20.5	.	.
6380	SONCHUS OLERACEUS	S	.	0	18.8	31.5	.	.
6381	SONCHUS OLERACEUS	L	265	0	27.9	4.4	44.1	5.9
6382	SONCHUS OLERACEUS	L	250	0	30.0	.	48.8	.
6383	SONCHUS SP.	L	265	0	27.9	4.4	44.1	5.9
6384	SOPHIA OCHROLEUCA	S	.	0	26.9	35.9	.	.
6385	SOPHORA ALOPECUROIDES	S	.	0	27.4	5.4	.	.
6386	SOPHORA GRIFFITHII	S	.	0	18.5	16.1	.	.
6387	SOPHORA JAPONICA	S	.	0	21.4	6.9	.	.
6388	SOPHORA JAPONICA	S	.	0	23.0	7.3	.	.
6389	SOPHORA JAPONICA	S	434	0	17.2	12.1	66.8	9.9
6390	SOPHORA RETICULATA	S	.	0	26.5	7.4	.	.
6391	SOPHORA SECUNDIFLORA	S	.	0	21.2	25.8	.	.
6392	SOPHORA SECUNDIFLORA	S	.	0	22.4	25.2	.	.
6393	SOPHORA SECUNDIFLORA	S	.	0	15.6	13.4	.	.
6394	SOPHORA SECUNDIFLORA	L	.	0	19.1	2.6	73.1	31.1
6395	SOPHORA SECUNDIFLORA	F	.	0	18.2	0.6	78.0	15.1
6396	SORBUS AMERICANA	S	.	0	31.9	23.9	.	.

OBS	ASH	CALCIUM	PHOSPHO	IRON	SODIUM	POTASS	CAROT	THIAMINE	RIBOFLAV	NIACIN	ASCORBIC	REFCO
6373	12.9	.	.	.	.	.	.	.	.	.	.	18
6374	4.1	69.7	.	24.6	.	.	.	0.20	0.08	4.10	4.1	1
6375	5.7	.	.	.	.	.	.	.	.	.	.	19
6376	.	.	.	.	.	.	.	.	.	.	.	21
6377	5.6	.	.	.	.	.	.	.	.	.	.	19
6378	7.1	1130.0	520.0	.	.	.	.	.	.	.	.	6
6379	.	.	.	.	.	.	.	.	.	.	.	21
6380	.	.	.	.	.	.	.	.	.	.	.	20
6381	23.5	1926.5	544.1	45.6	.	.	20955.9	1.91	11.76	.	0.0	1
6382	21.3	1162.5	437.5	.	.	.	54625.0	0.88	1.50	5.00	62.5	3
6383	23.5	1926.5	544.1	45.6	.	.	21029.4	.	1.91	11.76	0.0	1
6384	.	.	.	.	.	.	.	.	.	.	.	19
6385	.	.	.	.	.	.	.	.	.	.	.	21
6386	.	.	.	.	.	.	.	.	.	.	.	21
6387	3.7	.	.	.	.	.	.	.	.	.	.	19
6388	.	.	.	.	.	.	.	.	.	.	.	21
6389	3.9	264.9	272.3	.	.	1066	.	.	.	.	.	2
6390	.	.	.	.	.	.	.	.	.	.	.	21
6391	.	.	.	.	.	.	.	.	.	.	.	19
6392	.	.	.	.	.	.	.	.	.	.	.	20
6393	.	.	.	.	.	.	.	.	.	.	.	20
6394	5.2	1990.0	140.0	.	.	890	.	.	.	.	.	6
6395	3.2	500.0	140.0	.	.	1150	.	.	.	.	.	6
6396	.	.	.	.	.	.	.	.	.	.	.	19

TABLE 3 - DATA CONVERTED TO A ZERO-MOISTURE BASIS.

OBS	PLNA	PART	CALORIES	WATER	PROTEIN	FAT	TOTCARB	FIBER	ASH	CALCIUM	PHOSPHO	IRON	SODIUM	POTASS	CAROT	THIAMINE	RIBOFLAV	NIACIN	ASCORBIC	REFCO
6397	SORBUS AMERICANA	S	.	0	5.0	1.4	.	.	.	.	.	.	.	.	.	.	.	.	.	19
6398	SORBUS DOMESTICA	S	.	0	28.4	22.9	.	.	.	.	.	.	.	.	.	.	.	.	.	21
6399	SORBUS TORMINALIS	S	.	0	26.8	1.3	.	.	.	.	.	.	.	.	.	.	.	.	.	21
6400	SORGHASTRUM NUTANS	S	.	0	26.9	5.6	.	.	2.2	.	.	.	.	.	.	.	.	.	.	19
6401	SORGHASTRUM NUTANS	S	.	0	7.5	3.0	.	.	7.2	.	.	.	.	.	.	.	.	.	.	19
6402	SORGHASTRUM NUTANS	W	.	0	8.1	2.3	81.0	32.9	8.6	530.0	150.0	.	.	.	.	.	.	.	.	6
6403	SORGHUM ALMUM	W	.	0	11.3	3.8	75.5	29.4	9.4	.	.	.	.	.	.	.	.	.	.	18
6404	SORGHUM ALMUM	W	.	0	7.8	2.6	82.3	36.6	7.3	.	.	.	.	.	.	.	.	.	.	18
6405	SORGHUM ALMUM	S	.	0	8.1	4.0	.	.	5.9	.	.	.	.	.	.	.	.	.	.	19
6406	SORGHUM ALMUM	S	.	0	23.1	3.0	.	.	.	.	.	.	.	.	.	.	.	.	.	19
6407	SORGHUM ALMUM	S	.	0	18.8	5.3	.	.	.	.	.	.	.	.	.	.	.	.	.	20
6408	SORGHUM BICOLOR	W	.	0	8.2	1.5	84.5	22.6	5.8	.	.	.	.	.	.	.	.	.	.	18
6409	SORGHUM BICOLOR	H	.	0	4.2	0.8	87.1	32.1	7.9	.	.	.	.	.	.	.	.	.	.	18
6410	SORGHUM BICOLOR	S	.	0	11.2	3.0	.	.	.	.	.	.	.	.	.	.	.	.	.	20
6411	SORGHUM BICOLOR	S	.	0	16.9	5.0	.	.	.	.	.	.	.	.	.	.	.	.	.	20
6412	SORGHUM BICOLOR	W	.	0	9.4	2.5	79.5	29.3	8.6	380.0	180.0	.	.	1580.0	.	.	.	.	.	6
6413	SORGHUM BICOLOR	H	.	0	7.9	2.5	81.1	26.1	8.5	400.0	170.0	5.0	.	1410.0	.	.	.	.	.	6
6414	SORGHUM BICOLOR	S	385	0	9.6	3.5	85.0	3.4	1.9	25.3	490.6	4.3	.	.	.	0.39	0.13	3.74	.	1
6415	SORGHUM BICOLOR	S	413	0	8.9	1.3	89.1	0.2	0.8	12.5	269.1	3.4	.	.	0.00	0.24	0.09	1.25	.	1
6416	SORGHUM BICOLOR	S	381	0	16.7	3.3	77.7	2.4	2.3	29.8	459.2	13.8	.	.	.	.	.	.	.	1
6417	SORGHUM BICOLOR	S	391	0	18.3	5.5	73.9	1.3	2.3	19.8	629.3	8.1	.	.	.	.	.	.	.	1
6418	SORGHUM BICOLOR	S	379	0	12.5	2.3	83.2	1.5	2.0	19.9	430.5	3.8	.	.	.	.	.	.	.	1
6419	SORGHUM BICOLOR	S	391	0	14.2	5.2	78.4	1.6	2.2	20.0	432.9	5.0	.	.	.	.	.	.	.	1
6420	SORGHUM BICOLOR	S	390	0	12.8	4.7	80.4	1.4	2.1	35.3	432.7	4.6	.	.	.	0.58	0.08	2.87	.	1
6421	SORGHUM BICOLOR	S	391	0	15.4	5.0	77.6	1.0	2.0	19.7	459.0	6.6	.	.	.	.	.	.	.	1
6422	SORGHUM BICOLOR	S	382	0	16.6	3.3	78.1	2.8	2.0	19.8	529.7	15.0	.	.	.	.	.	.	.	1
6423	SORGHUM BICOLOR	S	390	0	15.1	4.8	78.0	2.2	2.1	19.8	449.3	5.2	.	.	.	.	.	.	.	1
6424	SORGHUM BICOLOR	S	379	0	17.2	2.6	77.9	2.1	2.2	19.9	449.2	4.7	.	.	.	.	.	.	.	1
6425	SORGHUM BICOLOR	S	389	0	11.4	4.2	82.6	2.5	1.7	25.0	275.0	4.3	9.1	50.0	0.00	0.38	0.20	4.43	0	2
6426	SORGHUM BICOLOR	S	380	0	9.8	3.6	84.8	2.6	1.9	21.1	332.2	4.1	.	.	22.22	0.46	0.13	3.56	0	3
6427	SORGHUM HALAPENSE	S	.	0	17.7	3.8	.	.	.	.	.	.	.	.	.	.	.	.	.	21
6428	SORGHUM HALEPENSE	W	.	0	16.1	2.8	70.0	29.6	11.1	.	.	.	.	.	.	.	.	.	.	18
6429	SORGHUM HALEPENSE	H	.	0	6.6	1.9	85.6	34.6	5.9	.	.	.	.	.	.	.	.	.	.	18
6430	SORGHUM HALEPENSE	S	.	0	9.4	4.6	.	.	6.3	.	.	.	.	.	.	.	.	.	.	19
6431	SORGHUM HALEPENSE	H	.	0	7.7	2.1	81.4	33.3	8.8	810.0	310.0	.	.	1350.0	.	.	.	.	.	6
6432	SORGHUM HALEPENSE	W	.	0	11.4	2.7	75.4	30.9	10.5	940.0	240.0	.	.	3120.0	.	.	.	.	.	6
6433	SORGHUM LANCEOLATUM	W	.	0	17.3	1.4	70.4	37.2	10.9	.	.	.	.	.	.	.	.	.	.	18
6434	SORGHUM SP.	S	.	0	21.8	4.5	.	.	2.5	.	.	.	.	.	.	.	.	.	.	19
6435	SORGHUM SP.	S	.	0	8.2	2.2	.	.	10.4	.	.	.	.	.	.	.	.	.	.	19
6436	SORGHUM SP.	S	373	0	12.4	3.7	82.0	1.9	1.9	31.5	322.5	4.9	.	393.3	0.00	0.43	0.17	4.38	0	4
6437	SORGHUM SUDANENSE	W	.	0	15.4	3.7	71.0	23.4	9.9	.	.	.	.	.	.	.	.	.	.	18
6438	SORGHUM SUDANENSE	H	.	0	7.3	2.0	81.8	35.7	8.9	.	.	.	.	.	.	.	.	.	.	18
6439	SORGHUM SUDANENSE	S	.	0	11.9	4.0	.	.	4.6	.	.	.	.	.	.	.	.	.	.	19
6440	SORGHUM SUDANENSE	W	.	0	11.2	2.5	76.5	32.2	9.8	430.0	410.0	21.0	.	2140.0	.	.	.	.	.	6
6441	SORGHUM VERTICILLIFLORUM	W	.	0	8.3	1.5	79.2	35.0	11.0	.	.	.	.	.	.	.	.	.	.	18

TABLE 3 - DATA CONVERTED TO A ZERO-MOISTURE BASIS.

OBS	PLNA	PART	CALORIES	WATER	PROTEIN	FAT	TOTLCARB	FIBER
6442	SORGHUM VULGARE	S	.	0	7.5	1.6	.	.
6443	SPARGANIUM RAMOSUM	S	.	0	6.5	0.8	.	.
6444	SPARGANIUM RAMOSUM	S	.	0	5.9	0.9	.	.
6445	SPARTINA PATENS(IMMATURE)	W	.	0	11.1	2.4	75.1	31.9
6446	SPARTINA PATENS(MATURE)	W	.	0	5.0	2.4	84.1	33.3
6447	SPARTINA PECTINATA	W	.	0	7.1	2.1	85.5	35.6
6448	SPARTINA SPARTINAE(IMMATURE)	W	.	0	11.1	2.5	76.4	32.1
6449	SPARTINA SPARTINAE(MATURE)	W	.	0	5.0	2.4	84.1	33.3
6450	SPARTIUM JUNCEUM	S	.	0	37.5	10.3	.	.
6451	SPARTIUM JUNCEUM	S	.	0	34.5	12.4	.	.
6452	SPARTIUM JUNCEUM	S	.	0	37.8	10.0	.	.
6453	SPARTIUM JUNCEUM	S	.	0	31.0	10.3	.	.
6454	SPARTIUM JUNCEUM	Y	.	0	11.0	4.6	80.0	39.7
6455	SPATHIPHYLLUM PHRYNIIFOLIUM	I	313	0	35.8	6.8	47.7	8.5
6456	SPERGULA ARVENSIS	S	.	0	16.1	14.7	.	.
6457	SPERGULA ARVENSIS	S	.	0	14.1	14.0	.	.
6458	SPERGULA ARVENSIS	L	.	0	19.5	4.0	61.7	20.3
6459	SPERMOLEPIS INERMIS	S	.	0	17.5	5.1	.	.
6460	SPHAERALCEA ANGUSTIFOLIA	W	.	0	20.4	2.4	65.5	23.2
6461	SPHAERALCEA ANGUSTIFOLIA VAR. CUSPIDATA	S	.	0	25.7	17.0	.	.
6462	SPHAERALCEA COCCINEA	S	.	0	12.9	7.4	.	.
6463	SPHENOPHOLIS OBTUSATA	W	.	0	6.7	5.1	78.3	34.5
6464	SPHENOSTYLIS STENOCARPA	R	366	0	10.8	0.6	86.4	1.1
6465	SPHENOSTYLIS STENOCARPA	S	386	0	21.2	1.2	74.4	5.7

OBS	ASH	CALCIUM	PHOSPHO	IRON	SODIUM	POTASS	CAROT	THIAMINE	RIBOFLAV	NIACIN	ASCORBIC	REFCO
6442	7.9	.	.	.	.	.	.	.	.	.	.	19
6443	.	.	.	.	.	.	.	.	.	.	.	21
6444	.	.	.	.	.	.	.	.	.	.	.	21
6445	11.4	.	.	.	.	.	.	.	.	.	.	6
6446	8.5	370.0	90.0	.	.	.	.	.	.	.	.	6
6447	5.3	250.0	110.0	.	.	.	.	.	.	.	.	6
6448	10.0	560.0	160.0	.	.	.	.	.	.	.	.	6
6449	8.5	320.0	100.0	.	.	.	.	.	.	.	.	6
6450	.	.	.	.	.	.	.	.	.	.	.	20
6451	.	.	.	.	.	.	.	.	.	.	.	21
6452	.	.	.	.	.	.	.	.	.	.	.	21
6453	.	.	.	.	.	.	.	.	.	.	.	21
6454	4.4	650.0	260.0	.	.	.	.	.	.	.	.	5
6455	9.7	2289.8	772.7	8.0	.	.	113.6	0.23	2.78	10.23	73.9	3
6456	.	.	.	.	.	.	.	.	.	.	.	21
6457	.	.	.	.	.	.	.	.	.	.	.	21
6458	14.8	1960.0	1810.0	18.4	200.0	.	.	.	.	.	.	5
6459	.	.	.	.	.	.	.	.	.	.	.	20
6460	11.7	3340.0	310.0	.	.	2690.0	.	.	.	.	.	6
6461	.	.	.	.	.	.	.	.	.	.	.	21
6462	3.8	.	.	.	.	.	.	.	.	.	.	19
6463	9.9	.	.	.	.	.	.	.	.	.	.	6
6464	2.3	28.4	227.3	.	.	.	.	.	.	.	.	1
6465	3.2	60.7	439.3	.	.	.	.	0.76	.	.	.	1

TABLE 3 - DATA CONVERTED TO A ZERO-MOISTURE BASIS.

OBS	PLANA	PART	CALORIES	WATER	PROTEIN	FAT	TOTLCARB	FIBER	ASH	CALCIUM	PHOSPHO	IRON	SODIUM	POTASS	CAROT	THIAMINE	RIBOFLAV	NIACIN	ASCORBIC	REFCO
6466	SPHINTOSPERMUM CONSTRICTUM	S	.	0	30.0	2.9	.	.	.	.	.	.	.	.	.	.	.	.	.	19
6467	SPILANTHES ACMELLA	L	291	0	17.3	2.7	64.5	.	15.5	1472.7	372.7	36.4	.	.	21363.6	0.27	.	.	181.8	2
6468	SPILANTHES OLERACEA	L	271	0	20.8	.	62.5	32.3	16.7	.	.	.	.	.	16250.0	.	.	.	.	1
6469	SPINACIA OLERACEA	S	.	0	15.2	4.2	.	.	2.6	.	.	.	.	.	.	.	.	.	.	19
6470	SPINACIA OLERACEA	S	.	0	14.8	3.9	.	.	6.9	.	.	.	.	.	.	.	.	.	.	19
6471	SPINACIA OLERACEA	S	.	0	24.1	7.0	.	.	4.0	.	.	.	.	.	.	.	.	.	.	19
6472	SPINACIA OLERACEA	S	.	0	10.9	1.9	.	.	4.7	.	.	.	.	.	.	.	.	.	.	19
6473	SPINACIA OLERACEA	S	.	0	20.0	6.9	.	.	.	.	.	.	.	.	.	.	.	.	.	20
6474	SPINACIA OLERACEA	S	.	0	25.6	7.5	.	.	.	.	.	.	.	.	.	.	.	.	.	20
6475	SPINACIA OLERACEA	S	.	0	15.0	3.1	.	.	.	.	.	.	.	.	.	.	.	.	.	20
6476	SPINACIA OLERACEA	L	.	0	31.0	4.8	33.1	8.5	31.1	.	.	.	.	.	.	.	.	.	.	6
6477	SPINACIA OLERACEA	L	277	0	22.3	2.1	56.4	8.5	19.1	648.9	489.4	18.1	.	.	.	0.32	2.87	.	489.4	1
6478	SPINACIA OLERACEA	L	271	0	34.3	5.7	40.0	10.0	20.0	885.7	557.1	55.7	542.9	6585.7	52000.0	0.86	3.14	10.00	800.0	2
6479	SPINACIA OLERACEA	L	294	0	27.5	6.9	48.0	6.9	17.6	588.2	294.1	31.4	.	.	22941.2	0.59	1.67	5.88	451.0	3
6480	SPINACIA OLERACEA	L	280	0	34.4	3.2	46.2	6.5	16.1	1000.0	548.4	33.3	763.4	5053.8	52258.1	1.08	2.15	6.45	548.4	4
6481	SPINACIA OLERACEA	L	.	0	25.3	8.9	44.3	7.6	21.5	924.1	265.8	138.0	734.2	2607.6	70632.9	0.38	0.89	6.33	354.4	5
6482	SPODIOPOGON BYRONIS	L	.	0	7.6	1.2	76.7	32.5	14.4	.	.	.	.	.	.	.	.	.	.	.
6483	SPONDIAS DULCIS	F	351	0	1.5	0.8	94.7	8.4	3.1	427.5	511.5	2.3	7.6	725.2	1564.9	0.38	0.15	10.69	274.8	2
6484	SPONDIAS MOMBIN	F	350	0	6.7	1.7	88.3	4.2	3.3	250.0	325.0	8.3	.	.	11666.7	0.33	0.25	11.67	133.3	1
6485	SPONDIAS MOMBIN	F	405	0	4.6	12.1	79.8	5.8	3.5	150.3	179.2	12.7	.	.	809.3	0.46	0.35	2.89	161.8	3
6486	SPONDIAS PINNATA	P	.	0	7.2	30.9	56.7	10.3	5.2	371.1	113.4	40.2	.	.	2783.5	0.21	0.21	3.09	216.5	5
6487	SPONDIAS PURPUREA	SH	307	0	37.7	7.0	44.7	14.0	10.5	105.3	640.4	24.6	.	.	.	0.70	1.05	13.16	201.8	2
6488	SPONDIAS PURPUREA	L	311	0	18.4	1.6	70.5	.	9.5	2842.1	431.6	32.6	.	.	9157.9	0.32	.	.	152.6	2
6489	SPONDIAS PURPUREA	F	355	0	4.1	1.2	92.1	2.1	2.5	57.9	144.6	3.7	8.3	1115.7	929.8	0.37	0.21	1.65	202.5	2
6490	SPONDIAS PURPUREA	F	349	0	3.8	0.4	92.4	1.7	3.4	92.4	168.1	2.5	.	.	252.1	0.29	0.13	4.20	189.1	3
6491	SPOROBOLUS AIROIDES	W	.	0	6.8	1.8	82.5	34.2	8.9	380.0	130.0	13.0	.	380.0	.	.	.	.	.	6
6492	SPOROBOLUS ASPER	W	.	0	4.2	1.9	90.7	37.9	3.2	290.0	110.0	.	.	.	.	.	.	.	.	6
6493	SPOROBOLUS CRYPTANDRUS	S	.	0	14.4	2.8	.	.	4.9	.	.	.	.	.	.	.	.	.	.	19
6494	SPOROBOLUS CRYPTANDRUS	W	.	0	7.4	1.5	82.5	36.6	8.6	380.0	110.0	.	.	1530.0	.	.	.	.	.	6
6495	SPOROBOLUS DIANDER	L	.	0	6.4	3.6	72.8	23.0	16.1	590.0	440.0	.	.	.	.	.	.	.	.	5
6496	SPOROBOLUS FLEXUOSUS	W	.	0	6.4	1.5	86.4	37.1	5.7	190.0	90.0	.	.	.	.	.	.	.	.	6
6497	SPOROBOLUS GLAUCIFOLIUS	L	.	0	4.2	1.0	86.2	26.5	8.7	340.0	420.0	.	.	.	.	.	.	.	.	5
6498	SPOROBOLUS HELVOLUS	W	.	0	16.8	4.0	69.1	28.0	10.1	.	.	.	.	.	.	.	.	.	.	18
6499	SPOROBOLUS HETEROLEPIS	W	.	0	5.7	1.4	87.8	33.5	5.1	.	.	.	.	.	.	.	.	.	.	6
6500	SPOROBOLUS LONGIBRACHIATUS	W	.	0	16.2	2.2	69.4	22.4	12.2	.	.	.	.	.	.	.	.	.	.	18
6501	SPOROBOLUS MARGINATUS	W	.	0	23.3	3.1	64.3	25.7	9.3	.	.	.	.	.	.	.	.	.	.	18
6502	SPOROBOLUS MARGINATUS	N	.	0	6.1	1.0	84.0	34.4	8.8	230.0	140.0	.	.	.	.	.	.	.	.	5
6503	SPOROBOLUS MARGINATUS	L	.	0	9.1	1.4	80.1	31.5	9.4	290.0	230.0	.	.	.	.	.	.	.	.	5
6504	SPOROBOLUS POIRETII	W	.	0	6.7	1.8	83.2	32.7	8.3	420.0	110.0	.	.	.	.	.	.	.	.	6
6505	SPOROBOLUS PYRAMIDALIS	W	.	0	7.7	1.8	84.3	35.9	6.2	.	.	.	.	.	.	.	.	.	.	18
6506	SPOROBOLUS PYRAMIDALIS	H	.	0	10.0	1.9	79.1	34.4	9.0	.	.	.	.	.	.	.	.	.	.	18
6507	SPOROBOLUS TREMULUS	L	.	0	6.1	0.9	81.6	24.6	11.4	.	.	.	.	.	.	.	.	.	.	5
6508	SPOROBOLUS WALLICHII	N	.	0	7.1	1.9	73.6	43.8	17.4	520.0	400.0	.	.	.	.	.	.	.	.	5
6509	STACHYS CRETICA	S	.	0	16.0	29.2	.	.	.	.	.	.	.	.	.	.	.	.	.	21
6510	STACHYS OLYMPICA	S	.	0	20.4	31.9	.	.	7.0	.	.	.	.	.	.	.	.	.	.	19

TABLE 3 - DATA CONVERTED TO A ZERO-MOISTURE BASIS.

OBS	PLNA	PART	CALORIES	WATER	PROTEIN	FAT	TOTLCARB	FIBER
6511	STACHYS VITICINA	S	.	0	13.8	30.9	.	.
6512	STAPHYLEA PINNATA	S	.	0	35.0	49.2	.	.
6513	STAPHYLEA PINNATA	S	.	0	33.6	42.9	.	.
6514	STELLARIA MEDIA	S	.	0	17.8	5.9	.	.
6515	STELLARIA MEDIA	L	.	0	.	.	.	.
6516	STELLARIA MEDIA	L	.	0	14.5	2.4	63.9	20.5
6517	STENOTAPHRUM SECUNDATUM	H	.	0	6.7	2.7	86.9	32.5
6518	STENOTAPHRUM SECUNDATUM	W	.	0	9.2	1.8	80.3	29.2
6519	STERCULIA APETELA	S	.	0	19.4	25.9	.	.
6520	STERCULIA FOETIDA	S	.	0	20.5	51.5	.	.
6521	STERCULIA FOETIDA	S	.	0	17.7	55.1	.	.
6522	STERCULIA MONOSPERMA	F	352	0	10.8	2.2	82.7	2.9
6523	STEREOSPERMUM PERSONATUM	L	.	0	9.8	2.1	.	25.5
6524	STEVIA RHOMBIFOLIA VAR. UNIARISTATA	S	.	0	28.8	16.4	.	.
6525	STEVIA SERRATA	S	.	0	24.4	18.3	.	.
6526	STEVIA SERRATA	S	.	0	26.2	16.4	.	.
6527	STEVIA SERRATA VAR. ARGUTA	S	.	0	28.6	16.2	.	.
6528	STEVIA SERRATA VAR. ARGUTA	S	.	0	29.4	20.7	.	.
6529	STEVIA SERRATA VAR. LINOIDES	S	.	0	33.8	25.1	.	.
6530	STEVIA SERRATA VAR. LINOIDES	S	.	0	22.5	11.9	.	.
6531	STEVIA SERRATA VAR. LINOIDES	S	.	0	21.0	14.3	.	.
6532	STEVIA SP.	S	.	0	27.5	24.0	.	.
6533	STEVIA VISCIDA	S	.	0	29.4	14.9	.	.
6534	STILLINGIA SP.	W	.	0	21.7	3.2	63.4	13.9

OBS	ASH	CALCIUM	PHOSPHO	IRON	SODIUM	POTASS	CAROT	THIAMINE	RIBOFLAV	NIACIN	ASCORBIC	REFCO
6511	.	.	.	.	.	.	.	.	.	.	.	21
6512	.	.	.	.	.	.	.	.	.	.	.	20
6513	.	.	.	.	.	.	.	.	.	.	.	21
6514	.	.	.	.	.	.	.	.	.	.	.	21
6515	.	.	.	.	.	.	32500.0	0.02	0.14	0.51	375.0	5
6516	19.3	.	.	.	.	.	7228.9	.	.	.	.	5
6517	3.7	.	.	.	.	.	.	.	.	.	.	18
6518	8.7	510.0	210.0	20.0	.	630.0	.	.	.	.	.	6
6519	.	.	.	.	.	.	.	.	.	.	.	20
6520	3.8	.	.	.	.	.	.	.	.	.	.	19
6521	3.7	51.2	644.4	2.6	.	802.8	.	0.09	0.12	1.71	7.8	5
6522	4.3	14.5	433.7	2.7	33.7	1855.4	2096.4	0.22	0.43	2.89	74.7	2
6523	12.5	1870.0	415.0	.	225.0	855.0	.	.	.	.	.	5
6524	.	.	.	.	.	.	.	.	.	.	.	20
6525	.	.	.	.	.	.	.	.	.	.	.	20
6526	.	.	.	.	.	.	.	.	.	.	.	20
6527	.	.	.	.	.	.	.	.	.	.	.	20
6528	.	.	.	.	.	.	.	.	.	.	.	20
6529	.	.	.	.	.	.	.	.	.	.	.	20
6530	.	.	.	.	.	.	.	.	.	.	.	20
6531	.	.	.	.	.	.	.	.	.	.	.	21
6532	.	.	.	.	.	.	.	.	.	.	.	20
6533	.	.	.	.	.	.	.	.	.	.	.	20
6534	11.7	310.0	240.0	.	.	2590.0	.	.	.	.	.	6

TABLE 3 - DATA CONVERTED TO A ZERO-MOISTURE BASIS.

OBS	PLNA	PART	CALORIES	WATER	PROTEIN	FAT	TOTLCARB	FIBER
6535	STIPA CAPILLATA	L	.	0	4.1	1.6	82.6	31.7
6536	STIPA CLARAZII	S	.	0	16.2	3.2	.	.
6537	STIPA CLARAZII	S	.	0	25.9	4.4	.	.
6538	STIPA COMATA	W	.	0	7.6	2.7	79.3	34.5
6539	STIPA LAGASCAE	S	.	0	21.5	2.6	.	.
6540	STIPA LETTERMANII	H	.	0	10.9	3.2	78.8	31.3
6541	STIPA LEUCOTRICHA	W	.	0	9.2	2.7	76.2	31.2
6542	STIPA MEGAPOTAMIA VAR. LATIFLORA	S	.	0	14.9	2.6	.	.
6543	STIPA ORIENTALIS	L	.	0	5.7	4.6	78.2	21.7
6544	STIPA SIBIRICA	L	.	0	7.7	2.3	81.7	31.1
6545	STIPA SPARTEA	W	.	0	5.6	0.9	85.0	41.8
6546	STIPA VIRIDULA	S	.	0	30.6	3.1	.	.
6547	STIPA VIRIDULA	W	.	0	7.6	1.1	84.4	40.7
6548	STIZOLOBIUM DEERINGIANUM	S	.	0	29.2	5.6	.	.
6549	STIZOLOBIUM SP.	S	.	0	28.1	4.2	.	.
6550	STIZOLOBIUM SP.	S	.	0	28.8	3.2	.	.
6551	STIZOLOBIUM SP.	H	.	0	17.7	3.3	71.0	29.6
6552	STIZOLOBIUM SP.	W	.	0	19.6	3.9	65.3	28.5
6553	STOCKSIA BRAHUICA	S	.	0	11.9	19.4	.	.
6554	STRELITZIA REGINAE	R	321	0	4.9	1.2	82.7	19.8
6555	STREPTOPUS AMPLEXIFOLIUS	S	.	0	17.5	12.8	.	.
6556	STROPHANTHUS KOMBE	S	.	0	27.0	35.4	.	.
6557	STROPHOSTYLES HELVEOLA	S	.	0	30.0	0.5	.	.
6558	STRYCHNOS COCCULOIDES	F	.	0	.	.	.	.

OBS	ASH	CALCIUM	PHOSPHO	IRON	SODIUM	POTASS	CAROT	THIAMINE	RIBOFLAV	NIACIN	ASCORBIC	REFCO
6535	10.8	680.0	170.0	.	.	.	.	.	.	.	.	5
6536	.	.	.	.	.	.	.	.	.	.	.	21
6537	.	.	.	.	.	.	.	.	.	.	.	21
6538	10.4	300.0	90.0	.	.	.	.	.	.	.	.	6
6539	.	.	.	.	.	.	.	.	.	.	.	21
6540	7.1	84.5	210.0	.	.	.	.	.	.	.	.	6
6541	11.9	560.0	150.0	.	.	1350	.	.	.	.	.	6
6542	.	.	.	.	.	.	.	.	.	.	.	21
6543	9.3	1630.0	530.0	.	.	.	.	.	.	.	.	5
6544	7.2	710.0	340.0	.	.	.	.	.	.	.	.	5
6545	8.5	.	.	.	.	.	.	.	.	.	.	6
6546	1.3	.	.	.	.	.	.	.	.	.	.	19
6547	6.9	.	.	.	.	.	.	.	.	.	.	6
6548	3.4	.	.	.	.	.	.	.	.	.	.	19
6549	3.8	.	.	.	.	.	.	.	.	.	.	19
6550	3.5	.	.	.	.	.	.	.	.	.	.	19
6551	8.0	.	.	.	.	.	.	.	.	.	.	6
6552	11.2	.	.	.	.	.	.	.	.	.	.	6
6553	.	.	.	.	.	.	.	.	.	.	.	21
6554	11.1	555.6	271.6	12.3	.	.	.	0.12	0.37	1.23	.	2
6555	.	.	.	.	.	.	.	.	.	.	.	20
6556	.	.	.	.	.	.	.	.	.	.	.	21
6557	3.7	.	.	.	.	.	.	.	.	.	.	19
6558	.	.	.	.	.	.	.	.	.	.	4.9	1

TABLE 3 - DATA CONVERTED TO A ZERO-MOISTURE BASIS.

OBS	PLNA	PART	CALORIES	WATER	PROTEIN	FAT	TOTLCARB	FIBER
6559	STRYCHNOS INNOCUA	F	.	0	.	.	.	.
6560	STRYCHNOS POTATORUM	S	.	0	6.5	2.8	.	.
6561	STRYCHNOS SPINOSA	S	.	0	10.6	4.0	.	.
6562	STRYCHNOS SPINOSA	F	355	0	7.9	3.0	84.2	3.0
6563	STYLOSANTHES BOJERI	W	.	0	10.1	1.2	80.1	38.4
6564	STYLOSANTHES GRACILIS	S	.	0	40.0	17.1	.	.
6565	STYLOSANTHES GUIANENSIS	W	.	0	18.1	2.1	71.5	26.8
6566	STYLOSANTHES GUIANENSIS	H	.	0	16.7	1.7	71.6	31.7
6567	STYLOSANTHES GUIANENSIS	L	.	0	16.7	1.7	71.7	31.7
6568	STYLOSANTHES HUMILIS	W	.	0	22.0	3.0	67.0	21.0
6569	STYLOSANTHES MUCRONATA	L	.	0	14.6	2.0	73.2	30.6
6570	STYRAX AMERICANA	S	.	0	18.8	42.7	.	.
6571	STYRAX GRANDIFOLIA	S	.	0	12.4	36.2	.	.
6572	STYRAX JAPONICA	S	.	0	13.8	50.0	.	.
6573	STYRAX OBESSIA	S	.	0	17.5	46.6	.	.
6574	STYRAX OFFICINALIS	S	.	0	16.5	50.8	.	.
6575	SUAEDA MARITIMA	L	.	0	18.7	3.7	42.1	8.4
6576	SUAEDA PROSTRATA	S	.	0	25.1	13.9	.	.
6577	SUAEDA SETIGERA	S	.	0	21.4	15.9	.	.
6578	SUAEDA SUFFRUTESCENS	S	.	0	.	36.4	.	.
6579	SUTHERLANDIA FRUTESCENS	S	.	0	29.2	2.8	.	.
6580	SYMPHORICARPOS ORBICULATUS	S	.	0	10.0	15.6	.	.
6581	SYMPHORICARPOS ROTUNDIFOLIUS	C	.	0	4.2	2.0	90.0	41.1
6582	SYMPHORICARPOS ROTUNDIFOLIUS(IMMATURE)	L	.	0	16.0	3.9	73.2	11.7

OBS	ASH	CALCIUM	PHOSPHO	IRON	SODIUM	POTASS	CAROT	THIAMINE	RIBOFLAV	NIACIN	ASCORBIC	REFCO
6559	.	.	.	.	.	.	.	.	.	.	47.6	1
6560	.	.	.	.	.	.	.	.	.	.	.	21
6561	1.4	.	.	.	.	.	.	.	.	.	.	19
6562	4.9	137.9	206.9	3.4	.	.	.	0.54	0.84	9.36	88.7	1
6563	8.6	1230.0	430.0	.	.	.	.	.	.	.	.	18
6564	4.1	.	.	.	.	.	.	.	.	.	.	19
6565	8.3	.	.	.	.	.	.	.	.	.	.	18
6566	10.0	1550.0	560.0	.	.	.	.	.	.	.	.	18
6567	10.0	2166.7	1291.7	.	.	.	.	.	.	.	.	5
6568	8.0	800.0	220.0	.	.	.	.	.	.	.	.	18
6569	10.1	2430.0	120.0	.	.	.	.	.	.	.	.	5
6570	2.8	.	.	.	.	.	.	.	.	.	.	19
6571	.	.	.	.	.	.	.	.	.	.	.	21
6572	.	.	.	.	.	.	.	.	.	.	.	21
6573	.	.	.	.	.	.	.	.	.	.	.	20
6574	.	.	.	.	.	.	.	.	.	.	.	21
6575	35.5	467.3	738.3	15.0	.	.	19626.2	0.09	0.47	1.87	121.5	5
6576	.	.	.	.	.	.	.	.	.	.	.	21
6577	.	.	.	.	.	.	.	.	.	.	.	21
6578	.	.	.	.	.	.	.	.	.	.	.	21
6579	.	.	.	.	.	.	.	.	.	.	.	21
6580	.	.	.	.	.	.	.	.	.	.	.	20
6581	3.8	.	.	.	.	.	.	.	.	.	.	6
6582	6.9	1220.0	380.0	.	.	.	.	.	.	.	.	6

TABLE 3 - DATA CONVERTED TO A ZERO-MOISTURE BASIS.

OBS	PLNA	PART	CALORIES	WATER	PROTEIN	FAT	TOTLCARB	FIBER
6583	SYMPHORICARPOS ROTUNDIFOLIUS(MATURE)	L	.	0	8.7	4.6	77.0	13.6
6584	SYMPHORICARPOS SP.	L	.	0	10.6	5.3	76.6	14.6
6585	SYMPHORICARPOS VACCINIOIDES	L	.	0	3.9	2.2	89.8	38.8
6586	SYMPHYTUM ASPERRINUM	W	.	0	19.8	2.5	59.2	13.6
6587	SYMPHYTUM PEREGRINUM	W	.	0	19.0	2.9	44.4	14.0
6588	SYMPHYTUM PEREGRINUM	W	.	0	20.1	3.3	55.7	11.3
6589	SYMPLOCOS PANICULATA	S	.	0	21.2	51.7	.	.
6590	SYMPLOCOS PANICULATA	S	.	0	21.0	52.0	.	.
6591	SYMPLOCOS RAMOSISSIMA	L	.	0	11.9	.	.	.
6592	SYNELCOSCIADIUM CARMELLI	S	.	0	15.8	8.7	.	.
6593	SYRINGA VULGARIS	S	.	0	20.0	18.8	.	.
6594	SYZYGIUM AQUEUM	F	354	0	4.6	1.5	90.8	6.9
6595	SYZYGIUM AROMATICUM	I	347	0	6.4	21.6	65.7	10.3
6596	SYZYGIUM AROMATICUM	I	.	0	7.0	11.9	74.2	12.7
6597	SYZYGIUM CUMINI	L	.	0	7.9	2.6	82.5	20.7
6598	SYZYGIUM CUMINI	F	347	0	4.0	0.6	91.3	1.7
6599	SYZYGIUM CUMINI	F	359	0	5.6	1.4	90.8	2.1
6600	SYZYGIUM CUMINI	F	.	0	4.3	1.8	91.4	5.5
6601	SYZYGIUM GUINEENSE	F	.	0	.	.	.	.
6602	SYZYGIUM JAMBOS	F	358	0	3.4	1.1	93.2	8.0
6603	SYZYGIUM MALACCENSE	F	349	0	5.8	1.2	88.4	11.6
6604	SYZYGIUM MALACCENSE	F	352	0	6.6	1.1	87.9	7.7
6605	SYZYGIUM MALACCENSE	F	.	0	1.5	6.2	90.8	33.8
6606	SYZYGIUM SAMARANGENSE	F	353	0	4.7	1.2	91.8	9.4

OBS	ASH	CALCIUM	PHOSPHO	IRON	SODIUM	POTASS	CAROT	THIAMINE	RIBOFLAV	NIACIN	ASCORBIC	REFCO
6583	9.7	2580.0	520.0	.	.	.	.	.	.	.	.	6
6584	7.5	1680.0	390.0	.	.	.	.	.	.	.	.	6
6585	4.1	770.0	140.0	.	.	.	.	.	.	.	.	6
6586	18.5	.	.	.	.	.	.	.	.	.	.	6
6587	33.7	.	.	.	.	.	.	.	.	.	.	18
6588	20.9	1460.0	410.0	.	.	.	.	.	.	.	.	18
6589	2.5	.	.	.	.	.	.	.	.	.	.	19
6590	.	.	.	.	.	.	.	.	.	.	.	5
6591	11.2	2950.0	.	.	.	.	.	.	.	.	.	5
6592	.	.	.	.	.	.	.	.	.	.	.	21
6593	.	.	.	.	.	.	.	.	.	.	.	20
6594	3.1	61.5	69.2	8.5	.	.	0.0	0.00	0.23	4.62	38.5	2
6595	6.3	693.9	112.8	9.3	261.0	1183.7	341.6	0.13	0.29	1.57	87.0	16
6596	7.0	989.3	133.7	6.6	.	.	338.2	0.11	0.17	2.02	.	5
6597	7.0	1380.0	120.0	.	.	.	.	.	.	.	.	18
6598	4.0	46.2	75.1	1.2	52.0	670.5	0.0	0.00	0.06	1.16	132.9	2
6599	2.1	56.3	112.7	11.3	.	.	0.0	0.07	0.07	2.11	42.3	3
6600	2.5	92.0	92.0	7.4	159.5	337.4	294.5	0.18	0.06	1.23	110.4	5
6601	.	.	.	.	.	.	.	.	.	.	0.0	1
6602	2.3	204.5	85.2	2.3	.	.	852.3	0.11	0.17	3.41	125.0	3
6603	4.7	209.3	139.5	4.7	23.3	1325.6	0.0	0.35	0.23	2.33	197.7	2
6604	4.4	65.9	175.8	4.4	.	.	0.0	0.33	0.33	3.30	142.9	3
6605	1.5	261.5	46.2	0.0	.	.	.	.	0.31	2.00	461.5	5
6606	2.4	200.0	82.4	3.5	23.5	1235.3	0.0	0.35	0.12	3.53	152.9	2

TABLE 3 - DATA CONVERTED TO A ZERO-MOISTURE BASIS.

OBS	PLNA	PART	CALORIES	WATER	PROTEIN	FAT	TOTCARB	FIBER	ASH	CALCIUM	PHOSPHO	IRON	SODIUM	POTASS	CAROT	THIAMINE	RIBOFLAV	NIACIN	ASCORBIC	REFCO
6607	TABEBUIA PALMERI	S	.	0	23.1	35.4	.	.	.	.	.	.	.	.	.	.	.	.	.	20
6608	TABEBUIA PALMERI	S	.	0	23.1	36.3	.	.	.	.	.	.	.	.	.	.	.	.	.	20
6609	TACCA LEONTOPETALODES	R	388	0	4.8	0.3	92.3	1.3	2.6	86.5	205.1	.	.	.	.	.	.	.	41.7	1
6610	TAGETES ERECTA	S	.	0	23.8	20.5	.	.	.	.	.	.	.	.	.	.	.	.	.	19
6611	TAGETES ERECTA	S	.	0	24.0	20.0	.	.	.	.	.	.	.	.	.	.	.	.	.	5
6612	TAGETES LUCIDA	S	.	0	18.8	9.4	.	.	.	.	.	.	.	.	.	.	.	.	.	20
6613	TAGETES LUCIDA	S	.	0	21.2	11.4	.	.	.	.	.	.	.	.	.	.	.	.	.	20
6614	TAGETES SUBULATA	S	.	0	33.8	18.8	.	.	.	.	.	.	.	.	.	.	.	.	.	20
6615	TALINUM TRIANGULARE	L	.	0	21.1	1.5	42.8	10.3	34.6	.	.	.	.	.	.	.	.	.	.	18
6616	TALINUM TRIANGULARE	L	272	0	26.1	4.3	47.8	10.9	21.7	1315.2	728.3	54.3	.	.	.	0.87	1.96	3.26	337.0	1
6617	TALINUM TRIANGULARE	L	303	0	25.0	5.3	53.9	7.9	15.8	1184.2	276.3	63.2	131.6	5157.9	39276.3	1.18	2.50	7.89	763.2	2
6618	TALINUM TRIANGULARE	L	268	0	27.8	6.2	42.3	8.2	23.7	1237.1	278.4	56.7	.	.	.	1.03	1.75	4.12	556.7	3
6619	TALINUM TRIANGULARE	L	.	0	21.6	8.0	48.9	.	21.6	1363.6	181.8	10.2	.	.	42272.7	1.48	2.73	5.23	659.1	5
6620	TAMARINDUS INDICA	L	.	0	15.8	9.6	66.7	23.7	7.9	2860.0	260.0	.	.	.	.	.	.	.	.	18
6621	TAMARINDUS INDICA	S	.	0	18.3	7.4	70.8	26.4	3.5	140.0	300.0	.	.	.	.	.	.	.	.	18
6622	TAMARINDUS INDICA	F	343	0	10.1	1.0	83.8	42.4	5.1	303.0	489.9	.	.	.	.	.	.	.	40.4	1
6623	TAMARINDUS INDICA	L	332	0	14.1	3.6	75.0	18.6	7.3	2300.0	100.0	.	.	.	.	.	.	.	.	1
6624	TAMARINDUS INDICA	L	342	0	22.4	4.4	70.6	5.7	2.6	105.3	228.1	8.8	35.1	1197.4	11008.8	0.44	0.48	6.58	26.3	2
6625	TAMARINDUS INDICA	I	375	0	12.5	9.0	75.0	6.0	3.5	265.0	220.0	7.0	25.0	1270.0	1025.0	0.40	0.60	6.00	60.0	2
6626	TAMARINDUS INDICA	P	349	0	3.8	0.3	92.5	3.1	3.4	132.1	140.3	2.1	4.9	929.9	16.3	0.36	0.13	1.79	4.9	2
6627	TAMARINDUS INDICA	G	346	0	11.7	0.5	83.9	3.9	3.9	282.9	141.5	3.4	14.6	1541.5	48.8	0.73	0.24	1.95	58.5	2
6628	TAMARINDUS INDICA	F	351	0	4.0	0.5	92.8	3.9	2.7	69.8	139.5	1.3	.	.	51.7	0.57	0.21	2.71	7.8	3
6629	TAMARINDUS INDICA	P	348	0	4.1	0.9	91.1	7.4	3.9	107.9	164.7	4.1	74.3	1138.5	26.2	0.50	0.20	1.75	2.9	4
6630	TAMARIX GALLICA	S	.	0	26.2	4.5	.	.	.	.	.	.	.	.	.	.	.	.	.	20
6631	TAMARIX GALLICA	S	.	0	36.9	17.4	.	.	.	.	.	.	.	.	.	.	.	.	.	20
6632	TAMUS COMMUNIS	S	.	0	12.1	11.5	.	.	.	.	.	.	.	.	.	.	.	.	.	21
6633	TANACETUM CAPITATUM	W	.	0	4.1	7.6	77.9	26.8	10.4	.	.	.	.	.	.	.	.	.	.	6
6634	TANACETUM MICROPHYLLUM	S	.	0	21.9	25.5	.	.	.	.	.	.	.	.	.	.	.	.	.	21
6635	TANACETUM VULGARE	S	.	0	28.9	26.5	.	.	.	.	.	.	.	.	.	.	.	.	.	21
6636	TARAXACUM OFFICINALE	S	.	0	30.0	26.7	.	.	.	.	.	.	.	.	.	.	.	.	.	20
6637	TARAXACUM OFFICINALE	W	.	0	23.0	4.8	59.3	11.1	12.9	.	440.0	62.0	.	.	.	.	.	.	.	6
6638	TARAXACUM OFFICINALE	L	313	0	18.8	4.9	63.9	11.1	12.5	1298.6	458.3	21.5	527.8	2756.9	58333.3	1.32	1.81	.	243.1	2
6639	TARAXACUM OFFICINALE	L	308	0	18.9	4.9	62.2	12.6	14.0	1049.0	.	21.7	.	.	34965.0	1.33	0.98	5.59	195.8	3
6640	TARAXACUM OFFICINALE	L	.	0	32.1	14.3	33.0	3.6	20.5	4223.2	526.8	29.5	.	.	35892.9	1.70	1.25	7.14	651.8	5
6641	TARCHONANTHUS CAMPHORATUS	SH	.	0	11.4	6.7	76.1	26.8	5.8	370.0	100.0	.	.	.	.	.	.	.	.	18
6642	TARCHONANTHUS MINOR	SH	.	0	12.1	11.6	71.6	25.6	6.7	1420.0	130.0	.	.	.	.	.	.	.	.	18
6643	TAXODIUM DISTICHUM	S	.	0	40.6	37.4	.	.	.	.	.	.	.	.	.	.	.	.	.	20
6644	TELANTHERA VERSICOLOR	L	324	0	10.7	1.8	79.6	6.2	8.0	1422.2	240.0	53.8	.	.	18271.1	0.04	0.67	4.44	124.4	2
6645	TELEPHIUM IMPERATI	S	.	0	14.9	3.7	.	.	.	.	.	.	.	.	.	.	.	.	.	21
6646	TELFAIRIA OCCIDENTALIS	L	.	0	21.2	12.9	52.0	12.8	13.9	.	.	.	.	.	.	.	.	.	.	18
6647	TELFAIRIA OCCIDENTALIS	L	346	0	21.3	13.2	51.5	12.5	14.0	.	.	.	.	.	.	.	.	.	.	1
6648	TELFAIRIA OCCIDENTALIS	S	579	0	21.9	48.0	25.1	2.3	5.1	89.6	609.8	.	.	.	.	.	.	.	.	1
6649	TELFAIRIA PEDATA	S	.	0	31.1	66.2	.	.	2.7	10.5	596.2	4.3	.	.	.	.	.	.	.	5
6650	TELOSMA MINOR	L	333	0	25.6	5.6	62.1	8.2	6.7	359.0	461.5	5.1	.	.	9692.3	0.51	0.62	7.69	230.8	2
6651	TEPHROSIA CANDIDA	S	.	0	40.8	10.4	.	.	.	.	.	.	.	.	.	.	.	.	.	21

TABLE 3 - DATA CONVERTED TO A ZERO-MOISTURE BASIS.

OBS	PLNA	PART	CALORIES	WATER	PROTEIN	FAT	TOTLCARB	FIBER	ASH
6652	TEPHROSIA GRANDIFLORA	S	.	0	38.2	13.4	.	.	.
6653	TEPHROSIA LEIOCARPA	S	.	0	43.8	11.4	.	.	.
6654	TEPHROSIA LONGIPES	S	.	0	33.9	14.1	.	.	.
6655	TEPHROSIA NOCTIFLORA	L	.	0	22.0	8.3	61.7	22.1	7.9
6656	TEPHROSIA NYIKENSIS	L	.	0	17.1	2.0	73.0	28.7	7.8
6657	TEPHROSIA PURPUREA	S	.	0	35.8	9.1	.	.	.
6658	TEPHROSIA PURPUREA	L	.	0	24.4	2.4	65.4	28.0	7.7
6659	TEPHROSIA PURPUREA VAR.PUMILA	L	.	0	14.7	2.7	74.8	27.9	7.9
6660	TEPHROSIA SP.	S	.	0	44.4	10.4	.	.	3.2
6661	TEPHROSIA SP.	W	.	0	8.6	2.3	84.0	42.6	5.1
6662	TEPHROSIA TENELLA	S	.	0	36.2	9.4	.	.	.
6663	TEPHROSIA VOGELII	S	.	0	43.0	10.9	.	.	.
6664	TERMINALIA BELLERICA	L	.	0	8.6	4.7	78.7	18.6	8.0
6665	TERMINALIA BELLERICA	S	.	0	40.9	43.8	.	.	.
6666	TERMINALIA BROWNEI	SH	.	0	10.3	2.9	81.2	16.7	5.6
6667	TERMINALIA BUCERAS	S	.	0	5.0	1.0	.	.	.
6668	TERMINALIA CATAPPA	S	634	0	20.3	59.6	15.9	1.9	4.3
6669	TERMINALIA CATAPPA	S	619	0	21.7	56.3	17.9	2.4	4.2
6670	TERMINALIA CATAPPA	S	624	0	23.7	56.1	17.7	9.5	2.5
6671	TERMINALIA CHEBULA	L	397	0	8.5	12.1	76.6	17.7	2.8
6672	TERMINALIA TOMENTOSA	L	.	0	8.9	4.9	75.9	21.8	10.3
6673	TERMINALIA TOMENTOSA	L	.	0	12.7	1.4	.	16.9	10.3
6674	TETRACARPIDIUM CONOPHORUM	S	456	0	31.2	19.9	44.5	3.0	4.4
6675	TETRACHNE DREGEI	S	.	0	30.6	4.2	.	.	2.9

OBS	CALCIUM	PHOSPHO	IRON	SODIUM	POTASS	CAROT	THIAMINE	RIBOFLAV	NIACIN	ASCORBIC	REFCO
6652	.	.	.	.	.	.	.	.	.	.	21
6653	.	.	.	.	.	.	.	.	.	.	20
6654	.	.	.	.	.	.	.	.	.	.	21
6655	1750.0	360.0	.	.	.	.	.	.	.	.	5
6656	1520.0	400.0	.	.	.	.	.	.	.	.	5
6657	.	.	.	.	.	.	.	.	.	.	21
6658	1650.0	520.0	.	.	.	.	.	.	.	.	5
6659	2020.0	260.0	.	.	.	.	.	.	.	.	5
6660	.	.	.	.	.	.	.	.	.	.	19
6661	2420.0	370.0	.	.	.	.	.	.	.	.	6
6662	.	.	.	.	.	.	.	.	.	.	19
6663	.	.	.	.	.	.	.	.	.	.	21
6664	2080.0	270.0	.	.	.	.	.	.	.	.	18
6665	.	.	.	.	.	.	.	.	.	.	21
6666	820.0	130.0	.	.	.	.	.	.	.	.	18
6667	.	.	.	.	.	.	.	.	.	.	20
6668	419.1	919.1	.	.	.	.	.	.	.	.	1
6669	33.3	821.9	9.6	72.9	816.7	.	0.33	0.08	0.63	0.0	2
6670	510.8	983.6	2.5	.	.	0.0	0.73	0.29	0.72	.	3
6671	127.7	127.7	0.0	.	.	2127.7	0.00	0.07	14.18	822.7	2
6672	3200.0	250.0	.	.	.	.	.	.	.	.	18
6673	3110.0	370.0	.	70.0	980.0	.	.	.	.	.	5
6674	.	.	.	.	.	.	.	.	.	.	1
6675	.	.	.	.	.	.	.	.	.	.	19

TABLE 3 - DATA CONVERTED TO A ZERO-MOISTURE BASIS.

OBS	PLNA	PART	CALORIES	WATER	PROTEIN	FAT	TOTLCARB	FIBER	ASH	CALCIUM	PHOSPHO	IRON	SODIUM	POTASS	CAROT	THIAMINE	RIBOFLAV	NIACIN	ASCORBIC	REFCO
6676	TETRACLEA COULTERI	S	.	0	16.9	8.4	.	.	.	.	.	.	.	.	.	.	.	.	.	19
6677	TETRADYMIA SPINOSA	W	.	0	8.7	6.3	79.6	36.8	5.4	940.0	250.0	.	.	.	.	.	.	.	.	6
6678	TETRAGONIA ARBUSCULA	S	.	0	12.2	6.0	.	.	3.2	.	.	.	.	.	.	.	.	.	.	19
6679	TETRAGONIA ARBUSCULA	S	.	0	13.5	2.6	.	.	7.8	.	.	.	.	.	.	.	.	.	.	19
6680	TETRAGONIA ARBUSCULA	S	.	0	12.8	6.1	.	.	4.0	.	.	.	.	.	.	.	.	.	.	19
6681	TETRAGONIA ARBUSCULA	S	.	0	15.1	1.5	.	.	9.2	.	.	.	.	.	.	.	.	.	.	19
6682	TETRAGONIA ARBUSCULA	S	.	0	13.1	5.8	.	.	3.5	.	.	.	.	.	.	.	.	.	.	19
6683	TETRAGONIA ARBUSCULA	S	.	0	15.0	3.2	.	.	7.8	.	.	.	.	.	.	.	.	.	.	19
6684	TETRAGONIA ARBUSCULA	S	.	0	13.1	5.3	.	.	3.4	.	.	.	.	.	.	.	.	.	.	19
6685	TETRAGONIA TETRAGONIOIDES	L	268	0	19.7	.	63.4	33.8	16.9	.	.	.	.	.	22816.9	.	.	.	.	1
6686	TETRAGONIA TETRAGONIOIDES	L	259	0	32.9	4.7	37.6	9.4	24.7	2094.1	717.6	44.7	1035.3	6529.4	41647.1	0.59	2.82	14.12	317.6	2
6687	TETRAGONIA TETRAGONIOIDES	L	278	0	30.4	5.1	44.3	8.9	20.3	822.8	430.4	43.0	.	.	27341.8	0.38	2.03	6.33	392.4	3
6688	TETRAGONIA TETRAGONIOIDES	L	257	0	29.7	4.1	41.9	9.5	24.3	783.8	621.6	35.1	2148.6	10743.2	34864.9	0.54	2.30	8.11	405.4	4
6689	TETRAGONIA TETRAGONIOIDES	L	.	0	36.0	4.4	.	8.8	19.3	3298.2	596.5	21.9	.	.	67017.5	0.26	1.67	13.07	315.8	5
6690	TETRAGONOTHECA TEXANA	S	.	0	13.8	8.2	.	.	.	.	.	.	.	.	.	.	.	.	.	19
6691	TETRAPOGON SPATHACEUS	W	.	0	12.6	1.7	77.0	34.6	8.7	.	.	.	.	.	.	.	.	.	.	18
6692	TEUCRIUM CRETICUM	S	.	0	36.8	52.1	.	.	.	.	.	.	.	.	.	.	.	.	.	21
6693	TEUCRIUM POLIUM	S	.	0	14.8	14.6	.	.	.	.	.	.	.	.	.	.	.	.	.	21
6694	THALIA GENICULATA	S	.	0	14.1	2.3	.	.	.	.	.	.	.	.	.	.	.	.	.	21
6695	THALICTRUM DIPTEROCARPUM	S	.	0	28.1	36.7	.	.	.	.	.	.	.	.	.	.	.	.	.	20
6696	THALICTRUM FENDLERI	W	.	0	8.9	3.1	.	.	9.6	.	.	.	.	.	.	.	.	.	.	6
6697	THALICTRUM FENDLERI	L	.	0	16.2	5.0	.	.	12.1	.	.	.	.	.	.	.	.	.	.	6
6698	THALICTRUM FENDLERI	Y	.	0	4.4	1.4	.	.	7.4	.	.	.	.	.	.	.	.	.	.	6
6699	THALICTRUM LUCIDUM	S	.	0	32.5	37.6	.	.	.	.	.	.	.	.	.	.	.	.	.	21
6700	THALICTRUM MINUS	S	.	0	29.2	41.1	.	.	.	.	.	.	.	.	.	.	.	.	.	21
6701	THALICTRUM POLYCARPUM	S	.	0	24.5	33.1	.	.	3.4	.	.	.	.	.	.	.	.	.	.	19
6702	THALICTRUM POLYCARPUM	S	.	0	9.9	7.4	.	.	6.6	.	.	.	.	.	.	.	.	.	.	19
6703	THALICTRUM POLYCARPUM	S	.	0	26.9	36.8	.	.	.	.	.	.	.	.	.	.	.	.	.	21
6704	THALICTRUM REVOLUTUM	S	.	0	27.5	38.4	.	.	3.1	.	.	.	.	.	.	.	.	.	.	19
6705	THEA SINENSIS	L	318	0	26.6	3.0	63.9	9.5	6.4	355.4	340.2	26.4	54.3	.	2934.8	0.08	0.87	8.26	9.8	2
6706	THEA SINENSIS	L	325	0	30.7	5.2	58.1	10.4	6.1	265.4	449.6	20.5	65.0	.	9100.8	0.41	1.34	4.98	249.2	2
6707	THEA SINENSIS	L	325	0	26.2	3.8	64.2	10.6	5.8	348.2	201.3	34.4	.	.	9140.4	0.08	0.86	7.94	92.5	2
6708	THELEPOGON ELEGANS	L	.	0	5.6	1.3	88.0	28.2	5.1	.	.	.	.	.	.	.	.	.	.	5
6709	THELESPERMA AMBIGUUM	S	.	0	12.3	9.1	.	.	.	.	.	.	.	.	.	.	.	.	.	20
6710	THELESPERMA AMBIGUUM	S	.	0	10.0	10.1	.	.	.	.	.	.	.	.	.	.	.	.	.	20
6711	THELESPERMA MEGAPOTAMICUM	S	.	0	15.6	13.7	.	.	.	.	.	.	.	.	.	.	.	.	.	20
6712	THELESPERMA SP.	L	.	0	10.0	1.6	83.2	33.6	5.2	1300.0	150.0	.	.	1600.0	.	.	.	.	.	6
6713	THELYPODIUM TEXANUM	S	.	0	27.5	35.6	.	.	.	.	.	.	.	.	.	.	.	.	.	20
6714	THEMEDA ARGUENS	N	.	0	4.0	.	.	31.2	3.6	720.0	210.0	490.0	.	.	.	.	.	.	.	5
6715	THEMEDA ARGUENS	L	.	0	5.5	1.2	80.7	34.2	12.5	.	.	.	.	.	.	.	.	.	.	5
6716	THEMEDA ARUNDINACEA	N	.	0	1.9	1.0	92.4	48.9	4.6	460.0	140.0	.	.	.	.	.	.	.	.	5
6717	THEMEDA CYMBARIA	L	.	0	4.6	2.1	87.1	37.8	9.1	320.0	200.0	.	.	.	.	.	.	.	.	5
6718	THEMEDA QUADRIVALVIS	L	.	0	3.0	3.0	.	26.2	12.5	330.0	140.0	.	.	330.0	.	.	.	.	.	5
6719	THEMEDA TREMULA	L	.	0	9.0	3.5	76.8	30.9	10.8	630.0	290.0	.	.	.	.	.	.	.	.	5
6720	THEMEDA TRIANDRA	W	.	0	6.9	3.0	78.1	32.3	12.0	.	.	.	.	.	.	.	.	.	.	18

TABLE 3 - DATA CONVERTED TO A ZERO-MOISTURE BASIS.

OBS	PLNA	PART	CALORIES	WATER	PROTEIN	FAT	TOTCARB	FIBER	ASH	CALCIUM	PHOSPHO	IRON	SODIUM	POTASS	CAROT	THIAMINE	RIBOFLAV	NIACIN	ASCORBIC	REFCO
6721	THEMEDA TRIANDRA	H	.	0	4.4	1.6	83.7	40.4	10.3	.	.	.	.	.	.	.	.	.	.	18
6722	THEMEDA TRIANDRA	S	.	0	23.8	5.5	.	.	.	.	.	.	.	.	.	.	.	.	.	20
6723	THEMEDA TRIANDRA	S	.	0	26.2	8.3	.	.	.	.	.	.	.	.	.	.	.	.	.	20
6724	THEMEDA TRIANDRA	S	.	0	21.2	6.7	.	.	.	.	.	.	.	.	.	.	.	.	.	20
6725	THEMEDA TRIANDRA	S	.	0	11.2	3.6	.	.	.	.	.	.	.	.	.	.	.	.	.	20
6726	THEMEDA TRIANDRA	S	.	0	11.2	2.6	.	.	.	.	.	.	.	.	.	.	.	.	.	20
6727	THEMEDA TRIANDRA	S	.	0	15.0	6.8	.	.	.	.	.	.	.	.	.	.	.	.	.	20
6728	THEMEDA TRIANDRA	S	.	0	11.9	4.5	.	.	.	.	.	.	.	.	.	.	.	.	.	20
6729	THEMEDA TRIANDRA	S	.	0	12.5	2.9	.	.	.	.	.	.	.	.	.	.	.	.	.	20
6730	THEMEDA TRIANDRA	S	.	0	11.2	3.7	.	.	.	.	.	.	.	.	.	.	.	.	.	20
6731	THEMEDA TRIANDRA	S	.	0	25.0	3.4	.	.	.	.	.	.	.	.	.	.	.	.	.	20
6732	THEMEDA TRIANDRA	S	.	0	13.8	4.0	.	.	.	.	.	.	.	.	.	.	.	.	.	20
6733	THEMEDA TRIANDRA	S	.	0	13.8	4.8	.	.	.	.	.	.	.	.	.	.	.	.	.	20
6734	THEMEDA TRIANDRA	S	.	0	13.1	10.6	.	.	.	.	.	.	.	.	.	.	.	.	.	20
6735	THEMEDA TRIANDRA	S	.	0	12.5	6.1	.	.	.	.	.	.	.	.	.	.	.	.	.	20
6736	THEMEDA TRIANDRA	S	.	0	15.0	5.7	.	.	.	.	.	.	.	.	.	.	.	.	.	20
6737	THEMEDA TRIANDRA	S	.	0	15.6	5.9	.	.	.	.	.	.	.	.	.	.	.	.	.	20
6738	THEMEDA TRIANDRA	S	.	0	11.9	9.6	.	.	.	.	.	.	.	.	.	.	.	.	.	20
6739	THEMEDA TRIANDRA	L	.	0	2.5	2.2	.	32.2	5.7	260.0	90.0	.	.	210.0	.	.	.	.	.	5
6740	THEOBROMA BICOLOR	P	341	0	13.5	1.4	79.3	5.3	5.8	28.8	197.1	3.4	.	.	913.46	8.65	0.72	15.38	101.0	3
6741	THEOBROMA BICOLOR	S	624	0	17.7	56.2	22.7	2.7	3.3	67.4	569.5	3.5	.	.	0.00	0.23	0.35	1.66	0.0	3
6742	THEOBROMA CACAO	S	.	0	15.1	37.1	44.5	5.9	3.3	80.0	360.0	.	.	.	.	.	.	.	.	18
6743	THEOBROMA CACAO	S	473	0	12.4	48.0	36.0	8.9	3.5	110.0	557.1	3.7	.	.	31.12	0.18	0.15	1.76	3.1	3
6744	THERMOPSIS DIVARICARPA	S	.	0	32.3	10.8	.	.	.	.	.	.	.	.	.	.	.	.	.	21
6745	THERMOPSIS RHOMBIFOLIA	S	.	0	34.9	11.6	.	.	3.1	.	.	.	.	.	.	.	.	.	.	19
6746	THEVETIA SP.	S	.	0	16.2	57.5	.	.	2.5	.	.	.	.	.	.	.	.	.	.	19
6747	THEVETIA THEVETIOIDES	S	.	0	17.5	62.2	.	.	.	.	.	.	.	.	.	.	.	.	.	20
6748	THLASPI ARVENSE	S	.	0	25.6	28.2	.	.	.	.	.	.	.	.	.	.	.	.	.	20
6749	THLASPI ARVENSE	S	.	0	23.8	32.0	.	.	.	.	.	.	.	.	.	.	.	.	.	20
6750	THLASPI ARVENSE	S	.	0	24.2	32.9	.	.	.	.	.	.	.	.	.	.	.	.	.	21
6751	THLASPI ARVENSE	S	.	0	26.0	28.9	.	.	.	.	.	.	.	.	.	.	.	.	.	21
6752	THLASPI ARVENSE	L	.	0	54.2	.	33.1	.	.	.	.	.	.	.	.	.	.	.	1901.4	5
6753	THLASPI PERFOLIATUM	S	.	0	23.8	30.8	.	.	.	.	.	.	.	.	.	.	.	.	.	20
6754	THYMBRA SPICATA	S	.	0	10.4	6.4	.	.	.	.	.	.	.	.	.	.	.	.	.	21
6755	THYMUS CAPITATUS	S	.	0	8.9	5.3	.	.	.	.	.	.	.	.	.	.	.	.	.	21
6756	THYMUS CHAUBARDII	S	.	0	33.7	37.7	.	.	.	.	.	.	.	.	.	.	.	.	.	21
6757	THYMUS SP.	L	313	0	19.8	3.7	65.0	25.3	11.5	258.1	207.4	42.4	.	.	.	.	.	.	.	1
6758	THYMUS VULGARIS	S	.	0	28.2	37.2	.	.	4.3	.	.	.	.	.	.	.	.	.	.	19
6759	THYMUS VULGARIS	S	.	0	25.6	38.9	.	.	.	.	.	.	.	.	.	.	.	.	.	19
6760	THYMUS VULGARIS	H	299	0	9.9	8.0	69.3	20.2	12.7	2049.9	218.0	134.1	59.7	882.9	2472.89	0.55	0.43	5.36	.	16
6761	THYRALLIS GLAUCA	S	.	0	20.6	1.2	.	.	.	.	.	.	.	.	.	.	.	.	.	19
6762	THYSANOCARPUS RADIANS	S	.	0	18.4	10.3	.	.	.	.	.	.	.	.	.	.	.	.	.	21
6763	TIDESTROMIA SP.	W	.	0	16.2	1.3	71.4	22.5	11.1	.	.	.	.	.	.	.	.	.	.	6
6764	TILIA X EUROPAEA	L	.	0	18.4	3.3	65.3	15.0	13.0	.	.	.	.	.	.	.	.	.	.	5
6765	TILIA X EUROPAEA	F	.	0	11.3	9.4	72.7	42.3	6.6	.	.	.	.	.	.	.	.	.	.	5

TABLE 3 - DATA CONVERTED TO A ZERO-MOISTURE BASIS.

OBS	PLNA	PART	CALORIES	WATER	PROTEIN	FAT	TOTLCARB	FIBER	ASH
6766	TILLANDSIA USNEOIDES	W	.	0	5.2	2.4	84.5	30.2	7.9
6767	TITHONIA CALVA	S	.	0	16.2	24.6	.	.	.
6768	TITHONIA CALVA	S	.	0	16.2	31.5	.	.	.
6769	TITHONIA SP.	S	.	0	21.2	18.8	.	.	.
6770	TITHONIA SPECIOSA	S	.	0	20.9	18.3	.	.	5.2
6771	TITHONIA THURBERI	S	.	0	22.5	22.2	.	.	.
6772	TITHONIA THURBERI	S	.	0	21.9	22.0	.	.	.
6773	TITHONIA TUBAEFORMIS	S	.	0	22.5	23.0	.	.	.
6774	TOFIELDIA OCCIDENTALIS	S	.	0	31.1	35.0	.	.	.
6775	TOONA SINENSIS	SH	311	0	36.6	6.1	48.2	7.9	9.1
6776	TORDYLIUM AEGYPTIACUM	S	.	0	13.6	10.6	.	.	.
6777	TORILIS ARVENSIS SUBSP. NEGLECTA	S	.	0	19.9	23.0	.	.	.
6778	TORILIS JAPONICA	S	.	0	21.2	16.9	.	.	.
6779	TORILIS LEPTOPHYLLA	S	.	0	16.6	10.6	.	.	.
6780	TORILIS LEPTOPHYLLA	S	.	0	16.7	14.4	.	.	.
6781	TORILIS NODOSA	S	.	0	20.6	11.2	.	.	.
6782	TORREYA NUCIFERA	S	623	0	11.9	54.8	30.2	6.1	3.1
6783	TORULARIA TORULOSA	S	.	0	37.4	34.3	.	.	.
6784	TRACHYPOGON MONTUFARII	W	.	0	3.5	2.3	88.2	38.0	6.0
6785	TRACHYPOGON SPICATUS	W	.	0	5.7	1.8	82.9	40.2	9.6
6786	TRACHYSPERMUM AMMI	F	363	0	12.5	7.0	64.1	.	16.4
6787	TRAGIA SP.	L	.	0	16.2	3.2	73.0	19.7	7.6
6788	TRAGOPOGON DUBIUS	S	.	0	16.3	12.1	.	.	.
6789	TRAGOPOGON DUBIUS	S	.	0	18.2	11.6	.	.	.

OBS	CALCIUM	PHOSPHO	IRON	SODIUM	POTASS	CAROT	THIAMINE	RIBOFLAV	NIACIN	ASCORBIC	REFCO
6766	.	.	.	.	.	.	.	.	.	.	6
6767	.	.	.	.	.	.	.	.	.	.	20
6768	.	.	.	.	.	.	.	.	.	.	20
6769	.	.	.	.	.	.	.	.	.	.	20
6770	.	.	.	.	.	.	.	.	.	.	19
6771	.	.	.	.	.	.	.	.	.	.	20
6772	.	.	.	.	.	.	.	.	.	.	20
6773	.	.	.	.	.	.	.	.	.	.	20
6774	.	.	.	.	.	.	.	.	.	.	20
6775	182.9	622.0	19.5	.	3512.2	.	.	.	.	.	2
6776	.	.	.	.	.	.	.	.	.	.	21
6777	.	.	.	.	.	.	.	.	.	.	21
6778	.	.	.	.	.	.	.	.	.	.	20
6779	.	.	.	.	.	.	.	.	.	.	21
6780	.	.	.	.	.	.	.	.	.	.	21
6781	.	.	.	.	.	.	.	.	.	.	19
6782	42.8	294.4	3.9	.	910.1	26.8	0.04	0.12	2.03	7.5	2
6783	.	.	.	.	.	.	.	.	.	.	21
6784	240.0	110.0	.	.	.	.	.	.	.	.	6
6785	.	.	.	.	.	.	.	.	.	.	18
6786	.	.	.	.	.	532.1	0.48	0.83	.	2.2	1
6787	.	.	.	.	.	.	.	.	.	.	6
6788	.	.	.	.	.	.	.	.	.	.	21
6789	.	.	.	.	.	.	.	.	.	.	21

TABLE 3 - DATA CONVERTED TO A ZERO-MOISTURE BASIS.

OBS	PLNA	PART	CALORIES	WATER	PROTEIN	FAT	TOTLCARB	FIBER	ASH	CALCIUM	PHOSPHO	IRON	SODIUM	POTASS	CAROT	THIAMINE	RIBOFLAV	NIACIN	ASCORBIC	REFCO
6790	TRAGOPOGON MAJOR	S	.	0	15.0	14.7	.	.	.	.	.	.	.	.	.	.	.	.	.	20
6791	TRAGOPOGON PORRIFOLIUS	S	.	0	17.7	22.0	.	.	.	.	.	.	.	.	.	.	.	.	.	21
6792	TRAGOPOGON PORRIFOLIUS	R	384	0	6.0	0.9	88.8	7.8	4.3	206.9	215.5	6.0	.	.	0.0	0.17	0.17	1.29	43.1	3
6793	TRAGOPOGON PORRIFOLIUS	R	214	0	12.9	2.7	80.4	8.0	4.0	209.8	294.6	6.7	.	1696.4	26.8	0.18	0.18	1.34	49.1	4
6794	TRAGOPOGON PORRIFOLIUS	R	.	0	6.0	0.9	88.8	7.8	4.3	206.9	215.5	6.0	.	.	.	0.17	0.17	1.29	43.1	5
6795	TRAGOPOGON PRATENSIS	S	.	0	18.8	11.4	.	.	4.2	.	.	.	.	.	.	.	.	.	.	19
6796	TRAGOPOGON PRATENSIS	S	.	0	17.5	19.5	.	.	.	.	.	.	.	.	.	.	.	.	.	20
6797	TRAGOPOGON PRATENSIS	S	.	0	18.5	9.5	.	.	.	.	.	.	.	.	.	.	.	.	.	20
6798	TRAGOPOGON PRATENSIS	S	.	0	15.9	15.7	.	.	.	.	.	.	.	.	.	.	.	.	.	21
6799	TRAPA BISPINOSA	F	348	0	12.2	1.2	82.7	2.4	3.9	160.7	339.3	3.6	62.5	1345.2	0.0	0.39	0.18	5.95	20.8	2
6800	TRAPA BISPINOSA	S	.	0	15.7	1.0	79.7	2.0	3.7	66.7	500.0	2.7	163.3	2166.7	.	0.17	0.23	2.00	30.0	5
6801	TRAPA NATANS	S	.	0	11.9	1.0	.	.	.	.	.	.	.	.	.	.	.	.	.	21
6802	TRECULIA AFRICANA	S	415	0	13.9	6.2	77.5	1.8	2.4	139.9	349.1	.	.	.	.	.	.	.	.	1
6803	TREWIA NUDIFLORA	F	.	0	7.5	1.1	84.9	.	5.9	161.3	322.6	9.7	.	.	.	.	.	.	.	5
6804	TRIANTHEMA PENTANDRA	W	.	0	16.5	2.5	63.1	23.1	17.9	1040.0	270.0	.	.	.	.	.	.	.	.	18
6805	TRIANTHEMA PORTULACASTRUM	L	256	0	28.9	1.7	47.9	9.9	21.5	1743.8	611.6	35.5	.	.	50909.1	0.33	0.99	8.26	247.9	1
6806	TRIANTHEMA PORTULACASTRUM	L	.	0	23.0	4.6	47.1	10.3	25.3	1149.4	344.8	442.5	.	.	26436.8	.	.	.	804.6	5
6807	TRIBULUS TERRESTRIS	S	.	0	46.2	29.2	.	.	.	.	.	.	.	.	.	.	.	.	.	21
6808	TRIBULUS TERRESTRIS	S	.	0	38.5	37.8	.	.	.	.	.	.	.	.	.	.	.	.	.	21
6809	TRIBULUS TERRESTRIS	W	.	0	18.9	2.7	66.6	25.7	11.8	3350.0	240.0	.	.	2840.0	.	.	.	.	.	6
6810	TRIBULUS TERRESTRIS	L	250	0	19.3	2.1	52.1	22.9	26.4	.	339.3	.	.	.	.	.	.	.	.	1
6811	TRIBULUS TERRESTRIS	L	.	0	34.4	2.4	.	.	22.0	7416.3	382.8	44.0	.	.	.	.	.	.	201.0	5
6812	TRICHACHNE CALIFORNICA	W	.	0	8.9	1.9	77.7	36.1	11.5	.	.	.	.	.	.	.	.	.	.	6
6813	TRICHILIA SP.	S	.	0	14.4	53.8	.	.	4.0	.	.	.	.	.	.	.	.	.	.	19
6814	TRICHILIA SP.	S	551	0	11.1	38.6	46.7	20.0	3.6	289.4	312.8	1.4	.	.	.	0.05	0.50	.	13.8	1
6815	TRICHOSANTHES ANGUINA	F	.	0	9.3	5.6	75.9	14.8	9.3	925.9	370.4	20.4	463.0	629.6	1777.8	0.74	1.11	5.56	0.0	5
6816	TRICHOSANTHES CUCUMERINA	S	.	0	33.8	56.3	.	.	.	.	.	.	.	.	.	.	.	.	.	21
6817	TRICHOSANTHES CUCUMERINA	F	316	0	15.8	0.0	78.9	8.8	5.3	333.3	596.5	17.5	.	2157.9	14210.5	0.70	0.88	12.28	105.3	2
6818	TRICHOSANTHES DIOICA	F	.	0	25.0	3.8	65.0	37.5	6.3	375.0	500.0	21.3	37.5	1037.5	1912.5	0.63	0.75	6.25	362.5	5
6819	TRICHOSANTHES DIOICA	L	.	0	27.7	5.6	51.3	21.5	15.4	2723.1	374.4	.	.	.	.	.	.	.	.	5
6820	TRICOSCYPHA SP.	F	367	0	4.2	3.0	91.0	.	1.8	253.0	361.4	12.0	.	.	.	.	.	.	451.8	1
6821	TRIFOLIUM AGRARIUM	W	.	0	16.7	3.7	73.2	20.5	6.4	1170.0	310.0	25.0	.	1850.0	.	.	.	.	.	6
6822	TRIFOLIUM ALEXANDRINUM	W	.	0	26.7	3.0	54.7	14.9	15.6	1480.0	310.0	.	.	.	.	.	.	.	.	18
6823	TRIFOLIUM ALEXANDRINUM	H	.	0	16.6	2.4	69.9	26.2	11.1	1310.0	210.0	.	.	.	.	.	.	.	.	18
6824	TRIFOLIUM ALEXANDRINUM	S	.	0	34.2	5.3	.	.	.	.	.	.	.	.	.	.	.	.	.	21
6825	TRIFOLIUM ALEXANDRINUM	W	.	0	19.2	3.7	60.2	22.1	16.9	3560.0	320.0	.	.	2390.0	.	.	.	.	.	6
6826	TRIFOLIUM ANGUSTIFOLIUM	S	.	0	45.8	7.9	.	.	.	.	.	.	.	.	.	.	.	.	.	21
6827	TRIFOLIUM ARMENIUM	S	.	0	39.8	5.6	.	.	.	.	.	.	.	.	.	.	.	.	.	21
6828	TRIFOLIUM ARVENSE	S	.	0	38.2	6.1	.	.	.	.	.	.	.	.	.	.	.	.	.	21
6829	TRIFOLIUM CHERLERI	S	.	0	35.8	5.6	.	.	.	.	.	.	.	.	.	.	.	.	.	21
6830	TRIFOLIUM CLYPEATUM	S	.	0	29.8	7.9	.	.	.	.	.	.	.	.	.	.	.	.	.	21
6831	TRIFOLIUM DUBIUM	S	.	0	39.5	9.8	.	.	.	.	.	.	.	.	.	.	.	.	.	21
6832	TRIFOLIUM FRAGIFERUM	S	.	0	31.4	7.9	.	.	3.5	.	.	.	.	.	.	.	.	.	.	19
6833	TRIFOLIUM FRAGIFERUM	S	.	0	32.4	7.4	.	.	.	.	.	.	.	.	.	.	.	.	.	21
6834	TRIFOLIUM GLOMERATUM	S	.	0	35.4	5.4	.	.	3.5	.	.	.	.	.	.	.	.	.	.	19

TABLE 3 - DATA CONVERTED TO A ZERO-MOISTURE BASIS.

OBS	PLNA	PART	CALORIES	WATER	PROTEIN	FAT	TOTCARB	FIBER	ASH	CALCIUM	PHOSPHO	IRON	SODIUM	POTASS	CAROT	THIAMINE	RIBOFLAV	NIACIN	ASCORBIC	REFCO
6835	TRIFOLIUM GLOMERATUM	S	.	0	24.0	9.6	.	.	9.3	.	.	.	.	.	.	.	.	.	.	19
6836	TRIFOLIUM HIRTUM	S	.	0	34.0	4.2	.	.	3.6	.	.	.	.	.	.	.	.	.	.	19
6837	TRIFOLIUM HIRTUM	S	.	0	37.5	4.8	.	.	.	.	.	.	.	.	.	.	.	.	.	21
6838	TRIFOLIUM HYBRIDUM	S	.	0	36.6	5.2	.	.	4.1	.	.	.	.	.	.	.	.	.	.	19
6839	TRIFOLIUM HYBRIDUM	S	.	0	35.4	5.5	.	.	4.4	.	.	.	.	.	.	.	.	.	.	19
6840	TRIFOLIUM HYBRIDUM	W	.	0	17.5	3.1	69.4	25.3	10.0	1270	300	43	.	2700	.	.	.	.	.	6
6841	TRIFOLIUM INCARNATUM	W	.	0	18.9	3.2	66.4	14.8	11.5	1320	200	.	.	.	.	.	.	.	.	18
6842	TRIFOLIUM INCARNATUM	S	.	0	43.5	3.9	.	.	4.1	.	.	.	.	.	.	.	.	.	.	19
6843	TRIFOLIUM INCARNATUM	S	.	0	38.4	4.7	.	.	4.1	.	.	.	.	.	.	.	.	.	.	19
6844	TRIFOLIUM INCARNATUM	L	.	0	4.2	0.3	14.9	5.0	1.6	160	70	.	.	.	.	.	.	.	.	5
6845	TRIFOLIUM INCARNATUM	W	.	0	16.7	3.1	70.9	28.3	9.3	1620	350	70	.	2410	.	.	.	.	.	6
6846	TRIFOLIUM LAPPACEUM	H	.	0	20.5	3.2	66.3	25.6	10.0	1340	230	63	.	.	.	.	.	.	.	6
6847	TRIFOLIUM LUPINASTER	S	.	0	32.3	6.0	.	.	4.3	.	.	.	.	.	.	.	.	.	.	19
6848	TRIFOLIUM LUPINASTER	S	.	0	30.8	6.3	.	.	4.0	.	.	.	.	.	.	.	.	.	.	19
6849	TRIFOLIUM MEDIUM	S	.	0	38.9	9.0	.	.	4.1	.	.	.	.	.	.	.	.	.	.	19
6850	TRIFOLIUM MEDIUM	S	.	0	39.1	9.1	.	.	4.0	.	.	.	.	.	.	.	.	.	.	19
6851	TRIFOLIUM MEDIUM	S	.	0	37.5	8.6	.	.	.	.	.	.	.	.	.	.	.	.	.	19
6852	TRIFOLIUM NIGRESCENS	S	.	0	31.6	6.8	.	.	11.3	.	.	.	.	.	.	.	.	.	.	19
6853	TRIFOLIUM NIGRESCENS SUBSP. PETRISAVII	S	.	0	34.3	5.0	.	.	.	.	.	.	.	.	.	.	.	.	.	21
6854	TRIFOLIUM PALAESTINUM	S	.	0	43.6	9.9	.	.	.	.	.	.	.	.	.	.	.	.	.	21
6855	TRIFOLIUM PILULARE	S	.	0	38.4	12.4	.	.	.	.	.	.	.	.	.	.	.	.	.	21
6856	TRIFOLIUM PRATENSE	W	.	0	20.7	5.3	63.0	22.5	11.0	1420	320	.	.	.	.	.	.	.	.	18
6857	TRIFOLIUM PRATENSE	H	.	0	9.8	2.9	81.7	33.8	5.6	980	130	.	.	.	.	.	.	.	.	18
6858	TRIFOLIUM PRATENSE	S	.	0	36.2	8.5	.	.	4.1	.	.	.	.	.	.	.	.	.	.	19
6859	TRIFOLIUM PRATENSE	S	.	0	38.1	8.6	.	.	.	.	.	.	.	.	.	.	.	.	.	21
6860	TRIFOLIUM PRATENSE	L	.	0	13.5	1.9	76.4	27.8	8.1	1270	220	.	.	.	.	.	.	.	.	5
6861	TRIFOLIUM PRATENSE	H	.	0	14.9	2.9	74.3	30.1	7.9	1610	220	13	.	1760	.	.	.	.	.	6
6862	TRIFOLIUM PROCUMBENS	S	.	0	39.1	7.5	.	.	3.6	.	.	.	.	.	.	.	.	.	.	19
6863	TRIFOLIUM PROCUMBENS	S	.	0	30.9	6.1	.	.	4.1	.	.	.	.	.	.	.	.	.	.	19
6864	TRIFOLIUM PURPUREUM	S	.	0	39.9	12.8	.	.	.	.	.	.	.	.	.	.	.	.	.	21
6865	TRIFOLIUM REPENS	W	.	0	21.0	1.4	68.1	23.8	9.5	510	350	.	.	.	.	.	.	.	.	18
6866	TRIFOLIUM REPENS	S	.	0	35.5	6.0	.	.	3.8	.	.	.	.	.	.	.	.	.	.	19
6867	TRIFOLIUM REPENS	S	.	0	33.9	6.2	.	.	4.5	.	.	.	.	.	.	.	.	.	.	19
6868	TRIFOLIUM REPENS	S	.	0	36.1	6.2	.	.	4.0	.	.	.	.	.	.	.	.	.	.	19
6869	TRIFOLIUM REPENS	S	.	0	34.8	6.9	.	.	.	.	.	.	.	.	.	.	.	.	.	21
6870	TRIFOLIUM REPENS	L	.	0	25.8	3.4	59.2	18.6	11.4	1850	340	.	.	.	.	.	.	.	.	5
6871	TRIFOLIUM REPENS	W	.	0	27.8	3.2	56.9	15.6	12.1	1480	460	34	.	2200	.	.	.	.	.	6
6872	TRIFOLIUM REPENS "LADINO"	H	.	0	23.1	3.4	64.0	19.2	9.5	1380	400	65	.	2170	.	.	.	.	.	6
6873	TRIFOLIUM REPENS "LADINO"	W	.	0	25.1	4.8	58.7	14.5	11.4	1330	370	37	.	2190	.	.	.	.	.	6
6874	TRIFOLIUM REPENS "WHITE"	H	.	0	22.0	2.6	65.3	23.3	10.1	1240	390	106	.	2030	.	.	.	.	.	6
6875	TRIFOLIUM RESUPINATUM	W	.	0	27.8	4.0	52.1	15.6	16.1	1830	430	.	.	.	.	.	.	.	.	18
6876	TRIFOLIUM RESUPINATUM	S	.	0	36.0	7.0	.	.	4.3	.	.	.	.	.	.	.	.	.	.	19
6877	TRIFOLIUM RESUPINATUM	S	.	0	36.8	7.1	.	.	.	.	.	.	.	.	.	.	.	.	.	21
6878	TRIFOLIUM RESUPINATUM	H	.	0	16.3	1.8	71.6	30.8	10.3	1660	230	.	.	.	.	.	.	.	.	6
6879	TRIFOLIUM SQUARROSUM	S	.	0	32.2	3.2	.	.	3.6	.	.	.	.	.	.	.	.	.	.	19

TABLE 3 - DATA CONVERTED TO A ZERO-MOISTURE BASIS.

OBS	PLANA	PART	CALORIES	WATER	PROTEIN	FAT	TOTLCARB	FIBER	ASH	CALCIUM	PHOSPHO	IRON	SODIUM	POTASS	CAROT	THIAMINE	RIBOFLAV	NIACIN	ASCORBIC	REFCO
6880	TRIFOLIUM STELLATUM	S	.	0	47.7	10.5	.	.	.	.	.	.	.	.	.	.	.	.	.	21
6881	TRIFOLIUM STRIATUM	S	.	0	28.5	1.6	.	.	3.6	.	.	.	.	.	.	.	.	.	.	19
6882	TRIFOLIUM SUBTERRANEUM	W	.	0	12.5	4.4	73.3	28.7	9.8	1490.0	330.0	.	.	.	.	.	.	.	.	18
6883	TRIFOLIUM SUBTERRANEUM	H	.	0	8.0	2.2	81.5	36.9	8.3	730.0	230.0	.	.	.	.	.	.	.	.	18
6884	TRIFOLIUM SUBTERRANEUM	S	.	0	36.4	15.2	.	.	4.9	.	.	.	.	.	.	.	.	.	.	19
6885	TRIFOLIUM SUBTERRANEUM	S	.	0	38.6	16.2	.	.	.	.	.	.	.	.	.	.	.	.	.	21
6886	TRIFOLIUM TRIDENTATUM	W	.	0	14.3	.	.	24.8	10.2	.	.	.	.	.	.	.	.	.	.	6
6887	TRIGLOCHIN BULBOSUM	S	.	0	37.0	28.7	.	.	.	.	.	.	.	.	.	.	.	.	.	21
6888	TRIGONELLA ARABICA	S	.	0	28.3	3.3	.	.	.	.	.	.	.	.	.	.	.	.	.	21
6889	TRIGONELLA BALANSAE	S	.	0	35.2	6.3	.	.	4.7	.	.	.	.	.	.	.	.	.	.	19
6890	TRIGONELLA COERULEA	S	.	0	27.8	6.5	.	.	4.4	.	.	.	.	.	.	.	.	.	.	19
6891	TRIGONELLA FOENUM-GRAECUM	W	.	0	15.7	2.1	73.4	31.1	8.8	1470.0	220.0	.	.	.	.	.	.	.	.	18
6892	TRIGONELLA FOENUM-GRAECUM	S	.	0	27.2	7.8	61.0	1.0	4.0	880.0	300.0	.	.	.	.	.	.	.	.	18
6893	TRIGONELLA FOENUM-GRAECUM	F	.	0	11.3	2.9	82.8	4.2	3.0	660.0	140.0	.	.	.	.	.	.	.	.	18
6894	TRIGONELLA FOENUM-GRAECUM	H	.	0	10.0	2.2	77.0	35.1	10.8	.	.	.	.	.	.	.	.	.	.	18
6895	TRIGONELLA FOENUM-GRAECUM	S	.	0	28.9	5.0	.	.	3.7	.	.	.	.	.	.	.	.	.	.	19
6896	TRIGONELLA FOENUM-GRAECUM	S	.	0	32.0	5.4	.	.	4.1	.	.	.	.	.	.	.	.	.	.	19
6897	TRIGONELLA FOENUM-GRAECUM	S	.	0	27.9	5.4	.	.	3.8	.	.	.	.	.	.	.	.	.	.	19
6898	TRIGONELLA FOENUM-GRAECUM	S	.	0	29.4	5.3	.	.	3.8	.	.	.	.	.	.	.	.	.	.	19
6899	TRIGONELLA FOENUM-GRAECUM	S	.	0	29.4	5.3	.	.	4.5	.	.	.	.	.	.	.	.	.	.	19
6900	TRIGONELLA FOENUM-GRAECUM	S	.	0	30.3	6.0	.	.	3.5	.	.	.	.	.	.	.	.	.	.	19
6901	TRIGONELLA FOENUM-GRAECUM	S	.	0	31.8	5.9	.	.	4.4	.	.	.	.	.	.	.	.	.	.	19
6902	TRIGONELLA FOENUM-GRAECUM	S	.	0	32.5	6.5	.	.	3.9	.	.	.	.	.	.	.	.	.	.	19
6903	TRIGONELLA FOENUM-GRAECUM	S	.	0	29.5	5.2	.	.	4.3	.	.	.	.	.	.	.	.	.	.	19
6904	TRIGONELLA FOENUM-GRAECUM	S	.	0	27.2	5.0	.	.	3.4	.	.	.	.	.	.	.	.	.	.	19
6905	TRIGONELLA FOENUM-GRAECUM	S	.	0	28.7	5.5	.	.	3.2	.	.	.	.	.	.	.	.	.	.	19
6906	TRIGONELLA FOENUM-GRAECUM	S	.	0	28.3	5.5	.	.	3.2	.	.	.	.	.	.	.	.	.	.	19
6907	TRIGONELLA FOENUM-GRAECUM	S	.	0	31.0	5.5	.	.	3.2	.	.	.	.	.	.	.	.	.	.	19
6908	TRIGONELLA FOENUM-GRAECUM	S	.	0	30.4	5.4	.	.	5.0	.	.	.	.	.	.	.	.	.	.	19
6909	TRIGONELLA FOENUM-GRAECUM	S	.	0	28.8	7.3	.	.	2.4	.	.	.	.	.	.	.	.	.	.	19
6910	TRIGONELLA FOENUM-GRAECUM	S	.	0	30.3	6.8	.	.	3.0	.	.	.	.	.	.	.	.	.	.	19
6911	TRIGONELLA FOENUM-GRAECUM	S	.	0	27.4	6.9	.	.	3.2	.	.	.	.	.	.	.	.	.	.	19
6912	TRIGONELLA FOENUM-GRAECUM	S	.	0	27.6	6.2	.	.	3.6	.	.	.	.	.	.	.	.	.	.	19
6913	TRIGONELLA FOENUM-GRAECUM	S	.	0	39.8	6.5	.	.	4.1	.	.	.	.	.	.	.	.	.	.	19
6914	TRIGONELLA FOENUM-GRAECUM	S	.	0	34.1	6.0	.	.	3.5	.	.	.	.	.	.	.	.	.	.	19
6915	TRIGONELLA FOENUM-GRAECUM	S	.	0	32.8	6.3	.	.	4.2	.	.	.	.	.	.	.	.	.	.	19
6916	TRIGONELLA FOENUM-GRAECUM	S	.	0	26.1	6.3	.	.	3.5	.	.	.	.	.	.	.	.	.	.	19
6917	TRIGONELLA FOENUM-GRAECUM	S	.	0	26.2	4.9	.	.	3.6	.	.	.	.	.	.	.	.	.	.	19
6918	TRIGONELLA FOENUM-GRAECUM	S	.	0	26.2	6.8	.	.	.	.	.	.	.	.	.	.	.	.	.	20
6919	TRIGONELLA FOENUM-GRAECUM	N	.	0	26.2	5.6	57.3	13.6	11.0	1290.0	460.0	.	.	.	.	.	.	.	.	5
6920	TRIGONELLA FOENUM-GRAECUM	S	354	0	25.2	7.0	64.0	11.1	3.7	193.0	324.6	36.7	73.5	844.3	.	0.35	0.41	1.80	3.3	16
6921	TRIGONELLA FOENUM-GRAECUM	S	.	0	30.4	6.7	59.4	8.3	3.5	185.4	428.7	16.3	22.0	614.1	111.2	0.39	0.34	1.27	.	5
6922	TRIGONELLA FOENUM-GRAECUM	L	.	0	31.7	6.5	51.1	7.9	10.8	2841.7	366.9	118.7	546.8	223.0	16834.5	0.29	2.23	5.76	374.1	5
6923	TRIGONELLA GLADIATA	S	.	0	26.4	3.3	.	.	.	.	.	.	.	.	.	.	.	.	.	21
6924	TRIGONELLA KOTSCHYI	S	.	0	27.6	5.1	.	.	7.4	.	.	.	.	.	.	.	.	.	.	19

TABLE 3 - DATA CONVERTED TO A ZERO-MOISTURE BASIS.

OBS	PLNA	PART	CALORIES	WATER	PROTEIN	FAT	TOTLCARB	FIBER	ASH	CALCIUM	PHOSPHO	IRON	SODIUM	POTASS	CAROT	THIAMINE	RIBOFLAV	NIACIN	ASCORBIC	REFCO
6925	TRIODIA ALBESCENS	W	.	0	9.7	2.0	78.1	33.7	10.2	520.0	200.0	.	.	1610.0	.	.	.	.	.	6
6926	TRIODIA FLAVA	S	.	0	36.4	3.1	.	.	.	.	.	.	.	.	.	.	.	.	.	21
6927	TRIODIA PULCHELLA	W	.	0	7.6	1.4	68.3	29.1	22.7	1190.0	60.0	.	.	.	.	.	.	.	.	6
6928	TRIODIA TEXANA	W	.	0	12.2	2.0	75.1	33.4	10.7	440.0	170.0	.	.	.	.	.	.	.	.	6
6929	TRIPSACUM DACTYLOIDES	W	.	0	8.5	2.0	79.3	30.2	10.2	620.0	310.0	.	.	.	.	.	.	.	.	6
6930	TRIPSACUM FASCICULATUM	W	.	0	15.9	2.8	71.7	31.4	9.6	.	.	.	.	.	.	.	.	.	.	18
6931	TRIPSACUM FASCICULATUM	L	.	0	6.1	3.4	85.2	40.4	5.3	.	.	.	.	.	.	.	.	.	.	18
6932	TRIPSACUM FASCICULATUM	S	.	0	4.6	1.9	87.1	36.1	6.4	.	.	.	.	.	.	.	.	.	.	18
6933	TRIPSACUM LATIFOLIUM	W	.	0	9.3	0.7	83.0	32.5	7.0	.	.	.	.	.	.	.	.	.	.	18
6934	TRIPSACUM LAXUM	W	.	0	6.4	1.1	80.1	32.8	12.4	.	.	.	.	.	.	.	.	.	.	6
6935	TRISETUM INTERRUPTUM	W	.	0	11.2	1.8	76.3	31.3	10.7	.	.	.	.	.	.	.	.	.	.	6
6936	TRITICUM AEGILOPOIDES	S	.	0	25.8	3.2	.	.	.	.	.	.	.	.	.	.	.	.	.	21
6937	TRITICUM AESTIVUM	W	.	0	18.1	3.7	70.6	15.5	7.6	.	.	.	.	.	.	.	.	.	.	18
6938	TRITICUM AESTIVUM	H	.	0	3.1	1.1	85.6	45.4	10.2	.	.	.	.	.	.	.	.	.	.	18
6939	TRITICUM AESTIVUM	S	.	0	13.5	2.1	81.1	3.3	3.3	30.0	480.0	.	.	.	.	.	.	.	.	18
6940	TRITICUM AESTIVUM	W	.	0	22.6	3.8	61.7	20.8	11.9	400.0	350.0	29.0	.	3370.0	.	.	.	.	.	6
6941	TRITICUM AESTIVUM	H	.	0	3.6	1.7	86.6	41.5	8.1	170.0	80.0	15.0	.	1110.0	.	.	.	.	.	6
6942	TRITICUM AESTIVUM	S	377	0	14.1	2.1	81.9	2.6	1.9	52.6	404.6	3.9	.	.	0.00	0.59	0.14	4.91	0.0	3
6943	TRITICUM AESTIVUM	S	379	0	16.1	2.5	79.4	2.6	2.0	41.4	440.2	3.6	.	.	0.00	0.66	0.14	4.94	0.0	3
6944	TRITICUM AESTIVUM	S	379	0	11.9	2.3	83.8	2.7	2.0	48.8	465.1	4.1	.	.	0.00	0.50	0.13	4.19	0.0	3
6945	TRITICUM AESTIVUM	S	379	0	16.1	2.5	79.4	2.6	2.0	41.4	440.2	3.6	3.4	425.3	0.00	0.66	0.14	4.94	0.0	4
6946	TRITICUM DICOCCON	S	380	0	14.3	2.7	81.0	3.1	2.1	43.3	410.5	5.4	.	.	0.00	0.44	0.15	5.36	.	1
6947	TRITICUM DICOCCUM	H	.	0	10.8	2.2	76.9	36.4	10.1	.	.	.	.	.	.	.	.	.	.	6
6948	TRITICUM DURUM	S	.	0	15.0	2.2	80.8	2.5	2.0	.	.	.	.	.	.	.	.	.	.	6
6949	TRITICUM DURUM	S	374	0	14.9	1.9	80.8	2.8	2.4	55.8	348.8	.	.	.	0.00	0.26	0.12	3.49	.	1
6950	TRITICUM DURUM	S	379	0	13.3	2.5	82.4	2.4	1.8	54.9	436.6	3.8	5.7	571.4	0.00	0.42	0.14	5.26	0.0	2
6951	TRITICUM SPELTA	S	.	0	13.3	2.2	80.6	9.1	3.9	.	.	.	.	.	.	.	.	.	.	6
6952	TRIUMFETTA BARTRAMIA	S	.	0	23.8	17.2	.	.	.	.	.	.	.	.	.	.	.	.	.	21
6953	TRIUMFETTA BREVIPES	S	.	0	30.6	22.8	.	.	.	.	.	.	.	.	.	.	.	.	.	20
6954	TRIUMFETTA POLYANDRA	S	.	0	16.8	9.8	.	.	.	.	.	.	.	.	.	.	.	.	.	21
6955	TRIUMFETTA SP.	L	315	0	19.4	1.9	70.4	15.7	8.3	1814.8	351.9	135.2	.	.	.	.	.	.	.	1
6956	TRIXIS RADIALIS	S	.	0	23.8	6.4	.	.	.	.	.	.	.	.	.	.	.	.	.	20
6957	TROPAEOLUM SP.	S	.	0	26.2	10.2	.	.	.	.	.	.	.	.	.	.	.	.	.	20
6958	TROPAEOLUM TUBEROSUM	R	371	0	11.4	4.3	78.6	5.7	5.7	50.0	300.0	8.6	.	.	214.29	0.43	0.57	4.29	478.6	3
6959	TULIPA SP.	S	.	0	20.8	3.8	.	.	.	.	.	.	.	.	.	.	.	.	.	21
6960	TURGENIA LATIFOLIA	S	.	0	14.5	19.5	.	.	.	.	.	.	.	.	.	.	.	.	.	21
6961	TURGENIA LATIFOLIA	S	.	0	11.6	5.3	.	.	.	.	.	.	.	.	.	.	.	.	.	21
6962	TYPHONIUM TRILOBATUM	R	.	0	4.7	0.3	89.7	3.3	5.3	116.3	66.4	4.3	29.9	787.4	259.14	0.23	.	2.33	.	5
6963	UAPACA GUINEENSIS	F	346	0	6.8	0.4	89.5	.	3.4	469.9	.	.	.	.	.	.	.	.	.	1
6964	UAPACA KIRKIANA	F	.	0	.	.	.	.	.	.	.	.	.	.	.	.	.	.	8.3	1
6965	ULEX EUROPAEUS	S	.	0	30.6	11.7	.	.	4.3	.	.	.	.	.	.	.	.	.	.	19
6966	ULEX EUROPAEUS	S	.	0	37.3	12.5	.	.	.	.	.	.	.	.	.	.	.	.	.	21
6967	ULEX EUROPAEUS	L	.	0	9.7	1.9	85.2	30.2	3.4	.	.	.	.	.	.	.	.	.	.	5
6968	ULLUCUS TUBEROSUS	R	362	0	7.1	0.0	88.7	4.3	4.3	21.3	248.2	5.7	.	.	0.00	0.28	0.14	2.13	163.1	3
6969	ULMUS AMERICANA	S	.	0	42.9	34.5	.	.	4.7	.	.	.	.	.	.	.	.	.	.	19

TABLE 3 - DATA CONVERTED TO A ZERO-MOISTURE BASIS.

OBS	PLANA	PART	CALORIES	WATER	PROTEIN	FAT	TOTLCARB	FIBER	ASH	CALCIUM	PHOSPHO	IRON	SODIUM	POTASS	CAROT	THIAMINE	RIBOFLAV	NIACIN	ASCORBIC	REFCO
6970	ULMUS AMERICANA	S	.	0	13.4	5.9	.	.	13.2	.	.	.	.	.	.	.	.	.	.	19
6971	ULMUS PUMILA	S	.	0	43.8	27.7	.	.	.	.	.	.	.	.	.	.	.	.	.	20
6972	UNGNADIA SPECIOSA	W	.	0	12.8	2.4	78.4	23.4	6.4	2420.0	170.0	.	.	1230.0	.	.	.	.	.	6
6973	URENA LOBATA	S	.	0	28.8	19.9	.	.	.	.	.	.	.	.	.	.	.	.	.	19
6974	URENA LOBATA	L	297	0	17.6	0.5	70.3	9.9	11.5	3065.9	368.1	.	.	.	.	.	.	.	340.7	1
6975	URERA OBOVATA	L	282	0	14.9	3.2	61.7	14.4	20.2	.	218.1	.	.	.	.	.	.	.	.	1
6976	UROCHLOA BOLBODES	W	.	0	13.7	2.8	72.9	30.0	10.6	.	.	.	.	.	.	.	.	.	.	18
6977	UROCHLOA BOLBODES	H	.	0	17.7	2.0	67.4	24.6	12.9	.	.	.	.	.	.	.	.	.	.	18
6978	UROCHLOA MOSAMBICENSIS	W	.	0	11.0	2.0	58.8	22.4	28.2	.	.	.	.	.	.	.	.	.	.	18
6979	UROCHLOA MOSAMBICENSIS	H	.	0	14.9	1.5	70.7	26.1	12.9	.	.	.	.	.	.	.	.	.	.	18
6980	UROCHLOA PANICOIDES	W	.	0	14.7	1.7	68.7	29.9	14.9	.	.	.	.	.	.	.	.	.	.	18
6981	UROCHLOA PANICOIDES	L	.	0	14.7	1.7	67.7	29.9	14.9	480.0	250.0	.	.	.	.	.	.	.	.	5
6982	UROCHLOA PULLULANS	L	.	0	5.8	1.4	79.7	33.4	13.2	510.0	310.0	.	.	.	.	.	.	.	.	5
6983	UROSPERMUM PICROIDES	S	.	0	13.8	16.9	.	.	.	.	.	.	.	.	.	.	.	.	.	21
6984	URSINIA ANETHEMOIDES	S	.	0	23.7	13.8	.	.	5.3	.	.	.	.	.	.	.	.	.	.	19
6985	URSINIA ANETHEMOIDES	S	.	0	13.4	7.5	.	.	5.6	.	.	.	.	.	.	.	.	.	.	19
6986	URSINIA ANETHEMOIDES	S	.	0	22.5	14.0	.	.	.	.	.	.	.	.	.	.	.	.	.	19
6987	URSINIA CAKILEFOLIA	S	.	0	24.8	13.6	.	.	.	.	.	.	.	.	.	.	.	.	.	21
6988	URTICA DIOICA	S	.	0	15.5	23.1	.	.	.	.	.	.	.	.	.	.	.	.	.	21
6989	URTICA DIOICA VAR. PROCERA	S	.	0	20.1	38.2	.	.	.	.	.	.	.	.	.	.	.	.	.	21
6990	UVARIA RUFA	F	357	0	7.1	4.1	82.7	15.3	6.1	234.7	153.1	6.1	.	.	0.00	0.71	0.31	7.14	122.4	2
6991	VACCARIA PYRAMIDATA	S	.	0	13.9	3.2	.	.	.	.	.	.	.	.	.	.	.	.	.	21
6992	VACCARIA PYRAMIDATA	S	.	0	13.8	1.6	.	.	.	.	.	.	.	.	.	.	.	.	.	21
6993	VACCARIA PYRAMIDATA	S	.	0	16.1	2.9	.	.	.	.	.	.	.	.	.	.	.	.	.	21
6994	VACCINIUM FLORIBUNDUM	F	373	0	3.9	3.9	90.7	12.7	1.5	98.0	78.4	4.4	.	.	343.14	0.05	0.05	3.43	39.2	3
6995	VACCINIUM MACROCARPON	F	380	0	3.3	5.8	89.3	11.6	1.7	115.7	82.6	4.1	16.5	677.7	206.61	0.25	0.17	0.83	90.9	2
6996	VACCINIUM OVATUM(DRY)	F	352	0	6.0	0.0	92.0	.	2.0	174.0	.	2.0	.	.	.	.	.	.	384.0	15
6997	VACCINIUM OVATUM(FRESH)	F	360	0	8.0	1.0	89.0	.	1.0	156.0	.	2.0	.	.	.	.	.	.	346.0	15
6998	VACCINIUM PARVIFOLIUM	F	394	0	16.0	9.0	72.0	.	2.0	216.0	.	4.0	.	.	.	.	.	.	354.0	15
6999	VACCINIUM RETICULATUM	F	364	0	4.0	2.0	91.9	13.1	3.0	70.7	101.0	1.0	.	.	5050.51	0.20	0.40	3.03	60.6	2
7000	VACCINIUM SP.	F	369	0	4.2	3.0	91.1	8.9	1.8	89.3	77.4	6.0	6.0	482.1	357.14	0.18	0.36	2.98	83.3	4
7001	VACCINIUM SYMPLOCIFOLIUM	F	.	0	3.9	2.9	91.7	35.6	1.5	97.6	48.8	6.8	.	.	234.15	.	.	.	.	5
7002	VALERIANA OFFICINALIS	S	.	0	19.9	30.0	.	.	5.8	.	.	.	.	.	.	.	.	.	.	19
7003	VALERIANA OFFICINALIS	S	.	0	19.4	34.4	.	.	.	.	.	.	.	.	.	.	.	.	.	19
7004	VALERIANELLA LOCUSTA	L	292	0	27.8	5.6	50.0	11.1	16.7	.	.	.	.	.	.	.	.	.	.	4
7005	VALERIANELLA RADIATA	S	.	0	15.6	21.9	.	.	.	.	.	.	.	.	.	.	.	.	.	20
7006	VALERIANELLA VESICARIA	S	.	0	9.6	10.1	.	.	.	.	.	.	.	.	.	.	.	.	.	21
7007	VALERIANOPSIS POLYSTACHYA	S	.	0	18.4	30.5	.	.	.	.	.	.	.	.	.	.	.	.	.	21
7008	VALLISNERIA SPIRALIS	L	.	0	.	.	.	.	14.1	1153.8	3205.1	141.0	.	.	.	.	.	.	.	5
7009	VANGUERIA INFAUSTA	F	.	0	.	.	.	.	.	.	.	.	.	.	.	.	.	.	0.0	1
7010	VANILLA SP.	S	.	0	10.0	21.2	62.8	26.3	6.0	1900.0	70.0	.	.	.	.	.	.	.	.	6
7011	VERBASCUM COROMANDELICUM	S	.	0	21.2	28.7	.	.	.	.	.	.	.	.	.	.	.	.	.	21
7012	VERBASCUM PULVERULENTUM	S	.	0	22.5	35.5	.	.	.	.	.	.	.	.	.	.	.	.	.	21
7013	VERBASCUM SINUATUM	S	.	0	19.4	38.3	.	.	.	.	.	.	.	.	.	.	.	.	.	21
7014	VERBASCUM SINUATUM	S	.	0	21.0	34.2	.	.	.	.	.	.	.	.	.	.	.	.	.	21

TABLE 3 - DATA CONVERTED TO A ZERO-MOISTURE BASIS.

OBS	PLANA	PART	CALORIES	WATER	PROTEIN	FAT	TOTLCARB	FIBER	ASH	CALCIUM	PHOSPHO	IRON	SODIUM	POTASS	CAROTE	THIAMIN	RIBOFLAV	NIACIN	ASCORBIC	REFCO
7015	VERBASCUM SINUATUM	S	.	0	21.1	38.4	.	.	.	.	.	.	.	.	.	.	.	.	.	21
7016	VERBASCUM SP.	S	.	0	20.0	30.1	.	.	.	.	.	.	.	.	.	.	.	.	.	20
7017	VERBASCUM THAPSUS	S	.	0	18.3	27.1	.	.	.	.	.	.	.	.	.	.	.	.	.	20
7018	VERBASCUM THAPSUS	S	.	0	17.8	26.8	.	.	.	.	.	.	.	.	.	.	.	.	.	21
7019	VERBENA BIPINNATIFIDA	S	.	0	15.2	17.5	.	.	.	.	.	.	.	.	.	.	.	.	.	20
7020	VERBENA BIPINNATIFIDA	L	.	0	11.5	1.8	78.3	18.1	8.4	1990.0	180.0	.	.	1600	.	.	.	.	.	6
7021	VERBENA BRACTEATA	S	.	0	9.8	16.9	.	.	.	.	.	.	.	.	.	.	.	.	.	21
7022	VERBENA BRASILIENSIS	S	.	0	12.9	19.9	.	.	.	.	.	.	.	.	.	.	.	.	.	21
7023	VERBENA CAROLINA	S	.	0	8.9	11.2	.	.	.	.	.	.	.	.	.	.	.	.	.	21
7024	VERBENA MACDOUGALII	S	.	0	11.6	20.0	.	.	.	.	.	.	.	.	.	.	.	.	.	21
7025	VERBENA RIGIDA	S	.	0	11.2	14.6	.	.	.	.	.	.	.	.	.	.	.	.	.	19
7026	VERBENA RIGIDA	S	.	0	10.6	16.2	.	.	.	.	.	.	.	.	.	.	.	.	.	20
7027	VERBENA SP.	L	.	0	11.5	3.0	75.7	15.6	9.8	2460.0	170.0	.	.	1590	.	.	.	.	.	6
7028	VERBENA STRICTA	S	.	0	8.9	13.8	.	.	5.1	.	.	.	.	.	.	.	.	.	.	19
7029	VERBESINA ENCELIOIDES	S	.	0	28.8	13.1	.	.	.	.	.	.	.	.	.	.	.	.	.	19
7030	VERBESINA ENCELIOIDES VAR. EXAURICULATA	S	.	0	30.0	24.5	.	.	.	.	.	.	.	.	.	.	.	.	.	20
7031	VERBESINA ENCELIOIDES VAR. EXAURICULATA	S	.	0	28.8	26.6	.	.	.	.	.	.	.	.	.	.	.	.	.	20
7032	VERBESINA ENCELIOIDES VAR. EXAURICULATA	S	.	0	27.9	29.3	.	.	.	.	.	.	.	.	.	.	.	.	.	21
7033	VERBESINA MYRIOCEPHALA	S	.	0	31.2	31.9	.	.	.	.	.	.	.	.	.	.	.	.	.	19
7034	VERBESINA ROTHROCKII	S	.	0	47.5	24.9	.	.	.	.	.	.	.	.	.	.	.	.	.	19
7035	VERBESINA SERRATA	S	.	0	33.8	14.4	.	.	.	.	.	.	.	.	.	.	.	.	.	20
7036	VERBESINA SERRATA VAR. PRINGLEI	S	.	0	32.5	15.9	.	.	.	.	.	.	.	.	.	.	.	.	.	20
7037	VERNONIA ALTISSIMA	S	.	0	23.7	22.7	.	.	.	.	.	.	.	.	.	.	.	.	.	21
7038	VERNONIA AMYGDALINA	L	299	0	30.5	2.3	57.5	8.6	9.8	833.3	385.1	28.7	.	.	.	.	.	.	293.1	1
7039	VERNONIA AMYGDALINA	L	267	0	44.8	2.1	39.4	.	13.8	144.6	738.3	7.6	.	.	.	.	.	.	27.5	1
7040	VERNONIA ANTHELMINTICA	S	.	0	18.1	26.5	.	.	5.9	.	.	.	.	.	.	.	.	.	.	19
7041	VERNONIA ANTHELMINTICA	S	.	0	16.9	26.1	.	.	.	.	.	.	.	.	.	.	.	.	.	20
7042	VERNONIA ANTHELMINTICA	S	.	0	18.8	25.6	.	.	.	.	.	.	.	.	.	.	.	.	.	20
7043	VERNONIA ANTHELMINTICA	S	.	0	15.8	24.3	.	.	.	.	.	.	.	.	.	.	.	.	.	21
7044	VERNONIA ANTHELMINTICA	S	.	0	14.6	24.0	.	.	.	.	.	.	.	.	.	.	.	.	.	21
7045	VERNONIA BALDWINI	S	.	0	23.1	29.2	.	.	6.2	.	.	.	.	.	.	.	.	.	.	19
7046	VERNONIA DEPPEANA	S	.	0	23.8	21.1	.	.	7.6	.	.	.	.	.	.	.	.	.	.	19
7047	VERNONIA DEPPEANA	S	.	0	20.0	10.0	.	.	.	.	.	.	.	.	.	.	.	.	.	20
7048	VERNONIA FASCICULATA	S	.	0	21.8	27.6	.	.	6.5	.	.	.	.	.	.	.	.	.	.	19
7049	VERNONIA LIATROIDES	S	.	0	14.4	11.8	.	.	.	.	.	.	.	.	.	.	.	.	.	20
7050	VERNONIA MISSURICA	S	.	0	23.4	21.5	.	.	4.8	.	.	.	.	.	.	.	.	.	.	19
7051	VERNONIA PALLENS	S	.	0	15.0	3.4	.	.	.	.	.	.	.	.	.	.	.	.	.	20
7052	VERNONIA SP.	S	.	0	28.1	18.5	.	.	5.7	.	.	.	.	.	.	.	.	.	.	19
7053	VERNONIA SP.	S	.	0	16.9	13.6	.	.	.	.	.	.	.	.	.	.	.	.	.	19
7054	VERONICA LONGIFOLIA	S	.	0	27.5	18.5	.	.	.	.	.	.	.	.	.	.	.	.	.	20
7055	VERONICA PEREGRINA	S	.	0	15.0	13.7	.	.	.	.	.	.	.	.	.	.	.	.	.	20
7056	VERONICA SALINA	S	.	0	21.4	15.8	.	.	.	.	.	.	.	.	.	.	.	.	.	21
7057	VERONICA SPICATA	S	.	0	26.2	26.6	.	.	6.0	.	.	.	.	.	.	.	.	.	.	19
7058	VERONICA SPICATA	S	.	0	26.2	35.4	.	.	.	.	.	.	.	.	.	.	.	.	.	20
7059	VIBURNUM ACERIFOLIUM	S	.	0	18.1	38.9	.	.	3.4	.	.	.	.	.	.	.	.	.	.	19

TABLE 3 - DATA CONVERTED TO A ZERO-MOISTURE BASIS.

OBS	PLNA	PART	CALORIES	WATER	PROTEIN	FAT	TOTLCARB	FIBER	ASH	CALCIUM	PHOSPHO	IRON	SODIUM	POTASS	CAROT	THIAMINE	RIBOFLAV	NIACIN	ASCORBIC	REFCO
7060	VIBURNUM ACERIFOLIUM	S	.	0	4.4	7.2	.	.	4.8	.	.	.	.	.	.	.	.	.	.	19
7061	VIBURNUM CASSINOIDES	S	.	0	19.4	30.4	.	.	.	.	.	.	.	.	.	.	.	.	.	19
7062	VIBURNUM DENTATUM	S	.	0	16.9	41.5	.	.	3.3	.	.	.	.	.	.	.	.	.	.	19
7063	VIBURNUM DENTATUM	S	.	0	5.1	23.3	.	.	3.0	.	.	.	.	.	.	.	.	.	.	19
7064	VIBURNUM LANTANA	S	.	0	4.6	10.4	.	.	.	.	.	.	.	.	.	.	.	.	.	21
7065	VIBURNUM ORIENTALE	S	.	0	4.8	7.9	.	.	.	.	.	.	.	.	.	.	.	.	.	21
7066	VIBURNUM TINUS	S	.	0	12.2	32.0	.	.	.	.	.	.	.	.	.	.	.	.	.	21
7067	VICIA AMERICANA(IMMATURE)	H	.	0	27.3	2.3	61.9	24.2	8.5	1300.0	390.0	.	.	.	.	.	.	.	.	6
7068	VICIA ANGUSTIFOLIA	S	.	0	47.8	6.2	.	.	3.1	.	.	.	.	.	.	.	.	.	.	19
7069	VICIA ANGUSTIFOLIA	S	.	0	29.5	2.2	.	.	.	.	.	.	.	.	.	.	.	.	.	20
7070	VICIA ANGUSTIFOLIA	S	.	0	31.9	1.0	.	.	.	.	.	.	.	.	.	.	.	.	.	20
7071	VICIA ANGUSTIFOLIA	S	.	0	31.2	0.7	.	.	.	.	.	.	.	.	.	.	.	.	.	21
7072	VICIA ATROPURPUREA	S	.	0	26.7	0.5	.	.	3.2	.	.	.	.	.	.	.	.	.	.	19
7073	VICIA BENGHALENSIS	W	.	0	23.6	3.0	65.3	28.8	8.1	8500.0	350.0	.	.	.	.	.	.	.	.	18
7074	VICIA BENGHALENSIS	S	.	0	27.2	0.6	.	.	3.5	.	.	.	.	.	.	.	.	.	.	19
7075	VICIA CASSUBICA	S	.	0	43.8	1.0	.	.	.	.	.	.	.	.	.	.	.	.	.	21
7076	VICIA CRACCA	S	.	0	34.9	0.6	.	.	3.4	.	.	.	.	.	.	.	.	.	.	19
7077	VICIA CRACCA	S	.	0	31.2	0.7	.	.	3.4	.	.	.	.	.	.	.	.	.	.	19
7078	VICIA DASYCARPA	S	.	0	33.1	1.1	.	.	3.6	.	.	.	.	.	.	.	.	.	.	19
7079	VICIA DASYCARPA	S	.	0	35.3	1.0	.	.	3.5	.	.	.	.	.	.	.	.	.	.	19
7080	VICIA DASYCARPA	S	.	0	31.2	1.7	.	.	.	.	.	.	.	.	.	.	.	.	.	20
7081	VICIA ERYILIA	S	.	0	22.8	1.0	.	.	.	.	.	.	.	.	.	.	.	.	.	21
7082	VICIA FABA	S	.	0	25.2	1.2	.	.	3.2	.	.	.	.	.	.	.	.	.	.	19
7083	VICIA FABA	W	.	0	20.5	2.7	67.3	22.7	9.5	.	.	.	.	.	.	.	.	.	.	6
7084	VICIA FABA	H	.	0	14.6	0.9	78.5	24.0	6.0	.	.	.	.	.	.	.	.	.	.	6
7085	VICIA FABA	X	300	0	38.3	1.7	53.6	2.0	6.3	8.6	910.7	17.9	253.6	2242.1	216.14	0.86	0.43	2.88	317.0	
7086	VICIA FABA	S	383	0	29.1	1.4	66.1	7.6	3.3	115.7	334.8	7.5	.	.	144.61	0.42	0.27	2.34	17.8	1
7087	VICIA FABA	IS	316	0	30.0	1.7	64.6	13.5	3.8	160.3	535.9	.	.	.	.	0.42	0.93	.	590.7	1
7088	VICIA FABA	X	313	0	30.7	0.8	64.5	15.8	3.9	.	.	.	.	.	.	.	.	.	.	1
7089	VICIA FABA	IS	377	0	33.6	2.0	60.2	7.0	4.1	106.6	606.6	7.4	16.4	1795.1	553.28	1.39	0.70	8.61	106.6	2
7090	VICIA FABA	IS	381	0	30.0	1.3	65.5	12.3	3.2	100.0	451.6	7.4	.	.	387.10	0.90	0.55	5.48	90.3	3
7091	VICIA FABA	S	388	0	27.5	2.5	66.6	6.8	3.4	88.1	427.9	7.2	.	.	68.65	0.61	0.34	2.86	6.9	3
7092	VICIA FABA	F	.	0	30.8	0.7	63.0	13.7	5.5	342.5	506.8	.	301.4	267.1	61.64	0.55	.	5.48	82.2	5
7093	VICIA FABA	L	317	0	25.0	1.3	67.9	16.5	5.8	495.5	665.2	.	.	.	.	.	.	.	.	5
7094	VICIA GIGANTEA	S	.	0	38.1	1.2	.	.	3.0	.	.	.	.	.	.	.	.	.	.	19
7095	VICIA HIRSUTA	S	.	0	27.6	0.6	.	.	3.8	.	.	.	.	.	.	.	.	.	.	19
7096	VICIA LEAVENWORTHII	S	.	0	33.2	0.7	.	.	3.2	.	.	.	.	.	.	.	.	.	.	19
7097	VICIA LEAVENWORTHII	S	.	0	33.1	0.9	.	.	.	.	.	.	.	.	.	.	.	.	.	20
7098	VICIA LUTEA	S	.	0	31.6	0.8	.	.	.	.	.	.	.	.	.	.	.	.	.	21
7099	VICIA MONATHA	S	.	0	27.5	0.8	.	.	.	.	.	.	.	.	.	.	.	.	.	21
7100	VICIA MONATHA	S	.	0	28.6	1.0	.	.	.	.	.	.	.	.	.	.	.	.	.	21
7101	VICIA MONATHA	S	.	0	26.4	1.0	.	.	.	.	.	.	.	.	.	.	.	.	.	21
7102	VICIA NARBONENSIS	W	.	0	24.7	3.2	59.9	22.7	12.2	590.0	360.0	.	.	.	.	.	.	.	.	18
7103	VICIA NARBONENSIS	S	.	0	27.8	0.7	.	.	.	.	.	.	.	.	.	.	.	.	.	21
7104	VICIA NARBONENSIS	S	.	0	25.6	1.2	.	.	.	.	.	.	.	.	.	.	.	.	.	21

TABLE 3 - DATA CONVERTED TO A ZERO-MOISTURE BASIS.

OBS	PLANA	PART	CALORIES	WATER	PROTEIN	FAT	TOTLCARB	FIBER	ASH	CALCIUM	PHOSPHO	IRON	SODIUM	POTASS	CAROT	THIAMINE	RIBOFLAV	NIACIN	ASCORBIC	REFCO
7105	VICIA NARBONENSIS	S	.	0	27.1	0.9	.	.	.	.	.	.	.	.	.	.	.	.	.	21
7106	VICIA NARBONENSIS	S	.	0	23.8	0.6	.	.	.	.	.	.	.	.	.	.	.	.	.	21
7107	VICIA NOEANA	S	.	0	28.2	2.4	.	.	.	.	.	.	.	.	.	.	.	.	.	20
7108	VICIA PANNONICA	S	.	0	27.1	1.7	.	.	3.2	.	.	.	.	.	.	.	.	.	.	19
7109	VICIA PEREGRINA	S	.	0	29.5	1.0	.	.	.	.	.	.	.	.	.	.	.	.	.	21
7110	VICIA PEREGRINA	S	.	0	28.8	0.8	.	.	.	.	.	.	.	.	.	.	.	.	.	21
7111	VICIA SATIVA	W	.	0	35.9	2.8	52.1	21.3	9.2	490.0	610.0	.	.	.	.	.	.	.	.	18
7112	VICIA SATIVA	H	.	0	19.0	2.4	69.5	28.5	9.1	1110.0	300.0	.	.	.	.	.	.	.	.	18
7113	VICIA SATIVA	S	.	0	26.1	0.8	.	.	3.2	.	.	.	.	.	.	.	.	.	.	19
7114	VICIA SATIVA	S	.	0	29.9	0.8	.	.	3.3	.	.	.	.	.	.	.	.	.	.	19
7115	VICIA SATIVA	S	.	0	30.2	1.0	.	.	.	.	.	.	.	.	.	.	.	.	.	20
7116	VICIA SATIVA	S	.	0	31.6	0.6	.	.	.	.	.	.	.	.	.	.	.	.	.	20
7117	VICIA SATIVA	S	.	0	31.6	1.0	.	.	.	.	.	.	.	.	.	.	.	.	.	21
7118	VICIA SATIVA	H	.	0	19.1	2.1	69.8	27.6	9.0	.	.	.	.	.	.	.	.	.	.	6
7119	VICIA SATIVA SUBSP. SATIVA	S	386	0	31.0	1.9	64.3	2.2	2.7	154.1	270.0	10.5	.	.	0.00	0.17	0.11	3.04	.	1
7120	VICIA SINENSIS	S	.	0	27.2	1.4	.	.	3.8	.	.	.	.	.	.	.	.	.	.	19
7121	VICIA SP.	S	.	0	30.0	1.2	.	.	.	.	.	.	.	.	.	.	.	.	.	19
7122	VICIA VILLOSA	W	.	0	30.5	2.4	53.8	24.2	13.3	640.0	460.0	.	.	.	.	.	.	.	.	18
7123	VICIA VILLOSA	S	.	0	25.6	0.9	.	.	3.3	.	.	.	.	.	.	.	.	.	.	19
7124	VICIA VILLOSA	S	.	0	30.2	0.7	.	.	.	.	.	.	.	.	.	.	.	.	.	20
7125	VICIA VILLOSA	W	.	0	23.1	2.7	62.1	27.5	12.1	1100.0	330.0	.	.	2250.0	.	.	.	.	.	6
7126	VICIA VILLOSA	H	.	0	22.6	3.2	64.3	28.7	9.9	.	.	.	.	.	.	.	.	.	.	6
7127	VIGNA ACONITIFOLIA	W	.	0	15.5	1.9	70.0	19.7	12.6	2370.0	540.0	.	.	.	.	.	.	.	.	18
7128	VIGNA ACONITIFOLIA	H	.	0	10.6	1.9	73.7	26.8	13.8	2340.0	180.0	.	.	.	.	.	.	.	.	18
7129	VIGNA ACONITIFOLIA	F	.	0	9.6	2.9	73.1	19.4	14.4	2010.0	250.0	.	.	.	.	.	.	.	.	18
7130	VIGNA ACONITIFOLIA	S	.	0	26.6	0.6	67.2	5.3	5.6	350.0	380.0	.	.	.	.	.	.	.	.	18
7131	VIGNA ANGULARIS	IS	386	0	27.4	0.3	71.1	1.7	1.1	68.6	228.6	6.0	.	.	0.00	0.03	0.03	0.57	0.0	2
7132	VIGNA ANGULARIS	S	374	0	29.1	0.7	65.7	6.6	4.5	290.0	364.8	8.7	.	.	11.51	0.66	0.21	3.68	.	3
7133	VIGNA ANGULARIS	S	.	0	29.1	0.7	65.6	6.6	4.5	289.7	.	8.7	.	.	10.34	0.66	0.21	3.68	.	5
7134	VIGNA CYLINDRICA	S	.	0	25.7	1.4	.	.	3.6	.	.	.	.	.	.	.	.	.	.	19
7135	VIGNA HOSEI	L	.	0	21.9	.	.	.	11.4	2139.3	895.5	.	.	1940.3	.	.	.	.	.	5
7136	VIGNA LUTEOLA	S	.	0	25.2	0.6	.	.	.	.	.	.	.	.	.	.	.	.	.	21
7137	VIGNA MUNGO	W	.	0	19.4	2.5	62.1	26.8	16.0	1970.0	240.0	.	.	.	.	.	.	.	.	18
7138	VIGNA MUNGO	H	.	0	8.9	2.8	75.7	28.6	12.6	1740.0	160.0	.	.	.	.	.	.	.	.	18
7139	VIGNA MUNGO	F	.	0	9.0	2.3	76.5	29.9	12.2	2710.0	190.0	.	.	.	.	.	.	.	.	18
7140	VIGNA MUNGO	S	.	0	26.8	0.9	66.7	5.3	5.6	220.0	390.0	.	.	.	.	.	.	.	.	18
7141	VIGNA MUNGO	S	371	0	26.7	1.2	66.0	5.4	6.1	122.2	424.4	9.9	.	.	50.00	0.53	0.23	2.56	0.0	1
7142	VIGNA MUNGO	S	385	0	23.5	1.8	70.9	4.9	3.8	123.0	389.3	9.4	.	191.3	22.37	0.65	0.22	2.57	2.2	2
7143	VIGNA MUNGO	S	389	0	26.9	1.6	67.9	1.0	3.6	172.8	432.1	10.2	44.9	897.9	42.65	0.47	0.22	2.24	.	5
7144	VIGNA RADIATA	W	.	0	13.0	3.7	71.9	21.0	11.4	2470.0	340.0	.	.	.	.	.	.	.	.	18
7145	VIGNA RADIATA	H	.	0	9.0	2.3	76.6	29.9	12.1	2710.0	200.0	.	.	.	.	.	.	.	.	18
7146	VIGNA RADIATA	F	.	0	8.9	2.8	75.2	28.1	13.1	2490.0	150.0	.	.	.	.	.	.	.	.	18
7147	VIGNA RADIATA	S	.	0	23.3	1.0	70.4	6.7	5.3	80.0	.	.	.	.	.	.	.	.	.	18
7148	VIGNA RADIATA	S	378	0	26.1	1.0	68.5	4.7	4.4	113.1	485.6	.	.	.	138.58	0.58	0.22	2.66	.	1
7149	VIGNA RADIATA	X	303	0	42.4	2.0	50.5	9.1	5.1	151.5	717.2	12.1	70.7	2242.4	202.02	1.11	1.01	8.08	181.8	2

TABLE 3 - DATA CONVERTED TO A ZERO-MOISTURE BASIS.

OBS	PLNA	PART	CALORIES	WATER	PROTEIN	FAT	TOTLCARB	FIBER
7150	VIGNA RADIATA	X	313	0	33.9	1.8	58.9	6.3
7151	VIGNA RADIATA	S	373	0	26.8	1.5	67.9	4.6
7152	VIGNA RADIATA	X	.	0	33.9	1.8	58.9	6.3
7153	VIGNA SINENSIS	S	.	0	25.7	1.4	.	.
7154	VIGNA SINENSIS	S	.	0	27.0	1.3	.	.
7155	VIGNA TRILOBATA	N	.	0	12.8	1.5	71.1	24.7
7156	VIGNA UMBELLATA	W	.	0	16.9	1.9	73.4	30.6
7157	VIGNA UMBELLATA	S	.	0	23.3	0.6	71.5	6.0
7158	VIGNA UMBELLATA	S	377	0	24.1	1.0	70.0	5.5
7159	VIGNA UMBELLATA	S	390	0	21.5	1.2	75.0	.
7160	VIGNA UMBELLATA	S	378	0	21.6	0.7	73.1	6.5
7161	VIGNA UMBELLATA	S	.	0	21.9	0.7	.	6.3
7162	VIGNA UMBELLATA	F	.	0	17.4	6.2	72.2	7.8
7163	VIGNA UMBELLATA	SH	.	0	16.9	1.9	73.4	30.6
7164	VIGNA UNGUICULATA	W	.	0	17.6	2.2	73.1	21.4
7165	VIGNA UNGUICULATA	H	.	0	10.6	1.8	74.5	16.2
7166	VIGNA UNGUICULATA	F	.	0	7.2	1.2	80.1	37.7
7167	VIGNA UNGUICULATA	S	.	0	23.8	1.1	68.4	6.4
7168	VIGNA UNGUICULATA	W	.	0	15.1	4.8	67.4	24.8
7169	VIGNA UNGUICULATA	H	.	0	18.4	2.9	67.1	27.3
7170	VIGNA UNGUICULATA	F	.	0	23.8	1.4	68.7	13.6
7171	VIGNA UNGUICULATA	L	345	0	30.9	6.4	48.2	10.9
7172	VIGNA UNGUICULATA SUBSP. CYLINDRICA	IS	299	0	29.3	2.4	57.9	12.8
7173	VIGNA UNGUICULATA SUBSP. CYLINDRICA	L	293	0	31.3	2.0	55.3	13.3

OBS	ASH	CALCIUM	PHOSPHO	IRON	SODIUM	POTASS	CAROT	THIAMINE	RIBOFLAV	NIACIN	ASCORBIC	REFCO
7150	5.4	169.6	571.4	11.6	44.6	1991.1	107.1	1.16	1.16	7.14	169.6	4
7151	3.9	138.4	363.8	8.1	31.3	940.8	104.9	0.52	0.30	2.34	0.0	5
7152	5.4	169.6	571.4	11.6	44.6	1991.1	.	.	.	.	.	5
7153	3.6	.	.	.	.	.	.	.	.	.	.	19
7154	3.5	.	.	.	.	.	.	.	.	.	.	19
7155	15.0	3012.3	447.9	.	.	.	.	.	.	.	.	5
7156	7.8	1030.0	260.0	.	.	.	.	.	.	.	.	18
7157	4.6	670.0	390.0	.	.	.	.	.	.	.	.	18
7158	4.8	230.7	449.8	12.6	.	.	.	0.57	0.24	2.77	.	1
7159	2.3	93.0	465.1	5.8	.	.	0.0	0.35	0.24	2.79	10.5	2
7160	4.7	243.6	349.7	8.0	.	.	22.3	0.65	0.09	2.46	.	3
7161	4.9	412.8	802.8	12.5	.	.	.	0.68	0.24	2.75	.	5
7162	4.3	352.0	264.0	8.8	.	.	.	0.52	0.23	2.42	.	5
7163	7.8	1437.5	593.8	.	.	.	.	.	.	.	.	5
7164	7.1	550.0	290.0	.	.	.	.	.	.	.	.	18
7165	13.1	1810.0	1180.0	.	.	.	.	.	.	.	.	18
7166	11.5	1660.0	130.0	.	.	.	.	.	.	.	.	18
7167	6.7	400.0	290.0	.	.	.	.	.	.	.	.	18
7168	12.7	.	.	.	.	.	.	.	.	.	.	6
7169	11.6	1340.0	320.0	90.0	.	1990.0	.	.	.	.	.	6
7170	6.1	489.8	401.4	17.0	.	.	3836.7	0.48	0.61	6.12	95.2	5
7171	14.5	2636.4	527.3	182.7	.	.	55200.0	0.45	1.64	5.45	36.4	5
7172	10.4	920.7	414.6	.	.	.	914.6	.	.	.	.	1
7173	11.3	1706.7	420.0	38.0	.	.	53133.3	1.33	2.47	14.00	373.3	1

TABLE 3 - DATA CONVERTED TO A ZERO-MOISTURE BASIS.

OBS	PLNA	PART	CALORIES	WATER	PROTEIN	FAT	TOTLCARB	FIBER
7174	VIGNA UNGUICULATA SUBSP. CYLINDRICA	S	383	0	25.9	1.6	68.8	5.4
7175	VIGNA UNGUICULATA SUBSP. SESQUIPEDALIS	F	316	0	25.6	1.7	67.5	13.7
7176	VIGNA UNGUICULATA SUBSP. SESQUIPEDALIS	L	293	0	36.2	3.4	50.0	14.7
7177	VIGNA UNGUICULATA SUBSP. SESQUIPEDALIS	F	309	0	30.0	0.9	62.7	11.8
7178	VIGNA UNGUICULATA SUBSP. UNGUICULATA	S	.	0	24.9	0.6	71.0	6.0
7179	VIGNA UNGUICULATA SUBSP. UNGUICULATA	S	381	0	25.4	1.6	68.8	6.1
7180	VIGNA UNGUICULATA SUBSP. UNGUICULATA	SH	273	0	43.6	2.7	40.0	.
7181	VIGNA UNGUICULATA SUBSP. UNGUICULATA	F	319	0	32.7	5.3	54.9	10.6
7182	VIGNA UNGUICULATA SUBSP. UNGUICULATA	F	320	0	24.5	2.0	68.0	12.2
7183	VIGNA UNGUICULATA SUBSP. UNGUICULATA	S	381	0	27.0	1.3	67.9	5.5
7184	VIGNA VEXILLATA	S	.	0	33.1	3.1	.	.
7185	VIGUIERA CILIATA	S	.	0	26.2	30.6	.	.
7186	VIGUIERA DECURRENS	S	.	0	19.4	11.2	.	.
7187	VIGUIERA DENTATA	S	.	0	25.0	24.9	.	.
7188	VIGUIERA EXCELSA	S	.	0	21.9	14.0	.	.
7189	VIGUIERA HYPARGYREA	S	.	0	29.4	19.8	.	.
7190	VIGUIERA LACINIATA	S	.	0	19.3	19.9	.	.
7191	VIGUIERA LINEARIS	S	.	0	27.5	24.8	.	.
7192	VIGUIERA MULTIFLORA	S	.	0	24.1	32.7	.	.
7193	VIGUIERA MULTIFLORA	S	.	0	24.0	35.9	.	.
7194	VIGUIERA SP.	S	.	0	31.9	18.7	.	.
7195	VIGUIERA STENOLOBA	S	.	0	20.6	19.4	.	.
7196	VIGUIERA STENOLOBA	S	.	0	20.4	20.4	.	.
7197	VINCA ROSEA	S	.	0	19.4	34.4	.	.

OBS	ASH	CALCIUM	PHOSPHO	IRON	SODIUM	POTASS	CAROT	THIAMINE	RIBOFLAV	NIACIN	ASCORBIC	REFCO
7174	3.7	113.2	429.4	8.5	.	.	78.5	0.84	0.20	2.80	1.1	1
7175	5.1	376.1	384.6	6.0	51.3	1991.5	1923.1	1.03	0.94	8.55	188.0	2
7176	10.3	931.0	913.8	40.5	215.5	3155.2	20689.7	2.41	2.07	10.34	301.7	2
7177	6.4	490.9	618.2	12.7	.	.	4272.7	0.82	1.09	10.00	236.4	3
7178	3.5	317.5	442.2	8.6	.	.	80.5	.	.	1.70	.	
7179	4.2	117.4	469.5	.	.	.	.	0.09	0.10	4.51	2.3	1
7180	16.4	663.6	963.6	20.0	.	.	.	3.18	1.64	10.00	327.3	2
7181	7.1	477.9	522.1	12.4	17.7	1946.9	4026.5	1.24	0.88	8.85	212.4	2
7182	5.4	306.1	353.7	8.2	.	.	2312.9	0.88	0.68	6.80	149.7	3
7183	3.8	86.1	469.8	8.1	.	.	22.4	0.97	0.26	2.13	3.4	3
7184	4.3	.	.	.	.	.	.	.	.	.	.	19
7185	.	.	.	.	.	.	.	.	.	.	.	20
7186	.	.	.	.	.	.	.	.	.	.	.	20
7187	.	.	.	.	.	.	.	.	.	.	.	20
7188	.	.	.	.	.	.	.	.	.	.	.	21
7189	.	.	.	.	.	.	.	.	.	.	.	20
7190	4.2	.	.	.	.	.	.	.	.	.	.	19
7191	.	.	.	.	.	.	.	.	.	.	.	20
7192	.	.	.	.	.	.	.	.	.	.	.	21
7193	.	.	.	.	.	.	.	.	.	.	.	21
7194	.	.	.	.	.	.	.	.	.	.	.	20
7195	.	.	.	.	.	.	.	.	.	.	.	20
7196	.	.	.	.	.	.	.	.	.	.	.	20
7197	.	.	.	.	.	.	.	.	.	.	.	20

TABLE 3 - DATA CONVERTED TO A ZERO-MOISTURE BASIS.

OBS	PLNA	PART	CALORIES	WATER	PROTEIN	FAT	TOTLCARB	FIBER
7198	VIROLA GUATEMALENSIS	S	.	0	11.20	70.70	.	.
7199	VITEX AGNUS-CASTUS	S	.	0	8.10	9.00	.	.
7200	VITEX AGNUS-CASTUS	S	.	0	6.10	6.50	.	.
7201	VITEX AGNUS-CASTUS	S	.	0	5.40	3.80	.	.
7202	VITEX DONIANA	F	354	0	2.40	1.40	93.20	4.40
7203	VITEX NEGUNDO	S	.	0	2.50	0.40	.	.
7204	VITIS AMURENSIS	F	354	0	5.10	3.10	86.70	17.90
7205	VITIS RIPARIA	S	.	0	10.60	17.40	.	.
7206	VITIS TRIFOLIA	S	.	0	11.30	15.80	.	.
7207	VITIS TRIFOLIA	F	354	0	12.40	2.50	80.10	21.70
7208	VITIS VINIFERA	L	.	0	15.80	13.40	62.20	9.20
7209	VITIS VINIFERA	F	358	0	3.50	2.30	91.30	11.60
7210	VITIS VINIFERA	F	357	0	3.60	2.10	91.40	6.40
7211	VITIS VINIFERA	F	370	0	3.30	3.80	90.80	2.70
7212	VITIS VINIFERA	F	353	0	3.10	0.40	94.20	1.30
7213	VITIS VINIFERA	F	375	0	7.10	5.40	85.30	3.30
7214	VITIS VINIFERA	F	360	0	3.20	1.60	93.00	2.70
7215	VITIS VULPINA	S	.	0	10.00	15.90	.	.
7216	VOANDZEIA SUBTERRANEA	H	.	0	15.90	1.80	74.80	31.70
7217	VOANDZEIA SUBTERRANEA	F	.	0	18.20	5.50	70.90	14.20
7218	VOANDZEIA SUBTERRANEA	S	.	0	21.30	7.50	67.90	6.70
7219	VOANDZEIA SUBTERRANEA	S	356	0	18.30	7.30	70.30	7.00
7220	VOANDZEIA SUBTERRANEA	S	409	0	21.00	6.90	68.30	5.40
7221	VOANDZEIA SUBTERRANEA	S	412	0	17.80	6.70	72.30	.

OBS	ASH	CALCIUM	PHOSPHO	IRON	SODIUM	POTASS	CAROT	THIAMINE	RIBOFLAV	NIACIN	ASCORBIC	REFCO
7198	.	.	.	.	.	.	.	.	.	.	.	20
7199	.	.	.	.	.	.	.	.	.	.	.	19
7200	.	.	.	.	.	.	.	.	.	.	.	21
7201	.	.	.	.	.	.	.	.	.	.	.	21
7202	3.1	115.6	159.9	9.2	.	.	.	0.07	.	.	30.6	1
7203	.	.	.	.	.	.	.	.	.	.	.	21
7204	5.1	374.4	51.3	8.7	.	.	0.0	0.26	0.15	2.56	41.0	2
7205	.	.	.	.	.	.	.	.	.	.	.	19
7206	.	.	.	.	.	.	.	.	.	.	.	21
7207	5.0	385.1	254.7	13.7	.	.	124.2	0.25	0.25	7.45	37.3	3
7208	8.6	2250.0	160.0	.	.	.	.	.	.	.	.	18
7209	2.9	121.4	109.8	4.6	.	.	289.0	.	.	.	.	1
7210	2.9	64.3	142.9	4.3	42.9	792.9	357.1	0.71	0.43	1.43	28.6	2
7211	2.2	65.2	81.5	4.9	.	.	0.0	0.27	0.22	2.72	16.3	3
7212	2.3	62.7	91.5	3.8	.	.	0.0	0.15	0.16	0.50	15.0	3
7213	2.2	87.0	65.2	2.2	16.3	858.7	326.1	0.27	0.16	1.63	21.7	4
7214	2.2	64.5	107.5	2.2	16.1	930.1	322.6	0.27	0.16	1.61	21.5	4
7215	3.8	.	.	.	.	.	.	.	.	.	.	19
7216	7.5	.	.	.	.	.	.	.	.	.	.	18
7217	5.4	.	.	.	.	.	.	.	.	.	.	18
7218	3.3	10.0	300.0	.	.	.	.	.	.	.	.	18
7219	4.2	32.8	604.2	2.8	.	.	.	.	.	.	.	1
7220	3.8	69.1	307.7	13.6	.	.	11.1	0.52	0.16	2.01	0.0	1
7221	3.3	94.5	293.7	4.7	.	.	0.0	0.20	.	.	0.0	2

TABLE 3 - DATA CONVERTED TO A ZERO-MOISTURE BASIS.

OBS	PLNA	PART	CALORIES	WATER	PROTEIN	FAT	TOTLCARB	FIBER
7222	WATSONIA MERIANA	S	.	0	14.60	2.70	.	.
7223	WATSONIA PYRAMIDATA	S	.	0	17.80	3.40	.	.
7224	WATSONIA SCHLECHTERI	S	.	0	17.10	9.70	.	.
7225	WATSONIA SP.	S	.	0	14.20	7.40	.	.
7226	WIEDEMANNIA ORIENTALIS	S	.	0	25.70	40.20	.	.
7227	WILDENOWIA STRIATA		.	0	3.92	1.06	92.42	34.41
7228	WILLARDIA MEXICANA	S	.	0	26.20	33.70	.	.
7229	WILLDENOWIA ARGENTEA	S	.	0	46.90	1.30	.	.
7230	WILLDENOWIA STRIATA	S	.	0	44.10	0.60	.	.
7231	WISLIZENIA REFRACTA	S	.	0	18.60	23.00	.	.
7232	WISTERIA FLORIBUNDA VAR. MULTIJUGA	S	.	0	36.20	11.00	.	.
7233	WITHANIA SOMNIFERA	S	.	0	18.40	17.50	.	.
7234	WITHANIA SOMNIFERA	S	.	0	18.50	19.60	.	.
7235	WYETHIA ANGUSTIFOLIA	S	.	0	24.10	5.40	.	.
7236	WYETHIA HELENIOIDES	S	.	0	21.20	15.20	.	.
7237	XANTHISMA TEXANUM	S	.	0	17.10	11.20	.	.
7238	XANTHIUM PENSYLVANICUM	S	.	0	41.40	37.80	.	.
7239	XANTHIUM SP.	S	.	0	43.80	34.30	.	.
7240	XANTHIUM SPECIOSUM	S	.	0	47.40	34.80	.	.
7241	XANTHIUM STRUMARIUM	S	.	0	49.10	35.70	.	.
7242	XANTHIUM STRUMARIUM	S	.	0	41.20	38.20	.	.
7243	XANTHOSOMA SAGITTIFOLIUM	L	.	0	22.90	7.20	55.30	10.00
7244	XANTHOSOMA SAGITTIFOLIUM	C	.	0	13.00	2.40	63.20	15.60
7245	XANTHOSOMA SAGITTIFOLIUM	R	.	0	6.90	0.20	88.70	2.60

OBS	ASH	CALCIUM	PHOSPHO	IRON	SODIUM	POTASS	CAROT	THIAMINE	RIBOFLAV	NIACIN	ASCORBIC	REFCO
7222	.	.	.	.	.	.	.	.	.	.	.	21
7223	.	.	.	.	.	.	.	.	.	.	.	21
7224	.	.	.	.	.	.	.	.	.	.	.	21
7225	.	.	.	.	.	.	.	.	.	.	.	21
7226	.	.	.	.	.	.	.	.	.	.	.	20
7227	2.6	150.0	60.0	6.8	.	.	.	.	.	.	.	8
7228	.	.	.	.	.	.	.	.	.	.	.	19
7229	.	.	.	.	.	.	.	.	.	.	.	21
7230	.	.	.	.	.	.	.	.	.	.	.	21
7231	.	.	.	.	.	.	.	.	.	.	.	21
7232	3.4	.	.	.	.	.	.	.	.	.	.	19
7233	.	.	.	.	.	.	.	.	.	.	.	21
7234	.	.	.	.	.	.	.	.	.	.	.	21
7235	5.6	.	.	.	.	.	.	.	.	.	.	19
7236	5.6	.	.	.	.	.	.	.	.	.	.	19
7237	3.3	.	.	.	.	.	.	.	.	.	.	19
7238	.	.	.	.	.	.	.	.	.	.	.	21
7239	.	.	.	.	.	.	.	.	.	.	.	19
7240	.	.	.	.	.	.	.	.	.	.	.	21
7241	.	.	.	.	.	.	.	.	.	.	.	21
7242	.	.	.	.	.	.	.	.	.	.	.	21
7243	14.6	1780.0	640.0	.	.	.	.	.	.	.	.	18
7244	21.4	.	.	.	.	.	.	.	.	.	.	18
7245	4.2	190.0	200.0	.	.	.	.	.	.	.	.	18

TABLE 3 - DATA CONVERTED TO A ZERO-MOISTURE BASIS.

OBS	PLNA	PART	CALORIES	WATER	PROTEIN	FAT	TOTLCARB	FIBER	ASH	CALCIUM	PHOSPHO	IRON	SODIUM	POTASS	CAROT	THIAMINE	RIBOFLAV	NIACIN	ASCORBIC	REFCO
7246	XANTHOSOMA SAGITTIFOLIUM	R	385	0	6.2	0.6	89.9	2.8	3.4	44.9	132.0	2.5	.	.	.	.	.	.	22.5	1
7247	XANTHOSOMA SAGITTIFOLIUM	L	337	0	24.8	9.9	52.5	20.8	12.9	940.6	3841.6	19.8	.	.	32673.3	.	.	.	366.3	1
7248	XANTHOSOMA SAGITTIFOLIUM	R	401	0	0.6	0.7	97.6	.	1.1	24.7	188.2	2.4	.	.	.	.	.	.	.	1
7249	XANTHOSOMA SAGITTIFOLIUM	SH	314	0	29.5	5.7	54.3	30.5	10.5	466.7	761.9	2.9	.	.	.	.	.	.	781.0	1
7250	XANTHOSOMA SAGITTIFOLIUM	R	387	0	5.0	0.9	90.6	1.8	3.5	41.1	164.2	2.3	.	.	58.6	0.38	0.09	2.05	14.7	3
7251	XANTHOXYLUM SP.	S	.	0	9.4	11.0	.	.	8.0	.	.	.	.	.	.	.	.	.	.	19
7252	XANTHOXYLUM SP.	S	.	0	12.5	2.3	.	.	5.5	.	.	.	.	.	.	.	.	.	.	19
7253	XERANTHEMUM ANNUUM	S	.	0	30.8	14.5	.	.	4.4	.	.	.	.	.	.	.	.	.	.	19
7254	XERANTHEMUM ANNUUM	S	.	0	24.4	9.3	.	.	7.3	.	.	.	.	.	.	.	.	.	.	19
7255	XERANTHEMUM ANNUUM	S	.	0	19.9	18.4	.	.	.	.	.	.	.	.	.	.	.	.	.	21
7256	XEROPHYLLUM SP.	L	.	0	7.3	2.3	83.0	42.0	7.4	.	.	.	.	.	.	.	.	.	.	6
7257	XIMENESIA ENCELIOIDES	S	.	0	35.6	30.8	.	.	4.6	.	.	.	.	.	.	.	.	.	.	19
7258	XIMENIA AMERICANA	S	.	0	20.0	62.4	.	.	.	.	.	.	.	.	.	.	.	.	.	19
7259	XIMENIA AMERICANA	S	703	0	18.4	63.9	13.7	.	4.0	.	.	.	.	.	.	.	.	.	9.5	5
7260	XIMENIA CAFFRA	F	.	0	.	.	.	.	.	.	.	.	.	.	.	.	.	.	145.8	1
7261	YUCCA ALOIFOLIA	I	330	0	31.0	2.0	60.0	.	7.0	470.0	730.0	5.0	.	.	200.0	1.40	0.90	6.00	.	3
7262	YUCCA ARIZONICA	S	.	0	14.4	28.4	.	.	.	.	.	.	.	.	.	.	.	.	.	19
7263	YUCCA BACCATA	S	.	0	14.4	27.2	.	.	.	.	.	.	.	.	.	.	.	.	.	19
7264	YUCCA CONSTRICTA	S	.	0	21.4	25.9	.	.	3.1	.	.	.	.	.	.	.	.	.	.	19
7265	YUCCA ELATA	S	.	0	20.6	29.1	.	.	3.0	.	.	.	.	.	.	.	.	.	.	19
7266	YUCCA ELATA	S	.	0	22.2	29.7	.	.	2.8	.	.	.	.	.	.	.	.	.	.	19
7267	YUCCA ELATA	S	.	0	21.2	31.7	.	.	.	.	.	.	.	.	.	.	.	.	.	19
7268	YUCCA ELATA	S	.	0	22.5	32.6	.	.	.	.	.	.	.	.	.	.	.	.	.	20
7269	YUCCA ELATA	Y	.	0	4.5	1.4	87.6	29.6	6.5	.	.	.	.	.	.	.	.	.	.	6
7270	YUCCA ELEPHANTIPES	L	306	0	12.2	3.1	65.3	11.2	19.4	3489.8	244.9	9.2	.	.	0.0	2.04	0.41	3.06	265.3	3
7271	YUCCA ELEPHANTIPES	I	363	0	11.9	1.8	81.5	5.4	4.8	202.4	410.7	8.3	.	.	119.0	0.95	0.89	8.93	2339.3	3
7272	YUCCA GLAUCA	S	.	0	20.9	26.6	.	.	3.4	.	.	.	.	.	.	.	.	.	.	19
7273	YUCCA GLAUCA	W	.	0	5.7	1.9	84.9	32.9	7.5	.	.	.	.	.	.	.	.	.	.	6
7274	YUCCA PENINSULARIS	S	.	0	15.0	27.5	.	.	.	.	.	.	.	.	.	.	.	.	.	19
7275	YUCCA SCHIDIGERA	S	.	0	11.9	34.5	.	.	.	.	.	.	.	.	.	.	.	.	.	20
7276	YUCCA SCHOTTII	S	.	0	12.5	29.4	.	.	.	.	.	.	.	.	.	.	.	.	.	20
7277	YUCCA SP.	S	.	0	13.8	28.3	.	.	.	.	.	.	.	.	.	.	.	.	.	19
7278	YUCCA TORREYI	L	.	0	12.0	1.9	75.6	36.1	10.5	860.0	140.0	12.0	.	860.0	.	.	.	.	.	6
7279	YUCCA TRECULEANA	L	.	0	4.2	2.1	86.8	30.0	6.9	3820.0	170.0	.	.	1780.0	.	.	.	.	.	6
7280	YUCCA WHIPPLEI	S	.	0	24.4	23.4	.	.	.	.	.	.	.	.	.	.	.	.	.	19
7281	ZALUZANIA DISCOIDEA	S	.	0	33.1	29.2	.	.	.	.	.	.	.	.	.	.	.	.	.	20
7282	ZALUZANIA SP.	S	.	0	32.5	26.5	.	.	.	.	.	.	.	.	.	.	.	.	.	20
7283	ZANTHOXYLUM PIPERITUM	S	.	0	15.8	31.8	.	.	.	.	.	.	.	.	.	.	.	.	.	21
7284	ZEA MAYS	W	.	0	10.3	2.4	77.0	28.6	10.3	.	.	.	.	.	.	.	.	.	.	18
7285	ZEA MAYS	C	.	0	7.7	0.8	83.0	46.2	8.5	.	.	.	.	.	.	.	.	.	.	18
7286	ZEA MAYS	H	.	0	7.0	1.4	84.7	27.0	6.9	.	.	.	.	.	.	.	.	.	.	18
7287	ZEA MAYS	H	.	0	8.9	2.4	82.0	25.9	6.7	500.0	250.0	10.0	.	930.0	.	.	.	.	.	6
7288	ZEA MAYS	W	.	0	8.2	2.7	83.4	25.3	5.7	270.0	220.0	9.0	.	1260.0	.	.	.	.	.	6
7289	ZEA MAYS SUBSP MEXICANA	S	380	0	24.5	2.8	71.7	0.5	0.9	10.2	220.5	3.8	.	.	0.0	0.16	0.08	0.80	0.0	3
7290	ZEA MAYS SUBSP MEXICANA	L	.	0	4.4	1.3	83.4	32.2	10.9	1062.5	375.0	.	.	2250.0	.	.	.	.	.	5

TABLE 3 - DATA CONVERTED TO A ZERO-MOISTURE BASIS.

OBS	PLANA	PART	CALORIES	WATER	PROTEIN	FAT	TOTCARB	FIBER	ASH	CALCIUM	PHOSPHO	IRON	SODIUM	POTASS	CAROT	THIAMINE	RIBOFLAV	NIACIN	ASCORBIC	REFCO
7291	ZEA MAYS SUBSP. MAYS	S	404	0	10.6	4.8	83.3	2.1	1.4	18.1	248.9	4.1	.	.	0.00	0.37	0.11	2.49	0.0	1
7292	ZEA MAYS SUBSP. MAYS	S	405	0	9.8	4.9	83.8	1.6	1.6	6.8	206.3	3.2	.	.	22.68	0.41	0.09	1.59	5.7	1
7293	ZEA MAYS SUBSP. MAYS	S	406	0	11.2	5.4	82.1	2.2	1.3	14.5	244.4	5.5	.	.	111.61	0.36	0.13	1.90	4.5	1
7294	ZEA MAYS SUBSP. MAYS	X	402	0	9.3	3.5	86.0	3.7	1.1	.	264.5	3.7	.	.	0.00	0.00	0.20	3.78	.	1
7295	ZEA MAYS SUBSP. MAYS	G	357	0	11.2	4.5	81.9	2.9	2.4	13.3	336.0	2.4	8.0	690.7	426.67	0.53	0.29	4.00	21.3	2
7296	ZEA MAYS SUBSP. MAYS	G	357	0	11.2	4.5	81.9	2.9	2.4	13.3	336.0	2.4	8.0	690.7	0.00	0.53	0.29	4.00	21.3	2
7297	ZEA MAYS SUBSP. MAYS	S	404	0	10.5	4.9	83.0	2.7	1.6	16.2	283.6	3.2	5.8	.	0.00	0.34	0.13	2.43	0.0	2
7298	ZEA MAYS SUBSP. MAYS	S	404	0	10.5	4.9	83.0	2.7	1.6	16.2	283.6	3.2	5.8	.	312.50	0.34	0.13	2.43	0.0	2
7299	ZEA MAYS SUBSP. MAYS	S	404	0	10.5	4.8	83.2	2.0	1.5	10.1	324.4	2.8	.	.	156.60	0.48	0.11	2.13	0.0	3
7300	ZEA MAYS SUBSP. MAYS	S	353	0	11.2	3.6	83.0	2.7	2.2	13.7	350.7	3.0	.	.	191.78	0.49	0.22	5.21	24.7	3
7301	ZEA MAYS SUBSP. MAYS	S	352	0	12.8	3.7	81.0	2.6	2.6	11.0	406.6	2.6	0.0	1025.6	879.12	0.55	0.44	6.23	44.0	4
7302	ZEA MAYS SUBSP. MAYS	S	401	0	13.2	5.2	79.9	2.3	1.7	11.1	292.7	2.8	3.3	.	.	0.43	0.12	2.33	0.0	4
7303	ZEA MAYS SUSP. MEXICANA	W	.	0	9.5	2.9	79.6	23.0	8.0	.	.	.	.	.	.	.	.	.	.	18
7304	ZELKOVA SERRATA	S	.	0	18.2	21.3	.	.	8.3	.	.	.	.	.	.	.	.	.	.	19
7305	ZEPHYRANTHES LONGIFOLIA	S	.	0	5.0	9.6	.	.	.	.	.	.	.	.	.	.	.	.	.	19
7306	ZEXMANIA HISPIDA(BUDDING)	W	.	0	17.5	3.2	63.8	21.3	15.5	3510.0	200.0	.	.	2480.0	.	.	.	.	.	6
7307	ZEXMENIA HISPIDA(FLOWERING)	W	.	0	12.2	1.3	70.0	25.7	16.5	2930.0	180.0	.	.	3150.0	.	.	.	.	.	6
7308	ZEXMENIA SEEMANNII	S	.	0	17.0	6.6	.	.	.	.	.	.	.	.	.	.	.	.	.	21
7309	ZINGIBER MIOGA	I	293	0	20.7	1.7	65.5	29.3	12.1	569.0	275.9	10.3	86.2	1724.1	258.62	1.21	1.03	8.62	86.2	2
7310	ZINGIBER OFFICINALE	R	335	0	8.5	3.2	80.6	.	7.7	200.4	.	.	.	.	133.63	0.18	0.30	9.35	0.0	1
7311	ZINGIBER OFFICINALE	R	383	0	10.0	6.6	78.1	6.5	5.3	128.0	163.4	12.7	35.3	1481.2	97.13	0.06	0.20	5.70	.	16
7312	ZINGIBER OFFICINALE	R	365	0	12.7	6.3	73.0	10.3	7.9	150.8	254.0	10.3	55.6	2507.9	436.51	0.08	0.24	13.49	31.7	2
7313	ZINGIBER OFFICINALE	SH	314	0	11.4	8.6	60.0	17.1	20.0	600.0	785.7	10.0	.	6785.7	.	.	.	.	.	2
7314	ZINGIBER OFFICINALE	R	379	0	12.9	6.5	72.6	7.3	8.1	354.8	532.3	14.5	.	.	0.00	0.16	0.48	5.65	16.1	3
7315	ZINGIBER OFFICINALE	R	377	0	10.8	7.7	73.1	8.5	8.5	176.9	276.9	16.2	46.2	2030.8	46.15	0.15	0.31	5.38	30.8	4
7316	ZINGIBER SP.?	R	.	0	12.0	4.7	77.0	12.6	6.3	104.7	314.1	13.6	.	.	.	0.31	0.16	3.14	31.4	5
7317	ZINNIA ELEGANS	S	.	0	37.5	28.3	.	.	.	.	.	.	.	.	.	.	.	.	.	19
7318	ZIZANIA AQUATICA	S	.	0	15.1	0.7	82.7	1.0	1.5	.	.	.	.	.	.	.	.	.	.	6
7319	ZIZANIA AQUATICA	S	386	0	15.4	0.8	82.3	1.1	1.5	20.8	370.5	4.6	7.7	240.4	0.00	0.49	0.69	6.78	0.0	4
7320	ZIZANIA PALUSTRIS	SH	337	0	18.6	2.3	70.9	12.8	8.1	93.0	651.2	11.6	.	3407.0	406.98	0.70	0.47	4.65	139.5	2
7321	ZIZIPHUS JUJUBA	L	.	0	11.8	4.3	75.3	14.3	8.6	1970.0	200.0	.	.	.	.	.	.	.	.	18
7322	ZIZIPHUS JUJUBA	SH	.	0	8.6	1.7	78.9	30.1	10.8	2160.0	230.0	.	.	.	.	.	.	.	.	18
7323	ZIZIPHUS JUJUBA	S	.	0	39.4	27.6	.	.	.	.	.	.	.	.	.	.	.	.	.	20
7324	ZIZIPHUS JUJUBA	S	.	0	36.9	33.5	.	.	.	.	.	.	.	.	.	.	.	.	.	20
7325	ZIZIPHUS JUJUBA	F	355	0	6.9	1.7	88.7	4.8	2.6	160.2	212.1	3.5	13.0	1203.5	173.16	0.13	0.13	3.03	199.1	2
7326	ZIZIPHUS JUJUBA	F	347	0	10.6	0.6	85.3	3.5	3.5	152.9	158.8	4.7	.	.	117.65	0.12	0.24	5.29	388.2	3
7327	ZIZIPHUS JUJUBA	F	352	0	4.0	0.7	92.6	4.7	2.7	97.3	124.2	2.3	10.1	902.7	80.54	0.07	0.13	3.02	231.5	4
7328	ZIZIPHUS JUJUBA	F	.	0	5.3	.	75.4	3.8	2.7	.	.	.	.	.	.	.	.	.	.	5
7329	ZIZIPHUS JUJUBA	S	.	0	2.9	0.5	95.0	2.8	1.7	.	.	.	.	.	.	.	.	.	.	5
7330	ZIZIPHUS MAURITIANA	F	340	0	6.7	0.0	88.4	7.4	4.9	178.9	70.2	.	.	.	.	.	.	.	203.5	1
7331	ZIZIPHUS MAURITIANA	F	347	0	5.3	0.1	91.0	4.0	3.5	245.7	92.5	4.5	.	.	0.00	0.03	.	3.03	50.6	1
7332	ZIZIPHUS MAURITIANA	F	338	0	13.7	0.7	79.9	6.5	5.8	230.2	237.4	4.3	.	1589.9	431.65	0.14	0.14	3.60	446.0	2
7333	ZIZIPHUS MAURITIANA	F	.	0	4.3	1.6	92.4	.	1.6	21.7	48.9	9.8	.	.	114.13	0.11	0.11	3.80	413.0	5
7334	ZIZIPHUS MUCRONATA	SH	.	0	14.3	2.6	73.7	8.4	9.4	4680.0	170.0	.	.	.	.	.	.	.	.	18
7335	ZIZIPHUS NUMMULARIA	L	.	0	11.5	1.6	80.7	33.8	6.2	1900.0	310.0	.	.	.	.	.	.	.	.	18

TABLE 3 - DATA CONVERTED TO A ZERO-MOISTURE BASIS.

OBS	PLNA	PART	CALORIES	WATER	PROTEIN	FAT	TOTLCARB	FIBER	ASH	CALCIUM	PHOSPHO	IRON	SODIUM	POTASS	CAROT	THIAMINE	RIBOFLAV	NIACIN	ASCORBIC	REFCO
7336	ZIZIPHUS RUGOSA	F	.	0	7.2	2.9	85.5	11.0	4.5	604.0	210.3	.	.	.	.	.	.	.	.	5
7337	ZIZIPHUS SPINA-CHRISTI	F	346	0	5.3	1.0	88.9	.	4.9	154.4	.	3.3	.	.	0	0.04	0.14	4.08	33.1	1
7338	ZORNIA DIPHYLLA	L	.	0	15.3	3.8	.	23.4	10.0	1020.0	180.0	.	.	530	.	.	.	.	.	5
7339	ZOYSIA JAPONICA	S	.	0	8.3	2.2	.	.	3.4	.	.	.	.	.	.	.	.	.	.	19
7340	ZOYSIA MATRELLA	L	.	0	5.0	1.3	86.0	35.3	7.6	.	.	.	.	.	.	.	.	.	.	5
7341	ZYGOPHYLLUM FABAGO	S	.	0	22.5	7.7	.	.	.	.	.	.	.	.	.	.	.	.	.	20
7342	ZYGOPHYLLUM SPINOSUM	S	.	0	19.1	8.0	.	.	.	.	.	.	.	.	.	.	.	.	.	21

REFERENCES

1. **Wu Leung, Woot-Tseun, Busson, F., and Jardin, C.,** Food Composition Table for Use in Africa, Food and Agriculture Organization and U.S. Department of Health, Education and Welfare, Washington, D.C., 1968, 306.
2. **Wu Leung, Woot-Tseun, Butrum, R. R., and Chang, F. H.,** Proximate composition mineral and vitamin contents of East Asian food, Part 1, In Food Composition Table for Use in East Asia, Food and Agriculture Organization and U.S. Department of Health, Education and Welfare, Washington, D.C., 1972, 334.
3. **Wu Leung, Woot-Tsuen and Flores, M.,** Tabla de Composicion de Alimentos Para Uso en America Latina, National Institutes of Health, Bethesda, Md., 1961, 132.
4. **Watt, B. K. and Merrill, A. L.,** Composition of Foods. Agriculture Handbook No. 8, Agricultural Research Service, U.S. Department of Agriculture, Washington, D.C., 1963, 189.
5. Council of Scientific and Industrial Research, The Wealth of India, 11 vols, New Delhi, 1948—1976.
6. **Miller, D. F.,** Composition of Cereal Grains and Forages, Publ. 585, National Academy of Sciences — National Research Council, Washington, D.C., 1958, 663.
7. **Jhang, J. J., Staba, E. J., and Kim, J. V.,** American and Korean ginseng tissue cultures: growth, chemical analysis, and plantlet production, *In Vitro,* 9(4), 253, 1974.
8. **Rutherford, M. C.,** Biomass structure and utilization of the natural vegetation in the winter rainfall region of South Africa, in *Mediterranean — Climate Regions — Basic and Applied Aspects,* Margaris, N. S., Ed., W. Junk, The Hague, 1981.
9. **Hilty, I. E., Peters, J. H., Benson, E. M., Edwards, M. A., and Miller, L. T.,** Nutritive Values of Native Foods of Warm Springs Indians, Extension Circular 809, Oregon State University, Corvallis, 1972.
10. **Duke, J. A.,** *Handbook of Legumes of World Economic Importance,* Plenum Press, New York, 1981, 345.
11. **Krikorian, A. D. and Getahun, A.,** Chat: coffee's rival from Harar, Ethiopia: II, chemical composition, *Econ. Bot.,* 27, 378, 1973.
12. **Oliveria, J. S. and Fidalgo de Carvalho, M.,** Nutritional value of some edible leaves used in Mozambique, *Econ. Bot.,* 29(3), 255, 1975.
13. **Wolverton, B. C. and McDonald, R. C.,** Nutritional composition of water hyacinths grown on domestic sewage, *Econ. Bot.,* 32(4), 363, 1978.
14. **Duke, J. A., Aulik, D., and Plowman, T.,** Nutritional Value of Coca, Bot. Mus. Leaflets, Harvard University, 24(6), 113, 1975.
15. **Norton, H. H.,** Plant use in Kaigani Haida culture: correction of an ethnohistorical oversight, *Econ. Bot.,* 35(4), 434, 1981.
16. USDA, Composition of Foods, Spices and Herbs. Raw-Processed-Prepared. Agriculture Handbook No. 8-2. Agricultural Research Service, U.S. Department of Agriculture, Washington, D.C. (looseleaf), 1977.
17. **Aoudia, M. O. and Gaudine, F.,** Composition and nutritive value of some foods of Algeria and wild plants of Kabylia, *Ann. Nut. et l'Aliment,* 24(6), 107, 1970.
18. **Gohl, B.,** Tropical Feeds. Feed Information Summaries and Nutritive Values, FAO Animal Production and Health Series No. 12, Food and Agriculture Organization, Rome, 1981, 529.
19. **Earle, F. R. and Jones, Q.,** Analyses of seed samples from 113 plant families, *Econ. Bot.,* 16(4), 221, 1962.
20. **Jones, Q. and Earle, F. R.,** Chemical analyses of seeds II: oil and protein content of 759 species, *Econ. Bot.,* 20(2), 127, 1966.
21. **Barclay, A. S. and Earle, F. R.,** Chemical analyses of seeds III: oil and protein content of 1253 species, *Econ. Bot.,* 28(2), 178, 1974.
22. **Duke, J. A.,** Vegetarian vitachart, *Quart. J. Crude Drug Res.,* 15, 45, 1977.